Fundamentals of Electrical Engineering

THE OXFORD SERIES IN ELECTRICAL AND COMPUTER ENGINEERING

M.E. VAN VALKENBURG   *Senior Consulting Editor*

ADEL S. SEDRA   *Series Editor, Electrical Engineering*

MICHAEL R. LIGHTNER   *Series Editor, Computer Engineering*

Allen and Holberg, *CMOS Analog Circuit Design*
Bobrow, *Elementary Linear Circuit Analysis, 2/e*
Bobrow, *Fundamentals of Electrical Engineering, 2/e*
Campbell, *The Science and Engineering of Microelectronic Fabrication*
Chen, *Linear System Theory and Design*
Chen, *System and Signal Analysis, 2/e*
Comer, *Digital Logic and State Machine Design, 3/e*
Comer, *Microprocessor Based System Design*
Cooper and McGillem, *Probabilistic Methods of Signal and System Analysis, 2/e*
Franco, *Electric Circuits Fundamentals*
Ghausi, *Electronic Devices and Circuits: Discrete and Integrated*
Houts, *Signal Analysis in Linear Systems*
Jones, *Introduction to Optical Fiber Communication Systems*
Kennedy, *Operational Amplifier Circuits: Theory and Application*
Kuo, *Digital Control Systems, 3/e*
Leventhal, *Microcomputer Experimentation with the IBM PC*
Leventhal, *Microcomputer Experimentation with the Intel SDK-86*
McGillem and Cooper, *Continuous and Discrete Signal and System Analysis, 3/e*
Miner, *Lines and Electromagnetic Fields for Engineers*
Navon, *Semiconductor Microdevices and Materials*
Papoulis, *Circuits and Systems: A Modern Approach*
Ramshaw and Van Heeswijk, *Energy Conversion*
Sadiku, *Elements of Electromagnetics, 2/e*
Schwarz, *Electromagnetics for Engineers*
Schwarz and Oldham, *Electrical Engineering: An Introduction, 2/e*
Sedra and Smith, *Microelectronic Circuits, 3/e*
Stefani, Savant, and Hostetter, *Design of Feedback Control Systems, 3/e*
Van Valkenburg, *Analog Filter Design*
Vranesic and Zaky, *Microcomputer Structures*
Warner and Grung, *Semiconductor Device Electronics*
Wolovich, *Automatic Control Systems*
Yariv, *Optical Electronics in Modern Communications, 5/e*

Fundamentals of

Electrical Engineering

Leonard S. Bobrow

Department of Electrical and Computer Engineering

University of Massachusetts, Amherst

New York   Oxford   OXFORD UNIVERSITY PRESS   1996

Oxford University Press

Oxford   New York
Athens   Auckland   Bangkok   Bombay
Calcutta   Cape Town   Dar es Salaam   Delhi
Florence   Hong Kong   Istanbul   Karachi
Kuala Lumpur   Madras   Madrid   Melbourne
Mexico City   Nairobi   Paris   Singapore
Taipei   Tokyo   Toronto

and associated companies in
Berlin   Ibadan

Bobrow, Leonard S.
     Fundamentals of electrical engineering / Leonard S. Bobrow — 2nd
ed.
          p.   cm. — (The Oxford series in electrical engineering)
     Includes bibliographical references and index.
     ISBN 0-19-510509-5 (cl)
     1. Electric engineering.   I. Title.   II. Series.
TK146.B693   1996
621.3—dc20   95-35576
                                                       CIP

9   8   7   6   5   4   3   2   1

Printed in the United States of America
on acid-free paper

In memory of my sister Adrienne

and my brother-in-law Jay,

who introduced me to electrical engineering

Contents

Preface

This book is intended to be the text for a first course in electrical engineering. It is partitioned into four parts: circuits, electronics, digital systems, and electromechanics. Although many topics are covered in each of these parts, this book is more than just a survey of the basics of electrical engineering. Even at this introductory level, material is usually covered in sufficient detail such that the reader may gain a good understanding of the fundamental principles on which modern electrical engineering is based.

The circuits portion of this book includes the traditional topics of Ohm's law, Kirchhoff's laws, resistive analysis techniques, various circuit theorems and principles, time-domain and frequency-domain analysis procedures, power, three-phase circuits, resonance, frequency response, and elementary system concepts. For this section of the book, it is assumed that the reader has had a course in college freshman-level (or higher) physics in which the concepts of the physics of electricity and magnetism (e.g., electric charge, electric potential, and magnetic fields) were introduced and studied. In addition, it is assumed that the reader has had one or more courses in elementary calculus in which the topics of differentiation and integration were covered.

The electronics portion of this book deals with both theory and applications of the major semiconductor devices: diodes and transistors—bipolar junction transistors (BJTs) and field-effect transistors (FETs)—in both discrete and integrated-circuit (IC) form. In addition to the coverage of the application of semiconductor devices to digital logic circuits, established analog topics such as small-signal, operational, and power amplifiers are included. Also covered are the concepts of feedback, oscillations, and elementary notions of communications via modulation. For a thorough understanding of the electronics section of the book, knowledge of a number of principles in the circuits section is necessary. When studying particular electronics topics, it should be obvious which circuits material is a prerequisite. The electronics material has been updated, and there is greater coverage of the 741 op amp and its analysis.

In the digital systems portion of this book, basic digital logic elements and logic design in both discrete and IC forms are covered. Included is sequential as well as combinational logic. In addition to the design of logic circuits which perform various types of operations, digital devices which count and store information are discussed. The digital systems section requires a minimal background from the circuits portion of the book. Familiarity with the electronics material, although illuminating as to the internal behavior of digital logic elements, is not a prerequisite for this part of the book.

The electromechanics portion of this book covers topics such as magnetic circuits, magnetic induction, and transformers on an elementary level. In addition, the basic principles of electromechanical devices such as transducers, meters, generators, and motors are studied. This section of the book employs numerous concepts from the circuits part of the book.

This edition of the book contains a chapter on the circuit analysis software SPICE. In particular, the PC version of SPICE, known as PSpice (from MicroSim Corp.), is discussed in detail. This subject matter, which is optional for the instructor, deals with topics covered in chapters 2, 3, 4, 5, 6, 7, 8, 9, 10, and 14.

In the study of electrical engineering I feel it is imperative that a student solve problems related to the material covered. This not only reinforces the understanding of the subject matter, but in some cases, it allows for the extension of various concepts discussed in the text. In certain situations, even new material is introduced via the problem sets.

Numerous drill exercises (243) have been added to this edition. Typically, each example is followed by an appropriate drill exercise so as to reinforce the important principles being discussed.

At the end of each chapter there is a set of problems. Answers to selected problems are located at the back of the book. Many of the end-of-chapter problems have been revised, and there are numerous new problems as well. This edition contains 1182 end-of-chapter problems—346 more problems than in the first edition.

This book can be used as the text for various types of electrical engineering courses. The entire book may be employed as the text for a two-semester or three-quarter sequence, which is an introduction to electrical engineering for EE majors or nonmajors or both. The circuits section (Chapters 1–5), the electromechanics section (Chapters 14 and 15), and the material on SPICE (Chapter 16) can be utilized as the text for a course on introductory circuits. Portions of the circuits section, the electronics section (Chapters 6—10), and the SPICE chapter can be used as the text for a first electronics course. Furthermore, Chapters 6, 7, 8, 11, 12, 13, and 16 can serve as the text for a course on digital electronics and logic design. I personally use portions of Chapters 1 to 10, 14, and 15 for a one-semester course in the fundamentals of electrical engineering for junior and senior nonmajors, and Chapters 6, 7, 8, and 16 for a second-semester sophomore course (34 lectures, 3 laboratory experiments, 3 review classes, 3 SPICE projects, and 3 midterm exams) in digital electronics.

I would like to thank the reviewers of the manuscript who made many helpful and important comments and suggestions. In particular, I received numerous useful remarks from Prof. Frank W. Smith. Finally, I would like to acknowledge my good friend and colleague, Prof. Donald E. Scott, who not only class tested material in the book, but whose friendship and support are greatly appreciated.

Amherst, Mass.   L.S.B.
http://www.ecs.umass.edu/ece/gradfac/bobrow.html
September 1995

Part I

Circuits

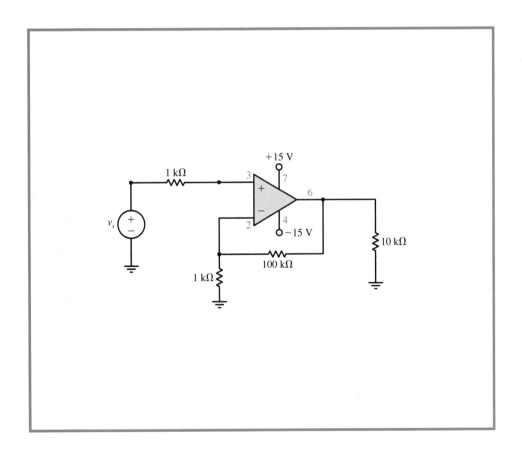

1

Basic Elements and Laws

INTRODUCTION

The study of electric circuits is fundamental in electrical engineering education and can be quite valuable in other disciplines as well. The skills acquired not only are useful in such electrical engineering areas as electronics, communications, microwaves, electromechanics, and control and power systems but also can be employed in other, seemingly different fields (e.g., acoustics and mechanics).

By an **electric circuit** or **network**, we mean a collection of electric components (e.g., voltage and current sources, resistors, inductors, capacitors, operational amplifiers, transformers, and transistors) that are interconnected in some manner. Our prime interest in this section will be in the process of determining the behavior of a given circuit, referred to as **circuit analysis**.

We begin our study by discussing some basic electric circuit elements and the laws that describe them. It is assumed that the reader has been introduced to the concepts of electric charge, potential, and current in various science and physics courses in high school and college.

1.1 Voltage Sources, Current Sources, and Resistors

Electric charge[1] is measured in **coulombs** (abbreviated C) in honor of the French scientist Charles de Coulomb (1736–1806); the unit of work or energy—the **joule**

---

[1] An electron has a charge of 1.6×10^{-19} C.

(J)—is named for the British physicist James P. Joule (1818–1889). Although for energy expended on electric charge the unit is joules per coulomb (J/C), we give it the special name **volt** (V) in honor of the Italian physicist Alessandro Volta (1745–1827), and we say that it is a measure of **electric potential difference,** or **voltage**. These units are part of the **Système International d'Unités** (International System of Units). Units of this system are referred to as **SI** units. Unless indicated to the contrary, SI units are the units used in this book. The SI unit for time is the second (s).

Voltage Sources

An **ideal voltage source,** which is represented in Fig. 1.1, is a device that produces a voltage or potential difference of *v* volts across its terminals *regardless of what is connected to it.*

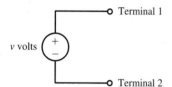

Fig. 1.1 Ideal voltage source.

For the device shown in Fig. 1.1, terminal 1 is marked plus (+) and terminal 2 is marked minus (−). This denotes that terminal 1 is at an electric potential that is *v* volts higher than that of terminal 2. (Alternatively, the electric potential of terminal 2 is *v* volts lower than that of terminal 1.)

The quantity *v* can have either a positive or a negative value. For the latter case, it is possible to obtain an equivalent source with a positive value. Suppose that $v = -5$ V for the voltage source shown in Fig. 1.1. Then the potential at terminal 1 is -5 V higher than that of terminal 2. However, this is equivalent to saying that terminal 1 is at a potential of $+5$ V lower than terminal 2. Consequently, the two ideal voltage sources shown in Fig. 1.2 are equivalent.

In the discussion so far, we may have implied that the value of an ideal voltage source is constant, that is, it does not change with time. Such a situation is plotted

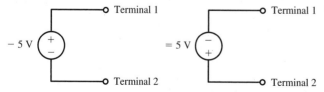

Fig. 1.2 Equivalent ideal voltage sources.

in Fig. 1.3 for the case that $v = 3$ V. For occasions such as this, an ideal voltage source is commonly represented by the equivalent notation shown in Fig. 1.4. We refer to such a device as an ideal **battery.** Although an actual battery is not ideal, there are many circumstances under which an ideal battery is a very good approximation to an actual battery. For example, the $1\frac{1}{2}$-V batteries that are used for portable transistor radios or cassette players roughly behave as ideal batteries. A 12-V automobile storage battery is another case in point. More generally, however, the voltage produced by an ideal voltage source will be a function of time. A few of the multitude of possible voltage waveforms are shown in Fig. 1.5.

Fig. 1.3 Constant voltage.

Fig. 1.4 Battery symbols.

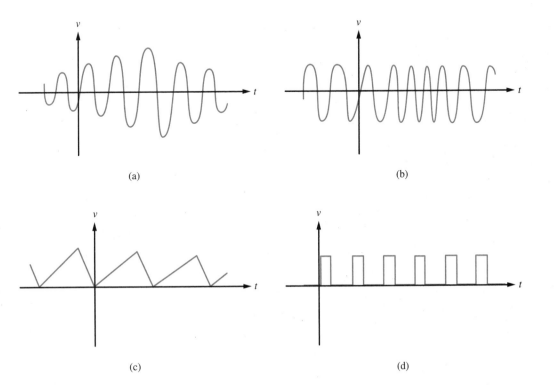

(a)

(b)

(c)

(d)

Fig. 1.5 Typical voltage waveforms.

Since the voltage produced by a source is, in general, a function of time, say $v(t)$, then the most general representation of an ideal voltage source is that shown in Fig. 1.6. There should be no confusion if the units "volts" are not included in the representation of the source. Thus the ideal voltage source in Fig. 1.7 is identical to the one in Fig. 1.6 with "volts" being understood.

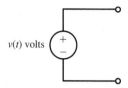

Fig. 1.6 Generalized ideal voltage source.

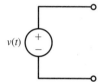

Fig. 1.7 Equivalent generalized ideal voltage source.

Example 1.1

Suppose that the voltage $v(t)$ produced by the ideal voltage source shown in Fig. 1.7 is described by $v(t) = 10e^{-t}$ V. Let us determine the value of this voltage at the instants of time $t = 0$ s, $t = 1$ s, $t = 2$ s, and $t = 3$ s.

At time $t = 0$ s, $v(t) = v(0) = 10e^{-0} = 10$ V. Furthermore, at time $t = 1$ s, $v(t) = v(1) = 10e^{-1} = 3.68$ V. Similarly, $v(2) = 10e^{-2} = 1.35$ V, and, finally, $v(3) = 10e^{-3} = 0.498$ V.

Drill Exercise 1.1

Suppose that the voltage $v(t)$ produced by the ideal voltage source shown in Fig. 1.7 is described by $v(t) = 5 - 10e^{-2t}$ V. Determine the value of this voltage at the instants of time $t = 0$ s, $t = 0.5$ s, $t = 1$ s, and $t = 1.5$ s.

ANSWER -5 V, 1.32 V, 3.65 V, 4.50 V

Note that due to the definition of an ideal voltage source, we will never allow two (or more) voltage sources to be connected to the same pair of terminals.

Current Sources

Placing an electric potential difference (voltage) across some material generally results in a flow of electric charge. Negative charge (in the form of electrons) flows from a given electric potential to a higher potential. Conversely, positive charge tends to flow from a given potential to a lower potential. Charge is usually denoted by q, and because this quantity is generally time dependent, the total amount of charge that is present in a given region is designated by $q(t)$.

We define **current,** denoted $i(t)$, to be the flow rate of the charge; that is,

$$i(t) = \frac{dq(t)}{dt}$$

is the current in the region containing $q(t)$. The units of current (coulombs per second, or C/s) are referred to as **amperes** (A), or **amps,** in honor of the French physicist André Ampère (1775–1836). Following the convention of Benjamin Franklin (a positive thinker), the direction of current has been chosen to be that direction in which positive charge would flow.

In the study of electric circuits, typically the important circuit variables are voltage and current, and the vast majority of circuit topics are described in terms of voltages and currents. As a consequence, electric charge is relegated to a minor role in the subject of circuit analysis.

An **ideal current source,** represented in Fig. 1.8, is a device that *when connected to anything,* will always move I amperes in the direction indicated by the arrow. As a consequence of the definition, it should be quite clear that the ideal current sources in Fig. 1.9 are equivalent.

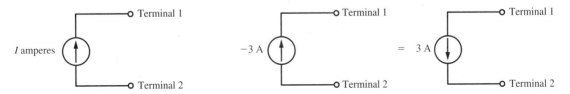

Fig. 1.8 Ideal current source.   **Fig. 1.9** Equivalent ideal current sources.

Again, in general, the amount of current produced by an ideal source will be a function of time. Thus, the general representation of an ideal current source is shown in Fig. 1.10, where the units ''amperes'' are understood.

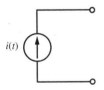

Fig. 1.10 Generalized ideal current source.

Although the current sources depicted in Figs. 1.8, 1.9, and 1.10 are not shown connected to any other circuit elements, in general, the implicit assumption is that a (nonzero) current source is always connected to something so that the current produced by the source has somewhere to go.

To summarize the properties of the two types of sources already discussed, **a voltage source places a constraint on the voltage across its terminals—there is no constraint on the current through a voltage source.** The current through a voltage source depends upon what is connected to that source. Conversely, **a current source places a constraint on the current through it—there is no constraint on the voltage across a current source.** The voltage across a current source depends upon what is connected to that source.

Resistors and Ohm's Law

Suppose that some material is connected to the terminals of an ideal voltage source $v(t)$ as shown in Fig. 1.11. Suppose that $v(t) = 1$ V. Then the electric potential at the top of the material is 1 V above the potential at the bottom. Since an electron has a negative charge, electrons in the material will tend to flow from bottom to top. Therefore, we say that current tends to go from top to bottom. Hence, for the given polarity, when $v(t)$ is a positive number, $i(t)$ will be a positive number with the direction indicated. If $v(t) = 2$ V, again the potential at the top is greater than at the bottom, so $i(t)$ will again be positive. However, because the potential is now twice as large as before, the current will be greater. (If the material is a "linear" element, the current will be twice as great.) Suppose now that $v(t) = 0$ V. Then the potentials at the top and the bottom of the material are the same. The result is no flow of electrons and, hence, no current. In this case, $i(t) = 0$ A. But suppose that $v(t) = -2$ V. Then the top of the material will be at a potential lower than at the bottom of the material. A current from bottom to top will result, and $i(t)$ will be a negative number. Note that the current $i(t)$ through the material must also go through the voltage source because there is nowhere else for it to go.

If in Fig. 1.11 the resulting current $i(t)$ is always directly proportional to the voltage for any function $v(t)$, then the material is called a **linear resistor**, or **resistor**, for short.

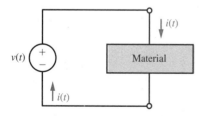

Fig. 1.11 Material with an applied voltage.

Since voltage and current are directly proportional for a resistor, there exists a proportionality constant R, called **resistance,** such that

$$v(t) = Ri(t)$$

In dividing both sides of this equation by $i(t)$, we obtain

$$R = \frac{v(t)}{i(t)}$$

The units of resistance (volts per ampere) are referred to as **ohms**[2] and are denoted by the capital Greek letter omega, Ω. The accepted circuit symbol for a resistor whose resistance is R ohms is shown in Fig. 1.12. A plot of voltage versus current for a (linear) resistor is given in Fig. 1.13.

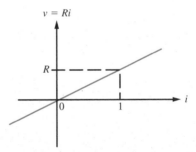

Fig. 1.12 Circuit symbol for a resistor.   **Fig. 1.13** Plot of voltage versus current for a resistor.

It was Ohm who discovered that if a resistor R has a voltage $v(t)$ *across* it and a current $i(t)$ *through* it, then if one is the cause, the other is the effect. Furthermore, if the polarity of the voltage and the direction of the current are as shown in Fig. 1.14, then it is true that

$$v(t) = Ri(t)$$

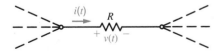

Fig. 1.14 Current and voltage convention for Ohm's law.

---

[2]Named for the German physicist Georg Ohm (1787–1854).

This equation is often called **Ohm's law.** From it, we may immediately deduce that

$$R = \frac{v(t)}{i(t)}$$

and

$$i(t) = \frac{v(t)}{R}$$

These last two equations may also be referred to as Ohm's law.

It should be pointed out here that directions of currents and polarities of voltages are crucial when writing Ohm's law—the accepted convention is given in Fig. 1.14. For example, in the situation depicted in Fig. 1.15, before writing Ohm's law, note that the current's direction is opposite to that dictated by convention. However, this difficulty is easily remedied by redrawing the figure in the equivalent form given in Fig. 1.16. (A current of 5 A going through the resistor to the left is the same as a current of -5 A going to the right.) Since the direction of the current and the polarity of the voltage now conform to convention, we use Ohm's law to write[3]

$$v_1(t) = R[-i_1(t)] \quad \Rightarrow \quad v_1(t) = -Ri_1(t)$$

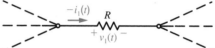

Fig. 1.15 Situation requiring negative sign for Ohm's law.

Fig. 1.16 Equivalent of resistor in Fig. 1.15.

Alternatively, we could have redrawn Fig. 1.15 in another equivalent form as shown in Fig. 1.17. Again we use Ohm's law to write

$$-v_1(t) = Ri_1(t) \quad \Rightarrow \quad v_1(t) = -Ri_1(t)$$

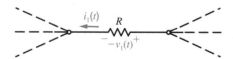

Fig. 1.17 Another equivalent of resistor in Fig. 1.15.

---

[3]The symbol $\Rightarrow$ means "implies."

Example 1.2

Consider the circuit shown in Fig. 1.18. We use the letter "k" to represent the prefix "kilo," which indicates a value of 10^3. Frequently used symbols include the following:

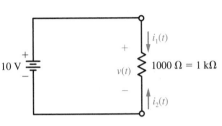

Value	Prefix	Symbol
10^{-12}	pico	p
10^{-9}	nano	n
10^{-6}	micro	μ
10^{-3}	milli	m
10^3	kilo	k
10^6	mega	M
10^9	giga	G

Fig. 1.18 Circuit with a battery.

For the circuit in Fig. 1.18, the voltage across the 1-kΩ resistor is, by the definition of an ideal voltage source, $v(t) = 10$ V. Thus, by Ohm's law, we get

$$i_1(t) = \frac{v(t)}{R} = \frac{10}{1000} = \frac{1}{100} = 0.01 = 10 \text{ mA}$$

and

$$i_2(t) = \frac{-v(t)}{R} = \frac{-10}{1000} = \frac{-1}{100} = -0.01 = -10 \text{ mA}$$

Note that $i_2(t) = -i_1(t)$ as expected.

For the circuit shown in Fig. 1.19, by the definition of an ideal current source, $i(t) = 25 \ \mu A = 25 \times 10^{-6}$ A. By Ohm's law, we have

$$v(t) = -Ri(t) = -(2 \times 10^6)(25 \times 10^{-6}) = -50 \text{ V}$$

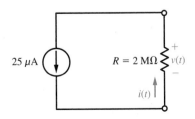

Fig. 1.19 Circuit with a current source.

For the circuit shown in Fig. 1.20, suppose that the voltage produced by the source is described by the sinusoid $v(t) = 170 \cos 120\pi t = 170 \cos 377t$ V and $R = 85 \ \Omega$. By Ohm's law, we have

$$i(t) = \frac{v(t)}{R} = \frac{170 \cos 120\ \pi t}{85} = 2 \cos 120\pi t = 2 \cos 377t \ \text{A}$$

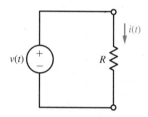

Fig. 1.20 Circuit with a nonconstant voltage source.

Specifically, at time $t = 0$ s, the voltage is $v(0) = 170 \cos 0 = 170$ V and the current is $i(0) = 2 \cos 0 = 2$ A. When $t = 1/240$ s, then $v(1/240) = 170 \cos \pi/2 = 0$ V and $i(1/240) = 2 \cos \pi/2 = 0$ A. When $t = 1/120$ s, then $v(1/120) = 170 \cos \pi = -170$ V and $i(1/120) = 2 \cos \pi = -2$ A.

Drill Exercise 1.2

(a) For the circuit shown in Fig. 1.19, what value of R will result in $v(t) = -2.5$ V? (b) For the circuit shown in Fig. 1.20, when $R = 4 \ \Omega$ what voltage will produce the current $i(t) = 3e^{-2t} - 7e^{-7t}$ A?

ANSWER   (a) 100 kΩ, (b) $12e^{-2t} - 28e^{-7t}$ V

Given a resistor R connected to an ideal voltage source $v(t)$ as shown in Fig. 1.21, we conclude the following. Since $i(t) = v(t)/R$ for any particular ideal source $v(t)$, the amount of current $i(t)$ that results can be made to be any finite value by choosing the appropriate value for R (e.g., to make $i(t)$ large, make R small). Thus we see that *the current through an ideal voltage source can be anything!* The current through a voltage source depends on what is connected to the voltage source—only the voltage across the terminals of the voltage source is constrained to be $v(t)$ volts.

When a resistor R is connected to an ideal current source as in Fig. 1.22, we know that $v(t) = Ri(t)$. Therefore, for a given current source $i(t)$, the voltage $v(t)$ that results

Fig. 1.21 Current through a voltage source.   **Fig. 1.22** Voltage across a current source.

can be made to be any finite value by appropriately choosing R (e.g., to make $v(t)$ large, make R large). Hence, we conclude that *the voltage across an ideal current source can be anything!* The voltage across the current source depends on what is connected to the current source—only the current through the current source is constrained to be $i(t)$ amperes.

Physical (nonideal) sources do not have the ability to produce unlimited currents and voltages. As a matter of fact, an actual source may approximate an ideal source only for a limited range of values.

Short Circuits and Open Circuits

Now consider a resistor $R = 0\ \Omega$. An equivalent representation, called a **short circuit**, of such a resistance is given in Fig. 1.23. By Ohm's law, we have

$$v(t) = Ri(t) = 0i(t) = 0 \text{ V}$$

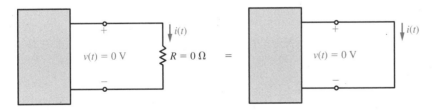

Fig. 1.23 Short-circuit equivalents.

Thus, no matter what (finite) value $i(t)$ has, $v(t)$ will be zero. Hence, we see that *a zero-ohm resistor is equivalent to an ideal voltage source whose value is zero volts* (provided that the current through it is finite). Therefore, for a zero resistance to be synonymous with a constraint of zero volts (and to avoid the unpleasantness of infinite currents), we will insist that we never be allowed to place a short circuit directly across a voltage source. In actuality, the reader will be spared a lot of grief by never attempting this in a laboratory or field situation.

Next consider a resistor having infinite resistance. An equivalent representation, called an **open circuit**, of such a situation is depicted in Fig. 1.24. By Ohm's law,

$$i(t) = \frac{v(t)}{R} = 0 \text{ A}$$

as long as $v(t)$ has a finite value. Thus, we may conclude that *an infinite resistance is equivalent to an ideal current source whose value is zero amperes*. Remember, we will always assume that an ideal current source has something connected to its terminals.

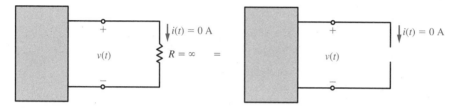

Fig. 1.24 Open-circuit equivalents.

1.2   Kirchhoff's Current Law (KCL)

It is a consequence of the work of the German physicist Gustav Kirchhoff (1824–1887) that enables us to analyze an interconnection of any number of elements (voltage sources, current sources, and resistors, as well as elements not yet discussed). We will refer to any such interconnection as a **circuit** or a **network**.

For a given circuit, a connection of two or more elements shall be called a **node**. The partial circuit shown in Fig. 1.25 depicts an example of a node. In addition to using a solid dot, we may also indicate a node by a hollow dot, as was done for a terminal. Conversely, we may use a solid dot for the terminal of a device.

We now present the first of Kirchhoff's two laws, Kirchhoff's current law (KCL), which is essentially the law of conservation of electric charge.

> KCL: At any node of a circuit, at every instant of time, the sum of the currents into the node is equal to the sum of the currents out of the node.

Specifically for the portion of the network shown in Fig. 1.25, by applying KCL we obtain the equation

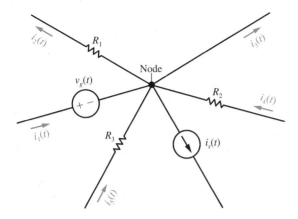

Fig. 1.25 Portion of a circuit.

$$i_1(t) + i_4(t) + i_5(t) = i_2(t) + i_3(t) + i_s(t)$$

Note that one of the elements (the one in which $i_3(t)$ flows) is a short circuit—KCL holds regardless of the kinds of elements in the circuit.

An alternative, but equivalent, form of KCL can be obtained by considering currents directed into a node to be positive in sense and currents directed out of a node to be negative in sense (or vice versa). Under this circumstance, the alternative form of KCL can be stated as follows:

KCL: At any node of a circuit, the currents algebraically sum to zero.

Applying this form of KCL to the node in Fig. 1.25 and considering currents directed in to be positive in sense, we get

$$i_1(t) - i_2(t) - i_3(t) + i_4(t) - i_s(t) + i_5(t) = 0$$

A close inspection of the last two equations, however, reveals that they are the same!

From this point on, we will simplify our notation somewhat by often abbreviating functions of time t such as $v(t)$ and $i(t)$ as v and i, respectively. For instance we may rewrite the last two equations, respectively, as

$$i_1 + i_4 + i_5 = i_2 + i_3 + i_s \qquad \text{and} \qquad i_1 - i_2 - i_3 + i_4 - i_s + i_5 = 0$$

It should always be understood, however, that lowercase letters such as v and i, in general, represent time-varying quantities.[4]

---

[4]A constant is a special case of a function of time.

Example 1.3

Let us find the voltage v in the two-node circuit given in Fig. 1.26 in which the directions of i_1, i_2, and i_3 and the polarity of v were chosen arbitrarily. (The directions of the 2-A and 13-A sources are given.)

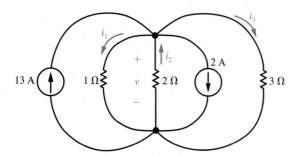

Fig. 1.26 Circuit for Example 1.3.

By KCL (at either of the two nodes), we have

$$13 - i_1 + i_2 - 2 - i_3 = 0 \quad \Rightarrow \quad i_1 - i_2 + i_3 = 11 \tag{1.1}$$

By Ohm's law,

$$i_1 = \frac{v}{1} \qquad i_2 = \frac{-v}{2} \qquad i_3 = \frac{v}{3}$$

Substituting these expressions into Eq. 1.1 yields

$$\left(\frac{v}{1}\right) - \left(\frac{-v}{2}\right) + \left(\frac{v}{3}\right) = 11 \quad \Rightarrow \quad v = 6 \text{ V}$$

Having solved for v, we now find that

$$i_1 = \frac{v}{1} = 6 \text{ A}, \qquad i_2 = \frac{-v}{2} = -3 \text{ A}, \qquad i_3 = \frac{v}{3} = 2 \text{ A}$$

Note that a reordering of the circuit elements, as shown in Fig. 1.27, will result in Eq. 1.1 when KCL is applied. Since Ohm's law remains unchanged, the same answers are obtained.

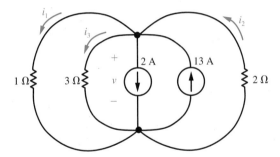

Fig. 1.27 Circuit equivalent to Fig. 1.26.

Drill Exercise 1.3

For the circuit shown in Fig. 1.26, change the value of the 13-A current source to -9 A. Find the resulting values for v, i_1, i_2, and i_3.

ANSWER -6 V, -6 A, 3 A, -2 A

General Form of KCL

Just as KCL applies to any node of a circuit (i.e., to satisfy the physical law of conservation of charge, the current going in must equal the current coming out), so must KCL hold for any closed region.

For the circuit shown in Fig. 1.28, three regions have been arbitrarily identified. Applying KCL to Region 1, we get

$$i_1 + i_4 + i_5 = i_2$$

Applying KCL to Region 2, we obtain

$$i_5 + i_6 + i_7 = 0$$

and by applying KCL to Region 3, we get

$$i_2 + i_7 = i_4 + i_g$$

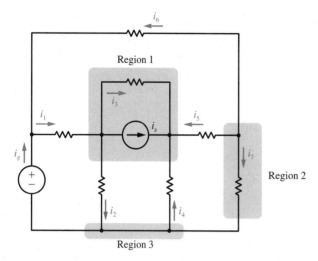

Fig. 1.28 Circuit with three arbitrarily selected regions.

Note that Region 3 apparently contains two nodes. However, since they are connected by a short circuit, there is no difference in voltage between these two points. Therefore, we can shrink the short circuit so as to coalesce the two points into a single node without affecting the behavior of the rest of the circuit. Applying KCL at the resulting node again yields $i_2 + i_7 = i_4 + i_g$.

The converse process of expanding a node into apparently different nodes interconnected by short circuits also does not affect the behavior of the remainder of a circuit. For example, the portion of a circuit shown in Fig. 1.25 has the equivalent form shown in Fig. 1.29. Applying KCL to the shaded region results in the same

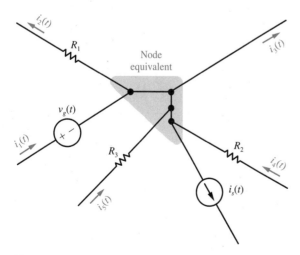

Fig. 1.29 Equivalent of circuit portion in Fig. 1.25.

equation as is obtained when KCL is applied to the node shown in Fig. 1.25. Thus, although it may appear that there are four distinct nodes in the shaded region depicted in Fig. 1.29, the region actually constitutes a single node.

It is because of the foregoing that we can redraw the circuit given in Fig. 1.26 in the equivalent form shown in Fig. 1.30. Even though it may not appear so at first glance, the circuit shown in Fig. 1.30 has just two distinct nodes—and these nodes are indicated by the shaded regions. With this fact in mind, from this point on we shall draw such a circuit without the shaded regions, as shown in Fig. 1.31.

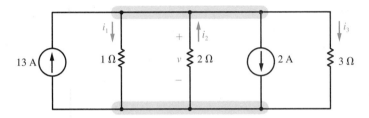

Fig. 1.30 Equivalent form of circuit in Fig. 1.26.

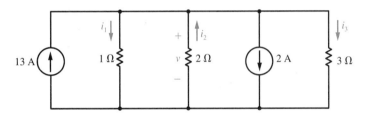

Fig. 1.31 Circuit in Fig. 1.30 with region indications omitted.

Resistors Connected in Parallel

Consider the circuit given in Fig. 1.32a. Let us express the current i in terms of R_1, R_2, and v. Note that the voltage across each resistor is v. By Ohm's law, we can write

$$i_1 = \frac{v}{R_1} \quad \text{and} \quad i_2 = \frac{v}{R_2}$$

Applying KCL (at either node), we get

$$i = i_1 + i_2 = \frac{v}{R_1} + \frac{v}{R_2} = v\left(\frac{1}{R_1} + \frac{1}{R_2}\right) \tag{1.2}$$

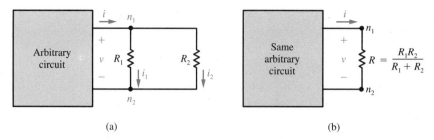

Fig. 1.32 (a) Parallel connection of resistors, and (b) equivalent resistance.

Note that resistors R_1 and R_2 are connected to the same pair of nodes (nodes n_1 and n_2). We call this a **parallel connection**. Specifically, we say that two elements are **connected in parallel** if they are connected to the same pair of nodes, regardless of what else is connected to those two nodes. This definition holds not only for two elements, but for three, four, or more, as well. As a consequence of the definition, we see that elements connected in parallel all have the same voltage across them.

As far as the arbitrary circuit in Fig. 1.32a is concerned, there is an equivalent resistance R between nodes n_1 and n_2 as depicted in Fig. 1.32b. But how is R related to R_1 and R_2? By Ohm's law, $R = v/i$. Combining this result with Eq. 1.2 we get

$$\frac{i}{v} = \frac{1}{R} = \frac{1}{R_1} + \frac{1}{R_2}$$

Thus, we see that the parallel connection of resistors R_1 and R_2 is equivalent to a single resistor R, provided that

$$\frac{1}{R} = \frac{1}{R_1} + \frac{1}{R_2} \qquad \Rightarrow \qquad R = \frac{R_1 R_2}{R_1 + R_2}$$

Using reasoning as in the previous discussion, we deduce that m resistors R_1, R_2, R_3, . . . , R_m connected in parallel are equivalent to a single resistor R, provided that

$$\frac{1}{R} = \frac{1}{R_1} + \frac{1}{R_2} + \cdots + \frac{1}{R_m}$$

In the previous discussion, the reciprocal of a resistance appeared a number of times. It is because of the frequent occurrence of the quantity $1/R$ that we denote it by a separate symbol. Given an R-ohm resistor, we define its **conductance** G to be

$$G = \frac{1}{R}$$

Furthermore, since the units of R are ohms, denoted Ω, the units[5] of G are often called **mhos** and are denoted by the symbol $\mho$. Figure 1.33 shows two equivalent ways of representing the same circuit element. For example, a 4 $\mho$ resistance is the same thing as a $\frac{1}{4}$-Ω conductance.

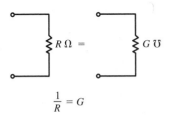

$$\frac{1}{R} = G$$

Fig. 1.33 Resistance and conductance.

In summary, the equivalent forms of Ohm's law are:

$$v = Ri = \frac{i}{G} \qquad i = \frac{v}{R} = Gv \qquad R = \frac{v}{i} = \frac{1}{G} \qquad G = \frac{i}{v} = \frac{1}{R}$$

Note that when combining resistors in parallel, to obtain an equivalent resistor, we add conductances. This fact is demonstrated in Fig. 1.34.

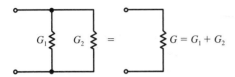

Fig. 1.34 Combining conductances in parallel.

---

[5]In terms of SI units, conductance is measured in **siemens** (S) in honor of the British inventor William Siemens (1823–1883). However, because it is in widespread use, we will use mhos.

Current Division

Now consider the circuit given in Fig. 1.35a. We ask the question that has been haunting scientists for a long time: "If you were an ampere, where would you go?" To answer this question, we simply replace the parallel connection of R_1 and R_2 by its equivalent resistance as shown in Fig. 1.35b. We then have

$$v = Ri = \frac{R_1 R_2}{R_1 + R_2} i \qquad\qquad (1.3)$$

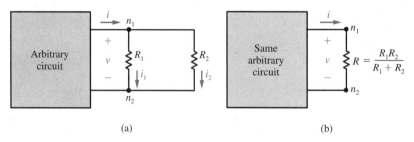

(a)   (b)

Fig. 1.35 (*a*) Current division, and (*b*) equivalent resistance of parallel connection.

Since this is the same v that appears in Fig. 1.35(a), we conclude that

$$i_1 = \frac{v}{R_1} \qquad \text{and} \qquad i_2 = \frac{v}{R_2}$$

Using the expression for v given in Eq. 1.3, we get

$$i_1 = \frac{R_2}{R_1 + R_2} i \qquad \text{and} \qquad i_2 = \frac{R_1}{R_1 + R_2} i$$

Since, by KCL, $i_1 + i_2 = i$, these two formulas describe how the current i is divided by the resistors. For this reason, a pair of resistors connected in parallel is often referred to as a **current divider**. Note that if R_1 and R_2 are both positive, and if R_1 is greater than R_2, then i_2 is greater than i_1. In other words, a larger amount of current will go through the smaller resistor—and amperes tend to take the path of least resistance!

For the circuit shown in Fig. 1.36, since $G_1 = 1/R_1$ and $G_2 = 1/R_2$, then by the current-divider formula for i_1, we have

$$i_1 = \frac{R_2}{R_1 + R_2} i = \frac{1/G_2}{(1/G_1) + (1/G_2)} i = \frac{1/G_2}{(G_1 + G_2)/G_1G_2} i = \frac{G_1}{G_1 + G_2} i$$

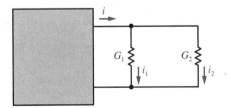

Fig. 1.36 Conductances connected in parallel.

Similarly,

$$i_2 = \frac{G_2}{G_1 + G_2} i$$

Thus we also have current-divider formulas in terms of conductances.

Example 1.4

For the circuit given in Fig. 1.37, let us determine the currents indicated by using current division.

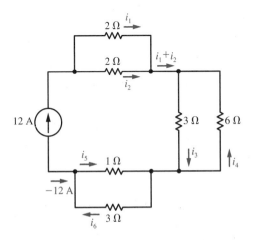

Fig. 1.37 Example of current division.

The current that is to be divided by the two 2-Ω resistors is 12 A. By the current-division formulas,

$$i_1 = i_2 = \frac{2}{2 + 2}(12) = 6 \text{ A}$$

The current that is to be divided by the 3-Ω and 6-Ω resistors is $i_1 + i_2 = 12$ A. By the current-division formulas, we have

$$i_3 = \frac{6}{3 + 6}(12) = 8 \text{ A}$$

$$-i_4 = \frac{3}{3 + 6}(12) = 4 \text{ A} \quad \Rightarrow \quad i_4 = -4 \text{ A}$$

Finally, the current that is to be divided by the 1-Ω and 3-Ω resistors is -12 A. Thus, by current division,

$$i_5 = \frac{3}{1 + 3}(-12) = -9 \text{ A}$$

$$-i_6 = \frac{1}{1 + 3}(-12) = -3 \text{ A} \quad \Rightarrow \quad i_6 = 3 \text{ A}$$

Drill Exercise 1.4

For the circuit shown in Fig 1.37, replace the 1-Ω resistance with a 1-℧ conductance, the 2-Ω resistances with 2-℧ conductances, the 3-Ω resistances with 3-℧ conductances, and the 6-Ω resistance with a 6-℧ conductance. Use the current-divider formulas to determine i_1, i_2, i_3, i_4, i_5, and i_6.

ANSWER   6 A, 6 A, 4 A, -8 A, -3 A, 9 A

1.3   Kirchhoff's Voltage Law (KVL)

We now present the second of Kirchhoff's laws—the voltage law. To do this, we must introduce the concept of a "loop."

Starting at any node n in a circuit, we form a **loop** by traversing through elements (open circuits included!) and returning to the starting node n, and never encountering any other node more than once. As an example, consider the partial circuit shown in Fig. 1.38. In this circuit, the 1-Ω, 2-Ω, 3-Ω, and 6-Ω resistors, along with the 2-A current source, constitute the loop a, b, c, e, f, a. A few (but not all) of the other loops are: (a) a, b, c, d, e, f, a; (2) c, d, e, c; (3) d, e, f, d; (4) a, b, c, e, d, f, a; (5) a, b, e, f, a; (6) b, c, e, b.

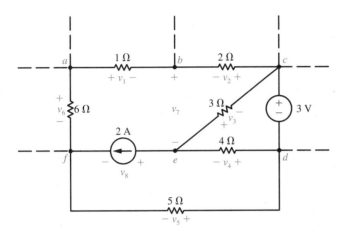

Fig. 1.38 Portion of a circuit.

We now can present Kirchhoff's voltage law (KVL):

KVL: In traversing any loop in any circuit, at every instant of time, the sum of the voltages having one polarity equals the sum of the voltages having the opposite polarity.[6]

For the partial circuit shown in Fig. 1.38, suppose that the voltages across the elements are as shown. Then by KVL around loop a, b, c, e, f, a, we get

$$v_1 + v_8 = v_2 + v_3 + v_6$$

and around loop b, c, d, e, b, we get

$$v_2 + v_7 = 3 + v_4$$

In this last loop, one of the elements traversed (the element between nodes b and e) is an open circuit—KVL holds regardless of the nature of the elements in the circuit.

Since voltage is energy (or work) per unit charge, then KVL is another way of stating the physical law of the conservation of energy.

---

[6]Another form of this statement is: The sum of the voltage rises equals the sum of the voltage drops, where traversing a voltage from minus to plus is a voltage rise and from plus to minus is a voltage drop.

An alternative statement of KVL can be obtained by considering voltages across elements that are traversed from plus to minus to be positive in sense and voltages across elements that are traversed from minus to plus to be negative in sense (or vice versa). Under this circumstance, KVL has the following alternative form:

KVL: Around any loop in a circuit, the voltages algebraically sum to zero.

By applying this form of KVL to the partial circuit shown in Fig. 1.38, and selecting a traversal from plus to minus to be positive in sense, around loop a, b, c, e, f, a, we get

$$v_1 - v_2 - v_3 + v_8 - v_6 = 0$$

and around loop b, c, d, e, b, we get

$$-v_2 + 3 + v_4 - v_7 = 0$$

These two equations are the same, respectively, as the preceding two equations.

Example 1.5

Let us find the current i for the circuit shown in Fig. 1.39a, where the polarities of v_1, v_2, v_3 and the direction of i were chosen arbitrarily.

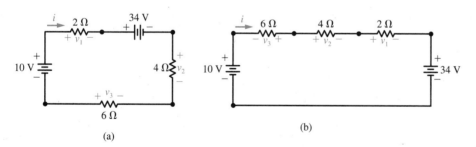

Fig. 1.39 (a) Single-loop circuit, and (b) circuit with elements reordered.

Applying KVL, we obtain

$$v_1 + 34 + v_2 - v_3 - 10 = 0 \quad \Rightarrow \quad v_1 + v_2 - v_3 = -24$$

By Ohm's law, $v_1 = 2i$, $v_2 = 4i$, and $v_3 = -6i$. Substituting these expressions into the preceding equation yields

$$2i + 4i - (-6i) = -24 \quad \Rightarrow \quad i = -2 \text{ A}$$

and using this value of i, the indicated voltages are

$$v_1 = 2i = 2(-2) = -4 \text{ V}, \qquad v_2 = 4i = 4(-2) = -8 \text{ V},$$

$$v_3 = -6i = -6(-2) = 12 \text{ V}$$

Note that a reordering of the circuit elements in Fig. 1.39a as shown in Fig. 1.39b will result in the same equation when KVL is applied. Since Ohm's law yields the same expressions for v_1, v_2, and v_3, the same values are obtained for the current and the voltages.

Drill Exercise 1.5

For the circuit shown in Fig. 1.39a, to what value should the 34-V battery be changed such that the resulting current is $i = 2$ A? What are the corresponding values of v_1, v_2, and v_3?

ANSWER -14 V, 4 V, 8 V, -12 V

Resistors Connected in Series

Now consider the circuit given in Fig. 1.40a. For this circuit, let us express the voltage v in terms of R_1, R_2, and i. By Ohm's law, we can write

$$v_1 = R_1 i \qquad \text{and} \qquad v_2 = R_2 i$$

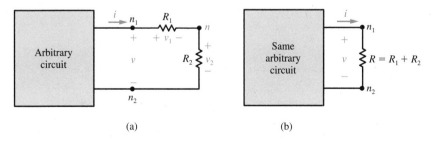

(a)   (b)

Fig. 1.40 (a) Series connection of resistors, and (b) equivalent resistance.

Applying KVL, we get

$$v = v_1 + v_2 = R_1 i + R_2 i = (R_1 + R_2)i \tag{1.4}$$

Note that resistors R_1 and R_2 have a node in common (node n), and no other element is connected to this node. This is known as a **series connection**. Specifically, we say that two elements are **connected in series** if they have a node in common and no other element is connected to this common node. In general, if m elements are connected together such that each resulting node joins no more than two of the elements, then these elements are connected in series. As a consequence of the definition, we see that elements connected in series all have the same current through them.

As far as the arbitrary circuit in Fig. 1.40a is concerned, there is an equivalent resistance R between nodes n_1 and n_2 as depicted in Fig. 1.40b. From Eq. 1.4, by Ohm's law, $R = v/i$. Combining this result with Eq. 1.4 we get

$$\frac{v}{i} = R = R_1 + R_2$$

Thus we see that the series connection of resistors R_1 and R_2 is equivalent to a single resistor R, provided that

$$\boxed{R = R_1 + R_2}$$

Using reasoning as above, we deduce that m resistors $R_1, R_2, R_3, \ldots, R_m$ connected in series are equivalent to a single resistor R, provided that

$$\boxed{R = R_1 + R_2 + R_3 + \cdots + R_m}$$

Example 1.6

Let us find i in the circuit shown in Fig. 1.41. To find i, we can replace series and parallel connections of resistors by their equivalent resistances. We begin by noting that the 1-Ω and 3-Ω resistors are connected in series. Combining them, we get the circuit in Fig. 1.42a. Note that in Fig. 1.42a it is not possible to display the voltage v_2 indicated in Fig. 1.41.

Since the two 4-Ω resistors in Fig. 1.42a are connected in parallel, we can simplify this circuit and obtain the circuit shown in Fig. 1.42b. Here the 5-Ω and 2-Ω resistors

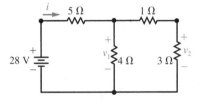

Fig. 1.41 Series-parallel circuit.

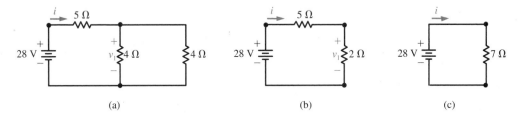

Fig. 1.42 Combining resistors connected in series and in parallel.

are connected in series, so we may combine them and obtain the circuit in Fig. 1.42c. In this circuit, by Ohm's law, we have

$$i = \frac{28}{7} = 4 \text{ A}$$

In this example, we see that the equivalent resistance "seen by the source" or "loading the source," is 7 Ω.

Drill Exercise 1.6

For the circuit shown in Fig. 1.41, change the value of the 1-Ω resistor to 9 Ω. Find the resistance loading the source and i.

ANSWER 8 Ω, 3.5 A

Voltage Division

Now consider the circuit given in Fig. 1.43a. To determine how the voltage v is divided between R_1 and R_2, we express i in terms of R_1, R_2, and v by replacing the

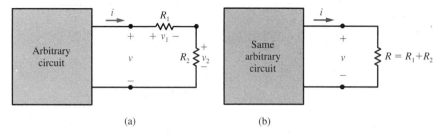

Fig. 1.43 (a) Voltage division, and (b) equivalent resistance of series connection.

series connection of the resistors by its equivalent resistance R as shown in Fig. 1.43b. By Ohm's law,

$$i = \frac{v}{R} = \frac{v}{R_1 + R_2} \qquad\qquad (1.5)$$

Thus for the circuit in Fig. 1.43a, we have

$$v_1 = R_1 i \qquad \text{and} \qquad v_2 = R_2 i$$

and substituting the expression for i given by Eq. 1.5 we get

$$\boxed{v_1 = \frac{R_1}{R_1 + R_2} v} \qquad \text{and} \qquad \boxed{v_2 = \frac{R_2}{R_1 + R_2} v}$$

and these two **voltage-divider formulas** describe how the voltage $v = v_1 + v_2$ is divided between the resistors. Because of this, a pair of resistors that are connected in series is often called a **voltage-divider network**, or **voltage divider.** Note that for positive-valued resistors, if R_1 is greater than R_2, then v_1 is greater than v_2. In other words, the larger voltage will be across the larger resistor.

Example 1.7

Let us find the voltage v_2 in the circuit given in Fig. 1.41 by using the voltage-divider formulas. Combining the series connection of the 1-Ω and 3-Ω resistors, we obtain the circuit shown in Fig. 1.42a. The resulting pair of 4-Ω resistors that are connected in parallel can be combined as shown in Fig. 1.42b. By voltage division, we have

$$v_1 = \frac{2}{2+5}(28) = \frac{56}{7} = 8 \text{ V}$$

Returning to the original circuit (Fig. 1.41) and employing the voltage-divider formula again yields

$$v_2 = \frac{3}{3+1}v_1 = \frac{3}{4}(8) = 6 \text{ V}$$

Drill Exercise 1.7

For the circuit shown in Fig. 1.41, change the value of both the 1-Ω and 5-Ω resistors to 9 Ω. Use voltage division to find v_1 and v_2.

ANSWER   7 V, 1.75 V

For a given circuit, there is generally more than one way to analyze it—as long as the different techniques are employed correctly, of course, the same results will be obtained. To demonstrate this, let us again consider the circuit given in Fig. 1.41 and take a slightly different approach.

Example 1.8

Consider the circuit shown in Fig. 1.44. By Ohm's law, $i_2 = v_2/3$. By KCL,

$$i_3 = i_2 = \frac{v_2}{3}$$

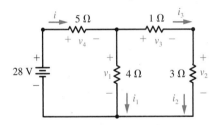

Fig. 1.44 Series-parallel circuit given in Fig. 1.41.

By Ohm's law, $v_3 = 1i_3 = v_2/3$. By KVL,

$$v_1 = v_2 + v_3 = v_2 + \frac{v_2}{3} = \frac{4v_2}{3}$$

By Ohm's law, $i_1 = v_1/4 = v_2/3$. By KCL,

$$i = i_1 + i_3 = \frac{v_2}{3} + \frac{v_2}{3} = \frac{2v_2}{3}$$

By Ohm's law, $v_4 = 5i = 10v_2/3$. By KVL,

$$28 = v_1 + v_4 = \frac{4}{3}v_2 + \frac{10}{3}v_2 = \frac{14}{3}v_2 \quad \Rightarrow \quad v_2 = 6 \text{ V}$$

Drill Exercise 1.8

For the circuit shown in Fig. 1.44, when the value of both the 1-Ω and 5-Ω resistors are changed to 9 Ω, then $v_2 = 1.75$ V. (a) Use the voltage-divider formula to find v_1. (b) Find i_1. (c) Use the current-divider formula to find i.

ANSWER (a) 7 V, (b) 1.75 A, (c) 7/3 A

1.4 Independent and Dependent Sources

Combining Sources

We mentioned earlier that we will not be allowed to connect ideal voltage sources in parallel. However, consider the series connection of two ideal voltage sources as shown in Fig. 1.45. By KVL, we have that $v = v_1 + v_2$, and by the definition of an ideal voltage source, this must always be the voltage between nodes a and b, regardless of what is connected to them. Thus, a series connection of two voltage sources as shown in Fig. 1.45 (remember, no other elements can be connected to node c!) is equivalent to the ideal voltage source shown in Fig. 1.46. Clearly, the obvious generalization to m voltage sources connected in series holds.

In a dual manner, consider a parallel connection of ideal current sources as shown in Fig. 1.47. By KCL, we have that $i = i_1 + i_2$, and by the definition of an ideal

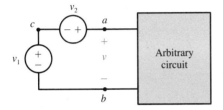

Fig. 1.45 Series connection of voltage sources.

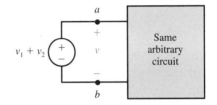

Fig. 1.46 Equivalent of voltage sources connected in series.

current source, this must always be the current into the arbitrary circuit. Therefore, a parallel connection of two current sources as shown in Fig. 1.47 (remember, other elements *can* be connected to nodes *a* and *b*!) is equivalent to the ideal current source shown in Fig. 1.48. Again, this result can be generalized to the case of *m* current sources connected in parallel.

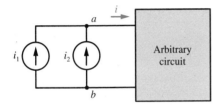

Fig. 1.47 Parallel connection of current sources.

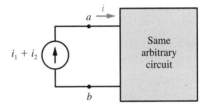

Fig. 1.48 Equivalent of current sources connected in parallel.

Example 1.9

Reconsider the circuit shown in Fig. 1.31 on p. 19. Combining the parallel connection of the current sources, we obtain the circuit given in Fig. 1.49. The equivalent resistance of the three resistors connected in parallel is R, where

$$\frac{1}{R} = \frac{1}{1} + \frac{1}{2} + \frac{1}{3} = \frac{6 + 3 + 2}{6} = \frac{11}{6} \quad \Rightarrow \quad R = \frac{6}{11} \, \Omega$$

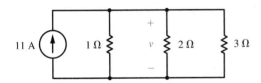

Fig. 1.49 Equivalent of circuit in Fig. 1.31.

In other words, the resistance loading the current source is $R = \frac{6}{11}\ \Omega$. By Ohm's law, we have that

$$v = \frac{6}{11}(11) = 6 \text{ V}$$

Drill Exercise 1.9

For the circuit shown Fig. 1.39*b* on p. 26, what is the value of the voltage obtained by combining the two batteries? What is the equivalent resistance loading this combined voltage source?

ANSWER -24 V, $12\ \Omega$

Dependent Sources

Up to this point, we have been considering voltage and current sources whose values are, in general, time dependent. However, these values were given to be independent of the behavior of the circuits to which the sources belonged. For this reason, we say that such sources are **independent**.

We now consider an ideal source, either voltage or current, whose value depends upon some variable (usually a voltage or current) in the circuit to which the source belongs. We call such an ideal source a **dependent** (or **controlled**) **source**, and represent it as shown in Fig. 1.50. Note that a dependent source is represented by a diamond-shaped symbol so as not to confuse it with an independent source.

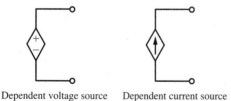

Dependent voltage source   Dependent current source

Fig. 1.50 Circuit symbols for dependent sources.

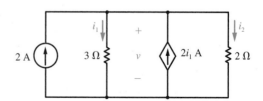

Fig. 1.51 Circuit with current-dependent current source.

Example 1.10

Consider the circuit shown in Fig. 1.51 in which the value of the dependent current source depends on the current i_1 through the 3-Ω resistor. Thus, we say that it is a **current-dependent current source**. Specifically, the value of the dependent source is $2i_1$, and the units are amperes, of course. (This implies that the constant 2 is dimensionless.) Hence, if the current though the 3-Ω resistor is $i_1 = 2$ A, then the dependent current source has a value of $2i_1 = 4$ A; if the current is $i_1 = -3$ A, then the dependent current source has a value of $2i_1 = -6$ A; and so on. To find the actual quantity i_1, we proceed as follows:

Applying KCL (at either node), we get

$$i_1 + i_2 = 2i_1 + 2 \quad \Rightarrow \quad -1i_1 + i_2 = 2 \tag{1.6}$$

By Ohm's law, $i_1 = v/3$ and $i_2 = v/2$. Substituting these expressions into Eq. 1.6, we get

$$-1\left(\frac{v}{3}\right) + \frac{v}{2} = 2 \quad \Rightarrow \quad v = 12 \text{ V}$$

Thus

$$i_1 = \frac{v}{3} = \frac{12}{3} = 4 \text{ A} \quad \text{and} \quad i_2 = \frac{v}{2} = \frac{12}{2} = 6 \text{ A}$$

and the value of the dependent current source is

$$2i_1 = 2(4) = 8 \text{ A}$$

Remember, if the independent source has a different value, then the quantity $2i_1$ will in general be different from 8 A.

Drill Exercise 1.10

For the circuit shown in Fig. 1.51, reverse the direction of the arrow in the independent current source. Find v, i_1, and i_2.

ANSWER -12 V, -4 A, -6 A

Example 1.11

Let us now consider the circuit given in Fig. 1.52. In this circuit, the value of the dependent current source is specified by a voltage! In other words, the value of the source is $0.5v$ amperes, where v is the *amount* of voltage across the 3-Ω resistor (also the 4-Ω resistor in this case). This implies that the constant 0.5 has as units A/V, or mhos—so it is a conductance. Such a device is a **voltage-dependent current source**.

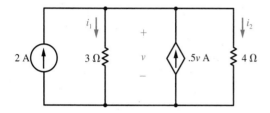

Fig. 1.52 Circuit with voltage-dependent current source.

To solve for v, we apply KCL and obtain

$$i_1 + i_2 = 0.5v + 2$$

Since $i_1 = v/3$ and $i_2 = v/4$, then

$$\frac{v}{3} + \frac{v}{4} = 0.5v + 2 \quad \Rightarrow \quad v = 24 \text{ V}$$

Consequently, the numerical value of the dependent current source is

$$0.5v = 0.5(24) = 12 \text{ A}$$

Again, we will point out that if the independent source has a value other than 2 A, then the dependent source's value of $0.5v$ amperes will be something other than 12 A.

The other variables in this circuit are

$$i_1 = \frac{v}{3} = \frac{24}{3} = 8 \text{ A} \quad \text{and} \quad i_2 = \frac{v}{4} = \frac{24}{4} = 6 \text{ A}$$

Drill Exercise 1.11

For the circuit given in Fig. 1.52, change the voltage-dependent current source to a current-dependent current source whose value is $3i_1/2$. (Do not change the direction of the arrow in the dependent source.) Find v, i_1, and i_2.

ANSWER 24 V, 8 A, 6 A

Example 1.12

The circuit shown in Fig. 1.53 contains a **current-dependent voltage source**. The value of this dependent voltage source is determined by the loop current i and is $2i$ V. In this case, the constant 2 has as its units V/A, or ohms—so it is a resistance.

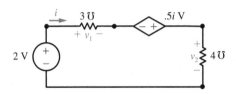

Fig. 1.53 Circuit with current-dependent voltage source.

Before we analyze this circuit, let us compare it with the circuit given in Fig. 1.52. What is a parallel connection there is a series connection here; what is a voltage (v) there is a current (i) here; what is a current there (2 A, i_1, i_2, $0.5v$ A) is a voltage here (2 V, v_1, v_2, $0.5i$ V, respectively); and what is a resistance there (3 Ω, 4 Ω) is a conductance here (3 ℧, 4 ℧, respectively). For this reason, we say that this circuit is the **dual** of the preceding circuit (and vice versa). Once a circuit has been analyzed, its dual is analyzed automatically. A more formal discussion of the concept of duality is given in Chapter 3. However, for the time being, let us verify it for this particular case.

By KVL,

$$v_1 - 0.5i + v_2 = 2$$

By Ohm's law, $v_1 = i/3$ and $v_2 = i/4$. Thus,

$$\frac{i}{3} - 0.5i + \frac{i}{4} = 2 \qquad \Rightarrow \qquad i = 24 \text{ A}$$

which is the numerical value for v in volts in the dual circuit (Fig. 1.52). Therefore, the value of the dependent source is

$$0.5i = 0.5(24) = 12 \text{ V}$$

Furthermore,

$$v_1 = \frac{i}{3} = \frac{24}{3} = 8 \text{ V} \quad \text{and} \quad v_2 = \frac{i}{4} = \frac{24}{4} = 6 \text{ V}$$

which are the respective values for i_1 and i_2 in the dual circuit.

Drill Exercise 1.12

For the circuit shown in Fig. 1.53, triple the value of the independent voltage source to 6 V. Find i, v_1, and v_2. How do these values compare to those obtained in Example 1.12?

ANSWER 72 A, 24 V, 18 V. The values are tripled.

The concept of dependent sources is not introduced simply for abstraction, but it is precisely this type of source that models the behavior of such important electronic devices as transistors and amplifiers.

Comments about Circuit Elements

Certain types of nonideal circuit elements, such as positive-valued resistors, come in a wide variety of values and are readily available in the form of discrete components. For many situations, the assumption that an actual resistor behaves as an ideal resistor is a reasonable one.

With a few exceptions, however, this is not the case with independent voltage and current sources. Although an actual battery often can be thought of as an ideal voltage source, other nonideal independent sources are approximated by a combination of circuit elements (some of which will be discussed later). Among these elements are the dependent sources, which are not discrete components, as are many resistors and batteries, but are in a sense part of electronic devices such as transistors and operational amplifiers (which consist of numerous transistors and resistors). But don't try to peel open a transistor's metal can (for those that are constructed that way) so that you can see a little diamond-shaped object. The dependent source is a theoretical element that is used to help describe or model the behavior of various electric devices.

In summary, although ideal circuit elements are not off-the-shelf circuit components, their importance lies in the fact that they can be interconnected (on paper or on a computer) to approximate actual circuits that are comprised of nonideal elements and assorted electrical components—thus allowing for the analysis of such circuits.

1.5 Instantaneous Power

Recall that electrons flow through a resistor from a given potential to a higher potential, and hence (positive-valued) current goes from a given potential to a lower one. Potential difference or voltage is a measure of work per unit charge (i.e., J/C). To obtain a current through an element, as shown in Fig. 1.54, it takes a certain amount of work or energy—and we say that this energy is absorbed by the element. By taking the product of voltage (energy per unit charge) and current (charge per unit time), we get a quantity that measures energy per unit time. Such a term is known as **power**. It is for this reason that we define $p(t)$, the **instantaneous power absorbed** by an element as in Fig. 1.54, as the product of voltage and current. That is,

$$p(t) = v(t)i(t)$$

The unit of power (J/s) is called the **watt**[7] (W). Given an expression for instantaneous power absorbed $p(t)$, to determine the power absorbed at time t_0, simply substitute the number t_0 into the expression.

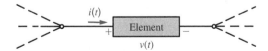

Fig. 1.54 Arbitrary element with conventional current direction and voltage polarity.

---

[7]Named for the Scottish inventor James Watt (1736–1819).

As with Ohm's law, when using the formula for power absorbed, we must always be conscious of both the voltage polarity and the current direction. For example, for the situation shown in Fig. 1.55, going back to the definition for instantaneous power absorbed, we have in this case

$$p_1(t) = -v_1(t)i_1(t)$$

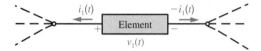

Fig. 1.55 Nonconventional current direction with respect to voltage polarity.

Since power absorbed in a given element can be either a positive or a negative quantity (depending on the relationship between voltage and current for the element), we can say that the element absorbs x watts, or, equivalently, that it **supplies** (or **delivers**) $-x$ watts.

Example 1.13

Consider the circuit in Fig. 1.56. Let us determine the (instantaneous) power absorbed by each of the elements for the case that $I = 10$ A.

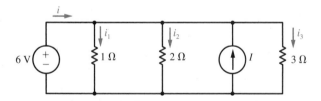

Fig. 1.56 Circuit containing a voltage and a current source.

Since all the elements are connected in parallel, then the voltage across each of the elements is 6 V. Therefore, by Ohm's law,

$$i_1 = \frac{6}{1} = 6 \text{ A}, \qquad i_2 = \frac{6}{2} = 3 \text{ A}, \qquad i_3 = \frac{6}{3} = 2 \text{ A}$$

and the powers absorbed by the 1-Ω, 2-Ω, and 3-Ω resistors are

$$p_1 = 6i_1 = 36 \text{ W}, \qquad p_2 = 6i_2 = 18 \text{ W}, \qquad p_3 = 6i_3 = 12 \text{ W}$$

respectively, for a total of

$$p_R = p_1 + p_2 + p_3 = 36 + 18 + 12 = 66 \text{ W}$$

absorbed by the resistors. By KCL,

$$i + 10 = i_1 + i_2 + i_3 = 6 + 3 + 2 = 11 \qquad \Rightarrow \qquad i = 1 \text{ A}$$

Thus the power absorbed by the voltage source is

$$p_V = -6i = -6 \text{ W}$$

in other words, the power supplied by the voltage source is 6 W. Also note that the power absorbed by the current source is

$$p_I = -6(10) = -60 \text{ W}$$

in other words, the power supplied by the current source is 60 W. Hence, the total power absorbed in the circuit is

$$p_R + p_V + p_I = 66 - 6 - 60 = 0 \text{ W}$$

Recalling from elementary physics that power is work (energy) per unit time, we see that the fact that the total power absorbed is zero is equivalent to saying that the principle of the conservation of energy is satisfied in this circuit (as it is in any circuit). In this case, the sources supplied all the power and the resistors absorbed it all.

Now, however, consider the situation in which $I = 12$ A. In this case

$$i_1 = 6 \text{ A}, \qquad i_2 = 3 \text{ A}, \qquad i_3 = 2 \text{ A}$$

and

$$p_1 = 36 \text{ W}, \qquad p_2 = 18 \text{ W}, \qquad p_3 = 12 \text{ W}$$
$$p_R = p_1 + p_2 + p_3 = 66 \text{ W}$$

as before. But by KCL,

$$i + 12 = i_1 + i_2 + i_3 \qquad \Rightarrow \qquad i = -1 \text{ A}$$

Therefore,

$$p_V = 6(1) = 6 \text{ W} \qquad \text{and} \qquad p_I = -6(12) = -72 \text{ W}$$

so that the total power absorbed is

$$p_R + p_V + p_I = 66 + 6 - 72 = 0 \text{ W}$$

and again power (energy) is conserved. However, in this case, not only do the resistors absorb power, but so does the voltage source.[8] It is the current source that supplies all the power absorbed in the circuit.

Drill Exercise 1.13

For the circuit shown in Fig. 1.56, determine the value of I for which $i = 0$ A. Find the power absorbed by each resistor, the voltage source, and the current source under this circumstance.

ANSWER 11 A, 36 W, 18 W, 12 W, 0 W; −66 W

Power Absorbed by Resistors

In all the examples worked so far, the reader may have noted that the power absorbed in every resistor was a nonnegative[9] number. As we now shall see, this is always the case.

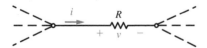

Fig. 1.57 Resistor with conventional current direction and voltage polarity.

Consider the resistor shown in Fig. 1.57. By definition, the power absorbed by the resistor is $p = vi$. But by Ohm's law, $v = Ri$. Thus, $p = (Ri)i$, or

---

[8]In practical terms, the voltage source (battery) is being charged.
[9]A **positive number** is a number that is greater than zero; a **nonnegative number** is a number that is greater than or equal to zero.

$$p = Ri^2$$

Also, $i = v/R$, so that $p = v(v/R)$, or

$$p = \frac{v^2}{R}$$

and both formulas for calculating the power absorbed by a resistor R demonstrate that p is always a nonnegative number when R is positive.

Example 1.14

For the circuit given in Fig. 1.56, the voltage across each resistor is 6 V. Therefore, regardless of the value of I, the power absorbed by the 1-Ω resistor is

$$p_1 = \frac{6^2}{1} = 36 \text{ W}$$

the power absorbed by the 2-Ω resistor is

$$p_2 = \frac{6^2}{2} = 18 \text{ W}$$

and the power absorbed by the 3-Ω resistor is

$$p_3 = \frac{6^2}{3} = 12 \text{ W}$$

A resistor always absorbs power. In a physical resistor, this power is dissipated as heat. In some types of resistors (e.g., an incandescent lamp or bulb, a toaster, or an electric space heater), this property is desirable in that the net result may be light or warmth. In other types of resistors, such as those in electronic circuits, the heat dissipated by a resistor may affect the operation of its circuit. In this case, heat dissipation cannot be ignored.

The common carbon resistor comes in values that range from less than 10 Ω to more than 10 MΩ. The physical size of such resistors will determine the amount of power that they can safely dissipate. These amounts are referred to as **power ratings**. Typical power ratings of electronic-circuit carbon resistors are $\frac{1}{8}, \frac{1}{4}, \frac{1}{2}$, 1, and 2 W. The dissipation of power that exceeds the rating of a resistor can damage the resistor physically. When an application requires the use of a resistor that must dissipate more than 2 W, another type—the wire-wound resistor—is often used. Metal-film resistors, although more expensive to construct, can have higher power ratings than carbon resistors and are more reliable and more stable. In addition to the lumped circuit elements such as carbon, wire-wound, and metal-film resistors, there are resistors in integrated-circuit (IC) form. These types of resistors, which dissipate only small amounts of power, are part of the countless ICs employed in present-day electronics.

S U M M A R Y

1. An ideal voltage source is a device that produces a specific (not necessarily constant with time) electric potential difference across its terminals regardless of what is connected to it.

2. An ideal current source is a device that produces a specific (not necessarily constant with time) current through it regardless of what is connected to it.

3. A (linear) resistor is a device in which the voltage across it is directly proportional to the current through it. If we write the equation (Ohm's law) describing this as $v = Ri$, then the constant of proportionality R is called resistance. If we write $i = Gv$, then G is known as conductance. Thus $R = 1/G$.

4. An open circuit is an infinite resistance and a short circuit is a zero resistance. The former is equivalent to an ideal current source whose value is zero; the latter is equivalent to an ideal voltage source whose value is zero.

5. At any node of a circuit, the currents algebraically sum to zero Kirchhoff's current law (KCL).

6. Around any loop in a circuit, the voltages algebraically sum to zero Kirchhoff's voltage law (KVL).

7. Current going into a parallel connection of resistors divides among them the smallest resistance has the most current through it.

8. Voltage across a series connection of resistors divides among them the largest resistance has the most voltage across it.

9. Resistances in a series behave as a single resistance whose value equals the sum of the individual resistances.

10. Conductances in parallel behave as a single conductance whose value equals the sum of the individual conductances.

11. Voltage sources in series can be combined into a single voltage source and current sources in parallel can be combined into a single current source.

12. Voltage and current sources can be dependent as well as independent.

13. A dependent source is a voltage or current source whose value depends on some other circuit variable.

14. The power absorbed by an element is the product of the current through it and the voltage across it.

Problems

1.1 An ideal voltage source is described by the function $v(t) = 10e^{-t}$ V. Find the value of this voltage source when (a) $t = 0$ s, (b) $t = 1$ s, (c) $t = 2$ s, (d) $t = 3$ s, (e) $t = 4$ s.

1.2 An ideal voltage source is described by the function $v(t) = 5 \sin (\pi/2)t$ V. Find the value of this voltage source when (a) $t = 0$ s, (b) $t = 1$ s, (c) $t = 2$ s, (d) $t = 3$ s, and (e) $t = 4$ s.

1.3 An ideal voltage source is described by the function $v(t) = 3 \cos (\pi/2)t$ V. Find the value of this voltage source when (a) $t = 0$ s, (b) $t = 1$ s, (c) $t = 2$ s, (d) $t = 3$ s, and (e) $t = 4$ s.

1.4 Find the current in a region when the total charge in the region is described by the function (a) $q(t) = 4e^{-2t}$ C, (b) $q(t) = 3 \sin \pi t$ C, (c) $q(t) = 6 \cos 2\pi t$ C, and (d) $q(t) = 5e^{-4t} \cos 3t$ C.

1.5 An ideal voltage source is described by the function shown in Fig. P1.5. Find the value of this voltage source when (a) $t = 0$ s, (b) $t = 1$ s, (c) $t = 2$ s, (d) $t = 3$ s, and (e) $t = 4$ s.

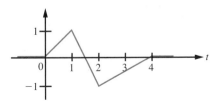

Fig. P1.5

1.6 The total charge $q(t)$ in some region is described by the function shown in Fig. P1.5. Sketch the current $i(t)$ in this region.

1.7 Consider the circuit shown in Fig. P1.7. (a) Given $i_1 = 4$ A, find v_1. (b) Given $i_2 = -2$ A, find v_2. (c) Given $i_3 = 2$ A, find v_3. (d) Given $i_4 = -2$ A, find v_4.

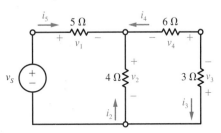

Fig. P1.7

1.8 Consider the circuit in Fig. P1.7. (a) Given $v_1 = 30$ V, find i_1. (b) Given $v_2 = 12$ V, find i_2. (c) Given $v_3 = -9$ V, find i_3. (d) Given $v_4 = -3$ V, find i_4.

1.9 Consider the circuit shown in Fig. P1.7. (a) Given $v_1 = -10$ V, find i_1. (b) Given $i_2 = 1$ A, find v_2. (c) Given $v_3 = 3$ V, find i_3. (d) Given $i_4 = 1$ A, find v_4.

1.10 Consider the circuit in Fig. P1.10. (a) Given $v_1 = -6$ V, find i_1. (b) Given $v_2 = 24$ V, find i_2. (c) Given $v_3 = 11$ V, find i_3. (d) Given $v_4 = 21$ V, find i_4. (e) Given $v_5 = -14$ V, find i_5.

1.11 Consider the circuit shown in Fig. P1.10. (a) Given $i_1 = 1.5$ A, find v_1. (b) Given $i_2 = -4$ A, find v_2. (c) Given $i_3 = 5.5$ A, find v_3. (d) Given $i_4 = 3.5$ A, find v_4. (e) Given $i_5 = 3.5$ A, find v_5.

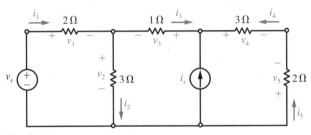

Fig. P1.10

1.12 Consider the circuit shown in Fig. P1.12. (a) Given $i_1 = -4$ A, find v_1. (b) Given $i_2 = 1$ A, find v_2. (c) Given $i_3 = 1$ A, find v_3. (d) Given $i_4 = 2$ A, find v_4.

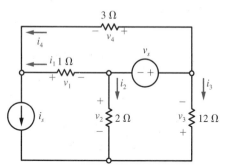

Fig.P1.12

1.13 Consider the circuit in Fig. P1.12. (a) Given $v_1 = -2$ V, find i_1. (b) Given $v_2 = -1$ V, find i_2. (c) Given $v_3 = -6$ V, find i_3. (d) Given $v_4 = 3$ V, find i_4.

1.14 Consider the circuit in Fig. P1.14. (a) Given $i_1 = 3$ A and $v_1 = 6$ V, find R_1. (b) Given $i_2 = 3$ A and $v_2 = -15$ V, find R_2. (c) Given $i_3 = -2$ A and $v_3 = 6$ V, find R_3. (d) Given $i_4 = -1$ A and $v_3 = 6$ V, find R_4.

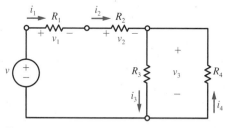

Fig. P1.14

1.15 Consider the circuit in Fig. P1.14. (a) Given $i_1 = 6$ A and $v_1 = 18$ V, find R_1. (b) Given $i_2 = 6$ A and $v_2 = -36$ V, find R_2. (c) Given $i_3 = 4$ A and $v_3 = 16$ V, find R_3. (d) Given $i_4 = -2$ A and $v_3 = 16$ V, find R_4.

1.16 For the circuit shown in Fig. P1.16, find v when (a) $i_s = 1$ A, (b) $i_s = 2$ A, (c) $i_s = 3$ A.

Fig. P1.16

1.17 For the circuit shown in Fig. P1.17, find i when (a) $v_s = 1$ V, (b) $v_s = 2$ V, (c) $v_s = 3$ V.

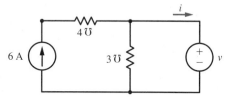

Fig. P1.17

1.18 For the circuit shown in Fig. P1.18, find v_4 when (a) $v_s = 2$ V, (b) $v_s = 4$ V, (c) $v_s = 6$ V.

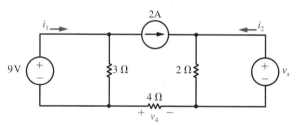

Fig. P1.18

1.19 For the circuit shown in Fig. P1.19, suppose that $i_1 = 6$ A. Use the current-divider formula to determine i_2, i_3, i_4, and i_5.

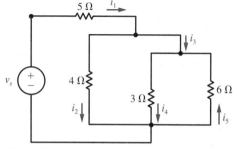

Fig. P1.19

1.20 For the circuit shown in Fig. P1.19, suppose that $i_4 = 4$ A. Use the current-divider formula to determine i_1, i_2, i_3, and i_5.

1.21 For the circuit shown in Fig. P1.19, suppose that $i_2 = -2$ A. Use the current-divider formula to determine i_1, i_3, i_4, and i_5.

1.22 For the circuit given in Fig. P1.19, suppose that $i_5 = 4$ A. Use the current-divider formula to determine i_1, i_2, i_3, and i_4.

1.23 For the circuit shown in Fig. P1.23, suppose that $i_1 = 2$ A. Find v for the case that (a) $i_2 = 1$ A, (b) $i_2 = 2$ A, and (c) $i_2 = 3$ A.

1.24 Consider the circuit shown in Fig. P1.23. Find v when (a) $i_1 = 12$ A and $i_2 = 6$ A, (b) $i_1 = 6$ A and $i_2 = 6$ A, (c) $i_1 = 6$ A and $i_2 = 12$ A.

1.25 Find the variables indicated for the circuits shown in Fig. P1.25.

1.26 Find the variables indicated for the circuits shown in Fig. P1.26. (See p. 48.)

1.27 Find the variables indicated for the circuits shown in Fig. P1.27. (See p. 48.)

1.28 For the circuit shown in Fig. P1.28, find the variables indicated when R is (a) 2 Ω, (b) 4 Ω, and (c) 6 Ω.

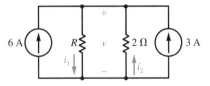

Fig. P1.28

1.29 For the circuit shown in Fig. P1.29, find the variables indicated when R is (a) 2 Ω, (b) 4 Ω, and (c) 6 Ω.

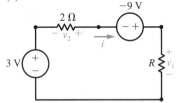

Fig. P1.29

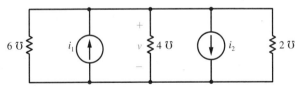

Fig. P1.23

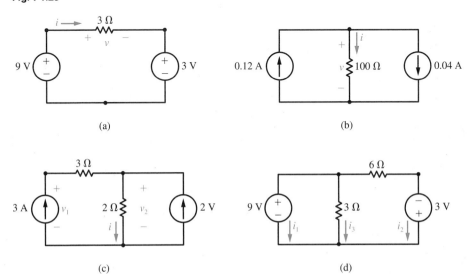

(a)

(b)

(c)

(d)

Fig. P1.25 *a-d*

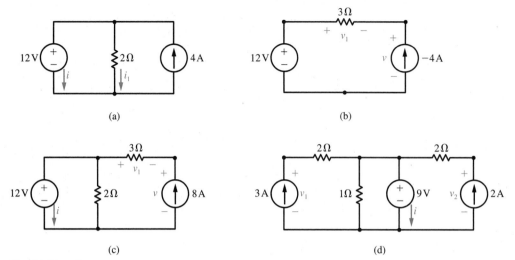

(a)

(b)

(c)

(d)

Fig. P1.26 a-d

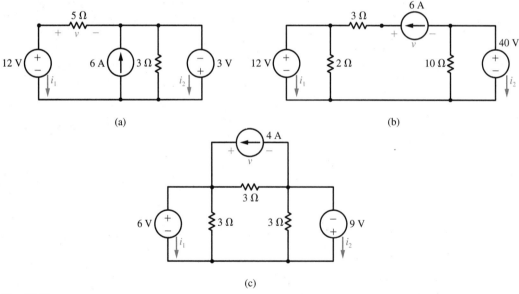

(a)

(b)

(c)

Fig. P1.27 a-c

1.30 Find v and i for the series-parallel circuit shown in Fig. P1.30.

1.31 Find v and i for the series-parallel circuit shown in Fig. P1.31.

1.32 Consider the circuit shown in Fig. P1.32. (a) Find i, v_1, v_2, and v_3. (b) Remove the short circuit between a and b (erase it), and find i, v_1, and v_2. (Don't try to find v_3—it can't be done!)

1.33 Consider the series-parallel circuit shown in Fig. P1.33. (a) Find V_s when $v_1 = 2$ V. (b) Find V_s when $i_3 = 3$ A. (c) Find V_s when $i_5 = 4$ A. (d) What is the resistance $R_{eq} = V_s/i$ loading the battery for part (a)? For part (b)? For part (c)?

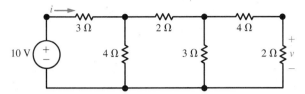

Fig. P1.30

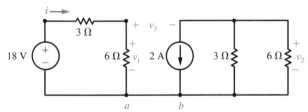

Fig. P1.31

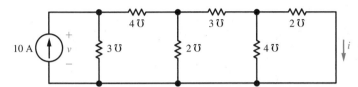

Fig. P1.32

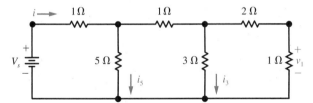

Fig. P1.33

1.34 Consider the nonseries-parallel circuit shown in Fig. P1.34. (a) When $R = \frac{1}{2}\ \Omega$, then $v_1 = 6$ V. Determine the resistance $R_{eq} = V_s/i$ loading the battery.

1.35 Consider the nonseries-parallel circuit shown in Fig. P1.34. When $R = 4\ \Omega$, then $v_1 = 4$ V. Determine the resistance $R_{eq} = V_s/i$ loading the battery.

1.36 Consider the nonseries-parallel circuit shown in Fig. P1.34. Determine R and the resistance $R_{eq} = V_s/i$ loading the battery when $v_1 = 3$ V.

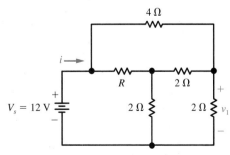

Fig. P1.34

1.37 The nonseries-parallel circuit shown in Fig. P1.37 is known as a **twin-T network**. (a) When $R_1 = 1\ \Omega$ and $R_2 = 3\ \Omega$, then $v_2 = 6$ V. Determine the resistance $R_{eq} = V_s/i$ loading the battery.

1.38 For the twin-T network shown in Fig. P1.37, suppose that $R_2 = \frac{3}{4}\ \Omega$ and $v_2 = 3$ V. Determine R_1 and the resistance $R_{eq} = V_s/i$ loading the battery.

1.39 Shown in Fig. P1.39 is a nonseries-parallel connection known as a **bridge circuit**. When $R_1 = 10\ \Omega$ and $R_2 = 1\ \Omega$, then $v_1 = 10$ V. Find v_2, i, v_3, and the resistance $R_{eq} = V_s/i$ loading the voltage source.

1.40 For the bridge circuit shown in Fig. P1.39, when $R_1 = 2\ \Omega$ and $R_2 = 4\ \Omega$, then $v_1 = 4$ V. Find v_2, i, v_3, and the resistance $R_{eq} = V_s/i$ loading the voltage source.

1.41 For the bridge circuit shown in Fig. P1.39, when the current $i = 0$ A, we say that the bridge is **balanced**. Under what condition (find an expression relating R_1 and R_2) will this bridge be balanced?

1.42 For the circuit shown in Fig. P1.42, find i_1 when (a) $K = 2$, (b) $K = 3$, and (c) $K = 4$.

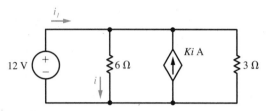

Fig. P1.42

1.43 The circuit shown in Fig. P1.43 contains a **voltage-dependent voltage source** as well as a current-dependent current source. Find i_1 when (a) $K = -3$, (b) $K = -1.5$, and (c) $K = 1.5$.

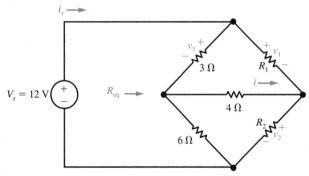

Fig. P1.37

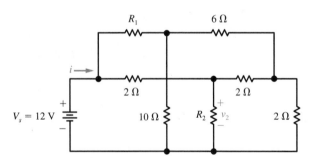

Fig. P1.39

1.44 Consider the circuit shown in Fig. P1.44. Find v when (a) $K = 2$, and (b) $K = 4$.

1.45 Consider the circuit shown in Fig. P1.45. Find i when (a) $K = 2$, and (b) $K = 4$.

1.46 Consider the circuit shown in Fig. P1.46. (a) Find the resistance $R_{eq} = v_1/i_1$. (b) Find the voltage v_2 in terms of the applied voltage v_1.

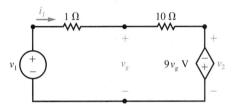

Fig. P1.46

1.47 Consider the circuit shown in Fig. P1.47. (a) Find the resistance $R_{eq} = v_1/i_1$. (b) Use voltage division to find v in terms of v_g. (c) Find the voltage v_2 in terms of the applied voltage v_1.

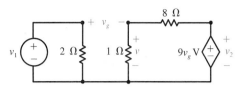

Fig. P1.47

1.48 For the circuit shown in Fig. P1.48, suppose that $R = 10\ \Omega$. Determine (a) v_s, and (b) $R_{eq} = v_s/i_s$.

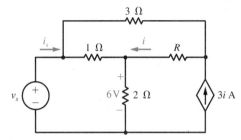

Fig. P1.48

1.49 For the circuit shown in Fig. P1.48, suppose that $R = 8\ \Omega$. Determine (a) v_s, and (b) $R_{eq} = v_s/i_s$.

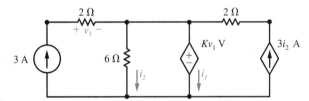

Fig. P1.43

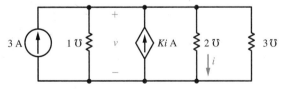

Fig. P1.44

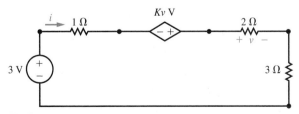

Fig. P1.45

1.50 For the circuit shown in Fig. P1.50, suppose that $R = 5 \, \Omega$. Determine (a) i_s, and (b) $R = v_s/i_s$.

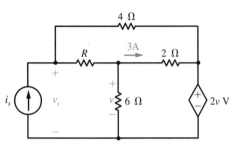

Fig. P1.50

1.51 For the circuit shown in Fig. P1.50, suppose that $R = 3 \, \Omega$. Determine (a) i_s, and (b) $R_{eq} = v_s/i_s$.

1.52 The circuit shown in Fig. P1.52 is a single field-effect transistor (FET) amplifier in which the input is v_1 and the output is v_2. The portion of the circuit in the shaded box is an approximate model of an FET. (a) Find v_{gs} in terms of v_1. (b) Find v_2 in terms of v_1. (c) Find v_2 when $v_1 = 0.1 \cos 120\pi t$ V.

1.53 The circuit shown in Fig. P1.53 is a single bipolar junction transistor (BJT) amplifier in which the input is v_1 and the output is v_2. The portion of the circuit in the shaded box is an approximate model of a BJT in the common-emitter configuration. (a) Find i_b in terms of the input voltage v_1. (b) Find the output voltage v_2 in terms of v_1. (c) Find v_2 when $v_1 = 0.1 \cos 120\pi t$ V.

1.54 The circuit shown in Fig. P1.54 is another single bipolar junction transistor (BJT) amplifier in which the input is v_1 and the output is v_2. The portion in the shaded box is an approximate model of a BJT in the common-base configuration. (a) Find i_e in terms of the input voltage v_1. (b) Find the output voltage v_2 in terms of v_1. (c) Find v_1 when $v_1 = 0.1 \cos 120\pi t$ V.

1.55 For the circuit given in Fig. 1.51 on p. 34, $v = 12$ V, $i_1 = 4$ A, and $i_2 = 6$ A. Determine the power absorbed by each element in the circuit.

1.56 For the circuit given in Fig. 1.52 on p. 36, $v = 24$ V. Determine the power absorbed by each element in the circuit.

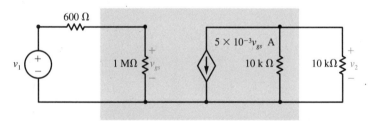

Fig. P1.52

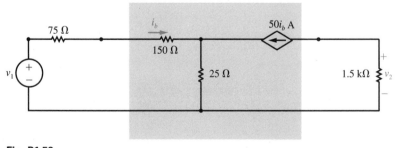

Fig. P1.53

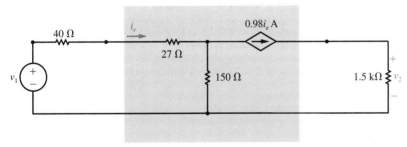

Fig. P1.54

1.57 For the circuit given in Fig. 1.53 on p. 37, $i = 24$ A. Determine the power absorbed by each element in the circuit.

1.58 For the circuit given in Fig. P1.42, determine the power absorbed by each element when (a) $K = 2$, and (b) $K = -2$.

1.59 For the circuit shown in Fig. P1.44, determine the power absorbed by each element given that (a) $K = 2$ and $v = 1.5$ V; (b) $K = 4$ and $v = -1.5$ V.

1.60 For the circuit shown in Fig. P1.45, determine the power absorbed by each element given that (a) $K = 2$ and $i = 1.5$ A; (b) $K = 4$ and $i = -1.5$ A.

2

Circuit Analysis Principles

INTRODUCTION

The process by which we determine a variable (either a voltage or a current) of a circuit is called **analysis**. Up to this point, we have been dealing with circuits that are relatively uncomplicated. Don't be fooled, though, because some very simple circuits can be quite useful. As part of the set of problems, we have already come across some simple single-stage amplifier circuits whose analysis was accomplished by applying the basic principles covered to date—Ohm's law, KCL, and KVL. Nonetheless, we will want to analyze more complicated circuits—circuits for which it is simply not possible to write and solve a single equation that has only one unknown. Instead, we will have to write and solve a set of linear algebraic equations. To obtain such equations, we again utilize Ohm's law, KCL, and KVL.

There are two distinct approaches that we will take. In one, we will write a set of simultaneous equations in which the variables are voltages, known as **nodal analysis**. In the other, we will write a set of simultaneous equations in which the variables are currents, known as **mesh analysis** (also called **loop analysis**). Although nodal analysis can be used for any circuit, mesh analysis is valid only for **planar networks**—that is, circuits that can be drawn in a two-dimensional plane in such a way that no element crosses over another.

A very important circuit element is the operational amplifier. Because of the evolution of integrated-circuit (IC) technology, the operational amplifier (or *op amp*) is both small in size and inexpensive. Its versatility and usefulness have made it extremely popular. In this chapter, we study the operational amplifier from the point of view of an ideal circuit element, and a number of applications are presented. A discussion of the operational amplifier from the electronics point of view is given in Chapter 10.

An important circuit concept is Thévenin's theorem. This result says, in essence, that an arbitrary circuit behaves as an appropriately valued voltage source in series with an appropriately valued resistance to the outside world. A major consequence of this fact is the determination of the maximum power that can be delivered to a load and the condition for which this occurs.

When a circuit contains more than one independent source, a response (a voltage or a current) of the circuit can be obtained by finding the response to each individual independent source and then summing these individual responses. This notion, known as the principle of superposition, is another important circuit concept that is frequently used.

Although in this chapter we consider **resistive circuits**—that is, circuits that contain only resistors and sources (both independent and dependent)—we will see later that the same techniques are applicable to networks that contain other types of elements as well.

2.1 Nodal Analysis

Given a circuit to be analyzed, the first step in employing nodal analysis is the arbitrary choice of one of the nodes of the circuit as the **reference** (or **datum**) **node**. Although this node can be indicated by any of the three symbols shown in Fig. 2.1, we will use the symbol depicted in Fig. 2.1*a* exclusively.

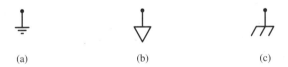

(a)   (b)   (c)

Fig. 2.1 Symbols used to indicate a reference node.

We can label any node in a circuit with its potential with respect to the reference node in the following manner.

Suppose that the voltages between nodes *a*, *c*, and the reference node are as shown in Fig. 2.2. Then, Fig. 2.3 indicates how the nodes are labeled. Of course, since the potential difference between the reference node and itself must be zero volts, we can mark the reference node "0 V" if we wish, but this is redundant.

It is possible and easy to express the voltage between any pair of nodes in terms of the potentials (with respect to the reference node) of those two nodes. Suppose

Fig. 2.2 Some node voltages.   **Fig. 2.3** Node voltage designations.

that we wish to determine the voltage v_{ac} between nodes a and c, with the plus at node a and the minus at node c, in terms of v_a and v_c. This situation is depicted in Fig. 2.4. Since the labeling in Fig. 2.5 is equivalent to Fig. 2.4, by KVL we have that $v_{ac} = v_a - v_c$. Similarly, the voltage v_{ca} between nodes a and c, with the plus at node c and the minus at node a, is $v_{ca} = v_c - v_a$.

Fig. 2.4 Voltage between nonreference nodes.   **Fig. 2.5** Determination of voltage between nonreference nodes.

With these ideas in mind, let us see how to employ nodal analysis for the circuit shown in Fig. 2.6. After choosing the reference node arbitrarily, the two nonreference nodes are labeled v_1 and v_2. The next step is to indicate currents (the names and directions are arbitrary) through elements that do not already have currents indicated through them. For this example, the values and directions of the current sources are given whereas i_1, i_2, and i_3 were chosen arbitrarily.

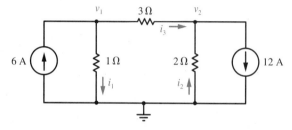

Fig. 2.6 Example of nodal analysis.

Now apply KCL at the node labeled v_1. The result is

$$i_1 + i_3 = 6 \tag{2.1}$$

By Ohm's law, $i_1 = v_1/1$ and $i_3 = (v_1 - v_2)/3$. Substituting these into Eq. 2.1 yields

$$\frac{v_1}{1} + \frac{v_1 - v_2}{3} = 6 \quad \Rightarrow \quad 4v_1 - v_2 = 18 \tag{2.2}$$

When we apply KCL at node v_2 (i.e., the node labeled v_2), we obtain

$$i_2 + i_3 = 12$$

and by Ohm's law $i_2 = -v_2/2$, this equation becomes

$$-\frac{v_2}{2} + \frac{v_1 - v_2}{3} = 12 \quad \Rightarrow \quad 2v_1 - 5v_2 = 72 \tag{2.3}$$

Equations 2.2 and 2.3 are simultaneous, linear equations in the unknowns v_1 and v_2. Multiplying Eq. 2.3 by 2 and subtracting the result from Eq. 2.2, we get

$$9v_2 = -126 \quad \Rightarrow \quad v_2 = -14 \text{ V}$$

Substituting this value of v_2 in Eq. 2.2, we obtain

$$4v_1 - (-14) = 18 \quad \Rightarrow \quad v_1 = 1 \text{ V}$$

Having determined the values of v_1 and v_2, it is a routine matter to determine i_1, i_2, and i_3. By Ohm's law,

$$i_1 = \frac{v_1}{1} = 1 \text{ A}, \qquad i_2 = \frac{-v_2}{2} = 7 \text{ A}, \qquad i_3 = \frac{v_1 - v_2}{3} = \frac{1 + 14}{3} = 5 \text{ A}$$

Now that we have seen an example of the application of nodal analysis, we can state the rules of nodal analysis for circuits that do not contain voltage sources. We will discuss the case of circuits with voltage sources immediately thereafter.

Nodal Analysis for Circuits with No Voltage Sources

Given a circuit with n nodes and no voltage sources, proceed as follows:

1. Select any node as the reference node.

2. Label the remaining $n - 1$ nodes (e.g., $v_1, v_2 \ldots, v_{n-1}$).

3. Arbitrarily assign currents to the elements in which no current is designated.

4. Apply KCL at each nonreference node.

5. Use Ohm's law to express the currents through resistors in terms of the node voltages, and substitute these expressions into the current equations obtained in Step 4.

6. Solve the resulting set of $n - 1$ simultaneous equations for the node voltages.

Having seen an example of nodal analysis for a circuit without a voltage source, let us now consider a circuit in which a voltage source is present.

Example 2.1

Figure 2.7 shows a circuit that contains a 3-V independent voltage source, as well as a dependent current source whose value depends on the voltage across the 6-Ω resistor which is drawn vertically. This circuit has four nodes—one is the reference node and the other three are labeled v_1, v_2, and v_3. The directions of the currents i_1, i_2, i_3, and i_4 through the resistors were chosen arbitrarily.

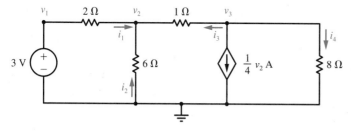

Fig. 2.7 Nodal analysis of a circuit with an independent voltage source.

Because there is a voltage source connected between node v_1 and the reference node, we do not have to apply KCL[1] at node v_1. By inspection we have that

---

[1]Since there is no constraint on the current through a voltage source, KCL should not be applied at node v_1 at this time.

$v_1 = 3$ V This is the equation that is obtained at node v_1. Thus, essentially, there are only two unknowns, v_2 and v_3. To determine these, we proceed as discussed previously.

At node v_2, by KCL,

$$i_1 + i_2 + i_3 = 0$$

By Ohm's law, this equation becomes

$$\frac{v_1 - v_2}{2} + \frac{-v_2}{6} + \frac{v_3 - v_2}{1} = 0 \quad \Rightarrow \quad -10v_2 + 6v_3 = -9 \qquad (2.4)$$

At node v_3, by KCL,

$$\frac{1}{4}v_2 + i_3 + i_4 = 0$$

and so by Ohm's law,

$$\frac{1}{4}v_2 + \frac{v_3 - v_2}{1} + \frac{v_3}{8} = 0 \quad \Rightarrow \quad -6v_2 + 9v_3 = 0 \qquad (2.5)$$

From Eq. 2.5, $v_3 = 2v_2/3$. Substituting this into Eq. 2.4, we get

$$-10v_2 + 6\left(\frac{2v_2}{3}\right) = -9 \quad \Rightarrow \quad v_2 = \frac{3}{2} = 1.5 \text{ V} \qquad (2.6)$$

and hence

$$v_3 = \frac{2v_2}{3} = \frac{2(1.5)}{3} = 1 \text{ V}$$

Having completed the nodal analysis for the circuit, that is having found all of the circuit's node voltages, we can now calculate the currents. By Ohm's law

$$i_1 = \frac{v_1 - v_2}{2} = \frac{3 - 1.5}{2} = 0.75 \text{ A}, \qquad i_2 = \frac{-v_2}{6} = \frac{-1.5}{6} = -0.25 \text{ A}$$

$$i_3 = \frac{v_3 - v_2}{1} = \frac{1 - 1.5}{1} = -0.5 \text{ A}, \qquad i_4 = \frac{v_3}{8} = \frac{1}{8} = 0.125 \text{ A}$$

Thus, the resistance loading the voltage source is $R = 3/i_1 = 3/0.75 = 4\ \Omega$.

Drill Exercise 2.1

For the circuit shown in Fig. 2.7, replace the dependent current source with a voltage-dependent voltage source (+ on top) whose value is $2v_2$ V. Use nodal analysis to find v_1, v_2, and v_3. What is the current, directed down, through the dependent source? What is the resistance loading the independent voltage source?

ANSWER   3 V, -4.5 V, -9 V, 5.63 A, 0.8 Ω

Voltage Sources Between Nonreference Nodes

Up to this point, when we employed nodal analysis, a voltage source in a circuit was connected between a nonreference node and the reference node. For the case that a voltage source is connected between two nonreference nodes, say v_a and v_b, and no other voltage source is connected to either of these two nodes, assign a current, say i, through the voltage source in question. Apply KCL at nodes v_a and v_b, and then combine the two equations to eliminate i. Thus, instead of getting two equations from nodes v_a and v_b, only one is obtained. However, we can get the other equation from KVL by expressing the value of the voltage source in terms of v_a and v_b.

Example 2.2

For the circuit shown in Fig. 2.8, the 4-V source is connected between two nonreference nodes.

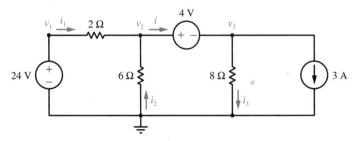

Fig. 2.8 Circuit with voltage source connected between two nonreference nodes.

By inspection, $v_1 = 24$ V. By KCL at node v_2,

$$i_1 + i_2 = i \qquad (2.7)$$

By Ohm's law, $i_1 = (24 - v_2)/2$ and $i_2 = -v_2/6$. However, there is no constraint on the current through a voltage source. Therefore, we cannot express i directly in terms of v_2 and v_3.

By KCL at node v_3,

$$i = i_3 + 3 \qquad (2.8)$$

where $i_3 = v_3/8$. Again, we cannot express i directly in terms of v_2 and v_3. However, by combining Eq. 2.7 and 2.8, we get

$$i_1 + i_2 = i_3 + 3$$

and by Ohm's law this equation becomes

$$\frac{24 - v_2}{2} + \frac{-v_2}{6} = \frac{v_3}{8} + 3 \quad \Rightarrow \quad 16v_2 + 3v_3 = 216 \qquad (2.9)$$

So far we have one equation and two unknowns. We need another equation in the unknowns v_2 and v_3. Such an equation can be obtained from the 4-V source itself. Specifically, by KVL,

$$v_2 - v_3 = 4 \qquad (2.10)$$

and this is the needed equation.

The solution to Eq. 2.9 and 2.10 is $v_2 = 12$ V and $v_3 = 8$ V. Using these voltages, we may now calculate the circuit's currents. In particular,

$$i_1 = \frac{24 - 12}{2} = 6 \text{ A}, \qquad i_2 = \frac{-12}{6} = -2 \text{ A}$$

$$i = i_1 + i_2 = 6 - 2 = 4 \text{ A}, \qquad i_3 = \frac{8}{8} = 1 \text{ A}$$

Drill Exercise 2.2

For the circuit shown in Fig. 2.8, change the value of the 4-V source to 23 V. Use nodal analysis to find v_1, v_2, and v_3. Determine the currents for the circuit.

ANSWER   24 V, 15 V, -8 V, 4.5 A, -2.5 A, 2 A, -1 A

Circuits with more than one voltage source connected between nonreference nodes are treated in a manner similar to that already discussed.

2.2 Determinants and Cramer's Rule

For the analysis of circuits previously encountered, we had situations in which we were required to solve two equations in two unknowns. Although the approaches taken were slightly different, there is yet another technique, known as **Cramer's rule**, that can be employed. To see how this is done, first consider the pair of simultaneous equations,

$$a_1x_1 + a_2x_2 = d_1$$
$$b_1x_1 + b_2x_2 = d_2$$

where the coefficients a_1, a_2, b_1, b_2 are given, as are d_1, d_2, and the unknowns x_1, x_2, are to be determined.

The **determinant** D of the coefficients a_1, a_2, b_1, b_2, is defined by

$$D = \begin{vmatrix} a_1 & a_2 \\ b_1 & b_2 \end{vmatrix} = a_1b_2 - b_1a_2$$

To use Cramer's rule, two more determinants D_1 and D_2 must be calculated. To find D_1, replace coefficients a_1 and b_1 by d_1 and d_2, respectively, that is,

$$D_1 = \begin{vmatrix} d_1 & a_2 \\ d_2 & b_2 \end{vmatrix} = d_1b_2 - d_2a_2$$

To find D_2, replace the coefficients a_2 and b_2 by d_1 and d_2 respectively. That is,

$$D_2 = \begin{vmatrix} a_1 & d_1 \\ b_1 & d_2 \end{vmatrix} = a_1d_2 - b_1d_1$$

By Cramer's rule,

$$x_1 = \frac{D_1}{D} \quad \text{and} \quad x_2 = \frac{D_2}{D}$$

Example 2.3

Consider Eq. 2.2 and 2.3, which are now repeated.

$$4v_1 - v_2 = 18 \tag{2.2}$$

$$2v_1 - 5v_2 = 72 \tag{2.3}$$

Forming determinants, we have

$$D = \begin{vmatrix} 4 & -1 \\ 2 & -5 \end{vmatrix} = (4)(-5) - (2)(-1) = -18$$

$$D_1 = \begin{vmatrix} 18 & -1 \\ 72 & -5 \end{vmatrix} = (18)(-5) - (72)(-1) = -18$$

$$D_2 = \begin{vmatrix} 4 & 18 \\ 2 & 72 \end{vmatrix} = (4)(72) - (2)(18) = 252$$

and by Cramer's rule,

$$v_1 = \frac{D_1}{D} = \frac{-18}{-18} = 1 \text{ V} \qquad \text{and} \qquad v_2 = \frac{D_2}{D} = \frac{252}{-18} = -14 \text{ V}$$

Drill Exercise 2.3

For the two equations $10v_1 - 6v_2 = 9$ and $2v_1 - v_2 = 0$, use Cramer's rule to find D, D_1, D_2, v_1, and v_2.

ANSWER 2, −9, −18, −4.5 V, −9 V

For the case of three equations and three unknowns

$$a_1x_1 + a_2x_2 + a_3x_3 = d_1$$
$$b_1x_1 + b_2x_2 + b_3x_3 = d_2$$
$$c_1x_1 + c_2x_2 + c_3x_3 = d_3$$

the determinant D is given by

$$D = \begin{vmatrix} a_1 & a_2 & a_3 \\ b_1 & b_2 & b_3 \\ c_1 & c_2 & c_3 \end{vmatrix} = a_1b_2c_3 + a_2b_3c_1 + a_3b_1c_2 - a_3b_2c_1 - a_1b_3c_2 - a_2b_1c_3$$

A simple way to remember this formula is to repeat the first two columns and then put in solid and dashed lines as follows:

$$\begin{vmatrix} a_1 & a_2 & a_3 & a_1 & a_2 \\ b_1 & b_2 & b_3 & b_1 & b_2 \\ c_1 & c_2 & c_3 & c_1 & c_2 \end{vmatrix}$$

The solid lines indicate the positive products in D and the dashed lines indicate the negative products.

To obtain D_1, replace the column consisting of a_1, b_1, c_1 with the column consisting of d_1, d_2, d_3 and the resulting determinant. Similarly, D_2 is obtained by replacing a_2, b_2, c_2 with d_1, d_2, d_3; and D_3 is obtained by replacing a_3, b_3, c_3 with d_1, d_2, d_3. Again, by Cramer's rule,

$$x_1 = \frac{D_1}{D} \qquad x_2 = \frac{D_2}{D} \qquad x_3 = \frac{D_3}{D}$$

Just as Cramer's rule can be applied to the cases of two simultaneous equations and three simultaneous equations, so can it be employed for the case of four or more simultaneous equations. However, this subject is beyond the scope of this book.

Example 2.4

Consider the circuit shown in Fig. 2.9, which contains two voltage sources—one independent and one dependent. In this circuit, the reference node and the node voltages have been labeled. Even though currents have not been explicitly indicated, let us go ahead and employ nodal analysis.

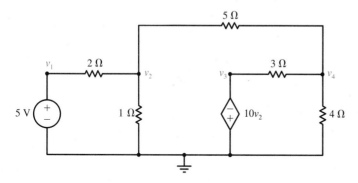

Fig. 2.9 Nodal analysis of a circuit with two voltage sources.

Since there is a voltage source between node v_1 and the reference node, by inspection we have

$$v_1 = 5 \text{ V}$$

Summing the currents out of node v_2 to zero, by KCL,

$$\frac{v_2 - 5}{2} + \frac{v_2}{1} + \frac{v_2 - v_4}{5} = 0 \qquad \Rightarrow \qquad 17v_2 - 2v_4 = 25 \qquad (2.11)$$

Since there is a voltage source (even though it is a dependent source) between node v_3 and the reference node, we again obtain, by inspection,

$$v_3 = -10v_2 \quad \Rightarrow \quad 10v_2 + v_3 = 0 \tag{2.12}$$

Summing the currents into node v_4 to zero, we get

$$\frac{v_2 - v_4}{5} + \frac{v_3 - v_4}{3} + \frac{-v_4}{4} = 0 \quad \Rightarrow \quad -12v_2 - 20v_3 + 47v_4 = 0 \tag{2.13}$$

The three equations in three unknowns that must be solved are given in Eq. 2.11, 2.12, and 2.13. Again, these equations are

$$17v_2 + 0v_3 - 2v_4 = 25$$
$$10v_2 + 1v_3 + 0v_4 = 0$$
$$-12v_2 - 20v_3 + 47v_4 = 0$$

Calculating determinants:

$$D = \begin{vmatrix} 17 & 0 & -2 \\ 10 & 1 & 0 \\ -12 & -20 & 47 \end{vmatrix} \begin{aligned} &= (17)(1)(47) + (0)(0)(-12) + (-2)(10)(-20) \\ &\quad - (-12)(1)(-2) - (-20)(0)(17) - (47)(0)(10) \\ &= 1175 \end{aligned}$$

$$D_2 = \begin{vmatrix} 25 & 0 & -2 \\ 0 & 1 & 0 \\ 0 & -20 & 47 \end{vmatrix} \begin{aligned} &= (25)(1)(47) + (0)(0)(0) + (-2)(0)(-20) \\ &\quad - (0)(1)(-2) - (-20)(0)(25) - (47)(0)(0) \\ &= 1175 \end{aligned}$$

$$D_3 = \begin{vmatrix} 17 & 25 & -2 \\ 10 & 0 & 0 \\ -12 & 0 & 47 \end{vmatrix} \begin{aligned} &= (17)(0)(47) + (25)(0)(-12) + (-2)(10)(0) \\ &\quad - (-12)(0)(-2) - (0)(0)(17) - (47)(25)(10) \\ &= -11{,}750 \end{aligned}$$

$$D_4 = \begin{vmatrix} 17 & 0 & 25 \\ 10 & 1 & 0 \\ -12 & -20 & 0 \end{vmatrix} \begin{aligned} &= (17)(1)(0) + (0)(0)(-12) + (25)(10)(-20) \\ &\quad - (-12)(1)(25) - (-20)(0)(17) - (0)(0)(10) \\ &= -4700 \end{aligned}$$

Thus by Cramer's rule, the remaining node voltages are

$$v_2 = \frac{D_2}{D} = \frac{1175}{1175} = 1 \text{ V}, \qquad v_3 = \frac{D_3}{D} = \frac{-11{,}750}{1175} = -10 \text{ V},$$

$$v_4 = \frac{D_4}{D} = \frac{-4700}{1175} = -4 \text{ V}$$

Having determined all the node voltages, by Ohm's law we have that

$$i_1 = \frac{v_1 - v_2}{2} = \frac{5 - 1}{2} = 2 \text{ A}$$

and therefore, the resistance loading the 5-V source is

$$R = \frac{5}{i_1} = \frac{5}{2} = 2.5 \ \Omega$$

Drill Exercise 2.4

For the circuit shown in Fig. 2.9, replace the voltage-dependent voltage source with a current-dependent voltage source (having the same polarity) whose value is $5i_1$ V. Determine the resulting node voltage v_1, v_2, v_3, and v_4.

ANSWER 5 V, 1 V, −10 V, −4 V

2.3 Mesh Analysis

As mentioned previously, mesh analysis can be used only for planar networks, that is, circuits that can be drawn in the plane in such a way that elements do not cross. As a result of its definition, we deduce that a planar network necessarily partitions the plane into regions or "windows" called **meshes**. One of the meshes, which surrounds the circuit, is infinite. The remaining meshes are finite.

The first step in employing mesh analysis is to visualize a current, called a **mesh current**, around each finite mesh (whether clockwise or counterclockwise is arbitrary). To be specific, consider the circuit shown in Fig. 2.10. The (finite) mesh

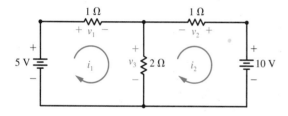

Fig. 2.10 Example of mesh analysis.

bounded by the 5-V source, the leftmost 1-Ω resistor, and the 2-Ω resistor has the clockwise mesh current i_1. The other finite mesh (the mesh bounded by the 10-V source, the 2-Ω resistor, and the rightmost 1-Ω resistor) has the clockwise mesh current i_2.

Although mesh currents are an invention, we can express the actual current through any element in a circuit in terms of the mesh currents. For the circuit in Fig. 2.10, the current going to the right through the leftmost 1-Ω resistor is i_1, and that going to the right through the other 1-Ω resistor is i_2. The current going down through the 2-Ω resistor is $i_1 - i_2$. What is the current through the 5-V source? What is it through the 10-V source?

Having selected mesh currents for the circuit in Fig. 2.10, we now apply KVL to each mesh. (The polarities for the voltages v_1, v_2, and v_3 were chosen arbitrarily.) For mesh i_1 (i.e., the mesh labeled with the mesh current i_1) by KVL,

$$v_1 + v_3 = 5$$

Using Ohm's law this equation becomes

$$1i_1 + 2(i_1 - i_2) = 5 \quad \Rightarrow \quad 3i_1 - 2i_2 = 5 \tag{2.14}$$

For mesh i_2, by KVL,

$$v_2 + v_3 = 10$$

Using Ohm's law this equation becomes

$$-1i_2 + 2(i_1 - i_2) = 10 \quad \Rightarrow \quad 2i_1 - 3i_2 = 10 \tag{2.15}$$

Solving Eq. 2.14 and 2.15, we get the values of the mesh currents:

$$i_1 = -1 \text{ A} \quad \text{and} \quad i_2 = -4 \text{ A}$$

Once we know the mesh currents, we can find any current or voltage in the circuit. For example, the current going down through the 2-Ω resistor is $i_1 - i_2 = 3\text{A}$. The voltages labeled in the circuit are

$$v_1 = 1i_1 = -1 \text{ V}, \quad v_2 = -1i_2 = 4 \text{ V}, \quad v_3 = 2(i_i - i_2) = 6 \text{ V}$$

Now that we have seen an example of the application of mesh analysis, we can state the rules for mesh analysis for planar circuits that do not contain current sources. We will discuss the case of circuits with current sources immediately thereafter.

Mesh Analysis for Circuits with No Current Sources

Given a planar circuit with m meshes and no current sources, proceed as follows:

1. Place mesh currents (e.g., $i_1, i_2, \ldots, i_m$) in the m (finite) meshes.

2. Arbitrarily assign voltages to the elements for which no voltage is designated.

3. Apply KVL to each of the m meshes.

4. Use Ohm's law to express the voltages across resistors in terms of the mesh currents, and substitute these expressions into the voltage equations obtained in Step 3.

5. Solve the resulting set of m simultaneous equations for the mesh currents.

Having seen an example of mesh analysis for a circuit without a current source, let us now consider a circuit in which a current source is present.

Example 2.5

Let us determine the clockwise mesh currents i_1, i_2, and i_3 indicated in the circuit shown in Fig. 2.11. (The polarities for the voltages v_1, v_2, v_3, and v_4 were chosen arbitrarily.)

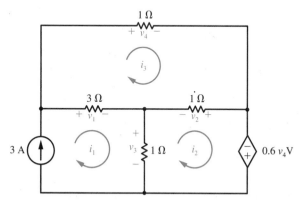

Fig. 2.11 Mesh analysis of a circuit with a current source.

Since the only mesh current that goes through the current source is i_1, we have by inspection that $i_1 = 3$ A. This is the equation obtained for mesh i_1. Thus, essentially, there are only two unknown mesh currents, i_2 and i_3. To determine these, we proceed as discussed previously.

For mesh i_2, by KVL,

$$-v_2 - v_3 - 0.6v_4 = 0$$

Using Ohm's law this equation becomes

$$-1(-i_2 + i_3) - 1(i_1 - i_2) - 0.6(1i_3) = 0 \quad \Rightarrow$$
$$10i_2 - 8i_3 = 5i_1 = 15 \quad (2.16)$$

For mesh i_3, by KVL,

$$-v_1 + v_2 + v_4 = 0$$

Using Ohm's law this equation becomes

$$-3(i_1 - i_3) + 1(-i_2 + i_3) + 1i_3 = 0 \quad \Rightarrow$$
$$-i_2 + 5i_3 = 3i_1 = 9 \quad (2.17)$$

Solving Eq. 2.16 and 2.17, we obtain

$$i_2 = 3.5 \text{ A} \qquad \text{and} \qquad i_3 = 2.5 \text{ A}$$

Since $i_1 = 3$ A, then by Ohm's law,

$$v_1 = 3(i_1 - i_3) = 1.5 \text{ V}, \qquad v_2 = 1(-i_2 + i_3) = -1 \text{ V}$$
$$v_3 = 1(i_1 - i_2) = -0.5 \text{ V}, \qquad v_4 = 1i_3 = 2.5 \text{ V}$$

Drill Exercise 2.5

For the circuit shown in Fig. 2.11, replace the dependent voltage source with a 3-Ω resistor. Use mesh analysis to find the three clockwise mesh currents for the circuit. Use these currents to determine the voltages v_1, v_2, v_3, and v_4.

ANSWER 3 A, 1 A, 2 A, 3 V, 1 V, 2 V, 2 V

In the preceding example, the circuit contained a current source that was common to a finite mesh and the infinite mesh. In such a case, we can express the finite mesh current in terms of the value of the current source by inspection. For the case that a current source is common to two finite meshes, say i_a and i_b, assign a voltage, say

v, across the current source in question. Apply KVL around meshes i_a and i_b, and then combine the two equations to eliminate v. Thus instead of getting two equations from meshes i_a and i_b, only one is obtained. However, we can get the other equation by expressing the value of the current source in terms of the mesh currents i_a and i_b.

Example 2.6

Let us find the clockwise mesh currents depicted in the circuit shown in Fig. 2.12.

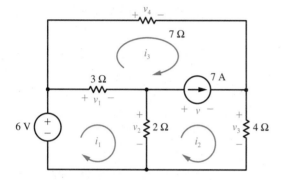

Fig. 2.12. Circuit with a current source common to two finite meshes.

For mesh i_1, by KVL,

$$v_1 + v_2 - 6 = 0$$

Substituting into this equation the fact that $v_1 = 3(i_1 - i_3)$ and $v_2 = 2(i_1 - i_2)$, then

$$3(i_1 - i_3) + 2(i_1 - i_2) - 6 = 0 \quad \Rightarrow \quad 5i_1 - 2i_2 - 3i_3 = 6 \quad (2.18)$$

For mesh i_2, by KVL,

$$v = v_2 - v_3 \quad (2.19)$$

Although $v_2 = 2(i_1 - i_2)$ and $v_3 = 4i_2$, there is no constraint upon the voltage v across the 7-A current source.

For mesh i_3, by KVL,

$$v = v_4 - v_1 \quad (2.20)$$

whereas $v_1 = 3(i_1 - i_3)$ and $v_4 = 7i_3$.

Combining Eq. 2.19 and 2.20 we get

$$v_2 - v_3 = v_4 - v_1$$

from which

$$2(i_1 - i_2) - 4i_2 = 7i_3 - 3(i_1 - i_3) \quad \Rightarrow \quad 5i_1 - 6i_2 - 10i_3 = 0 \quad (2.21)$$

Therefore, we have eliminated the need to express v in terms if the mesh currents. However, so far we have only two equations—Eq. 2.18 and 2.21—in terms of the three mesh currents.

The third equation is obtained from the 7-A current source. Although there is no constraint on the voltage v across it, there is a constrint of 7 A through it. In terms of mesh currents, we have that

$$i_2 - i_3 = 7 \quad (2.22)$$

and this is the third equation.

The solution to the simultaneous Eq. 2.18, 2.21, and 2.22 is

$$i_1 = 2 \text{ A}, \quad i_2 = 5 \text{ A}, \quad i_3 = -2 \text{ A}$$

Having found the three mesh currents and, therefore, completed mesh analysis, we can now determine the resistor voltages by using Ohm's law. Specifically,

$$v_1 = 3(i_1 - i_3) = 3(2 + 2) = 12 \text{ V}, \qquad v_2 = 2(i_1 - i_2) = 2(2 - 5) = -6 \text{ V}$$
$$v_3 = 4i_2 = 4(5) = 20 \text{ V}, \qquad v_4 = 7i_3 = 7(-2) = -14 \text{ V}$$

Finally, we may now calculate v by using KVL. From Eq. 2.19,

$$v = v_2 - v_3 = -6 - 20 = -26 \text{ V}$$

or from Eq. 2.20,

$$v = v_4 - v_1 = -14 - 12 = -26 \text{ V}$$

Drill Exercise 2.6

For the circuit shown in Fig. 2.12, replace the 7-A current source with a 1-Ω resistor and the 2-Ω resistor with a 14-A current source directed up. Determine the three clockwise mesh currents for the circuit.

ANSWER -9 A, 5 A, -2 A

2.4 Ideal Amplifiers

Of fundamental importance in the study of electric circuits is the **ideal voltage amplifier**. Such a device, in general, has two inputs, v_1 and v_2, and one output, v_o. The relationship between the output and the inputs is given by $v_o = A(v_2 - v_1)$, where A is called the **gain** of the amplifier. The ideal amplifier is modeled by the circuit shown in Fig. 2.13a, which contains a dependent voltage source. The resistance R is the **input resistance** of the amplifier. Note that since the input resistance $R = \infty$ Ω for an ideal voltage amplifier, when such an amplifier is connected to any circuit, no current will go into the input terminals. (In general, however, there will be a current going into or coming out of the output terminal.) Also, since the output v_o is the voltage across an ideal source, we have that $v_o = A(v_2 - v_1)$ regardless of what is connected to the output. For the sake of simplicity, the ideal amplifier having gain A is often represented as shown in Fig. 2.13b. (Note that the reference node is not explicitly displayed in this figure.) We refer to the input terminal labeled ''$-$'' as the **inverting input** and the input terminal labeled ''$+$'' as the **noninverting input**.

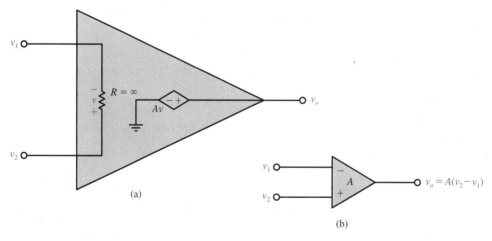

Fig. 2.13 Model and circuit symbol of an ideal amplifier having gain A.

Example 2.7

Let us find v_o for the ideal amplifier circuit shown in Fig. 2.14a. The explicit form of this circuit is shown in Fig. 2.14b.

In the circuit given in Fig. 2.14, the noninverting input is at the reference potential, that is, $v_2 = 0$ V. Furthermore, since node v_s and node v_o are constrained by voltage sources (independent and dependent, respectively), in using nodal analysis we sum

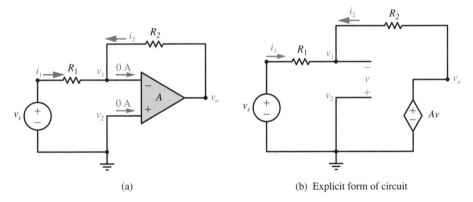

(a)   (b) Explicit form of circuit

Fig. 2.14 (*a*) Ideal-amplifier circuit, and (*b*) explicit form of the circuit.

currents only at node v_1 (the inverting input). Since the amplifier inputs draw no current, by KCL,

$$i_1 + i_2 = 0$$

By Ohm's law

$$\frac{v_s - v_1}{R_1} + \frac{v_o - v_1}{R_2} = 0 \quad \Rightarrow \quad R_2 v_s = (R_1 + R_2)v_1 - R_1 v_o \qquad (2.23)$$

But due to the amplifier, $v_o = A(v_2 - v_1) = -Av_1$, so

$$v_1 = \frac{-v_o}{A} \qquad (2.24)$$

and substituting this into Eq. 2.23, we get

$$R_2 v_s = (R_1 + R_2)\frac{-v_o}{A} - R_1 v_o = -\left[\frac{1}{A}(R_1 + R_2) + R_1\right]v_o$$

Thus,

$$v_o = \frac{-R_2 v_s}{R_1 + (1/A)(R_1 + R_2)} \qquad (2.25)$$

Drill Exercise 2.7

For the ideal-amplifier circuit given in Fig. 2.14, suppose that $R_1 = 1$ kΩ, $R_2 = 10$ kΩ, $A = 100{,}000$, and $v_s = 1$ V. (a) Find v_o, v_1, i_1, and i_2. (b) Find the power absorbed by each resistor, the independent voltage source, and the ideal amplifier (i.e., the dependent voltage source).

ANSWER (a) $-$ 10.0 V, 0.10 mV, 1.0 mA, -1.0 mA;
(b) 1.0 mW, 10.0 mW, -1.0 mW, -10.0 mW

For a circuit such as that in Fig. 2.14, let us consider the case that the gain A becomes arbitrarily large. When $A \to \infty$, from Eq. 2.25 we have that

$$v_o = -\frac{R_2}{R_1} v_s$$

We see that although the gain of the amplifier is infinite, for a finite input voltage v_s, the output voltage v_o is finite (provided, of course, that $R_1 \neq 0$). Inspection of Eq. 2.24 indicates why the output voltage remains finite—as $A \to \infty$, then $v_1 = -v_o/A \to 0$ V. This result occurs because there is a resistor connected between the output and the negative input. Such a connection is called **negative feedback**.

The Operational Amplifier

An ideal amplifier having gain $A = \infty$ is known as an **operational amplifier**, or **op amp.** In an op-amp circuit, because of the infinite gain property, we must have a feedback resistor and must not connect a voltage source directly between the amplifier's input terminals. For the circuit given in Fig. 2.14, the corresponding op-amp circuit is usually drawn as shown in Fig. 2.15. Using the fact that $v_1 = 0$ V, and summing the currents at the inverting input (node v_1), we get

$$\frac{v_s}{R_1} = -\frac{v_o}{R_2} \quad \Rightarrow \quad v_o = -\frac{R_2}{R_1} v_s \tag{2.26}$$

and the gain of the overall circuit is

$$\frac{v_o}{v_s} = -\frac{R_2}{R_1} \tag{2.27}$$

This circuit is called an **inverting amplifier**.

Notice how simple the analysis of the op-amp circuit in Fig. 2.15 is when we use the fact that $v_1 = 0$ V. Although this result was originally deduced from Eq. 2.25,

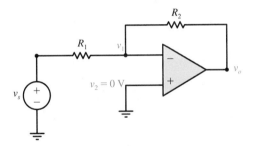

Fig. 2.15 Op-amp circuit—an inverting amplifier.

the combination of infinite gain and feedback constrains the voltage applied to the op-amp (between terminals v_1 and v_2) to be 0 V. In other words, we must have that $v_1 = v_2$.

Example 2.8

Consider the op-amp circuit with feedback in Fig. 2.16a. Again, the inputs of the amplifier draw no current, and so in applying KCL at node v_1, we have

$$i_1 + i_2 = 0 \quad \Rightarrow \quad \frac{v_1}{R_1} + \frac{v_1 - v_o}{R_2} = 0$$

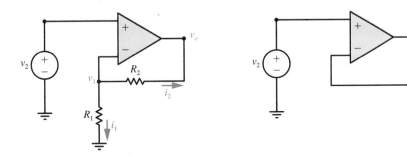

(a) Noninverting amplifier

(b) Voltage follower

Fig. 2.16 (a) Noninverting amplifier, and (b) voltage follower.

Since the input voltage to the amplifier is 0 V, or equivalently, since both input terminals must be at the same potential, then $v_1 = v_2$, and substituting this fact into the last equation, we obtain

$$\frac{v_2}{R_1} + \frac{v_2 - v_o}{R_2} = 0 \quad \Rightarrow \quad R_2 v_2 + R_1 v_2 - R_1 v_o = 0$$

from which

$$v_o = \frac{R_1 + R_2}{R_1} v_2 = \left(1 + \frac{R_2}{R_1}\right) v_2$$

$$\frac{v_o}{v_2} = \frac{R_1 + R_2}{R_1} = 1 + \frac{R_2}{R_1} \tag{2.28}$$

and this is overall gain of this circuit, which is called a **noninverting amplifier**.

Note that if $R_2 = 0\ \Omega$, then $v_o = v_2$. Under the circumstance, R_1 is superfluous and may be removed. The resulting op-amp circuit shown in Fig. 2.16b is known as a **voltage follower**. Such a configuration is used to isolate or buffer one circuit from another.

Drill Exercise 2.8

For the noninverting amplifier given in Fig. 2.16a, suppose that $R_1 = 1\ \text{k}\Omega$, $R_2 = 9\ \text{k}\Omega$, and $v_2 = 1\ \text{V}$. (a) Find v_o, i_1, and i_2. (b) Find the power absorbed by each resistor, the independent voltage source, and the op amp.

ANSWER (a) 10 V, 1 mA, -1 mA; (b) 1 mW, 9 mW, 0 W, -10 mW

Operational-amplifier circuits can be quite useful in some situations in which there is more than a single input.

Example 2.9

Let us determine the output voltage v_o in terms of the input voltages v_a and v_b for the op-amp circuit shown in Fig. 2.17. Since $v_2 = 0\ \text{V}$, then $v_1 = 0\ \text{V}$. By KCL, at the inverting input,

$$i_1 + i_2 = i_3$$

from which we get

$$\frac{v_a}{R_1} + \frac{v_b}{R_1} = -\frac{v_o}{R_2} \quad \Rightarrow \quad v_o = -\frac{R_2}{R_1}(v_a + v_b) \tag{2.29}$$

Since the output v_o is the sum of the input voltages v_a and v_b (multiplied by the constant $-R_2/R_1$), this circuit is called an **adder** (or **summer**). A common applica-

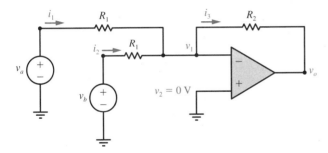

Fig. 2.17 Op-amp adder.

tion of such a circuit is as an "audio mixer," where, for example, the inputs are voltages attributable to two separate microphones and the output voltage is their sum.

Drill Exercise 2.9

For the op-amp adder shown in Fig. 2.17, suppose that $R_1 = 1$ kΩ, $R_2 = 10$ kΩ, $v_a = -0.2 \cos 2000\pi t$ V, and $v_b = 0.3 \cos 4000\pi t$ V. Find v_o, i_1, i_2, and i_3.

ANSWER $2 \cos 2000\pi t - 3 \cos 4000\pi t$ V, $-0.2 \cos 2000\pi t$ mA, $0.3 \cos 4000\pi t$ mA, $-0.2 \cos 2000\pi t + 0.3 \cos 4000\pi t$ mA

Example 2.10

Let us find the output voltage v_o of the op-amp circuit with two inputs is shown in Fig. 2.18.

Using the fact that $v_1 = v_2 = v$, by KCL at the inverting input

$$i_1 = i_3 \quad \Rightarrow \quad \frac{v_a - v}{R_1} = \frac{v - v_o}{R_2}$$

from which

$$R_1 v_o + R_2 v_a = (R_1 + R_2)v \qquad (2.30)$$

Applying KCL at the noninverting input, we get

$$i_2 = i_4 \quad \Rightarrow \quad \frac{v_b - v}{R_1} = \frac{v}{R_2}$$

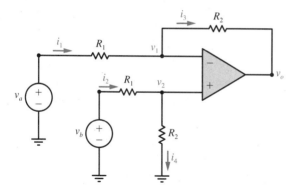

Fig. 2.18 Difference amplifier.

from which

$$R_2 v_b = (R_1 + R_2)v \tag{2.31}$$

Combining Eq. 2.30 and 2.31 results in

$$R_1 v_o + R_2 v_a = R_2 v_b \quad \Rightarrow \quad v_o = \frac{R_2}{R_1}(v_b - v_a) \tag{2.32}$$

Since the output is the difference of the inputs (multiplied by the constant R_2/R_1), such a circuit is called a **difference amplifier** or **differential amplifier**.

Drill Exercise 2.10

For the difference amplifier shown in Fig. 2.18, suppose that $R_1 = 1$ kΩ, $R_2 = 9$ kΩ, $v_a = -0.2 \cos 2000\pi t$ V, and $v_b = 0.3 \cos 4000\pi t$ V. Find v_o, v, i_1, i_2, i_3, and i_4.

ANSWER $1.8 \cos 2000\pi t + 2.7 \cos 4000\pi t$ V, $0.27 \cos 4000\pi t$ V,
$\quad\quad\quad -0.2 \cos 2000\pi t - 0.27 \cos 4000\pi t$ mA, $0.03 \cos 4000\pi t$ mA,
$\quad\quad\quad -0.2 \cos 2000\pi t - 0.27 \cos 4000\pi t$ mA, $0.03 \cos 4000\pi t$ mA

For a physical operational amplifier, the gain and the input resistance are large, but not infinite. A more practical model of an actual op amp is to have a resistance R_o (called the **output resistance** of the op amp) connected in series with the dependent voltage source. Yet, assuming that an op amp is ideal often yields a simple analysis with very accurate results.

Although its usefulness is great, the physical size of an op amp typically is small. This is a result of the fact that op amps are commonly available on IC chips, which normally contain between one and four op amps. In our previous discussions, we depicted the op amp as a three-terminal device. In actuality, a package containing one or more op amps typically has between 8 and 14 terminals. As we have seen, three of the terminals are the inverting input, the noninverting input, and the output. However, there are also terminals which are used for applying constant voltages (called "power-supply voltages") that operate or "bias" the numerous transistors comprising an op amp, for frequency compensation, and for other details to be discussed in Chapter 10. One of the important practical electronic considerations for proper op-amp operation is that when feedback is between the output and only one input terminal, that input terminal should be the inverting input.

Example 2.11

Shown in Fig. 2.19 is a practical amplifier that uses the popular 741 op amp. For this circuit, terminal (pin) 2 is the inverting input, terminal 3 is the noninverting input, and terminal 6 is the output. Furthermore, pin 4 has a constant voltage of -15 V applied to it and pin 7 has a constant voltage of $+15$ V applied to it.[2] The 741 op amp typically has a gain of $A = 200{,}000$, and input resistance of $R = 2$ MΩ, and an output resistance of $R_o = 75$ Ω.

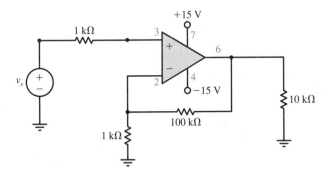

Fig. 2.19 Practical op-amp circuit.

Suppose that v_j is the voltage at terminal j, where $j = 1, 2, 3, \ldots$ Assuming that the op amp is ideal, then $v_3 = v_s$ since the inputs of an (ideal) op amp

---

[2]These constant voltages supply power to the actual components forming the op amp and are ignored when the op-amp circuit is analyzed.

draw no current. Also, due to feedback, $v_2 = v_3 = v_s$, and by KCL at terminal 2,

$$\frac{v_6 - v_s}{100,000} = \frac{v_s}{1000} \qquad \Rightarrow \qquad v_6 = 101 v_s$$

We can get a more accurate result by not assuming that the op amp is ideal. Doing this, which requires a much greater analysis effort (see Drill Exercise 2.11), yields

$$v_6 = 100.95 v_s$$

which is not significantly different from the simple approach taken above, and is even the same when rounded off to three significant digits.

Drill Exercise 2.11

For the circuit given in Fig. 2.19, suppose that the amplifier has input resistance $R = 2$ MΩ, output resistance $R_o = 75$ Ω (which is connected in series with the dependent source Av), and gain $A = 200,000$. Use nodal analysis to get an accurate value for the output voltage v_6.

ANSWER $100.95 v_s$

2.5 Thévenin's and Norton's Theorems

Suppose that a resistor R_L, called a **load resistor**, is connected to an arbitrary (in the sense that it contains only elements discussed previously) circuit as shown in Fig. 2.20. What value of the load resistor R_L will absorb the maximum amount of power? Knowing the particular circuit, we can use nodal or mesh analysis to obtain an expression for the power absorbed by R_L, then take the derivative of this expression to determine what value of R_L results in the maximum power absorbed by R_L. The effort required for such an approach can be quite great. Fortunately, though, a very important circuit theory concept states that as far as R_L is concerned, the arbitrary circuit shown in Fig. 2.20 behaves as though it is a single independent voltage source connected in series with a single resistance. Once we determine the values of this source and this resistance, we simply apply the results on maximum power transfer (which follows shortly) to find the value of R_L that results in the maximum power transfer to the load.

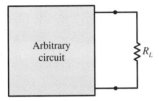

Fig. 2.20 Arbitrary circuit and its associated load.

Thévenin's Theorem

Suppose we are given an arbitrary circuit containing any or all of the following elements: resistors, voltage sources, current sources. (The sources can be dependent as well as independent.) Let us identify a pair of nodes, say node *a* and node *b*, such that the circuit can be partitioned into two parts as shown in Fig. 2.21. Furthermore, suppose that circuit A contains no dependent source that is dependent on a variable in circuit B, and vice versa. Then we can model circuit A by an appropriate independent voltage source, call it v_{oc}, that is connected in series with an appropriate resistance, call it R_o. This series combination of a voltage source and a resistance is called the **Thévenin equivalent** of circuit A. In other words, circuit A in Fig. 2.21 and the circuit in the shaded box in Fig. 2.22 have the same effect on circuit B. This result is known as **Thévenin's theorem,**[3] and is one of the more useful and significant concepts in circuit theory.

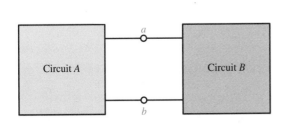

Fig. 2.21 Circuit partitioned into two parts.

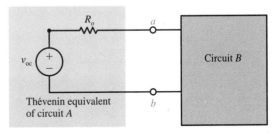

Fig. 2.22 Application of Thévenin's theorem.

To obtain the voltage v_{oc}—called the **open-circuit voltage**—remove circuit B from circuit A, and determine the voltage between nodes *a* and *b* (where the + is at node *a*). This voltage, as shown in Fig. 2.23a, is v_{oc}.

To obtain the resistance R_o—called the **Thévenin-equivalent resistance** or the **output resistance** of circuit A—again remove circuit B from circuit A. Next, set all

---

[3]Named for the French engineer M. L. Thévenin (1857–1926).

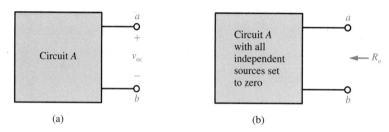

Fig. 2.23 Determination of (*a*) open-circuit voltage, and (*b*) Thévenin-equivalent (output) resistance.

independent sources in circuit *A* to zero. (A zero voltage source is equivalent to a short circuit, and a zero current source is equivalent to an open circuit.) ***Leave the dependent sources as is!*** Then determine the resistance between nodes *a* and *b*—this is R_o—as shown in Fig. 2.23*b*.

If circuit *A* contains no dependent sources, when all independent sources are set to zero, the result may just be a series-parallel resistive circuit. In this case, R_o can be found by appropriately combining resistors that are connected in series and/or parallel. In general, however, R_o can be found by applying an independent source between nodes *a* and *b* and then by taking the ratio of voltage to current. This procedure is depicted in Fig. 2.24.

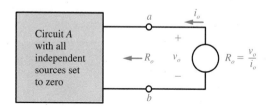

Fig. 2.24 Determination of output resistance.

In applying Thévenin's theorem, circuit *B* (which is often called a **load**) may consist of many circuit elements, a single element (e.g., a load resistor), or no elements (i.e., circuit *B* already may be an open circuit).

Example 2.12

Let us find the voltage v_L across the 3-Ω load resistor for the circuit shown in Fig. 2.25 by replacing circuit *A* by its Thévenin equivalent.

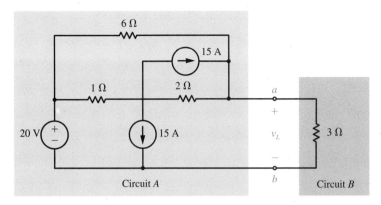

Fig. 2.25 Example of Thévenin's theorem.

First, we remove the 3-Ω load resistor and calculate v_{oc} from the circuit shown in Fig. 2.26. Summing the currents out of node v, by KCL, we get

$$\frac{v - 20}{1} + \frac{v - v_{oc}}{2} + 15 + 15 = 0$$

Summing the currents into node v_{oc}, by KCL, we get

$$\frac{v - v_{oc}}{2} + \frac{20 - v_{oc}}{6} + 15 = 0$$

Solving these equations for v_{oc} yields $v_{oc} = 30$ V.

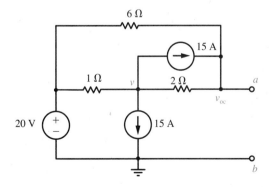

Fig. 2.26 Determination of open-circuit voltage.

To find the Thévenin equivalent resistance R_o, set the independent sources in circuit A to zero. The resulting circuit is shown in Fig. 2.27. Since the 1-Ω and 2-Ω resistors are connected in series, their combined resistance is 3 Ω. But this resulting 3-Ω resistance is connected in parallel with 6 Ω. Thus,

$$R_o = \frac{(3)(6)}{3 + 6} = 2 \; \Omega$$

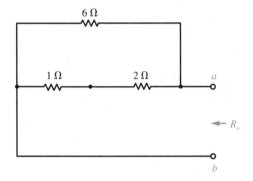

Fig. 2.27 Determination of output resistance.

Replacing circuit A by its Thévenin equivalent, we get the circuit shown in Fig. 2.28. By voltage division, we have that

$$v_L = \frac{3}{3 + 2}(30) = 18 \text{ V}$$

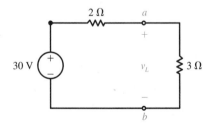

Fig. 2.28 Use of the Thévenin-equivalent circuit.

Direct analysis of the circuit given in Fig. 2.25, of course, yields the same result for v_L (see Problem 2.6).

Drill Exercise 2.12

For the circuit given in Fig. 2.25, replace the lower current source with a 1-Ω resistor and the upper current source with a 2-Ω resistor. Find the Thévenin equivalent of the resulting circuit A, and use this Thévenin-equivalent circuit to determine v_L.

ANSWER 12 V, 1.2 Ω, 8.57 V

Having done an example of finding the Thévenin equivalent of a circuit that does not contain a dependent source, let us now consider circuits with dependent sources. In particular, since an ideal op amp is modeled with a dependent source (see Fig. 2.13), let us consider an op-amp circuit.

Example 2.13

Let us find the Thévenin equivalent of the op-amp circuit (with no load) shown in Fig. 2.29.

We have already seen (Eq. 2.26 on p. 74) that

$$v_{oc} = -\frac{R_2}{R_1} v_s$$

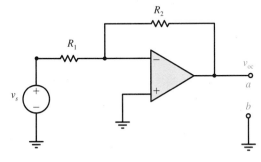

Fig. 2.29 Thévenin's theorem applied to an op-amp circuit.

To find the output resistance R_o, we set v_s to zero and take the ratio $R_o = v_o/i_o$ as depicted in Fig. 2.30. By KCL,

$$i_1 + i_2 = 0 \quad \Rightarrow \quad \frac{0}{R_1} + \frac{v_o}{R_2} = 0$$

from which

$$v_o = 0 \text{ V} \quad \Rightarrow \quad R_o = \frac{v_o}{i_o} = 0 \text{ }\Omega$$

and the Thévenin equivalent of the op-amp circuit given in Fig. 2.29 is the (ideal) voltage source shown in Fig. 2.31.

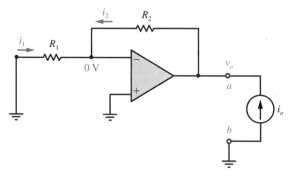

Fig. 2.30 Determination of output resistance.   **Fig. 2.31** Thévenin-equivalent circuit.

Drill Exercise 2.13

For the circuit given in Fig. 2.9, on p. 64, consider the 4-Ω resistor to be the load, and find the Thévenin equivalent of the remainder of the circuit. Use this Thévenin-equivalent circuit to find v_4.

ANSWER -5.11 V, 1.11 Ω, -4 V

Norton's Theorem

Let us find the Thévenin equivalent of the circuit shown in Fig. 2.32a. Clearly, $v_{oc} = Ri$ and $R_o = R$. Thus the Thévenin equivalent of the circuit given in Fig. 2.32a is the circuit shown in Fig. 2.32b. Furthermore, $i = v_{oc}/R = v_{oc}/R_o$. But placing a short circuit between nodes a and b results in a current of $i_{sc} = v_{oc}/R_o = Ri/R = i$ through the short circuit directed from node a and node b. Since the circuits in Figs. 2.32a and 2.32b are equivalent, we deduce that the circuits

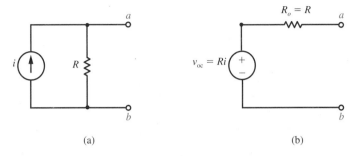

Fig. 2.32 (*a*) Current source connected in parallel with a resistor, and (*b*) its Thévenin-equivalent circuit.

shown in Figs. 2.33*a* and 2.33*b* are equivalent. The relationships between v_{oc}, i_{sc}, and R_o are

$$ v_{oc} = R_o i_{sc} \qquad i_{sc} = \frac{v_{oc}}{R_o} \qquad R_o = \frac{v_{oc}}{i_{sc}} $$

If the circuit given in Fig. 2.33*a* is the Thévenin equivalent of some circuit *A*, then the circuit shown in Fig. 2.33*b* is known as the **Norton equivalent**[4] of the circuit.

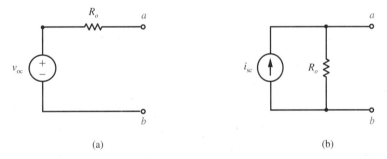

Fig. 2.33 (*a*) Voltage source connected in series with a resistor, and (*b*) its Norton-equivalent circuits.

To obtain the Norton equivalent of some circuit *A* without first finding the Thévenin equivalent, determine the **short-circuit current** i_{sc} by placing a short circuit between nodes *a* and *b* and then calculate the resulting current (directed from *a* to *b*) through the short circuit. The output resistance R_o can be found as was done for the Thévenin-equivalent circuit. Alternatively, v_{oc} can be determined as was done for the Thévenin-equivalent circuit, and then the formula $R_o = v_{oc}/i_{sc}$ can be employed.

---

[4]Named for the American engineer Edward L. Norton.

Example 2.14

For the circuit shown in Fig. 2.34, let us find the voltage across the 8-Ω resistor by replacing the remainder of the circuit by its Norton equivalent.

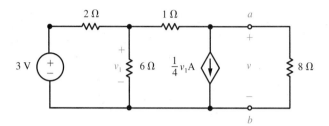

Fig. 2.34 Example of the use of the Norton-equivalent circuit.

First, replace the 8-Ω resistor that is connected between nodes a and b with a short circuit, and calculate i_{sc} from the circuit shown in Fig. 2.35. Summing the currents out of node v_1, by KCL we have that

$$\frac{v_1 - 3}{2} + \frac{v_1}{6} + \frac{v_1}{1} = 0$$

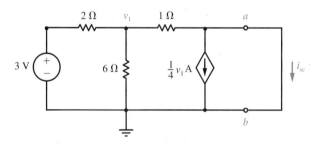

Fig. 2.35 Determination of short-circuit current.

Solving this equation for v_1 yields $v_1 = 0.9$ V. By KCL,

$$i_{sc} = \frac{v_1}{1} - \frac{v_1}{4} = \frac{3}{4}v_1 = \left(\frac{3}{4}\right)\left(\frac{9}{10}\right) = \frac{27}{40} = 0.675 \text{ A}$$

To find R_o, we remove the 8-Ω load resistor, set the independent source to zero, and calculate $R_o = v_o/i_o$ between nodes a and b as depicted in Fig. 2.36. (Be careful not to use the value of v_1 that was calculated for the circuit in Fig. 2.35; the circuit in Fig. 2.36 is a different circuit.) Summing the currents out of node v_1, we get

$$\frac{v_1}{2} + \frac{v_1}{6} + \frac{v_1 - v_o}{1} = 0 \quad \Rightarrow \quad v_1 = 0.6v_o$$

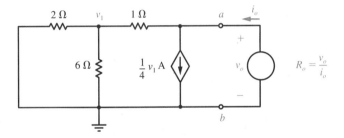

Fig. 2.36 Determination of output resistance.

By KCL,

$$i_o = \frac{v_o - v_1}{1} + \frac{1}{4}v_1 = \frac{v_o}{1} - \frac{3}{4}v_1 = v_o - (0.75)(0.6v_o) = 0.55v_o$$

from which

$$R_o = \frac{v_o}{i_o} = \frac{1}{0.55} = \frac{20}{11} = 1.82 \ \Omega$$

An alternative procedure for finding R_o is to remove the 8-Ω resistor from the circuit given in Fig. 2.34, and calculate the open-circuit voltage v_{oc} between nodes a and b. Doing so results in $v_{oc} = 27/22 = 1.23$ V. Thus, we also have that

$$R_o = \frac{v_{oc}}{i_{sc}} = \frac{27/22}{27/40} = \frac{20}{11} = 1.82 \ \Omega$$

as was determined above.

To calculate the voltage v indicated in Fig. 2.34, we may model the portion of the circuit to the left of nodes a and b by its Norton equivalent. This is illustrated by Fig. 2.37. Using the current-divider formula, we get

$$i = \frac{1.82}{1.82 + 8}(0.675) = 0.125 \ \text{A} \qquad \Rightarrow \qquad v = 8(0.125) = 1 \ \text{V}$$

Compare this result with that obtained in Example 2.1.

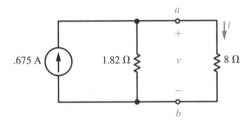

Fig. 2.37 Use of the Norton-equivalent circuit.

In the preceding example, we mentioned that R_o can be determined by finding v_{oc} and i_{sc} and then using $R_o = v_{oc}/i_{sc}$ rather than taking the direct approach shown in Fig. 2.36. However, $R_o = v_{oc}/i_{sc}$ cannot be employed if neither v_{oc} nor i_{sc} exists. For instance, the op-amp circuit given in Fig. 2.29 is equivalent to an ideal voltage source (Fig. 2.31). This means that i_{sc} does not exist for this circuit. In other words, an ideal voltage source cannot be equivalent to (modeled by) a current source connected in parallel with a resistance. Nor can an ideal current source be equivalent to (modeled by) a voltage source connected in series with a resistance. This means that for a circuit that is equivalent to an ideal current source, v_{oc} does not exist.

There are circuits whose Thévenin and Norton equivalents are identical. For example, in general, a circuit that contains no independent sources will have $v_{oc} = 0$ V and $i_{sc} = 0$ A. This means that the Thévenin- and Norton-equivalent circuits consist only of the output resistance R_o (e.g., see Problem 2.33). Of course, in such a case, the formula $R_o = v_{oc}/i_{sc}$ cannot be used to find R_o. However, there are the exceptional circuits with nonzero independent sources that also have Thévenin- and Norton-equivalent circuits comprised only of a resistance (e.g., see Problem 2.40).

Maximum Power Transfer

We may now answer the question that was posed at the beginning of this section. For the situation depicted in Fig. 2.20 on p. 81, we wish to determine the value of the load R_L that will absorb the maximum amount of power. By Thévenin's theorem, we can model the arbitrary circuit by its Thévenin equivalent and the effect on R_L will be the same. Doing so, we obtain the circuit shown in Fig. 2.38. Since the power absorbed by the load is $p = R_L i^2$, it may be tempting to believe that to increase the power absorbed, we simply increase the load resistance R_L. This reasoning, however, is invalid since an increase in R_L will result in a decrease in i. To find the value for which maximum power is delivered to the load, we proceed as follows:

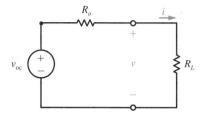

Fig. 2.38 Determination of maximum power transfer.

By Ohm's law, $i = v_{oc}/(R_o + R_L)$. The (instantaneous) power absorbed by the load is

$$p = R_L i^2 = R_L \left(\frac{v_{oc}}{R_o + R_L}\right)^2 = \frac{R_L v_{oc}^2}{(R_o + R_L)^2} \tag{2.33}$$

Thus we have an expression for the power in terms of the variable R_L. For what value of R_L is the power a maximum? From introductory calculus we know that if we have a function $f(x)$ of the real variable x, then maxima occur for the values of x where the derivative of $f(x)$ is equal to zero. Thus let us take the derivative of p with respect to the variable R_L. Using the fact that

$$\frac{d}{dx}\left[\frac{f(x)}{g(x)}\right] = \frac{g(x)\, df(x)/dx - f(x)\, dg(x)/dx}{[g(x)]^2}$$

we obtain

$$\frac{dp}{dR_L} = \frac{(R_o + R_L)^2\, d(R_L v_{oc}^2)/dR_L - R_L v_{oc}^2\, d(R_o + R_L)^2/dR_L}{[(R_o + R_L)^2]^2}$$

Simplifying this expression yields

$$\frac{dp}{dR_L} = \frac{(R_o - R_L)v_{oc}^2}{(R_o + R_L)^3}$$

Clearly $dp/dR_L = 0$ when the numerator $(R_o - R_L)v_{oc}^2 = 0$. Furthermore, given that v_{oc} is a nonzero source, then $dp/dR_L = 0$ when

$$R_o - R_L = 0 \quad \Rightarrow \quad \boxed{R_o = R_L}$$

Therefore, when $R_o = R_L$, the power absorbed by the load (Eq. 2.33) becomes

$$p_m = \frac{v_{oc}^2}{4R_o} = \frac{v_{oc}^2}{4R_L} \tag{2.34}$$

and this quantity is either a maximum or a minimum. However, since we can make the expression for power absorbed by the load arbitrarily small by making R_L arbitrarily small, the quantity $v_{oc}^2/4R_o$ cannot be a minimum. We may also verify that $p_m = v_{oc}^2/4R_o$ is maximum by taking the second derivative of p, setting $R_o = R_L$, and obtaining a negative quantity.

In summary, for the situation depicted in Fig. 2.20 on p. 81 (or Fig. 2.38), the maximum power that can be absorbed by (or delivered to) a resistive load R_L is obtained when $R_L = R_o$, and that power is given by Eq. 2.34.

Example 2.15

For the circuit given in Fig. 2.25 on p. 83, let us determine to what value the 3-Ω load resistor should be changed so that it will absorb the maximum amount of power.

In Example 2.12 we have already seen that for the given circuit $v_L = 18$ V. Therefore, the power absorbed by the 3-Ω load is

$$p = \frac{v^2}{3} = \frac{18^2}{3} = 108 \text{ W}$$

But since the Thévenin equivalent of circuit A has $v_{oc} = 30$ V and $R_o = 2$ Ω, then the load resistance R_L, which absorbs maximum power, is $R_L = R_o = 2$ Ω. Thus, if we replace the 3-Ω resistor with a 2-Ω resistor, from Eq. 2.34, this resistor will absorb

$$p_m = \frac{v_{oc}^2}{4R_o} = \frac{30^2}{4(2)} = 112.5 \text{ W}$$

and this is the maximum power that can be absorbed by any load resistance R_L.

Drill Exercise 2.15

For the circuit given in Fig. 2.34, on p. 88, to what value should the 8-Ω load resistor be changed such that the resulting load will absorb maximum power? What is the maximum power absorbed?

ANSWER 1.82 Ω, 0.207 W

2.6 Linearity and Superposition

Given a circuit that contains two or more independent sources (voltage or current or both), as we have seen, one way to determine a specific variable (either voltage or current) is by the direct use of nodal or mesh analysis. An alternative, however, is to find that portion of the variable attributable to each independent source, and then sum these up. This concept, known as the **principle of superposition**, is a direct consequence of the linearity property of circuits.

Linear Circuits

Reconsider the relationship between voltage and current for a resistor (i.e., Ohm's law). Suppose that a current i_1 (the **excitation** or **input**) is applied to a resistor, R. Then the resulting voltage v_1 (the **response** or **output**) is $v_1 = Ri_1$. Similarly, if i_2 is applied to R, then $v_2 = Ri_2$ results. But if $i = i_1 + i_2$ is applied, then the response is

$$v = Ri = R(i_1 + i_2) = Ri_1 + Ri_2 = v_1 + v_2$$

In other words, the response to a sum of inputs is equal to the sum of the individual responses (Condition 1).

In addition, if v is the response to i (i.e., $v = Ri$), then the response to Ki is

$$R(Ki) = K(Ri) = Kv$$

In other words, if the excitation is scaled by the constant K, then the response is also scaled by K, (Condition 2).

Because Conditions 1 and 2 are satisfied, we say that the relationship between current (input) and voltage (output) is **linear** for a resistor. Similarly, by using the alternative form of Ohm's law $i = v/R$, we can show that the relationship between voltage (excitation) and current (response) is also linear for a resistor.

Although the relationships between voltage and current for a resistor are linear, the power relationships $p = Ri^2$ and $p = v^2/R$ are not. For instance, if the current through a resistor is i_1, then the power absorbed by the resistor R is $p_1 = Ri_1^2$, whereas if the current is i_2, then the power absorbed is $p_2 = Ri_2^2$. However, the power absorbed due to the current $i_1 + i_2$ is

$$p_3 = R(i_1 + i_2)^2 = Ri_1^2 + Ri_2^2 + 2Ri_1i_2 \neq p_1 + p_2$$

Hence the relationship $p = Ri^2$ is **nonlinear**.

Since the relationships between voltage and current are linear for resistors, we say that a resistor is a **linear element**. A dependent source (either current or voltage) whose value is directly proportional to some voltage or current is also a linear element. Because of this, we say that a circuit consisting of independent sources, resistors, and linear dependent sources (as well as other linear elements to be introduced later) is a **linear circuit**. All of the circuits dealt with so far have been linear circuits.

The Principle of Superposition

We now return to the simple case of a resistor R connected to a voltage source v. If the voltage applied across R is v_1, then the resulting current through R is $i_1 = v_1/R$, whereas if v_2 is applied, then $i_2 = v_2/R$ results. By linearity, we have that for the excitation $v_1 + v_2$, the response is $i_1 + i_2$. However, we can represent this situation as shown in Fig. 2.39. But each component (i_1 and i_2) of the response can be determined as shown in Fig. 2.40. This is a very special case showing how it is possible to obtain the response due to two independent voltage sources by calculating the response to each one separately and then summing the responses. However, for any linear circuit, we can take the same approach: Find the response to each independent source (both voltage and current) separately, and then sum the responses. A formal general proof is much more involved, but utilizes analogous ideas.

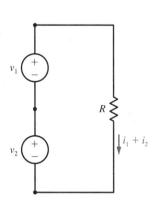

Fig. 2.39 Circuit with two independent voltage sources.

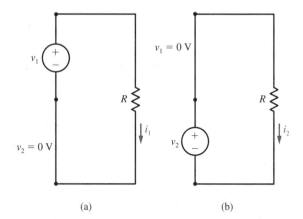

(a)   (b)

Fig. 2.40 Treating each independent source separately.

For demonstration purposes, consider the linear circuit shown in Fig. 2.41, which contains three independent sources. By KCL, at node v,

$$\frac{v}{6} + \frac{v - v_g}{3} + i_y - i_x = 0 \qquad \Rightarrow \qquad 3v = 2v_g + 6i_x - 6i_y \qquad (2.35)$$

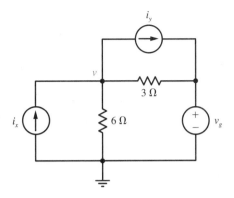

Fig. 2.41 Circuit with three independent voltage sources.

Suppose that when $i_x = i_y = 0$ A, the resulting node voltage is $v = v_1$. Thus we have the equality

$$3v_1 = 2v_g + 6(0) - 6(0) \tag{2.36}$$

For the case that $v_g = 0$ V and $i_y = 0$ A, let the resulting node voltage be $v = v_2$. This yields the equality

$$3v_2 = 2(0) + 6i_x - 6(0) \tag{2.37}$$

Finally, let $v = v_3$ be the node voltage that results when $v_g = 0$ V and $i_x = 0$ A. This gives us the equality

$$3v_3 = 2(0) + 6(0) - 6i_y \tag{2.38}$$

Adding the three equalities from Eq. 2.36, 2.37, and 2.38 produces the equality

$$3(v_1 + v_2 + v_3) = 2v_g + 6i_x - 6i_y$$

which means that $v_1 + v_2 + v_3$ is the solution to Eq. 2.35.

We now formally state the principle of superposition:

Given a linear circuit with independent sources $s_1, s_2, \ldots, s_n$, let r be the response (either voltage or current) of this circuit. If r_i is the response of the circuit to source s_i with all other independent sources set to zero (dependent sources are left as is), then $r = r_1 + r_2 + \cdots + r_n$.

Example 2.16

Let us find the voltage v indicated in the circuit shown in Fig. 2.42 by employing the principle of superposition.

This circuit contains three independent sources. To find the voltage v by using the principle of superposition, we first find that portion of v, call it v_a, which is due to the independent voltage source. To do this, we set the two independent current sources to zero as shown in Fig. 2.43. In this circuit note that the 1-Ω and 2-Ω resistors are connected in series, and the resulting equivalent 3-Ω resistance is connected in parallel with the 6-Ω resistor. These three resistors, therefore, are equivalent to a 2-Ω resistance. By voltage division, we have that

$$v_a = \frac{3}{3+2}(20) = 12 \text{ V}$$

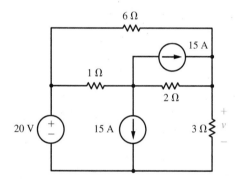

Fig. 2.42 Example of the principle of superposition.

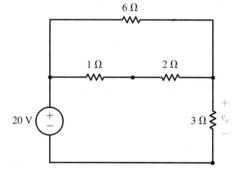

Fig. 2.43 Determination of the response due to 20-V source.

To find the portion of the voltage v, call it v_b, that is due to the lower 15-A current source, set the other current source and the voltage source to zero as shown in Fig. 2.44. In this circuit, the 6-Ω resistor is connected in parallel with the 3-Ω resistor. We can redraw the circuit as shown in Fig. 2.45. The 3-Ω and 6-Ω resistors connected in parallel are equivalent to a 2-Ω resistor—which is connected in series with another 2-Ω resistor. The combination is effectively a 4-Ω resistance that is connected in parallel with the 1-Ω resistor. By current division,

$$i_2 = \frac{1}{1+4}(-15) = -3 \text{ A}$$

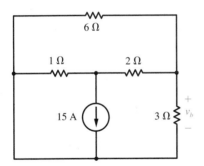

Fig. 2.44 Determination of response due to the lower 15-A source.

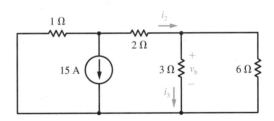

Fig. 2.45 Circuit for calculating v_b.

Having determined i_2, we can use current division again to get

$$i_3 = \frac{6}{6+3}i_2 = -2 \text{ A} \quad \Rightarrow \quad v_b = 3i_3 = -6 \text{ V}$$

To find the voltage v_c due to the upper 15-A current source, set the other two independent sources to zero. The resulting circuit is shown in Fig. 2.46. Since the 6-Ω and 3-Ω resistors are connected in parallel, we can redraw the circuit as is done in Fig. 2.47. Suppose for this circuit that we apply Thévenin's theorem to the 15-A source connected in parallel with the 2-Ω resistor. The result is a $15(2) = 30$-V source (with the plus toward node n) connected in series with a 2-Ω resistor. Since the 3-Ω and 6-Ω resistors connected in parallel are an equivalent 2-Ω resistance, by voltage division we have that

$$v_c = \frac{2}{2+1+2}(30) = 12 \text{ V}$$

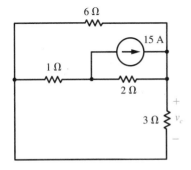

Fig. 2.46 Determination of response due to the upper 15-A source.

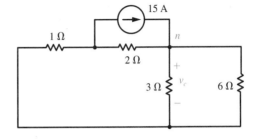

Fig. 2.47 Equivalent circuit for calculating v_c.

By the principle of superposition, we have that

$$v = v_a + v_b + v_c = 12 - 6 + 12 = 18 \text{ V}$$

Drill Exercise 2.16

For the circuit given in Fig. 2.42, replace the lower current source with an independent voltage source (+ on the bottom) having value $\frac{2}{3}$ V. (a) Find the portion of v that is due to the 20-V voltage source. (b) Find the portion of v that is due to the $\frac{2}{3}$-V voltage source. (c) Find the portion of v that is due to the 15-A current source. (d) Find v by using the principle of superposition.

ANSWER (a) 10/3 V, (b) −1/3 V, (c) 15 V, (d) 18 V

In Example 2.16, all of the sources in the circuit were independent sources. For the case in which a circuit also contains dependent sources, the dependent sources are left as is—only independent sources are set to zero when applying the principle of superposition.

S U M M A R Y

1. In nodal analysis, the circuit variables that are determined are the node voltages (with respect to the reference node).

2. In mesh analysis, the circuit variables that are determined are the mesh currents. Mesh analysis can be employed only for planar networks.

3. An operational amplifier is an ideal amplifier with infinite gain. When used in conjunction with feedback, the input voltage to an op amp is constrained to be zero volts. The input terminals of an op amp draw no current.

4. The effect of an arbitrary circuit on a load is equivalent to an appropriate voltage source in series with an appropriate resistance (Thévenin's theorem) or an appropriate current source in parallel with that resistance (Norton's theorem).

5. An arbitrary circuit delivers maximum power to a resistive load R_L when $R_L = R_o$, where R_o is the Thévenin-equivalent (output) resistance of the arbitrary circuit.

6. The response of a circuit having n independent sources equals the sum of the n responses to each individual independent source (the principle of superposition).

Problems

2.1 For the circuit shown in Fig. P2.1, select node *d* as the reference node. (a) Use nodal analysis to find the node voltages. (b) Use the node voltages to determine i_1, i_2, i_3, and i_4.

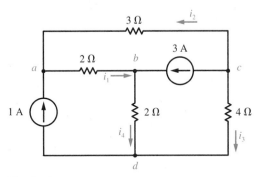

Fig. P2.1

2.2 For the circuit shown in Fig. P2.1, select node *c* as the reference node. (a) Use nodal analysis to find the node voltages. (b) Use the node voltages to determine i_1, i_2, i_3, and i_4.

2.3 For the circuit shown in Fig. P2.1, select node *b* as the reference node. (a) Use nodal analysis to find the node voltages. (b) Use the node voltages to determine i_1, i_2, i_3, and i_4.

2.4 For the circuit shown in Fig. P2.1, select node *a* as the reference node. (a) Use nodal analysis to find the node voltages. (b) Use the node voltages to determine i_1, i_2, i_3, and i_4.

2.5 Find the node voltages for the circuit shown in Fig. P2.5.

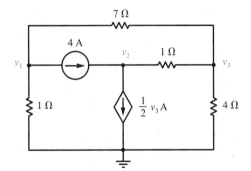

Fig. P2.5

2.6 Find the node voltages for the circuit shown in Fig. P2.6.

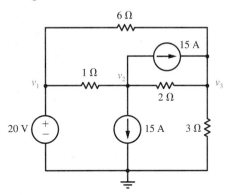

Fig. P2.6

2.7 Find the node voltages for the circuit shown in Fig. P2.7. (See p. 100.)

2.8 Find the node voltages for the circuit shown in Fig. P2.8.

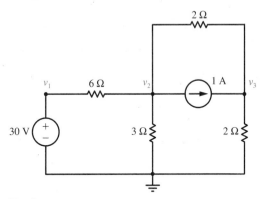

Fig. P2.8

2.9 Find the node voltages for the circuit shown in Fig. P2.9.

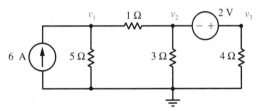

Fig. P2.9

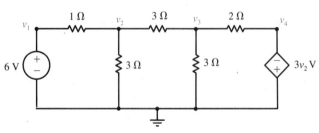

Fig. P2.7

2.10 Find the node voltages for the circuit shown in Fig. P2.10.

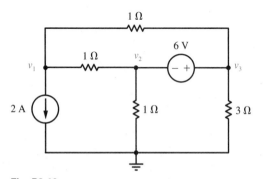

Fig. P2.10

2.11 Fig. P2.11 shows a single transistor amplifier circuit where the portion in the shaded box is the **hybrid-** or **h-parameter model** of a bipolar junction transistor (BJT). Note that h_i is a resistance and h_o is a conductance. Suppose that $h_i = 1$ kΩ, $h_r = 2.5 \times 10^{-4}$, $h_f = 50$, and $h_o = 25$ $\mu\mho$. (a) Use nodal analysis to find the voltage gain v_2/v_1 of this amplifier. (b) Determine the input resistance v_1/i_1 of this amplifier.

2.12 Fig. P2.11 shows a single transistor amplifier circuit where the portion in the shaded box is the hybrid- or h-parameter model of a BJT. Note that h_i is a resistance and h_o is a conductance. Use nodal analysis to show that the voltage gain v_2/v_1 of this amplifier is

$$\frac{v_2}{v_1} = \frac{-h_f R_L}{h_i + (h_i h_o - h_f h_r)R_L}$$

2.13 Fig. P2.11 shows a single transistor amplifier circuit where the portion in the shaded box is the hybrid- or h-parameter model of a BJT. Note that h_i is a resistance and h_o is a conductance. Use the result given in Problem 2.12 to show that the input resistance v_1/v_1 of this amplifier is

$$\frac{v_1}{i_1} = h_i - \frac{h_f h_r}{h_o + 1/R_L}$$

2.14 The circuit shown in Fig. P2.14 is a single BJT amplifier with "feedback." The portion of the circuit in the shaded box is an approximate T-model of a transistor in the common-emitter configuration. (a) Use nodal analysis to find the voltage gain

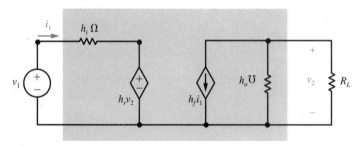

Fig. P2.11

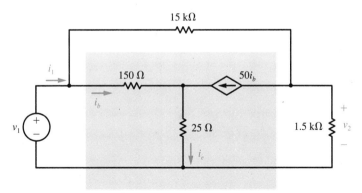

Fig. P2.14

v_2/v_1 of the amplifier. (b) Use the results of part (a) to determine the input resistance v_1/i_1 of the amplifier.

2.15 The circuit shown in Fig. P2.15 is a single BJT amplifier. The portion of the circuit in the shaded box is an approximate T-model of a transistor in the common-base configuration. (a) Use nodal analysis to find the voltage gain v_2/v_1 of the amplifier. (b) Use the results of part (a) to determine the input resistance v_1/i_e of the amplifier.

2.16 For the circuit given in Fig. P2.16, use nodal analysis to determine the conductance $G = i/v$ loading the source.

2.17 Suppose that a circuit contains the shaded box in Fig. P2.17a. Without affecting the remainder of the circuit, the box in Fig. P2.17a can be replaced by the box in Fig. P2.17b, provided that

$$G_{AB} = \frac{G_A G_B}{G_A + G_B + G_C} \quad G_{AC} = \frac{G_A G_C}{G_A + G_B + G_C}$$

$$G_{BC} = \frac{G_B G_C}{G_A + G_B + G_C}$$

Such a process is called a **Y-Δ (wye-delta) transformation.** The circuit in Fig. P2.17c is identical to the circuit given in Fig. P2.16. Use a Y-Δ transformation on the 1-℧, 2-℧, and 5-℧ conductances, and then combine elements in series and parallel to determine $G = i/v$. (See p. 102.)

2.18 Suppose that a circuit contains the shaded box in Fig. P2.17b. Without affecting the remainder of the circuit, the box in Fig. P2.17b can be replaced by the box in Fig. 2.17a, provided that

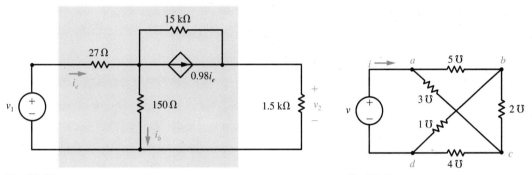

Fig. P2.15   **Fig. P2.16**

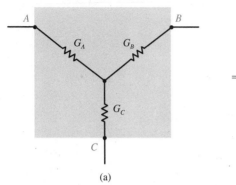

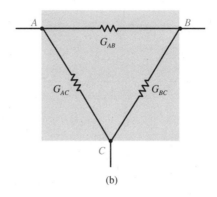

Fig. P2.17 *a,b*

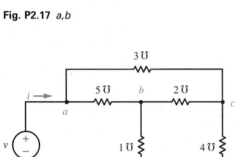

Fig. P2.17 *c*

$$R_A = \frac{R_{AB}R_{AC}}{R_{AB} + R_{AC} + R_{BC}} \qquad R_B = \frac{R_{AB}R_{BC}}{R_{AB} + R_{AC} + R_{BC}}$$

$$R_C = \frac{R_{AC}R_{BC}}{R_{AB} + R_{AC} + R_{BC}}$$

where $R = 1/G$. Such a process is called a **Δ-Y (delta-wye) transformation.**

The circuit shown in Fig. P2.18 is identical to the circuit given in Fig. P2.16. Use a Δ-Y transformation on the 2-℧, 3-℧, and 5-℧ conductances, and then combine elements in series and parallel to determine $G = i/v$.

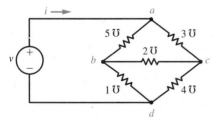

Fig. P2.18

2.19 Find the mesh currents for the circuit shown in Fig. P2.19.

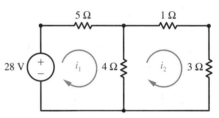

Fig. P2.19

2.20 Assume clockwise mesh currents for the circuit shown in Fig. 2.9 on p. 64. Use mesh analysis to find these mesh currents.

2.21 Assume clockwise mesh currents for the circuit shown in Fig. P2.7. Use mesh analysis to find these mesh currents.

2.22 Assume clockwise mesh currents for the circuit shown in Fig. P2.9. Use mesh analysis to find these mesh currents.

2.23 Assume clockwise mesh currents for the circuit shown in Fig. P2.10. Use mesh analysis to find these mesh currents.

2.24 Use mesh analysis to find the conductance $G = i/v$ for the circuit given in Fig. P2.18.

2.25 Assume clockwise mesh currents for the circuit shown in Fig. P2.8. Use mesh analysis to find these mesh currents.

2.26 Assume clockwise mesh currents for the circuit shown in Fig. P2.26 (below). Use mesh analysis to find these mesh currents.

2.27 For the circuit shown in Fig. P2.27, find v_o when the ideal amplifier (a) is an op amp, and (b) has finite gain A.

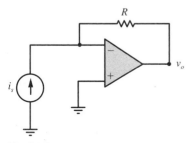

Fig. P2.27

2.28 For the op-amp circuit shown in Fig. P2.28, find (a) v_o, and (b) i_o.

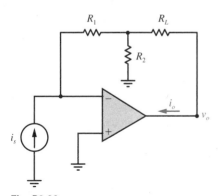

Fig. P2.28

2.29 For the op-amp circuit shown in Fig. P2.29, find (a) v_o, and (b) i_o.

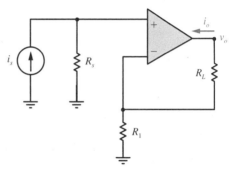

Fig. P2.29

2.30 The op-amp circuit shown in Fig. P2.30 is known as a **negative-impedance converter.** For this circuit, find (a) v_o, and (b) the resistance v_s/i_s.

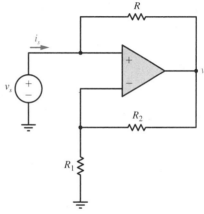

Fig. P2.30

2.31 For the op-amp circuit shown in Fig. P2.31, find (a) v_o, and (b) the resistance v_s/i_s. (See p. 104.)

2.32 For the op-amp circuit shown in Fig. P2.31, interchange the 1-Ω and 2-Ω resistors, and find (a) v_o, and (b) the resistance v_s/i_s. (See p. 104.)

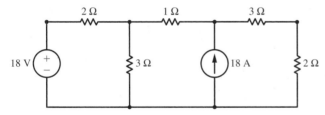

Fig. P2.26

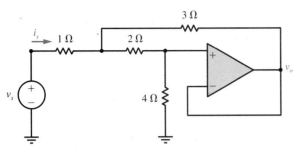

Fig. P2.31

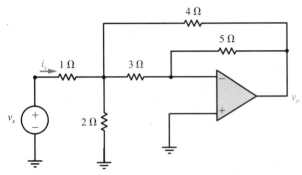

Fig. P2.33

2.33 For the op-amp circuit shown in Fig. P2.33, find (a) v_o, and (b) the resistance v_s/i_s.

2.34 For the op-amp circuit shown in Fig. P2.34, find (a) v_o, and (b) the resistance v_s/i_s. (See p. 105.)

2.35 For the op-amp circuit shown in Fig. P2.35, find v_o.

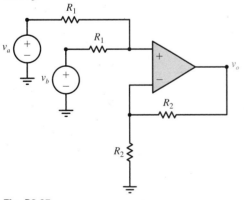

Fig. P2.35

2.36 For the op-amp circuit shown in Fig. P2.36, find v_o. (See p. 105.)

2.37 Consider the circuit shown in Fig. P2.37. (a) Find the Thévenin equivalent of the circuit to the left of terminals a and b. (b) Use the Thévenin-equivalent circuit to find the power absorbed by $R_L = 2 \, \Omega$. (c) Determine the value of R_L, which absorbs the maximum amount of power, and find this power.

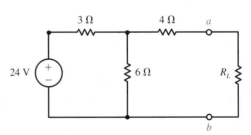

Fig. P2.37

2.38 For the circuit shown in Fig. P2.37, connect a 12-Ω resistor between terminal a and the positive terminal of the voltage source. (a) Find the Thévenin equivalent of the resulting circuit to the left of ter-

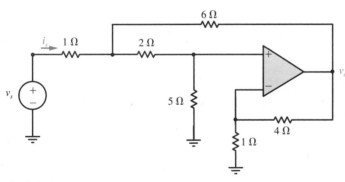

Fig. P2.34

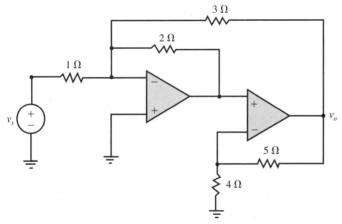

Fig. P2.36

minals a and b. (b) Use the Thévenin-equivalent circuit to find the power absorbed by $R_L = 2\ \Omega$. (c) Determine the value of R_L which absorbs the maximum amount of power, and find this power.

2.39   Consider the circuit shown in Fig. P2.39. (a) Find the Thévenin equivalent of the circuit to the left of terminals a and b. (b) Use the Thévenin-equivalent circuit to find i and the power absorbed by R_L when $R_L = 6\ \Omega$. (c) Determine the value of R_L, which absorbs the maximum amount of power, and find this power. (See p. 106.)

2.40   Consider the circuit shown in Fig. P2.40. (a) Find the Thévenin equivalent of the circuit to the left of terminals a and b. (b) Use the Thévenin-equivalent circuit to find v and the power absorbed by R_L when $R_L = 3\ \Omega$. (c) Determine the value of R_L, which

absorbs the maximum amount of power, and find this power. (See p. 106.)

2.41   For the circuit given in Fig. P2.41, determine the value of R_L, which absorbs the maximum amount of power, and find this power.

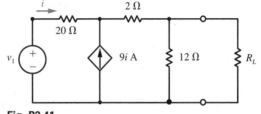

Fig. P2.41

2.42   Find the Norton equivalent of the circuit to the left of terminals a and b for the circuit shown in Fig. P2.42. Use this result to find i.

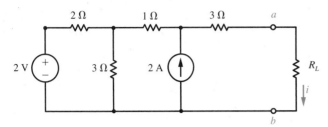

Fig. P2.39

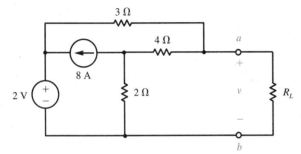

Fig. P2.40

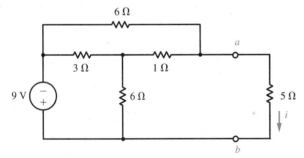

Fig. P2.42

2.43 Find the Norton equivalent of the circuit shown in Fig. P2.43.

2.44 Find the Thévenin equivalent of the circuit shown in Fig. P2.44.

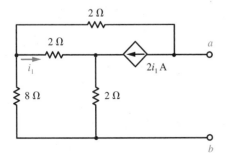

Fig. P2.43

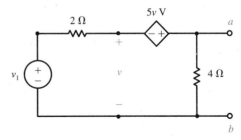

Fig. P2.44

2.45 Find the Thévenin equivalent of the op-amp circuit shown in Fig. P2.45. (*Hint:* To find R_o, apply a current source i_o and calculate the resulting voltage v_o.)

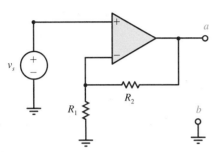

Fig. P2.45

2.46 Find the Thévenin equivalent of the op-amp circuit shown in Fig. P2.46. (*Hint:* To find $2R_o$, apply a current source i_o and calculate the resulting voltage v_o.)

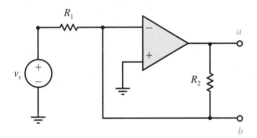

Fig. P2.46

2.47 Show that the Norton equivalent of the circuit shown in Fig. P2.47 is an ideal current source. (*Hint:* To find R_o, apply a voltage source v_o and calculate the resulting current i_o.)

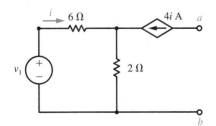

Fig. P2.47

2.48 For the circuit shown in Fig. P2.48, find (a) the Norton-equivalent circuit, and (b) the Thévenin-equivalent circuit.

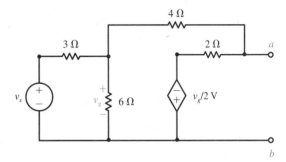

Fig. P2.48

2.49 Figure P2.49 demonstrates the concept of a **source transformation.** Specifically, the voltage source v_s connected in series with a resistance R_s is equivalent to a current source v_s/R_s connected in parallel with R_s. Without using Thévenin's or Norton's theorem, confirm the equivalence of Fig. P2.49*a* and *b* by writing expressions relating i and v.

2.50 Figure P2.50 also demonstrates the concept of a **source transformation.** Specifically, the current source i_g connected in parallel with R_g is equivalent to a voltage source $R_g i_g$ connected in series with a resistance R_g. Without using Thévenin's or Norton's theorem, confirm the equivalence of Fig. P2.50*a* and *b* by writing expressions relating i and v.

2.51 Use source transformations as described in Problems 2.49 and 2.50 to reduce the circuit given in Fig. P2.8 to a circuit having one mesh. Calculate v_3 from this reduced circuit.

2.52 Use source transformations as described in Problems 2.49 and 2.50 to reduce the circuit given in Fig. P2.9 to a circuit having one mesh. Calculate v_3 from this reduced circuit.

2.53 Use source transformations as described in Problems 2.49 and 2.50 and combine independent sources to reduce the circuit given in Fig. P2.39 to a circuit having one mesh. Calculate i from this reduced circuit when $R_L = 6 \ \Omega$.

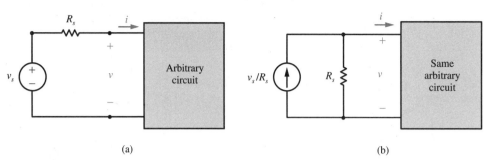

(a)   (b)

Fig. P2.49 *a,b*

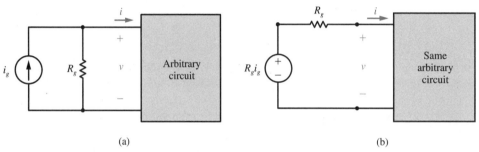

(a)   (b)

Fig. P2.50 *a,b*

2.54 Confirm that the source transformations described in Problems 2.49 and 2.50 can be applied to dependent sources, as well as independent sources, by reducing the circuit given in Fig. P2.7 to a circuit with one independent and one dependent current source, and then determining v_2.

2.55 Consider the circuit shown in Fig. 2.6 on p. 56. (a) Find the portion of i_3 that is due to the 6-A current source. (b) Find the portion of i_3 that is due to the 12-A current source. (c) Find i_3.

2.56 Consider the circuit shown in Fig. 2.10 on p. 66. (a) Find the portion of v_3 that is due to the 5-V voltage source. (b) Find the portion of v_3 that is due to the 10-V voltage source. (c) Find v_3.

2.57 Consider the circuit shown in Fig. P2.39, where $R_L = 6\ \Omega$. (a) Find the portion of i that is due to the 2-V voltage source. (b) Find the portion of i that is due to the 2-A current source. (c) Find i.

2.58 Consider the circuit shown in Fig. P2.40, where $R_L = 3\ \Omega$. (a) Find the portion of v that is due to the 2-V voltage source. (b) Find the portion of v that is due to the 8-A current source. (c) Find v.

2.59 Consider the circuit shown in Fig. P2.59. (a) Find the portion of i and the portion of v that are due to the 6-V voltage source. (b) Find the portion of i and the portion of v that are due to the 2-A current source. (c) Find i and v.

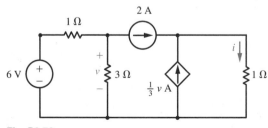

Fig. P2.59

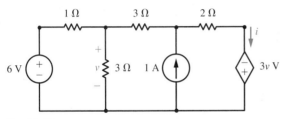

Fig. P2.60

2.60 Consider the circuit shown in Fig. P2.60. (a) Find the portion of i and the portion of v that are due to the 6-V voltage source. (b) Find the portion of i and the portion of v that are due to the 1-A current source. (c) Find i and v.

2.61 Consider the circuit shown in Fig. P2.61. (a) Find the portion of i and the portion of v that are due to the 2-A current source. (b) Find the portion of i and the portion of v that are due to the 6-V voltage source. (c) Find the portion of i and the portion of v that are due to the 4-V voltage source. (d) Find i and v.

2.62 Consider the circuit shown in Fig. P2.62. (a) Find the portion of i and the portion of v that are due to the 12-V voltage source. (b) Find the portion of i and the portion of v that are due to the 6-V voltage source. (c) Find the portion of i and the portion of v that are due to the 6-A current source. (d) Find i and v.

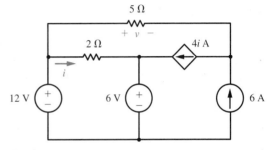

Fig. P2.62

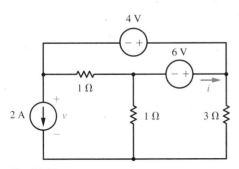

Fig. P2.61

3

Time-Domain Circuit Analysis

INTRODUCTION

The applications of electric circuits consisting solely of sources and resistors are quite limited. Furthermore, sources that produce voltages or currents that vary with time can be extremely useful. Because of this, in the present chapter, we introduce two circuit elements, the inductor and the capacitor, which have voltage-current relationships that are not just direct proportionalities. Instead, the behavior of each element can be described by either a differential or an integral relationship.

Resistors dissipate power (or energy), but inductors and capacitors store energy. The relationships that describe this property are developed in this chapter. We shall also see how to write node and mesh equations for circuits containing inductors and capacitors. By using the concept of duality, the analysis of a planar circuit results in the automatic analysis of a second circuit.

To analyze a resistive circuit, we can write a set of node or mesh equations—which are algebraic equations—and solve them. However, for circuits with inductors and/or capacitors, not all the equations will be algebraic. To see how to analyze such circuits, we begin by considering simple circuits, which, in addition to resistors, contain a single inductor or capacitor. Such a network, called a **first-order circuit,** is described by a first-order, linear differential equation. We first consider the situation in which an initial condition is present but an independent source is not (the natural response), and then the situation in which an independent source is present (the complete response). The analysis of such circuits requires the solution of "simple" differential equations (a topic that is part of our study). Once a simple circuit has been analyzed, we will see that the use of various principles such as linearity and time-invariance will enable us to handle more complicated situations without having to start from scratch.

After dealing with circuits containing either a single inductor or capacitor, we consider the case of two energy-storage elements in a circuit. Most such networks, called **second-order circuits,** are described by second-order, linear differential equations, the solution of which takes three forms. As with first-order circuits, we start by considering natural responses, and follow that with the study of complete responses. We shall see that the complexity of analysis increases significantly as compared with that encountered with first-order circuits. It is because of this fact that we do not extend such an approach to higher-order circuits, but take a different approach to circuit analysis in the following chapters.

3.1 Inductors and Capacitors

In our high school science courses (or earlier), we learned that a current going through a conductor, such as a wire, produces a magnetic field around that conductor. Furthermore, winding such a conductor into a coil strengthens the magnetic field. For the resulting element, known as an **inductor,** the voltage across it is directly proportional to the time rate of change of the current through it. This fact is credited to the independent work of the American inventor Joseph Henry (1797–1878) and the English physicist Michael Faraday (1791–1867).

To signify a coil of wire, we represent an **ideal inductor** as shown in Fig. 3.1. The relationship between voltage and current (for the polarity and direction, respectively indicated) for this element is given by

$$v = L\frac{di}{dt}$$

(3.1)

where L is the **inductance** of the element, and its unit is the **henry** (abbreviated H) in honor of Joseph Henry. Remember that in the differential relationship just given, v and i represent the functions of time $v(t)$ and $i(t)$, respectively.

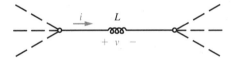

Fig. 3.1 Inductor.

For the special case that $i(t)$ is a constant current—called a **direct current** (abbreviated **dc**)—i.e., for the special case that $i(t) = K$, from Eq. 3.1 we have that

$$v = L\frac{di}{dt} = L\frac{dK}{dt} = 0 \text{ V}$$

In other words, for the dc case, the voltage across an inductor is 0 V. Thus, *an inductor behaves as a short circuit to dc.*

Given the current through an inductor, for the case that $i(t)$ is not constant with time, the determination of $v(t)$ involves taking the derivative of $i(t)$.

Example 3.1

For the inductor shown in Fig. 3.1, suppose that $L = 2$ H and the current $i(t)$ is described by the function of time given in Fig. 3.2a. Since

$$v = L\frac{di}{dt} = 2\frac{di}{dt}$$

then a sketch of $v(t)$ is as shown in Fig. 3.2b.

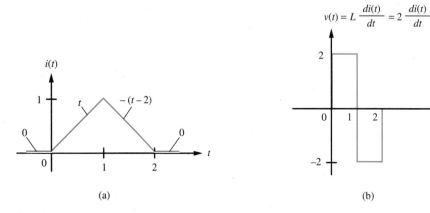

Fig. 3.2 (*a*) Nonconstant current, and (*b*) resulting inductor voltage.

We know that a resistor always absorbs power and that the energy absorbed is dissipated as heat. But how about an inductor? For the inductor in question, the instantaneous power $p(t) = v(t)i(t)$ absorbed by the inductor is as shown in Fig. 3.3. We now see that the power absorbed by the inductor is 0 W for $t \le 0$ s and $t \ge 2$ s. For $0 < t < 1$ s, since $p(t) = 2t$ is a positive quantity, the inductor absorbs power (which is supplied by the source). However, for $1 < t < 2$ s, since $p(t) = 2(t - 2)$

is a negative quantity, the inductor actually supplies power (to the source of the current)!

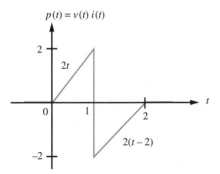

Fig. 3.3 Power absorbed by the inductor.

Drill Exercise 3.1

For the inductor shown in Fig. 3.1, suppose that $L = 3$ H and the current through the inductor is described by

$$i(t) = \begin{cases} 0 \text{ A} & \text{for } t \le \text{s} \\ 2t \text{ A} & \text{for } 0 < t \le 1 \text{ s} \\ 2 \text{ A} & \text{for } t > 1 \text{ s} \end{cases}$$

Find the voltage $v(t)$ across the inductor and the power $p(t)$ absorbed by the inductor. Sketch these functions.

ANSWER 0 V for $t \le 0$ s, 6 V for $0 < t \le 1$ s, 0 V for $t > 1$ s;
0 W for $t \le 0$ s, $12t$ W for $0 < t \le 1$ s, 0 W for $t > 1$ s

Energy Stored by an Inductor

Since energy is the product of power and time, to get the energy absorbed by an inductor, we just integrate the power absorbed over time.

Given an expression for $p(t)$, the instantaneous power absorbed by an inductor, we can obtain an expression for the energy absorbed by the inductor—call it $w_L(t)$—

we just sum up the power over time. In other words, the energy absorbed by an inductor at time t is

$$w_L(t) = \int_{-\infty}^{t} p(t)\, dt$$

However, rather than find the particular expression $w_L(t)$ for a specific example, let us derive a formula for the energy absorbed by an inductor. We proceed as follows:

For an inductor (see Fig. 3.1), since

$$p(t) = i(t)v(t) \qquad \text{and} \qquad v(t) = L\frac{di(t)}{dt}$$

then

$$p(t) = i(t)\left[L\frac{di(t)}{dt}\right] = Li(t)\frac{di(t)}{dt}$$

Thus the energy absorbed by the inductor at time t is $w_L(t)$, where

$$w_L(t) = \int_{-\infty}^{t} p(t)\, dt = \int_{-\infty}^{t} Li(t)\frac{di(t)}{dt}\, dt$$

By using the chain rule of calculus, the variable of integration can be changed from t to $i(y)$. Changing the limits of integration appropriately results in

$$w_L(t) = \int_{i(-\infty)}^{i(t)} Li(t)\, di(t) = L\frac{i^2(t)}{2}\bigg|_{i(-\infty)}^{i(t)} = \frac{1}{2}L[i^2(t) - i^2(-\infty)]$$

Assuming that we can go back far enough in time, we will eventually return to that time when there was no current through the inductor. Thus we adopt the convention that $i(-\infty) = 0$ A. The result is the following expression for the energy absorbed at time t by an inductor whose value is L henries[1]:

$$w_L(t) = \tfrac{1}{2}Li^2(t)$$

---

[1]The plural of the unit "henry" is spelled either "henries" or "henrys."

Example 3.2

Let us determine the energy absorbed by the inductor described in Example 3.1.

Since the current through the 2-H inductor is described by Fig. 3.2a, then the energy absorbed by the inductor is depicted in Fig. 3.4.

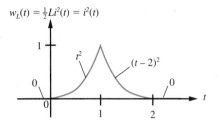

Fig. 3.4 Energy stored by the inductor.

In this example, the energy absorbed by the inductor increases from 0 to $\frac{1}{2}(1)(2) = 1$ J as time goes from $t = 0$ to $t = 1$ s. However, from $t = 1$ s to $t = 2$ s, the inductor supplies energy, such that at time $t = 2$ s, and thereafter, the energy absorbed by the inductor is 0 J. Since the energy absorbed by the inductor is not dissipated, but is eventually returned, we say that the inductor **stores** energy. The energy is stored in the magnetic field that surrounds the inductor.

Drill Exercise 3.2

Determine the energy stored by the inductor described in Drill Exercise 3.1. Sketch this function.

ANSWER 0 J for $t \leq 0$ s, $6t^2$ J for $0 < t \leq 1$ s, 6 J for $t > 1$ s

Unlike an ideal inductor, an actual inductor (an automobile ignition coil is an example) has some resistance associated with it. This arises from the fact that a physical inductor is constructed from real wire that may have a very small, but still nonzero resistance. For this reason, we model a physical inductor as an ideal inductor in series with a resistance. Practical inductors (also called ''coils'' and ''chokes'') range in value from about 10^{-6} H = 1 μH all the way up to around 100 H. To construct inductors having large values requires many turns of wire and iron cores, and consequently results in larger-value series resistances. The series resistance is typically in the range from a fraction of an ohm to several hundred ohms.

The Capacitor

Another extremely important circuit element is obtained when two conducting surfaces (called **plates**) are placed in proximity to one another, and between them is a nonconducting material called a **dielectric**. For the resulting element, known as a **capacitor** (formerly **condenser**), a voltage across the plates results in an electric field between them and the current through the capacitor is directly proportional to the time rate of change of the voltage across it. The **ideal capacitor** is depicted in Fig. 3.5 and the relationship between current and voltage (for the direction and polarity respectively indicated) is given by[2]

$$i = C\frac{dv}{dt} \qquad\qquad (3.2)$$

where C is the **capacitance** of the element, and its unit is the **farad** (abbreviated F) in honor of Michael Faraday.

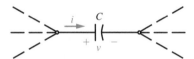

Fig. 3.5 Capacitor.

For the special case that the voltage across a capacitor is constant (i.e., the dc case), then $v = K$, and we have that

$$i = C\frac{dv}{dt} = C\frac{dK}{dt} = 0 \text{ A}$$

and we deduce, therefore, that *a capacitor behaves as an open circuit to dc*.

Perhaps the fact that a capacitor acts as an open circuit to dc does not seem surprising since such an element has a nonconducting dielectric (material that will not allow the flow of charge) between its (conducting) plates. How, then, can there ever be any current through a capacitor? For the case that v is a constant, then indeed $i = 0$ A (i.e., there is no current through the capacitor). However, there is a voltage v across the capacitor and a charge q on the plates. From elementary physics, we know that $q = Cv$. For the case that the voltage is not constant with time (let's

---

[2]For those who remember from elementary physics that $q = Cv$, where q is charge, C is capacitance, and v is voltage, by taking the derivative, we get

$$C\frac{dv}{dt} = \frac{dq}{dt} = i$$

explicitly write $v(t)$), the charge will vary with time, that is, $q(t) = Cv(t)$. Since the net charge on the plates fluctuates, and no charge crosses the dielectric, there must be a transfer of charge throughout the remaining portion of the circuit. Hence when $v(t)$ is nonconstant, there is a nonzero current $i(t) = C\,dv(t)/dt$. In summary, even though there is no flow of charge across the dielectric of a capacitor, the effect of alternate charging and discharging due to a nonconstant voltage results in an actual current outside of the dielectric.

Example 3.3

For the capacitor shown in Fig. 3.5, suppose that $C = 2$ F and the voltage v is described by the function given in Fig. 3.6a. Then the current through the capacitor is as given in Fig. 3.6b. (Compare these results with those given in Figs. 3.2a and 3.2b.) Calculating the instantaneous power absorbed by the capacitor in this example results in the same function $p(t)$ that was obtained in Fig. 3.3. As a consequence, a similar subsequent discussion leads us to the conclusion that a capacitor, like an inductor, is an energy-storage element.

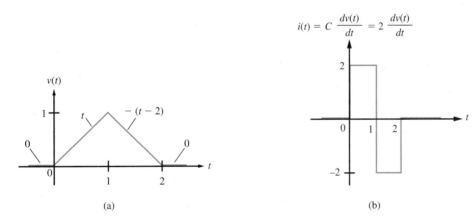

$$i(t) = C\,\frac{dv(t)}{dt} = 2\,\frac{dv(t)}{dt}$$

(a)   (b)

Fig. 3.6 (a) Nonconstant voltage, and (b) resulting capacitor current.

For a capacitor, energy is stored in the electric field that exists between its plates. To obtain an expression for the energy stored in a capacitor, we proceed in a manner analogous to the case of the inductor. The result is that the energy absorbed by a capacitor of C farads at time t is

$$w_C(t) = \tfrac{1}{2}Cv^2(t)$$

Drill Exercise 3.3

Determine the energy stored by the capacitor described in Example 3.3. Sketch this function.

ANSWER   0 J for $t \leq 0$ s, t^2 J for $0 < t \leq 1$ s, $(t - 2)^2$ J for $1 < t \leq 2$ s, 0 J for $t > 2$ s, see Fig. 3.5

Since no physical dielectric is perfect, an actual capacitor will allow a certain amount of current (perhaps extremely small), called **leakage current,** through it. For this reason, we can model a physical capacitor as an ideal capacitor in parallel with a resistance. This leakage resistance is inversely proportional to the capacitance. Depending upon the fabrication of a capacitor, its value can range from a few picofarads (10^{-12} F) up to 10,000 μF and more. The product of leakage resistance and capacitance typically lies between 10 and 10^6 Ω-F. The **working voltage** of a capacitor—that is, the maximum voltage that can be applied to the capacitor without damaging it or breaking down the dielectric—can be anything from a few volts to hundreds, or even thousands, of volts.

The capacitor, like the operational amplifier and the resistor, has the valuable property of miniaturization—it can be made part of integrated circuits. On the other hand, because semiconductors do not possess the appropriate magnetic properties, and because of size limitations, inductors are not, in general, readily adaptable to IC form. This, however, does not diminish the importance of the inductor—your radio, television set, and various kinds of telecommunications equipment employ many of them.

Example 3.4

Let us find the output voltage v_o in terms of the input voltage v for the op-amp circuit shown in Fig. 3.7a for the case that the input voltage $v(t)$ is described by Fig. 3.6a.

Because the inverting input is at a potential of 0 V, then $v_C = v$. Thus

$$i = C\frac{dv_C}{dt} = 2\frac{dv}{dt}$$

Since the inputs of an op amp draw no current, $i_R = i$, and

$$v_R = \frac{1}{2}i_R = \frac{1}{2}i = \frac{1}{2}\left(2\frac{dv}{dt}\right) = \frac{dv}{dt}$$

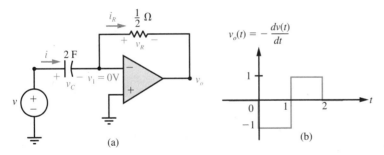

Fig. 3.7 (a) Differentiator circuit, and (b) its output voltage.

Finally, by KVL we have that $v_o = -v_R$. Hence

$$v_o = -\frac{dv}{dt}$$

In other words, the output voltage $v_o(t)$ is the derivative of the input voltage $v(t)$ (multiplied by the constant -1). Because of this, we call such a circuit a **differentiator**.

For the case that $v(t)$ is the function given in Fig. 3.6a, the output voltage $v_o(t)$ is described by the function shown in Fig. 3.7b.

Drill Exercise 3.4

For the op-amp circuit shown in Fig. 3.7a, suppose that the input voltage is described by

$$v(t) = \begin{cases} 0 \text{ V} & \text{for } t \le 0 \text{ s} \\ 2t \text{ V} & \text{for } 0 < t \le 1 \text{ s} \\ 2 \text{ V} & \text{for } t > 1 \text{ s} \end{cases}$$

Find the output voltage $v_o(t)$ and sketch this function.

ANSWER 0 V for $t \le 0$ s, 2 V for $0 < t \le 1$ s, 0 V for $t > 1$ s

---

3.2 Integral Relationships for Inductors and Capacitors

Given any fixed value t_0, the integral of the real function $f(t)$ is

$$\int_{-\infty}^{t_0} f(t)\, dt$$

which is a real number that represents the net area under the curve $f(t)$ for the segment of the t-axis that extends from $-\infty$ to t_0. Since t_0 can be any real number between $-\infty$ and ∞, let us instead use for the upper limit on the integral the variable t. In this way, the integral will not result in a specific number, but rather will result in a function of the variable t. Let us call this function $g(t)$—that is, let us define

$$g(t) = \int_{-\infty}^{t} f(t) \, dt$$

Specifically then, the area under the curve $f(t)$ between $-\infty$ and t_0 is

$$g(t_0) = \int_{-\infty}^{t_0} f(t) \, dt$$

Therefore, for all $t > t_0$, we have that

$$g(t) = \int_{-\infty}^{t} f(t) \, dt = \int_{-\infty}^{t_0} f(t) \, dt + \int_{t_0}^{t} f(t) \, dt = g(t_0) + \int_{t_0}^{t} f(t) \, dt \qquad (3.3)$$

Inductor Integral Relationships

Now let us return to the inductor. With reference to Fig. 3.1, we know that

$$v = L \frac{di}{dt}$$

and this is the differential relationship between voltage and current for an inductor. From this, we may now obtain an integral relationship between voltage and current for an inductor. To do this, first integrate both sides of this equation with respect to time, choosing t_0 and t as the lower and upper limits respectively. Doing this we get

$$\int_{t_0}^{t} v(t) \, dt = \int_{t_0}^{t} L \frac{di(t)}{dt} \, dt$$

By the chain rule of calculus, we can change the variable of integration for the term on the right. Since $t = t_0$ implies that $i(t) = i(t_0)$, we have

$$\int_{t_0}^{t} v(t) \, dt = \int_{i(t_0)}^{i(t)} L \, di(t) = Li(t) \Big|_{i(t_0)}^{i(t)} = L[i(t) - i(t_0)]$$

Hence

$$i(t) - i(t_0) = \frac{1}{L}\int_{t_0}^{t} v(t)\,dt \quad \Rightarrow \quad \boxed{i(t) = i(t_0) + \frac{1}{L}\int_{t_0}^{t} v(t)\,dt} \tag{3.4}$$

But by Eq. 3.3, for $t > t_0$, we can write Eq. 3.4 as

$$\boxed{i(t) = \frac{1}{L}\int_{-\infty}^{t} v(t)\,dt} \tag{3.5}$$

Since Eq. 3.4 and 3.5 are equivalent, either one is referred to as the integral relationship between voltage and current for an inductor.

Example 3.5

For the circuit shown in Fig. 3.8*a*, let us find the inductor current i given that the voltage v is as described in Fig. 3.8*b*.

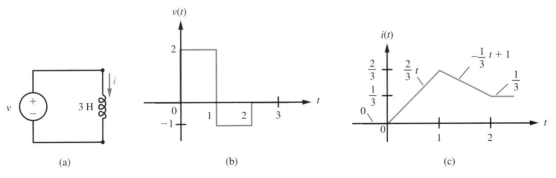

Fig. 3.8 (*a*) Circuit with an inductor, (*b*) waveform of applied voltage, and (*c*) resulting inductor current.

For $t \leq 0$ s, we have that

$$i(t) = \frac{1}{L}\int_{-\infty}^{t} v(t)\,dt = \frac{1}{3}\int_{-\infty}^{t} 0\,dt = 0 \text{ A}$$

For $0 < t \leq 1$ s,

$$i(t) = \frac{1}{L}\int_{-\infty}^{t} v(t)\,dt = i(t_0) + \frac{1}{L}\int_{t_0}^{t} v(t)\,dt = i(0) + \frac{1}{L}\int_{0}^{t} v(t)\,dt$$

and since $i(t) = 0$ A for $t \le 0$ s, then

$$i(t) = 0 + \frac{1}{3}\int_0^t 2\, dt = \frac{1}{3}(2t)\Big|_0^t = \frac{2}{3}t\ \text{A}$$

For $1 < t \le 2$ s,

$$i(t) = \frac{1}{L}\int_{-\infty}^t v(t)\, dt = i(t_0) + \frac{1}{L}\int_{t_0}^t v(t)\, dt = i(1) + \frac{1}{L}\int_1^t v(t)\, dt$$

and since $i(t) = \frac{2}{3}t$ for $0 < t \le 1$ s, then

$$i(t) = \frac{2}{3}(1) + \frac{1}{3}\int_1^t (-1)\, dt = \frac{2}{3} - \frac{1}{3}t\Big|_1^t = \frac{2}{3} - \frac{1}{3}(t-1) = -\frac{1}{3}t + 1\ \text{A}$$

For $t > 2$ s,

$$i(t) = i(t_0) + \frac{1}{L}\int_{t_0}^t v(t)\, dt = i(2) + \frac{1}{L}\int_2^t v(t)\, dt$$

and since $i(t) = -\frac{1}{3}t + 1$ for $1 < t \le 2$ s, then

$$i(t) = \left[-\frac{1}{3}(2) + 1\right] + \frac{1}{3}\int_2^t 0\, dt = \frac{1}{3}\ \text{A}$$

Thus in summary, we have that the current through the inductor is described by Fig. 3.8c.

Drill Exercise 3.5

For the circuit shown in Fig. 3.8a, suppose that $v(t)$ is

$$v(t) = \begin{cases} 0\ \text{V} & \text{for } t \le 0\ \text{s} \\ -3\ \text{V} & \text{for } 0 < t \le 1\ \text{s} \\ 3\ \text{V} & \text{for } 1 < t \le 2\ \text{s} \\ 0\ \text{V} & \text{for } t > 2\ \text{s} \end{cases}$$

Determine $i(t)$ and sketch this function.

ANSWER 0 A for $t \leq 0$ s, $-t$ A for $0 < t \leq 1$ s, $t - 2$ A for $1 < t \leq 2$ s, 0 A for $t > 2$ s

For Example 3.5, we see that for time $t \leq 0$ s, the voltage applied to the inductor is 0 V, and as we would expect, the resulting current through the inductor is 0 A. For time $0 < t \leq 2$ s, the voltage is nonzero, and as our analysis tells us, so is the current. But look—for time $t > 2$ s, the voltage applied is again zero, but the current is not. It remains constant forever! Why?

For time $t > 2$ s, the voltage source has a value of 0 V, and thus behaves as a short circuit. Unlike a resistor, an inductor can have zero volts across it and, at the same time, have a nonzero current through it (remember that an inductor acts as a short circuit to dc). So there is no paradox.

The energy stored by the inductor is shown in Fig. 3.9a. We see that for $t \leq 0$ s, since the current is 0 A, the energy stored is 0 J. When the input voltage is positive, for $0 < t \leq 1$ s, then the energy stored in the inductor increases with time. When the input voltage is negative, for $1 < t \leq 2$ s, the energy stored decreases until time $t = 2$ s, where it is $\frac{1}{6}$ J. For $t > 2$ s, however, when the voltage applied is 0 V, we have the equivalent situation shown in Fig. 3.9b. Since the inductor and the voltage source are ideal, no energy is dissipated, and the energy stays constant for $t > 2$ s. This causes the current to stay constant for $t > 2$ s.

$$w_L(t) = \tfrac{1}{2} L i^2(t) = \tfrac{3}{2} i^2(t)$$

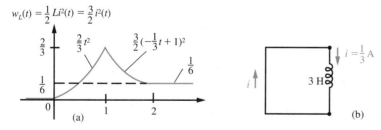

Fig. 3.9 (a) Energy stored in the inductor, and (b) equivalent circuit for $t > 2$ s.

Since all physical inductors and voltage sources have some associated resistance, in a corresponding actual circuit the current $i(t)$ and the energy stored $w_L(t)$ would eventually become zero. The more resistance, the sooner zero values would be reached.

Capacitor Integral Relationships

Returning to the capacitor with the voltage and current indicated as shown in Fig. 3.5, we know that

$$i = C\frac{dv}{dt}$$

and this is the differential relationship between current and voltage for a capacitor. In a manner identical to that discussed for the inductor, we can obtain the equivalent integral relationships between voltage and current for a capacitor. These are

$$v(t) = v(t_0) + \frac{1}{C}\int_{t_0}^{t} i(t)\,dt$$

and

$$v(t) = \frac{1}{C}\int_{-\infty}^{t} i(t)\,dt \qquad (3.6)$$

Example 3.6

For the circuit shown in Fig. 3.10a, let us determine v given that i is described by Fig. 3.10b. In this case, let us consider three intervals of time.

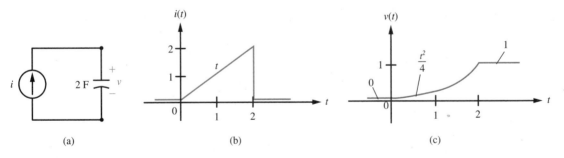

Fig. 3.10 (a) Circuit with a capacitor, (b) waveform of applied current, and (c) resulting capacitor voltage.

For $t \leq 0$ s,

$$v(t) = \frac{1}{C}\int_{-\infty}^{t} i(t)\,dt = \frac{1}{2}\int_{-\infty}^{t} 0\,dt = 0 \text{ V}$$

For $0 < t \leq 2$ s,

$$v(t) = \frac{1}{C} \int_{-\infty}^{t} i(t) \, dt = v(0) + \frac{1}{C} \int_{0}^{t} i(t) \, dt$$

$$= 0 + \frac{1}{2} \int_{0}^{t} t \, dt = \frac{1}{2} \left(\frac{t^2}{2} \right) \Big|_{0}^{t} = \frac{1}{4} t^2 \text{ V}$$

For $t > 2$ s,

$$v(t) = \frac{1}{C} \int_{-\infty}^{t} i(t) \, dt = v(2) + \frac{1}{2} \int_{2}^{t} i(t) \, dt$$

and since $v(t) = \frac{1}{4} t^2$ V for $0 < t \leq 2$ s, then

$$v(t) = \frac{2^2}{4} + \frac{1}{2} \int_{2}^{t} 0 \, dt = 1 \text{ V}$$

In summary, then, the voltage $v(t)$ is as shown in Fig. 3.10c.

Drill Exercise 3.6

For the circuit shown in Fig. 3.10a, find $v(t)$ and sketch this function when $i(t)$ is

$$i(t) = \begin{cases} 0 \text{ A} & \text{for } t \leq 0 \text{ s} \\ 1 \text{ A} & \text{for } 0 < t \leq 1 \text{ s} \\ -2 \text{ A} & \text{for } 1 < t \leq 2 \text{ s} \\ 0 \text{ A} & \text{for } t > 2 \text{ s} \end{cases}$$

ANSWER 0 V for $t \leq 0$ s, $\frac{1}{2} t$ V for $0 < t \leq 1$ s, $-t + \frac{3}{2}$ V for $1 < t \leq 2$ s, $-\frac{1}{2}$ V for $t > 2$ s

---

In Example 3.6, for time $t \leq 0$ s, the current applied to the capacitor is 0 A, and the resulting voltage across it is, therefore, 0 V. For $0 < t \leq 2$ s when the current is nonzero, so is the voltage. For $t > 2$ s, the current is again 0 A, but the voltage remains nonzero—in particular, it stays constant at 1 V.

The energy stored in the capacitor is

$$w_C(t) = \frac{1}{2} C v^2(t) = v^2(t) = \begin{cases} 0 \text{ J} & \text{for } t \leq 0 \text{ s} \\ t^4/16 \text{ J} & \text{for } 0 < t \leq 2 \text{ s} \\ 1 \text{ J} & \text{for } t > 2 \text{ s} \end{cases}$$

Thus the energy stored in the capacitor increases from 0 to 1 J in the time interval $0 < t \leq 2$ s. But for $t > 2$ s, since the current source has a value of 0 A, the capacitor behaves as an open circuit. Therefore, there is no way for the capacitor to discharge—that is, there is no element that can absorb the energy stored in the capacitor. Consequently, the energy stored and the voltage across the capacitor remain constant for $t > 2$ s.

Since a physical capacitor has some associated leakage resistance (it may be very large), and since a practical current source has an internal resistance, then in a corresponding actual circuit the energy stored by and the voltage across the capacitor would eventually become zero—the smaller the effective resistance, the sooner zero is reached.

Example 3.7

Let us find the output voltage v_o in terms of the input voltage v for the op-amp circuit shown in Fig. 3.11.

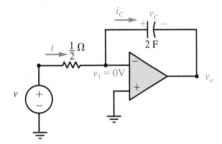

Fig. 3.11 Integrator circuit.

Since $v_1 = 0$ V, then $i = 2v$. Since the input terminals of the op amp draw no current, then $i_C = i = 2v$. Thus

$$v_C = \frac{1}{C}\int_{-\infty}^{t} i_C(t)\, dt = \frac{1}{2}\int_{-\infty}^{t} 2v(t)\, dt = \int_{-\infty}^{t} v(t)\, dt$$

By KVL, $v_o = -v_C$. Hence

$$v_o(t) = -\int_{-\infty}^{t} v(t)\, dt$$

and the output voltage is the integral of the input voltage (multiplied by the constant -1). For this reason such a circuit is known as an **integrator**. This type of circuit is extremely useful because it is the backbone of the analog computer.

Drill Exercise 3.7

For the op-amp integrator circuit shown in Fig. 3.11, suppose that the input is the voltage $v(t)$ given in Fig. 3.6a on p. 117. Determine the output voltage $v_o(t)$. Sketch this function.

ANSWER 0 V for $t \leq 0$ s, $-\frac{1}{2} t^2$ V for $0 < t \leq 1$ s,
$\frac{1}{2}(t^2 - 4t + 2)$ V for $1 < t \leq 2$ s, -1 V for $t > 2$ s

Duality

Let us formalize our discussion of duality, which was mentioned briefly earlier (see Examples 1.11 and 1.12).

Consider the two circuits shown in Fig. 3.12. Using mesh analysis for circuit A and nodal analysis for circuit B, we obtain the following results.

Circuit A   *Circuit B*

$v_s = v_1 + v_2$ $i_s = i_1 + i_2$

$v_s = Ri + L\dfrac{di}{dt}$ $i_s = Gv + C\dfrac{dv}{dt}$

In other words, both circuits are described by the same pair of equations:

$$x_s = x_1 + x_2 \quad \text{and} \quad x_s = a_1 y + a_2 \frac{dy}{dt}$$

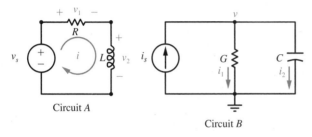

Circuit A

Circuit B

Fig. 3.12 Dual circuits.

Note that a variable that is a current in one circuit is a voltage in the other, and vice versa. For these two circuits, this result is not a coincidence, but rather is due to the concept known as **duality**.

In actuality, circuits A and B are **dual circuits** when the values $R = G$ and $L = C$ and the functions $v_s = i_s$, since the following roles are interchanged:

series ↔ parallel   resistance ↔ conductance

voltage ↔ current   capacitance ↔ inductance

In summary, the relationships between a circuit and its dual are given in the following table.

Circuit	*Dual Circuit*
Series connection	Parallel connection
Parallel connection	Series connection
Voltage of x volts	Current of x amperes
Current of x amperes	Voltage of x volts
Resistance of x ohms	Conductance of x mhos
Conductance of x mhos	Resistance of x ohms
Capacitance of x farads	Inductance of x henries
Inductance of x henries	Capacitance of x farads

The usefulness of duality lies in the fact that once a circuit is analyzed, its dual is, in essence, analyzed also—two for the price of one! It may even be advantageous to construct the dual of a given circuit rather than work with the given circuit itself. Note that if circuit B is the dual of circuit A, then taking the dual of circuit B in essence results in circuit A. Not every circuit, however, has a dual. A circuit has a dual if, and only if, it is a planar network.

Given a series-parallel network, its dual can be found by inspection using the above table. However, a circuit may not be series-parallel. We now present the rules for obtaining the dual of a planar circuit, regardless of whether or not it is a series-parallel network.

Rule 1. Inside of each mesh, including the infinite region surrounding the circuit, place a node.

Rule 2. Suppose two of these nodes, say nodes a and b, are in adjacent meshes. Then there is at least one element in the boundary common to these two meshes. Place the dual of each common element between nodes a and b.

As an example of the above rules, the dual of the circuit drawn in black in Fig. 3.12 is the circuit drawn in color, and this dual circuit is redrawn as shown in Fig. 3.13.

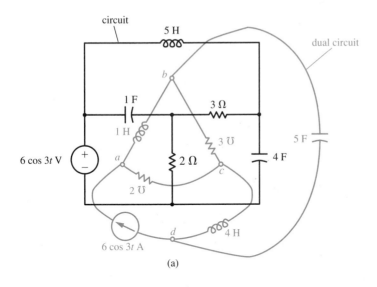

(a)

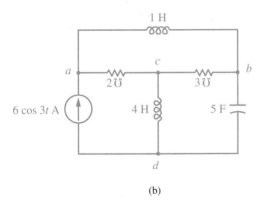

(b)

Fig. 3.13 *(a)* Determination of dual circuit, and *(b)* dual circuit redrawn.

Instantaneous Changes in Current and Voltage

Again consider the inductor shown in Fig. 3.1. Since

$$i(t) = \frac{1}{L} \int_{-\infty}^{t} v(t) \, dt$$

then the current through the inductor at time $t = a$ is

$$i(a) = \frac{1}{L} \int_{-\infty}^{a} v(t) \, dt$$

For any real number $\epsilon > 0$, the current through the inductor at time $t = a + \epsilon$ is

$$i(a + \epsilon) = \frac{1}{L}\int_{-\infty}^{a+\epsilon} v(t)\,dt = \frac{1}{L}\int_{-\infty}^{a} v(t)\,dt + \frac{1}{L}\int_{a}^{a+\epsilon} v(t)\,dt$$

$$= i(a) + \frac{1}{L}\int_{a}^{a+\epsilon} v(t)\,dt$$

If ϵ gets arbitrarily small, then $1/L\int_{a}^{a+\epsilon} v(t)\,dt$ gets arbitrarily small and $i(a + \epsilon)$ gets arbitrarily close to $i(a)$. We conclude, therefore, that **the current through an inductor cannot change instantaneously**.

Since for a capacitor, depicted in Fig. 3.5, it is true that

$$v(t) = \frac{1}{C}\int_{-\infty}^{t} i(t)\,dt$$

proceeding in a manner as above, we can deduce that **the voltage across a capacitor cannot change instantaneously**.

3.3 First-Order Circuits—The Natural Response

Consider the RC circuit shown in Fig. 3.14a. This circuit has an **ideal switch** S, which is **closed** (a short circuit) for time $t < 0$ s. Since for $t < 0$ s this circuit is a dc circuit, the capacitor behaves as an open circuit. Thus by voltage division,

$$v(t) = \frac{RV}{R + R_g} \qquad \text{for } t < 0 \text{ s} \tag{3.7}$$

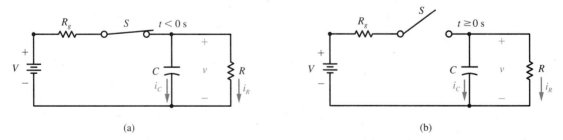

Fig. 3.14 RC circuit (a) with a closed switch, and (b) with the switch opened.

Now suppose that the switch S **opens** (becomes an open circuit) instantaneously at time $t = 0$ s and remains open for $t > 0$ s as shown in Fig. 3.14b. Since the voltage across the capacitor cannot change instantaneously, then the voltage remains the same at $t = 0$ s, that is,

$$v(0) = \frac{RV}{R + R_g} \tag{3.8}$$

But what is $v(t)$ for $t > 0$ s? To answer this question, refer to Fig. 3.14b. By KCL, $i_C + i_R = 0$ or

$$C\frac{dv}{dt} + \frac{v}{R} = 0 \tag{3.9}$$

In Eq. 3.9 there appears the variable $v(t)$ we wish to determine, as well as its first derivative $dv(t)/dt$. For this reason, we say that this equation is a **first-order differential equation**. Since the circuit in Fig. 3.14b is described by a first-order differential equation, we call it a **first-order circuit**. Since every nonzero term in Eq. 3.9 is of the first degree in the dependent variable v and its derivative, we call it an **homogeneous, linear** differential equation. Note that the coefficients of v and its derivative are constants. To solve Eq. 3.9, that is, to find the function $v(t)$ that satisfies the equation and the initial (or boundary) condition of $v(0)$ given by Eq. 3.8, we proceed as follows.

The Natural Response

From the differential Eq. 3.9, we have that

$$\frac{dv}{dt} = -\frac{1}{RC}v$$

Dividing both sides by v (called **separating variables**), we get

$$\frac{1}{v}\frac{dv}{dt} = -\frac{1}{RC} \quad \Rightarrow \quad \int \frac{1}{v}\frac{dv}{dt}dt = \int -\frac{1}{RC}dt$$

after integrating both sides of this equation with respect to time. By the chain rule of calculus, we can write

$$\int \frac{1}{v}\frac{dv}{dt}dt = \int \frac{dv}{v} = \int -\frac{1}{RC}dt \quad \Rightarrow \quad \ln v(t) = -\frac{t}{RC} + K$$

where K is a constant of integration. From this equation, taking powers of e, the result is

$$v(t) = e^{-t/RC + K} = e^{-t/RC}e^{K} \tag{3.10}$$

Setting $t = 0$ s, yields $v(0) = e^{0}e^{K} = e^{K}$, and substituting this value of e^{K} into Eq. 3.10 yields

$$v(t) = v(0)e^{-t/RC} \qquad \text{for } t \geq 0 \text{ s} \tag{3.11}$$

Thus this is the solution to the differential Eq. 3.9 subject to the constraint of the initial condition—that is, the voltage across the capacitor at time $t = 0$ s is $v(0) = RV/(R + R_g)$. A sketch of the function $v(t)$ versus t for $t \geq 0$ s is the color curve shown in Fig. 3.15.

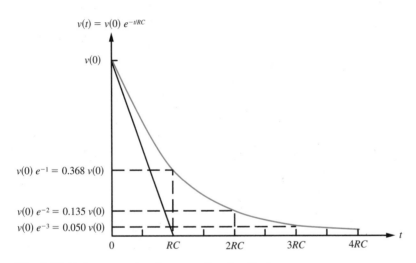

Fig. 3.15 Voltage across the capacitor for $t \geq 0$ s.

Of course, a variable is a variable is a variable—so even if we change the variable's name from $v(t)$ to $x(t)$, by going through the same routine, we get the same results. In other words, the solution to the homogeneous, first-order, linear differential equation (with constant coefficients)

$$\frac{dx(t)}{dt} + ax(t) = 0 \tag{3.12}$$

subject to the initial (or boundary) condition $x(0)$ is

$$x(t) = x(0)e^{-at} \qquad \text{for } t \geq 0 \text{ s} \tag{3.13}$$

Note that the resistor current can now be determined for $t \geq 0$ s since

$$i_R(t) = \frac{v(t)}{R} = \frac{v(0)}{R} e^{-t/RC}$$

Furthermore, the current through the capacitor for $t \geq 0$ s is

$$i_C(t) = -i_R(t) = -\frac{v(0)}{R} e^{-t/RC}$$

or alternatively,

$$i_C(t) = C\frac{dv(t)}{dt} = C\frac{d}{dt}[v(0)e^{-t/RC}] = C\left(-\frac{1}{RC}\right)v(0)e^{-t/RC} = -\frac{v(0)}{R} e^{-t/RC}$$

From these results, we see that initially (at $t = 0$ s) the capacitor is charged to $v(0)$ volts, and for $t > 0$ s, the capacitor discharges through the resistor exponentially. From the formula

$$w_C(t) = \tfrac{1}{2} Cv^2(t)$$

we see that the energy stored in the capacitor at $t = 0$ s is

$$w_C(0) = \tfrac{1}{2} Cv^2(0) \text{ J}$$

Since the voltage goes to zero as time goes to infinity, the energy stored in the capacitor also goes to zero as time goes to infinity. But where does this energy go? The power absorbed by the resistor is

$$p_R(t) = Ri_R^2(t) = R\left(\frac{v(0)}{R} e^{-t/RC}\right)^2 = \frac{v^2(0)}{R} e^{-2t/RC}$$

Therefore, the total energy absorbed by the resistor is

$$w_R = \int_0^\infty p_R(t)\, dt = \int_0^\infty \frac{v^2(0)}{R} e^{-2t/RC} dt = -\frac{RC}{2} \frac{v^2(0)}{R} e^{-2t/RC}\Big|_0^\infty$$

$$= -\tfrac{1}{2} Cv^2(0)[0 - 1] = \tfrac{1}{2} Cv^2(0) = w_C(0)$$

Thus, we see that the energy initially stored in the capacitor is eventually dissipated as heat by the resistor.

Time Constant

We now see that the voltage and current in the circuit shown in Fig. 3.14b have the form

$$f(t) = Ke^{-t/\tau}$$

where K and τ are constants. The constant τ is called the **time constant** of the circuit and, since the power of e should be a dimensionless number, the units of τ, therefore, are seconds. For the voltage and current expressions above, the time constant is $\tau = RC$. Thus we see that the product of resistance and capacitance has seconds as units, that is, ohms $\times$ farads $=$ seconds. This fact can be verified from $R = v/i$ and $C = q/v$, which have the respective units volts per ampere (V/A) $=$ volts per coulomb per second (V/C/s) and coulombs per volt (C/V). Hence the product RC has as its unit the second.

From Fig. 3.15, we see that as t goes to infinity, $v(t)$ goes to zero. Although $v(t)$ never identically equals zero for any finite value of t, when $t = 3\tau$, the value of $v(t)$ is down to only about 5% of the value of $v(t)$ at $t = 0$ s. Thus after just a few time constants, the value of the function $v(t)$ is practically zero. Also note how the initial slope (i.e., the slope immediately after time 0, designated $t = 0^+$) of $v(t)$ is related to τ. Specifically,

$$\frac{dv(t)}{dt} = -\frac{K}{\tau}e^{-t/\tau} \qquad \Rightarrow \qquad \left.\frac{dv(t)}{dt}\right|_{t=0+} = \frac{-K}{\tau}$$

Hence the corresponding tangent line intersects the horizontal axis at $t = \tau$ as shown in Fig. 3.15.

For the circuit given in Fig. 3.14b, for $t \geq 0$ s, the independent source (battery) has no effect on the circuit—it is disconnected. Only the initial condition $v(0)$ has an effect on the circuit for $t \geq 0$ s. For this reason, we say that the voltage and current that we determined for $t \geq 0$ s are **natural responses**.

Example 3.8

Let us find v_C and i_C for the parallel RC circuit shown in Fig. 3.16a for the case that the applied current i_s is described by Fig. 3.16b.

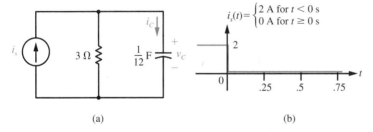

(a)   (b)

Fig. 3.16 (a) Parallel *RC* circuit, and (b) applied current.

For $t < 0$ s, the applied current is $i_s = 2$ A—a constant. Thus for $t < 0$ s we have the dc case. Since the capacitor behaves as an open circuit, then

$$i_C = 0 \text{ A} \quad \text{and} \quad v_C = 3i_s = 3(2) = 6 \text{ V}$$

At time $t = 0$ s, the applied current changes instantaneously (from 2 to 0 A). However, the voltage across the capacitor cannot change instantaneously. Therefore, $v_C(0) = 6$ V, and this is the initial condition.

For $t \geq 0$ s, the circuit given in Fig. 3.16a is not a dc circuit (because the applied current is not constant for all time from $-\infty$ to t). For this circuit, by KCL,

$$i_s = \frac{v_C}{3} + \frac{1}{12}\frac{dv_C}{dt} \quad \Rightarrow \quad \frac{dv_C}{dt} + 4v_C = i_s = 0$$

since $i_s = 0$ A (i.e., the current source is equivalent to an open circuit). As discussed previously, the solution to this differential equation is

$$v_C(t) = v_C(0)e^{-4t} = 6e^{-4t} \text{ V} \qquad \text{for } t \geq 0 \text{ s}$$

and this is the voltage across the capacitor. The current through the capacitor can be found by using the fact that

$$i_C(t) = -\frac{v_C(t)}{3} = \frac{-6e^{-4t}}{3} = -2e^{-4t} \text{ A} \qquad \text{for } t \geq 0 \text{ s}$$

or

$$i_C(t) = \frac{1}{12}\frac{dv_C}{dt} = \frac{1}{12}\frac{d}{dt}(6e^{-4t}) = \frac{1}{12}(-24e^{-4t}) = -2e^{-4t} \text{ A} \qquad \text{for } t \geq 0 \text{ s}$$

In summary, the capacitor voltage and current for the circuit given in Fig. 3.16a are

$$v_C(t) = \begin{cases} 6\text{ V} & \text{for } t < 0 \text{ s} \\ 6e^{-4t}\text{ V} & \text{for } t \geq 0 \text{ s} \end{cases} \quad \text{and}$$

$$i_C(t) = \begin{cases} 0\text{ A} & \text{for } t < 0 \text{ s} \\ -2e^{-4t}\text{ A} & \text{for } t \geq 0 \text{ s} \end{cases}$$

Sketches of $v_C(t)$ and $i_C(t)$ for all time are shown in Fig. 3.17. Note that although the voltage across the capacitor does not change instantaneously, the current through the capacitor does change instantaneously at time $t = 0$ s.

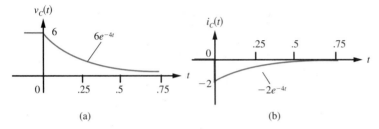

(a) (b)

Fig. 3.17 (a) Voltage across capacitor, and (b) current through capacitor.

Drill Exercise 3.8

For the circuit given in Fig. 3.16a, connect a 1-Ω resistor in series with the capacitor, and determine the capacitor voltage and current for the resulting circuit for the case that the applied current i_s is described by Fig. 3.16b.

ANSWER 6 V for $t < 0$ s, $6e^{-3t}$ V for $t \geq 0$ s; 0 A for $t < 0$ s, $-1.5e^{-3t}$ A for $t \geq 0$ s

A First-Order RL Circuit

Now consider the circuit shown in Fig. 3.18a. In this circuit, the switch S is closed for $t < 0$ s. Since an inductor behaves as a short circuit to dc, the voltage across the inductor is $v = 0$ V. Since this is the voltage across the resistor R, then the current through R is 0 A. Thus

$$i(t) = \frac{V}{R_g} \quad \text{for } t < 0 \text{ s}$$

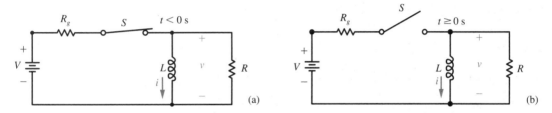

Fig. 3.18 *RL* circuit with (*a*) a switch that is closed, and (*b*) after the switch is opened.

If the switch *S* opens at time $t = 0$ s and remains open for $t > 0$ s as shown in Fig. 3.18*b*, since the current through an inductor cannot change instantaneously, then $i(0) = V/R_g$. For $t \geq 0$ s, by KVL,

$$L\frac{di}{dt} + Ri = 0 \qquad \Rightarrow \qquad \frac{di}{dt} + \frac{R}{L}i = 0 \qquad (3.14)$$

Since this equation has the form of Eq. 3.12, then the solution of Eq. 3.14 has the form of Eq. 3.13. In other words, the solution of differential Eq. 3.14 is

$$i(t) = i(0)e^{-Rt/L} \qquad \text{for } t \geq 0 \text{ s}$$

A sketch of $i(t)$ versus t for $t \geq 0$ s is shown in Fig. 3.19. Note that in this case the time constant is $\tau = L/R$. Thus, dividing henries by ohms yields seconds.

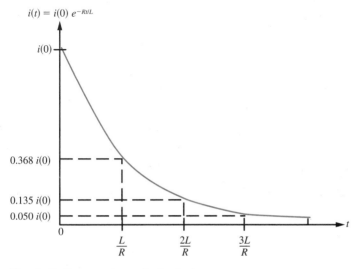

Fig. 3.19 Current through the inductor.

For the *RL* circuit given in Fig. 3.18b, the voltage $v(t)$ is

$$v(t) = L\frac{di(t)}{dt} = L\frac{d}{dt}[i(0)e^{-Rt/L}] = -Ri(0)e^{-Rt/L} \qquad \text{for } t \geq 0 \text{ s}$$

Alternatively, we have that

$$v(t) = -Ri(t) = -Ri(0)e^{-Rt/L} \qquad \text{for } t \geq 0 \text{ s}$$

The energy stored in the inductor initially (at $t = 0$ s) is determined from $w_L(t) = \frac{1}{2}Li^2(t)$ to be

$$w_L(0) = \frac{1}{2}Li^2(0) \text{ J}$$

and as time goes to infinity, since the current in the inductor goes to zero exponentially, the energy stored in the inductor also goes to zero.

The instantaneous power absorbed by the resistor is

$$p_R(t) = Ri^2(t) = R[i(0)e^{-Rt/L}]^2 = Ri^2(0)e^{-2Rt/L}$$

and, therefore, the total energy absorbed by the resistor is

$$w_R = \int_0^\infty p_R(t)\, dt = \int_0^\infty Ri^2(0)e^{-2Rt/L}\, dt = -\frac{L}{2R}Ri^2(0)e^{-2Rt/L}\Big|_0^\infty$$

$$= -\left(\frac{L}{2}\right)i^2(0)[0 - 1] = \frac{1}{2}Li^2(0) = w_L(0)$$

Thus we see that the energy initially stored in the inductor is eventually dissipated as heat by the resistor.

Example 3.9

Let us find i_L and v_L for the series *RL* circuit shown in Fig. 3.20a for the case that the applied voltage v_s is described by Fig. 3.20b.

For $t < 0$ s, the applied voltage is $v_s = 2$ V—a constant. Thus for $t < 0$ s we have the dc case. Since the inductor behaves as a short circuit to dc, then

$$v_L = 0 \text{ V} \qquad \text{and} \qquad i_L = \frac{v_s}{\frac{1}{3}} = \frac{2}{\frac{1}{3}} = 6 \text{ A}$$

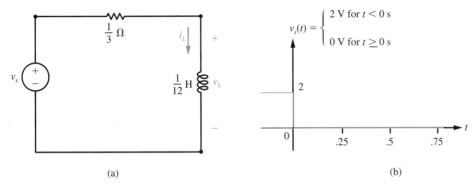

Fig. 3.20 (a) Series *RL* circuit, and (b) applied voltage.

At time $t = 0$ s, the applied voltage changes instantaneously (from 2 to 0 V). However, the current through the inductor cannot change instantaneously. Therefore, $i_L(0) = 6$ A, and this is the initial condition.

For $t \geq 0$ s, the circuit given in Fig. 3.20a is not a dc circuit (because the applied voltage is not constant for all time from $-\infty$ to t). For this circuit, by KVL,

$$v_s = \frac{1}{3}i_L + \frac{1}{12}\frac{di_L}{dt} \quad \Rightarrow \quad \frac{di_L}{dt} + 4i_L = v_s = 0$$

since $v_s = 0$ V (i.e., the voltage source is equivalent to a short circuit). As discussed previously, the solution to this differential equation is

$$i_L(t) = i_L(0)e^{-4t} = 6e^{-4t} \text{ A} \qquad \text{for } t \geq 0 \text{ s}$$

and this is the current through the inductor. The voltage across the inductor can be found by using the fact that

$$v_L(t) = -\tfrac{1}{3}i_L(t) = -\tfrac{1}{3}(6e^{-4t}) = -2e^{-4t} \text{ V}$$

or

$$v_L(t) = \frac{1}{12}\frac{di_L}{dt} = \frac{1}{12}\frac{d}{dt}(6e^{-4t}) = \frac{1}{12}(-24e^{-4t}) = -2e^{-4t} \text{ V}$$

In summary, the inductor current and voltage for the circuit given in Fig. 3.20a are

$$i_L(t) = \begin{cases} 6 \text{ A} & \text{for } t < 0 \text{ s} \\ 6e^{-4t} \text{ A} & \text{for } t \geq 0 \text{ s} \end{cases} \quad \text{and} \quad v_L(t) = \begin{cases} 0 \text{ V} & \text{for } t < 0 \text{ s} \\ -2e^{-4t} \text{ V} & \text{for } t \geq 0 \text{ s} \end{cases}$$

Sketches of $i_L(t)$ and $v_L(t)$ for all time are shown in Fig. 3.21. Note that although the current through the inductor does not change instantaneously, the voltage across the inductor does change instantaneously at time $t = 0$ s.

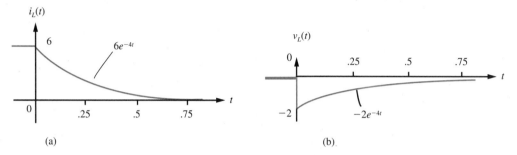

Fig. 3.21 (a) Current through inductor, and (b) voltage across inductor.

Also note that Example 3.9 is the dual of Example 3.8.

Drill Exercise 3.9

For the circuit given in Fig. 3.20a, connect a 1-Ω resistor in parallel with the inductor, and determine the inductor current and voltage for the resulting circuit for the case that the applied voltage v_s is described by Fig. 3.20b. How is this drill exercise related to the preceding one?

ANSWER   6 A for $t < 0$ s, $6e^{-3t}$ A for $t \geq 0$ s; 0 V for $t < 0$ s, $-1.5e^{-3t}$ V for $t \geq 0$ s. It is the dual.

3.4   First-Order Circuits—The Complete Response

Now that we have discussed the responses of first-order circuits that have no applied inputs and nonzero initial (or boundary) conditions, let us consider the case of a response due to a nonzero input.

For the RC circuit shown in Fig. 3.22, the switch S moves (instantaneously) from a to b at time $t = 0$ s. We see that the voltage $v_s(t) = 0$ V for $t < 0$ s, and $v_s(t) = V$ for $t \geq 0$ s. Thus let us define a function $u(t)$, called the **unit step function,** by

$$u(t) = \begin{cases} 0 & \text{for } t < 0 \text{ s} \\ 1 & \text{for } t \geq 0 \text{ s} \end{cases}$$

Hence we can write $v_s(t)$ as the product $v_s(t) = Vu(t)$. This function is depicted in Fig. 3.23.

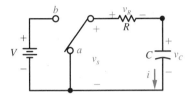

Fig. 3.22 *RC* circuit with an applied step of voltage.

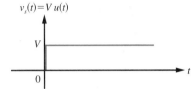

Fig. 3.23 Step function.

From Fig. 3.22 we see that the voltage $v_s(t) = Vu(t)$ is applied to a series *RC* combination. Thus we can call the resulting voltage $v(t)$ and current $i(t)$ **step responses**.

Before we determine a step response, say $v(t)$, mathematically, let us try to predict what will happen. The capacitor is originally uncharged and the input voltage $v_s(t)$ is zero for $t < 0$ s. Since there is no voltage to produce a current that will charge the capacitor, it should be obvious that $v(t) = 0$ V for $t < 0$ s. At time $t = 0$ s, when the input voltage goes from 0 to V volts, since the voltage across the capacitor cannot change instantaneously (the capacitor acts as a short circuit to an instantaneous change of voltage), $v(0) = 0$ V. By KVL, $v_R(0) = V - 0 = V$ volts, and $i(0) = v_R(0)/R = V/R$. So the voltage across the resistor changes instantaneously, as does the current through the resistor, which is the current through the capacitor. The nonzero current through the capacitor begins to charge it to some nonzero value. Since the input voltage remains constant for $t \geq 0$ s, by KVL, an increase in the capacitor voltage means a decrease in the resistor voltage. This, in turn, means a reduction in the resistor current—which is the capacitor current—so that the capacitor charges at a slower rate than before. The charging process continues until the capacitor is completely charged. After a long, long time, the input voltage seems more and more like a constant (dc), so as time goes to infinity, the capacitor behaves as an open circuit. Therefore, the final voltage across the capacitor is V volts. For the case of a natural response, we have seen that a capacitor discharges exponentially. Thus we can take an educated guess that it will charge up exponentially as well.

To summarize, for $t < 0$ s and $t = 0$ s, the voltage across the capacitor is 0 V. For $t > 0$ s, the voltage across the capacitor increases at a slower and slower rate until it is completely charged to V volts. By KVL, the voltage across the resistor is 0 V for $t < 0$ s, it jumps to V volts instantaneously at $t = 0$ s, and decreases exponentially to zero as time increases. Furthermore, the current through the resistor—which equals the current through the capacitor—is proportional to the voltage across the resistor.

Let us now confirm our intuition by a mathematical analysis. Using nodal analysis, by KCL,

$$\frac{Vu(t) - v_C}{R} = C\frac{dv_C}{dt} \quad \Rightarrow \quad \frac{dv_C}{dt} + \frac{1}{RC}v_C = \frac{V}{RC}u(t) \tag{3.15}$$

Since $u(t) = 0$ for $t < 0$ s, and $u(t) = 1$ for $t \geq 0$ s, let us consider differential Eq. 3.15 for the two time intervals $t < 0$ s and $t \geq 0$ s.

First, for $t < 0$ s, the right-hand side of the differential equation becomes zero. Since we want to find the response to the input under the circumstance that there is no initial voltage across the capacitor, and since the excitation is zero before time $t = 0$ s, the solution is $v_C(t) = 0$ V for $t < 0$ s.

Next, for $t \geq 0$ s, differential Eq. 3.15 becomes

$$\frac{dv_C}{dt} + \frac{1}{RC}v_C = \frac{V}{RC} \tag{3.16}$$

Therefore, we are now required to find the solution to this equation, i.e., we wish to determine that function $v_C(t)$ which satisfies Eq. 3.16.

Solution of a Nonhomogeneous First-Order Differential Equation

Eq. 3.16 is a nonhomogeneous first-order differential equation which has the more general form

$$\frac{dx(t)}{dt} + ax(t) = f(t) \tag{3.17}$$

where $x(t) = v_C(t)$, $a = 1/RC$, and $f(t) = V/RC$. To find the solution of this equation, multiply both sides by e^{at}. Therefore, we have

$$e^{at}\frac{dx(t)}{dt} + e^{at}ax(t) = e^{at}f(t)$$

Then

$$\frac{d}{dt}[e^{at}x(t)] = e^{at}\frac{dx(t)}{dt} + e^{at}ax(t) = e^{at}f(t)$$

Integrating both sides of this equation with respect to t, we get

$$\int \frac{d[e^{at}x(t)]}{dt} \, dt = \int e^{at}f(t) \, dt \quad \Rightarrow \quad e^{at}x(t) = \int e^{at}f(t) \, dt + A$$

where A is a constant of integration. Multiplying both sides of this equation by e^{-at}, we obtain the solution to Eq. 3.17, which is

$$x(t) = e^{-at} \int e^{at}f(t) \, dt + Ae^{-at} = x_f(t) + x_n(t)$$

where the constant A is determined from the initial condition. Thus we see that, in general, the solution of differential Eq. 3.17, which is called the **complete response,** consists of the sum of two parts. The first part is

$$x_f(t) = e^{-at} \int e^{at}f(t) \, dt$$

which basically is determined by the function $f(t)$. Since $f(t)$ results from the excitation of a circuit, we call this part of the solution the **forced response** (also known as the **steady-state response**), and we refer to $f(t)$ as the **forcing function**. The other part of the solution is

$$x_n(t) = Ae^{-at}$$

which we recognize has the form of the **natural response**. (It is also called the **transient response**).

Constant Forcing Functions

Now let us be specific and consider the case that the forcing function is a constant, say b. Then differential Eq. 3.17 becomes

$$\frac{dx(t)}{dt} + ax(t) = b$$

The forced response is

$$x_f(t) = e^{-at} \int e^{at}b \, dt = e^{-at}\left(\frac{1}{a}e^{at}\right)b = \frac{b}{a}$$

which is a constant, while the natural response is $x_n(t) = Ae^{-at}$. Thus, the complete response is

$$x(t) = x_f(t) + x_n(t) = \frac{b}{a} + Ae^{-at}$$

where the constant A is determined from the initial condition.

We now have the solution to Eq. 3.16:

$$v_C(t) = \frac{V/RC}{1/RC} + Ae^{-t/RC} = V + Ae^{-t/RC} \tag{3.18}$$

Since $v_C(t) = 0$ V for $t < 0$ s and the voltage across a capacitor cannot change instantaneously, then $v_C(0) = 0$ V. Setting $t = 0$ s in Eq. 3.18, we get

$$v_C(0) = V + Ae^{-0} = 0 \quad \Rightarrow \quad A = -V$$

Thus the voltage across the capacitor is

$$v_C(t) = V - Ve^{-t/RC} = V(1 - e^{-t/RC}) \qquad \text{for } t \geq 0 \text{ s}$$

Combining this result with the fact that $v_C(t) = 0$ V for $t < 0$ s, we can express $v_C(t)$ for all t as the product

$$v_C(t) = V(1 - e^{-t/RC})u(t) \tag{3.19}$$

Having determined the capacitor voltage, we can now find the capacitor current (which is equal to the resistor current in this case). For $t < 0$ s, $i(t) = C\,dv_C(t)/dt = 0$ A.

For $t \geq 0$ s,

$$i(t) = C\frac{d}{dt}[V(1 - e^{-t/RC})] = CV(-e^{-t/RC})\left(-\frac{1}{RC}\right) = \frac{V}{R}e^{-t/RC}$$

Thus we can write the single expression for all time (i.e., $-\infty < t < \infty$):

$$i(t) = \frac{V}{R}e^{-t/RC}u(t) \tag{3.20}$$

Sketches of the step responses $v_C(t)$ and $i(t)$ are shown in Fig. 3.24.

Fig. 3.24 Waveforms of the resulting capacitor voltage and current.

Example 3.10

Let us find the step responses $i_L(t)$ and $v_L(t)$ for the series *RL* circuit shown in Fig. 3.25a.

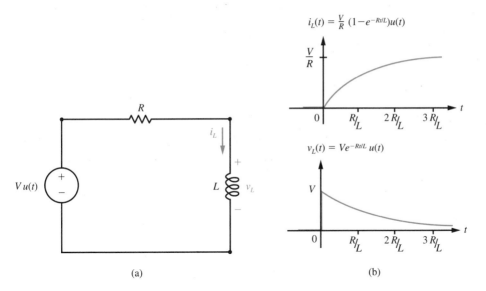

(a)   (b)

Fig. 3.25 (*a*) Series *RL* circuit, and (*b*) step responses.

By KVL, we have that

$$Vu(t) = Ri_L + L\frac{di_L}{dt} \quad \Rightarrow \quad \frac{di_L}{dt} + \frac{R}{L}i_L = \frac{V}{L}u(t) \qquad (3.21)$$

For $t < 0$ s, the excitation is zero, and therefore, the solution to this differential that we seek is $i_L(t) = 0$ A. Since the current through an inductor cannot change instantaneously, we also have that $i_L(0) = 0$ A.

For $t \geq 0$ s, Eq. 3.21 becomes

$$\frac{di_L}{dt} + \frac{R}{L}i_L = \frac{V}{L} \quad \Rightarrow \quad i_L(t) = \frac{V/L}{R/L} + Ae^{-Rt/L} = \frac{V}{R} + Ae^{-Rt/L} \quad (3.22)$$

Since $i_L(t) = 0$ A for $t < 0$ s and the current through an inductor cannot change instantaneously, then $i_L(0) = 0$ A. Setting $t = 0$ s in Eq. 3.22, we get

$$i_L(0) = V/R + Ae^{-0} = 0 \quad \Rightarrow \quad A = -V/R$$

Thus the current through the inductor is

$$i_L(t) = \frac{V}{R} - \frac{V}{R}e^{-Rt/L} = \frac{V}{R}(1 - e^{-Rt/L}) \quad \text{for } t \geq 0 \text{ s}$$

Combining this result with the fact that $i_L(t) = 0$ A for $t < 0$ s, we can express $i_L(t)$ for all t (i.e., $-\infty < t < \infty$) as the product

$$i_L(t) = \frac{V}{R}(1 - e^{-Rt/L})u(t) \quad (3.23)$$

Having determined the inductor current, we can now find the inductor voltage. For $t < 0$ s, $v_L(t) = L \, di_L(t)/dt = 0$ V. For $t \geq 0$ s,

$$v_L(t) = L\frac{d}{dt}\left[\frac{V}{R}(1 - e^{-Rt/L})\right] = L\frac{V}{R}(-e^{-Rt/L})\left(-\frac{R}{L}\right) = Ve^{-Rt/L}$$

Thus, we can write the single expression for all time ($-\infty < t < \infty$):

$$v_L(t) = Ve^{-Rt/L}u(t) \quad (3.24)$$

Sketches of the step responses $i_L(t)$ and $v_C(t)$ are shown in Fig. 3.25b on p. 145.

Drill Exercise 3.10

For the circuit shown in Fig. 3.25a, connected a second resistor R in parallel with the inductor L. Find the step responses $i_L(t)$ and $v_L(t)$ for the resulting circuit.

ANSWER $(V/R)(1 - e^{-Rt/2L})u(t)$, $\frac{1}{2}Ve^{-Rt/2L}u(t)$

Linearity and Time Invariance

Again consider the circuit in Fig. 3.22, but suppose that the switch S returns to position a at time $t = 1$ s. Under this circumstance, the voltage $v_s(t)$ is the voltage pulse shown in Fig. 3.26. To determine the voltage $v_C(t)$ across the capacitor, for $t < 1$ s, we proceed as above. In other words, $v_C(t) = 0$ V for $t < 0$ s, and $v_C(t) = V(1 - e^{-t/RC})$ for $0 \le t < 1$ s. Since the voltage across a capacitor cannot change instantaneously, $v_C(1) = V(1 - e^{-1/RC})$. But for $t \ge 1$ s, we have to determine a natural response. In other words, for $t \ge 1$ s, as discussed in the section on natural responses, we can get Eq. 3.9. A similar approach to the solution of this equation as taken previously results in

$$v_C(t) = v_C(1)e^{-(t-1)/RC} = V(1 - e^{-1/RC})e^{-(t-1)/RC} \qquad \text{for } t \ge 1 \text{ s}$$

A sketch of the voltage $v_C(t)$ is shown in Fig. 3.27.

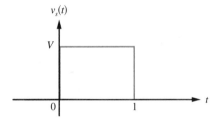

Fig. 3.26 Voltage pulse.

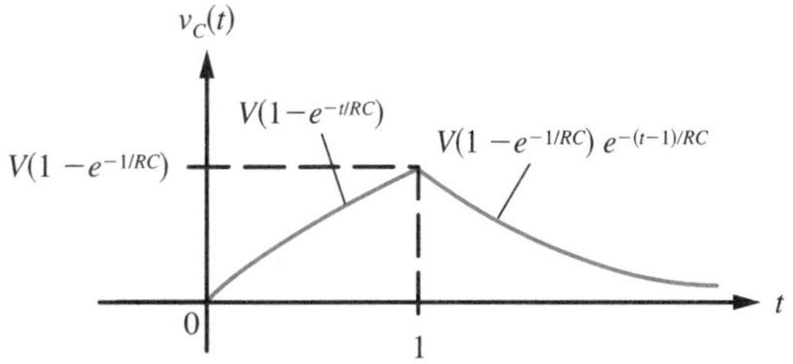

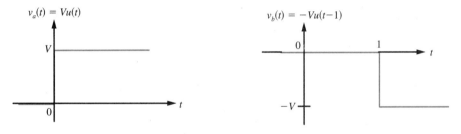

Fig. 3.27 Voltage response to applied pulse.

Although this last problem employed a complete response (for $t < 1$ s) and a natural response (for $t \ge 1$ s), we can use an alternative approach. Since inductors and capacitors are linear circuit elements, we can use the properties of linearity on circuits that contain such elements. In particular, we can express the pulse $v_s(t)$ as the sum of the two voltage steps $v_a(t) = Vu(t)$ and $v_b(t) = -Vu(t - 1)$ as shown in Fig. 3.28. By linearity, the response to $v_s(t) = v_a(t) + v_b(t)$ is the sum of the response

Fig. 3.28 Components of the given voltage pulse.

to $v_a(t)$ and the response to $v_b(t)$. But we have already seen that the response to $v_a(t)$ is

$$v_{Ca}(t) = V(1 - e^{-t/RC})u(t)$$

Also, by linearity, if the input is scaled by a constant, the output is also scaled by the same constant. In particular, the response to $-Vu(t)$ is $-V(1 - e^{-t/RC})u(t)$.

We are now in a position to use another property, known as **time invariance,** which is attributable to the fact that the differential equations describing our circuits have constant coefficients. (Such circuits are called **time-invariant** circuits.) This property says that delaying an excitation by a certain amount of time delays the response by the same amount of time. In particular, the response to $v_b(t) = -Vu(t - 1)$, and this is the input $-Vu(t)$ delayed by 1 s, is

$$v_{Cb}(t) = -V(1 - e^{-(t-1)/RC})u(t - 1)$$

Thus by linearity, we have that the response to $v_s(t) = v_a(t) + v_b(t)$ is

$$v_C(t) = v_{Ca}(t) + v_{Cb}(t) = V(1 - e^{-t/RC})u(t) - V(1 - e^{-(t-1)/RC})u(t - 1)$$

To compare this answer with our previous results, note that when $t < 0$ s, then $u(t) = u(t - 1) = 0$, so $v_C(t) = 0$ V. Also, when $0 \le t < 1$ s, then $u(t) = 1$ and $u(t - 1) = 0$, so $v_C(t) = V(1 - e^{-t/RC})$. Finally, when $t \ge 1$ s, then $u(t) = u(t - 1) = 1$, and we have that

$$v_C(t) = V(1 - e^{-t/RC}) - V(1 - e^{-(t-1)/RC})$$
$$= V(1 - e^{-1/RC})e^{-(t-1)/RC} \qquad \text{for } t \ge 1 \text{ s}$$

which agrees with our previous analysis (see Fig. 3.27).

From Eq. 3.20, the current due to $v_a(t)$ is

$$i_a(t) = \frac{V}{R}e^{-t/RC}u(t)$$

By the linearity and time-invariance properties, the current due to $v_b(t)$ is

$$i_b(t) = -\frac{V}{R}e^{-(t-1)/RC}u(t - 1)$$

Thus by linearity, the current due to $v_s(t) = v_a(t) + v_b(t)$ is

$$i(t) = i_a(t) + i_b(t) = \frac{V}{R}e^{-t/RC}u(t) - \frac{V}{R}e^{-(t-1)/RC}u(t-1)$$

When $t < 0$ s, then $i(t) = 0$ A, whereas when $0 \le t < 1$ s, then $i(t) = (V/R)e^{-t/RC}$. Finally, when $t \ge 1$ s, then

$$i(t) = \frac{V}{R}e^{-t/RC} - \frac{V}{R}e^{-(t-1)/RC} = \frac{V}{R}(e^{-1/RC} - 1)e^{-(t-1)/RC}$$

A sketch of $i(t)$ is shown in Fig. 3.29.

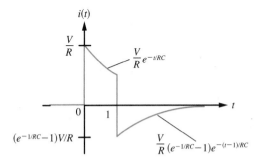

Fig. 3.29 Current response to applied voltage pulse.

Example 3.11

Let us find the current and voltage responses for the series RL circuit given in Fig. 3.25a to the voltage pulse shown in Fig. 3.26 for the case that $R = 1\ \Omega$, $L = 1$ H, and $V = 1$ V.

Since we can write $v_s(t) = v_a(t) + v_b(t)$, where $v_a(t) = u(t)$ V and $v_b(t) = -u(t-1)$ V, then as in Example 3.10 (see Eq. 3.23), the current response to $v_a(t) = u(t)$ V is

$$i_{La}(t) = (1 - e^{-t})u(t)\ \text{A}$$

By the linearity and time-invariance properties, the current response to $v_b(t) = -u(t-1)$ V is

$$i_{Lb}(t) = -(1 - e^{-(t-1)})u(t-1)\ \text{A}$$

Therefore, the current response to $v_s(t) = v_a(t) + v_b(t)$ is

$$i_L(t) = i_{La}(t) + i_{Lb}(t) = (1 - e^{-t})u(t) - (1 - e^{-(t-1)})u(t-1)\ \text{A}$$

The voltage response to $v_a(t) = u(t)$ V is (see Eq. 3.24)

$$v_{La}(t) = e^{-t}u(t) \text{ V}$$

whereas by the linearity and time-invariance properties, the voltage response to $v_b(t) = -u(t - 1)$ V is

$$v_{Lb}(t) = -e^{-(t-1)}u(t - 1) \text{ V}$$

Therefore, the voltage response to $v_s(t) = v_a(t) + v_b(t)$ is

$$v_L(t) = v_{La}(t) + v_{Lb}(t) = e^{-t}u(t) - e^{-(t-1)}u(t - 1) \text{ V}$$

Sketches of the current and voltage responses are shown in Fig. 3.30.

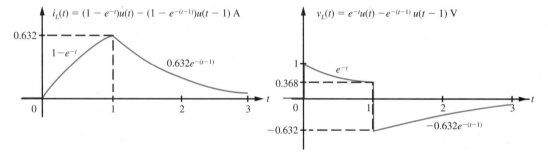

Fig. 3.30 Responses of series *RL* circuit to applied voltage pulse.

Drill Exercise 3.11

Find the voltage $v_C(t)$ and current $i_C(t)$ for the parallel *RC* circuit given in Fig. 3.16a on p. 135 for the case that $R = 1$ Ω, $C = 1$ F, and the applied current is the pulse $i_s(t) = u(t) - u(t - 1)$ A.

ANSWER $(1 - e^{-t})u(t) - (1 - e^{-(t-1)})u(t - 1)$ V,
$e^{-t}u(t) - e^{-(t-1)}u(t - 1)$ A

3.5 Second-Order Circuits—The Natural Response

Having dealt with circuits containing either a single inductor or capacitor, we will now consider the case of two energy-storage elements in a circuit. Most such networks, called **second-order circuits,** are described by second-order, linear differential equations. We shall see that the complexity of analysis increases significantly when compared with that encountered with first-order circuits. It is because of this fact that we do not extend such an approach to higher-order circuits, but instead take a different point of view in the following chapters.

The Series RLC Circuit

Suppose that for the series *RLC* circuit shown in Fig. 3.31, the initial current through the inductor is $i(0)$ A and the initial voltage across the capacitor is $v(0)$ V. By KVL,

$$v + Ri + L\frac{di}{dt} = 0 \tag{3.25}$$

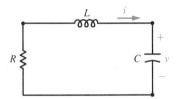

Fig. 3.31 Series *RLC* circuit.

Substituting the fact that $i = C\,dv/dt$ into Eq. 3.25, we get

$$v + RC\frac{dv}{dt} + LC\frac{d^2v}{dt^2} = 0 \quad \Rightarrow \quad \frac{d^2v}{dt^2} + \frac{R}{L}\frac{dv}{dt} + \frac{1}{LC}v = 0 \tag{3.26}$$

This equation has the following general form:

$$\boxed{\frac{d^2y(t)}{dt^2} + 2\alpha\frac{dy(t)}{dt} + \omega_n^2\,y(t) = 0} \tag{3.27}$$

We have seen that the solution to a first-order homogeneous differential equation is $y(t) = Ae^{st}$, where A and s are constants. However, is this a solution to Eq. 3.27? To find out, let us substitute Ae^{st} into Eq. 3.27. The result is

$$\frac{d^2}{dt^2}(Ae^{st}) + 2\alpha\frac{d}{dt}(Ae^{st}) + \omega_n^2(Ae^{st}) = 0 \tag{3.28}$$

from which

$$s^2 A e^{st} + 2\alpha s A e^{st} + \omega_n^2 A e^{st} = 0$$

Dividing both sides by $A e^{st}$ yields

$$s^2 + 2\alpha s + \omega_n^2 = 0 \qquad (3.29)$$

If this equality holds, then so does the one given in Eq. 3.28. By the quadratic formula, this equality holds when

$$s = \frac{-2\alpha \pm \sqrt{4\alpha^2 - 4\omega_n^2}}{2} = -\alpha \pm \sqrt{\alpha^2 - \omega_n^2}$$

In other words, the two values of s that satisfy Eq. 3.29 are

$$s_1 = -\alpha - \sqrt{\alpha^2 - \omega_n^2} \quad \text{and} \quad s_2 = -\alpha + \sqrt{\alpha^2 - \omega_n^2}$$

Thus $A_1 e^{s_1 t}$ and $A_2 e^{s_2 t}$ satisfy Eq. 3.22. As a consequence of this fact, it follows that their sum

$$y(t) = A_1 e^{s_1 t} + A_2 e^{s_2 t} \qquad (3.30)$$

also satisfies Eq. 3.27. Since this is the most general expression (it contains both $A_1 e^{s_1 t}$ and $A_2 e^{s_2 t}$) that satisfies Eq. 3.27, it is the solution. In this expression, A_1 and A_2 are arbitrary constants that are determined from initial conditions.

Note that, depending on the relative values of α and ω_n, the values of s_1 and s_2 can be either real or complex numbers. If $\alpha > \omega_n$, then $\alpha^2 - \omega_n^2 > 0$, so s_1 and s_2 are real numbers. This condition is referred to as the **overdamped** case. If $\alpha < \omega_n$, then $\alpha^2 - \omega_n^2 < 0$ (equivalently, $\omega_n^2 - \alpha^2 > 0$), so that s_1 and s_2 are complex numbers. In particular,

$$s_1 = -\alpha - \sqrt{\alpha^2 - \omega_n^2} = -\alpha - \sqrt{-(\omega_n^2 - \alpha^2)} = -\alpha - j\omega_d$$

$$s_2 = -\alpha + \sqrt{\alpha^2 - \omega_n^2} = -\alpha + \sqrt{-(\omega_n^2 - \alpha^2)} = -\alpha + j\omega_d$$

where $j = \sqrt{-1}$ and $\omega_d = \sqrt{\omega_n^2 - \alpha^2}$. This condition is referred to as the **underdamped** case. Finally, the special case that $\alpha = \omega_n$, called the **critically damped** case, results in $s_1 = s_2 = -\alpha$.

The Overdamped Case

Let us discuss the first of the three possibilities for the form of the natural response by means of an example.

Example 3.12

Let us determine the natural responses $v(t)$ and $i(t)$ for $t \geq 0$ s for the series RLC circuit shown in Fig. 3.31 for the case that $R = 5\ \Omega$, $L = \frac{1}{2}$ H, and $C = \frac{1}{8}$ F given that the initial conditions are $v(0) = 2$ V and $i(0) = 1$ A.

First we must determine which type of damping results due to the given values of R, L, C. From the second-order differential Eq. 3.26, we deduce that

$$2\alpha = \frac{R}{L} \quad \Rightarrow \quad \alpha = \frac{R}{2L} \quad \text{and} \quad \omega_n^2 = \frac{1}{LC} \quad \Rightarrow \quad \omega_n = \frac{1}{\sqrt{LC}}$$

and consequently, $\alpha = R/2L = 5 > \omega_n = 1/\sqrt{LC} = 4$, so the circuit is overdamped. Since

$$s_1 = -\alpha - \sqrt{\alpha^2 - \omega_n^2} = -8 \quad \text{and} \quad s_2 = -\alpha + \sqrt{\alpha^2 - \omega_n^2} = -2$$

then the expression for the voltage $v(t)$ has the form

$$v(t) = A_1 e^{-8t} + A_2 e^{-2t} \tag{3.31}$$

Setting $t = 0$ s and using the initial condition $v(0) = 2$ V, we get the equation

$$2 = A_1 + A_2 \tag{3.32}$$

Taking the derivative of $v(t)$ given in Eq. 3.31 yields

$$\frac{dv(t)}{dt} = -8A_1 e^{-8t} - 2A_2 e^{-2t} \tag{3.33}$$

The second initial condition given is $i(0) = 1$ A. In order to use this condition, note that $i(t)$ is not only the current through the inductor, it is also the current through the capacitor. Thus

$$i(t) = C\frac{dv(t)}{dt} = \frac{1}{8}\left(-8A_1 e^{-8t} - 2A_2 e^{-2t}\right) = -A_1 e^{-8t} - \frac{1}{4}A_2 e^{-2t} \tag{3.34}$$

Setting $t = 0$ s in Eq. 3.34 yields

$$i(0) = -A_1 e^{-8(0)} - 2A_2 e^{-2(0)} \quad \Rightarrow \quad 1 = -A_1 - \tfrac{1}{4} A_2 \quad (3.35)$$

The solution of the simultaneous equations (3.32 and 3.35) is

$$A_1 = -2 \quad \text{and} \quad A_2 = 4$$

Thus the solution of Eq. 3.26 is

$$v(t) = -2e^{-8t} + 4e^{-2t} \text{ V} \quad \text{for } t \geq 0 \text{ s}$$

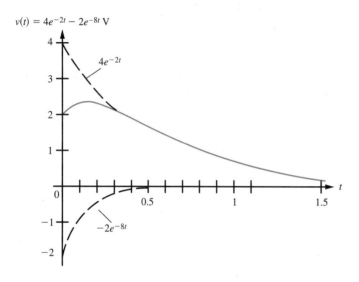

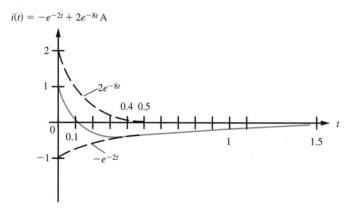

Fig. 3.32 Overdamped natural responses for series *RLC* circuit.

We then have that

$$i(t) = C\frac{dv(t)}{dt} = \frac{1}{8}\frac{d}{dt}(-2e^{-8t} + 4e^{-2t}) = 2e^{-8t} - e^{-2t} \text{ A} \qquad \text{for } t \geq 0 \text{ s}$$

From our expressions for $i(t)$ and $v(t)$, setting $t = 0$ s, we get

$$v(0) = -2e^{-0} + 4e^{-0} = -2 + 4 = 2 \text{ V} \qquad \text{and}$$

$$i(0) = 2e^{-0} - e^{-0} = 2 - 1 = 1 \text{ A}$$

as required. Plots of $v(t)$ and $i(t)$ for $t \geq 0$ s are shown in Fig. 3.32.

Drill Exercise 3.12

A parallel *RLC* circuit has $R = \frac{1}{2}\,\Omega$, $L = \frac{1}{3}\,H$, and $C = \frac{1}{4}\,F$. The current through the inductor is $i(t)$, the voltage across the capacitor is $v(t)$, the circuit is described by the differential equation

$$\frac{d^2i(t)}{dt^2} + 8\frac{di(t)}{dt} + 12i(t) = 0$$

and the initial conditions are $i(0) = 0$ A and $v(0) = -4$ V. Use the fact that $v(t) = \frac{1}{3}\,di(t)/dt$ to find $i(t)$ and $v(t)$ for $t \geq 0$ s.

ANSWER $3e^{-6t} - 3e^{-2t}$ A, $-6e^{-6t} + 2e^{-2t}$ V

The Underdamped Case

In the second of the three cases (underdamping) to be considered, we will see quite different waveforms than are seen in Fig. 3.32. Again we proceed by means of an example.

Example 3.13

In order to obtain the underdamped case ($\alpha < \omega_n$) for the series *RLC* circuit shown in Fig. 3.31, we must change the values of R, L, and C from those given in Example 3.12—the initial conditions $v(0)$ and $i(0)$, however, have no effect on the damping of a circuit.

Let us determine the natural responses $v(t)$ and $i(t)$ for $t \geq 0$ s for the series RLC circuit shown in Fig. 3.31 for the case that $R = 2\ \Omega$, $L = 1$ H, and $C = 1/50$ F given that the initial conditions are $v(0) = 2$ V and $i(0) = -0.32$ A.

Since $\alpha = R/2L = 1 < \omega_n = 1/\sqrt{LC} = \sqrt{50}$, then the circuit is underdamped. Thus

$$\omega_d = \sqrt{\omega_n^2 - \alpha^2} = \sqrt{50 - 1} = 7$$

so that we can write s_1 and s_2 in the form

$$s_1 = -\alpha - j\omega_d = -1 - j7 \qquad \text{and} \qquad s_2 = -\alpha + j\omega_d = -1 + j7$$

and the expression for the voltage $v(t)$ has the form

$$v(t) = A_1 e^{(-1-j7)t} + A_2 e^{(-1+j7)t} = A_1 e^{-1t} e^{-j7t} + A_2 e^{-1t} e^{j7t}$$
$$= e^{-t}(A_1 e^{-j7t} + A_2 e^{j7t})$$

Using **Euler's formula** (we will derive this very important result later),

$$\boxed{e^{j\theta} = \cos\theta + j\sin\theta} \qquad (3.36)$$

we get

$$v(t) = e^{-t}(A_1 \cos 7t - jA_1 \sin 7t + A_2 \cos 7t + jA_2 \sin 7t)$$
$$= e^{-t}[(A_1 + A_2) \cos 7t + j(-A_1 + A_2) \sin 7t]$$
$$= e^{-t}(B_1 \cos 7t + B_2 \sin 7t) \qquad (3.37)$$

Letting $t = 0$ s, we get

$$v(0) = e^{-0}(B_1 \cos 0 + B_2 \sin 0) \qquad \Rightarrow \qquad 2 = B_1$$

Thus so far we have that $v(t) = e^{-t}(2 \cos 7t + B_2 \sin 7t)$. Since

$$i(t) = C\frac{dv(t)}{dt} = \frac{1}{50}\frac{d}{dt}[e^{-t}(2 \cos 7t + B_2 \sin 7t)]$$

then

$$i(t) = \tfrac{1}{50}[e^{-t}(-14 \sin 7t + 7B_2 \cos 7t) + (2 \cos 7t + B_2 \sin 7t)(-e^{-t})] \quad (3.38)$$

Setting $t = 0$ s in this expression yields

$$i(0) = \tfrac{1}{50}[e^{-0}(-14 \sin 0 + 7B_2 \cos 0) + (2 \cos 0 + B_2 \sin 0)(-e^{-0})]$$

from which

$$-0.32 = 0.02[7B_2 - 2] \quad \Rightarrow \quad B_2 = -2$$

Substituting the values $B_1 = 2$ and $B_2 = -2$ into Eq. 3.37 yields the expression for the voltage, which is

$$v(t) = e^{-t}(2 \cos 7t - 2 \sin 7t) = 2e^{-t}(\cos 7t - \sin 7t) \text{ V} \qquad \text{for } t \geq 0 \text{ s}$$

Furthermore, substituting the values $B_1 = 2$ and $B_2 = -2$ into Eq. 3.38 yields the expression for the current, which is

$$i(t) = e^{-t}(-0.32 \cos 7t - 0.24 \sin 7t) \text{ A} \qquad \text{for } t \geq 0 \text{ s}$$

Drill Exercise 3.13

A parallel RLC circuit has $R = 1\ \Omega$, $L = \tfrac{1}{2}$ H, and $C = \tfrac{1}{4}$ F. The current through the inductor is $i(t)$, the voltage across the capacitor is $v(t)$, the circuit is described by the differential equation

$$\frac{d^2 i(t)}{dt^2} + 4\frac{di(t)}{dt} + 8i(t) = 0$$

and the initial conditions are $i(0) = 4$ A and $v(0) = 0$ V. Use the fact that $v(t) = \tfrac{1}{2}\, di(t)/dt$ to find $i(t)$ and $v(t)$ for $t \geq 0$ s.

ANSWER $4e^{-2t}(\cos 2t + \sin 2t)$ A, $-8e^{-2t}\sin 2t$ V

Combining a Cosine and a Sine

In Example 3.13, no attempt was made to sketch the functions $v(t)$ and $i(t)$. Before attempting to sketch the functions obtained in that example, let us derive alternative forms for those types of expressions.

For the underdamped case, the solution to Eq. 3.27 has the form

$$y(t) = A_1 e^{s_1 t} + A_2 e^{s_2 t} = A_1 e^{-\alpha t} e^{-j\omega_d t} + A_2 e^{-\alpha t} e^{j\omega_d t}$$
$$= e^{-\alpha t}(A_1 e^{-j\omega_d t} + A_2 e^{j\omega_d t})$$

where $\omega_d = \sqrt{\omega_n^2 - \alpha^2}$. Using Euler's formula and combining constants, we have just seen an example of the fact that for the underdamped case, we can rewrite $y(t)$ in the form

$$y(t) = e^{-\alpha t}(B_1 \cos \omega_d t + B_2 \sin \omega_d t) \tag{3.39}$$

Even so, it is possible further to combine the sum $B_1 \cos \omega_d t + B_2 \sin \omega_d t$ into the single sinusoid $B \cos(\omega_d t - \phi)$, where we call B the **amplitude,** ω_d the **frequency,** and $\theta = -\phi$ the **phase angle** of the sinusoid. Since the angle of a sinusoid is measured in radians and the unit of time is the second, the unit of ω_d is radians per second (rad/s).

Now we want to determine under what conditions the following equality holds:

$$B_1 \cos \omega_d t + B_2 \sin \omega_d t = B \cos(\omega_d t - \phi) \tag{3.40}$$

Using the trigonometric identity for the cosine of the difference of two angles, Eq. 3.40 becomes

$$B_1 \cos \omega_d t + B_2 \sin \omega_d t = B(\cos \omega_d t \cos \phi + \sin \omega_d t \sin \phi)$$
$$= B \cos \phi \cos \omega_d t + B \sin \phi \sin \omega_d t$$

and equality holds when

$$B_1 = B \cos \phi \quad \text{and} \quad B_2 = B \sin \phi$$

Thus

$$\frac{B_2}{B_1} = \frac{B \sin \phi}{B \cos \phi} = \tan \phi \quad \Rightarrow \quad \phi = \tan^{-1}\left(\frac{B_2}{B_1}\right)$$

The angle ϕ can be depicted by the right triangle shown in Fig. 3.33. We then have that

$$B_1 = B \cos \phi \quad \Rightarrow \quad B = \frac{B_1}{\cos \phi} = \frac{B_1}{B_1/\sqrt{B_1^2 + B_2^2}} = \sqrt{B_1^2 + B_2^2}$$

In summary, the conditions for which Eq. 3.40 holds are

$$B = \sqrt{B_1^2 + B_2^2} \quad \text{and} \quad \phi = \tan^{-1}\left(\frac{B_2}{B_1}\right)$$

and the expression given by Eq. 3.39 can be rewritten as

$$y(t) = Be^{-\alpha t} \cos(\omega_d t - \phi) \tag{3.41}$$

where $\omega_d = \sqrt{\omega_n^2 - \alpha^2}$. A typical sketch of $y(t)$ for $t \geq 0$ s is shown in Fig. 3.34. Such a curve is called a **damped sinusoid**.

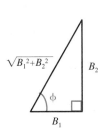

Fig. 3.33 Relationship between B_1, B_2, and ϕ.

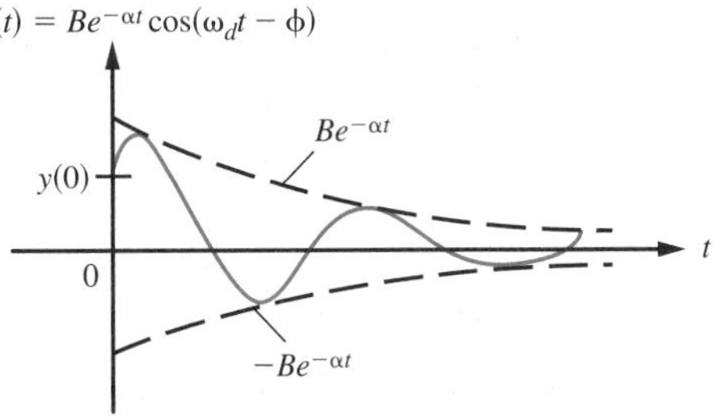

$$y(t) = Be^{-\alpha t} \cos(\omega_d t - \phi)$$

Fig. 3.34 Damped sinusoid.

As a consequence of this discussion, we call α the **damping factor,** since its value determines how fast the sinusoid's amplitude diminishes. We call ω_d the **damped natural frequency,** or **damped frequency,** for short. For the case that the damping factor $\alpha = 0$, we have that $\omega_d = \sqrt{\omega_n^2 - \alpha^2} = \omega_n$, and we refer to ω_n as the **undamped natural frequency,** or **undamped frequency,**[3] for short.

Example 3.14

In Example 3.13, we determined that

$$v(t) = e^{-t}(2 \cos 7t - 2 \sin 7t) = e^{-t}(B_1 \cos 7t + B_2 \sin 7t)$$

---

[3]Although some texts refer to ω_n as the "resonance frequency," except for circuits such as the series and parallel *RLC* circuits, the resonance frequency (to be discussed in Chapter 4) of the circuit is generally different from the undamped frequency.

so

$$B = \sqrt{B_1^2 + B_2^2} = 2\sqrt{2} \quad \text{and} \quad \phi = \tan^{-1}\left(\frac{B_2}{B_1}\right) = \tan^{-1}\left(\frac{-2}{2}\right) = -\frac{\pi}{4} \text{ rad}$$

Thus we can also write $v(t)$ as

$$v(t) = 2\sqrt{2}e^{-t}\cos(7t + \pi/4) \text{ V} \quad \text{for } t \geq 0 \text{ s}$$

Since it was also determined that $i(t) = e^{-t}(-0.32 \cos 7t - 0.24 \sin 7t)$ A for $t \geq 0$ s, then in this case

$$B = \sqrt{(-0.32)^2 + (-0.24)^2} = 0.4 \quad \text{and}$$

$$\phi = \tan^{-1}\left(\frac{-0.24}{-0.32}\right) = 3.79 = -2.5 \text{ rad}$$

so

$$i(t) = 0.4e^{-t}\cos(7t + 2.5) \text{ A} \quad \text{for } t \geq 0 \text{ s}$$

What's that? Your calculator says that $\tan^{-1}(-0.24/-0.32)$ is equal to 0.643 rad (36.9 degrees)? If it does, be careful. Your calculator cannot distinguish between $\tan^{-1}(-0.24/-0.32)$ and $\tan^{-1}(0.24/0.32)$—these are different quantities! And it is the former we require in this case. By inspection of Fig. 3.35 we can see that $\tan^{-1}(0.24/0.32) = 0.644$ rad and $\tan^{-1}(-0.24/-0.32) = 3.79 = -2.5$ rad.

Plots of $v(t)$ and $i(t)$ for $t \geq 0$ s are shown in Fig. 3.36.

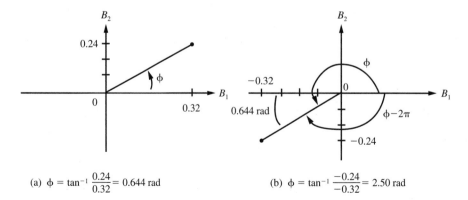

(a) $\phi = \tan^{-1} \dfrac{0.24}{0.32} = 0.644$ rad

(b) $\phi = \tan^{-1} \dfrac{-0.24}{-0.32} = 2.50$ rad

Fig. 3.35 Determination of arctangent.

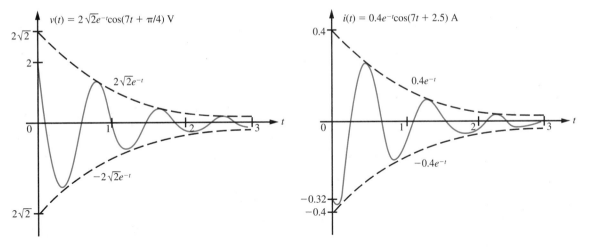

Fig. 3.36 Underdamped natural responses.

Drill Exercise 3.14

Express the current and the voltage for $t \geq 0$ s in the form of Eq. 3.41 for $i(t)$ and $v(t)$ determined in Drill Exercise 3.13.

ANSWER $4\sqrt{2}e^{-2t}\cos(2t - \pi/4)$ A for $t \geq 0$ s,
$8e^{-2t}\cos(2t + \pi/2)$ V for $t \geq 0$ s

What happens in the underdamped case is that the energy stored in the inductor and capacitor eventually gets dissipated by the resistor. However, the energy does not simply go from the L and C directly to the R. Instead, it is transferred back and forth between the two energy storage elements, with the resistance taking its toll during the process. The result of this is a response that is oscillatory (changes sign more than once). In particular, an underdamped response is the product of a real exponential and a sinusoid, that is, a damped sinusoid.

In the Examples 3.13 and 3.14, although oscillatory, the responses (current and voltage) become negligible before too many oscillations occurred. This is due to the fact that ω_n is not much larger than α (i.e., the circuit is slightly underdamped). By reducing the damping (making α smaller), the circuit becomes more underdamped, and more oscillations occur before the responses get close to zero. A sketch of an underdamped response where α is relatively small is shown in Fig. 3.37. For a series RLC circuit when $R = 0$ Ω or a parallel RLC circuit when $R = \infty$, we have the underdamped case with no damping ($\alpha = 0$). A consequence of this is a response that will be perfectly sinusoidal, and thus will never decrease in amplitude. Such a

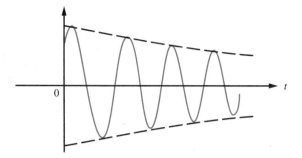

Fig. 3.37 Damped sinusoid having a relatively small damping factor.

situation cannot be physically constructed with ordinary capacitors and inductors. There are, however, electronic ways for producing such a result. A device that accomplishes this is known as an **oscillator**.

The Critically Damped Case

Having considered the case of overdamping ($\alpha > \omega_n$) and underdamping ($\alpha < \omega_n$), it is now time to investigate the condition of critical damping ($\alpha = \omega_n$). For the case that $\alpha = \omega_n$, we have that $s_1 = s_2 = -\alpha$, and Eq. 3.30 becomes

$$y(t) = A_1 e^{s_1 t} + A_2 e^{s_2 t} = A_1 e^{-\alpha t} + A_2 e^{-\alpha t} = A e^{-\alpha t}$$

However, this is not the solution to Eq. 3.27 in general. This is so because two initial conditions cannot be satisfied with the one arbitrary constant A. To determine what the solution is, let us return to Eq. 3.27. Using the critically damped condition $\alpha = \omega_n$, Eq. 3.27 becomes

$$\frac{d^2 y(t)}{dt^2} + 2\alpha \frac{dy(t)}{dt} + \alpha^2 y(t) = 0$$

which can be rewritten as

$$\frac{d}{dt}\left[\frac{dy(t)}{dt} + \alpha y(t)\right] + \alpha \left[\frac{dy(t)}{dt} + \alpha y(t)\right] = 0 \tag{3.42}$$

If we define the function $f(t) = dy(t)/dt + \alpha y(t)$, Eq. 3.42 becomes

$$\frac{df(t)}{dt} + \alpha f(t) = 0$$

which is a first-order differential equation whose solution is

$$f(t) = A_1 e^{-\alpha t} \quad \Rightarrow \quad \frac{dy(t)}{dt} + \alpha y(t) = A_1 e^{-\alpha t} \tag{3.43}$$

due to the definition of $f(t)$. Multiplying both sides of Eq. 3.43 by $e^{\alpha t}$ yields

$$e^{\alpha t} \frac{dy(t)}{dt} + e^{\alpha t} \alpha y(t) = A_1 \quad \text{or} \quad \frac{d}{dt}[e^{\alpha t} y(t)] = A_1 \tag{3.44}$$

Integrating both sides of Eq. 3.44 with respect to t results in

$$\int \frac{d[e^{\alpha t} y(t)]}{dt} dt = \int A_1 \, dt \quad \Rightarrow \quad e^{\alpha t} y(t) = A_1 t + A_2 \tag{3.45}$$

where A_2 is a constant of integration. Multiplying both sides of Eq. 3.45 by $e^{-\alpha t}$ gives

$$y(t) = A_1 t e^{-\alpha t} + A_2 e^{-\alpha t} = (A_1 t + A_2)e^{-\alpha t} \tag{3.46}$$

and this is the solution of Eq. 3.27 for the case of critical damping.

Example 3.15

Let us determine the natural responses $v(t)$ and $i(t)$ for $t \geq 0$ s for the series RLC circuit shown in Fig. 3.31 on p. 151 for the case that $R = 4 \, \Omega$, $L = 1$ H, and $C = \frac{1}{4}$ F given that the initial conditions are $v(0) = 2$ V and $i(0) = 1$ A.

Since $\alpha = R/2L = 2 = \omega_n = 1/\sqrt{LC} = 2$, then the circuit is critically damped. Thus the expression for $v(t)$ is

$$v(t) = A_1 t e^{-2t} + A_2 e^{-2t}$$

Setting $t = 0$ s yields

$$v(0) = 0 + A_2 = 2 \quad \Rightarrow \quad A_2 = 2$$

Since

$$i(t) = C \frac{dv(t)}{dt} = \frac{1}{4} \frac{d}{dt}[A_1 t e^{-2t} + 2e^{-2t}] = -\frac{1}{2} A_1 t e^{-2t} + \left(\frac{1}{4} A_1 - 1\right)e^{-2t}$$

Setting $t = 0$ s yields

$$i(0) = -0 + (\tfrac{1}{4} A_1 - 1)e^{-0} = 1 \quad \Rightarrow \quad A_1 = 8$$

Therefore, the expression for $v(t)$ is

$$v(t) = 8te^{-2t} + 2e^{-2t} = (8t + 2)e^{-2t} \text{ V} \qquad \text{for } t \geq 0 \text{ s}$$

In addition, from the expression for $i(t)$ above,

$$i(t) = -\tfrac{1}{2}(8)te^{-2t} + (\tfrac{1}{4}[8] - 1)e^{-2t} = -4te^{-2t} + e^{-2t} \text{ A} \quad \text{for } t \geq 0 \text{ s}$$

Sketches of $v(t)$ and $i(t)$ for $t \geq 0$ s are shown in Fig. 3.38.

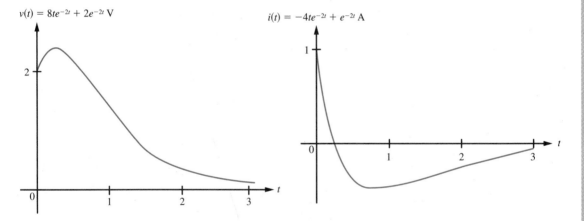

Fig. 3.38 Critically damped natural responses.

Drill Exercise 3.15

A parallel *RLC* circuit has $R = 1$ Ω, $L = 4$ H, and $C = 1$ F. The current through the inductor is $i(t)$, the voltage across the capacitor is $v(t)$, the circuit is described by the differential equation

$$\frac{d^2 i(t)}{dt^2} + \frac{di(t)}{dt} + \frac{1}{4}i(t) = 0$$

and the initial conditions are $i(0) = -3$ A and $v(0) = 14$ V. Use the fact that $v(t) = 4\, di(t)/dt$ to find $i(t)$ and $v(t)$ for $t \geq 0$ s.

ANSWER $2te^{-t/2} - 3e^{-t/2}$ A, $-4te^{-t/2} + 14e^{-t/2}$ V

3.6 Second-Order Circuits—The Complete Response

Having considered the natural response for some second-order circuits, let us now look at second-order circuits with nonzero inputs. We will limit our discussion to constant forcing functions. Other types of forcing functions will be dealt with in the next two chapters when additional concepts are introduced.

Consider the parallel *RLC* circuit shown in Fig. 3.39, where the excitation is a step function of current. By *KCL,*

$$Iu(t) = \frac{v}{R} + i + C\frac{dv}{dt}$$

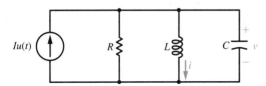

Fig. 3.39 Step of current applied to a parallel *RLC* circuit.

Substituting $v = L\, di/dt$ into this equation, we get

$$\frac{d^2i}{dt^2} + \frac{1}{RC}\frac{di}{dt} + \frac{1}{LC}i = \frac{I}{LC}u(t) \qquad (3.47)$$

For $t < 0$ s, the right-hand side of this equation is zero, and for zero initial conditions the solution is $i(t) = 0$ A. However, for $t \geq 0$ s, Eq. 3.47 becomes

$$\frac{d^2i}{dt^2} + \frac{1}{RC}\frac{di}{dt} + \frac{1}{LC}i = \frac{I}{LC} \qquad (3.48)$$

As in the case of first-order differential equations, the solution $i(t)$ to Eq. 3.48 consists of two parts—the forced response $i_f(t)$ and the natural response $i_n(t)$—that is, the solution has the form

$$i(t) = i_f(t) + i_n(t)$$

where $i_n(t)$ is the solution to

$$\frac{d^2i}{dt^2} + 2\alpha\frac{di}{dt} + \omega_n^2 i = 0$$

where $\alpha = 1/2RC$ and $\omega_n = 1/\sqrt{LC}$.

Since the right side of Eq. 3.48 is a constant, the forced response is a constant—call it K. If the natural response $v_n(t)$ is substituted into the left side of Eq. 3.48, then the left side of the equation becomes zero. Thus, the forced response $v_f(t) = K$ must satisfy Eq. 3.48, that is,

$$\frac{d^2K}{dt^2} + \frac{1}{RC}\frac{dK}{dt} + \frac{1}{LC}K = \frac{I}{LC} \quad \Rightarrow \quad \frac{1}{LC}K = \frac{I}{LC} \quad \Rightarrow \quad K = I$$

Hence the solution to Eq. 3.48 is as follows:

For the overdamped case: $i(t) = I + A_1 e^{s_1 t} + A_2 e^{s_2 t}$

For the underdamped case: $i(t) = I + e^{-\alpha t}(A_1 \cos \omega_d t + A_2 \sin \omega_d t)$

For the critically damped case: $i(t) = I + A_1 t e^{-\alpha t} + A_2 e^{-\alpha t}$

where the constants A_1 and A_2 are determined from the circuit topology and the initial conditions $i(0) = 0$ A and $v(0) = 0$ V.

For the circuit in Fig. 3.39, since no energy initially is stored in the inductor and capacitor, when the applied current is zero, the current $i(t)$ and the voltage $v(t)$ will both be zero. At $t = 0$ s, when the applied current become I amperes, since the current through an inductor and the voltage across a capacitor cannot change instantaneously, we conclude that $i(0) = 0$ A and $v(0) = 0$ V. After a long time, the excitation acts as a constant, the inductor behaves as a short circuit, and the capacitor behaves as an open circuit. Thus, eventually, the current through the inductor will be I amperes and the voltage across the capacitor will be 0 V. The shape of the current and voltage waveforms between their initial and final values will depend upon whether the circuit is overdamped, underdamped, or critically damped.

The Overdamped Case

First we consider the responses to a step function for an overdamped circuit.

Example 3.16

Let us find the step responses $i(t)$ and $v(t)$ for the parallel RLC circuit shown in Fig. 3.39 when $R = 6\ \Omega$, $L = 7$ H, $C = 1/42$ F, and $Iu(t) = 6u(t)$ A. From Eq. 3.48 we get

$$\frac{d^2i}{dt^2} + 7\frac{di}{dt} + 6i = 36\ u(t) \tag{3.49}$$

For $t < 0$ s, the right-hand side of this equation is zero, and for zero initial conditions the solution is $i(t) = 0$ A. However, for $t \geq 0$ s, Eq. 3.49 becomes

$$\frac{d^2i}{dt^2} + 7\frac{di}{dt} + 6i = 36 \tag{3.50}$$

Since $\alpha = \frac{7}{2} > \omega_n = \sqrt{6}$, then the circuit is overdamped, and

$$s = -\alpha \pm \sqrt{\alpha^2 - \omega_n^2} = -\frac{7}{2} \pm \sqrt{49/4 - 6} = -\frac{7}{2} \pm \frac{5}{2} = -6, -1$$

Thus $s_1 = -6$ and $s_2 = -1$. Therefore, the natural response has the form

$$i_n(t) = A_1e^{-6t} + A_2e^{-t}$$

For $t \geq 0$ s, the forcing function is a constant, so the forced response is a constant K that can be obtained by substituting K into the differential Eq. 3.50. Doing so, we get

$$\frac{d^2K}{dt^2} + 7\frac{dK}{dt} + 6K = 36 \quad \Rightarrow \quad 0 + 0 + 6K = 36 \quad \Rightarrow \quad K = 6$$

(Alternatively, from the circuit, for the case that the applied current is a constant of value 6 A, then the inductor current is $i_f(t) = 6$ A since the inductor behaves as a short circuit to dc.) Hence the complete response has the form

$$i(t) = 6 + A_1e^{-6t} + A_2e^{-t} \tag{3.51}$$

Setting $t = 0$ s, we get

$$i(0) = 6 + A_1 + A_2 = 0 \quad \Rightarrow \quad A_1 + A_2 = -6 \tag{3.52}$$

However, from the circuit,

$$v(t) = L\frac{di(t)}{dt} = 7\frac{d}{dt}[6 + A_1e^{-6t} + A_2e^{-t}] = -42A_1e^{-6t} - 7A_2e^{-t} \tag{3.53}$$

Setting $t = 0$ s, we obtain

$$v(0) = -42A_1 - 7A_2 = 0 \tag{3.54}$$

The solution to Eq. 3.52 and 3.54 is $A_1 = 1.2$ and $A_2 = -7.2$. Substituting these values into Eq. 3.51 yields the complete response

$$i(t) = 6 + 1.2e^{-6t} - 7.2e^{-t} \text{ A} \qquad \text{for } t \geq 0 \text{ s}$$

Furthermore, from Eq. 3.53, we have that

$$v(t) = -42(1.2)e^{-6t} - 7(-7.2)e^{-t} = -50.4e^{-6t} + 50.4e^{-t} \text{ V} \qquad \text{for } t \geq 0 \text{ s}$$

Combining these responses for $t \geq 0$ s with the fact that $i(t) = 0$ A for $t < 0$ s and $v(t) = 0$ V for $t < 0$ s, we deduce that the step responses for the given circuit are:

$$i(t) = (6 + 1.2e^{-6t} - 7.2e^{-t})u(t) \text{ A} \qquad \text{and}$$
$$v(t) = (-50.4e^{-6t} + 50.4e^{-t})u(t) \text{ V}$$

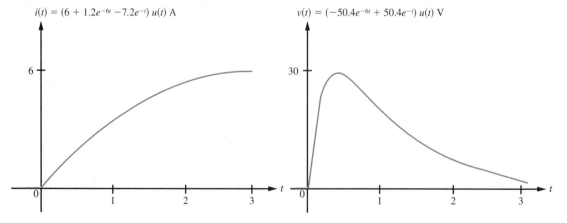

Fig. 3.40 Step responses for overdamped parallel RLC circuit.

Sketches of these step responses are shown in Fig. 3.40.

Drill Exercise 3.16

For the parallel *RLC* circuit given in Fig. 3.39, suppose that $R = 0.4 \ \Omega$, $L = 0.25$ H, and $C = 0.25$ F. Find the responses $i(t)$ and $v(t)$ to a unit step function, i.e., the responses when the applied current is $u(t)$ A.

ANSWER $(1 + \frac{1}{3} e^{-8t} - \frac{4}{3} e^{-2t})u(t)$ A, $(-\frac{2}{3} e^{-8t} + \frac{2}{3} e^{-2t})u(t)$ V

The Underdamped Case

Having considered the parallel *RLC* circuit, let us now look at the series *RLC* circuit shown in Fig. 3.41. By KVL,

$$Vu(t) = Ri + L\frac{di}{dt} + v \tag{3.55}$$

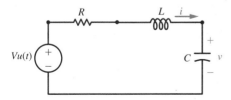

Fig. 3.41 Step of voltage applied to a series *RLC* circuit.

Substituting $i = C \ dv/dt$ into Eq. 3.55 results in

$$\frac{d^2v}{dt^2} + \frac{R}{L}\frac{dv}{dt} + \frac{1}{LC}v = \frac{V}{LC}u(t) \tag{3.56}$$

For $t < 0$ s, the right-hand side of this equation becomes zero, and for zero initial conditions the solution is $v(t) = 0$ V. However, for $t \geq 0$ s, Eq. 3.56 becomes

$$\frac{d^2v}{dt^2} + \frac{R}{L}\frac{dv}{dt} + \frac{1}{LC}v = \frac{V}{LC} \tag{3.57}$$

Thus we have that $\alpha = R/2L$ and $\omega_n = 1/\sqrt{LC}$.

Having seen an example of an overdamped step response, let us now consider the underdamped case.

Example 3.17

Let us find the step responses $v(t)$ and $i(t)$ for the series RLC circuit shown in Fig. 3.41 for the case that $R = 2\ \Omega$, $L = 1$ H, $C = 0.02$ F, and $V = 14$.

For $t < 0$ s, the solution to Eq. 3.56 is $v(t) = 0$ V. For $t \geq 0$ s, substituting $v(t) = K$ into Eq. 3.57, we get the forced response $v_f(t) = V = 14$. Since $\alpha = R/2L = 2/2(1) = 1 < \omega_n = 1/\sqrt{LC} = 1/\sqrt{(1)(0.02)} = \sqrt{50}$, then the form of the natural response is described by the underdamped case with $\omega_d = \sqrt{\omega_n^2 - \alpha^2} = 7$ rad/s. That is, $v_n(t) = e^{-t}(A_1 \cos 7t + A_2 \sin 7t)$. Therefore, the complete response $v(t) = v_f(t) + v_n(t)$ is

$$v(t) = 14 + e^{-t}(A_1 \cos 7t + A_2 \sin 7t)$$

Since we use zero initial conditions for step responses, setting $t = 0$ s yields

$$v(0) = 14 + A_1 = 0 \quad \Rightarrow \quad A_1 = -14$$

Furthermore, since $i(t) = C\ dv(t)/dt$, then

$$i(t) = 0.02\ [e^{-t}(-7A_1 \sin 7t + 7A_2 \cos 7t) \tag{3.58}$$
$$- e^{-t}(A_1 \cos 7t + A_2 \sin 7t)]$$

from which

$$i(0) = 0.02\ [7A_2 - A_1] = 0 \quad \Rightarrow \quad A_2 = A_1/7 = -14/7 = -2$$

Therefore, for $t \geq 0$ s

$$v(t) = 14 + e^{-t}(-14 \cos 7t - 2 \sin 7t)$$
$$= 14 - 10\sqrt{2}e^{-t} \cos(7t - 0.142)\ \text{V}$$

Combining this with the fact that $v(t) = 0$ V for $t < 0$ s results in the unit voltage step response,

$$v(t) = [14 - 10\sqrt{2}e^{-t} \cos(7t - 0.142)]u(t)\ \text{V}$$

Furthermore, from Eq. 3.58, the current is

$$i(t) = 0.02 \left[e^{-t}(98 \sin 7t - 14 \cos 7t) - e^{-t}(-14 \cos 7t - 2 \sin 7t) \right]$$
$$= 2e^{-t} \sin 7t = 2e^{-t} \cos(7t - \pi/2) \text{ A} \qquad \text{for } t \geq 0 \text{ s}$$

Thus the current step response is

$$i(t) = 2e^{-t} \sin 7t \, u(t) \text{ A}$$

Sketches of $v(t)$ and $i(t)$ are shown in Fig. 3.42.

$v(t) = [14 - 10\sqrt{2} \, e^{-t} \cos(7t - 0.142)] \, u(t)$ V

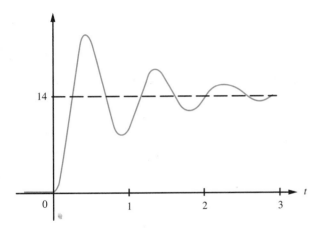

$i(t) = 2e^{-t} \sin 7t \, u(t)$ A

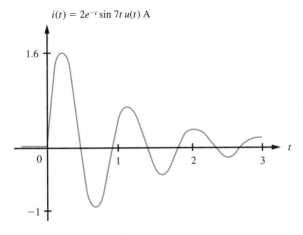

Fig. 3.42 Step responses of an underdamped series *RLC* circuit.

The Critically Damped Case

Consider the *RLC* circuit shown in Fig. 3.43. By KCL,

$$\frac{Vu(t) - v}{R} = i + C\frac{dv}{dt} \tag{3.59}$$

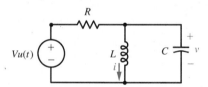

Fig. 3.43 Another *RLC* circuit.

Substituting $v = L\ di/dt$ into Eq. 3.59 and simplifying results in

$$\frac{d^2i}{dt^2} + \frac{1}{RC}\frac{di}{dt} + \frac{1}{LC}i = \frac{V}{RLC}u(t) \tag{3.60}$$

For $t < 0$ s, the right-hand side of this equation becomes zero, and for zero initial conditions the solution is $i(t) = 0$ A. However, for $t \geq 0$ s, Eq. 3.59 becomes

$$\frac{d^2i}{dt^2} + \frac{1}{RC}\frac{di}{dt} + \frac{1}{LC}i = \frac{V}{RLC} \tag{3.61}$$

Thus we have that $\alpha = 1/2RC$ and $\omega_n = 1/\sqrt{LC}$.

We finally consider the critically damped case.

Example 3.18

Let us find the unit ($V = 1$) step responses $i(t)$ and $v(t)$ for the RLC circuit shown in Fig. 3.43 when $R = 1\ \Omega$, $L = 4$ H, and $C = 1$ F.

For $t < 0$ s, the solution to Eq. 3.60 is $i(t) = 0$ A. For $t \geq 0$ s, substituting $i(t) = K$ into Eq. 3.61, we get the forced response $i_f(t) = V/R = 1$. Since $\alpha = \frac{1}{2} = \omega_n = \frac{1}{2}$, then the form of the natural response is described by the critically damped case. That is, $i_n(t) = A_1 t e^{-t/2} + A_2 e^{-t/2}$. Therefore, the complete response $i(t) = i_f(t) + i_n(t)$ is

$$i(t) = 1 + A_1 t e^{-t/2} + A_2 e^{-t/2} \tag{3.62}$$

Since we use zero initial conditions for step responses, setting $t = 0$ s yields

$$i(0) = 1 + 0 + A_2 = 0 \quad \Rightarrow \quad A_2 = -1$$

Furthermore, since $v(t) = L\, di(t)/dt$, then

$$v(t) = 4\frac{di(t)}{dt} = 4(-\tfrac{1}{2}A_1 t e^{-t/2} + A_1 e^{-t/2} + \tfrac{1}{2}e^{-t/2}) \tag{3.63}$$

from which

$$v(0) = 4(0 + A_1 + \tfrac{1}{2}) = 0 \quad \Rightarrow \quad A_1 = -\tfrac{1}{2}$$

Therefore, from Eq. 3.62,

$$i(t) = 1 - \tfrac{1}{2}t e^{-t/2} - e^{-t/2} \text{ A} \qquad \text{for } t \geq 0 \text{ s}$$

Combining this with the fact that $i(t) = 0$ A for $t < 0$ s results in the current step response,

$$i(t) = (1 - \tfrac{1}{2}t e^{-t/2} - e^{-t/2})u(t) \text{ A}$$

Furthermore, from Eq. 3.63, the voltage is

$$v(t) = 4(\tfrac{1}{4}t e^{-t/2} - \tfrac{1}{2}e^{-t/2} + \tfrac{1}{2}e^{-t/2}) = t e^{-t/2} \text{ V} \qquad \text{for } t \geq 0 \text{ s}$$

Thus the voltage step response is

$$v(t) = t e^{-t/2} u(t) \text{ V}$$

Sketches of $i(t)$ and $v(t)$ are shown in Fig. 3.44.

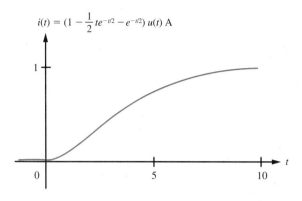

$$i(t) = (1 - \frac{1}{2} te^{-t/2} - e^{-t/2})\, u(t) \text{ A}$$

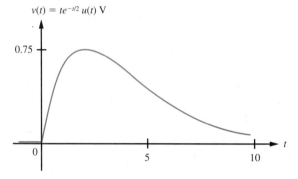

$$v(t) = te^{-t/2}\, u(t) \text{ V}$$

Fig. 3.44 Unit step responses for a critically damped *RLC* circuit.

Drill Exercise 3.18

Find the unit step responses $i(t)$ and $v(t)$ for the *RLC* circuit given in Fig. 3.43 when $R = \frac{1}{4}\,\Omega$, $L = \frac{1}{4}\,$H, and $C = 1\,$F.

ANSWER $i(t) = (1 - 2te^{-2t} - e^{-2t})u(t)$ A, $te^{-2t}u(t)$ V

SUMMARY

1. The voltage across an inductor is directly proportional to the derivative of the current through it.

2. An inductor behaves as a short circuit to dc.

3. The current through a capacitor is directly proportional to the derivative of the voltage across it.

4. A capacitor behaves as an open circuit to dc.

5. The current through an inductor is directly proportional to the integral of the voltage across it.

6. The voltage across a capacitor is directly proportional to the integral of the current through it.

7. The energy stored in an inductor is $\frac{1}{2}Li^2$ joules, whereas the energy stored in a capacitor is $\frac{1}{2}Cv^2$ joules.

8. The current through an inductor cannot change instantaneously. The voltage across a capacitor cannot change instantaneously.

9. A planar circuit and its dual are, in essence, described by the same equations.

10. Writing node and mesh equations for circuits containing inductors and capacitors is done as was done for resistive circuits. Except for simple circuits, the solutions of such equations will be avoided.

11. A circuit containing either a single inductor or a single capacitor is described by a first-order differential equation.

12. If no independent source is present, the differential equation describing a circuit is homogeneous and the solution is the natural response.

13. The complete response is the sum of two terms—one is the forced response and the other has the form of the natural response.

14. The natural response of a second-order circuit is either overdamped, underdamped, or critically damped.

15. An overdamped response is the sum of two (real) exponentials.

16. An underdamped response is a damped sinusoid.

17. A critically damped response is the sum of an exponential and another exponential (with the same time constant) multiplied by time.

18. As with the case of first-order circuits, the complete response of a second-order circuit is the sum of two terms—one is the forced response and the other has the form of the natural response.

Problems

3.1 For the circuit shown in Fig. P3.1a, suppose that $i(t)$ is described by the function given in Fig. P3.1b. Sketch (a) $v(t)$, (b) $w_L(t)$, (c) $p_R(t)$, (d) $v_R(t)$, and (e) $v_s(t)$.

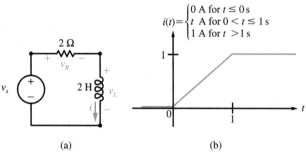

$$i(t) = \begin{cases} 0 \text{ A for } t \le 0 \text{ s} \\ t \text{ A for } 0 < t \le 1 \text{ s} \\ 1 \text{ A for } t > 1 \text{ s} \end{cases}$$

(a)   (b)

Fig. P3.1

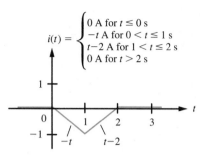

$$i(t) = \begin{cases} 0 \text{ A for } t \le 0 \text{ s} \\ -t \text{ A for } 0 < t \le 1 \text{ s} \\ t-2 \text{ A for } 1 < t \le 2 \text{ s} \\ 0 \text{ A for } t > 2 \text{ s} \end{cases}$$

Fig. P3.2

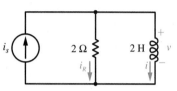

Fig. P3.3

3.2 For the circuit shown in Fig. P3.1a, suppose that $i(t)$ is described by the function given in Fig. P3.2. Sketch (a) $v(t)$, (b) $w_L(t)$, (c) $p_R(t)$, (d) $v_R(t)$, and (e) $v_s(t)$.

3.3 For the circuit shown in Fig. P3.3, suppose that $i(t)$ is described by the function given in Fig. P3.1b. Sketch (a) $v(t)$, (b) $w_L(t)$, (c) $p_R(t)$, (d) $i_R(t)$, and (e) $i_s(t)$.

3.4 For the circuit shown in Fig. P3.3, suppose that $i(t)$ is described by the function given in Fig. P3.2. Sketch (a) $v(t)$, (b) $w_L(t)$, (c) $p_R(t)$, (d) $i_R(t)$, and (e) $i_s(t)$.

3.5 For the circuit shown in Fig. P3.5, suppose that $i(t)$ is described by the function given in Fig. P3.1b. Sketch (a) $v_R(t)$, (b) $v_L(t)$, and (c) $v(t)$.

3.6 For the circuit shown in Fig. P3.5, suppose that $i(t)$ is described by the function given in Fig. P3.2. Sketch (a) $v_R(t)$, (b) $v_L(t)$, and (c) $v(t)$.

3.7 For the circuit shown in Fig. P3.7a, suppose that $v(t)$ is described by the function given in Fig. P3.7b. Sketch (a) $i(t)$, (b) $w_C(t)$, (c) $p_R(t)$, (d) $v_R(t)$, and (e) $v_s(t)$.

3.8 For the circuit shown in Fig. P3.8, suppose that $v(t)$ is described by the function given in Fig. P3.7b.

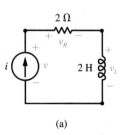

(a)

Fig. P3.5

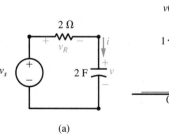

(a)

Fig. P3.7

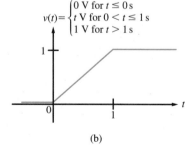

$$v(t) = \begin{cases} 0 \text{ V for } t \le 0 \text{ s} \\ t \text{ V for } 0 < t \le 1 \text{ s} \\ 1 \text{ V for } t > 1 \text{ s} \end{cases}$$

(b)

Sketch (a) $i(t)$, (b) $w_C(t)$, (c) $p_R(t)$, (d) $i_R(t)$, and (e) $i_s(t)$.

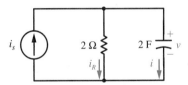

Fig. P3.8

3.9 For the op-amp circuit shown in Fig. P3.9, suppose that $v(t)$ is described by the function given in Fig. P3.7b. Sketch (a) $i(t)$, (b) $i_R(t)$, (c) $v_R(t)$, (d) $v_s(t)$, and (e) $v_o(t)$.

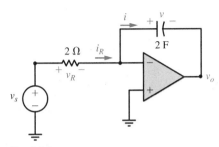

Fig. P3.9

3.10 For the op-amp circuit shown in Fig. P3.9, connect an additional 2-Ω resistor in parallel with the capacitor. Suppose that $v(t)$ is described by the function given in Fig. P3.7b. Sketch (a) $i(t)$, (b) $i_R(t)$, (c) $v_R(t)$, (d) $v_s(t)$, and (e) $v_o(t)$.

3.11 For the op-amp circuit shown in Fig. P3.11, suppose that $v(t)$ is described by the function given in Fig. P3.7b. Sketch (a) $i(t)$, (b) $i_R(t)$, (c) $v_R(t)$, (d) $v_s(t)$, and (e) $v_o(t)$.

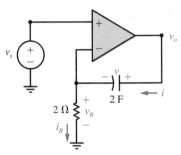

Fig. P3.11

3.12 For the op-amp circuit given in Fig. P3.11, connect an additional 2-Ω resistor in parallel with the capacitor. Suppose that $v(t)$ is described by the function given in Fig. P3.7b. Sketch (a) $i(t)$, (b) $i_R(t)$, (c) $v_R(t)$, (d) $v_s(t)$, and (e) $v_o(t)$.

3.13 For the op-amp circuit shown in Fig. P3.13, suppose that $v(t)$ is described by the function given in Fig. P3.7b. Sketch (a) $i(t)$, (b) $i_R(t)$, (c) $v_R(t)$, (d) $v_s(t)$, and (e) $v_o(t)$.

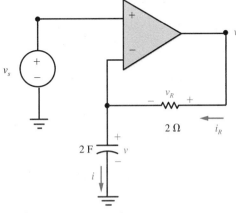

Fig. P3.13

3.14 For the op-amp circuit shown in Fig. P3.13, connect an additional 2-Ω resistor in parallel with the capacitor. Suppose that $v(t)$ is described by the function given in Fig. P3.7b. Sketch (a) $i(t)$, (b) $i_R(t)$, (c) $v_R(t)$, (d) $v_s(t)$, and (e) $v_o(t)$.

3.15 Show the following: (See p. 178.)
(a) Inductors connected in series can be combined as depicted in Fig. P3.15a.
(b) Inductors connected in parallel can be combined as depicted in Fig. P3.15b.
(c) Capacitors connected in parallel can be combined as depicted in Fig. P3.15c.
(d) Capacitors connected in series can be combined as depicted in Fig. P3.15d.

3.16 For the circuit shown in Fig. P3.1a, suppose that $v(t)$ is described by the function given in Fig. P3.16. Sketch (a) $i(t)$, (b) $v_R(t)$, and (c) $v_s(t)$. (See p. 178.)

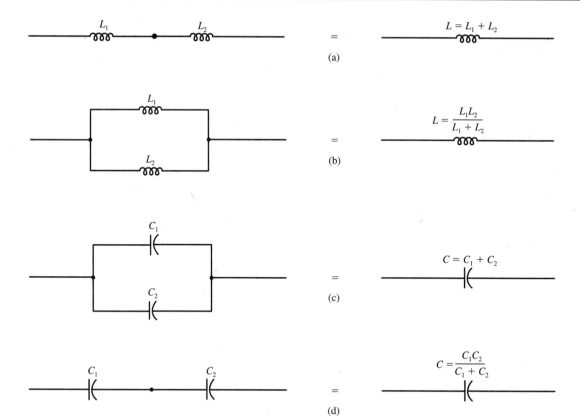

Fig. P3.15

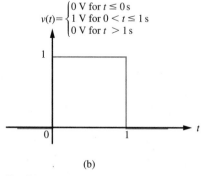

(b)

Fig. P3.16

Fig. P3.17

3.18 For the op-amp circuit shown in Fig. P3.18, suppose that $v(t)$ is described by the function given in Fig. P3.16. Sketch (a) $i_R(t)$, (b) $v_C(t)$, and (c) $v_o(t)$.

3.17 For the circuit shown in Fig. P3.17, suppose that $v(t)$ is described by the function given in Fig. P3.16. Sketch (a) $i_L(t)$, (b) $i_R(t)$, and (c) $i_s(t)$.

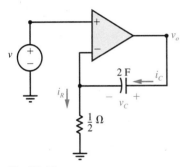

Fig. P3.18

3.19 For the circuit shown in Fig. P3.7a, suppose that $i(t)$ is described by the function given in Fig. P3.19. Sketch (a) $v(t)$, (b) $v_R(t)$, and (c) $v_s(t)$.

$$i(t) = \begin{cases} 0 \text{ A for } t \le 0 \text{ s} \\ 1 \text{ A for } 0 < t \le 1 \text{ s} \\ 0 \text{ A for } t > 1 \text{ s} \end{cases}$$

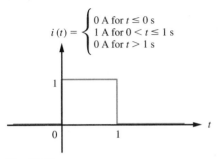

Fig. P3.19

3.20 For the circuit shown in Fig. P3.8, suppose that $i(t)$ is described by the function given in Fig. P3.19. Sketch (a) $v(t)$, (b) $i_R(t)$, and (c) $i_s(t)$.

3.21 For the op-amp circuit shown in Fig. P3.9, suppose that $i(t)$ is described by the function given in Fig. P3.19. Sketch (a) $v(t)$, (b) $v_R(t)$, (c) $v_s(t)$, and (d) $v_o(t)$.

3.22 For the op-amp circuit shown in Fig. P3.9, interchange the resistor and the capacitor. Suppose that $i(t)$ is described by the function given in Fig. P3.19. Sketch (a) $v(t)$, (b) $v_C(t)$, (c) $v_s(t)$, and (d) $v_o(t)$.

3.23 For the op-amp circuit shown in Fig. P3.11, suppose that $i(t)$ is described by the function given in Fig. P3.19. Sketch (a) $v(t)$, (b) $v_R(t)$, (c) $v_s(t)$, and (d) $v_o(t)$.

3.24 For the op-amp circuit shown in Fig. P3.13, suppose that $i(t)$ is described by the function given in Fig. P3.19. Sketch (a) $v(t)$, (b) $v_R(t)$, (c) $v_s(t)$, and (d) $v_o(t)$.

3.25 Find the dual of the circuit given in Fig. P3.25.

3.26 Find the dual of the circuit given in Fig. P3.26. (See p. 180.)

3.27 Find the dual of the circuit given in Fig. P3.27. (See p. 180.)

3.28 For the circuit shown in Fig. P3.28, the switch opens at time $t = 0$ s. Write a differential equation in $v(t)$ for ≥ 0 s. Find $v(t)$ and $i(t)$ for all time and sketch these functions.

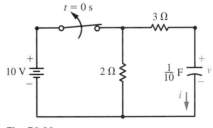

Fig. P3.28

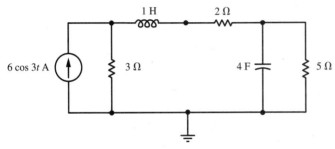

Fig. P3.25

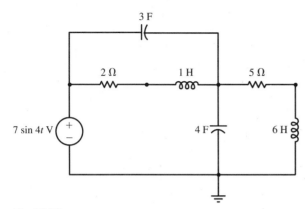

Fig. P3.26

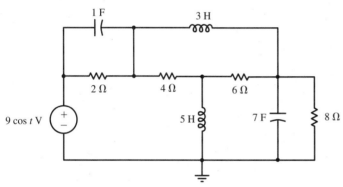

Fig. P3.27

3.29 For the circuit shown in Fig. P3.28, replace the capacitor with a 5-H inductor. For the resulting circuit, the switch opens at time $t = 0$ s. Write a differential equation in $i(t)$ for $t \geq 0$ s. Find $i(t)$ and $v(t)$ for all time and sketch these functions.

3.30 For the circuit shown in Fig. P3.30, suppose that $i_s(t) = 10$ A for $t < 0$ s and $i_s(t) = 0$ A for $t \geq 0$ s. Write a differential equation in $i(t)$ for $t \geq 0$ s. Find $i(t)$ and $v(t)$ for all time and sketch these functions.

3.31 For the circuit shown in Fig. P3.30, replace the inductor with a 0.1-F capacitor. Suppose that $i_s(t) = 10$ A for $t < 0$ s and $i_s(t) = 0$ A for $t \geq 0$ s. Write a differential equation in $v(t)$ for $t \geq 0$ s. Find $v(t)$ and $i(t)$ for all time and sketch these functions.

3.32 For the circuit shown in Fig. P3.32, suppose that $v_s(t) = 18$ V for $t < 0$ s and $v_s(t) = 0$ V for $t \geq 0$ s. Write a differential equation in $i(t)$ for $t \geq 0$ s. Find $i(t)$ and $v(t)$ for all time and sketch these functions.

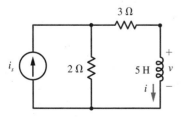

Fig. P3.30

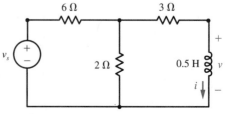

Fig. P3.32

3.33 For the circuit shown in Fig. P3.32, replace the inductor with a $\frac{1}{9}$-F capacitor. Suppose that $v_s(t) = 18$ V for $t < 0$ s and $v_s(t) = 0$ V for $t \geq 0$ s. Write a differential equation in $v(t)$ for $t \geq 0$ s. Find $v(t)$ and $i(t)$ for all time and sketch these functions.

3.34 For the circuit shown in Fig. P3.34, suppose that $v_s(t) = 12$ V for $t < 0$ s and $v_s(t) = 0$ V for $t \geq 0$ s. Write a differential equation in $v(t)$ for $t \geq 0$ s. Find $v(t)$ and $i(t)$ for all time and sketch these functions.

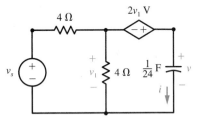

Fig. P3.34

3.35 For the circuit shown in Fig. P3.34, replace the capacitor with a 3-H inductor. Suppose that

$v_s(t) = 12$ V for $t < 0$ s and $v_s(t) = 0$ V for $t \geq 0$ s. Write a differential equation in $i(t)$ for $t \geq 0$ s. Find $i(t)$ and $v(t)$ for all time and sketch these functions.

3.36 For the circuit shown in Fig. P3.36, the switch opens at time $t = 0$ s. Write a differential equation in $i(t)$ for $t \geq 0$ s. Find $i(t)$ and $v(t)$ for all time and sketch these functions.

3.37 For the circuit shown in Fig. P3.36, replace the inductor with a $\frac{1}{8}$-F capacitor. For the resulting circuit, the switch opens at time $t = 0$ s. Write a differential equation in $v(t)$ for $t \geq 0$ s. Find $v(t)$ and $i(t)$ for all time and sketch these functions.

3.38 For the circuit shown in Fig. P3.38, the switch opens at time $t = 0$ s. Find $v_1(t)$, $v_2(t)$, $i_1(t)$, $i_2(t)$, and $v(t)$ for all time.

3.39 For the circuit shown in Fig. P3.38, change the value of the 2-Ω resistor to 1 Ω. The switch in the circuit opens at time $t = 0$ s. Find $v_1(t)$, $v_2(t)$, $i_1(t)$, $i_2(t)$, and $v(t)$ for all time.

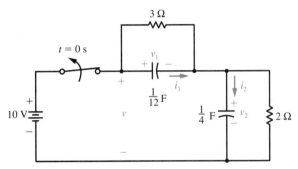

Fig. P3.36

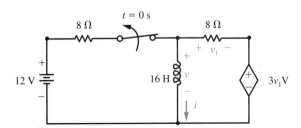

Fig. P3.38

3.40 For the parallel RC circuit given in Fig. P3.8, suppose that $i_s(t) = 6u(t)$ A. Find the step responses $v(t)$ and $i(t)$, and sketch these functions.

3.41 For the parallel RL circuit given in Fig. P3.17, find the step responses $i_L(t)$ and $v(t)$, and sketch these functions.

3.42 For the circuit shown in Fig. P3.42, suppose that $i_s(t) = 6u(t)$ A. Find the step responses $v(t)$ and $i(t)$, and sketch these functions.

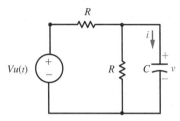

Fig. P3.42

3.43 For the circuit given in Fig. P3.30, suppose that $i_s(t) = 10u(t)$ A. Use Thévenin's theorem to find the step responses $i(t)$ and $v(t)$, and sketch these functions.

3.44 For the circuit given in Fig. P3.30, replace the inductor with a 0.1-F capacitor. Suppose that $i_s(t) = 10u(t)$ A. Use Thévenin's theorem to find the step responses $v(t)$ and $i(t)$, and sketch these functions.

3.45 For the circuit given in Fig. P3.34, suppose that $v_s(t) = 12u(t)$ V. Find the step responses $v(t)$ and $i(t)$, and sketch these functions.

3.46 For the circuit given in Fig. P3.34, replace the capacitor with a 3-H inductor. Suppose that $i_s(t) = 12u(t)$ V. Find the step responses $i(t)$ and $v(t)$, and sketch these functions.

3.47 The step responses $v_C(t)$ and $i(t)$ for the series RC circuit shown in Fig. P3.47a are given by Eq. 3.19 and Eq. 3.20, respectively. Use duality to determine the step responses $i_L(t)$ and $v(t)$ for the parallel GL circuit shown in Fig. P3.47b.

3.48 Find the step response $v_o(t)$ for the op-amp circuit shown in Fig. P3.48.

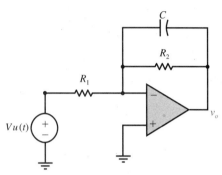

Fig. P3.48

3.49 Find the step responses $v(t)$ and $v_o(t)$ for the op-amp circuit shown in Fig. P3.49.

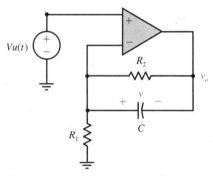

Fig. P3.49

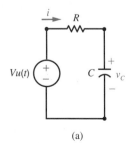

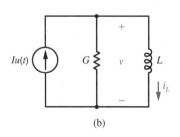

(a)   (b)

Fig. P3.47

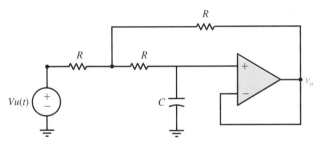

Fig. P3.50

3.50 Find the step response $v_o(t)$ for the op-amp circuit shown in Fig. P3.50.

3.51 For the series RC circuit given in Fig. P3.7a, suppose that $v_s(t) = 12e^{-t/2}u(t)$ V. Find the responses $v(t)$ and $i(t)$.

3.52 For the series RC circuit given in Fig. P3.7a, suppose that $v_s(t) = 12e^{-t/4}u(t)$ V. Find the responses $v(t)$ and $i(t)$.

3.53 For the series RL circuit given in Fig. P3.1a, suppose that $v_s(t) = 12e^{-2t}u(t)$ V. Find the responses $i(t)$ and $v(t)$.

3.54 For the series RL circuit given in Fig. P3.1a, suppose that $v_s(t) = 12e^{-t}u(t)$ V. Find the responses $i(t)$ and $v(t)$.

3.55 For the circuit shown in Fig. P3.30, when $i_s(t) = 10u(t)$ A, then $i(t) = 4(1 - e^{-t})u(t)$ A and $v(t) = 20e^{-t}u(t)$ V. Find $i(t)$ and $v(t)$ when $i_s(t) = 5u(t) - 5u(t - 1)$ A.

3.56 For the circuit shown in Fig. P3.34, when $v_s(t) = 12u(t)$ V, then $v(t) = 18(1 - e^{-4t})u(t)$ V and $i(t) = 3e^{-4t}u(t)$ A. Find $v(t)$ and $i(t)$ when $v_s(t) = 4u(t) - 4u(t - 2)$ V.

3.57 For the circuit shown in Fig. P3.57, the switch opens at time $t = 0$ s. Find $v(t)$ and $i(t)$ for all time.

3.58 For the circuit shown in Fig. P3.57, change the value of the capacitor to $\frac{3}{5}$ F. For the resulting circuit, the switch opens at time $t = 0$ s. Find $v(t)$ and $i(t)$ for all time.

3.59 For the circuit shown in Fig. P3.57, change the value of the capacitor to 3 F. For the resulting circuit, the switch opens at time $t = 0$ s. Find $v(t)$ and $i(t)$ for all time.

3.60 For the circuit shown in Fig. P3.60, the switch opens at time $t = 0$ s. Find $i(t)$ and $v(t)$ for all time. (See p. 184.)

3.61 For the circuit shown in Fig. P3.60, change the value of the resistor to $\frac{1}{2}$ Ω. For the resulting circuit, the switch opens at time $t = 0$ s. Find $i(t)$ and $v(t)$ for all time. (See p. 184.)

3.62 For the circuit shown in Fig. P3.60, change the value of the inductor to $\frac{2}{9}$ H. For the resulting circuit, the switch opens at time $t = 0$ s. Find $v(t)$ and $i(t)$ for all time. (See p. 184.)

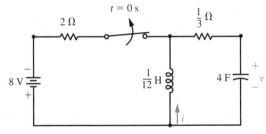

Fig. P3.57

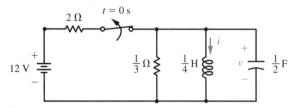

Fig. P3.60

3.63 For the series *RLC* circuit shown in Fig. P3.63, suppose that $R = 7\ \Omega$, $L = 1$ H, $C = 0.1$ F, $v_s(t) = 12$ V for $t < 0$ s and $v_s(t) = 0$ V for $t \geq 0$ s. Find $v(t)$ and $i(t)$ for all time.

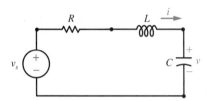

Fig. P3.63

3.64 For the series *RLC* circuit shown in Fig. P3.63, suppose that $R = 2\ \Omega$, $L = 0.25$ H, $C = 0.2$ F, $v_s(t) = 10$ V for $t < 0$ s and $v_s(t) = 0$ V for $t \geq 0$ s. Find $v(t)$ and $i(t)$ for all time.

3.65 For the series *RLC* circuit shown in Fig. P3.63, suppose that $R = 2\ \Omega$, L $= 1$ H, $C = 1$ F, $v_s(t) = 6$ V for $t < 0$ s and $v_s(t) = 0$ V for $t \geq 0$ s. Find $v(t)$ and $i(t)$ for all time.

3.66 For the circuit shown in Fig. P3.66, suppose that $v_s(t) = 6$ V for $t < 0$ s and $v_s(t) = 0$ V for $t \geq 0$ s. Find $v_2(t)$ and $v_1(t)$ for all time.

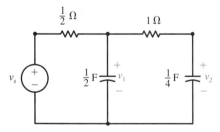

Fig. P3.66

3.67 For the circuit shown in Fig. P3.67, suppose that $v_s(t) = 6$ V for $t < 0$ s and $v_s(t) = 0$ V for $t \geq 0$ s. Find $i(t)$ and $v(t)$ for all time.

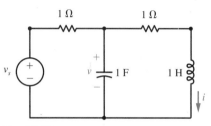

Fig. P3.67

3.68 For the circuit shown in Fig. P3.67, interchange the inductor and the capacitor. Suppose that $v_s(t) = 6$ V for $t < 0$ s and $v_s(t) = 0$ V for $t \geq 0$ s. Find the capacitor voltage $v(t)$ and the inductor current $i(t)$ for all time.

3.69 For the parallel RLC circuit shown in Fig. P3.69, suppose that $R = 0.5\ \Omega$, $L = 0.2$ H, $C = 0.25$ F, and $i_s(t) = 2u(t)$ A. Find the step responses $i(t)$ and $v(t)$.

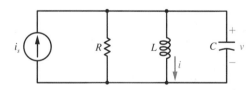

Fig. P3.69

3.70 For the parallel *RLC* circuit shown in Fig. P3.69, suppose that $R = 3\ \Omega$, L $= 3$ H, $C = \frac{1}{12}$ F, and $i_s(t) = 2u(t)$ A. Find the step responses $i(t)$ and $v(t)$.

3.71 For the series *RLC* circuit shown in Fig. P3.63, suppose that $R = 7\ \Omega$, $L = 1$ H, $C = 0.1$ F, and $v_s(t) = 12u(t)$ V. Find the step responses $v(t)$ and $i(t)$.

3.72 For the series *RLC* circuit shown in Fig. P3.63, suppose that $R = 2\ \Omega$, $L = 1$ H, $C = 1$ F, and $v_s(t) = 12u(t)$ V. Find the step responses $v(t)$ and $i(t)$.

3.73 For the *RLC* circuit shown in Fig. 3.43 on p. 172, suppose that $R = \frac{1}{2}$ Ω, $L = \frac{1}{3}$ H, $C = \frac{1}{4}$ F, and $V = 1$ V. Find the unit step responses $i(t)$ and $v(t)$.

3.74 For the *RLC* circuit shown in Fig. 3.43 on p. 172, suppose that $R = \frac{1}{2}$ Ω, $L = \frac{1}{4}$ H, $C = \frac{1}{2}$ F, and $V = 1$ V. Find the unit step responses $i(t)$ and $v(t)$.

3.75 For the circuit shown in Fig. P3.66, suppose that $v_s(t) = 9u(t)$ V. Find the step response $v_2(t)$.

3.76 For the circuit shown in Fig. P3.67, suppose that $v_s(t) = 6u(t)$ V. Find the step responses $i(t)$ and $v(t)$.

3.77 Find the step response $v_o(t)$ for the op-amp circuit shown in Fig. P3.77 when $C = \frac{1}{3}$ F and $v_s(t) = 4u(t)$ V.

3.78 Find the step response $v_o(t)$ for the op-amp circuit shown in Fig. P3.77 when $C = \frac{1}{8}$ F and $v_s(t) = 8u(t)$ V.

3.79 Find the step response $v_o(t)$ for the op-amp circuit shown in Fig. P3.77 when $C = \frac{1}{4}$ F and $v_s(t) = 6u(t)$ V.

3.80 Find the step response $v_o(t)$ for the op-amp circuit shown in Fig. P3.80 when $C = \frac{4}{3}$ F and $v_s(t) = 4u(t)$ V.

3.81 Find the step response $v_o(t)$ for the op-amp circuit shown in Fig. P3.80 when $C = 1$ F and $v_s(t) = 3u(t)$ V.

3.82 Find the step response $v_o(t)$ for the op-amp circuit shown in Fig. P3.80 when $C = \frac{1}{5}$ F and $v_s(t) = 2u(t)$ V.

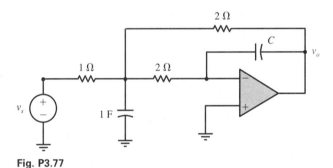

Fig. P3.77

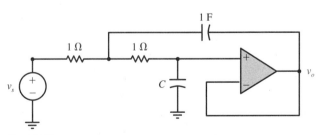

Fig. P3.80

4

AC Analysis

INTRODUCTION

Step functions are useful in determining the responses of circuits when they are first turned on or when sudden or irregular changes occur in the input. However, to see how a circuit responds to a regular or repetitive input, the function that is by far the most useful is the sinusoid.

The sinusoid is an extremely important and ubiquitous function. The shape of ordinary household voltage is sinusoidal. Broadcast radio transmissions are either AM, in which the **a**mplitude of a sinusoid is changed or **m**odulated according to some information signal, or FM, in which the **f**requency of a sinusoid is **m**odulated. (Broadcast television uses AM for the picture [video] and FM for the sound [audio].) For these reasons and more, sinusoidal analysis—also called **ac**[1] **analysis**—is a fundamental topic in the study of electric circuits.

In this chapter, we shall see that although we can analyze ac circuits with the techniques discussed previously, by generalizing the sinusoid, we can perform ac analyses in a more simplified manner that avoids the direct solution of differential equations.

In our previous analyses, our major concern was with determining voltages and currents. In many applications (e.g., electric utilities), the energy or power supplied

---

[1]ac is the abbreviation of the term "alternating current."

and that absorbed are extremely important quantities. Knowing the instantaneous power is useful, because many electric and electronic devices have maximum instantaneous or "peak" power ratings, which, for satisfactory operation, should not be exceeded. By averaging instantaneous power, we obtain the average rate at which power is supplied or absorbed. Power usages vary from a fraction of a watt for small electronic circuits to millions of watts for large electric utilities.

4.1 Time-Domain Analysis

Of all the functions encountered in electrical engineering, perhaps the most important is the sinusoid.

From trigonometry, recall the plot of $\cos x$ versus the angle x (whose units are radians) shown in Fig. 4.1a. By changing the angle from x to ωt, we get the plot shown in Fig. 4.1b, where t is time in seconds, so that ω must have the units radians per second (rad/s).[2] We say that ω is the **radian** or **angular frequency** of $\cos \omega t$. To plot $\cos \omega t$ versus time t, divide the numbers on the horizontal axis (abscissa) by ω. The result is shown in Fig. 4.1c. From this plot, it is easy to see that this function goes through a complete cycle in π/ω seconds. We call the time to complete one cycle the **period** of the sinusoid and denote it by T, that is, the number of seconds per cycle is

$$T = \frac{2\pi}{\omega} \quad \text{seconds}$$

The inverse of this quantity is the number of cycles that occur in one second. That is, the number of cycles per second is

$$\frac{1}{T} = \frac{\omega}{2\pi} \quad \frac{1}{\text{seconds}}$$

The number of cycles per second is designated by f and is called the **cyclical** or **ordinary frequency** of the sinusoid. The units of ordinary frequency, formerly called "cycles per second," are **hertz** (Hz).[3] Thus the relationships between ordinary and angular frequencies are

$$f = \frac{\omega}{2\pi} \quad \text{Hz} \qquad \text{and} \qquad \omega = 2\pi f \quad \text{rad}/\text{s}$$

---

[2]It has been suggested that the term "radians per second" be replaced by "metz"—abbreviated Mz—to honor the American electrical engineer Charles P. Steinmetz (1865–1923).

[3]Named for the German physicist Heinrich Hertz (1857–1894).

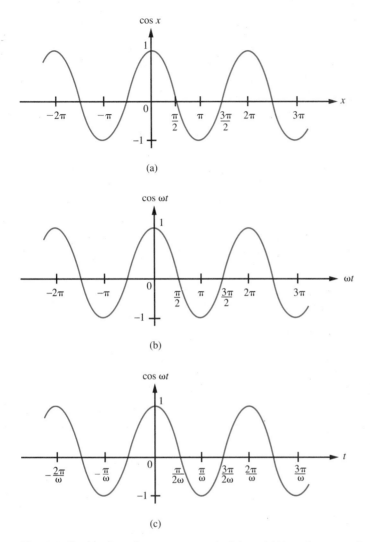

(a)

(b)

(c)

Fig. 4.1 Cosine function versus angle (*a*) and (*b*) and versus time (*c*).

To obtain a plot of the general sinusoid $\cos(\omega t + \phi)$ versus t, reconsider the plot of $\cos \omega t$ versus ωt shown in Fig. 4.1*b*. Replacing ωt by $\omega t + \phi$, where ϕ is a positive quantity, corresponds to translating (or shifting) the sinusoid to the left by the amount ϕ. Figure 4.2 shows a plot of $\cos(\omega t + \phi)$ versus ωt. Dividing the numbers on the horizontal axis by ω yields the plot shown in Fig. 4.3. (Of course, if ϕ is a negative number, the shift is to the right.) We call ϕ the **phase angle** or simply **angle** of the sinusoid. Note that when $\phi = -\pi/2$ rad (or $\phi = -90$ degrees), then

$$\cos(\omega t - \pi/2) = \sin \omega t$$

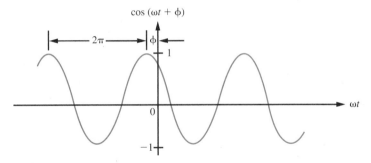

Fig. 4.2 Sinusoid versus angle.

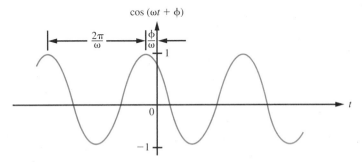

Fig. 4.3 Sinusoid versus time.

Similarly,

$$\sin(\omega t + \pi/2) = \cos \omega t$$

For the case of a circuit with a sinusoidal input, in general, the response consists of the natural response and the forced response. In particular, let us concentrate on the forced response. We know that the integral or derivative of a sinusoid that has a given frequency is again a sinusoid having the same frequency. For a resistor, the voltage and the current differ by a constant, whereas for an inductor and a capacitor, the relationships between voltage and current are given in terms of integrals and derivatives. Thus for each of these elements, a sinusoidal input produces a sinusoidal response. Because adding sinusoids having frequency ω results in a single sinusoid of frequency ω (e.g., see Eq. 3.40 on p. 158), we can deduce that the forced response of any *RLC* circuit whose input is a sinusoid of frequency ω is also a sinusoid of frequency ω.

Example 4.1

Consider the series *RC* circuit shown in Fig. 4.4a. Since

$$v_s = 1i + v_o \qquad \text{and} \qquad i = \frac{1}{2}\frac{dv_o}{dt}$$

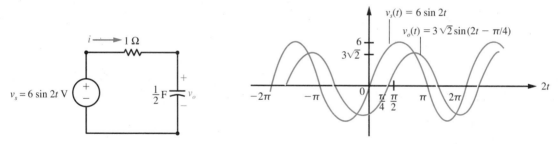

Fig. 4.4 (a) Series *RC* circuit, (b) output voltage $v_o(t)$ lags input voltage $v_s(t)$ by 45°.

then

$$\frac{dv_o}{dt} + 2v_o = 12 \sin 2t \qquad (4.1)$$

Because the forced response is a sinusoid of frequency $\omega = 2$ rad/s, it has the general form $A \cos(2t + \theta)$. That is, the forced response portion of the solution to the differential Eq. 4.1 has the form $v_o(t) = A \cos(2t + \theta)$. However, to avoid using the trigonometric identity for the cosine of the sum of two angles, let us use the alternative form (see Eq. 3.40 on p. 158):

$$v_o(t) = A_1 \cos 2t + A_2 \sin 2t$$

Substituting this into Eq. 4.1 results in

$$(-2A_1 + 2A_2) \sin 2t + (2A_1 + 2A_2) \cos 2t = 12 \sin 2t$$

Equating coefficients yields the simultaneous equations

$$-2A_1 + 2A_2 = 12 \qquad \text{and} \qquad 2A_1 + 2A_2 = 0$$

the solution of which is $A_1 = -3$ and $A_2 = 3$. Hence, the forced response is

$$v_o(t) = -3 \cos 2t + 3 \sin 2t = 3\sqrt{2} \cos\left(2t - \frac{3\pi}{4}\right) \text{V}$$

by Eq. 3.29. Alternatively,

$$v_o(t) = 3\sqrt{2} \cos\left(2t - \frac{\pi}{2} - \frac{\pi}{4}\right) = 3\sqrt{2} \sin\left(2t - \frac{\pi}{4}\right) \text{V}$$

A plot of the functions $v_s(t)$ and $v_o(t)$ versus ωt is shown in Fig. 4.4b. Because $v_s(t)$ reaches its peak before $v_o(t)$, we can say that $v_s(t)$ **leads** $v_o(t)$ by $\frac{\pi}{4}$ rad (45 degrees) or that $v_o(t)$ **lags** $v_s(t)$ by $\frac{\pi}{4}$ rad. Since the output of the circuit lags the input, the given circuit is an example of what is called a **lag network**.

The current in this circuit is

$$i(t) = C\frac{dv_o(t)}{dt} = \frac{1}{2}\frac{dv_o(t)}{dt} = \frac{1}{2}(6\sin 2t + 6\cos 2t)$$

$$= 3\cos 2t + 3\sin 2t = 3\sqrt{2}\cos\left(2t - \frac{\pi}{4}\right)$$

$$= 3\sqrt{2}\cos\left(2t - \frac{\pi}{2} + \frac{\pi}{4}\right) = 3\sqrt{2}\sin\left(2t + \frac{\pi}{4}\right)\text{ A}$$

Thus, the current $i(t)$ leads the voltage $v_s(t)$ by 45 degrees, and $i(t)$ leads the voltage $v_o(t)$ by 90 degrees.

Drill Exercise 4.1

For the circuit shown in Fig. 4.4a, replace the $\frac{1}{2}$-F capacitor with a $\frac{1}{2}$-H inductor, and find the voltage $v_o(t)$ across the inductor. Does $v_o(t)$ lag or lead $v_s(t)$? By how much?

ANSWER $3\cos 2t + 3\sin 2t = 3\sqrt{2}\cos(2t - \frac{\pi}{4}) = 3\sqrt{2}\sin(2t + \frac{\pi}{4})$ V, $v_o(t)$ leads $v_s(t)$ by $\frac{\pi}{4}$ rad (or 45 degrees)

As indicated by the preceding example and drill exercise, sinusoidal analysis is straightforward. However, the amount of arithmetic required can be extremely tedious. A lot of drudgery involved in analyzing sinusoidal circuits can be eliminated if we utilize the concept of a complex sinusoid and the property of linearity. Such an approach will not only be numerically advantageous but gives rise to concepts that will prove invaluable in the future. To proceed, though, it will be necessary to study complex numbers.

4.2 Complex Numbers

As mentioned previously, we designate the constant $\sqrt{-1}$ by j, that is, $j = \sqrt{-1}$. Thus taking powers of j, we get

$$j^2 = (\sqrt{-1})^2 = -1, \quad j^3 = j^2 j = -j, \quad j^4 = j^2 j^2 = (-1)(-1) = 1, \quad \text{etc.}$$

A **complex number A** is a number that can be written in the form

$$\mathbf{A} = a + jb$$

where a and b are real numbers. We call a the **real part** of **A**, and we call b the **imaginary part** of **A**. These are denoted

$$\text{Re}(\mathbf{A}) = a \quad \text{and} \quad \text{Im}(\mathbf{A}) = b$$

respectively. If $a = 0$, we say that **A** is **imaginary**, and if $b = 0$, we say that **A** is **real**.

Two complex numbers $\mathbf{A}_1 = a_1 + jb_1$ and $\mathbf{A}_2 = a_2 + jb_2$ are **equal** if both

$$a_1 = a_2 \quad \text{and} \quad b_1 = b_2$$

The **sum** of two complex numbers is

$$\mathbf{A}_1 + \mathbf{A}_2 = (a_1 + jb_1) + (a_2 + jb_2) = (a_1 + a_2) + j(b_1 + b_2)$$

and the **difference** is

$$\mathbf{A}_1 - \mathbf{A}_2 = (a_1 + jb_1) - (a_2 + jb_2) = (a_1 - a_2) + j(b_1 - b_2)$$

The **complex conjugate** of $\mathbf{A} = a + jb$ is denoted by $\mathbf{A}^*$ and is defined as

$$\mathbf{A}^* = a - jb$$

In other words, to obtain the complex conjugate of a complex number, simply change the sign of the imaginary part of the complex number.

The **product** of two complex numbers is

$$\mathbf{A}_1\mathbf{A}_2 = (a_1 + jb_1)(a_2 + jb_2) = a_1 a_2 + ja_1 b_2 + ja_2 b_1 + j^2 b_1 b_2$$

$$= (a_1 a_2 - b_1 b_2) + j(a_1 b_2 + a_2 b_1)$$

and the **quotient** is

$$\frac{\mathbf{A}_1}{\mathbf{A}_2} = \frac{a_1 + jb_1}{a_2 + jb_2} = \frac{(a_1 + jb_1)(a_2 - jb_2)}{(a_2 + jb_2)(a_2 - jb_2)} = \frac{a_1 a_2 + b_1 b_2}{a_2^2 + b_2^2} + j\frac{b_1 a_2 - a_1 b_2}{a_2^2 + b_2^2}$$

Although addition and subtraction of complex numbers are simple operations, multiplication and division are a little more complicated. This is a consequence of one form of a complex number—specifically, a complex number **A** can be written as $\mathbf{A} = a + jb$. We call this the **rectangular form** of the complex number **A**. There is an alternative form for a complex number that lends itself nicely to multiplication and division, although not to addition and subtraction. To develop this form, we first derive Euler's formula,[4] which was mentioned earlier.

Euler's Formula

Let θ be a real variable. Define the function $f(\theta)$ by

$$f(\theta) = \cos\theta + j\sin\theta$$

Taking the derivative with respect to θ yields

$$\frac{df(\theta)}{d\theta} = -\sin\theta + j\cos\theta = j^2\sin\theta + j\cos\theta$$

$$= j(\cos\theta + j\sin\theta) = jf(\theta) \quad\Rightarrow\quad j = \frac{1}{f(\theta)}\frac{df(\theta)}{d\theta}$$

Integrating both sides of the last expression with respect to θ results in

$$\int j\,d\theta = \int \frac{1}{f(\theta)}\frac{df(\theta)}{d\theta}d\theta = \int \frac{df(\theta)}{f(\theta)}$$

Integrating yields

$$j\theta + K = \ln f(\theta) = \ln(\cos\theta + j\sin\theta)$$

where K is a constant of integration. Setting $\theta = 0$ gives

$$K = \ln 1 = 0 \quad\Rightarrow\quad j\theta = \ln(\cos\theta + j\sin\theta)$$

Taking powers of e, we get

$$\boxed{e^{j\theta} = \cos\theta + j\sin\theta} \qquad\qquad (4.2)$$

---

[4]Named for the Swiss mathematician Leonhard Euler (1707–1783).

and this result is known as **Euler's formula**. Replacing θ by $-\theta$, we obtain another version:

$$e^{-j\theta} = \cos \theta - j \sin \theta \qquad (4.3)$$

We now see that $e^{j\theta}$ is in fact a complex number for which

$$\text{Re}(e^{j\theta}) = \cos \theta \qquad \text{and} \qquad \text{Im}(e^{j\theta}) = \sin \theta$$

By adding and subtracting Eq. 4.2 and 4.3, we can get, respectively,

$$\cos \theta = \frac{1}{2}(e^{j\theta} + e^{-j\theta}) \qquad \text{and} \qquad \sin \theta = \frac{1}{2j}(e^{j\theta} - e^{-j\theta}) \qquad (4.4)$$

If M is a real number, since

$$Me^{j\theta} = M(\cos \theta + j \sin \theta) = M \cos \theta + jM \sin \theta$$

then $Me^{j\theta}$ is a complex number for which

$$\text{Re}(Me^{j\theta}) = M \cos \theta \qquad \text{and} \qquad \text{Im}(Me^{j\theta}) = M \sin \theta$$

Exponential Form of a Complex Number

Given a (nonzero) complex number $\mathbf{A} = a + jb$, is there a complex number of the form $Me^{j\theta}$, where $M > 0$, that equals $\mathbf{A}$? If there is, then

$$a + jb = M \cos \theta + jM \sin \theta \qquad \Rightarrow$$
$$a = M \cos \theta \qquad \text{and} \qquad b = M \sin \theta$$

Thus

$$\cos \theta = aM \qquad \text{and} \qquad \sin \theta = bM$$

from which

$$\tan \theta = \frac{\sin \theta}{\cos \theta} = \frac{b/M}{a/M} = \frac{b}{a} \qquad \Rightarrow \qquad \theta = \tan^{-1}\left(\frac{b}{a}\right)$$

Pictorially, we have the right triangle shown in Fig. 4.5. Thus

$$a = M \cos \theta \quad \Rightarrow \quad M = \frac{a}{\cos \theta} = \frac{a}{\dfrac{a}{\sqrt{a^2 + b^2}}} = \sqrt{a^2 + b^2}$$

Fig. 4.5 Right triangle describing angle θ.

In summary, we can write a complex number as

$$a + jb = Me^{j\theta}$$

where the real number $M > 0$ and

$$\begin{aligned} M &= \sqrt{a^2 + b^2} \\ \theta &= \tan^{-1}(b/a) \end{aligned} \qquad \text{or} \qquad \begin{aligned} a &= M \cos \theta \\ b &= M \sin \theta \end{aligned}$$

We say that $Me^{j\theta}$ is the **exponential** (or **polar**) **form** of a complex number, where the positive number M is called the **magnitude** and θ is called the **angle** (or **argument**) of the complex number. We denote the magnitude of $\mathbf{A}$ by $|\mathbf{A}|$ and the angle of $\mathbf{A}$ by $\text{ang}(\mathbf{A})$.

A complex number $\mathbf{A} = a + jb$ can be represented by the point (a, b) in the **complex-number plane** shown in Fig. 4.6. In this plane, the abscissa is the real axis and the ordinate is the imaginary axis. Note that a straight line connecting the origin $(0, 0)$ and the point (a, b) has length $M = \sqrt{a^2 + b^2}$ and the angle formed by this line and the positive real axis is $\theta = \tan^{-1}(b/a)$—and these are precisely the magnitude and the angle of the complex number $\mathbf{A}$, respectively.

Note that by using the exponential forms of the complex numbers $\mathbf{A}_1 = M_1 e^{j\theta_1}$ and $\mathbf{A}_2 = M_2 e^{j\theta_2}$, the product is

$$\mathbf{A}_1 \mathbf{A}_2 = (M_1 e^{j\theta_1})(M_2 e^{j\theta_2}) = M_1 M_2 e^{j(\theta_1 + \theta_2)}$$

Thus

$$|\mathbf{A}_1 \mathbf{A}_2| = M_1 M_2 = |\mathbf{A}_1||\mathbf{A}_2|$$

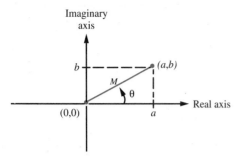

Fig. 4.6 Complex-number plane.

that is, the magnitude of a product of complex numbers is equal to the product of the magnitudes of the complex numbers, and

$$\text{ang}(\mathbf{A}_1\mathbf{A}_2) = \theta_1 + \theta_2 = \text{ang}(\mathbf{A}_1) + \text{ang}(\mathbf{A}_2)$$

that is, the angle of a product of complex numbers is equal to the sum of the angles of the complex numbers.

The quotient of the complex numbers $\mathbf{A}_1$ and $\mathbf{A}_2$ is

$$\frac{\mathbf{A}_1}{\mathbf{A}_2} = \frac{M_1 e^{j\theta_1}}{M_2 e^{j\theta_2}} = \frac{M_1}{M_2} e^{j(\theta_1 - \theta_2)}$$

Thus

$$\left|\frac{\mathbf{A}_1}{\mathbf{A}_2}\right| = \frac{M_1}{M_2} = \frac{|\mathbf{A}_1|}{|\mathbf{A}_2|}$$

that is, the magnitude of a quotient of complex numbers is equal to the quotient of the magnitudes of the complex numbers, and

$$\text{ang}\left(\frac{\mathbf{A}_1}{\mathbf{A}_2}\right) = \theta_1 - \theta_2 = \text{ang}(\mathbf{A}_1) - \text{ang}(\mathbf{A}_2)$$

that is, the angle of a quotient of complex numbers is equal to the difference of the angles of the complex numbers.

Note that the complex conjugate of $\mathbf{A} = M e^{j\theta}$ is

$$(M e^{j\theta})^* = (M \cos \theta + jM \sin \theta)^* = M \cos \theta - jM \sin \theta = M e^{-j\theta}$$

Therefore, to obtain the (complex) conjugate of a complex number in exponential form, simply change the sign of its angle.

In addition, $\mathbf{A}_1 = \mathbf{A}_2$ if and only if $M_1 = M_2$ and $\theta_1 = \theta_2$.

Example 4.2

Suppose that $\mathbf{A}_1 = -3 + j4 = M_1 e^{j\theta_1}$ and $\mathbf{A}_2 = -5 - j12 = M_2 e^{j\theta_2}$. Then we have that

$$|\mathbf{A}_1| = M_1 = \sqrt{(-3)^2 + (4)^2} = 5 \quad \text{and}$$

$$\text{ang}(\mathbf{A}_1) = \theta_1 = \tan^{-1}(4/-3) = 127°$$

and

$$|\mathbf{A}_2| = M_2 = \sqrt{(-5)^2 + (-12)^2} = 13 \quad \text{and}$$

$$\text{ang}(\mathbf{A}_2) = \theta_2 = \tan^{-1}(-12/-5) = -113°$$

Thus, the exponential form of the product is

$$\mathbf{A}_1 \mathbf{A}_2 = (5e^{j127°})(13e^{-j113°}) = (5)(13)e^{j(127°-113°)} = 65e^{j14°}$$

whereas the rectangular form of the product is

$$\mathbf{A}_1 \mathbf{A}_2 = 65 \cos(14°) + j65 \sin(14°) = 63.1 + j15.7$$

The exponential form of the quotient is

$$\frac{\mathbf{A}_1}{\mathbf{A}_2} = \frac{5e^{j127°}}{13e^{-j113°}} = \frac{5}{13}e^{j(127°+113°)} = 0.385e^{j240°} = 0.385e^{-j120°}$$

while the rectangular form of the quotient is

$$\mathbf{A}_1/\mathbf{A}_2 = 0.385 \cos(-120°) + j0.385 \sin(-120°) = -0.193 - j0.333$$

Drill Exercise 4.2

Suppose that $\mathbf{A}_1 = 1 - j\sqrt{3}$ and $\mathbf{A}_2 = -2 + j2$. Find the exponential and polar forms of (a) $\mathbf{A}_1 \mathbf{A}_2$ and (b) $\mathbf{A}_1/\mathbf{A}_2$.

ANSWER (a) $4\sqrt{2}e^{j75°}$, $1.46 + j5.46$; (b) $(\sqrt{2}/2)e^{j165°}$, $-0.683 + j0.183$

Suppose that $\mathbf{A}_1 = 10e^{j36.9°}$ and $\mathbf{A}_2 = 13e^{-j67.4°}$. Find the rectangular and exponential forms of $\mathbf{A}_1 + \mathbf{A}_2$.

ANSWER $13.0 - j6.0$, $14.3e^{-j24.8°}$

4.3 Frequency-Domain Analysis

Given an *RLC* circuit whose input is the sinusoid $A \cos(\omega t + \theta)$, the forced response is a sinusoid of frequency ω, say $B \cos(\omega t + \phi)$. If the input is shifted by 90 degrees, that is, if the input is $A \sin(\omega t + \theta)$, then by the time-invariance property, the corresponding response is $B \sin(\omega t + \phi)$. Furthermore, if the input is scaled by a constant K, then by the property of linearity, so will be the response; that is, the response to $KA \sin(\omega t + \phi)$ is $KB \sin(\omega t + \phi)$. Again, by linearity, the response to

$$A \cos(\omega t + \theta) + KA \sin(\omega t + \theta) \quad \text{is} \quad B \cos(\omega t + \phi) + KB \sin(\omega t + \phi)$$

Consider now the special case that the constant $K = \sqrt{-1} = j$. Thus the response to

$$A \cos(\omega t + \theta) + jA \sin(\omega t + \theta) = Ae^{j(\omega t + \theta)} \quad \text{is}$$

$$B \cos(\omega t + \theta) + jB \sin(\omega t + \theta) = Be^{j(\omega t + \theta)}$$

Therefore, we see that whether we wish to find the response to $A \cos(\omega t + \theta)$ or the response to the **complex sinusoid** $Ae^{j(\omega t + \theta)}$, we must determine the same two constants B and ϕ.

Although it may be somewhat easier computationally to find the response to a (real) sinusoid by considering the corresponding complex sinusoidal case instead, the determination of the appropriate differential equation is still required. By introducing the appropriate concepts, however, we will be able to eliminate this requirement, and the result will be the simplification of sinusoidal analysis.

Phasors

Just as the sinusoid $A \cos(\omega t + \theta)$ has frequency ω, amplitude A, and phase angle θ, we can say that the complex sinusoid

$$Ae^{j(\omega t + \theta)} = Ae^{j\theta}e^{j\omega t}$$

also has frequency ω, amplitude A, and phase angle θ. Let us denote the complex number

$$\mathbf{A} = Ae^{j\theta} \quad \text{by} \quad \mathbf{A} = A\underline{/\theta}$$

which is known as the **phasor** representation of the (real or complex) sinusoid.

Why do we do this? We begin to answer this question by considering a resistor. If the current $i(t)$ through the resistor is a complex sinusoid, then so is the voltage $v(t)$ across the resistor, and vice versa. Suppose that

$$i(t) = Ie^{j(\omega t + \theta)} \quad \text{and} \quad v(t) = Ve^{j(\omega t + \phi)}$$

From Ohm's law, $v(t) = Ri(t)$. Thus

$$Ve^{j(\omega t + \phi)} = RIe^{j(\omega t + \theta)} \quad \Rightarrow \quad Ve^{j\omega t}e^{j\phi} = RIe^{j\omega t}e^{j\theta}$$

and dividing both sides of the last expression by $e^{j\omega t}$ yields

$$Ve^{j\phi} = RIe^{j\theta}$$

From this equation, we deduce that $V = RI$ and $\phi = \theta$. Using the phasor forms of the voltage and current, we can write the phasor equation

$$V\underline{/\phi} = RI\underline{/\theta} \quad \text{or} \quad \boxed{\mathbf{V} = R\mathbf{I}}$$

where $\mathbf{V} = V\underline{/\phi}$ and $\mathbf{I} = I\underline{/\theta}$. Because $\phi = \theta$, the voltage and the current are "in phase" for a resistor.

For an inductor with complex sinusoidal current $i(t) = Ie^{j(\omega t + \theta)}$ and voltage $v(t) = Ve^{j(\omega t + \phi)}$, since

$$v(t) = L\frac{di(t)}{dt} \quad \Rightarrow \quad Ve^{j(\omega t + \phi)} = L\frac{d}{dt}(Ie^{j(\omega t + \theta)})$$

then

$$Ve^{j\omega t}e^{j\phi} = j\omega LIe^{j\omega t}e^{j\theta} \quad \Rightarrow \quad Ve^{j\phi} = j\omega LIe^{j\theta}$$

and using the phasor form of voltage and current, we get the phasor equation

$$V\underline{/\phi} = j\omega LI\underline{/\theta} \quad \text{or} \quad \boxed{\mathbf{V} = j\omega L\,\mathbf{I}}$$

where $\mathbf{V} = V\underline{/\phi}$ and $\mathbf{I} = I\underline{/\theta}$. Note that

$$\text{ang}(\mathbf{V}) = \text{ang}(j\omega L\mathbf{I}) = \text{ang}(j\omega L) + \text{ang}(\mathbf{I}) \quad \Rightarrow \quad \phi = 90° + \theta$$

Because the voltage angle is 90 degrees greater than the current angle, then for an inductor the voltage leads the current by 90 degrees, or equivalently, the current lags the voltage by 90 degrees.

For a capacitor with complex sinusoidal current $i(t) = Ie^{j(\omega t + \theta)}$ and voltage $v(t) = Ve^{j(\omega t + \phi)}$, since

$$i(t) = C\frac{dv(t)}{dt} \quad \Rightarrow \quad Ie^{j(\omega t + \theta)} = C\frac{d}{dt}(Ve^{j(\omega t + \phi)})$$

then

$$Ie^{j\omega t}e^{j\theta} = j\omega CVe^{j\omega t}e^{j\phi} \quad \Rightarrow \quad Ie^{j\theta} = j\omega CVe^{j\phi}$$

and using the phasor form of voltage and current

$$I\underline{/\theta} = j\omega CV\underline{/\phi} \quad \Rightarrow \quad \mathbf{I} = j\omega C\mathbf{V} \quad \Rightarrow \quad \boxed{\mathbf{V} = \frac{1}{j\omega C}\mathbf{I}}$$

where $\mathbf{V} = V\underline{/\phi}$ and $\mathbf{I} = I\underline{/\theta}$. Note that since

$$\text{ang}(\mathbf{I}) = \text{ang}(j\omega L\mathbf{V}) = \text{ang}(j\omega L) + \text{ang}(\mathbf{V}) \quad \Rightarrow \quad \theta = 90° + \phi$$

we see that for a capacitor, the current leads the voltage by 90 degrees. (If you would like an easy way to remember whether current leads or lags voltage for a capacitor or an inductor, just think of the word "capacitor" as the key. Let the first three letters stand for "**c**urrent **a**nd **p**otential," thus indicating that current comes before (leads) voltage for a capacitor. This, of course, implies that current lags voltage for an inductor.)

Because a phasor is a complex number, we can represent it as a point in the complex-number plane, and therefore, specify it by an arrow directed from the origin to the point. For example, for a resistor R having a voltage phasor $\mathbf{V} = V\underline{/\phi}$ and a current phasor $\mathbf{I} = I\underline{/\theta}$, a diagram characterizing the relationship between these two phasors is as shown in Fig. 4.7a. We refer to this as a **phasor diagram**. Note that because $\phi = \theta$ for a resistor (i.e., the voltage and the current are in phase), the voltage and current phasors are collinear.

The phasor diagram for an L-henry inductor is shown in Fig. 4.7b. Given the voltage phasor $\mathbf{V} = V\underline{/\phi}$ and the current phasor $\mathbf{I} = I\underline{/\theta}$, the phasor diagram dis-

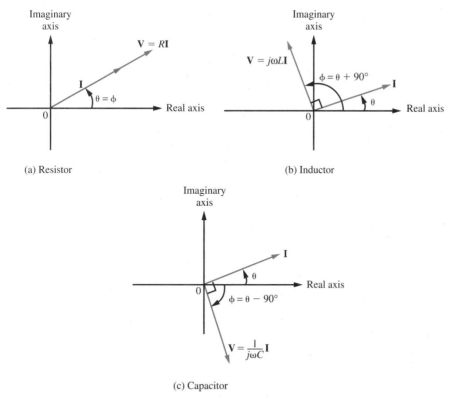

Fig. 4.7 Phasor diagrams for the (*a*) resistor, (*b*) inductor, and (*c*) capacitor.

plays the fact that the voltage leads the current by 90 degrees—this is indicated by the voltage phasor having an angle that is 90 degrees greater than the current phasor.

Figure 4.7*c* shows the phasor diagram for a *C*-farad capacitor whose voltage phasor is $\mathbf{V} = V\underline{/\phi}$ and whose current phasor is $\mathbf{I} = I\underline{/\theta}$. In this case, the current angle is 90 degrees greater than the voltage angle (i.e., $\theta = 90° + \phi$), so the current leads the voltage by 90 degrees.

Impedance and Admittance

We have seen that for the resistor, inductor, and capacitor, we get a phasor equation of the form

$$\mathbf{V} = \mathbf{Z}\mathbf{I}$$

The quantity **Z** is called the **impedance** of the element. In other words, the impedance of an R-ohm resistor, L-henry inductor, and C-farad capacitor are, respectively:

$$\mathbf{Z}_R = R \qquad \mathbf{Z}_L = j\omega L \qquad \mathbf{Z}_C = 1/j\omega C$$

The equation $\mathbf{V} = \mathbf{Z}\mathbf{I}$ is the phasor version of Ohm's law. From this equation, we see that impedance is the ratio of voltage phasor to current phasor, that is,

$$\mathbf{Z} = \mathbf{V}/\mathbf{I}$$

It follows that,

$$\mathbf{I} = \mathbf{V}/\mathbf{Z}$$

In this expression, the reciprocal of impedance appears. As this can occur frequently, it is useful to give it a name. We call the ratio of current phasor to voltage phasor **admittance** and denote it by the symbol **Y,** that is,

$$\mathbf{Y} = \mathbf{I}/\mathbf{V}$$

Thus the admittances of an R-ohm resistor, L-henry inductor, and C-farad capacitor are, respectively:

$$\mathbf{Y}_R = \frac{1}{R} \qquad \mathbf{Y}_L = \frac{1}{j\omega L} \qquad \mathbf{Y}_C = j\omega C$$

Alternative forms of the phasor version of Ohm's law are:

$$\mathbf{I} = \mathbf{Y}\mathbf{V} \qquad \text{and} \qquad \mathbf{V} = \mathbf{I}/\mathbf{Y}$$

As for the resistive case, the units of impedance are ohms and those of admittance are mhos (or siemens).

Example 4.3

For the case that $\omega = 100 \text{ rad}/\text{s}$, the impedance of a 25-$\Omega$ resistor is $\mathbf{Z}_R = 25 \ \Omega$ whereas its admittance is $\mathbf{Y}_R = 1/25 = 0.04 \ \mho$. (Actually, these are the values for the impedance and the admittance regardless of the value of ω.)

The impedance of a 50-mH inductor is $\mathbf{Z}_L = j\omega L = j(100)(0.05) = j5 \ \Omega$, whereas its admittance is $\mathbf{Y}_L = 1/j5 = j/j^2 5 = -j(0.2) \ \mho$.

The impedance of a 0.02-F capacitor is $\mathbf{Z}_C = 1/j\omega C = 1/j(100)(0.02) = -j(0.5)\ \Omega$, whereas its its admittance is $\mathbf{Y}_C = 1/-j(0.5) = j2\ \mho$.

Drill Exercise 4.3

Find the impedance and the admittance of (a) a 1-kΩ resistor, (b) a 1-μF capacitor, and (c) a 50-mH inductor for the case that the frequency is 10 kHz.

ANSWER (a) 1 kΩ, 1 m$\mho$; (b) $-j15.9\ \Omega$, $j62.8$ m$\mho$;
(c) $j3.14$ kΩ, $-j318\ \mu\mho$

AC Analysis

Suppose that for *RLC* circuit, writing KVL around a loop results in the equation

$$v_1(t) + v_2(t) + v_3(t) + \cdots + v_n(t) = 0$$

For the case that voltages are (real) sinusoids, this equation takes the form.

$$V_1 \cos(\omega t + \theta_1) + V_2 \cos(\omega t + \theta_2) + \cdots + V_n \cos(\omega t + \theta_n) = 0$$

whereas for the case of complex sinusoids

$$V_1 e^{j(\omega t + \theta_1)} + V_2 e^{j(\omega t + \theta_2)} + \cdots + V_n e^{j(\omega t + \theta_n)} = 0$$

Dividing both sides of the preceding equation by $e^{j\omega t}$ yields

$$V_1 e^{j\theta_1} + V_2 e^{j\theta_2} + \cdots + V_n e^{j\theta_n} = 0$$

or using phasor notation

$$V_1 \underline{/\theta_1} + V_2 \underline{/\theta_2} + \cdots + V_n \underline{/\theta_n} = 0 \qquad \Rightarrow \qquad \mathbf{V}_1 + \mathbf{V}_2 + \cdots + \mathbf{V}_n = 0$$

where $\mathbf{V}_i = V_i \underline{/\theta_i}$. Thus we see that in addition to voltage functions of time, KVL holds for voltage phasors. We may also deduce that KCL holds for current phasors as well as for functions of time.

For inductors and capacitors, as well as resistors, using impedance, the phasor relationships between (phasor) voltage and (phasor) current are

$$\mathbf{V} = \mathbf{Z}\,\mathbf{I}, \qquad \mathbf{I} = \mathbf{V}/\mathbf{Z}, \qquad \mathbf{Z} = \mathbf{V}/\mathbf{I}$$

That is, these are the phasor versions of Ohm's law. By using KVL and KCL, we can analyze an ac circuit without resorting to writing and solving differential equations—we simply look at the circuit from the phasor point of view and then proceed as was done for resistive circuits using nodal or mesh analysis. For resistive circuits, we have to manipulate real numbers; for ac circuits, complex-number arithmetic is required.

If we analyze a circuit by using time functions and the various differential and integral relationships between voltage and current—and, hence, are required to write and solve differential equations—we say that we are working in the **time domain**. However, if we have an ac circuit, we can use the phasor representations of the time functions and the impedances (or admittances) of the various (nonsource) elements or combinations of elements. This is said to be the **frequency-domain** approach.

Making the transformation from the time to the frequency domain is a simple matter. Just express a sinusoidal function of time by its phasor representation and label resistors, inductors, and capacitors (or combinations of them) by their impedances or admittances. Then proceed as was done for resistive circuits to find the phasor representations of the desired responses. The transformation back to the time domain is done by inspection.

Example 4.4

Reconsider the circuit given in Fig. 4.4a on p. 190. This circuit [note that $6 \sin 2t = 6 \cos(2t - 90°)$] is shown in the frequency domain in Fig. 4.8. By KVL,

$$\mathbf{V}_s = 1\mathbf{I} - j\mathbf{I}$$

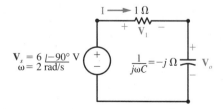

Fig. 4.8 Frequency-domain representation of circuit in Fig. 4.4a.

from which

$$6\underline{/-90°} = (1 - j)\mathbf{I} = \sqrt{2}\underline{/-45°}\ \mathbf{I} \quad \Rightarrow$$

$$\mathbf{I} = \frac{6\underline{/-90°}}{\sqrt{2}\underline{/-45°}} = 3\sqrt{2}\underline{/-45°}\ \text{A}$$

Also,

$$\mathbf{V}_o = -j\mathbf{I} = (1\underline{/-90°})(3\sqrt{2}\underline{/-45°}) = 3\sqrt{2}\underline{/-135°}\ \text{V}$$

The phasor diagram for this circuit is shown in Fig. 4.9. From this diagram, we note that the capacitor voltage lags applied voltage by 45 degrees. Therefore, if the capacitor voltage is considered to be the output voltage, then the circuit is a lag network. Note, however, that the current leads the applied voltage by 45 degrees. Since $\mathbf{V}_1 = 1\mathbf{I} = \mathbf{I}$, then the voltage across the resistor leads the applied voltage by 45 degrees. Consequently, if the voltage across the resistor is considered to be the output, then the given circuit is a **lead network**.

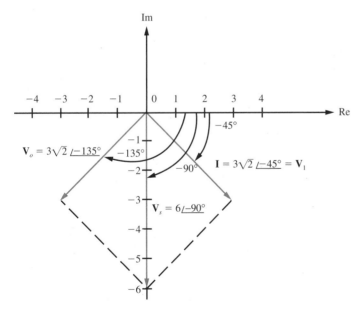

Fig. 4.9 Phasor diagram.

For this circuit, by KVL,

$$\mathbf{V}_s = \mathbf{V}_1 + \mathbf{V}_o$$

The phasor diagram shows that, indeed, $\mathbf{V}_s$ is the (complex-number) sum of $\mathbf{V}_o$ and $\mathbf{V}_1$.

Having found the voltage phasor $\mathbf{V}_o$ and the current phasor $\mathbf{I}$, we can write the corresponding time functions, respectively,

$$v_o(t) = 3\sqrt{2}\cos(2t - 135°)\ \text{V} \qquad \text{and} \qquad i(t) = 3\sqrt{2}\cos(2t - 45°)\ \text{A}$$

Note that we can instead obtain $\mathbf{V}_o$ by voltage division as follows:

$$\mathbf{V}_o = \frac{\mathbf{Z}_C}{\mathbf{Z}_C + \mathbf{Z}_R}\mathbf{V}_s = \frac{-j}{-j + 1}(6\underline{/-90°})$$

$$= \frac{1\underline{/-90°}}{\sqrt{2}\underline{/-45°}}(6\underline{/-90°}) = 3\sqrt{2}\underline{/-135°}\ \text{V}$$

Drill Exercise 4.4

For the circuit given in Fig. 4.4a, on p. 190, replace the $\frac{1}{2}$-F capacitor with a $\frac{1}{2}$-H inductor. (a) Determine the resulting voltage phasor $\mathbf{V}_o$ across the inductor. What is the corresponding function of time $v_o(t)$? (b) Determine the resulting current phasor $\mathbf{I}$ through the inductor. What is the corresponding function of time $i(t)$?

ANSWER (a) $3\sqrt{2}\underline{/-45°}$ V, $3\sqrt{2}\cos(2t - 45°)$ V;

(b) $3\sqrt{2}\underline{/-135°}$ A, $3\sqrt{2}\cos(2t - 135°)$ A

Series and Parallel Connections of Impedances

The voltage-divider formula used in Example 4.4 can be derived as was done for resistive circuits by considering the situation depicted in Fig. 4.10. By KVL,

$$\mathbf{V} = \mathbf{Z}_1\mathbf{I} + \mathbf{Z}_2\mathbf{I} = (\mathbf{Z}_1 + \mathbf{Z}_2)\mathbf{I} \qquad \Rightarrow \qquad \mathbf{I} = \frac{\mathbf{V}}{\mathbf{Z}_1 + \mathbf{Z}_2} \qquad (4.5)$$

Since $\mathbf{V}_2 = \mathbf{Z}_2\mathbf{I}$, then replacing $\mathbf{I}$ with the expression given in Eq. 4.5, we get the voltage-divider formula

$$\mathbf{V}_2 = \frac{\mathbf{Z}_2}{\mathbf{Z}_1 + \mathbf{Z}_2}\mathbf{V}$$

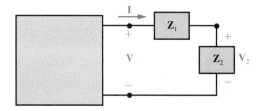

Fig. 4.10 Voltage division.

Note also that the current through the series connection of $\mathbf{Z}_1$ and $\mathbf{Z}_2$ is given by Eq. 4.5. That is, the series combination behaves as a single impedance $\mathbf{Z}$, where $\mathbf{Z} = \mathbf{Z}_1 + \mathbf{Z}_2$. This is shown in Fig. 4.11.

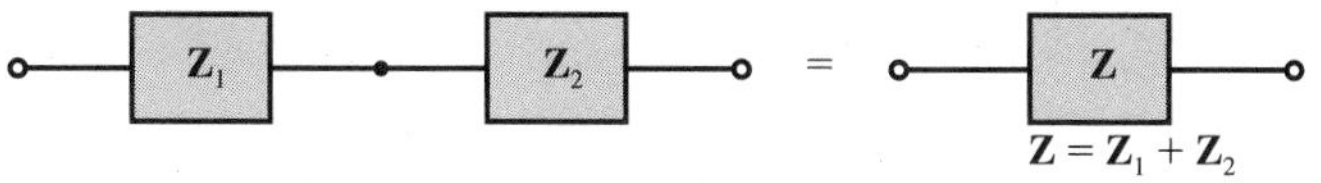

$$\mathbf{Z}_1 \quad \mathbf{Z}_2 \quad = \quad \mathbf{Z} \qquad \mathbf{Z} = \mathbf{Z}_1 + \mathbf{Z}_2$$

Fig. 4.11 Combining impedances in series.

Consider now the case of current division depicted in Fig. 4.12. By KCL,

$$\mathbf{I} = \frac{\mathbf{V}}{\mathbf{Z}_1} + \frac{\mathbf{V}}{\mathbf{Z}_2} = \left(\frac{1}{\mathbf{Z}_1} + \frac{1}{\mathbf{Z}_2}\right)\mathbf{V} \quad \Rightarrow$$

$$\mathbf{V} = \frac{\mathbf{I}}{(1/\mathbf{Z}_1) + (1/\mathbf{Z}_2)} = \frac{\mathbf{Z}_1\mathbf{Z}_2\mathbf{I}}{\mathbf{Z}_1 + \mathbf{Z}_2} \tag{4.6}$$

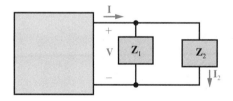

Fig. 4.12 Current division.

Since $\mathbf{I}_2 = \mathbf{V}/\mathbf{Z}_2$, substituting $\mathbf{V}$ from Eq. 4.6 into this expression, we obtain

$$\mathbf{I}_2 = \frac{\mathbf{Z}_1}{\mathbf{Z}_1 + \mathbf{Z}_2}\mathbf{I}$$

and this is the current-divider formula. Also from Eq. 4.6, we have that the impedance of the parallel connection of $\mathbf{Z}_1$ and $\mathbf{Z}_2$ is

$$\mathbf{Z} = \frac{\mathbf{V}}{\mathbf{I}} = \frac{1}{(1/\mathbf{Z}_1) + (1/\mathbf{Z}_2)} = \frac{\mathbf{Z}_1\mathbf{Z}_2}{\mathbf{Z}_1 + \mathbf{Z}_2}$$

Thus two impedances connected in parallel behave as a single impedance $\mathbf{Z}$, where $1/\mathbf{Z} = 1/\mathbf{Z}_1 + 1/\mathbf{Z}_2$. This is shown pictorially in Fig. 4.13. We now conclude that impedances in series and parallel combine as resistances do. Using dual arguments, we may deduce that admittances in parallel and series combine as conductances do.

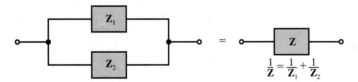

Fig. 4.13 Combining impedances in parallel.

Thévenin's Theorem

In addition to voltage and current division, other resistive network concepts can be employed in ac circuits, if such circuits are described in the frequency domain, that is, in terms of phasors and impedances (or admittances). One very important example is Thévenin's theorem, which says, that (with respect to a pair of terminals) a circuit can be modeled by an appropriate voltage source connected in series with an appropriate impedance. This situation is depicted in Fig. 4.14.

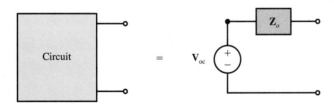

Fig. 4.14 Thévenin's theorem.

Example 4.5

For the circuit shown in Fig. 4.15, let us determine the effect of the circuit on the load impedance $\mathbf{Z}_L$ by employing Thévenin's theorem.

We first determine $\mathbf{V}_{oc}$ from the circuit in Fig. 4.16 with nodal analysis. By KCL at the node labeled $\mathbf{V}_{oc}$,

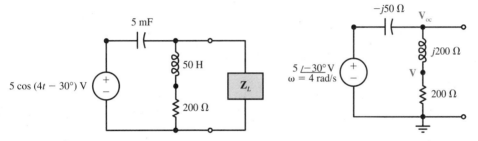

Fig. 4.15 Given ac circuit. **Fig. 4.16** Determination of open-circuit voltage.

$$\frac{(5\underline{/-30°}) - \mathbf{V}_{oc}}{-j50} = \frac{\mathbf{V}_{oc} - \mathbf{V}}{j200} \quad \Rightarrow \quad 3\mathbf{V}_{oc} + \mathbf{V} = 20\underline{/-30°} \qquad (4.7)$$

At the node labeled $\mathbf{V}$, we get

$$\frac{\mathbf{V}_{oc} - \mathbf{V}}{j200} = \frac{\mathbf{V}}{200} \quad \Rightarrow \quad \mathbf{V} = \frac{1}{1 + j}\mathbf{V}_{oc}$$

Substituting this expression for $\mathbf{V}$ into Eq. 4.7 yields

$$3\mathbf{V}_{oc} + \frac{1}{1 + j}\mathbf{V}_{oc} = 20\underline{/-30°}$$

from which

$$\mathbf{V}_{oc} = \frac{1 + j}{4 + j3}(20\underline{/-30°}) = \frac{\sqrt{2}\underline{/45°}}{5\underline{/36.9°}}(20\underline{/-30°}) = 4\sqrt{2}\underline{/-21.9°} \text{ V}$$

Thus,

$$v_{oc}(t) = 4\sqrt{2}\cos(4t - 21.9°) \text{ V}$$

To find Thévenin-equivalent (output) impedance $\mathbf{Z}_o$, set the independent source to zero. The resulting circuit is shown in Fig. 4.17. We can find $\mathbf{Z}_o$ by combining impedances in series and parallel. In particular (the symbol ‖ means "in parallel with"),

$$\mathbf{Z}_o = (\mathbf{Z}_L + \mathbf{Z}_R) \parallel \mathbf{Z}_C = \frac{(j200 + 200)(-j50)}{j200 + 200 - j50} = \frac{200\,(1 - j)}{(4 + j3)}$$

$$= 8 - j56 = 40\sqrt{2}\underline{/-81.9°} \text{ } \Omega$$

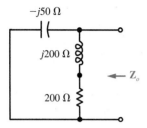

Fig. 4.17 Determination of the Thévenin-equivalent (output) impedance.

In the frequency domain, we have, therefore, the Thévenin-equivalent circuit shown in Fig. 4.18.

In representing this circuit in the time domain, we can use the *RLC* combination from which we determined $\mathbf{Z}_o$, thus yielding the circuit in Fig. 4.19*a*. Alternatively, we can use the fact that $\mathbf{Z}_o = 8 - j56\ \Omega$ has the form of a resistor in series with a capacitor, that is,

$$\mathbf{Z}_o = R + \frac{1}{j\omega C} = R - \frac{j}{\omega C}$$

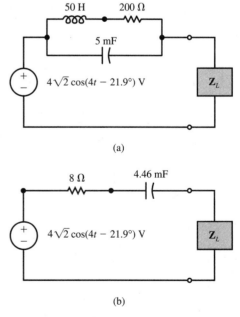

(a)

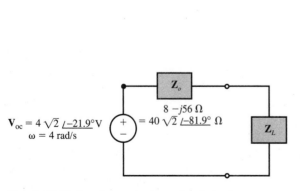

Fig. 4.18 Thévenin-equivalent circuit in the frequency domain.

(b)

Fig. 4.19 Thévenin-equivalent circuits in the time domain.

Thus,

$$R = 8 \ \Omega \quad \text{and} \quad \frac{1}{\omega C} = 56 \quad \Rightarrow \quad C = \frac{1}{224} = 4.46 \text{ mF}$$

and an alternative equivalent circuit is as shown in Fig. 4.19*b*.

Drill Exercise 4.5

Consider the circuit given in Fig. 4.15. (a) If $\mathbf{Z}_L$ is a 48-Ω resistor, what is the voltage across $\mathbf{Z}_L$? (b) What impedance $\mathbf{Z}_L$ will have a voltage of cos 4*t* V across it?

ANSWER   (a) 3.43 cos (4*t* + 23.1°) V, (b) a series connection of a 6.74-Ω resistor and a 25.5-mF capacitor.

Example 4.6

Let us now give an example of how we can apply mesh analysis and Cramer's rule to ac circuits. First consider the ac circuit shown in Fig. 4.20*a*, which contains a dependent source. By making the transformation to the frequency domain and arbitrarily assigning clockwise mesh currents, we obtain the circuit shown in Fig. 4.20*b*. By applying KVL to mesh $\mathbf{I}_1$, we have that

$$6\mathbf{I}_1 - j3(\mathbf{I}_1 - \mathbf{I}_2) = 9 \quad \Rightarrow \quad (2 - j)\mathbf{I}_1 + j\mathbf{I}_2 = 3$$

For mesh $\mathbf{I}_2$, by KVL,

$$-\mathbf{V}_1 - 2\mathbf{V}_1 + 3\mathbf{I}_2 = 0 \quad \Rightarrow \quad j3\mathbf{I}_1 + (1 - j3)\mathbf{I}_2 = 0$$

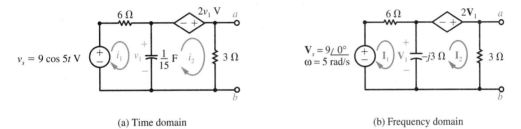

(a) Time domain   (b) Frequency domain

Fig. 4.20 Example of mesh analysis.

since $V_1 = -j3(I_1 - I_2)$. Calculating determinants yields

$$D = \begin{vmatrix} 2 - j & j \\ j3 & 1 - j3 \end{vmatrix} = (2 - j)(1 - j3) - (j3)(j) = 2 - j7$$

$$D_1 = \begin{vmatrix} 3 & j \\ 0 & 1 - j3 \end{vmatrix} = 3(1 - j3), \qquad D_2 = \begin{vmatrix} 2 - j & 3 \\ j3 & 0 \end{vmatrix} = -j9$$

By Cramer's rule

$$I_1 = \frac{D_1}{D} = \frac{3(1 - j3)}{2 - j7} = \frac{3(23 + j)}{53} = 1.30\underline{/2.49°} \text{ A}$$

$$I_2 = \frac{D_2}{D} = \frac{-j9}{2 - j7} = \frac{-j9(2 + j7)}{(2 - j7)(2 + j7)} = \frac{9(7 - j2)}{53} = 1.24\underline{/-15.9°} \text{ A}$$

Therefore,

$$i_1(t) = 1.30 \cos(5t + 2.49°) \text{ A} \qquad \text{and} \qquad i_2(t) = 1.24 \cos(5t - 15.9°) \text{ A}$$

Drill Exercise 4.6

For the circuit given in Fig. 4.15, on p. 209, suppose that Z_L is a 48-Ω resistor. Use mesh analysis to find the two clockwise mesh currents.

ANSWER $0.0805 \cos(4t + 17.0°)$ A, $0.0714 \cos(4t + 23.1°)$ A

4.4 Power

We previously defined power to be the product of voltage and current. For the case that voltage and current are constants (the dc case), the instantaneous power is equal to the average value of the power. This result is a consequence of the fact that the instantaneous and average values of a constant are equal. A sinusoid, however, has an average value of zero. This does not mean that power due to a sinusoid is zero. We shall see what effect sinusoids have in terms of power.

Although our prime interest will be in the sinusoid, we shall also be able to discuss average power for the case of repetitive or periodic nonsinusoidal waveforms by using the notion of effective values. Yet it is the sinusoid that receives the greatest attention, because it is in this form that electric power is produced and distributed.

Average Power

Recall that for the arbitrary element (or an interconnection of elements) shown in Fig. 4.21 the instantaneous power absorbed by the element is $p(t) = v(t)i(t)$. For the case that the voltage is a sinusoid, say $v(t) = V \cos(\omega t + \phi_1)$, the current will also be sinusoidal, say $i(t) = I \cos(\omega t + \phi_2)$. Then the instantaneous power absorbed by the element is

$$p(t) = [V \cos(\omega t + \phi_1)][I \cos(\omega t + \phi_2)]$$

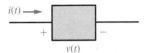

Fig. 4.21 Arbitrary element.

Using the trigonometric identity $(\cos \alpha)(\cos \beta) = \frac{1}{2} \cos(\alpha - \beta) + \frac{1}{2} \cos(\alpha + \beta)$, we get

$$p(t) = VI[\tfrac{1}{2} \cos(\phi_1 - \phi_2) + \tfrac{1}{2} \cos(2\omega t + \phi_1 + \phi_2)]$$

$$= \tfrac{1}{2}VI \cos(\phi_1 - \phi_2) + \tfrac{1}{2}VI \cos(2\omega t + \phi_1 + \phi_2)$$

If this is the instantaneous power absorbed by the element, what is the average power P absorbed by the element? The second term on the right in the last equation is a sinusoid of radian frequency 2ω. We know that the average value of a sinusoid is zero—this can be verified with calculus. Thus the **average power** absorbed by the element is

$$P = \tfrac{1}{2}VI \cos (\phi_1 - \phi_2) = \tfrac{1}{2}VI \cos \theta$$

where $\theta = \phi_1 - \phi_2$ is the difference in phase angles between voltage and current.

For the special case that the element in Fig. 4.21 is a resistor R, since $v(t) = Ri(t)$, then

$$V \cos(\omega t + \phi_1) = RI \cos(\omega t + \phi_2) \quad \Rightarrow \quad V = RI \quad \text{and} \quad \phi_1 = \phi_2$$

Thus, the average power absorbed by the resistor is

$$P_R = \tfrac{1}{2}VI \cos 0 = \tfrac{1}{2}VI$$

From the fact that $V = RI$, and hence $I = V/R$, we get that the average power absorbed by a resistor is

$$P_R = \tfrac{1}{2}VI = \tfrac{1}{2}RI^2 = \tfrac{1}{2}V^2/R$$

Note that P_R cannot be a negative number (assuming a positive value for R, of course).

Inductive and Capacitive Reactance

Since an impedance $\mathbf{Z}$ is a complex number, it can be written in the rectangular form $\mathbf{Z} = a + jb$. We call the real part of the impedance a, the **resistance** and the imaginary part b, the **reactance**. For the case that b is negative, we say that the reactance is **capacitive** and denote it by X_C. If b is positive, we say that the reactance is **inductive** and denote it by X_L. Consequently, impedance is often expressed in the form $\mathbf{Z} = R + jX$.

Suppose that an impedance is purely imaginary (reactive), that is, $\mathbf{Z} = jX$, where X is either a positive or a negative number, as depicted in Fig. 4.22. Let us find the average power absorbed by such an element. Since

$$\mathbf{Z} = \frac{\mathbf{V}}{\mathbf{I}} = \frac{V\underline{/\phi_1}}{I\underline{/\phi_2}} = \frac{V}{I}\underline{/(\phi_1 - \phi_2)}$$

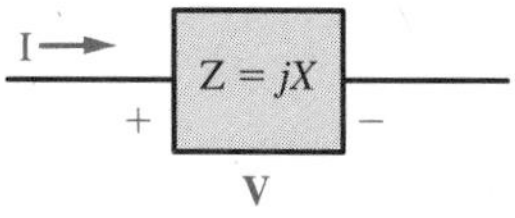

Fig. 4.22 Arbitrary reactance.

then for positive X we have that $jX = X\underline{/90°}$, whereas for negative X we have that $jX = -j(-X) = -X\underline{/-90°}$. Thus

$$\mathbf{Z} = |X|\ \underline{/\pm90°} = \frac{V}{I}\ \underline{/(\phi_1 - \phi_2)}$$

and the average power P_X absorbed by this impedance is

$$P_X = \tfrac{1}{2}VI \cos(\phi_1 - \phi_2) = \tfrac{1}{2}VI \cos(\pm 90°) = 0 \text{ W}$$

In other words, purely reactive impedances, such as (ideal) inductors and capacitors or combinations of the two, absorb zero average power, and are referred to as **lossless elements**.

Example 4.7

Let us calculate the power absorbed by each element in the circuit given in Example 4.6.

Figure 4.23 shows the circuit in the frequency domain, where the currents $I_1 = 1.30\underline{/2.49°}$ A and $I_2 = 1.24\underline{/-15.9°}$ A were previously determined.

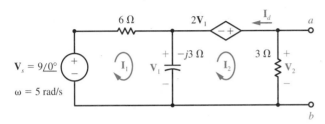

Fig. 4.23 Circuit from Example 4.6.

Since $|I_2| = 1.24$ A, the average power absorbed by the 3-Ω resistor is

$$P_3 = \tfrac{1}{2}R_3|I_2|^2 = \tfrac{1}{2}(3)(1.24)^2 = 2.31 \text{ W}$$

By Ohm's law, $V_2 = 3I_2 = 3(1.24\underline{/-15.9°}) = 3.72\underline{/-15.9°}$ V. By KVL,

$$V_2 = 2V_1 + V_1 = 3V_1 \quad \Rightarrow \quad V_1 = \frac{V_2}{3} = 1.24\underline{/-15.9°} \text{ V}$$

The average power absorbed by the dependent voltage source is

$$P_d = \tfrac{1}{2}|2V_1|\,|I_d|\cos[\text{ang}(2V_1) - \text{ang}(I_d)]$$

Since $2V_1 = 2(1.24\underline{/-15.9°}) = 2.48\underline{/-15.9°}$ V, then $|2V_1| = 2.48$ and $\text{ang}(2V_1) = -15.9°$. Also, since $I_d = -I_2 = (-1)I_2 = (1\underline{/180°})(1.24\underline{/-15.9°}) = 1.24\underline{/164°})$ A, then $|I_d| = 1.24$ A and $\text{ang}(I_d) = 164°$. Thus

$$P_d = \tfrac{1}{2}(2.48)(1.24)\cos(-15.9° - 164°) = -1.54 \text{ W}$$

The current through the capacitor is

$$\mathbf{I}_3 = \frac{\mathbf{V}_1}{-j3} = \frac{1.24/\!-15.9°}{3/\!-90°} = 0.413/\!74.1° \text{ A}$$

However, we need not calculate the average power P_c absorbed by the capacitor—it must be 0 W. (Confirm this fact!)

Since $|\mathbf{I}_1| = 1.30$ A, the average power absorbed by the 6-Ω resistor is

$$P_6 = \tfrac{1}{2}R_6|\mathbf{I}_1|^2 = \tfrac{1}{2}(6)(1.30)^2 = 5.07 \text{ W}$$

Furthermore, $\mathbf{I}_s = -\mathbf{I}_1 = (-1)\mathbf{I}_1 = (1/\!-180°)(1.30/\!2.49°) = 1.30/\!-177.5°$ A, so $|\mathbf{I}_s| = 1.30$ A and $\text{ang}(\mathbf{I}_s) = -177.5°$. Thus the average power absorbed by the independent voltage source $\mathbf{V}_s$ is

$$P_s = \tfrac{1}{2}|\mathbf{V}_s|\,|\mathbf{I}_s|\cos[\text{ang}(\mathbf{V}_s) - \text{ang}(\mathbf{I}_s)] = \tfrac{1}{2}(9)(1.30)\cos(0° + 177.5°) = -5.84 \text{ W}$$

Summing the average power absorbed by all the elements in the circuit, we get

$$P_3 + P_d + P_C + P_6 + P_s = 2.31 - 1.54 + 0 + 5.07 - 5.84 = 0 \text{ W}$$

Hence we see that average power is conserved, and this is true for any circuit.

Drill Exercise 4.7

For the circuit given in Fig. 4.15, on p. 209, when $\mathbf{Z}_L$ is a 48-Ω resistor, the resulting clockwise mesh currents (in the frequency domain) are $0.0805/\!17.0°$ A and $0.0714/\!23.1°$ A for the left and right meshes, respectively. Find the average power absorbed by each element in the circuit.

ANSWER 0.122 W, 0.015 W, 0 W, 0 W, −0.137 W

Maximum Power Transfer

Given a circuit with a load impedance $\mathbf{Z}_L = R_L + jX_L$, as depicted in Fig. 4.24, what impedance $\mathbf{Z}_L$ will absorb the maximum amount of power? In a manner similar

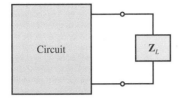

Fig. 4.24 Circuit and its associated load.

to that for resistive circuits (but mathematically slightly more involved), it can be shown that the load that absorbs the maximum power is

$$\mathbf{Z}_L = R_L + jX_L = R_o - jX_o = \mathbf{Z}_o^*$$

where $\mathbf{Z}_o = R_o + jX_o$ is the Thévenin-equivalent (output) impedance of the circuit.

For the case that $\mathbf{Z}_L$ is restricted to be real (resistive), it can be shown that the value of $\mathbf{Z}_L = R_L$ that absorbs the maximum amount of power (among all resistive loads) is

$$R_L = \sqrt{R_o^2 + X_o^2} = |\mathbf{Z}_o|$$

Example 4.8

For the ac circuit given in Fig. 4.15, on p. 209, we have already seen in Example 4.5 that $\mathbf{Z}_o = 8 - j56 \ \Omega$. Thus the impedance that absorbs maximum power is

$$\mathbf{Z}_L = \mathbf{Z}_o^* = 8 + j56 \ \Omega$$

We can realize this impedance as a resistor R_o connected in series with an inductor L, where $R_o = 8 \ \Omega$ and $\omega L = 56 \Rightarrow L = 56/\omega = 56/4 = 14$ H. From Fig. 4.18, on p. 210, we see that the current through $\mathbf{Z}_L$ is

$$\mathbf{I}_L = \frac{\mathbf{V}_{oc}}{\mathbf{Z}_o + \mathbf{Z}_L} = \frac{4\sqrt{2}\underline{/-21.9°}}{8 - j56 + 8 + j56} = \frac{\sqrt{2}}{4}\underline{/-21.9°} \text{ A}$$

But, the average power absorbed by $\mathbf{Z}_L$ is equal to the average power absorbed by the real part of $\mathbf{Z}_L$ since the average power absorbed by a reactance is 0 W. Thus the average power absorbed by $\mathbf{Z}_L$ is

$$P_L = \tfrac{1}{2}R_o|\mathbf{I}_L|^2 = \tfrac{1}{2}(8)(\sqrt{2}/4)^2 = 0.5 \text{ W}$$

If the load is restricted to be a resistance $\mathbf{Z}_L = R_L$, then the resistance that absorbs the maximum average power is

$$R_L = |\mathbf{Z}_o| = \sqrt{8^2 + (-56)^2} = 40\sqrt{2} = 56.6 \ \Omega$$

The current through this load is

$$\mathbf{I}_L = \frac{\mathbf{V}_{oc}}{\mathbf{Z}_o + R_L} = \frac{4\sqrt{2}\underline{/-21.9°}}{8 - j56 + 56.6} = \frac{5.66\underline{/-21.9°}}{85.5\underline{/-40.9°}} = 0.0662\underline{/19°} \ \text{A}$$

Thus the average power absorbed by R_L is

$$P_L = \tfrac{1}{2}R_L|\mathbf{I}_L|^2 = \tfrac{1}{2}(56.6)(0.0662)^2 = 0.124 \ \text{W}$$

Drill Exercise 4.8

The Thévenin equivalent of an ac circuit has $\mathbf{V}_{oc} = 12\underline{/-45°}$ V and $\mathbf{Z}_o = 8 + j8 \ \Omega$. (a) What load impedance $\mathbf{Z}_L$ absorbs maximum average power, and what is the value of this power? (b) What load resistance $\mathbf{Z}_L = R_L$ absorbs maximum average power when $\mathbf{Z}_L$ is restricted to be a resistance, and what is the value of this power?

ANSWER (a) $8 - j8 \ \Omega$, 2.25 W; (b) 11.3 Ω, 1.86 W

Effective Value of a Sinusoid

We have seen that for a resistor R with a sinusoidal current through it, say, $i(t) = I \cos(\omega t + \phi)$, the average power absorbed by the resistor is $\tfrac{1}{2}RI^2$. What dc current I_e would yield the same power absorption by R? Since the power absorbed by R due to a direct current I_e is RI_e^2, we get the equality

$$RI_e^2 = \tfrac{1}{2}RI^2 \qquad \Rightarrow \qquad \boxed{I_e = \frac{I}{\sqrt{2}} \approx 0.707I}$$

Thus the constant current $I_e = I/\sqrt{2}$ and the sinusoidal current $i(t) = I\cos(\omega t + \phi)$ result in the same average power absorption by R. For this reason, we could say that $I/\sqrt{2}$ is the **effective value** of the sinusoid whose amplitude is I.

Because the voltage across R is

$$v(t) = Ri(t) = RI\cos(\omega t + \phi) = V\cos(\omega t + \phi)$$

then the effective value of $v(t)$ is $V_e = V/\sqrt{2}$. Furthermore, the average power absorbed by R is

$$P_R = \frac{VI}{2} = \frac{V}{\sqrt{2}}\frac{I}{\sqrt{2}} = V_e I_e$$

Alternatively, since $V = RI$, then

$$P_R = \frac{RI^2}{2} = R\frac{I}{\sqrt{2}}\frac{I}{\sqrt{2}} = RI_e^2 \qquad \text{and} \qquad P_R = \frac{V^2}{2R} = \frac{1}{R}\frac{V}{\sqrt{2}}\frac{V}{\sqrt{2}} = \frac{V_e^2}{R}$$

In summary, given a resistor R having current $i(t) = I\cos(\omega t + \phi)$ through it and voltage $v(t) = V\cos(\omega t + \phi)$ across it, then the average power absorbed by the resistor is

$$\boxed{P_R = V_e I_e = RI_e^2 = \frac{V_e^2}{R}}$$

where $V_e = V/\sqrt{2}$ and $I_e = I/\sqrt{2}$ are the effective values of the voltage and the current, respectively.

For an arbitrary element

$$\boxed{P = \tfrac{1}{2}VI\cos\theta = V_e I_e \cos\theta} \tag{4.8}$$

where θ is the difference in phase between the voltage and the current.

Ordinary household voltage is designated typically as 115 V ac, 60 Hz. The term "ac," which stands for **alternating current**, indicates a sinusoid, and 60 Hz is the cyclical frequency of the sinusoid. However, the 115 V does not refer to the amplitude of the sinusoid, but rather to its effective value. Thus, the amplitude of the sinusoid is $115\sqrt{2} \approx 163$ V.

RMS Values

Consider for a resistor R, the case that the current (or the voltage) is nonsinusoidal. Since the instantaneous power absorbed by the resistor is $p(t) = v(t)i(t)$, and since by Ohm's law $v(t) = Ri(t)$, then

$$p(t) = Ri^2(t)$$

Summing the instantaneous power from time t_1 to t_2 and dividing by the width of the interval $t_2 - t_1$ yields the average power absorbed by the resistor in the interval, that is,

$$P_R = \frac{1}{t_2 - t_1} \int_{t_1}^{t_2} p(t)\ dt = \frac{1}{t_2 - t_1} \int_{t_1}^{t_2} Ri^2(t)\ dt$$

Suppose that the current (and hence the voltage) is "repetitive" or "periodic," and the period is T seconds. Then the average power absorbed in one period is

$$P_R = \frac{1}{T} \int_{t_o}^{t_o+T} p(t)\ dt = \frac{1}{T} \int_{t_o}^{t_o+T} Ri^2(t)\ dt$$

where t_o, is an arbitrary value. Again, let I_e be the dc current that results in the same amount of average power absorption by the resistor. Thus

$$RI_e^2 = \frac{1}{T} \int_{t_o}^{t_o+T} Ri^2(t)\ dt$$

from which the expression for I_e, the effective value of $i(t)$, is

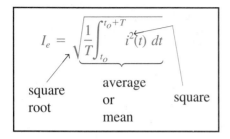

square root — average or mean — square

and for this reason the effective value is also known as the **root-mean-square** or **rms** value.

It is a routine exercise in integral calculus to confirm that the rms value of the sinusoid $A \cos(\omega t + \phi)$, whose period is $T = 2\pi/\omega$, is $A/\sqrt{2}$.

Example 4.9

The function $f(t)$ shown in Fig. 4.25 is periodic, where the period is $T = 4$ s. Although the average or mean value of this function is zero (why?), the rms value of $f(t)$ is

$$\sqrt{\frac{1}{T} \int_{-2}^{2} f^2(t)\, dt} = \sqrt{\frac{1}{4}\left[\int_{-2}^{-1} 1^2\, dt + \int_{-1}^{1} (-t)^2\, dt + \int_{1}^{2} (-1)^2\, dt \right]}$$

$$= \sqrt{\frac{1}{4}\left[t\Big|_{-2}^{-1} + \frac{t^3}{3}\Big|_{-1}^{1} + t\Big|_{1}^{2} \right]} = \sqrt{\frac{1}{4}\left(1 + \frac{2}{3} + 1\right)} = \sqrt{\frac{2}{3}}$$

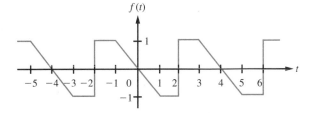

Fig. 4.25 Nonsinusoidal, periodic waveform.

Therefore, if the current through a 1-Ω resistor is $i(t) = f(t)$, then the average power absorbed by that resistor is

$$P_R = R I_e^2 = 1\left(\sqrt{\frac{2}{3}}\right)^2 = \frac{2}{3}\ \text{W}$$

Drill Exercise 4.9

Find the rms values of the waveforms shown in Fig. DE4.9 on p. 222.

ANSWER (a) $\sqrt{6}$, (b) $\sqrt{6}$, (c) $\sqrt{6}$

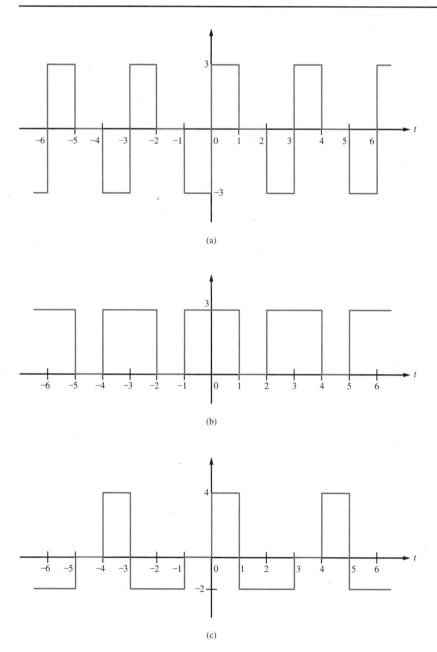

(a)

(b)

(c)

Fig. DE4.9

4.5 Important Power Concepts

Consider the arbitrary load, as depicted in Fig. 4.26, where $v(t) = V \cos(\omega t + \phi_1)$ and $i(t) = I \cos(\omega t + \phi_2)$. Then the average power absorbed by the load is, repeating Eq. 4.8,

$$P = \tfrac{1}{2}VI \cos \theta = V_e I_e \cos \theta \tag{4.8}$$

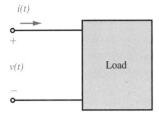

Fig. 4.26 Arbitrary load.

where $\theta = \phi_1 - \phi_2$. If the load is purely resistive, then the average power it absorbs is $P_R = V_e I_e$. However, if the load is not purely resistive, even though it might "appear" at first glance that the average power absorbed is $V_e I_e$, the average power absorbed is given by Eq. 4.8. We call $V_e I_e$ the **apparent power** and use the unit **volt-ampere** (VA) to distinguish this quantity from the actual power (the average power) that is absorbed by the load. From Eq. 4.8 we see that the quantity $\cos \theta$, called the **power factor** (pf), is the ratio of average power to apparent power, that is,

$$\text{pf} = \cos \theta = \frac{P}{V_e I_e} = \frac{\text{average power}}{\text{apparent power}}$$

The angle θ is referred to as the **power-factor angle**.

The impedance of the load shown in Fig. 4.26 is

$$\mathbf{Z} = \frac{\mathbf{V}}{\mathbf{I}} = \frac{V\underline{/\phi_1}}{I\underline{/\phi_2}} = \frac{V}{I}\underline{/(\phi_1 - \phi_2)} = \frac{V}{I}\underline{/\theta}$$

so that the angle of the load is the pf angle. The general form of a load impedance is $\mathbf{Z} = R + jX$. If $X > 0$, we say that the load is **inductive**; if $X < 0$, we say it is **capacitive.** Assuming a positive resistance R, then for an inductive load

$$\theta = \text{ang}(\mathbf{Z}) = \tan^{-1}\left(\frac{X}{R}\right) > 0$$

that is, the pf angle is positive; whereas for a capacitive load

$$\theta = \text{ang}(\mathbf{Z}) = \tan^{-1}\left(\frac{X}{R}\right) < 0$$

that is, the pf angle is negative. The former case ($\phi_1 > \phi_2$) implies that the current lags the voltage, whereas the latter case ($\phi_1 < \phi_2$) implies that the current leads the voltage. In the vernacular of the power industry, we refer to the pf as being **lagging** or **leading**, respectively. In other words, if the current lags the voltage for a load (the inductive case), the result is a lagging pf; if the current leads the voltage (the capacitive case), the result is a leading pf.

You may ask, "Why talk about lagging and leading pf at all? Why not just talk about the impedance of the load?" One reason is that, in some applications (e.g., power systems), a load may not consist simply of ordinary inductors, capacitors, and resistors. In the following example, the load is an electric motor.

Example 4.10

A 1000-W electric motor is connected to a 200-V rms ac, 60-Hz source, and the result is a lagging pf of 0.8. Since the pf is lagging, the angle of the current is less than the voltage angle by $\cos^{-1}(0.8) = 36.9°$. Arbitrarily selecting the voltage angle to be zero, we can depict the given situation as shown in Fig. 4.27. From Eq. 4.8, we have

$$I_e = \frac{P}{V_e \cos \theta} = \frac{1000}{200(0.8)} = \frac{25}{4} = 6.25 \text{ A} \quad \Rightarrow$$

$$\mathbf{I}_m = 6.25\underline{/-36.9°} = 5 - j3.75 \text{ A}$$

$\mathbf{V}_e = 200 \underline{/0°}$
$f = 60 \text{ Hz}$
$\omega = 2\pi f$
$\quad = 120\pi \text{ rad/s}$

$\mathbf{I}_m = I_e \underline{/-36.9°} \text{ A}$

Motor

Fig. 4.27 Example of a lagging pf.

Of course, because the motor absorbs 1000 W of average power, the source supplies this power. But suppose that a 28-μF capacitor is placed in parallel with the motor. Since the current through the motor does not change, we have the situation shown in Fig. 4.28. Since

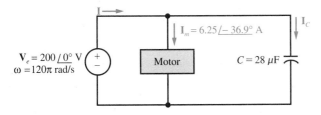

Fig. 4.28 Power-factor correction.

$$\mathbf{I}_C = \frac{\mathbf{V}_e}{\mathbf{Z}_C} = j\omega C\mathbf{V}_e = (1\underline{/90°})(120\pi)(28 \times 10^{-6})(200\underline{/0°})$$

$$= 2.11\underline{/90°} = j2.11 \text{ A}$$

then

$$\mathbf{I} = \mathbf{I}_m + \mathbf{I}_C = 5 - j1.64 = 5.26\underline{/-18.1°} \text{ A}$$

and the pf for the new load (the parallel connection of the motor and the capacitor) is $\cos(18.1°) = 0.95$, and since $\text{ang}(\mathbf{I}) = -18.1° < \text{ang}(\mathbf{V}_e) = 0°$, then the pf is lagging. Figure 4.29 shows a phasor diagram of the currents $\mathbf{I}_C$, $\mathbf{I}_m$, and $\mathbf{I}$—and this pictorially depicts that $\mathbf{I}$ is the phasor (complex-number) sum of $\mathbf{I}_C$ and $\mathbf{I}_m$.

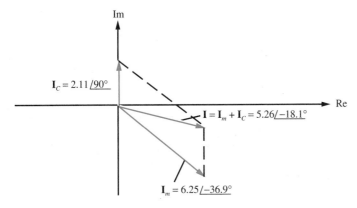

Fig. 4.29 Phasor diagram.

We now see that placing a capacitor in parallel with the motor increases the lagging pf from 0.8 to 0.95. This is an example of what is known as **power-factor correction**. Since capacitors absorb zero average power, and since the motor absorbs 1000 W in either case, the voltage source supplies 1000 W in both cases. Note, however, that originally 6.25 A of rms current was drawn from the voltage source, whereas after the capacitor is connected in parallel with the motor, the rms current through the voltage source is 5.26 A—a decrease of about 16%.

Drill Exercise 4.10

Consider the load consisting of the motor connected in parallel with the capacitor shown in Fig. 4.28. What value of C will cause this load to have a unity pf, i.e., $\cos \theta = 1$?

ANSWER 49.7 μF

There are situations for which current changes such as was done in Example 4.10 have significant implications. One important case is when the voltage source is an electric utility company generator and the load is some demand by a consumer. Between the two is the power transmission line, which is not ideal and has some associated nonzero resistance.

Example 4.11

Suppose that a large consumer of electricity requires 10 kW of power by using 230 V rms at a pf angle of 60° lagging, that is, a pf of 0.5 lagging. We conclude that the current drawn by the load has an effective value of

$$I_e = \frac{P}{V_e \cos \theta} = \frac{10,000}{230(0.5)} = 87.0 \text{ A}$$

If the transmission-line resistance is $R = 0.1 \ \Omega$, then, as indicated in Fig. 4.30,

$$\mathbf{V}_e = 0.1 \ \mathbf{I}_L + \mathbf{V}_L = (8.7 \underline{/-60°}) + (230 \underline{/0°}) = 234 \underline{/-1.84°} \text{ V}$$

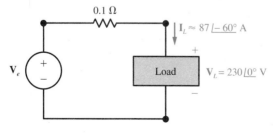

Fig. 4.30 Inclusion of transmission-line resistance.

Therefore, the voltage that must be generated by the power company must have an effective value of $V_e = 234$ V. Also, the line loss is $RI_e^2 = (0.1)(87)^2 = 757$ W. Thus to supply 10 kW of power to the load, the utility has to produce 10,757 W of power.

If the consumer can correct the pf from 0.5 to 0.9 lagging, the load current will become $I_e = P/(V_e \cos \theta) = 10,000/(230)(0.9) = 48.3$ A rms. The consequence of this is that the line loss will be $RI_e^2 = (0.1)(48.3)^2 = 233$ W—which is only about 30% of the loss than for the lower pf. It is for this reason that an electric utility takes pf into consideration when dealing with large consumers.

Drill Exercise 4.11

For the power consumption described in Example 4.11, what is the line loss when the pf is corrected to unity?

ANSWER   189 W

An average household uses relatively small amounts of power and typically has a reasonably high (close to unity) pf. Thus pf is not a consideration in ordinary home electricity usage.

Complex Power

Previously, we found it computationally convenient to introduce the abstract concept of a complex sinusoid—a real sinusoid is just a special case. So, too, we now discuss and apply the notion of "complex power."

Again consider the arbitrary load depicted in Fig. 4.31, where $\mathbf{V} = V/\underline{\phi_1}$, $\mathbf{I} = I/\underline{\phi_2}$, $V_e = V/\sqrt{2}$, $I_e = I/\sqrt{2}$. Since $P = V_e I_e \cos \theta$, and $\cos \theta = \text{Re}[e^{j\theta}]$, then

$$P = V_e I_e(\text{Re}[e^{j\theta}] = V_e I_e(\text{Re}[e^{j(\phi_1 - \phi_2)}]) = V_e I_e(\text{Re}[e^{j\phi_1} e^{-j\phi_2}])$$

$$= \text{Re}[V_e I_e e^{j\phi_1} e^{-j\phi_2}] = \text{Re}[(V_e e^{j\phi_1})(I_e e^{-j\phi_2})] = \text{Re}[(V_e e^{j\phi_1})(I_e e^{j\phi_2})^*]$$

Thus we see that the average power absorbed by the load is equal to the real part of some complex quantity. Defining $\mathbf{V}_{rms}$ and $\mathbf{I}_{rms}$, respectively, by $\mathbf{V}_{rms} = V_e e^{j\phi_1}$ and $\mathbf{I}_{rms} = I_e e^{j\phi_2}$, then

$$P = \text{Re}[\mathbf{V}_{rms} \mathbf{I}_{rms}^*]$$

In other words, the average power absorbed by the load is the real part of the quantity

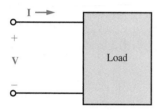

Fig. 4.31 Arbitrary load.

$$S = V_{rms}I_{rms}^*$$

and **S** is called **complex power**[5] absorbed by the load. We may also write **S** as

$$S = (V_e e^{j\phi_1})(I_e e^{-j\phi_2}) = V_e I_e e^{j(\phi_1 - \phi_2)} = V_e I_e e^{j\theta}$$

Since $|S| = V_e I_e$, we see that the magnitude of the complex power is the apparent power, and the angle of the complex power is the pf angle. By Euler's formula, we can write complex power in rectangular form as follows:

$$S = V_e I_e(\cos \theta + j \sin \theta) = V_e I_e \cos \theta + jV_e I_e \sin \theta = P + jQ$$

where $P = V_e I_e \cos \theta$ is the average power—also called the **real power**—absorbed by the load, and $Q = V_e I_e \sin \theta$ is known as the **reactive power** absorbed by the load. As mentioned before, the unit of P is the watt (W). The unit of Q is the **var** (or VAR)—short for **v**olt-**a**mpere **r**eactive. The unit of complex power **S**, like its magnitude $|S|$ (i.e., apparent power), is the volt-ampere (VA). Thus, we have

$$V_e I_e = |S| = \sqrt{P^2 + Q^2}$$

Furthermore, the angle of **S** is the pf angle θ, so

$$\theta = \tan^{-1}\left(\frac{Q}{P}\right)$$

From the triangle—called a **power triangle**—shown in Fig. 4.32, we deduce that the pf is

$$\cos \theta = \frac{P}{\sqrt{P^2 + Q^2}} = \frac{P}{|S|}$$

---

[5]In terms of ordinary (non-rms) phasors **V** and **I**, the definition of complex power is

$$S = \tfrac{1}{2}VI^*$$

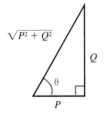

Fig. 4.32 Power triangle.

Example 4.12

Let us find the complex power absorbed by each element in the circuit shown in Fig. 4.33, where the voltage source is given in terms of its rms phasor representation.

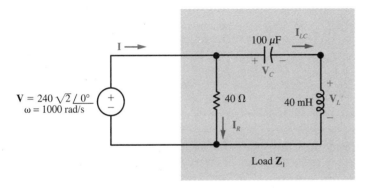

Fig. 4.33 Voltage source and its associated load.

The current through the resistor is

$$\mathbf{I}_R = \mathbf{V}/40 = (240\underline{/0°})/40 = 6\underline{/0°} = 6 \text{ A (rms)}$$

and the complex power absorbed by the resistor is

$$\mathbf{S}_R = \mathbf{V}\mathbf{I}_R^* = (240\underline{/0°})(6\underline{/0°}) = 1440 \text{ VA}$$

which, of course, is purely real. The current through the series LC combination is

$$\mathbf{I}_{LC} = \frac{\mathbf{V}}{(1/j\omega C) + j\omega L} = \frac{240\underline{/0°}}{(10/j) + j40} = -j8 = 8\underline{/-90°} \text{ A}$$

The voltage across the capacitor is

$$\mathbf{V}_C = \frac{1}{j\omega C}\mathbf{I}_{LC} = \frac{-j8}{j10^{-1}} = -80 = 80\underline{/180°} \text{ V}$$

and the voltage across the inductor is

$$\mathbf{V}_L = j\omega L\,\mathbf{I}_{LC} = (j40)(-j8) = 320 \text{ V}$$

The complex power absorbed by the capacitor is

$$\mathbf{S}_C = \mathbf{V}_C\mathbf{I}_{LC}^* = (80\underline{/180°})(8\underline{/90°}) = 640\underline{/-90°} = -j640 \text{ VA}$$

and the complex power absorbed by the inductor is

$$\mathbf{S}_L = \mathbf{V}_L\mathbf{I}_{LC}^* = (320\underline{/0°})(8\underline{/90°}) = 2560\underline{/-90°} = j2560 \text{ VA}$$

The total complex power absorbed by the load $\mathbf{Z}_1$ consisting of the resistor, capacitor, and inductor is

$$\mathbf{S} = \mathbf{S}_R + \mathbf{S}_C + \mathbf{S}_L = 1440 - j640 + j2560$$

$$= 1440 + j1920 = 2400\underline{/53.1°} \text{ VA}$$

Thus the apparent power absorbed by the load is $|\mathbf{S}| = 2400$ VA, the pf angle is ang($\mathbf{S}$) = 53.1°, and thus the pf is $\cos(53.1°) = 0.6$ lagging. The real power absorbed by the load is $P = 1440$ W and the reactive power absorbed by the load is $Q = 1920$ VAR.

Since $\mathbf{I} = \mathbf{I}_R + \mathbf{I}_{LC} = 6 - j8 = 10\underline{/-53.1°}$ A, then the load has impedance

$$\mathbf{Z}_1 = \frac{\mathbf{V}}{\mathbf{I}} = \frac{240\underline{/0°}}{10\underline{/-53.1°}} = 2.4\underline{/53.1°} \ \Omega$$

The complex power supplied[6] by the source is

$$\mathbf{VI}^* = (240\underline{/0°})(10\underline{/53.1°}) = 2400\underline{/53.1°} = 1440 + j1920 \text{ VA}$$

Note that this exactly equals the complex power that is absorbed by the load, and this result is a consequence of the fact that, as with the case of real power, complex power must be conserved.

---

[6]For power supplied, the direction of current is opposite to that used for power absorbed.

Drill Exercise 4.12

For the circuit given in Fig. 4.28, on p. 225, the 1000-W motor has a lagging pf of 0.8. Find the complex power absorbed by (a) the capacitor, (b) the motor, and (c) the voltage source.

ANSWER   (a) $-j422$ VA, (b) $1000 + j750$ VA, (c) $-1000 - j328$ VA

Since a pf that is too small may result in additional expense, let us try to increase the pf for the load described in Fig. 4.33. Since pf $= \cos\theta = P/|\mathbf{S}|$, one way to increase the pf is to increase the real power P. But real power is costly. Instead, let us connect another load $\mathbf{Z}_2$ in parallel with $\mathbf{Z}_1$. This will not affect the voltage and current for $\mathbf{Z}_1$, but it will have an effect on the current through the voltage source. The resulting situation is depicted in Fig. 4.34, where $\mathbf{Z}_1 = 24\underline{/53.1°}\ \Omega$, $\mathbf{I}_1 = 10\underline{/-53.1°}$ A, and $\mathbf{S}_1 = 2400\underline{/53.1°} = 1440 + j1920$ VA.

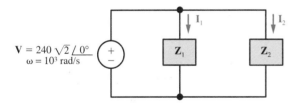

Fig. 4.34 Power-factor correction for $\mathbf{Z}_1$.

Suppose we wish to increase the lagging pf to 0.9. Since $\cos^{-1}(0.9) = 25.8°$, let us call the current directed out of the source $\mathbf{I} = I\underline{/-25.8°}$ A. The resulting complex power supplied by the source is

$$\mathbf{S} = \mathbf{V}\mathbf{I}^* = (240\underline{/0°})(I\underline{/25.8°}) = 240I\underline{/25.8°}\ \text{VA}$$

As we wish the real power to remain the same as before, we have that

$$P = |\mathbf{S}|\cos\theta = (240I)(0.9) = 1440\ \text{W}$$

from which

$$I = \frac{1440}{(240)(0.9)} = 6.67 \quad \Rightarrow \quad \mathbf{I} = 6.67\underline{/-25.8°}\ \text{A}$$

By KCL,

$$\mathbf{I}_2 = \mathbf{I} - \mathbf{I}_1 = (6.67\underline{/-25.8°}) - (10\underline{/-53.1°}) = 5.1\underline{/90°} \text{ A}$$

Thus,

$$\mathbf{Z}_2 = \frac{\mathbf{V}}{\mathbf{I}_2} = \frac{240\underline{/0°}}{5.1\underline{/90°}} = 47.1\underline{/-90°} = -j47.1 \ \Omega$$

and this impedance is purely reactive—which might have been expected because of the constraint that there be no additional real power supplied by the source. The impedance $\mathbf{Z}_2$ can be realized with a capacitor as follows:

$$\frac{1}{j\omega C} = \frac{-j}{\omega C} = -j47.1 \quad \Rightarrow \quad C = \frac{1}{47.1\omega} = \frac{1}{47.1 \times 10^3} = 21.2 \ \mu\text{F}$$

This concludes the demonstration of how complex-power concepts can be used for purposes of power-factor correction.

4.6 Polyphase Circuits

Previously, we mentioned that ordinary household voltage is sinusoidal, having an approximate rms value of 115 V and a frequency of 60 Hz. Let us discuss the subject further.

Single-Phase Three-Wire Circuits

Coming from a power line into a home's service panel or fuse box are three wires. One wire is bare—it has no insulation. This is the **neutral** or **ground wire**, so called since it must be connected to a water pipe or a rod (or both) that goes into the ground. (The earth is the reference potential of zero volts.) Typically, another wire is red and a third is black. The voltage between the red and neutral wires (plus sign at the red wire) is $v_{rn}(t) = 115\sqrt{2} \cos(120\pi t + \phi)$ V—that is, 115 V (rms) ac, 60 Hz. The voltage $v_{bn}(t)$ between the black and the neutral wires (plus at the black wire) is $v_{bn}(t) = -v_{rn}(t)$. This means that the voltage between the neutral and black wires (with the plus at the neutral wire) is $v_{nb}(t) = -v_{bn}(t) = v_{rn}(t)$, and, therefore, the two voltages $v_{rn}(t)$ and $v_{nb}(t)$ have the same phase angle. As these voltages remain fairly constant for a wide variety of practical loads (appliances), we can reasonably approximate the **single-phase, three-wire source** just described as shown in Fig. 4.35, where $v_{rn}(t) = v_{nb}(t) = v_s(t) = 115\sqrt{2}\cos(120\pi t + \phi)$ V. By KVL, we have that $v_{rb}(t) = 230\sqrt{2}\cos(120\pi t + \phi)$ V. Thus, in addition to two 115-V ac, 60-Hz sources, there is a supply of 230 V (rms) ac, 60 Hz. This higher voltage is useful for higher power appliances, since for a given power requirement, a higher voltage requires less current than a lower voltage. The consequence of this is that a higher voltage results in a lower line loss.

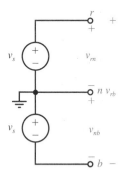

Fig. 4.35 Single-phase, three-wire source.

The single-phase, three-wire circuit shown in Fig. 4.36 (which is represented in the frequency domain) consists of a single-phase, three-wire source and three load impedances. The short circuits between nodes a and A and between nodes b and B are the **lines** of the circuit. The short circuit between nodes n and N is the **neutral wire** of the circuit. It may be that the lines have a nonzero impedance $\mathbf{Z}_g$ and the neutral wire has a nonzero impedance $\mathbf{Z}_n$ (see Problem 4.59).

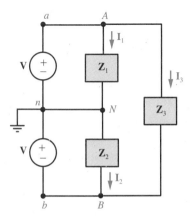

Fig. 4.36 Single-phase, three-wire circuit.

Example 4.13

For the single-phase, three-wire circuit given in Fig. 4.36, suppose that $Z_1 = 60 \; \Omega$, $Z_2 = 80 \; \Omega$, $Z_3 = 20 + j15 \; \Omega$, and $V_s = 120 \underline{/0°}$ V is an rms phasor. Let us find the average power absorbed by each load impedance.

Since

$$I_1 = \frac{V_s}{Z_1} = \frac{120 \underline{/0°}}{60} = 2 \underline{/0°} \; A$$

$$I_2 = \frac{V_s}{Z_2} = \frac{120 \underline{/0°}}{80} = 1.5 \underline{/0°} \; A$$

$$I_3 = \frac{2V_s}{Z_3} = \frac{240 \underline{/0°}}{20 + j15} = \frac{240 \underline{/0°}}{25 \underline{/36.9°}} = 9.6 \underline{/-36.9°} \; A$$

then the power absorbed by the combined load consisting of Z_1, Z_2, and Z_3 is $P_1 + P_2 + P_3$, where

$$P_1 = R_1 |I_1|^2 = 60(2)^2 = 240 \; W$$

$$P_2 = R_2 |I_2|^2 = 80(1.5)^2 = 180 \; W$$

$$P_3 = R_3 |I_3|^2 = 20(9.6)^2 = 1843 \; W$$

Thus $P_1 + P_2 + P_3 = 2263$ W. Since

$$I_1 + I_3 = 2 \underline{/0°} + 9.6 \underline{/-36.9°} = 11.3 \underline{/-30.8°} \; A$$

$$I_2 + I_3 = 1.5 \underline{/0°} + 9.6 \underline{/-36.9°} = 10.8 \underline{/-32.1°} \; A$$

then, as a check, the power supplied by the upper source is

$$P_{s1} = |V_s| |I_1 + I_3| \cos \theta_1 = (120)(11.3) \cos(30.8°) = 1165 \; W$$

and the power supplied by the lower source is

$$P_{s2} = |V_s| |I_2 + I_3| \cos \theta_2 = (120)(10.8) \cos(32.1°) = 1098 \; W$$

for a total of $P_{s1} + P_{s2} = 2263$ W.

Drill Exercise 4.13

Suppose that the single-phase, three-wire circuit given in Fig. 4.36 has $\mathbf{Z}_1 = \mathbf{Z}_2$ (this is known as a **balanced load**). Specifically, suppose that $\mathbf{V}_s = 115\underline{/0°}$ V rms, $\mathbf{Z}_1 = \mathbf{Z}_2 = 100$ Ω, and $\mathbf{Z}_3 = 15 + j20$ Ω. Find the power absorbed by each impedance, and determine the current in the neutral wire.

ANSWER 132 W, 132 W, 1270 W, 0 A

Three-Phase Circuits (Systems)

A single-phase, three-wire circuit contains a source that produces two sinusoidal voltages that have the same amplitude and the same angle. A circuit that contains a source that produces (sinusoidal) voltages with different phases is called a **polyphase system**. The importance of this concept lies in the fact that most of the generation and distribution of electric power in the United States is accomplished with polyphase systems. The most common polyphase system is the balanced three-phase system, which has the property that it supplies constant instantaneous power. This results in less vibration of the rotating machinery used to generate electric power.

A **Y (wye)-connected three-phase source** represented in the frequency domain is shown in Fig. 4.37a. Terminals a, b, and c are called the **line terminals** and n is called the **neutral terminal**. The source is said to be **balanced** if the voltages $\mathbf{V}_{an}$, $\mathbf{V}_{bn}$, and $\mathbf{V}_{cn}$, called **phase voltages**, have the same magnitude and sum to zero, that is, if

$$|\mathbf{V}_{an}| = |\mathbf{V}_{bn}| = |\mathbf{V}_{cn}| \quad \text{and} \quad \mathbf{V}_{an} + \mathbf{V}_{bn} + \mathbf{V}_{cn} = 0$$

Suppose that the amplitude of the sinusoids is V_m. If we arbitrarily select the angle of $\mathbf{V}_{an}$ to be zero, that is, if $\mathbf{V}_{an} = V_m\underline{/0°}$, then the two situations that result in a balanced source are as follows:

Case 1
$$\mathbf{V}_{an} = V_m\underline{/0°}$$
$$\mathbf{V}_{bn} = V_m\underline{/-120°}$$
$$\mathbf{V}_{cn} = V_m\underline{/-240°} = V_m\underline{/120°}$$

Case 2
$$\mathbf{V}_{an} = V_m\underline{/0°}$$
$$\mathbf{V}_{bn} = V_m\underline{/120°} = V_m\underline{/-240°}$$
$$\mathbf{V}_{cn} = V_m\underline{/240°} = V_m\underline{/-120°}$$

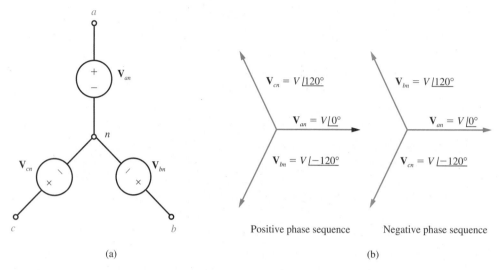

Fig. 4.37 (*a*) A Y-connected three-phase source, and (*b*) phasor diagrams.

For Case 1, $v_{an}(t)$ leads $v_{bn}(t)$ by 120 degrees, and $v_{bn}(t)$ leads $v_{cn}(t)$ by 120 degrees. Case 1 is, therefore, called a **positive phase sequence** (or **abc phase sequence**). Similarly, Case 2 is called a **negative phase sequence** (or **acb phase sequence**). Phasor diagrams for the positive phase and negative phase sequences are shown in Fig. 4.37*b*. Clearly, a negative phase sequence can be converted to a positive phase sequence simply by relabeling the terminals. Thus we need only consider positive phase sequences.

It is a simple matter to determine the voltages between the line terminals—called **line voltages**. Let us denote the voltage between line x and line y, where the $+$ is at line x, by $\mathbf{V}_{xy}$. By KVL,

$$\mathbf{V}_{ab} = \mathbf{V}_{an} - \mathbf{V}_{bn} = V_m\underline{/0°} - V_m\underline{/-120°} = \sqrt{3}V_m\underline{/30°}$$

Similarly,

$$\mathbf{V}_{bc} = \sqrt{3}V_m\underline{/-90°} \quad \text{and} \quad \mathbf{V}_{ca} = \sqrt{3}V_m\underline{/-210°} = \sqrt{3}V_m\underline{/150°}$$

A phasor diagram showing the relationships between the phase voltages and the line voltages is shown in Fig. 4.38.

Let us now connect our balanced source to a **balanced Y-connected three-phase load** as shown in Fig. 4.39. The short circuits between nodes a and A, between nodes

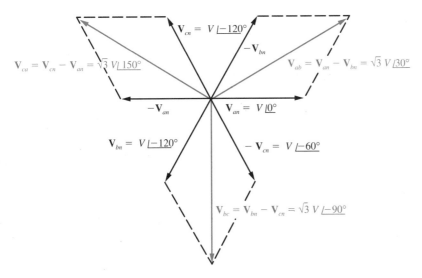

Fig. 4.38 Phasor diagram relating phase and line voltages.

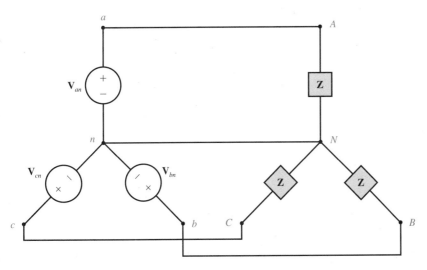

Fig. 4.39 Balanced Y-Y-connected three-phase circuit.

b and B, and between nodes c and C are the **lines** of the system. If we denote the current through $\mathbf{Z}$ from terminal x to y by $\mathbf{I}_{xy}$, then the line currents are

$$\mathbf{I}_{AN} = \frac{\mathbf{V}_{an}}{\mathbf{Z}} \qquad \mathbf{I}_{BN} = \frac{\mathbf{V}_{bn}}{\mathbf{Z}} \qquad \mathbf{I}_{CN} = \frac{\mathbf{V}_{cn}}{\mathbf{Z}}$$

By KCL,

$$\mathbf{I}_{Nn} = \mathbf{I}_{AN} + \mathbf{I}_{BN} + \mathbf{I}_{CN} = \frac{\mathbf{V}_{an}}{\mathbf{Z}} + \frac{\mathbf{V}_{bn}}{\mathbf{Z}} + \frac{\mathbf{V}_{cn}}{\mathbf{Z}} = \frac{\mathbf{V}_{an} + \mathbf{V}_{bn} + \mathbf{V}_{cn}}{\mathbf{Z}} = 0 \text{ A}$$

so we see that there is no current in the "neutral wire" (the short circuit between terminals n and N). We deduce that the neutral wire can be removed without affecting the voltages and currents.

   If the lines all have the same impedance, the effective load is still balanced and so the neutral current is zero, and the neutral wire can again be removed. If the current in the neutral wire is zero when the neutral wire's impedance is both zero and infinity, we may deduce that the current is also zero for any value in between. For the case of a balanced source, balanced load, and balanced (equal) line impedances (called a **balanced system**), whether or not there is a neutral wire connecting terminals n and N, it is convenient to think of these two points as being connected by a short circuit. In this way, we can analyze the system on a "per-phase" basis, that is, as three separate single-phase problems. This can greatly simplify the analysis of the system.

Example 4.14

For the balanced three-phase circuit given in Fig. 4.39, suppose that $\mathbf{V}_{an} = 120\underline{/0°}$ V is an rms phasor. Assuming a positive phase sequence, then $\mathbf{V}_{bn} = 120\underline{/-120°}$ V and $\mathbf{V}_{cn} = 120\underline{/120°}$ V. Thus, the rms line voltage phasors are

$$\mathbf{V}_{ab} = 120\sqrt{3}\underline{/30°} \text{ V}, \qquad \mathbf{V}_{bc} = 120\sqrt{3}\underline{/-90°} \text{ V},$$

$$\mathbf{V}_{ca} = 120\sqrt{3}\underline{/150°}$$

and the line voltage is $120\sqrt{3} = 208$ V rms.
   If $\mathbf{Z} = 30 + j40 = 50\underline{/53.1°}$ Ω, then the rms line current $\mathbf{I}_{aA} = \mathbf{I}_{AN}$ is

$$\mathbf{I}_{aA} = \frac{\mathbf{V}_{an}}{\mathbf{Z}} = \frac{120\underline{/0°}}{50\underline{/53.1°}} = 2.4\underline{/-53.1°} \text{ A}$$

and the power absorbed by the single load impedance $\mathbf{Z}$ connected between terminals A and N is[7]

---

[7] An alternative approach is to use $P_{AN} = R_{AN} |\mathbf{I}_{AN}|^2 = 30(2.4)^2 = 172.8$ W.

$$P_{AN} = |\mathbf{V}_{an}| \, |\mathbf{I}_{AN}| \cos (53.1°) = 120(2.4)(0.6) = 172.8$$

so that the total power absorbed by the three-phase load is 3(172.8) = 518.4 W.

Drill Exercise 4.14

For the balanced three-phase circuit depicted in Fig. 4.39, the line voltage is 250 V rms and the total power absorbed by the load is 600 W at a pf of 0.866 leading. Analyze the circuit on a per-phase basis to determine (a) the rms per-phase voltage, (b) the rms line current, and (c) the impedance $\mathbf{Z}$.

ANSWERS (a) 144 V rms, (b) 1.6 A rms, (c) $90\underline{/-30°}$ Ω

The powers mentioned in Example 4.14 are average powers. But, how do they compare with the instantaneous powers for that example?

Since the rms value for a phase voltage is 120 V, the corresponding sinusoidal amplitude is $120\sqrt{2}$ V. Also, since the rms value for a line current is 2.4 A, the corresponding sinusoidal amplitude is $2.4\sqrt{2}$ A. Thus, for the single load between A and N, the instantaneous power absorbed is

$$P_{AN}(t) = v_{an}(t) \, i_{AN}(t) = [120\sqrt{2} \cos \omega t][2.4\sqrt{2} \cos(\omega t - 53.1°)]$$

$$= 576 \cos \omega t \cos(\omega t - 53.1°) \text{ W}$$

which is a time-varying quantity. However, the total instantaneous power absorbed by the three-phase load is

$$p(t) = p_{AN}(t) + p_{BN}(t) + p_{CN}(t)$$

where

$$p_{BN}(t) = v_{bn}(t) \, i_{BN}(t)$$

$$= [120\sqrt{2} \cos(\omega t - 120°)][2.4\sqrt{2} \cos(\omega t - 53.1° - 120°)] \text{ W}$$

and

$$p_{CN}(t) = v_{cn}(t) \, i_{CN}(t)$$

$$= [120\sqrt{2} \cos(\omega t - 240°)][2.4\sqrt{2} \cos(\omega t - 53.1° - 240°)] \text{ W}$$

Using the trigonometric identity

$$\cos \alpha \cos \beta = \tfrac{1}{2}\cos(\alpha + \beta) + \tfrac{1}{2}\cos(\alpha - \beta)$$

we get

$$p(t) = 288[\cos(2\omega t - 53°) + \cos 53.1°] + 288[\cos(2\omega t - 53.1° - 240°)$$
$$+ \cos 53.1°] + 288[\cos(2\omega t - 53.1° - 120°) + \cos 53.1°] \text{ W}$$

It is a routine matter to show that since the above three sinusoids are spaced 120 degrees apart, they sum to zero. Thus

$$p(t) = 288[3 \cos 53.1°] = 518.4 \text{ W}$$

so we see that the instantaneous power absorbed by the load (supplied by the source) is constant! This result holds for any balanced Y-Y-connected three-phase system, which is one reason for its importance.

Delta Connections

More common than a balanced Y-connected three-phase load is a Δ (**delta**)-**connected** load. A Y-connected source with a balanced Δ-connected load is shown in Fig. 4.40. We see that the individual loads are connected directly between the lines, and, consequently, it is relatively easier to add or remove one of the components of a Δ-connected load than for a Y-connected load. As a matter of fact, it may not even be possible to do so for a Y-connected load since the neutral terminal may not be accessible.

Suppose that the voltage between terminals x and y, where the $+$ is at terminal x, is denoted by $\mathbf{V}_{xy}$. For a balanced source

$$\mathbf{V}_{an} = V_m\underline{/0°}, \qquad \mathbf{V}_{bn} = V_m\underline{/-120°}, \qquad \mathbf{V}_{cn} = V_m\underline{/-240°}$$

and as was shown previously, the line voltages are

$$\mathbf{V}_{ab} = \sqrt{3}V_m\underline{/30°}, \qquad \mathbf{V}_{bc} = \sqrt{3}V_m\underline{/-90°}, \qquad \mathbf{V}_{ca} = \sqrt{3}V_m\underline{/-210°}$$

which are the phase voltages for a Δ-connected load. The phase currents for the load are

$$\mathbf{I}_{AB} = \frac{\mathbf{V}_{ab}}{\mathbf{Z}} \qquad \mathbf{I}_{BC} = \frac{\mathbf{V}_{bc}}{\mathbf{Z}} \qquad \mathbf{I}_{CA} = \frac{\mathbf{V}_{ca}}{\mathbf{Z}}$$

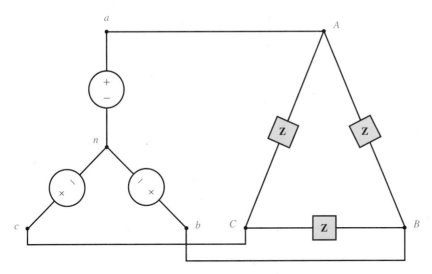

Fig. 4.40 Three-phase system with a balanced Δ-connected load.

Thus by KCL, the line current $\mathbf{I}_{aA}$ is

$$\mathbf{I}_{aA} = \mathbf{I}_{AB} - \mathbf{I}_{CA} = \frac{\mathbf{V}_{ab}}{\mathbf{Z}} - \frac{\mathbf{V}_{ca}}{\mathbf{Z}} = \frac{(\mathbf{V}_{ab} - \mathbf{V}_{ca})}{\mathbf{Z}}$$

$$= \frac{\sqrt{3}V_m\underline{/30°} - \sqrt{3}V_m\underline{/-210°}}{\mathbf{Z}} = \frac{3V_m\underline{/0°}}{\mathbf{Z}}$$

Similarly,

$$\mathbf{I}_{bB} = \frac{3V_m\underline{/-120°}}{\mathbf{Z}} \qquad \text{and} \qquad \mathbf{I}_{cC} = \frac{3V_m\underline{/-240°}}{\mathbf{Z}}$$

Since the magnitude of a load phase current, for example $\mathbf{I}_{AB}$, is $|\mathbf{I}_{AB}| = \dfrac{|\mathbf{V}_{ab}|}{|\mathbf{Z}|} = \dfrac{\sqrt{3}V_m}{|\mathbf{Z}|}$, and the magnitude of a line current, for example $\mathbf{I}_{aA}$, is $|\mathbf{I}_{aA}| = \dfrac{3V_m}{|\mathbf{Z}|}$, then

$$|\mathbf{I}_{aA}| = \sqrt{3}\left(\frac{\sqrt{3}V_m}{|\mathbf{Z}|}\right) = \sqrt{3}\,|\mathbf{I}_{AB}|$$

That is, a line current is $\sqrt{3}$ times as great as a load phase current. Furthermore, since the load phase currents are 120 degrees apart, they are a balanced set—as are the line currents.

Example 4.15

For the balanced three-phase circuit given in Fig. 4.40, let us determine the impedance $\mathbf{Z}$ for the case that the line voltage is 250 V rms and the total power absorbed by the Δ-connected load is 600 W at a pf of 0.866 leading.

Since the per-phase power absorbed is $600/3 = 200$ W, from the fact that $P_{AB} = |\mathbf{V}_{ab}|\,|\mathbf{I}_{AB}| \cos \theta$, we have that the rms load phase current, for example $\mathbf{I}_{AB}$, has magnitude

$$|\mathbf{I}_{AB}| = \frac{P_{AB}}{|\mathbf{V}_{ab}| \cos \theta} = \frac{200}{250(0.866)} = 0.924 \text{ A}$$

whereas the rms line current, for example $\mathbf{I}_{aA}$, has magnitude

$$|\mathbf{I}_{aA}| = \sqrt{3}\,|\mathbf{I}_{AB}| = (1.73)(0.924) = 1.6 \text{ A}$$

The per-phase load impedance has magnitude

$$|\mathbf{Z}| = \frac{|\mathbf{V}_{ab}|}{|\mathbf{I}_{AB}|} = \frac{250}{0.924} = 271 \text{ } \Omega$$

Since the pf is leading, the current leads to voltage. Also, $\cos^{-1}(0.866) = 30°$. Thus, $\text{ang}(\mathbf{I}_{AB}) > \text{ang}(\mathbf{V}_{ab})$. Hence

$$\mathbf{Z} = 271\underline{/-30°} = 235 - j136 \text{ } \Omega$$

Drill Exercise 4.15

For the balanced three-phase circuit shown in Fig. 4.40, the line voltage is 250 V rms and the line current is 10 A rms. If the load has a pf of 0.866 lagging, determine the total average power it absorbs.

ANSWER 3750 W

Suppose now that a balanced three-phase load is Y-connected, the magnitude of the line voltage is $V_L = |\mathbf{V}_{ab}|$ volts rms, and the line current is $I_L = |\mathbf{I}_{aA}| = |\mathbf{I}_{AN}|$ amperes rms. Then we know that the average power absorbed per phase is $P_{AN} = |\mathbf{V}_{an}|\,|\mathbf{I}_{AN}| \cos \theta$, where the pf angle θ is the angle of $\mathbf{Z}$. We have already seen that $V_L = |\mathbf{V}_{ab}| = \sqrt{3}\,|\mathbf{V}_{an}|$. Thus, $|\mathbf{V}_{an}| = |\mathbf{V}_{ab}|/\sqrt{3} = V_L/\sqrt{3}$, so

$$P_{AN} = |\mathbf{V}_{an}| \, |\mathbf{I}_{AN}| \cos \theta = (V_L/\sqrt{3})I_L \cos \theta$$

and, therefore, the total power absorbed by the load is

$$P = 3P_{AN} = \sqrt{3}V_L I_L \cos \theta$$

For the case of a balanced Δ-connected load, $P_{AB} = |\mathbf{V}_{ab}| \, |\mathbf{I}_{AB}| \cos \theta$. But we have already seen that $I_L = |\mathbf{I}_{aA}| = \sqrt{3}\,|\mathbf{I}_{AB}|$. Therefore, $|\mathbf{I}_{AB}| = I_L/\sqrt{3}$, and

$$P_{AB} = |\mathbf{V}_{ab}| \, |\mathbf{I}_{AB}| \cos \theta = V_L(I_L/\sqrt{3}) \cos \theta \quad \Rightarrow$$

$$P = 3P_{AB} = \sqrt{3}V_L I_L \cos \theta$$

Thus we see that regardless of whether a balanced load is Y-connected or Δ-connected, in terms of the rms line voltage V_L, the rms line current I_L, and angle θ of the load impedance (i.e., the pf angle), we can use the same formula for the total power absorbed by the load

$$\boxed{P = \sqrt{3}V_L I_L \cos \theta}$$

4.7   Three-Phase Loads

In our discussions, we have assumed that the source is balanced and Y-connected. Although the same conclusions can be obtained for a balanced Δ-connected source, for practical reasons, such sources are uncommon. If a Δ-connected source is not balanced exactly, a large current can circulate around the loop formed by the elements comprising the delta. The result is the reduction of the source's current-delivering capacity and an increase in the system's losses.

Δ-Y and Y-Δ Transformations

One might feel that a Y connection is equivalent to a Δ connection, and vice versa. But under what conditions is equivalence obtained? To answer this question, consider the Δ and Y connections shown in Fig. 4.41. For the Δ connection, the mesh equations are

$$\mathbf{V}_1 = \mathbf{Z}_{AC}\mathbf{I}_1 - \mathbf{Z}_{AC}\mathbf{I}_3 \tag{4.9}$$

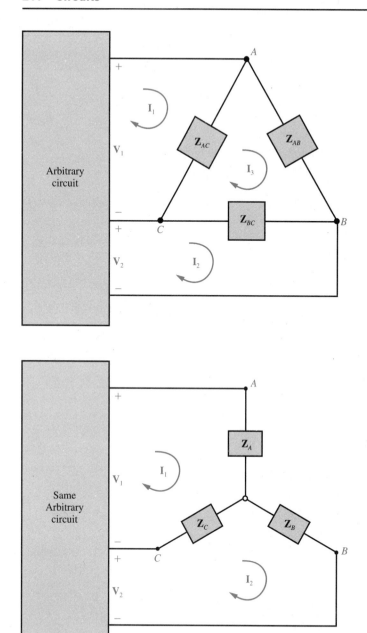

Fig. 4.41 Equivalence of Δ and Y connections.

$$\mathbf{V}_2 = \mathbf{Z}_{BC}\mathbf{I}_2 - \mathbf{Z}_{BC}\mathbf{I}_3 \qquad (4.10)$$

$$0 = -\mathbf{Z}_{AC}\mathbf{I}_1 - \mathbf{Z}_{BC}\mathbf{I}_2 + (\mathbf{Z}_{AB} + \mathbf{Z}_{BC} + \mathbf{Z}_{AC})\mathbf{I}_3 \qquad (4.11)$$

From Eq. 4.11,

$$\mathbf{I}_3 = \frac{\mathbf{Z}_{AC}\mathbf{I}_1}{\mathbf{Z}_{AB} + \mathbf{Z}_{BC} + \mathbf{Z}_{AC}} + \frac{\mathbf{Z}_{BC}\mathbf{I}_2}{\mathbf{Z}_{AB} + \mathbf{Z}_{BC} + \mathbf{Z}_{AC}}$$

and substituting this expression for $\mathbf{I}_3$ into Eq. 4.9 and Eq. 4.10, we get, respectively,

$$\mathbf{V}_1 = \frac{\mathbf{Z}_{AB}\mathbf{Z}_{AC} + \mathbf{Z}_{AC}\mathbf{Z}_{BC}}{\mathbf{Z}_{AB} + \mathbf{Z}_{BC} + \mathbf{Z}_{AC}}\mathbf{I}_1 - \frac{\mathbf{Z}_{AC}\mathbf{Z}_{BC}}{\mathbf{Z}_{AB} + \mathbf{Z}_{BC} + \mathbf{Z}_{AC}}\mathbf{I}_2$$

and

$$\mathbf{V}_2 = \frac{-\mathbf{Z}_{AC}\mathbf{Z}_{BC}}{\mathbf{Z}_{AB} + \mathbf{Z}_{BC} + \mathbf{Z}_{AC}}\mathbf{I}_1 + \frac{\mathbf{Z}_{AB}\mathbf{Z}_{BC} + \mathbf{Z}_{AC}\mathbf{Z}_{BC}}{\mathbf{Z}_{AB} + \mathbf{Z}_{BC} + \mathbf{Z}_{AC}}\mathbf{I}_2$$

The mesh equations for the Y connection are

$$\mathbf{V}_1 = (\mathbf{Z}_A + \mathbf{Z}_C)\mathbf{I}_1 - \mathbf{Z}_C\mathbf{I}_2 \qquad \text{and} \qquad \mathbf{V}_2 = -\mathbf{Z}_C\mathbf{I}_1 + (\mathbf{Z}_B + \mathbf{Z}_C)\mathbf{I}_2$$

Since the two connections are equivalent when the values of $\mathbf{V}_1$ and $\mathbf{V}_2$ and $\mathbf{I}_1$ and $\mathbf{I}_2$ are the same for both cases, by equating coefficients we get

$$\mathbf{Z}_A = \frac{\mathbf{Z}_{AB}\mathbf{Z}_{AC}}{\mathbf{Z}_{AB} + \mathbf{Z}_{BC} + \mathbf{Z}_{AC}} \qquad \mathbf{Z}_B = \frac{\mathbf{Z}_{AB}\mathbf{Z}_{BC}}{\mathbf{Z}_{AB} + \mathbf{Z}_{BC} + \mathbf{Z}_{AC}}$$

$$\mathbf{Z}_C = \frac{\mathbf{Z}_{AC}\mathbf{Z}_{BC}}{\mathbf{Z}_{AB} + \mathbf{Z}_{BC} + \mathbf{Z}_{AC}}$$

and these are the conditions for finding the equivalent Y connection given a Δ connection.

By using dual arguments, we can derive the conditions for finding the equivalent Δ connection given a Y connection. (In this case, we can use admittances instead of impedances.) They are

$$Y_{AB} = \frac{Y_A Y_B}{Y_A + Y_B + Y_C} \qquad Y_{BC} = \frac{Y_B Y_C}{Y_A + Y_B + Y_C}$$

$$Y_{AC} = \frac{Y_A Y_C}{Y_A + Y_B + Y_C}$$

Alternatively, it is a simple matter to express these conditions in terms of impedances as follows:

$$Z_{AB} = \frac{Z_A Z_B + Z_B Z_C + Z_A Z_C}{Z_C} \qquad Z_{BC} = \frac{Z_A Z_B + Z_B Z_C + Z_A Z_C}{Z_A}$$

$$Z_{AC} = \frac{Z_A Z_B + Z_B Z_C + Z_A Z_C}{Z_B}$$

Replacing a Δ connection with its equivalent Y connection, or vice versa, is called a **Δ-Y (delta-wye) transformation**, or **Y-Δ (wye-delta) transformation,** respectively. Since the equivalence is with respect to the three terminals A, B, and C, the equivalence holds for any pair of them also.

For the special case that a Δ-connected load is balanced, that is, $Z_{AB} = Z_{AC} = Z_{BC} = Z_\Delta$, then by performing a Δ-Y transformation, we get a Y-connected balanced load with

$$Z_A = Z_B = Z_C = \frac{Z_\Delta}{3}$$

Conversely, for a balanced Y-connected load having $Z_A = Z_B = Z_C = Z_Y$, then by performing a Y-Δ transformation, we obtain a balanced Δ-connected load having

$$Z_{AB} = Z_{AC} = Z_{BC} = 3Z_Y$$

Example 4.16

Suppose that for the balanced three-phase load shown in Fig. 4.42a, $Z_Y = 10 - j10$ Ω and $Z_\Delta = 40 + j40$ Ω. Let us find an equivalent single Δ-connected load.

Performing a Y-Δ transformation on the Y-connected portion of the load, we obtain a Δ connection having impedances

$$Z_\Delta' = 3Z_Y = 30 - j30 \ \Omega$$

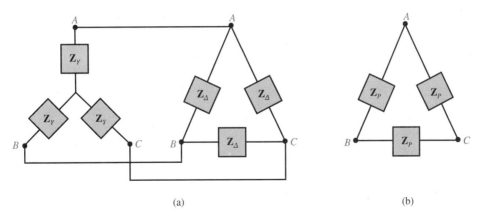

(a)   (b)

Fig. 4.42 (*a*) Parallel-connected three-phase load, and (*b*) equivalent three-phase load.

Since each impedance $\mathbf{Z}'_\Delta$ is connected in parallel with an impedance $\mathbf{Z}_\Delta$, the impedance of the parallel combination is

$$\mathbf{Z}_P = \mathbf{Z}'_\Delta \| \mathbf{Z}_\Delta = \frac{\mathbf{Z}'_\Delta \mathbf{Z}_\Delta}{\mathbf{Z}'_\Delta + \mathbf{Z}_\Delta} = \frac{(30 - j30)(40 + j40)}{(30 - j30) + (40 + j40)} = 33.6 - j4.8 \ \Omega$$

and we have the equivalent Δ-connected load shown in Fig. 4.42*b*.

If a Δ-Y transformation is performed on the Δ-connected load shown in Fig. 4.42*b*, the result would be a balanced Y-connected load, where each impedance in the Y connection would be $\mathbf{Z} = \mathbf{Z}_P/3 = (33.6 - j4.8)/3 = 11.2 - j1.6 \ \Omega$.

Drill Exercise 4.16

For the parallel-connected three-phase load depicted in Fig. 4.42*a*, suppose that $\mathbf{Z}_Y = 10 - j10 \ \Omega$ and $\mathbf{Z}_\Delta = 40 + j40 \ \Omega$. (a) Replace the Δ-connected portion with an equivalent Y connection having impedances $\mathbf{Z}'_Y$. Determine the impedance $\mathbf{Z}'_Y$. (b) Place a short circuit between the nodes in the centers of the Y connections so that each impedance $\mathbf{Z}_Y$ is connected in parallel with an impedance $\mathbf{Z}'_Y$. Determine the impedance $\mathbf{Z}$ of each resulting parallel combination. (c) How does this result compare with the impedance $\mathbf{Z}$ obtained in Example 4.16?

ANSWER   (a) $40/3 + j40/3 \ \Omega$, (b) $11.2 + j1.6 \ \Omega$, (c) same result

Power Measurements

The **wattmeter** is a device that measures the average power absorbed by a two-terminal load. Such a device is depicted in Fig. 4.43. The wattmeter, which is used typically at frequencies between a few hertz and a few hundred hertz, has two coils. One, the current coil, has a very low impedance (ideally zero), and the other, the voltage coil, has a very high impedance (ideally infinite). The current coil is connected in series with the load and the voltage coil is connected in parallel with the load. When the current and the voltage shown for the load are both positive valued, the wattmeter has an upscale deflection and reads the average power absorbed by the load. The same result occurs when both the current and voltage are negative valued.

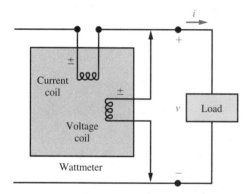

Fig. 4.43 Wattmeter connected to a load.

Having seen how to measure the power absorbed by a single load, how can we use the wattmeter to measure the power absorbed by a three-phase load? For a balanced load, if we can connect a wattmeter to one phase of the load as shown in Fig. 4.43, then the total power absorbed is three times the reading. For an unbalanced load, if three separate wattmeters can be connected, then the total power absorbed is the sum of the three readings. However, a Y-connected load may not have its neutral terminal accessible. In such a case, although the current coil can be connected properly, the voltage coil cannot. On the other hand, for a Δ-connected load such as a three-phase rotating machine, only the three terminals of the load are accessible. This means that the voltage coil can be connected appropriately, but the current coil cannot. Despite these apparent difficulties, there is a way to measure the power absorbed by a three-phase load, whether balanced or unbalanced, with the use of wattmeters. To see how, consider the three wattmeters and Δ-connected load shown in Fig. 4.44.

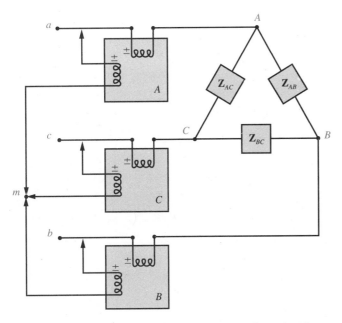

. **Fig. 4.44** Power measurement for a three-phase load.

In this situation, the current coils of the wattmeters are connected in series with the lines aA, bB, and cC. The voltage coils are connected between a line and some arbitrary point m. The average power reading on wattmeter A is

$$P_A = \frac{1}{T}\int_0^T v_{am}i_{aA}\,dt$$

and similar expressions are obtained for wattmeters B and C. Summing these average powers, we get

$$P = P_A + P_B + P_C = \frac{1}{T}\int_0^T v_{am}i_{aA}\,dt + \frac{1}{T}\int_0^T v_{bm}i_{bB}\,dt + \frac{1}{T}\int_0^T v_{cm}i_{cC}\,dt$$

$$= \frac{1}{T}\int_0^T (v_{am}i_{aA} + v_{bm}i_{bB} + v_{cm}i_{cC})\,dt$$

But, by KCL,

$$i_{aA} = i_{AB} + i_{AC} \qquad i_{bB} = -i_{AB} + i_{BC} \qquad i_{cC} = -i_{AC} - i_{BC}$$

Substituting these expressions for the line currents into the preceding integral, we obtain

$$P = \frac{1}{T}\int_0^T [(v_{am} - v_{bm})i_{AB} + (v_{bm} - v_{cm})i_{BC} + (v_{am} - v_{cm})i_{AC}]\, dt$$

and, by KVL,

$$P = \frac{1}{T}\int_0^T (v_{ab}i_{AB} + v_{bc}i_{BC} + v_{ac}i_{AC})\, dt$$

$$= \frac{1}{T}\int_0^T v_{ab}i_{AB}\, dt + \frac{1}{T}\int_0^T v_{bc}i_{BC}\, dt + \frac{1}{T}\int_0^T v_{ac}i_{AC}\, dt$$

which is the total power absorbed by the three-phase load.

Since point m was chosen arbitrarily, we can place it anywhere without affecting the end result—the sum of the three wattmeter readings is the power absorbed by the three-phase load. The judicious choice of point m to be on line cC means that wattmeter C will have zero volts across its voltage coil and, hence, its reading will be 0 W. For such a case, wattmeter C is superfluous, and the sum of the readings of wattmeters A and B is equal to the total power absorbed by the three-phase load. The resulting configuration, shown in Fig. 4.45, is an example of what is known as the **two-wattmeter method** for measuring power. It can also be applied when a three-phase load is Y-connected, and as for the case of a Δ connection, the load can be unbalanced as well as balanced. Furthermore, it is immaterial whether or not the source is balanced—the variables utilized are line voltages and line currents.

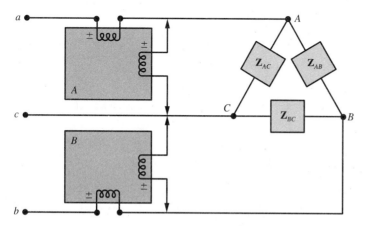

Fig. 4.45 Two-wattmeter method for measuring power.

Example 4.17

Let us find the wattmeter readings for the circuit shown in Fig. 4.45 when $\mathbf{Z}_{AB} = \mathbf{Z}_{BC} = \mathbf{Z}_{AC} = 271\underline{/-30°}$ Ω and the line voltage is 250 V rms, that is, $\mathbf{V}_{ab} = 250\underline{/30°}$ V, $\mathbf{V}_{bc} = 250\underline{/-90°}$ V, $\mathbf{V}_{ca} = 250\underline{/-210°}$ V.

Since the voltage across the voltage coil for wattmeter A is $\mathbf{V}_{ac}$ and the current through the current coil is $\mathbf{I}_{aA}$, then the reading for wattmeter A is

$$P_A = |\mathbf{V}_{ac}|\,|\mathbf{I}_{aA}|\,\cos\theta_A$$

where $\theta_A = \text{ang}(\mathbf{V}_{ac}) - \text{ang}(\mathbf{I}_{aA})$. Since $\mathbf{V}_{ac} = -\mathbf{V}_{ca} = (-1)(\mathbf{V}_{ca}) = (1\underline{/180°})(250\underline{/-210°}) = 250\underline{/-30°}$ V and

$$\mathbf{I}_{aA} = \mathbf{I}_{AB} - \mathbf{I}_{CA} = \frac{\mathbf{V}_{ab}}{\mathbf{Z}} - \frac{\mathbf{V}_{ca}}{\mathbf{Z}} = \frac{\mathbf{V}_{ab} - \mathbf{V}_{ca}}{\mathbf{Z}}$$

$$= \frac{250\underline{/30°} - 250\underline{/-210°}}{271\underline{/-30°}} = 1.6\underline{/30°}\ \text{A}$$

then $P_A = (250)(1.6)\cos(-30° - 30°) = 200$ W.

Since the voltage across the voltage coil for wattmeter B is $\mathbf{V}_{bc}$ and the current through the current coil is $\mathbf{I}_{bB}$, then the reading for wattmeter B is

$$P_B = |\mathbf{V}_{bc}|\,|\mathbf{I}_{bB}|\,\cos\theta_B$$

where $\theta_B = \text{ang}(\mathbf{V}_{bc}) - \text{ang}(\mathbf{I}_{bB})$. Since

$$\mathbf{I}_{bB} = \mathbf{I}_{BC} - \mathbf{I}_{AB} = \frac{\mathbf{V}_{bc}}{\mathbf{Z}} - \frac{\mathbf{V}_{ab}}{\mathbf{Z}} = \frac{\mathbf{V}_{bc} - \mathbf{V}_{ab}}{\mathbf{Z}}$$

$$= \frac{250\underline{/-90°} - 250\underline{/30°}}{271\underline{/-30°}}$$

$$= 1.6\underline{/-90°}\ \text{A}$$

then $P_B = (250)(1.6)\cos(-90° + 90°) = 400$ W.

Hence the total power absorbed by the Δ-connected load is $P_A + P_B = 200 + 400 = 600$ W. (See Example 4.15 on p. 242.)

Drill Exercise 4.17

Replace the Δ-connected load shown in Fig. 4.45 with a balanced Y-connected load having a per-phase impedance of $\mathbf{Z} = 30 + j40\ \Omega$. Find the wattmeter readings given the following rms voltages:

$$\text{phase voltages: } \mathbf{V}_{an} = 120\underline{/0°}\ \text{V},\ \mathbf{V}_{bn} = 120\underline{/-120°}\ \text{V},$$

$$\mathbf{V}_{cn} = 120\underline{/-240°}\ \text{V}$$

$$\text{line voltages: } \mathbf{V}_{ab} = 120\sqrt{3}\underline{/30°},\ \mathbf{V}_{bc} = 120\sqrt{3}\underline{/-90°}\ \text{V},$$

$$\mathbf{V}_{ca} = 120\sqrt{3}\underline{/150°}\ \text{V}$$

ANSWER 459 W, 59.9 W

For the case that a load is balanced, it is also possible to determine the pf angle by using the two-wattmeter method. To see how this is done, again consider the Δ-connected load shown in Fig. 4.45, where $\mathbf{Z}_{AB} = \mathbf{Z}_{BC} = \mathbf{Z}_{AC} = \mathbf{Z} = |\mathbf{Z}|\underline{/\theta}$.

For the balanced source

$$\mathbf{V}_{an} = V_m\underline{/0°}, \qquad \mathbf{V}_{bn} = V_m\underline{/-120°}, \qquad \mathbf{V}_{cn} = V_m\underline{/-240°}$$

we have previously seen that line voltages are

$$\mathbf{V}_{ab} = \sqrt{3}V_m\underline{/30°}, \qquad \mathbf{V}_{bc} = \sqrt{3}V_m\underline{/-90°}, \qquad \mathbf{V}_{ca} = \sqrt{3}V_m\underline{/-210°}$$

and the line currents are

$$\mathbf{I}_{aA} = (3V_m\underline{/0°})/\mathbf{Z}, \qquad \mathbf{I}_{bB} = (3V_m\underline{/-120°})/\mathbf{Z}, \qquad \mathbf{I}_{cC} = (3V_m\underline{/-240°})/\mathbf{Z}$$

Thus the rms line voltage is $V_L = \sqrt{3}V_m$ and the rms line current is $I_L = 3V_m$.

Since $\text{ang}(\mathbf{I}_{aA}) = -\text{ang}(\mathbf{Z}) = -\theta$, then the reading for wattmeter A is

$$P_A = |\mathbf{V}_{ac}|\,|\mathbf{I}_{aA}|\cos[\text{ang}(\mathbf{V}_{ac}) - \text{ang}(\mathbf{I}_{aA})]$$

$$= V_L I_L \cos(-30° + \theta) = V_L I_L \cos(30° - \theta)$$

and since $\text{ang}(\mathbf{I}_{bB}) = -120° - \text{ang}(\mathbf{Z}) = -120° - \theta$, then the reading for wattmeter B is

$$P_B = |\mathbf{V}_{bc}| \, |\mathbf{I}_{bB}| \cos[\text{ang}(\mathbf{V}_{bc}) - \text{ang}(\mathbf{I}_{bB})]$$

$$= V_L I_L \cos(-90° - [-120° - \theta]) = V_L I_L \cos(30° + \theta)$$

Taking the ratio of P_A to P_B, we get

$$\frac{P_A}{P_B} = \frac{\cos(30° - \theta)}{\cos(30° + \theta)}$$

Using the trigonometric identity $\cos(\alpha - \beta) = \cos \alpha \cos \beta + \sin \alpha \sin \beta$, this ratio becomes

$$\frac{P_A}{P_B} = \frac{\cos 30° \cos \theta + \sin 30° \sin \theta}{\cos 30° \cos \theta - \sin 30° \sin \theta} = \frac{(\sqrt{3}/2)\cos \theta + (1/2)\sin \theta}{(\sqrt{3}/2)\cos \theta - (1/2)\sin \theta}$$

from which

$$\sqrt{3}(P_A - P_B) \cos \theta = (P_A + P_B) \sin \theta \quad \Rightarrow$$

$$\frac{\sin \theta}{\cos \theta} = \frac{\sqrt{3}(P_A - P_B)}{P_A + P_B} = \tan \theta$$

Thus the pf angle is

$$\boxed{\theta = \tan^{-1} \frac{\sqrt{3}(P_A - P_B)}{P_A + P_B}} \tag{4.12}$$

Example 4.18

Suppose that the Δ-connected load shown in Fig. 4.45 is balanced, that is, $\mathbf{Z}_{AB} = \mathbf{Z}_{BC} = \mathbf{Z}_{AC} = \mathbf{Z} = |\mathbf{Z}|\underline{/\theta}$. Let us determine $\mathbf{Z}$ for the case that the line voltage is 250 V rms, and the wattmeter readings are $P_A = 200$ W and $P_B = 400$ W.

From Eq. 4.12,

$$\theta = \tan^{-1} \frac{\sqrt{3}(200 - 400)}{200 + 400} = \tan^{-1}\left(\frac{-1}{\sqrt{3}}\right) = -30°$$

Since the total power absorbed by the load is $200 + 400 = 600$ W, then each impedance absorbs $600/3 = 200$ W. In particular, for the impedance connected between terminals A and B,

$$P_{AB} = \left| \mathbf{V}_{ab} \right| \left| \mathbf{I}_{AB} \right| \cos \theta$$

$$200 = (250) \left| \mathbf{I}_{AB} \right| \cos(-30°) \quad \Rightarrow \quad \left| \mathbf{I}_{AB} \right| = \frac{200}{250 \cos(-30°)} = 0.924 \text{ A}$$

However,

$$\left| \mathbf{Z} \right| = \frac{\left| \mathbf{V}_{ab} \right|}{\left| \mathbf{I}_{AB} \right|} = \frac{250}{0.924} = 271 \ \Omega$$

Therefore, $\mathbf{Z} = \left| \mathbf{Z} \right| \underline{/\theta} = 271 \underline{/-30°} \ \Omega$.

Drill Exercise 4.18

Suppose that the Δ-connected load shown in Fig. 4.45 is balanced, that is, $\mathbf{Z}_{AB} = \mathbf{Z}_{BC} = \mathbf{Z}_{AC} = \mathbf{Z} = \left| \mathbf{Z} \right| \underline{/\theta}$. Determine $\mathbf{Z}$ for the case that the line voltage is 208 V rms, and the wattmeter readings are $P_A = 459$ W and $P_B = 60$ W.

ANSWER $150 \underline{/53.1°} \ \Omega$

SUMMARY

1. If the input of a linear, time-invariant circuit is a sinusoid, the response is a sinusoid of the same frequency.

2. A complex number $\mathbf{c}$ can be written in rectangular form as $\mathbf{c} = a + jb$, where a is the real part of $\mathbf{c}$ and b is the imaginary part of $\mathbf{c}$. It can also be written in polar form as $\mathbf{c} = Me^{j\theta}$, where M (also denoted $|\mathbf{c}|$) is a positive number called the magnitude of $\mathbf{c}$ and θ is the angle of $\mathbf{c}$.

3. Finding the amplitude and phase angle of a sinusoidal steady-state response can be accomplished with either real or complex sinusoids.

4. Using the concepts of phasors and impedance (or admittance), sinusoidal circuits can be analyzed in the frequency domain in a manner analogous to that for resistive circuits by using phasor versions of KCL, KVL, and nodal or mesh analysis.

5. Important circuit concepts such as the principle of superposition and Thévenin's theorem are also applicable in the frequency domain.

6. The instantaneous power absorbed by an element is equal to the product of the voltage across it and the current through it.

7. The average power absorbed by a resistance R having a sinusoidal current of amplitude I and voltage of amplitude V is

$$P_R = \tfrac{1}{2}VI = \tfrac{1}{2}RI^2 = \tfrac{1}{2}\frac{V^2}{R}$$

8. The average power absorbed by a capacitance or an inductance is zero.

9. A circuit whose Thévenin-equivalent (output) impedance is $\mathbf{Z}_o$ transfers maximum power to a load $\mathbf{Z}_L$ when $\mathbf{Z}_L$ is equal to the complex conjugate of $\mathbf{Z}_o$.

10. For the case in which $\mathbf{Z}_L$ is restricted to be purely resistive, maximum power is transferred when $\mathbf{Z}_L$ equals the magnitude of $\mathbf{Z}_o$.

11. The effective or rms value of a sinusoid of amplitude A is $A/\sqrt{2}$.

12. The average power absorbed by a resistance R having a current whose effective value is I_e and a voltage whose effective value is V_e is

$$P_R = V_e I_e = RI_e^2 = \frac{V_e^2}{R}$$

13. The power factor (pf) is the ratio of average power to apparent power.

14. If current lags voltage, the pf is lagging. If current leads voltage, the pf is leading.

15. Average or real power can be generalized with the notion of complex power.

16. The ordinary household uses a single-phase, three-wire electrical system.

17. The most common polyphase electrical system is the balanced three-phase system.

18. Three-phase sources are generally Y connected, and three-phase loads are generally Δ connected.

19. The device commonly used to measure power is the wattmeter.

20. Three-phase load power measurements can be taken with the two-wattmeter method.

Problems

4.1 Find the exponential form of the following complex numbers given in rectangular form: (a) $4 + j7$, (b) $3 - j5$, (c) $-2 + j3$, (d) $-1 - j6$, (e) 4, (f) -5, (g) $j7$, (h) $-j2$.

4.2 Find the rectangular form of the following complex numbers given in exponential form:
(a) $3e^{j70°}$, (b) $2e^{j120°}$, (c) $5e^{-j60°}$, (d) $4e^{-j150°}$, (e) $6e^{j90°}$, (f) $e^{-j90°}$, (g) $2e^{j180°}$, (h) $2e^{-j180°}$.

4.3 Find the rectangular form of the product $\mathbf{A}_1\mathbf{A}_2$ given that: (a) $\mathbf{A}_1 = 3e^{j30°}$, $\mathbf{A}_2 = 4e^{j60°}$; (b) $\mathbf{A}_1 = 3e^{j30°}$, $\mathbf{A}_2 = 4e^{j30°}$; (c) $\mathbf{A}_1 = 5e^{-j60°}$, $\mathbf{A}_2 = 2e^{j120°}$; (d) $\mathbf{A}_1 = 4e^{j45°}$, $\mathbf{A}_2 = 2e^{-j90°}$.

4.4 Find the rectangular form of the quotient $\mathbf{A}_1/\mathbf{A}_2$ for $\mathbf{A}_1$ and $\mathbf{A}_2$ given in Problem 4.3.

4.5 Find the rectangular form of the sum $\mathbf{A}_1 + \mathbf{A}_2$ for $\mathbf{A}_1$ and $\mathbf{A}_2$ given in Problem 4.3.

4.6 For the ac circuit shown in Fig. P4.6, suppose that $v_s(t) = 13\cos(2t - 22.6°)$ V. Find $v_o(t)$ by using voltage division. Draw a phasor diagram. Is this circuit a lag network or a lead network?

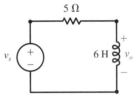

Fig. P4.6

4.7 Connect a 5-Ω resistor in parallel with the inductor in the circuit shown in Fig. P4.6. Suppose that $v_s(t) = 13 \cos(2t - 22.6°)$ V. Find the voltage $v_o(t)$ across the inductor by using voltage division. Draw a phasor diagram. Is this circuit a lag network or a lead network?

4.8 Connect a 5-Ω resistor in parallel with the inductor in the circuit shown in Fig. P4.6. Suppose that $v_s(t) = 13 \cos(2t - 22.6°)$ V. Find the voltage $v_o(t)$ across the inductor by using nodal analysis. Draw a phasor diagram. Is this circuit a lag network or a lead network?

4.9 For the circuit given in Fig. P4.9, suppose that $i_s(t) = 5 \cos 3t$ A. Find $v_o(t)$ and $v_s(t)$ by using current division.

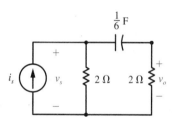

Fig. P4.9

4.10 For the circuit given in Fig. P4.9, suppose that $i_s(t) = 5 \cos 3t$ A. Find $v_o(t)$ and $v_s(t)$ by using nodal analysis.

4.11 A voltage of $v_s(t) = 10 \cos \omega t$ V is applied to a series *RLC* circuit. If $R = 5$ Ω, $L = \frac{1}{5}$ H, and $C = \frac{1}{5}$ F, by how many degrees does $v_C(t)$ lead or lag $v_s(t)$ when (a) $\omega = 1$ rad/s, (b) $\omega = 5$ rad/s, and (c) $\omega = 10$ rad/s?

4.12 A voltage of $v_s(t) = 10 \cos \omega t$ V is applied to a series *RLC* circuit. If $R = 5$ Ω, $L = \frac{1}{5}$ H, and $C = \frac{1}{5}$ F, by how many degrees does $v_R(t)$ lead or lag $v_s(t)$ when (a) $\omega = 1$ rad/s, (b) $\omega = 5$ rad/s, and (c) $\omega = 10$ rad/s?

4.13 For the *RLC* connection given in Fig. P4.13, find the impedance **Z** when ω is (a) 2, (b) 4, and (c) 8 rad/s.

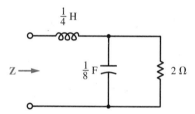

Fig. P4.13

4.14 For the *RLC* connection shown in Fig. P4.14, find the admittance **Y** when ω is: (a) 1, (b) 3, and (c) 7 rad/s.

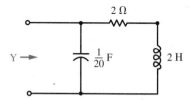

Fig. P4.14

4.15 Show that a general expression for the impedance **Z** depicted in Fig. P4.13 is

$$\mathbf{Z} = \frac{32}{\omega^2 + 16} + j\frac{\omega(\omega^2 - 16)}{4(\omega^2 + 16)}$$

4.16 Show that a general expression for the admittance **Y** depicted in Fig. P4.14 is

$$\mathbf{Y} = \frac{1}{2(\omega^2 + 1)} + j\frac{\omega(\omega^2 - 9)}{20(\omega^2 + 1)}$$

4.17 For the circuit shown in Fig. P4.17, find the Thévenin equivalent of the circuit in the shaded box when $v_s(t) = 4 \cos(4t - 60°)$ V. Use this to determine $v_o(t)$.

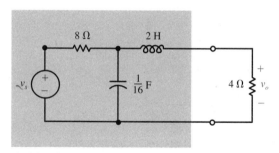

Fig. P4.17

4.18 For the circuit shown in Fig. P4.17, find the Thévenin equivalent of the circuit in the shaded box when $v_s(t) = 4 \cos(2t - 60°)$ V. Use this to determine $v_o(t)$.

4.19 Find the frequency-domain Thévenin equivalent (to the left of terminals a and b) of the circuit shown in Fig. 4.20 on p. 211. (*Hint:* Use the fact that $\mathbf{Z}_o = \mathbf{V}_{oc}/\mathbf{I}_{sc}$.)

4.20 The frequency-domain Thévenin equivalent of a circuit having $\omega = 5$ rad/s has $\mathbf{V}_{oc} = 3.71\underline{/-15.9°}$ V and $\mathbf{Z}_o = 2.38 - j0.667$ Ω. Determine a corresponding time-domain Thévenin-equivalent circuit.

4.21 For the op-amp circuit shown in Fig. P4.21, find $v_o(t)$ when $v_s(t) = 6 \sin 2t$ V.

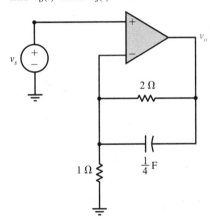

Fig. P4.21

4.22 For the op-amp circuit given in Fig. P4.22, find $v_o(t)$ when $v_s(t) = 3 \cos 2t$ V.

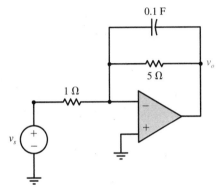

Fig. P4.22

4.23 For the op-amp circuit shown in Fig. P4.23, find $v_o(t)$ when $v_s(t) = 4 \cos(2t - 30°)$ V. (See p. 258.)

4.24 For the circuit shown in Fig. P4.24, find the currents $\mathbf{I}_1$ and $\mathbf{I}_2$ when $\mathbf{V}_{s1} = 250\sqrt{2}\underline{/-30°}$ V, $\mathbf{V}_{s2} = 250\sqrt{2}\underline{/-90°}$ V, and $\mathbf{Z} = 78 - j45$ Ω.

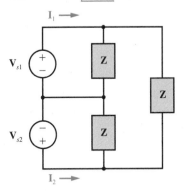

Fig. P4.24

4.25 Use mesh analysis to find $\mathbf{I}_1$ and $\mathbf{I}_2$ for the circuit given in Fig. P4.25 when $\mathbf{V}_{s1} = 250\sqrt{2}\underline{/-30°}$ V, $\mathbf{V}_{s2} = 250\sqrt{2}\underline{/-90°}$ V, and $\mathbf{Z} = 26 - j15$ Ω.

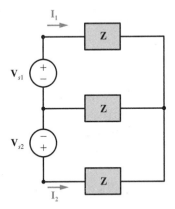

Fig. P4.25

4.26 For the circuit shown in Fig. P4.9, when $i_s(t) = 5 \cos 3t$ A then $v_o(t) = 4.47 \cos(3t + 26.6°)$ V. Find the average power absorbed by each element in the circuit.

4.27 For the circuit shown in Fig. P4.17, when $v_s(t) = 10 \cos 4t$ V, then the Thévenin equivalent of the portion of the circuit in the shaded box is $\mathbf{V}_{oc} =$

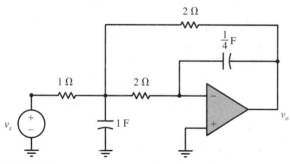

Fig. P4.23

4.47 $\underline{/-63.4°}$ V and $\mathbf{Z}_o = 1.6 + j4.8$ Ω. (a) Replace the 4-Ω load resistor by an impedance $\mathbf{Z}_L$ that absorbs the maximum average power, and determine this maximum power. (b) Replace the 4-Ω load resistor with a resistance R_L that absorbs the maximum power for resistive loads, and determine this power.

4.28 For the *RLC* circuit shown in Fig. P4.28, suppose that $v_s(t) = 10 \cos 3t$ V. Find the average power absorbed by the 4-Ω resistor for the case that (a) $C = \frac{1}{6}$ F; (b) $C = \frac{1}{18}$ F; (c) $C = \frac{1}{30}$ F.

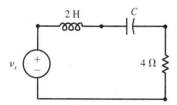

Fig. P4.28

4.29 For the circuit shown in Fig. P4.29, suppose that $v_s(t) = 8 \cos 2t$ V. Find the average power absorbed by each element in the circuit for the case that $\mathbf{Z}_L = 1$ Ω.

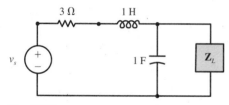

Fig. P4.29

4.30 For the circuit shown in Fig. P4.29, change the value of the resistor to 2 Ω and the value of the capacitor to $\frac{1}{4}$ F. Suppose that $v_s(t) = 8 \cos 2t$ V. (a) Find the load impedance $\mathbf{Z}_L$ that absorbs the maximum average power, and determine this power. (b) Find the load resistance R_L that absorbs the maximum power for resistive loads, and determine this power.

4.31 For the op-amp circuit given in Fig. P4.21, when $v_s(t) = 6 \sin 2t$ V, then the output voltage $v_o(t) = 13.4 \cos(2t - 117°)$ V. Find the average power absorbed by each element.

4.32 For the op-amp circuit given in Fig. P4.22, when $v_s(t) = 3 \cos 2t$ V, then the output voltage $v_o(t) = 10.6 \cos(2t + 135°)$ V. Find the average power absorbed by each element.

4.33 For the op-amp circuit given in Fig. P4.23, when $v_s(t) = 4 \cos(2t - 30°)$ V, then $v_1(t) = 1.6\cos(2t - 66.9°)$ V and $v_o(t) = 1.6\cos(2t + 23.1°)$ V. Find the average power absorbed by each element.

4.34 For the circuit given in Fig. P4.24, when $\mathbf{V}_{s1} = 250\sqrt{2}\underline{/-30°}$ V, $\mathbf{V}_{s2} = 250\sqrt{2}\underline{/-90°}$ V, and $\mathbf{Z} = 78 - j45$ Ω, then $\mathbf{I}_1 = 6.8\underline{/30°}$ A and $\mathbf{I}_2 = 6.8\underline{/-90°}$ A. (a) Find the average power absorbed by each impedance. (b) Find the average power supplied by each source.

4.35 For the circuit given in Fig. P4.25, when $\mathbf{V}_{s1} = 250\sqrt{2}\underline{/-30°}$ V, $\mathbf{V}_{s2} = 250\sqrt{2}\underline{/-90°}$ V, and $\mathbf{Z} = 26 - j15$ Ω, then $\mathbf{I}_1 = 6.8\underline{/30°}$ A and $\mathbf{I}_2 = 6.8\underline{/-90°}$ A. (a) Find the average power absorbed by each impedance. (b) Find the average power supplied by each source.

4.36 For the op-amp circuit shown in Fig. P4.36, find the average power absorbed by each element for the case that $v_s(t) = \cos \omega t$ V.

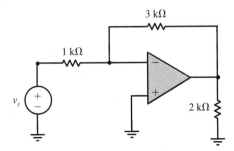

Fig. P4.36

4.37 For the op-amp circuit shown in Fig. P4.37, find the average power absorbed by each element for the case that $v_s(t) = \cos \omega t$ V.

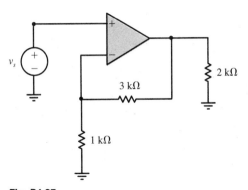

Fig. P4.37

4.38 Find the rms value of each function given in Fig. P4.38. (See p. 260.)

4.39 Find the rms value of the "half-wave rectified" sine wave that is shown in Fig. P4.39. [*Hint:* $\sin^2 x = \frac{1}{2}(1 - \cos 2x)$.]

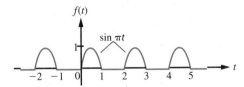

Fig. P4.39

4.40 Find the rms value of the "full-wave rectified" sine wave that is shown in Fig. P4.40. [*Hint:* $\sin^2 x = \frac{1}{2}(1 - \cos 2x)$.]

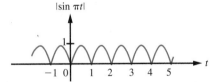

Fig. P4.40

4.41 The load shown in Fig. P4.41 operates at 60 Hz. (a) What are the pf and the pf angle of this load? (b) Is the pf leading or lagging? (c) To what value should the capacitor be changed to get a unity pf (pf = 1)?

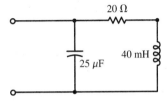

Fig. P4.41

4.42 A 115-V rms, 60-Hz electric hair dryer absorbs 500 W at a lagging pf of 0.95. What is the rms value of the current drawn by this dryer?

4.43 An electric motor which operates at 220 V rms, 20 A rms, 60 Hz, absorbs 2200 W. (a) What is the pf of the motor. (b) For the case that the pf is lagging, what value capacitor should be connected in parallel with the motor such that the resulting combination has a unity pf (pf = 1)?

4.44 An electric motor operating at 220 V rms, 60 Hz, draws a current of 20 A rms at a pf of 0.75 lagging. (a) What is the average power absorbed by the motor? (b) What value capacitor should be connected in parallel with the motor such that the resulting combination has a unity pf (pf = 1)?

4.45 Two loads, which are connected in parallel, operate at 230 V rms. One load absorbs 500 W at a pf of 0.8 lagging, and the other absorbs 1000 W at

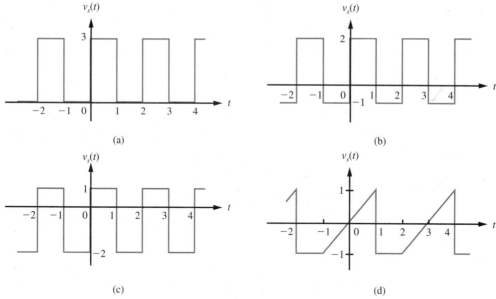

$v_s(t)$

(a)

$v_s(t)$

(b)

$v_s(t)$

(c)

$v_s(t)$

(d)

Fig. P4.38

a pf of 0.9 lagging. Find the pf of the combined load. Is this pf leading or lagging?

4.46 Three loads, which are connected in parallel, operate at 230 V rms. One load absorbs 500 W at a pf of 0.8 lagging. The second absorbs 1000 W at a pf of 0.9 lagging. The third absorbs 1500 W at a pf of 0.9 leading. Find the pf of the combined load. Is this pf leading or lagging?

4.47 The parallel connection of two 115-V rms loads absorbs 2000 W at a lagging pf of 0.95. Suppose that one load absorbs 1200 W at a pf of 0.8 lagging. What are the power absorbed and the pf of the second load?

4.48 A load, which operates at 220 V rms, draws 5 A rms at a lagging pf of 0.95. (a) Find the complex power absorbed by the load. (b) Find the average power absorbed by the load. (c) Find the reactive power absorbed by the load. (d) Find the apparent power absorbed by the load. (e) Find the impedance of the load.

4.49 Consider the circuit shown in Fig. P4.28. Suppose that $v_s(t) = 12\sqrt{2} \cos 3t$ V and $C = \frac{1}{6}$ F.

Find the complex power absorbed by each element. Is complex power conserved?

4.50 Consider the circuit shown in Fig. P4.28. Suppose that $v_s(t) = 12\sqrt{2} \cos 3t$ V and $C = \frac{1}{6}$ F. Find the apparent power absorbed by each element. Is apparent power conserved?

4.51 Consider the circuit shown in Fig. P4.28. Suppose that $v_s(t) = 12\sqrt{2} \cos 3t$ V and $C = \frac{1}{6}$ F. Find the reactive power absorbed by each element. Is reactive power conserved?

4.52 For the circuit given in Fig. P4.24, when $V_{s1} = 250\sqrt{2}\underline{/-30°}$ V, $V_{s2} = 250\sqrt{2}\underline{/-90°}$ V, and $Z = 78 - j45$ Ω, then $I_1 = 6.8\underline{/30°}$ A and $I_2 = 6.8\underline{/-90°}$ A. (a) Find the complex power absorbed by each impedance. (b) Find the complex power supplied by each source.

4.53 For the circuit given in Fig. P4.24, when $V_{s1} = 250\sqrt{2}\underline{/-30°}$ V, $V_{s2} = 250\sqrt{2}\underline{/-90°}$ V, and $Z = 78 - j45$ Ω, then $I_1 = 6.8\underline{/30°}$ A and $I_2 = 6.8\underline{/-90°}$ A. (a) Find the apparent power absorbed by each impedance. (b) Find the apparent power supplied by each source.

4.54 For the circuit given in Fig. P4.24, when $\mathbf{V}_{s1} = 250\sqrt{2}\underline{/-30°}$ V, $\mathbf{V}_{s2} = 250\sqrt{2}\underline{/-90°}$ V, and $\mathbf{Z} = 78 - j45$ Ω, then $\mathbf{I}_1 = 6.8\underline{/30°}$ A and $\mathbf{I}_2 = 6.8\underline{/-90°}$ A. (a) Find the reactive power absorbed by each impedance. (b) Find the reactive power supplied by each source.

4.55 An R-ohm resistor has the voltage $v(t) = V\cos(\omega t + \phi_1)$ across it and it has the current $i(t) = I\cos(\omega t + \phi_2)$ through it. Show that the complex power absorbed by the resistor is given by

$$\mathbf{S}_R = \tfrac{1}{2}RI^2 = \tfrac{1}{2}V^2/R$$

4.56 An L-henry inductor has the voltage $v(t) = V\cos(\omega t + \phi_1)$ across it and it has the current $i(t) = I\cos(\omega t + \phi_2)$ through it. Show that the complex power absorbed by the inductor is given by

$$\mathbf{S}_L = \frac{j\omega LI^2}{2} = \frac{jV^2}{2\omega L}$$

4.57 An C-farad capacitor has the voltage $v(t) = V\cos(\omega t + \phi_1)$ across it and it has the current $i(t) = I\cos(\omega t + \phi_2)$ through it. Show that the complex power absorbed by the capacitor is given by

$$\mathbf{S}_C = \frac{-jI^2}{2\omega C} = \frac{-j\omega CV^2}{2}$$

4.58 For the single-phase, three-wire circuit shown in Fig. P4.58, suppose that $\mathbf{V}_s = 120\underline{/0°}$ V rms. Find the average power supplied by each source if $\mathbf{Z}_1 = 60$ Ω, $\mathbf{Z}_2 = 80$ Ω, $\mathbf{Z}_3 = 40$ Ω, and $R_g = R_n = 0$ Ω.

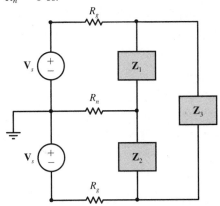

Fig. P4.58

4.59 For the single-phase, three-wire circuit shown in Fig. P4.58, suppose that $\mathbf{V}_s = 115\underline{/0°}$ V rms. Find the average power supplied by each source if $\mathbf{Z}_1 = 60$ Ω, $\mathbf{Z}_2 = 80$ Ω, $\mathbf{Z}_3 = 40$ Ω, $R_g = 1$ Ω, and $R_n = 2$ Ω.

4.60 For the single-phase, three-wire circuit shown in Fig. P4.58, suppose that $R_g = R_n = 0$ Ω. For the case that $\mathbf{Z}_1$ absorbs 500 W at a lagging pf of 0.8, $\mathbf{Z}_2$ absorbs 1000 W at a lagging pf of 0.9, and $\mathbf{Z}_3$ absorbs 1500 W at a leading pf of 0.95, find the average power supplied by each source.

4.61 A balanced Y-Y three-phase circuit has 130-V rms phase voltages and a per-phase impedance of $\mathbf{Z} = 12 + j12$ Ω. Find the line currents and the total power absorbed by the load.

4.62 A balanced Y-Y three-phase circuit has 210-V rms, 60-Hz line voltages. Suppose that the load absorbs a total of 3 kW of power at a lagging pf of 0.85. (a) Find the per-phase impedance. (b) What value capacitors should be connected in parallel with the per-phase impedances to result in a unity pf (pf = 1)?

4.63 A balanced, three-phase Y-connected source, whose phase voltages are 115 V rms, has the unbalanced Y-connected load $\mathbf{Z}_{AN} = 3 + j4$ Ω, $\mathbf{Z}_{BN} = 10$ Ω, and $\mathbf{Z}_{CN} = 5 + j12$ Ω. Find the line currents and the total power absorbed by the load for the case that there is a neutral wire.

4.64 A balanced, three-phase Y-connected source, whose phase voltages are 120 V rms, has the unbalanced Y-connected load $\mathbf{Z}_{AN} = 10$ Ω, $\mathbf{Z}_{BN} = 20$ Ω, and $\mathbf{Z}_{CN} = 60$ Ω. Find the line currents and the total power absorbed by the load for the case that there is no neutral wire.

4.65 Suppose that the balanced Y-Δ three-phase circuit shown in Fig. 4.40 on p. 241 has a line voltage of 130 V rms and $\mathbf{Z} = 4\sqrt{2}\underline{/45°}$ Ω. Find the line currents and the total power absorbed by the load.

4.66 A balanced, three-phase Y-connected source with 230-V rms line voltages has an unbalanced Δ-connected load whose impedances are $\mathbf{Z}_{AB} = 8$ Ω, $\mathbf{Z}_{BC} = 4 + j3$ Ω, and $\mathbf{Z}_{AC} = 12 - j5$ Ω. Find the line currents and the total power absorbed by the load.

4.67 Suppose that the balanced Y-Δ three-phase circuit shown in Fig. 4.40 on p. 241 has 210-V rms, 60-Hz line voltages. If the load absorbs a total of 3 kW of power at a lagging pf of 0.85, (a) find the per-phase impedance, and (b) what value capacitors should be connected in parallel with the per-phase impedances to result in a unity pf (pf = 1)?

4.68 A Δ-connected load has impedances $Z_{AB} = 2$ Ω, $Z_{BC} = 3$ Ω, and $Z_{AC} = 5$ Ω. Find the equivalent Y-connected load.

4.69 A Δ-connected load has impedances $Z_{AB} = 2 + j2$ Ω, $Z_{BC} = 3 - j3$ Ω, and $Z_{AC} = 1 + j1$ Ω. Find the equivalent Y-connected load.

4.70 A Y-connected load has impedances $Z_A = 1$ Ω, $Z_B = 0.6$ Ω, and $Z_C = 1.5$ Ω. Find the equivalent Δ-connected load.

4.71 A Y-connected load has impedances $Z_A = j2/3$ Ω, $Z_B = 2$ Ω, and $Z_C = 1$ Ω. Find the equivalent Δ-connected load.

4.72 A Y-connected load has admittances $Y_A = 2\,\mho$, $Y_B = 3\,\mho$, and $Y_C = 5\,\mho$. Find the equivalent Δ-connected load.

4.73 A Y-connected load has admittances $Y_A = 2 + j2\,\mho$, $Y_B = 3 - j3\,\mho$, and $Y_C = 1 + j1\,\mho$. Find the equivalent Δ-connected load.

4.74 For the load shown in Fig. 4.42a on p. 247, $Z_Y = 3 + j4$ Ω and $Z_\Delta = 5 - j12$ Ω. Find the pf and the total power absorbed by the load for the case that the line voltage is 120 V rms.

4.75 A balanced, three-phase Δ-connected load has a per-phase impedance of $Z = 36 + j36$ Ω. Find the readings for the two-wattmeter connection shown in Fig. 4.45 on p. 250 for the case that the line voltage of $130\sqrt{3}$ V rms is produced by a balanced Y-connected source.

4.76 A balanced Y-Y three-phase circuit has a line voltage of 210 V rms, a line current of 9.72 A rms,

and a pf of 0.95 lagging. Find the wattmeter readings for the two-wattmeter method.

4.77 An unbalanced Y-connected load has rms line voltages $V_{ab} = 208\underline{/30°}$ V, $V_{bc} = 208\underline{/-90°}$ V, $V_{ca} = 208\underline{/-210°}$ V, rms line currents $I_{aA} = 7.49\underline{/16.1°}$ A, $I_{bB} = 6.82\underline{/-142°}$ A, $I_{cC} = 2.75\underline{/131°}$ A. Find the wattmeter readings for the two-wattmeter method.

4.78 A balanced Y-Δ three-phase circuit has a line voltage of 225 V rms, a line current of 39.8 A rms, and a pf of 0.707 lagging. Find the wattmeter readings for the two-wattmeter method.

4.79 A balanced Y-Δ three-phase circuit has a line voltage of 210 V rms, a line current of 9.7 A rms, and a pf of 0.85 lagging. Find the wattmeter readings for the two-wattmeter method.

4.80 An unbalanced Δ-connected load has rms line voltages $V_{ab} = 230\underline{/30°}$ V, $V_{bc} = 230\underline{/-90°}$ V, $V_{ca} = 230\underline{/-210°}$ V, rms line currents $I_{aA} = 44.2\underline{/16°}$ A, $I_{bB} = 73.3\underline{/-136°}$ A, $I_{cC} = 40.4\underline{/75.7°}$ A. Find the wattmeter readings for the two-wattmeter method.

4.81 The unbalanced Δ-connected load shown in Fig. 4.45 on p. 250 has rms line voltages $V_{ab} = 220\underline{/30°}$ V, $V_{bc} = 220\underline{/-90°}$ V, $V_{ca} = 220\underline{/-210°}$ V, rms phase currents $I_{AB} = 11\underline{/30°}$ A, $I_{BC} = 5.5\underline{/-135°}$ A, $I_{CA} = 22\underline{/60°}$ A. (a) Find the impedances Z_{AB}, Z_{BC}, and Z_{AC}. (b) Find the power absorbed by each impedance.

4.82 The unbalanced Δ-connected load shown in Fig. 4.45 on p. 250 has rms line voltages $V_{ab} = 220\underline{/30°}$ V, $V_{bc} = 220\underline{/-90°}$ V, $V_{ca} = 220\underline{/-210°}$ V, rms phase currents $I_{AB} = 11\underline{/30°}$ A, $I_{BC} = 5.5\underline{/-135°}$ A, $I_{CA} = 22\underline{/60°}$ A. Find the wattmeter readings for the two-wattmeter method.

4.83 The unbalanced Δ-connected load shown in Fig. 4.45 on p. 250 has impedances $Z_{AB} = 12 - j5$ Ω, $Z_{BC} = 26$ Ω, $Z_{AC} = 8 - j6$ Ω, and line voltages of 130 V rms that are produced by a balanced Y-connected source. Find the wattmeter readings for the two-wattmeter method.

4.84 Find the wattmeter readings for the circuit shown in Fig. P4.84.

4.85 Find the wattmeter readings for the circuit shown in Fig. P4.85.

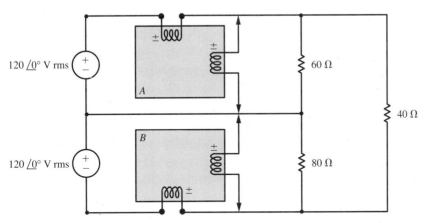

Fig. P4.84

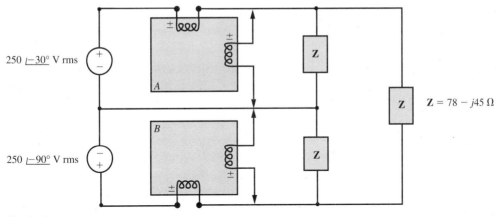

Fig. P4.85

4.86 Find the wattmeter readings for the circuit shown in Fig. P4.86.

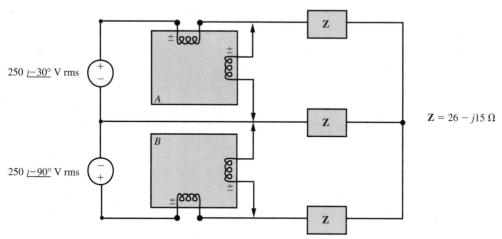

250 $\underline{/-30°}$ V rms

250 $\underline{/-90°}$ V rms

A

B

$\mathbf{Z} = 26 - j15 \ \Omega$

Fig. P4.86

5

Important Circuit and System Concepts

INTRODUCTION

As the impedances of capacitors and inductors are frequency dependent, so is the impedance of an arbitrary *RLC* connection. This dependence, which can be depicted graphically by plots of magnitude and phase angle versus frequency, constitutes the frequency response of the impedance. Impedance, which is the ratio of voltage to current (phasors), is only one network function whose frequency response can be displayed. It is the frequency response of a circuit that describes the circuit's behavior to all sinusoidal forcing functions, and nonsinusoidal forcing functions that are comprised of sinusoids.

An important frequency is that for which a network function reaches a maximum value. In certain basic circuits, this occurs when an impedance or admittance is purely real—a condition known as resonance. In such a situation, we can quantitatively describe the frequency selectivity of the circuit with a figure of merit known as the quality factor. We can also talk about the notion of bandwidth.

Just as the concept of complex sinusoids allowed us to treat ac circuits in a relatively simple manner, the concept of complex frequency does a similar job for circuits that have damped-sinusoidal excitations. This generalization not only allows for the direct determination of forced responses due to sinusoidal and damped sinusoidal forcing functions, but indicates the form of the natural response as well.

A **system** is an interconnection of components or devices that perform some objective. The components of a system can be electrical, mechanical, thermal, chemical, et cetera. Electrical systems are commonly used for communications, computation, control, and instrumentation. An electric circuit is an example of an electrical system, in particular, a linear circuit is an example of a **linear system**—which is a system that is described by a linear differential equation.

The description of a linear circuit or system resulting from a damped-sinusoidal excitation may first appear to be of limited usefulness. However, with the use of a very important mathematical tool—the Laplace transform—such circuit and system descriptions can be used to obtain complete responses to many types of common excitations. By using Laplace transforms, functions of time are transformed to functions of frequency. Consequently, frequency-domain analysis can be performed, thereby eliminating the necessity to solve differential equations.

5.1 Frequency Response

For the parallel connection of a resistor R and capacitor C shown in Fig. 5.1, the impedance $\mathbf{Z}$ of the combination is given by

$$\mathbf{Z} = \frac{\mathbf{Z}_R \mathbf{Z}_C}{\mathbf{Z}_R + \mathbf{Z}_C} = \frac{R\left(\dfrac{1}{j\omega C}\right)}{R + \dfrac{1}{j\omega C}} = \frac{R}{1 + j\omega RC}$$

$$= \frac{R}{\sqrt{1 + (\omega RC)^2}} \underline{/-\tan^{-1}(\omega RC)} = |\mathbf{Z}|\underline{/\theta}$$

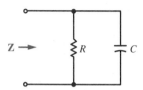

Fig. 5.1 Parallel *RC* impedance.

Since $\mathbf{Z}$ is a complex function—the notation $\mathbf{Z}(j\omega)$ is often used—and, therefore, can be expressed in terms of a magnitude and an angle as $\mathbf{Z} = |\mathbf{Z}|\underline{/\theta}$, both the magnitude $|\mathbf{Z}|$ and the angle θ are, in general, functions of ω. We, therefore, may plot $|\mathbf{Z}|$ versus ω and θ versus ω. In obtaining the former, note that when $\omega = 0$ rad/s, the magnitude is $|\mathbf{Z}| = R$, whereas when $\omega = 1/RC$, then $|\mathbf{Z}| = R/\sqrt{2} \approx 0.707R$. Also, when $\omega \to \infty$, then $|\mathbf{Z}| \to 0$. Thus we have the plot shown in Fig. 5.2, which we call the **amplitude response** of $\mathbf{Z}$. The plot of θ versus ω is known as

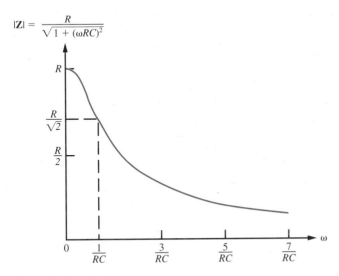

Fig. 5.2 Amplitude response.

the **phase response** of $\mathbf{Z}$, and for this impedance, it is shown in Fig. 5.3. In this plot, we see that when $\omega = 0$ rad/s, then $\theta = -\tan^{-1} \omega RC = 0°$, and when $\omega = 1/RC$, then $\theta = -\tan^{-1} \omega RC = -45°$. Furthermore, when $\omega \to \infty$, then $\theta \to -90°$. The amplitude and phase responses comprise the **frequency response** of the impe-

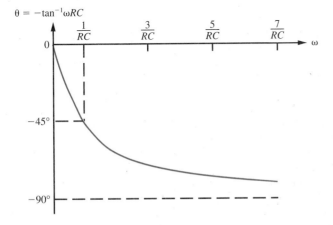

Fig. 5.3 Phase response.

dance $\mathbf{Z}$. Although a frequency response consists of both an amplitude response and a phase response, quite often the amplitude response alone is referred to as the frequency response.

For the circuit shown in Fig. 5.4, $\mathbf{V} = \mathbf{Z}\mathbf{I}$ where $\mathbf{Z}$ is the impedance of the parallel RC combination. From complex-number arithmetic,

$$|\mathbf{V}| = |\mathbf{Z}\mathbf{I}| = |\mathbf{Z}||\mathbf{I}|$$

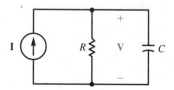

Fig. 5.4 Current applied to parallel RC impedance.

Given the current $\mathbf{I}$, then $|\mathbf{V}|$ is maximum when $|\mathbf{Z}|$ is maximum. This occurs when $\omega = 0$ rad/s, where $|\mathbf{Z}| = R$. For this case, $|\mathbf{V}| = R|\mathbf{I}|$, and the average power absorbed by the resistor is

$$P_0 = \frac{1}{2}\frac{|\mathbf{V}|^2}{R} = \frac{1}{2}\frac{R^2|\mathbf{I}|^2}{R} = \frac{1}{2}R|\mathbf{I}|^2$$

However, when $\omega = 1/RC = \omega_c$, then $|\mathbf{Z}| = R/\sqrt{2}$. For this case, the power absorbed by the resistor is

$$P_1 = \frac{1}{2}\frac{(R/\sqrt{2})^2|\mathbf{I}|^2}{R} = \frac{1}{2}\left(\frac{R|\mathbf{I}|^2}{2}\right) = \frac{1}{2}P_0$$

That is, P_1 is half of the maximum power P_0. For this reason, we say that $\omega_c = 1/RC$ is the **half-power frequency** or **cutoff frequency**.

In general, given an amplitude response whose maximum value is M, the half-power frequencies are those frequencies for which the magnitude of the amplitude response is $M/\sqrt{2}$.

Example 5.1

Let us sketch the amplitude response for the RC circuit shown in Fig. 5.5.

Since the input voltage (phasor) is $\mathbf{V}_1$ and the output voltage is $\mathbf{V}_2$, by voltage division,

$$\mathbf{V}_2 = \frac{R}{R + (1/j\omega C)}\mathbf{V}_1 \quad \Rightarrow \quad \frac{\mathbf{V}_2}{\mathbf{V}_1} = \frac{j\omega RC}{1 + j\omega RC}$$

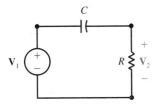

Fig. 5.5 *RC* circuit.

Denoting the ratio of output voltage to input voltage by $\mathbf{H}(j\omega)$, then

$$\left|\mathbf{H}(j\omega)\right| = \left|\frac{\mathbf{V}_2}{\mathbf{V}_1}\right| = \frac{|\mathbf{V}_2|}{|\mathbf{V}_1|} = \frac{\omega RC}{\sqrt{1 + (\omega RC)^2}}$$

For $\omega = 0$ rad/s, we have that $\left|\mathbf{H}(j\omega)\right| = \left|\mathbf{H}(j0)\right| = 0$. To determine the value of $\left|\mathbf{H}(j\omega)\right|$ as $\omega \to \infty$, note that

$$\left|\mathbf{H}(j\omega)\right| = \sqrt{\frac{(\omega RC)^2}{1 + (\omega RC)^2}} = \frac{1}{\sqrt{1 + 1/(\omega RC)^2}}$$

Thus $\left|\mathbf{H}(j\omega)\right| \to 1$ as $\omega \to \infty$. Also, note that for $\omega = 1/RC$, $\left|\mathbf{H}(j\omega)\right| = 1/\sqrt{2}$. The amplitude response of the function $\mathbf{H}(j\omega)$—called a **system function** or a **transfer function**—is shown in Fig. 5.6. Clearly, $\omega_c = 1/RC$ is the half-power frequency.

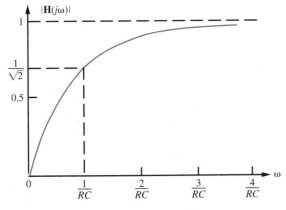

Fig. 5.6 Amplitude response of *RC* circuit.

Drill Exercise 5.1

For the circuit shown in Fig. 5.5, replace the capacitor C with an inductor L. (a) Find an expression for $|\mathbf{H}(j\omega)| = |\mathbf{V}_2/\mathbf{V}_1|$, (b) sketch the amplitude response, and (c) determine the half-power frequency.

ANSWER (a) $1/\sqrt{1 + (\omega L/R)^2}$,
(b) $|\mathbf{H}(j0)| = 1$, $|\mathbf{H}(jR/L)| = 1/\sqrt{2}$, $|\mathbf{H}(j\infty)| = 0$, (c) R/L

It is possible to intuitively obtain a rough sketch of the amplitude response for the circuit shown in Fig. 5.5 without writing an analytical expression for the transfer function $\mathbf{H}(j\omega)$. In particular, for $\omega = 0$ rad/s—the dc case—the capacitor behaves as an open circuit, so $\mathbf{V}_2 = 0$, and consequently, $|\mathbf{H}(j\omega)| = 0$. For small values of ω, the impedance of the capacitor is still large relative to R, so by voltage division $\mathbf{V}_2$ is relatively small, as is $|\mathbf{H}(j\omega)|$. However, as frequency (ω) increases, the impedance of the capacitor decreases, so that the values of $\mathbf{V}_2$ and $|\mathbf{H}(j\omega)|$ increase. In the limit, as frequency becomes infinite, the impedance of the capacitor becomes zero, so $\mathbf{V}_2 \rightarrow \mathbf{V}_1 \Rightarrow |\mathbf{H}(j\omega)| \rightarrow 1$.

From the amplitude response of $\mathbf{H}(j\omega)$ shown in Fig. 5.5, we see that if the input voltage is a sinusoid, then the amplitude of the output sinusoid will be relatively small for low frequencies (i.e., frequencies well below ω_c). Conversely, for high frequencies (i.e., frequencies well above ω_c), the amplitude of the output sinusoid approximately equals the amplitude of the input sinusoid. In other words, the high frequencies are "passed" by the RC circuit, whereas the low frequencies are "stopped." For this reason, we say that the circuit is an example of a simple **high-pass filter**. The amplitude response shown in Fig. 5.7a is indicative of a **low-pass filter**, and the one depicted in Fig. 5.7b describes a **bandpass filter**. From Fig. 5.7a

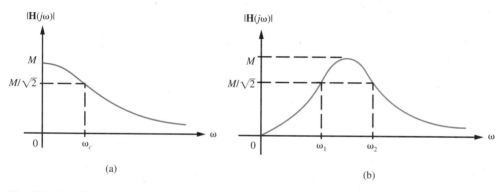

Fig. 5.7 Amplitude responses of (a) low-pass filter, (b) bandpass filter.

we see that the low-pass filter has a single half-power frequency, whereas from Fig. 5.7b we see that the bandpass filter has two half-power frequencies.

Bode Plots

The magnitude of a voltage transfer function, that is, $|\mathbf{H}(j\omega)| = |\mathbf{V}_2/\mathbf{V}_1|$, which in general is a function of frequency, is also referred to as **voltage gain**. Although we can, as discussed above, directly plot gain versus frequency in order to obtain the amplitude response, in practice it is convenient to use logarithmic scales rather than linear scales for gain and frequency. As a consequence of this, it is often possible to obtain an approximate frequency response without having to perform many calculations.

Suppose that P_1 and P_2 are the input and output (load) powers, respectively, of a circuit or system. We define the power gain to be $\log_{10} P_2/P_1$, where the unit is the **bel** (B), named for Alexander Graham Bell (1847–1922). Since gain in bels is often a relatively small number in practice, the accepted convention is to multiply the value by 10 and express gain in **decibels** (dB). In other words,

$$\text{gain} = 10 \log_{10} \frac{P_2}{P_1} \text{ dB}$$

If the input resistance is R_1 and the load resistance is R_2, then $P_1 = \frac{1}{2}V_1^2/R_1$ and $P_2 = \frac{1}{2}V_2^2/R_2$, where V_1 and V_2 are, respectively, the input and output voltage magnitudes. For the case that $R_1 = R_2$, we have that

$$\text{gain} = 10 \log_{10} \frac{V_2^2}{V_1^2} = 20 \log_{10} \frac{V_2}{V_1} \text{ dB}$$

A plot of gain in decibels versus frequency on a logarithmic scale (the amplitude response), and a plot of the angle (on a linear scale) versus frequency, again on a logarithmic scale (the phase response), constitute a **Bode plot** or **diagram**.[1]

Example 5.2

Let us sketch the Bode plot for the low-pass filter shown in Fig. 5.8. Since

$$\frac{\mathbf{V}_2}{\mathbf{V}_1} = \frac{1/j\omega C}{(1/j\omega C) + R} = \frac{1}{1 + j\omega RC} = \frac{1}{\sqrt{1 + (\omega RC)^2}} \underline{/-\tan^{-1} \omega RC}$$

---

[1]Named for the American electrical engineer Hendrik Bode.

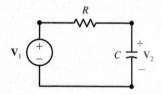

Fig. 5.8 Low-pass filter.

the gain in decibels is given by

$$20 \log_{10}\left|\frac{\mathbf{V}_2}{\mathbf{V}_1}\right| = 20 \log_{10}\frac{1}{\sqrt{1 + (\omega RC)^2}} = -20 \log_{10}\sqrt{1 + (\omega RC)^2}$$

For low frequencies ($\omega \ll 1/RC$), the gain is approximately constant and approximately equal to $-20 \log_{10}\sqrt{1} = 0$ dB. For high frequencies ($\omega \gg 1/RC$), the gain is approximately given by $-20 \log_{10} \omega RC$ dB.

The two straight lines 0 and $-20 \log_{10} \omega RC$ are called **asymptotes**—the former is the low-frequency asymptote and the latter is the high-frequency asymptote—of the amplitude response. These asymptotes are indicated in Fig. 5.9.

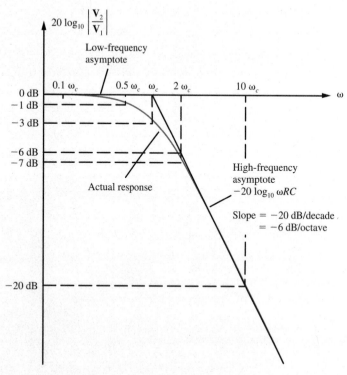

Fig. 5.9 Amplitude response of low-pass filter.

The frequency at which the asymptotes intersect is called the **corner frequency** or **break frequency**. In this case,

$$-20 \log_{10} \omega RC = 0 \quad \Rightarrow \quad \omega = \frac{1}{RC}$$

and this is the corner frequency.

As indicated in Fig. 5.9, a plot of $-20 \log_{10} \omega RC$ versus ω on a logarithmic scale is a straight line. When $\omega = 1/RC$, then $-20 \log_{10} \omega RC = 0$; whereas when $\omega = 10/RC$, then $-20 \log_{10} \omega RC = -20 \log_{10} 10 = -20$ dB. Since a change in frequency by a factor of 10 is referred to as a **decade**, the straight line $-20 \log_{10} \omega RC$ has a slope of -20 dB/decade. Alternatively, when $\omega = 2/RC$, then $-20 \log_{10} \omega RC = -20 \log_{10} 2 = -6.02 \approx -6$ dB. Since a change in frequency by a factor of 2 is referred to as an **octave**, we can also say that the straight line $-20 \log_{10} \omega RC$ has a slope of -6 dB/octave.

The low-frequency and high-frequency asymptotes are good approximations of the actual amplitude response for low frequencies ($\omega << 1/RC$) and high frequencies ($\omega >> 1/RC$), respectively. However, at the corner frequency $\omega = 1/RC$,

$$20 \log_{10} |V_2/V_1| = -20 \log_{10} \sqrt{1 + (\omega RC)^2}$$
$$= -20 \log_{10} \sqrt{2} = -3.01 \approx -3 \text{ dB}$$

Thus the actual amplitude response is 3 dB down from the intersection of the asymptotes. At $\omega = 0.5/RC$, that is, at one octave below the corner frequency,

$$20 \log_{10} \left| \frac{V_2}{V_1} \right| = -20 \log_{10} \sqrt{1 + (\omega RC)^2}$$
$$= -20 \log_{10} \sqrt{5/4} = -0.969 \approx -1 \text{ dB}$$

whereas at $\omega = 2/RC$, that is, at one octave above the corner frequency,

$$20 \log_{10} \left| \frac{V_2}{V_1} \right| = -20 \log_{10} \sqrt{1 + (\omega RC)^2}$$
$$= -20 \log_{10} \sqrt{5} = -6.99 \approx -7 \text{ dB}$$

Therefore, one octave below and above the corner frequency, the actual amplitude response is 1 dB below the approximate amplitude response formed by the asymptotes. One decade below and above the corner frequency, the actual amplitude response is only 0.043 dB below the asymptotes. (Verify this fact.)

For the given transfer function, note that for the corner frequency $\omega = 1/RC$, we have

$$\left|\frac{\mathbf{V}_2}{\mathbf{V}_1}\right| = \frac{1}{\sqrt{1 + (\omega RC)^2}} = \frac{1}{\sqrt{2}}$$

Since the maximum value of $|\mathbf{V}_2/\mathbf{V}_1|$ occurs when $\omega = 0$ rad/s and is $|\mathbf{V}_2/\mathbf{V}_1| = 1$, we see that the corner frequency $\omega = 1/RC$ is also the half-power frequency.

Since $\text{ang}(\mathbf{V}_2/\mathbf{V}_1) = -\tan^{-1} \omega RC$, when $\omega \ll 1/RC$, then $\text{ang}(\mathbf{V}_2/\mathbf{V}_1) \approx 0°$. Furthermore, when $\omega \gg 1/RC$, then $\text{ang}(\mathbf{V}_2/\mathbf{V}_1) \approx -90°$. In addition, when $\omega = 1/RC$, then $\text{ang}(\mathbf{V}_2/\mathbf{V}_1) = -\tan^{-1} 1 = -45°$. Therefore, let us approximate the phase response of $\mathbf{V}_2/\mathbf{V}_1$ by using three straight-line segments as shown in Fig. 5.10.

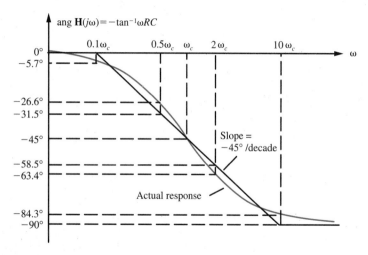

Fig. 5.10 Phase response.

The actual phase response is also indicated in Fig. 5.10. In this case, at the corner frequency, the approximation gives the actual value of the phase (-45 degrees). However, one decade below the corner frequency at $\omega = 0.1/RC$, the actual phase is

$$\text{ang}\left(\frac{\mathbf{V}_2}{\mathbf{V}_1}\right) = -\tan^{-1}\left(\frac{0.1}{RC}\right)RC = -5.7°$$

whereas one decade above the corner frequency at $\omega = 10/RC$, the actual phase is

$$\text{ang}\left(\frac{\mathbf{V}_2}{\mathbf{V}_1}\right) = -\tan^{-1}\left(\frac{10}{RC}\right)RC = -84.3°$$

Drill Exercise 5.2

For the circuit given in Fig. 5.8, replace the capacitor C with an inductor L. Sketch the amplitude-response portion of the Bode plot for V_2/V_1. (a) What is the expression for the low-frequency asymptote? (b) What is the expression for the high-frequency asymptote? (c) What is the corner frequency?

ANSWER   (a) $20 \log_{10} \omega L/R$, (b) 0, (c) R/L

A major motivation for using Bode plots is the fact that the response of a product of transfer functions is equal to the sum of the individual transfer-function responses. Suppose that we can express a transfer function $H(j\omega)$ as the product of two transfer functions $H_1(j\omega)$ and $H_2(j\omega)$. In other words, suppose that

$$H(j\omega) = H_1(j\omega)H_2(j\omega)$$

Since $|H(j\omega)| = |H_1(j\omega)H_2(j\omega)| = |H_1(j\omega)||H_2(j\omega)|$, then

$$20 \log_{10}|H(j\omega)| = 20 \log_{10}|H_1(j\omega)| + 20 \log_{10}|H_2(j\omega)|$$

Therefore, the addition of the amplitude responses for $H_1(j\omega)$ and $H_2(j\omega)$ results in the amplitude response for $H(j\omega)$. Furthermore, since

$$\text{ang}[H(j\omega)] = \text{ang}[H_1(j\omega)H_2(j\omega)] = \text{ang}[H_1(j\omega)] + \text{ang}[H_2(j\omega)]$$

then the addition of the phase responses for $H_1(j\omega)$ and $H_2(j\omega)$ results in the phase response for $H(j\omega)$.

Applications of these important properties are left as a series of exercises for the reader (see Problems 5.12 to 5.19 at the end of this chapter).

5.2 Resonance

For the parallel RLC circuit shown in Fig. 5.11, the admittance loading the current source is $Y = I/V$, where

$$\mathbf{Y} = \mathbf{Y}_R + \mathbf{Y}_L + \mathbf{Y}_C = \frac{1}{R} + \frac{1}{j\omega L} + j\omega C = \frac{1}{R} + j\left(\omega C - \frac{1}{\omega L}\right)$$

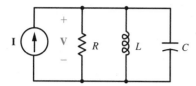

Fig. 5.11 Parallel *RLC* circuit.

From this expression, we can see that there is some frequency ω_r for which the imaginary part of **Y** will be zero, that is,

$$\omega_r C - \frac{1}{\omega_r L} = 0 \quad \Rightarrow \quad \omega_r = \frac{1}{\sqrt{LC}}$$

We say that a circuit with at least one capacitor and one inductor is in **resonance**, or is **resonant**, when the imaginary part of its admittance (or impedance) is equal to zero. The circuit given in Fig. 5.11 is in resonance when $\omega = 1/\sqrt{LC}$, and this, therefore, is the **resonance frequency** ω_r.

At the resonance frequency, $\mathbf{Y} = 1/R$, and therefore, $|\mathbf{Y}|$ is minimum. Thus, for a given **I**, since

$$|\mathbf{V}| = \left| \frac{\mathbf{I}}{\mathbf{Y}} \right| = \frac{|\mathbf{I}|}{|\mathbf{Y}|}$$

then $|\mathbf{V}|$ is maximum. Hence, the maximum value for $|\mathbf{V}|$ is

$$|\mathbf{V}| = \frac{|\mathbf{I}|}{1/R} = R|\mathbf{I}|$$

For $\omega = 0$ rad/s, the impedance of the inductor is zero; thus, so is the voltage **V**. As $\omega \to \infty$, the impedance of the capacitor $1/j\omega C \to 0$, and again $\mathbf{V} \to 0$. A sketch of $|\mathbf{V}|$ versus ω is shown in Fig. 5.12 where ω_1 and ω_2 are the "lower" and "upper" half-power frequencies respectively.

Since, at resonance, the parallel *RLC* combination acts simply as the resistance R, the parallel *LC* combination—known as a **tank circuit**—behaves as an open circuit.

We define the **bandwidth** (*BW*) of the resonant circuit in terms of the half-power frequencies by

$$BW = \omega_2 - \omega_1$$

Note that the smaller the bandwidth *BW* is, the "sharper" or "narrower" is the amplitude response.

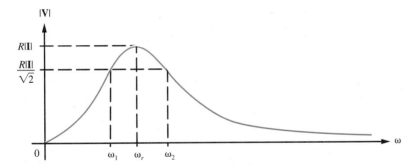

Fig. 5.12 Voltage magnitude versus frequency.

Quality Factor

Another conventional quantity that describes the sharpness of an amplitude response is the **quality factor**, designated Q, which is defined at the resonance frequency to be

$$Q = 2\pi \left(\frac{\text{maximum energy stored}}{\text{total energy lost in a period}} \right)$$

Since energy is stored in capacitors and inductors, the maximum energy stored is designated $[w_C(t) + w_L(t)]_{\text{max}}$. Since energy is dissipated by resistors, the energy lost in a period T is $P_R T$, where P_R is the total power absorbed by all the resistors. Therefore,

$$Q = \frac{2\pi[w_C(t) + w_L(t)]_{\text{max}}}{P_R T}$$

Now let us determine the quality factor for the parallel RLC circuit given in Fig. 5.11. Assume that the current source is given by $i(t) = I \cos \omega_r t$ A, where $\omega_r = 1/\sqrt{LC}$. At resonance, $\mathbf{Y} = 1/R$, so

$$v(t) = Ri(t) = RI \cos \omega_r t \text{ V}$$

Then the energy stored in the capacitor is

$$w_C(t) = \tfrac{1}{2}Cv^2(t) = \tfrac{1}{2}C(RI \cos \omega_r t)^2 = \tfrac{1}{2}CR^2I^2 \cos^2 \omega_r t$$

The (phasor) inductor current $\mathbf{I}_L$ (directed down) is $\mathbf{I}_L = \mathbf{V}/j\omega L$. At resonance, since $v(t) = RI \cos \omega_r t$, then $\mathbf{V} = RI\underline{/0°}$, and therefore,

$$\mathbf{I}_L = \frac{RI\underline{/0°}}{\omega_r L \underline{/90°}} = \frac{RI}{\omega_r L}\underline{/-90°} \quad \Rightarrow \quad i_L(t) = \frac{RI}{\omega_r L}\cos(\omega_r t - 90°)$$

$$= \frac{RI}{\omega_r L}\sin \omega_r t$$

The fact that the *LC* tank portion of the circuit behaves as an open circuit at resonance does not mean that there are no currents in the inductor and capacitor. As a matter of fact, the current $i_L(t)$ circulates around the loop formed by the inductor and capacitor.

At resonance, the energy stored in the inductor is

$$w_L(t) = \frac{1}{2}Li_L^2(t) = \frac{1}{2}\left(\frac{R^2 I^2}{\omega_r^2 L}\right)\sin^2 \omega_r t = \frac{1}{2}CR^2 I^2 \sin^2 \omega_r t$$

since $\omega_r^2 = 1/LC$. Therefore, the total energy stored is

$$w_C(t) + w_L(t) = \frac{1}{2}CR^2 I^2(\cos^2 \omega_r t + \sin^2 \omega_r t) = \frac{1}{2}CR^2 I^2$$

which is a constant. Since the power absorbed by a resistor is $P_R = \frac{1}{2}RI^2$, then the energy lost in a period is

$$P_R T = \frac{1}{2}RI^2 T = \frac{1}{2}RI^2\left(\frac{2\pi}{\omega_r}\right) = \frac{\pi RI^2}{\omega_r}$$

Thus the quality factor of the given parallel *RLC* circuit is

$$Q = \frac{2\pi(\frac{1}{2}CR^2 I^2)}{\pi RI^2/\omega_r} = \omega_r RC$$

Since $\omega_r C = 1/\omega_r L$, alternatively, we can write

$$Q = \omega_r RC = \frac{R}{\omega_r L} = R\sqrt{\frac{C}{L}}$$

Relationship Between Q and Bandwidth

Again considering the expression for the admittance of the circuit in Fig. 5.11, we have that

$$
\begin{aligned}
\mathbf{Y} &= \frac{1}{R} + j\left(\omega C - \frac{1}{\omega L}\right) = \frac{1}{R} + j\left(\frac{\omega C \omega_r R}{\omega_r R} - \frac{\omega_r R}{\omega L \omega_r R}\right) \\
&= \frac{1}{R} + j\frac{1}{R}\left(\frac{\omega}{\omega_r}Q - \frac{\omega_r}{\omega}Q\right) = \frac{1}{R}\left[1 + jQ\left(\frac{\omega}{\omega_r} - \frac{\omega_r}{\omega}\right)\right]
\end{aligned}
$$

At the half-power frequencies ω_1 and ω_2,

$$
|\mathbf{V}| = \frac{R|\mathbf{I}|}{\sqrt{2}} = \frac{|\mathbf{I}|}{|\mathbf{Y}|} \quad \Rightarrow \quad |\mathbf{Y}| = \frac{\sqrt{2}}{R}
$$

This condition occurs when

$$
Q\left(\frac{\omega}{\omega_r} - \frac{\omega_r}{\omega}\right) = \pm 1 \tag{5.1}
$$

For the case that the right-hand side of Eq. 5.1 is $+1$, then

$$
Q\left(\frac{\omega^2 - \omega_r^2}{\omega_r \omega}\right) = 1 \quad \Rightarrow \quad Q\omega^2 - \omega_r\omega - Q\omega_r^2 = 0
$$

By the quadratic formula, we get a positive value and a negative value of ω that satisfy this equation. Since the half-power frequencies are positive quantities, we select the positive value. Thus one of the half-power frequencies is

$$
\frac{\omega_r}{2Q} + \omega_r\sqrt{\left(\frac{1}{2Q}\right)^2 + 1}
$$

For the case that the right-hand side of Eq. 5.1 is -1, as above we can show that the resulting half-power frequency is

$$
\frac{-\omega_r}{2Q} + \omega_r\sqrt{\left(\frac{1}{2Q}\right)^2 + 1}
$$

Thus we see that the latter half-power frequency is ω_1 whereas the former is ω_2; consequently, the bandwidth for the parallel RLC circuit is

$$
BW = \omega_2 - \omega_1 = \frac{\omega_r}{Q} \tag{5.2}
$$

We also have that

$$BW = \frac{\omega_r}{\omega_r RC} = \frac{1}{RC}$$

Thus we see that when the Q of the circuit is relatively small, the bandwidth BW is relatively large; whereas if the Q is relatively large, then the BW is relatively small. The latter case indicates a sharper, or more frequency-selective, amplitude response.

From Eq. 5.2, we have that $Q = \omega_r/BW$ for the parallel RLC circuit. This ratio of resonance frequency to bandwidth is sometimes defined to be the **selectivity** of a circuit. For a parallel RLC circuit, we see that the circuit's quality factor and selectivity are equal. In general, though, the selectivity and the quality factor of a circuit are not the same.

Example 5.3

For the parallel RLC circuit shown in Fig. 5.11 on p. 276, suppose that $R = 10$ kΩ, $L = 50.7$ μH, and $C = 500$ pF. Then the resonance frequency is

$$\omega_r = \frac{1}{\sqrt{LC}} = 6.28 \text{ Mrad/s} \qquad (1 \text{ MHz})$$

the quality factor is

$$Q = R\sqrt{\frac{C}{L}} = 31.4$$

and the bandwidth is

$$BW = \frac{1}{RC} = 200 \text{ krad/s} \qquad (31.8 \text{ kHz})$$

The lower half-power frequency is

$$\omega_1 = -\frac{\omega_r}{2Q} + \omega_r\sqrt{\left(\frac{1}{2Q}\right)^2 + 1} = 6.18 \text{ Mrad/s} \qquad (984 \text{ kHz})$$

and the upper half-power frequency is

$$\omega_2 = \frac{\omega_r}{2Q} + \omega_r \sqrt{\left(\frac{1}{2Q}\right)^2 + 1} = 6.38 \text{ Mrad/s} \qquad (1.016 \text{ MHz})$$

Drill Exercise 5.3

For the parallel *RLC* circuit shown in Fig. 5.11 on p. 276, find expressions for the half-power frequencies in terms of *R*, *L*, and *C*.

ANSWER $-1/2RC + \sqrt{(1/2RC)^2 + 1/LC}$, $1/2RC + \sqrt{(1/2RC)^2 + 1/LC}$

Series Resonance

Let us now consider the case of a series *RLC* circuit as shown in Fig. 5.13*a*. The impedance $\mathbf{Z} = \mathbf{V}_1/\mathbf{I}$ is given by

$$\mathbf{Z} = R + j\omega L + \frac{1}{j\omega C} = R + j\left(\omega L - \frac{1}{\omega C}\right)$$

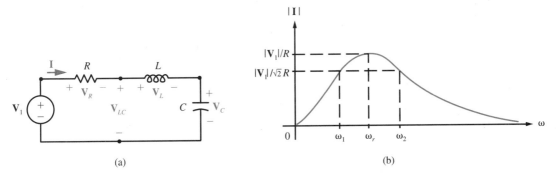

Fig. 5.13 (*a*) Series *RLC* circuit, and (*b*) current magnitude versus frequency.

so the resonance frequency ω_r is that frequency for which $\omega_r L - 1/\omega_r C = 0$. Thus

$$\omega_r = \frac{1}{\sqrt{LC}}$$

as it was for the parallel *RLC* circuit. In this case, the amplitude response is the plot of $|\mathbf{I}|$ versus ω as sketched in Fig. 5.13*b*. Since $\mathbf{V}_R = R\mathbf{I}$, then $|\mathbf{V}_R| = R|\mathbf{I}|$ and a plot of $|\mathbf{V}_R|$ versus ω has the same shape as the plot of $|\mathbf{I}|$ versus ω—the difference being the scale of the vertical axis.

By duality, we have that the quality factor of the series *RLC* circuit is

$$Q = \frac{\omega_r L}{R} = \frac{1}{\omega_r RC} = \frac{1}{R}\sqrt{\frac{L}{C}}$$

since $\omega_r L = 1/\omega_r C$.

In a process similar to that for the parallel *RLC* circuit, it can be shown that the bandwidth for the series *RLC* circuit is

$$BW = \frac{\omega_r}{Q} = \frac{R}{L}$$

so that the selectivity and the quality factor are again equal.

A derivation of the formulas for the half-power frequencies for the series *RLC* circuit can be accomplished as was done for the parallel *RLC* circuit. What results are the same expressions. Thus the lower half-power frequency is

$$\omega_1 = -\frac{\omega_r}{2Q} + \omega_r \sqrt{\left(\frac{1}{2Q}\right)^2 + 1}$$

and the upper half-power frequency is

$$\omega_2 = \frac{\omega_r}{2Q} + \omega_r \sqrt{\left(\frac{1}{2Q}\right)^2 + 1}$$

Example 5.4

For the series *RLC* circuit shown in Fig. 5.13, the voltage transfer function $\mathbf{H}_C(j\omega) = \mathbf{V}_C/\mathbf{V}_1$ is

$$\mathbf{H}_C(j\omega) = \frac{\mathbf{V}_C}{\mathbf{V}_1} = \frac{1/j\omega C}{R + j(\omega L - 1/\omega C)}$$

At the resonance frequency $\omega_r = 1/\sqrt{LC}$, this transfer function is

$$\mathbf{H}_C(j\omega_r) = \frac{1/j\omega_r C}{R} = \frac{1}{j\omega_r RC} \quad \Rightarrow \quad |\mathbf{H}_C(j\omega_r)| = \frac{1}{\omega_r RC} = Q$$

Therefore, if the circuit has a Q that is large (or high), then the gain $|\mathbf{H}_C(j\omega)|$ at the resonance frequency will be large, and the bandwidth $BW = \omega_r/Q$ will be small.

The impedance of the series LC combination is

$$\mathbf{Z}_{LC} = \mathbf{Z}_L + \mathbf{Z}_C = j\omega L + \frac{1}{j\omega C} = j\omega L - \frac{j}{\omega C} = j\left(\omega L - \frac{1}{\omega C}\right)$$

and at the resonance frequency, $\mathbf{Z}_{LC} = 0\ \Omega$. Therefore, a series connection of an inductor L and a capacitor C behaves as a short circuit at the resonance frequency $\omega_r = 1/\sqrt{LC}$. Yet the voltage across each element is nonzero. In particular, at resonance,

$$\mathbf{V}_{LC} = \mathbf{Z}_{LC}\mathbf{I} = 0\ \text{V} \qquad \Rightarrow \qquad \mathbf{I} = \frac{\mathbf{V}_1}{R}$$

Consequently,

$$\mathbf{V}_C = \mathbf{Z}_C\mathbf{I} = \frac{\mathbf{I}}{j\omega C} \qquad \text{and} \qquad \mathbf{V}_L = \mathbf{Z}_L\mathbf{I} = j\omega L\mathbf{I}$$

Since $|\mathbf{V}_C| = |\mathbf{I}|/\omega C$, $|\mathbf{V}_L| = \omega L|\mathbf{I}|$, and $\omega L = 1/\omega C$ at resonance, then $|\mathbf{V}_C| = |\mathbf{V}_L|$. But $\text{ang}(\mathbf{V}_C) = \text{ang}(\mathbf{I}) - 90°$, whereas $\text{ang}(\mathbf{V}_L) = 90° + \text{ang}(\mathbf{I})$. Hence $\mathbf{V}_C$ and $\mathbf{V}_L$ have the same magnitude, but their angles differ by 180 degrees. Consequently, $\mathbf{V}_L + \mathbf{V}_C = 0$ V.

Drill Exercise 5.4

For the series RLC circuit shown in Fig. 5.13, show that at the resonance frequency the gain $|\mathbf{H}_L(j\omega)| = |\mathbf{V}_L/\mathbf{V}_1| = Q$.

Series and Parallel Reactances

Just as we defined the quality factor Q for a resonant circuit, we can define it for a reactance in series or parallel with a resistance. To demonstrate this fact, consider the series case for an inductor L as shown in Fig. 5.14. Let us assume that $\mathbf{I} = I\underline{/0°}$. Then $i(t) = I\cos\omega t$ and the energy stored in the inductor is

$$w(t) = \tfrac{1}{2}Li^2(t) = \tfrac{1}{2}LI^2\cos^2\omega t$$

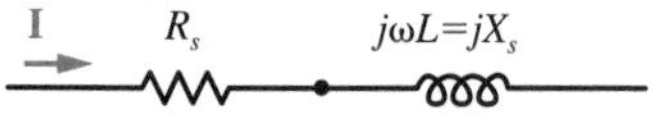

Fig. 5.14 Series inductive reactance.

which has the maximum value $w_m = \frac{1}{2}LI^2$. Since $X_s = \omega L \Rightarrow L = X_s/\omega$, then

$$w_m = \frac{1}{2}\left(\frac{X_s}{\omega}\right)I^2$$

The energy lost per period is

$$P_sT = \left(\frac{1}{2}R_sI^2\right)\left(\frac{2\pi}{\omega}\right) = \frac{\pi R_sI^2}{\omega}$$

Therefore, the quality factor is

$$Q = 2\pi\frac{\frac{1}{2}(X_s/\omega)I^2}{\pi R_sI^2/\omega} = \frac{X_s}{R_s}$$

For the case of a resistor in series with a capacitor C as shown in Fig. 5.15, proceeding as above, we may obtain the fact that $Q = |X_s|/R_s$. Hence, in general, for a resistance in series with a reactance X_s, the quality factor is

$$Q = \frac{|X_s|}{R_s}$$

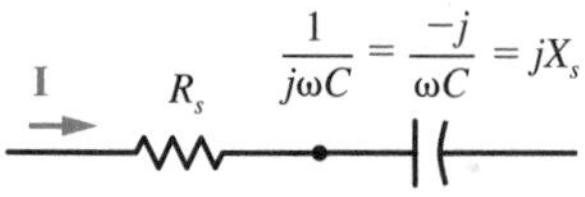

Fig. 5.15 Series capacitive reactance.

For the case of a resistance in parallel with a reactance X_p as shown in Fig. 5.16, it can be shown that the quality factor is

$$Q = \frac{R_p}{|X_p|}$$

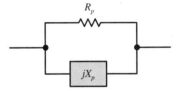

Fig. 5.16 Parallel reactance.

High-Q Coils

Consider the admittance $\mathbf{Y}$ shown in Fig. 5.17, where

$$\mathbf{Y} = j\omega C + \frac{1}{R_s + j\omega L}$$

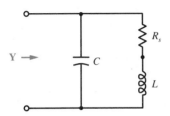

Fig. 5.17 High-Q resonant circuit.

Suppose that the series connection consisting of the inductor L and resistor R_s has a large (at least 20) quality factor. (We call such a combination a **high-Q coil**). Then

$$\frac{X_s}{R_s} \gg 1 \quad \Rightarrow \quad \frac{\omega L}{R_s} \gg 1 \quad \Rightarrow \quad \omega L \gg R_s$$

Under this circumstance, the admittance $\mathbf{Y}$ is approximated by

$$\mathbf{Y} \approx j\omega C + \frac{1}{j\omega L} \tag{5.3}$$

Thus the resonance frequency is approximated by $\omega_r \approx 1/\sqrt{LC}$.
  The impedance of the connection shown in Fig. 5.17 is

$$\mathbf{Z} = \frac{(R_s + j\omega L)(1/j\omega C)}{R_s + j\omega L + 1/j\omega C}$$

$$\approx \frac{(j\omega L)(1/j\omega C)}{R_s + j\omega L + 1/j\omega C} = \frac{L/C}{R_s + j\omega L + 1/j\omega C} \tag{5.4}$$

since $R_s + j\omega L \approx j\omega L$ for a high-Q coil. Thus, the admittance shown in Fig. 5.17 is approximated by[2]

$$\mathbf{Y} \approx \frac{C}{L}\left(R_s + j\omega L + \frac{1}{j\omega C}\right) = \frac{R_s C}{L} + j\omega C + \frac{1}{j\omega L}$$

which is the admittance of a parallel RLC connection, where $R = L/R_s C$. Hence, for a high-Q coil, the two admittances $\mathbf{Y}_1$ and $\mathbf{Y}_2$ shown in Fig. 5.18 are approximately equal.

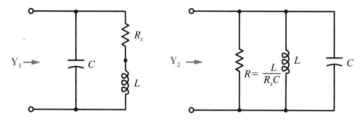

Fig. 5.18 Approximate equivalent circuits.

Example 5.5

The practical tank circuit shown in Fig. 5.19 is to be in resonance at 1 MHz, that is, $\omega_r = 2\pi \times 10^6$ rad/s. Assuming a high-Q coil, $\omega_r \approx 1/\sqrt{LC}$, from which

$$L \approx \frac{1}{\omega_r^2 C} = \frac{1}{(4\pi^2 \times 10^{12})(500 \times 10^{-12})} = 50.7 \ \mu\text{H}$$

The quality factor of the high-Q coil at the resonance frequency is

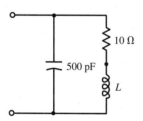

Fig. 5.19 Practical tank circuit.

---

[2]The approximation given in Eq. 5.4 is more accurate than the approximation given by Eq. 5.3. In particular, if $R_s + j\omega L$ in the denominator of Eq. 5.4 is replaced by $j\omega L$, then the approximation given by Eq. 5.3 is obtained.

$$\frac{X_s}{R_s} = \frac{\omega_r L}{10} = \frac{(2\pi \times 10^6)(50.7 \times 10^{-6})}{10} = 31.9$$

which is much greater than one, as was assumed. Since

$$\frac{L}{R_s C} = \frac{50.7 \times 10^{-6}}{(10)(500 \times 10^{-12})} \approx 10,000$$

then the tank circuit given in Fig. 5.19 can be approximated by the parallel RLC connection shown in Fig. 5.20. The quality factor of this parallel RLC circuit is

$$Q = \omega_r RC = (2\pi \times 10^6)(10 \times 10^3)(500 \times 10^{-12}) = 31.4$$

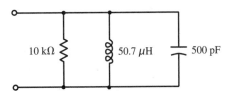

Fig. 5.20 Equivalent of practical tank circuit.

Drill Exercise 5.5

For the practical tank circuit given in Fig. 5.19, suppose that $L = 10$ μH. (a) To what value should the capacitor be changed such that the resonance frequency will be approximately 4 MHz? (b) What is the Q of the coil? (c) What are the values for the equivalent parallel RLC connection shown in Fig. 5.20? (d) What is the quality factor of this equivalent parallel RLC connection?

ANSWER (a) 158 pF, (b) 25.1, (c) 6.33 kΩ, 10 μH, 158 pF, (d) 25.1

5.3   Complex Frequency

In the preceding chapter, our major preoccupation was with the sinusoid. By generalizing real sinusoids to complex sinusoids, we developed the notion of phasors with which we were able to find forced responses of ac circuits without having to deal with differential equations. As a consequence, we were able to perform analyses in the frequency domain much more easily than in the time domain.

We are now in a position to go one step further. By extending the above-mentioned concepts, we can employ frequency-domain analysis for circuits with damped sinusoidal forcing functions of the form $Ae^{\sigma t}\cos(\omega t + \theta)$.

For the damped sinusoidal function $Ae^{\sigma t}\cos(\omega t + \theta)$, we know that the frequency ω has radians per second as its unit. Since the power of the natural logarithm base e is to be dimensionless, the units of the (usually nonpositive) damping factor σ have been designated **nepers per second**, where the **neper**[3] (Np) is a dimensionless unit. For this reason, σ is referred to as the **neper frequency**.

Now suppose that the input to an *RLC* circuit has the form $Ae^{\sigma t}\cos(\omega t + \theta)$. Using an argument similar to that for the real sinusoidal case, the forced response will have the form $Be^{\sigma t}\cos(\omega t + \phi)$. Thus by the property of linearity, the response to

$$Ae^{\sigma t}\cos(\omega t + \theta) + jAe^{\sigma t}\sin(\omega t + \theta) = Ae^{\sigma t}e^{j(\omega t + \theta)}$$
$$= Ae^{j\theta}e^{(\sigma + j\omega)t} = Ae^{j\theta}e^{st}$$
$$= \mathbf{V}_{in}e^{st}$$

where $\mathbf{V}_{in} = Ae^{j\theta}$ and $s = \sigma + j\omega$, which is called **complex frequency**, has the form

$$Be^{\sigma t}\cos(\omega t + \phi) + jBe^{\sigma t}\sin(\omega t = \phi) = Be^{\sigma t}e^{j(\omega t + \phi)}$$
$$= Be^{j\phi}e^{(\sigma + j\omega)t} = Be^{j\phi}e^{st}$$
$$= \mathbf{V}_{o}e^{st}$$

where $\mathbf{V}_{o} = Be^{j\phi}$. Hence, knowing the response to $Ae^{j\theta}e^{st}$ is tantamount to knowing the response to $Ae^{\sigma t}\cos(\omega t + \theta)$. That is, we can determine B and ϕ by finding the response either to $Ae^{j\theta}e^{st}$ or to $Ae^{\sigma t}\cos(\omega t + \theta)$.

Just as we associated the real sinusoid $A\cos(\omega t + \theta)$ or the complex sinusoid $Ae^{j(\omega t + \theta)}$ with the phasor $\mathbf{A} = Ae^{j\theta}$, so we can associate the damped sinusoid $Ae^{\sigma t}\cos(\omega t + \theta)$ or the damped complex sinusoid

$$Ae^{\sigma t}e^{j(\omega t + \theta)} = Ae^{j\theta}e^{st}$$

with the phasor $\mathbf{A} = Ae^{j\theta}$ also. However, for the latter case, $s = \sigma + j\omega$ must be implicit, whereas only ω need be for the former.

---

[3]Named for the Scottish mathematician John Napier (1550–1617).

Impedance

For a resistor R having voltage v across it and current i through it, if the current is a damped complex sinusoid, say $i = Ie^{\sigma t}e^{j(\omega t + \theta)} = Ie^{j\theta}e^{st}$, then the voltage is also a damped complex sinusoid, say $v = Ve^{\sigma t}e^{j(\omega t + \phi)} = Ve^{j\phi}e^{st}$. Since $v = Ri$, then

$$Ve^{j\phi}e^{st} = RIe^{j\theta}e^{st}$$

and dividing by e^{st} yields

$$Ve^{j\phi} = RIe^{j\theta} \tag{5.5}$$

from which we make the identities $V = RI$ and $\phi = \theta$. Using phasor notation, Eq. 5.5 can be written as

$$V\underline{/\phi} = RI\underline{/\theta} \quad \Rightarrow \quad \mathbf{V} = R\mathbf{I}$$

where $\mathbf{V} = V\underline{/\phi}$ is the voltage phasor and $\mathbf{I} = I\underline{/\theta}$ is the current phasor. Defining the **impedance** $\mathbf{Z}_R$ of the resistance to be the ratio of voltage phasor to current phasor, we have

$$\boxed{\mathbf{Z}_R = \frac{\mathbf{V}}{\mathbf{I}} = R}$$

For an inductor L, if $i = Ie^{j\theta}e^{st}$ and $v = Ve^{j\phi}e^{st}$, since $v = L\,di/dt$, then

$$Ve^{j\phi}e^{st} = L\frac{d}{dt}(Ie^{j\theta}e^{st}) = LsIe^{j\theta}e^{st}$$

Dividing this by e^{st} gives

$$Ve^{j\phi} = LsIe^{j\theta} \quad \text{or} \quad V\underline{/\phi} = LsI\underline{/\theta} \quad \Rightarrow \quad \mathbf{V} = Ls\mathbf{I}$$

Thus the impedance $\mathbf{Z}_L$ of an inductor is

$$\boxed{\mathbf{Z}_L = \frac{\mathbf{V}}{\mathbf{I}} = Ls}$$

For a capacitor C, if $i = Ie^{j\theta}e^{st}$ and $v = Ve^{j\phi}e^{st}$, since $i = C\,dv/dt$, then

$$Ie^{j\theta}e^{st} = C\frac{d}{dt}(Ve^{j\phi}e^{st}) = CsVe^{j\phi}e^{st}$$

from which

$$Ie^{j\theta} = CsVe^{j\phi} \quad \text{or} \quad I\underline{/\theta} = CsV\underline{/\phi} \quad \Rightarrow \quad \mathbf{I} = Cs\mathbf{V}$$

The impedance of a capacitor is, therefore,

$$\mathbf{Z}_C = \frac{\mathbf{V}}{\mathbf{I}} = \frac{1}{Cs}$$

From the definition of impedance, we see that we have the equations

$$\mathbf{V} = \mathbf{Z}\mathbf{I} \quad \text{and} \quad \mathbf{I} = \frac{\mathbf{V}}{\mathbf{Z}}$$

This latter equation suggests the usefulness of the reciprocal or impedance. We, therefore, define the ratio of current phasor to voltage phasor as **admittance**, denoted **Y**. Thus we have the element admittances

$$\mathbf{Y}_R = \frac{1}{R} \qquad \mathbf{Y}_L = \frac{1}{Ls} \qquad \mathbf{Y}_C = Cs$$

and the alternative forms of Ohm's law:

$$\mathbf{Y} = \frac{\mathbf{I}}{\mathbf{V}} \qquad \mathbf{I} = \mathbf{Y}\mathbf{V} \qquad \mathbf{V} = \frac{\mathbf{I}}{\mathbf{Y}}$$

Just as we analyzed ac circuits with the use of phasors and impedance, so can we analyze circuits with damped-sinusoidal excitations. For the former case, impedance is a function of ω—denoted $\mathbf{Z}(j\omega)$, and for the latter, it is a function of σ and ω—denoted $\mathbf{Z}(s)$.

Example 5.6

For the circuit given in Fig. 5.21, let us determine the voltage $v_C(t)$.

Figure 5.22 shows the frequency-domain representation of the given circuit. For this circuit, by KCL, we have that

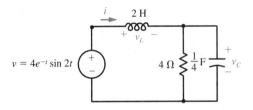

Fig. 5.21 Circuit with damped-sinusoidal excitation.

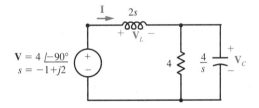

Fig. 5.22 Frequency-domain representation.

$$\frac{\mathbf{V}_C - \mathbf{V}}{2s} + \frac{\mathbf{V}_C}{4} + \frac{\mathbf{V}_C}{4/s} = 0 \quad \Rightarrow \quad 2(\mathbf{V}_C - \mathbf{V}) + s\mathbf{V}_C + s^2\mathbf{V}_C = 0$$

from which

$$(s^2 + s + 2)\mathbf{V}_C = 2\mathbf{V} \quad \Rightarrow \quad \mathbf{V}_C = \frac{2\mathbf{V}}{s^2 + s + 2}$$

One alternative procedure for finding $\mathbf{V}_C$ consists of first expressing the parallel *RC* combination as the impedance

$$\mathbf{Z}_{RC} = \frac{4(4/s)}{4 + 4/s} = \frac{4}{s + 1}$$

and then using the voltage-divider formula to obtain

$$\mathbf{V}_C = \frac{\mathbf{Z}_{RC}}{\mathbf{Z}_{RC} + \mathbf{Z}_L}\mathbf{V} = \frac{4/(s + 1)}{4/(s + 1) + 2s}\mathbf{V} = \frac{2\mathbf{V}}{s^2 + s + 2}$$

From the fact that $\mathbf{V} = 4\underline{/-90°}$ and $s = -1 + j2$, we get

$$\mathbf{V}_C = \frac{2(4\underline{/-90°})}{(-1 + j2)^2 + (-1 + j2) + 2} = 2\sqrt{2}\underline{/45°} \text{ V}$$

Thus we deduce that for the circuit shown in Fig. 5.21,

$$v_C(t) = 2\sqrt{2}\, e^{-t} \cos(2t + 45°) \text{ V}$$

Drill Exercise 5.6

Find $v_L(t)$ for the circuit shown in Fig. 5.21.

ANSWER $2\sqrt{10}\, e^{-t} \cos (2t - 108°)$ V

Circuit concepts such as Thévenin's theorem are also valid for circuits having damped-sinusoidal excitations. In particular, for the circuit shown in Fig. 5.22, consider the load to be the capacitor.

We can readily obtain $\mathbf{V}_{oc}$ by removing the capacitor, and for the resulting circuit shown in Fig. 5.23a, employing voltage division. Doing so yields

$$\mathbf{V}_{oc} = \frac{4\mathbf{V}}{2s + 4} = \frac{2\mathbf{V}}{s + 2} \tag{5.6}$$

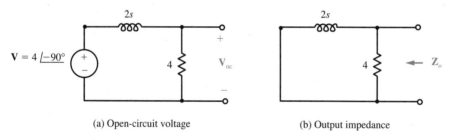

(a) Open-circuit voltage　　　　(b) Output impedance

Fig. 5.23 Determination of Thévenin-equivalent circuit.

To find $\mathbf{Z}_o$, set the voltage source in Fig. 5.23a to zero, and determine $\mathbf{Z}_o$ as indicated in Fig. 5.23b. The consequence of doing this is

$$\mathbf{Z}_o = \frac{2s(4)}{2s + 4} = \frac{4s}{s + 2} \tag{5.7}$$

The frequency-domain Thévenin-equivalent circuit is shown in Fig. 5.24a. From this circuit, by voltage division,

$$\mathbf{V}_C = \frac{(4/s)\mathbf{V}_{oc}}{4/s + 4s/(s + 2)} = \frac{(s + 2)\mathbf{V}_{oc}}{(s + 2) + s^2} = \frac{2\mathbf{V}}{s^2 + s + 2}$$

as was obtained in Example 5.6.

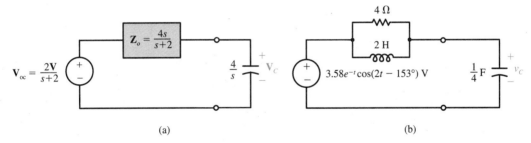

Fig. 5.24 (a) Frequency-domain Thévenin-equivalent circuit, and (b) time-domain Thévenin-equivalent circuit.

The specific value for $\mathbf{V}_{oc}$ can be obtained by substituting $s = -1 + j2$ and $\mathbf{V} = 4\underline{/-90°}$ into Eq. 5.6. Similarly, the specific value for $\mathbf{Z}_o$ can be obtained from Eq. 5.7. What result are $\mathbf{V}_{oc} = 3.58\underline{/-153°}$ V and $\mathbf{Z}_o = 2.4 + j3.2 = 4\underline{/53.1°}$ Ω.

In order to obtain the time-domain version of the Thévenin-equivalent circuit, from the fact that $\mathbf{V}_{oc} = 3.58\underline{/-153°}$ V, we conclude that $v_{oc}(t) = 3.58e^{-t}\cos(2t - 153°)$ V. Furthermore, $\mathbf{Z}_o$ is realized in the time-domain by the parallel connection of a 4-Ω resistor and a 2-H inductor as suggested in Fig. 5.23b, Consequently, the time-domain Thévenin-equivalent circuit is depicted in Fig. 5.24b.

Poles and Zeros

For the circuit given in Fig. 5.22, if we define the voltage transfer function $\mathbf{H}_C(s)$ to be the ratio of $\mathbf{V}_C$ to $\mathbf{V}$, then from Example 5.6 we have that

$$\mathbf{H}_C(s) = \frac{\mathbf{V}_C}{\mathbf{V}} = \frac{2}{s^2 + s + 2} = \frac{2}{(s + 1/2 - j\sqrt{7}/2)(s + 1/2 + j\sqrt{7}/2)}$$

Thus we see for the special case that

$$s = -\frac{1}{2} + j\frac{\sqrt{7}}{2} \qquad \text{or} \qquad s = -\frac{1}{2} - j\frac{\sqrt{7}}{2}$$

the transfer function becomes infinite—which implies that the forced response is infinite. We say that the function $\mathbf{H}_C(s)$ has a **pole** at each of these two complex frequencies.

To find the voltage transfer function $\mathbf{H}_L(s) = \mathbf{V}_L/\mathbf{V}$ for the circuit in Fig. 5.22, we can again use voltage division. Doing so yields

$$V_L = \frac{2sV}{2s + 4/(s + 1)} = \frac{2s(s + 1)V}{2s(s + 1) + 4} \qquad \Rightarrow$$

$$H_L(s) = \frac{V_L}{V} = \frac{s(s + 1)}{s^2 + s + 2}$$

Therefore, $H_L(s)$ has the same poles as $H_C(s)$. However, there are two frequencies, $s = 0$ and $s = -1$, for which $H_L(s) = 0$. For this reason, we say that the network function $H_L(s)$ has **zeros** at $s = 0$ and $s = -1$.

Since $s = \sigma + j\omega$ is a complex quantity, we can indicate any value of s as a point in a complex-number plane. If we label the horizontal axis σ and the vertical axis $j\omega$, the result is known as the **s plane**. If we denote a pole by the symbol "$\times$" and a zero by the symbol "$\circ$," then depicting the poles and zeros of a network function in the s plane is referred to as a **pole-zero plot**. The pole-zero plots for $H_C(s)$ and $H_L(s)$ are shown in Fig. 5.25.

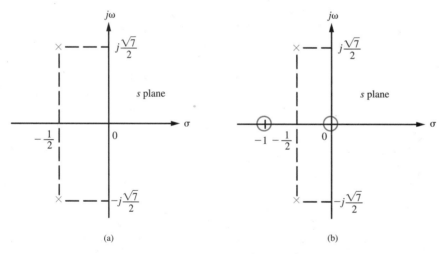

Fig. 5.25 Pole-zero plot for (a) $H_C(s) = V_C/V$, and (b) $H_L(s) = V_L/V$.

The Natural Response

It is a routine exercise to show that the circuit given in Fig. 5.21 is described by the differential equation

$$\frac{d^2 v_C(t)}{dt^2} + \frac{d v_C(t)}{dt} + 2v_C(t) = 2v(t)$$

When $v(t) = 0$ V, the (nontrivial) solution to this equation is the natural response. As was done in Chapter 3, we get the natural response by assuming a solution of the form Ae^{st}. Doing so, we obtain two values for s that satisfy the equation

$$s^2 + s + 2 = 0$$

Using the quadratic formula, these values of s are

$$s = \frac{-1 \pm \sqrt{1-8}}{2} = -\frac{1}{2} \pm j\frac{\sqrt{7}}{2}$$

In other words, the natural response $v_n(t)$ has the form

$$v_n(t) = A_1 e^{s_1 t} + A_2 e^{s_2 t}$$

where

$$s_1 = -1/2 + j\sqrt{7}/2 \quad \text{and} \quad s_2 = -1/2 - j\sqrt{7}/2$$

But note, these are precisely the poles of the voltage transfer function $\mathbf{H}_C(s) = \mathbf{V}_C/\mathbf{V}$. Thus, and this is true in general, the poles of the transfer function specify the natural response—provided that there was no cancellation of a common pole and zero. In this example, the complex values for s_1 and s_2, that is, the complex poles, indicate that this second-order circuit is underdamped. Thus the natural response has the more common alternative forms

$$v_n(t) = B_1 e^{-\alpha t} \cos \omega_d t + B_2 e^{-\alpha t} \sin \omega_d t = Be^{-\alpha t} \cos(\omega_d t + \phi)$$

From the above differential equation, $\alpha = \frac{1}{2}$ and $\omega_n = \sqrt{2}$. Thus, $\omega_d = \sqrt{\omega_n^2 - \alpha^2} = \sqrt{2 - 1/4} = \sqrt{7}/2$. Hence for an input of

$$v(t) = 4e^{-t} \sin 2t \, u(t) \text{ V}$$

the complete response has the form (see Example 5.6 for the forced response)

$$v_C(t) = Be^{-t/2} \cos(\sqrt{7}t/2 + \phi) + 2\sqrt{2}e^{-t} \cos(2t + 45°) \text{ V}$$

where the constants B and ϕ (or B_1 and B_2) are determined from the initial conditions $i(0)$ and $v_C(0)$.

If in the previous discussion, the poles of $H_C(s)$ were real, say $s_1 = -a_1$ and $s_2 = -a_2$, instead of complex, then the form of the natural response would be that of the overdamped case, that is,

$$v_n(t) = A_1 e^{-a_1 t} + A_2 e^{-a_2 t}$$

Finally, if the two poles of $H_C(s)$ were real and equal (called a **double pole**), say $s_1 = s_2 = -\alpha$, then the form of the natural response would be that of the critically damped case, that is,

$$v_n(t) = A_1 e^{-\alpha t} + A_2 t e^{-\alpha t}$$

Natural Frequencies

Again let us return to the circuit given in Fig. 5.22 and consider yet another network function for the same input variable. This time, suppose that the output variable is the inductor current $\mathbf{I}$. Since the impedance loading the voltage source is

$$\mathbf{Z} = \mathbf{Z}_L + \mathbf{Z}_{RC} = 2s + \frac{4}{s + 1} = \frac{2(s^2 + s + 2)}{s + 1}$$

and since $\mathbf{Z} = \mathbf{V}/\mathbf{I}$, then the transfer function of interest $\mathbf{I}/\mathbf{V}$ is just the admittance $\mathbf{Y}(s)$ loading the source, that is,

$$\mathbf{Y}(s) = \frac{\mathbf{I}}{\mathbf{V}} = \frac{\frac{1}{2}(s + 1)}{s^2 + s + 2}$$

and again the poles of $\mathbf{Y}(s)$ are the same as the poles of $H_C(s)$ and $H_L(s)$. The pole-zero plot for $\mathbf{Y}(s)$ can be obtained from the pole-zero plot for $H_L(s)$ by removing the zero at the origin of the s plane.

It is not just coincidence that the poles for the network functions $\mathbf{V}_C/\mathbf{V}$, $\mathbf{V}_L/\mathbf{V}$, and $\mathbf{I}/\mathbf{V}$ determined above are the same. Provided that one portion of a circuit is not physically separated from the rest, each transfer function (defined as the ratio of an output to a given input) will have the same poles regardless of which voltage or current is chosen as the output variable. This should not be surprising though, for the poles are those values of s, called **natural frequencies**, that determine the natural response. And since the form of the natural response is the same throughout a (non-separated) circuit, the poles of any transfer function (with respect to a given input variable) will be the same.

5.4   Introduction to Systems

As was demonstrated in Section 5.3, a linear circuit (or any system described by a linear differential equation with constant coefficients, for that matter) having an input, say $x(t) = \mathbf{X}e^{st}$, will have a forced response of the form $y(t) = \mathbf{Y}e^{st}$. Substituting these terms into the general form of the differential equation describing the circuit (or system)

$$a_n\frac{d^n y(t)}{dt^n} + a_{n-1}\frac{d^{n-1}y(t)}{dt^{n-1}} + \cdots + a_1\frac{dy(t)}{dt} + a_0 y(t)$$

$$= b_m\frac{d^m x(t)}{dt^m} + b_{m-1}\frac{d^{m-1}x(t)}{dt^{m-1}} + \cdots + b_1\frac{dx(t)}{dt} + b_0 x(t) \quad (5.8)$$

results in

$$(a_n s^n + a_{n-1}s^{n-1} + \cdots + a_1 s + a_0)\mathbf{Y}e^{st}$$

$$= (b_m s^m + b_{m-1}s^{m-1} + \cdots + b_1 s + b_0)\mathbf{X}e^{st}$$

From this expression, we find that the general form of the transfer function $\mathbf{H}(s)$ is

$$\mathbf{H}(s) = \frac{\mathbf{Y}}{\mathbf{X}} = \frac{b_m s^m + b_{m-1}s^{m-1} + \cdots + b_1 s + b_0}{a_n s^n + a_{n-1}s^{n-1} + \cdots + a_1 s + a_0} \qquad (5.9)$$

That is, $\mathbf{Y}/\mathbf{X}$ is a ratio of polynomials in s with real coefficients.

A result from mathematics theory is that a polynomial with real coefficients can be factored into a product of quadratics of the form $as^2 + bs + c$, where a, b, and c are real numbers. From the quadratic formula, such a term can further be factored into the product $(s - s_1)(s - s_2)$, where

$$s_1 = \frac{-b + \sqrt{b^2 - 4ac}}{2a} \quad \text{and} \quad s_2 = \frac{-b - \sqrt{b^2 - 4ac}}{2a}$$

For the case that $b^2 \geq 4ac$, the numbers s_1 and s_2 are real. However, if $b_2 < 4ac$, then s_1 and s_2 are complex numbers and are conjugates. As a consequence of this, we can express the transfer function, or any other network function for that matter, in the form

$$\mathbf{H}(s) = \frac{K(s - z_1)(s - z_2) \cdots (s - z_m)}{(s - p_1)(s - p_2) \cdots (s - p_n)} \qquad (5.10)$$

where $z_1, z_2, \ldots, z_m$ are the zeros and $p_1, p_2, \ldots, p_n$ are the poles of $\mathbf{H}(s)$. The zeros and poles can be either real or complex, but complex zeros or poles occur in conjugate pairs.

For the case of distinct poles, the natural response $y_n(t)$ has the form

$$y_n(t) = A_1 e^{p_1 t} + A_2 e^{p_2 t} + \cdots + A_n e^{p_n t}$$

If pole p_i is at the origin of the s plane, that is, if $p_i = 0$, then the term in the natural response corresponding to it is $A_i e^0 = A_i$, a constant. If p_i is on the negative real axis or the positive real axis of the s plane, then $A_i e^{p_i t}$ is a decaying exponential or increasing exponential, respectively, as shown in Fig. 5.26. If p_i is a complex number, then there is a pole $p_j = p_i^*$. As we saw in the discussion on second-order circuits, when $p_i = \sigma + j\omega$, the corresponding terms in the natural response can be combined in the form $A e^{\sigma t} \cos(\omega t + \theta)$. If the conjugate pair of poles is in the left half of the s plane (i.e., $\sigma < 0$), this term is a damped sinusoid. If the pair is on the $j\omega$ axis ($\sigma = 0$), the term is a sinusoid. If the pair is in the right half of the s plane ($\sigma > 0$), the term is an increasing sinusoid. These three cases are depicted in Fig. 5.27.

As may be surmised from this discussion, a circuit that has a "stable" operation does not have a network function with poles in the right half of the s plane. In some

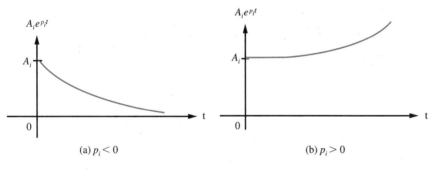

(a) $p_i < 0$ (b) $p_i > 0$

Fig. 5.26 Natural-response forms for a real pole.

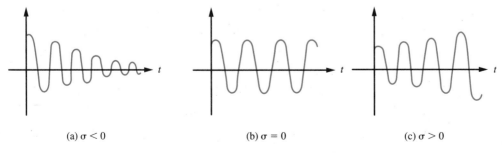

(a) $\sigma < 0$ (b) $\sigma = 0$ (c) $\sigma > 0$

Fig. 5.27 Natural-response forms for complex-conjugate poles.

applications, poles on the imaginary axis are undesirable, and only poles in the left half of the s plane are allowable.

Block Diagrams

Consider a linear system (i.e., a system described by a linear differential equation) whose input is **X** and whose output is **Y** as depicted by the "block" in Fig. 5.28. The transfer function of the system is

$$\mathbf{H}(s) = \frac{\mathbf{Y}}{\mathbf{X}}$$

Suppose that the transfer function of the system is $\mathbf{H}(s) = 1/s$. Then from Eq. 5.4, we have that $a_1 = b_0 = 1$ and $a_0 = a_2 = a_3 = \cdots = a_n = b_1 = b_2 = \cdots = b_m = 0$. Thus the differential equation describing the system is

$$\frac{dy(t)}{dt} = x(t)$$

Taking the integral of both sides of this expression with respect to time yields,

$$\int \frac{dy(t)}{dt}\, dt = \int x(t)\, dt = y(t)$$

In other words, the output of the system is the integral of the input. For this reason, we call this simple system an **integrator** and can represent a device that performs integration as in Fig. 5.29.

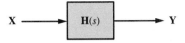

Fig. 5.28 Linear system.   **Fig. 5.29** Integrator.

Another system is shown in Fig. 5.30. In this case, the output is just the sum of the two inputs $\mathbf{X}_1$ and $\mathbf{X}_2$. For this reason, this device is called an **adder** or **summer**. Although the adder in Fig. 5.30 has two inputs, an adder can have three or more inputs as well.

For the system shown in Fig. 5.31, the output equals the input scaled by the constant A. Thus we call this device a **scaler** or **amplifier**.

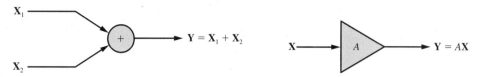

Fig. 5.30 Two-input adder.　　　　　　　　**Fig. 5.31** Scaler.

Using the three types of devices described (integrators, adders, and scalers), we now can "simulate" or "build" a linear system that is described by a transfer function having the form of Eq. 5.9, provided that $n \geq m$.

Example 5.7

Let us construct a system whose transfer function is

$$\mathbf{H}(s) = \frac{\mathbf{Y}}{\mathbf{X}} = \frac{s + 2}{3s^2 + 4s + 5} \tag{5.11}$$

From this expression, we obtain

$$(3s^2 + 4s + 5)\mathbf{Y} = (s + 2)\mathbf{X}$$

$$3s^2\mathbf{Y} = s\mathbf{X} + 2\mathbf{X} - 4s\mathbf{Y} - 5\mathbf{Y}$$

$$\mathbf{Y} = \frac{1}{3s}\mathbf{X} + \frac{2}{3s^2}\mathbf{X} - \frac{4}{3s}\mathbf{Y} - \frac{5}{3s^2}\mathbf{Y}$$

This last equation can be realized as shown in Fig. 5.32. We deduce, therefore, that the system given in Fig. 5.32 has the transfer function specified by Eq. 5.11.

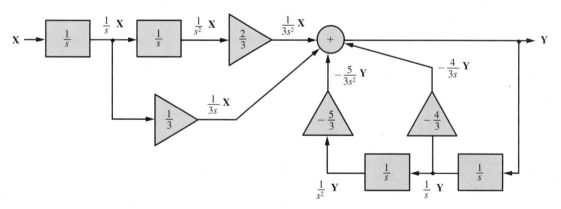

Fig. 5.32 Realization (simulation) of given transfer function.

We did not indicate what type of system (electrical, mechanical, etc.) was described by Eq. 5.11. However, regardless of what type of system it is, we can simulate it with the system shown in Fig. 5.32. Thus if the original system is cumbersome or not easily accessible, we can test or study its responses to various inputs by applying these inputs to the simulated system in Fig. 5.32. This is precisely what the analog computer is all about. An **analog computer** is an electronic system that contains integrators, adders, and scalers, and on which we can simulate linear systems.

In our discussion of operational amplifiers, we saw how we could use op amps and other circuit elements to construct integrators, adders, and scalers. Yet, as in Fig. 5.32, for example, we do not need to indicate specifically how these devices are fabricated, and instead we merely represent them by appropriately labeled "blocks" (or circles or triangles). For this reason, we can refer to Fig. 5.32 as a **block diagram** of the system described by Eq. 5.11.

In general, there may be more than one way to represent a system in terms of various building blocks, and blocks other than integrators, adders, and scalers can be utilized. For instance, it should be relatively easy by now to show that the low-pass filter given in Fig. 5.33a has the voltage transfer function

$$\mathbf{H}(s) = \frac{\mathbf{V}_2}{\mathbf{V}_1} = \frac{1}{RCs + 1}$$

and we can represent this system by the single block shown in Fig. 5.33b.

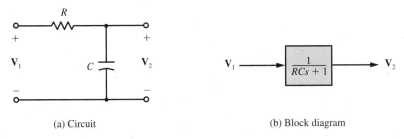

(a) Circuit   (b) Block diagram

Fig. 5.33 Low-pass filter.

Example 5.8

Consider the system with input **X** and output **Y** indicated by the block diagram given in Fig. 5.34. Since the output of the block labeled **H** is **HY**, then the output of the adder is **X − HY**.

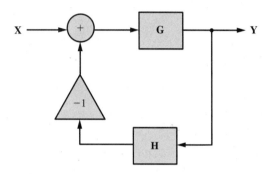

Fig. 5.34 System with negative feedback.

Hence an expression for the overall system output **Y** is

$$\mathbf{Y} = \mathbf{G}(\mathbf{X} - \mathbf{HY}) \qquad \Rightarrow \qquad \mathbf{Y} = \frac{\mathbf{G}}{1 + \mathbf{GH}}\mathbf{X}$$

and, therefore, the system transfer function is

$$\frac{\mathbf{Y}}{\mathbf{X}} = \frac{\mathbf{G}}{1 + \mathbf{GH}}$$

Drill Exercise 5.8

For the system shown in Fig. 5.34, the desired transfer function is $\mathbf{Y}/\mathbf{X} = (3s + 12)/(s^2 + 9s + 8)$. Determine **G** for the case that $\mathbf{H} = s/(s + 4)$.

ANSWER $3/(s + 2)$

Feedback

Note that, in the preceding example, a portion (**HY**) of the output **Y** is "fed back" and subtracted from the input **X** and applied to **G**. (The output of **G** is the overall system output **Y**.) This, then, is an example of **feedback**, and since a portion of the output is subtracted from the input, the feedback is **negative**. If a portion of the output were added to the input, we would have **positive** feedback.

There are two major applications of feedback in electrical systems. One is to change the characteristics or performance of a system to obtain a desirable effect, and the other is to control a system (such a system is called an **automatic control system**, or **control system,** for short).

Example 5.9

To illustrate how negative feedback can be used to improve system performance, consider an amplifier whose transfer function (gain) **G** = 1100. Suppose that, because of a temperature change, aging of components, faulty repair, or a combination of these and other factors, the gain drops to **G′** = 900. This is a change of 18 percent of the original value.

Now, however, suppose that the amplifier **G** = 1100 is part of a feedback system as depicted in Fig. 5.34. Further, suppose that the feedback element **H** is a simple resistive voltage-divider network with voltage transfer function **H** = $1/110$. Then the gain of the overall system (amplifier) is

$$\mathbf{G}_f = \frac{\mathbf{G}}{1 + \mathbf{GH}} = \frac{1100}{1 + 1100(1/110)} = 100$$

If the gain of **G** drops to **G′** = 900, the overall system gain is

$$\mathbf{G}'_f = \frac{\mathbf{G}'}{1 + \mathbf{G'H}} = \frac{900}{1 + 900(1/110)} = 98$$

and this is a change of only 2 percent. Thus we see that although the amplifier **G** = 1100 without feedback has more gain than the feedback amplifier **G**$_f$ = 100, it is the latter whose gain is much less sensitive to possible parameter variations. The fact that the feedback amplifier has less gain than the amplifier without feedback is not usually considered a problem as further stages of amplification (with feedback) can be added if needed.

In the preceding example, we saw how negative feedback reduced the sensitivity of amplifier gain by reducing its dependence on various device parameters. Negative feedback can offer other advantages as well. It can also reduce the sensitivity of amplifier gain to frequency-dependent elements such as inductors and capacitors. In addition, negative feedback can reduce the effect of disturbances resulting from distortion, external interference, or internal noise. Network parameters such as circuit time constants and input and output impedance also can be modified with the use of negative feedback.

Not only are there advantages to negative feedback, but positive feedback also has useful applications, as is illustrated in the following example.

Example 5.10

Consider the positive-feedback system shown in Fig. 5.35, which contains a high-pass filter $\mathbf{G}(s) = 2s/(s + 1)$ and whose feedback element is a low-pass filter $\mathbf{H}(s) = 1/(s + 1)$. As was done for the feedback system in Fig. 5.34, we can easily obtain

$$\mathbf{G}_f(s) = \frac{\mathbf{Y}}{\mathbf{X}} = \frac{\mathbf{G}(s)}{1 - \mathbf{G}(s)\mathbf{H}(s)}$$

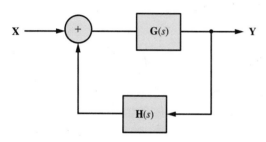

Fig. 5.35 System with positive feedback.

from which

$$G_f(s) = \frac{2s/(s+1)}{1 - [2s/(s+1)][1/(s+1)]} = \frac{2s(s+1)}{s^2+1} = \frac{2s(s+1)}{(s-j)(s+j)}$$

Hence the transfer function for this feedback system has poles on the imaginary axis of the s plane (at $s = j$ and $s = -j$). From our discussion at the beginning of this section (see Fig. 5.27), we see that this system will produce a sinusoid of frequency $\omega = 1$ rad/s. Such a device is called an **oscillator**, and there are a number of electronic circuits for constructing one (see Section 10.4).

Drill Exercise 5.10

Suppose that for the positive-feedback connection given in Fig. 5.35, $G(s) = 4s/(s+2)$. What is the transfer function $H(s)$ that will result in $G_f(s) = (4s^2 + 8s)/(s^2 + 4)$? What is the corresponding frequency of oscillation?

ANSWER $1/(s+2)$, 2 rad/s

So far we have seen two specific examples demonstrating how feedback can be used to obtain desirable performance from a system. We shall now describe how feedback can be employed for automatic control. The feedback system (albeit non-linear) that will be discussed is the phase-locked loop—a device that, in recent years, has seen increasing use in FM demodulation.

A **phase-locked loop** (PLL) is an electronic feedback control system that can be described by the block diagram shown in Fig. 5.36. The PLL contains a **voltage-controlled oscillator** (VCO), which is an oscillator whose frequency is controlled by a dc voltage coming from the low-pass filter. When an input is applied to the PLL, its frequency and phase are compared with the frequency and phase of the VCO by the **phase detector**. The output of the phase detector is an ''error voltage'' that is proportional to the difference between the frequencies and phases of the input

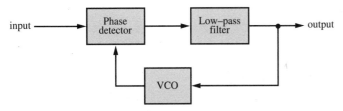

Fig. 5.36 Phase-locked loop (PLL).

and VCO. The output of the phase detector goes to a low-pass filter to remove higher-frequency components. The output of the low-pass filter, which is the output of the system, changes the frequency of the VCO such that the difference between the input and the VCO is reduced. This process continues until the input and the VCO have the same frequency. At this point, we say that the PLL is **locked** or is in **phase lock**. There is a difference between the phases of the input and VCO that produces an error voltage that keeps the PLL locked.

If the input to the PLL shown in Fig. 5.36 is an FM signal, then the output is the demodulated signal. In addition to FM demodulation, the PLL has numerous other practical applications. For some of these, the output of the VCO is the output of the system. Under this circumstance, the connection between the VCO and the phase detector is the feedback element.

Although the concept of the PLL was originated by British scientists in 1932, its utilization for most applications was economically unfeasible until its appearance as an integrated-circuit (IC) package in the 1970s.

5.5 The Laplace Transform

In Section 5.4 we saw that the transfer function $\mathbf{H}(s)$ for a linear circuit or system is a ratio of polynomials in the complex-frequency variable s [see Equations (5.9) and (5.10)]. Such a circuit or system characterization was obtained by considering forced responses to damped sinusoids. Although this may seem like a very restrictive characterization, we will now see how we can generalize it so that complete responses to arbitrary inputs can be obtained without the need for writing and solving differential equations.

In order to accomplish this, the transformation of circuits from the time domain to the frequency domain will not be done by using phasors (for the sinusoidal case or the damped-sinusoidal case), but instead will be done with the use of a more sophisticated mathematical transformation—the Laplace transform.

Definition of the Laplace Transform

Given a function of time $f(t)$, we define its **Laplace transform**[4]—designated either $\mathcal{L}[f(t)]$ or $\mathbf{F}(s)$—to be

$$\mathcal{L}[f(t)] = \mathbf{F}(s) = \int_0^\infty f(t)e^{-st}\, dt \qquad (5.12)$$

---

[4]Named for the French mathematician Marquis Pierre Simon de Laplace (1749–1827).

where $s = \sigma + j\omega$. Because the lower limit of the integral in the definition of the Laplace transform is zero, the Laplace transform treats a function $f(t)$ as if $f(t) = 0$ for $t < 0$ s. Consequently, we will consider only such functions.

Example 5.11

Recall the unit step function $u(t)$ defined by

$$u(t) = \begin{cases} 0 \text{ for } & t < 0 \text{ s} \\ 1 \text{ for } & t \geq 0 \text{ s} \end{cases}$$

Then the Laplace transform of a unit step function is

$$\mathcal{L}[u(t)] = \int_0^\infty u(t)e^{-st}\, dt = \int_0^\infty e^{-st}\, dt = -\frac{1}{s}e^{-st}\Big|_0^\infty = -\frac{1}{s}(0 - 1) = \frac{1}{s}$$

For the decaying exponential $e^{-at}u(t)$, where $a > 0$, we have that

$$\mathcal{L}[e^{-at}u(t)] = \int_0^\infty e^{-at}u(t)e^{-st}\, dt = \int_0^\infty e^{-(s+a)t}\, dt$$

$$= -\frac{1}{s+a}e^{-(s+a)t}\Big|_0^\infty = -\frac{1}{s+a}(0 - 1) = \frac{1}{s+a}$$

Drill Exercise 5.11

Determine the Laplace transform of the function $f(t) = (1 - e^{-at})u(t)$.

ANSWER $a/(s^2 + as)$

Properties of the Laplace Transform

Although the Laplace transform of a function $f(t)$ may be obtained by using the defining integral Eq. 5.12, sometimes it is more convenient to use some of the properties of this transform. We will now derive some of the more useful properties.

If $f(t) = f_1(t) + f_2(t)$, then

$$\mathcal{L}[f(t)] = \int_0^\infty f(t)e^{-st}\,dt = \int_0^\infty [f_1(t) + f_2(t)]e^{-st}\,dt$$

$$= \int_0^\infty [f_1(t)e^{-st} + f_2(t)e^{-st}]\,dt = \int_0^\infty f_1(t)e^{-st}\,dt + \int_0^\infty f_2(t)e^{-st}\,dt$$

$$= \mathcal{L}[f_1(t)] + \mathcal{L}[f_2(t)] \tag{5.13}$$

In other words, the Laplace transform of a sum of functions is equal to the sum of the transforms of the individual functions.

If K is a constant, then

$$\mathcal{L}[Kf(t)] = \int_0^\infty Kf(t)e^{-st}\,dt = K\int_0^\infty f(t)e^{-st}\,dt = K\mathcal{L}[f(t)] \tag{5.14}$$

In other words, if a function is scaled by a constant, then the Laplace transform of the function is scaled by the same constant.

The properties of the Laplace transform given by Eq. 5.13 and Eq. 5.14 collectively are referred to as the **linearity property** of the Laplace transform. We also say that the Laplace transform is a **linear transformation**.

Example 5.12

Let us find the Laplace transform of $f(t) = 3(1 - e^{-2t})u(t)$.
Since we can express this function in the form

$$f(t) = (3 - 3e^{-2t})u(t) = 3u(t) - 3e^{-2t}u(t) = f_1(t) + f_2(t)$$

where $f_1(t) = 3u(t)$ and $f_2(t) = -3e^{-2t}u(t)$. Then by the linearity property of the Laplace transform

$$\mathcal{L}[f(t)] = \mathcal{L}[3u(t)] + \mathcal{L}[-3e^{-2t}u(t)] = 3\mathcal{L}[u(t)] - 3\mathcal{L}[e^{-2t}u(t)]$$

$$= 3\left(\frac{1}{s}\right) - 3\left(\frac{1}{s+2}\right) = \frac{3(s+2) - 3s}{s(s+2)} = \frac{6}{s(s+2)}$$

Drill Exercise 5.12

Find the Laplace transform of $e^{-a(t-1)}u(t)$.

ANSWER $e^{a}/(s + a)$

Now recall Euler's formula (see p. 193)

$$e^{j\theta} = \cos \theta + j \sin \theta \qquad (5.15)$$

Replacing θ by $-\theta$, we get

$$e^{-j\theta} = \cos(-\theta) + j \sin(-\theta)$$

Since $\cos(-\theta) = \cos \theta$ and $\sin(-\theta) = -\sin \theta$, then

$$e^{-j\theta} = \cos \theta - j \sin \theta \qquad (5.16)$$

By adding Eq. 5.15 and Eq. 5.16, we get

$$\cos \theta = (e^{j\theta} + e^{-j\theta})/2 \qquad (5.17)$$

and by subtracting Eq. 5.16 from Eq. 5.15, we obtain

$$\sin \theta = (e^{j\theta} - e^{-j\theta})/j2 \qquad (5.18)$$

As we will now see, Eq. 5.17 and Eq. 5.18 can be used find the Laplace transforms of sinusoids.

Example 5.13

For the case that $f(t) = \cos \beta t\, u(t)$, then

$$\mathcal{L}[\cos \beta t\, u(t)] = \mathcal{L}\left[\frac{1}{2}(e^{j\beta t} + e^{-j\beta t})u(t)\right] = \frac{1}{2}\mathcal{L}[e^{j\beta t}u(t) + e^{-j\beta t}u(t)]$$

$$= \frac{1}{2}\mathcal{L}[e^{j\beta t}u(t)] + \frac{1}{2}\mathcal{L}[e^{-j\beta t}u(t)]$$

Just as $\mathcal{L}[e^{-at}u(t)] = 1/(s + a)$, so too $\mathcal{L}[e^{-jat}u(t)] = 1/(s + ja)$. Therefore,

$$\mathcal{L}[\cos \beta t \; u(t)] = \frac{1}{2}\left(\frac{1}{s - j\beta}\right) + \frac{1}{2}\left(\frac{1}{s + j\beta}\right) = \frac{1}{2}\frac{s + j\beta + s - j\beta}{(s - j\beta)(s + j\beta)}$$

$$= \frac{s}{s^2 + \beta^2}$$

Drill Exercise 5.13

Find the Laplace transform of $f(t) = \sin \beta t \; u(t)$.

ANSWER $\beta/(s^2 + \beta^2)$

Differentiation

Another property of the Laplace transform involves the derivative of a function. Specifically,

$$\mathcal{L}\left[\frac{df(t)}{dt}\right] = \int_0^\infty \frac{df(t)}{dt}e^{-st}\,dt = \int_0^\infty e^{-st}\frac{df(t)}{dt}\,dt$$

We may employ the formula for integration by parts:

$$\int_a^b u\,dv = uv\Big|_a^b - \int_a^b v\,du$$

By selecting $u = e^{-st}$ and $dv = [df(t)/dt]dt = df(t)$, we have that

$$du = -se^{-st}\,dt \qquad \text{and} \qquad v = f(t)$$

Thus,

$$\mathcal{L}\left[\frac{df(t)}{dt}\right] = e^{-at}f(t)\Big|_0^\infty - \int_0^\infty f(t)[-se^{-st}]\,dt = 0 - f(0) + s\int_0^\infty f(t)e^{-st}\,dt$$

$$= -f(0) + s\mathcal{L}[f(t)] = -f(0) + s\mathbf{F}(s)$$

This result is known as the **differentiation property** of the Laplace transform.

Example 5.14

Let us find the Laplace transform of $\sin \beta t\, u(t)$ by using the differentiation property. Since

$$\frac{d[\sin \beta t\, u(t)]}{dt} = \beta \cos \beta t\, u(t)$$

then taking the Laplace transform of both sides of this expression, we obtain

$$\mathcal{L}\left(\frac{d[\sin \beta t\, u(t)]}{dt}\right) = \beta\mathcal{L}[\cos \beta t\, u(t)]$$

Therefore,

$$-\sin 0\, u(0) + s\mathcal{L}[\sin \beta t\, u(t)] = \beta \frac{s}{s^2 + \beta^2}$$

and hence

$$\mathcal{L}[\sin \beta t\, u(t)] = \frac{\beta}{s^2 + \beta^2}$$

Drill Exercise 5.14

Define the unit ramp function $r(t)$ by $r(t) = t\, u(t)$. Use the fact that $dr(t)/dt = u(t)$ to determine the Laplace transform of $r(t)$.

ANSWER $1/s^2$

In order to find $\mathcal{L}[d^2 f(t)/dt^2]$, we can apply the differentiation property twice to obtain

$$\mathcal{L}\left[\frac{d^2 f(t)}{dt^2}\right] = -\frac{df(0)}{dt} - sf(0) + s^2\mathcal{L}[f(t)]$$

where $df(0)/dt$ is the derivative of $f(t)$ evaluated at $t = 0$ s.

Formulas for the Laplace transforms of higher-order derivatives can be obtained by repeated applications of the differentiation property.

Complex Translation

Another important property of the Laplace transform is obtained as follows: Suppose that $\mathbf{F}(s) = \mathcal{L}[f(t)]$. Then

$$\mathcal{L}[e^{-at}f(t)] = \int_0^\infty e^{-at}f(t)e^{-st}\,dt = \int_0^\infty f(t)e^{-(s+a)t}\,dt = \mathbf{F}(s + a)$$

In other words, $\mathcal{L}[e^{-at}f(t)]$ can be obtained from $\mathcal{L}[f(t)]$—simply replace each s in $\mathcal{L}[f(t)]$ by $s + a$. This result is referred to as the **complex-translation property** of the Laplace transform.

Example 5.15

Since $\mathcal{L}[u(t)] = \mathbf{F}(s) = 1/s$, then by the complex-translation property

$$\mathcal{L}[e^{-at}u(t)] = \mathbf{F}(s + a) = \frac{1}{s + a}$$

Furthermore,

$$\mathcal{L}[\cos \beta t\, u(t)] = \frac{s}{s^2 + \beta^2} \quad\Rightarrow\quad \mathcal{L}[e^{-\alpha t}\cos \beta t\, u(t)] = \frac{s + \alpha}{(s + \alpha)^2 + \beta^2}$$

Also,

$$\mathcal{L}[\sin \beta t\, u(t)] = \frac{\beta}{s^2 + \beta^2} \quad\Rightarrow\quad \mathcal{L}[e^{-\alpha t}\sin \beta t\, u(t)] = \frac{\beta}{(s + \alpha)^2 + \beta^2}$$

Drill Exercise 5.15

Use the complex-translation property and the result from Drill Exercise 5.14 to determine the Laplace transform of $te^{-at}u(t)$.

ANSWER $1/(s + a)^2$

Complex Differentiation

Given that $\mathcal{L}[f(t)] = \mathbf{F}(s)$, then

$$\frac{d\,\mathbf{F}(s)}{ds} = \frac{d}{ds}\int_0^\infty f(t)e^{-st}\,dt = \int_0^\infty f(t)\frac{d[e^{-st}]}{ds}\,dt$$

$$= \int_0^\infty f(t)[-te^{-st}]\,dt = -\int_0^\infty tf(t)e^{-st}\,dt = -\mathcal{L}[tf(t)]$$

Thus,

$$\mathcal{L}[tf(t)] = -\frac{d\mathbf{F}(s)}{ds} = -\frac{d}{ds}\{\mathcal{L}[f(t)]\}$$

and this is known as the **complex-differentiation property** of the Laplace transform.

Example 5.16

Since $\mathcal{L}[e^{-at}u(t)] = 1/(s + a)$, then

$$\mathcal{L}[te^{-at}u(t)] = -\frac{d}{ds}\left(\frac{1}{s + a}\right) = \frac{1}{(s + a)^2}$$

Setting $a = 0$, we get

$$\mathcal{L}[t\,u(t)] = 1/s^2$$

Drill Exercise 5.16

Use the complex-differentiation property to determine the Laplace transform of $t^2e^{-at}u(t)$.

ANSWER $2/(s + a)^3$

A summary of some of the properties of the Laplace transform, as well as the transforms of some important functions, is given in Table 5.1.

Table 5.1 Table of Laplace Transforms

$f(t)$	Property	$F(s)$
$f(t)$	Definition	$\int_0^\infty f(t)e^{-st}\, dt$
$f_1(t) + f_2(t)$	Linearity	$\mathbf{F}_1(s) + \mathbf{F}_2(s)$
$Kf(t)$	Linearity	$K\mathbf{F}(s)$
$\dfrac{df(t)}{dt}$	Differentiation	$s\mathbf{F}(s) - f(0)$
$\dfrac{d^2f(t)}{dt^2}$	Differentiation	$s^2\mathbf{F}(s) - sf(0) - \dfrac{df(0)}{dt}$
$\displaystyle\int_0^t f(t)\, dt$	Integration	$\dfrac{1}{s}\mathbf{F}(s)$
$tf(t)$	Complex differentiation	$-\dfrac{d\mathbf{F}(s)}{ds}$
$e^{-at}f(t)$	Complex translation	$\mathbf{F}(s + a)$
$f(t - a)\, u(t - a)$	Real translation	$e^{-as}\mathbf{F}(s)$
$u(t)$		$\dfrac{1}{s}$
$e^{-at}u(t)$		$\dfrac{1}{s + a}$
$\cos \beta t\, u(t)$		$\dfrac{s}{s^2 + \beta^2}$
$\sin \beta t\, u(t)$		$\dfrac{\beta}{s^2 + \beta^2}$
$e^{-\alpha t} \cos \beta t\, u(t)$		$\dfrac{s + \alpha}{(s + \alpha)^2 + \beta^2}$
$e^{-\alpha t} \sin \beta t\, u(t)$		$\dfrac{\beta}{(s + \alpha)^2 + \beta^2}$
$t\, u(t)$		$\dfrac{1}{s^2}$
$te^{-at}u(t)$		$\dfrac{1}{(s + a)^2}$

5.6 Inverse Laplace Transforms

Given a function $\mathbf{F}(s)$, in order to determine the function $f(t)$ such that $\mathscr{L}[f(t)] = \mathbf{F}(s)$, we must take the **inverse Laplace transform** of $\mathbf{F}(s)$—which is denoted as $\mathscr{L}^{-1}[\mathbf{F}(s)] = f(t)$. Although this can be done with the use of a mathematical formula, such an approach requires the use of advanced mathematics. Therefore, we will take inverse Laplace transforms instead by inspecting the table of Laplace transforms (see Table 5.1) to see what function $f(t)$ has the Laplace transform $\mathbf{F}(s)$. If $\mathbf{F}(s)$ is not in the table, we will decompose it into functions that are in the table or are readily obtainable by using the properties of Laplace transforms.

Just as the Laplace transform is a linear transformation, so too the inverse Laplace transform is a linear transformation. Specifically, if $\mathbf{F}(s) = \mathbf{F}_1(s) + \mathbf{F}_2(s)$, then

$$\mathscr{L}^{-1}[\mathbf{F}(s)] = \mathscr{L}^{-1}[\mathbf{F}_1(s) + \mathbf{F}_2(s)] = \mathscr{L}^{-1}[\mathbf{F}_1(s)] + \mathscr{L}^{-1}[\mathbf{F}_2(s)]$$

Furthermore,

$$\mathscr{L}^{-1}[K\mathbf{F}(s)] = K\,\mathscr{L}^{-1}[\mathbf{F}(s)]$$

Example 5.17

Let us determine $f_1(t)$ given that its Laplace transform is

$$\mathscr{L}[f_1(t)] = \mathbf{F}_1(s) = \frac{14s + 23}{s^2 + 4s + 5}$$

By completing the square for the denominator, this function can be written in the form

$$\mathbf{F}_1(s) = \frac{14s + 23}{(s + 2)^2 + 1^2} = \frac{14(s + 2) - 5}{(s + 2)^2 + 1^2}$$

$$= \frac{14(s + 2)}{(s + 2)^2 + 1^2} - \frac{5(1)}{(s + 2)^2 + 1^2}$$

Since the inverse Laplace transform of a sum is equal to the sum of the individual inverse Laplace transforms, by using Table 5.1, we get

$$f_1(t) = 14e^{-2t} \cos t\, u(t) - 5e^{-2t} \sin t\, u(t) = e^{-2t}(14 \cos t - 5 \sin t)u(t)$$

Next let us find $f_2(t)$ given that its Laplace transform is

$$\mathcal{L}[f_2(t)] = \mathbf{F}_2(s) = \frac{14s + 23}{s^2 + 5s + 4}$$

In this case the denominator cannot be put into the form $(s + \alpha)^2 + \beta^2$, where α and β are real numbers. However, it is true that—and we will see why shortly—

$$\mathbf{F}_2(s) = \frac{14s + 23}{s^2 + 5s + 4} = \frac{3}{s + 1} + \frac{11}{s + 4} \tag{5.19}$$

Therefore, from Table 5.1, the inverse Laplace transform of $\mathbf{F}_2(s)$ is

$$f_2(t) = 3e^{-t}u(t) + 11e^{-4t}u(t) = (3e^{-t} + 11e^{-4t})u(t)$$

Drill Exercise 5.17

Determine the inverse Laplace transform of

$$\frac{2s + 26}{s^2 + 6s + 25}$$

ANSWER $e^{-3t}(2 \cos 4t + 5 \sin 4t)u(t)$

Partial-Fraction Expansions

In Example 5.17, Eq. 5.19 indicates that $\mathbf{F}_2(s)$ can be expressed as the sum of two functions, each of which is in the form of a function in Table 5.1. There is a systematic method for decomposing a function into a sum of simpler functions—such a decomposition is called a **partial-fraction expansion**, and we now describe a procedure for obtaining it.

Suppose we are given a function $\mathbf{F}(s) = \mathbf{N}(s)/\mathbf{D}(s)$, where $\mathbf{N}(s)$ and $\mathbf{D}(s)$ are polynomials in s with real coefficients. If the roots of $\mathbf{D}(s)$ are $s_1, s_2, s_3, \ldots s_n$, we can write $\mathbf{F}(s)$ in the form

$$F(s) = \frac{N(s)}{D(s)} = \frac{N(s)}{(s - s_1)(s - s_2)(s - s_3) \cdots (s - s_n)} \tag{5.20}$$

If the degree of $D(s)$ is greater than the degree of $N(s)$, and if the roots of $D(s)$— that is, the poles of $F(s)$—are distinct, then we may write

$$F(s) = \frac{K_1}{s - s_1} + \frac{K_2}{s - s_2} + \frac{K_3}{s - s_3} + \cdots + \frac{K_n}{s - s_n} \tag{5.21}$$

To find K_1, first multiply both sides of Eq. 5.21 by $s - s_1$. This yields

$$(s - s_1)F(s) = K_1 + \frac{K_2(s - s_1)}{s - s_2} + \frac{K_3(s - s_1)}{s - s_3} + \cdots + \frac{K_n(s - s_1)}{s - s_n}$$

If we set $s = s_1$, then this equation becomes

$$(s - s_1)F(s)\Big|_{s=s_1} = K_1$$

Similarly,

$$(s - s_2)F(s)\Big|_{s=s_2} = K_2$$

et cetera.

Example 5.18

Let us take the partial-fraction expansion of

$$F(s) = \frac{2s^2 + 11s + 19}{(s + 1)(s + 2)(s + 3)} = \frac{K_1}{s + 1} + \frac{K_2}{s + 2} + \frac{K_3}{s + 3}$$

Multiplying this expression by $s + 1$ and then setting $s = -1$, we get

$$K_1 = \frac{2s^2 + 11s + 19}{(s + 2)(s + 3)}\Big|_{s=-1} = \frac{2(-1)^2 + 11(-1) + 19}{(-1 + 2)(-1 + 3)} = 5$$

Multiplying $\mathbf{F}(s)$ by $s + 2$ and then setting $s = -2$, we obtain

$$K_2 = \frac{2s^2 + 11s + 19}{(s + 1)(s + 3)}\bigg|_{s=-2} = \frac{2(-2)^2 + 11(-2) + 19}{(-2 + 1)(-2 + 3)} = -5$$

Finally,

$$K_3 = \frac{2s^2 + 11s + 19}{(s + 1)(s + 2)}\bigg|_{s=-3} = \frac{2(-3)^2 + 11(-3) + 19}{(-3 + 1)(-3 + 2)} = 2$$

Hence

$$\mathbf{F}(s) = \frac{5}{s + 1} - \frac{5}{s + 2} + \frac{2}{s + 3}$$

and from Table 5.1, the inverse Laplace transform of $\mathbf{F}(s)$ is

$$f(t) = 5e^{-t}u(t) - 5e^{-2t}u(t) + 2e^{-3t}u(t) = (5e^{-t} - 5e^{-2t} + 2e^{-3t})u(t)$$

Drill Exercise 5.18

Determine the inverse Laplace transform of

$$\frac{600}{s^3 + 40s^2 + 300s}$$

ANSWER $(2 - 3e^{-10t} + e^{-30t})u(t)$

Multiple and Complex Poles

In Example 5.18, the function $\mathbf{F}(s)$ has poles that are both distinct and real. In general, however, the poles of $\mathbf{F}(s)$ can be nondistinct or complex—occurring in conjugate pairs. Although there are formal procedures for taking partial-fraction expansions for these cases as well, we will study only circuits or systems whose complexity is such that we may deal with these cases by utilizing the technique for distinct, real poles.

Example 5.19

Consider the following function $\mathbf{F}_1(s)$ with a double pole at $s = -1$:

$$\mathbf{F}_1(s) = \frac{s^2 + 5s + 6}{s(s + 1)^2}$$

Let us determine the term K_0 associated with the pole at the origin of the s plane by writing

$$\mathbf{F}_1(s) = \frac{K_0}{s} + \mathbf{G}_1(s)$$

Then

$$K_0 = s\mathbf{F}_1(s)\bigg|_{s=0} = \frac{s^2 + 5s + 6}{(s + 1)^2}\bigg|_{s=0} = 6$$

Thus, we can write

$$\mathbf{G}_1(s) = \mathbf{F}_1(s) - \frac{K_0}{s} = \frac{s^2 + 5s + 6}{s(s + 1)^2} - \frac{6}{s} = \frac{-5s - 7}{(s + 1)^2} = \frac{-5(s + 1 - 1) - 7}{(s + 1)^2}$$

$$= \frac{-5(s + 1) - 2}{(s + 1)^2} = -\frac{5}{s + 1} - \frac{2}{(s + 1)^2}$$

Hence

$$\mathbf{F}_1(s) = \frac{6}{s} - \frac{5}{s + 1} - \frac{2}{(s + 1)^2}$$

and from Table 5.1, we have that

$$f_1(t) = 6u(t) - 5e^{-t}u(t) - 2te^{-t}u(t) = (6 - 5e^{-t} - 2te^{-t})u(t)$$

Drill Exercise 5.19

Determine the inverse Laplace transform of

$$\frac{s^2 - 3s - 6}{(s + 1)(s + 2)^2}$$

ANSWER $(-2e^{-t} + 3e^{-2t} - 4te^{-2t})u(t)$

Having considered the case of multiple poles, let us now look at a case of complex poles.

Example 5.20

The following function $F_2(s)$ has complex poles at $s = -1 - j2$ and $s = -1 + j2$.

$$F_2(s) = \frac{5s^2 + 30s - 15}{s(s^2 + 2s + 5)} = \frac{5s^2 + 30s - 15}{s(s + 1 + j2)(s + 1 - j2)}$$

However, we need not replace $s^2 + 2s + 5$ with $(s + 1 + j2)(s + 1 - j2)$ as indicated above. Instead, we can write

$$F_2(s) = \frac{K_0}{s} + G_2(s)$$

where

$$K_0 = sF_2(s)\Big|_{s=0} = \frac{5s^2 + 30s - 15}{s^2 + 2s + 5}\Big|_{s=0} = -3$$

Thus

$$G_2(s) = F_2(s) - \frac{K_0}{s} = \frac{5s^2 + 30s - 15}{s(s^2 + 2s + 5)} + \frac{3}{s} = \frac{8s + 36}{s^2 + 2s + 5}$$

$$= \frac{8s + 36}{(s + 1)^2 + 2^2} = \frac{8(s + 1) + 14(2)}{(s + 1)^2 + 2^2}$$

Therefore,

$$F_2(s) = -\frac{3}{s} + \frac{8(s + 1)}{(s + 1)^2 + 2} + \frac{14(2)}{(s + 1)^2 + 2^2}$$

and from Table 5.1, we get that

$$f_2(t) = -3u(t) + 8e^{-t} \cos 2t \, u(t) + 14e^{-t} \sin 2t \, u(t)$$

Drill Exercise 5.20

Determine the inverse Laplace transform of

$$\frac{14s - 50}{s^3 + 6s^2 + 25s}$$

ANSWER $-2u(t) + 2e^{-3t} \cos 4t \, u(t) + 5e^{-3t} \sin 4t \, u(t)$

5.7 Application of the Laplace Transform

One very important application of the Laplace transform is to the solution of differential equations.

Example 5.21

Suppose that we wish to solve the linear, second-order differential equation

$$\frac{d^2x(t)}{dt^2} + 3\frac{dx(t)}{dt} + 2x(t) = 4e^{-3t}u(t) \tag{5.22}$$

subject to the initial conditions $x(0) = 2$ and $dx(0)/dt = -1$. Taking the Laplace transform of this differential equation, if we let $\mathbf{X}(s) = \mathcal{L}[x(t)]$, then

$$\left[s^2\mathbf{X}(s) - sx(0) - \frac{dx(0)}{dt} \right] + 3[s\mathbf{X}(s) - x(0)] + 2\mathbf{X}(s) = \frac{4}{s + 3}$$

from which

$$s^2\mathbf{X}(s) - 2s + 1 + 3s\mathbf{X}(s) - 6 + 2\mathbf{X}(s) = \frac{4}{s + 3}$$

Therefore,

$$(s^2 + 3s + 2)\mathbf{X}(s) = \frac{4}{s + 3} + 2s + 5 = \frac{2s^2 + 11s + 19}{s + 3}$$

and hence

$$\mathbf{X}(s) = \frac{2s^2 + 11s + 19}{(s + 1)(s + 2)(s + 3)}$$

By Example 5.18, the solution to Eq. 5.22 subject to the given initial conditions is

$$x(t) = (5e^{-t} - 5e^{-2t} + 2e^{-3t})u(t)$$

Drill Exercise 5.21

Find the solution of the differential equation

$$\frac{d^2x(t)}{dt^2} + 8\frac{dx(t)}{dt} + 12x(t) = 0$$

subject to the initial conditions $x(0) = 0$ and $dx(0)/dt = -12$.

ANSWER $(3e^{-6t} - 3e^{-2t})u(t)$ (See Drill Exercise 3.12 on p. 155.)

Application to Circuit Analysis

Of course, we can write the differential equation or equations that characterize a circuit and then use Laplace transforms to solve such equations. However, we can avoid writing differential equations if we employ Laplace-transform (frequency-domain) concepts directly. Let's see how this is done.

For a resistor having a value of R ohms, we know that

$$v(t) = Ri(t)$$

Taking the Laplace transform of both sides of this equation results in

$$\mathbf{V}(s) = R\mathbf{I}(s)$$

where $\mathbf{V}(s) = \mathcal{L}[v(t)]$ and $\mathbf{I}(s) = \mathcal{L}[i(t)]$. Defining the impedance $\mathbf{Z}_R(s)$ of the resistor to be the ratio of voltage transform to current transform, we get

$$\boxed{\mathbf{Z}_R(s) = \frac{\mathbf{V}(s)}{\mathbf{I}(s)} = R}$$

The circuit symbols of a resistor in the time domain and in the frequency (transform) domain are shown in Fig. 5.37a and b, respectively.

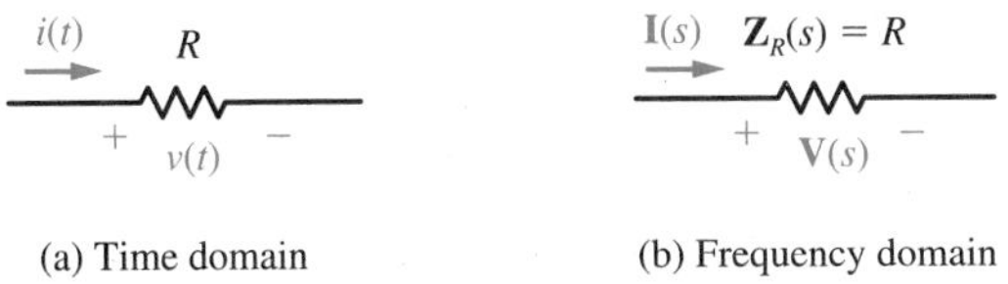

(a) Time domain   (b) Frequency domain

Fig. 5.37 Resistor circuit symbols.

For an inductor having a value of L henries,

$$v(t) = L\frac{di(t)}{dt}$$

Taking the Laplace transform, we get

$$\mathbf{V}(s) = L[s\mathbf{I}(s) - i(0)] = Ls\mathbf{I}(s) - Li(0) \qquad (5.23)$$

from which

$$\mathbf{I}(s) = \frac{1}{Ls}\mathbf{V}(s) + \frac{i(0)}{s} \qquad (5.24)$$

For the case of zero initial conditions, $i(0) = 0$ A, and thus

$$\mathbf{V}(s) = Ls\mathbf{I}(s)$$

We then define the impedance $\mathbf{Z}_L(s)$ of the inductor to be the ratio of voltage transform to current transform when the initial current is zero. Thus

$$Z_L(s) = \frac{V(s)}{I(s)} = Ls$$

The time-domain circuit symbol for an inductor is shown in Fig. 5.38a, whereas Fig. 5.38b shows the circuit symbol in the frequency domain when the initial current is zero. For the case that the initial current is not necessarily zero, the parallel connection shown in Fig. 5.38c models the frequency-domain description given by Eq. 5.24. Note that if $i(0) = 0$ A, then the model in Fig. 5.38c is equivalent to the one in Fig. 5.38b.

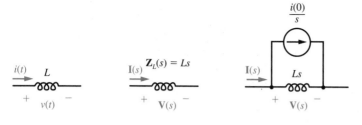

(a) Time domain   (b) Frequency domain—   (c) Frequency domain—
               zero initial current   nonzero initial current

Fig. 5.38 Inductor circuit symbols.

For a capacitor having a value of C farads,

$$i(t) = C\frac{dv(t)}{dt}$$

Taking the Laplace transform, we get

$$I(s) = C[sV(s) - v(0)] = CsV(s) - Cv(0) \tag{5.25}$$

from which

$$V(s) = \frac{1}{Cs}I(s) + \frac{v(0)}{s} \tag{5.26}$$

The impedance $Z_C(s)$ of the capacitor (for zero initial voltage) is

$$Z_C(s) = \frac{V(s)}{I(s)} = \frac{1}{Cs}$$

The circuit symbols for a capacitor in the time domain and the frequency domain, with zero and nonzero initial conditions, are shown in Fig. 5.39.

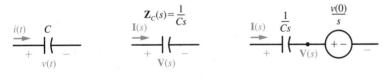

(a) Time domain

(b) Frequency domain— zero initial voltage

(c) Frequency domain— nonzero initial voltage

Fig. 5.39 Capacitor circuit symbols.

Since we have the same expressions for the impedances of resistors, inductors, and capacitors as we had for the damped-sinusoidal case, we can use the same circuit analysis techniques—the difference being that we use the Laplace transforms of time functions rather than their phasor representations.

Example 5.22

Suppose that the series *RLC* circuit shown in Fig. 5.40*a* has zero initial conditions. Let us find the step response $v(t)$.

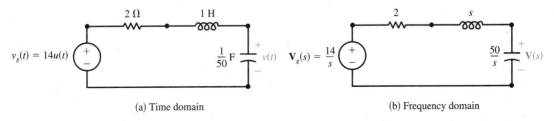

(a) Time domain

(b) Frequency domain

Fig. 5.40 A series *RLC* circuit (*a*) in the time domain, and (*b*) in the frequency domain.

Figure 5.40*b* shows the circuit in the frequency domain. Of course this circuit can be analyzed by using either mesh analysis or nodal analysis. However, even more simply, by voltage division we can write

$$\mathbf{V}(s) = \frac{50/s}{50/s + s + 2}\mathbf{V}_g(s) = \frac{50}{s^2 + 2s + 50}\left(\frac{14}{s}\right)$$

or

$$V(s) = \frac{700}{s(s^2 + 2s + 50)} = \frac{K_0}{s} + F(s)$$

where

$$K_0 = sV(s)\Big|_{s=0} = \frac{700}{s^2 + 2s + 50}\Big|_{s=0} = 14$$

Thus

$$F(s) = V(s) - \frac{K_0}{s} = \frac{700}{s(s^2 + 2s + 50)} - \frac{14}{s} = \frac{-14s - 28}{s^2 + 2s + 50}$$

Hence

$$V(s) = \frac{14}{s} - \frac{14s + 28}{s^2 + 2s + 50} = \frac{14}{s} - \frac{14s + 28}{(s + 1)^2 + 7^2}$$

$$= \frac{14}{s} - \frac{14(s + 1)}{(s + 1)^2 + 7^2} - \frac{(2)(7)}{(s + 1)^2 + 7^2}$$

Therefore, from Table 5.1, the inverse Laplace transform of $V(s)$ is

$$v(t) = 14u(t) - 14e^{-t} \cos 7t\, u(t) - 2e^{-t} \sin 7t\, u(t)$$

$$= [14 - e^{-t}(14 \cos 7t + 2 \sin 4t)]u(t) \text{ V}$$

and this is the complete response—i.e., forced and natural responses—for an underdamped series *RLC* circuit. (See Example 3.17 on p. 170.)

Drill Exercise 5.22

For the series *RLC* circuit shown in Fig. 5.40, change the value of the resistor to 16 Ω, change the value of the inductor to 2 H, and find the step response $v(t)$ when $v_g(t) = 6u(t)$ V. (See Drill Exercise 3.17 on p. 172.)

ANSWER $[6 - e^{-4t}(6 \cos 3t + 8 \sin 3t)]u(t)$ V

Nonzero Initial Conditions

Having analyzed a circuit with zero initial conditions, let us now consider the case of a circuit with nonzero initial conditions. In the following examples, we will simplify the notation by replacing $\mathbf{V}_1(s)$, $\mathbf{V}_2(s)$, and $\mathbf{V}(s)$ with $\mathbf{V}_1$, $\mathbf{V}_2$, and $\mathbf{V}$, respectively.

Example 5.23

Suppose that we wish to find $v(t)$ for the series *RLC* circuit shown in Fig. 5.41*a* subject to the initial conditions $v(0) = 2$ V and $i(0) = 1$ A.

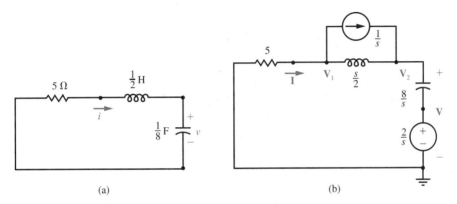

(a)   (b)

Fig. 5.41 A series *RLC* circuit (*a*) in the time domain, and (*b*) in the frequency domain.

In the frequency domain, the circuit is as shown in Fig. 5.41*b*. Here, the initial voltage across the capacitor is modeled with an independent voltage source, and the initial current through the inductor is modeled with an independent current source. Note that the voltage (transform) $\mathbf{V}$ that is to be determined is the voltage across the series combination of the $\frac{1}{8}$-F capacitor and a voltage source having a value of $v(0)/s = 2/s$. Although two node voltages $\mathbf{V}_1$ and $\mathbf{V}_2$ are indicated in Fig. 5.41*b*, note that $\mathbf{V}_2 = \mathbf{V}$.

Summing the currents directed out of the node labeled $\mathbf{V}_1$, by KCL, we get

$$\frac{\mathbf{V}_1}{5} + \frac{\mathbf{V}_1 - \mathbf{V}_2}{s/2} + \frac{1}{s} = 0 \quad \Rightarrow \quad (s + 10)\mathbf{V}_1 - 10\mathbf{V}_2 = -5 \quad (5.27)$$

By KCL at the node labeled $\mathbf{V}_2$, we obtain

$$\frac{V_2 - V_1}{s/2} + \frac{V_2 - 2/s}{8/s} = \frac{1}{s} \quad \Rightarrow$$

$$-16V_1 - (s^2 + 16)V_2 = 2s + 8 \tag{5.28}$$

Using Cramer's rule to solve simultaneous Eq. 5.27 and Eq. 5.28 results in

$$V^2 = \frac{\begin{vmatrix} s + 10 & -5 \\ -16 & 2s + 8 \end{vmatrix}}{\begin{vmatrix} s + 10 & -5 \\ -16 & s^2 + 16 \end{vmatrix}}$$

$$= \frac{(s + 10)(2s + 8) - 80}{(s + 10)(s^2 + 16) - 160} = \frac{4}{s + 2} + \frac{-2}{s + 8}$$

Since $V = V_2$, then

$$v(t) = 4e^{-t}u(t) - 2e^{-8t}u(t) = (4e^{-2t} - 2e^{-8t})u(t) \text{ V}$$

and this is an example of an overdamped natural response (see Example 3.12 on p. 153).

Drill Exercise 5.23

Determine $v(t)$ for the series *RLC* circuit shown in Fig. 5.41 when $R = 4\ \Omega$, $L = 1$ H, and $C = \frac{1}{4}$ F subject to the initial conditions $v(0) = 2$ V and $i(0) = 1$ A. (See Example 3.15 on p. 163.)

ANSWER $(2e^{-2t} + 8te^{-2t})u(t)$ V

Circuits with Sinusoidal Sources

In Chapter 4 we saw how to find forced sinusoidal responses by using phasors. We will now see an example of how to determine the complete response to a sinusoidal source (which is zero for $t < 0$ s) by using Laplace-transform techniques.

Example 5.24

Let us determine the responses $i(t)$ and $v_o(t)$ to a sinusoidal excitation (which begins at time $t = 0$ s) for the circuit shown in Fig. 5.42a. The frequency-domain representation of this circuit is shown in Fig. 5.42b.

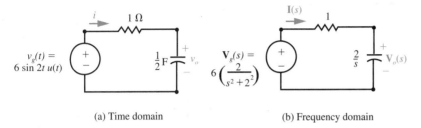

(a) Time domain   (b) Frequency domain

Fig. 5.42 A series *RC* circuit (*a*) in the time domain, and (*b*) in the frequency domain.

By KVL,

$$\mathbf{V}_g = 1\mathbf{I} + \frac{2}{s}\mathbf{I} = \frac{s + 2}{s}\mathbf{I} \quad \Rightarrow \quad \mathbf{I} = \frac{s}{s + 2}\mathbf{V}_g = \frac{12s}{(s + 2)(s^2 + 4)}$$

However, we can express $\mathbf{I}$ in the form

$$\mathbf{I} = \frac{12s}{(s + 2)(s^2 + 4)} = \frac{K_1}{s + 2} + \mathbf{F}_1(s) \tag{5.29}$$

where

$$K_1 = \frac{12s}{s^2 + 4}\bigg|_{s=-2} = -3$$

From Eq. 5.29,

$$\mathbf{F}_1(s) = \mathbf{I} - \frac{K_1}{s + 2} = \frac{12s}{(s + 2)(s^2 + 4)} - \frac{-3}{s + 2} = \frac{3s + 6}{s^2 + 4}$$

Substituting the above values of K_1 and $\mathbf{F}_1(s)$ into Eq. 5.29 yields

$$\mathbf{I} = \frac{-3}{s + 2} + \frac{3s}{s^2 + 4} + \frac{6}{s^2 + 4} = \frac{-3}{s + 2} + \frac{3s}{s^2 + 2^2} + \frac{3(2)}{s^2 + 2^2}$$

Taking the inverse Laplace transform of **I** gives us

$$i(t) = -3e^{-2t}u(t) + 3\cos 2t\, u(t) + 3\sin 2t\, u(t) \text{ A}$$

But note that this response is a complete response—the natural response is $i_n(t) = -3e^{-2t}u(t)$ A and the forced response is $i_f(t) = 3\cos 2t\, u(t) + 3\sin 2t\, u(t) = 3\sqrt{2}\cos(2t - 45°)\, u(t)$ A.

To find $v_o(t)$, we first calculate $\mathbf{V}_o$ from

$$\mathbf{V}_o = \mathbf{Z}_C\mathbf{I} = \frac{2}{s}\frac{12s}{(s+2)(s^2+4)} = \frac{24}{(s+2)(s^2+4)}$$

$$= \frac{K_2}{s+2} + \mathbf{F}_2(s) \tag{5.30}$$

where

$$K_2 = \frac{24}{s^2+4}\Bigg|_{s=-2} = 3$$

and

$$\mathbf{F}_2(s) = \mathbf{V}_o - \frac{K_2}{s+2} = \frac{24}{(s+2)(s^2+4)} - \frac{3}{s+2)}$$

$$= \frac{-3(s^2-4)}{(s+2)(s^2+4)} = \frac{-3(s-2)}{\cdot\ s^2+4}$$

Substituting the above values of K_2 and $\mathbf{F}_2(s)$ into Eq. 5.30 yields

$$\mathbf{V}_o = \frac{3}{s+2} + \frac{-3s}{s^2+4} + \frac{6}{s^2+4} = \frac{3}{s+2} - \frac{3s}{s^2+2^2} + \frac{3(2)}{s^2+2^2}$$

Taking the inverse Laplace transform of $\mathbf{V}_o$ gives us

$$v_o(t) = 3e^{-2t}u(t) - 3\cos 2t\, u(t) + 3\sin 2t\, u(t) \text{ V}$$

This response again is a complete response—the natural response is $v_n(t) = 3e^{-2t}u(t)$ V and the forced response is $v_f(t) = -3\cos 2t\, u(t) + 3\sin 2t\, u(t) = 3\sqrt{2}\cos(2t - 135°)\, u(t)$ V.

If only the forced response of an ac circuit is of interest,[5] Laplace transform techniques can be avoided by taking the simpler phasor-analysis approach discussed in Chapter 4. In particular, for this circuit see Example 4.4 on p. 204.

Drill Exercise 5.24

For the circuit given in Fig. 5.42, replace the $\frac{1}{2}$-F capacitor with a $\frac{1}{2}$-H inductor. Use Laplace transform techniques to determine $i(t)$ and $v_o(t)$.

ANSWER $[3e^{-2t} - 3 \cos 2t + 3 \sin 2t]u(t)$ A,
$[-3e^{-2t} + 3 \cos 2t + 3 \sin 2t]u(t)$ V

Application to Linear Systems

Suppose that the input to a linear system has a Laplace transform of $X(s)$, and suppose that the Laplace transform of the output, given that all the initial conditions are zero, is $Y(s)$. Then the **transfer function H(s)** of the system is defined to be

$$H(s) = \frac{Y(s)}{X(s)}$$

If the transfer function of a linear system is known, then when the input is specified, the output transform can be determined from the equation

$$Y(s) = H(s)X(s)$$

Taking the inverse Laplace transform of $Y(s)$ yields the corresponding output $y(t)$ in the time domain.

Example 5.25

Consider a low-pass filter with a voltage transfer function of

$$H(s) = \frac{V_2(s)}{V_1(s)} = \frac{3}{s + 3}$$

---

[5]The natural response is negligible after just a few time constants—in this example the time constant is $\frac{1}{3}$ s.

Let us find the output voltage $v_2(t)$ for the case that the input voltage is $v_1(t) = 2e^{-3t}u(t)$ V.

Since

$$\mathbf{V}_1(s) = \mathscr{L}[2e^{-3t}u(t)] = \frac{2}{s + 3}$$

then

$$\mathbf{V}_2(s) = \mathbf{H}(s)\mathbf{V}_1(s) = \left(\frac{3}{s + 3}\right)\left(\frac{2}{s + 3}\right) = \frac{6}{(s + 3)^2}$$

and

$$v_2(t) = 6te^{-3t}u(t) \text{ V}$$

In this case, the forced and natural responses combine into the single term $6te^{-3t}u(t)$. This is a consequence of exciting the system at its pole; that is, the pole of $\mathbf{V}_1(s)$ is the same as the pole of $\mathbf{H}(s)$.

Drill Exercise 5.25

A high-pass filter has the voltage transfer function $\mathbf{H}(s) = 3s/(s + 3)$. Find the output voltage $v_2(t)$ when the input voltage is $v_1(t) = 2e^{-3t}u(t)$ V.

ANSWER $(6e^{-3t} - 18te^{-3t})u(t)$ V

SUMMARY

1. The frequency response of a circuit consists of the amplitude response and the phase response.

2. The frequencies at which the amplitude response drops to $1/\sqrt{2}$ of its maximum value are the half-power frequencies.

3. The frequencies at which an impedance (or admittance) is purely real are the resonance frequencies of the impedance (or admittance).

4. The bandwidth and the quality factor are measures of the sharpness of an amplitude response.

5. An impedance can be expressed as a function of the complex frequency s, as can other parameters like the voltage transfer function.

6. The poles and zeros of a ratio of polynomials in s can be depicted in the s plane with a pole-zero plot.

7. If there are no cancellations of common poles and zeros, the poles of a network function indicate the form of the natural response.

8. A linear system can be simulated with integrators, adders, and scalers (i.e., with an analog computer).

9. Systems are often represented by block diagrams.

10. Feedback can improve system performance and can be used for purposes of control.

11. The Laplace transform is a linear transformation that can be used to solve linear differential equations or analyze linear circuits.

12. The inverse Laplace transform can be found by using a table of transforms and various transform properties, as well as partial-fraction expansions.

13. The impedance of an R-ohm resistor is R, of an L-henry inductor is Ls, and of a C-farad capacitor is $1/Cs$.

14. An inductor (or a capacitor) with a nonzero initial condition can be modeled by an independent source and an inductor (or capacitor) with a zero initial condition.

15. Circuit analysis using Laplace transforms results in complete (both forced and natural) responses.

Problems

5.1 Sketch the phase response ang($\mathbf{V}_2/\mathbf{V}_1$) versus ω for the high-pass filter given in Fig. 5.5 on p. 269.

5.2 For the circuit given in Fig. 5.5 on p. 269, replace the capacitor C with an inductor L, and sketch the phase response ang($\mathbf{V}_2/\mathbf{V}_1$) versus ω for the resulting low-pass filter.

5.3 Sketch the amplitude response of $\mathbf{V}_2/\mathbf{V}_1$ for the op-amp circuit shown in Fig. P5.3. Determine the half-power frequency. What type of filter is this circuit?

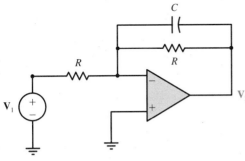

Fig. P5.3

5.4 Show that for the circuit given in Fig. P5.4 the voltage transfer function is

$$\mathbf{H}(j\omega) = \frac{\mathbf{V}_2}{\mathbf{V}_1} = \frac{R_2(1 + j\omega R_1 C_1)}{(R_1 + R_2) + j\omega R_1 R_2(C_1 + C_2)}$$

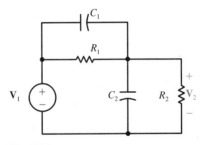

Fig. P5.4

5.5 For the circuit shown in Fig. P5.4, suppose that $R_1 = R_2 = R$ and $C_1 = C_2 = C$. Sketch the amplitude response and the phase response of $\mathbf{V}_2/\mathbf{V}_1$.

5.6 For the circuit shown in Fig. P5.4, suppose that $R_1 = R_2 = R$, $C_1 = C$ and $C_2 = 0$ F. Sketch the amplitude response of $\mathbf{V}_2/\mathbf{V}_1$. What is the half-power frequency?

5.7 For the circuit shown in Fig. P5.4, suppose that $R_1 = R_2 = R$, $C_1 = 0$ F and $C_2 = C$. Sketch the amplitude response of $\mathbf{V}_2/\mathbf{V}_1$. What is the half-power frequency?

5.8 For the op-amp circuit shown in Fig. P5.8, sketch the amplitude response of $\mathbf{V}_2/\mathbf{V}_1$, indicating the half-power frequency. What type of filter is this circuit?

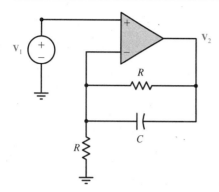

Fig. P5.8

5.9 Sketch the Bode plot—both the amplitude and phase responses—for (a) $\mathbf{H}(j\omega) = K$, where $K > 0$, (b) $\mathbf{H}(j\omega) = j\omega$, and (c) $\mathbf{H}(j\omega) = 1/j\omega$.

5.10 Use only the straight-line asymptotes to sketch the Bode plot—both the amplitude and phase responses—for

$$\mathbf{H}(j\omega) = \frac{(a + j\omega)}{a}$$

where $a > 0$. What is the corner frequency?

5.11 Use only the straight-line asymptotes to sketch the Bode plot—both the amplitude and phase responses—for

$$\mathbf{H}(j\omega) = \frac{a}{a + j\omega} = \frac{1}{1 + j\omega/a}$$

where $a > 0$. What is the corner frequency? What type of filter is this?

5.12 The transfer function

$$\mathbf{H}(j\omega) = \frac{j\omega/a}{1 + j\omega/a} = \frac{j\omega}{a + j\omega}$$

where $a > 0$, can be expressed as the product $\mathbf{H}(j\omega) = \mathbf{H}_1(j\omega)\mathbf{H}_2(j\omega)$, where $\mathbf{H}_1(j\omega) = j\omega/a$ and $\mathbf{H}_2(j\omega) = 1/(1 + j\omega/a)$. Use only the straight-line asymptotes to sketch the Bode plot—both the amplitude and phase responses—for $\mathbf{H}(j\omega)$ by adding the Bode plots for $\mathbf{H}_1(j\omega)$ and $\mathbf{H}_2(j\omega)$. What type of filter is this?

5.13 The transfer function

$$\mathbf{H}(j\omega) = \frac{1}{1 + j\omega 11 + (j\omega)^2 10}$$

$$= \left(\frac{1}{1 + j\omega}\right)\left(\frac{1}{1 + j\omega 10}\right)$$

is expressed as the product $\mathbf{H}(j\omega) = \mathbf{H}_1(j\omega)\mathbf{H}_2(j\omega)$, where $\mathbf{H}_1(j\omega) = 1/(1 + j\omega)$ and $\mathbf{H}_2(j\omega) = 1/(1 + j\omega 10)$. Use only the straight-line asymptotes to sketch the Bode plot—both the amplitude and phase responses—for $\mathbf{H}(j\omega)$ by adding the Bode plots for $\mathbf{H}_1(j\omega)$ and $\mathbf{H}_2(j\omega)$. What type of filter is this?

5.14 The transfer function

$$\mathbf{H}(j\omega) = \frac{j\omega}{10 + j\omega 11 + (j\omega)^2}$$

$$= \left(\frac{1}{1 + j\omega}\right)\left(\frac{j\omega}{10 + j\omega}\right)$$

is expressed as the product $\mathbf{H}(j\omega) = \mathbf{H}_1(j\omega)\mathbf{H}_2(j\omega)$, where $\mathbf{H}_1(j\omega) = 1/(1 + j\omega)$ and $\mathbf{H}_2(j\omega) = j\omega/(10 + j\omega)$. Use only the straight-line asymptotes to sketch the Bode plot—both the amplitude and phase responses—for $\mathbf{H}(j\omega)$ by adding the Bode plots for $\mathbf{H}_1(j\omega)$ and $\mathbf{H}_2(j\omega)$. What type of filter is this?

5.15 For the circuit given in Fig. P5.15, show that:

(a) $\mathbf{H}_C(j\omega) = \dfrac{\mathbf{V}_C}{\mathbf{V}_1} = \dfrac{1/LC}{1/LC + j\omega R/L + (j\omega)^2}$

(b) $\mathbf{H}_L(j\omega) = \dfrac{\mathbf{V}_L}{\mathbf{V}_1} = \dfrac{(j\omega)^2}{1/LC + j\omega R/L + (j\omega)^2}$

(c) $\mathbf{H}_R(j\omega) = \dfrac{\mathbf{V}_R}{\mathbf{V}_1} = \dfrac{j\omega L/R}{1/LC + j\omega R/L + (j\omega)^2}$

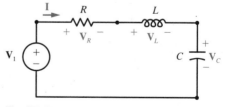

Fig. P5.15

5.16 For the circuit shown in Fig. P5.15, when $R = 11\ \Omega$, $L = 1$ H, and $C = 0.1$ F, then

$$\mathbf{H}_C(j\omega) = \frac{\mathbf{V}_C}{\mathbf{V}_1} = \frac{10}{10 + j\omega 11 + (j\omega)^2}$$

$$= \left(\frac{1}{1 + j\omega}\right)\left(\frac{10}{10 + j\omega}\right)$$

Thus $\mathbf{H}_C(j\omega)$ can be expressed as the product $\mathbf{H}_C(j\omega) = \mathbf{H}_1(j\omega)\mathbf{H}_2(j\omega)$, where $\mathbf{H}_1(j\omega) = 1/(1 + j\omega)$ and $\mathbf{H}_2(j\omega) = 10/(10 + j\omega)$. Use only the straight-line asymptotes to sketch the Bode plot—both the amplitude and phase responses—for $\mathbf{H}_C(j\omega)$ by adding the Bode plots for $\mathbf{H}_1(j\omega)$ and $\mathbf{H}_2(j\omega)$. What type of filter is this?

5.17 For the circuit shown in Fig. P5.15, when $R = 2\ \Omega$, $L = 1$ H, and $C = 1$ F, then

$$\mathbf{H}_C(j\omega) = \frac{\mathbf{V}_C}{\mathbf{V}_1} = \frac{1}{1 + j\omega 2 + (j\omega)^2}$$

$$= \left(\frac{1}{1 + j\omega}\right)\left(\frac{1}{1 + j\omega}\right)$$

Thus $\mathbf{H}_C(j\omega)$ can be expressed as the product $\mathbf{H}_C(j\omega) = \mathbf{H}_1(j\omega)\mathbf{H}_2(j\omega)$, where $\mathbf{H}_1(j\omega) = \mathbf{H}_2(j\omega) = 1/(1 + j\omega)$. Use only the straight-line asymptotes to sketch the Bode plot—both the amplitude and phase responses—for $\mathbf{H}_C(j\omega)$ by adding the Bode plots for $\mathbf{H}_1(j\omega)$ and $\mathbf{H}_2(j\omega)$. What type of filter is this?

5.18 For the circuit shown in Fig. P5.15, when $R = 2\ \Omega$, $L = 1$ H, and $C = 1$ F, then

$$\mathbf{H}_L(j\omega) = \frac{\mathbf{V}_L}{\mathbf{V}_1} = \frac{(j\omega)^2}{1 + j\omega 2 + (j\omega)^2}$$

$$= \left(\frac{j\omega}{1 + j\omega}\right)\left(\frac{j\omega}{1 + j\omega}\right)$$

Thus $\mathbf{H}_L(j\omega)$ can be expressed as the product $\mathbf{H}_L(j\omega) = \mathbf{H}_1(j\omega)\mathbf{H}_2(j\omega)$, where $\mathbf{H}_1(j\omega) = \mathbf{H}_2(j\omega) = j\omega/(1 + j\omega)$. Use only the straight-line asymptotes to sketch the Bode plot—both the amplitude and phase responses—for $\mathbf{H}_L(j\omega)$ by adding the Bode plots for $\mathbf{H}_1(j\omega)$ and $\mathbf{H}_2(j\omega)$. What type of filter is this?

5.19 For the circuit shown in Fig. P5.15, when $R = 2\ \Omega$, $L = 1$ H, and $C = 1$ F, then

$$\mathbf{H}_R(j\omega) = \frac{\mathbf{V}_R}{\mathbf{V}_1} = \frac{j\omega 2}{1 + j\omega 2 + (j\omega)^2}$$

$$= j\omega 2\left(\frac{1}{1 + j\omega}\right)\left(\frac{1}{1 + j\omega}\right)$$

Thus $\mathbf{H}_R(j\omega)$ can be expressed as the product $\mathbf{H}_R(j\omega) = \mathbf{H}_1(j\omega)\mathbf{H}_2(j\omega)\mathbf{H}_3(j\omega)$, where $\mathbf{H}_1(j\omega) = j\omega 2$ and $\mathbf{H}_2(j\omega) = \mathbf{H}_3(j\omega) = 1/(1 + j\omega)$. Use only the straight-line asymptotes to sketch the Bode plot—both the amplitude and phase responses—for $\mathbf{H}_R(j\omega)$ by adding the Bode plots for $\mathbf{H}_1(j\omega)$, $\mathbf{H}_2(j\omega)$, and $\mathbf{H}_3(j\omega)$. What type of filter is this?

5.20 Suppose that a filter has gain $|\mathbf{H}(j\omega)| = 16\omega/(\omega^2 + 4)$. (a) For what value of ω is the gain $|\mathbf{H}(j\omega)|$ maximum? (b) What is the maximum value M of the gain $|\mathbf{H}(j\omega)|$? (c) What are the half-power frequencies? (d) What is the bandwidth of the filter?

5.21 For the admittance shown in Fig. P5.21, suppose that $R = 1\ \Omega$, $L = 1$ H, and $C = \frac{1}{5}$ F. Find the resonance frequency.

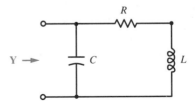

Fig. P5.21

5.22 Find a formula for the resonance frequency of the admittance shown in Fig. P5.21.

5.23 For the admittance shown in Fig. P5.23, suppose that $R = 1\ \Omega$, $L = 10$ H, and $C = 1$ F. Find the resonance frequency.

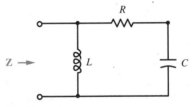

Fig. P5.23

5.24 Find a formula for the resonance frequency of the admittance shown in Fig. P5.23.

5.25 For the impedance shown in Fig. P5.25, suppose that $R = 1\ \Omega$, $L = 1$ H, and $C = 10$ F. Find the resonance frequency.

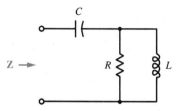

Fig. P5.25

5.26 Find a formula for the resonance frequency of the impedance shown in Fig. P5.25.

5.27 For the impedance shown in Fig. P5.27, suppose that $R = 1\ \Omega$, $L = \frac{1}{5}$ H, and $C = 1$ F. Find the resonance frequency.

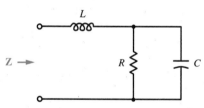

Fig. P5.27

5.28 Find a formula for the resonance frequency of the impedance shown in Fig. P5.27.

5.29 When $R = 6\ \Omega$, $L = 2$ H, and $C = \frac{1}{36}$ F, then the resonance frequency of the admittance shown in Fig. P5.21 is $\omega_r = 3$ rad/s. Apply a voltage source $v_1(t) = \cos 3t$ V to the admittance and calculate the following: (a) the energy stored by the inductor, (b) the energy stored by the capacitor, (c) the maximum total energy stored, (d) the power absorbed by the resistor, (e) the energy dissipated by the resistor in one period, and (f) the quality factor of the circuit.

5.30 When $R = 1\ \Omega$, $L = 1$ H, and $C = 2$ F, then the resonance frequency of the impedance shown in Fig. P5.25 is $\omega_r = 1$ rad/s. Apply a current source $i_1(t) = \cos t$ A to the impedance and calculate the following: (a) the energy stored by the inductor, (b) the energy stored by the capacitor, (c) the maximum total energy stored, (d) the power absorbed by the resistor, (e) the energy dissipated by the resistor in one period, and (f) the quality factor of the circuit.

5.31 Show that expressions for the half-power frequencies for the series RLC circuit given in Fig. 5.13a on p. 281 are:

$$\omega_1 = \frac{-R}{2L} + \sqrt{\left(\frac{R}{2L}\right)^2 + \frac{1}{LC}}$$

$$\omega_2 = \frac{R}{2L} + \sqrt{\left(\frac{R}{2L}\right)^2 + \frac{1}{LC}}$$

5.32 For the series RLC circuit given in Fig. 5.13a on p. 281, suppose that $R = 2\ \Omega$, $L = 1$ H, and $C = 1$ F. Find (a) the resonance frequency, (b) the quality factor, (c) the bandwidth, and (d) the lower and upper half-power frequencies.

5.33 For the series RLC circuit given in Fig. 5.13a on p. 281, suppose that $R = 20\ \Omega$, $L = 10$ H, and $C = 0.001$ F. Find (a) the resonance frequency, (b) the quality factor, (c) the bandwidth, and (d) the lower and upper half-power frequencies.

5.34 For the series RLC circuit given in Fig. 5.13a on p. 281, suppose that $R = 2\ \Omega$, $L = 1$ H, and $C = 1$ F. (a) Find an expression for the gain $|H_R(j\omega)| = |V_R/V_1|$. (b) For what value of ω is $|H_R(j\omega)|$ maximum? (c) What are the half-power frequencies? (d) What is the bandwidth?

5.35 A 10-Ω resistor and a 2-H inductor are connected in series, and $\omega = 50$ rad/s. (a) What is the Q of this series connection? (b) What parallel RL connection has the same admittance as the series connection? (c) What is the Q of this parallel connection?

5.36 A 10-Ω resistor and a 2-H inductor are connected in parallel, and $\omega = 50$ rad/s. (a) What is the Q of this parallel connection? (b) What series RL connection has the same impedance as the parallel connection? (c) What is the Q of this series connection?

5.37 Consider the practical tank circuit shown in Fig. 5.17 on p. 285. Suppose that $R_s = 25\ \Omega$, $L =$

0.62 mH, and $C = 20$ pF. Approximate this admittance by a parallel RLC connection. What is the quality factor of this parallel connection?

5.38 Consider the practical tank circuit shown in Fig. 5.17 on p. 285. Suppose that $R_s = 50\ \Omega$, $L = 50$ mH, and $C = 0.005\ \mu$F. Approximate this admittance by a parallel RLC connection. What is the quality factor of this parallel connection?

5.39 An AM radio has a parallel RLC circuit for tuning in stations. Suppose that $R = 1.24\ \text{M}\Omega$, $L = 0.62$ mH, and $C = 20$ pF. (a) What is the resonance frequency in hertz? (b) What is the quality factor of the circuit? (c) What is the bandwidth in hertz? (d) If the current into the parallel connection due to a station at 1430 kHz is 5 μA rms, what is the resulting voltage at that frequency? (e) If the current into the parallel connection due to a station at 1450 kHz is 5 μA rms, what is the resulting voltage at that frequency?

5.40 For the series RLC circuit shown in Fig. P5.15, suppose that $R = \frac{5}{3}\ \Omega$, $L = 5$ H, $C = \frac{1}{25}$ F, and $v_1(t) = 20e^{-6t}\cos 3t$ V. Find (a) $v_R(t)$, (b) $v_L(t)$, and (c) $v_C(t)$.

5.41 For the series RLC circuit shown in Fig. P5.15, suppose that $R = 2\ \Omega$, $L = \frac{1}{2}$ H, $C = 2$ F, and $v_1(t) = 20e^{-6t}\cos 3t$ V. Find (a) $v_R(t)$, (b) $v_L(t)$, and (c) $v_C(t)$.

5.42 For the series RLC circuit shown in Fig. P5.15, suppose that $R = 2\ \Omega$, $L = 2$ H, $C = 2$ F, and $v_1(t) = 20e^{-6t}\cos 3t$ V. Find (a) $v_R(t)$, (b) $v_L(t)$, and (c) $v_C(t)$.

5.43 For the series-parallel RLC circuit shown in Fig. P5.43, suppose that $R = \frac{5}{3}\ \Omega$, $L = 5$ H, $C = \frac{1}{25}$ F, and $v_1(t) = 20e^{-6t}\cos 3t$ V. Find (a) $v_R(t)$, (b) $v_L(t)$, and (c) $v_C(t)$.

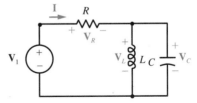

Fig. P5.43

5.44 For the series RLC circuit shown in Fig. P5.15, suppose that $R = \frac{5}{3}\ \Omega$, $L = 5$ H, and $C = \frac{1}{25}$ F. Draw a pole-zero plot for (a) $\mathbf{I}/\mathbf{V}_1$, (b) $\mathbf{V}_R/\mathbf{V}_1$, (c) $\mathbf{V}_L/\mathbf{V}_1$, and (d) $\mathbf{V}_C/\mathbf{V}_1$.

5.45 For the series RLC circuit shown in Fig. P5.15, suppose that $R = 2\ \Omega$, $L = \frac{1}{2}$ H, and $C = 2$ F. Draw a pole-zero plot for (a) $\mathbf{I}/\mathbf{V}_1$, (b) $\mathbf{V}_R/\mathbf{V}_1$, (c) $\mathbf{V}_L/\mathbf{V}_1$, and (d) $\mathbf{V}_C/\mathbf{V}_1$.

5.46 For the series RLC circuit shown in Fig. P5.15, suppose that $R = 2\ \Omega$, $L = 2$ H, and $C = 2$ F. Draw a pole-zero plot for (a) $\mathbf{I}/\mathbf{V}_1$, (b) $\mathbf{V}_R/\mathbf{V}_1$, (c) $\mathbf{V}_L/\mathbf{V}_1$, and (d) $\mathbf{V}_C/\mathbf{V}_1$.

5.47 For the series-parallel RLC circuit shown in Fig. P5.43, suppose that $R = \frac{5}{3}\ \Omega$, $L = 5$ H, and $C = \frac{1}{25}$ F. Draw a pole-zero plot for (a) $\mathbf{I}/\mathbf{V}_1$, (b) $\mathbf{V}_R/\mathbf{V}_1$, (c) $\mathbf{V}_L/\mathbf{V}_1$, and (d) $\mathbf{V}_C/\mathbf{V}_1$.

5.48 For the op-amp circuit shown in Fig. P5.48, draw the pole-zero plot of $\mathbf{H}(s) = \mathbf{V}_2/\mathbf{V}_1$ for the case that C is (a) 1 F, (b) $\frac{1}{4}$ F, and (c) $\frac{1}{16}$ F.

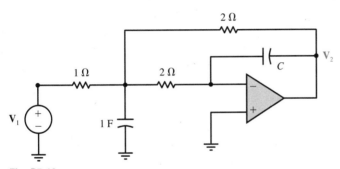

Fig. P5.48

5.49 For the op-amp circuit shown in Fig. P5.49, draw the pole-zero plot of $\mathbf{H}(s) = \mathbf{V}_2/\mathbf{V}_1$ for the case that C is (a) $\frac{1}{2}$ F, (b) 1 F, and (c) 2 F.

5.50 For the series-parallel RLC circuit given in Fig. P5.43, consider the capacitor to be the load. Find the Thévenin equivalent of the voltage source, resistor, and inductor combination. Use this Thévenin-equivalent circuit to determine the voltage $v_C(t)$ across the load for the case that $v_1(t) = 20e^{-6t}\cos 3t$ V.

5.51 Use integrators, adders, and scalers to simulate the transfer function

$$\mathbf{H}(s) = \frac{4s}{(s^2 + 2s + 3)}$$

5.52 Use integrators, adders, and scalers to simulate the transfer function

$$\mathbf{H}(s) = \frac{(s^2 + 2)}{(s^2 + 3s + 4)}$$

5.53 Find the transfer function $\mathbf{H}(s) = \mathbf{Y}/\mathbf{X}$ of the system shown in Fig. P5.53.

5.54 Find the transfer function $\mathbf{H}(s) = \mathbf{Y}/\mathbf{X}$ of the system shown in Fig. P5.54.

5.55 For the feedback system shown in Fig. 5.35 on p. 304, find the transfer function $\mathbf{Y}/\mathbf{X}$ when $\mathbf{G}(s) = (s + 1)/(s + 2)$ and $\mathbf{H}(s) = 1/(s + 3)$.

5.56 For the feedback system shown in Fig. 5.35 on p. 304, suppose that $\mathbf{G}(s) = (s + 1)/(s + 2)$. Determine $\mathbf{H}(s)$ such that the resulting transfer function is $\mathbf{Y}/\mathbf{X} = (s + 1)(s + 5)/(s + 3)^2$.

5.57 For the feedback system given in Fig. 5.35 on p. 304, suppose that $\mathbf{H}(s) = (s + 1)/(s + 2)$. Determine $\mathbf{G}(s)$ such that the resulting transfer function is $\mathbf{Y}/\mathbf{X} = (s + 2)^2/(s + 1)(s + 4)$.

5.58 For the feedback system given in Fig. 5.35 on p. 304, suppose that $\mathbf{G}(s) = 4s(s + 1)/(s + 2)^2$ and $\mathbf{H}(s) = 1/(s + 1)$. At what frequency will the system oscillate?

5.59 Find the Laplace transform of (a) $(2e^{-8t} - e^{-2t})u(t)$, (b) $(6 + 2e^{-6t} - 12e^{-t})u(t)$, (c) $(2 + 3t)e^{-2t}u(t)$, and (d) $e^{-3t}(\cos 4t - \sin 4t)u(t)$.

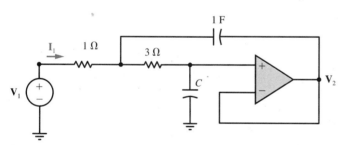

Fig. P5.49

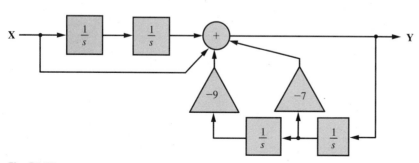

Fig. P5.53

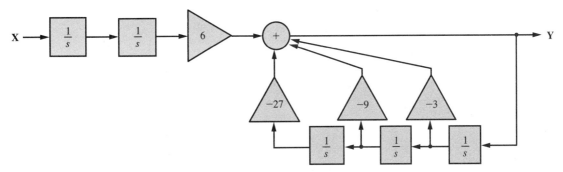

Fig. P5.54

5.60 Find the Laplace transform of (a) $\sin(\beta t - \phi)u(t)$, (b) $\cos(\beta t - \phi)u(t)$, (c) $e^{-\alpha t}\sin(\beta t - \phi)u(t)$, and (d) $e^{-\alpha t}\cos(\beta t - \phi)u(t)$.

5.61 Find the inverse Laplace transform of each of the following functions:

(a) $\dfrac{600}{s(s + 10)(s + 30)}$ (b) $\dfrac{60(s + 4)}{s(s + 2)(s + 12)}$

5.62 Find the inverse Laplace transform of each of the following functions:

(a) $\dfrac{12s}{(s + 3)(s^2 + 9)}$ (b) $\dfrac{4(s^2 + 1)}{s(s^2 + 4)}$

5.63 Find the inverse Laplace transform of each of the following functions:

(a) $\dfrac{(s + 2)(s + 3)}{s(s + 1)^2}$ (b) $\dfrac{10s + 80}{s^2 + 8s + 20}$

5.64 Find the solution to the differential equation

$$\frac{d^2x(t)}{dt^2} + 7\frac{dx(t)}{dt} + 6x(t) = 36u(t)$$

subject to the initial conditions $dx(0)/dt = 0$ and $x(0) = -4$.

5.65 Find the solution to the differential equation

$$\frac{d^2x(t)}{dt^2} + 3\frac{dx(t)}{dt} + 2x(t) = 20 \cos 2t \, u(t)$$

subject to the zero initial conditions $dx(0)/dt = x(0) = 0$.

5.66 For the series RC circuit shown in Fig. P5.66, suppose that $R = 5 \, \Omega$ and $C = 0.1$ F. Find the step responses $v(t)$ and $i(t)$ when $v_s(t) = 20u(t)$ V.

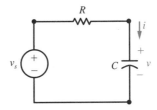

Fig. P5.66

5.67 For the series RC circuit shown in Fig. P5.66, suppose that $R = 2 \, \Omega$ and $C = 2$ F. Find $v(t)$ and $i(t)$ when $v_s(t) = 12e^{-t/2}u(t)$ V.

5.68 For the series RC circuit shown in Fig. P5.66, suppose that $R = 2 \, \Omega$ and $C = 2$ F. Find $v(t)$ and $i(t)$ when $v_s(t) = 12e^{-t/4}u(t)$ V.

5.69 For the series RL circuit shown in Fig. P5.69, suppose that $R = 5 \, \Omega$ and $L = 5$ H. Find the step responses $i(t)$ and $v(t)$ when $v_s(t) = 20u(t)$ V.

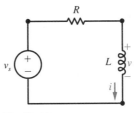

Fig. P5.69

5.70 For the series *RL* circuit shown in Fig. P5.69, suppose that $R = 2\ \Omega$ and $L = 2$ H. Find $i(t)$ and $v(t)$ when $v_s(t) = 12e^{-2t}u(t)$ V.

5.71 For the series *RL* circuit shown in Fig. P5.69, suppose that $R = 2\ \Omega$ and $L = 2$ H. Find $i(t)$ and $v(t)$ when $v_s(t) = 12e^{-t}u(t)$ V.

5.72 Find the step responses $v(t)$ and $i(t)$ for the circuit shown in Fig. P5.72 when $v_s(t) = 12u(t)$ V.

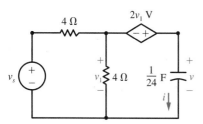

Fig. P5.72

5.73 For the circuit shown in Fig. P5.72, replace the capacitor with a 3-H inductor, and find the step responses $v(t)$ and $i(t)$ when $v_s(t) = 20u(t)$ V.

5.74 For the op-amp circuit shown in Fig. P5.3, suppose that $R = 2\ \Omega$ and $C = \frac{1}{8}$ F. Find the step response $v_2(t)$ when $v_1(t) = 3u(t)$ V.

5.75 For the op-amp circuit shown in Fig. P5.3, suppose that $R = 2\ \Omega$ and $C = \frac{1}{8}$ F. Find $v_2(t)$ when $v_1(t) = 3e^{-2t}u(t)$ V.

5.76 For the op-amp circuit shown in Fig. P5.3, suppose that $R = 2\ \Omega$ and $C = \frac{1}{8}$ F. Find $v_2(t)$ when $v_1(t) = 3e^{-4t}u(t)$ V.

5.77 For the op-amp circuit shown in Fig. P5.8, suppose that $R = 2\ \Omega$ and $C = \frac{1}{8}$ F. Find the step response $v_2(t)$ when $v_1(t) = 3u(t)$ V.

5.78 For the op-amp circuit shown in Fig. P5.8, suppose that $R = 2\ \Omega$ and $C = \frac{1}{8}$ F. Find $v_2(t)$ when $v_1(t) = 3e^{-2t}u(t)$ V.

5.79 For the op-amp circuit shown in Fig. P5.8, suppose that $R = 2\ \Omega$ and $C = \frac{1}{8}$ F. Find $v_2(t)$ when $v_1(t) = 3e^{-4t}u(t)$ V.

5.80 For the series *RLC* circuit shown in Fig. P5.80, suppose that $R = \frac{1}{3}\ \Omega$, $L = \frac{1}{12}$ H, $C = 3$ F,

and $v_s(t) = 0$ V. Find $v(t)$ and $i(t)$ when $i(0) = 4$ A and $v(0) = 0$ V.

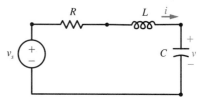

Fig. P5.80

5.81 For the series *RLC* circuit shown in Fig. P5.80, suppose that $R = \frac{1}{3}\ \Omega$, $L = \frac{1}{12}$ H, $C = \frac{3}{5}$ F, and $v_s(t) = 0$ V. Find $v(t)$ and $i(t)$ when $i(0) = 4$ A and $v(0) = 0$ V.

5.82 For the series *RLC* circuit shown in Fig. P5.80, suppose that $R = \frac{1}{3}\ \Omega$, $L = \frac{1}{12}$ H, $C = 4$ F, and $v_s(t) = 0$ V. Find $v(t)$ and $i(t)$ when $i(0) = 4$ A and $v(0) = 0$ V.

5.83 For the parallel *RLC* circuit shown in Fig. P5.83, suppose that $R = \frac{1}{2}\ \Omega$, $L = \frac{1}{4}$ H, $C = \frac{1}{2}$ F, and $i_s(t) = 0$ A. Find $v(t)$ and $i(t)$ when $i(0) = 6$ A and $v(0) = 0$ V.

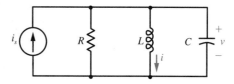

Fig. P5.83

5.84 For the parallel *RLC* circuit shown in Fig. P5.83, suppose that $R = \frac{1}{3}\ \Omega$, $L = \frac{1}{4}$ H, $C = \frac{1}{2}$ F, and $i_s(t) = 0$ A. Find $v(t)$ and $i(t)$ when $i(0) = 6$ A and $v(0) = 0$ V.

5.85 For the parallel *RLC* circuit shown in Fig. P5.83, suppose that $R = \frac{1}{3}\ \Omega$, $L = \frac{2}{9}$ H, $C = \frac{1}{2}$ F, and $i_s(t) = 0$ A. Find $v(t)$ and $i(t)$ when $i(0) = 6$ A and $v(0) = 0$ V.

5.86 For the series *RLC* circuit shown in Fig. P5.80, suppose that $R = 7\ \Omega$, $L = 1$ H, $C = 0.1$ F, and $v_s(t) = 0$ V. Find $i(t)$ and $v(t)$ when $v(0) = 12$ V and $i(0) = 0$ A.

5.87 For the series *RLC* circuit shown in Fig. P5.80, suppose that $R = 2\ \Omega$, $L = \frac{1}{4}$ H, $C = 0.2$ F, and $v_s(t) = 0$ V. Find $i(t)$ and $v(t)$ when $v(0) = 10$ V and $i(0) = 0$ A.

5.88 For the series *RLC* circuit shown in Fig. P5.80, suppose that $R = 2\ \Omega$, $L = 1$ H, $C = 1$ F, and $v_s(t) = 0$ V. Find $i(t)$ and $v(t)$ when $v(0) = 6$ V and $i(0) = 0$ A.

5.89 For the circuit shown in Fig. P5.89, find $v_2(t)$ when $v_s(t) = 0$ V and $v_1(0) = v_2(0) = 6$ V.

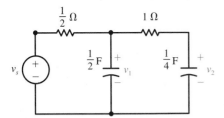

Fig. P5.89

5.90 For the circuit shown in Fig. P5.90, find $v(t)$ when $v_s(t) = 0$ V, $v(0) = 3$ V, and $i(0) = 3$ A.

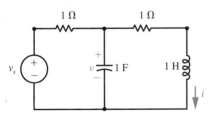

Fig. P5.90

5.91 For the circuit shown in Fig. P5.90, interchange the inductor and the capacitor. Find the capacitor voltage $v(t)$ and the inductor current $i(t)$ when $v_s(t) = 0$ V, $v(0) = 0$ V and $i(0) = 6$ A.

5.92 For the parallel *RLC* circuit shown in Fig. P5.83, suppose that $R = \frac{1}{2}\ \Omega$, $L = \frac{1}{5}$ H, and $C = \frac{1}{4}$ F. Find the step responses $v(t)$ and $i(t)$ when $i_s(t) = 2u(t)$ A.

5.93 For the parallel *RLC* circuit shown in Fig. P5.83, suppose that $R = 3\ \Omega$, $L = 3$ H, and $C = \frac{1}{12}$ F. Find the step responses $v(t)$ and $i(t)$ when $i_s(t) = 4u(t)$ A.

5.94 For the series *RLC* circuit shown in Fig. P5.80, suppose that $R = 7\ \Omega$, $L = 1$ H, and $C = 0.1$ F. Find the step responses $v(t)$ and $i(t)$ when $v_s(t) = 12u(t)$ V.

5.95 For the series *RLC* circuit shown in Fig. P5.80, suppose that $R = 2\ \Omega$, $L = 1$ H, and $C = 1$ F. Find the step responses $v(t)$ and $i(t)$ when $v_s(t) = 12u(t)$ V.

5.96 For the *RLC* circuit shown in Fig. P5.96, suppose that $R = \frac{1}{2}\ \Omega$, $L = \frac{1}{3}$ H, and $C = \frac{1}{4}$ F. Find the unit step responses $v(t)$ and $i(t)$ when $v_s(t) = u(t)$ V.

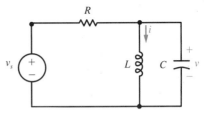

Fig. P5.96

5.97 For the *RLC* circuit shown in Fig. P5.96, suppose that $R = \frac{1}{2}\ \Omega$, $L = \frac{1}{4}$ H, and $C = \frac{1}{2}$ F. Find the unit step responses $v(t)$ and $i(t)$ when $v_s(t) = u(t)$ V.

5.98 For the circuit shown in Fig. P5.89, find the step response $v_2(t)$ when $v_s(t) = 9u(t)$ V.

5.99 For the circuit shown in Fig. P5.90, find the step response $v(t)$ when $v_s(t) = 6u(t)$ V.

5.100 For the op-amp circuit shown in Fig. P5.48, suppose that $C = \frac{1}{3}$ F. Find the step response $v_2(t)$ when $v_1(t) = 4u(t)$ V.

5.101 For the op-amp circuit shown in Fig. P5.48, suppose that $C = \frac{1}{8}$ F. Find the step response $v_2(t)$ when $v_1(t) = 8u(t)$ V.

5.102 For the op-amp circuit shown in Fig. P5.48, suppose that $C = \frac{1}{4}$ F. Find the step response $v_2(t)$ when $v_1(t) = 6u(t)$ V.

5.103 For the op-amp circuit shown in Fig. P5.49, suppose that $C = 1$ F. Find the step response $v_2(t)$ when $v_1(t) = 3u(t)$ V.

5.104 For the op-amp circuit shown in Fig. P5.49, suppose that $C = \frac{4}{3}$ F. Find the step response $v_2(t)$ when $v_1(t) = 4u(t)$ V.

5.105 For the op-amp circuit shown in Fig. P5.49, suppose that $C = \frac{1}{5}$ F. Find the step response $v_2(t)$ when $v_1(t) = 2u(t)$ V.

5.106 For the parallel RLC circuit shown in Fig. P5.83, suppose that $R = 6 \; \Omega$, $L = 7$ H, and $C = \frac{1}{42}$ F. Find $v(t)$ and $i(t)$ when $i_s(t) = 6u(t)$ A, $i(0) = -4$ A, and $v(0) = 0$ V.

5.107 For the series RLC circuit shown in Fig. P5.80, suppose that $R = 2 \; \Omega$, $L = \frac{1}{4}$ H, and $C = \frac{1}{5}$ F. Find $i(t)$ and $v(t)$ when $v_s(t) = 2u(t)$ V, $i(0) = 0$ A, and $v(0) = -2$ V.

5.108 For the RLC circuit shown in Fig. P5.96, suppose that $R = 3 \; \Omega$, $L = 4$ H, and $C = \frac{1}{12}$ F. Find $v(t)$ and $i(t)$ when $v_s(t) = -12u(t)$ V, $i(0) = 4$ A, and $v(0) = 0$ V.

5.109 Given that the transfer function of a linear system is $\mathbf{H}(s) = 1/(s + 2)$, find the output $y(t)$ when the input $x(t)$ is (a) $u(t)$, (b) $e^{-t}u(t)$, (c) $(1 - e^{-t})u(t)$, and (d) $e^{-2t}u(t)$.

5.110 Given that the transfer function of a linear system is $\mathbf{H}(s) = s/(s + 2)$, find the output $y(t)$ when the input $x(t)$ is (a) $u(t)$, (b) $e^{-t}u(t)$, (c) $(1 - e^{-t})u(t)$, and (d) $e^{-2t}u(t)$.

5.111 Given that the transfer function of a linear system is $\mathbf{H}(s) = (s - 1)/(s + 10)$, find the input $x(t)$ when the output $y(t)$ is (a) $(-1 + 2e^{-t})u(t)$, (b) $(-2e^{-t} + 3e^{-2t})u(t)$, (c) $(1 - 11t)e^{-10t}u(t)$, and (d) $(1 - 2t)e^{-t}u(t)$.

5.112 For the case that the input to a linear system is $x(t) = e^{-t}u(t)$, find the transfer function $\mathbf{H}(s)$ when the output $y(t)$ is (a) $e^{-2t}u(t)$, (b) $\sin t \, u(t)$, (c) $e^{-t} \sin t \, u(t)$, (d) $te^{-t}u(t)$, and (e) $(e^{-t} - e^{-2t})u(t)$.

5.113 For the case that the input to a linear system is $x(t) = \cos t \, u(t)$, find the transfer function $\mathbf{H}(s)$ when the output $y(t)$ is (a) $e^{-2t}u(t)$, (b) $\sin t \, u(t)$, (c) $e^{-t} \sin t \, u(t)$, (d) $te^{-t}u(t)$, and (e) $(e^{-t} - e^{-2t})u(t)$.

Part II

Electronics

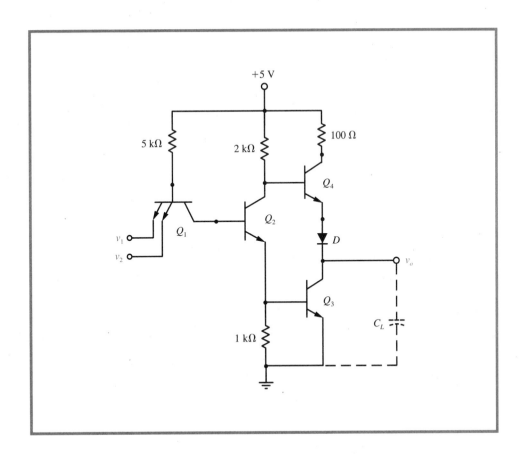

6

Diodes

INTRODUCTION

We begin our study of electronics with a discussion of semiconductor diodes. This topic naturally evolves into the subject of the bipolar junction transistor (BJT), which will be the focus of Chapter 7. Although a BJT is inherently a current-controlled device, a field-effect transistor (FET), to be studied in Chapter 8, is inherently a voltage-controlled device. Throughout the chapters on diodes, BJTs, and FETs, numerous examples and applications are presented. Chapters 9 and 10 consider applications not covered in the chapters preceding them.

Unlike elements encountered previously in this book, the diode is a nonlinear element. In terms of analysis, nonlinear elements tend to make circuits more complicated. However, with the judicious use of approximations and assumptions (or educated guesses), we will be able to deal with nonlinear circuits with not too much more effort than is required for linear circuits. This is not a great price to pay considering that innumerable important devices employ nonlinear elements, and these devices would be impossible to fabricate using only linear elements. Even though we will be dealing with nonlinear circuits, we will utilize some of the linear analysis techniques studied in earlier chapters.

Unlike with electric circuits, an in-depth study of electronics requires more emphasis on physical principles. It is for this reason that we begin this chapter with a discussion of the electric properties of some types of materials.

6.1 Semiconductors

The flow of charge (current) in a metal results from the movement of electrons. An electron is a negatively charged particle having charge magnitude $q = 1.60 \times 10^{-19}$ C. In a metal, the atoms consist of outer, or valence, electrons that are free to move, and positively charged ions, each of which consists of the nucleus and the tightly bound inner electrons of the atom. Figure 6.1 is a two-dimensional depiction of the ions and the valence, or free electrons, of a metal. Because of thermal energy, the free electrons are continually in motion. The direction of the motion of an electron changes after each inelastic collision with a heavy (essentially stationary) ion. We call the average distance between collisions the **mean free path.** If no voltage is applied to a metal, then the random motion of the electrons results in zero average current.

The application of a voltage to a metal will result in an **electric field** $\mathscr{E}$ measured in volts per meter (V/m). Consider the case that $\mathscr{E}$ is a constant. Then on the average, electrons will go from a given potential to a higher potential at a speed, denoted u, called the **drift speed.** This quantity is directly proportional to the electric field, that is,

$$u = \mu\mathscr{E} \tag{6.1}$$

where the proportionality constant μ is known as the **mobility** of the electrons. Since the units of u and $\mathscr{E}$ are m/s and V/m, respectively, the unit of μ is $\mathrm{m^2/V\text{-}s}$.

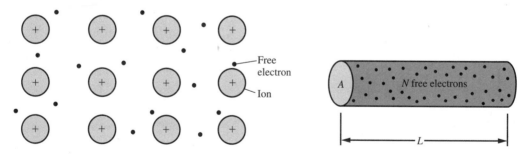

Fig. 6.1 Positive ions and free electrons of a metal.

Fig. 6.2 Section of a metallic conductor.

Figure 6.2 shows a length-L section of a metallic conductor that has a cross-sectional area of A square meters. Let N be the number of free electrons in this section of the conductor, and suppose that these N electrons move to the left with a drift speed of u. If it takes T seconds for an electron to travel L meters, then from the drift speed $u = L/T$, we deduce that $T = L/u$. But a flow to the left of N electrons in T seconds means a current i going to the right, where

$$i = \frac{Nq}{T} = \frac{Nq}{L/u} = \frac{Nqu}{L}$$

We define **current density** J by

$$J = \frac{i}{A} \tag{6.2}$$

Thus we have that

$$J = \frac{i}{A} = \frac{Nqu}{LA} \tag{6.3}$$

Since the volume of the conductor section shown in Fig. 6.2 is LA, and since the number of free electrons in this volume is N, then the density of free electrons, called the **free-electron concentration,** is $n = N/LA$. Since N is a number, then the units of n are $1/m^3$ or m^{-3}. Hence Eq. 6.3 can be rewritten

$$J = nqu \tag{6.4}$$

The product nq, known as **charge density,** has as units C/m^3.

From Eq. 6.1 and Eq. 6.4, we have that

$$J = nq\mu\mathscr{E} = \sigma\mathscr{E} \tag{6.5}$$

where $\sigma = nq\mu$ is known as the **conductivity** of the metal. The units of σ are $(\Omega\text{-m})^{-1}$ or $\mho/m$. The reciprocal of conductivity is called **resistivity,** denoted by ρ, and has units of Ω-m. In other words, $\rho = 1/\sigma$.

For the conductor depicted in Fig. 6.2, from Eq. 6.2

$$i = JA = \sigma\mathscr{E}A = \frac{\sigma\mathscr{E}AL}{L} = \frac{\sigma A}{L}\mathscr{E}L$$

As the unit of $\mathscr{E}L$ is $(V/m)(m) = V$, then $v = \mathscr{E}L$ is the voltage across the length-L conductor. Then, by Ohm's law,

$$i = \frac{\sigma A}{L}v = \frac{1}{R}v$$

where the resistance of the conductor is $R = L/\sigma A = \rho L/A$.

Even though the SI unit for temperature is the **kelvin**[1] (K), because manufacturers typically use **degrees Celsius**[2] (°C), we shall use both temperature measurements. If T_K is the temperature in K and T_C is the temperature in °C, then the relationship between the two is given simply by

$$T_K = T_C + 273$$

Example 6.1

A length of copper wire with a cross-sectional area of 1 square millimeter (mm^2 or 10^{-6} m^2) carries a current of 0.4 A. The conductivity of copper at 20°C (293 K) is 5.78×10^7 $\mho/m$, and the free-electron concentration is 8.43×10^{28} m^{-3}. Since $J = nqu$, we have that the drift speed of free electrons is

$$u = \frac{J}{nq} = \frac{i}{nqA}$$

$$= \frac{0.4}{(8.43 \times 10^{28})(1.6 \times 10^{-19})(10^{-6})} = 2.97 \times 10^{-5} \ m/s$$

From the fact that $\sigma = nq\mu$, the mobility of the free electrons is

$$\mu = \frac{\sigma}{nq} = \frac{5.78 \times 10^7}{(8.43 \times 10^{28})(1.6 \times 10^{-19})} = 4.29 \times 10^{-3} \ m^2/V\text{-}s$$

Since $u = \mu\mathscr{E}$, the electric field strength is

$$\mathscr{E} = \frac{u}{\mu} = \frac{2.97 \times 10^{-5}}{4.29 \times 10^{-3}} = 6.92 \times 10^{-3} \ V/m$$

From the formula obtained on the preceding page, we have that the resistance of a 1-meter length of copper wire is

$$R = \frac{L}{\sigma A} = \frac{1}{(5.78 \times 10^7)(10^{-6})} = 17.3 \ m\Omega$$

---

[1]Formerly "degrees Kelvin" (°K), named for the British physicist William Thompson, first Baron Kelvin (1824–1907).

[2]Named for the Swedish astronomer Anders Celsius (1701–1744), it is also known as "degrees centigrade."

Drill Exercise 6.1

A piece of aluminum wire has a cross-sectional area of 10^{-6} m^2. The aluminum has a free-electron concentration of 1.81×10^{29} m^{-3}, and the mobility of the free electrons is 10^{-3} m^2/V-s. (a) Find the conductivity of the aluminum wire. (b) Find the resistivity of the wire. (c) Find the resistance of a 1-meter length of wire.

ANSWER (a) 2.9×10^7 ℧/m; (b) 3.45×10^{-8} Ω-m; (c) 34.5 mΩ

From the equation $\sigma = nq\mu$, we see that conductivity σ is proportional to the concentration of free electrons n. A good conductor, such as copper, has a value of n on the order of 10^{28} to 10^{29}. Conversely, an **insulator** has a value of n on the order of 10^7. Materials such as silicon (Si) and germanium (Ge) have conductivities between these values and are known as **semiconductors.** At 300 K, for silicon $n = 1.5 \times 10^{16}$ m^{-3} and for germanium $n = 2.5 \times 10^{19}$ m^{-3}.

A semiconductor such as silicon or germanium has a crystal lattice structure as illustrated two-dimensionally in Fig. 6.3. An atom has four valence electrons—each of which is shared with an adjacent atom. This is called a **covalent bond.** In Fig. 6.3, the nucleus and inner electrons of an atom are designated by a circle labeled "+4" since this is a positively charged ion, and a valence electron is designated by a solid dot. Since the valence electrons are shared with adjacent atoms, these electrons are tightly bound to the atoms—and the crystal has a low conductivity.

At low temperatures, this type of material acts as an insulator. However, at room temperatures, thermal energy can break the covalent bond by dislodging a valence

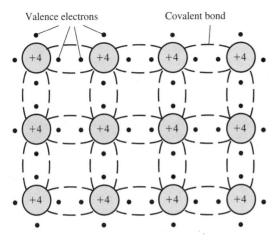

Fig. 6.3 Crystal lattice structure of a semiconductor.

electron. The resulting vacated space, designated by a hollow dot in Fig. 6.4, is called a **hole.** The combination of the hole and the resulting **free electron** is known as a **hole-electron pair.**

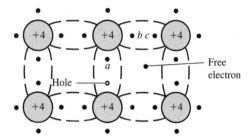

Fig. 6.4 Hole-electron pair in a semiconductor.

Relative to an electron, 1 joule (J) is a very large amount of energy. For this reason, we define the **electron volt** (eV) by

$$1 \text{ eV} = 1.60 \times 10^{-19} \text{ J}$$

The energy required to form a hole-electron pair is 1.1 eV for silicon and 0.72 eV for germanium at room temperature.

Once a hole has been created, another valence electron can fill in this hole and, thus, produce a different hole. This process can continue, giving the effect of a hole in motion. For example, if the electron at point a in Fig. 6.4 fills in the hole, there will be a hole at point a. If the electron at b then fills the hole at a, the result is a hole at b. If the electron at c fills the hole at b, there is now a hole at c. Thus, even though it is electrons that are moving, we can think of the original hole as moving to a, then to b, and then to c. Consequently, we can think of this phenomenon in terms of a flow of positive charge, called a **hole current.**

In a pure (**intrinsic**) semiconductor, the number of holes is equal to the number of free electrons. Thus the free-electron concentration n and the **hole concentration,** designated p, must be equal, that is $n = p$. We call this number the **intrinsic concentration** of the semiconductor and designate it by n_i. That is, for an intrinsic (pure) semiconductor,

$$n_i = n = p$$

Recombination occurs when a free electron fills in a hole. As a result of thermal energy, new hole-electron pairs are continually being generated, and old ones recombine at the same rate.

In a metal, current is conducted only by negative charges (electrons). For this reason, we say that a metal is **unipolar.** However, a semiconductor such as silicon or germanium is said to be **bipolar** since current can be conducted by both negative charges (electrons) and positive charges (holes).

Repeating Eq. 6.5, we have for a metal that the current density J is given by

$$J = nq\mu\mathscr{E} \tag{6.5}$$

where n is the free-electron concentration, μ is the free-electron mobility, $q = 1.60 \times 10^{-19}$ C is the charge, σ is the conductivity of the metal, and $\mathscr{E}$ is the electric field.

For a semiconductor, let μ_n denote the mobility of the free electrons and μ_p denote the mobility of the holes. Under the influence of an electric field, electrons and holes will move in opposite directions. The consequence of this is electron and hole currents in the same direction. Generalizing Eq. 6.5, the resulting current density J is given by

$$J = (n\mu_n + p\mu_p)q\mathscr{E} = \sigma\mathscr{E} \tag{6.6}$$

where $\sigma = (n\mu_n + p\mu_p)q$ is the conductivity of the semiconductor, n is the free-electron concentration, p is the hole concentration, and $q = 1.6 \times 10^{-19}$ C is the charge. (Remember that for an intrinsic semiconductor, $p = n = n_i$.)

Example 6.2

At 300 K, the intrinsic concentration of silicon (Si) is 1.5×10^{16} m^{-3}. Furthermore, the free-electron mobility is 0.13 m^2/V-s and the hole mobility is 0.05 m^2/V-s. Thus the (intrinsic) conductivity of Si is

$$\sigma = (n\mu_n + p\mu_p)q = (n_i\mu_n + n_i\mu_p)q = n_i(\mu_n + \mu_p)q$$
$$= (1.5 \times 10^{16})(0.13 + 0.05)(1.6 \times 10^{-19}) = 4.32 \times 10^{-4}\ \mho/m$$

and the resistivity of Si is

$$\rho = 1/\sigma = 2.31 \times 10^3 = 2.31\ k\Omega\text{-m}$$

In Example 6.2, we determined the conductivity of silicon (at 300 K) to be 4.32×10^{-4} ℧/m. A metal such as copper has a much higher conductivity—about 5.8×10^7 ℧/m. For a metal, as temperature increases, random thermal motion of the free electrons increases, the time between collisions decreases, and electron mobility decreases. As a consequence, the conductivity of a metal decreases as temperature increases. However, for a semiconductor, as temperature increases, more hole-electron pairs are generated so that the free-electron and hole concentrations increase. The result is an increase in the conductivity of the semiconductor.

6.2   Doped Semiconductors

Not only does the conductivity of silicon or germanium increase as temperature increases, but the conductivity of semiconductors can be dramatically increased by adding small amounts of appropriate impurities. Such a process is referred to as **doping,** and we can say that the resulting semiconductor is **doped.**

To see the consequences of doping, first consider the case that a small amount of an element, called an **impurity,** with five valence electrons (e.g., arsenic or phosphorus) is added to a semiconductor such as silicon or germanium. As shown in Fig. 6.5, what results is a doped semiconductor with a covalent bonding. However, because each impurity atom has five valence electrons, only four of which are used for the covalent bond—there is an excess free electron for each impurity atom. Since such an impurity "donates" free electrons (negative charges), we call the impurity a **donor** and refer to the impure semiconductor as an *n*-type **semiconductor**—the "*n*" signifying donor or negative.

For the case in which silicon or germanium is doped with an impurity having three valence electrons (e.g., boron or gallium), what results is a doped semiconductor, as illustrated in Fig. 6.6, which again has a covalent bonding. In this case, however, because it has only three valence electrons, each impurity atom produces an excess hole. Since a hole will "accept" a free electron, we call the impurity an **acceptor** and refer to the doped semiconductor as a *p*-type **semiconductor**—the "*p*" signifying acceptor or positive.

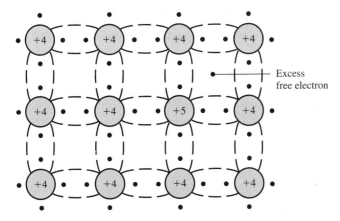

Fig. 6.5 An *n*-type semiconductor.

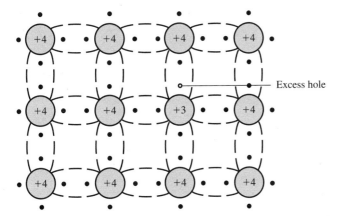

Fig. 6.6 A *p*-type semiconductor.

In a doped semiconductor, the current will be carried predominantly either by (free) electrons or by holes. In an *n*-type semiconductor, electrons are the **majority carriers** and holes are the **minority carriers.** The converse is true for a *p*-type semiconductor. Both holes and free electrons are said to be **mobile carriers.**

To see how doping affects the conductivity of a semiconductor, we now present the **mass-action law.** This law states that regardless of the amount of doping for a semiconductor, under thermal equilibrium

$$np = n_i^2 \qquad (6.7)$$

where n is the free-electron concentration, p is the hole concentration, and n_i is the intrinsic concentration of the semiconductor.

As was seen in Fig. 6.5, when a semiconductor is doped with donor atoms, we get the situation shown in Fig. 6.7. We can consider the boxed region around the donor atom shown to be an immobile positive ion. If the concentration of donor atoms is N_D then at ordinary temperatures, the concentration of immobile positive ions will be N_D also. Similarly, if a semiconductor is doped with acceptor atoms having a concentration of N_A then the resulting concentration of immobile negative ions will be N_A as well. In general, for a semiconductor with n-type and p-type impurities, since the semiconductor must be electrically neutral,

$$N_D + p = N_A + n \tag{6.8}$$

where p is the hole concentration and n is the (free) electron concentration.

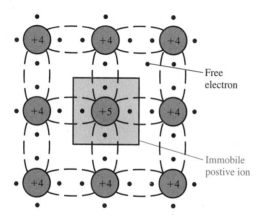

Free electron

Immobile postive ion

Fig. 6.7 Immobile positive ion.

For an n-type semiconductor, $N_A = 0$, and Equation (6.8) becomes $N_D + p = n$. Also, for an n-type semiconductor, there are many more free electrons than holes ($n \gg p$). Thus we deduce that $N_D = n - p \approx n$, and by the mass-action law (Eq. 6.7),

$$p = \frac{n_i^2}{n} \approx \frac{n_i^2}{N_D}$$

In summary, for an n-type semiconductor

$$\boxed{n \approx N_D} \qquad \text{and} \qquad \boxed{p \approx \frac{n_i^2}{N_D}} \tag{6.9}$$

Similarly, for a *p*-type semiconductor

$$p \approx N_A$$ and $$n \approx \frac{n_i^2}{N_A}$$ (6.10)

We are now in a position to see an example of how doping a semiconductor affects its conductivity.

Example 6.3

At 300 K, the intrinsic concentration of Si is 1.5×10^{16} m^{-3}. Furthermore, the free-electron mobility is 0.13 m^2/V-s and the hole mobility is 0.05 m^2/V-s. In Example 6.2, we calculated the conductivity to be 4.32×10^{-4} ℧/m. Let us find the conductivity of Si that is doped with two parts per 10^8 of a donor impurity.

Since there are 5×10^{28} Si atoms/m^3, then the concentration of donor atoms is

$$N_D = (5 \times 10^{28})\left(\frac{2}{10^8}\right) = 10^{21} \text{ m}^{-3}$$

From Eq. 6.9,

$$n \approx N_D = 10^{21} \text{ m}^{-3}   \text{and}   p \approx \frac{n_i^2}{N_D} = \frac{(1.5 \times 10^{16})^2}{10^{21}}$$

$$= 2.25 \times 10^{11} \text{ m}^{-3}$$

Thus the conductivity of the doped silicon is

$$\sigma = (n\mu_n + p\mu_p)q \approx [(10^{21})(0.13) + (2.25 \times 10^{11})(0.05)](1.6 \times 10^{-19})$$

$$= 20.8 \text{ ℧/m}$$

Hence by doping the semiconductor with two parts per 100 million, the conductivity increases by a factor of $20.8/(4.32 \times 10^{-4}) \approx 48,000$.

Previously, we stated that for an *n*-type semiconductor, typically, $n \gg p$, and this was numerically demonstrated in Example 6.3. Since the mobilities of holes and free electrons have the same order of magnitude, a good approximation for the conductivity of an *n*-type semiconductor is given by

$$\sigma \approx n\mu_n q$$

Similarly, a good approximation for the conductivity of p-type semiconductor is

$$\sigma \approx p\mu_p q$$

Drill Exercise 6.3

Silicon is doped with 1 part per billion of an impurity. Find the conductivity at 300 K of the resulting doped semiconductor when the impurity is (a) a donor impurity and (b) an acceptor impurity.

ANSWER (a) 1.04 $\mho/\text{m}$; (b) 0.4 $\mho/\text{m}$

Graded Semiconductors

We have seen that a current in a semiconductor can be obtained as a result of the influence of an electric field (Eq. 6.6). Yet there is another way in which electric charge can be transported in a semiconductor. If there is a nonuniform concentration of electric charge, then the charge will move from a region of a given concentration to one of lower concentration. This phenomenon is known as **diffusion.** (It is analogous to the diffusion of gases.) The movement of charge is not due to the repulsion of like charges, but rather is a result of a statistical phenomenon.

In general, the concentration of holes p in a semiconductor is not constant but rather is a function of distance x. The hole-concentration rate of change dp/dx is known as the **concentration gradient** for holes.

The hole current density J_p due to diffusion is proportional to the concentration gradient. In particular,

$$J_p = -qD_p\frac{dp}{dx} \tag{6.11}$$

where D_p, called the **diffusion constant** for holes, has square meters per second (m^2/s) as units. Similarly, for the case of (free) electrons, the electron current density J_n is given by

$$J_n = qD_n\frac{dn}{dx} \tag{6.12}$$

where dn/dx is the concentration gradient for electrons, and the diffusion constant for electrons D_n also has as units m^2/s.

Diffusion and mobility are not independent of each other, and in fact are related by the **Einstein equation,**[3]

$$\frac{D_n}{\mu_n} = \frac{D_p}{\mu_p} = \frac{kT}{q} \tag{6.13}$$

where $k = 1.38 \times 10^{-23}$ J/K is the **Boltzmann constant,**[4] T is the temperature (in kelvins), and $q = 1.60 \times 10^{-19}$ C. Since the units of kT/q are J/C = V, we define the quantity V_T by

$$V_T = \frac{kT}{q} = \frac{T}{11,586} \approx \frac{T}{11,600} \tag{6.14}$$

and refer to V_T as the "volt equivalent of temperature."

Given a semiconductor with hole and electron concentrations p and n, respectively, if an electric field $\mathscr{E}$ is also present, then the total current density for holes is (see Eq. 6.5, Eq. 6.6, and Eq. 6.11).

$$J_p = p\mu_p q\mathscr{E} - qD_p \frac{dp}{dx} \tag{6.15}$$

whereas for electrons the total current density is (see Eq. 6.5, Eq. 6.6, and Eq. 6.12).

$$J_n = n\mu_n q\mathscr{E} + qD_n \frac{dn}{dx} \tag{6.16}$$

Consider a doped semiconductor with a nonuniform hole concentration p. Such a semiconductor is said to be **graded.** Suppose that there are no external connections to the semiconductor. Since p is nonuniform, there is a hole diffusion current. But nothing is connected to the semiconductor, so in order to cancel the diffusion current, there must be a current due to hole drift that is equal and opposite to the diffusion current. However, the existence of a drift current means there is an electric field; otherwise the drift is random and averages to zero.

From Eq. 6.15, since the net current is zero, we have

$$0 = p\mu_p q\mathscr{E} - qD_p \frac{dp}{dx}$$

---

[3]Named for the American (German-born) physicist Albert Einstein (1879–1955).
[4]Named for the Austrian physicist Ludwig Boltzmann (1844–1906).

from which

$$\mathscr{E} = \frac{D_p}{p\mu_p} \frac{dp}{dx}$$

From Eq. 6.13 and Eq. 6.14, this equation becomes

$$\mathscr{E} = \frac{V_T}{p} \frac{dp}{dx}$$

However, there is a relationship between potential v and electric field $\mathscr{E}$, that being $\mathscr{E} = -dv/dx$. Using this equation, we have that

$$-\frac{dv}{dx} = \frac{V_T}{p} \frac{dp}{dx}$$

Integrating both sides of this expression with respect to distance x between points x_1 and x_2, we obtain

$$\int_{x_1}^{x_2} -\frac{dv}{dx} \, dx = \int_{x_1}^{x_2} \frac{V_T}{p} \frac{dp}{dx} \, dx$$

Applying the chain rule of calculus, this integral equation becomes

$$\int_{v_1}^{v_2} -dv = \int_{p_1}^{p_2} V_T \frac{dp}{p}$$

where v_1 and v_2 are the potentials at x_1 and x_2, respectively; and p_1 and p_2 are the hole concentrations at x_1 and x_2, respectively. Evaluating these integrals, we find that

$$v_2 - v_1 = V_T (\ln p_1 - \ln p_2)$$

and if we define $v_{21} = v_2 - v_1$, then

$$v_{21} = V_T \ln (p_1/p_2) \tag{6.17}$$

Note, therefore, that the difference in potential between points x_1 and x_2 depends on the hole concentrations at the two points and not on the distance between the points. From Eq. 6.17, we have

$$\ln \frac{p_1}{p_2} = \frac{v_{21}}{V_T} \qquad \Rightarrow \qquad \frac{p_1}{p_2} = e^{v_{21}/V_T}$$

from which

$$p_1 = p_2 e^{v_{21}/V_T} \tag{6.18}$$

Similarly, for the case of electrons

$$n_1 = n_2 e^{-v_{21}/V_T} \tag{6.19}$$

Taking the product of Eq. 6.18 and Eq. 6.19 results in

$$n_1 p_1 = (n_2 e^{-v_{21}/V_T})(p_2 e^{v_{21}/V_T}) = n_2 p_2$$

Thus we see that (under thermal equilibrium) the product of n and p is always the same and is independent of position and the amount of doping. Since $n = p = n_i$ for an intrinsic semiconductor, the product of n and p must be $np = n_i^2$—which is the mass-action law (Eq. 6.7).

The pn Junction

Now consider the case that one side of a semiconductor is p-type material and the other side is n-type material as depicted in Fig. 6.8. We shall deal with the situation that there are no external connections (i.e., the open-circuit case). The boundary between the two sides of the semiconductor is called a ***pn* junction,** and we say that the semiconductor is **step graded.**

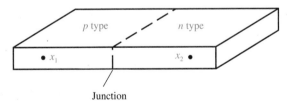

Fig. 6.8 A *pn* junction.

Suppose that for the p-type material the hole concentration p_p is uniform. From Eq. 6.10, then

$$p_p \approx N_A \qquad \text{and} \qquad n_p \approx \frac{n_i^2}{N_A}$$

In addition, suppose that for the n-type material, the free-electron concentration n_n is uniform. From Eq. 6.9, then

$$n_n \approx N_D \quad \text{and} \quad p_n \approx \frac{n_i^2}{N_D}$$

In general, there is a difference in concentrations between the two sides with the concentration at the junction zero. From Eq. 6.17, the potential difference v_0, called the **barrier potential** or **built-in voltage,** of the pn junction is

$$v_0 = V_T \ln \frac{p_1}{p_2} = V_T \ln \frac{p_p}{p_n} = V_T \ln \frac{N_A}{n_i^2/N_D} = V_T \ln \frac{N_A N_D}{n_i^2} \qquad (6.20)$$

Use of a similar argument for free-electron concentrations will result in the same expression for v_0. The order of magnitude for v_0 typically is a few tenths of a volt (see Problems 6.13–6.18 at the end of this chapter).

The barrier potential of a pn junction cannot cause an external current. If external connections are made, what results is the creation of contact potentials that negate the barrier potential.

Example 6.4

The p side of a germanium pn junction has a hole concentration of 2×10^{21} m^{-3}, whereas the n side has a free-electron concentration of 5×10^{22} m^{-3}. Therefore, from Eq. 6.20, the barrier potential at 300 K of this pn junction is

$$v_0 = V_T \ln \frac{N_A N_D}{n_i^2} = \frac{300}{11{,}586} \ln \frac{(2 \times 10^{21})(5 \times 10^{22})}{(2.5 \times 10^{19})^2} = 0.31 \text{ V}$$

Drill Exercise 6.4

A silicon pn junction is to have a barrier potential of 0.7 V at 300 K. Given that the p side has an acceptor-atom concentration of N_A and the n side has a donor-atom concentration of N_D, determine these concentrations for the case that $N_A = N_D = N$.

ANSWER 1.09×10^{22} m^{-3}

6.3 The Junction Diode

Let us now look at the pn junction in further detail. Figure 6.9 shows a two-dimensional depiction of a pn junction, where the p-type material is on the left-hand

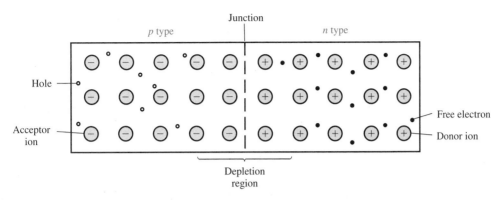

Fig. 6.9 A *pn* junction.

side of the semiconductor and the *n*-type material is on the right-hand side. Since there is a concentration gradient across the junction, diffusion results—holes diffuse to the right and free electrons diffuse to the left. The net effect is a diffusion current directed to the right. When they meet, in the vicinity of the junction, holes and free electrons combine and are neutralized. This results in uncovered ions in the neighborhood of the junction. Since this region is depleted of mobile carriers (holes and free electrons), it is called the **depletion region.** It is also referred to as the **space-charge region** and the **transition region.** (The width of this region typically is of the order of 5×10^{-7} m.)

The ions in the depletion region establish an electric field directed from right to left. This, in turn, produces a drift to the left of holes, and a drift to the right of free electrons—that is, a drift current directed to the left. If there are no external connections to the junction (i.e., the open-circuit case), the diffusion current directed to the right and the drift current directed to the left must sum to a net current of zero.

If we connect **metal** (or **ohmic**) **contacts** to a *pn* junction as illustrated in Fig. 6.10*a*, we have a circuit element known as a **junction diode.** The circuit symbol for this element is given in Fig. 6. 10*b*. The *p* side of the junction diode is called the **anode** and the *n* side is called the **cathode.**

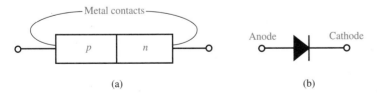

Fig. 6.10 Junction diode.

Let us connect a battery (dc voltage source) to a junction diode as shown in Fig. 6.11. Here the positive terminal of the battery is connected to the cathode (*n* side)

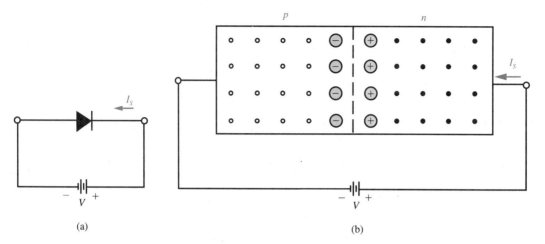

Fig. 6.11 Reverse-biased junction diode.

and the negative terminal is connected to the anode (p side). Under this condition, known as **reverse bias,** holes have a tendency to diffuse to the right and drift to the left. Conversely, electrons tend to diffuse to the left and drift to the right. The result is a widened depletion region, and a potential across the junction of $v_0 + V$, where v_0 is the barrier potential (Eq. 6.20) and V is the battery potential.

In the n-type material, holes are minority carriers. There are relatively few holes, and they are continually being thermally generated and recombining with free electrons. Some of these holes drift to the left into the depletion region, where they are pushed across by the electric field. A similar situation exists in the p-type material where free electrons are the minority carriers. Here the electron drift is to the right. The net effect is a small current I_S, called the **reverse saturation current,** or **saturation current,** through the diode directed to the left. For the most part, this current is not dependent on the value of V, but rather depends on temperature. A higher temperature means that more hole-electron pairs are generated, and more minority carriers produce a greater value for I_S. (Typically, at 300 K, I_S is in the microampere (μA), or 10^{-6} A, range for germanium, and in the nanoampere (nA), or 10^{-9} A, range for silicon.) However, if V is made sufficiently large, the junction will "break down" (see Section 6.6) and a large current can result.

Now suppose that the polarity of the battery shown in Fig. 6.11 is reversed. In other words, suppose that the positive terminal of the battery is connected to the anode (p side) and the negative terminal is connected to the cathode (n side). Under this condition, known as **forward bias,** holes tend to diffuse to the right. Furthermore, free electrons tend to diffuse to the left. If V is large enough (a few tenths of a volt), diffusion overcomes the opposing tendency to drift, and holes (majority carriers) from the p-type material cross the junction and are injected into the n-type material, where they recombine with free electrons. In addition, free electrons (majority carriers) from the *n*-type material are injected into the p-type material, where they recombine with holes. The net result of this is a current directed to the right that is sustained by the continual thermal generation of majority carriers.

Although the applied voltage *V* narrows the depletion region, the depletion region never disappears. A narrower depletion region means more current, but the current is limited by the "bulk" resistance of the semiconductor crystal material and the resistance of the metal contacts.

Shown in Fig. 6.12 is a (junction) diode that has a voltage *v* across it and a current *i* through it. The polarity of *v* and the direction of *i* are conventional. The relationship between *i* and *v* is given by

$$i = I_S(e^{v/\eta V_T} - 1) \tag{6.21}$$

where I_S is the reverse saturation current, $V_T = T/11{,}586$ is the volt-equivalent of temperature, and the constant η (eta), called the **emission coefficient,** is 1 for Ge and 2 for Si. A typical plot of current *i* versus voltage *v*, called the *i-v* **characteristic curve,** for a diode is shown in Fig. 6.13.

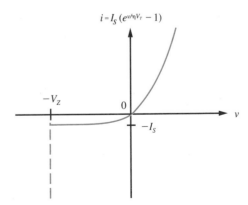

Fig. 6.12 Diode.

Fig. 6.13 The *i-v* characteristic curve of a diode.

If the diode is forward biased and *v* > 0.2 V, then at room temperature a good approximation (for both silicon and germanium) of Eq. 6.21 is

$$i \approx I_S e^{v/\eta V_T} \tag{6.22}$$

If the diode is reverse biased and $-V_Z < v < -0.2$ V, then a good approximation is

$$i \approx -I_S \tag{6.23}$$

If the voltage across the diode is made negative enough, that is, for a large enough reverse bias, the diode will break down and allow negative current (more negative than $-I_S$) through it. From the i-v characteristic given in Fig. 6.13, we see that when $v = -V_Z$, the diode breaks down, and the dashed line indicates that a negative (reverse) current results. Note that when $i < -I_S$, the voltage across the diode is essentially constant ($v = V_Z$). We will discuss the phenomenon of diode breakdown further in Section 6.6.

In Eq. 6.21, v is the independent variable and i is the dependent variable. That is, given a value for v, it is a simple matter to determine the corresponding value for i. From Eq. 6.21, we have that

$$i + I_S = I_S e^{v/\eta V_T}$$

so

$$e^{v/\eta V_T} = \frac{i + I_S}{I_S} = \frac{i}{I_S} + 1 \quad \Rightarrow \quad \frac{v}{\eta V_T} = \ln\left(\frac{i}{I_S} + 1\right)$$

Thus,

$$\boxed{v = \eta V_T \ln\left(\frac{i}{I_S} + 1\right)} \tag{6.24}$$

In this expression, i is the independent variable and v is the dependent variable. However, since the natural logarithm exists only for positive numbers, Eq. 6.24 is valid only for the case that $i > -I_S$.

Example 6.5

At 300 K, a 1N4153 silicon diode has a saturation current of $I_S = 10$ nA. If the voltage across the diode is $v = 0.6$ V, then by Eq. 6.21, the current through it is

$$i = I_S(e^{v/\eta V_T} - 1) = (10 \times 10^{-9})(e^{(0.6)(11,586)/2(300)} - 1) = 1.08 \text{ mA}$$

whereas if $v = 0.7$ V, then

$$i = I_S(e^{v/\eta V_T} - 1) = (10 \times 10^{-9})(e^{(0.7)(11,586)/2(300)} - 1) = 7.42 \text{ mA}$$

Conversely, if the current through the diode is $i = 5$ mA, then by Eq. 6.24, the voltage across it is

$$v = \eta V_T \ln\left(\frac{i}{I_S} + 1\right) = 2\left(\frac{300}{11{,}586}\right) \ln\left(\frac{5 \times 10^{-3}}{10 \times 10^{-9}} + 1\right) = 0.68 \text{ V}$$

Drill Exercise 6.5

A silicon diode has a reverse saturation current of 1 nA at 300 K. With reference to Fig. 6.12, find i when v is (a) 0.7 V, (b) 0.1 V, (c) 0 V, (d) -0.1 V, (e) -0.7 V.

ANSWER   (a) 0.742 mA; (b) 5.90 nA; (c) 0 A; (d) -0.855 nA; (e) -1.0 nA

A silicon diode has a reverse saturation current of 1 nA at 300 K. With reference to Fig. 6.12, find v when i is (a) -0.5 nA, (b) 0 A, (c) 0.5 nA, (d) 50 nA, (e) 50 μA.

ANSWER   (a) -0.036 V; (b) 0 V; (c) 0.021 V; (d) 0.204 V; (e) 0.56 V

Diode Behavior

Now let us inspect the *i-v* characteristic (see Fig. 6.13) of a diode more closely. For the case that the diode is forward biased ($i > 0$ A, $v > 0$ V), Fig. 6.14a shows the *i-v* characteristic for a typical germanium diode and Fig. 6.14b shows the *i-v* characteristic for a typical silicon diode. As can be seen in Fig. 6.14, for a given voltage, a germanium diode has a much larger current than does a silicon diode. Alternatively, for a given current, a germanium diode has a smaller voltage than does a silicon diode.

From Fig. 6.14a, note that the current through the germanium diode is very small[5] for voltage values of less than 0.2 V. In other words, there is no significant current through a germanium diode until the voltage across it exceeds 0.2 V. For this reason, we say that 0.2 V is the **cut-in voltage** V_γ of a germanium diode. Furthermore, inspection of Fig. 6.14b reveals that the approximate value for the cut-in voltage for a silicon diode is about $V_\gamma = 0.5$ V.

It has been determined experimentally that for a temperature increase of 1°C (an increase of 1 K), in both germanium and silicon diodes the saturation current increases by about 7 percent. This means that for an increase in temperature of 10°C,

---

[5]By "very small" we mean a value of less than 1 percent of the maximum current rating of the diode.

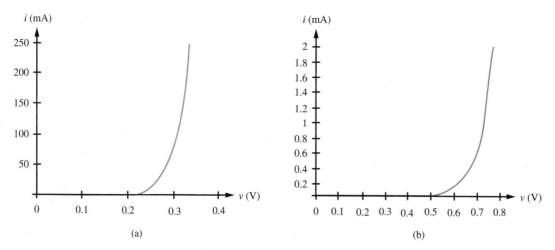

Fig. 6.14 Typical *i-v* characteristic curve for (*a*) a Ge diode, and (*b*) a Si diode.

the saturation current increases by a factor of $(1.07)^{10} \approx 2$. In other words, I_S is a function of temperature T, and I_S approximately doubles for every rise of 10°C. To express this fact in mathematical form, if $I_S(T_a)$ is the saturation current at temperature T_a, then the saturation current at temperature T_b is given approximately by the equation

$$I_S(T_b) = 2^{(T_b - T_a)/10} I_S(T_a) \qquad (6.25)$$

Note that in Eq. 6.25, we can use either °C or K (but not both together) as the unit for temperature.

From the fact that the saturation current of a diode is a function of temperature, it is apparent that, for a given voltage, the diode current is also temperature dependent. Specifically, if temperature increases, then I_S increases. Even though the exponent in Eq. 6.21 decreases as T increases, I_S increases at a higher rate. The result is that for a given voltage v, the diode current i increases as T increases. This phenomenon is illustrated in Fig. 6.15, where the solid curve is the *i-v* characteristic of a diode (either germanium or silicon) at temperature T_a, and the dashed curve is the *i-v* characteristic of the same diode at temperature $T_b > T_a$. From Fig. 6.15, we see that if the diode voltage is $v = v_2$, then the diode current is $i = i_1$ for temperature T_a, and $i = i_2$ for $T_b > T_a$. To maintain the current $i = i_1$ when the temperature increases from T_a to T_b, the diode voltage should be lowered from $v = v_2$ to $v = v_1$. In other words, by decreasing the voltage appropriately, we can compensate for the increase in temperature. Specifically, at room temperature (for either germanium or silicon), a constant current can be

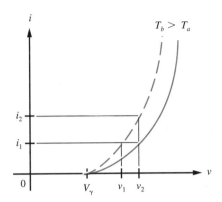

Fig. 6.15 Temperature dependence of diode *i-v* characteristic.

maintained if the voltage is decreased by approximately 2.5 mV for each 1°C increase in temperature. In mathematical terms,

$$\frac{dv}{dT} \approx -2.5 \text{ mV}/°C \qquad (6.26)$$

Thus, for Fig. 6.15, we have when $i = i_1$ that $v_2 - v_1 \approx -2.5(T_a - T_b)$ mV, or

$$v_2 - v_1 \approx 2.5(T_b - T_a) \times 10^{-3} \text{ V} \qquad (6.27)$$

Example 6.6

For a 1N4153 silicon diode, the saturation current at 300 K is $I_S(300) = 10$ nA. From Eq. 6.25, at 316 K the saturation current is approximately

$$I_S(316) = 2^{(316-300)/10}I_S(300) = 2^{1.6}(10 \times 10^{-9}) = 30.3 \text{ nA}$$

In Example 6.5, we determined that at 300 K when $v = 0.7$ V, then $i = 7.42$ mA. To have the same current at 316 K, by Eq. 6.27, the voltage must be reduced by

$$v_2 - v_1 \approx 2.5(316 - 300) \times 10^{-3} = 0.04 \text{ V}$$

that is, to get $i = 7.42$ mA at 316 K, the voltage should be $v \approx 0.7 - 0.04 = 0.66$ V.

As a check, note that

$$v = \eta V_T \ln\left(\frac{i}{I_S} + 1\right) = 2\left(\frac{316}{11,586}\right) \ln\left(\frac{7.42 \times 10^{-3}}{30.3 \times 10^{-9}} + 1\right)$$

$$= 0.677 \approx 0.66 \text{ V}$$

Drill Exercise 6.6

Given that the current through a 1N4153 silicon diode is 10 mA, find the voltage across the diode when the temperature is (a) 290 K, (b) 310 K, (c) 320 K.

ANSWER (a) 0.726 V; (b) 0.70 V; (c) 0.687 V

Diode Circuits

Now let us look at the diode circuit shown in Fig. 6.16a. We can attempt to analyze this circuit by writing some equations, and begin with the equation relating the current and voltage for the diode:

$$i = I_S(e^{v/\eta V_T} - 1) \tag{6.21}$$

where it is assumed that I_S, η, and T are given. By KVL, we can also write

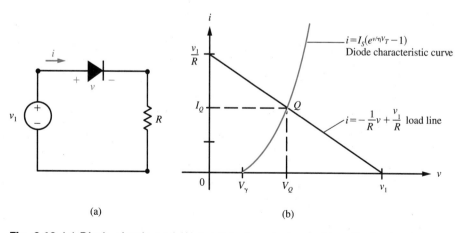

(a)

(b)

Fig. 6.16 (a) Diode circuit, and (b) graphical analysis of given diode circuit.

$$v_1 = v + Ri \qquad (6.28)$$

where it is assumed that v_1 and R are given. Thus we have two equations, (Eq. 6.21 and 6.28), and two unknowns, i and v. However, Eq. 6.21 is nonlinear, so that the techniques used for linear, algebraic simultaneous equations cannot be employed. For example, substituting Eq. 6.21 into Eq. 6.28 results in a transcendental equation that has no analytical solution.

Fortunately, though, there are other analysis techniques that can be utilized. Suppose that v_1 is a positive constant. Since current goes through a resistor from a given potential to a lower potential, we should recognize for Fig. 6.16a that i and hence v, are positive; that is, the diode is forward biased. Next, by using Eq. 6.21 and a calculator, we can determine enough points to sketch an i-v characteristic curve for the diode. But from Eq. 6.28 we have that

$$i = -\frac{1}{R}v + \frac{v_1}{R} \qquad (6.29)$$

This is the equation of a straight line of slope $-1/R$ and vertical-axis intercept v_1/R. This straight line, called the **load line**, can be plotted using the same axes as for the diode curve as is shown in Fig. 6.16b. Since any point on the diode curve satisfies Eq. 6.21, and any point on the load line satisfies Eq. 6.29, the intersection of both plots satisfies both equations simultaneously. The point of intersection is designated Q and is called the **quiescent operating point** (or **operating point,** or ***Q* point,** for short). Thus the solution to the equations, which is the current I_Q and the voltage V_Q, has been obtained graphically instead of analytically.

As an alternative to a graphical approach, with the aid of a hand-held calculator, a simple numerical method can be employed. The technique is an iterative process that begins by assuming a partial numerical solution, which is used to calculate the remainder of the solution. The result, in turn, can be used to refine the original assumed value. The process is repeated until a desired numerical accuracy is obtained.

Example 6.7

For the diode circuit given in Fig. 6.16a, $v_1 = 3$ V, $R = 1$ kΩ, and the silicon diode has a saturation current of 10 nA at 300 K. Let us assume that $v = 0.6$ V. Then, by Ohm's law,

$$i = \frac{-v + v_1}{R} = \frac{-0.6 + 3}{1 \times 10^3} = 2.4 \text{ mA}$$

Now let us use this value of current to refine the value of the voltage. By Eq. 6.24

$$v = \eta V_T \ln\left(\frac{i}{I_S} + 1\right) = 2\left(\frac{300}{11,586}\right) \ln\left(\frac{2.4 \times 10^{-3}}{10 \times 10^{-9}} + 1\right) = 0.6416 \text{ V}$$

and using this value of the voltage, we can refine the value of the current from

$$i = \frac{-v + v_1}{R} = \frac{-0.6416 + 3}{1 \times 10^3} = 2.358 \text{ mA}$$

Performing one more iteration,

$$v = \eta V_T \ln\left(\frac{i}{I_S} + 1\right) = 2\left(\frac{300}{11,586}\right) \ln\left(\frac{2.358 \times 10^{-3}}{10 \times 10^{-9}} + 1\right) = 0.6406 \text{ V}$$

and

$$i = \frac{-v + v_1}{R} = \frac{-0.6406 + 3}{1 \times 10^3} = 2.359 \text{ mA}$$

Since this value of the current changed only in the fourth significant digit, from a practical point of view, we need not proceed any further.

Hence we conclude that $I_Q = 2.36$ mA and $V_Q = 0.641$ V.

Drill Exercise 6.7

Suppose that for the diode circuit shown in Fig. 6.16a, $v_1 = 9$ V, $R = 2$ kΩ, and the silicon diode has a saturation current of 5 nA at 300 K. Determine i and v by using the numerical technique described in Example 6.7. Begin by assuming that $v = 0.5$ V.

ANSWER 4.15 mA, 0.706 V

For some diode circuits, it may be possible to avoid graphical or numerical analysis techniques.

Example 6.8

For the diode circuit shown in Fig 6.17, suppose that both diodes have the same saturation current. The 6-V source will forward bias D_1 and reverse bias D_2. Consequently, we deduce that $i_2 = -I_S$. Therefore, $i_1 = -i_2 = I_S$ and

$$v_1 = \eta V_T \ln\left(\frac{i_1}{I_S} + 1\right) = 2 \frac{300}{11{,}586} \ln\left(\frac{I_S}{I_S} + 1\right) = 0.036 \text{ V}$$

so that $v_2 = -6 + v_1 = -6 + 0.036 = -5.964$ V. This value of v_2 indicates that D_2 is sufficiently reverse biased so that indeed $i_2 = -I_S$, and $i_1 = -i_2 = I_S > 0$ A indicates that D_1 is indeed forward biased.

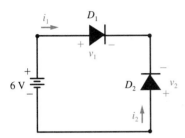

Fig. 6.17 Diode circuit for Example 6.8.

Drill Exercise 6.8

For the diode circuit given in Fig. 6.17, suppose that the silicon diodes D_1 and D_2 have saturation currents of 1 nA and 20 nA, respectively, at 300 K. Find i_1, i_2, v_1, and v_2.

ANSWER   20 nA; -20 nA; 0.16 V; -5.84 V

6.4   The Ideal Diode

Figure 6.13 shows a typical i-v characteristic curve for a diode. The relationships between current and voltage given in Eq. 6.21 and Eq. 6.24 are clearly nonlinear and analytically cumbersome. There are many situations in which a simplified description for a diode can be used for analysis purposes without sacrificing reasonable accuracy. Because of this, we define an **ideal diode** to be a two-terminal device whose i-v characteristic is that given in Fig. 6.18. The circuit symbol for the ideal diode shown in Fig. 6.19a differs slightly (it is hollow, not solid) from the nonideal diode (see Fig. 6.12). For the case that the current through an ideal diode is in the forward direction (i.e., for forward bias), the voltage across the diode is zero—that is, the diode behaves as a short circuit. For the case that the voltage across an ideal

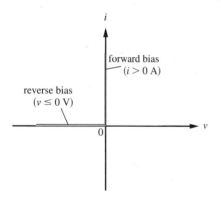

Fig. 6.18 Ideal diode *i-v* characteristic.

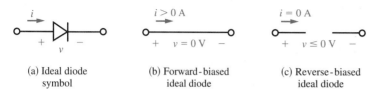

(a) Ideal diode symbol   (b) Forward-biased ideal diode   (c) Reverse-biased ideal diode

Fig. 6.19 Ideal-diode behavior.

diode is nonpositive (i.e., for reverse bias), the current through the diode is zero—that is, the diode behaves as an open circuit. These situations are depicted in Figs. 6.19*b* and *c*, respectively.

Since an ideal diode, as well as a nonideal diode, is a nonlinear element, we cannot immediately apply linear circuit analysis techniques to circuits containing diodes. Instead, what we shall do is assume ideal diodes to be forward biased or reverse biased. For the former case, we say that a diode is **ON,** whereas for the latter case, we say that a diode is **OFF.** After assuming that diodes are ON or OFF, we replace them by short circuits or open circuits, respectively. Next, we can apply known analysis techniques to the resulting linear circuits. If the current through an ON diode is calculated to be nonpositive, the assumption about its state was incorrect, and it must be assumed to be OFF. After changing assumptions, the analysis must be repeated from the beginning. Similarly, if the voltage across an OFF diode is calculated to be positive, its assumption about being OFF was incorrect, and it should be assumed to be ON. Again, the analysis must begin anew. If all diodes assumed to be ON are found to have positive currents, and if all diodes assumed to be OFF are found to have nonpositive voltages, then the assumptions about them were right, and the circuit currents and voltages calculated are correct. Diode circuit analysis often is shortened by using intuition or educated guessing as to whether a diode should be assumed to be OFF or ON.

Example 6.9

Given the ideal-diode circuit shown in Fig. 6.20, let us first consider the case of a positive input voltage, that is, $v_S > 0$ V. Since current goes through a resistor from a given potential to a lower potential, let us assume that D is ON. Replacing D by a short circuit, we get the circuit shown in Fig. 6.21. By Ohm's law, we have that $i = v_S/R$. Since $v_S > 0$ V, then $i > 0$ A, and the assumption that D is ON is confirmed.

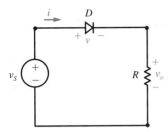

Fig. 6.20 An ideal-diode circuit.

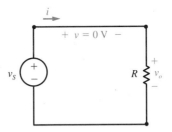

Fig. 6.21 Ideal-diode circuit with D assumed ON.

Next consider the case that $v_S \leq 0$ V. For this situation, let us assume that D is OFF. Replacing D by an open circuit yields the circuit shown in Fig. 6.22. Since $i = 0$ A, by KVL, $v = v_S - Ri = v_S$. Since $v_S \leq 0$ V, then $v \leq 0$ V and our assumption that D is OFF is confirmed.

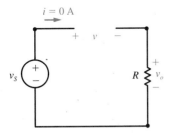

Fig. 6.22 Diode circuit with D assumed OFF.

From this discussion, $v_S > 0$ V, then the output voltage is $v_o = v_S$. When $v_S \leq 0$ V, then $v_o = Ri = 0$ V. Therefore, the output voltage v_o is given by

$$v_o = \begin{cases} 0 \text{ V} & \text{for } v_S \leq 0 \text{ V} \\ v_S & \text{for } v_S > 0 \text{ V} \end{cases}$$

Consider the case that the input voltage v_S is a sinusoid as shown in Fig. 6.23a. Then the resulting output voltage v_o is the **half-wave rectified** sinusoid shown in Fig. 6.23b. We call the diode circuit shown in Fig. 6.20 a **half-wave rectifier.**

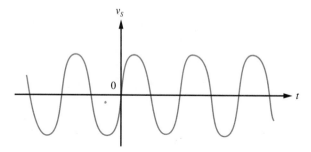

(a) Input to half-wave rectifier

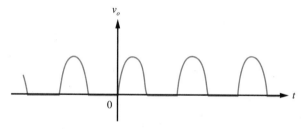

(b) Output of half-wave rectifier

Fig. 6.23 Input and output for half-wave rectifier circuit.

Drill Exercise 6.9

For the ideal-diode circuit given in Fig. 6.20, reverse the diode. Find v_o given that v_S is the sinusoid shown in Fig. 6.23a.

ANSWER v_S for $v_S < 0$ V, 0 V for $v_S \geq 0$ V

Having seen one application of diodes, let us now look at another.

Example 6.10

In addition to an ideal diode, the circuit shown in Fig. 6.24 contains two independent voltage sources. If the input voltage is $v_S > 0$ V, then v_S tends to forward bias D.

However, the 3-V source tends to reverse bias D. Let us find out which voltage source dominates.

We begin by assuming that D is ON. From the resulting circuit shown in Fig. 6.25, by Ohm's law,

$$i_1 = \frac{v_S - 3}{1000} \quad \text{and} \quad i_2 = \frac{3}{2000} = 1.5 \text{ mA}$$

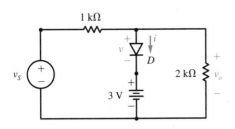

Fig. 6.24 Diode clipper circuit.

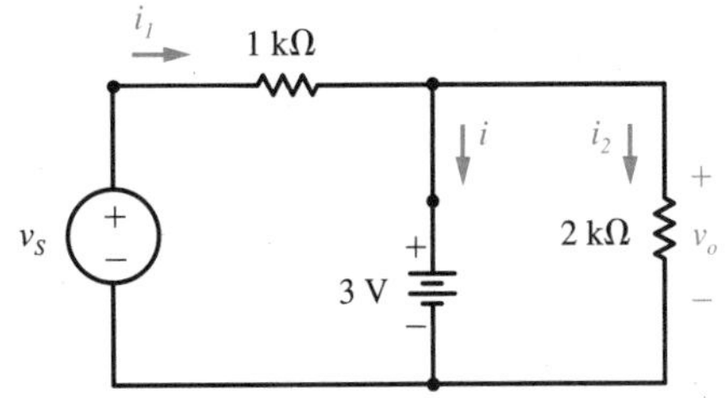

Fig. 6.25 Circuit with D ON.

and by KCL,

$$i = i_1 - i_2 = \frac{v_S}{1000} - 3 \times 10^{-3} - 1.5 \times 10^{-3} = \frac{v_S}{1000} - 4.5 \times 10^{-3}$$

Since the diode is ON when $i > 0$ A, then D is ON when

$$i = \frac{v_S}{1000} - 4.5 \times 10^{-3} > 0 \quad \Rightarrow \quad v_S > 4.5 \text{ V}$$

We may, therefore, deduce that D will be OFF for $v_S \le 4.5$ V.

When D is ON, by KVL, we have that the output voltage is $v_o = 3$ V. To determine the output voltage when D is OFF, we need to analyze the circuit shown in Fig. 6.26. By voltage division, we have that the output voltage is

$$v_o = \frac{2000}{1000 + 2000} v_S = \frac{2}{3} v_S$$

As a matter of redundancy, D is OFF when $v \le 0$ V. Since $v = v_o - 3$, then D is OFF when

$$v = v_o - 3 = \tfrac{2}{3} v_S - 3 \le 0 \quad \Rightarrow \quad v_S \le 4.5 \text{ V}$$

as was previously deduced.

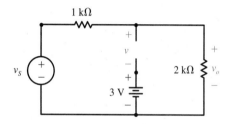

Fig. 6.26 Circuit with *D* OFF.

In summary, for the diode circuit given in Fig. 6.24, the output voltage v_o is given by

$$v_o = \begin{cases} \frac{2}{3}v_S & \text{for } v_S \le 4.5 \text{ V} \\ 3 \text{ V} & \text{for } v_S > 4.5 \text{ V} \end{cases}$$

Suppose now that the input voltage v_S is a sinusoid with an amplitude of 6 V as shown in Fig. 6.27a. Then the output voltage v_o is the "clipped" sinusoid shown in Fig. 6.27b. Because of this, we say that the circuit shown in Fig. 6.24 is an example of a diode **clipper circuit**.

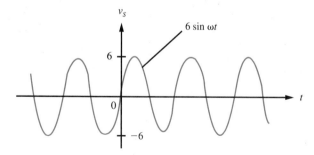

(a) Input to clipper

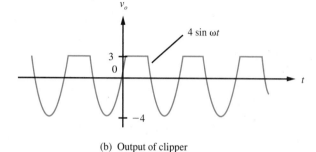

(b) Output of clipper

Fig. 6.27 Input and output of diode clipper circuit.

Application to Digital Logic Circuits

A major use of electronics is for digital logic circuits that are employed in computers, control systems, data processing systems, and digital communication systems. Typically, a digital circuit has an input and an output that are equal to one of only two possible voltages—a low voltage and a high voltage. Because there are only two possible states, one can be designated by a 0 and the other by a 1. If we use the 0 to represent the low voltage and the 1 to represent the high voltage, the result is known as **positive logic.** Conversely, if 0 represents the high voltage and 1 represents the low voltage, then we have **negative logic.** The importance and applications of digital logic to digital systems will be studied in detail in the portion of this book on digital systems (Chapters 11, 12, and 13). Now, however, we will see how ideal diodes can be used to fabricate some types of digital logic circuits.

We begin by considering the two-input diode circuit shown Fig. 6.28, where v_1 and v_2 are the inputs and v_o is the output. Let us choose the case of positive logic— the low voltage (logic 0) is 0 V and the high voltage (logic 1) is V_H volts, where $V_H > 0$ V.

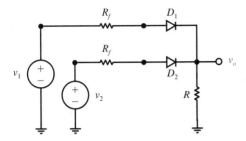

Fig. 6.28 Diode OR gate.

Case 1: $v_1 = v_2 = 0$ V. Since there is no applied voltage, no current is produced, and therefore, D_1 and D_2 are OFF, and $v_o = 0$ V.

Case 2: $v_1 = 0$ V, $v_2 = V_H$. Let us assume that D_1 is OFF and D_2 is ON. This means that we must analyze the circuit shown in Fig. 6.29. By KVL,

$$V_H = R_f i_2 + R i_2 = (R_f + R)i_2 \quad \Rightarrow \quad i_2 = \frac{V_H}{R_f + R}$$

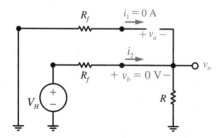

Fig. 6.29 D_1 is OFF, D_2 is ON.

Since $R > 0 \ \Omega$ and $R_f > 0 \ \Omega$, then $i_2 > 0$ A, and this confirms that D_2 is indeed ON. Furthermore, by voltage division,

$$v_o = \frac{R}{R_f + R} V_H > 0 \text{ V} \tag{6.30}$$

By KVL, $v_a = -v_o < 0$ V, and this confirms that D_1 is indeed OFF. For the case that $R >> R_f$, from Eq. 6.30, we have that $v_o \approx V_H$.

Case 3: $v_1 = V_H$, $v_2 = 0$ V. Similar to Case 2, we deduce that D_1 is ON, D_2 is OFF, $v_o = RV_H / (R_f + R)$, and for $R >> R_f$, we have that $v_o \approx V_H$.

Case 4: $v_1 = v_2 = V_H$. Let us assume that both D_1 and D_2 are ON. Thus we need to analyze the circuit shown in Fig. 6.30. By KCL at the output,

$$\frac{V_H - v_o}{R_f} + \frac{V_H - v_o}{R_f} = \frac{v_o}{R}$$

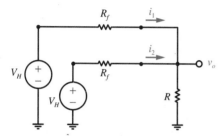

Fig. 6.30 D_1 and D_2 are ON.

Solving this equation for v_o results in

$$v_o = \frac{2R}{2R + R_f} V_H = \frac{R}{R + R_f/2} V_H$$

and for $R >> R_f$, we have that $v_o \approx V_H$.

By Ohm's law,

$$i_1 = i_2 = \frac{V_H - v_o}{R_f} = \frac{V_H - 2RV_H/(2R + R_f)}{R_f} = \frac{V_H}{R_f + R}$$

and since $i_1 = i_2 > 0$ A, then D_1 and D_2 are confirmed to be ON.

In summary, for the case that $R >> R_f$, the circuit given in Fig. 6.28 yields the truth table shown in Table 6.1. The operation described in this table is called the **OR operation** since the output is 1 if either one input OR the other is 1. For this reason, the circuit shown in Fig. 6.28 is known as an **OR gate,** and it is an example of a **diode logic** (DL) circuit.

Table 6.1 OR Truth Table

v_1	v_2	OR
0	0	0
0	1	1
1	0	1
1	1	1

Although Fig. 6.28 shows a two-input DL OR gate, an OR gate with more than two inputs is easily obtained—for each new input, simply place another series connection of a resistor R_f and a diode between that new input and the output.

A DL AND Gate

A second important DL gate is shown in Fig. 6.31. Again we deal with positive logic, where the low voltage is 0 V and the high voltage is $V_H > 0$ V.

Case 1: $v_1 = v_2 = 0$ V. Let us assume that D_1 and D_2 are ON. Therefore, from the resulting circuit in Fig. 6.32, by KCL at the output,

$$\frac{v_o}{R_f} + \frac{v_o}{R_f} + \frac{v_o - V_H}{R} = 0$$

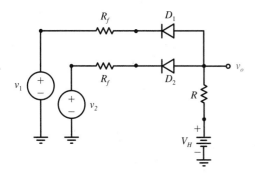

Fig. 6.31 A DL AND gate.   **Fig. 6.32** D_1 and D_2 are ON.

Solving this equation for v_o results in

$$v_o = \frac{R_f}{R_f + 2R} V_H = \frac{R_f/2}{R_f/2 + R} V_H$$

and for $R \gg R_f$, then $v_o \approx 0$ V.
   By Ohm's law,

$$i_1 = i_2 = \frac{v_o}{R_f} = \frac{V_H}{R_f + 2R}$$

and since $i_1 = i_2 > 0$ A, then D_1 and D_2 are confirmed to be ON.
   Case 2: $v_1 = 0$ V, $v_2 = V_H$. Let us assume that D_1 is ON and D_2 is OFF. From the resulting circuit shown in Fig. 6. 33, by KCL,

$$\frac{v_o}{R_f} + \frac{v_o - V_H}{R} = 0 \quad \Rightarrow \quad v_o = \frac{R_f}{R_f + R} V_H$$

so for $R \gg R_f$, then $v_o \approx 0$ V.
   Since $i_1 = v_o/R_f = V_H/(R_f + R) > 0$ A, then D_1 is confirmed to indeed be ON. Furthermore,

$$v_b = v_o - V_H = \frac{R_f}{R_f + R} V_H - V_H = \frac{-R}{R_f + R} V_H < 0 \text{ V}$$

and this confirms that D_2 is indeed OFF.
   Case 3: $v_1 = V_H$, $v_2 = 0$ V. Similar to Case 2, we deduce that D_1 is OFF, D_2 is ON, $v_o = R_f V_H/(R_f + R)$, and for $R \gg R_f$, we have that $v_o \approx 0$ V.
   Case 4: $v_1 = v_2 = V_H$. Let us assume that both D_1 and D_2 are OFF. From Fig. 6.34, since $i_1 = i_2 = 0$ A, then

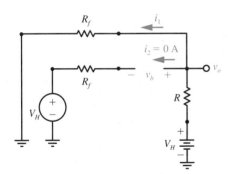

Fig. 6.33 D_1 is ON, D_2 is OFF.

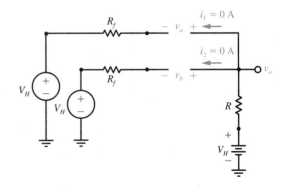

Fig. 6.34 D_1 and D_2 are OFF.

$$v_o = -R(i_1 + i_2) + V_H = V_H$$

Since $v_a = v_b = v_o - V_H = V_H - V_H = 0$ V, then this confirms that D_1 and D_2 are indeed OFF.

In summary, for the case that $R \gg R_f$, the circuit given in Fig. 6.31 yields the truth table shown in Table 6.2. The operation described by this table is called the **AND operation** since the output is 1 only if one input AND the other are both 1. For this reason, the circuit shown in Fig. 6.31 is known as a DL **AND gate**.

Table 6.2 AND Truth Table

v_1	v_2	AND
0	0	0
0	1	0
1	0	0
1	1	1

6.5 Nonideal-Diode Models

Although the ideal model of a diode can be used with relatively good accuracy in the analysis and design of different types of electronic circuits, there may be occasions when a more accurate model of a diode is needed. Therefore, we now discuss how to model nonideal diodes.

First consider the circuit shown in Fig. 6.35a. For the case that v_S is a positive dc voltage source, say $v_S = V_Q$, call the resulting current I_Q. This situation is depicted in Fig. 6.35b. In this case the diode is said to have a **dc (or static) resistance** R of

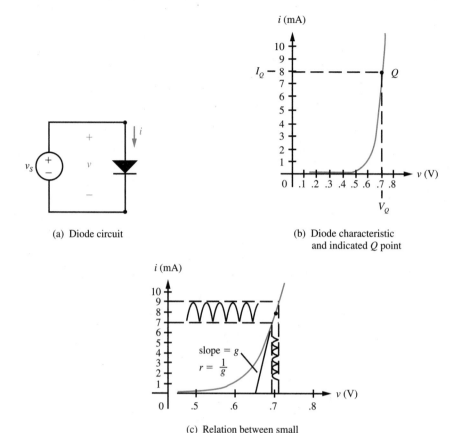

(a) Diode circuit

(b) Diode characteristic and indicated Q point

(c) Relation between small ac voltage and current.

Fig. 6.35 Diode with dc and ac components.

$$R = \frac{V_Q}{I_Q}$$

Thus we see that the dc resistance of a diode depends upon where the diode is operating (or is biased)—i.e., the dc resistance of a diode depends on its dc current I_Q and voltage V_Q.

As an example, a silicon diode having $I_S = 10$ nA at 300 K that operates (is biased) at $V_Q = 0.62$ V, $I_Q = 1.58$ mA has a dc resistance of $R = V_Q/I_Q = 392$ Ω. If the diode operates (is biased) at $V_Q = 0.68$ V, $I_Q = 5.04$ mA, it has a dc resistance of $R = V_Q/I_Q = 135$ Ω.

AC Resistance

Suppose that the voltage source v_S shown in Fig. 6.35(a) has an ac component v_{ac} in addition to its dc component V_Q. That is, suppose that

$$v_S = V_Q + v_{ac}$$

where v_{ac} is an ac voltage (e.g., a sinusoid). This means that the resulting current through the diode will also have dc and ac components, that is

$$i = I_Q + i_{ac}$$

At any instant of time, the diode's voltage and current must correspond to a point on the diode's i-v characteristic shown in Fig. 6.35b. If the ac voltage v_{ac} is small in magnitude, the portion of the i-v characteristic, which relates the voltage variations to the current variations, is approximately a straight line. The slope of this straight line, depicted in Fig. 6.35c, is equal to the derivative of the i-v characteristic curve at the Q point. Since the units of di/dv are A/V $=$ ℧, we say that at a particular operating point the **ac** (or **incremental** or **dynamic**) **conductance** of the diode is

$$g = \frac{di}{dv}$$

and the **ac resistance** of the diode at a particular operating point is

$$r = \frac{1}{g} = \frac{dv}{di}$$

Figure 6.35c demonstrates that the variation in current (i.e., the ac current) is essentially linearly related to the variation in voltage (i.e., the ac voltage) for the diode.
    For a diode, we know that

$$i = I_S(e^{v/\eta V_T} - 1) = I_S e^{v/\eta V_T} - I_S \tag{6.31}$$

Thus taking the derivative with respect to v yields

$$\frac{di}{dv} = \frac{1}{\eta V_T} I_S e^{v/\eta V_T} \tag{6.32}$$

However, from Eq. 6.31, $I_S e^{v/\eta V_T} = i + I_S$. Substituting this fact into Eq. 6.32 yields the ac conductance g and ac resistance r of a diode:

$$g = \frac{di}{dv} = \frac{i + I_S}{\eta V_T} \quad \Rightarrow \quad r = \frac{dv}{di} = \frac{\eta V_T}{i + I_S}$$

Thus the ac resistance of a diode operating at the point $i = I_Q$, $v = V_Q$ is

$$r = \frac{\eta V_T}{I_Q + I_S}$$

For the typical case that $I_Q \gg I_S$, a good approximation for the ac resistance of a diode at 300 K (27°C or 80.6°F) is

$$r = \frac{\eta V_T}{I_Q} = \begin{cases} 0.026/I_Q \text{ for Ge} \\ 0.052/I_Q \text{ for Si} \end{cases}$$

Example 6.11

Let us find the voltage v_o for the circuit shown in Fig. 6.36a, where the silicon diode has a saturation current of 10 nA at 300 K.

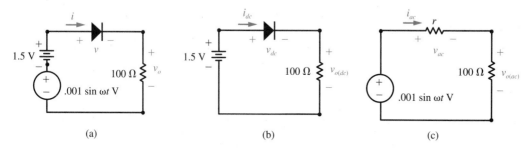

(a) (b) (c)

Fig. 6.36 Diode circuit for Example 6.11.

Since there are two independent voltage sources in the circuit—one dc and one ac—the voltage v_o will have a dc component and an ac component. That is, we can write v_o in the form $v_o = v_{o(dc)} + v_{o(ac)}$.

We find $v_{o(dc)}$ from the dc equivalent circuit shown in Fig. 6.36b. For this circuit we can use either the numerical or the graphical technique described earlier to find that $i_{dc} = I_Q = 7.96$ mA and $v_{dc} = V_Q = 0.704$ V. Thus we have that $v_{o(dc)} = 100 i_{dc} = 100(7.96 \times 10^{-3}) = 0.796$ V. Alternatively, since in this circuit the diode behaves as a resistance of $R = V_Q/I_Q = 0.704/(7.96 \times 10^{-3}) = 88.4 \,\Omega$, then by using the voltage-divider formula, we get

$$v_{o(dc)} = \frac{100}{88.4 + 100}(1.5) = 0.796 \text{ V}$$

We find $v_{o(ac)}$ from the ac equivalent circuit shown in Fig. 6.36c. For this circuit we replaced the diode with its ac resistance. The corresponding approximate value is $r = 0.052/I_Q = 0.052/(7.96 \times 10^{-3}) = 6.5 \,\Omega$. Using the voltage-divider formula, in this case we get

$$v_{o(ac)} = \frac{100}{6.5 + 100}(0.001 \sin \omega t) = 0.00094 \sin \omega t \text{ V}$$

Hence

$$v_o = v_{o(dc)} + v_{o(ac)} = 0.796 + 0.00094 \sin \omega t \text{ V}$$

Drill Exercise 6.11

Find the dc and ac resistances of a silicon diode with a saturation current of 10 nA at 300 K for the case that I_Q is (a) 1.2 mA, (b) 5 mA.

ANSWER (a) 505 Ω, 43.3 Ω; (b) 136 Ω, 10.4 Ω

This last example has demonstrated that the ac resistance of a diode depends on the current through the diode.

Modeling with Ideal Diodes

The dashed curve shown in Fig. 6.37 is a typical *i-v* characteristic of a diode. Since the current is very small when the diode is reverse biased (if the actual scale were used, the dashed curve in the third quadrant would be practically coincident with the negative-voltage axis), as for the ideal case, we will assume that the diode behaves as an open circuit (this situation corresponds to the heavy solid line drawn on the negative-voltage axis). Now, however, for the case that the diode is forward biased, let us approximate the dashed curve in the first quadrant

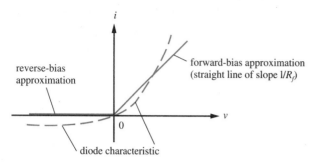

Fig. 6.37 Approximation of diode *i-v* characteristic.

by a straight line of slope $1/R_f$ as shown. (Since the slope of the straight line has units amperes per volt (A/V), the slope is designated by the reciprocal of resistance.) The value of this **forward resistance** R_f is a typical ac resistance of the forward-biased diode.

For the ideal case, a reverse-biased (OFF) diode behaves as an open circuit when $v \leq 0$ V, and a forward-biased (ON) diode behaves as a short circuit when $i > 0$ A. According to the above discussion, a better approximation is to treat a forward-biased diode ($i > 0$ A) as a resistance R_f. We still treat the reverse-biased diode ($v \leq 0$ V) as an open circuit. Consequently, we can model the diode shown in Fig. 6.38a simply by the series connection of an ideal diode and a resistance R_f as shown in Fig. 6.38b—and this model is more accurate than just an ideal diode. Of course, because they are in series, we can interchange the positions of the ideal diode and R_f in Fig. 6.38b without changing the effect of the model.

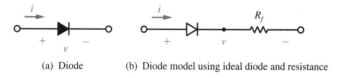

(a) Diode   (b) Diode model using ideal diode and resistance

Fig. 6.38 A diode and its simple model.

As an example, consider the OR gate shown in Fig. 6.39a. With the diode model depicted in Fig. 6.38, the OR gate can be analyzed using the ideal-diode circuit shown in Fig. 6.39b. However, in that circuit, we can combine R_s and R_f into a single resistance. The resulting circuit, therefore, can be treated identically to the OR gate given in Fig. 6.28, and this further demonstrates the usefulness of the concept of the ideal diode.

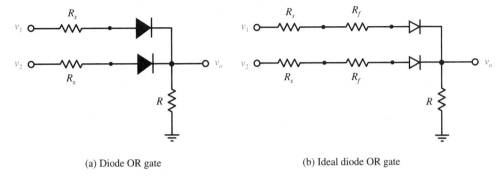

(a) Diode OR gate   (b) Ideal diode OR gate

Fig. 6.39 Use of simple diode model.

More Accurate Models

We now can go one step further and get an even better (but still relatively simple) approximation of a diode. In order to do this, we will enlarge the *i-v* characteristic curve of a forward-biased diode so that the cut-in voltage V_γ becomes apparent. Such a situation is shown in Fig. 6.40, where the *i-v* characteristic is the dashed curve. Let us approximate this dashed curve by a straight line of slope $1/R_f$ as shown. This line intersects the horizontal axis at $v = V_\gamma$ and is extended, in dotted form, to the vertical axis, where it intersects at some point $i = -a$. (This is done for computa-

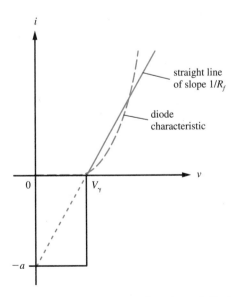

Fig. 6.40 Further approximation of diode *i-v* characteristic.

tional purposes.) The equation of the solid line has the form $i = mv + b$, where m is the slope (i.e., $m = 1/R_f$) and b is the vertical-axis intercept (i.e., $b = -a$). Thus we have

$$i = \frac{1}{R_f}v - a \tag{6.33}$$

However, from Fig. 6.40, we see that the slope is $1/R_f = a/V_\gamma$. Thus, $a = V_\gamma/R_f$, and substituting this fact into Eq. 6.33 yields

$$i = \frac{1}{R_f}v - \frac{V_\gamma}{R_f} \quad\Rightarrow\quad v = V_\gamma + R_f i \tag{6.34}$$

It is easy to see that these equivalent relationships between current i and voltage v are realized by the series connection shown in Fig. 6.41. In other words, a forward-biased diode behaves approximately as a series connection of a battery V_γ and a resistance R_f, where V_γ is the cut-in voltage of the diode and R_f is a typical ac resistance of the diode. If the diode is reverse biased, it behaves approximately as an open circuit. Thus a diode model more accurate than the one shown in Fig. 6.38b is the one shown in Fig. 6.42b—it consists of an ideal diode, a battery V_γ, and a resistance R_f. For this situation, however, the diode is OFF if $v \leq V_\gamma$. As before, though, the diode is ON if $i > 0$ A. (For an even more accurate model, see Problem 6.54 at the end of this chapter.)

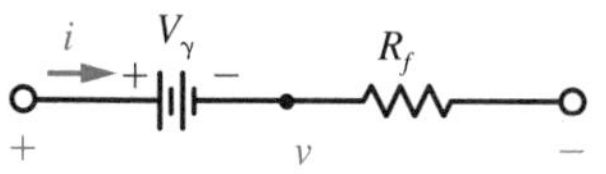

Fig. 6.41 Realization of $v = V_\gamma + R_f i$.

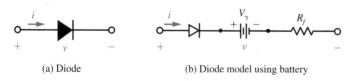

(a) Diode   (b) Diode model using battery

Fig. 6.42 A diode and an accurate model.

Example 6.12

Let us demonstrate that the i-v characteristic of the ideal-diode connection shown in Fig. 6.43a, where $V_Z > 0$ V, is depicted in Fig. 6.43b. We shall solve this problem by considering all possibilities for the states of the diodes.

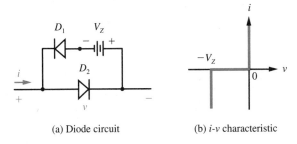

(a) Diode circuit   (b) *i-v* characteristic

Fig. 6.43 Determining an *i-v* characteristic.

Case 1: D_1 is OFF and D_2 is ON. This situation is depicted in Fig. 6.44a. By KVL, $v_1 = -V_Z \le 0$ V as is required, and $v = 0$ V. Furthermore, by KCL, $i = i_2$. Since D_2 is ON, we have that $i = i_2 > 0$ A. In summary, for this case, when $i > 0$ A, then $v = 0$ V. The corresponding *i-v* characteristic is shown in Fig. 6.44b.

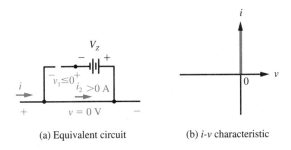

(a) Equivalent circuit   (b) *i-v* characteristic

Fig. 6.44 D_1 is OFF, D_2 is ON.

Case 2: D_1 is ON and D_2 is OFF. This situation is depicted in Fig. 6.45a. By KVL, $v_2 = -V_Z \le 0$ V as is required, and $v = -V_Z$. Furthermore, by KCL, $i = -i_1$. Since D_1 is ON, then $i_1 > 0$ A $\Rightarrow -i_1 < 0$ A. Therefore, $i = -i_1 < 0$ A. In summary, for this case, when $i < 0$ A, then $v = -V_Z$. The corresponding *i-v* characteristic is shown in Fig. 6.45b.

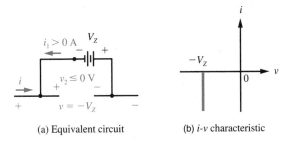

(a) Equivalent circuit   (b) *i-v* characteristic

Fig. 6.45 D_1 is ON, D_2 is OFF.

Case 3: D_1 and D_2 are OFF. This situation is depicted in Fig. 6.46*a*. By KCL, $i = 0$ A. Since D_2 is OFF, then $v_2 \leq 0$ V, and by KVL, $v = v_2 \leq 0$ V. Also, since D_1 is OFF, then $v_1 \leq 0$ V, and by KVL, $v_1 = -V_Z - v \leq 0$ V. From this inequality, $-V_Z \leq v$. In summary, when $-V_Z \leq v \leq 0$ V, then $i = 0$ A. The corresponding *i-v* characteristic is shown in Fig. 6.46*b*.

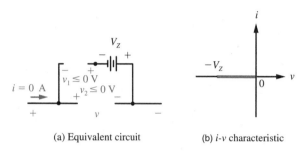

(a) Equivalent circuit   (b) *i-v* characteristic

Fig. 6.46 D_1 and D_2 are OFF.

Case 4: D_1 and D_2 are ON. This case is not possible because this situation would result in the paradoxical condition that both $v = 0$ V and $v = -V_Z$.

Combining the individual *i-v* characteristics obtained for all the possible cases results in the overall *i-v* characteristic shown in Fig. 6.43*b*.

Drill Exercise 6.12

For the diode circuit given in Fig. 6.43*a*, replace the voltage source (battery) with a resistor R_1, and place a resistor R_2 in series with ideal diode D_2. Find the *i-v* characteristic of the resulting connection.

ANSWER $i = v/R_1$ for $i > 0$ A, $i = v/R_2$ for $i < 0$ A, $v = 0$ V for $i = 0$ A

Another model of a diode that is frequently used in the analysis of digital logic circuits is portrayed in Fig. 6.47. The dashed curve in Fig. 6.47*a* is the forward-biased *i-v* characteristic of the diode. The positive portion of this curve is approximated by the vertical line $v = V_{ON}$, where $V_{ON} > V_\gamma$ is a typical forward-bias voltage across the diode. For a silicon diode at room temperature, the value that is normally used is $V_{ON} = 0.7$ V. Figure 6.47*c* shows how we can obtain this characteristic by connecting an ideal diode in series with a voltage source (battery) V_{ON}. Suppose that the ideal diode is ON. Then $i > 0$ A and $v = V_{ON}$. When the ideal diode is OFF, then $i = 0$ A, and since $v_D = v - V_{ON} \leq 0$ V, then $v \leq V_{ON}$.

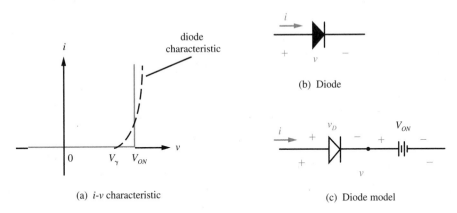

(a) i-v characteristic

(b) Diode

(c) Diode model

Fig. 6.47 Digital-logic characterization of a diode.

Example 6.13

Let us demonstrate that the DL circuit shown in Fig. 6.48 is an AND gate whose low voltage (logic 0) is 2 V and whose high voltage (logic 1) is 8 V, when $R_1 = 1$ kΩ and $R_2 = 10$ kΩ.

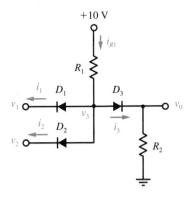

Fig. 6.48 A DL AND gate.

Case 1: $v_1 = v_2 = 2$ V. Since the 10-V supply is larger than both input voltages, assume that D_1 and D_2 are ON. Then, by KVL,

$$v_3 = v_{D1} + v_1 = 0.7 + 2 = 2.7 \text{ V}$$

This value of v_3 is enough to forward bias D_3, so assume that D_3 is ON. Thus the output voltage is

$$v_o = -v_{D3} + v_3 = -0.7 + 2.7 = 2 \text{ V}$$

Since $i_{R1} = (10 - v_3)/R_1 = (10 - 2.7)/1000 = 7.3$ mA and $i_3 = v_o/R_2 = 2/10{,}000 = 0.2$ mA, then by KCL, $i_1 + i_2 = i_{R1} - i_3 = 7.3 \times 10^{-3} - 2 \times 10^{-4} = 7.1$ mA. By symmetry, $i_1 = i_2 = (7.1 \times 10^{-3})/ = 3.55$ mA > 0, and this confirms that D_1 and D_2 are ON, whereas $i_3 = 0.2$ mA > 0 confirms that D_3 is ON.

Case 2: $v_1 = 2$ V, $v_2 = 8$ V. Although the 10-V supply is larger than both input voltages, there is a larger potential difference between the supply and v_1. Therefore, assume that D_1 in ON. Then, by KVL,

$$v_3 = v_{D1} + v_1 = 0.7 + 2 = 2.7 \text{ V}$$

This means that $v_{D2} = v_3 - v_2 = 2.7 - 8 = -5.3$ V. Thus assume that D_2 is OFF. The value of v_3 is enough to forward bias D_3, so assume that D_3 is ON. Thus the output voltage is

$$v_o = -v_{D3} + v_3 = -0.7 + 2.7 = 2 \text{ V}$$

Since $i_{R1} = (10 - v_3)/R_1 = (10 - 2.7)/1000 = 7.3$ mA and $i_3 = v_o/R_2 = 2/10{,}000 = 0.2$ mA, then by KCL, $i_1 + i_2 = i_{R1} - i_3 = 7.3 \times 10^{-3} - 2 \times 10^{-4} = 7.1$ mA $= i_1$, since D_2 is OFF and, hence, $i_2 = 0$ A . Since $i_1 > 0$ A and $i_3 > 0$ A, then D_1 and D_3 are confirmed to be ON, whereas $v_{D2} < 0$ V confirms that D_2 is OFF.

Case 3: $v_1 = 8$ V, $v_2 = 2$ V. Similar to Case 2, D_1 is OFF, D_2 and D_3 are ON, and the output voltage is $v_o = 2$ V.

Case 4: $v_1 = v_2 = 8$ V. Since the 10-V supply is larger than both input voltages, assume that D_1 and D_2 are ON. Then, by KVL,

$$v_3 = v_{D1} + v_1 = 0.7 + 8 = 8.7 \text{ V}$$

This value of v_3 is enough to forward bias D_3, so assume that D_3 is ON. Thus the output voltage is

$$v_o = -v_{D3} + v_3 = -0.7 + 8.7 = 8 \text{ V}$$

Since $i_{R1} = (10 - v_3)/R_1 = (10 - 8.7)/1000 = 1.3$ mA and $i_3 = v_o/R_2 = 8/10{,}000 = 0.8$ mA, then by KCL, $i_1 + i_2 = i_{R1} - i_3 = 1.3 \times 10^{-3} - 8 \times 10^{-4} = 0.5$ mA. By symmetry, $i_1 = i_2 = (5 \times 10^{-4})/2 = 0.25$ mA > 0, and this confirms that D_1 and D_2 are ON, whereas $i_3 = 0.8$ mA > 0 confirms that D_3 is ON.

This verifies that the circuit given in Fig. 6.48 is a DL AND gate.

6.6   Zener Diodes

As was mentioned in Section 6.3, if a diode is sufficiently reverse biased, that is, if the voltage is made negative enough, the diode will allow the conduction of a wide range of currents in the reverse direction. Furthermore, the diode voltage remains approximately constant at some value, say $-V_Z$, over this current range. This property, known as **breakdown,** is illustrated in the reverse-bias i-v diode characteristic shown in Fig. 6.49a. The value V_Z is called the (**reverse**) **breakdown voltage** of the diode.

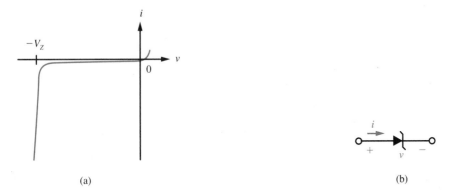

(a)   (b)

Fig. 6.49 (*a*) Reverse-bias portion of diode *i-v* characteristic, and (*b*) Zener diode circuit symbol.

The phenomenon of diode breakdown is attributable to two types of mechanisms:

1. In a semiconductor diode, holes and free electrons are continually being thermally generated. For the case of reverse bias, a small reverse saturation current is due to minority carriers—and this current is, in essence, independent of applied voltage (before the breakdown voltage is reached). But increasing the reverse-bias voltage causes the depletion region to widen and makes the junction potential greater. Minority carriers crossing the junction, therefore, acquire greater energy. If this energy is large enough,

the collision of a carrier with an ion will disrupt a covalent bond, thereby generating a new hole-electron pair. The new carriers produced in this manner can have sufficient energy, as a result of further collisions, to generate additional carriers. This cumulative process, which allows large reverse currents, is known as the **avalanche effect** or **avalanche breakdown**, and V_Z is called the **avalanche (breakdown) voltage**.

2. When the strength of an electric field across the junction of a semiconductor becomes sufficiently large (about 2×10^7 V/m), electrons are pulled from their covalent bonds, thus creating hole-electron pairs. As with the case of the avalanche effect, the increase in carriers can result in large reverse currents. However, unlike the avalanche effect, the carrier increase is not the result of collisions, but of electric field strength. This is known as the **Zener**[6] **effect** or **Zener breakdown**. Consequently, V_Z is also referred to as the **Zener breakdown voltage**.

For heavily doped diodes, Zener breakdown occurs at about 6 V or less, whereas for lightly doped diodes the breakdown voltage, due predominantly to the avalanche effect, is higher. Diodes that are designed to be used in the breakdown state are called **avalanche**, **breakdown**, or **Zener diodes**—the last being the most common terminology. This term is used even in cases where the avalanche effect predominates. Silicon Zener diodes have breakdown voltages in the range of a few volts to a few hundred volts. Furthermore, they are able to dissipate up to 50 W of power. The circuit symbol for a Zener diode is shown in Fig. 6.49b.

Example 6.14

The circuit shown in Fig. 6.50a contains two silicon Zener diodes D_1 and D_2 with saturation currents of 10 nA and 20 nA, respectively, at 300 K, and both diodes have breakdown voltages of 10 V. The i-v characteristics (in the third quadrant) for these diodes are shown in Fig. 6.50b. Let us find the current i and the voltages v_1 and v_2 for this circuit when V_S is (a) 6 V and (b) 12 V.

(a) For the case that $V_S = 6$ V, since the voltage source appears to produce a negative current, let's assume so. If both D_1 and D_2 are each in the breakdown state, then by KVL, we have that $V_S = -v_1 - v_2 \Rightarrow 6 = 10 + 10$ V—a contradiction. Thus both diodes cannot be in the breakdown state.

Next assume that D_2 is reverse biased, but is not in the breakdown state, i.e., $i = -20$ nA and $-10 \leq v_2 \leq -0.2$ V. This implies that D_1 is in the breakdown state, and hence, $v_1 = -10$ V. By KVL, $v_2 = -V_S - v_1 = -6 + 10 = 4$ V—a contradiction. Therefore, the condition that D_2 is reverse biased, but is not in the breakdown state, is not possible.

Now assume that D_1 is reverse biased, but is not in the breakdown state, i.e., $i = -10$ nA and $-10 \leq v_1 \leq -0.2$ V. From Eq. 6.24 for D_2,

---

[6]Named for the American physicist Clarence Zener.

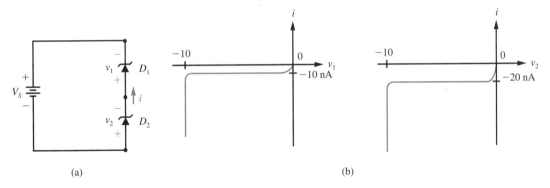

Fig. 6.50 (a) Circuit with two Zener diodes, and (b) diode *i-v* characteristics.

$$v_2 = \eta V_T \ln\left(\frac{i}{I_{S2}} + 1\right) = 2\left(\frac{300}{11,586}\right) \ln\left(\frac{-10 \times 10^{-9}}{20 \times 10^{-9}} + 1\right) = -0.036 \text{ V}$$

By KVL, $v_1 = -v_2 - V_S = 0.036 - 6 = -5.96$ V. These results indicate that both D_1 and D_2 are reverse biased and not in the breakdown state.

In summary, for the case that $V_S = 6$ V, then $i = -10$ nA, $v_1 = -5.96$ V, and $v_2 = -0.036$ V.

(b) For the case that $V_S = 12$ V, since the voltage source appears to produce a negative current, let us assume so. If both D_1 and D_2 are in the breakdown state, then by KVL, we have that $V_S = -v_1 - v_2 \Rightarrow 12 = 10 + 10$ V—a contradiction. Thus both diodes cannot be in the breakdown state.

Next assume that D_2 is reverse biased, but is not in the breakdown state, that is, $i = -20$ nA and $-10 \le v_2 \le -0.2$ V. This implies that D_1 is in the breakdown state and $v_1 = -10$ V. By KVL, $v_2 = -V_S - v_1 = -12 + 10 = -2$ V. This is not a contradiction, and it satisfies the condition on v_2 for D_2 to be reverse biased but not in the breakdown state. Consequently, for the case that $V_S = 12$ V, then $i = -20$ nA, $v_1 = -10$ V, and $v_2 = -2$ V.

Drill Exercise 6.14

For the Zener-diode circuit given in Fig. 6.50, D_1 has a saturation current of 10 nA and a breakdown voltage of 5 V whereas D_2 has a saturation current of 20 nA and a breakdown voltage of 10 V. Find i, v_1, and v_2 for the case that V_S is (a) 6 V, and (b) 12 V.

ANSWER (a) -20 nA, -5 V, -1 V; (b) -20 nA, -5 V, -7 V

Zener diodes are commonly used in "voltage regulators"—circuits that keep load voltages essentially constant despite variations in supply voltage and/or load resistance. In such applications, to ensure operation in the breakdown state, the minimum (reverse) current is often chosen to be 10 percent of the maximum (reverse) current.

Example 6.15

A Zener diode has a breakdown voltage of 6 V and a maximum current rating of 50 mA. This diode is used in the voltage-regulator circuit shown in Fig. 6.51a. Here v_S is the supply voltage and R_L is the load resistance. Since the maximum (reverse) current is 50 mA, the minimum (reverse) current will be chosen to be $0.1(50 \times 10^{-3}) = 5$ mA.

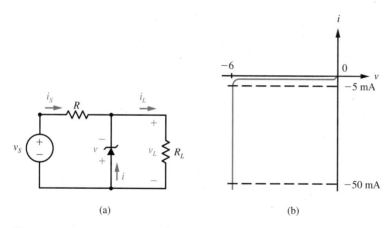

(a)   (b)

Fig. 6.51 (a) Zener-diode voltage-regulator circuit, and (b) desired range of operation.

(a) Suppose that we are given that $v_S = 24$ V and $R = 300\ \Omega$. Let us determine the range of values for R_L for which the diode will operate in the breakdown state, that is, $v = -6$ V (i.e., $v_L = 6$ V) and -50 mA $\leq i \leq -5$ mA as depicted in Fig. 6.51b. By Ohm's law, we have that

$$i_S = \frac{24 - 6}{300} = \frac{18}{300} = 60 \text{ mA}$$

But $i_L = v_L/R_L = 6/R_L$. Since $i = i_L - i_S = 6/R_L - 60 \times 10^{-3}$, then

$$-50 \times 10^{-3} \leq 6/R_L - 60 \times 10^{-3} \leq -5 \times 10^{-3}$$

In this expression there are two inequalities. From the first,

$$-50 \times 10^{-3} \le 6/R_L - 60 \times 10^{-3} \qquad \Rightarrow \qquad R_L \le 600 \ \Omega$$

whereas from the second,

$$6/R_L - 60 \times 10^{-3} \le -5 \times 10^{-3} \qquad \Rightarrow \qquad R_L \ge 109 \ \Omega$$

Combining these two results, we see that the voltage across the load resistance R_L will be regulated at $v_L = 6$ V as long as R_L is in the range $109 \le R_L \le 600 \ \Omega$.

(b) Now suppose that we are given that $R_L = 1$ kΩ and $R = 300 \ \Omega$. Let us determine the range of values for v_S for which the diode will operate in the breakdown state and -50 mA $\le i \le -5$ mA. By Ohm's law, we have that $i_L = v_L/R_L = 6/(1 \times 10^3) = 6$ mA and $i_S = (v_S - 6)/300$. Since $i = i_L - i_S = 6 \times 10^{-3} - (v_S - 6)/300$, then

$$-50 \times 10^{-3} \le 6 \times 10^{-3} - \frac{(v_S - 6)}{300} \le 5 \times 10^{-3}$$

From the first inequality,

$$-50 \times 10^{-3} \le 6 \times 10^{-3} - \frac{(v_S - 6)}{300} \qquad \Rightarrow \qquad v_S \le 22.8 \ \text{V}$$

whereas from the second

$$6 \times 10^{-3} - \frac{(v_S - 6)}{300} \le -5 \times 10^{-3} \qquad \Rightarrow \qquad v_S \ge 9.3 \ \text{V}$$

Combining these two results, we see that the voltage across the load resistance $R_L = 1$ kΩ will be regulated at $v_L = 6$ V as long as v_S is in the range $9.3 \le v_S \le 22.8$ V.

Drill Exercise 6.15

For the voltage-regulator circuit given in Fig. 6.51, suppose that $R_L = 1$ kΩ and $v_S = 24$ V. Determine the range of values for R for which the diode will operate in the breakdown state and -50 mA $\le i \le -5$ mA.

ANSWER $321 \le R \le 1.64$ kΩ

Zener Diode Characteristics

In the preceding two examples, we assumed that the voltage across a Zener diode operating in the breakdown state was a constant value and did not depend upon the diode current. Such a situation can be modeled by a battery V_Z. However, a more accurate model of a Zener diode in the breakdown state is a battery V_Z in series with a resistance R_Z. These ideal and nonideal models of a Zener diode in the breakdown state are shown in Fig. 6.52. The resistance R_Z normally ranges from several ohms to several hundred ohms. Generally, Zener diodes with breakdown voltages between 6 and 10 V have relatively small values of R_Z, whereas those with breakdown voltages below 6 V and above 10 V have relatively large values of R_Z.

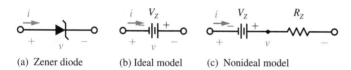

(a) Zener diode   (b) Ideal model   (c) Nonideal model

Fig. 6.52 Models of a Zener diode in the breakdown state.

When a Zener diode is not operating in the breakdown state, it behaves as an ordinary diode, and we can model it as was done previously for the diode. In particular, an **ideal Zener diode** (symbolized in Fig. 6.53a—note that the symbol is hollow, not solid), when in the breakdown state, behaves as the battery shown in Fig. 6.52b, and when not in the breakdown state, behaves as the ideal diode depicted in Fig. 6.19. The i-v characteristic corresponding to an ideal Zener diode, therefore, is as shown in Fig. 6.53b. (See Example 6.12 on p. 388.)

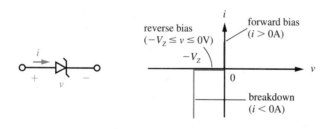

(a) Ideal Zener diode model   (b) Ideal Zener diode i-v characteristic

Fig. 6.53 Symbol and i-v characteristic for the ideal Zener diode.

Example 6.16

Let us find the output voltage v_o of the ideal Zener-diode circuit shown in Fig. 6.54 when the input voltage is $v_S = 24 \sin \omega t$ V as depicted in Fig. 6.55a. Suppose that the Zener diodes, which are connected in series, have a breakdown voltage of 12 V.

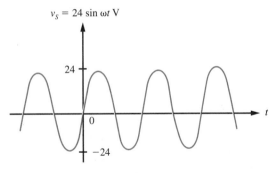

(a) Sinusoidal input voltage

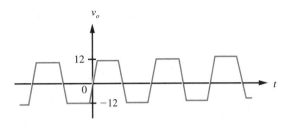

(b) Nonsinusoidal output voltage

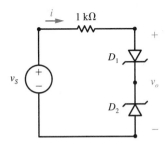

Fig. 6.54 Circuit with series-connected ideal Zener diodes.

Fig. 6.55 (*a*) Input voltage, and (*b*) output voltage for given Zener-diode circuit.

Case 1: $i > 0$ A. In this case, D_1 is forward biased and D_2 is in breakdown. Thus $v_{D1} = 0$ V and $v_{D2} = -12$ V. Thus $v_o = v_{D1} - v_{D2} = 0 - (-12) = 12$ V, and

$$i = \frac{v_S - v_o}{1000} = \frac{v_S - 12}{1000} > 0 \quad \Rightarrow \quad v_S > 12 \text{ V}$$

In other words, when $v_S > 12$ V, then $v_o = 12$ V.

Case 2: $i < 0$ A. In this case, D_2 is forward biased and D_1 is in breakdown. Thus $v_{D2} = 0$ V and $v_{D1} = -12$ V. Thus $v_o = v_{D1} - v_{D2} = -12 - 0 = -12$ V, and

$$i = \frac{v_S - v_o}{1000} = \frac{v_S + 12}{1000} < 0 \quad \Rightarrow \quad v_S < -12 \text{ V}$$

In other words, when $v_S < -12$ V, then $v_o = -12$ V.

Case 3: $i = 0$ A. In this case, $v_o = -1000i + v_S = v_S$. Since $i = 0$ A, then

$$-12 \leq v_{D1} \leq 0 \text{ V} \quad \text{and} \quad -12 \leq v_{D2} \leq 0 \text{ V} \quad \Rightarrow \quad 0 \leq -v_{D2} \leq 12 \text{ V}$$

Since $v_S = v_o = v_{D1} - v_{D2}$, then $-12 \leq v_S \leq 12$ V. In other words, when $-12 \leq v_S \leq 12$ V, then $v_o = v_S$.

In summary, when the input is the sinusoidal voltage given in Fig. 6.55a, then the output voltage is the clipped sinusoid shown in Fig. 6.55b.

Drill Exercise 6.16

For the circuit given in Fig. 6.51 on p. 396, replace the Zener diode with a 6-V ideal Zener diode having the same orientation. Given that $R_L = R = 1$ kΩ, find v_L when $v_S = 24 \sin \omega t$ V.

ANSWER 0 V for $v_S < 0$ V; 12 sin ωt V for $0 \leq v_S \leq 12$ V; 6 V for $v_S > 12$ V

As with other diode characteristics, the breakdown voltage V_Z of a diode is also temperature dependent. The variation in V_Z is specified by its percentage change per degree Celsius. Such a **temperature coefficient** of a Zener diode typically lies in the range of ± 0.1 percent/°C. If diode breakdown is due predominantly to the avalanche effect ($V_Z > 6$ V), the temperature coefficient is positive. Conversely, if breakdown involves the Zener effect ($V_Z < 6$ V), the temperature coefficient is negative.

Example 6.17

A Zener diode with a breakdown voltage of 12 V at 25°C has a temperature coefficient of +0.075 percent/°C. Therefore, the increase in the breakdown voltage per degree Celsius is

$$\left(\frac{0.075}{100}\right)(12) = 0.009 \text{ V}/°\text{C}$$

Thus, if the temperature of the diode rises to 45°C, then the breakdown voltage increases by $(0.009)(45 - 25) = 0.18$ V. Hence at 45°C, the Zener diode has a breakdown voltage of

$$V_Z = 12 + 0.18 = 12.18 \text{ V}$$

6.7 Effects of Capacitance

A common diode application is to rectify a sinusoid as depicted in Fig. 6.23 on p. 374. A reason for this rectification can be illustrated by placing a capacitor C in parallel with the resistor R in the half-wave rectifier circuit shown in Fig. 6.20 on p. 373. The resulting circuit is shown in Fig. 6.56. Suppose that for $t < 0$ s, the input voltage $v_S = 0$ V, and for $t \geq 0$ s, $v_S = A \sin \omega t$ V, as shown in Fig. 6.57a. Clearly, for $t \leq 0$ s, since the input voltage $v_S = 0$ V, then $v_o = 0$ V. For $0 < t \leq \pi/2\omega$, then $v_S > 0$ V and the (ideal) diode is forward biased. Thus, in this interval of time, $v_o = v_S$. This means that when $t = \pi/2\omega$, the capacitor is charged up to A volts. After time $t = \pi/2\omega$, the input voltage v_S decreases and becomes less than A volts. However, since the voltage across the capacitor is $v_o = A$ volts, then the diode is reverse biased. The voltage across capacitor C, though, will discharge (exponentially, of course) through the resistor R. If the time constant is much larger than the period of the sinusoid, that is, if $RC \gg T = 2\pi/\omega$, then while the diode is reverse biased, the voltage v_o will be approximately equal to A volts. The diode will remain reverse biased until, somewhere near $t = 5\pi/2\omega$, the voltage v_S increases enough to forward bias the diode. Then, as v_S increases to its maximum value, the capacitor will again charge up to A volts at time $t = 5\pi/2\omega$. This quickly charging and slowly discharging process is continually repeated, and the result is the output

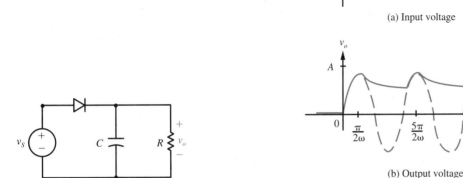

(a) Input voltage

(b) Output voltage

Fig. 6.56 Half-wave rectifier with added capacitor.

Fig. 6.57 Input and output of peak detector.

voltage v_o shown in Fig. 6.57*b*. Since the time constant RC is not infinite, the voltage v_o fluctuates with time. This voltage variation, known as **ripple,** can be made smaller by making the time constant larger.

As we have now seen, the circuit given in Fig. 6.56 converts a sinusoidal voltage to a nearly constant voltage whose value is approximately equal to the peak value of the input voltage. Because of this, we refer to such a circuit as a **peak detector.** Although this circuit can be used as part of a dc power supply, the same type of circuit can also be employed for demodulating or detecting AM signals (e.g., in ordinary AM radio receivers or in the video sections of television sets). For this reason, a peak detector is also known as an **envelope detector.**

Now consider the diode circuit shown in Fig. 6.58, where the applied (input) voltage v_S is the sinusoid $A \sin \omega t$ V shown in Fig. 6.59*a*. The first time that the input voltage goes negative, the diode will be forward biased and the capacitor will charge to A volts with the polarity indicated. Since the voltage across the capacitor cannot discharge through the diode, it is possible only for it to discharge through the resistor R. If $RC \gg T = 2\pi/\omega$, then the voltage across the capacitor will discharge only slightly before it again fully charges to A volts with the polarity

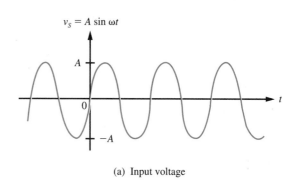

$v_S = A \sin \omega t$

(a) Input voltage

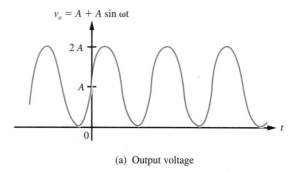

$v_o = A + A \sin \omega t$

(a) Output voltage

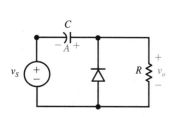

Fig. 6.58 Clamping circuit.

Fig. 6.59 Input and output of clamping circuit.

shown. Hence, for a large time constant RC, the voltage across the capacitor will remain, in essence, constant at a value of A volts. By KVL, the resulting output voltage is $v_o = A + v_S = A + A \sin \omega t$ V. From a plot of this voltage as shown in Fig. 6.59b, we see that the output has its minimum value "clamped" to 0 V. For this reason, the circuit given in Fig. 6.58 is called a **clamping circuit.** Note that the clamping circuit took a purely ac voltage (the input v_S) and added a constant (A) to it. In some types of applications (e.g., television sets), such a circuit is called a **dc restorer.**

Drill Exercise 6.17

For the clamping circuit shown in Fig. 6.58, reverse the direction of the diode. Determine the approximate output voltage v_o when $RC \gg T = 2\pi/\omega$ and the input voltage is $v_S = A \sin \omega t$ V.

ANSWER $-A + A \sin \omega t$ V

By combining the peak detector of Fig. 6.56 with the clamping circuit given in Fig. 6.58, we obtain the circuit shown in Fig. 6.60. Suppose that the input to this circuit is $v_S = A \sin \omega t$ V as depicted in Fig. 6.59a. For the case that $RC \gg T = 2\pi/\omega$, the clamping portion of the circuit will produce a voltage whose waveform is as shown in Fig. 6.59b. As a consequence, the peak-detecting portion of the circuit will produce an output voltage of $v_o \approx 2A$ V. Since this is the value of the voltage between the two peaks of the input waveform, the circuit is called a **peak-to-peak detector.** Furthermore, since the output voltage is twice as large as the output of a peak detector, the circuit also is sometimes called a **voltage doubler.**

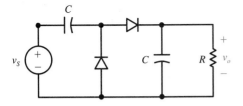

Fig. 6.60 Peak-to-peak detector.

Diode Capacitance

Having considered some external effects of capacitance, we now turn our attention to inherent diode capacitances. By definition, current i is related to charge q by $i = dq/dt$. Since the relationship between current i and voltage v for a capacitor C is

$i = C \, dv/dt$, we have that $C \, dv/dt = dq/dt$. Using the chain rule of calculus, we deduce that $C = dq/dv$.

We know that the depletion region of a *pn* junction has charge and a potential difference across it. From elementary physics, we know that a pair of parallel conducting plates with area A that are separated by a distance d has a capacitance $C = \epsilon A/d$, where ϵ is the **permittivity** (measured in F/m) of the material between the plates. So, too, a diode has a capacitance C_T, called the **transition capacitance** (or **depletion-region capacitance,** or **space-charge capacitance**) given by

$$C_T = \frac{\epsilon A}{W} \tag{6.35}$$

where ϵ is the permittivity of the semiconductor material, A is the area of the junction, and W is the width of the depletion region.

Since the depletion-region width changes with applied voltage (the larger the reverse bias, the wider the depletion region), the value of the transition capacitance depends on the diode voltage. For the case of reverse bias, values of C_T generally range between 5 and 500 pF. Diodes that are constructed specifically for reverse-bias use as voltage-variable capacitors are called **varactors** or **varicaps.** These devices are used, for example, for electronically tuning television sets.

Diffusion Capacitance

For the case of forward bias the depletion region is narrower than for the case of reverse bias, and, therefore, C_T is larger for forward bias than for reverse bias. However, in the forward-bias condition, the transition capacitance is negligible compared with another type of diode capacitance. Therefore, let us consider a forward-biased diode in some detail.

When a diode is forward biased, holes (majority carriers) go from the p side to the n side (and vice versa for free electrons). When they enter the n side, holes (minority carriers) encounter many free electrons with which they can recombine. The average time that a hole exists before it recombines with a free electron is the **mean lifetime** of the hole and is denoted by τ_p (τ_n for free electrons). Since the flow rate of charge (mobile carriers) is current, then the flow of charge q having average lifetime τ yields a diode current of

$$i = q \big/ \tau = I_S(e^{v/\eta V_T} - 1) \approx I_S \, e^{v/\eta V_T} \tag{6.36}$$

where the approximation is quite good for $v \geq 0.2$ V. Thus we have that

$$q = \tau I_S \, e^{v/\eta V_T}$$

and the resulting capacitance C_D is

$$C_D = \frac{dq}{dv} \approx \frac{\tau I_S}{\eta V_T} e^{v/\eta V_T}$$

Substituting the approximation given in Eq. 6.36 into the preceding approximation, we have that

$$C_D \approx \frac{\tau i}{\eta V_T}$$

Since, for the case of forward bias, the current through a diode is due almost entirely to diffusion, (i.e., drift current is negligible), the capacitance C_D is called the **diffusion capacitance** (also known as the **storage capacitance**) of the diode.

For example, a forward-biased silicon diode having a current of 30 mA and a mobile carrier mean lifetime of 25 μs has a diffusion capacitance of

$$C_D \approx \frac{\tau i}{\eta V_T} = \frac{(25 \times 10^{-6})(30 \times 10^{-3})}{2(300)/11{,}586} = 14.5 \ \mu\text{F}$$

at 300 K. Although this typical value of capacitance is relatively large (much larger than the transition capacitance), recall that the forward resistance R_f of a diode is typically small. Consequently, depending upon the circuit application of the diode, the time constant $R_f C_D$ may be small enough that the diffusion capacitance can be ignored.

Diode Switching Times

Because of its inherent capacitance, a (nonideal) diode cannot be driven from a reverse-biased condition to the forward-biased state (or vice versa) instantaneously— a certain nonzero time is required. For instance, consider the simple diode circuit shown in Fig. 6.61a, where the applied voltage v_S is depicted in Fig. 6.61b. Initially, the applied voltage is $v_S = -V_R$ and the diode is reverse biased. Thus the current is $i = -I_S$. At time $t = t_1$, the applied voltage instantaneously changes to $v_S = V_F$, and this will then forward bias the diode. If $R_L \gg R_f$ and $V_F \gg V_\gamma$, then the current increases from $i = -I_S$ to $i \approx V_F/R_L$ as depicted in Fig. 6.61c. The time required for the current to go from about 10 percent of V_F/R_L to 90 percent of V_F/R_L is called the **forward recovery time** t_{fr} of the diode. This time is dependent upon the relatively small transition capacitance and is usually negligible compared

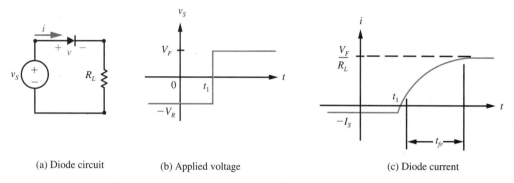

(a) Diode circuit   (b) Applied voltage   (c) Diode current

Fig. 6.61 Diode switching from reverse to forward bias.

with the time required for the diode to switch from forward to reverse bias. We now discuss this case.

Suppose that an applied voltage results in a forward-biased diode. Although the situation we discuss is for holes, a similar description applies to free electrons. Holes (majority carriers) in the p-type material cross the junction into the n-type material, where they are minority carriers. The original hole concentration in the n side is attributable to thermal energy. The holes injected into the n-type material, therefore, increase the hole concentration. (The deeper a hole goes into the n side, the more chance there is of recombination. Thus the hole concentration is greater closer to the junction.) When the applied voltage changes polarity so as to reverse bias the diode, the excess (injected) holes in the n side return to the p side. Hence, until these excess holes (minority carriers) are removed from the n-type material, there is a significant current in the reverse direction. After the elimination of the excess minority carriers, the reverse current decays to its final value of I_S.

To see the consequences of this discussion, suppose that the diode circuit given in Fig. 6.61a has the applied voltage shown in Fig. 6.62a. Initially, the applied voltage is $v_S = V_F$ and the diode is forward biased. If $R_L \gg R_f$ and $V_F \gg V_\gamma$, then the current is $i \approx V_F/R_L$. At time $t = t_1$, the applied voltage instantaneously changes to $v_S = -V_R$, and this will reverse bias the diode. However, due to the excess minority carriers, the current becomes $i = -V_R/R_L$, and this current remains until time $t = t_2$, when the excess minority carriers are removed. After this has occurred, the current will increase until time $t = t_3$ when $i = -I_S$. This behavior is summarized in Fig. 6.62b. The time difference $t_2 - t_1$ is called the **storage time** t_s of the diode (it is the time during which the excess holes are "stored" in the n side and the excess free electrons are "stored" in the p side). The time difference $t_3 - t_2$, called the **transition time** t_t, is dependent upon the relatively large diffusion capacitance. The sum $t_s + t_t$ is called the **reverse recovery time** t_{rr} of the diode. Diode reverse recovery times typically range from 1 ns to 1 μs—generally the larger the current-carrying capability of a diode, the larger is the value of t_{rr}.

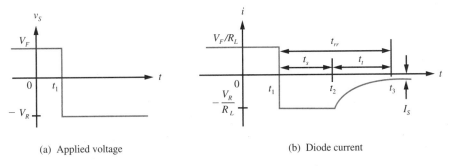

Fig. 6.62 Diode switching from forward to reverse bias.

SUMMARY

1. Semiconductors such as silicon and germanium have covalent bonds. Thermal energy can disrupt covalent bonds, thus creating hole-electron pairs.

2. A metal is unipolar—current is conducted by only one type of charge (electrons). A semiconductor such as germanium or silicon is bipolar—current is conducted by two types of charges (electrons and holes).

3. For a metal, conductivity decreases as temperature increases. For a semiconductor, conductivity increases as temperature increases.

4. Doping an intrinsic semiconductor with donor atoms produces an *n*-type semiconductor. Doping with acceptor atoms produces a *p*-type semiconductor.

5. In an *n*-type semiconductor, electrons are the majority carriers and holes are the minority carriers. Conversely, in a *p*-type semiconductor, the majority carriers are holes and the minority carriers are electrons.

6. Doping can significantly increase the conductivity of a semiconductor.

7. Current in a semiconductor is due to diffusion as well as drift.

8. A *pn* junction with appropriate metal contacts forms a junction diode.

9. A forward-biased (junction) diode can easily pass current, whereas a reverse-biased diode allows only a very small current.

10. The cut-in voltage of a diode is approximately 0.2 V for germanium and 0.5 V for silicon.

11. Diode characteristics are temperature sensitive; for example, the reverse saturation current approximately doubles for each 10-degree increase in temperature.

12. Some diode circuits can be analyzed either graphically or numerically.

13. A forward-biased ideal diode behaves as a short circuit; a reverse-biased ideal diode acts as an open circuit.

14. A nonideal diode can be accurately modeled with an ideal diode, a battery, and a resistor.

15. A diode that is excessively reverse biased will break down. Diode breakdown is due to the avalanche effect and the Zener effect. Zener diodes are fabricated to be used in the breakdown state.

16. The capacitance of a *pn* junction is predominantly the transition capacitance when a diode is reverse biased, and essentially the diffusion capacitance

when a diode is forward biased. Diode switching times are dependent upon these capacitances.

17. Varactor diodes are voltage-variable capacitors.

18. Diodes can be used to design such circuits as rectifiers, clippers, voltage regulators, peak or enve-

lope detectors, clampers, voltage doublers, electronic tuners, and digital logic circuits such as AND gates and OR gates.

Problems

6.1 Ordinary household copper wire has a cross-sectional area of 3.32×10^{-6} m^2. Given that the free-electron concentration of copper is 8.43×10^{28} m^{-3}, find the current corresponding to a typical drift speed of 4×10^{-4} m/s.

6.2 A 1-meter length of copper wire 1 millimeter (mm) in diameter has a resistance of 2.2×10^{-2} Ω and a current of 2 A. Given that the free-electron concentration of copper is 8.43×10^{28} m^{-3}, determine (a) the current density, (b) the drift speed of the free electrons, (c) the mobility of the free electrons, and (d) the conductivity of the copper.

6.3 An aluminum wire 2 mm in diameter has a conductivity of 2.9×10^7 ℧/m and a current of 50 mA. Find the voltage across a 1-m length of this wire.

6.4 A bar of pure silicon has a length of 1 cm (10^{-2} m) and cross-sectional area of 1 mm^2. At 300 K, the intrinsic concentration of Si is 1.5×10^{16} m^{-3}. The free-electron and hole mobilities are 0.13 m^2/V-s and 0.05 m^2/V-s, respectively. Find (a) the conductivity, and (b) the resistance of the bar.

6.5 A bar of pure silicon has a cross-sectional area of 1 mm^2. At 300 K, the intrinsic concentration of Si is 1.5×10^{16} m^{-3}. The free-electron and hole mobilities are 0.13 m^2/V-s and 0.05 m^2/V-s, respectively. Find (a) the conductivity, and (b) the length of a bar whose resistance is 50 kΩ.

6.6 A bar of pure germanium has a length of 1 cm (10^{-2} m) and a cross-sectional area of 1 mm^2. At 300 K, the intrinsic concentration of Ge is 2.5×10^{19} m^{-3}. The free-electron and hole mobilities are

0.38 m^2/V-s and 0.18 m^2/V-s, respectively. Find (a) the conductivity, and (b) the resistance of the bar.

6.7 A bar of silicon has a length of 1 cm (10^{-2} m) and cross-sectional area of 1 mm^2. At 300 K, Si has 5×10^{28} atoms/m^3 and an intrinsic concentration of 1.5×10^{16} m^{-3}. The free-electron and hole mobilities are 0.13 m^2/V-s and 0.05 m^2/V-s, respectively. The bar is doped with two parts per 10^8 of an acceptor impurity. Determine (a) the conductivity, and (b) the resistance of the resulting p-type semiconductor bar.

6.8 At 300 K, Si has 5×10^{28} atoms/m^3 and an intrinsic concentration of 1.5×10^{16} m^{-3}. The free-electron and hole mobilities are 0.13 m^2/V-s and 0.05 m^2/V-s, respectively. Determine the concentration of acceptor impurity required to dope silicon such that the resulting conductivity is 20.8 ℧/m. How many parts per 10^8 is this doping?

6.9 A bar of germanium has a length of 1 cm (10^{-2} m) and a cross-sectional area of 1 mm^2. At 300 K, the intrinsic concentration of Ge is 2.5×10^{19} m^{-3}. The free-electron and hole mobilities are 0.38 m^2/V-s and 0.18 m^2/V-s, respectively. This bar is doped with one part per 10^8 of an acceptor impurity. Given that there are 4.4×10^{28} Ge atoms/m^3, determine (a) the conductivity, and (b) the resistance of the resulting p-type semiconductor bar.

6.10 A bar of germanium has a length of 1 cm (10^{-2} m) and a cross-sectional area of 1 mm^2. At 300 K, the intrinsic concentration of Ge is 2.5×10^{19} m^{-3}. The free-electron and hole mobilities are 0.38 m^2/V-s and 0.18 m^2/V-s, respectively. This bar is doped with one part per 10^8 of a donor impurity.

Given that there are 4.4×10^{28} Ge atoms/m^3, determine (a) the conductivity, and (b) the resistance of the resulting *n*-type semiconductor bar.

6.11 A bar of germanium has a length of 1 cm (10^{-2} m) and a cross-sectional area of 1 mm^2. At 300 K, the intrinsic concentration of Ge is 2.5×10^{19} m^{-3}. The free-electron and hole mobilities are 0.38 m^2/V-s and 0.18 m^2/V-s, respectively. Given that there are 4.4×10^{28} Ge atoms/m^3, determine the parts per 10^9 of *p*-type doping that results in a bar resistance of 1 kΩ.

6.12 The free-electron and hole mobilities for Si are 0.13 m^2/V-s and 0.05 m^2/V-s, respectively. The free-electron and hole mobilities for Ge are 0.38 m^2/V-s and 0.18 m^2/V-s, respectively. Find the diffusion constants for holes and electrons at 300 K for (a) silicon and (b) germanium.

6.13 A piece of silicon has 5×10^{28} atoms/m^3. If one side is doped with one part per 10^8 of an acceptor impurity, and the other side is doped with two parts per 10^7 of a donor impurity, find the barrier potential across the resulting *pn* junction at 300 K, given that the intrinsic concentration of Si is 1.5×10^{16} m^{-3}.

6.14 A piece of germanium has 4.4×10^{28} atoms/m^3. If one side is doped with one part per 10^8 of an acceptor impurity, and the other side is doped with two parts per 10^7 of a donor impurity, find the barrier potential across the resulting *pn* junction at 300 K, given that the intrinsic concentration of Ge is 2.5×10^{19} m^{-3}.

6.15 The *p* side of a silicon *pn* junction has a conductivity of 50 ℧/m, whereas the *n* side has a conductivity of 100 ℧/m. Determine the barrier potential across the junction at 300 K, given that the electron and hole mobilities are 0.13 m^2/V-s and 0.05 m^2/V-s, respectively, and the intrinsic concentration of Si is 1.5×10^{16} m^{-3}.

6.16 The *p* side of a germanium *pn* junction has a conductivity of 50℧/m, whereas the *n* side has a conductivity of 100℧/m. Determine the barrier potential across the junction at 300 K, given that the

electron and hole mobilities are 0.38 m^2/V-s and 0.18 m^2/V-s, respectively, and the intrinsic concentration of Ge is 2.5×10^{19} m^{-3}.

6.17 A piece of silicon has 5×10^{28} atoms/m^3 and an intrinsic concentration of 1.5×10^{16} m^{-3} at 300 K. If one side is doped with one part per 10^8 of an acceptor impurity, how many parts per million of a donor impurity should the other side be doped such that the barrier potential across the resulting *pn* junction 0.7 V?

6.18 A piece of germanium has 4.4×10^{28} atoms/m^3 and has an intrinsic concentration of 2.5×10^{19} m^{-3} at 300 K. If one side is doped with one part per 10^8 of an acceptor impurity, how many parts per million of a donor impurity should the other side be doped such that the barrier potential across the resulting *pn* junction 0.3 V?

6.19 A silicon diode has a saturation current of 5 nA at 300 K.
(a) Find the current for the case that the forward-bias voltage is 0.7 V.
(b) Find the forward-bias voltage that results in a current of 15 mA.
(c) Suppose that this diode is used in the circuit shown in Fig. 6.16a on p. 368. If $v_1 = 6$ V, what value of *R* results in a current of 15 mA?

6.20 A germanium diode has a saturation current of 10 μA at 300 K.
(a) Find the current for the case that the forward-bias voltage is 0.3 V.
(b) Find the forward-bias voltage that results in a current of 0.5 A.
(c) Suppose that this diode is used in the circuit shown in Fig. 6.16a on p. 368. If $v_1 = 12$ V, what value of *R* results in a current of 0.5 A?

6.21 A silicon diode is forward biased at 20 mA and 0.8 V at 300 K. Find the saturation current of the diode for the case that the temperature is (a) 300 K, (b) 310 K, (c) 316 K, and (d) 284 K.

6.22 A silicon diode with a saturation current of 4 nA at 300 K has a forward-bias current of 20 mA. Find the voltage across the diode for the case that the temperature is (a) 310 K, (b) 316 K, and (c) 284 K.

6.23 A germanium diode is forward biased at 0.5 A and 0.3 V at 300 K. Find the saturation current of the diode for the case that the temperature is (a) 300 K, (b) 310 K, (c) 324 K, and (d) 276 K.

6.24 A germanium diode with a saturation current of 4 μA at 300 K has a forward-bias current of 0.5 A. Find the voltage across the diode for the case that the temperature is (a) 310 K, (b) 324 K, and (c) 276 K.

6.25 For the diode circuit shown in Fig. P6.25, $V_S = 2$ V and the silicon diode has a saturation current of 1 nA at 300 K. Given that $v = 0.7$ V, find (a) R_2 when $R_1 = 1$ kΩ, and (b) R_1 when $R_2 = 1$ kΩ.

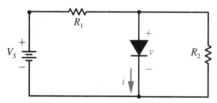

Fig. P6.25

6.26 For the diode circuit shown in Fig. P6.25, $V_S = 2$ V and the silicon diode has a saturation current of 1 nA at 300 K. Given that $i = 0.5$ mA, find (a) R_2 when $R_1 = 1$ kΩ, and (b) R_1 when $R_2 = 1$ kΩ.

6.27 For the diode circuit shown in Fig. P6.27, $V_S = 1.5$ V and the silicon diodes have a temperature of 300 K. Given that $v_2 = 0.7$ V, find (a) I_{S1} when $R = 100$ Ω and $I_{S2} = 1$ nA, and (b) R when $I_{S1} = I_{S2} = 1$ nA.

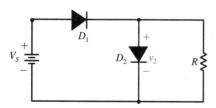

Fig. P6.27

6.28 For the diode circuit shown in Fig. P6.28, $V_S = 3$ V and the silicon diodes have a saturation current of 10 nA at 300 K. (a) Find R_1 for the case that $R_2 = 850$ Ω and $i_2 = 2$ mA. (b) Find R_2 for the case that $R_1 = 1$ kΩ and $i_1 = 5$ mA.

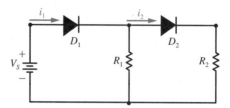

Fig. P6.28

6.29 For the diode circuit shown in Fig. P6.29, $V_S = 6$ V and the silicon diodes have a saturation current of 1 nA at 300 K. (a) Find R_2 for the case that $R_1 = 10$ kΩ and $v_1 = 0.66$ V. (b) Find R_1 for the case that $R_2 = 100$ Ω and $v_2 = 0.66$ V. (c) Find R_1 and R_2 for the case that $v_1 = 0.68$ V and $v_2 = 0.66$ V.

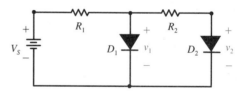

Fig. P6.29

6.30 For the diode circuit given in Fig. 6.17, on p. 371, D_1 is silicon with a saturation current of 5 nA and D_2 is germanium with a saturation current of 10 μA. Find v_1 and v_2 at 300 K.

6.31 For the diode circuit shown in Fig. P6.31, D_1 and D_2 are silicon diodes with saturation currents of 5 nA and 10 nA, respectively, at 300 K. Given that both diodes are reverse biased, find v_1 and v_2.

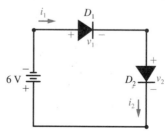

Fig. P6.31

6.32 For the diode circuit shown in Fig. P6.32, D_1 and D_2 are silicon diodes having saturation currents of 5 nA and 10 nA, respectively, at 300 K. Given that both diodes are forward biased, find the value of R for which the current is 15 mA.

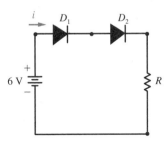

Fig. P6.32

6.33 A germanium diode has a saturation current of 20 μA at 300 K.
(a) Plot an i-v characteristic curve on a piece of graph paper.
(b) Suppose that this diode is used in the circuit given in Fig. 6.16a, on p. 368, where $v_1 = 1$ V and $R = 0.5$ Ω. Use the i-v curve obtained in part (a) to graphically determine i and v.

6.34 A germanium diode has a saturation current of 20 μA at 300 K. Suppose that this diode is used in the circuit given in Fig. 6.16a on p. 368, where $V_1 = 1$ V and $R = 0.5$ Ω. Use the numerical method described in Example 6.7 on p. 369 to find i and v. (Begin with $v = 0.2$ V.)

6.35 A silicon diode has a saturation current of 10 nA at 300 K.
(a) Plot an i-v characteristic curve on a piece of graph paper.
(b) Suppose that this diode is used in the circuit given in Fig. 6.16a, on p. 368, where $v_1 = 1$ V and $R = 100$ Ω. Use the i-v curve obtained in part (a) to determine graphically i and v.

6.36 A silicon diode has a saturation current of 10 nA at 300 K. Suppose that this diode is used in the circuit given in Fig. 6.16a, on p. 368, where $v_1 = 1$ V and $R = 100$ Ω. Use the numerical method described in Example 6.7 on p. 369 to find i and v. (Begin with $v = 0.5$ V.)

6.37 Given that the input voltage is $v_S = A \sin \omega t$ V, sketch the resulting output voltage v_o for the ideal-diode circuit shown in Fig. P6.37.

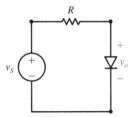

Fig. P6.37

6.38 The input voltage to the clipper circuit shown in Fig. P6.38 is $v_S = 12 \sin \omega t$ V. Determine the output voltage v_o and sketch this function.

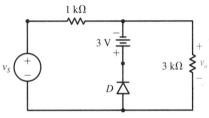

Fig. P6.38

6.39 For the ideal-diode circuit given in Fig. P6.38, reverse the polarity of the 3-V source. The input voltage to the resulting clipper circuit is $v_S = 12 \sin \omega t$ V. Determine the output voltage v_o and sketch this function.

6.40 The input voltage to the clipper circuit shown in Fig. P6.40 is $v_S = 6 \sin \omega t$ V. Determine the output voltage v_o and sketch this function.

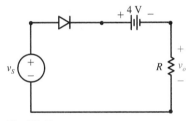

Fig. P6.40

6.41 The input voltage to the clipper circuit shown in Fig. P6.41 is $v_S = 12 \sin \omega t$ V. Determine the output voltage v_o and sketch this function.

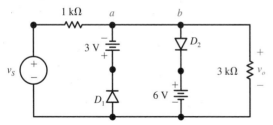

Fig. P6.41

6.42 For the ideal-diode circuit given in Fig. P6.41, reverse the polarity of the 3-V source. The input voltage to the resulting clipper circuit is $v_S = 12 \sin \omega t$ V. Determine the output voltage v_o and sketch this function.

6.43 For the ideal-diode circuit given in Fig. P6.41, remove the short circuit between points a and b, and replace it with a 2-kΩ resistor. Find the range of values of v_S for which (a) D_1 and D_2 are OFF; (b) D_1 is ON and D_2 is OFF; and (c) D_1 is OFF and D_2 is ON. Is it possible for both D_1 and D_2 to be ON?

6.44 The ideal-diode circuit shown in Fig. P6.44 is known as a **full-wave** (or **bridge**) **rectifier circuit.** Determine which diodes are ON and which are OFF when (a) $v_S > 0$ V, and (b) $v_S < 0$ V. Sketch the output voltage v_o for the case that the input voltage is $v_S = A \sin \omega t$ V.

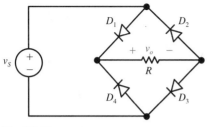

Fig. P6.44

6.45 The ideal-diode circuit shown in Fig. P6.45 is also an example of a full-wave rectifier. Determine which diodes are ON and which are OFF when (a) $v_S > 0$ V, and (b) $v_S < 0$ V. Sketch the output voltage v_o when the input voltage is $v_S = A \sin \omega t$ V.

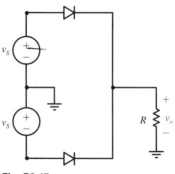

Fig. P6.45

6.46 When a diode is used in a rectifier circuit, the maximum reverse-bias voltage across that diode is known as the **peak inverse voltage** (PIV) of the diode. (Of course, the PIV must be less than the breakdown voltage of the diode.) If $v_S = A \sin \omega t$ V, find the PIV of the diodes in the rectifier circuit given in (a) Fig. 6.20, on p. 373, (b) Fig. P6.44, and (c) Fig. P6.45.

6.47 For the ideal-diode circuit given in Fig. 6.28 on p. 377, let $R_f = 1$ kΩ and $R = 9$ kΩ. Find the output voltage v_o when the input voltages are (a) $v_1 = v_2 = 0$ V; (b) $v_1 = 0$ V, $v_2 = 10$ V; (c) $v_1 = v_2 = 10$ V.

6.48 For the ideal-diode circuit given in Fig. 6.31, on p. 380, let $R_f = 1$ kΩ, $R = 9$ kΩ, and $V_H = 10$ V. Find the output voltage v_o when the input voltages are (a) $v_1 = v_2 = 0$ V; (b) $v_1 = 0$ V, $v_2 = 10$ V; (c) $v_1 = v_2 = 10$ V.

6.49 A silicon diode with a saturation current of 5 nA at 300 K has a forward-bias voltage of 0.7 V. (a) Find the dc resistance of the diode. (b) Find the ac resistance of the diode. (c) Find the dc resistance of the diode for 310 K. (d) Find the ac resistance of the diode for 310 K.

6.50 A germanium diode with a saturation current of 10 μA at 300 K has a forward-bias voltage of 0.3 V. (a) Find the dc resistance of the diode. (b) Find the ac resistance of the diode. (c) Find the dc resistance of the diode for 310 K. (d) Find the ac resistance of the diode for 310 K.

6.51   Sketch the *i-v* characteristic of the ideal-diode circuit shown in Fig. P6.51 for the case that that $I_S > 0$ A, $R_f = 0$ Ω, and $V_\gamma = 0$ V.

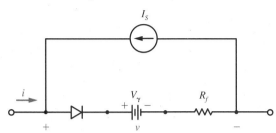

Fig. P6.51

6.52   Sketch the *i-v* characteristic of the ideal-diode circuit shown in Fig. P6.51 for the case that $I_S > 0$ A, $R_f > 0$ Ω, and $V_\gamma = 0$ V.

6.53   Sketch the *i-v* characteristic of the ideal-diode circuit shown in Fig. P6.51 for the case that $I_S > 0$ A, $R_f = 0$ Ω, and $V_\gamma > 0$ V.

6.54   Sketch the *i-v* characteristic of the ideal-diode circuit shown in Fig. P6.51 for the case that $I_S > 0$ A, $R_f > 0$ Ω, and $V_\gamma > 0$ V.

6.55   Sketch the *i-v* characteristic of the ideal-diode circuit shown in Fig. P6.55 for the case that $V_Z > 0$ V and $R_Z = 0$ Ω.

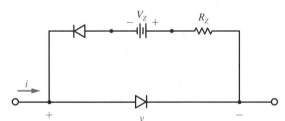

Fig. P6.55

6.56   Sketch the *i-v* characteristic of the ideal-diode circuit shown in Fig. P6.55 for the case that $V_Z > 0$ V and $R_Z > 0$ Ω.

6.57   For the DL AND gate given in Fig. 6.48, on p. 391, suppose that $R_2 = 10$ kΩ, the low voltage is 2 V, and the high voltage is 8 V. Assume that a forward-biased diode has a voltage of 0.7 V, and determine the range of values of R_1 required for proper operation.

6.58   The circuit shown in Fig. P6.58 is a DL OR gate. Suppose that $R_2 = 2$ kΩ, the low voltage is 2 V, and the high voltage is 8 V. Assume that a forward-biased diode has a voltage of 0.7 V, and determine the range of values of R_1 required for proper operation.

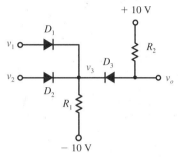

Fig. P6.58

6.59   The circuit shown in Fig. P6.58 is a DL OR gate. Suppose that $R_1 = 2$ kΩ, the low voltage is 2 V, and the high voltage is 8 V. Assume that a forward-biased diode has a voltage of 0.7 V, and determine the range of values of R_2 required for proper operation.

6.60   For the half-wave rectifier circuit given in Fig. 6.20, on p. 373, replace the ideal diode with a diode having $R_f = 20$ Ω and $V_\gamma = 0.6$ V (see Fig. 6.42 on p. 388). Sketch the output voltage v_o, labeling it sufficiently, for the case that the input voltage is $v_S = 6 \sin \omega t$ V and $R = 100$ Ω.

6.61   For the diode circuit shown in Fig. P6.61, D_1 is germanium with $R_f = 10$ Ω and $V_\gamma = 0.2$ V whereas D_2 is silicon with $R_f = 10$ Ω and $V_\gamma = 0.6$ V. Find the range of values of v_S for which (a) D_1 and D_2 are both OFF; (b) D_1 and D_2 are both ON; and (c) D_1 is ON and D_2 is OFF.

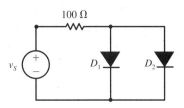

Fig. P6.61

6.62 For the circuit given in Fig. 6.50, on p. 395, reverse D_1. The identical silicon Zener diodes have saturation currents of 10 nA at 300 K. Find i, v_1, and v_2 when $V_S = 6$ V and the breakdown voltage of the diodes is (a) 9 V and (b) 5 V.

6.63 For the Zener diode circuit shown in Fig. P6.63, at 300 K the germanium diodes D_1 and D_2 have saturation currents of 1 μA and 2 μA, respectively. If both diodes have a breakdown voltage of 10 V, find v_1 and v_2 when V_S is (a) 6 V and (b) 12 V.

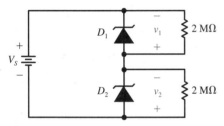

Fig. P6.63

6.64 The Zener diode in the voltage-regulator circuit given in Fig. 6.51 on p. 396 has a breakdown voltage of 9 V and is to operate with a reverse current between 10 and 100 mA. Given that $R = 200$ Ω,
(a) Find the range of the load resistance R_L that results in a 9-V load voltage when $v_S = 24$ V.
(b) Find the range of the supply voltage v_S that results in a 9-V load voltage when $R_L = 600$ Ω.

6.65 The Zener diode in the voltage-regulator circuit given in Fig. 6.51 on p. 396 has a breakdown voltage of 12 V. Suppose that the diode (reverse) current is to be 10 mA.
(a) If $v_S = 24$ V and $R_L = 2$ kΩ, find R.
(b) If $R = 2$ kΩ and $R_L = 2$ kΩ, find v_S.
(c) If $v_S = 24$ V and $R = 300$ Ω, find R_L.

6.66 Repeat Problem 6.65 using the Zener diode model given in Fig. 6.52c, on p. 398, where $R_Z = 200$ Ω.

6.67 The circuit shown in Fig. P6.67 contains a 6-V Zener diode, that is, a diode with a breakdown voltage of 6 V. Suppose that the supply voltage v_S is a half-wave rectified sine wave (see Fig. 6.23b on p. 374) with an amplitude of 12 V. (a) Use the ideal Zener diode model given in Fig. 6.52b on p. 398 to obtain a sketch of v_L. (b) Use the model given in Fig. 6.52c with $R_Z = 100$ Ω to obtain a sketch of v_L.

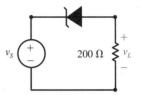

Fig. P6.67

6.68 The Zener diode in the voltage-regulator circuit given in Fig. 6.51 on p. 396 has a breakdown voltage of 9 V. Suppose that the diode has a resistance of 100 Ω in the breakdown state and the diode is to operate with a reverse current between 10 and 100 mA. Find the range of v_S for the case that $R = 200$ Ω and $R_L = 600$ Ω.

6.69 The Zener diode in the voltage-regulator circuit given in Fig. 6.51 on p. 396 has a breakdown voltage of 9 V. Suppose that the diode has a resistance of 100 Ω in the breakdown state and the diode is to operate with a reverse current between 10 and 100 mA. Find the range of R_L for the case that $R = 200$ Ω and $v_S = 24$ V.

6.70 Sketch the i-v characteristic for each connection of elements shown in Fig. P6.70, where the breakdown voltage of an ideal Zener diode is V_Z. (See p. 415.)

6.71 For the circuit shown in Fig. 6.54, on p. 399, replace the ideal Zener diode D_1 with an (ordinary) ideal diode having the same orientation. Given that the breakdown voltage for D_2 is 12 V, sketch v_o for the case that $v_S = 24 \sin \omega t$ V.

6.72 For the half-wave rectifier given in Fig. 6.56, on p. 401, reverse the diode. Sketch the output voltage v_o of the resulting circuit when the input is $v_S = A \sin \omega t \, u(t)$ V and $RC \gg T = 2\pi/\omega$.

6.73 For the full-wave rectifier given in Fig. P6.44, connect a capacitor C in parallel with the

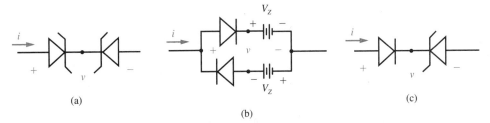

Fig. P6.70

resistor R. Sketch the resulting output voltage v_o when the input voltage is $v_S = A \sin \omega t \, u(t)$ V and $RC \gg T = 2\pi/\omega$.

6.74 For the full-wave rectifier given in Fig. P6.45, connect a capacitor C in parallel with the resistor R. Sketch the resulting output voltage v_o when the input is $v_S = A \sin \omega t \, u(t)$ V and $RC \gg T = 2\pi/\omega$.

6.75 For the clamping circuit given in Fig. 6.58, on p. 402, sketch the output voltage v_o for the case

$RC \gg T$ and the input v_S is each voltage shown in Fig. P6.75.

6.76 Reverse the diode in the clamping circuit given in Fig. 6.58 on p. 402. Sketch the output voltage v_o when $RC \gg T$ and the input v_S is each voltage shown in Fig. P6.75.

6.77 Consider the clamping circuit shown in Fig. P6.77. Given that the input voltage is $v_S = A \sin \omega t$ V and $RC \gg T = 2\pi/\omega$, sketch the output voltage v_o for the case that (a) $V_1 = A/2$, and (b) $V_1 = -A/2$.

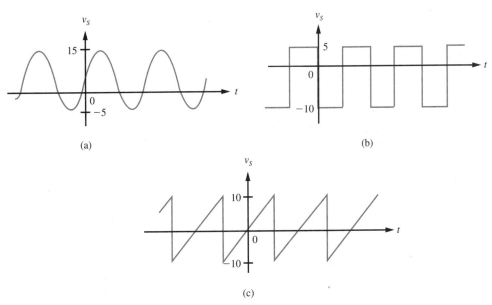

Fig. P6.75

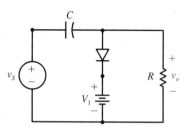

Fig. P6.77

6.78 Reverse the diode in the clamping circuit shown in Fig. P6.77. For the resulting circuit, given that the input voltage is $v_S = A \sin \omega t$ V and that $RC >> T = 2\pi/\omega$, sketch the output voltage v_o for the case that (a) $V_1 = A/2$, and (b) $V_1 = -A/2$.

6.79 For the clamping circuit shown in Fig. P6.79, the Zener diode has a breakdown voltage of 6 V. Sketch the output voltage v_o when the input voltage is $v_S = 24 \sin \omega t$ V and $RC >> T = 2\pi/\omega$.

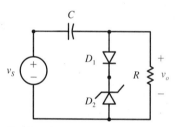

Fig. P6.79

6.80 Reverse the directions of both diodes in the clamping circuit shown in Fig. P6.79, and sketch the output voltage v_o when the input voltage is $v_S = 24 \sin \omega t$ V, $RC >> T = 2\pi/\omega$, and the Zener diode has a breakdown voltage of 6 V.

6.81 For the ideal-diode circuit shown in Fig. P6.81, sketch the output voltage v_o when the input is $v_S = A \sin \omega t$ V and $RC >> T = 2\pi/\omega$.

6.82 A silicon diode that has a permittivity of 1.04×10^{-10} F/m is reverse biased such that the depletion region has a width of 4×10^{-6} m. (a) If the cross section of the diode is a square measuring 1 mm on a side, what is the value of the transition capacitance? (b) What cross-sectional area would result in a transition capacitance of 15 pF?

6.83 A silicon diode with a saturation current of 16 nA at 300 K has a forward-bias voltage of 0.7 V and a mean carrier lifetime of 25 μs. (a) Find the diffusion capacitance. (b) Find the voltage that will result in a diffusion capacitance of 10 μF.

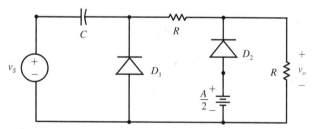

Fig. P6.81

7

Bipolar Junction Transistors (BJTs)

INTRODUCTION

In the preceding chapter, we studied the diode (a *pn* junction) and discussed many of its practical applications. By sandwiching a narrow section of *n*-type semiconductor between two sections of *p*-type semiconductor (or vice versa), we obtain a bipolar junction transistor (BJT), or transistor for short. Such a device does not act simply as two *pn* junctions, but instead displays a more elaborate behavior that has far-reaching consequences. In this chapter, we study the various characteristics and behavior of such a device. We also see how transistors can be utilized to design digital logic circuits in both discrete and integrated-circuit (IC) form. The importance and applications of digital logic circuits are covered in Chapters 11, 12, and 13.

7.1 The *pnp* Transistor

By sandwiching a narrow section of *n*-type material between two sections of *p*-type material as illustrated in Fig. 7.1*a,* we obtain a *pnp* **bipolar junction transistor** (BJT). The three sections of the BJT are known as the **emitter** (E), the **base** (B), and the **collector** (C). The circuit symbol for a *pnp* BJT is shown in Fig. 7.1*b.* If the *p*-type semiconductor is replaced by *n*-type semiconductor material, and vice

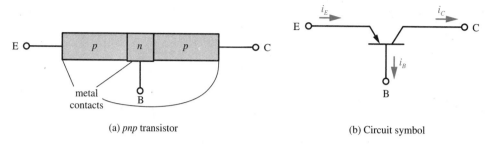

(a) *pnp* transistor (b) Circuit symbol

Fig. 7.1 A *pnp* BJT and its circuit symbol.

versa, what results is an *npn* BJT as depicted in Fig. 7.2*a*. The reason why the current directions are as shown will be discussed shortly. We denote the voltage between terminals *a* and *b* (with the plus at terminal *a*) by v_{ab}. For example, the voltage between the emitter and the base (with the plus at the emitter) is v_{EB}, and also, $v_{EB} = -v_{BE}$. In addition to the directions of the currents, *pnp* and *npn* BJT circuit symbols differ in the direction of the arrowhead on the emitter. As is the case for any *pn* junction, the arrowhead points from the *p*-type to the *n*-type material.

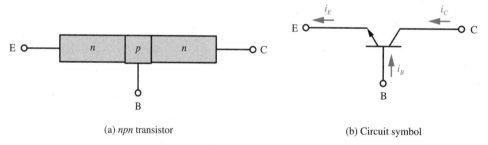

(a) *npn* transistor (b) Circuit symbol

Fig. 7.2 An *npn* BJT and its circuit symbol.

Some of the important operating characteristics of a *pnp* transistor are seen when the emitter-base junction (or E-B junction, for short) is forward biased and the collector-base junction (or C-B junction) is reverse biased. This situation is depicted in Fig. 7.3. Here the supply voltage V_{EE} forward biases the E-B junction and produces an emitter current i_E, while V_{CC} reverse biases the C-B junction. As mentioned in the preceding chapter, the current through a *pn* junction is due almost entirely to diffusion—drift current being negligible. Thus, the emitter current i_E has two components—a hole current i_{pE} due to the diffusion of holes from emitter to base, and an electron current i_{nE} due to the diffusion of electrons from base to emitter. That is, $i_E = i_{pE} + i_{nE}$. However, because the emitter in a BJT is much more heavily doped than the base, the emitter current is due almost entirely to the diffusion of holes, that is, $i_E \approx i_{pE}$. Holes are ''emitted'' from the emitter and are injected into

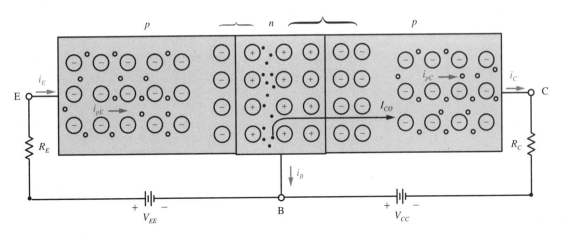

Fig. 7.3 A *pnp* BJT with E-B junction forward biased and C-B junction reverse biased.

the *n*-type base. Although some recombine with electrons, the vast majority of these holes diffuse across the very narrow base region to the C-B junction where they are then swept across the C-B junction (as a drift current) by its electric field. The holes are then "collected" by the collector. Let us denote the hole current from base to collector by i_{pC}. Further, let us denote that fraction of the emitted holes that makes it to the collector by α. In other words, the collector hole current i_{pC} is a fraction α of the emitter hole current $i_{pE} = i_E$ (for simplicity, we have changed the approximation to an equality), that is,

$$i_{pC} = \alpha i_E$$

Just as a reverse-biased *pn* junction has a reverse saturation current going from the *n*-type to the *p*-type semiconductor, there is a reverse-bias current going from base to collector. This current is attributable mainly to thermally generated minority carriers (holes going from base to collector and free electrons going from collector to base). Some additional carriers may be produced in the C-B depletion region as a result of collisions, as is the situation for the avalanche effect. Also, there is a small current due to charge that moves along the surface of the device known as **(surface) leakage current.** It is proportional to the voltage across the junction. The total reverse current is denoted I_{CO}. Typically, at 300 K, I_{CO} is on the order of nanoamperes for silicon and, as was the case for the reverse saturation current of a junction diode, I_{CO} approximately doubles for every 10-K (°C) rise in temperature.

With the E-B junction forward biased and the C-B junction reverse biased, therefore, we have that the collector current i_C is given by

$$i_C = \alpha i_E + I_{CO} \tag{7.1}$$

from which

$$\alpha i_E = i_C - I_{CO} \tag{7.2}$$

Thus,

$$\alpha = \frac{i_C - I_{CO}}{i_E} \tag{7.3}$$

This quantity α (the fraction of emitted holes that gets collected) is called the **large-signal current gain** of the transistor, and typically ranges between 0.95 and 0.998. Furthermore, the value of α depends upon such variables as the collector-base voltage v_{CB} and the temperature T.

Active-Region Operation

For the situation just described, where the E-B junction is forward biased and the C-B junction is reverse biased, we say that the transistor is **biased** in, or operates in, the **active region** (also called the **linear region**) or is in the **active mode.** That is, for active-region operation, Eq. 7.1 through 7.3 are valid. In particular, even though it can change slightly with v_{CB}, the quantity α is close to unity. Further, because I_{CO} is normally very small, since $i_C = \alpha i_E + I_{CO}$, we see that the collector current i_C depends on the emitter current i_E and is, in essence, independent of v_{CB}. Therefore, a reasonable approximation relating the collector current to the emitter current is

$$i_C \approx \alpha i_E \tag{7.4}$$

We now see why the directions for the currents in a *pnp* BJT were defined as indicated in Fig. 7.1*b*. Doing so means that when a BJT is operating in the active region, its terminal (dc) currents i_E, $i_C \approx \alpha i_E$, and $i_B = i_E - i_C = i_E - \alpha i_E = (1 - \alpha)i_E$ all will be positive.

Example 7.1

The circuit shown in Fig. 7.4*a* is the circuit given in Fig. 7.3 using the circuit symbol of a silicon *pnp* BJT. Suppose that $R_E = 500 \ \Omega$, $R_C = 1.2 \ \text{k}\Omega$, $V_{EE} = 3 \ \text{V}$, and

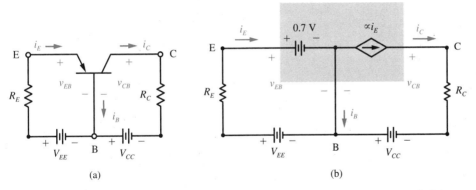

Fig. 7.4 (a) A *pnp* BJT biased in the active region, and (b) active-region model of BJT.

$V_{CC} = 6$ V. Let us analyze this circuit to confirm that the BJT is biased (operating) in the active region.

Assuming active-region operation, the E-B junction is forward biased and the C-B junction is reverse biased. Since a typical value for a forward-biased silicon *pn* junction is 0.7 V, we will use $v_{EB} \approx 0.7$ V. By KVL around the loop on the left,

$$v_{EB} - V_{EE} + R_E i_E = 0$$

so

$$0.7 - 3 + 500 i_E = 0 \quad \Rightarrow \quad i_E = (3 - 0.7)/500 = 4.6 \text{ mA}$$

By KVL around the loop on the right,

$$v_{CB} + V_{CC} - R_C i_C = 0 \tag{7.5}$$

Since for the active region, $i_C \approx \alpha i_E$, and since $\alpha \approx 1$, then $i_C \approx i_E$, and Eq. 7.5 becomes

$$v_{CB} + 6 - (4.6 \times 10^{-3})(1.2 \times 10^3) = 0 \quad \Rightarrow \quad v_{CB} = -6 + 5.52$$

$$= -0.48 \text{ V}$$

Since $i_E > 0$ A confirms that the E-B junction is forward biased and $v_{CB} = -0.48$ V $< V_\gamma = 0.5$ V (which is a typical silicon *pn* junction cut-in voltage) confirms that the C-B junction is reverse biased, the BJT is indeed biased in the active region.

By using the two active-region approximations, $v_{EB} = 0.7$ V and $i_C = \alpha i_E$, we are in effect modeling the behavior of a BJT operating in the active region as shown in Fig. 7.4*b*. The fact that the voltage $v_{EB} = 0.7$ V is a constant is modeled by a 0.7-V battery, and the fact that the collector current i_C is dependent (or controlled) by the emitter current i_E is modeled by a current-dependent (or controlled) current source labeled αi_E.

Drill Exercise 7.1

For the BJT circuit given in Fig. 7.4*a*, suppose that $V_{CC} = V_{EE} = 5$ V. Find R_E and R_C such that the BJT is biased in the active region at $i_E = 5$ mA and $v_{CB} = -2$ V.

ANSWER 860 Ω, 600 Ω

Example 7.1 demonstrated how a BJT transfers a current (i_E), which flows through a 500-Ω resistor (R_E), to a current ($i_C \approx i_E$), which flows through a 1.2-kΩ resistor (R_C). It was because such a device behaved as a "**trans**fer re**sistor**," that it became known as a **transistor.**

For the circuit in Example 7.1, consider the emitter as the input and the collector as the output. Since the base is common to both the input and output portions of the circuit, we refer to this as a **common-base** (CB) configuration. Also, note that since the collector current (approximately) equals the emitter current, and since the collector resistor R_C is larger than the emitter resistor R_E, the voltage across R_C (the output voltage) is larger than the voltage across R_E (the input voltage)—specifically, $R_C = 1.2$ kΩ and $R_E = 500$ Ω means that the output voltage is 2.4 times greater than the input voltage. This illustrates how a BJT can produce a voltage gain.

Transistor Characteristics

For a BJT operating in the active region, $i_C = \alpha i_E \approx i_E$. Since the E-B junction is a forward biased *pn* junction, the emitter current, and hence, the collector current, is essentially related to the emitter-base voltage by the usual junction equation

$$i_C = I_S(e^{v_{EB}/V_T} - 1) \tag{7.6}$$

where I_S is the **saturation current** of the transistor. In this expression, the emission coefficient η is not needed since silicon BJTs are fabricated so that typically $\eta = 1$. Also, I_S usually ranges from 10^{-12} to 10^{-15} A, and it approximately doubles for each temperature increase of 5 K (5°C).

Fig. 7.5*a* shows plots of i_C versus v_{EB} for various values of v_{CB} for a typical silicon *pnp* BJT. (Note that a typical emitter-base voltage is in the vicinity of 0.7 V.) Although Eq. 7.6 indicates that i_C does not depend on v_{CB}, in actual devices it does, and this is indicated by these **collector-emitter static characteristics.** But why should this characteristic change with v_{CB}? Recall that, as the reverse bias of a junction increases, its depletion region increases. Thus, as the reverse bias of the C-B junction increases, the depletion region widens and, in effect, narrows the base. Hence the hole concentration gradient between the emitter and the base increases, meaning a larger emitter current (and, hence, collector current) for the same value of v_{EB}. Fig. 7.5*b* shows the **emitter static characteristics** for the same BJT. Although the curves for these latter characteristics are essentially the same as for the former characteristics, they also include the case that the collector is open ($i_C = 0$ A), which cannot be included with the collector-emitter characteristics. When the collector is open, the corresponding *i-v* characteristic is that of a *pn* junction.

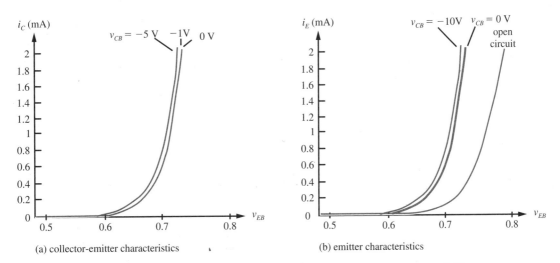

(a) collector-emitter characteristics   (b) emitter characteristics

Fig. 7.5 (*a*) Collector-emitter, and (*b*) emitter characteristics for a silicon *pnp* BJT.

Another illuminating transistor characteristic consists of a family of curves of collector current i_C versus base-collector voltage v_{BC} for various values of emitter current i_E. The resulting (CB) **collector static characteristics** of a typical *pnp* BJT are shown in Fig. 7.6. We can obtain the case that $i_E = 0$ A by opening the emitter. When $v_{BC} > 0$ V (or $v_{CB} < 0$ V), the C-B junction is reverse biased, and the resulting collector current is $-I_{CBO}$, where I_{CBO} is the current from collector to base with the emitter open. Since this reverse current is very small (on the order of nanoamperes), the characteristic curve i_C versus v_{BC} for $i_E = 0$ A appears to be coincident with the horizontal axis. The value of I_{CBO} depends on v_{BC}, and I_{CBO} approximately doubles for each increase in temperature of 10°C.

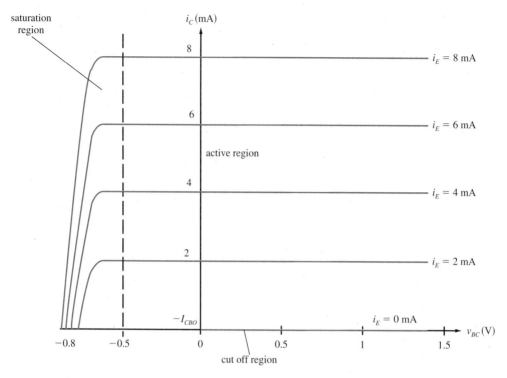

Fig. 7.6 Collector characteristics for a silicon *pnp* BJT.

For the case that $v_{BC} > -0.5$ V (or $v_{CB} < 0.5$ V), the C-B junction is reverse biased. In addition, if $i_E > 0$ A, then the E-B junction is forward biased. Thus the region to the right of the dashed line $v_{BC} = -0.5$ V and above the curve $i_E = 0$ A is the active region.

For the case that $v_{BC} < -0.5$ V (or $v_{CB} > 0.5$ V), the C-B junction is forward biased. For the region where $i_E > 0$ A and $-0.8 < v_{BC} < -0.5$ V, both the E-B and C-B junctions are forward biased—this is known as the **saturation region.** Finally, the region on and below the curve $i_E = 0$ A corresponds to the case that the E-B and C-B junctions are both reverse biased—and this is called the **cutoff region.** The very important notions of transistor saturation and cutoff will be discussed in detail later.

7.2 The *npn* Transistor

Having discussed the common-base (CB) configuration for a *pnp* transistor, let us consider the *npn* BJT in the **common-emitter** (CE) configuration shown in Fig. 7.7*a*. An equivalent form of the circuit is shown in Fig. 7.7*b*. As was mentioned previously, a transistor is biased in the active region when the B-E junction is forward

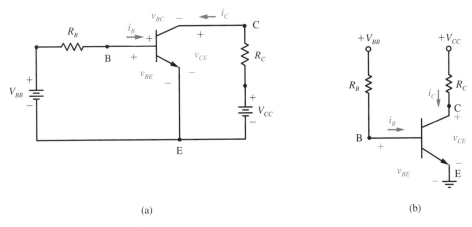

(a)

(b)

Fig. 7.7 An *npn* CE transistor circuit and its alternative representation.

biased and the B-C junction is reverse biased. Clearly, the former is accomplished with a large enough supply voltage V_{BB}. But when is the latter condition satisfied?

For an *npn* transistor, the B-C junction is reverse biased when $v_{BC} < V_\gamma$ (typically 0.5 V for silicon). By KVL, $v_{BC} = v_{BE} - v_{CE}$. Thus the B-C junction is reverse biased when

$$v_{BE} - v_{CE} < V_\gamma \qquad \Rightarrow \qquad v_{CE} > v_{BE} - V_\gamma$$

For the CB configuration, the collector (output) characteristics are plots of collector (output) current i_C versus base-collector (output) voltage v_{BC} for different values of emitter (input) current i_E. For the CE configuration, the collector (output) characteristics are plots of collector (output) current i_C versus collector-emitter (output) voltage v_{CE} for different values of base (input) current i_B. The collector (static) characteristics for a typical silicon *npn* BJT in the CE configuration are shown in Fig. 7.8. When an *npn* BJT is in the active region, as indicated above, the B-C junction is reverse biased when $v_{CE} > v_{BE} - V_\gamma > 0.7 - 0.5 = 0.2$ V. Since the B-E junction is forward biased when $i_B > 0$ A, the portion of the characteristics above the curve $i_B = 0$ A and to the right of the dashed line $v_{CE} = 0.2$ V is the active region. The area to the left of the dashed line $v_{CE} = 0.2$ V (this corresponds to the case that the B-C junction is forward biased) and above the curve $i_B = 0$ A (this corresponds to the case that the B-E junction is forward biased) is the saturation region. Finally, for the case that $i_B \le 0$ A, the B-E junction is reverse biased, so the region that is bounded above by the curve $i_B = 0$ A is the cutoff region. For the case that $i_B = 0$ A, the resulting collector current is I_{CEO}, since by definition I_{CEO} is the current that goes from collector to emitter with the base open ($i_B = 0$ A). As was the case for I_{CBO}, the value of I_{CEO} is very small so the curve $i_B = 0$ A appears to coincide with the horizontal axis.

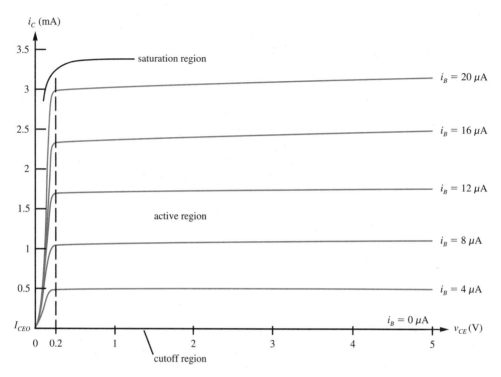

Fig. 7.8 CE collector characteristics for a silicon *npn* BJT.

For a BJT operating in the active region, by Eq. 7.1, we have that

$$i_C = \alpha i_E + I_{CO}$$

However, by KCL, $i_E = i_B + i_C$. Therefore,

$$i_C = \alpha(i_B + i_C) + I_{CO} = \alpha i_B + \alpha i_C + I_{CO}$$

from which

$$i_C = \frac{\alpha}{1 - \alpha} i_B + \frac{1}{1 - \alpha} I_{CO} \tag{7.7}$$

We now define the **large-signal CE current gain**[1] β by

$$\beta = \frac{\alpha}{1 - \alpha} \tag{7.8}$$

---

[1]α is the large-signal CB current gain.

From this definition, we can write

$$\frac{1}{1-\alpha} = \frac{1-\alpha+\alpha}{1-\alpha} = \frac{1-\alpha}{1-\alpha} + \frac{\alpha}{1-\alpha} = 1 + \beta \tag{7.9}$$

Substituting Eq. 7.8 and 7.9 into Eq. 7.7 results in

$$i_C = \beta i_B + (1 + \beta)I_{CO} \tag{7.10}$$

To get an idea of some typical values of β, consider the case that $\alpha = 0.98$. Then by Eq. 7.8,

$$\beta = \frac{\alpha}{1-\alpha} = \frac{0.98}{1-0.98} = 49$$

As another example, if $\alpha = 0.99$, then

$$\beta = \frac{\alpha}{1-\alpha} = \frac{0.99}{1-0.99} = 99$$

As has been indicated here, typically, $\beta \gg 1$, so a reasonable approximation is $1 + \beta \approx \beta$. Furthermore, for a transistor biased in the active region, generally $i_B \gg I_{CO}$. For these reasons, Eq. 7.10 yields

$$\boxed{i_C \approx \beta i_B} \qquad \text{or} \qquad \boxed{\beta \approx i_C/i_B}$$

For example, for a transistor that has the collector characteristics given in Fig. 7.8, when $i_B = 20 \ \mu A$ and $v_{CE} = 2$ V, then the approximate collector current is $i_C = 3$ mA. Under this condition,

$$\beta \approx i_C/i_B = (3 \times 10^{-3})/(20 \times 10^{-6}) = 150$$

Again from Eq. 7.1, we know that

$$i_C = \alpha i_E + I_{CO} \approx \alpha i_E$$

Since α changes only slightly as v_{CB} changes (remember, the effective base width changes), then for a given value of i_E, there will be only a small change in the value of i_C as v_{CB} changes. Because of this, CB collector characteristic curves (see Fig. 7.6) are nearly horizontal. However, since

$$i_C \approx \beta i_B \qquad \text{and} \qquad \beta = \frac{\alpha}{1-\alpha}$$

then a slight change in α due to a change in v_{CB}—which, in turn, means a change in $v_{CE} = v_{CB} + v_{BE}$—can cause a large change in β. In the numerical example above, an approximately 1 percent change in α from 0.98 to 0.99 corresponds to more than a 100 percent change in β from 49 to 99. It is because of this that CE collector characteristic curves (see Fig. 7.8) are not horizontal.

For the CE configuration, the **base (input) characteristics** are plots of base (input) current i_B versus base-emitter (input) voltage v_{BE} for different values of collector-emitter (output) voltage v_{CE}. The base (static) characteristics of a typical silicon *npn* BJT in the CE configuration are shown in Fig. 7.9a. When there is a short circuit connected between the collector and emitter ($v_{CE} = 0$ V) as shown in Fig. 7.9b, the resulting i_B-v_{BE} curve is typical of a forward-biased diode. Such a connection is an example of a diode-connected transistor.

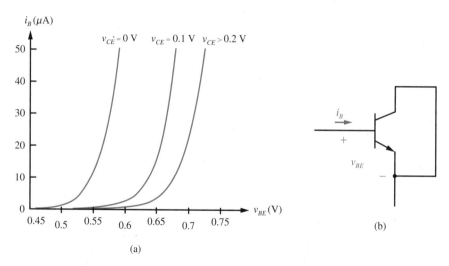

(a)

(b)

Fig. 7.9 (*a*) Base characteristics for a silicon *npn* CE transistor, and (*b*) diode-connected BJT.

By holding v_{BE} constant, if v_{CE} increases, then $v_{CB} = v_{CE} - v_{BE}$ increases. This, in turn, decreases the effective base width. The result is less hole-electron recombination in the base, and less base current. In other words, as v_{CE} increases, an i_B-v_{BE} curve moves to the right.

Since the characteristic curve labeled $v_{CE} > 0.2$ V corresponds to active-region operation, we again see that a typical value for the base-emitter voltage is $v_{BE} \approx 0.7$ V.

Example 7.2

For the circuit shown in Fig. 7.7a, suppose that $R_B = 330$ kΩ, $R_C = 2.7$ kΩ, $V_{BB} = V_{CC} = 10$ V, and the BJT is a silicon *npn* transistor having $\beta = 100$. The resulting circuit is shown in Fig. 7.10a. Since $V_{BB} = V_{CC} = 10$ V, the two batteries shown in Fig. 7.7b can be replaced by the single power supply labeled $+10$ V as indicated in Fig. 7.10b.

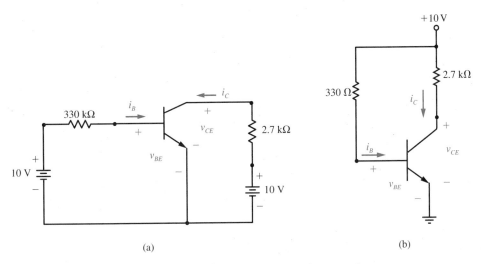

(a)   (b)

Fig. 7.10 (a) A CE transistor circuit, and (b) an equivalent version.

If we assume active-region operation, we can employ the two conditions that $v_{BE} \approx 0.7$ V and $i_C \approx \beta i_B = 100 i_B$. By KVL and Ohm's law,

$$i_B = \frac{10 - v_{BE}}{330 \times 10^3} = \frac{10 - 0.7}{330 \times 10^3} = 2.82 \times 10^{-5} = 28.2 \ \mu A$$

and

$$i_C = \beta i_B = 100(2.82 \times 10^{-5}) = 2.82 \ mA$$

The fact that $i_B > 0$ A indicates that the B-E junction is forward biased.

To check that the collector junction is reverse biased, by KVL,

$$v_{CE} = -(2.7 \times 10^3)i_C + 10 \tag{7.11}$$

$$= -(2.7 \times 10^3)(2.82 \times 10^{-3}) + 10 = 2.39 \ V$$

Since $v_{CE} > 0.2$ V, the B-C junction is indeed reverse biased. A simple confirmation is obtained by KVL from

$$v_{BC} = v_{BE} - v_{CE} = 0.7 - 2.39 = -1.69 < V_\gamma = 0.5 \ V$$

Hence the transistor is indeed operating in the active region.

If the collector characteristics of the transistor are available, a graphical analysis can be employed. Plot the load line corresponding to Eq. 7.11 on the collector characteristics. This line intersects the curve $i_B = 28.2$ μA (and such a curve probably will have to be interpolated) at the operating point whose ordinate is the value for i_C and whose abscissa is the value for v_{CE}.

Now consider the case that R_C is changed to a 4.7 kΩ. Proceeding as above, we again get that $i_C \approx \beta i_B = 2.82$ mA. However, for this situation

$$v_{CE} = -(4.7 \times 10^3)i_C + 10 = -(4.7 \times 10^3)(2.82 \times 10^{-3}) + 10$$

$$= -3.25 \text{ V}$$

and

$$v_{BC} = v_{BE} - v_{CE} = 0.7 + 3.25 = 3.95 \text{ V}$$

Using either the fact that $v_{CE} < 0.2$ V or $v_{BC} > V_\gamma = 0.5$ V, we see that the B-C junction is not reverse biased, and, therefore, the transistor is not in the active region.

Drill Exercise 7.2

For the circuit given in Fig. 7.7, suppose that $R_B = 270$ kΩ, $R_C = 1.5$ kΩ, $V_{BB} = V_{CC} = 6$ V, and $\beta = 120$. Assume that the BJT operates in the active region and find i_B, i_C, and v_{CE}. Is the assumption about active-region operation correct? Explain why. Does the BJT operate in the active region if the value of R_C is changed to 3 kΩ? Explain why.

ANSWER 19.6 μA, 2.35 mA, 2.47 V, yes, $i_B > 0$ A and $v_{CE} > 0.2$ V, no, $v_{CE} < 0.2$ V

Example 7.3

Suppose that for the circuit described in Example 7.2, a 1-kΩ resistor is placed in series with the emitter. The resulting circuit is shown in Fig. 7.11a. Assuming that the BJT is operating in the active region, again we can employ the two approximations $v_{BE} = 0.7$ V and $i_C = \beta i_B = 100i_B$. This situation is explicitly indicated in Fig. 7.11b, where the BJT has been replaced by a model that realizes these conditions.

By KVL, we have

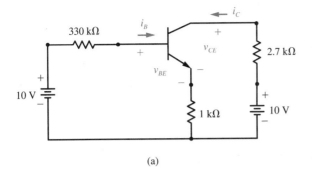

(a)

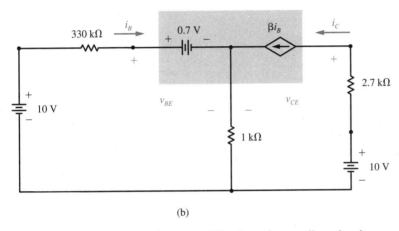

(b)

Fig. 7.11 Circuit resulting from modification of preceding circuit.

$$10 = (330 \times 10^3)i_B + v_{BE} + 1000(i_B + i_C)$$
$$= (330 \times 10^3)i_B + 0.7 + 1000(i_B + 100i_B)$$

Solving this expression for i_B results in $i_B = 21.6 \ \mu A$. Thus

$$i_C = \beta i_B = 100(21.6 \times 10^{-6}) = 2.16 \text{ mA}$$

Furthermore,

$$v_{CE} = -(2.7 \times 10^3)i_C + 10 - 1000(i_B + i_C)$$
$$= -(2.7 \times 10^3)(2.16 \times 10^{-3})$$
$$\quad + 10 - 1000(21.6 \times 10^{-6} + 2.16 \times 10^{-3})$$
$$= 1.99 \text{ V}$$

Since $i_B > 0$ A, the B-E junction is indeed forward biased, and since $v_{CE} > 0.2$ V, the B-C junction is indeed reverse biased. This confirms that the transistor is in the active region.

Drill Exercise 7.3

For the circuit given in Fig. 7.7, suppose that $R_B = 270$ kΩ, $R_C = 1.5$ kΩ, $V_{BB} = V_{CC} = 6$ V, and $\beta = 120$. Place a 500-Ω resistor in series with the emitter, and assume that the BJT operates in the active region. Find i_B, i_C, and v_{CE}. Is the assumption about active-region operation correct? Explain why. Does the BJT operate in the active region if the value of R_C is changed to 3 kΩ? Explain why.

ANSWER 16 μA, 1.92 mA, 2.15 V, yes, $i_B > 0$ A and $v_{CE} > 0.2$ V, no, $v_{CE} < 0.2$ V

Example 7.4

The circuit shown in Fig. 7.12a has two *npn* transistors labeled Q_1 and Q_2. Suppose that $R_1 = 150$ kΩ, $R_2 = 75$ kΩ, $R_{C1} = 2$ kΩ, $R_{C2} = 1$ kΩ, $V_{CC} = 6$ V, and $\beta = 100$ for both BJTs. Let us analyze this circuit assuming active-region operation for both transistors.

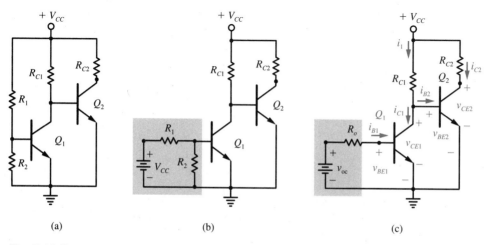

(a)　　　　　(b)　　　　　(c)

Fig. 7.12 Two-transistor circuit.

First note that the circuit given in Fig. 7.12b is equivalent to the circuit shown in Fig. 7.12a. For the circuit shown in Fig. 7.12b, we can apply Thévenin's theorem to the portion of the circuit to the left of transistor Q_1. The resulting circuit is shown in Fig. 7.12c, where

$$v_{oc} = \frac{R_2}{R_1 + R_2} v_1 = \frac{75 \times 10^3}{150 \times 10^3 + 75 \times 10^3}(6) = 2 \text{ V}$$

and

$$R_o = \frac{(150 \times 10^3)(75 \times 10^3)}{150 \times 10^3 + 75 \times 10^3} = 50 \text{ k}\Omega$$

Since Q_1 and Q_2 are assumed to be in the active region, we will use the approximations

$$v_{BE1} = v_{BE2} = 0.7 \text{ V}, \qquad i_{C1} = \beta i_{B1} = 100 i_{B1}, \qquad i_{C2} = \beta i_{B2} = 100 i_{B2}$$

By KVL and Ohm's law, we have that

$$i_{B1} = \frac{2 - 0.7}{50 \times 10^3} = 26 \text{ }\mu\text{A} \qquad \Rightarrow \qquad i_{C1} = 100(26 \times 10^{-6}) = 2.6 \text{ mA}$$

Since

$$i_1 = \frac{6 - 0.7}{2 \times 10^3} = 2.65 \text{ mA}$$

then

$$i_{B2} = i_1 - i_{C1}$$

$$= 2.65 \times 10^{-3} - 2.6 \times 10^{-3}$$

$$= 50 \text{ }\mu\text{A}$$

$$i_{C2} = 100 i_{B2}$$

$$= 100(50 \times 10^{-6})$$

$$= 5 \text{ mA}$$

Finally,

$$v_{CE1} = v_{BE2} = 0.7 \text{ V} \qquad \text{and} \qquad v_{CE2} = -1000(5 \times 10^{-3}) + 6 = 1 \text{ V}$$

Since $i_{B1} > 0$ A, $i_{B2} > 0$ A, $v_{CE1} > 0.2$ V, and $v_{CE2} > 0.2$ V, both BJTs are confirmed to be operating in the active region.

Drill Exercise 7.4

For the circuit described in Example 7.4, place a 100-Ω resistor in series with the emitter of Q_2. Find the resulting values for i_{B1}, i_{C1}, i_{B2}, i_{C2}, v_{CE1}, and v_{CE2}.

ANSWER 26 μA, 2.6 mA, 8.26 μA, 0.826 mA, 0.78 V, 5.09 V

7.3 Cutoff and Saturation

Previously we saw that a transistor is in the active mode when the B-E junction is forward biased and the B-C junction is reverse biased. We mentioned that a transistor whose B-C junction is reverse biased, and whose B-E junction is also reverse biased, is in the cutoff region—we also say that the transistor is **cut off.** When a transistor is not cut off, we say that it **conducts** (current). A transistor whose B-E junction is forward biased and whose B-C junction is also forward biased is in the saturation region—we also say that the transistor is **saturated.** For linear applications such as amplification, a transistor operates in the active region. Conversely, transistors often operate mainly in cutoff and saturation for such digital-circuit applications as logic gates.

Consider the CE *npn* transistor circuit shown in Fig. 7.13. For this circuit, we have that

$$i_C = \frac{V_{CC} - v_{CE}}{R_C} = -\frac{1}{R_C}v_{CE} + \frac{V_{CC}}{R_C}$$

A plot of this equation on the collector characteristics of the transistor gives us the load line shown in Fig. 7.14. With reference to the input characteristics shown in Fig. 7.9, for there to be a significant base current, we should have $v_{BE} > 0.5$ V. For the case that $v_{BE} < 0.5$ V, we can consider the base current to be $i_B = 0$ A and the B-E junction to be reverse biased. As a consequence of this, from the collector (output) characteristics given in Fig. 7.14, we see that, in essence, $i_C = 0$ A and $v_{CE} = V_{CC}$. Furthermore, by KCL, $i_E = i_B + i_C = 0$ A. When $v_{BE} < 0.5$ V, then

$$v_{BC} = v_{BE} - v_{CE} < 0.5 - V_{CC}$$

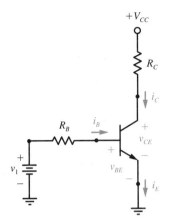

Fig. 7.13 CE *npn* transistor circuit.

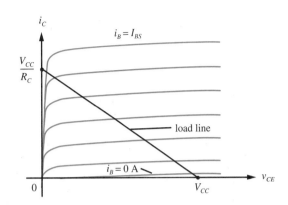

Fig. 7.14 Output characteristics and load line.

and so the B-C junction is reverse biased as well. In summary, if the applied voltage v_1 is such that $v_{BE} < 0.5$ V, then the silicon transistor in Fig. 7.13 will be cut off ($i_B = i_C = i_E = 0$ A and $v_{CE} = V_{CC}$).

Perhaps an overzealous individual may want to make v_1 very negative to guarantee that the BJT is cut off. However, there is a limit as to how much the B-E junction can be reverse biased. The maximum reverse-bias voltage, called the **emitter-base breakdown voltage,** is designated BV_{EBO} since it is measured from the emitter to the base with the collector open ($i_C = 0$ A). Values of BV_{EBO} typically range from volts to tens of volts. Exceeding BV_{EBO} can result in excessive current (and damage to the transistor) and undesirable operating characteristics.

Another type of breakdown occurs when v_{CE} is made too large. Specifically, the **collector-emitter breakdown voltage,** designated BV_{CEO}, is the maximum voltage that can be applied from the collector to the emitter with the base open ($i_B = 0$ A). If this value of voltage is exceeded, then excessive collector current can result, as indicated by the collector characteristics shown in Fig. 7.15. Inspection of the curve for $i_B = 0$ A reveals that $BV_{CEO} \approx 40$ V. For the case of a transistor in the CB configuration, the output breakdown voltage is BV_{CBO}—the maximum voltage from collector to base with the emitter open ($i_E = 0$ A). Generally, BV_{CBO} is greater than 50 V, and usually BV_{CEO} is about half of BV_{CBO}.

The excessive current resulting from transistor breakdown can be due to either the avalanche effect or to another phenomenon known as **reach-through** (or **punch-through**). This mechanism is a result of a widened B-C junction depletion region. If this junction is sufficiently reverse biased, then the depletion region can completely spread across the narrow base and "reach through" to the B-E junction. The consequence of this can be a large current. The breakdown-voltage limitation of a particular transistor is determined by the phenomenon, either the avalanche effect or reach-through, that occurs at the lower voltage.

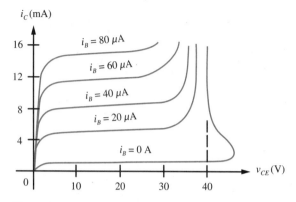

Fig. 7.15 Collector (output) characteristics, including the effects of breakdown.

DC Current Gain

In Section 7.2, we saw how, for a BJT operating in the active region, the large-signal (CE) current gain β was reasonably approximated by $\beta \approx i_C/i_B$. Therefore, let us now designate this current ratio by its own notation. Specifically, we define the **dc current gain** h_{FE} (also called the **dc beta,** and also denoted β_{dc}) by

$$h_{FE} = \frac{i_C}{i_B} \tag{7.12}$$

and, as a consequence of this definition, we have that $h_{FE} \approx \beta$.

Not only does h_{FE} change with temperature (it increases as temperature increases), but for a given temperature, h_{FE} depends on the collector current i_C. A typical variation in h_{FE} is displayed in Fig. 7.16. Despite this characteristic, for the sake of simplicity, in our calculations we shall treat h_{FE} as a constant.

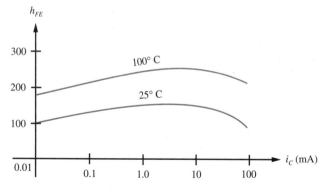

Fig. 7.16 Dependence of h_{FE} on i_C and on temperature. (Note the logarithmic horizontal axis.)

Again consider the circuit given in Fig. 7.13 and the associated output characteristics and load line shown in Fig. 7.14. If $v_1 > 0.5$ V, then this voltage will forward bias the B-E junction and make $i_B > 0$ A. This means that the operation of the transistor goes (along the load line) from cutoff into the active region. By further increasing v_1, and hence i_B, the transistor eventually will enter the saturation region. In particular, when i_B increases beyond I_{BS}, the collector current ($i_C \approx V_{CC}/R_C$) can no longer increase—the collector current is saturated (hence the name for the region). Thus we see that when there is sufficient base current, the transistor will be put into saturation. For this case, the collector current is $i_C \approx V_{CC}/R_C$ and the corresponding collector-emitter voltage is designated $v_{CE(sat)}$—typically 0.2 V for silicon. Therefore, the collector saturation current $i_{C(sat)} = (V_{CC} - v_{CE(sat)})/R_C = (V_{CC} - 0.2)/R_C \approx V_{CC}/R_C$, when $V_{CC} >> 0.2$ V.

When there is sufficient base current to saturate a transistor, typically the corresponding base-emitter voltage, designated $v_{BE(sat)}$, is about 0.8 V for silicon. For a silicon transistor in saturation, $i_B > 0$ A and the B-E junction is forward biased. Thus, $v_{BC} = v_{BE(sat)} - v_{CE(sat)} = 0.8 - 0.2 = 0.6 > V_\gamma = 0.5$ V, and this means the B-C junction is also forward biased.

When a transistor is in saturation, we shall say that it is **ON;** when it is in cutoff, we shall say that it is **OFF.** In summary, then, when a transistor is ON, it conducts current ($\approx V_{CC}/R_C$) from collector to emitter, and the voltage between collector and emitter (≈ 0.2 V) is close to 0 V. In other words, the BJT behaves approximately as a short circuit between collector and emitter. Conversely, when a transistor is OFF, it will not conduct any significant current between collector and emitter, and, therefore, it acts essentially as an open circuit. Hence we see that a transistor operating in saturation and cutoff behaves as an electronic switch that is turned ON and OFF, respectively, by base current.

Example 7.5

Suppose that for the CE circuit shown in Fig. 7.17a, the silicon transistor has $h_{FE} = 100$. Let us assume that the transistor is in the active region. For this case, we use $v_{BE} = 0.7$ V, so we have that

$$i_B = \frac{10 - 0.7}{100 \times 10^3} = 93 \ \mu\text{A}$$

and

$$i_C = h_{FE}i_B = 100(93 \times 10^{-6}) = 9.3 \ \text{mA}$$

Since

$$v_{CE} = -(2 \times 10^3)i_C + 10 = -8.6 < 0.2 \ \text{V}$$

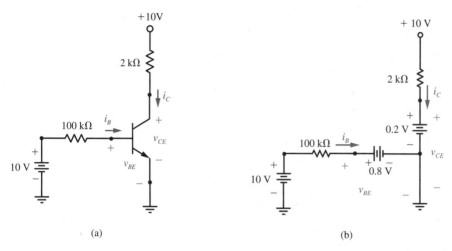

Fig. 7.17 (*a*) CE transistor circuit, and (*b*) transistor replaced by model for a BJT in saturation.

our assumption that the transistor was biased in the active region was incorrect. Therefore, let us assume that the transistor is operating in the saturation region, that is, $v_{BE} = v_{BE(sat)} = 0.8$ V and $v_{CE} = v_{CE(sat)} = 0.2$ V. Fig. 7.17*b* shows how we may model a BJT in saturation so as to result in the appropriate saturation voltages. By Ohm's law, the saturation currents are

$$i_B = \frac{10 - 0.8}{100 \times 10^3} = 92 \ \mu\text{A} \qquad \text{and} \qquad i_C = \frac{10 - 0.2}{2 \times 10^3} = 4.9 \ \text{mA}$$

On the border between the active and saturation regions, we have that $h_{FE} = i_C/i_B$ (Eq. 7.12). Thus if i_C is the saturation current, then the minimum current required for the BJT to be on the border of the saturation region is $i_B = i_C/h_{FE}$. In other words, for the transistor to go into saturation, if i_C is the saturation current, then we must have that

$$i_B \geq \frac{i_C}{h_{FE}} \tag{7.13}$$

For this example,

$$i_B = 92 \ \mu\text{A} > \frac{i_C}{h_{FE}} = \frac{4.9 \times 10^{-3}}{100} = 49 \ \mu\text{A}$$

so the transistor is indeed in saturation.

Suppose now that h_{FE} for the transistor was not specified. What is the range of values for h_{FE} that will result in saturation-region operation? Again assuming that the transistor is in saturation, we get $i_B = 92 \ \mu A$ and $i_C = 4.9$ mA. From Inequality 7.13 for saturation-region operation, we must have that

$$h_{FE} \geq \frac{i_C}{i_B} = \frac{4.9 \times 10^{-3}}{92 \times 10^{-6}} = 53.3 \approx 53$$

Drill Exercise 7.5

For the circuit shown in Fig. 7.17a, place a 2-kΩ resistor in series with the emitter. Find the minimum value of h_{FE} required for the BJT to be in saturation. (*Hint:* Write simultaneous equations in the variables i_B and i_C.)

ANSWER 57

Example 7.6

Suppose that for the circuit shown in Fig. 7.18a, the BJTs Q_1 and Q_2 have dc betas of $h_{FE1} = h_{FE2} = 100$. Let us analyze this circuit for the case that $R_1 = 4$ kΩ, $R_2 = 2.5$ kΩ, $R_E = 1$ kΩ, and (a) $V_{BB} = 3$ V, and (b) $V_{BB} = 1$ V.

(a) Since 3 V is applied directly to the base of Q_1, it should be sufficient to strongly forward bias the B-E junction of Q_1. Therefore, let us assume that Q_1 is ON. Under

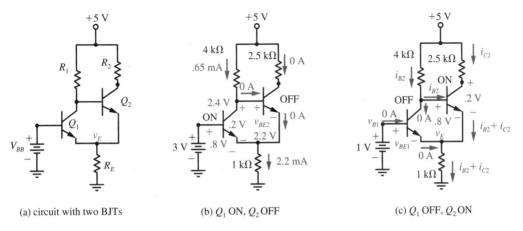

(a) circuit with two BJTs   (b) Q_1 ON, Q_2 OFF   (c) Q_1 OFF, Q_2 ON

Fig. 7.18 Emitter-coupled Schmitt trigger.

this assumption, $v_{BE1} = 0.8$ V and $v_{CE1} = 0.2$ V. However, $v_{BE2} = v_{CE1} = 0.2 < 0.5$ V indicates that Q_2 is OFF. Consequently, $i_{B2} = i_{C2} = i_{E2} = 0$ A as indicated in Fig. 7.18b. By KVL, $v_E = -v_{BE1} + V_{BB} = -0.8 + 3 = 2.2$ V, so the current going down through the 1-kΩ resistor is $i_{B1} + i_{C1} = 2.2/1000 = 2.2$ mA. By KVL, $v_{C1} = v_{CE1} + v_E = 0.2 + 2.2 = 2.4$ V. Thus, $i_{C1} = (5 - 2.4)/4000 = 0.65$ mA. Since $i_{B1} + i_{C1} = 2.2$ mA, then $i_{B1} = 2.2 \times 10^{-3} - i_{C1} = 2.2 \times 10^{-3} - 0.65 \times 10^{-3} = 1.55$ mA.

For Q_1 to indeed be ON, we must have that $i_{B1} \geq i_{C1}/h_{FE1}$. Since $i_{B1} = 1.55$ mA $> i_{C1}/h_{FE1} = (0.65 \times 10^{-3})/100 = 6.5$ μA, this confirms that Q_1 is ON. Furthermore, $v_{BE2} = v_{CE1} = 0.2 < 0.5$ V, and this confirms that Q_2 is OFF.

The condition that $i_{B1} \geq i_{C1}/h_{FE1}$ for Q_1 to be ON is equivalent to the condition that $h_{FE1} \geq i_{C1}/i_{B1}$. This indicates that $h_{FE1} \geq (0.65 \times 10^{-3})/(1.55 \times 10^{-3}) = 0.42$ will be sufficient to turn ON Q_1.

(b) Let us assume that $V_{BB} = 1$ V is not sufficient to forward bias the B-E junction of Q_1. In other words, suppose that Q_1 is OFF. Then, $i_{B1} = i_{C1} = i_{E1} = 0$ A as indicated in Fig. 7.18c. Since the 5-V supply voltage seems sufficient to forward bias the B-E junction of Q_2, let us assume that Q_2 is ON. Therefore, $v_{BE2} = 0.8$ V and $v_{CE2} = 0.2$ V. By KVL, we have that

$$5 = (4 \times 10^3)i_{B2} + 0.8 + (1 \times 10^3)(i_{B2} + i_{C2}) \quad \Rightarrow$$

$$4.2 = (5 \times 10^3)i_{B2} + (1 \times 10^3)i_{C2}$$

and

$$5 = (2.5 \times 10^3)i_{C2} + 0.2 + (1 \times 10^3)(i_{B2} + i_{C2}) \quad \Rightarrow$$

$$4.8 = (1 \times 10^3)i_{B2} + (3.5 \times 10^3)i_{C2}$$

Solving these two equations results in $i_{B2} = 0.6$ mA and $i_{C2} = 1.2$ mA. For Q_2 to indeed be ON, we must have that $i_{B2} \geq i_{C2}/h_{FE2}$. Since $i_{B2} = 0.6$ mA $= 600$ μA $> i_{C2}/h_{FE2} = (1.2 \times 10^{-3})/100 = 12$ μA, this confirms that Q_2 is ON. Furthermore, $v_E = (1 \times 10^3)(i_{B2} + i_{C2}) = (1 \times 10^3)(0.6 \times 10^{-3} + 1.2 \times 10^{-3}) = 1.8$ V. Therefore, $v_{BE1} = v_{B1} - v_E = 1 - 1.8 = -0.8 < 0.5$ V, and this confirms that Q_1 is OFF.

The condition that $i_{B2} \geq i_{C2}/h_{FE2}$ for Q_2 to be ON is equivalent to the condition that $h_{FE2} \geq i_{C2}/i_{B2}$. This indicates that $h_{FE2} \geq (1.2 \times 10^{-3})/(0.6 \times 10^{-3}) = 2$ will be sufficient to turn ON Q_2.

The circuit given in Fig. 7.18a is an example of an emitter-coupled **Schmitt trigger.** See Section 10.5 for a detailed discussion on Schmitt triggers.

Transistor Switching Times

Since an actual transistor is a physical device, it cannot change state (go from cutoff to saturation, or vice versa) instantaneously. To discuss what happens, therefore, let us again consider the CE transistor circuit, shown in Fig. 7.19, where $V_{CC} \gg$

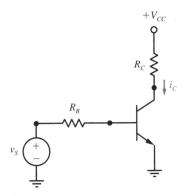

Fig. 7.19 A CE transistor circuit.

0.2 V and the applied voltage v_S is as indicated in Fig. 7.20. For time $t < 0$ s, since $v_S = -V_1 < 0$ V, the transistor is cut off and $i_C = 0$ A. At $t = 0$ s, the applied voltage instantaneously changes to $v_S = V_2$. Suppose that V_2 is sufficiently large to saturate the transistor. When the transistor is in saturation, $i_C = i_{C(\text{sat})} \approx V_{CC}/R_C$. However, the transistor will not go from cutoff to saturation instantaneously. Instead, the current will increase from $i_C = 0$ A to $i_C = i_{C(\text{sat})}$ as depicted in Fig. 7.21. The time it takes for the collector current to increase from $i_C = 0$ A to $i_C = 0.1i_{C(\text{sat})}$ is called the **delay time** t_d. The time it takes for this current to rise from $i_C = 0.1i_{C(\text{sat})}$ to $i_C = 0.9i_{C(\text{sat})}$ is known as the **rise time** t_r. We then define the **turn-on time** t_{ON} to be $t_{ON} = t_d + t_r$.

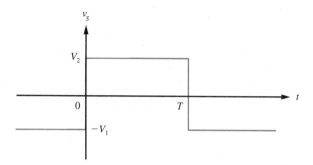

Fig. 7.20 Voltage applied to CE transistor circuit.

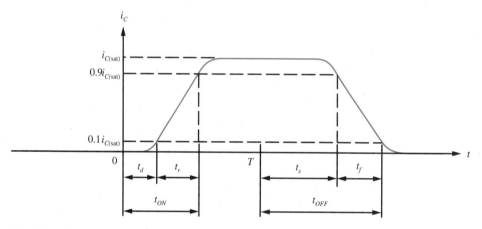

Fig. 7.21 Transistor switching times.

At time $t = T$, the applied voltage instantaneously changes from $v_S = V_2$ to $v_S = -V_1$. As a consequence of this, the transistor will go from saturation to cut-off—but not instantaneously. The time it takes for the collector current to drop from $i_C = i_{C(sat)}$ to $i_C = 0.9i_{C(sat)}$ is called the **storage time** (or **saturation time**) t_s, whereas the time it takes this current to fall from $i_C = 0.9i_{C(sat)}$ to $i_C = 0.1i_{C(sat)}$ is known as the **fall time** t_f. We then define the **turn-off time** t_{OFF} to be $t_{OFF} = t_s + t_f$.

When the transistor is cut off, both the B-E and B-C junctions are reverse biased. For the B-E junction to be forward biased, first a certain amount of time is required for the B-E junction transition capacitance to be charged up. Time is also required for minority carriers to diffuse across the base and then enter the collector. This produces the delay time t_d (on the order of nanoseconds). Before it can go into saturation, the transistor goes (via the load line—see Fig. 7.14) into the active region. The time required for the transistor to go through the active region from cutoff to saturation accounts for the rise time t_r (on the order of nanoseconds to tens of nanoseconds).

When a transistor is in saturation, both junctions are forward biased and conducting current. As a consequence of this, an excess of minority carriers is stored in the base. For the B-C junction to be reverse biased, this large amount of minority-carrier charge must first be removed. The time required for the removal of the excess charge determines the storage time t_s (on the order of hundreds of nanoseconds). Before it can be cut off, the transistor goes into the active region. The time required for the transistor to go through the active region from saturation to cutoff accounts for the fall time t_f (on the order of tens of nanoseconds).

7.4 Applications to Digital Logic Circuits

An important digital logic circuit is the **inverter** or **NOT gate.** The output of such a device is the logic inverse of the input. In other words, if the input is a logic 0, then the output is a logic 1; if the input is a logic 1, then the output is a logic 0.

Example 7.7

Fig. 7.22 shows a BJT inverter, where the low voltage is 0.2 V (the typical collector-emitter saturation voltage) and the high voltage is 5 V. Let us analyze this digital logic circuit.

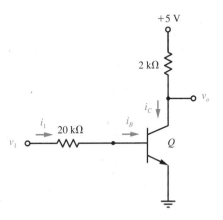

Fig. 7.22 A BJT inverter.

For the case that the input is the low voltage $v_1 = 0.2$ V, since $v_1 < 0.5$ V, the input tends to turn OFF the transistor. Therefore, let us assume that the transistor is OFF. Thus $i_B = 0$ A and, hence, $v_{BE} = v_1 = 0.2 < 0.5$ V, so the BJT is indeed cut off. Since $i_C = 0$ A, the resulting output voltage is

$$v_o = -2000i_C + 5 = 5 \text{ V}$$

For the case that the input voltage is $v_1 = 5$ V, the input tends to turn ON the transistor. Therefore, let us assume that the transistor is ON. Since $v_{BE} = 0.8$ V and $v_{CE} = 0.2$ V, then

$$i_B = \frac{5 - 0.8}{20 \times 10^3} = 0.21 \text{ mA} \quad \text{and} \quad i_C = \frac{5 - 0.2}{2 \times 10^3} = 2.4 \text{ mA}$$

Hence for the transistor to be in saturation, the transistor must have

$$h_{FE} \geq \frac{i_C}{i_B} = \frac{2.4 \times 10^{-3}}{0.21 \times 10^{-3}} = 11.4$$

When this inequality is satisfied, the transistor is indeed in saturation, and the resulting output voltage is

$$v_o = v_{CE(\text{sat})} = 0.2 \text{ V}$$

We have now confirmed that the given circuit is an inverter. Therefore, we say that the output is the complement of the input, and denote this by writing $v_o = \bar{v}_1$.

Fig. 7.23 shows a plot of the output voltage v_o versus the input voltage v_1 for the inverter given in Fig. 7.22. Such a plot is called the **voltage transfer characteristic** of the inverter, and this curve depicts how the output varies with the input. Since the desired output voltage of the inverter is to be either 0.2 V (the low voltage) or 5 V (the high voltage), we designate these voltages by $V_{OL} = 0.2$ V and $V_{OH} = 5$ V. These values are indicated on the vertical (output voltage) axis of the transfer characteristic.

When the input voltage v_1 to the given inverter is 0.2 V (or less), then Q is OFF, and the output voltage is 5 V. If v_1 increases, Q will still be OFF until v_1 exceeds approximately 0.5 V. When $v_1 > 0.5$ V, then Q conducts (has nonzero) current, so $v_o = -(2 \times 10^3)i_C + 5 < 5$ V, and Q is in the active region. We designate the input voltage $v_1 = 0.5$ V—the value below which Q is OFF and above which Q is in the active region—by V_{IL}. This is the largest input voltage that still results in the same output produced by the low input voltage.

As v_1 becomes larger, the currents in Q get larger, and the output gets smaller. If v_1 is large enough, then Q goes into saturation, and the output becomes 0.2 V. We designate the smallest input voltage, which results in the same output that is produced by the high input voltage, by V_{IH}.

We can determine V_{IH} as follows: Suppose that we are given that $h_{FE} = 100$. When Q is ON, then $v_{BE} = 0.8$ V, $v_{CE} = 0.2$ V, and

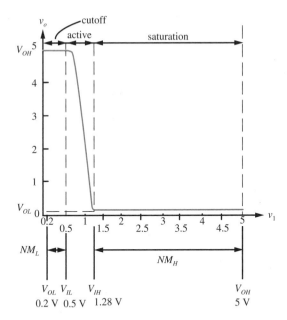

Fig. 7.23 Transfer characteristic for given inverter.

$$i_B = \frac{v_1 - v_{BE}}{20 \times 10^3} = \frac{v_1 - 0.8}{20 \times 10^3}$$

and

$$i_C = \frac{5 - v_{CE}}{2 \times 10^3} = \frac{5 - 0.2}{2 \times 10^3} = 2.4 \times 10^{-3}$$

For Q to be ON, we require that $i_B \geq i_C / h_{FE}$. Therefore,

$$i_B = \frac{v_1 - 0.8}{20 \times 10^3} \geq \frac{2.4 \times 10^{-3}}{100} \qquad \Rightarrow \qquad v_1 \geq 1.28 \text{ V}$$

Hence $V_{IH} = 1.28$ V.

Drill Exercise 7.7

For the inverter given in Fig. 7.22, connect a 50-kΩ resistor between the base of the BJT and a second power supply of −5 V. (The additional power supply will result in Q turning OFF faster after the input goes low by producing a negative base current, which removes excess stored minority carriers from the base region of the BJT.) Find the minimum value of h_{FE} required for the resulting inverter.

ANSWER 25.5

Noise Margin

Noise voltages in circuits are unwanted voltages that can be caused by such things as power-supply voltage spikes, transients resulting from switching, or unwanted capacitive or inductive coupling. The minimum noise voltage at the input of a digital logic gate that causes the output to deviate from its proper value is known as the **noise margin** of the gate. Specifically, NM_L denotes the minimum amount of noise that produces an output other than that which results from a low input. Similarly, NM_H denotes the minimum amount of noise that produces an output other than that which results from a high input.

Example 7.8

For the inverter discussed in Example 7.7, the desired outputs are $V_{OL} = 0.2$ V and $V_{OH} = 5$ V. When the input is low ($v_1 = V_{OL} = 0.2$ V), then Q is OFF and $v_o = V_{OH} = 5$ V. When v_1 exceeds $V_{IL} = 0.5$ V, then Q goes into the active region, and $v_o < V_{OH} = 5$ V. Therefore, $NM_L = V_{IL} - V_{OL} = 0.5 - 0.2 = 0.3$ V. When the input is high ($v_1 = V_{OH} = 5$ V), then Q is ON and $v_o = V_{OL} = 0.2$ V. When v_1 falls below $V_{IH} = 1.28$ V, then Q goes into the active region, and $v_o > V_{OL} = 0.2$ V. Therefore, $NM_H = V_{OH} - V_{IH} = 5 - 1.28 = 3.72$ V. These results are indicated on the transfer characteristic for the inverter (Fig. 7.23).

Drill Exercise 7.8

For the inverter given in Fig. 7.22, connect a 50-kΩ resistor between the base of the BJT and a a second power supply of -5 V. Suppose that the BJT has $h_{FE} = 100$. Find the noise margins NM_L and NM_H for the resulting inverter. (*Hint:* Find V_{IL} by determining the value of v_1 for which $v_{BE} = 0.5$ V, and find V_{IH} by determining the value of v_1 for which $v_{BE} = 0.8$ V.)

ANSWER 2.5 V, 1.4 V

Propagation Delay

Fig. 7.24 shows input and output voltages (as functions of time) for an inverter. When the input goes from low to high, the output will go from high to low, and vice versa. However, as depicted in Fig. 7.24, the response to an input is not instantaneous. The **turn-off delay time** t_{PLH} and the **turn-on delay time** t_{PHL} for the inverter are the times indicated. The **propagation delay** t_p of the gate is defined to be

$$t_p = \frac{1}{2}(t_{PLH} + t_{PHL})$$

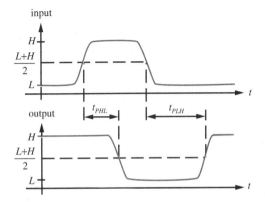

t_{PHL}: high-to-low propagation delay

t_{PLH}: low-to-high propagation delay

propagation delay: $t_p = \frac{1}{2}(t_{PHL} + t_{PLH})$

Fig. 7.24 Definition of propagation delay.

Diode-Transistor Logic (DTL)

If we invert the output of an AND gate, the resulting combination of an AND gate followed by a NOT gate is called a **NAND gate**—NAND being a contraction of NOT-AND. This very important digital logic circuit can be constructed by connecting a transistor NOT gate to a diode AND gate. One such NAND gate, shown in Fig. 7.25, which utilizes diodes and a transistor, is an example of **diode-transistor logic** (DTL).

Example 7.9

Let us analyze the NAND gate given in Fig. 7.25. Table 7.1 is the truth table for AND/NAND operation.

In the analysis of this gate, we will assume that a reverse-biased diode is an open circuit, and a forward-biased diode has a voltage of 0.7 V across it (see Fig. 6.47 on p. 391). Furthermore, the low voltage is 0.2 V and the high voltage is 6 V.

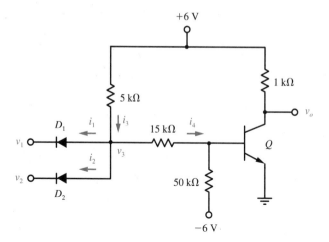

Fig. 7.25 A DTL NAND gate.

Table 7.1 AND/NAND Truth Table

V_1	V_2	AND	NAND
0	0	0	1
0	1	0	1
1	0	0	1
1	1	1	0

For the case that $v_1 = v_2 = 0.2$ V, the positive supply voltage (6 V) tends to forward bias the diodes. Although the negative supply voltage (-6 V) tends to reverse bias the diodes, the resistance in series with the negative supply voltage is greater than that for the positive supply voltage. Therefore, let us assume that D_1 and D_2 are ON. Since the voltage across a forward-biased diode is 0.7 V, then

$$v_3 = 0.7 + 0.2 = 0.9 \text{ V}$$

Although this voltage tends to turn ON the transistor Q, the negative supply voltage tends to turn Q OFF. Therefore, let us assume that Q is OFF. Thus, $i_B = 0$ A and

$$i_4 = \frac{0.9 - (-6)}{15 \times 10^3 + 50 \times 10^3} = 0.106 \text{ mA}$$

Hence

$$v_{BE} = (50 \times 10^3)(0.106 \times 10^{-3}) - 6 = -0.7 \text{ V}$$

and Q is indeed cut off. Also, the total diode current is

$$i_1 + i_2 = i_3 - i_4 = \frac{6 - 0.9}{5 \times 10^3} - 0.106 \times 10^{-3} = 0.914 \text{ mA}$$

By symmetry, for identical diodes, $i_1 = i_2 = (0.914 \times 10^{-3})/2 = 0.457$ mA, and since $i_1 > 0$ A and $i_2 > 0$ A, then D_1 and D_2 are indeed ON. Finally, since Q is OFF, then $i_C = 0$ A and $v_o = 6$ V.

For the case that $v_1 = 0.2$ V and $v_2 = 6$ V, as previously shown, let us assume that D_1 is ON. Therefore, again $v_3 = 0.7 + 0.2 = 0.9$ V. This means that the voltage across D_2 is $v_3 - v_2 = 0.9 - 6 = -5.1 < 0.7$ V, so when D_1 is ON, then D_2 is OFF. As before, we again find that Q is OFF, and $v_o = 6$ V. Now, however, $i_1 = 0.914$ mA > 0 A so D_1 is indeed ON and D_2 is indeed OFF.

By symmetry, for the case that $v_1 = 6$ V and $v_2 = 0.2$ V, we deduce that D_1 is OFF, D_2 is ON, Q is OFF, and $v_o = 6$ V.

Finally, consider the case that $v_1 = v_2 = 6$ V. Since the supply voltages are not higher in potential than either input, let us assume that D_1 and D_2 are OFF ($i_1 = i_2 = 0$ A). Since Q tends to be turned ON by $+6$ V applied to 5 kΩ + 15 kΩ = 20 kΩ, and tends to be turned OFF by -6 V applied to 50 kΩ, let us assume that Q is ON. Since $i_1 = i_2 = 0$ A and $v_{BE} = 0.8$ V, then

$$i_3 = i_4 = \frac{6 - 0.8}{5 \times 10^3 + 15 \times 10^3} = 0.26 \text{ mA}$$

and

$$i_B = 0.26 \times 10^{-3} + \frac{-6 - 0.8}{50 \times 10^3} = 0.124 \text{ mA}$$

When Q is ON, the collector current is

$$i_C = \frac{6 - 0.2}{1 \times 10^3} = 5.8 \text{ mA}$$

Hence Q is indeed in saturation when

$$h_{FE} \geq \frac{i_C}{i_B} = \frac{5.8 \times 10^{-3}}{0.124 \times 10^{-3}} = 46.8$$

Furthermore,

$$v_3 = -(5 \times 10^3)i_3 + 6 = -(5 \times 10^3)(0.26 \times 10^{-3}) + 6 = 4.7 \text{ V}$$

so that the voltage across each diode is $v_3 - 6 = 4.7 - 6 = -1.3$ V. Thus D_1 and D_2 are indeed OFF, and when $h_{FE} \geq 46.8$, Q is in saturation and $v_o = 0.2$ V.

As indicated by the AND/NAND truth table given in Table 7.1, this then confirms that the DTL circuit given in Fig. 7.25 is a NAND gate.

Drill Exercise 7.9

For the NAND gate given in Fig. 7.25, suppose that the BJT has $h_{FE} = 100$. Find the noise margins NM_L and NM_H for this gate. (*Hint:* Find V_{IL} by determining the value of v_1 for which D_1 is ON and $v_{BE} = 0.5$ V, and find V_{IH} by determining the value of v_1 for which D_1 is ON and $v_{BE} = 0.8$ V.)

ANSWER   1.55 V, 2.99 V

7.5 DTL Integrated-Circuit Logic

Up to this point, we have been treating diodes, transistors, and resistors as distinct, separate elements, that is, as **discrete components.** However, a circuit having such elements can be fabricated on a single, small[2] silicon chip, and what results is called an **integrated circuit** (IC). The same processes used to fabricate discrete diodes and transistors are used to make ICs.

The number of gates on a digital IC determines the category in which such an IC is placed. Thirty or fewer gates on a chip is known as **small-scale integration** (SSI). Between 30 and 300 gates is called **medium-scale integration** (MSI). **Large-scale integration** (LSI) refers to between 300 and 3000 gates, whereas **very-large-scale integration** (VLSI) denotes more than 3000 gates.

DTL gates commonly appear in IC form. Under this circumstance, because of economics, resistance values are limited to 30 kΩ and capacitance values to 100 pF. On the other hand, diodes and transistors can be fabricated inexpensively. Because of this, large resistive values, such as the 50 kΩ in the NAND gate given in Fig. 7.25, are to be avoided for IC designs.

---

[2]A typical cross section is 50 by 50 mils, where 1 mil = 0.001 inch = 0.0254 mm.

Example 7.10

Let us verify that the DTL circuit shown in Fig. 7.26 is a NAND gate when h_{FE} is sufficiently large. Note, too, that only a single 5-V supply voltage is used. In this case, therefore, although the low voltage is still 0.2 V, the high voltage is 5 V.

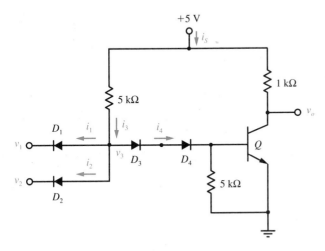

Fig. 7.26 An IC DTL NAND gate.

Case 1. $v_1 = v_2 = 0.2$ V

Since the 5-V supply voltage tends to forward bias D_1 and D_2, let us assume that D_1 and D_2 are ON. Thus, $v_3 = 0.7 + 0.2 = 0.9$ V. However, this voltage is not large enough to turn ON D_3 and D_4, and so we shall assume that D_3 and D_4 are OFF. Under this assumption,

$$i_1 + i_2 = i_3 = \frac{5 - 0.9}{5 \times 10^3} = 0.82 \text{ mA}$$

and by symmetry, for identical diodes, $i_1 = i_2 = 0.41$ mA > 0. Hence, D_1 and D_2 are indeed ON. Since $i_4 = 0$ A, then $i_B = 0$ A. Consequently, $v_{BE} = 0$ V and the voltage across the series connection of D_3 and D_4 is

$$v_3 - v_{BE} = 0.9 < 0.7 + 0.7 = 1.4 \text{ V}$$

and so D_3 and D_4 indeed are OFF. Furthermore, $v_{BE} = 0 < 0.5$ V means that Q is OFF, and hence, $v_o = 5$ V.

Case 2. $v_1 = 0.2$ V, $v_2 = 5$ V

We again assume that D_1 is ON. Again, $v_3 = 0.7 + 0.2 = 0.9$ V. As in Case 1, this means we assume that D_3 and D_4 are OFF. Furthermore, the voltage across D_2 is $v_3 - v_2 = 0.9 - 5 = -4.1 < 0.7$ V, so we assume that D_2 is OFF. This means that $i_2 = i_4 = 0$ A, and

$$i_1 = i_3 = \frac{5 - 0.9}{5 \times 10^3} = 0.82 \text{ mA} > 0$$

Thus D_1 is indeed ON, and D_2, D_3, D_4 are indeed OFF. Again, since $i_4 = 0$ A, we deduce that Q is OFF and $v_o = 5$ V.

Case 3. $v_1 = 5$ V, $v_2 = 0.2$ V

Just as in Case 2, we find that D_2 is ON; D_1, D_3, D_4, and Q are OFF; and $v_o = 5$ V.

Case 4. $v_1 = v_2 = 5$ V

Since there is no potential difference between either input and the supply voltage, let us assume that D_1 and D_2 are OFF ($i_1 = i_2 = 0$ A). This means that the supply voltage has a tendency to turn ON D_3, D_4, and Q. Therefore, let us make these assumptions. By KVL, we have that

$$v_3 = 0.7 + 0.7 + 0.8 = 2.2 \text{ V}$$

and since $v_3 - 5 = 2.2 - 5 = -2.8 < 0.7$ V, then D_1 and D_2 are indeed OFF. Furthermore,

$$i_3 = \frac{5 - 2.2}{5 \times 10^3} = 0.56 \text{ mA}$$

Since $i_4 = i_3 > 0$ A, then D_3 and D_4 are indeed ON. In addition, if Q is ON, then

$$i_B = i_4 - \frac{v_{BE}}{5 \times 10^3} = 0.56 \times 10^{-3} - \frac{0.8}{5 \times 10^3} = 0.4 \text{ mA}$$

and

$$i_C = \frac{5 - 0.2}{1 \times 10^3} = 4.8 \text{ mA}$$

Thus for Q indeed to be ON, we require that

$$h_{FE} \geq \frac{i_C}{i_B} = \frac{4.8 \times 10^{-3}}{0.4 \times 10^{-3}} = 12$$

and when Q is in saturation, $v_o = 0.2$ V. This confirms that the DTL circuit given in Fig. 7.26 is a NAND gate.

Drill Exercise 7.10

Find the noise margins NM_L and NM_H for the NAND gate given in Fig. 7.26. (*Hint:* Find V_{IL} by determining the value of v_1 for which D_1 is ON and $v_{BE} = 0.5$ V, and find V_{IH} by determining the value of v_1 for which D_1 is ON and $v_{BE} = 0.8$ V.) What are the noise margins NM_L and NM_H when D_4 is replaced by a short circuit?

ANSWER 1.0 V, 3.5 V, 0.3 V, 4.2 V

Power Dissipation

An important property of a logic gate is the power dissipated by the gate.

Example 7.11

Let us now determine the power dissipated by the NAND gate given in Fig. 7.26. To do this, let us find the power supplied by the 5-V source.

For the case that the output is 0.2 V (logic 0), D_1 and D_2 are OFF and Q is ON. Using the results from Example 7.10, the current directed out of the 5-V power supply is

$$i_S = i_3 + i_C = 0.56 \times 10^{-3} + 4.8 \times 10^{-3} = 5.36 \text{ mA}$$

and the power supplied by the 5-V source, therefore, is

$$P_0 = 5(5.36 \times 10^{-3}) = 26.8 \text{ mW}$$

For the case that the output is 5 V (logic 1), D_1 and/or D_2 is ON and Q is OFF. Again, from Example 7.10,

$$i_S = i_3 + i_C = 0.82 \times 10^{-3} + 0 = 0.82 \text{ mA}$$

and the power supplied by the 5-V source, therefore, is

$$P_1 = 5(0.82 \times 10^{-3}) = 4.1 \text{ mW}$$

Assuming that the NAND gate spends the same amount of time in the 0 state that it does in the 1 state, the average power supplied by the 5-V source is

$$P = \frac{P_0 + P_1}{2} = \frac{26.8 \times 10^{-3} + 4.1 \times 10^{-3}}{2} = 15.45 \text{ mW}$$

If we ignore the small amount of power (on the order of 0.1 mW) absorbed by the sources for v_1 and v_2 (see Problem 7.55 at the end of this chapter), then the power supplied by the 5-V source is equal to the power dissipated by the gate. Thus the average power dissipated by the NAND gate given in Fig. 7.26 is 15.45 mW.

Drill Exercise 7.11

Determine the power dissipated by the NAND gate given in Fig. 7.25.

ANSWER 22.0 mW

Fan-out

In our discussions so far, we have considered the situation in which a digital logic circuit was unloaded—that is, nothing was connected to the output of the circuit. Let us now investigate the condition that a gate is loaded with N similar gates.

Example 7.12

Reconsider the NAND gate given in Fig. 7.26. Shown in Fig. 7.27 is such a gate, and only the input portion of one of the N gates that loads it is explicitly depicted.

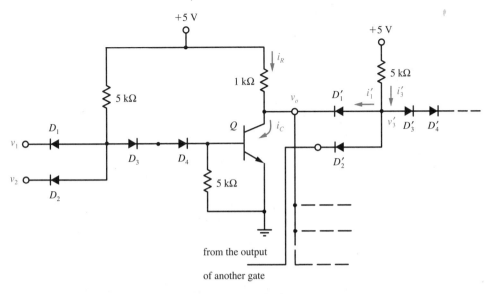

Fig. 7.27 Two-input NAND gate loaded by N similar gates.

If $v_1 = 0.2$ V or $v_2 = 0.2$ V, then Q will be OFF and v_o will tend to be 5 V. This, in turn, means that D_1' is OFF. Similarly, the other diodes also connected to v_o will be OFF. Thus, $v_o = 5$ V.

Now, however, consider the case that $v_1 = v_2 = 5$ V. Under this circumstance, Q is ON and $v_o = 0.2$ V. This, in turn, means that D_1' is ON and $v_3' = 0.7 + 0.2 = 0.9$ V. As a result,

$$i_3' = \frac{5 - 0.9}{5 \times 10^3} = 0.82 \text{ mA}$$

We know that for Q to be ON, $h_{FE} \geq i_C/i_B$. The base current is the same value as was obtained in Example 7.10—that being $i_B = 0.4$ mA. However, the collector current is the sum of i_R and the current coming from the load. This load current is maximum when, for each of the N output stages, only one of the diodes conducts; i.e., for the first stage shown, the maximum contribution to the collector current is $i_1' = i_3' = 0.82$ mA. This current is called the **standard load.** Since there are N stages, the maximum collector current is

$$i_C = i_R + Ni_1' = \frac{5 - 0.2}{1 \times 10^3} + N(0.82 \times 10^{-3})$$

$$= 4.8 \times 10^{-3} + N(0.82 \times 10^{-3})$$

To guarantee that Q is ON, we must have that $h_{FE} \geq i_C/i_B$ or $h_{FE}i_B \geq i_C$. Specifically, if we are given that $h_{FE} = 40$, then for proper operation

$$h_{FE}i_B \geq i_C \quad \Rightarrow$$

$$40(0.4 \times 10^{-3}) \geq 4.8 \times 10^{-3} + N(0.82 \times 10^{-3})$$

from which

$$\frac{16 - 4.8}{0.82} \geq N \quad \Rightarrow \quad N \leq 13.7$$

The maximum number of similar output gates that can load a given gate, and still have that gate function properly, is called the **fan-out** of the gate. Since fan-out must be an integer, for the circuit under consideration, the fan-out is 13. The number of inputs to a gate is known as **fan-in.** In this case, the fan-in is 2.

Drill Exercise 7.12

Find the fan-out of the NAND gate given in Fig. 7.25 for the case that $h_{FE} = 100$.

ANSWER 7

Example 7.13

One way to significantly increase the fan-out of the NAND gate given in Fig. 7.26 is to replace diode D_3 with a transistor Q_1, as shown in Fig. 7.28. Also shown is a portion of the first gate of the fan-out.

For the case that at least one input is 0.2 V, the corresponding diode is ON and $v_3 = 0.7 + 0.2 = 0.9$ V. Since this potential is not high enough to turn ON Q_1, D_4, and Q, then Q is OFF and $v_o = 5$ V. (The diodes in the load that are connected to v_o also are OFF.)

For the case that $v_1 = v_2 = 5$ V, diodes D_1 and D_2 are OFF ($i_1 = i_2 = 0$ A). Since the 5-V supply tends to turn ON Q_1, let us assume that D_4 and Q are ON. We will not assume that Q_1 is in saturation because a positive current i_3 results in

$$v_{BC1} = -2000i_3 < 0.5 \text{ V}$$

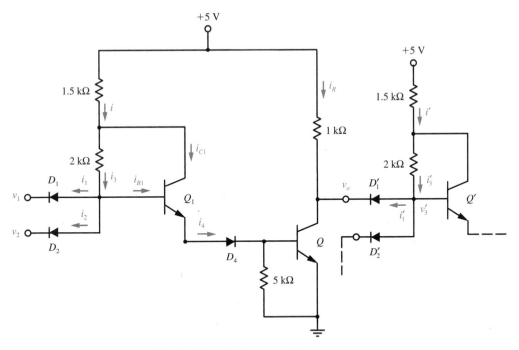

Fig. 7.28 A NAND gate with significantly increased fan-out.

and this means that the B-C junction of Q_1 is reverse biased. Thus let us assume that Q_1 is in the active region. Under these assumptions,

$$v_3 = 0.7 + 0.7 + 0.8 = 2.2 \text{ V}$$

Since $i_1 = i_2 = 0$ A, then $i_{B1} = i_3$. Furthermore, since Q_1 is assumed to be in the active region,

$$i = i_{B1} + i_{C1} = i_{B1} + h_{FE}i_{B1} = (1 + h_{FE})i_{B1} = i_4$$

Specifically, for the case that $h_{FE} = 40$, by KVL,

$$5 = 1500i + 2000i_{B1} + v_3 = 1500(41i_{B1}) + 2000i_{B1} + 2.2$$

from which

$$i_{B1} = \frac{5 - 2.2}{61.5 \times 10^3 + 2 \times 10^3} = 44 \text{ } \mu\text{A}$$

Since $i_{B1} > 0$ A and $v_{BC1} = -2000i_{B1} < 0.5$ V, then Q_1 is indeed in the active region. Also,

$$i_{C1} = h_{FE}i_{B1} = 40(44 \times 10^{-6}) = 1.76 \text{ mA}$$

Since the current $i_4 = i_{B1} + i_{C1} > 0$ A, then D_4 is indeed ON. We also have that the base current for Q is

$$i_B = i_{B1} + i_{C1} - \frac{v_{BE}}{5 \times 10^3} = 44 \times 10^{-6} + 1.76 \times 10^{-3} - \frac{0.8}{5 \times 10^3}$$

$$= 1.64 \text{ mA}$$

If Q is ON, then $v_o = 0.2$ V and D_1' is ON. This, in turn, means that $v_3' = 0.7 + 0.2 = 0.9$ V and Q' is OFF. Thus the standard-load current is

$$i_1' = \frac{5 - 0.9}{1.5 \times 10^3 + 2 \times 10^3} = 1.17 \text{ mA}$$

and the collector current for a fan-out of N is $i_C = i_R + Ni_1'$. Given that $h_{FE} = 40$, for Q indeed to be in saturation, we must have that

$$h_{FE}i_B \geq i_C = i_R + Ni_1'$$

or

$$40(1.64 \times 10^{-3}) \geq \frac{5 - 0.2}{1000} + N(1.17 \times 10^{-3})$$

from which

$$\frac{65.6 - 4.8}{1.17} \geq N \quad \Rightarrow \quad N \leq 51.97$$

Hence the fan-out for the NAND gate given in Fig. 7.28 is 51, and a dramatic increase in fan-out has been achieved.

Drill Exercise 7.13

For the NAND gate given in Fig. 7.28, remove diode D_4. Replace the diode with an *npn* BJT Q_2 as follows: (1) connect the base of Q_2 to the emitter of

Q_1, (2) connect the emitter of Q_2 to the base of Q_3, and (3) connect a 2-kΩ resistor between the collector of Q_2 and the 5-V supply. Find the fan-out of the resulting NAND gate given that each BJT has $h_{FE} = 40$.

ANSWER 118

7.6 Transistor-Transistor Logic (TTL)

A simplified version of an IC-chip *npn* transistor is shown in Fig. 7.29a and its circuit symbol is shown in Fig. 7.29b. Here we see that the device is fabricated on a *p*-type **substrate** (or **body**) in a vertical manner. By embedding more than one *n*-type region into the *p*-type base, we obtain a multiple-emitter *npn* transistor as depicted in Fig. 7.30a for the case of two emitters. The corresponding circuit symbol for such a device is shown in Fig. 7.30b.

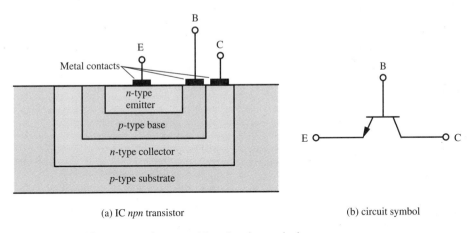

(a) IC *npn* transistor   (b) circuit symbol

Fig. 7.29 An IC *npn* transistor and its circuit symbol.

By using a multiple-emitter transistor to replace the input diodes, we can construct an IC NAND gate that uses only transistors (and resistors). Such a circuit is an example of **transistor-transistor logic,** abbreviated TTL (or T²L). This type of logic is faster (smaller switching times) than DTL. Let us now examine TTL gates in detail.

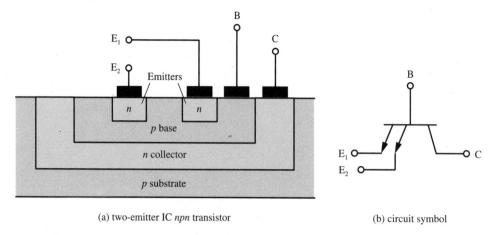

(a) two-emitter IC *npn* transistor

(b) circuit symbol

Fig. 7.30 A two-emitter IC *npn* transistor and its circuit symbol.

Example 7.14

For the TTL NAND gate shown in Fig. 7.31a, if either input is 0.2 V, then the corresponding B-E junction will be forward biased by the 5-V supply voltage. If we assume active region operation for Q_1, then $v_{B1} = 0.7 + 0.2 = 0.9$ V, whereas if we assume operation in the saturation region, then $v_{B1} = 0.8 + 0.2 = 1$ V. In either

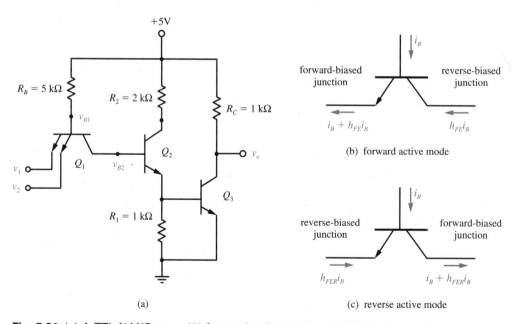

(a)

(b) forward active mode

(c) reverse active mode

Fig. 7.31 (a) A TTL NAND gate, (b) forward active mode, and (c) reverse active mode.

case, v_{B1} is not large enough to forward bias the B-E junctions of both Q_2 and Q_3—to do this, we must have that $v_{BE2} + v_{BE3} > 0.5 + 0.5 = 1$ V. Thus assume that Q_2 and Q_3 are OFF. When Q_2 is OFF, there is a very small reverse current I_{CO2} (on the order of a nanoampere) going from the collector to the base of Q_2, so

$$I_{CO2} = -i_{B2} = i_{C1}$$

Depending on whether Q_1 is in the active or saturation region,

$$i_{B1} = \frac{5 - v_{B1}}{R_B} = \frac{5 - 0.9}{5 \times 10^3} = 0.82 \text{ mA}$$

or

$$i_{B1} = \frac{5 - v_{B1}}{R_B} = \frac{5 - 1}{5 \times 10^3} = 0.8 \text{ mA}$$

respectively. However, if $h_{FE1} \geq i_{C1}/i_{B1}$, then Q_1 will be in saturation. Since $i_{C1} = I_{CO2}$ is very small compared with either value of i_{B1}, then indeed Q_1 is ON as only a very small value of h_{FE1} is required. Thus for the case that, say, $v_1 = 0.2$ V, we have

$$v_{B2} = v_{CE1} + v_1 = 0.2 + 0.2 = 0.4 \text{ V}$$

which confirms that Q_2 and Q_3 are OFF. Hence $v_o = 5$ V.

Now consider the case that $v_1 = v_2 = 5$ V. Since the supply voltage and the input voltages are equal, the B-E junctions of Q_1 are reverse biased. However, the 5-V supply voltage is large enough to forward bias the collector junction of Q_1. But this is a transistor operation that we have not yet considered—B-E junction reverse biased and B-C junction forward biased. Since this is like active-region operation with the emitter and collector reversing roles, it is known as **reverse (or inverted-mode) active-region operation** or **reverse active mode**. Because a transistor is fabricated to have a large h_{FE} for normal or forward active-region operation, when used in the reverse active mode, it has a small (reverse) dc beta h_{FER} (also denoted β_R) which is typically less than unity. Figure 7.31b and 7.31c respectively show a BJT in the (forward) active mode and the reverse active mode.

As indicated in Fig. 7.31c, when Q_1 operates in the reverse active mode, if its base current is i_{B1}, then the total current going into the emitter is $h_{FER}i_{B1}$. Thus the current coming out of the collector of Q_1 is $i_{B1} + h_{FER}i_{B1} = (1 + h_{FER})i_{B1}$ and the current going into the base of Q_2 is

$$i_{B2} = (1 + h_{FER})i_{B1}$$

Since this current (due to the 5-V supply) tends to turn ON Q_2 and Q_3, let us assume that they are in saturation. Then

$$v_{B1} = v_{BC1} + v_{BE2} + v_{BE3} = 0.7 + 0.8 + 0.8 = 2.3 \text{ V}$$

and

$$i_{B1} = \frac{5 - 2.3}{5 \times 10^3} = 0.54 \text{ mA}$$

Even neglecting the reverse beta ($h_{FER} = 0$), we have that at a minimum $i_{B2} = i_{B1} = 0.54$ mA. Since

$$i_{C2} = \frac{5 - (0.2 + 0.8)}{2 \times 10^3} = 2 \text{ mA}$$

then for Q_2 to be ON, we require that

$$h_{FE2} \geq \frac{i_{C2}}{i_{B2}} = \frac{2 \times 10^{-3}}{0.54 \times 10^{-3}} = 3.7$$

Furthermore, since

$$i_{B3} = i_{B2} + i_{C2} - \frac{0.8}{1000} = (0.54 + 2 - 0.8) \times 10^{-3} = 1.74 \text{ mA}$$

and

$$i_{C3} = \frac{5 - 0.2}{1000} = 4.8 \text{ mA}$$

then for Q_3 to be ON, we require that

$$h_{FE3} \geq \frac{i_{C3}}{i_{B3}} = \frac{4.8 \times 10^{-3}}{1.74 \times 10^{-3}} = 2.76$$

Hence when the above conditions for h_{FE2} and h_{FE3} are satisfied, Q_2 and Q_3 will be ON, and $v_o = 0.2$ V.

Hence we have verified that the TTL circuit given in Fig. 7.31a is a NAND gate.

Drill Exercise 7.14

Find the power dissipation of the TTL NAND gate in Fig. 7.31a given that $h_{FE} = 50$ and $h_{FER} = 0.5$.

ANSWER 20.35 mW

In calculating the fan-out for preceding NAND gates, we needed only to consider the case that the output was low. Under this circumstance, current flows from the load into the collector of the output transistor. We only considered this case because when the output was high, the input diodes of the NAND gates in the fan-out were reverse biased, and consequently, there was no current through them. However, for the TTL NAND gate given in Fig. 7.31a, there is an output current when the gate's output is high as well as low. When the output is low, again current flows from the load into the collector of the output transistor. However, when the output is high (Q_3 is OFF), some of the multiple-emitter input transistors in the fan-out may be operating in the reverse active mode. In this case, current must be supplied to the load by the gate. Since Q_3 is OFF, therefore, the output (high) voltage must be less than the supply voltage. After a value for the high output voltage is selected, the fan-out can be calculated by assuming the worst-case situation that all of the input transistors in the load are in the reverse active mode.

The fan-out of a TTL gate such as the one given in Fig. 7.31a can be relatively small because the gate has to supply current to the load when the output is high. This problem can be overcome, however, by using a more sophisticated circuit. We will discuss this subject next.

Totem-Pole Output Stage

Recall that when a transistor goes from saturation to cutoff, there is a significant storage time t_s that is attributable to the storage of minority carriers in the base. For the TTL NAND gate given in Fig. 7.31a, when $v_1 = v_2 = 5$ V, Q_1 is in reverse active mode while Q_2 and Q_3 are ON. Under these conditions,

$$v_{B2} = v_{BE2} + v_{BE3} = 0.8 + 0.8 = 1.6 \text{ V}$$

When an input becomes 0.2 V, then the corresponding B-E junction of Q_1 becomes forward biased, and $v_{B1} = 0.7 + 0.2 = 0.9$ V. This means that the voltage across the B-C junction of Q_1 (because Q_2 is still ON during the storage time) is

$$v_{B1} - v_{B2} = 0.9 - 1.6 = -0.7 < V_\gamma = 0.5 \text{ V}$$

so the B-C junction is reverse biased. Thus Q_1 is in the active region and its collector current is

$$i_{C1} = h_{FE1}i_{B1} = h_{FE1}\left(\frac{5 - 0.9}{5 \times 10^3}\right) = (8.2 \times 10^{-4})h_{FE1} = -i_{B2}$$

For the case that $h_{FE1} = 50$, then $i_{B2} = -41$ mA. This relatively large current quickly removes stored charge from Q_2 and Q_3 and, subsequently, Q_1 will turn ON while Q_2 and Q_3 will turn OFF. Now we can see why, for saturating logic, TTL has the highest speed.

For the TTL NAND gate given in Fig. 7.31a, the output transistor Q_3 has a collector resistance of $R_C = 1$ kΩ. When Q_3 goes from ON to OFF, v_o goes from 0.2 to 5 V. However, the time required for this transition is affected in the following manner.

There is an effective capacitance C_L loading Q_3, which is, in essence, due to the junctions that are loading the gate and the capacitance of the connecting wires. Although this capacitance charges to 5 V with a time constant of R_CC_L, the resulting charge time may be longer than desired. The time constant can be reduced by decreasing R_C, but as a consequence, there is an increase in collector current, and, hence, power dissipation—an undesirable effect. As we will see shortly, by modifying the given NAND gate as shown in Fig. 7.32, the charge time can be reduced.

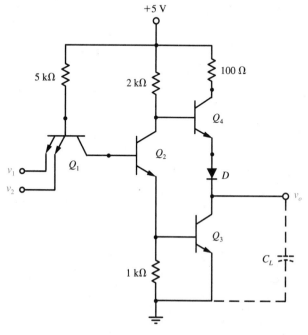

Fig. 7.32 A TTL NAND gate with totem-pole output stage.

The TTL NAND gate shown in Fig. 7.32, which utilizes a **totem-pole** connection of transistors Q_3 and Q_4, has the transfer characteristic shown in Fig. 7.33. This characteristic is obtained by keeping one input (say, v_2) high, and letting the other input (v_1) vary from low to high so that the change in the output from high to low can be observed. For $0 \leq v_1 \leq 0.5$ V, Q_1 is in saturation, Q_2 and Q_3 are cut off, Q_4 and D conduct very small currents, and the output is high. For $0.5 < v_1 \leq 1.2$ V, Q_1 is in saturation, Q_2 is in the active region, Q_3 is cut off, and Q_4 and D conduct very small currents. For $1.2 < v_1 \leq 1.4$ V, Q_1 is in saturation, Q_2, Q_3 and Q_4 are in the active region, and D is ON. For $v_1 > 1.4$ V, Q_1 is in reverse active mode, Q_2 and Q_3 are in saturation, Q_4 and D are OFF, and the output is low.

When both input voltages are high, as discussed earlier, Q_1 is in the reverse active mode while Q_2 and Q_3 are ON. Suppose that the capacitance C_L has already discharged to $v_o = 0.2$ V. Then the collector voltage of Q_2 is

$$v_{C2} = v_{CE2} + v_{BE3} = 0.2 + 0.8 = 1 \text{ V} = v_{B4}$$

and this is the voltage at the base of Q_4. Thus the voltage across the combination of the B-E junction of Q_4 and the diode D is

$$v_{B4} - v_{CE3} = 1 - 0.2 = 0.8 \text{ V}$$

Because this voltage is less than the sum of their cut-in voltages, Q_4 and D are OFF. (If D was not present, then Q_4 would be ON, and its subsequent collector current of 46 mA would result in unnecessary higher power dissipation.)

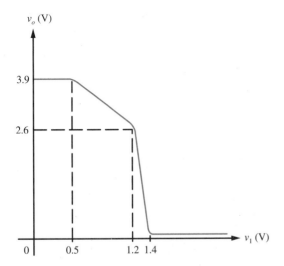

Fig. 7.33 Transfer characteristic for TTL NAND gate.

When either input changes to 0.2 V, as discussed previously, Q_1 goes into saturation while Q_2 and Q_3 go OFF. However, because of the capacitance C_L, the output voltage does not instantaneously rise to the high voltage. While the output voltage is still $v_o = 0.2$ V, since Q_2 is OFF, then $v_{B4} = v_{C2} = 5$ V. This will tend to turn ON Q_4 and D. Assuming this, we have that

$$v_{B4} = v_{BE4} + v_D + v_o = 0.8 + 0.7 + 0.2 = 1.7 \text{ V}$$

Since Q_2 is OFF, then

$$i_{B4} = \frac{5 - v_{C2}}{2000} = \frac{5 - v_{B4}}{2000} = \frac{5 - 1.7}{2000} = 1.65 \text{ mA}$$

and since Q_3 is OFF, then

$$i_{C4} = \frac{5 - v_{C4}}{100} = \frac{5 - (v_{CE4} + v_D + v_o)}{100}$$

$$= \frac{5 - (0.2 + 0.7 + 0.2)}{100} = 39 \text{ mA}$$

Thus, for Q_4 indeed to be in saturation, we require that

$$h_{FE4} \geq \frac{i_{C4}}{i_{B4}} = \frac{39 \times 10^{-3}}{1.65 \times 10^{-3}} = 23.6$$

and the fact that $i_{B4} + i_{C4} = 1.65 \times 10^{-3} + 39 \times 10^{-3} = 40.65$ mA > 0 confirms that D is ON. Since Q_3 is OFF ($i_{C3} = 0$ A), the current $i_{B4} + i_{C4}$ rapidly charges the capacitance C_L. The time constant is RC_L, where R is the sum of the 100-Ω resistance connected in series with the collector of Q_4, the saturation resistance of Q_4 (i.e., the resistance between the collector and emitter of Q_4 when it is ON—typically tens of ohms or less), and the forward resistance of the diode. As v_o increases,

$$i_{C4} = \frac{5 - (v_{CE4} + v_D + v_o)}{100}$$

decreases. When v_o becomes sufficiently large, then $v_{B4} - v_o = 5 - v_o$ will not be large enough to forward bias the B-E junction of Q_4 and D. This occurs when

$$5 - v_o = v_{BE4\text{(cut-in)}} + v_{D\text{(cut-in)}} = 0.5 + 0.6 = 1.1 \text{ V}$$

from which

$$v_o = 5 - 1.1 = 3.9 \text{ V}$$

and this is the high voltage for the NAND gate.

Now suppose that either input is the low voltage 0.2 V and the output voltage is $v_o = 3.9$ V (Q_3 is OFF). If the inputs become $v_1 = v_2 = 3.9$ V, since the 5-V supply voltage is large enough to forward bias the B-C junction of Q_1 and turn ON Q_2 and Q_3, then the voltage at the base of Q_1 is

$$v_{B1} = v_{BC1} + v_{BE2} + v_{BE3} = 0.7 + 0.8 + 0.8 = 2.3 \text{ V}$$

Since $v_{B1} - 3.9 = 2.3 - 3.9 = -1.6 < 0.5$ V, the B-E junctions are reverse biased and Q_1 is in the reverse active mode. As discussed above, under these conditions, D and Q_4 will turn OFF. In the process of turning ON, Q_3 has a large collector current, which results in C_L discharging quickly.

The transistor Q_4 in the TTL NAND gate given in Fig. 7.32 is known as an **active pull-up.** The 1-kΩ resistor in series with the collector of Q_3 in the NAND gate given in Fig. 7.31a is called a **passive pull-up.** The term ''pull-up'' is used because the capacitance C_L is charged up, or its voltage is pulled up, by current going through these elements.

For the NAND gate given in Fig. 7.32, the diode D is connected in series with the emitter of Q_4. If this diode instead is placed in series with the base of Q_4, the operation of the TTL gate will remain unchanged. But why is such a gate referred to as TTL if it contains a diode? It is because such a diode can be realized by the B-E junction of some other transistor—i.e., a diode-connected BJT.

Example 7.15

Let us determine the fan-out for the TTL NAND gate given in Fig. 7.32 for the case that $h_{FE} = 40$, $h_{FER} = 0.1$, and the high voltage is 3.5 V. To do this, consider the circuit shown in Fig. 7.34.

First consider the case that the output is high. A high output is obtained when either or both inputs are low (0.2 V). So suppose that $v_1 = 0.2$ V. Then, Q_1 is in saturation, while Q_2 and Q_3 are OFF. Since the supply voltage is 5 V, the 1.5-V difference between the power supply and the output is large enough to forward bias the B-E junction of Q_4 and D. Therefore, let us assume that Q_4 is in the active mode and D is ON. As a consequence of this, $v_{B4} = v_{BE4} + v_D + v_o = 0.7 + 0.7 + 3.5 = 4.9$ V, and $i_{B4} = (5 - 4.9)/(2 \times 10^3) = 50$ μA. Thus, $i_{C4} = h_{FE}i_{B4} = 40(50 \times 10^{-6}) = 2$ mA, and, $i_{E4} = i_{B4} + i_{C4} = 50 \times 10^{-6} + 2 \times 10^{-3} = 2.05$ mA, and since Q_3 is OFF, this is the output current.

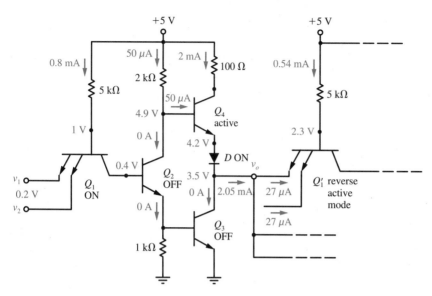

Fig. 7.34 TTL NAND gate with output high.

By KVL, $v_{C4} = -100i_{C4} + 5 = -100(2 \times 10^{-3}) + 5 = 4.8$ V and $v_{E4} = v_D + v_o = 0.7 + 3.5 = 4.2$ V. Thus, $v_{CE4} = v_{C4} - v_{E4} = 4.8 - 4.2 = 0.6 > 0.2$ V. This result along with the fact that $i_{B4} = 50$ μA > 0 confirms that Q_4 is in the active mode, and $i_{E4} > 0$ A confirms that D is ON.

Since the output is high, each input transistor in the load may be in the reverse active mode. When it is, as we have seen before, its base voltage is 2.3 V, its base current is $(5 - 2.3)/(5 \times 10^3) = 0.54$ mA, and the total current going into its two emitters is $h_{FER}i_{B1} = 0.1(0.54 \times 10^{-3}) = 54$ μA, and so 27 μA is the maximum current per gate in the fan-out drawn by the load. Since the output current is 2.05 mA, then for a load with N gates,

$$2.05 \times 10^{-3} \geq N(27 \times 10^{-6}) \quad \Rightarrow \quad N \leq 75.9$$

Therefore, the fan-out is 75 for the case that the output is high.

Next we must consider the case that the output is low (0.2 V). In this case, current goes from the load into the collector of Q_3. We leave the details of this case to the reader to show that the resulting fan-out is 89 (Drill Exercise 7.15).

Considering the results of the two possibilities—either the output is high or it is low—we deduce that the fan-out for the given TTL NAND gate is the smaller of the results for the two cases, that is, 75.

Drill Exercise 7.15

Find the fan-out for the TTL NAND gate given in Fig. 7.32 for the case that $h_{FE} = 40$, $h_{FER} = 0.1$, and the output voltage is low (0.2 V).

ANSWER 89

Schottky TTL

When a metal is connected to a semiconductor, the resulting contact can be either **ohmic (nonrectifying)** or **rectifying.** In previous discussions, the implicit assumption was that a metal-semiconductor junction was ohmic. However, the formation of such a contact requires an appropriate fabrication process. As an example, for attaching aluminum to n-type silicon, the region of the silicon adjacent to the aluminum is made highly conductive by doping it heavily—denoted as n^+. This results in an ohmic contact.

Aluminum in contact with n-type silicon that is not heavily doped acts as a p-type impurity. As a consequence of this fact, such a metal-semiconductor contact is rectifying and behaves in a manner similar to that of an ordinary pn junction. This type of junction, however, is known as a **Schottky**[3] (or **Schottky-barrier) diode.** The circuit symbol for this device is shown in Fig. 7.35. The i-v characteristic for a Schottky diode is like that of an ordinary pn junction diode. A major distinction, however, is that the cut-in voltage for a Schottky diode is about 0.3 V as compared with 0.6 V for a silicon junction diode. Furthermore, a typical forward-bias voltage for the former is about 0.4 V, whereas it is about 0.7 V for the latter. Typical Schottky-diode parameters are $\eta = 1$ and $I_S = 1$ nA.

For an aluminum n-type silicon Schottky diode, a current in the forward direction is essentially due to the flow of electrons from the n-type silicon to the metal. In essence, therefore, a Schottky diode is a majority-carrier device and there is no storage of minority carriers as takes place with an ordinary junction diode. Conse-

i

$+$ v $-$

Fig. 7.35 Schottky diode.

---

[3]Named for the Swiss physicist Walter Schottky.

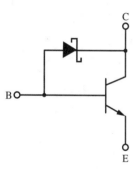

Fig. 7.36 A BJT with a Schottky-diode clamp.

quently, the storage time of a Schottky diode is negligible; hence Schottky diodes have higher switching speeds than do junction diodes.

Consider the situation that a Schottky diode is connected between the base and the collector of a BJT as shown in Fig. 7.36. Suppose that an attempt is made to saturate the transistor. As the base current increases, the base-emitter voltage increases and the collector-emitter voltage decreases. Eventually, the Schottky diode will be forward biased and the base-collector voltage will be ''clamped'' at about $v_{BC} = 0.4$ V. Assuming that the BJT is in saturation, we get

$$v_{CE} = -v_{BC} + v_{BE} = -0.4 + 0.8 = 0.4 > 0.2 \text{ V}$$

which is a contradiction to our assumption, and, therefore, the transistor is not in saturation. Hence the Schottky-diode clamp prevents the BJT from going into saturation.

A BJT, along with a Schottky-diode clamp, can be fabricated as shown in Fig. 7.37a. Such a device is known as a **Schottky** (or **Schottky-clamped**) **transistor,** and its circuit symbol is given in Fig. 7.37b.

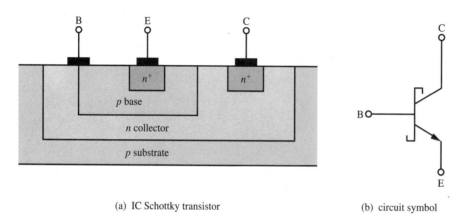

(a) IC Schottky transistor

(b) circuit symbol

Fig. 7.37 (a) An IC Schottky transistor, and (b) its circuit symbol.

Example 7.16

Fig. 7.38a shows a Schottky BJT inverter, and Fig. 7.38b shows the same inverter with the ordinary BJT Q and Schottky-diode clamp explicitly displayed. Suppose that $R_B = 20$ kΩ, $R_C = 1$ kΩ, $V_{CC} = 5$ V, and $h_{FE} = 50$ for Q.

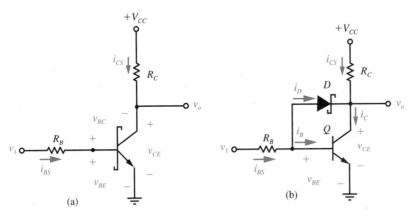

Fig. 7.38 (a) Schottky inverter, and (b) equivalent circuit.

Consider the case that $v_1 = 5$ V is the high voltage. Since Q cannot go into saturation, let us assume that it is in the active region and that the Schottky diode D is ON. Under these assumptions, $v_{BE} = 0.7$ V, $v_{BC} = 0.4$ V, and by KVL, $v_{CE} = -v_{BC} + v_{BE} = -0.4 + 0.7 = 0.3$ V, and this is the low voltage. Furthermore,

$$i_{BS} = \frac{v_1 - v_{BE}}{R_B} = \frac{5 - 0.7}{20 \times 10^3} = 215 \ \mu\text{A} \qquad \text{and}$$

$$i_{CS} = \frac{5 - v_{CE}}{R_C} = \frac{5 - 0.3}{1 \times 10^3} = 4.7 \ \text{mA}$$

By KCL,

$$i_{BS} = i_B + i_D = 215 \ \mu\text{A} \quad \text{and} \quad i_{CS} = i_C - i_D = 4.7 \ \text{mA}$$

Since Q is assumed to be in the active region, $i_C = h_{FE}i_B = 50i_B$. Solving these equations for i_B, i_C, and i_D, we get that $i_B = 96.4 \ \mu\text{A}$, $i_C = 4.82$ mA, and $i_D = 119 \ \mu\text{A}$.

Since $i_D > 0$ A, then D is confirmed to be ON, and since $i_B > 0$ A and $v_{CE} = 0.3 > 0.2$ V, then Q is confirmed to be in the active mode.

Drill Exercise 7.16

For the circuit given in Fig. 7.38, find i_B, i_C, and i_D for the case that $R_B = 20$ kΩ, $R_C = 500$ Ω, $V_{CC} = 5$ V, $h_{FE} = 50$, and $v_1 = 5$ V.

ANSWER 189 μA, 9.45 mA, 26 μA

A Schottky BJT is said to be ON when the diode between the base and the collector of Q is forward biased (i.e., $i_D > 0$ A) and Q is in the active region. Under this circumstance, $v_{BE} = 0.7$ V, $v_{BC} = 0.4$ V, and $v_{CE} = 0.3$ V. With reference to Fig. 7.38, if the Schottky BJT is ON, then

$$i_{BS} = i_B + i_D \quad \Rightarrow \quad h_{FE}i_{BS} = h_{FE}i_B + h_{FE}i_D \tag{7.14}$$

$$i_{CS} = i_C - i_D \quad \Rightarrow \quad i_{CS} = h_{FE}i_B - i_D \tag{7.15}$$

Subtracting Eq. 7.15 from Eq. 7.14 yields

$$h_{FE}i_{BS} - i_{CS} = h_{FE}i_D + i_D = (h_{FE} + 1)i_D$$

from which

$$i_D = \frac{h_{FE}i_{BS} - i_{CS}}{h_{FE} + 1} > 0 \quad \Rightarrow \quad h_{FE}i_{BS} > i_{CS} \quad \Rightarrow \quad h_{FE} > \frac{i_{CS}}{i_{BS}}$$

and this is the range for h_{FE} for which a Schottky BJT is ON.

Example 7.17

For the Schottky BJT circuit given in Fig. 7.38, suppose that $R_B = 20$ kΩ, $R_C = 1$ kΩ, and $V_{CC} = 5$ V. Assume that when $v_1 = 5$ V, the Schottky BJT is ON. Then

$$i_{BS} = \frac{v_1 - v_{BE}}{R_B} = \frac{5 - 0.7}{20 \times 10^3} = 215 \ \mu\text{A} \quad \text{and}$$

$$i_{CS} = \frac{5 - v_{CE}}{R_C} = \frac{5 - 0.3}{1 \times 10^3} = 4.7 \ \text{mA}$$

Therefore, the Schottky BJT will be ON when

$$h_{FE} > \frac{i_{CS}}{i_{BS}} = \frac{4.7 \times 10^{-3}}{215 \times 10^{-6}} = 21.9 \approx 22$$

Drill Exercise 7.17

For the Schottky BJT circuit given in Fig. 7.38, suppose that $R_B = 20$ kΩ, $R_C = 1$ kΩ, $V_{CC} = 5$ V, and $h_{FE} = 100$. What is the minimum value of v_1 for which the Schottky BJT will be ON?

ANSWER 1.64 V

Since a Schottky transistor cannot go into saturation, its storage time is reduced significantly (to about 50 ps) and, therefore, the turn-off time is shortened consid-

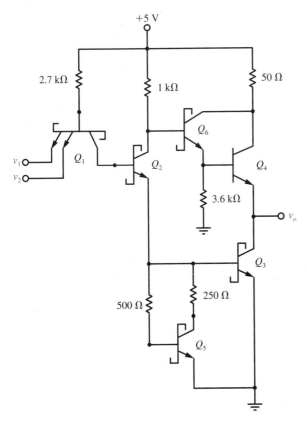

Fig. 7.39 Schottky TTL NAND gate.

erably. A consequence of this is that Schottky TTL has a higher switching speed than does standard TTL. A Schottky TTL version of the TTL NAND gate given in Fig. 7.32 is shown in Fig. 7.39. As mentioned previously, the diode D in Fig. 7.32 can be placed in series with the base of Q_4. For the NAND gate in Fig. 7.39, the diode is replaced by the Schottky transistor Q_6. Furthermore, since $v_{CE4} = v_{BE4} + v_{CE6}$, transistor Q_4 in Fig. 7.39 cannot saturate, so an ordinary BJT can be used.

Although the speed of the Schottky TTL NAND gate shown in Fig. 7.39 is about three times as fast as the TTL NAND gate given in Fig. 7.32, the former dissipates about twice as much power as does the latter. Where power dissipation has priority over speed, **low-power Schottky TTL** can be employed. For such a NAND gate, shown in Fig. 7.40, the resistance values are significantly larger. Although the speed of such a gate is about the same as for standard TTL, the power dissipation for the gate in Fig. 7.40 is approximately 20 percent of that for the gate in Fig. 7.32.

Despite the fact that Schottky TTL is fast logic, another type of nonsaturating logic (ECL) is even faster. This topic will be discussed next.

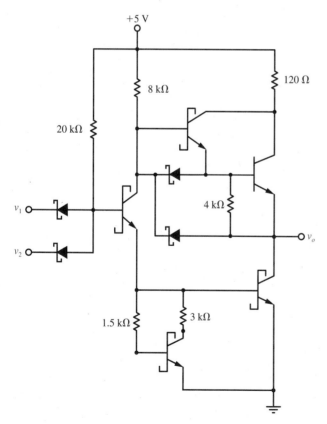

Fig. 7.40 Low-power Schottky TTL NAND gate.

7.7 Other IC Logic Families

Consider the emitter-coupled BJT circuit shown in Fig. 7.41, where the voltage applied to the base of Q_2 is some reference voltage V_R. If v_1 has a value in the interval $-V_{EE} + 0.7 < v_1 < V_{CC} + 0.5$ V, then v_1 will tend to forward bias the B-E junction and reverse bias the B-C junction of Q_1, that is, put Q_1 in the active region.

Suppose that the value of v_1 is such that Q_1 is in the active region and v_1 exceeds V_R by 0.2 V, that is, $v_1 > V_R + 0.2$. From this inequality,

$$V_R - v_1 < -0.2 \text{ V}$$

Since

$$v_{BE2} = V_R - v_E = V_R - (-0.7 + v_1) = 0.7 + V_R - v_1$$

from the above inequality,

$$v_{BE2} = 0.7 + V_R - v_1 < 0.7 - 0.2 = 0.5 \text{ V}$$

and this means that Q_2 is cut off. On the other hand, if V_R has a value in the range $-v_{EE} + 0.7 < V_R < V_{CC} + 0.5$ V, then V_R will tend to put Q_2 in the active region.

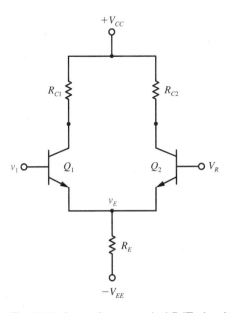

Fig. 7.41 An emitter-coupled BJT circuit.

Furthermore, if V_R exceeds v_1 by 0.2 V, then, using reasoning as above, when Q_2 is in the active region, Q_1 will be cut off.

Emitter-Coupled Logic (ECL)

We now describe another type of nonsaturating logic, called **emitter-coupled logic** (ECL) or **current-mode logic** (CML), which is based on the circuit shown in Fig. 7.41. With ECL, as described above, when one transistor is cut off, the other transistor is not in saturation, but is in the active region. Because of this kind of operation, the problem of excess minority-carrier charge storage in the base of a transistor in saturation is eliminated. The consequence of this is that ECL is the fastest type of logic.

Let us use the basic circuit given in Fig. 7.41 to try to construct an OR gate. We will do this by adding another transistor, as shown in Fig. 7.42. For the case that either input voltage is high, the corresponding transistor should be in the active region and Q_3 will be OFF. Then, $v_o = -500i_{C3} + V_{CC} = 0 + 0 = 0$ V should be the high voltage. Conversely, if v_1 and v_2 are low, then Q_3 should be in the active region whereas Q_1 and Q_2 will be OFF. Then the output $v_o = -500i_{C3}$ is the low voltage. Such a behavior would result in OR-gate operation (see Table 7.2). Note that the voltage at the collectors of Q_1 and Q_2 is the complement of v_o—that is why it is denoted $\bar{v}_o$. Therefore, if v_o is the output of an OR gate, then $\bar{v}_o$ would be the output of a NOR gate. Let us now determine whether or not the circuit given in Fig. 7.42 is indeed an OR/NOR gate.

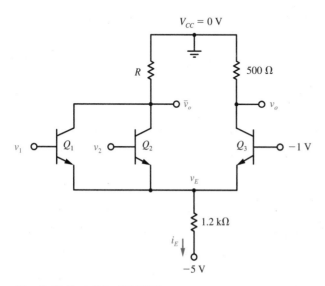

Fig. 7.42 Possible OR/NOR gate.

Table 7.2 OR/NOR Truth Table

v_1	v_2	OR	NOR
0	0	0	1
0	1	1	0
1	0	1	0
1	1	1	0

If both input voltages are low, assume that Q_1 and Q_2 are OFF and Q_3 is in the active region. Then

$$v_E = -v_{BE3} - 1 = -1.7 \text{ V}$$

and

$$i_E = \frac{v_E - (-5)}{1.2 \times 10^3} = \frac{-1.7 + 5}{1.2 \times 10^3} = 2.75 \text{ mA}$$

Since

$$i_{E3} = i_{C3} + i_{B3} = h_{FE3}i_{B3} + i_{B3} \approx h_{FE3}i_{B3} = i_{C3}$$

then

$$v_o = -500i_{C3} \approx -500(2.75 \times 10^{-3}) = -1.375 \text{ V}$$

and this is the low voltage. Since

$$v_{BE1} = v_{BE2} = v_1 - v_E = -1.375 - (-1.7) = 0.325 < 0.5 \text{ V}$$

then Q_1 and Q_2 are indeed cut off. Furthermore,

$$v_{BC3} = -1 - (-1.375) = 0.375 < 0.5 \text{ V}$$

so that the B-C junction of Q_3 is reverse biased, and Q_3 indeed is in the active region.

Now consider the case that at least one input voltage, say v_1, is high. Assume that Q_1 is in the active region and Q_3 is cut off. Under these circumstances, $v_o = -500i_{C3} + 0 = 0$ V is the high voltage. Also,

$$v_E = -v_{BE1} + v_1 = -0.7 + 0 = -0.7 \text{ V}$$

Since

$$v_{BE3} = -1 - v_E = -1 + 0.7 = -0.3 < 0.5 \text{ V}$$

then Q_3 is indeed cut off. Furthermore,

$$i_E = \frac{v_E - (-5)}{1.2 \times 10^3} = \frac{-0.7 + 5}{1.2 \times 10^3} = 3.58 \text{ mA}$$

Since $i_{C1} + i_{C2} \gg i_{B1} + i_{B2}$, then $i_{C1} + i_{C2} \approx i_E$. Because we wish to have $\bar{v}_o = -1.375$ V (the low voltage), then

$$\bar{v}_o = -R(i_{C1} + i_{C2}) \approx -Ri_E$$

from which

$$R \approx \frac{-\bar{v}_o}{i_E} = \frac{1.375}{3.58 \times 10^{-3}} = 384 \text{ } \Omega$$

But note that

$$v_{BC1} = v_1 - \bar{v}_o = 0 - (-1.375) = 1.375 > 0.5 \text{ V}$$

This means that the B-C junction of Q_1 is not reverse biased, and, therefore, contrary to our assumption, Q_1 is not in the active region. Because of this kind of problem,

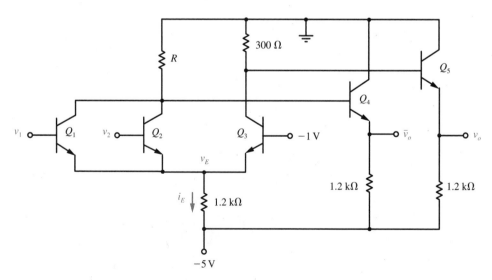

Fig. 7.43 An ECL OR/NOR gate.

we will change the value of the collector resistance for Q_3 and modify the given circuit by adding two transistors as shown in Fig. 7.43.

Example 7.18

Let us demonstrate that the circuit given in Fig. 7.43 is indeed an OR/NOR gate.

Suppose that both input voltages are low, and assume that Q_1 and Q_2 are OFF and Q_3 is in the active region. As was the case for the preceding circuit, $v_E = -1.7$ V and $i_E = 2.75$ mA. Also, $i_{C3} \approx i_E$. Ignoring i_{B5}, we have

$$v_{C3} = -300 i_{C3} \approx -300 i_E = -300(2.75 \times 10^{-3}) = -0.825 \text{ V}$$

This voltage is large enough (relative to -5 V) to forward bias the B-E junction of Q_5. Since

$$v_{BC5} = v_{C3} = -0.825 < 0.5 \text{ V}$$

then the B-C junction of Q_5 is reverse biased. Thus Q_5 is in the active region, and, therefore,

$$v_o = -v_{BE5} + v_{C3} = -0.7 - 0.825 = \textbf{-1.525 V is the low voltage}$$

For active-region operation, typically $i_{C5} \gg i_{B5}$, and, therefore,

$$v_{C5} \approx i_{B5} + i_{C5} = \frac{v_o - (-5)}{1.2 \times 10^3} = \frac{-1.525 + 5}{1.2 \times 10^3} = 2.9 \text{ mA}$$

Since $i_{C5} = 2.9$ mA $\gg i_{B5}$, then also $i_E = 2.75$ mA $\gg i_{B5}$, and ignoring i_{B5} in the calculation of v_{C3} above is a reasonable thing to do. Since

$$v_{BE1} = v_{BE2} = v_1 - v_E = -1.525 - (-1.7) = 0.175 < 0.5 \text{ V}$$

then Q_1 and Q_2 are indeed OFF. Also, since

$$v_{BC3} = -1 - v_{C3} = -1 - (0.825) = -0.175 < 0.5 \text{ V}$$

then the B-C junction of Q_3 is reverse biased, and Q_3 indeed is in the active region.

Now consider the case that at least one input voltage, say v_1, is high. Assume that Q_1 is in the active region and Q_3 is cut off. Ignoring i_{B5}, we have that

$$v_{C3} = -300 i_{C3} = 0 \text{ V}$$

This voltage is large enough (relative to -5 V) to forward bias the B-E junction of Q_5. Since

$$v_{BC5} = v_{C3} = 0 < 0.5 \text{ V}$$

then the B-C junction of Q_5 is reverse biased. Thus Q_5 is in the active region, and, therefore,

$$v_o = v_{BE5} + v_{C3} = -0.7 + 0 = \textbf{-0.7 V is the high voltage}$$

Note that

$$i_{C5} \approx i_{B5} + i_{C5} = \frac{v_o - (-5)}{1.2 \times 10^3} = \frac{-0.7 + 5}{1.2 \times 10^3} = 3.58 \text{ mA} \gg i_{B5}$$

For a typical value such as $h_{FE5} = 50$,

$$i_{B5} \geq \frac{i_{C5}}{h_{FE5}} = \frac{3.58 \times 10^{-3}}{50} = 71.6 \text{ μA}$$

So even if we did not ignore i_{B5} in calculating v_{C3}, we would obtain $v_{C3} = 300(i_{C3} + i_{B5}) = -300(71.6 \times 10^{-6}) = -0.02 \approx 0$ V. Thus ignoring i_{B5} is justified. We also have that

$$v_E = -v_{BE1} + v_1 = -0.7 - 0.7 = -1.4 \text{ V}$$

Since

$$v_{BE3} = -1 - v_E = -1 + 1.4 = 0.4 < 0.5 \text{ V}$$

then Q_3 is indeed cut off. Furthermore,

$$i_E = \frac{v_E - (-5)}{1.2 \times 10^3} = \frac{-1.4 + 5}{1.2 \times 10^3} = 3 \text{ mA}$$

Previously, we determined that when Q_3 was in the active region, we had $v_{C3} = -0.825$ V. Therefore, now that Q_1 is in the active region, we wish to have $v_{C1} = -0.825$ V. Thus ignoring base currents

$$v_{C1} = -Ri_E$$

from which

$$R = -\frac{v_{C1}}{i_E} = \frac{0.825}{3 \times 10^{-3}} = 275 \ \Omega$$

Furthermore, since

$$v_{BC1} = v_1 - v_{C1} = -0.7 - (-0.825) = 0.125 < 0.5 \ \text{V}$$

the B-C junction of Q_1 is reverse biased, and Q_1 is indeed in the active region. Note too that Q_4, like Q_5, is always in the active region.

This concludes our demonstration that, for the ECL gate given in Fig. 7.43, v_o is the output of an OR gate, whereas $\bar{v}_o$ is the output of a NOR gate.

Drill Exercise 7.18

For the ECL OR/NOR gate given in Fig. 7.43, replace the 300-Ω resistor with a 400-Ω resistor, and change the reference voltage from -1 V to -1.3 V. Ignore base currents and determine the gate's low voltage, high voltage, and the value of the resistor R.

ANSWER -1.7 V, -0.7 V, 333 Ω

Resistor-Transistor Logic (RTL)

Another realization of a NOR gate is the **resistor-transistor logic** (RTL) circuit shown in Fig. 7.44.

For the case that $v_1 = v_2 = 0.2$ V, both Q_1 and Q_2 must be OFF. Thus, $i_{C1} = i_{C2} = 0$ A and $i_R = i_{C1} + i_{C2} = 0$ A, so

$$v_o = -Ri_R + 3 = 3 \ \text{V}$$

If $v_1 = 0.2$ V and $v_2 = 3$ V, then Q_1 is OFF. Assume that Q_2 is ON. Then

$$i_{B2} = \frac{v_2 - v_{BE2}}{R_B} = \frac{3 - 0.8}{500} = 4.4 \ \text{mA}$$

and

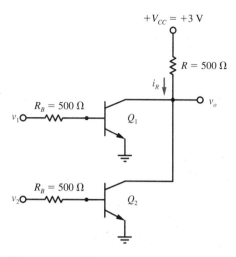

Fig. 7.44 An RTL NOR gate.

$$i_{C2} = i_R = \frac{3 - 0.2}{500} = 5.6 \text{ mA}$$

For Q_2 to be in saturation, therefore, we must have that

$$h_{FE2} \geq \frac{i_{C2}}{i_{B2}} = \frac{5.6 \times 10^{-3}}{4.4 \times 10^{-3}} = 1.27$$

Similarly, when $v_1 = 3$ V and $v_2 = 0.2$ V, for Q_1 to be ON and Q_2 to be OFF, we require that $h_{FE1} \geq 1.27$.

Finally, for the case that $v_1 = v_2 = 3$ V, assume that Q_1 and Q_2 are ON. Under this circumstance, as above, $i_{B1} = i_{B2} = 4.4$ mA. Furthermore, if Q_1 and Q_2 are identical, then, as above, $i_R = 5.6$ mA, and $i_{C1} = i_{C2} = i_R/2 = 2.8$ mA. Thus Q_1 and Q_2 are in saturation when

$$h_{FE1} = h_{FE2} \geq \frac{i_{C1}}{i_{B1}} = \frac{2.8 \times 10^{-3}}{4.4 \times 10^{-3}} = 0.64$$

Even if Q_1 and Q_2 are not identical, the maximum collector current is $i_R = 5.6$ mA. Thus to guarantee that Q_1 and Q_2 are ON, we should have for Q_1 and Q_2 that

$$h_{FE} \geq \frac{5.6 \times 10^{-3}}{4.4 \times 10^{-3}} = 1.27$$

Hence we see that the RTL circuit given in Fig. 7.44 is a NOR gate.

Now consider the situation in which the RTL NOR gate given in Fig. 7.44 is loaded by 10 identical gates. When either v_1 or v_2 is high, $v_o = 0.2$ V and the 10 load transistors Q'_1 through Q'_{10} will be OFF. However, when $v_1 = v_2 = 0.2$ V, both Q_1 and Q_2 will be OFF and v_o will be high. Consequently, Q'_1 through Q'_{10} are to be ON. Thus, by KCL,

$$i_R = i_{C1} + i_{C2} + i_{B1} + i_{B2} + \cdots + i_{B10}$$

or

$$\frac{3 - v_o}{500} = 0 + 0 + \frac{v_o - 0.8}{500} + \frac{v_o - 0.8}{500} \cdots + \frac{v_o - 0.8}{500}$$

$$= 10(\frac{v_o - 0.8}{500})$$

from which

$$3 - v_o = 10v_o - 8 \quad \Rightarrow \quad v_o = 1 \text{ V}$$

Thus the high voltage is 1 V. Furthermore, for each transistor in saturation

$$i_B = \frac{v_o - v_{BE}}{R_B} = \frac{1 - 0.8}{500} = 0.4 \text{ mA}$$

and

$$i_C = \frac{3 - 0.2}{500} = 5.6 \text{ mA}$$

Hence we require for each transistor that

$$h_{FE} \geq \frac{i_C}{i_B} = \frac{5.6 \times 10^{-3}}{0.4 \times 10^{-3}} = 14$$

Direct-Coupled Transistor Logic (DCTL)

Since TTL is superior to RTL in switching speed and fan-out, the former is the overwhelming choice in practice and the latter sees little application. However, RTL can be modified to increase fan-out. This is done by making the base resistance

$R_B = 0 \; \Omega$. What results is known as **direct-coupled transistor logic** (DCTL). An example of this type of logic is illustrated by the DCTL NOR gate and its fan-out shown in Fig. 7.45.

For the case that $v_1 = v_2 = 0.2$ V (the low voltage), Q_1 and Q_2 are OFF. Under this circumstance, the output voltage v_o will be the high voltage. This high voltage, then, will turn ON the fan-out transistors Q_1' through Q_N'. But, because of the direct coupling, we have that the high voltage is $v_o = 0.8$ V. Thus we see that the high voltage is independent of V_{CC} (when $V_{CC} > 0.8$ V).

For the case that an input is high, say $v_1 = 0.8$ V, Q_1 is ON and $v_o = 0.2$ V. Under this condition, the fan-out transistors will be OFF. We have now confirmed NOR-gate operation.

The advantages of DCTL are that low-voltage supplies (as low as $V_{CC} = 1.5$ V) and transistors with low breakdown voltages can be used, and the power dissipation is low. Unfortunately, DCTL has numerous disadvantages. To begin with, since $R_B = 0 \; \Omega$, there are large base currents, and hence, large amounts of stored charge in the bases of the transistors—this means a reduction in switching speed. Since the difference between the high and low voltages of DCTL is only 0.6 V, the noise margins are low, and such logic is susceptible to noise. Another problem can arise when the fan-out transistors do not have identical characteristics. If one fan-out transistor has a lower value of base-emitter saturation voltage $v_{BE(\text{sat})}$ than another, the former will draw more base current than the latter, and there may be insufficient current for the latter to go into saturation. In such a situation, we say that the former transistor **hogs** the current. Needless to say, current hogging can result in improper gate operation. Although there are other disadvantages, these are usually sufficient to preclude the use of DCTL that employ BJTs. However, as we see in the next

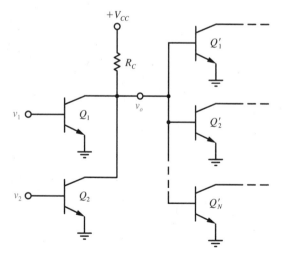

Fig. 7.45 A DCTL NOR gate and its fan-out.

chapter, this type of logic sees important applications with field-effect transistors (FETs). However, we will now see how BJT DCTL evolved into a practical BJT logic family.

Integrated-Injection Logic (I^2L)

The problem of current hogging in DCTL was eliminated with the development of another type of logic known as **integrated-injection logic** (I^2L or IIL). To see how I^2L can evolve from DCTL, let us first consider the DCTL NOR gate and its two-transistor load shown in Fig. 7.46 (compare this with Fig. 7.45). This circuit is to be fabricated on a single IC chip.

For the given circuit, all four BJTs are *npn* transistors. But, since the bases of Q_3 and Q_4 are connected together, they can be merged by using the same *p*-type region for both transistors. Similarly, both emitters of Q_3 and Q_4 can be merged into a single *n*-type region. In essence, what results is a two-collector transistor. (Such a device can be fabricated in a manner similar to the two-emitter transistor illustrated in Fig. 7.30.) Of course, *m*-collector (*m* greater than two) transistors can also be fabricated. Because of the space-saving construction of such devices, a typical value for h_{FE} (total collector current divided by the base current) is 5, whereas 50 is more typical for a regular BJT.

We have now seen how we can reduce the space requirements of the circuit given in Fig. 7.46 by merging regions for the load transistors. Because IC resistors require

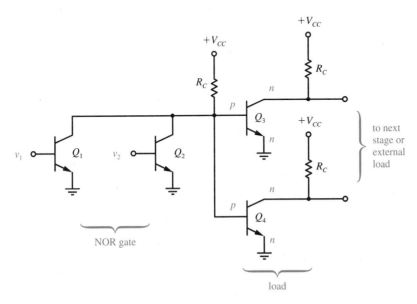

Fig. 7.46 A DCTL NOR gate and its load.

much more space than do IC transistors, let us now concentrate on the collector resistor R_C. Either of the driver transistors, Q_1 or Q_2, will saturate when its respective input is high and, in either case, both load transistors, Q_3, and Q_4, will be cut off. However, when v_1 and v_2 are low, Q_1 and Q_2 will be OFF. Consequently, Q_3 and Q_4 will be ON. The current required to turn ON Q_3 and Q_4 goes through R_C. Thus we can eliminate R_C if we replace it with some other supplier of current. For this reason, let us replace R_C by the *pnp* transistor shown in Fig. 7.47. Such a transistor is called a **current injector.** Only the *pnp* transistor shown in Fig. 7.47 is to replace the resistor R_C. The resistor R and the supply voltage V_{CC} are external to the chip. In other words, even though there originally may be many resistors labeled R_C, each one of them is replaced by a *pnp* transistor, and all such transistors use the same external resistor R and supply voltage V_{CC}. This supply voltage will forward bias the emitter junction of the pnp transistor, producing an emitter current of

$$i_E = \frac{V_{CC} - v_{EB}}{R}$$

When a driver BJT is ON, the collector of the associated *pnp* current injector is at 0.2 V. Hence the C-B junction is reverse biased, so the *pnp* BJT is in the active mode. The resulting current directed out of the collector is αi_E, where $i_E = (V_{CC} - 0.7)/R$. When a load BJT is ON, the collector of the current injector is at 0.8 V. Hence, the C-B junction is forward biased, so the *pnp* BJT is in saturation, and the current directed out of the collector is roughly $(V_{CC} - 0.8)/R$. Therefore, when connected to a load, the resulting current coming out of the collector of the current injector is approximately some constant I.

Let us now redraw the circuit given in Fig. 7.46, replacing Q_3 and Q_4 with a multiple (two)-collector transistor and the resistors labeled R_C with *pnp* current injectors. The resulting circuit is shown in Fig. 7.48. But note that the collector of the leftmost *pnp* transistor and the base of the multiple-collector transistor are both

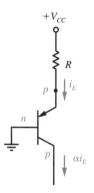

Fig. 7.47 Current injector.

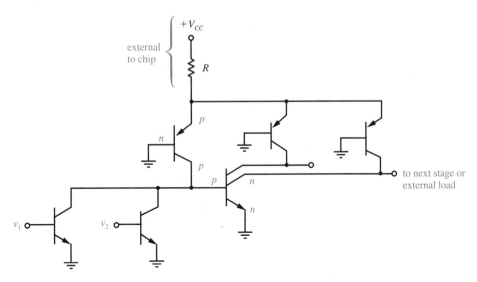

Fig. 7.48 Circuit obtained after substitution of multiple-collector transistor and current injectors.

p-type regions and, hence can be merged. In addition, the base of the *pnp* transistor and the emitter of the multiple-collector transistor can be merged into the same n-type region. Consequently, even more space can be saved. It is because of the merging of regions that I^2L is sometimes referred to as **merged-transistor logic** (MTL). Because of its small space requirements, I^2L has application in LSI and VLSI circuits such as calculators, digital wristwatches, and microprocessors.

The basic I^2L unit (building block) consists of a multiple-collector *npn* transistor to whose base is connected a *pnp* current-injector transistor as shown in Fig. 7.49a

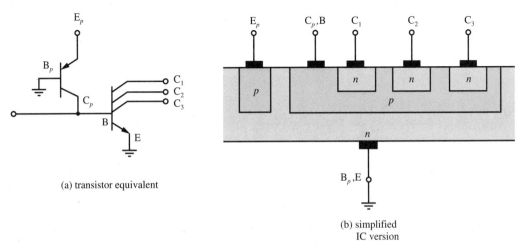

(a) transistor equivalent

(b) simplified
IC version

Fig. 7.49 Basic I^2L unit.

for the case of three collectors. A simplified IC version of this building block is shown in Fig. 7.49*b*. For the sake of simplicity, however, as shown in Fig. 7.50, the circuit symbol of this basic I²L gate uses an independent current source *I* to represent the current injector.

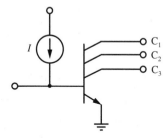

Fig. 7.50 Circuit symbol for the basic I²L gate.

To form an I²L inverter, we require only single-collector *npn* transistors. Specifically, Fig. 7.51 shows an I²L inverter and its load. When the input voltage is $v_1 = 0.2$ V, then Q_1 is OFF—and the leftmost injected current *I* goes through the source of v_1 (a transistor in saturation). Since Q_1 is OFF, the next injected current *I* turns ON Q_2. Thus $v_o = v_{BE2} = 0.8$ V. Conversely, when the source of v_1 is a transistor in cutoff, then assume that the injected current $I = i_{B1}$ turns ON Q_1 and the input voltage is $v_1 = v_{BE1} = 0.8$ V. When Q_1 is in saturation, $v_o = 0.2$ V. This implies that Q_2 is OFF, so $i_{C1} = I$. Therefore, for Q_1 to be in saturation, we require that

$$h_{FE1} \geq \frac{i_C}{i_B} = \frac{I}{I} = 1$$

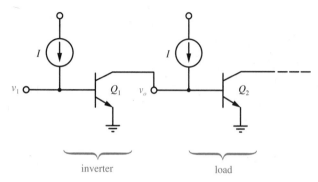

Fig. 7.51 An I²L inverter.

When the output voltage of an inverter changes from low voltage to high, the input capacitance of the following stage must charge up correspondingly. Conversely, this capacitance must discharge when the inverter changes from high voltage to low. This charging and discharging, which determine the switching times (logic speed), depend on the injection current I. Increasing I results in faster logic, but at a price— more power is required.

Example 7.19

Consider the I^2L circuit shown in Fig. 7.52. For the case that $v_1 = v_2 = 0.2$ V, then Q_1 and Q_2 are OFF. Since $i_{C1} = i_{C2} = 0$ A, then $i_{B3} = I$ and Q_3 is ON. Thus, $v_o = v_{BE3} = 0.8$ V.

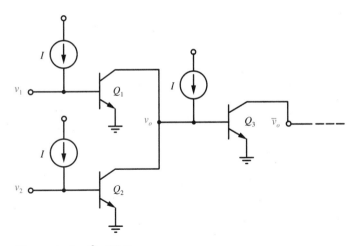

Fig. 7.52 An I^2L NOR gate.

When $v_1 = 0.2$ V and $v_2 = 0.8$ V, then Q_1 is OFF and Q_2 is ON. Thus $v_o = 0.2$ V (and Q_3 is OFF). Due to the symmetry of the circuit, when $v_1 = 0.8$ V and $v_2 = 0.2$ V, then Q_1 is ON and Q_2 is OFF. Again, $v_o = 0.2$ V (and Q_3 is OFF). Finally, when $v_1 = v_2 = 0.8$ V, then Q_1 and Q_2 are ON and $v_o = 0.2$ V (and Q_3 is OFF).

From these results, we now see that the I^2L circuit given in Fig. 7.52 is a NOR gate when v_o is taken as the output. Since the following stage (Q_3) is an inverter, if $\bar{v}_o$ is taken as the output, an OR gate is obtained.

Drill Exercise 7.19

Consider the I^2L circuit shown in Fig. DE7.19. Determine the type of logic gate this circuit is when the output is taken to be (a) v_o and (b) $\bar{v}_o$.

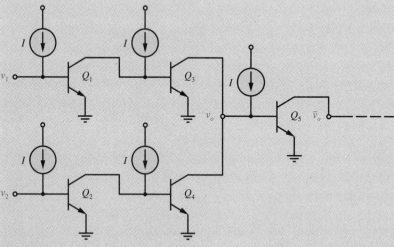

Fig. DE7.19

ANSWER (a) AND gate, (b) NAND gate

S U M M A R Y

1. Sandwiching a narrow section of n-type semiconductor between two sections of p-type semiconductor yields a *pnp* bipolar junction transistor (BJT). If the n- and p-type materials are interchanged, an *npn* BJT results.

2. The terminals of a BJT are the base (B), collector (C), and emitter (E).

3. The common-base (CB) large-signal current gain α of a BJT typically is in the range from 0.95 to 0.998. The common-emitter (CE) large-signal current gain β of a BJT often lies between 50 and 200.

4. A BJT is in the active region or active mode when its B-E junction is forward biased and its B-C junction is reverse biased. In this case, $i_C \approx \alpha i_E$ and $i_C \approx \beta i_B$.

5. A BJT is cut off (OFF) when its B-E and B-C junctions are reverse biased. In this case, $i_B = i_C = i_E = 0$ A.

6. A BJT is in saturation (ON) when its B-E and B-C junctions are forward biased.

7. A BJT is in the reverse (or inverted) active mode when its B-E junction is reverse biased and its B-C junction is forward biased.

8. The CB collector characteristics of a BJT are plots of collector current versus collector-base voltage for different values of emitter current. The CE collector characteristics of a BJT are plots of collector current versus collector-emitter voltage for different values of base current.

9. For a silicon *npn* BJT in saturation, $v_{BE} \approx$ 0.8 V and $v_{CE} \approx$ 0.2 V. A silicon *npn* BJT is cut off when $v_{BE} < 0.5$ V and $v_{BC} < 0.5$ V.

10. The conditions for the regions of operation for a typical *npn* silicon BJT are summarized by Table 7.3.

11. The dc current gain h_{FE} of a BJT is the ratio of collector current to base current ($h_{FE} \approx \beta$). For a BJT to be in saturation, a necessary condition is that $h_{FE}i_B \geq i_C$.

12. The sum of the delay time t_d and the rise time t_r of a BJT is the turn-on time t_{ON}. The sum of the storage time t_s and the fall time t_f is the turn-off time t_{OFF}. These times determine the switching speed of the BJT.

13. Transistors and diodes can be used to design such digital logic circuits as NOT, NAND, NOR, AND, and OR gates.

14. Noise margin is the input noise voltage that causes an improper output voltage. $NM_L = V_{IL} - V_{OL}$ and $NM_H = V_{OH} - V_{IH}$.

15. Fan-out is the maximum number of similar gates that can load a given gate. Fan-in is the number of inputs to a gate.

16. Diode-transistor logic (DTL) gates employ both diodes and transistors, whereas transistor-transistor logic (TTL) utilizes transistors in place of diodes.

17. Schottky TTL and emitter-coupled logic (ECL) do not require transistors to saturate. ECL is the fastest type of logic.

18. Although resistor-transistor logic (RTL) and BJT direct-coupled transistor logic (DCTL) are little used, their study is useful for understanding other types of logic (e.g., I^2L).

19. Integrated-injection logic (I^2L) saves considerable space by employing current-injecting transistors in place of resistors. Because of this, I^2L is suitable for large-scale integration (LSI) and very-large-scale integration (VLSI) circuits.

Table 7.3 Table of *npn* BJT Values

Region of Operation	v_{BE}	v_{BC}	v_{CE}	i_B	i_C	h_{FE}
Active	0.7 V	< 0.5 V	> 0.2 V	> 0 A	$h_{FE}i_B$	i_C/i_B
Cutoff	< 0.5 V	< 0.5 V	—	0 A	0 A	—
Saturation	0.8 V	0.6 V	0.2 V	> 0 A	> 0 A	> i_C/i_B
Schottky ON ($i_D > 0$ A)	0.7 V	0.4 V	0.3 V	> 0 A	$h_{FE}i_B$	i_C/i_B

Problems

7.1 For the circuit shown in Fig. 7.4, (p. 421), $R_C = R_E = 1$ kΩ and $V_{CC} = V_{EE} = 5$ V. Assume that the transistor is in the active region and that $\alpha \approx 1$. Find (a) i_E and (b) v_{CB}.

7.2 For the circuit given in Fig. 7.4, (p. 421), $V_{CC} = V_{EE} = 5$ V. Assume that the transistor is in the active region and that $\alpha \approx 1$. (a) Find R_E such that $i_E = 2$ mA. (b) Given that $i_E = 2$ mA, find R_C such that $v_{CB} = -1$ V. (c) Given that $i_E = 2$ mA, find the maximum value of R_C such that the transistor is in the active region.

7.3 For the circuit given in Fig. 7.4, (p. 421), $R_C = R_E = 1$ kΩ. Assume that the transistor is in the active region and that $\alpha \approx 1$. (a) Find V_{EE} such that $i_E = 5.3$ mA. (b) Given that $i_E = 5.3$ mA, find V_{CC} such that $v_{CB} = -3.7$ V. (c) Given that $i_E = 5.3$ mA, find the minimum value of V_{CC} such that the transistor is in the active region.

7.4 For the circuit given in Fig. 7.4, (p. 421), $R_C = 1.8$ kΩ, $R_E = 180$ Ω, and $V_{EE} = 1.4$ V. Assume that the transistor is in the active region and that $\alpha \approx 1$. (a) Find v_{CB} when $V_{CC} = 12$ V. (b) Find the minimum value of V_{CC} such that the transistor is in the active region.

7.5 For the circuit shown in Fig. P7.5, $R_E = 2$ kΩ and $V_{EE} = 5$ V. Assume that the transistor is in the active region and that $\alpha \approx 1$. Find i_E, v_{CB}, and v_{EC} for the case that (a) $R_C = 0$ Ω, $V_{CC} = 0$ V; (b) $R_C = 0$ Ω, $V_{CC} = 5$ V; (c) $R_C = 2$ kΩ, $V_{CC} = 5$ V.

7.6 For the circuit shown in Fig. P7.5, assume that $V_{CC} = V_{EE} = 10$ V, the transistor is in the active region, and $\alpha \approx 1$. Find R_C and R_E for the case that (a) $i_E = 3$ mA, $v_{CB} = -1.9$ V; and (b) $i_E = 5$ mA, $v_{EC} = 2.7$ V.

7.7 For the circuit shown in Fig. P7.7, $V_{BB} = 5$ V and $V_{EE} = 10$ V. Assume that the transistor is in the active region and that $\alpha \approx 1$. (a) Find i_E, v_{CB}, and v_{EC} for the case that $R_C = 1$ kΩ and $R_E = 2$ kΩ. (b) Find R_C and R_E for the case that $i_E = 5$ mA and $v_{CB} = -2.5$ V.

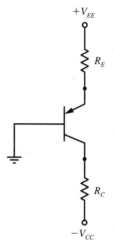

Fig. P7.5

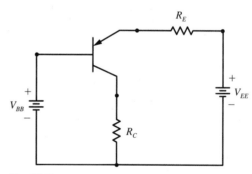

Fig. P7.7

7.8 Suppose that the circuit given in Fig. 7.7 (p. 425), has $V_{BB} = 3$ V, $V_{CC} = 10$ V, and the BJT has $\beta = 100$. Find R_B and R_C such that the transistor is biased in the active region at $i_C = 2.3$ mA and $v_{CE} = 5.4$ V.

7.9 The circuit shown in Fig. 7.7 (p. 425) has $V_{BB} = V_{CC} = 10$ V. Assuming that the transistor has $\beta = 100$, (a) find the maximum value of R_C for active-region operation when $R_B = 330$ kΩ, and (b) find the minimum value of R_B for active-region operation when $R_C = 2.7$ kΩ.

7.10 The BJT in the circuit shown in Fig. P7.10 has $\beta = 100$. Suppose that $R_1 = 90$ kΩ, $R_2 = 10$ kΩ, $R_C = 5$ kΩ, $R_E = 1$ kΩ, and $V_{CC} = 20$ V.

Verify that the transistor is biased in the active region by finding (a) i_B, (b) i_C, and (c) v_{CE}.

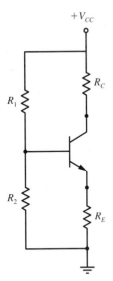

Fig. P7.10

7.11 The BJT in the circuit shown in Fig. P7.10 has $\beta = 100$. Suppose that $R_1 = 60$ kΩ, $R_2 = 30$ kΩ, $R_C = 1$ kΩ, $R_E = 100$ Ω, and $V_{CC} = 6$ V. Verify that the transistor is biased in the active region by finding (a) i_B, (b) i_C, and (c) v_{CE}.

7.12 The BJT in the circuit shown in Fig. P7.10 has $\beta = 100$. Suppose that $R_1 = 60$ kΩ, $R_2 = 30$ kΩ, $R_C = 1$ kΩ, and $V_{CC} = 6$ V. Find the minimum value of R_E such that the transistor will be in the active mode.

7.13 The BJT in the circuit shown in Fig. P7.10 has $\beta = 100$. Suppose that $R_1 = 60$ kΩ, $R_2 = 30$ kΩ, and $V_{CC} = 6$ V. Find R_E and R_C such that the transistor is biased in the active region at $i_C = 2$ mA and $v_{CE} = 2.1$ V.

7.14 The BJT in the circuit shown in Fig. P7.10 has $\beta = 100$. Suppose that $R_1 = 90$ kΩ, $R_2 = 10$ kΩ, and $V_{CC} = 20$ V. Find R_E and R_C such that the transistor is biased in the active region at $i_C = 2$ mA and $v_{CE} = 14.5$ V.

7.15 The BJT in the circuit shown in Fig. P7.10 has $\beta = 100$. Suppose that $R_1 = 150$ kΩ, $R_2 = \infty$, $R_C = R_E = 1$ kΩ, and $V_{CC} = 12$ V. Verify that the transistor is biased in the active region by finding (a) i_B, (b) i_C, and (c) v_{CE}.

7.16 The BJT in the circuit shown in Fig. P7.10 has $\beta = 100$. Suppose that $R_1 = 300$ kΩ, $R_2 = \infty$, and $V_{CC} = 12$ V. Find R_E and R_C such that the transistor is biased in the active region at $i_C = 2$ mA and $v_{CE} = 2.5$ V.

7.17 The BJT in the circuit shown in Fig. P7.10 has $\beta = 100$. Suppose that $R_1 = 120$ kΩ, $R_2 = \infty$, $R_C = 0$ Ω, $R_E = 1$ kΩ, and $V_{CC} = 6$ V. Verify that the transistor is biased in the active region by finding (a) i_B, (b) i_C, and (c) v_{CE}.

7.18 The BJT in the circuit shown in Fig. P7.10 has $\beta = 100$. Suppose that $R_C = 0$ Ω, $R_2 = \infty$, and $V_{CC} = 6$ V. Find R_E and R_1 such that the transistor is biased in the active region at $i_C = 2$ mA and $v_{CE} = 2$ V.

7.19 The BJT in the circuit shown in Fig. P7.19 has $\beta = 100$. Suppose that $R_B = 22$ kΩ, $R_C = 0$ Ω, $R_E = 2$ kΩ, $V_{BB} = V_{CC} = 0$ V, and $V_{EE} = 5$ V. Verify that the transistor is biased in the active region by finding (a) i_B, (b) i_C, and (c) v_{CE}.

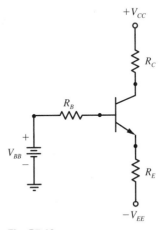

Fig. P7.19

7.20 The BJT in the circuit shown in Fig. P7.19 has $\beta = 100$. Suppose that $R_C = 0$ Ω, $V_{BB} = V_{CC} = 0$ V, and $V_{EE} = 5$ V. Find R_B and R_E such that the transistor is biased in the active region at $i_C = 2$ mA and $v_{CE} = 2.7$ V.

7.21 For the circuit shown in Fig. P7.19, suppose that R_B = 180 kΩ, R_C = R_E = 1 kΩ, V_{BB} = 0 V, and V_{CC} = V_{EE} = 5 V. Given that the BJT has β = 100, verify that the transistor is in the active region by finding (a) i_B, (b) i_C, and (c) v_{CE}.

7.22 For the circuit shown in Fig. P7.19 suppose that R_B = 165 kΩ, R_C = R_E = 1 kΩ, and V_{CC} = V_{EE} = 5 V. Given that the BJT is in the active region such that i_B = 50 μA and i_C = 4.25 mA, find (a) β, (b) V_{BB}, and (c) v_{CE}.

7.23 The BJT in the circuit shown in Fig. P7.19 has β = 100. Given that R_B = 500 kΩ, R_E = 5 kΩ, V_{BB} = 0 V, and V_{CC} = V_{EE} = 10 V, determine the value of R_C such that the transistor is in the active region and v_{CE} = 10 V.

7.24 For the circuit shown in Fig. P7.24, suppose that R_B = 250 kΩ, R_{C1} = 2.2 kΩ, R_{C2} = 100 Ω, R_E = 0 Ω, and V_{BB} = V_{CC} = 5 V. Given that the BJTs have β = 100, verify that the transistors are in the active region by finding i_{C1}, v_{CE1}, i_{C2}, and v_{CE2}.

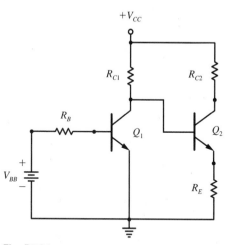

Fig. P7.24

7.25 For the circuit shown in Fig. P7.24, suppose that R_B = 230 kΩ, R_{C1} = 1 kΩ, R_{C2} = 0 Ω, R_E = 2 kΩ, V_{BB} = 3 V, and V_{CC} = 6 V. Given that the BJTs have β = 100, verify that the transistors are in the active region by finding i_{C1}, v_{CE1}, i_{C2}, and v_{CE2}.

7.26 The BJTs in the circuit shown in Fig. P7.24 have β = 100. Suppose that V_{BB} = 4 V and V_{CC} =

6 V. Find R_B, R_{C1}, R_E, and R_{C2} such that the transistors are biased in the active region at i_{C1} = 1 mA, v_{CE1} = 3.5 V, i_{C2} = 4 mA, and v_{CE2} = 2 V.

7.27 For the circuit shown in Fig. P7.27, suppose that R_B = 230 kΩ, R_{E1} = 400 Ω, R_{E2} = 0 Ω, R_C = 100 Ω, and V_{CC} = 6 V. Given that the BJTs have β = 100, verify that the transistors are in the active region by finding i_{C1}, v_{CE1}, i_{C2}, and v_{CE2}.

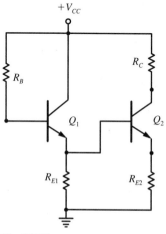

Fig. P7.27

7.28 For the circuit shown in Fig. P7.27, suppose that R_B = 65 kΩ, R_{E1} = 500 Ω, R_{E2} = 1 kΩ, R_C = 1 kΩ, and V_{CC} = 6 V. Given that the BJTs have β = 100, verify that the transistors are in the active region by finding i_{C1}, v_{CE1}, i_{C2}, and v_{CE2}.

7.29 The BJTs in the circuit shown in Fig. P7.27 have β = 100. Suppose that V_{CC} = 6 V. Find R_B, R_{E1}, R_{E2}, and R_C such that the transistors are biased in the active region at i_{C1} = 1 mA, v_{CE1} = 2.6 V, i_{C2} = 13.5 mA, and v_{CE2} = 1.3 V.

7.30 For the circuit given in Fig. 7.13, (p. 435), suppose that V_{CC} = 5 V. (a) Find the minimum value of h_{FE} required to saturate the BJT for the case that R_B = 50 kΩ, R_C = 1 kΩ, and v_1 = 5 V. (b) Find the minimum value of R_C required to saturate the BJT for the case that h_{FE} = 100, R_B = 50 kΩ, and v_1 = 5 V. (c) Find the maximum value of R_B required to saturate the BJT for the case that h_{FE} = 100, R_C =

1 kΩ, and $v_1 = 5$ V. (d) Find the minimum value of v_1 required to saturate the BJT for the case that $h_{FE} = 100$, $R_B = 50$ kΩ, and $R_C = 1$ kΩ.

7.31 For the circuit shown in Fig. P7.19, suppose that $V_{BB} = 5$ V, $V_{CC} = 10$ V, $V_{EE} = 0$ V, $R_B = 56$ kΩ, and $R_C = 2.7$ kΩ. Determine the minimum value of h_{FE} required for saturation when (a) $R_E = 0$ Ω and (b) $R_E = 1.5$ kΩ.

7.32 For the circuit shown in Fig. P7.19, suppose that $V_{BB} = 2$ V, $V_{CC} = 6$ V, $V_{EE} = 0$ V, $R_B = 20$ kΩ, and $R_C = 1$ kΩ. Determine the minimum value of h_{FE} required for saturation when (a) $R_E = 0$ Ω, and (b) $R_E = 100$ Ω.

7.33 For the circuit shown in Fig. P7.19, suppose that $V_{BB} = 2.5$ V, $V_{CC} = V_{EE} = 5$ V, $R_B = 10$ kΩ, and $R_C = R_E = 100$ Ω. Determine the minimum value of h_{FE} required for saturation.

7.34 For the circuit shown in Fig. P7.34, (a) verify that the BJT is OFF when the switch is closed, and (b) find the minimum value of h_{FE} for which the BJT is ON when the switch is open.

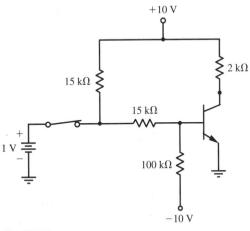

Fig. P7.34

7.35 The BJTs Q_1 and Q_2 in the circuit shown in Fig. P7.35 have dc betas of h_{FE1} and h_{FE2}, respectively. Suppose that $R_B = 68$ kΩ, $R_{C1} = 1$ kΩ, $R_{C2} = 500$ Ω, $R_E = 210$ Ω, and $V_{CC} = 5$ V. Find the minimum values of h_{FE1} and h_{FE2} such that Q_1 and Q_2 are in saturation.

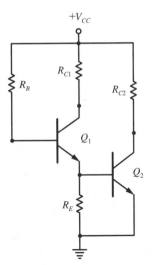

Fig. P7.35

7.36 For the circuit shown in Fig. P7.36, suppose that $R_B = 2$ kΩ, $R_{C1} = 1$ kΩ, $R_{C2} = 500$ Ω, $R_E = 250$ Ω, and $V_{CC} = 5$ V. Given that Q_1 operates in the active region and has a dc beta of $h_{FE1} = 100$, find the minimum value of h_{FE2} for which Q_2 will saturate.

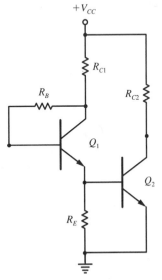

Fig. P7.36

7.37 For the circuit shown in Fig. P7.36, suppose that $R_B = 0$ Ω, $R_{C1} = 2.7$ kΩ, $R_{C2} = 500$ Ω 750 Ω, and $V_{CC} = 5$ V. Given that Q_1 oper the active region and has a dc beta of $h_{FE1} = 100$, find the minimum value of h_{FE2} for which Q_2 will saturate.

7.38 For the circuit shown in Fig. P7.36, suppose that $R_B = 50$ kΩ, $R_{C1} = 0$ Ω, $R_{C2} = 300$ Ω, $R_E = 125$ Ω, and $V_{CC} = 5$ V. Given that Q_1 operates in the active region and has a dc beta of $h_{FE1} = 100$, find the minimum value of h_{FE2} for which Q_2 will saturate.

7.39 For the circuit shown in Fig. 7.18a, (p. 439), suppose that $R_1 = R_2 = R_E = 2$ kΩ. When $V_{BB} = 4$ V, then Q_1 is ON and Q_2 is OFF. Find (a) v_E, (b) i_{C1}, (c) i_{B1}, (d) v_{BE2}, (e) v_{CE2}, and (f) the minimum value of h_{FE1}.

7.40 For the circuit shown in Fig. 7.18a, (p. 439), suppose that $R_1 = R_2 = R_E = 2$ kΩ. When $V_{BB} = 2$ V, then Q_1 is OFF and Q_2 is ON. Find (a) i_{B2}, (b) i_{C2}, (c) v_E, (d) v_{BE1}, (e) v_{CE1}, and (f) the minimum value of h_{FE2}.

7.41 For the inverter circuit shown in Fig. P7.41, $R_B = 50$ kΩ, the low voltage is 0.2 V, and the high voltage is 5 V. Determine the minimum value of h_{FE} required for proper operation if the collector resistor R_C has a value of (a) 1 kΩ and (b) 2 kΩ.

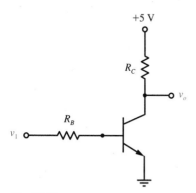

Fig. P7.41

7.42 For the inverter circuit shown in Fig. P7.41, $R_B = 50$ kΩ, the low voltage is 0.2 V, and the high voltage is 5 V. Determine the minimum value of the collector resistor R_C required for proper operation when h_{FE} is (a) 100 and (b) 80.

7.43 For the inverter circuit shown in Fig. P7.41, $R_C = 1$ kΩ, the low voltage is 0.2 V, and the high voltage is 5 V. Determine the maximum value of the base resistor R_B required for proper operation when h_{FE} is (a) 100 and (b) 80.

7.44 For the inverter circuit shown in Fig. P7.41, $R_B = 50$ kΩ, $h_{FE} = 100$, the low voltage is 0.2 V, and the high voltage is 5 V. Determine the noise margin NM_H if the collector resistor R_C has a value of (a) 1 kΩ and (b) 10 kΩ.

7.45 For the inverter circuit shown in Fig. P7.45, $R_1 = 15$ kΩ, $R_2 = 50$ kΩ, the low voltage is 0.2 V, and the high voltage is 6 V. Determine the minimum value of h_{FE} required for proper operation if the collector resistor R_C has a value of (a) 500 Ω and (b) 1 kΩ.

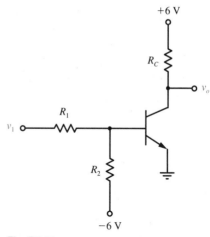

Fig. P7.45

7.46 For the inverter circuit shown in Fig. P7.45, $R_1 = 15$ kΩ, $R_2 = 50$ kΩ, the low voltage is 0.2 V, and the high voltage is 6 V. Determine the minimum value of R_C required for proper operation when h_{FE} is (a) 80 and (b) 100.

7.47 For the inverter circuit shown in Fig. P7.45, $h_{FE} = 100$, $R_1 = 15$ kΩ, $R_2 = 50$ kΩ, $R_C = 500$ Ω, the low voltage is 0.2 V, and the high voltage is 6 V. Determine the noise margins NM_L and NM_H.

7.48 For the inverter circuit shown in Fig. P7.45, $h_{FE} = 100$, $R_2 = 50$ kΩ, $R_C = 500$ Ω, the low voltage is 0.2 V, and the high voltage is 6 V. Determine the value of R_1 for which $NM_L = NM_H$, and find the values of these noise margins.

7.49 For the inverter circuit shown in Fig. P7.45, $h_{FE} = 100$, $R_1 = 15$ kΩ, $R_C = 500$ Ω, the low voltage is 0.2 V, and the high voltage is 6 V. Determine the value of R_2 for which $NM_L = NM_H$, and find the values of these noise margins.

7.50 For the inverter circuit shown in Fig. P7.45, $h_{FE} = 100$, $R_1 = 15$ kΩ, $R_2 = 50$ kΩ, the low voltage is 0.2 V, and the high voltage is 6 V. Determine the value of R_C for which $NM_L = NM_H$, and find the values of these noise margins.

7.51 For the inverter shown in Fig. P7.41, the low voltage is 0.2 V. Suppose that $R_B = 10$ kΩ, $R_C = 1$ kΩ, $V_{CC} = 5$ V, and the gate is loaded by 10 identical inverters. (a) Find the high voltage; that is, with the load connected, find v_o when $v_1 = 0.2$ V. (b) Find the minimum value of h_{FE} required for proper operation. (c) Find the fan-out when $h_{FE} = 50$.

7.52 The DTL circuit shown in Fig. P7.52 is a NOR gate. Suppose that the voltage across a forward-biased diode is 0.7 V. If the low voltage is 0.2 V and the high voltage is 5 V, find (a) the minimum value of h_{FE} required for proper operation, and (b) the noise margins for the case that $h_{FE} = 100$.

7.53 Suppose that the NOR gate given in Fig. P7.52, which has a low voltage of 0.2 V, is loaded by 20 identical gates. (a) Find the high voltage; that is, with the load connected, find v_o when $v_1 = v_2 = 0.2$ V. (b) Find the minimum value of h_{FE} required for proper operation. (c) Find the fan-out when $h_{FE} = 50$.

7.54 In addition to the DTL gate given in Fig. P7.52, a NOR gate can be constructed by connecting the collectors of two inverters together as shown in Fig. P7.54. This is an example of **wired logic** (or **collector logic**), so called because the collectors of Q_1 and Q_2 are wired together. (Note that this NOR gate requires no diodes.) Find the minimum value of h_{FE} required for proper operation.

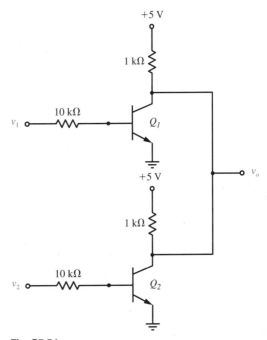

Fig. P7.54

7.55 For the NAND gate given in Fig. 7.26, (p. 451), the sources of v_1 and v_2 absorb relatively small amounts of power. Find the power absorbed by the sources of v_1 and v_2 when (a) $v_1 = v_2 = 0.2$ V; (b) $v_1 = 0.2$ V, $v_2 = 5$ V; and (c) $v_1 = v_2 = 5$ V.

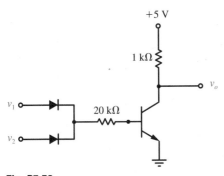

Fig. P7.52

7.56 Find the power dissipated by the NAND gate shown in Fig. 7.28 (p. 457) given that the BJTs have $h_{FE} = 40$.

7.57 Find the noise margins NM_L and NM_H for the NAND gate shown in Fig. 7.28 (p. 457) given that the BJTs have $h_{FE} = 40$.

7.58 By modifying the NAND gate given in Fig. 7.28 (p. 457) as shown in Fig. P7.58, higher noise margins are obtained. [This is desirable for operation in high noise environments and is an example of a **high-threshold logic** (HTL) gate.] The low voltage for the gate is 0.2 V and the high voltage is 15 V. The breakdown voltage of the Zener diode is 6 V. Find the noise margins NM_L and NM_H for this gate.

7.59 For the NAND gate shown in Fig. P7.58, suppose that $h_{FE1} = 50$. The low voltage for the gate is 0.2 V and the high voltage is 15 V. The breakdown voltage of the Zener diode is 6 V. What is the minimum value for h_{FE2}?

7.60 Find the power dissipated by the NAND gate shown in Fig. P7.58 given that $h_{FE1} = 50$, the low voltage for the gate is 0.2 V, the high voltage is 15 V, and the breakdown voltage of the Zener diode is 6 V.

7.61 For the NAND gate shown in Fig. P7.58, $h_{FE} = 50$ for Q_1 and Q_2. Given that the low voltage for the gate is 0.2 V, the high voltage is 15 V, and the breakdown voltage of the Zener diode is 6 V, find the fan-out of the gate.

7.62 For the NAND gate shown in Fig. P7.62, the low voltage for the gate is 0.2 V and the high voltage is 5 V. Find the minimum value of h_{FE} for (a) Q_1, (b) Q_2, and (c) Q_3.

7.63 For the NAND gate shown in Fig. P7.62, the low voltage for the gate is 0.2 V and the high voltage is 5 V. Find the power dissipated by the gate.

7.64 For the NAND gate shown in Fig. P7.62, the low voltage for the gate is 0.2 V and the high voltage is 5 V. Find the noise margins NM_L and NM_H for the gate.

7.65 For the NAND gate shown in Fig. P7.62, the low voltage for the gate is 0.2 V and the high voltage is 5 V. Find the fan-out for the gate given that $h_{FE3} = 100$.

7.66 Find the fan-out for the TTL NAND gate given in Fig. 7.31a on p. 460 for the case that $h_{FER} = 0.5$, $h_{FE} = 50$, and the high voltage is 3 V.

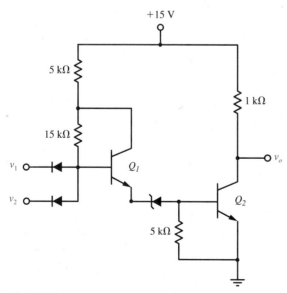

Fig. P7.58

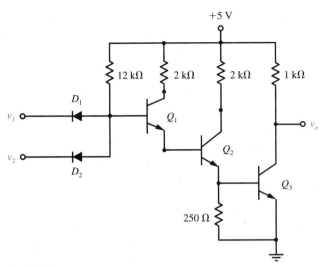

+5 V

12 kΩ   2 kΩ   2 kΩ   1 kΩ

D_1

v_1

Q_1

v_o

v_2

D_2

Q_2

Q_3

250 Ω

Fig. P7.62

7.67 Find the noise margins NM_L and NM_H for the TTL NAND gate given in Fig. 7.31a on p. 460.

7.68 For the TTL NAND gate shown in Fig. 7.31a, on p. 460, suppose that $R_B = 2.7$ kΩ, $R_C = 500$ Ω, $R_1 = 100$ Ω, and $R_2 = 500$ Ω. Given that $h_{FER} = 0.6$ for Q_1, find the minimum values of h_{FE} for Q_2 and Q_3.

7.69 For the TTL NAND gate shown in Fig. 7.31a, on p. 460, suppose that $R_B = 2.7$ kΩ, $R_C = 500$ Ω, $R_1 = 100$ Ω, and $R_2 = 500$ Ω. Find the power dissipated by the gate.

7.70 For the TTL NAND gate shown in Fig. 7.31a, on p. 460, suppose that $R_B = 2.7$ kΩ, $R_C = 500$ Ω, $R_1 = 100$ Ω, and $R_2 = 500$ Ω. Given that $h_{FER} = 0.6$ for Q_1, find the fan-out for the case that $h_{FE} = 50$ and the high voltage is 3 V.

7.71 For the TTL NAND gate given in Fig. 7.32 on p. 464, suppose that $h_{FER} = 0.5$. Find the minimum values of h_{FE} for Q_2 and Q_3.

7.72 Find the power dissipated by the TTL NAND gate shown in Fig. 7.32 on p. 464.

7.73 For the circuit given in Fig. 7.38 on p. 471, the Schottky BJT has $h_{FE} = 50$, $R_B = 20$ kΩ, $R_C = 1$ kΩ, and $V_{CC} = 5$ V. Find i_B, i_C, and i_D when $v_1 = 5$ V.

7.74 For the inverter given in Fig. 7.38 on p. 471, the Schottky BJT has $h_{FE} = 50$, $R_B = 20$ kΩ, $R_C = 1$ kΩ, and $V_{CC} = 5$ V. Find the noise margins NM_L and NM_H.

7.75 For the Schottky inverter shown in Fig. 7.38 on p. 471, $R_B = 20$ kΩ, $R_C = 1$ kΩ, and $V_{CC} = 5$ V. Suppose that the gate is loaded by 10 identical inverters. (a) Find the high voltage. (b) Find the minimum value of h_{FE} required for proper operation.

7.76 For the Schottky inverter shown in Fig. 7.38 on p. 471, $R_B = 20$ kΩ, $R_C = 1$ kΩ, and $V_{CC} = 5$ V. Find the fan-out when $h_{FE} = 50$.

7.77 For the NAND gate shown in Fig. P7.77, find the minimum values of h_{FE} for Q_1 and Q_2. (See p. 500.)

7.78 For the NAND gate shown in Fig. P7.77, find the fan-out for the case that $h_{FE} = 50$.

7.79 Find the power dissipated by the NAND gate shown in Fig. P7.77. (See p. 500.)

7.80 For the Schottky TTL NAND gate shown in Fig. P7.80, suppose that $h_{FER} = 0.5$ for Q_1. Find the minimum values of h_{FE} for Q_2 and Q_3. (See p. 500.)

7.81 Find the power dissipated by the Schottky TTL NAND gate shown in Fig. P7.80.

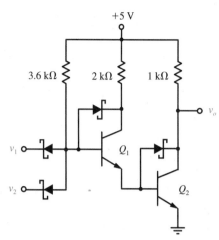

Fig. P7.77

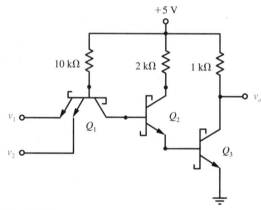

Fig. P7.80

7.82 For the Schottky TTL NAND gate shown in Fig. P7.80, suppose $h_{FER} = 0.5$ for Q_1. Find the fanout for the case that $h_{FE} = 50$ and the high voltage is 3 V.

7.83 Find the power dissipated by the ECL gate given in Fig. 7.43 on p. 478 (see Example 7.18).

7.84 Determine the noise margins NM_L and NM_H for the ECL OR/NOR gate given in Fig. 7.43 on p. 478 (see Example 7.18).

7.85 For the ECL OR/NOR gate shown in Fig. P7.85, suppose that $R_2 = 200$ Ω, $V_R = 4$ V, $V_{CC} = 5$ V, and $V_{EE} = 0$ V. Neglect base currents and determine (a) the low voltage, (b) the high voltage, and (c) the value of R_1. (See p. 501.)

7.86 For the ECL OR/NOR gate shown in Fig. P7.85, suppose that $R_1 = 250$ Ω, $V_R = 3.7$ V, $V_{CC} = 5$ V, and $V_{EE} = 0$ V. Neglect base currents and determine (a) the high voltage, (b) the low voltage, and (c) the value of R_2.

7.87 For the ECL OR/NOR gate shown in Fig. P7.85, suppose that $R_2 = 300$ Ω, $V_R = 1.5$ V, $V_{CC} = V_{EE} = 2.5$ V. Neglect base currents and determine (a) the low voltage, (b) the high voltage, and (c) the value of R_1.

7.88 For the ECL OR/NOR gate shown in Fig. P7.85, suppose that the low voltage is -1.6 V, $V_R = -1.3$ V, $V_{CC} = 0$ V, and $V_{EE} = 5$ V. Neglect base currents and determine the values of R_1 and R_2.

7.89 Determine the fan-out of the RTL NOR gate in Fig. 7.44 on p. 482 for the case that the transistors have $h_{FE} = 50$.

7.90 For the RTL NOR gate given in Fig. 7.44 on p. 482, let $R_B = 20$ kΩ, $R = 5$ kΩ, and $V_{CC} = 5$ V. Suppose that this gate is loaded by 10 identical gates. Determine (a) the high voltage, and (b) the minimum value of h_{FE} for the transistors.

7.91 The RTL circuit shown in Fig. P7.91 is a NAND gate. The low voltage is 0.4 V and the high voltage is 1.2 V. Determine whether each transistor is OFF or ON, and find the minimum values of h_{FE} for (a) $v_1 = v_2 = 0.4$ V; (b) $v_1 = 0.4$ V, $v_2 = 1.2$ V; (c) $v_1 = 1.2$ V, $v_2 = 0.4$ V; and (d) $v_1 = v_2 = 1.2$ V.

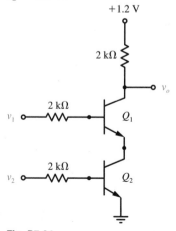

Fig. P7.91

7.92 For the sake of simplicity, the current injectors for the I^2L circuit shown in Fig. P7.92 are not shown. Construct the truth table corresponding to this circuit when the output is taken to be (a) v_o and (b) $\bar{v}_o$.

7.93 For the sake of simplicity, the current injectors for the I^2L circuit shown in Fig. P7.93 are not shown. Construct the truth table corresponding to this circuit when the output is taken to be (a) v_o and (b) $\bar{v}_o$. (See p. 502.)

7.94 For the sake of simplicity, the current injectors for the I^2L circuit shown in Fig. P7.94 are not shown. Construct the truth table corresponding to this circuit when the output is taken to be (a) v_o and (b) $\bar{v}_o$. (See p. 502.)

7.95 For the sake of simplicity, the current injectors for the I^2L circuit shown in Fig. P7.95 are not shown. This circuit is known as a **majority-logic gate**. Determine the truth table corresponding to this circuit when the output is taken to be v_o.

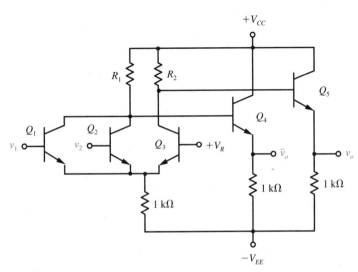

Fig. P7.85

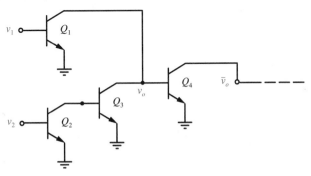

Fig. P7.92

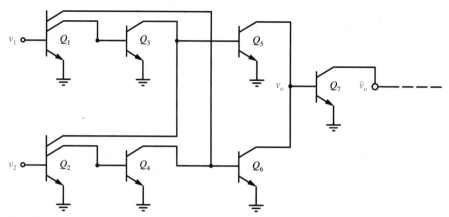

Fig. P7.93

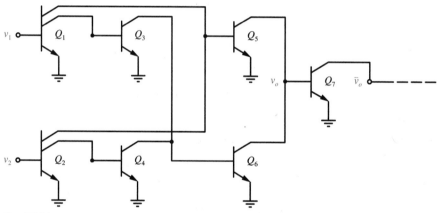

Fig. P7.94

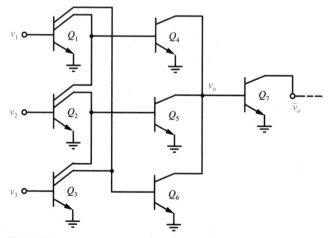

Fig. P7.95

8

Field-Effect Transistors (FETs)

8.1 The Junction Field-Effect Transistor (JFET)

8.2 Metal-Oxide-Semiconductor Field-Effect Transistors (MOSFETs)

8.3 MOSFET Logic Gates

8.4 Complementary MOSFETs (CMOS)

INTRODUCTION

Unlike the bipolar junction transistor (BJT) studied in Chapter 7, another type of transistor—the field-effect transistor (FET)—is a unipolar device; that is, current results from the flow of majority carriers only. There are two types of FETS—the junction field-effect transistor (JFET) and the metal-oxide-semiconductor field-effect transistor (MSOFET).

Whereas the BJT is a current-controlled device, the FET is a voltage-controlled device. A voltage controls an electric field (hence, the terminology "field effect"), which, in turn, controls a current. An FET inherently possesses a high input resistance; therefore, in digital circuit applications, FET logic generally has a higher fan-out than BJT logic. Although BJTs operate at higher speeds than FETS, the FET is inherently less noisy than the BJT.

FETs, especially MOSFETs, are simpler to fabricate and require less space than do BJTs. Also, MOSFET logic does not involve the need for resistors and diodes. As a consequence of this, MOSFETs find widespread use in large-scale integration (LSI) and very-large-scale integration (VLSI) applications, such as handheld calculators, digital wristwatches, microprocessors, and memories. For more general-purpose logic design, complementary MOSFETs (CMOS) offer greater speed and lower power consumption.

8.1 The Junction Field-Effect Transistor (JFET)

A side view of a simplified version of a **junction field-effect transistor** (JFET) is shown in Fig. 8.1*a*. In particular, this device is an ***n*-channel** JFET. The corresponding circuit symbol is given in Fig. 8.1*b*. Interchanging the *n*- and *p*-type semiconductors results in a *p*-channel JFET, and its circuit symbol is obtained from the one in Fig. 8.1*b* simply by reversing the direction of the arrow. In essence, an *n*-channel JFET is made by taking a bar of *n*-type semiconductor (silicon) and forming heavily doped *p*-type regions, called the **gate** (G), on both sides. This results in two *pn* junctions. Metal contacts are made to the gate as well as to the ends of the *n*-type semiconductor. One end of this *n*-type semiconductor, known as the **channel,** is called the **drain** (D) and the other end is called the **source** (S) of the JFET. Because of its physical symmetry, either end of the channel can be the drain and the other end the source. However, the end in which majority carriers (electrons for an *n*-type semiconductor) enter the channel is designated the source, and the end from which they leave the channel is the drain. Unlike the BJT, the JFET is a unipolar device—current is due to majority carriers only. As we shall see, current going from drain to source (i.e., electrons going from source to drain) through the channel can be controlled by a voltage applied between gate and source.

To understand the behavior of a JFET, consider the circuit shown in Fig. 8.2, where all of the JFET's currents and voltages are indicated. For the case that

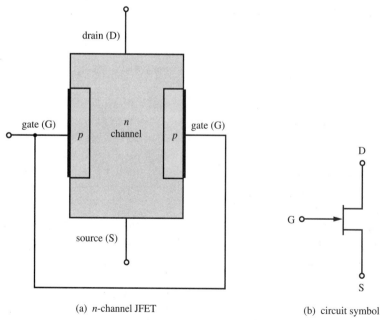

(a) *n*-channel JFET   (b) circuit symbol

Fig. 8.1 An *n*-channel JFET and its circuit symbol.

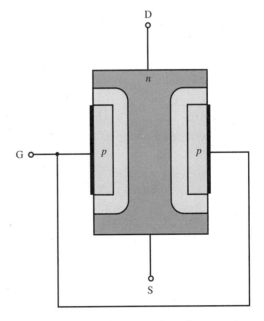

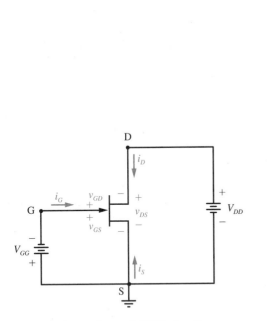

Fig. 8.2 An *n*-channel JFET circuit.

Fig. 8.3 The depletion regions in an *n*-channel JFET for small values of v_{DS}.

$V_{DD} = V_{GG} = 0$ V, all external currents and voltages, of course, will be zero. However, there are two *pn* junctions with their associated depletion regions as depicted in Fig. 8.3. For the case that $V_{DD} = v_{DS}$ is a small (positive) number, a small (positive) drain current i_D results. Since $v_{DS} = V_{DD}$ is small, the drain voltage v_D (with respect to the reference) is approximately equal to the source voltage v_S, that is, $v_D \approx v_S = 0$ V. However, if V_{GG} is increased, since $v_{GS} = -V_{GG}$, then the *pn* junctions become more reverse biased. This means that the depletion regions get wider, and because $v_D \approx v_S = 0$ V, they essentially are symmetrical. This widening of the depletion regions, in effect, narrows the channel. By sufficiently increasing V_{GG}, the depletion regions will widen enough to eliminate the channel. The value of $v_{GS} = -V_{GG}$ for which such a situation occurs is called the **pinchoff voltage** V_p of the JFET. Note that for an *n*-channel JFET, V_p is a negative number. The magnitude of V_p typically lies in the range from 2 to 6 V.

Let us now again consider the case that $v_{GS} = 0$ V ($V_{GG} = 0$ V). As just mentioned, if $V_{DD} = v_{DS} = 0$ V, then $i_D = 0$ A and $v_D = v_S = 0$ V. Thus the voltage across each *pn* junction is zero, which means that they are reverse biased. (We say that the gate is reverse biased.) If $V_{DD} > 0$ V, but is still small, then the drain current i_D is small and $v_D \approx v_S = 0$ V. Thus the gate remains reverse biased, so $i_G = 0$ A. As long as $V_{DD} = v_{DS} = v_D - v_S$ remains small (i.e., $v_D \approx v_S$), the depletion regions will be essentially symmetrical and i_D will be proportional to v_{DS}—that is, the channel will behave as a linear resistor. However, if $V_{DD} = v_{DS} = v_D - v_S$ gets larger, then i_D gets larger and v_D becomes greater than v_S. Since, when $v_{GS} = 0$ V,

$$v_{GD} = v_{GS} - v_{DS} = 0 - v_{DS} = -v_{DS}$$

the gate-drain voltage is negative whereas the gate-source voltage is zero. Thus the portion of the *pn* junction between the gate and drain is more reverse biased than the portion between the gate and source. As a consequence of this, the channel is narrower at the drain end than at the source end. This situation is illustrated in Fig. 8.4 for the condition that $0 < v_{DS} < |V_P|$. As v_{DS} increases, i_D increases, the channel narrows, and the resistance of the channel increases. Because of the manner in which the channel configuration changes, the nature of the channel resistance is nonlinear.

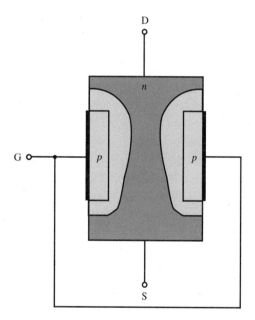

Fig. 8.4 The depletion regions in an *n*-channel JFET for $0 < v_{DS} < |V_p|$.

If v_{DS} is made large enough, the reverse-bias gate-drain voltage will be sufficient to pinch off the channel. By definition, this occurs when $v_{GD} = V_p$, that is, when

$$v_{GD} = v_{GS} - v_{DS} = 0 - v_{DS} = V_p \tag{8.1}$$

or $v_{DS} = -V_p$. For this condition, as illustrated in Fig. 8.5, even though it is technically "pinched off," the channel is not completely eliminated, but instead is very narrow. This allows a nonzero drain current i_D. If $v_{DS} > -V_p$, then the

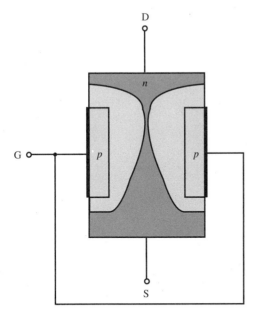

Fig. 8.5 The depletion regions in an *n*-channel JFET for $v_{DS} = -V_p$.

channel is still pinched off (very narrow), but it is narrow over a greater distance. As v_{DS} increases, the drain current i_D tends to increase. However, the increasing resistance due to the reduction in the width of the channel tends to decrease the drain current. The net effect is that for $v_{DS} \geq -V_p$, the drain current i_D is approximately constant. Because of this, the drain current is said to **saturate.** (The term "saturation" when used for a JFET has a different meaning than when it is used for a BJT.)

JFET Characteristics

The previous discussion can be summarized graphically by the plot of i_D versus v_{DS} (for the case that $v_{GS} = 0$ V) shown in Fig. 8.6. From this plot, we see that for small values of v_{DS}, the (channel of the) device behaves as a linear resistor, and as v_{DS} increases, the channel resistance becomes nonlinear. When $v_{DS} \geq -V_p$ and the channel is pinched off, the drain current remains essentially constant as v_{DS} varies. The value of this constant (saturation) current is designated I_{DSS} to indicate the current from drain to source when there is a short circuit connected between the gate and the source (i.e., when $v_{GS} = 0$ V). For a JFET, the current I_{DSS} can range from tenths of a milliampere to hundreds of milliamperes.

Throughout the previous discussion, due to reverse-biased *pn* junctions, we had that $i_G = 0$ A. However, practically speaking, we know that a *pn* junction will break down if too large a reverse-bias voltage is applied. As indicated in Fig. 8.6, if v_{DS}

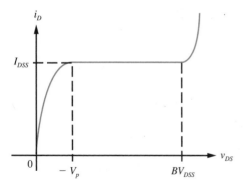

Fig. 8.6 Typical plot of i_D versus v_{DS} when $v_{GS} = 0$ V for an n-channel JFET.

increases too much, an excessive reverse bias will occur and junction breakdown can result in a large drain current (going from drain to gate). The value of v_{DS} for which this occurs is designated BV_{DSS} to indicate the drain-source breakdown voltage when there is a short circuit between the gate and the source (i.e., when $v_{GS} = 0$ V). This voltage typically ranges from 20 to 50 V.

Let us return to the JFET circuit given in Fig. 8.2. We have already seen (Eq. 8.1) that when $v_{GS} = 0$ V, the channel is pinched off when $v_{GD} = -v_{DS} = V_p$ or $v_{DS} = -V_p$. However, when $V_{GG} > 0$ V, that is, $v_{GS} < 0$ V, then the channel is pinched off when

$$v_{GD} = v_{GS} - v_{DS} = V_p$$

or when

$$v_{DS} = v_{GS} - V_p$$

Thus by decreasing v_{GS}, the value of v_{DS} required to pinch off the channel decreases. In particular, Fig. 8.7 shows a set of curves of i_D versus v_{DS} for different values of v_{GS} for an n-channel JFET. We refer to these curves as the **drain (static) characteristics** or **output characteristics** of the JFET. For this device, $I_{DSS} = 12$ mA and the pinchoff voltage is $V_p = -3$ V. Thus if $v_{GS} = 0$ V, the channel pinches off when $v_{DS} = -V_p = 3$ V. However, if $v_{GS} = -1$ V, then the channel pinches off when $v_{DS} = v_{GS} - V_p = -1 + 3 = 2$ V. Similarly, if $v_{GS} = -2$ V, pinchoff occurs when $v_{DS} = -2 + 3 = 1$ V. The dashed curve shown in Fig. 8.7 corresponds to the equation $v_{DS} = v_{GS} - V_p$. To the right of this curve ($v_{DS} > v_{GS} - V_p$) the channel of the JFET is pinched off, and this region is called the **pinchoff region** or **saturation region,** or more preferably, the **active region.** To the left of this curve ($v_{DS} < v_{GS} - V_p$) the channel of the JFET is

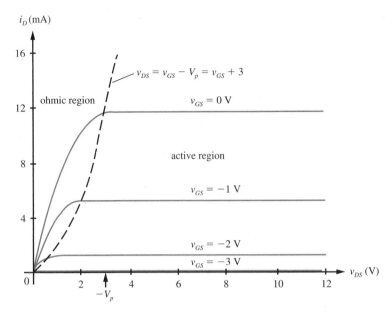

Fig. 8.7 Drain characteristics of an *n*-channel JFET.

not pinched off, and this region is known as the **ohmic region** (or **nonsaturation region**[1]).

Note that when the gate is sufficiently reverse biased, the channel will be eliminated—this, of course, occurs when $v_{GS} \leq V_p$. Under this circumstance, increasing v_{DS} to any positive value (below the breakdown voltage) will not be sufficient to produce a drain current. Thus we have that $i_D = 0$ A and we say that the JFET is **cut off** or is in **cut-off**. (In an actual device, due to leakage, the drain current of a cutoff JFET is small, but nonzero. However, we will use the approximation $i_D = 0$ A.)

In these discussions, we were concerned only with the situation in which the gate is reverse biased. Ideally, then, $i_G = 0$ A. However, practically speaking, there is a small gate current even when the JFET is cut off. Although such a current is of the order of a few nanoamperes or less at 25°C, it approximately doubles for every 10°C rise in temperature. For the sake of simplicity, we will assume the ideal case that $i_G = 0$ A.

JFET Equations

It can be shown that, for a JFET operating in the ohmic region,

---

[1]This terminology is unfortunate since the corresponding region for a BJT is called the saturation region. To avoid confusion, we will use the term ''ohmic region.''

$$i_D = I_{DSS}\left[2\left(1 - \frac{v_{GS}}{V_p}\right)\frac{v_{DS}}{-V_p} - \left(\frac{v_{DS}}{V_p}\right)^2\right] \tag{8.2}$$

For small values of v_{DS}, we have $-V_p \gg v_{DS}$. This in turn, means that $(v_{DS}/V_p)^2$ is small. Thus

$$i_D \approx I_{DSS}\left[2\left(1 - \frac{v_{GS}}{V_p}\right)\frac{v_{DS}}{-V_p}\right] = \frac{2I_{DSS}}{-V_p}\left(1 - \frac{v_{GS}}{V_p}\right)v_{DS}$$

from which

$$\frac{i_D}{v_{DS}} \approx \frac{2I_{DSS}}{-V_P}\left(1 - \frac{v_{GS}}{V_p}\right) = \frac{2I_{DSS}(v_{GS} - V_p)}{V_p^2}$$

Therefore, for small values of v_{DS}, when the channel behaves as a linear resistor r_{DS}, the approximate value of the resistance of the channel is

$$r_{DS} = \frac{v_{DS}}{i_D} \approx \frac{V_p^2}{2I_{DSS}(v_{GS} - V_p)} \tag{8.3}$$

and the smaller the value of v_{DS}, the better the approximation. Typically, the resistance r_{DS} ranges from several ohms to several hundred ohms.

It can also be shown, that when a JFET is in the active region,

$$i_D = I_{DSS}\left(1 - \frac{v_{GS}}{V_p}\right)^2 \tag{8.4}$$

The corresponding plot of i_D versus v_{GS} shown in Fig. 8.8 is known as the **transfer characteristic** of the JFET. For an n-channel JFET in the active region, $v_{GS} \leq 0$ V. However, when $v_{GS} \leq V_p$, the JFET is cut off and $i_D = 0$ A. For this reason, V_p is also known as the **gate-source cutoff voltage,** designated $V_{GS(OFF)}$. Ideally, the transfer characteristic of a JFET is independent of the value of v_{DS}, but practically speaking, this is not necessarily the case.

As shown in the JFET drain characteristics given in Fig. 8.7, the boundary between the active and ohmic regions corresponds to the case that

$$v_{DS} = v_{GS} - V_p \quad \Rightarrow \quad v_{GS} = v_{DS} + V_p$$

Substituting this expression for v_{GS} into either Eq. 8.3 or Eq. 8.4—they both must satisfy the boundary condition—we find that

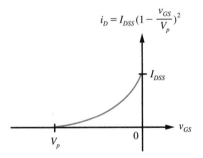

Fig. 8.8 An *n*-channel JFET transfer characteristic.

$$i_D = \frac{I_{DSS}}{V_p^2} \, v_{DS}^2$$

which is the equation of a parabola.

JFET Circuits

Let us now analyze some JFET circuits.

Example 8.1

For the JFET circuit shown in Fig. 8.9, the JFET has the parameters $I_{DSS} = 8$ mA and $V_p = -4$ V. Let us find i_D and v_{DS}.

Since there is a short circuit connected between the gate and the source, $v_{GS} = 0$ V. By Ohm's law, we have that

$$i_D = \frac{18 - v_{DS}}{500} = -\frac{1}{500} \, v_{DS} + 36 \times 10^{-3}$$

The load line corresponding to this equation can be plotted on the drain characteristics of the JFET, if available. The intersection of the load line and the curve $v_{GS} = 0$ V will graphically yield the values of i_D and v_{DS}. Furthermore, the region of operation of the JFET can be determined graphically. However, let us suppose that the drain characteristics are not available. As a consequence, let us assume that the JFET is in the active region. Then from Eq. 8.4 we have

$$i_D = I_{DSS} \left(1 - \frac{v_{GS}}{V_p} \right)^2 = (8 \times 10^{-3}) \left(1 - \frac{0}{-4} \right)^2 = 8 \text{ mA} = I_{DSS}$$

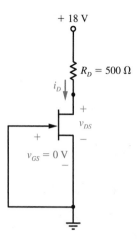

Fig. 8.9 An *n*-channel JFET circuit.

By KVL,

$$v_{DS} = -500(8 \times 10^{-3}) + 18 = 14 \text{ V}$$

Since $v_{DS} = 14 > v_{GS} - V_p = 0 - (-4) = 4$ V, the JFET is indeed in the active region.

Now suppose that the resistance R_D is changed to 2 kΩ. Again, if we assume that the JFET is in the active region, we get $i_D = 8$ mA and

$$v_{DS} = -2000(8 \times 10^{-3}) + 18 = 2 \text{ V}$$

But since $v_{DS} = 2 < v_{GS} - V_p = 4$ V, our assumption that the JFET is in the active region is contradicted. Therefore, let us assume that the JFET is in the ohmic region. From Eq. 8.2 we have that

$$i_D = (8 \times 10^{-3}) \left[2 \left(1 - \frac{0}{-4} \right) \frac{v_{DS}}{4} - \left(\frac{v_{DS}}{-4} \right)^2 \right]$$
$$= (4 \times 10^{-3}) v_{DS} - (0.5 \times 10^{-3}) v_{DS}^2$$

And from the circuit, we have that

$$i_D = \frac{18 - v_{DS}}{2000} = 9 \times 10^{-3} - (0.5 \times 10^{-3}) v_{DS}$$

Combining these two equations for i_D, we get

$$4v_{DS} - 0.5v_{DS}^2 = 9 - 0.5v_{DS} \quad \Rightarrow \quad v_{DS}^2 - 9v_{DS} + 18 = 0$$

Solving this quadratic equation, we get the two solutions $v_{DS1} = 3$ V and $v_{DS2} = 6$ V. However, for the JFET to be in the ohmic region, we require that $v_{DS} \leq v_{GS} - V_p = 0 - (-4) = 4$ V. Thus we select as the solution $v_{DS} = 3$ V, and this confirms that the JFET is in the ohmic region. Furthermore, we have that the drain current is

$$i_D = \frac{18 - 3}{2000} = 7.5 \text{ mA}$$

We may also obtain this value for i_D by substituting the appropriate values into Eq. 8.2.

Drill Exercise 8.1

For the circuit given in Fig. 8.9, suppose that the JFET has the parameters $I_{DSS} = 8$ mA and $V_p = -4$ V. Determine the value of R_D for which the JFET operates on the border between the active and the ohmic regions.

ANSWER 1.75 kΩ

Example 8.2

Suppose that for the circuit given in Fig. 8.9, we place a 1-kΩ resistor in series with the source of the JFET as shown in Fig. 8.10. Let us find v_{GS}, i_D, and v_{DS} given that $I_{DSS} = 8$ mA and $V_p = -4$ V.

Assuming that the gate is reverse biased ($v_{GS} \leq 0$ V), then $i_G = 0$ A. Thus

$$i_D = -\frac{v_{GS}}{1000} \tag{8.5}$$

If the transfer characteristic of the JFET is available, the line described by this equation can be plotted using the same axes. The intersection of this line and the transfer-characteristic curve gives the graphical solution of the values for i_D and v_{GS}. But suppose that the transfer characteristic is not given.

Assuming that the JFET is in the active region, from Equation (8.4) we have

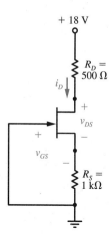

Fig. 8.10 Preceding circuit with added source resistance.

$$i_D = (8 \times 10^{-3})\left(1 - \frac{V_{GS}}{-4}\right)^2 = (8 \times 10^{-3})\left(1 + \frac{v_{GS}}{2} + \frac{v_{GS}^2}{16}\right)$$

Combining this with Eq. 8.5, we get

$$-v_{GS} = 8\left(1 + \frac{v_{GS}}{2} + \frac{v_{GS}^2}{16}\right) = 8 + 4v_{GS} + \frac{1}{2}v_{GS}^2$$

from which

$$v_{GS}^2 + 10v_{GS} + 16 = 0$$

The two solutions to this quadratic equation are $v_{GS1} = -2$ V and $v_{GS2} = -8$ V. If the JFET is to be in the active region, then $V_p = -4 < v_{GS} \le 0$ V. Thus the solution we seek is $v_{GS} = -2$ V, and the gate is reverse biased as assumed. Substituting this value into Eq. 8.5 yields $i_D = 2$ mA. Furthermore, by KVL,

$$v_{DS} = -500i_D + 18 - 1000i_D = -1500(2 \times 10^{-3}) + 18 = 15 \text{ V}$$

and since $v_{DS} = 15 > v_{GS} - V_p = -2 + 4 = 2$ V, then the JFET is indeed in the active region.

An alternative approach to the analysis of the circuit can be taken. By KVL, or from Eq. 8.5, we have

$$v_{GS} = -1000i_D$$

Substituting this fact into Eq. 8.4 yields

$$i_D = (8 \times 10^{-3})\left(1 - \frac{-1000i_D}{-4}\right)^2 = (8 \times 10^{-3})(1 - 500i_D + 62,500i_D^2)$$

from which

$$i_D^2 - (10 \times 10^{-3})i_D + 16 \times 10^{-6} = 0$$

The solutions to this equation are $i_{D1} = 2 \times 10^{-3}$ A and $i_{D2} = 8 \times 10^{-3}$ A. For the JFET to be in the active region, $0 < i_D \le I_{DSS} = 8$ mA. However, if we select $i_D = 8$ mA, then

$$v_{GS} = -1000i_D = -1000(8 \times 10^{-3}) = -8 \text{ V}$$

Since $v_{GS} = -8 < V_p = -4$ V, the JFET is not in the active region—a contradiction. Therefore, we choose $i_D = 2$ mA. This value of drain current results in

$$v_{GS} = -1000i_D = -1000(2 \times 10^{-3}) = -2 \text{ V}$$

and

$$v_{DS} = -500i_D + 18 - 1000i_D = 15 \text{ V}$$

Since $V_p = -4 < v_{GS} = -2 < 0$ V and $v_{DS} = 15 > v_{GS} - V_p = 2$ V, the JFET is indeed in the active region.

Drill Exercise 8.2

For the circuit given in Fig. 8.10, suppose that the JFET has the parameters $I_{DSS} = 8$ mA and $V_p = -4$ V. Determine the value of R_D for which the JFET operates on the border between the active and the ohmic regions.

ANSWER 7 kΩ

Although this section has discussed only n-channel JFETs, p-channel JFETs are treated in a similar manner. Figure 8.11 shows the circuit symbol for a p-channel JFET. Note that the currents and voltages are designated just as they are for an n-channel JFET. This means that the currents and voltages for a p-channel JFET are opposite in sign to those of the corresponding n-channel JFET. For example, for an *n*-channel JFET, the drain characteristics are plots of positive-valued drain current (i_D) versus positive-valued drain-source voltage (v_{DS}) for different nonpositive values of gate-source voltage (v_{GS}). For a *p*-channel JFET, though, the drain characteristics have negative values for i_D and v_{DS}, whereas v_{GS} takes on nonnegative values. Figure 8.12 shows typical drain characteristics for a *p*-channel JFET.

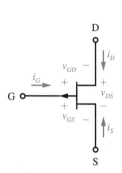

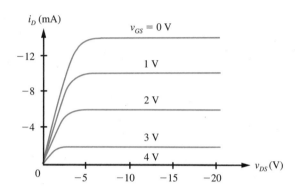

Fig. 8.11 Circuit symbol for a *p*-channel JFET.

Fig. 8.12 Drain characteristics for a *p*-channel JFET.

8.2 Metal-Oxide-Semiconductor Field-Effect Transistors (MOSFETs)

To function properly, the channel of a JFET is narrowed or depleted of majority charge carriers. Because of this fact, we say that a JFET operates in a **depletion mode,** and we refer to it as a **depletion-type** FET. The JFET, however, is not the only type of depletion FET.

The Depletion MOSFET

To see how another kind of depletion FET can be fabricated, we begin with a body or foundation of *p*-type silicon, called a **substrate,** or **body** and form two heavily-doped *n*-type regions, denoted n^+. An *n*-type channel, labeled *n,* is then formed between the two n^+ regions, as shown in Fig. 8.13*a*. In addition to the substrate (B), metal contacts are made to the two n^+ regions—these are the drain (D) and the source (S). Next, the channel is covered with a thin (about 0.1 μm) insulating oxide layer of silicon dioxide (SiO_2). The gate (G) is obtained by depositing a metal on the oxide layer as is illustrated. Such a device is known as a **metal-oxide-semiconductor field-effect transistor** (MOSFET). In particular, the transistor shown in Fig. 8.13*a* is an ***n*-channel** MOSFET (also denoted **NMOS**), and its circuit symbol is given in Fig. 8.13*b*. If the *n*- and *p*-type materials are interchanged, a ***p*-channel** MOSFET (also denoted **PMOS**) results. (To obtain the circuit symbol for a *p*-channel

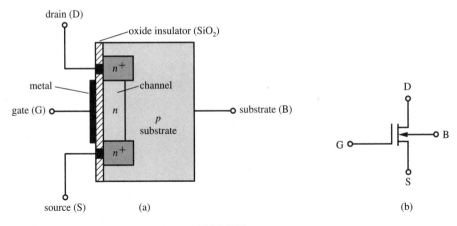

Fig. 8.13 An *n*-channel depletion MOSFET.

MOSFET, reverse the direction of the arrow in Fig. 8.13*b*. Because there is a layer of insulation between the gate and the channel, a MOSFET is also known as an **insulated-gate field-effect transistor** (IGFET).

For normal JFET operation, the gate is reverse biased, and the resistance between the gate and source is typically many megohms. However, for a MOSFET, because of the insulation between the gate and the substrate, the resistance between the gate and the source is extremely high—typically ranging from 10^{10} to 10^{15} Ω. Therefore, as we did for a JFET, we will take as a given that $i_G = 0$ A for a MOSFET.

Typically, the substrate and the source of a MOSFET are connected. If v_{GS} is made negative, then positive charges are induced in the *n*-type channel, thereby effectively narrowing it, as is the case for a JFET. In addition, for varying values of v_{DS}, the behavior of the device is similar to that of a JFET. Since the MOSFET depicted in Fig. 8.13 can operate in the depletion mode, it is called a **depletion MOSFET.**

When operating in the depletion mode, a depletion MOSFET has characteristics like those of a JFET. However, because of the insulation between the gate and the semiconductor, positive values of v_{GS} are allowable. When v_{GS} is made positive, additional negative charges are induced in the *n*-type channel, thereby effectively widening it. When the channel is widened by or enhanced with majority charge carriers, we say that it operates in an **enhancement mode.** Operation of a depletion MOSFET in the enhancement mode is similar to that in the depletion mode. Consequently, even though in the depletion mode it behaves similarly to a JFET, a depletion MOSFET's characteristics are extended versions of those of a JFET (because positive values of v_{GS} are included). This is demonstrated by the typical *n*-channel depletion MOSFET drain and transfer characteristics shown in Fig. 8.14. Eq. 8.2 and Eq. 8.4 on p. 510, which describe operation in the ohmic and active regions, respectively, again are valid for depletion MOSFETs.

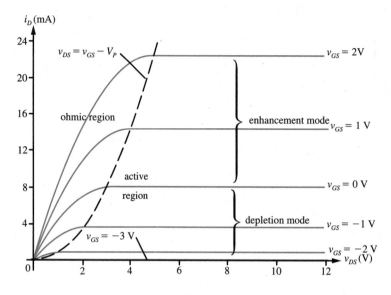

(a) drain characteristics

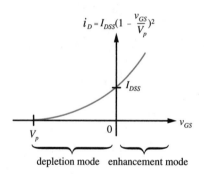

(b) transfer characteristic

Fig. 8.14 Characteristics of a typical *n*-channel depletion MOSFET.

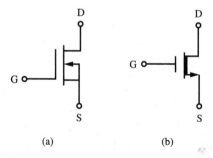

Fig. 8.15 Circuit symbols for an *n*-channel depletion MOSFET.

The typical situation that the substrate of an *n*-channel depletion MOSFET is connected to the source is depicted in Fig. 8.15*a*. For this case, we will use the simplified circuit symbol shown in Fig. 8.15*b*. By reversing the direction of the arrow in either circuit symbol, we obtain the circuit symbol for a *p*-channel depletion MOSFET.

Example 8.3

Consider the MOSFET circuit shown in Fig. 8.16, where the depletion NMOS transistor has $I_{DSS} = 8$ mA and $V_p = -2$ V. Let us find v_{GS}, i_D, and v_{DS}.

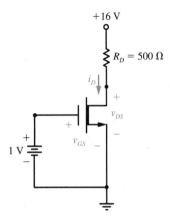

Fig. 8.16 MOSFET circuit for Example 8.3.

By inspection, $v_{GS} = 1$ V. If the MOSFET is in the active region, then from Eq. 8.4

$$i_D = I_{DSS}\left(1 - \frac{v_{GS}}{V_p}\right)^2 = (8 \times 10^{-3})\left(1 - \frac{1}{-2}\right)^2 = 18 \text{ mA}$$

Thus,

$$v_{DS} = -500(18 \times 10^{-3}) + 16 = 7 \text{ V}$$

and since $v_{DS} = 7 > v_{GS} - V_p = 1 + 2 = 3$ V, the MOSFET is indeed in the active region.

Suppose that R_D is changed to 750 Ω. If we again assume that the MOSFET is in the active region, then, as before, $i_D = 18$ mA and

$$v_{DS} = -750(18 \times 10^{-3}) + 16 = 2.5 \text{ V}$$

However, since $v_{DS} = 2.5 < v_{GS} - V_p = 3$ V, then, contrary to our assumption, the MOSFET is not in the active region. Therefore, let us assume that it is in the ohmic region. From Eq. 8.2, we have

$$i_D = (8 \times 10^{-3})\left[2\left((1 - \frac{1}{-2}\right)\frac{v_{DS}}{2} - \left(\frac{v_{DS}}{-2}\right)^2\right]$$

Using this expression and Ohm's law, we get

$$i_D = (12 \times 10^{-3})v_{DS} - (2 \times 10^{-3})v_{DS}^2 = \frac{16 - v_{DS}}{750}$$

This yields the quadratic equation

$$3v_{DS}^2 - 20v_{DS} + 32 = 0$$

the solutions of which are $v_{DS1} = \frac{8}{3}$ V and $v_{DS2} = 4$ V. Since $v_{DS1} = \frac{8}{3} < v_{GS} - V_p = 3$ V implies ohmic-region operation, and $v_{DS2} = 4 > v_{GS} - V_p = 3$ V implies active-region operation, we have that $v_{DS} = \frac{8}{3}$ V. The corresponding drain current, therefore, is

$$i_D = \frac{16 - \frac{8}{3}}{750} = 17.8 \text{ mA}$$

Drill Exercise 8.3

For the circuit given in Fig. 8.16, place a 125-Ω resistor in series with the source of the MOSFET. Suppose that the MOSFET has parameters $I_{DSS} = 8$ mA and $V_p = -2$ V. Find v_{GS}, i_D, and v_{DS}.

ANSWER   0 V, 8 mA, 11 V

On an IC chip, the space required for a resistor is generally much larger than the space required for a transistor (either BJT or FET). Consequently, transistors are often used in place of resistors.

Example 8.4

The circuit shown in Fig. 8.17 uses a depletion MOSFET M_2 in place of a resistor. Let us find the voltages and the currents for this circuit for the case that $I_{DSS} = 16$ mA and $V_p = -4$ V for both MOSFETs.

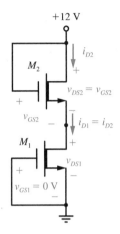

Fig. 8.17 MOSFET circuit without resistors.

Since there is a short circuit connected between the gate and the drain of transistor M_2, then $v_{GD2} = 0$ V. By KVL, therefore, $v_{DS2} = v_{GS2}$. But $v_{DS2} = v_{GS2} < v_{GS2} - V_p = v_{GS2} + 4$ means that M_2 will operate only in the ohmic region (when $i_{D2} > 0$ A). Let us assume that M_1 operates in the active region.

Since $v_{GS1} = 0$ V, then $i_{D1} = I_{DSS} = 16$ mA, and since the gate currents are zero, then $i_{D1} = i_{D2} = 16$ mA. But M_2 will only operate in the ohmic region. Therefore, from Eq. 8.2,

$$16 \times 10^{-3} = (16 \times 10^{-3}) \left[2 \left(1 - \frac{v_{GS2}}{-4} \right) \frac{v_{GS2}}{4} - \left(\frac{v_{GS2}}{-4} \right)^2 \right]$$

Simplifying this expression yields

$$v_{GS2}^2 + 8 v_{GS2} - 16 = 0 \quad \Rightarrow \quad v_{GS2} = -9.66 \text{ V}, \ 1.66 \text{ V}$$

Since $i_{D2} > 0$ A when $v_{GS2} > V_p = -4$ V, then $v_{GS2} = 1.66$ V $= v_{DS2}$. By KVL, $v_{DS1} = -v_{DS2} + 12 = -1.66 + 12 = 10.34$ V. Since, $v_{DS1} = 10.34 > v_{GS1} - V_p = 0 + 4 = 4$ V, then M_1 is confirmed to be operating in the active region.

> ### Drill Exercise 8.4
>
> For the circuit given in Fig. 8.17, if the short circuit between the gate and the source of M_1 is removed and instead placed between the gate and the drain of M_1, then both MOSFETs will operate only in the ohmic region. Find the resulting values for v_{DS1}, v_{DS2}, v_{GS1}, v_{GS2}, i_{D1}, i_{D2}, and the channel resistance r_{DS} for each MOSFET given that $I_{DSS} = 16$ mA and $V_p = 4$ V. (*Hint:* Use the fact that the MOSFETS have identical parameters and are identically connected.)
>
> **ANSWER** 6 V, 6 V, 6 V, 6 V, 84 mA, 84 mA, 71.4 Ω, 71.4 Ω

The Enhancement MOSFET

Consider again the MOSFET depicted in Fig. 8.13. Let us remove the *n*-type material that constitutes the channel between the two n^+ regions. The resulting device is shown in Fig. 8.18*a,* and the circuit symbol for this device is shown in Fig. 8.18*b.* Suppose again that the substrate is connected to the source, and suppose that $v_{GS} = 0$ V. To have a (positive) current go from drain to source, we require that the drain voltage be greater than the source voltage. But this means that the *pn* junction formed by the drain n^+ region and the *p*-type substrate is reverse biased. Consequently, there is no drain current.

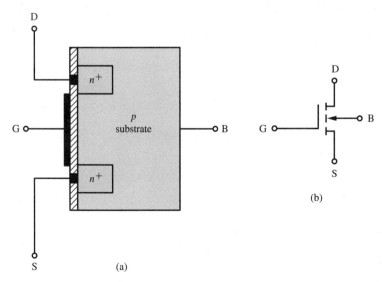

Fig. 8.18 (*a*) An *n*-channel enhancement MOSFET, and (*b*) its circuit symbol.

Next, let us consider the case that v_{GS} is made positive. This will induce negative charges near the p-type semiconductor (substrate) surface beneath the gate. The more positive v_{GS} is, the more induced negative charges there are. When v_{GS} is made sufficiently large, an ''inversion layer'' of electrons results and this, in effect, forms an n-type channel. The voltage required for this to happen is called the **gate-source threshold voltage** and is designated by V_t. Typically, V_t ranges between 1 and 3 V.

As v_{GS} is increased beyond V_t, the channel is widened or enhanced. Once a channel has been established (i.e., for $v_{GS} > V_t$), the behavior of such a device, called an **enhancement MOSFET,** is similar to that discussed previously. The drain characteristics of a typical n-channel enhancement MOSFET are shown in Fig. 8.19a. For this enhancement NMOS transistor, the (gate-source) threshold voltage is $V_t = 2$ V. Thus, if $v_{GS} \leq 2$ V, then the gate-source voltage is not sufficient to produce a channel, and the MOSFET is cut off. If $v_{GS} > 2$ V, then the gate-source voltage is large enough to establish a channel. Under this circumstance, however, the value of v_{DS} will determine whether or not the channel is pinched off, and, therefore, whether the MOSFET is in the active or ohmic region.

For the typical situation that the substrate of an n-channel enhancement MOSFET is connected to the source, we may use either of the circuit symbols shown in Fig. 8.19b.

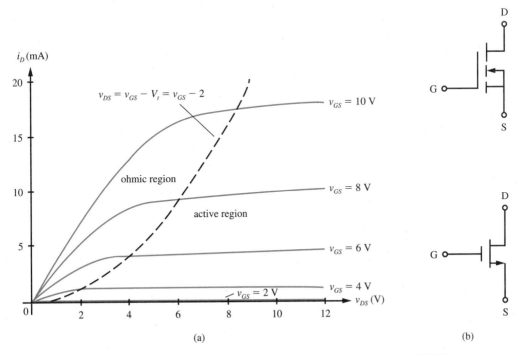

Fig. 8.19 (a) Drain characteristics of a typical n-channel enhancement MOSFET, and (b) circuit symbols for an n-channel enhancement MOSFET.

Again, by reversing the direction of the arrow in the circuit symbol for an n-channel enhancement MOSFET, we obtain the circuit symbol for a p-channel enhancement MOSFET.

Suppose that a channel has been established ($v_{GS} > V_t$) in an enhancement NMOS transistor. Because of the symmetry of the device, this channel will be pinched off when $v_{GD} \leq V_t$. Since $v_{GD} = v_{GS} - v_{DS}$, pinchoff occurs when

$$v_{GS} - v_{DS} \leq V_t \qquad \Rightarrow \qquad v_{DS} \geq v_{GS} - V_t$$

Under this condition, therefore, the MOSFET is in the active region. Of course, this means that it is in the ohmic region when $v_{DS} < v_{GS} - V_t$. For the drain characteristics given in Fig. 8.19a, the dashed curve described by

$$v_{DS} = v_{GS} - V_t$$

therefore, is the border between the active and the ohmic regions.

Unlike the JFET and the depletion MOSFET, an enhancement MOSFET operating in the active region satisfies the equation[2]

$$\boxed{i_D = K(v_{GS} - V_t)^2} \qquad (8.6)$$

where the value of the constant K depends on the fabrication parameters of the device. A plot of the curve described by this equation is the transfer characteristic of the MOSFET. Such a typical i_D-v_{GS} plot is shown in Fig. 8.20. If we know V_t and the values of current and voltage for any point on this curve (for $v_{GS} > V_t$), we can determine K easily by using Eq. 8.6. As an example of an alternative method for finding K, note that from the drain characteristics given in Fig. 8.19a, when the MOSFET is in the active region and $v_{GS} = 6$ V, then $i_D \approx 4$ mA. Thus from Eq. 8.6, we have the approximation

$$K = \frac{i_D}{(v_{GS} - V_t)^2} = \frac{4 \times 10^{-3}}{(6 - 2)^2} = 2.5 \times 10^{-4} = 0.25 \text{ mA}/\text{V}^2$$

Ideally, this value of K is the same throughout the active region. However, if we consider the case that $v_{GS} = 8$ V, then $i_D \approx 10$ mA, and from Eq. 8.6 we get the approximation

$$K = \frac{i_D}{(v_{GS} - V_t)^2} = \frac{10 \times 10^{-3}}{(8 - 2)^2} = 2.78 \times 10^{-4} = 0.278 \text{ mA}/\text{V}^2$$

---

[2]In practice, a better expression is $i_D = K(v_{GS} - V_t)^2(1 + \lambda v_{DS})$, where typically $0.01 \leq \lambda \leq 0.1$ V^{-1}. This expression indicates that i_D has some dependence on v_{DS}.

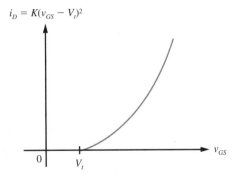

$i_D = K(v_{GS} - V_t)^2$

0

V_t

v_{GS}

Fig. 8.20 Transfer characteristic of an enhancement MOSFET.

Therefore, for the sake of simplicity in computation, it is reasonable to use the approximate value $K = 2.6 \times 10^{-4} = 0.26 \text{ mA}/\text{V}^2$ throughout the active region.

Previously we mentioned that the dashed curve in Fig. 8.19a, which separates the active and ohmic regions, corresponds to $v_{DS} = v_{GS} - V_t$. Substituting this fact into Eq. 8.6, we get the equation of this dashed curve, which is

$$i_D = K v_{DS}^2$$

and, therefore, this curve is a parabola.

For the case that an enhancement MOSFET is in the ohmic region, it can be shown that

$$i_D = K[2(v_{GS} - V_t)v_{DS} - v_{DS}^2] \tag{8.7}$$

where the constant K (ideally) has the same value as for active-region operation. For the case that v_{DS} is small, we can approximate Eq. 8.7 by

$$i_D \approx K[2(v_{GS} - V_t)v_{DS}] \qquad \Rightarrow \qquad \frac{i_D}{v_{DS}} \approx 2K(v_{GS} - V_t)$$

Thus for small values of $v_{DS,}$ when the MOSFET behaves as a linear resistor, the approximate value of drain-source resistance r_{DS} is given by

$$r_{DS} = \frac{v_{DS}}{i_D} \approx \frac{1}{2K(v_{GS} - V_t)} \tag{8.8}$$

For example, for the n-channel enhancement MOSFET just discussed, from Eq. 8.8, when $v_{GS} = 8 \text{ V}$, we have that for small values of v_{DS},

$$r_{DS} \approx \frac{1}{2(2.6 \times 10^{-4})(8 - 2)} = 321 \ \Omega$$

whereas when $v_{GS} = 6$ V,

$$r_{DS} \approx \frac{1}{2(2.6 \times 10^{-4})(6 - 2)} = 481 \ \Omega$$

Example 8.5

Consider the circuit shown in Fig. 8.21. From the characteristics of the n-channel enhancement MOSFET, suppose it is determined that, in the active region, when $v_{GS} = 8$ V, then $i_D \approx 9$ mA. Furthermore, suppose it is given that $V_t = 2$ V and $V_{GG} = 10$ V. Let us find i_D and v_{DS} for the case that (a) $R_D = 250 \ \Omega$, and (b) $R_D = 1 \ k\Omega$.

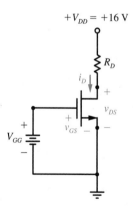

Fig. 8.21 Enhancement NMOS circuit.

From Eq. 8.6, we have that the approximate value of K is

$$K = \frac{i_D}{(v_{GS} - V_t)^2} = \frac{9 \times 10^3}{(8 - 2)^2} = 2.5 \times 10^{-4} = 0.25 \ \mathrm{mA/V^2}$$

(a) For the case that $R_D = 250 \ \Omega$, let us assume active-region operation. Since $v_{GS} = V_{GG} = 10$ V, from Eq. 8.6 we have

$$i_D = K(v_{GS} - V_t)^2 = (2.5 \times 10^{-4})(10 - 2)^2 = 16 \ \mathrm{mA}$$

Thus,

$$v_{DS} = 250(16 \times 10^{-3}) + 16 = 12 \ \mathrm{V}$$

Since $v_{DS} = 12 > v_{GS} - V_t = 10 - 2 = 8$ V, then active-region operation is confirmed.

(b) For the case that $R_D = 1$ kΩ, again assume active-region operation. As in Part (a), $i_D = 16$ mA. Thus,

$$v_{DS} = -1000(16 \times 10^{-3}) + 16 = 0 \text{ V}$$

But since $v_{DS} = 0 < v_{GS} - V_t = 8$ V, this is a contradiction to operation in the active region. Therefore, let us assume ohmic-region operation. From Eq. 8.7,

$$i_D = K[2(v_{GS} - V_t)v_{DS} - v_{DS}^2] = (2.5 \times 10^{-4})[2(10 - 2)v_{DS} - v_{DS}^2]$$

By Ohm's law, $i_D = (16 - v_{DS})/1000$. Thus

$$i_D = (2.5 \times 10^{-4})[2(10 - 2)v_{DS} - v_{DS}^2] = \frac{16 - v_{DS}}{1000}$$

Simplifying this expression, we get the quadratic equation

$$v_{DS}^2 - 20v_{DS} + 64 = 0$$

the solutions of which are $v_{DS1} = 4$ V and $v_{DS2} = 16$ V. However, as stated above, for ohmic-region operation, we require that $v_{DS} < 8$ V. Thus we choose the solution $v_{DS} = 4$ V. As a result of this

$$i_D = \frac{16 - 4}{1000} = 12 \text{ mA}$$

Drill Exercise 8.5

For the circuit given in Fig. 8.21, suppose that $R_D = 250$ Ω and the MOSFET has parameters $K = 0.25$ mA/V^2 and $V_t = 2$ V. Find V_{GG} such that $i_D = 4$ mA.

ANSWER 6 V

Example 8.6

Let us find the voltages and the currents for the circuit shown in Fig. 8.22 for the case that both enhancement MOSFETs have $K = 0.25 \text{ mA}/\text{V}^2$ and $V_t = 2$ V when (a) $V_{GG} = 2$ V, and (b) $V_{GG} = 12$ V.

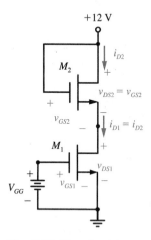

Fig. 8.22 Circuit with two enhancement MOSFETs.

Since there is a short circuit connected between the gate and the drain of M_2, then $v_{GD2} = 0$ V, and by KVL, $v_{DS2} = v_{GS2}$. But $v_{DS2} = v_{GS2} > v_{GS2} - V_t = v_{GS2} - 2$ means that M_2 can operate only in the active region (when $i_{D2} > 0$ A).

(a) By inspection, we see that $v_{GS1} = V_{GG} = 2$ V $= V_t$. Thus M_1 is cut off, and $i_{D1} = 0$ A. Since the gate currents are zero, $i_{D1} = i_{D2} = 0$ A. Since M_2 operates only in the active region,[3] then

$$i_{D2} = K(v_{GS2} - V_t)^2 = (2.5 \times 10^{-4})(v_{GS2} - 2)^2 = 0 \quad \Rightarrow \quad v_{GS2} = 2 \text{ V}$$

and by KVL,

$$v_{DS1} = -v_{DS2} + 12 = -2 + 12 = 10 \text{ V}$$

(b) For the case that $V_{GG} = 12$ V, by inspection $v_{GS1} = V_{GG} = 12$ V. Since M_2 operates only in the active region,

$$i_{D2} = K(v_{GS2} - V_t)^2 = (2.5 \times 10^{-4})(v_{GS2} - 2)^2 \tag{8.9}$$

---

[3]Even though M_1 is cut off, it has a very small (negligible) drain current. Thus, M_2 is in the active region—albeit very close to cutoff.

If M_1 is in the active region, then $i_{D1} = K(v_{GS1} - V_t)^2 = (2.5 \times 10^{-4})(12 - 2)^2 = 25$ mA. But $i_{D1} = i_{D2} = 25$ mA $\Rightarrow$ $v_{GS2} = 12$ V $= v_{DS2}$, and $v_{DS1} = -v_{DS2} + 12 = -12 + 12 = 0 < v_{GS1} - V_t = 12 - 2 = 10$ V—a contradiction. Therefore, M_1 must be in the ohmic region.

We thus have that

$$i_{D1} = K[2(v_{GS1} - V_t)v_{DS1} - v_{DS1}^2]$$
$$= (2.5 \times 10^{-4})[2(12 - 2)v_{DS1} - v_{DS1}^2] \tag{8.10}$$

Combining Eq. 8.9 and Eq. 8.10, along with the facts that $i_{D1} = i_{D2}$ and $v_{GS2} = v_{DS2}$, we get

$$(2.5 \times 10^{-4})(v_{DS2} - 2)^2 = (2.5 \times 10^{-4})[2(12 - 2)v_{DS1} - v_{DS1}^2]$$

Substituting $v_{DS2} = 12 - v_{DS1}$ into this equation and simplifying yields the quadratic equation

$$v_{DS1}^2 - 20v_{DS1} + 50 = 0 \quad \Rightarrow \quad v_{DS1} = 2.93 \text{ V}, 17.1 \text{ V}$$

Since M_1 is in the ohmic region, $v_{DS1} < v_{GS1} - V_t = 12 - 2 = 10$ V $\Rightarrow$ $v_{DS1} = 2.93$ V. Therefore, $v_{GS2} = v_{DS2} = 12 - v_{DS1} = 12 - 2.93 = 9.07$ V, and $i_{D1} = i_{D2} = K(v_{GS2} - V_t)^2 = (2.5 \times 10^{-4})(9.07 - 2)^2 = 12.5$ mA.

Drill Exercise 8.6

For the circuit shown in Fig. 8.22, suppose that both enhancement MOSFETs have $K = 0.25$ mA$/$V^2 and $V_t = 2$ V. When $V_{GG} = 6$ V, then M_1 operates in the active region. Find i_{D1}, i_{D2}, v_{DS1}, and v_{DS2}.

ANSWER 4 mA, 4 mA, 6 V, 6 V

8.3 MOSFET Logic Gates

Let us now employ a graphical approach to the analysis of MOSFET circuits. To be specific, consider the circuit shown in Fig. 8.23a, where the enhancement NMOS transistor has a threshold voltage of $V_t = 2$ V and its drain (output) characteristics are as depicted in Fig. 8.23b. For the given circuit,

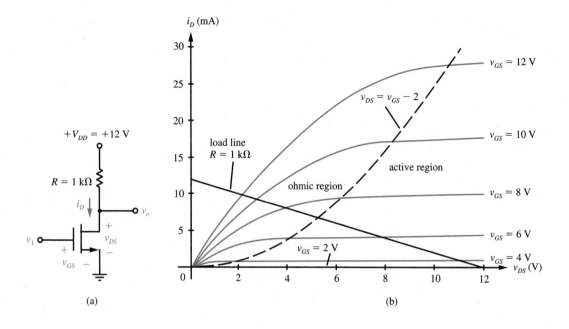

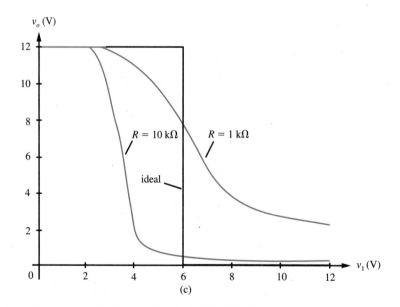

Fig. 8.23 (*a*) An enhancement-MOSFET circuit, (*b*) MOSFET drain characteristics, and (*c*) transfer characteristics.

$$i_D = \frac{V_{DD} - v_{DS}}{R} = \frac{12 - v_{DS}}{1 \times 10^3} = -(1 \times 10^{-3})v_{DS} + 12 \times 10^{-3}$$

The corresponding load line is plotted on the drain characteristics as indicated.

We are now in a position to determine graphically, and quite easily, the output voltage $v_o = v_{DS}$ that results when an input voltage $v_1 = v_{GS}$ is applied. For example, if the input is a high voltage of $v_1 = 12$ V, then from the intersection of the load line and the curve labeled $v_{GS} = 12$ V, we have approximately that $i_D = 10$ mA and $v_{DS} = 2$ V. Thus when the input is $v_1 = 12$ V, the output is $v_o = v_{DS} = 2$ V. Conversely, if the input is a low voltage of $v_1 = 2$ V, then from the intersection of the load line and the curve labeled $v_{GS} = 2$ V (which coincides with the horizontal axis), we have that $i_D = 0$ A and $v_{DS} = 12$ V. Thus, when the input is $v_1 = 2$ V, the output is $v_o = v_{DS} = 12$ V. Hence, choosing the low and high voltages to be 2 V and 12 V, respectively, the circuit given in Fig. 8.23(a) is an example of an inverter (NOT gate). Note that when the input is low, the MOSFET is in cutoff (OFF), and when the input is high, the MOSFET is in the ohmic region.

An ideal inverter has an output that is either high or low, and the transition between these two outputs occurs when the input is midway between the high and the low values. (This results in maximum noise margins.) Furthermore, a low voltage of 0 V and a high voltage of $+V_{DD}$ would result in the maximum separation between the low and the high voltages. The transfer characteristic of an ideal inverter is indicated in Fig. 8.23c. Also shown in Fig. 8.23c is the transfer characteristic for the inverter given in Fig. 8.23a for the case that $R = 1$ kΩ, and as we can see, this characteristic is far from being ideal. However, if we change R from 1 kΩ to 10 kΩ, then the result is a transfer characteristic, indicated in Fig. 8.23c, which is closer to being ideal.

Inverter with Active-Region Load

For the inverter circuit given in Fig. 8.23a, the load is a resistor R. In an IC, however, a resistor can occupy 20 times more space than a MOSFET does. Therefore, let us use a MOSFET as a resistive load, albeit nonlinear, in place of the linear resistance R. One way to do this is to connect a short circuit between the gate and the drain of an n-channel enhancement MOSFET as depicted in Fig. 8.24a. Suppose $V_t = 2$ V and the drain characteristics for this transistor are as given in Fig. 8.24b.

Since $v_{DS} = v_{GS} > v_{GS} - V_t = v_{GS} - 2$, the MOSFET will operate only in the active region (when $i_D > 0$ A). That is, the MOSFET will operate at some point on the curve labeled $v_{DS} = v_{GS}$ shown in Fig. 8.24b. By inspection, we see that the curve $v_{DS} = v_{GS}$ indicates that there is a nonlinear relationship between i_D and v_{DS}. In other words, between the drain and the source, the MOSFET has a nonlinear resistance.

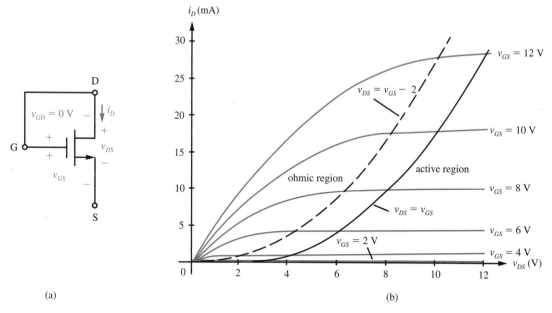

Fig. 8.24 (*a*) An *n*-channel enhancement MOSFET, and (*b*) drain characteristics.

Example 8.7

Let us analyze the inverter shown in Fig. 8.25*a* for the case that both enhancement MOSFETs have $K = 0.25$ mA$/$V^2 and $V_t = 2$ V.

Since it operates only in the active region, the enhancement NMOS transistor M_2 is called an **active-region load**, and the enhancement NMOS transistor M_1 is known as the **driver**. The drain characteristics for M_1 (M_2 has identical drain characteristics) are shown in Fig. 8.25*b*. Note, too, that since a MOSFET has no gate current, $i_{D1} = i_{D2} = i_D$.

Since it operates in the active region, for the load we have that

$$i_D = K_2(v_{GS2} - V_t)^2 = K_2(V_{DD} - v_{DS1} - V_t)^2 = K_2(12 - v_{DS1} - 2)^2$$
$$= K_2(10 - v_{DS1})^2$$

from which

$$i_D = (2.5 \times 10^{-4})(v_{DS1} - 10)^2 \tag{8.11}$$

A plot of this equation, called the **load curve**, is indicated on the drain characteristics in Fig. 8.25*b*, and the corresponding transfer characteristic is shown in Fig. 8.26.

Note that this transfer characteristic is far from being ideal. As we have seen, we can improve a transfer characteristic by increasing the resistance of the load. Since

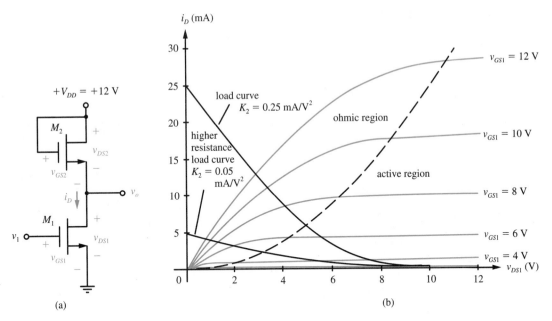

Fig. 8.25 (*a*) Enhancement NMOS inverter, and (*b*) drain characteristics.

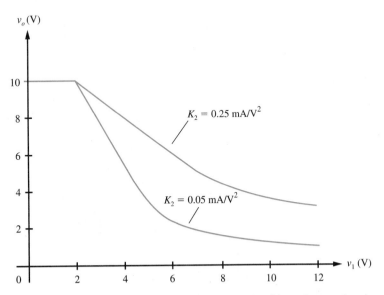

Fig. 8.26 Transfer characteristics for inverter with active-region load.

an increase in resistance corresponds to a decrease in current, let us decrease the value of K for the load MOSFET. In particular, consider the case that $K_2 = 0.05 \ \text{mA/V}^2 = 50 \ \mu\text{A/V}^2$. The resulting load curve, which is described by the equation

$$i_D = K_2(v_{DS1} + V_t - V_{DD})^2 = (5 \times 10^{-5})(10 - v_{DS1})^2$$

$$= (5 \times 10^{-5})(v_{DS1} - 10)^2$$

is also depicted in Fig. 8.25b. The corresponding transfer characteristic, indicated in Fig. 8.26, is seen to be closer to the ideal case.

Drill Exercise 8.7

For the inverter given in Fig. 8.25a, suppose that both MOSFETs have $V_t = 2$ V. Given that $K_1 = 0.25$ mA/V^2, analytically determine v_o for the case that $v_1 = 12$ V and $K_2 = 0.05$ mA/V^2. (*Hint:* M_1 operates in the ohmic region.)

ANSWER 0.87 V

Inverter with Ohmic-Region Load

For the enhancement-MOSFET inverter with an active-region load shown in Fig. 8.25, the highest output voltage is $V_{DD} - V_t = 12 - 2 = 10$ V. We can increase the highest output voltage for the inverter by slightly modifying the load.

Example 8.8

Let us analyze the NMOS inverter shown in Fig. 8.27a for the case that both of the MOSFETs have $K = 0.25$ mA/V^2 and $V_t = 2$ V, and the corresponding drain characteristics are given in Fig. 8.27b.

If $V_{GG} = V_{DD}$, then the circuit in Fig. 8.27a is identical to the one in Fig. 8.25a. However, if $V_{GG} = V_{DD} + V_t + a$, where $a > 0$, then

$$v_{DS2} = V_{DD} - V_{GG} + v_{GS2} \tag{8.12}$$

$$= V_{DD} - (V_{DD} + V_t + a) + v_{GS2}$$

$$= v_{GS2} - V_t - a < v_{GS2} - V_t$$

which means that M_2 operates in the ohmic region.

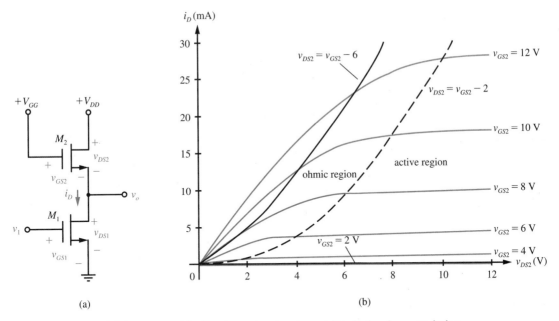

Fig. 8.27 (a) NMOS inverter with ohmic-region load, and (b) drain characteristics.

Suppose that, as for the preceding inverter, $V_{DD} = 12$ V. Let us choose $a = 4$. Then

$$V_{GG} = V_{DD} + V_t + a = 12 + 2 + 4 = 18 \text{ V}$$

From Eq. 8.12,

$$v_{DS2} = V_{DD} - V_{GG} + v_{GS2} = 12 - 18 + v_{GS2} = v_{GS2} - 6$$

A plot of this relationship is shown on the drain characteristics given in Fig. 8.27b. The resulting curve, which is nearly a straight line, also describes the relationship between $i_D = i_{D2}$ and v_{DS2}. We can use this nearly linear relationship, however, to sketch a plot of $i_D = i_{D2}$ versus v_{DS1}—this is the load curve and is is done as follows.

Since it operates in the ohmic region, for the load, we have that

$$i_D = K_2[2(v_{GS2} - V_t)v_{DS2} - v_{DS2}^2] = K_2[2(V_{GG} - v_{DS1} - V_t)v_{DS2} - v_{DS2}^2]$$

But, $V_{GG} = 18$ V and $v_{DS2} = 12 - v_{DS1}$. Thus,

$$i_D = K_2[2(18 - v_{DS1} - 2)(12 - v_{DS1}) - (12 - v_{DS1})^2]$$

$$= (2.5 \times 10^{-4})(v_{DS1}^2 - 32v_{DS1} + 240)$$

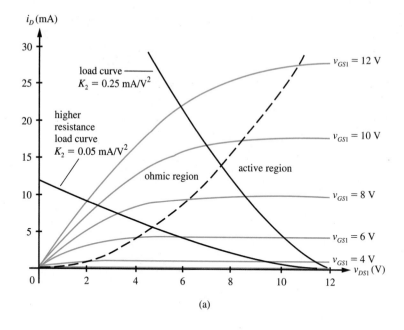

(a)

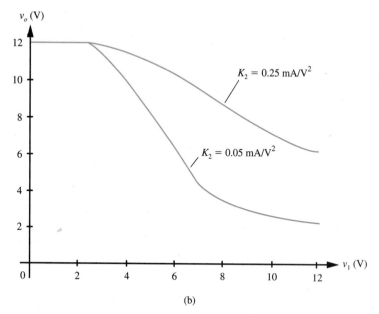

(b)

Fig. 8.28 (a) Load curves for ohmic-load inverter, and (b) transfer characteristics.

The corresponding load curve and transfer characteristic are shown in Fig. 8.28a and b, respectively.

Once again, we can improve the inverter's transfer characteristic by using a load MOSFET with a greater channel resistance (smaller K). Specifically, for the case that $K_2 = 0.05$ mA/$V^2 = 50$ μA/V^2, the equation for the resulting load curve is

$$i_D = (5 \times 10^{-5})(v_{DS1}^2 - 32v_{DS1} + 240)$$

and the load curve is indicated in Fig. 8.28a, whereas the transfer characteristic is indicated in Fig. 8.28b.

Drill Exercise 8.8

For the inverter given in Fig. 8.27a, suppose that $V_{GG} = 18$ V, $V_{DD} = 12$ V, and both MOSFETs have $V_t = 2$ V. Given that $K_1 = 0.25$ mA/V^2, analytically determine v_o for the case that $v_1 = 12$ V and K_2 is (a) 0.25 mA/V^2, and (b) 0.05 mA/V^2. (*Hint:* M_1 operates in the ohmic region.)

ANSWER   (a) 6 V, (b) 2 V

Note that the highest output voltage for the inverter with the active-region load (Fig. 8.25a) is $V_{DD} - V_t = 12 - 2 = 10$ V, whereas for the inverter with the ohmic-region load (Fig. 8.27a) it is $V_{DD} = 12$ V. The latter property, however, has to be paid for with the use of an additional supply voltage V_{GG}.

Inverter with Depletion-MOSFET Load

For the two types of inverters just discussed, one type had an active-region load and the other an ohmic-region load. In both cases, however, the load MOSFETs were enhancement NMOS transistors (as were the drivers). In practice, though, depletion MOSFETs are frequently used as load transistors.

Example 8.9

Let us analyze the NMOS inverter shown in Fig. 8.29a for the case that the enhancement MOSFET M_1 has $K = 0.25$ mA/V^2, $V_t = 2$ V, and the depletion

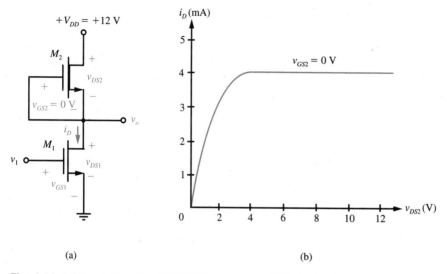

Fig. 8.29 (a) Depletion-load NMOS inverter, and (b) drain characteristics for M_2.

MOSFET M_2 has I_{DSS} = 4 mA, V_p = −4 V. The drain characteristics for M_2 are shown in Fig. 8.29b.

When M_2 is in the ohmic region,

$$I_D = I_{DSS}\left[2\left(1 - \frac{v_{GS2}}{V_p}\right)\frac{v_{DS2}}{-V_p} - \left(\frac{v_{DS2}}{V_p}\right)^2\right] = I_{DSS}\left[2\frac{v_{DS2}}{-V_p} - \frac{v_{DS2}^2}{V_p^2}\right]$$

Substituting $v_{DS2} = V_{DD} - v_{DS1} = 12 - v_{DS1}$, I_{DSS} = 4 mA, and V_p = −4 V into this equation yields

$$i_D = -(2.5 \times 10^{-4})(v_{DS1} - 4)(v_{DS1} - 12)$$

$$= -(2.5 \times 10^{-4})(v_{DS1}^2 - 16v_{DS1} + 48)$$

and this expression for the load curve is valid when M_2 is in the ohmic region, that is, when

$$v_{DS2} = V_{DD} - v_{DS1} \le v_{GS2} - V_p = -V_p \quad \Rightarrow$$

$$v_{DS1} \ge V_{DD} + V_p = 12 - 4 = 8 \text{ V}$$

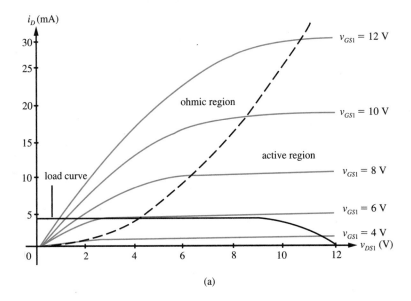

(a)

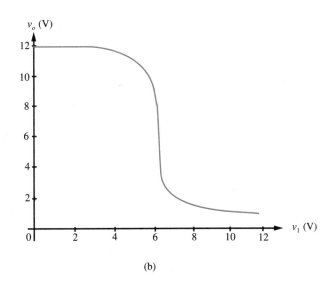

(b)

Fig. 8.30 (*a*) Load curve for depletion-load inverter, and (*b*) inverter transfer characteristic.

When M_2 is in the active region, i.e., when $v_{DS1} \leq 8$ V, then $i_D = I_{DSS} = 4$ mA. The corresponding load curve and transfer characteristic are shown in Fig. 8.30*a* and (b), respectively. Note how the transfer characteristic for the depletion-load inverter is closer to ideal than are those for the enhancement active- and ohmic-load inverters.

Drill Exercise 8.9

For the inverter given in Fig. 8.29a, suppose that the enhancement MOSFETs have $V_t = 2$ V and $K_1 = 0.25$ mA$/$V^2. Analytically determine v_o for the case that $v_1 = 12$ V given that the depletion MOSFET M_2 has parameters $I_{DSS} = 4$ mA and $V_p = -4$ V. (*Hint:* M_1 operates in the ohmic region and M_2 operates in the active region.)

ANSWER 0.835 V

An NMOS NOR Gate

Although NMOS logic gates can employ either active loads or ohmic loads, as we have already seen, depletion loads give superior results.

Example 8.10

Consider the MOSFET circuit shown in Fig. 8.31a, where M_1 and M_2 are enhancement NMOS transistors having a threshold voltage of V_t, and M_3 is a depletion MOSFET. Suppose that $V_{DD} > V_t > 0$ V and that both driver M_1 and driver M_2 have drain characteristics such as those shown in Fig. 8.31b, whereas the load M_3 is a MOSFET with high resistance (small value for I_{DSS}). Let us assume the ideal case that the low voltage is 0 V and the high voltage is V_{DD}.

Case 1. $v_1 = v_2 = 0$ V

Since $v_{GS1} = v_1 = v_{GS2} = v_2 = 0 < V_t$, then M_1 and M_2 are cut off (**OFF**). Thus $i_{D1} = i_{D2} = 0$ A, and, therefore, $i_{D3} = i_{D1} + i_{D2} = 0$ A. Since M_3 must operate on the characteristic curve $v_{GS3} = 0$ V (e.g., see Fig. 8.29b and change the subscripts from 2 to 3), it will operate at the point $i_{D3} = 0$ A and $v_{DS3} = 0$ V. Thus we have that $v_o = -v_{DS3} + V_{DD} = V_{DD}$.

Case 2. $v_1 = 0$ V, $v_2 = V_{DD}$

Since $v_{GS1} = v_1 = 0 < V_t$, then M_1 is OFF and $i_{D1} = 0$ A. However, $v_{GS2} = v_2 = V_{DD} > V_t$ means that M_2 can conduct current ($i_{D2} = i_{D3} \geq 0$ A), so we say that M_2 is **ON**. The intersection of the curve $v_{GS2} = V_{DD}$ and the load curve that results when M_1 is OFF yields the values of the drain current and output voltage v_o. From

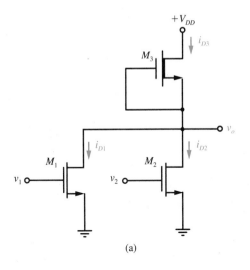

(a)

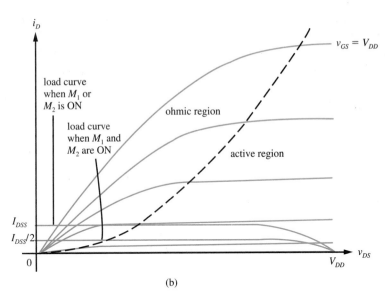

(b)

Fig. 8.31 (a) An NMOS NOR gate, and (b) drain characteristics for M_1 and M_2.

Fig. 8.31b, we see that the output voltage is small—let us approximate it by $v_o = 0$ V.

Case 3. $v_1 = V_{DD}$, $v_2 = 0$ V

Using reasoning as for Case 2, we deduce that M_1 is ON, M_2 is OFF, and the output voltage is approximately $v_o = 0$ V.

Case 4. $v_1 = v_2 = V_{DD}$

Since $v_{GS1} = v_1 = V_{DD} > V_t$ and $v_{GS2} = v_2 = V_{DD} > V_t$, then M_1 and M_2 are ON. Since M_1 and M_2 have the same characteristics, $i_{D1} = i_{D2} = i_{D3}/2$. Therefore, we can think of the depletion load M_3 as being a parallel connection of two equal depletion loads M_{3a} and M_{3b}, each of which has a drain current of $i_{D3}/2$. Thus, if M_3 has the parameter I_{DSS}, then M_{3a} and M_{3b} have parameters $I_{DSS3a} = I_{DSS3b} = I_{DSS}/2$. The corresponding load curve for each combination of an enhancement-MOSFET driver and a depletion-MOSFET load is given by the indicated trace in Fig. 8.31b. The intersection of this load curve and the characteristic curve $v_{GS} = V_{DD}$ demonstrates that again the output voltage is small—ideally, $v_o = 0$ V.

Hence, we have confirmed that the NMOS circuit given in Fig. 8.31a is a NOR gate.

Drill Exercise 8.10

For the NMOS NOR gate given in Fig. 8.31a, suppose that $V_{DD} = 12$ V, both enhancement MOSFETs have $V_t = 2$ V and $K = 0.25$ mA/V^2, and the depletion MOSFET has $I_{DSS} = 4$ mA and $V_p = -4$ V. Find v_o when $v_1 = v_2 = 12$ V. (*Hint:* M_1 and M_2 are in the ohmic region, and M_3 is in the active region.)

ANSWER 0.408 V

An NMOS NAND Gate

For the NMOS NOR gate discussed above, the driver MOSFETs were connected in parallel. By instead connecting them in series, we obtain a NAND gate.

Example 8.11

Consider the MOSFET circuit shown in Fig. 8.32, where M_1 and M_2 are enhancement NMOS transistors having a threshold voltage of V_t, and M_3 is a depletion MOSFET. Suppose that $V_{DD} > V_t > 0$ V and that both driver M_1 and driver M_2 have drain characteristics such as those shown in Fig. 8.31b, whereas the load M_3 is a MOSFET with a high resistance (small I_{DSS}). Let us assume the ideal case that the low voltage is 0 V and the high voltage is V_{DD}.

If $v_1 = 0$ V, then $v_{GS1} = v_1 = 0 < V_t$. Thus, M_1 is OFF and $i_{D1} = i_{D2} = i_{D3} = 0$ A. Since M_3 operates on the characteristic curve $v_{GS3} = 0$ V (e.g., see Fig. 8.29b

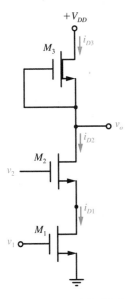

Fig. 8.32 An NMOS NAND gate.

and change the subscripts from 2 to 3), and since $i_{D3} = 0$ A, we must have that $v_{DS3} = 0$ V. Thus the output voltage is $v_o = -v_{DS3} + V_{DD} = V_{DD}$.

Case I: $v_1 = v_2 = 0$ V

As mentioned above, M_1 is OFF, $i_{D1} = i_{D2} = i_{D3} = 0$ A, and $v_o = V_{DD}$. Since the power-supply voltage is V_{DD}, we may surmise that $0 \leq v_{DS1} \leq V_{DD}$. Doing so, we conclude that $v_{GS2} = v_2 - v_{DS1} = -v_{DS1} \leq 0$ V and, therefore, M_2 is OFF also. Since M_1 and M_2 are assumed to have identical characteristics, the voltage across the series connection of their channels will be divided equally. Thus $v_{DS1} = v_{DS2} = v_o/2 = V_{DD}/2$. As a consequence of this, $v_{GS2} = -V_{DD}/2 < V_t$—which confirms that M_2 is OFF.

Case 2. $v_1 = 0$ V, $v_2 = V_{DD}$

As in Case 1, M_1 is OFF, $i_{D1} = i_{D2} = i_{D3} = 0$ A, $v_o = V_{DD}$, and we conclude that M_2 is OFF. Again, the tendency is to have $v_{DS1} = v_{DS2} = v_o/2 = V_{DD}/2$. The consequence of this is that $v_{GS2} = v_2 - v_{DS1} = V_{DD} - V_{DD}/2 = V_{DD}/2$. However, this value may be greater than V_t. Thus the voltage will divide across the channels of M_1 and M_2 such that M_2 will remain OFF. The maximum value for v_{GS2} for which M_2 is OFF is $v_{GS2} = V_t$. Thus, $v_{GS2} = v_2 - v_{DS1} = V_t \Rightarrow v_{DS1} = v_2 - V_t = V_{DD} - V_t$. Furthermore, $v_{DS2} = v_o - v_{DS1} = V_{DD} - (V_{DD} - V_t) = V_t$.

Case 3. $v_1 = V_{DD}$, $v_2 = 0$ V

Since $v_{GS1} = v_1 = V_{DD} > V_t$, then M_1 is ON and $v_{DS1} \geq 0$ V. Thus $v_{GS2} = v_2 - v_{DS1} = -v_{DS1} \leq 0$ V, so M_2 is OFF and $i_D = 0$ A. As described above, we therefore have that $v_o = V_{DD}$. Since $v_{GS1} = V_{DD}$, M_1 operates on the characteristic curve $v_{GS1} = V_{DD}$ (e.g., see Fig. 8.31b). Because the current is zero, M_1 will operate at the point $i_D = 0$ A and $v_{DS1} = 0$ V. Thus $v_{GS2} = v_2 - v_{DS1} = 0 - 0 = 0 < V_t$, and this confirms that M_2 is OFF. Furthermore, $v_{DS2} = v_o - v_{DS1} = V_{DD} - 0 = V_{DD}$.

Case 4. $v_1 = v_2 = V_{DD}$

Since $v_{GS1} = v_1 = V_{DD} > V_t$, then M_1 is ON. For a moment, ignore M_2. We have seen for the combination of a high-resistance depletion load and an enhancement MOSFET driver (Fig. 8.29a) that the resulting output voltage is low (ideally 0 V) when the input is high. Placing an enhancement MOSFET in series with the depletion MOSFET results in the driver having a load with a greater resistance than it would have with just a depletion load. The consequence of this is that v_{DS1} is a low voltage (ideally 0 V). This, in turn, means that $v_{GS2} = v_2 - v_{DS1} = V_{DD}$, so that M_2 is ON. But, the combination of M_2 and a high-resistance depletion load (Fig. 8.29a) again results in a low voltage (ideally 0 V) for v_{DS2}. Thus $v_o = v_{DS2} + v_{DS1}$ is a low voltage (ideally 0 V).

Hence we have confirmed that the NMOS circuit given in Fig. 8.32 is a NAND gate.

Drill Exercise 8.11

For the NMOS NAND gate given in Fig. 8.32, suppose that $V_{DD} = 12$ V, both enhancement MOSFETs have $V_t = 2$ V and $K = 0.25$ mA/V^2, and the depletion MOSFET has $I_{DSS} = 1$ mA and $V_p = -4$ V. Find v_o when $v_1 = v_2 = 12$ V. (*Hint:* M_1 and M_2 are in the ohmic region, and M_3 is in the active region.)

ANSWER 0.406 V

A MOSFET Switch

Having now seen some MOSFET applications to digital circuits, specifically logic gates, let us now present an example of the use of MOSFETs for analog purposes.

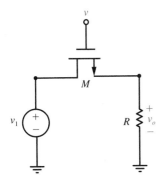

Fig. 8.33 MOSFET used as a switch.

Specifically, consider the *n*-channel enhancement MOSFET circuit shown in Fig. 8.33. We will demonstrate that in this context, the MOSFET *M* can be used as a switch.

Suppose that the applied voltage v_1 varies between 0 and the positive value V_m. Depending upon the condition of the MOSFET, the output voltage v_o will be some value in the interval $0 \le v_o \le V_m$. For example, if *M* is OFF, then $i_D = 0$ A, and as a consequence, $v_o = Ri_D = 0$ V. Conversely, if *M* is sufficiently ON and the result is $v_{DS} \approx 0$ V, then $v_o \approx v_1$.

Since *M* is OFF when $v_{GS} \le V_t$, then *M* is OFF when $v - v_o \le V_t$ or $v \le V_t + v_o$. Since the minimum value of the output voltage is 0 V, then to guarantee that *M* is OFF, we should use $v \le V_t$. For this condition, $v_o = 0$ V and *M* behaves as an open circuit (an open switch).

To get *M* to behave as a short circuit (a closed switch), we should make *v* large enough such that $v_{DS} \approx 0$ V. This can be done by making $v_{GS} \gg V_t$, that is, $v - v_o \gg V_t$ or $v \gg V_t + v_o$. Since the maximum value of the output voltage is V_m, we want to have $v \gg V_t + V_m$. Although a large value of *v* would guarantee a small value of v_{DS}, a smaller value of *v* can suffice if the value of *R* is sufficiently large (just recall the load line plotted on the drain characteristics of a MOSFET in Fig. 8.22*b*).

Example 8.12

Suppose that the MOSFET *M* in Fig. 8.33 has a threshold voltage of 2 V and its drain characteristics are as shown in Fig. 8.34. Also suppose that $R = 5$ kΩ and the applied voltage is in the range $0 \le v_1 \le 5$ V. We can turn OFF *M* by setting $v = 0$ V because $v_{GS} = v - v_o = -v_o \le 0$ V. Conversely, if *M* is ON, we wish to have $v_{DS} \approx 0$ V. When $v_1 = 5$ V (the maximum applied voltage), we see that the intersection of the corresponding load line and the curve $v_{GS} = 12$ V will yield $v_{DS} \approx 0$ V. Thus we can achieve reasonable accuracy by having $v_{GS} = v - v_o = 12$ V, or

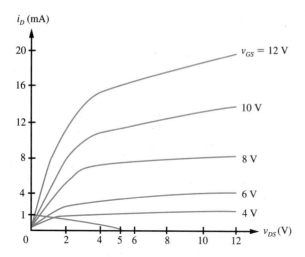

Fig. 8.34 MOSFET drain characteristics.

$v = 12 + v_o = 12 + 5 = 17$ V. In summary, therefore, for the circuit given in Fig. 8.33, the MOSFET M (whose drain characteristics are given in Fig. 8.34) can be used as a switch (for applied voltages in the range $0 \le v_1 \le 5$ V), which is open when $v = 0$ V and closed when $v = 17$ V.

Drill Exercise 8.12

For the circuit given in Fig. 8.33, the MOSFET has $V_t = 2$ V. Suppose that $R = 5$ kΩ, $v_1 = 5$ V, and $v = 17$ V. Analytically determine v_o for the case that (a) $K = 0.2$ mA$/$V^2, and (b) $K = 0.4$ mA$/$V^2. (*Hint: M* operates in the ohmic region.)

ANSWER (a) 4.77 V, (b) 4.88 V

8.4 COMPLEMENTARY MOSFETs (CMOS)

If two enhancement MOSFETs, one NMOS and one PMOS, have identical characteristics (differing only by a minus sign), we say that they possess **complementary symmetry.** This property can be obtained by constructing the MOSFETs on the same IC chip. The resulting devices are known as **complementary MOSFETs,** or **CMOS** for short. (They are also referred to as COSMOS by RCA, and as McMOS by Motorola.[4])

CMOS IC chips not only appear in small-scale integration (SSI) and medium-scale integration (MSI) form, but as large-scale integration (LSI) and very-large-scale integration (VLSI) packages, as well. SSI and MSI logic circuits are often used in conventional digital system design, whereas LSI CMOS circuits are used in electronic watches and calculators. VLSI CMOS has application for microprocessors and for memory circuits. Currently, CMOS is the most popular technology for digital circuits.

Figure 8.35 illustrates the construction of individual *n*-channel and *p*-channel enhancement MOSFETs (ignoring the substrate terminal). Figure 8.36 shows an example of CMOS, where the drains are connected together, as are the gates. Note that to fabricate an NMOS transistor, a *p*-type region, or "well," must be formed in the *n*-type substrate.

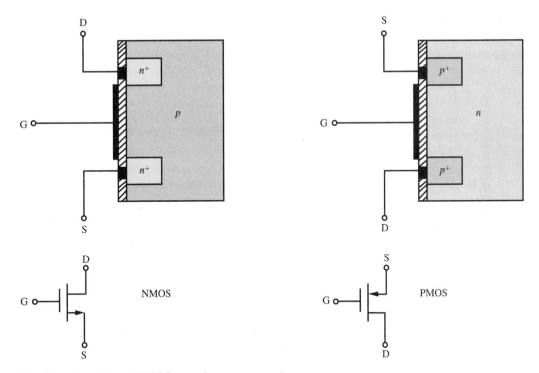

Fig. 8.35 NMOS and PMOS transistors.

---

[4]And by McDonald's.

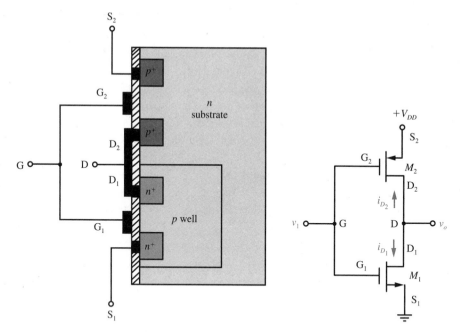

Fig. 8.36 Complementary MOSFETs (CMOS). **Fig. 8.37** A CMOS inverter.

A CMOS Inverter

Let us use the device depicted in Fig. 8.36 to form the CMOS circuit shown in Fig. 8.37, where the threshold voltage for the NMOS transistor is $V_{t1} = a > 0$ V, and the threshold voltage for the PMOS transistor is $V_{t2} = -a < 0$ V. Typical drain characteristics for M_1 and M_2 are shown in Figs. 8.38a and b respectively. Let us assume that $V_{DD} > a$. For the NMOS transistor, if $v_{GS} \leq a$, then it is OFF, and if $v_{GS} > a$, then it is ON. For the PMOS transistor, if $v_{GS} \geq -a$, then it is OFF, and if $v_{GS} < -a$, then it is ON. We will now see that the circuit given in Fig. 8.37 is an inverter.

If $v_1 = 0$ V, then $v_{GS1} = v_1 = 0 < a$. Thus, M_1 is OFF and $i_{D1} = 0$ A. But $v_{GS2} = v_1 - V_{DD} = -V_{DD} < -a$. Therefore, M_2 is ON and tends to conduct current. Specifically, the operation of M_2 is described by the characteristic curve $v_{GS2} = -V_{DD}$ given in Fig. 8.38b. However, the only point on the curve $v_{GS2} = -V_{DD}$ that satisfies the condition that $i_{D2} = i_{D1} = 0$ A is the origin. Thus $v_{DS2} = 0$ V, and hence, $v_o = v_{DS2} + V_{DD} = V_{DD}$.

If $v_1 = V_{DD}$, then $v_{GS1} = v_1 = V_{DD} > a$. Thus M_1 is ON and tends to conduct current. Specifically, the operation of M_1 is described by the curve $v_{GS1} = V_{DD}$ given in Fig. 8.38a. But $v_{GS2} = v_1 - V_{DD} = 0 > -a$. Therefore, M_2 is OFF and $i_{D2} = 0$ A. However, the only point on the curve $v_{GS1} = V_{DD}$ that satisfies the condition that $i_{D1} = i_{D2} = 0$ A is the origin. Thus $v_{DS1} = 0$ V, and hence, $v_o = v_{DS1} = 0$ V.

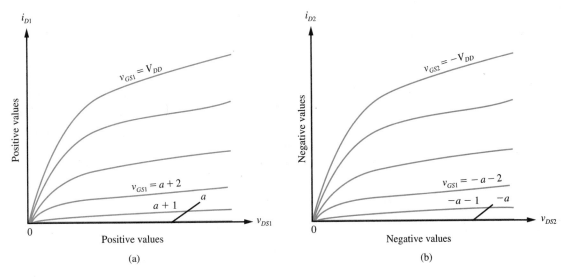

Fig. 8.38 Drain characteristics for (a) M_1 and (b) M_2.

This discussion confirms that the CMOS circuit given in Fig. 8.37 is an inverter. Note, too, that when the input voltage is either high or low, the drain current is essentially zero. Even though the drain current is nonzero when the inverter changes state, on the whole, such a device consumes a relatively small amount of power.

The transfer characteristic for the CMOS inverter given in Fig. 8.37 for the case that $a = 2$ V, $K = 0.25$ mA$/$V^2, and $V_{DD} = 12$ V is shown in Fig. 8.39. Note how much closer this characteristic is to ideal than for the inverters discussed previously.

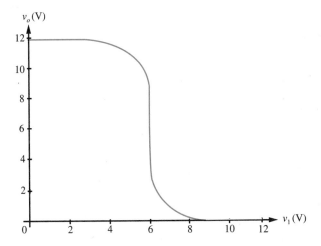

Fig. 8.39 Transfer characteristic for CMOS inverter.

Example 8.13

Let us find the output v_o for the CMOS inverter shown in Fig. 8.37 when $v_1 = 4$ V given that $V_{DD} = 10$ V, $K = 0.25$ mA/V^2 and the threshold voltages for M_1 and M_2 are $V_{t1} = 1$ V and $V_{t2} = -1$ V, respectively.

From the drain characteristics for M_1 given in Fig. 8.38a, when v_{GS1} is slightly larger than the threshold voltage, the resulting drain current will be relatively small. Since this current also goes through M_2, when v_{GS2} has a relatively large magnitude, v_{DS2} will have a relatively small magnitude. This suggests that M_2 will operate in the ohmic region and M_1 will operate in the active region. Therefore, let us make these assumptions.

For a MOSFET, the drain current is defined to be directed from the drain to the source. Since the drain current in a PMOS transistor actually goes in the opposite direction, we must modify the expressions for the drain current in such a device. Specifically, for a p-channel enhancement MOSFET, we have the following relations.

$$\text{active region:} \quad -i_D = K(v_{GS} - V_t)^2 \quad \text{for} \quad v_{DS} \le v_{GS} - V_t$$

$$\text{ohmic region:} \quad -i_D = K[2(v_{GS} - V_t)v_{DS} - v_{DS}^2] \quad \text{for} \quad v_{DS} \ge v_{GS} - V_t$$

(Note the opposite inequalities than for the NMOS case due to the use of negative values for the voltage variables.)

Since M_1 is assumed to be in the active region, we have that

$$i_{D1} = K(v_{GS1} - V_{t1})^2 = (2.5 \times 10^{-4})(4 - 1)^2 = 2.25 \text{ mA}$$

Since M_2 is assumed to be in the ohmic region, we also have that

$$-i_{D2} = K[2(v_{GS2} - V_{t2})v_{DS2} - v_{DS2}^2]$$

But, $i_{D2} = -i_{D1}$, and $v_{GS2} = v_1 - V_{DD} = 4 - 10 = 6$ V. Thus substituting into this last equation, we get

$$2.25 \times 10^{-3} = (2.5 \times 10^{-4})[2(-6 + 1)v_{DS2} - v_{DS2}^2]$$

$$= (2.5 \times 10^{-4})[-10v_{DS2} - v_{DS2}^2]$$

Solving this expression for v_{DS2} results in $v_{DS2} = -1$ V, -9 V. For ohmic region operation we require that $v_{DS2} \ge v_{GS2} - V_{t2} = -6 - (-1) = -5$ V. Thus $v_{DS2} = -1$ V. Consequently, $v_{DS1} = v_{DS2} + V_{DD} = -1 + 10 = 9$ V, and since $v_{DS1} = 9 > v_{GS1} - V_{t1} = 4 - 1 = 3$ V, we have confirmed that M_1 is operating in the active region.

Finally, $v_o = v_{DS1} = 9$ V.

A CMOS NAND Gate

Having discussed the CMOS inverter (NOT gate) given in Fig. 8.37, let us look at a more complicated CMOS logic circuit. Consider the CMOS circuit shown in Fig. 8.40. In essence, the driver (*n*-channel) enhancement MOSFETs M_1 and M_2 are connected in series, whereas the load (*p*-channel) enhancement MOSFETs M_3 and M_4 are connected in parallel. Suppose that M_1 and M_2 have a threshold voltage of $V_t = a > 0$ V, and M_3 and M_4 have a threshold voltage of $V_t = -a < 0$ V. Further, assume that $V_{DD} > a$.

If $v_1 = 0$ V, then $v_{GS1} = v_1 = 0 < a$. Thus M_1 is OFF and $i_{D1} = 0$ A. In addition, $v_{GS3} = v_1 - V_{DD} = -V_{DD} < -a$, so M_3 is ON. But $i_{D2} = i_{D1} = 0$ A regardless of the value of v_2, and, therefore, $i_{D3} + i_{D4} = -i_{D2} = 0$ A. Since $i_{D3} \leq 0$ A and $i_{D4} \leq 0$ A, we deduce that $i_{D3} = i_{D4} = 0$ A. However, as was discussed for the CMOS

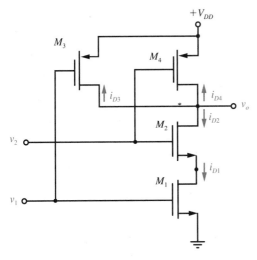

Fig. 8.40 A CMOS NAND gate.

inverter, if M_3 is ON and $i_{D3} = 0$ A, then we must have that $v_{DS3} = 0$ V. Hence $v_o = v_{DS3} + V_{DD} = V_{DD}$.

Case 1. $v_1 = v_2 = 0$ V

From above, when $v_1 = 0$ V, then M_1 is OFF, M_3 is ON, $i_{D1} = i_{D2} = i_{D3} = i_{D4} = 0$ A, $v_o = V_{DD}$. Since $v_2 = 0$ V, then $v_{GS4} = -V_{DD} < -a$, so M_4 is ON. Since $v_{DS1} \geq 0$ V, then $v_{GS2} = v_2 - v_{DS1} = -v_{DS1} \leq 0$ V, and M_2 is OFF. Since M_1 and M_2 have identical characteristics, $v_{DS1} = v_{DS2} = v_o/2 = V_{DD}/2$, and $v_{GS2} = v_2 - v_{DS1} = -V_{DD}/2 < a$—confirming that M_2 is OFF.

Case 2. $v_1 = 0$ V, $v_2 = V_{DD}$

From above, when $v_1 = 0$ V, then M_1 is OFF, M_3 is ON, $i_{D1} = i_{D2} = i_{D3} = i_{D4} = 0$ A, and $v_o = V_{DD}$. Since $v_2 = V_{DD}$, then $v_{GS4} = v_2 - V_{DD} = 0$ V $> -a$, so M_4 is OFF. Now assume that M_2 is ON. Since $i_{D2} = 0$ A, then $v_{DS2} = 0$ V, so $v_{DS1} = v_o = V_{DD}$. Thus $v_{GS2} = v_2 - v_{DS1} = V_{DD} - V_{DD} = 0 < a$—a contradiction to the assumption that M_2 is ON. Therefore, M_2 is OFF. As above, the tendency is to have $v_{DS1} = v_{DS2} = v_o/2 = V_{DD}/2$. However, in this case, $v_{GS2} = v_2 - v_{DS1} = V_{DD} - V_{DD}/2 = V_{DD}/2$ may be larger than a, thereby contradicting the fact that M_2 is OFF. The largest value that v_{GS2} can be such that M_2 is OFF is $v_{GS2} = a$. Under this circumstance, $v_{DS1} = -v_{GS2} + v_2 = -a + V_{DD}$ and $v_{DS2} = v_o - v_{DS1} = V_{DD} - (-a + V_{DD}) = a$.

Case 3. $v_1 = V_{DD}$, $v_2 = 0$ V

Since $v_{GS1} = v_1 = V_{DD} > a$, then M_1 is ON. Also, $v_{GS3} = v_1 - V_{DD} = 0$ V $> -a$ so M_3 is OFF and $i_{D3} = 0$ A. Furthermore, $v_{GS4} = v_2 - V_{DD} = -V_{DD} < -a$, so M_4 is ON. Since $v_{DS1} \geq 0$ V, then $v_{GS2} = v_2 - v_{DS1} = -v_{DS1} \leq 0$ V, so M_2 is OFF and $i_{D2} = 0$ A. Since $i_{D1} = i_{D2} = 0$ A and M_1 is ON, then $v_{DS1} = 0$ V. Thus $v_{GS2} = v_2 - v_{DS1} = 0$ V $< a$—confirming that M_2 is OFF. Since $i_{D4} = -i_{D3} - i_{D2} = 0$ A and M_4 is ON, then $v_{DS4} = 0$ V and $v_o = v_{DS4} + V_{DD} = V_{DD}$.

Case 4. $v_1 = v_2 = V_{DD}$

Since $v_{GS1} = v_1 = V_{DD} > a$, then M_1 is ON. Also, $v_{GS3} = v_1 - V_{DD} = 0$ V $> -a$ and $v_{GS4} = v_2 - V_{DD} = 0$ V $> -a$. Thus M_3 and M_4 are OFF, so $i_{D3} = i_{D4} = 0$ A. Since M_1 is ON and $i_{D1} = i_{D2} = -i_{D3} - i_{D4} = 0$ A, then $v_{DS1} = 0$ V. Consequently, $v_{GS2} = v_2 - v_{DS1} = V_{DD} > a$, so M_2 is ON. Since $i_{D2} = 0$ A, then $v_{DS2} = 0$ V, and $v_o = v_{DS2} + v_{DS1} = 0$ V.

Hence we have confirmed that the CMOS circuit given in Fig. 8.40 is a NAND gate.

Example 8.14

Let us find the output v_o for the CMOS NAND gate shown in Fig. 8.40 when $v_1 = 5$ V and $v_2 = 10$ V given that $V_{DD} = 10$ V, $K = 0.25$ mA/V^2, and the threshold voltages for M_1 and M_2 are $V_t = 1$ V, whereas the threshold voltages for M_3 and M_4 are $V_t = -1$ V. Furthermore, suppose we know that M_1 and M_3 are in the active region, M_2 is in the ohmic region, and M_4 is cut off.

Since $v_{GS1} = v_1 = 5$ V and M_1 is in the active region, then $i_{D1} = K(v_{GS1} - V_{t1})^2 = (2.5 \times 10^{-4})(5 - 1)^2 = 4$ mA. Since M_2 is in the ohmic region, then

$$i_{D2} = K[2(v_{GS2} - V_{t2})v_{DS2} - v_{DS2}^2]$$

But, $i_{D2} = i_{D1} = 4$ mA. Since $v_{GS2} = 10 - v_{DS1}$, this last equation becomes

$$4 \times 10^{-3} = (2.5 \times 10^{-4})[2(9 - v_{DS1})v_{DS2} - v_{DS2}^2] \tag{8.13}$$

By KVL,

$$V_{DD} = 10 = -v_{DS3} + v_{DS2} + v_{DS1} \tag{8.14}$$

Since M_3 is in the active region with $v_{GS3} = v_1 - V_{DD} = 5 - 10 = -5$ V, $-i_{D3} = K(v_{GS3} - V_{t1})^2 = (2.5 \times 10^{-4})(-5 + 1)^2 = 4$ mA, and since M_3 is the complement of M_1, we deduce that $v_{DS3} = -v_{DS1}$. Thus Eq. 8.14 becomes

$$10 = v_{DS1} + v_{DS2} + v_{DS1} = 2v_{DS1} + v_{DS2} \quad \Rightarrow \quad v_{DS2} = 10 - 2v_{DS1}$$

Substituting this into Eq. 8.13 yields

$$4 \times 10^{-3} = (2.5 \times 10^{-4})[2(9 - v_{DS1})(10 - 2v_{DS1}) - (10 - 2v_{DS1})^2] \quad \Rightarrow$$

$$v_{DS1} = 4 \text{ V}$$

Therefore, $v_{DS2} = 10 - 2v_{DS1} = 10 - 2(4) = 2$ V, and $v_o = v_{DS1} + v_{DS2} = 4 + 2 = 6$ V.

Since $v_{DS1} = 4 = v_{GS1} - V_{t1} = 5 - 1 = 4$ V, then M_1 is actually on the border between the active and ohmic regions, so either the active-region equation or the ohmic-region equation is valid. Since $v_{DS2} = 2 < v_{GS2} - V_{t2} = v_2 - v_{DS1} - V_{t2} = 10 - 4 - 1 = 5$ V, then M_2 is confirmed to be in the ohmic region. Furthermore, since $v_{DS3} = v_o - V_{DD} = 6 - 10 = -4 = v_{GS3} - V_{t3} = -5 + 1 = -4$ V, then M_3 is also on the border between the active and ohmic regions—which we may have deduced from M_1. Finally, $v_{GS4} = v_2 - V_{DD} = 10 - 10 = 0 > V_{t4} = -1$ V, so M_4 is confirmed to be OFF.

A CMOS NOR Gate

Another CMOS logic circuit is shown in Fig. 8.41. Conversely to the NAND gate given in Fig. 8.40, the driver MOSFETs M_1 and M_2 are connected in parallel, whereas the load MOSFETs M_3 and M_4 are connected in series. Again suppose that the NMOS transistors M_1 and M_2 have a threshold voltage of $V_t = a > 0$ V, whereas the PMOS transistors M_3 and M_4 have a threshold voltage of $V_t = -a < 0$ V. Also, assume that $V_{DD} > a$.

Case 1. $v_1 = v_2 = 0$ V

Since $v_{GS1} = v_{GS2} = 0$ V, then M_1 and M_2 are OFF, and $i_{D1} = i_{D2} = 0$ A. It follows that $i_{D3} = i_{D4} = -i_{D1} - i_{D2} = 0$ A. Since $v_{GS3} = v_1 - V_{DD} = -V_{DD} < -a$,

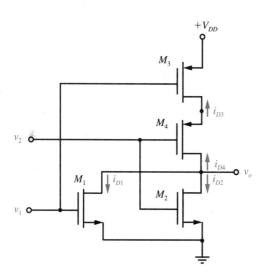

Fig. 8.41 A CMOS NOR gate.

then M_3 is ON. But since $i_{D3} = 0$ A, then $v_{DS3} = 0$ V. Therefore, $v_{GS4} = v_2 - (v_{DS3} + V_{DD}) = -V_{DD} < -a$, so M_4 is ON. Again, since $i_{D4} = 0$ A, then $v_{DS4} = 0$ V. Hence $v_o = v_{DS4} + v_{DS3} + V_{DD} = V_{DD}$.

Case 2. $v_1 = 0$ V, $v_2 = V_{DD}$

Since $v_{GS1} = 0$ V, then M_1 is OFF and $i_{D1} = 0$ A. Also, $v_{GS2} = V_{DD} > a$, and, hence M_2 is ON. Since $v_{GS3} = v_1 - V_{DD} = -V_{DD} < -a$, then M_3 is ON as well. But since $0 \leq v_{D3} \leq V_{DD}$ and $v_{GS4} = v_2 - v_{D3} = V_{DD} - v_{D3}$, then $v_{GS4} \geq 0 > -a$. Thus M_4 is OFF and $i_{D4} = 0$ A. Since $i_{D3} = i_{D4} = 0$ A and M_3 is ON, then $v_{DS3} = 0$ V, so $v_{D3} = v_{DS3} + V_{DD} = V_{DD}$ and $v_{GS4} = v_2 - v_{D3} = V_{DD} - V_{DD} = 0 > -a$—confirming that M_4 is OFF. Since $i_{D2} = -i_{D1} - i_{D4} = 0$ A, and since M_2 is ON, we have that $v_{DS2} = 0$ V. Hence $v_o = v_{DS2} = 0$ V.

Case 3. $v_1 = V_{DD}$, $v_2 = 0$ V

Since $v_{GS1} = V_{DD} > a$, then M_1 is ON. Since $v_{GS2} = 0$ V, then M_2 is OFF, and $i_{D2} = 0$ A. Since $v_{GS3} = v_1 - V_{DD} = 0$ V, then M_3 is OFF and $i_{D3} = i_{D4} = 0$ A. Thus $i_{D1} = -i_{D2} - i_{D4} = 0$ A, and since M_1 is ON, then $v_{DS1} = 0$ V. Hence $v_o = v_{DS1} = 0$ V.

Assume that M_4 is ON. Since $i_{D4} = 0$ A, then $v_{DS4} = 0$ V, so $v_{GS4} = v_2 - (-v_{DS4} + v_o) = 0 > -a$—a contradiction to the assumption that M_4 is ON. Therefore, M_4 is OFF. Since M_3 and M_4 have identical characteristics and they are OFF, the tendency is to have $v_{D3} = V_{DD}/2$. In this case, however, $v_{GS4} = v_2 - v_{D3} = -V_{DD}/2$ may be less than $-a$—thereby contradicting the fact that M_4 is OFF. The largest value that v_{D3} can be and still have M_4 OFF is $v_{D3} = a$. (This corresponds to $v_{GS4} = -a$.) As a consequence, $v_{DS4} = v_o - v_{D3} = -a$ and $v_{DS3} = v_{D3} - V_{DD} = a - V_{DD}$.

Case 4. $v_1 = v_2 = V_{DD}$

Since $v_{GS1} = v_{GS2} = V_{DD} > a$, then M_1 and M_2 are ON. However, $v_{GS3} = v_1 - V_{DD} = 0 > -a$, so M_3 is OFF and $i_{D3} = 0$ A. Since $i_{D1} \geq 0$ A, $i_{D2} \geq 0$ A, and $i_{D1} + i_{D2} = -i_{D4} = -i_{D3} = 0$ A, then we deduce that $i_{D1} = i_{D2} = 0$ A. But M_1 and M_2 are ON. Thus $v_{DS1} = v_{DS2} = 0$ V, and hence $v_o = 0$ V.

Assume that M_4 is ON. Since $i_{D4} = 0$ A, then $v_{DS4} = 0$ V, so $v_{D3} = -v_{DS4} + v_o = 0$ V. As a consequence, $v_{GS4} = v_2 - v_{D3} = 0 > -a$—a contradiction to the assumption that M_4 is ON. Therefore, M_4 is OFF. Since M_3 is also OFF, because M_3 and M_4 have identical characteristics, $v_{D3} = V_{DD}/2$, and hence $v_{DS3} = v_{DS4} = -V_{DD}/2$.

We now see that the CMOS circuit given in Fig. 8.41 is a NOR gate.

Example 8.15

Let us find the output v_o for the CMOS NOR gate shown in Fig. 8.41 when $v_1 = 4$ V and $v_2 = 0$ V given that $V_{DD} = 10$ V, $K = 0.25$ mA/V^2, and the threshold voltages for M_1 and M_2 are $V_t = 1$ V, whereas the threshold voltages for M_3 and M_4 are $V_t = -1$ V. Furthermore, suppose we know that M_1 is in the active region, M_3 and M_4 are in the ohmic region, and M_2 is cut off.

Since $v_{GS1} = v_1 = 4$ V and M_1 is in the active region, then $i_{D1} = K(v_{GS1} - V_{t1})^2 = (2.5 \times 10^{-4})(4 - 1)^2 = 2.25$ mA. Since M_3 is in the ohmic region, then

$$-i_{D3} = K[2(v_{GS3} - V_{t3})v_{DS3} - v_{DS3}^2]$$

But $v_{GS3} = v_1 - V_{DD} = 4 - 10 = -6$ V, so the preceding equation becomes

$$2.25 \times 10^{-3} = (2.5 \times 10^{-4})[-10v_{DS3} - v_{DS3}^2] \quad \Rightarrow \quad v_{DS3} = -1 \text{ V}$$

Thus $v_{D3} = v_{DS3} + V_{DD} = -1 + 10 = 9$ V. Since M_4 is in the ohmic region, then

$$-i_{D4} = K[2(v_{GS4} - V_{t4})v_{DS4} - v_{DS4}^2]$$

But $v_{GS4} = v_2 - v_{D3} = 0 - 9 = -9$ V, so the preceding equation becomes

$$2.25 \times 10^{-3} = (2.5 \times 10^{-4})[-16v_{DS4} - v_{DS4}^2] \quad \Rightarrow$$

$$v_{DS4} = -0.584 \text{ V}, -15.4 \text{ V}$$

Since M_4 is in the ohmic region, we must have that $v_{DS4} \geq v_{GS4} - V_{t4} = -9 + 1 = -8$ V. Therefore, $v_{DS4} = -0.584$ V, and so $v_o = v_{DS4} + v_{DS3} + V_{DD} = -0.584 - 1 + 10 = 8.42$ V.

Since $v_{DS1} = v_o = 8.42 > v_{GS1} - V_{t1} = 4 - 1 = 3$ V, then M_1 is confirmed to be in the active region. Since $v_{GS2} = v_2 = 0 < V_{t1} = 1$ V, then M_2 is confirmed to be OFF. Since $v_{DS3} = -1 > v_{GS3} - V_{t3} = -6 + 1 = -5$ V, then M_3 is confirmed to be in the ohmic region. Finally, $v_{DS4} = -0.584 > v_{GS4} - V_{t4} = -9 + 1 = -8$ V, so M_4 is confirmed to be in the ohmic region.

Drill Exercise 8.15

Find the output v_o for the CMOS NOR gate shown in Fig. 8.41 when $v_1 = 5$ V and $v_2 = 0$ V given that $V_{DD} = 10$ V, $K = 0.25$ mA/V^2, and the threshold voltages for M_1 and M_2 are $V_t = 1$ V, whereas the threshold voltages for M_3

and M_4 are $V_t = -1$ V. (*Hint:* M_1 and M_3 are in the active region, M_4 is in the ohmic region, and M_2 is cut off.)

ANSWER 4 V

A CMOS Transmission Gate

In the chapter on BJTs (Chapter 7), we saw how a BJT can be used as an electronic switch that is turned ON (saturated) and OFF (cut off) by base current. We have also seen that a MOSFET can be used as a switch that is turned ON (ohmic-region operation) and OFF (cut off) by an applied gate voltage. (A JFET may be used as a switch, too.) However, a more sophisticated device, known as a **transmission gate,** can be realized by the CMOS circuit shown in Fig. 8.42a. For this circuit, there is a low voltage, say 0 V, and a high voltage, say $V_H > a > 0$ V, where $V_{t1} = a$ and $V_{t2} = -a$ are the threshold voltages of M_1 and M_2, respectively. To operate this gate, we apply the high voltage to the gate of M_1 and the low voltage to the gate of M_2, or vice versa.

First consider the case that $v = V_H$ and $\bar{v} = 0$ V. If $v_1 = 0$ V, then $v_{GS1} = v - v_1 = V_H - 0 = V_H > a$. Thus M_1 is ON. Furthermore, $v_{GS2} = \bar{v} - v_1 = 0 - 0 = 0 > -a$, so M_2 is OFF and $i_{D2} = 0$ A. Assuming that no load or that the gate(s) of another MOSFET(s) are connected to the output, we must have that

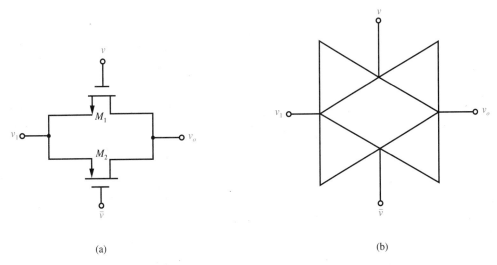

(a)   (b)

Fig. 8.42 (*a*) A transmission gate, and (*b*) circuit symbol for a transmission gate.

$i_{D2} = 0$ A. But since M_1 is ON, then $v_{DS1} = 0$ V and $v_o = v_1 = 0$ V. Conversely, if $v_1 = V_H$, then $v_{GS1} = v - v_1 = V_H - V_H = 0 < a$. Thus M_1 is OFF and $i_{D1} = 0$ A. In addition, $v_{GS2} = \bar{v} - v_1 = 0 - V_H = -V_H < -a$, so M_2 is ON. But since $i_{D2} = 0$ A, then $v_{DS2} = 0$ V and $v_o = v_1 = V_H$. In summary, therefore, when $v = V_H$ and $\bar{v} = 0$, the gate is "closed" ($v_o = v_1$) for either input $v_1 = 0$ V or $v_1 = V_H$.

Now consider the case that $v = 0$ V and $\bar{v} = V_H$. If $v_1 = 0$ V, then $v_{GS1} = v - v_1 = 0 - 0 = 0 < a$ and M_1 is OFF. Furthermore, $v_{GS2} = \bar{v} - v_1 = V_H - 0 = V_H > -a$ and M_2 is OFF. Thus the device behaves as an open circuit between output and input. If $v_1 = V_H$, then $v_{GS1} = v - v_1 = 0 - V_H = -V_H < a$ and M_1 is OFF. In addition, $v_{GS2} = \bar{v} - v_1 = V_H - V_H = 0 > -a$ and M_2 is OFF. Therefore, effectively there is again an open circuit between output and input. In summary, when $v = 0$ V and $\bar{v} = V_H$, for either input $v_1 = 0$ V or $v_1 = V_H$, the gate is "open."

We have just seen that the transmission gate given in Fig. 8.42a is closed when $v = V_H$ and $\bar{v} = 0$ V, whereas it is open when $v = 0$ V and $\bar{v} = V_H$. Even though the statements were verified for the cases that $v_1 = 0$ V and $v_1 = V_H$, it can be shown that they also hold for values of v_1 in the interval $0 \le v_1 \le V_H$ (see Problem 8.76 at the end of this chapter). In other words, a transmission gate can be used for analog as well as digital applications. The circuit symbol for the transmission gate given in Fig. 8.42a is shown in Fig. 8.42b.

Example 8.16

Let us verify transmission-gate operation for the CMOS circuit shown in Fig. 8.42a, where we are given that $v_H = 10$ V, $K = 0.25$ mA/V^2, $V_{t1} = 2$ V, $V_{t2} = -2$ V, and $0 \le v_1 \le 10$ V.

Case 1. $v = 10$ V and $\bar{v} = 0$ V (the gate is closed)

If $v_1 = 0$ V, then $v_{GS1} = v - v_1 = 10 - 0 = 10 > V_{t1} = 2$ V $\Rightarrow M_1$ is ON. Also, $v_{GS2} = \bar{v} - v_1 = 0 - 0 = 0 > V_{t2} = -2$ V $\Rightarrow M_2$ is OFF and $i_{D2} = 0$ A. Assuming no load current, $i_{D1} = -i_{D2} = 0$ A. But since M_1 is ON, then $v_{DS1} = 0$ V and $v_o = v_{DS1} + v_1 = 0 + 0 = 0$ V.

If $v_1 = 10$ V, then $v_{GS1} = v - v_1 = 10 - 10 = 0 < V_{t1} = 2$ V $\Rightarrow M_1$ is OFF and $i_{D1} = 0$ A. Also, $v_{GS2} = \bar{v} - v_1 = 0 - 10 = -10 < V_{t2} = -2$ V $\Rightarrow M_2$ is ON. Again, assuming no load current, $i_{D2} = -i_{D1} = 0$ A. But since M_2 is ON, then $v_{DS2} = 0$ V and $v_o = v_{DS2} + v_1 = 0 + 10 = 10$ V.

If $v_1 = 5$ V, then $v_{GS1} = v - v_1 = 10 - 5 = 5 > V_{t1} = 2$ V $\Rightarrow M_1$ is ON. Also, $v_{GS2} = \bar{v} - v_1 = 0 - 5 = -5 < V_{t2} = -2$ V $\Rightarrow M_2$ is ON. Since $v_{DS1} \ge 0$ V, and $v_{DS2} \le 0$ V, we deduce that $v_{DS1} = v_{DS2} = 0$ V, and therefore, $v_o = v_{DS1} + v_1 = 0 + 5 = 5$ V.

Case 2. $v = 0$ V and $\bar{v} = 10$ V (the gate is open)

If $v_1 = 0$ V, then $v_{GS1} = v - v_1 = 0 - 0 = 0 < V_{t1} = 2$ V $\Rightarrow M_1$ is OFF. Also, $v_{GS2} = \bar{v} - v_1 = 10 - 0 = 10 > V_{t2} = -2$ V $\Rightarrow M_2$ is OFF and $i_{D2} = 0$ A. Since both MOSFETs are OFF, the gate behaves as an open circuit between the output v_o and the input v_1.

If $v_1 = 10$ V, then $v_{GS1} = v - v_1 = 0 - 10 = -10 < V_{t1} = 2$ V $\Rightarrow M_1$ is OFF. Also, $v_{GS2} = \bar{v} - v_1 = 10 - 10 = 0 > V_{t2} = -2$ V $\Rightarrow M_2$ is OFF. Again, both MOSFETs are OFF, and the gate behaves as an open circuit between the output v_o and the input v_1.

If $v_1 = 5$ V, then $v_{GS1} = v - v_1 = 0 - 5 = -5 < V_{t1} = 2$ V $\Rightarrow M_1$ is OFF. Also, $v_{GS2} = \bar{v} - v_1 = 10 - 5 = 5 > V_{t2} = -2$ V $\Rightarrow M_2$ is OFF. Once more, both MOSFETs are OFF, and the gate behaves as an open circuit between the output v_o and the input v_1.

Drill Exercise 8.16

For the transmission gate given in Fig. 8.42a, suppose that $v = 0$ V and $\bar{v} = 10$ V. Given that $V_{t1} = 2$ V and $V_{t2} = -2$ V, for what range of values of v_1 will (a) M_1 and M_2 be OFF, (b) M_1 be ON and M_2 be OFF, and (c) M_1 be OFF and M_2 be ON?

ANSWER (a) $-2 \le v_1 \le 12$ V, (b) $v_1 < -2$ V, (c) $v_1 > 12$ V

S U M M A R Y

1. The terminals of a field-effect transistor (FET) are the gate (G), the drain (D), and the source (S). Unlike the BJT, the FET is a unipolar device—majority charge carriers go from source to drain.

2. The junction field-effect transistor (JFET) operates only in the depletion mode.

3. The voltage required to close a channel is the pinchoff voltage.

4. The drain characteristics of an FET are plots of drain current versus drain-source voltage for different values of gate-source voltage.

5. The gate current of an FET is zero.

6. When an FET is in cutoff, its drain current is zero.

7. A JFET is in the active region when its drain current is nonzero and its channel is pinched off.

8. A JFET is in the ohmic region when its drain current is nonzero and its channel is not pinched off.

9. For small values of drain-source voltage, an FET behaves as a linear resistor.

10. The transfer curve of an FET is a plot of drain current versus gate-source voltage.

11. A depletion metal-oxide-semiconductor FET (MOSFET) has a layer of insulation between the gate and the channel. A depletion MOSFET can operate in the enhancement mode as well as the depletion mode.

12. An enhancement MOSFET is fabricated with no channel, but a channel results from an applied gate-source voltage. The value of gate-source voltage required to induce a channel is the threshold voltage.

13. Simplified circuit symbols are often used (when the substrate is connected to the source) for depletion and enhancement MOSFETs.

14. An IC MOSFET inverter uses MOSFETs as both the driver and the load. Although the former is an enhancement type, the latter usually is a depletion MOSFET. An enhancement load can operate entirely in either the active region or the ohmic region.

15. Enhancement MOSFETs can be used to fabricate gates and switches.

16. Complementary MOSFETs (CMOS) are well suited for IC logic.

Problems

8.1 An n-channel JFET has $I_{DSS} = 10$ mA and $V_p = -2$ V. For small values of v_{DS}, determine the approximate drain-source resistance r_{DS} when (a) $v_{GS} = 0$ V and (b) $v_{GS} = -1$ V.

8.2 For the circuit shown in Fig. P8.2, the JFET has $I_{DSS} = 12$ mA and $V_p = -4$ V. Given that $R_D = 3$ kΩ, $R_S = 0$ Ω, $V_{DD} = 15$ V, and $V_{GG} = -2$ V, find (a) i_D and (b) v_{DS}.

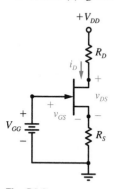

Fig. P8.2

8.3 For the circuit given in Fig. P8.2, the JFET has $I_{DSS} = 12$ mA and $V_p = -4$ V. Given that $R_S = 0$ Ω and $V_{DD} = 15$ V, find the value of R_D for which $v_{DS} = 0.1$ V when $V_{GG} = 0$ V.

8.4 For the circuit shown in Fig. P8.2, the JFET has $I_{DSS} = 12$ mA and $V_p = -4$ V. Given that $R_D = 1$ kΩ, $R_S = 0$ Ω, $V_{DD} = 15$ V, and $V_{GG} = 0$ V, find (a) i_D, (b) v_{DS}, and (c) the region of operation for the JFET.

8.5 For the circuit given in Fig. P8.2, the JFET is in the active region and has $I_{DSS} = 12$ mA and $V_p = -4$ V. Suppose that $V_{DD} = 12$ V and $V_{GG} = 0$ V. (a) Find i_D when $v_{GS} = -2$ V. (b) Find v_{GS} when $i_D = 9$ mA. (c) Find R_S when $v_{GS} = -3$ V. (d) Find R_D when $v_{GS} = -3$ V and $v_{DS} = 4.5$ V. (e) Find v_{DS} when $R_D = 2$ kΩ and $v_{GS} = -3$ V.

8.6 For the circuit given in Fig. P8.2, the JFET has $I_{DSS} = 16$ mA and $V_p = -4$ V. Given that $V_{DD} = 18$ V, $V_{GG} = 0$ V, and $R_D = R_S = 500$ Ω, determine (a) v_{GS}, (b) i_D, (c) v_{DS}, and (d) the region of operation for the JFET.

8.7 For the circuit given in Fig. P8.2, the JFET has $I_{DSS} = 8$ mA and $V_p = -4$ V. Given that $V_{DD} = 18$ V, $V_{GG} = 0$ V, $R_D = 8$ kΩ, $R_S = 1$ kΩ, and the JFET operates in the ohmic region, determine (a) v_{GS}, (b) i_D, and (c) v_{DS}.

8.8 Show that the equation for a JFET operating in the ohmic region (Equation 8.2 on p. 510) reduces to Eq. 8.4 for the case of operation on the border between the ohmic and active regions, that is, for $v_{DS} = v_{GS} - V_p$.

8.9 For the circuit given in Fig. P8.2, suppose that $R_D = R_S = R$, $V_{DD} = 15$ V, and the JFET has $V_p = -3$ V. (a) Assuming active-region operation, $V_{GG} = 5$ V, and $I_{DSS} = 10$ mA, determine the value of R for which $v_{GS} = 0$ V. (b) Assuming active-region operation, find I_{DSS} given that $R = 400$ Ω, $V_{GG} = 5$ V, and $v_{GS} = 0$ V. (c) If I_{DSS}, R, and the region of operation are not known, what is v_{GS} when $V_{GG} = 5$ V and $v_{DS} = 1$ V? (d) If R is not known, what is the region of operation when $v_{DS} = 2$ V and $v_{GS} = -1.5$ V?

8.10 For the circuit shown in Fig. P8.10, the JFET has $I_{DSS} = 9$ mA and $V_p = -3$ V. Given that $R_1 = 150$ kΩ, $R_2 = 50$ kΩ, $R_D = R_S = 7$ kΩ, and $V_{DD} = 20$ V, assume active-region operation and determine (a) v_{GS}, (b) i_D, and (c) v_{DS}. (*Hint:* Use Thévenin's theorem.)

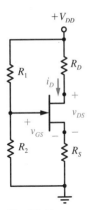

Fig. P8.10

8.11 For the circuit shown in Fig. P8.10, the JFET has $I_{DSS} = 12$ mA and $V_p = -4$ V. Given that $R_1 = 300$ kΩ, $R_2 = 100$ kΩ, $R_D = R_S = R$, $V_{DD} = 12$ V, and $v_{DS} = 6$ V, determine (a) v_{GS}, (b) i_D, and (c) R. (*Hint:* Use Thévenin's theorem.)

8.12 For the circuit shown in Fig. P8.10, the JFET has $I_{DSS} = 12$ mA and $V_p = -4$ V. Given that $R_1 = 300$ kΩ, $R_2 = 100$ kΩ, $R_D = R_S = 1$ kΩ, and $V_{DD} = 12$ V, determine (a) v_{GS}, (b) i_D, (c) v_{DS}, and (d) the region of operation for the JFET. (*Hint:* Use Thévenin's theorem.)

8.13 For the circuit shown in Fig. P8.10, the JFET has $I_{DSS} = 12$ mA and $V_p = -4$ V. Given that $R_1 = 300$ kΩ, $R_2 = 100$ kΩ, $R_D = R_S = 2$ kΩ, and $V_{DD} = 12$ V, determine (a) v_{GS}, (b) i_D, (c) v_{DS}, and (d) the region of operation for the JFET. (*Hint:* Use Thévenin's theorem.)

8.14 Both JFETs in the circuit shown in Fig. P8.14 have $I_{DSS} = 4$ mA and $V_p = -2$ V. Given that both JFETs are in the active region, $R_S = 0$ Ω and $V_{DD} = V_{SS} = 10$ V, find (a) i_D, (b) v_{GS}, (c) v_{DS}, and (d) confirm the regions of operation.

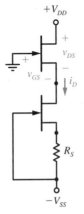

Fig. P8.14

8.15 Both JFETs in the circuit shown in Fig. P8.14 have $I_{DSS} = 4$ mA and $V_p = -2$ V. Place a 1-kΩ resistor in series with the drain of the lower JFET. Given that both JFETs are in the active region, $R_S = 1$ kΩ and $V_{DD} = V_{SS} = 10$ V, find (a) i_D, (b) v_{GS}, (c) v_{DS}, and (d) confirm the regions of operation.

8.16 Both JFETs in the circuit shown in Fig. P8.14 have $I_{DSS} = 16$ mA and $V_p = -4$ V. Given that $R_S = 0$ Ω, $V_{DD} = 12$ V, $V_{SS} = 0$ V, the upper JFET is in the active region, and the lower JFET is in the ohmic region, find (a) v_{GS}, (b) i_D, and (c) v_{DS}.

8.17 Both JFETs in the circuit shown in Fig. P8.14 have $I_{DSS} = 16$ mA and $V_p = -4$ V. Instead of connecting the gate of the upper JFET to the reference as shown, connect a short circuit between the gate of the upper JFET and the drain of the lower JFET. Given that $R_S = 1$ kΩ, $V_{DD} = 6$ V, $V_{SS} = $

0 V, the lower JFET is in the active region, and the upper JFET is in the ohmic region, find (a) i_D and (b) v_{DS}.

8.18 An *n*-channel depletion MOSFET has $I_{DSS} = 10$ mA and $V_p = -2$ V. For small values of v_{DS}, determine the approximate drain-source resistance r_{DS} when v_{GS} is (a) 1 V and (b) 2 V.

8.19 For the circuit given in Fig. P8.19, the depletion MOSFET has $I_{DSS} = 8$ mA and $V_p = -2$ V. Given that $R_S = 0$ Ω, $V_{DD} = 16$ V, and $V_{GG} = 1$ V, determine the value of R_D for which the depletion MOSFET will operate on the border between the active and the ohmic regions.

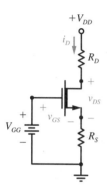

Fig. P8.19

8.20 For the MOSFET circuit shown in Fig. P8.19, the depletion NMOS transistor has $I_{DSS} = 8$ mA and $V_p = -2$ V. Given that $R_D = R_S = 250$ Ω, $V_{DD} = 16$ V, and $V_{GG} = 2$ V, find (a) v_{GS}, (b) i_D, (c) v_{DS}, and (d) the region of operation for the MOSFET.

8.21 For the MOSFET circuit shown in Fig. P8.19, the depletion NMOS transistor has $I_{DSS} = 8$ mA and $V_p = -2$ V. Suppose that $V_{DD} = 16$ V and $V_{GG} = 2$ V. (a) If $R_D = R_S = R$, then what value of R results in active-region operation and $v_{GS} = 1$ V? (b) Find R_D and R_S such that $v_{DS} = v_{GS} = 1$ V.

8.22 For the circuit shown in Fig. P8.19, the depletion MOSFET has $I_{DSS} = 10$ mA and $V_p = -4$ V. Given that $R_D = 1$ kΩ, $R_S = 0$ Ω, and $V_{DD} = 15$ V, find v_{DS} when V_{GG} is (a) 0 V and (b) 5 V.

8.23 For the MOSFET circuit shown in Fig. P8.23, the depletion NMOS transistor has $I_{DSS} = 4$ mA and $V_p = -4$ V. Given that $V_{DD} = 10$ V and $R_D = 1$ kΩ, find (a) the region of operation, (b) v_{GS}, (c) v_{DS}, and (d) i_D.

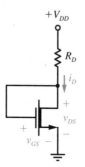

Fig. P8.23

8.24 For the MOSFET circuit shown in Fig. P8.23, the depletion NMOS transistor has $I_{DSS} = 4$ mA and $V_p = -4$ V. Given that $V_{DD} = 10$ V and $v_{DS} = 6$ V, find (a) v_{GS}, (b) the region of operation, (c) i_D, and (d) R_D.

8.25 For the MOSFET circuit shown in Fig. P8.23, the depletion NMOS transistor has $I_{DSS} = 4$ mA and $V_p = -4$ V. Given that $V_{DD} = 10$ V, and $i_D = 16$ mA, find (a) the region of operation, (b) v_{GS}, (c) v_{DS}, and (d) R_D.

8.26 For the depletion-MOSFET circuit shown in Fig. P8.26, M_1 has $I_{DSS} = 8$ mA and $V_p = -4$ V, whereas M_2 has $I_{DSS} = 16$ mA and $V_p = -4$ V. When $V_{DD} = 11$ V and $V_{GG} = 0$ V, then M_1 is in the active region and M_2 is in the ohmic region. Find (a) v_{DS2}, (b) v_{DS1}, and (c) confirm the regions of operation.

8.27 For the depletion-MOSFET circuit shown in Fig. P8.26, M_1 has $I_{DSS} = 8$ mA and $V_p = -4$ V, whereas M_2 has $I_{DSS} = 16$ mA and $V_p = -4$ V. When $V_{DD} = 11$ V and $V_{GG} = 1$ V, then M_1 is in the active region and M_2 is in the ohmic region. Find (a) i_D, (b) v_{DS2}, (c) v_{DS1}, and (d) confirm the regions of operation.

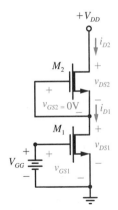

Fig. P8.26

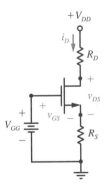

Fig. P8.32

8.28 For the depletion-MOSFET circuit shown in Fig. P8.26, M_1 has $I_{DSS} = 8$ mA and $V_p = -4$ V, whereas M_2 has $I_{DSS} = 16$ mA and $V_p = -4$ V. Given that $V_{DD} = 11$ V, find V_{GG} such that M_1 is in the active region and M_2 is on the border between the active and the ohmic regions.

8.29 For the depletion-MOSFET circuit shown in Fig. P8.26, M_1 has $I_{DSS} = 8$ mA and $V_p = -4$ V, whereas M_2 has $I_{DSS} = 16$ mA and $V_p = -4$ V. When $V_{DD} = 11$ V and $V_{GG} = 2$ V, then M_1 is in the ohmic region and M_2 is in the active region. Find (a) i_D, (b) v_{DS1}, (c) v_{DS2}, and (d) confirm the regions of operation.

8.30 For the depletion-MOSFET circuit shown in Fig. P8.26, M_1 has $I_{DSS} = 8$ mA and $V_p = -4$ V, whereas M_2 has $I_{DSS} = 16$ mA and $V_p = -4$ V. When $V_{DD} = 11$ V and $V_{GG} = 10$ V, then M_1 is in the ohmic region and M_2 is in the active region. Find (a) i_D, (b) v_{DS1}, (c) v_{DS2}, and (d) confirm the regions of operation.

8.31 An n-channel enhancement MOSFET has $K = 0.25$ mA/V^2 and $V_t = 2$ V. For small values of v_{DS}, determine the approximate drain-source resistance r_{DS} when v_{GS} is (a) 4 V, (b) 6 V, and (c) 10 V.

8.32 For the circuit given in Fig. P8.32, the enhancement MOSFET has $K = 0.15$ mA/V^2, $V_t = 2$ V, and operates in the active region. Suppose that $V_{DD} = 12$ V. (a) Find V_{GG} when $R_S = 0$ Ω and $i_D = 5.4$ mA. (b) Find i_D when $v_{GS} = 4$ V. (c) Find v_{GS} when $i_D = 2.4$ mA. (d) Find R_S when $v_{GS} = 3$ V and $V_{GG} = 4.5$ V. (e) Find R_D when $v_{GS} = 3$ V, $v_{DS} = 7.5$ V, and $V_{GG} = 4.5$ V. (f) Find v_{DS} when $R_D = 4$ kΩ, $v_{GS} = 3$ V, and $V_{GG} = 4.5$ V.

8.33 For the circuit given in Fig. P8.32, the enhancement MOSFET has $K = 0.25$ mA/V^2 and $V_t = 2$ V. Given that $R_S = 0$ Ω and $V_{DD} = 16$ V, determine the value of R_D for which the enhancement MOSFET will operate on the border between the active and the ohmic regions when V_{GG} is (a) 4 V and (b) 10 V.

8.34 For the circuit given in Fig. P8.32, the enhancement MOSFET has $K = 0.25$ mA/V^2 and $V_t = 2$ V. Given that $R_D = 1$ kΩ, $R_S = 0$ Ω, $V_{DD} = 16$ V, and $V_{GG} = 4$ V, find (a) i_D, (b) v_{DS}, and (c) the region of operation for the MOSFET.

8.35 For the circuit given in Fig. P8.32, the enhancement MOSFET has $K = 0.25$ mA/V^2 and $V_t = 2$ V. Given that $R_D = R_S = 1$ kΩ, $V_{DD} = 20$ V, and $V_{GG} = 10$ V, find (a) v_{GS}, (b) i_D, (c) v_{DS}, and (d) the region of operation for the MOSFET.

8.36 For the circuit given in Fig. P8.32, the enhancement MOSFET has $K = 0.25$ mA/V^2 and $V_t = 2$ V. Given that $R_D = R_S = 1$ kΩ, $V_{DD} = 9$ V, and $V_{GG} = 10$ V, find (a) v_{GS}, (b) i_D, (c) v_{DS}, and (d) the region of operation for the MOSFET.

8.37 For the MOSFET circuit shown in Fig. P8.37, the enhancement NMOS transistor has $K = 0.2$ mA/V^2 and $V_t = 1$ V. Given that $V_{DD} = 10$ V and $v_{DS} = 6$ V, find (a) v_{GS}, (b) the region of operation, (c) i_D, and (d) R_D.

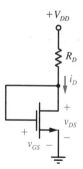

Fig. P8.37

8.38 For the MOSFET circuit shown in Fig. P8.37, the enhancement NMOS transistor has $K = 0.2$ mA/V² and $V_t = 1$ V. Given that $V_{DD} = 10$ V and $i_D = 3.2$ mA, find (a) the region of operation, (b) v_{GS}, (c) v_{DS}, and (d) R_D.

8.39 For the MOSFET circuit shown in Fig. P8.37, the enhancement NMOS transistor has $K = 0.2$ mA/V² and $V_t = 1$ V. Given that $V_{DD} = 10$ V and $R_D = 1$ kΩ, find (a) the region of operation, (b) v_{GS}, (c) v_{DS}, and (d) i_D.

8.40 For the circuit shown in Fig. P8.40, the enhancement MOSFET has $K = 0.25$ mA/V² and $V_t = 1$ V. Given that $R_1 = 300$ kΩ, $R_2 = 100$ kΩ, $R_D = R_S = R$, $V_{DD} = 12$ V, and $v_{DS} = 10$ V, determine (a) v_{GS}, (b) i_D, and (c) R. (*Hint:* Use Thévenin's theorem.)

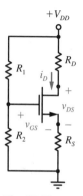

Fig. P8.40

8.41 For the circuit shown in Fig. P8.40, the enhancement MOSFET has $K = 0.25$ mA/V² and $V_t = 1$ V. Given that $R_1 = 300$ kΩ, $R_2 = 100$ kΩ, $R_D = R_S = 1$ kΩ, and $V_{DD} = 12$ V, find (a) v_{GS}, (b) i_D, (c) v_{DS}, and (d) the region of operation for the MOSFET. (*Hint:* Use Thévenin's theorem.)

8.42 For the MOSFET circuit given in Fig. P8.37, replace the short circuit that is connected between the gate and the drain with a 1-MΩ resistor. In addition, place another 1-MΩ resistor between the gate and the reference. Given that $K = 0.25$ mA/V², $V_t = 2$ V, $R_D = 500$ Ω, and $V_{DD} = 12$ V, find (a) the region of operation for the MOSFET, (b) v_{GS}, (c) v_{DS}, and (d) i_D.

8.43 The NMOS circuit shown in Fig. P8.43 is known as a **current mirror** since the current on one side equals (mirrors) the current on the other side. Given that all of the enhancement MOSFETs have $K = 0.1$ mA/V² and $V_t = 1$ V, $V_{DD} = 10$ V, and M_1 is in the active region, find all the currents and voltages for the circuit.

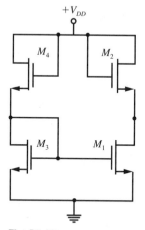

Fig. P8.43

8.44 For the NMOS circuit given in Fig. 8.25*a*, on p. 533, suppose that $V_{DD} = 12$ V. When $v_1 = 12$ V, then M_1 is in the ohmic region. Given that $V_t = 2$ V for both MOSFETs and $K_1 = K_2$, find v_o.

8.45 For the NMOS circuit given in Fig. 8.25a, on p. 533, suppose that $V_{DD} = 12$ V. When $v_1 = 3$ V, then M_1 is in the active region. Given that $V_t = 2$ V for both MOSFETs and $K_1 = K_2$, find v_o.

8.46 Consider the NMOS circuit given in Fig. 8.25a on p. 533. Given that $V_{t1} = V_{t2}$ and $K_1 = K_2$, find the range of values for v_1 such that M_1 is in the active region. What is the corresponding value of v_o?

8.47 For the NMOS circuit given in Fig. 8.29a, on p. 538, suppose that $V_{DD} = 12$ V, the enhancement MOSFET has $V_t = 1$ V and $K = 0.25$ mA$/$V^2, and the depletion MOSFET has $I_{DSS} = 4$ mA and $V_p = -4$ V. When $v_1 = 2$ V, then M_1 is in the active region and M_2 is in the ohmic region. Find v_o and confirm the regions of operation.

8.48 For the NMOS circuit given in Fig. 8.29a, on p. 538, suppose that $V_{DD} = 12$ V, the enhancement MOSFET has $V_t = 1$ V and $K = 0.25$ mA$/$V^2, and the depletion MOSFET has $I_{DSS} = 4$ mA and $V_p = -4$ V. When $v_1 = 12$ V, then M_1 is in the ohmic region and M_2 is in the active region. Find v_o and confirm the regions of operation.

8.49 For the NMOS circuit given in Fig. 8.29a, on p. 538, suppose that $V_{DD} = 8$ V, the enhancement MOSFET has $V_t = 1$ V and $K = 0.25$ mA$/$V^2, and the depletion MOSFET has $I_{DSS} = 4$ mA and $V_p = -4$ V. (a) For what value of v_1 will both MOSFETs be in the active region? (b) What is the corresponding current? (c) What is the corresponding value of v_o?

8.50 For the NMOS circuit given in Fig. 8.29a, on p. 538, suppose that $V_{DD} = 6$ V, the enhancement MOSFET has $V_t = 1$ V and $K = 0.25$ mA$/$V^2, and the depletion MOSFET has $I_{DSS} = 4$ mA and $V_p = -4$ V. When $v_1 = 5$ V, then M_1 and M_2 are in the ohmic region. Find v_o and confirm the regions of operation.

8.51 For the NMOS circuit given in Fig. P8.51, suppose that $V_{DD} = 12$ V, the enhancement MOSFET has $V_t = 2$ V and $K = 0.25$ mA$/$V^2, and the depletion MOSFET has $I_{DSS} = 4$ mA and $V_p = -4$ V. When $v_1 = 0$ V, then M_1 is in the active region. Find v_o.

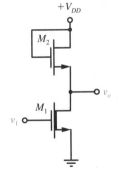

Fig. P8.51

8.52 For the NMOS circuit given in Fig. P8.51, suppose that $V_{DD} = 12$ V, the enhancement MOSFET has $V_t = 2$ V and $K = 0.25$ mA$/$V^2, and the depletion MOSFET has $I_{DSS} = 4$ mA and $V_p = -4$ V. Find the range of values of v_1 for which M_1 is in the active region.

8.53 For the NMOS circuit given in Fig. P8.51, suppose that $V_{DD} = 12$ V, the enhancement MOSFET has $V_t = 2$ V and $K = 0.25$ mA$/$V^2, and the depletion MOSFET has $I_{DSS} = 4$ mA and $V_p = -4$ V. When $v_1 = 12$ V, then M_1 is in the ohmic region. Find v_o.

8.54 For the NMOS NOR gate given in Fig. 8.31a, on p. 541, suppose that $V_{DD} = 10$ V, both enhancement MOSFETs have $V_t = 1$ V and $K = 0.25$ mA$/$V^2, and the depletion MOSFET has $I_{DSS} = 1$ mA and $V_p = -4$ V. Find v_o when $v_1 = v_2 = 0$ V.

8.55 For the NMOS NOR gate given in Fig. 8.31a, on p. 541, suppose that $V_{DD} = 10$ V, both enhancement MOSFETs have $V_t = 1$ V and $K = 0.25$ mA$/$V^2, and the depletion MOSFET has $I_{DSS} = 1$ mA and $V_p = -4$ V. Find v_o when $v_1 = 0$ V and $v_2 = 10$ V. (*Hint:* M_1 is OFF, M_2 is in the ohmic region, and M_3 is in the active region.)

8.56 For the NMOS NOR gate given in Fig. 8.31a, on p. 541, suppose that $V_{DD} = 10$ V, both enhancement MOSFETs have $V_t = 1$ V and $K = 0.25$ mA$/$V^2, and the depletion MOSFET has $I_{DSS} = 1$ mA and $V_p = -4$ V. Find v_o when $v_1 = v_2 = $

10 V. (*Hint:* M_1 and M_2 are in the ohmic region, and M_3 is in the active region.)

8.57 For the NMOS NAND gate given in Fig. 8.32, on p. 543, suppose that $V_{DD} = 12$ V, both enhancement MOSFETs have $V_t = 2$ V and $K = 0.25$ mA/V^2, and the depletion MOSFET has $I_{DSS} = 0.25$ mA and $V_p = -4$ V. Find v_o when $v_1 = v_2 = 12$ V. (*Hint:* M_1 and M_2 are in the ohmic region, and M_3 is in the active region.)

8.58 For the circuit shown in Fig. P8.58, the depletion MOSFET loads have a high resistance (small I_{DSS}). Assume that all enhancement MOSFETs have the threshold voltage V_t and that $V_{DD} \gg V_t$. What type of logic gate is this NMOS circuit?

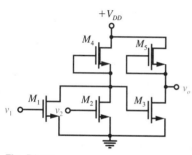

Fig. P8.58

8.59 For the circuit shown in Fig. P8.59, the depletion MOSFET loads have a high resistance (small I_{DSS}). Assume that all enhancement MOSFETs have the threshold voltage V_t and that $V_{DD} \gg V_t$. What type of logic gate is this NMOS circuit?

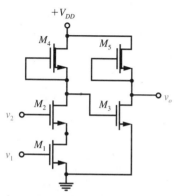

Fig. P8.59

8.60 For the circuit shown in Fig. P8.60, the depletion MOSFET loads have a high resistance (small I_{DSS}). Assume that all enhancement MOSFETs have the threshold voltage V_t and that $V_{DD} \gg V_t$. What type of logic gate is this NMOS circuit?

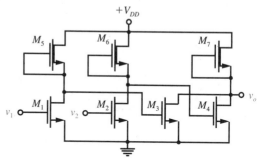

Fig. P8.60

8.61 For the circuit shown in Fig. P8.61, the depletion MOSFET loads have a high resistance (small I_{DSS}). Assume that all enhancement MOSFETs have the threshold voltage V_t and that $V_{DD} \gg V_t$. What type of logic gate is this NMOS circuit?

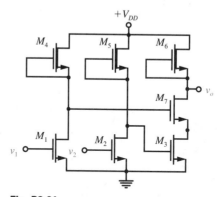

Fig. P8.61

8.62 For the CMOS circuit given in Fig. 8.37, on p. 548, suppose that $V_{DD} = 10$ V, $K = 0.25$ mA/V^2, $V_{t1} = 1$ V, and $V_{t2} = -1$ V. When $v_1 = 2$ V, then M_1 is in the active region and M_2 is in the ohmic region. Find v_o and confirm the regions of operation.

8.63 For the CMOS circuit given in Fig. 8.37, on p. 548, suppose that $V_{DD} = 10$ V, $K = 0.25$ mA/V^2, $V_{t1} = 1$ V, and $V_{t2} = -1$ V. For what value of v_1

are both MOSFETs in the active region? What is i_{D1}, i_{D2}, and v_o under this circumstance?

8.64 For the CMOS circuit shown in Fig. P8.64, M_1 and M_2 are enhancement MOSFETs with threshold voltages $V_{t1} = a > 0$ V and $V_{t2} = -a < 0$ V, respectively. Assuming that $V_{DD} > a$, determine the output voltage v_o when the input voltage v_1 is (a) 0 V and (b) $-V_{DD}$.

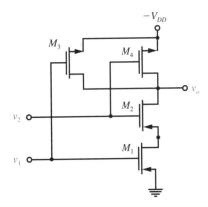

Fig. P8.66

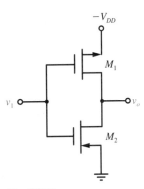

Fig. P8.64

8.65 Find the output v_o for the CMOS NAND gate shown in Fig. 8.40 on p. 551 when $v_1 = 3$ V and $v_2 = 10$ V given that $V_{DD} = 10$ V, $K = 0.25$ mA$/$V^2, and the threshold voltages for M_1 and M_2 are $V_t = 1$ V, whereas the threshold voltages for M_3 and M_4 are $V_t = -1$ V. (*Hint:* M_1 and M_2 are in the active region, M_3 is in the ohmic region, and M_4 is cut off.)

8.66 Consider the CMOS circuit shown in Fig. P8.66, where $V_{DD} = 10$ V, the NMOS transistors M_3 and M_4 have the threshold voltage $V_t = 1$ V, whereas the PMOS transistors M_1 and M_2 have the threshold voltage $V_t = -1$ V. Complete Table 8.1 for the case that $v_1 = v_2 = 0$ V.

8.67 Consider the CMOS circuit shown in Fig. P8.66, where $V_{DD} = 10$ V, the NMOS transistors M_3 and M_4 have the threshold voltage $V_t = 1$ V, whereas the PMOS transistors M_1 and M_2 have the threshold voltage $V_t = -1$ V. Complete Table 8.1 for the case that $v_1 = 0$ V and $v_2 = -10$ V.

8.68 Consider the CMOS circuit shown in Fig. P8.66, where $V_{DD} = 10$ V, the NMOS transistors M_3

Table 8.1

	v_{GS}	OFF/ON	i_D	v_{DS}
M_1				
M_2				
M_3				
M_4				

and M_4 have the threshold voltage $V_t = 1$ V, whereas the PMOS transistors M_1 and M_2 have the threshold voltage $V_t = -1$ V. Complete Table 8.1 for the case that $v_1 = -10$ V and $v_2 = 0$ V.

8.69 Consider the CMOS circuit shown in Fig. P8.66, where $V_{DD} = 10$ V, the NMOS transistors M_3 and M_4 have the threshold voltage $V_t = 1$ V, whereas the PMOS transistors M_1 and M_2 have the threshold voltage $V_t = -1$ V. Complete Table 8.1 for the case that $v_1 = v_2 = -10$ V.

8.70 Consider the CMOS circuit shown in Fig. P8.70, where $V_{DD} = 10$ V, the NMOS transistors M_3 and M_4 have the threshold voltage $V_t = 1$ V, whereas the PMOS transistors M_1 and M_2 have the threshold voltage $V_t = -1$ V. Complete Table 8.1 for the case that $v_1 = v_2 = 0$ V. (See p. 568.)

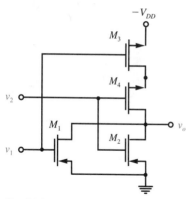

Fig. P8.70

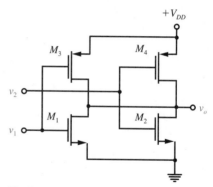

Fig. P8.74

8.71 Consider the CMOS circuit shown in Fig. P8.70, where $V_{DD} = 10$ V, the NMOS transistors M_3 and M_4 have the threshold voltage $V_t = 1$ V, whereas the PMOS transistors M_1 and M_2 have the threshold voltage $V_t = -1$ V. Complete Table 8.1 for the case that $v_1 = 0$ V and $v_2 = -10$ V.

8.72 Consider the CMOS circuit shown in Fig. P8.70, where $V_{DD} = 10$ V, the NMOS transistors M_3 and M_4 have the threshold voltage $V_t = 1$ V, whereas the PMOS transistors M_1 and M_2 have the threshold voltage $V_t = -1$ V. Complete Table 8.1 for the case that $v_1 = -10$ V and $v_2 = 0$ V.

8.73 Consider the CMOS circuit shown in Fig. P8.70, where $V_{DD} = 10$ V, the NMOS transistors M_3 and M_4 have the threshold voltage $V_t = 1$ V, whereas the PMOS transistors M_1 and M_2 have the threshold voltage $V_t = -1$ V. Complete Table 8.1 for the case that $v_1 = v_2 = -10$ V.

8.74 The CMOS circuit shown in Fig. P8.74 is a "NO-DAMN-GOOD" gate. The NMOS transistors M_1 and M_2 have threshold voltage $V_t = a$, whereas the PMOS transistors M_3 and M_4 have threshold voltage $V_t = -a$. Assume that, except for a difference in sign, all MOSFETs have the same characteristics. Given that $V_{DD} > a$, determine v_o when (a) $v_1 = v_2 = 0$ V, (b) $v_1 = v_2 = V_{DD}$, and (c) $v_1 = V_{DD}$, $v_2 = 0$ V.

8.75 The CMOS circuit shown in Fig. P8.75 is another "NO-DAMN-GOOD" gate. The NMOS transistors M_2 and M_4 have threshold voltage $V_t = a$,

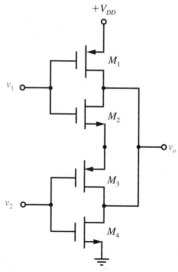

Fig. P8.75

whereas the PMOS transistors M_1 and M_3 have threshold voltage $-a$. Assume that, except for a difference in sign, all MOSFETs have the same characteristics. Given that $V_{DD}/2 > a$, determine v_o when (a) $v_1 = v_2 = 0$ V, (b) $v_1 = v_2 = V_{DD}$, and (c) $v_1 = V_{DD}$, $v_2 = 0$ V.

8.76 For the transmission gate given in Fig. 8.42a, on p. 557, suppose that $V_H = 5$ V, $V_{t1} = 2$ V, and $V_{t2} = -2$ V. Verify transmission-gate operation when v_1 is (a) 1 V, (b) 2.5 V, and (c) 4 V.

8.77 For the transmission gate given in Fig. 8.42a, on p. 557, $V_{t1} = 2$ V and $V_{t2} = -2$ V, the low voltage is -5 V and the high voltage is 5 V. Given that $v = 5$ V, $\bar{v} = -5$ V, and the input voltage v_1 varies between -5 and 5 V, determine the range of values of v_1 for which (a) M_1 is ON and M_2 is OFF, (b) M_1 and M_2 are ON, and (c) M_1 is OFF and M_2 is ON.

8.78 For the CMOS transmission gate given in Fig. 8.42a, on p. 557, suppose that $v = 10$ V, $\bar{v} = 0$ V, $V_{t1} = 2$ V, and $V_{t2} = -2$ V. Find the range of values of v_1 for which (a) M_1 and M_2 are ON, (b) M_1 is ON and M_2 is OFF, and (c) M_1 is OFF and M_2 is ON.

9

Transistor Amplifiers

9.1 Bipolar Junction Transistor Amplifiers

9.3 Frequency Response

9.2 Field-Effect Transistor Amplifiers

9.4 Power Amplifiers

INTRODUCTION

Amplification is an important, yet fundamental, signal-processing task. In many situations, important signals arise, but are much too small in magnitude to be directly useful. For us to utilize their information content, such signals must be enlarged or amplified—hence the need for amplifiers. Transistor amplifiers can be designed to result in voltage, current, and/or power gain of signals.

When desired signals are weak, typically, they are first amplified with small-signal amplifiers. Such devices can be constructed using bipolar junction transistors (BJTs), field-effect transistors (FETs), or a combination of both types of components. These amplifiers can be constructed in either discrete or integrated-circuit (IC) form. The range of signal frequencies that can be amplified depends upon whether or not capacitors are employed in the design of the amplifiers, and upon the inherent capacitances of the transistors themselves.

After small ac signals have been sufficiently amplified to produce large ac signals, they may be amplified further by power (or large-signal) amplifiers so that there is sufficient power to drive the final load, for example, a loudspeaker, a motor, a recording instrument, or a cathode-ray tube (CRT).

9.1 Bipolar Junction Transistor Amplifiers

In digital-circuit applications, transistors typically operate either in cutoff or in saturation. (An exception is nonsaturating logic such as ECL where transistors are either

570

in cutoff or in the active region.) However, if we operate a transistor strictly in the active region, also known as the linear region, then it can exhibit amplification properties. To see how this happens, consider the collector characteristics shown in Fig. 9.1. Included with these characteristics is the load line corresponding to some associated circuit. As an example, suppose that the base current of the BJT is 40 μA. The intersection of curve $i_B = 40$ μA and the load line is the point labeled Q. This point is called the **quiescent operating point**, or **operating point**, or **Q point** for short. From Fig. 9.1, therefore, we see that the resulting collector current is $i_C = I_{CQ} = 4$ mA and the resulting collector-emitter voltage is $v_{CE} = V_{CEQ} = 6$ V. In other words, the transistor is **biased** (clearly in the active region) at $i_B = I_{BQ} = 40$ μA, $i_C = I_{CQ} = 4$ mA, and $v_{CE} = V_{CEQ} = 6$ V.

Now suppose that we add an ac component—in particular, a sinusoid—to the base current so that it varies between 30 and 50 μA as illustrated in Fig. 9.1. Specifically,

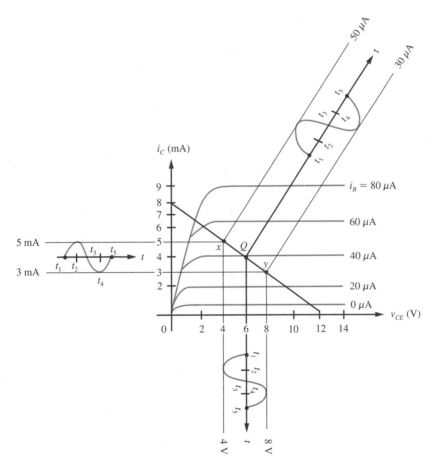

Fig. 9.1 Collector characteristics and load line of a BJT biased in the active region.

if $t_1 = 0$ s, then the total base current i_B is given by $i_B = I_{BQ} + i_b = 40 \times 10^{-6} +$ $(10 \times 10^{-6}) \sin \omega t$ A. (A lowercase subscript indicates an ac quantity.) Since the corresponding operation of the transistor is along the load line between points x and y, the resulting total collector current is $i_C = I_{CQ} + i_c = 4 \times 10^{-3} +$ $(1 \times 10^{-3}) \sin \omega t$ A, and the total collector-emitter voltage is $v_{CE} = V_{CEQ} + v_{ce} =$ $6 - 2 \sin \omega t$ V. Note that although the ac component of the base current has an amplitude of 10 μA, the ac component of the resulting collector current has an amplitude of 1 mA. In other words, the ac collector current is 100 times as large as the ac base current. Thus the given transistor amplifies alternating current by a factor of 100—or we say that it has an ac current gain of 100.

In general, we define the **ac current gain** h_{fe} (also called the **ac beta** β_{ac}) of a BJT to be the ratio of the ac collector current i_c (or variation in collector current Δi_C) to the ac base current i_b (or variation in base current Δi_B). That is,

$$h_{fe} = \frac{i_c}{i_b} = \frac{\Delta i_C}{\Delta i_B}$$

From Fig. 9.1, for example, we have that

$$h_{fe} = \frac{i_{Cx} - i_{Cy}}{i_{Bx} - i_{By}} = \frac{5 \times 10^{-3} - 3 \times 10^{-3}}{50 \times 10^{-6} - 30 \times 10^{-6}} = 100$$

Compare this with the dc beta h_{FE} (uppercase subscripts), which is the ratio of dc (quiescent) collector current to dc (quiescent) base current. In other words, the dc beta is

$$h_{FE} = \frac{I_{CQ}}{I_{BQ}} = \frac{4 \times 10^{-3}}{40 \times 10^{-6}} = 100$$

so in this case we have that $h_{FE} = h_{fe}$. In general, however, h_{FE} and h_{fe} need not be identical.

Note that if the Q point is located too close either to cutoff or to saturation, the ac component of the base current will, respectively, either cut off or saturate the transistor. Even when the Q point is centrally located between cutoff and saturation, if this ac component becomes too large, then the transistor can be driven into cutoff or saturation. Regardless of the cause, if the transistor is cut off or saturated, the undesirable result will be that the output voltage v_{CE} and output current i_C will be distorted (''clipped'').

Biasing the BJT

We have just seen that a BJT can be used as an amplifier when it is biased in the active region. Therefore, let us now concentrate on the biasing of BJTs. Shortly, we will see how to add an ac component that can be amplified.

Consider the dc *npn* CE transistor circuit shown in Fig. 9.2. For this circuit, we have that

$$i_B = \frac{V_{CC} - v_{BE}}{R_B} \tag{9.1}$$

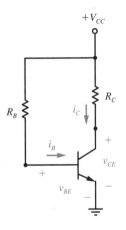

Fig. 9.2 Fixed bias.

If the transistor is in the active region, then $v_{BE} \approx 0.7$ V for silicon (0.3 V for germanium). Once R_B and V_{CC} have been selected, therefore, the base current is, in essence, constant or fixed. For this reason, Fig. 9.2 is an example of a **fixed-bias** circuit.

Figure 9.3 shows a BJT circuit with a **voltage-divider bias**. For such a situation, we can employ Thévenin's theorem. Doing so, we can erase the voltage-divider resistors R_1 and R_2, and then place a resistor $R_B = R_1 \parallel R_2 = R_1R_2/(R_1 + R_2)$ and a voltage source $V_{BB} = R_2V_{CC}/(R_1 + R_2)$ in series with the base as shown in Fig. 9.4. It is now apparent that the voltage-divider bias in Fig. 9.3 is really just another form of fixed bias.

For the case of fixed bias, when the transistor is in the active region, we have that

$$i_C = h_{FE}i_B \quad \text{and} \quad v_{CE} = -R_Ci_C + V_{CC} = -R_Ch_{FE}i_B + V_{CC}$$

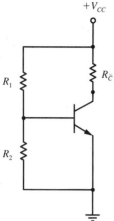

Fig. 9.3 Voltage-divider bias.

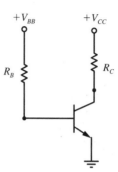

Fig. 9.4 Voltage-divider bias after application of Thévenin's theorem.

Thus we see that the operating point Q is dependent on the value of h_{FE}. If a fixed-bias circuit is duplicated using another transistor of the same type as the original, the value of h_{FE} can differ significantly. Even a temperature change will alter the value of h_{FE}. In either case, the Q point can wind up at an undesirable location. For these reasons, we conclude that a fixed-bias circuit does not have good stability.

There is another important reason why a fixed-bias circuit may result in unstable operation. Recall that the reverse current I_{CO} approximately doubles for every 10°C rise in temperature. A current i_C going across the C-B junction will cause a rise in its temperature. This, in turn, increases I_{CO} (and beta, too). From Eq. 7.10 on p. 427, we see that there will be an increase in collector current, and, thus, an even higher junction temperature. The result can be a cumulative process, known as **thermal runaway**, that will produce excessive heat and destroy the transistor. Even if such an extreme situation does not occur, it may be that the Q point moves close to or actually goes into saturation when the temperature increases.

A common way to improve bias stability is to put a resistor R_E in series with the emitter of the transistor as shown in Fig. 9.5. For analysis purposes, it is convenient to replace the voltage-division portion of the circuit by its Thévenin equivalent. Doing so, we obtain the equivalent circuit shown in Fig. 9.6, where $R_B = R_1 \parallel R_2 = R_1R_2/(R_1 + R_2)$ and $V_{BB} = R_2V_{CC}/(R_1 + R_2)$. By KVL, we have that

$$V_{BB} = R_B i_B + v_{BE} + R_E(i_B + i_C) \tag{9.2}$$

For active-region operation, $i_C = h_{FE}i_B$ and $v_{BE} = 0.7$ V (for silicon). Thus Eq. 9.2 becomes

$$R_B i_B + R_E(i_B + h_{FE}i_B) = V_{BB} - v_{BE}$$

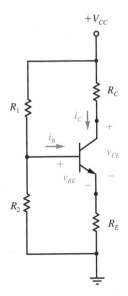

Fig. 9.5 A BJT self-biasing circuit.

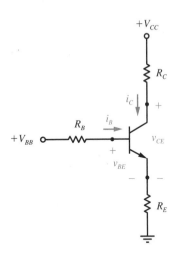

Fig. 9.6 Thévenin-equivalent of self-biasing circuit.

from which

$$i_B = \frac{V_{BB} - v_{BE}}{R_B + (1 + h_{FE})R_E} \qquad (9.3)$$

Unlike the case of fixed bias, from Eq. 9.3, we see that the base current is dependent on h_{FE}. However, this is actually beneficial. Suppose that I_{CO} and h_{FE} increase due to a rise in temperature. As mentioned above, this tends to increase i_C. But from Eq. 9.3 we see that an increase in h_{FE} means a decrease in i_B, and this, in turn, tends to decrease i_C. Hence the presence of R_E means that the overall increase in i_C is less than it would be if R_E were not present in the circuit. This results in more stable operation. By inserting the resistance R_E in series with the emitter of the transistor as shown in Fig. 9.5, we obtain what is known as a **self-biasing circuit**.

Example 9.1

For the self-biasing circuit given in Fig. 9.5, suppose that $R_1 = 20$ kΩ, $R_2 = 12$ kΩ, $R_C = R_E = 1$ kΩ, $V_{CC} = 10$ V, and $h_{FE} = 100$. Then, for the equivalent circuit given in Fig. 9.6, we have that

$$R_B = \frac{R_1 R_2}{R_1 + R_2} = \frac{(20 \times 10^3)(12 \times 10^3)}{20 \times 10^3 + 12 \times 10^3} = 7.5 \text{ kΩ}$$

and

$$V_{BB} = \frac{R_2}{R_1 + R_2} V_{CC} = \frac{12 \times 10^3}{20 \times 10^3 + 12 \times 10^3} (10) = 3.75 \text{ V}$$

Therefore, for active-region operation, Eq. 9.3 yields

$$i_B = \frac{3.75 - 0.7}{7500 + (1 + 100)(1000)} = 28.1 \text{ μA}$$

Thus

$$i_C = h_{FE}i_B = 100(28.1 \times 10^{-6}) = 2.81 \text{ mA} \tag{9.4}$$

and

$$\begin{aligned}
v_{CE} &= -R_C i_C + V_{CC} - R_E(i_B + i_C) \\
&= -1000(2.81 \times 10^{-3}) + 10 - 1000(28.1 \times 10^{-6} + 2.81 \times 10^{-3}) \\
&= 4.35 \text{ V}
\end{aligned}$$

Since $i_B > 0$ A and v_{CE} is well above the saturation voltage of 0.2 V, this confirms that the transistor is biased in the active region, and the coordinates of the operating point are $I_{CQ} = 2.81$ mA and $V_{CEQ} = 4.35$ V.

Drill Exercise 9.1

For the self-biasing circuit given in Fig. 9.5, suppose that $h_{FE} = 100$, $R_1 = R_2 = 26$ kΩ, and $V_{CC} = 10$ V. Find the values of R_E and R_C such that the BJT is biased in the active region at $I_{CQ} = 2$ mA and $V_{CEQ} = 4$ V.

ANSWER 2 kΩ, 980 Ω

Small-Signal AC Models

Given a transistor biased in the active region, let us see how ac components can be added to the existing quiescent operating (dc) conditions. One common way to do this is to capacitively couple an ac source (an ideal voltage source v_s connected in series with a resistance R_s) to the base of a transistor by using a capacitor, called a **coupling capacitor,** as shown in Fig. 9.7 for a CE amplifier. Since the capacitor C is an open circuit to direct current, the addition of the ac source will have no effect on the Q point (dc conditions) of the transistor. However, let us assume that the capacitance C is large enough that, in essence, it behaves as a short circuit for the ac signals produced by voltage source v_s. In this circumstance, therefore, the ac source will add ac components to the transistor's voltages and currents. One way to describe the relationships between these ac components is to do so graphically as illustrated in Fig. 9.1. But this is not the only approach to the ac analysis of transistor amplifiers.

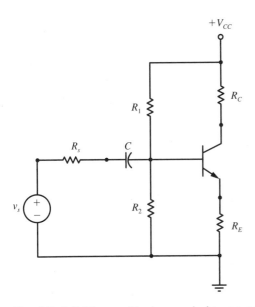

Fig. 9.7 A BJT capacitively coupled to an ac source.

When the transistor in the circuit given in Fig. 9.7 is biased in the active (linear) region, we can apply the principle of superposition to the circuit. For the case that $v_s = 0$ V and $V_{CC} \neq 0$ V (the dc case), the ac source is replaced by a short circuit and the capacitor is replaced by an open circuit. The analysis then proceeds as was discussed in the preceding section ("Biasing the BJT"). On the other hand, for the case that $v_s \neq 0$ V and $V_{CC} = 0$ V, the dc supply (implicit by the notation $+V_{CC}$) is replaced by a short circuit—and, as mentioned above, so too is the capacitor. The resulting **ac equivalent** of the circuit given in Fig. 9.7 is shown in Fig. 9.8. Note that since a short circuit is placed between the node labeled $+V_{CC}$ and the reference, in the ac-equivalent circuit, R_1 and R_2 are connected in parallel. For the ac-equivalent circuit given in Fig. 9.8, the transistor's alternating currents and voltages are denoted

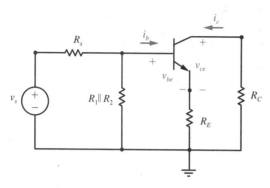

Fig. 9.8 The ac-equivalent circuit of a transistor amplifier.

by using lowercase letter subscripts (e.g., i_b, i_c, v_{be}, v_{ce}), and these are the ac components of the corresponding total currents and voltages.

Now, however, the question arises as to how we can analyze an ac-equivalent transistor circuit such as the one given in Fig. 9.8. That is, other than a graphical approach such as illustrated by Fig. 9.1, is there some description of a transistor that relates its alternating currents and voltages? The answer is yes. We can model the ac behavior of a transistor in a number of ways. We will begin, however, by considering the simpler models. It is always our assumption, though, that the ac components under consideration are small enough that the transistor will operate in the linear (active) region at all times.

Another assumption will be made as well. For the sake of simplified numerical calculations, the collector characteristics given in Fig. 9.1 were so drawn that the curves appear parallel and equally spaced. In actuality, however, this is not the case. As a consequence, a sinusoidal variation in the base current will produce somewhat distorted sinusoids as the collector current and the collector-emitter voltage. Even an applied sinusoidal base-emitter voltage will produce a somewhat distorted sinusoidal base current because of the nonlinear characteristic of the B-E junction of a BJT (see Fig. 7.9 on p. 488). However, if the ac signals are small, these nonlinear distortions become insignificant, and we can model the transistor's ac behavior with a linear circuit. It is because of this that we refer to such linear circuits as **small-signal (ac) models**.

Given the transistor shown in Fig. 9.9 with some of its associated alternating currents and voltages, the first small-signal model we present is shown in Fig. 9.10. The ac base resistance r_π is also designated by h_{ie}. Since $i_b = v_{be}/r_\pi$, the current-dependent current source can be relabeled as

$$h_{fe}i_b = \frac{h_{fe}}{r_\pi}v_{be} = g_m v_{be}$$

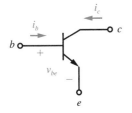

Fig. 9.9 A BJT.

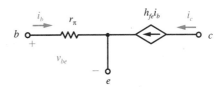

Fig. 9.10 Small-signal ac model of a BJT.

where $g_m = h_{fe}/r_\pi$ is known as the **transconductance** of the BJT. With this definition of transconductance, the model given in Fig. 9.10 is equivalent to the one shown in Fig. 9.11. For the latter case, though, the current source is dependent upon a voltage (v_{be}) instead of a current (i_b). Still another small-signal model of a BJT is shown in Fig. 9.12. This model makes use of the ac emitter resistance r_e instead of r_π. Note that

$$r_e = \frac{v_{be}}{i_b + h_{fe}i_b} = \frac{v_{be}}{(1 + h_{fe})i_b} = \frac{r_\pi}{1 + h_{fe}} \approx \frac{r_\pi}{h_{fe}} = \frac{1}{g_m} \tag{9.5}$$

since typically $h_{fe} \gg 1$. Thus,

$$r_\pi = (1 + h_{fe})r_e \approx h_{fe}r_e \tag{9.6}$$

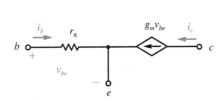

Fig. 9.11 Second small-signal model.

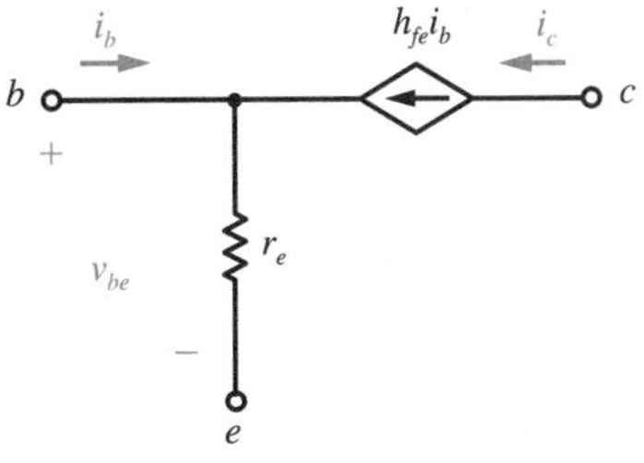

Fig. 9.12 Another small-signal model.

For an *npn* transistor operating in the active region (see Eq. 7.6 on p. 422),

$$i_C = I_S(e^{v_{BE}/V_T} - 1) \approx I_S e^{v_{BE}/V_T}$$

from which

$$di_C = d[I_S(e^{v_{BE}/V_T} - 1)] = \frac{1}{V_T}(I_S e^{v_{BE}/V_T})\, dv_{BE} \approx \frac{i_C}{V_T} dv_{BE}$$

Since $g_m v_{be} = h_{fe} i_b = i_c$, then

$$g_m = \frac{i_c}{v_{be}} = \frac{di_C}{dv_{BE}} = \frac{i_C}{V_T}$$

Hence for a BJT biased at $i_C = I_{CQ}$, we have that

$$\boxed{g_m = I_{CQ}/V_T} \tag{9.7}$$

where $V_T = T/11{,}600$ (Eq. 6.14) and T is the temperature in kelvins. At 27°C (300 K), we have that $V_T \approx 26$ mV, so a good approximation for transconductance is

$$\boxed{g_m = \frac{I_{CQ}}{0.026}} \tag{9.8}$$

From Eq. 9.5, therefore, we have the approximations

$$\boxed{r_e = \frac{1}{g_m} = \frac{0.026}{I_{CQ}}} \quad \text{and} \quad \boxed{r_\pi = h_{fe} r_e = \frac{0.026 h_{fe}}{I_{CQ}}} \tag{9.9}$$

For the case of a *pnp* transistor, replace I_{CQ} by $|I_{CQ}|$ in Eq. 9.7, Eq. 9.8, and Eq. 9.9. It should also be pointed out that a given small-signal ac model can be used for either an *npn* transistor, or a *pnp* transistor—there is no need to be concerned with signs as in the dc case.

Example 9.2

To demonstrate the use of a small-signal model, let us replace the transistor in the ac circuit shown in Fig. 9.8 by the model given in Fig. 9.12. The result is shown in Fig. 9.13, where $R_p = R_1 \parallel R_2 = R_1 R_2/(R_1 + R_2)$. Of course, this circuit can be analyzed by nodal or mesh analysis. However, let us take an alternative approach, thereby defining some new terms along the way.

The resistance between the base and the reference, looking into the base of the transistor, is $R_b = v_b/i_b$, where v_b is the base voltage (with respect to the reference). Since

$$v_b = r_e(i_b + h_{fe} i_b) + R_E(i_b + h_{fe} i_b)$$

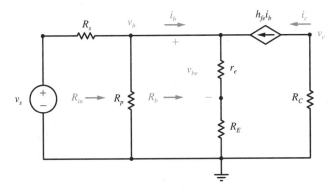

Fig. 9.13 Transistor amplifier ac-equivalent circuit incorporating small-signal model.

then

$$R_b = \frac{v_b}{i_b} = (1 + h_{fe})(r_e + R_E) \approx h_{fe}(r_e + R_E) \tag{9.10}$$

The resistance between the base and the reference, looking into the amplifier, is called the (ac) **input resistance** R_{in} of the amplifier. We have, therefore, that

$$R_{in} = R_p \parallel R_b = \frac{R_p R_b}{R_p + R_b} \tag{9.11}$$

The (ac) **voltage gain** A_v from base (amplifier input) to collector (amplifier output) is the ratio of collector voltage v_c to base voltage v_b, that is, $A_v = v_c/v_b$. For the circuit given in Fig. 9.13, the voltage gain is

$$A_v = \frac{v_c}{v_b} = \frac{-R_C h_{fe} i_b}{R_b i_b} = \frac{-h_{fe} R_C}{(1 + h_{fe})(r_e + R_E)} \approx \frac{-R_C}{r_e + R_E} \tag{9.12}$$

Since $v_c = A_v v_b$ and A_v is a negative number, then when the source voltage v_s is a sinusoid, the input v_b is a sinusoid, and the output voltage v_c is a sinusoid that is 180 degrees out of phase with the input v_b.

From Eq. 9.12, we see that if we decrease the denominator $r_e + R_E$, we increase the voltage gain A_v. The resistance r_e is determined by the transistor's Q point (Eq. 9.9). However, R_E is external to the transistor. Although setting R_E to zero will increase the gain, it will also change the operating point and cause a decrease in the stability of operation. Suppose, therefore, that we place in parallel with R_E a capacitor whose impedance to the ac signals produced by v_s is negligible compared to R_E. The addition of such a capacitor will not affect the dc conditions (quiescent operating point) of the circuit, so r_e remains unchanged. However, since for ac signals R_E is short-circuited or bypassed by the capacitor, in the ac circuit we replace R_E by a short circuit. Thus by adding such an element, called a **bypass capacitor**, to the amplifier, we get a voltage gain of

$$A_v = \frac{-h_{fe}R_C}{(1 + h_{fe})r_e} \approx \frac{-R_C}{r_e} \tag{9.13}$$

The voltage gain A_{vs} from the ac source to the amplifier output (collector) is the ratio of collector voltage v_c to source voltage v_s, that is, $A_{vs} = v_c/v_s$. For the circuit given in Fig. 9.13, by voltage division

$$v_b = \frac{R_{in}}{R_{in} + R_s} v_s$$

Thus

$$A_{vs} = \frac{v_c}{v_s} = \frac{v_c}{v_b}\frac{v_b}{v_s} = \frac{A_v R_{in}}{R_{in} + R_s}$$

where R_{in} is given by Eq. 9.11 and A_v is given by Eq. 9.12.

Drill Exercise 9.2

For the self-biasing circuit given in Fig. 9.7 on p. 577, suppose that $h_{FE} = h_{fe} = 100$, $R_1 = R_2 = 26$ kΩ, $R_C = 980$ Ω, $R_E = 2$ kΩ, and $V_{CC} = 10$ V. Find (a) g_m, (b) r_e, (c) R_{in}, and (d) A_v.

ANSWER (a) 76.9 m$\mho$, (b) 13 Ω, (c) 12.2 kΩ, (d) -0.482

Additional Amplifier Principles

Although the CE amplifier given in Fig. 9.7 is not externally loaded, in general, it may have some load resistance R_L. Such a situation, in which R_L is capacitively coupled to the amplifier, is depicted in Fig. 9.14. Note that for this circuit, in addition

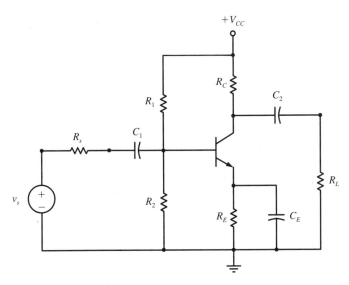

Fig. 9.14 A CE amplifier with a capacitively coupled external load.

to the coupling capacitors C_1 and C_2, there is also a bypass capacitor C_E. The dc equivalent of this circuit is given in Fig. 9.5 on p. 575.

Figure 9.15 shows typical collector characteristics for the transistor. Also shown are the quiescent operating point Q and the load line obtained for the dc case (C_1, C_2, and C_E behave as open circuits). Since, typically $h_{FE} \gg 1$, then $i_B + i_C = i_C/h_{FE} + i_C \approx i_C$. Thus, by KVL,

$$V_{CC} = R_C i_C + v_{CE} + R_E(i_B + i_C) \approx R_C i_C + v_{CE} + R_E i_C$$

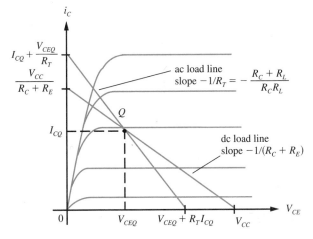

Fig. 9.15 Collector characteristics with ac and dc load lines.

from which we obtain the approximation

$$i_C = -\frac{1}{R_C + R_E} v_{CE} + \frac{V_{CC}}{R_C + R_E}$$

and the straight line described by this equation is called the **dc load line**. It passes through the Q point and has a slope of $-1/(R_C + R_E)$.

For the ac case, we assume that C_1, C_2, and C_E behave as short circuits. Therefore, the resulting ac-equivalent circuit is shown in Fig. 9.16, where $R_p = R_1 \parallel R_2 = R_1 R_2/(R_1 + R_2)$ and $R_T = R_C \parallel R_L = R_C R_L/(R_C + R_L)$. In this case, however, the transistor operates on a line that passes through the Q point but has a slope of $-1/R_T$. This line, known as the **ac load line**, is also indicated in Fig. 9.15. [If the bypass capacitor C_E is omitted, the slope of the ac load line is $-1/(R_T + R_E)$.]

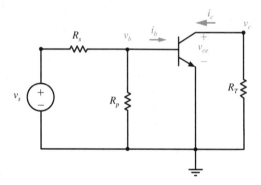

Fig. 9.16 The ac-equivalent circuit of the given transistor amplifier.

In order to obtain the equation of the ac load line, note that

$$v_{CE} = V_{CEQ} + v_{ce} \qquad \text{and} \qquad v_{ce} = -R_T i_c$$

whereas

$$i_C = I_{CQ} + i_c \qquad \Rightarrow \qquad i_c = i_C - I_{CQ}$$

Thus

$$v_{CE} = V_{CEQ} + (-R_T i_c) = V_{CEQ} - R_T(i_C - I_{CQ}) = V_{CEQ} - R_T i_C + R_T I_{CQ}$$

from which

$$i_C = -\frac{1}{R_T} v_{CE} + \left(I_{CQ} + \frac{V_{CEQ}}{R_T} \right)$$

and this is the equation for the ac load line.

If we replace the transistor in the ac-equivalent circuit of Fig. 9.16 by the model given in Fig. 9.12, we get the circuit shown in Fig. 9.17a. As for the case of the circuit in Fig. 9.13, we may determine that $R_b = (1 + h_{fe})r_e \approx h_{fe}r_e$, $R_{in} = R_p \| R_b = R_pR_b/(R_p + R_b)$, and $A_v = -h_{fe}R_T/(1 + h_{fe})r_e \approx -R_T/r_e$. Because there is an external load R_L (that is why R_T is explicitly shown as the parallel combination of R_C and R_L), another network function is the **current gain**. For the given transistor amplifier, the (ac) current gain A_i from the base to the load is $A_i = i_L/i_b$. From the circuit given in Fig. 9.17a, by current division we immediately have that

$$A_i = \frac{i_L}{i_b} = \frac{-h_{fe}R_C}{R_C + R_L} \qquad (9.14)$$

The current gain A_{is} from the source to the load is defined to be $A_{is} = i_L/i_s$. For the given amplifier, by current division, we know that $i_b = R_pi_s/(R_p + R_b)$. Thus we have that

$$A_{is} = \frac{i_L}{i_s} = \frac{i_L}{i_b}\frac{i_b}{i_s} = \frac{A_iR_p}{R_p + R_b}$$

where $R_b = (1 + h_{fe})r_e$ and A_i is given by Eq. 9.14.

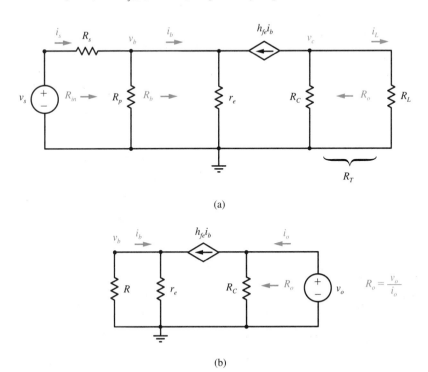

(a)

(b)

Fig. 9.17 (a) Transistor amplifier incorporating small-signal model, and (b) circuit for determining R_o.

Another important parameter of the amplifier is its **output** (Thévenin-equivalent) **resistance** R_o. As we know from our study of electric circuits, this is the resistance between the collector and the reference (with the load removed) when the input v_s is set to zero. To calculate R_o, we set $v_s = 0$ V and apply a voltage source v_o to the output as depicted in Fig. 9.17b. In this circuit, R represents the parallel combination of R_s and R_p obtained when v_s is replaced by a short circuit, i.e., $R = R_s R_p / (R_s + R_p)$. The output resistance is $R_o = v_o/i_o$.

By KCL,

$$i_b + h_{fe}i_b = \frac{v_b}{r_e} \qquad \Rightarrow \qquad i_b = \frac{v_b}{(1 + h_{fe})r_e}$$

Since $v_b = -Ri_b$, then $i_b = -Ri_b/(1 + h_{fe})r_e \Rightarrow i_b = 0$ A. Thus, by KCL,

$$i_o = h_{fe}i_b + \frac{v_o}{R_C} = \frac{v_o}{R_C} \qquad \Rightarrow \qquad R_o = \frac{v_o}{i_o} = R_C$$

is the output resistance.

Example 9.3

As another example of the use of a small-signal model, consider the BJT circuit shown in Fig. 9.18, where the resistor values are such that the transistor is biased in

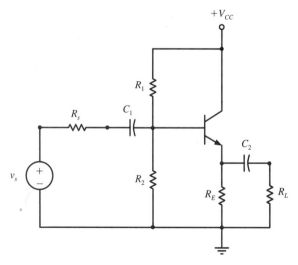

Fig. 9.18 Common-collector amplifier (emitter follower).

the active region. Figure 9.19 shows the ac-equivalent circuit for the case that the transistor is replaced by the model given in Fig. 9.10, and where $R_p = R_1R_2/(R_1 + R_2)$. Note that since R_L is the load, the emitter voltage v_e is the output voltage and the collector is common to both input and output. Therefore, such a circuit is known as a **common-collector amplifier**.

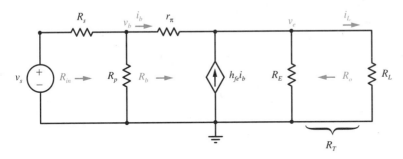

Fig. 9.19 Emitter-follower ac-equivalent circuit.

Let us designate the parallel connection of R_E and R_L by R_T. Thus

$$v_e = R_T(i_b + h_{fe}i_b) = (1 + h_{fe})R_Ti_b$$

and

$$v_b = r_\pi i_b + v_e = [r_\pi + (1 + h_{fe})R_T]i_b \qquad (9.15)$$

Hence the voltage gain A_v from base to emitter is

$$A_v = \frac{v_e}{v_b} = \frac{(1 + h_{fe})R_T}{r_\pi + (1 + h_{fe})R_T} \approx \frac{h_{fe}R_T}{r_\pi + h_{fe}R_T} = \frac{R_T}{r_\pi/h_{fe} + R_T} = \frac{R_T}{r_e + R_T} \qquad (9.16)$$

For the case that $h_{fe}R_T \gg r_\pi$ (or $R_T \gg r_e$), we have that $A_v \approx 1$. In other words, $v_e \approx v_b$ and the emitter voltage "follows" the base voltage. It is for this reason that the common-collector circuit is usually called an **emitter follower**.

The fact that the output (emitter) voltage approximates the base voltage is not the only important property of an emitter follower. The resistance between the base and the reference looking into the base is, from Eq. 9.15,

$$R_b = \frac{v_b}{i_b} = r_\pi + (1 + h_{fe})R_T \approx r_\pi + h_{fe}R_T \qquad (9.17)$$

and for large values of h_{fe} and R_T, then R_b can be made very large. The input resistance of the emitter follower, therefore, is

$$R_{in} = R_p \parallel R_b = \frac{R_p R_b}{R_p + R_b}$$

where $R_p = R_1 R_2/(R_1 + R_2)$. If R_1 and R_2 are selected to be large, the input resistance can be made large.

The output resistance R_o of the emitter follower can be determined from the circuit shown in Fig. 9.20. Since $R_o = R_E \parallel R'$, let us first find R'. For the sake of simplicity, denote the parallel connection of R_s and R_p by R. By voltage division, we have that

$$v_b = \frac{R}{r_\pi + R} v_o$$

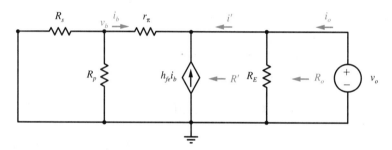

Fig. 9.20 Circuit for determining R_o for an emitter follower.

Since $i_b = -v_b/R \Rightarrow i_b = -v_o/(r_\pi + R)$, then

$$i' = -(1 + h_{fe})i_b = -(1 + h_{fe}) \frac{-v_o}{r_\pi + R}$$

from which

$$R' = \frac{v_o}{i'} = \frac{r_\pi + R}{1 + h_{fe}} \approx \frac{r_\pi + R}{h_{fe}} \tag{9.18}$$

where $R = R_s R_p/(R_s + R_p)$, and the output resistance is

$$R_o = R_E \parallel R'$$

Dividing the value $r_\pi + R$ by $1 + h_{fe}$ as indicated in Eq. 9.18 means that R' can be made relatively small and, consequently, so can the output resistance R_o.

The properties of approximately unity gain, a high input resistance, and a low output resistance allow the emitter follower to be used as a buffer or isolation amplifier for connecting a low-resistance load to a high-resistance source.

Drill Exercise 9.3

For the emitter follower given in Fig. 9.18, suppose that $h_{FE} = h_{fe} = 100$, $R_1 = R_2 = 26$ kΩ, $R_E = R_L = 2$ kΩ, $R_s = 1$ kΩ, and $V_{CC} = 10$ V. Find (a) v_e/v_b, (b) R_{in}, and (c) R_o.

ANSWER (a) 0.987, (b) 11.5 kΩ, (c) 22.0 Ω

The h Parameters

Having considered some simple small-signal ac models of a BJT biased in the active region, let us now discuss a model that yields greater accuracy. Shown in Fig. 9.21 is a BJT and its (CE) **hybrid-parameter** or **h-parameter** model. The four h parameters used in this model can be determined as follows:

(1) If we place a short circuit between the collector and the emitter ($v_{ce} = 0$ V), then from Fig. 9.21b we see that $v_{be} = h_{ie}i_b$. Thus

$$h_{ie} = \left.\frac{v_{be}}{i_b}\right|_{v_{ce}=0\,\text{V}} \qquad \text{(ohms)}$$

and h_{ie} is called the **short-circuit input resistance**.

(2) If the base is open-circuited ($i_b = 0$ A), then $v_{be} = h_{re}v_{ce}$. Thus

$$h_{re} = \left.\frac{v_{be}}{v_{ce}}\right|_{i_b=0\,\text{A}}$$

and h_{re} is called the **open-circuit reverse voltage gain**.

(3) If $v_{ce} = 0$ V, then there is no current through h_{oe} and $i_c = h_{fe}i_b$. Thus

$$h_{fe} = \left.\frac{i_c}{i_b}\right|_{v_{ce}=0\,\text{V}}$$

and h_{fe} is called the **short-circuit forward current gain**.

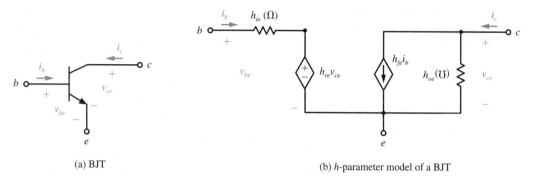

(a) BJT

(b) *h*-parameter model of a BJT

Fig. 9.21 (*a*) A BJT and (*b*) its *h*-parameter model.

(4) If $i_b = 0$ A, then $v_{ce} = i_c/h_{oe}$. Thus

$$h_{oe} = \left.\frac{i_c}{v_{ce}}\right|_{i_b = 0\,\text{A}} \qquad \text{(mhos)}$$

and h_{oe} is called the **open-circuit output conductance**.

Typical numerical values for *h* parameters are:

$$h_{ie} = 1.5 \text{ k}\Omega, \qquad h_{re} = 10^{-4}, \qquad h_{fe} = 100, \qquad h_{oe} = 10^{-5} \text{ ℧}$$

For the case that we ignore (set to zero) h_{re} and h_{oe} (which are typically small values), the *h*-parameter model given in Fig. 9.21*b* is reduced to the small-signal model shown in Fig. 9.10 on p. 579. Now you know why the ac beta of a BJT is designated "h_{fe}."

If we replace the BJT in the ac circuit given in Fig. 9.16 on p. 584 by the *h*-parameter model given in Fig. 9.21*b*, we get the ac circuit shown in Fig. 9.22. Note that, in this case, since $v_e = 0$ V, then $v_{be} = v_b$ and $v_{ce} = v_c$.

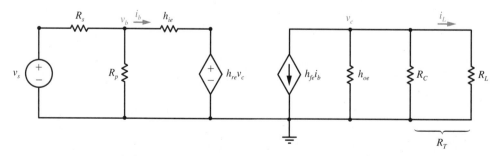

Fig. 9.22 Transistor amplifier incorporating the *h*-parameter BJT model.

Example 9.4

Let us calculate the voltage gain $A_v = v_c/v_b$ for the circuit given in Fig. 9.22 when $R_C = R_L = 2 \text{ k}\Omega$ and the BJT has the h parameters: $h_{ie} = 1.3 \text{ k}\Omega$, $h_{re} = 10^{-4}$, $h_{fe} = 100$, $h_{oe} = 10^{-5} \text{℧}$.

By Ohm's law,

$$i_b = \frac{v_b - h_{re}v_c}{h_{ie}} = \frac{v_b - 10^{-4}v_c}{1.3 \times 10^3}$$

Since $R_T = R_C \| R_L = 2 \text{ k}\Omega \| 2 \text{ k}\Omega = 1 \text{ k}\Omega$ and $r_{oe} = 1/h_{oe} = 1/10^{-5} = 100 \text{ k}\Omega$, then

$$v_c = -(r_{oe} \| R_T)h_{fe}i_b = -(100 \text{ k}\Omega \| 1 \text{ k}\Omega)100i_b = -(99 \times 10^3)i_b$$

and therefore,

$$v_c = -(99 \times 10^3)\left(\frac{v_b - 10^{-4}v_c}{1.3 \times 10^3}\right) \quad \Rightarrow \quad \frac{v_c}{v_b} = -76.7$$

Drill Exercise 9.4

For the ac circuit given in Fig. 9.22 suppose that $R_C = R_L = 2 \text{ k}\Omega$, $h_{ie} = 1.3 \text{ k}\Omega$, $h_{re} = 0$, $h_{fe} = 100$, and $h_{oe} = 0 \text{℧}$. Find the ac voltage gain v_c/v_b.

ANSWER -76.9

Drill Exercise 9.4 demonstrates that the simple model depicted in Fig. 9.10 on p. 579 yields virtually the same voltage gain that was obtained in Example 9.4 by using the accurate h-parameter model of the BJT shown in Fig. 9.21b.

Although the h-parameter model of a BJT is an accurate model, it only externally describes the ac behavior of a BJT, that is, it describes the transistor's behavior from the point of view of its terminals. An accurate small-signal ac model that describes the behavior of a BJT both internally as well as externally is the **hybrid-π model** of a BJT (see Problem 9.24 at the end of this chapter). Although the use of this model, which is a generalization of the one given in Fig. 9.11 on p. 579, gives more accurate results, generally they are not significantly different from those obtained with the simpler model. Therefore, when performing circuit analysis with hand calculations, the simpler models are usually employed.

9.2 Field-Effect Transistor Amplifiers

Just as a BJT biased in the active region can be used for amplification purposes, so too can an FET biased in the active region be employed as a small-signal ac amplifier. We begin our discussion of small-signal FET amplifiers by considering the case of depletion-type FETs, that is, JFETs and depletion MOSFETs.

Figure 9.23 shows an n-channel JFET circuit and Fig. 9.24 shows the transfer characteristic of the JFET. Since $i_G = 0$ A, then $v_{GS} = -V_{GG}$. Thus the intersection of the vertical line, whose equation is $v_{GS} = -V_{GG}$, and the transfer characteristic curve at point Q graphically determine the quiescent drain current I_{DQ}. Analytically, the active-region equation $i_D = I_{DSS}(1 - v_{GS}/V_p)^2$ can be used. From a realistic point of view, however, a JFET's transfer characteristic varies with temperature and from device to device. Suppose that such a variation range corresponds to the shaded region indicated in Fig. 9.24. In other words, let us allow the possibility that the transfer characteristic of the JFET could be anywhere within the shaded region. For the extreme case, for instance, that the transfer characteristic corresponds to the lower boundary, since v_{GS} is "fixed" at $v_{GS} = -V_{GG}$, the JFET actually will be cut off ($i_D = 0$ A). Because of such a possibility, this type of bias, called **fixed bias**, is undesirable.

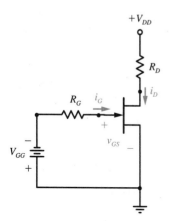

Fig. 9.23 A fixed-bias JFET circuit.

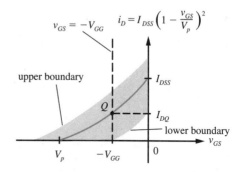

Fig. 9.24 JFET transfer characteristic and its range of variation.

An improvement on fixed bias is the arrangement shown in Fig. 9.25. In this case (since $i_G = 0$ A), we have that $i_D = -v_{GS}/R_S$ and the line described by this equation is plotted along with the JFET's transfer characteristic (and range of variation) in Fig. 9.26. Note, however, that even for the extreme case the transfer characteristic is at the lower boundary, the JFET will not be cut off. Thus, this type of bias is superior to fixed bias. Since $v_{GS} = -R_S i_D$, the value of v_{GS} depends upon one of the JFET's own variables (i_D) and, therefore, is an example of what is known as **self-**

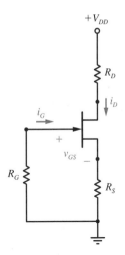

Fig. 9.25 Self-bias JFET circuit.

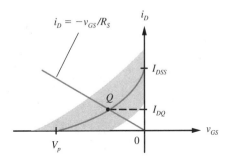

Fig. 9.26 JFET transfer characteristic and its range of variation.

bias. Note that if the transfer characteristic curve changes position, then, in general, so will the value of the quiescent drain current I_{DQ}. The possible variation in I_{DQ} can be reduced by making the line $i_D = -v_{DS}/R_S$ more nearly horizontal (i.e., by increasing R_S). Doing this, though, will put the Q point closer to cutoff. There is, however, an alternative way to decrease the dependence of I_{DQ} on the variation of the JFET's characteristics.

To accomplish this, we combine self-bias and fixed bias as shown in Fig. 9.27. In this case, we have that $i_D = (-v_{GS} + V_{GG})/R_S = -v_{GS}/R_S + V_{GG}/R_S$, and the line described by this equation is shown in Fig. 9.28. Of course, a separate second

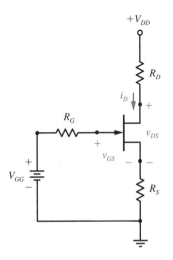

Fig. 9.27 Combination of fixed and self-bias.

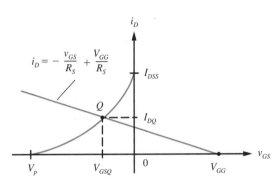

Fig. 9.28 Quiescent operating point for fixed and self-bias.

supply voltage V_{GG} need not be used since the series connection of V_{GG} and R_G can represent the Thévenin equivalent of a voltage divider connected to the supply voltage V_{DD}.

Let us now see why this type of bias results in a more stable operation than either fixed bias or self-bias does individually. Suppose that i_D tends to increase (e.g., due to a change in temperature). As a consequence, the voltage at the source of the JFET ($v_S = R_S i_D$) will tend to increase. However, since $i_G = 0$ A and the voltage at the gate of the JFET ($v_G = -R_G i_G + V_{GG} = V_{GG}$) is a constant, then $v_{GS} = v_G - v_S$ will tend to decrease. This, in turn, will tend to decrease i_D, thereby opposing the original tendency for an increase in i_D.

Example 9.5

For the amplifier circuit given in Fig. 9.29, suppose that the JFET has $I_{DSS} = 10$ mA and $V_p = -4$ V. Given that $R_1 = 500$ kΩ, $R_2 = 250$ kΩ, and $V_{DD} = 15$ V, let us determine the values of R_S and R_D such that the JFET is biased in the active region at $v_{DS} = 5.5$ V $= V_{DSQ}$ and $v_{GS} = -2$ V $= V_{GSQ}$. (Since $V_p = -4 < v_{GS} = -2 < 0$ V and $v_{DS} = 5.5 > v_{GS} - V_p = -2 - (-4) = 2$ V, then active-region operation is assured.)

For the dc-equivalent circuit shown in Fig. 9.27, by Thévenin's theorem we have that $R_G = R_1 \| R_2 = 167$ kΩ and $V_{GG} = R_2 V_{DD}/(R_1 + R_2) = 5$ V. For active-region operation, then

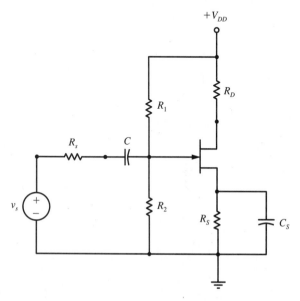

Fig. 9.29 A JFET amplifier.

$$i_D = I_{DSS}\left(1 - \frac{v_{GS}}{V_p}\right)^2 = (10 \times 10^{-3})\left(1 - \frac{-2}{-4}\right)^2 = 2.5 \text{ mA} = I_{DQ}$$

Therefore,

$$R_S = \frac{-v_{GS} + V_{GG}}{i_D} = \frac{-(-2) + 5}{2.5 \times 10^{-3}} = 2.8 \text{ k}\Omega$$

Since $v_S = -v_{GS} + V_{GG} = -(-2) + 5 = 7$ V, or
$v_S = R_S i_D = (2.8 \times 10^3)(2.5 \times 10^{-3}) = 7$ V, then

$$R_D = \frac{V_{DD} - (v_{DS} + v_S)}{i_D} = \frac{15 - (5.5 + 7)}{2.5 \times 10^{-3}} = 1 \text{ k}\Omega$$

Drill Exercise 9.5

For the amplifier given in Fig. 9.29, the JFET has $I_{DSS} = 16$ mA and $V_p = -3$ V. Given that $R_1 = 600$ kΩ, $R_2 = 360$ kΩ, and $V_{DD} = 12$ V, determine the values of R_S and R_D for which the JFET will be biased in the active region at $I_{DQ} = 4$ mA and $V_{DSQ} = 4$ V, and confirm active-region operation.

ANSWER 1.5 kΩ, 500 Ω

Biasing Enhancement MOSFETS

Let us now consider the case of an enhancement-type FET. Figure 9.30 shows an n-channel enhancement MOSFET circuit, and Fig. 9.31 shows the transfer characteristic curve for the MOSFET. Again, by Thévenin's theorem, we can replace the voltage divider consisting of R_1 and R_2 by a voltage source $V_{GG} = R_2 V_{DD}/(R_1 + R_2)$ and a resistance $R_G = R_1 \parallel R_2 = R_1 R_2/(R_1 + R_2)$ connected in series with the gate of the MOSFET (e.g., see Fig. 9.27). For the circumstance that $R_S = 0$ Ω, then $v_{GS} = V_{GG}$. However, as for the fixed-bias case of a JFET, a variation in the MOSFET's characteristics can result in a significant change in the quiescent drain current. For this reason, we should have $R_S > 0$ Ω.

In general, $i_D = (-v_{GS} + V_{GG})/R_S = -v_{GS}/R_S + V_{GG}/R_S$, and the intersection of the line described by this equation and the MOSFET transfer characteristic de-

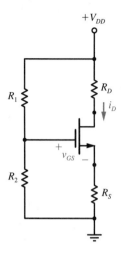

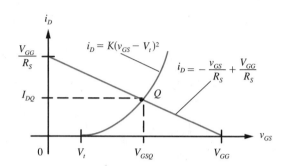

Fig. 9.30 Bias configuration for an enhancement MOSFET.

Fig. 9.31 Enhancement MOSFET transfer characteristic.

termines the Q point. Such a situation is illustrated in Fig. 9.31. As was the case for the JFET, the resistance R_S produces a biasing stabilizing effect.

Example 9.6

For the amplifier circuit shown in Fig. 9.32, suppose that the enhancement MOSFET has $V_t = 2$ V and $K = 0.25$ mA$/$V^2. Given that $R_1 = 1$ MΩ, $R_2 = 2$ MΩ, and $V_{DD} = 12$ V, let us determine the values of R_S and R_D for which the MOSFET will be biased in the active region at $I_{DQ} = 4$ mA and $V_{DSQ} = 6$ V.

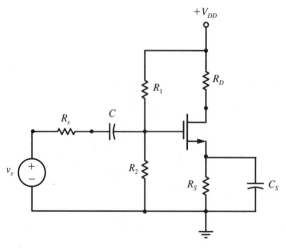

Fig. 9.32 A MOSFET amplifier.

For the dc-equivalent circuit shown in Fig. 9.33, $R_G = R_1 \| R_2 = 667$ kΩ and $V_{GG} = R_2 V_{DD}/(R_1 + R_2) = 8$ V. Since we wish to have $i_D = 4$ mA $= I_{DQ}$, then from $i_D = K(v_{GS} - V_t)^2$ we have that

$$v_{GS} = V_t + \sqrt{\frac{i_d}{K}} = 2 + \sqrt{\frac{4 \times 10^{-3}}{2.5 \times 10^{-4}}} = 6 \text{ V} = V_{GSQ}$$

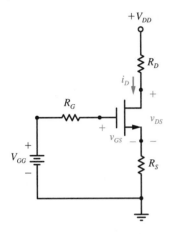

Fig. 9.33 The dc-equivalent circuit.

Therefore,

$$R_S = \frac{-v_{GS} + V_{GG}}{i_D} = \frac{-6 + 8}{4 \times 10^{-3}} = 500 \ \Omega$$

Since $v_S = -v_{GS} + V_{GG} = -6 + 8 = 2$ V, or $v_S = R_S i_D = (500)(4 \times 10^{-3}) = 2$ V, then

$$R_D = \frac{V_{DD} - (v_{DS} + v_S)}{i_D} = \frac{12 - (6 + 2)}{4 \times 10^{-3}} = 1 \text{ k}\Omega$$

Finally, since $v_{GS} = 6 > V_t = 2$ V and $v_{DS} = 6 > v_{GS} - V_t = 6 - 2 = 4$ V, then active-region operation is confirmed.

Drill Exercise 9.6

For the amplifier given in Fig. 9.32, the enhancement MOSFET has $K = 0.2$ mA/V^2 and $V_t = 2$ V. Given that $R_1 = R_2 = 1$ MΩ and $V_{DD} = 12$ V, determine the values of R_S and R_D for which the MOSFET will be biased at $V_{DSQ} = V_{GSQ} = 4$ V.

ANSWER 2.5 kΩ, 7.5 kΩ

An alternative popular biasing scheme for an enhancement MOSFET is **feedback bias**. Depicted in Fig. 9.34a is a feedback amplifier and Fig. 9.34b the dc-equivalent circuit of this amplifier. For this circuit, there is a "feedback" resistance R_F connected between the drain and the gate of the MOSFET. We know that an enhancement MOSFET is in the active region when $v_{DS} \geq v_{GS} - V_t$ and $v_{GS} > V_t$. Since there is no gate current, $v_{GS} = v_{DS}$. Thus $v_{DS} = v_{GS} > v_{GS} - V_t$, and the feedback bias guarantees active-region operation (when $v_{GS} > V_t$).

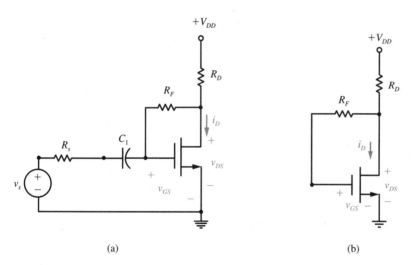

Fig. 9.34 (a) MOSFET feedback amplifier, and (b) dc-equivalent circuit.

Suppose that for this type of bias, i_D tends to increase. As a consequence of this, $v_{GS} = v_{DS} = -R_D i_D + V_{DD}$ tends to decrease. However, a tendency of v_{GS} to decrease means that i_D has a tendency to decrease—and this opposes the original tendency for i_D to increase. Thus we see that the feedback amplifier shown in Fig. 9.34 possesses operational stability.

Example 9.7

The amplifier circuit given in Fig. 9.34 has $K = 0.25$ mA/V^2 and $V_t = 2$ V. Suppose that $R_D = 1$ kΩ, $R_F = 1$ MΩ, and $V_{DD} = 10$ V. Since

$$i_D = K(v_{GS} - V_t)^2 \quad \text{and} \quad i_D = \frac{V_{DD} - v_{DS}}{R_D} = \frac{V_{DD} - v_{GS}}{R_D}$$

then

$$K(v_{GS} - V_t)^2 = \frac{V_{DD} - v_{GS}}{R_D}$$

from which

$$(2.5 \times 10^{-4})(v_{GS} - 2)^2 = \frac{10 - v_{GS}}{1000} \quad \Rightarrow \quad v_{GS} = \pm 6 \text{ V}$$

The solution we seek is the value of v_{GS} that is greater than $V_t = 2$ V. Thus $v_{GS} = 6$ V $= V_{GSQ}$, and by using this value, we get $i_D = 4$ mA $= I_{DQ}$.

Drill Exercise 9.7

For the amplifier given in Fig. 9.34, the enhancement MOSFET has $K = 0.25$ mA/V^2 and $V_t = 2$ V. Given that $R_D = 500$ Ω, $R_F = 1$ MΩ, and $V_{DD} = 10$ V, place a resistor $R_S = 500$ Ω in series with the source terminal of the MOSFET and determine V_{GSQ}, V_{DSQ}, and I_{DQ}.

ANSWER 6 V, 6 V, 4 mA

Small-Signal AC Models

Just as we modeled the small-signal ac behavior of a BJT, so too we can use small-signal (ac) models for FETs. Figure 9.35 depicts a generic circuit symbol of an FET, that is, it can represent either a JFET, a depletion MOSFET, or an enhancement MOSFET. Some of the FET's associated ac variables are indicated. A simple small-signal model of the given FET is presented in Fig. 9.36a, where the ac resistance between gate and source is $r = \infty$. (Practically speaking, $r > 10^{10}$ Ω.) Instead of using the circuit symbol for a resistance, we can indicate the infinite resistance by an open circuit and, consequently, the model in Fig. 9.36a can be redrawn in the equivalent form shown in Fig. 9.36b.

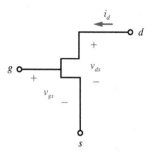

Fig. 9.35 A JFET, a depletion MOSFET, or an enhancement MOSFET.

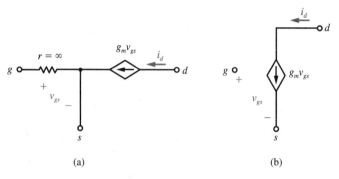

(a)   (b)

Fig. 9.36 Simple small-signal ac model of an FET.

As can be seen from the drain characteristics of a typical FET (e.g., see Fig. 8.7 on p. 509), the drain current is dependent upon the gate-source voltage. For the small-signal (linear) model of the FET given in Fig. 9.36, the ac drain current is directly proportional to the ac gate-source voltage, that is, $i_d = g_m v_{gs}$, where the proportionality constant g_m is the **transconductance** of the FET. Thus we can write

$$g_m = \frac{i_d}{v_{gs}} = \frac{di_D}{dv_{GS}}$$

since g_m is the ratio of the ac drain current i_d (that is, the variation in i_D) to the ac gate-source voltage v_{gs} (that is, the variation in v_{GS}).

For the case of a (*n*-channel) JFET or depletion MOSFET operating in the active region, recall that

$$i_D = I_{DSS} \left(1 - \frac{v_{GS}}{V_p} \right)^2 \tag{9.19}$$

Thus

$$g_m = \frac{di_D}{dv_{GS}} = -\frac{2I_{DSS}}{V_p}\left(1 - \frac{v_{GS}}{V_p}\right)$$

(9.20)

From Eq. 9.19, we have that $1 - v_{GS}/V_p = \sqrt{i_D/I_{DSS}}$. Substituting this fact into Eq. 9.20 yields the alternative expression for transconductance:

$$g_m = -\frac{2I_{DSS}}{V_p}\sqrt{\frac{i_D}{I_{DSS}}} = -\frac{2}{V_p}\sqrt{i_D I_{DSS}}$$

(9.21)

For the special case that $v_{GS} = 0$ V, the transconductance is designated g_{m0}. From Eq. 9.20, setting $v_{GS} = 0$ V, we get

$$g_{m0} = -\frac{2I_{DSS}}{V_p}$$

(9.22)

Practically speaking, for an FET, g_{m0} and I_{DSS} can be measured much more accurately than can be V_p. Thus by measuring g_{m0} and I_{DSS}, Eq. 9.22 can be utilized to accurately determine the pinchoff voltage V_p. Furthermore, by using Eq. 9.22, we have the additional formulas for the transconductance:

$$g_m = g_{m0}\left(1 - \frac{v_{GS}}{V_p}\right) = g_{m0}\sqrt{\frac{i_D}{I_{DSS}}}$$

For the case of an (n-channel) enhancement MOSFET operating in the active region, we know that

$$i_D = K(v_{GS} - V_t)^2$$

Therefore, such an FET has a transconductance of

$$g_m = \frac{di_D}{dv_{GS}} = 2K(v_{GS} - V_t)$$

(9.23)

Since $i_D = K(v_{GS} - V_t)^2 \Rightarrow v_{GS} - V_t = \sqrt{i_D/K}$, then substituting this fact into Eq. 9.23 yields another expression for the transconductance of an enhancement MOSFET:

$$g_m = 2\sqrt{Ki_D}$$

Alternatively, since $i_D = K(v_{GS} - V_t)^2 \Rightarrow K = i_D/(v_{GS} - V_t)^2$, then substituting this fact into Eq. 9.23 yields yet another expression for the transconductance of an enhancement MOSFET:

$$g_m = \frac{2i_D}{v_{GS} - V_t}$$

The value of g_m for an FET typically lies in the range between 10^{-4} and 2×10^{-2} ℧.

A more accurate small-signal model of an FET is shown in Fig. 9.37, where r_d is called the **drain** or **output resistance** of the FET. Since

$$i_d = g_m v_{gs} + \frac{v_{ds}}{r_d}$$

then when $v_{ds} = 0$ V (i.e., for fixed v_{DS}), we have that $g_m = i_d/v_{gs}$ as before. For the case that $v_{gs} = 0$ V (i.e., for fixed v_{GS}), though, we see that

$$r_d = \frac{v_{ds}}{i_d} = \frac{dv_{DS}}{di_D}$$

A typical value of r_d for a JFET is 100 kΩ, whereas for a MOSFET it is 10 kΩ.

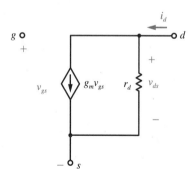

Fig. 9.37 More accurate small-signal model of an FET.

Example 9.8

For the JFET circuit given in Fig. 9.29 on p. 594, suppose that the capacitors C and C_S are large enough so that they behave as short circuits for the ac signals produced by v_s. The corresponding ac-equivalent circuit is shown in Fig. 9.38, where $R_p = R_1 \parallel R_2 = R_1 R_2 / (R_1 + R_2)$. Since the source of the JFET is common to both the input and the output of the amplifier, this configuration is referred to as a **common-source amplifier**.

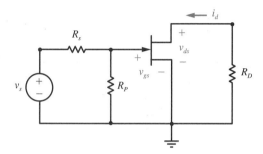

Fig. 9.38 The ac-equivalent circuit of the JFET amplifier given in Fig. 9.29.

By replacing the JFET with the small-signal model given in Fig. 9.36, we get the circuit shown in Fig. 9.39. Since $v_{ds} = -R_D i_d = -R_D g_m v_{gs}$, we have that the voltage gain A_v from gate to drain is

$$A_v = \frac{v_d}{v_g} = \frac{v_{ds}}{v_{gs}} = -g_m R_D \tag{9.24}$$

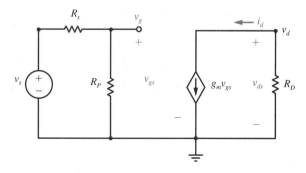

Fig. 9.39 The ac-equivalent circuit incorporating small-signal model.

If, instead of using the small-signal model given in Fig. 9.36, we had used the model in Fig. 9.37, then in the expression given by Eq. 9.24 we would replace R_D by $r_d \parallel R_D = r_d R_D / (r_d + R_D)$.

For the JFET amplifier given in Fig. 9.29 on p. 594, when $I_{DSS} = 10$ mA, $V_p = -4$ V, $R_1 = 500$ kΩ, $R_2 = 250$ kΩ, $R_S = 2.8$ kΩ, $R_D = 1$ kΩ, and $V_{DD} = 15$ V, then $v_{GS} = -2$ V $= V_{GSQ}$ and $i_D = 2.5$ mA $= I_{DQ}$ (see Example 9.5). Under these circumstances, from Eq. 9.20, we have that the transconductance is

$$g_m = -\frac{2I_{DSS}}{V_p}\left(1 - \frac{v_{GS}}{V_p}\right) = -\frac{2(10 \times 10^{-3})}{-4}\left(1 - \frac{-2}{-4}\right) = 2.5 \text{ m℧}$$

and, when r_d is ignored (i.e., $r_d = \infty$), the voltage gain is

$$A_v = -g_m R_D = -(2.5 \times 10^{-3})(1000) = -2.5$$

However, for the case that $r_d = 100$ kΩ, since $r_d \parallel R_D = 100$ kΩ $\parallel 1$ kΩ $= 990$ Ω, then the gain from gate to drain is

$$A_v = -g_m(r_d \parallel R_D) = -(2.5 \times 10^{-3})(990) = -2.48$$

Drill Exercise 9.8

For the amplifier given in Fig. 9.29 on p. 594, suppose that the JFET has $I_{DSS} = 16$ mA and $V_p = -3$ V. In addition, suppose that $R_1 = 600$ kΩ, $R_2 = 360$ kΩ, $R_D = 500$ Ω, $R_S = 1.5$ kΩ, and $V_{DD} = 12$ V. Find the voltage gain from gate to drain for the case that (a) $r_d = \infty$, and (b) $r_d = 100$ kΩ.

ANSWER (a) -2.67, (b) -2.65

The Source Follower

When discussing BJT amplifiers, we saw how an emitter follower (e.g., see Fig. 9.18 on p. 586) can be used as an isolation amplifier. The FET version of such an amplifier is known as a "source follower."

Example 9.9

Consider the MOSFET amplifier shown in Fig. 9.40. Replacing the MOSFET by the model given in Fig. 9.36, we obtain the ac-equivalent circuit shown in Fig. 9.41. In this case, the amplifier input is the gate voltage v_g and the output is the source voltage v_s. Note, too, that in ac terms, the drain is common to both the input and the output. It is because of this that the circuit is called a **common-drain amplifier**.

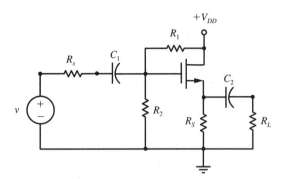

Fig. 9.40 A common-drain amplifier.

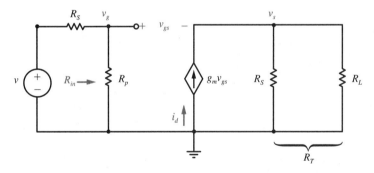

Fig. 9.41 Amplifier ac-equivalent circuit.

To determine the voltage gain $A_v = v_s/v_g$, let us use the fact that

$$v_s = R_T g_m v_{gs} = R_T g_m (v_g - v_s)$$

where $R_T = R_S \parallel R_L = \dfrac{R_S R_L}{R_S + R_L}$. From this expression, we get

$$A_v = \frac{v_s}{v_g} = \frac{g_m R_T}{1 + g_m R_T} = \frac{R_T}{1/g_m + R_T} \tag{9.25}$$

For the case that $R_T \gg 1/g_m$, then $A_v \approx 1$. In other words, $v_s \approx v_g$ and the source voltage (output) "follows" the gate voltage (input), and we refer to such an amplifier as a **source follower**.

Since the resistance between the gate and the reference looking into the gate is infinite, then clearly the input resistance of the source follower is

$$R_{in} = R_p = R_1 \| R_2$$

Thus if R_1 and R_2 are selected to be large, then so is the amplifier input resistance R_{in}. For a source follower, it is desirable to have this resistance large.

The output resistance R_o of the amplifier is equal to the ratio v_o/i_o for the circuit shown in Fig. 9.42. Since the current into the parallel combination of R_s and R_p must be zero (just apply KCL at the node labeled v_g), then by Ohm's law, $v_g = 0$ V. Since $v_s = v_o$, then

$$v_{gs} = v_g - v_s = -v_s = -v_o$$

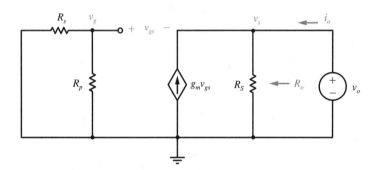

Fig. 9.42 Circuit for determining the output resistance.

Also, by KCL,

$$i_o = \frac{v_s}{R_S} - g_m v_{gs} = \frac{v_o}{R_S} - g_m(-v_o) = \left(\frac{1}{R_S} + g_m\right)v_o$$

Thus the output resistance is

$$R_o = \frac{v_o}{i_o} = \frac{1}{1/R_S + g_m} = \frac{R_S}{1 + g_m R_S} = \frac{(1/g_m)R_S}{1/g_m + R_S} = \frac{1}{g_m} \| R_S$$

and, for a source follower, it is desirable for the output resistance to be small.

For the amplifier given in Fig. 9.40, if we use the model shown in Fig. 9.37 instead of the model in Fig. 9.36, we only need replace the term R_S by $r_d \parallel R_S = r_d R_S / (r_d + R_S)$ in the above results.

In summary, because a source follower has approximately unity gain, high input resistance, and low output resistance, as for an emitter follower, it sees application as an isolation amplifier.

Drill Exercise 9.9

For the common-drain amplifier given in Fig. 9.40, suppose that the MOSFET has $K = 0.25$ mA/V^2 and $V_t = 2$ V. In addition, suppose that $R_1 = R_2 = 10$ MΩ, $R_L = R_S = 1$ kΩ, and $V_{DD} = 10$ V. Find the voltage gain from gate to source, the input resistance, and the output resistance for the case that (a) $r_d = \infty$, and (b) $r_d = 10$ kΩ.

ANSWER   (a) 0.333, 5 MΩ, 500 Ω; (b) 0.322, 5 MΩ, 476 Ω

A MOSFET Feedback Amplifier

Shown in Fig. 9.43a is a more general form of the feedback amplifier given in Fig. 9.34 on p. 598.

Example 9.10

Figure 9.43b shows the ac-equivalent circuit of the enhancement MOSFET amplifier in Fig. 9.43a. Let us define $R_T = R_D \parallel R_L = R_D R_L / (R_D + R_L)$. Since $v_g = v_{gs}$, then, by KCL,

$$g_m v_g + \frac{v_d}{R_T} + \frac{v_d - v_g}{R_F} = 0$$

Solving this equation for the voltage gain $A_v = v_d / v_g$, we get

$$A_v = \frac{v_d}{v_g} = \frac{R_T(1 - g_m R_F)}{R_F + R_T} = \frac{R_T[(1/g_m) - R_F]}{(1/g_m)(R_F + R_T)} \tag{9.26}$$

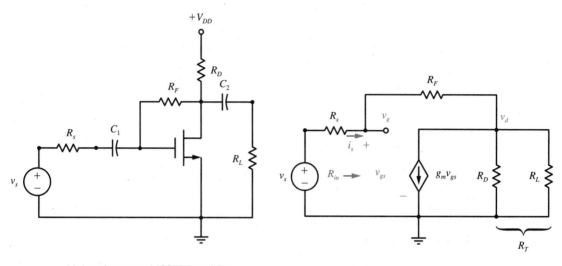

(a) An enhancement MOSFET amplifier

(b) The ac-equivalent circuit of the enhancement MOSFET amplifier

Fig. 9.43 (a) A MOSFET feedback amplifier, and (b) the ac-equivalent circuit.

The input resistance of the amplifier is $R_{in} = v_g/i_s$. We can use Eq. 9.26 to express v_d in terms of v_g and then substitute the result into the fact that $i_s = (v_g - v_d)/R_F$. Doing so, we obtain

$$i_s = \frac{v_g - [R_T(1 - g_m R_F)/(R_F + R_T)]v_g}{R_F}$$

and simplifying this expression results in

$$R_{in} = \frac{R_F + R_T}{1 + g_m R_T} = \frac{(1/g_m)(R_F + R_T)}{(1/g_m) + R_T} \tag{9.27}$$

We leave it as an exercise for the reader (see Problem 9.46 at the end of this chapter) to demonstrate that the output resistance R_o (seen by the external load R_L) of the amplifier is given by the expression

$$\frac{1}{R_o} = \frac{1}{R_D} + \frac{1 + g_m R_s}{R_F + R_s}$$

Drill Exercise 9.10

For the feedback amplifier given in Fig. 9.43, suppose that the MOSFET has $K = 0.25$ mA$/$V^2 and $V_t = 2$ V. In addition, suppose that $R_D = R_s = 1$ kΩ, $R_F = 1$ MΩ, $R_L = 4$ kΩ, and $V_{DD} = 10$ V. Find the voltage gain from gate to drain, the input resistance, and the output resistance for the case that (a) $r_d = \infty$, and (b) $r_d = 10$ kΩ.

ANSWER   (a) -1.6, 385 kΩ, 997 Ω; (b) -1.48, 403 kΩ, 907 Ω

A typical value of g_m for an FET is 5 m$\mho$, whereas for a BJT it is 40 m$\mho$. This means that for small-signal applications, when it comes to gain, the BJT is superior to the FET. In addition to this fact, the transfer characteristic of a BJT is much closer to being linear than that of an FET. Consequently, nonlinear distortion for a BJT is less than for an FET. With regard to input resistance, however, the FET is superior to the BJT in that the FET has a much higher resistance than the BJT—the FET draws negligible current.

9.3   Frequency Response

An important characteristic of an amplifier is its frequency response. The amplitude-response portion of the frequency response describes how the amplifier gain varies with frequency. Previously we have ignored any capacitive effects and have considered the gain of an amplifier to be independent of frequency. In general, however, gain is frequency dependent.

For many types of amplifiers, the gain is approximately constant over a frequency range—called the **mid-frequency range** or **midband** of the amplifier—and it falls off for very low and very high frequencies as depicted in Fig. 9.44, where the lower and upper cutoff (3-dB) frequencies of the amplitude response are designated ω_L and ω_H, respectively. In general, the lower cutoff frequency ω_L is determined by the coupling and bypass capacitors of the amplifier, whereas the upper cutoff frequency is determined by the internal capacitances of the amplifier's transistors.

In the preceding discussions of BJT and FET small-signal amplifiers, we assumed that the coupling and bypass capacitors behaved as short circuits to the ac signals that were being amplified. Since the impedance $\mathbf{Z}_C$ of a capacitor C is $\mathbf{Z}_C = 1/j\omega C$, when the frequency of an ac signal becomes sufficiently low, the impedances of coupling and bypass capacitors no longer can be ignored. Therefore, we will now investigate the effects that coupling and bypass capacitors have on the frequency responses of amplifiers. We first consider the case of a BJT amplifier, and then study a typical FET amplifier.

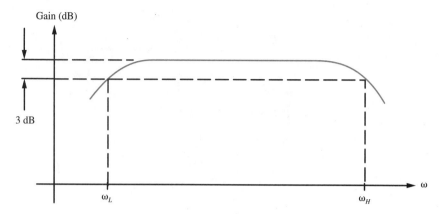

Fig. 9.44 Amplitude response of an amplifier.

Before we proceed, by way of a review, consider the *RC* circuit shown in Fig. 9.45a. By voltage division, we can write

$$\mathbf{V}_2 = \frac{\mathbf{Z}_R}{\mathbf{Z}_C + \mathbf{Z}_R}\mathbf{V}_1 = \frac{R}{1/j\omega C + R}\mathbf{V}_1 = \frac{j\omega RC}{1 + j\omega RC}\mathbf{V}_1$$

Thus

$$\mathrm{ang}(\mathbf{V}_2) = \mathrm{ang}(j\omega RC) - \mathrm{ang}(1 + j\omega RC) + \mathrm{ang}(\mathbf{V}_1)$$

$$= 90° - \tan^{-1}(\omega RC/1) + \mathrm{ang}(\mathbf{V}_1)$$

Since $0 < \tan^{-1} \omega RC < 90°$, then $\mathrm{ang}(\mathbf{V}_2) > \mathrm{ang}(\mathbf{V}_1)$. Thus if the input voltage v_1 is a sinusoid, then the output sinusoid v_2 leads v_1. Consequently, such a circuit is referred to as a **lead network**. It should be obvious that for the circuit shown in Fig. 9.45b, v_2 again leads v_1. Since v_3 is directly proportional to v_2 (by voltage division), then v_3 leads v_1. Therefore, the circuit shown in Fig. 9.45b is also a lead network.

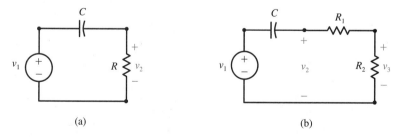

(a) (b)

Fig. 9.45 (*a*) Simple lead network, and (*b*) another lead network.

The lead network shown in Fig. 9.45a is, in fact, the high-pass filter given in Fig. 5.5 on p. 269, and as discussed in Section 5.1, it has a cutoff frequency of $\omega_c = 1/RC$. So, too, the high-pass filter shown in Fig. 9.45b has a cutoff frequency of $\omega_c = 1/RC$, where $R = R_1 + R_2$.

Consider now a BJT amplifier such as the one given in Fig. 9.14 on p. 583. To begin with, let us concentrate on the effects of the coupling capacitors C_1 and C_2. Treating the bypass capacitor C_E as a short circuit, but not ignoring C_1 and C_2, the ac-equivalent circuit of the amplifier is as shown in Fig. 9.46, where $R_p = R_1 \parallel R_2$.

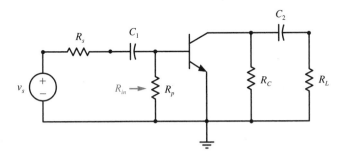

Fig. 9.46 Coupling capacitors included in the ac-equivalent circuit of the amplifier given in Fig. 9.14.

To find the effect due solely to C_1 let us ignore C_2. As was pointed out in the discussion in conjunction with Fig. 9.14, the amplifier input resistance R_{in} is independent of R_C and R_L, and is given by

$$R_{in} = R_p \parallel R_b$$

where $R_p = R_1R_2/(R_1 + R_2)$, $R_b = (1 + h_{fe})r_e$, and $r_e = 0.026/I_{CQ}$. Consequently, the ac input voltage v_b to the amplifier can be determined from the equivalent circuit shown in Fig. 9.47. This base-equivalent lead network, therefore, has a cutoff frequency of

$$\omega_1 = \frac{1}{(R_s + R_{in})C_1} \tag{9.28}$$

To determine the effect on the amplifier due solely to C_2, we ignore C_1 and replace the portion of the circuit (given in Fig. 9.46) to the left of C_2 by its Thévenin equivalent. The result is given by Fig. 9.48. We have already seen (see Fig. 9.17 and its associated discussion) that $R_o = R_C$. It should be clear that the output voltage v_L will not be affected if the circuit given in Fig. 9.48 is cast into the form of Fig.

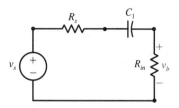

Fig. 9.47 Base-equivalent lead network.

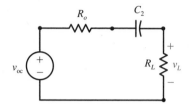

Fig. 9.48 Collector-equivalent lead network.

9.45*b*. Thus the collector-equivalent lead network shown in Fig. 9.48 has a cutoff frequency of

$$\omega_2 = \frac{1}{(R_C + R_L)C_2} \tag{9.29}$$

Effect of the Bypass Capacitor

Having determined the cutoff frequencies due to the coupling capacitors C_1 and C_2, let us now deal with the bypass capacitor C_E. For the amplifier given in Fig. 9.14 on p. 583, treating C_1 and C_2 as short circuits, we get the ac-equivalent circuit shown in Fig. 9.49, where $R_p = R_1 \parallel R_2$ and $R_T = R_C \parallel R_L$. To calculate the effective resistance R_{oE} that is in parallel with C_E, we replace the transistor by its small-signal model and set v_s to zero as shown in Fig. 9.50. From the resulting circuit, we can find $R_{oE} = v_o/i_o$.

Let us define $R = R_s \parallel R_p$. Applying KCL to node v_b, we get

$$\frac{v_b}{R} + \frac{v_b - v_o}{r_e} = h_{fe}i_b \tag{9.30}$$

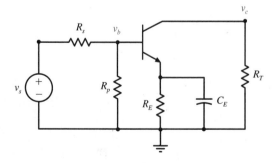

Fig. 9.49 Bypass capacitor included in the ac-equivalent circuit of the amplifier given in Fig. 9.14.

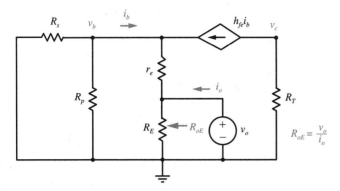

Fig. 9.50 Circuit for determining the effective resistance R_{oE} in parallel with C_E.

Since $i_b + h_{fe}i_b = (v_b - v_o)/r_e$, then $i_b = (v_b - v_o)/(1 + h_{fe})r_e$. Substituting this fact into Eq. 9.30, we obtain

$$\frac{v_b}{R} + \frac{v_b - v_o}{r_e} = \frac{h_{fe}(v_b - v_o)}{(1 + h_{fe})r_e}$$

Manipulating this equation results in

$$v_b = \frac{R/(1 + h_{fe})}{r_e + R/(1 + h_{fe})}\, v_o$$

Substituting this expression into the fact that $i_o = v_o/R_E + (v_o - v_b)/r_e$, we obtain

$$\frac{1}{R_{oE}} = \frac{i_o}{v_o} = \frac{1}{r_e + R/(1 + h_{fe})} + \frac{1}{R_E}$$

In other words,

$$R_{oE} = R_E \,\|\, \left(r_e + \frac{R}{1 + h_{fe}}\right) \tag{9.31}$$

For the ac-equivalent amplifier circuit shown in Fig. 9.49, for the case that R_E is unbypassed (C_E is an open circuit), just as for the derivation of Eq. 9.12 on p. 581, we have that

$$A_v = \frac{v_c}{v_b} \approx -\frac{R_T}{r_e + R_E}$$

However, if R_E is bypassed (C_E is a short circuit), then

$$A_v = \frac{v_c}{v_b} \approx -\frac{R_T}{r_e}$$

In other words, as frequency decreases and C_E behaves less and less as a short circuit, the voltage gain decreases. For the coupling capacitors C_1 and C_2, the critical frequencies are given by Eq. 9.28 and Eq. 9.29, respectively. For the bypass capacitor C_E, though, the critical frequency is

$$\omega_3 = \frac{1}{R_{oE}C_E} \tag{9.32}$$

where R_{oE} is given by Eq. 9.31. A more formal demonstration of this fact is presented in Problems 9.47 through 9.49 at the end of this chapter.

Assuming that the values of the frequencies given in Eq. 9.28, Eq. 9.29, and Eq. 9.32 are sufficiently different (at least a decade apart), the lower cutoff frequency ω_L for the BJT amplifier given in Fig. 9.14 on p. 583 is equal to the highest of those three values. In practice, quite often ω_3 has the highest value, and, therefore, it determines the lower cutoff frequency of the amplifier. If the three frequencies are not sufficiently different, hand calculations become impractical, and a computer-aided solution with a software package such as SPICE should be employed.

An FET Amplifier

For an FET amplifier, such as the one shown in Fig. 9.51, the corresponding ac-equivalent circuit, including the coupling and bypass capacitors, is shown in Fig. 9.52, where $R_p = R_1 \parallel R_2$. Similar to the case of the BJT amplifier just discussed, the cutoff frequency due to C_1 is given by Eq. 9.28. In this case, however, we have that $R_{in} = R_p$. Therefore, the gate-equivalent lead network has a cutoff frequency of

$$\omega_1 = \frac{1}{(R_s + R_p)C_1} \tag{9.33}$$

Again similar to the BJT amplifier discussed previously, the cutoff frequency of the drain-equivalent lead network is

$$\omega_2 = \frac{1}{(R_D + R_L)C_2} \tag{9.34}$$

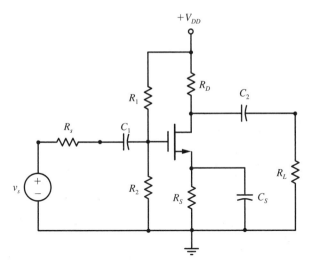

Fig. 9.51 An FET amplifier.

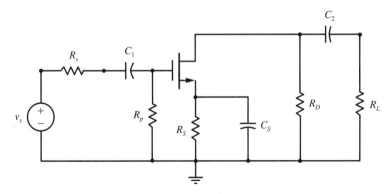

Fig. 9.52 The ac-equivalent circuit.

when r_d is ignored ($r_d = \infty$). If r_d is not ignored ($r_d < \infty$), then replace R_D by $r_d \parallel R_D$ in Eq. 9.34. Finally, to determine the effect of the bypass capacitor C_S, we replace C_1, C_2, and v_s by short circuits and determine $R_{oS} = v_o/i_o$ from the circuit shown in Fig. 9.53, where $R_T = R_D \parallel R_L$ when r_d is ignored or $R_T = r_d \parallel R_D \parallel R_L$ when r_d is not ignored.

Since, by KCL, there is no current going into the parallel combination of R_p and R_s, then $v_g = 0$ V. Consequently, by KVL, $v_{gs} = -v_o$. Since

$$i_o = \frac{v_o}{R_S} - g_m v_{gs} = \frac{v_o}{R_S} + g_m v_{gs}$$

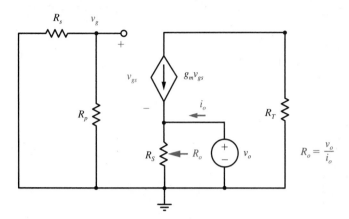

Fig. 9.53 Circuit for determining R_{oS}.

then

$$\frac{1}{R_{oS}} = \frac{i_o}{v_o} = \frac{1}{R_S} + g_m \quad \Rightarrow \quad R_{oS} = R_S \parallel \frac{1}{g_m}$$

Thus we have

$$\omega_3 = \frac{1}{R_{oS} C_S} \tag{9.35}$$

when r_d is ignored. For the case that r_d is not ignored, then $R_{oS} = R_S \parallel r_d \parallel 1/g_m$.

As was the situation for the BJT amplifier discussed earlier, for the case of sufficiently different values, the lower cutoff frequency for the FET amplifier shown in Fig. 9.51 is given by the highest of the values determined by Eq. 9.33, Eq. 9.34, and Eq. 9.35.

Miller's Theorem

Having discussed how to determine the lower cutoff frequency for some transistor amplifiers, we now consider the problem of finding the upper cutoff frequency. To simplify our analysis, we use an important result known as "Miller's theorem." Since the following discussion uses the concept of impedance, we shall use phasor notation for voltages and currents.

We begin by considering the situation depicted in Fig. 9.54. Here, the box represents an amplifier whose (phasor) input voltage is $\mathbf{V}_1$ and whose output voltage is $\mathbf{V}_2$. There is a feedback element, an impedance $\mathbf{Z}_F$ connected between the output

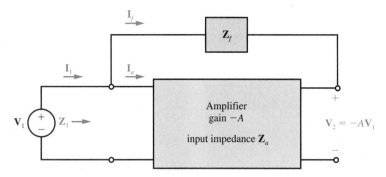

Fig. 9.54 Amplifier with feedback.

and the input. With the feedback impedance $\mathbf{Z}_F$ in place, the gain of the amplifier is $\mathbf{V}_2/\mathbf{V}_1 = -A$. Further, the input impedance of the amplifier is $\mathbf{V}_1/\mathbf{I}_a = \mathbf{Z}_a$. By KCL, we have that

$$\mathbf{I}_1 = \mathbf{I}_a + \mathbf{I}_f = \frac{\mathbf{V}_1}{\mathbf{Z}_a} + \frac{\mathbf{V}_1 - \mathbf{V}_2}{\mathbf{Z}_F} = \frac{\mathbf{V}_1}{\mathbf{Z}_a} + \frac{(1 + A)\mathbf{V}_1}{\mathbf{Z}_F} \qquad (9.36)$$

and

$$\frac{\mathbf{I}_1}{\mathbf{V}_1} = \frac{1}{\mathbf{Z}_a} + \frac{1}{\mathbf{Z}_F/(1 + A)}$$

Thus the input impedance $\mathbf{Z}_1$ to the overall feedback amplifier is

$$\mathbf{Z}_1 = \frac{\mathbf{V}_1}{\mathbf{I}_1} = \mathbf{Z}_a \, \| \, \frac{\mathbf{Z}_F}{1 + A} \qquad (9.37)$$

We also have that

$$\mathbf{I}_f = \frac{\mathbf{V}_1 - \mathbf{V}_2}{\mathbf{Z}_F} = \frac{-\mathbf{V}_2/A - \mathbf{V}_2}{\mathbf{Z}_F} = \frac{-\mathbf{V}_2 - A\mathbf{V}_2}{A\mathbf{Z}_F} = \frac{-\mathbf{V}_2}{A\mathbf{Z}_F/(1 + A)} \qquad (9.38)$$

Now note that Eq. 9.36, Eq. 9.37, and Eq. 9.38 are also satisfied for the configuration shown in Fig. 9.55. In other words, the effect on the amplifier in the box is the same as that for the circuit in Fig. 9.54, and the impedance $\mathbf{Z}_1$ is also the same. This result is known as **Miller's theorem,** and the equivalent manner in which the feedback element $\mathbf{Z}_F$ is reflected across the input as the impedance $\mathbf{Z}_F/(1 + A)$ and across the output as $A\mathbf{Z}_F/(1 + A)$ is referred to as the **Miller effect.**

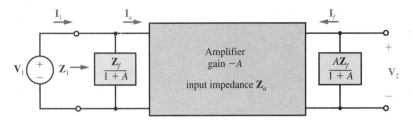

Fig. 9.55 Miller equivalent of the given amplifier with feedback.

Before we see how Miller's theorem is used to determine how transistor amplifiers behave at high frequencies, again by way of a review, we mention that the circuit shown in Fig. 9.56 is an example of a **lag network.** By voltage division, we can write

$$V_2 = \frac{Z_C}{Z_C + Z_R}V_1 = \frac{1/j\omega C}{1/j\omega C + R}V_1 = \frac{1}{1 + j\omega RC}V_1$$

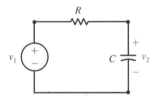

Fig. 9.56 Simple lag network.

Thus

$$\text{ang}(V_2) = \text{ang}(V_1) - \text{ang}(1 + j\omega RC) = \text{ang}(V_1) - \tan^{-1}(\omega RC/1)$$

Since $0 < \tan^{-1}\omega RC < 90°$, then $\text{ang}(V_2) < \text{ang}(V_1)$. Thus if the input voltage v_1 is a sinusoid, then the output sinusoid v_2 lags v_1. The lag network shown in Fig. 9.56 is, in fact, the low-pass filter given in Fig. 5.8 on p. 272, and it has a cutoff frequency of $\omega_c = 1/RC$.

High-Frequency Small-Signal BJT Model

The small-signal ac models of BJTs and FETs encountered previously are valid for low and middle frequencies. However, for high frequencies, junction capacitances become significant factors. Redrawing the small-signal BJT model given in Fig. 9.11

on p. 579, and including the B-E capacitance C_e (also denoted C_π) and the B-C capacitance C_c (also denoted C_μ), we get the high-frequency BJT small-signal model shown in Fig. 9.57. In amplification applications, the B-E junction of a BJT is forward biased and the B-C junction is reverse biased. Consequently, C_e is comprised of a transition capacitance and a diffusion capacitance (see Section 6.7), whereas C_c is strictly a transition capacitance. A typical value for C_e is 100 pF, whereas a typical value for C_c is 5 pF.

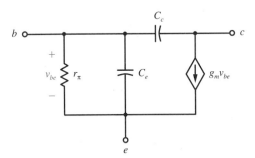

Fig. 9.57 High-frequency BJT model.

Again consider the BJT amplifier given in Fig. 9.14 on p. 583 and its ac-equivalent circuit shown in Fig. 9.16 on p. 584, where $R_p = R_1 \parallel R_2$ and $R_T = R_C \parallel R_L$. Replacing the BJT in Fig. 9.16 by the model given in Fig. 9.57, we get the circuit shown in Fig. 9.58. For middle frequencies when C_c and C_e can be ignored (treated as open circuits) and the coupling and bypass capacitors can be treated as short circuits, we have already seen in the discussion associated with Fig. 9.16 that the corresponding voltage gain, called the **mid-band gain**, is $A_v = v_c/v_b \approx -R_T/r_e = -g_m R_T = -A$.

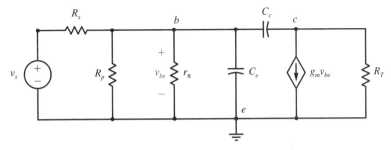

Fig. 9.58 Amplifier ac-equivalent circuit using high-frequency BJT model.

Now, by Miller's theorem, the capacitance C_c can be reflected across the input (between the base and emitter) and the output (between the collector and emitter) by appropriately valued capacitances. Specifically, since $\mathbf{Z}_F = 1/j\omega C_c$, then the reflected input impedance is

$$\frac{\mathbf{Z}_F}{1 + A} = \frac{1}{j\omega C_c(1 + A)} \qquad (9.39)$$

which is the impedance of a capacitance of value $(1 + A)C_c$, whereas the reflected output impedance is

$$\frac{A\mathbf{Z}_F}{1 + A} = \frac{1}{j\omega C_c(1 + A)/A} \qquad (9.40)$$

which is the impedance of a capacitance of value $(1 + A)C_c/A$. The use of Miller's theorem, therefore, yields the equivalent circuit shown in Fig. 9.59.

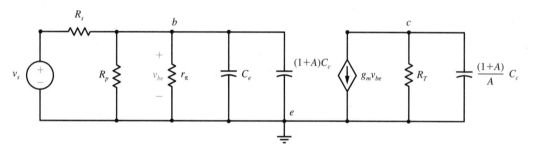

Fig. 9.59 Amplifier ac-equivalent circuit resulting from Miller's theorem.

Since C_e and $(1 + A)C_c$ are connected in parallel, they can be combined into a single capacitance $C_g = C_e + (1 + A)C_c$. Furthermore, we can apply Thévenin's theorem to the remainder of the base portion of the circuit and represent it as a single voltage source v_{oc} connected in series with a single resistance $R_g = R_s \parallel R_p \parallel r_\pi$. In addition, the current source $g_m v_{be}$ that is connected in parallel with R_T is equivalent to a voltage source $R_T g_m v_{be}$ connected in series with R_T (see Problem 9.58 at the end of this chapter). Therefore, the circuit given in Fig. 9.59 can be cast into the form shown in Fig. 9.60, which explicitly exhibits base and collector lag networks. We deduce that the cutoff frequency resulting from the base lag network is

$$\omega_g = \frac{1}{R_g C_g} \qquad (9.41)$$

where $R_g = R_s \parallel R_p \parallel r_\pi$ and $C_g = C_e + (1 + A)C_c$. The cutoff frequency due to the collector lag network is

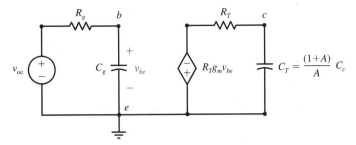

Fig. 9.60 Equivalent circuit showing base and collector lag networks.

$$\omega_T = \frac{1}{R_T C_T} \tag{9.42}$$

where $R_T = R_C \parallel R_L$ and $C_T = (1 + A)C_c/A$.

Assuming that ω_g and ω_T have sufficiently different values, the upper cutoff frequency ω_H for the BJT amplifier given in Fig. 9.14 on p. 583 is equal to the smaller of these two values.

Having determined the upper cutoff frequency ω_H of the amplifier as just described and the lower cutoff frequency ω_L as previously discussed, we can define the **bandwidth** *BW* of the amplifier to be

$$\boxed{BW = \omega_H - \omega_L} \tag{9.43}$$

The capacitance C_c, which typically lies in the range 1–20 pF, is often designated on BJT specification sheets as C_{ob}. On the other hand, C_e, which typically lies in the range 25–500 pF, does not have a similar designation. Instead, C_e can be determined from another parameter called the **current-gain/bandwidth product** f_T. The parameter f_T is the frequency (in hertz) for which the effect of the capacitances has become so significant that the short-circuit ($v_{ce} = 0$ V) current gain $|\mathbf{I}_c/\mathbf{I}_b|$ equals unity. The range of f_T typically extends from 100 to more than 1000 MHz. It can be shown (see Problems 9.51 and 9.52 at the end of this chapter) that

$$f_T \approx \frac{g_m}{2\pi(C_c + C_e)} \quad \Rightarrow \quad C_c + C_e \approx \frac{g_m}{2\pi f_T} = \frac{1}{2\pi f_T r_e} \tag{9.44}$$

High-Frequency Small-Signal FET Model

By adding capacitances C_{gs}, C_{gd}, and C_{ds} to the small-signal models of an FET shown in Fig. 9.36 and Fig. 9.37, we obtain the high-frequency FET small-signal models shown in Fig. 9.61. These models are valid for JFETs and MOSFETs—both deple-

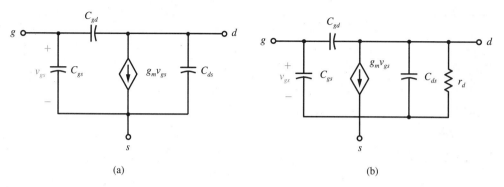

Fig. 9.61 High-frequency FET model for (a) $r_d = \infty$, and (b) $r_d < \infty$.

tion and enhancement types. Capacitances C_{gs} and C_{gd} typically range between 1 and 3 pF, whereas C_{ds} is usually 1 pF or less. Similar to the case of the BJT, these capacitances produce equivalent lag networks (low-pass filters) that establish the upper cutoff frequencies for FET amplifiers.

In particular, for the FET amplifier given in Fig. 9.51 on p. 615, at higher frequencies when the coupling and bypass capacitors behave as short circuits, using the model given in Fig. 9.61a we get the ac-equivalent circuit shown in Fig. 9.62, where $R_p = R_1 \| R_2$ and $R_T = R_D \| R_L$ when r_d is ignored, and $R_T = r_d \| R_D \| R_L$ when r_d is not ignored. For low and middle frequencies when C_{gs}, C_{gd}, and C_{ds} can be treated as open circuits, just as in the derivation of Eq. 9.23, we have that $A_v = -g_m R_T = -A$. By the Miller effect, C_{gd} is reflected between the gate and the source as $(1 + A)C_{gd}$ and between the drain and the source as $(1 + A)C_{gd}/A$. Combining

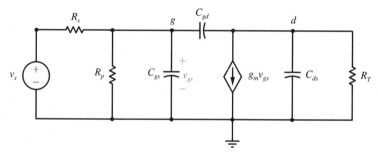

Fig. 9.62 Amplifier ac-equivalent circuit using high-frequency FET model.

(adding) capacitances in parallel and using Thévenin's theorem, we obtain the equivalent circuit shown in Fig. 9.63, where $C_g = C_{gs} + (1 + A)C_{gd}$, $C_T = C_{ds} + (1 + A)C_{gd}/A$, and $R_g = R_s \| R_p$. It is from this circuit that we notice that the gate lag network has a cutoff frequency given by Eq. 9.41 and the drain lag network has a cutoff frequency given by Eq. 9.42. If sufficiently different, the smaller

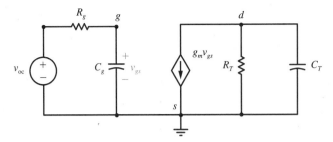

Fig. 9.63 Equivalent circuit showing gate and drain lag networks.

of these two values is the upper cutoff frequency of the FET amplifier. As with the case of the BJT amplifier, if the upper cutoff frequency of the FET amplifier is ω_H and the lower cutoff frequency is ω_L, then the bandwidth is $BW = \omega_H - \omega_L$.

9.4 Power Amplifiers

In our discussion of transistor amplifiers thus far, we have been preoccupied with small-signal analysis. In many applications, small ac signals are amplified sufficiently to produce large ac signals. The resulting ac signals are then amplified by **large-signal amplifiers** or **power amplifiers,** which, in turn, provide sufficient power to drive devices such as loudspeakers, motors, recording instruments, and cathode-ray tubes (CRTs).

In small-signal applications, transistors dissipate a fraction of watt, whereas when used for power amplification, they may have to be capable of dissipating greater amounts of power. It is because of this that a special class of transistors, referred to as **power transistors,** is required. Since such devices deal with large ac signals, it is desirable that their characteristics be more linear than was the case for small-signal transistors. The more linear a transistor's characteristics, the less nonlinear distortion will result.

Class-A Amplifiers

Class-A operation is the term used to indicate that a transistor operates entirely in the active (linear) region. In our previous discussions of small-signal amplifiers, transistors were always biased in the active region, and it was assumed that the ac signals were sufficiently small always to have class-A operation.

Let us now consider a typical ac (capacitively coupled) BJT amplifier as shown in Fig. 9.64. To obtain more general results, we have not used a bypass capacitor for R_E. Suppose that the BJT is biased in the active region, the quiescent collector current is I_{CQ}, and the quiescent collector-emitter voltage is V_{CEQ}. Figure 9.65 shows

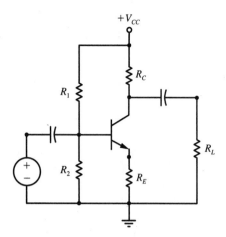

Fig. 9.64 A typical ac amplifier.

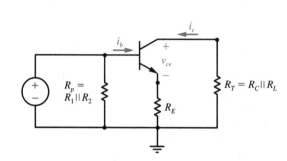

Fig. 9.65 Amplifier ac-equivalent circuit.

the ac-equivalent circuit of the amplifier, where $R_p = R_1 \parallel R_2$ and $R_T = R_C \parallel R_L$. The voltage and currents indicated in Fig. 9.65 are ac quantities and, therefore, lowercase subscripts are used. Thus the total collector current i_C is given by $i_C = I_{CQ} + i_c$ and the total collector-emitter voltage v_{CE} by $v_{CE} = V_{CEQ} + v_{ce}$.

From Fig. 9.65, since typically $h_{fe} \gg 1$ and therefore $i_b + i_c = i_b + h_{fe}i_b \approx h_{fe}i_b = i_c$, by KVL we have that

$$v_{ce} = -R_T i_c - R_E(i_b + i_c) \approx -R_T i_c - R_E i_c = -(R_T + R_E)i_c$$

from which we get the approximation

$$i_c = \frac{-v_{ce}}{R_T + R_E}$$

Thus the (total) collector current is

$$i_C = I_{CQ} + i_c = I_{CQ} - \frac{v_{ce}}{R_T + R_E} \tag{9.45}$$

But $v_{CE} = V_{CEQ} + v_{ce}$ implies that $v_{ce} = v_{CE} - V_{CEQ}$. Substituting this fact into Eq. 9.45 yields

$$i_C = I_{CQ} - \frac{v_{CE} - V_{CEQ}}{R_T + R_E} = -\frac{v_{CE}}{R_T + R_E} + I_{CQ} + \frac{V_{CEQ}}{R_T + R_E} \tag{9.46}$$

This is the equation of a straight line in the i_C-v_{CE} plane that has a slope of $-1/(R_T + R_E)$ and y-axis intercept $I_{CQ} + V_{CEQ}/(R_T + R_E)$. Setting $v_{CE} = V_{CEQ}$,

Eq. 9.46 yields $i_C = I_{CQ}$. Thus this line passes through the Q point $i_C = I_{CQ}$, $v_{CE} = V_{CEQ}$. Furthermore, setting $i_C = 0$ A in Eq. 9.46, we obtain the horizontal-axis intercept $V_{CEQ} + (R_T + R_E)I_{CQ}$. The line described by Eq. 9.46 is depicted in Fig. 9.66 and is precisely the amplifier's ac load line (see Fig. 9.15 on p. 583 and the associated discussion). For a silicon BJT, we know that the cutoff current is $i_{C(\text{off})} \approx 0$ A and the saturation voltage is $v_{CE(\text{sat})} \approx 0.2$ V. However, for the sake of simplicity, let us use $i_{C(\text{off})} = 0$ A and $v_{CE(\text{sat})} = 0$ V. Then the saturation current is $i_{C(\text{sat})} = I_{CQ} + V_{CEQ}/(R_T + R_E)$ and the cutoff voltage is $v_{CE(\text{off})} = V_{CEQ} + (R_T + R_E)I_{CQ}$.

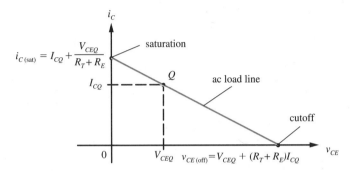

Fig. 9.66 The ac load line for the given amplifier.

For class-A operation, the transistor must not go into saturation or cutoff. Since the transistor operates along the load line, to allow for the largest possible ac signal we should have the Q point at the center of the load line. When the Q point is centered on the load line, then $i_{C(\text{sat})} = 2I_{CQ}$. Under this circumstance, therefore,

$$i_{C(\text{sat})} = I_{CQ} + \frac{V_{CEQ}}{R_T + R_E} = 2I_{CQ}$$

From this equation,

$$\boxed{\frac{V_{CEQ}}{I_{CQ}} = R_T + R_E} \qquad (9.47)$$

and this is the condition for a centered Q point.

Example 9.11

For the amplifier given in Fig. 9.64, suppose that $h_{FE} = 30$, $R_1 = 600$ Ω, $R_2 = 300$ Ω, $R_C = R_E = R_L = 100$ Ω, and $V_{CC} = 30$ V. By routine dc BJT analysis (see

Drill Exercise 9.11), we can find that $I_{BQ} = 2.82$ mA, $I_{CQ} = 84.6$ mA, and $V_{CEQ} = 12.8$ V. Therefore,

$$\frac{V_{CEQ}}{I_{CQ}} = \frac{12.8}{84.6 \times 10^{-3}} = 151\ \Omega$$

Since $R_T = R_C \parallel R_L = 100\ \Omega \parallel 100\ \Omega = 50\ \Omega$, then $R_T + R_E = 50 + 100 = 150\ \Omega \approx V_{CEQ}/I_{CQ} = 151\ \Omega$, and hence, the Q point is approximately centered on the load line.

Drill Exercise 9.11

Confirm that the BJT in the amplifier described in Example 9.11 is biased in the active region at $I_{CQ} = 84.6$ mA and $V_{CEQ} = 12.8$ V.

Power Terminology

Returning to the class-A amplifier given in Fig. 9.64 and its ac-equivalent circuit shown in Fig. 9.65, suppose that the Q point is centered on the load line. It follows that the largest possible sinusoidal collector current has amplitude I_{CQ}. The corresponding rms value is $I_{CQ}/\sqrt{2}$. The resulting average ac (signal) power absorbed by R_T is

$$P_T = R_T \left(\frac{I_{CQ}}{\sqrt{2}}\right)^2 = \frac{1}{2} R_T I_{CQ}^2$$

Since typically $i_b + i_c \approx i_c$, then the (average) signal power absorbed by R_E is

$$P_E = R_E \left(\frac{I_{CQ}}{\sqrt{2}}\right)^2 = \frac{1}{2} R_E I_{CQ}^2$$

Thus the total signal power (or output power) P_o absorbed by R_T and R_E is

$$P_o = P_T + P_E = \frac{1}{2}(R_T + R_E)I_{CQ}^2 \tag{9.48}$$

and this is the maximum output signal power since it was assumed that the sinusoidal collector current had maximum amplitude. Using Eq. 9.47—the condition for a centered Q point—we can substitute $R_T + R_E = V_{CEQ}/I_{CQ}$ into Eq. 9.48 to get the maximum output signal power

$$P_o = \frac{1}{2}V_{CEQ}I_{CQ} \tag{9.49}$$

For instance, the amplifier described in Example 9.11 has a maximum output signal power of

$$P_o = \frac{1}{2}(12.8)(84.6 \times 10^{-3}) = 541 \text{ mW} \approx 0.54 \text{ W}$$

Since the dc collector current of a BJT is typically much greater than the dc base current, we define the **instantaneous power dissipation** p of a BJT to be

$$p = v_{CE}i_C \tag{9.50}$$

We know that a BJT operates somewhere on its associated load line. But what point on the load line corresponds to maximum instantaneous power dissipation? With reference to Fig. 9.66, for the sake of simplicity, let us define the symbols $a = v_{CE(off)}$ and $b = i_{C(sat)}$. Then the equation of the load line can be written as

$$i_C = -\frac{b}{a}v_{CE} + b \tag{9.51}$$

Substituting this into Eq. 9.50 yields

$$p_C = v_{CE}\left(-\frac{b}{a}v_{CE} + b\right) = -\frac{b}{a}v_{CE}^2 + b\,v_{CE}$$

To determine what value of v_{CE} results in a maximum value of p, let us take the derivative of p with respect to v_{CE} and set the resulting expression to zero. Doing so we get

$$\frac{dp}{dv_{CE}} = -\frac{2b}{a}v_{CE} + b = 0 \qquad \Rightarrow \qquad v_{CE} = \frac{a}{2} = \frac{v_{CE(off)}}{2}$$

Substituting this value of v_{CE} into Eq. 9.51 yields

$$i_C = -\frac{b}{a}\left(\frac{a}{2}\right) + b = \frac{b}{2} = \frac{i_{C(\text{sat})}}{2}$$

Since $d^2p/dv_{CE}^2 = -2b/a < 0$, then the point $i_C = i_{C(\text{sat})}/2$, $v_{CE} = v_{CE(\text{off})}/2$ gives the maximum value of p. In other words, the instantaneous power dissipation of a BJT is maximum when the BJT is operating at the center of the load line.

The **quiescent power dissipation** P_{DQ} of a BJT is defined to be

$$\boxed{P_{DQ} = V_{CEQ}I_{CQ}} \tag{9.52}$$

This is the power dissipated by the transistor when no ac signal is present. According to the discussion above, P_{DQ} is maximum when the Q point is centered on the load line. A centered Q point, however, is required for maximum ac output power. Under this circumstance, though, the average power P_D dissipated by the transistor will be less than P_{DQ}. Thus given a centered Q point, the value of the quiescent power dissipation $P_{DQ} = V_{CEQ}I_{CQ}$ is the upper limit for the average power that is dissipated by the transistor.

The maximum average power that can be safely dissipated by a BJT is denoted $P_{D(\text{max})}$ and is found on a transistor's specification sheet. From the preceding discussion, $P_{D(\text{max})}$ should be chosen to be greater than or equal to P_{DQ} corresponding to a centered Q point. That is,

$$\boxed{P_{D(\text{max})} \geq P_{DQ} = V_{CEQ}I_{CQ}}$$

for a centered Q point. For instance, for the BJT used in the amplifier described in Example 9.11, $P_{DQ} = V_{CEQ}I_{CQ} = (12.8)(84.6 \times 10^{-3}) = 1.08$ W. Therefore, the transistor should have $P_{D(\text{max})} \geq 1.08$ W.

From Eq. 9.49 and Eq. 9.52, note that the maximum output signal power of a class-A amplifier is only half of the maximum quiescent power dissipation of the BJT. Thus, for example, if a class-A amplifier is to have a maximum output signal power of 5 W, then the transistor used must be rated at $P_{D(\text{max})} \geq 10$ W.

For a BJT, typically the collector current is much greater than the base current. For this reason, let us define the dc power P_{dc} delivered by the supply V_{CC} to be $P_{dc} = V_{CC}I_{CQ}$. In addition, let us define the **efficiency** η of an amplifier to be the ratio of the output signal power P_o to the dc power P_{dc} delivered by the supply, that is.

$$\eta = \frac{P_o}{P_{dc}}$$

It can be shown (see Problem 9.74 at the end of this chapter) that the maximum efficiency for a class-A amplifier of the type shown in Fig. 9.64 is $\eta = \frac{1}{4}$.

The power absorbed by a transistor is dissipated as heat. For a BJT, typically $P_{D(max)}$ is rated at 25°C. If the ambient (surrounding) temperature is greater than 25°C, the value of $P_{D(max)}$ must be **derated** (reduced) appropriately. This is done by using the **power derating factor** given on the transistor's specification sheet. For instance, a 2N1711 silicon power BJT is rated at $P_{D(max)} = 3$ W at 25°C. Its power derating factor is 17 mW/°C. Therefore, at 75°C, $P_{D(max)}$ must be derated by $(75 - 25)(17 \times 10^{-3}) = 0.85$ W. In other words, at 75°C,

$$P_{D(max)} = 3 - 0.85 = 2.15 \text{ W}$$

The value of $P_{D(max)}$ at which a BJT is rated can be increased by connecting the transistor's case to a relatively large metallic surface with good heat-radiating properties. Such a structure, called a **heat sink,** is often used with power transistors.

Class-B Amplifiers

Although class-A operation is both simple and stable, the $P_{D(max)}$ rating of a transistor must be at least twice the maximum ac output power. Also, for the case of no ac signal, a centered Q point means that the current drain on the voltage supply is $i_{C(sat)}/2$. For low-power applications, these properties may not be serious drawbacks. However, there are situations in which the use of class-A amplifiers yields results that are too inefficient. Fortunately, though, there are efficient alternatives to class-A operation. Such an alternative will now be discussed.

Class-B operation is the term used to indicate that a transistor is biased at cutoff, and for an applied ac signal, the transistor remains cut off for half of the cycle and operates in the active region for the other half. Since a transistor is biased at cutoff for class-B operation, then necessarily $I_{CQ} = 0$ A.

Suppose the circuit shown in Fig. 9.67 is the ac-equivalent circuit of an emitter follower that is biased at cutoff. For the time being, let us assume the ideal case that when the B-E junction is forward biased, the voltage is $v_{BE} = 0$ V. Furthermore, let

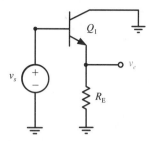

Fig. 9.67 Emitter follower biased at cutoff.

us assume that the cut-in voltage is $V_\gamma = 0$ V. Thus when the ac signal voltage $v_s > 0$ V, the B-E junction will be forward biased, and, therefore, $v_e = v_s$. Conversely, when $v_s \leq 0$ V, the B-E junction remains reverse biased, and, therefore, $v_e = 0$ V.

Figure 9.68a shows the ac-equivalent circuit of an emitter follower that uses a *pnp* transistor, which is again biased at cutoff. This circuit is redrawn in Fig. 9.68b. Proceeding as was done for the case of the emitter follower given in Fig. 9.67, we deduce that when $v_s < 0$ V, then $v_e = v_s$ and when $v_s \geq 0$ V, then $v_e = 0$ V.

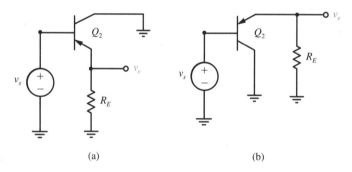

(a)   (b)

Fig. 9.68 A *pnp* emitter follower biased at cutoff.

If we combine the emitter followers given in Figs. 9.67 and 9.68 as shown in Fig. 9.69, when $v_s > 0$ V, then Q_1 conducts and Q_2 is cut off, and $v_o = v_s$. Similarly, when $v_s < 0$ V, then Q_2 conducts and Q_2 is cut off, and $v_o = v_s$. (Of course, when $v_s = 0$ V, then $v_o = 0$ V.) Thus, for an ac signal v_s (which is not large enough to saturate either Q_1 or Q_2), we have that $v_o = v_s$. The circuit given in Fig. 9.69 is an example of a **push-pull circuit**—specifically, it is a push-pull emitter follower. Although the output voltage v_o (ideally) equals the input voltage v_s, a BJT has a significant current gain (h_{fe}). Thus an emitter follower can produce a significant power gain.

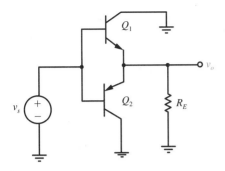

Fig. 9.69 Push-pull emitter follower.

In this latest discussion, we assumed the ideal situation that a forward-biased base-emitter junction had a voltage of $v_{BE} = 0$ V. Of course we know in actuality, for a silicon BJT, such a voltage is about 0.7 V. Furthermore, for the circuit given in Fig. 9.69, even when $v_s > 0$ V, there will be essentially no current in Q_1 until v_s exceeds the typical cut-in voltage $V_\gamma = 0.5$ V. Thus, $v_o \approx 0$ V for $0 \le v_s \le 0.5$ V. By symmetry, we deduce that $v_o \approx 0$ V for $-0.5 \le v_s \le 0$ V. Consequently, if v_s is a sinusoid, for example, the resulting output voltage v_o will be nonsinusoidal and will be a function like the one shown in Fig. 9.70. The resulting distortion is known as **crossover distortion.**

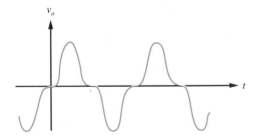

Fig. 9.70 Crossover distortion.

To minimize crossover distortion, we slightly forward bias the B-E junctions of the transistors. This is referred to as a **trickle bias.** As a consequence of having a trickle bias, I_{CQ} is nonzero and has a value that typically is between 1 percent and 5 percent of

$$i_{C(\text{sat})} = I_{CQ} + \frac{V_{CEQ}}{R_T + R_E} \approx \frac{V_{CEQ}}{R_E} \tag{9.53}$$

When $I_{CQ} > 0$ A, strictly speaking, we no longer have class-B operation. Technically, operation between class A and class B is called **class-AB operation.** However, since

I_{CQ} is a relatively small quantity for a trickle bias, we still say that we have class-B operation.

An example of an ac-coupled, class-B push-pull emitter follower is shown in Fig. 9.71*a*. The transistors Q_1 and Q_2 are assumed to be **complementary**—that is, they have identical characteristics that differ only in sign (polarity) since one is an *npn* BJT and the other is *pnp*. The dc-equivalent circuit of the emitter follower is shown in Fig. 9.71*b* where the quiescent currents and voltages are indicated.

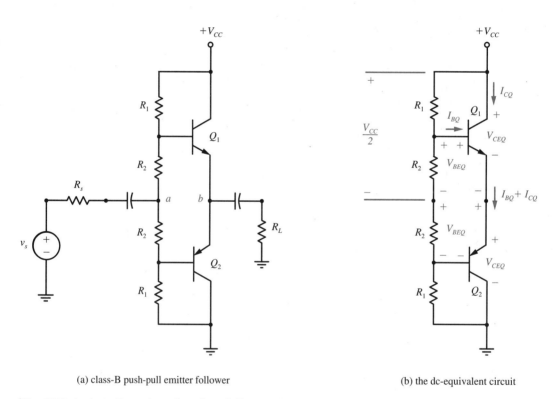

(a) class-B push-pull emitter follower

(b) the dc-equivalent circuit

Fig. 9.71 A class-B, push-pull emitter follower: (*a*) complete circuit, (*b*) dc-equivalent circuit.

By symmetry, we have that $V_{CEQ} = V_{CC}/2$, as is the voltage across R_1 and R_2. For a trickle bias, I_{CQ} is small, as is I_{BQ}. For the ideal case that $I_{BQ} = 0$ A, by voltage division we have

$$V_{BEQ} = \frac{R_2}{R_1 + R_2}\left(\frac{V_{CC}}{2}\right) \qquad (9.54)$$

Since $I_{BQ} \neq 0$ A, Eq. 9.54 is an approximation, but it is a good approximation if the current through R_2 is much greater than I_{BQ}. In other words, Eq. 9.54 is accurate when $V_{BEQ}/R_2 \gg I_{BQ}$.

Example 9.12

For the circuit given in Fig. 9.71, suppose that for the BJTs, $h_{FE} = 25$ and $I_S = 1.3 \times 10^{-14}$ A, whereas $R_2 = 180$ Ω, $R_L = 100$ Ω, and $V_{CC} = 20$ V. Let us find the value of R_1 required to obtain a trickle bias of $I_{CQ} = 0.02i_{C(sat)}$.

Since $V_{CEQ} = V_{CC}/2 = 20/2 = 10$ V, $R_E = R_L = 100$ Ω, and $R_T = 0$ Ω, from Eq. 9.53

$$i_{C(sat)} = 0.02i_{C(sat)} + \frac{10}{100} \quad \Rightarrow \quad i_{C(sat)} = 102 \text{ mA}$$

Thus

$$I_{CQ} = 0.02i_{C(sat)} = 0.02(102 \times 10^{-3}) = 2.04 \text{ mA}$$

and

$$I_{BQ} = \frac{I_{CQ}}{h_{FE}} = \frac{2.04 \times 10^{-3}}{25} = 81.6 \text{ } \mu A$$

Since $i_C = I_S(e^{v_{BE}/V_T} - 1) \approx I_S e^{v_{BE}/V_T} \Rightarrow v_{BE} = V_T \ln(i_C/I_S)$, then at 300 K, $V_T = 26$ mV and

$$v_{BE} = (0.026) \ln \frac{2.04 \times 10^{-3}}{1.3 \times 10^{-14}} = 0.67 \text{ V} = V_{BEQ}$$

From Eq. 9.54 we have

$$V_{BEQ} = \frac{180}{R_1 + 180}\left(\frac{20}{2}\right) = 0.67 \text{ V}$$

Solving this equation for R_1 we get $R_1 = 2.51$ k$\Omega \approx 2.5$ kΩ, and since

$$\frac{V_{BEQ}}{R_2} = \frac{0.67}{180} = 3.72 \text{ mA} \gg I_{BQ} = 81.6 \text{ } \mu A$$

then Eq. 9.54 yields accurate results.

The need for the coupling capacitors shown in Fig. 9.71a can be eliminated if the point where the collector of Q_2 and the lower resistor R_1 meet is connected to a supply voltage of $-V_{CC}$ instead of the reference. Under such a circumstance, due to the symmetry of the circuit, the dc potentials at nodes a and b would be 0 V. As a consequence of no dc voltage across R_L and R_s (assuming that v_s is purely an ac voltage source), the direct current through them is zero, so no coupling capacitors are required. The resulting dc, class-B push-pull emitter follower is shown in Fig. 9.72. For such a circuit, Eq. 9.54 becomes

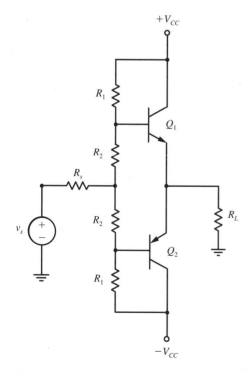

Fig. 9.72 A dc class-B, push-pull emitter follower.

$$V_{BEQ} = \frac{R_2}{R_1 + R_2} V_{CC} \qquad (9.55)$$

and, as was the case for Eq. 9.54, this equation is accurate when $V_{BEQ}/R_2 \gg I_{BQ}$.

Push-pull emitter followers are indeed important—the output stage of many IC amplifiers is a push-pull emitter follower.

Class-B push-pull amplifiers exist in other than emitter-follower configurations. For example, the circuit shown in Fig. 9.73 is a CE push-pull amplifier. In addition to power gain, an amplifier such as this can provide a voltage gain greater than unity. Because of the nonlinear characteristic of the B-E junction of a BJT, emitter resistors labeled R_E are included as part of the amplifier to reduce nonlinear distortion. Unfortunately, these resistors also reduce the voltage gain.

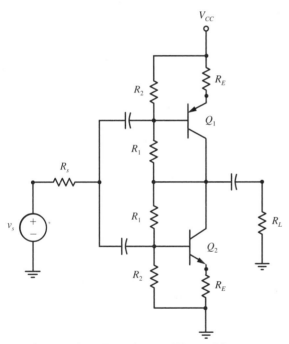

Fig. 9.73 A class-B push-pull CE amplifier.

Maximum Output Power

Now consider the case that a class-B push-pull emitter follower or amplifier is driven with a sinusoidal signal such that the maximum output results. Specifically, Fig. 9.74 shows the load line for the *npn* transistor along with the associated waveforms for i_C and v_{CE}. For the sake of notational simplicity, let us define $I_m = i_{C(\text{sat})} = V_{CEQ}/(R_T + R_E)$ and $V_m = v_{CE(\text{off})} = V_{CEQ}$. The waveforms shown in Fig. 9.74

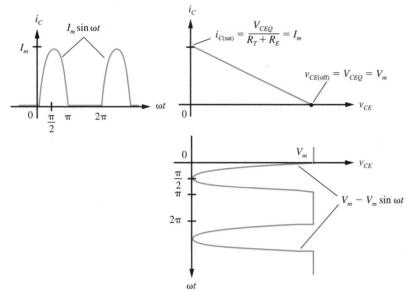

Fig. 9.74 Load line and waveforms of i_C and v_{CE} for *npn* transistor when maximum output is achieved.

are the current and the voltage associated with one (the *npn* BJT) of the two complementary transistors. The ac output current and voltage are sinusoids with amplitudes I_m and V_m, respectively. As a consequence, the maximum output signal power P_o is

$$P_o = \left(\frac{V_m}{\sqrt{2}}\right)\left(\frac{I_m}{\sqrt{2}}\right) = \frac{1}{2}\,V_m I_m = \frac{1}{2}V_{CEQ}i_{C(\text{sat})} \qquad (9.56)$$

Since class-B operation means that transistors are biased at cutoff, then the quiescent power dissipation P_{DQ} of the transistors is (ideally) zero, as is the dc power P_{dc} delivered by the supply (also called the **no-signal power drain**). However, the average power P_D dissipated by each transistor is not zero, and with a little bit of effort, it can be calculated.

The instantaneous power dissipated by a BJT is, from Eq. 9.50, $p = v_{CE}i_C$. With reference to Fig. 9.74, for the maximum ac-signal case, the average power P_D dissipated by the corresponding transistor is

$$P_D = \frac{1}{2\pi}\int_0^{2\pi} p\,d(\omega t) = \frac{1}{2\pi}\int_0^{\pi}(V_m - V_m \sin \omega t)(I_m \sin \omega t)\,d(\omega t) \qquad (9.57)$$

It can be shown (see Problem 9.82 at the end of this chapter) that from this integral we get

$$P_D = \frac{4 - \pi}{4\pi} V_m I_m = 0.0683 \, V_{CEQ} i_{C(\text{sat})} \tag{9.58}$$

This, however, is not the maximum average power that is dissipated by a transistor. In particular, if the ac signal is less than maximum and the entire load line is not utilized, then a larger value of P_D can be obtained. Specifically, suppose that the sinusoidal portions of i_C and v_{CE} have amplitudes kI_m and kV_m, where $0 < k < 1$. Then the average power dissipated by each transistor is

$$P_D = \frac{1}{2\pi} \int_0^\pi (V_m - kV_m \sin \omega t)(kI_m \sin \omega t) \, d(\omega t) \tag{9.59}$$

It can be shown (see Problem 9.83 at the end of this chapter) that from this integral we get

$$P_D = \frac{V_m I_m}{2\pi} \left(2k - \frac{\pi}{2} k^2 \right) \tag{9.60}$$

To find the value of k that yields the maximum value of P_D, let us take the derivative of P_D with respect to k and set the result equal to zero. Doing so, we get

$$\frac{dP_D}{dk} = \frac{V_m I_m}{2\pi} (2 - \pi k) = 0$$

from which we deduce that P_D is maximum when $k = 2/\pi$. Substituting this value of k into Eq. 9.60, we find that the maximum average power P_{DM} dissipated by each transistor is

$$P_{DM} = \frac{V_m I_m}{\pi^2} = 0.1 V_{CEQ} i_{C(\text{sat})} \tag{9.61}$$

Thus for each BJT we require that

$$\boxed{P_{D(\text{max})} \geq 0.1 V_{CEQ} i_{C(\text{sat})}} \tag{9.62}$$

From Eq. 9.56, we know that the maximum output signal power is $P_o = 0.5 V_{CEQ} i_{C(\text{sat})}$. Thus the maximum output signal power of a class-B amplifier is five times as large as the maximum power dissipated by each BJT. For example, if a class-B amplifier is to have a maximum output signal power of 5 W, then each transistor used must be rated at $P_{D(\text{max})} \geq 1$ W. Recall that for the corresponding class-A amplifier, $P_{D(\text{max})} \geq 10$ W. From an alternative point of view, as another example, given BJTs with $P_{D(\text{max})} = 4$ W, a class-A amplifier will safely yield an output signal power of at most 2 W, whereas a class-B amplifier can produce an output signal power of up to 20 W. This much higher efficiency is a major reason for the selection of class-B over class-A amplifiers.

With regard to efficiency, reconsider the class-B push-pull emitter follower given in Fig. 9.72. Since the transistors are (ideally) biased at cutoff, then the dc power P_{dc} delivered by the supply is zero. However, there is a current going through the supply. For the case of maximum output signal power, this current is (approximately) equal to the collector current i_C shown in Fig. 9.74. Thus, while Q_1 is conducting, the instantaneous power delivered by the supply is $V_{CC}I_m \sin \omega t$. Therefore, the average power P_d delivered by the supply while Q_1 is conducting is

$$P_D = \frac{1}{\pi}\int_0^\pi V_{CC}I_m \sin \omega t \; d(\omega t) = \frac{2}{\pi}V_{CC}I_m = \frac{2}{\pi}V_m I_m$$

since $V_m = V_{CEQ} = V_{CC}$ for the given circuit. By symmetry, this is also the average power delivered while Q_2 is conducting. From Eq. 9.56, the maximum output signal power is $\frac{1}{2}V_m I_m$. Hence, the maximum efficiency for the push-pull emitter follower under consideration is

$$\eta = \frac{\frac{1}{2}V_m I_m}{(2/\pi)V_m I_m} = \frac{\pi}{4}$$

The fact that class-B amplifiers have a maximum efficiency of $\eta = \pi/4$ (78.5%), whereas class-A amplifiers have a maximum efficiency of $\eta = 1/4$ (25%), affirms the practicality of class-B amplifiers.

An important concern with class-B amplifiers is their sensitivity to temperature changes. As we have seen, a push-pull configuration is trickle biased by using a voltage divider consisting of resistors R_1 and R_2 (e.g., see Fig. 9.72). The voltage across R_2 establishes V_{BEQ}, which, in turn, determines the trickle-bias collector current I_{CQ}. This current should be between 1 percent and 5 percent of $i_{C(\text{sat})}$. However, for a given value of V_{BEQ}, an increase in temperature can result in an excessive amount of current (see Fig. 6.15 on p. 367). We can compensate for this by replacing the resistors labeled R_2 with diodes that are directed downward (see Problem 9.85 at the end of this chapter). We should select diodes whose characteristics closely match those of the B-E junctions of the BJTs. This is not easy to do for discrete components, but it is readily accomplished on the same IC chip. Since $V_{BEQ} \approx 0.7$ V, then typically $R_1 \gg R_2$, and the current through R_2 is determined by the value of R_1. When R_2 is replaced by a diode, the current through the diode is still determined by the value of R_1. If the temperature increases, as the diode current remains the same, the diode voltage will decrease (see Fig. 6.15). This, in turn, means that V_{BEQ} decreases, and when the characteristics of the diode and the B-E junction are matched, I_{CQ} remains unchanged.

Power FETs

Although our discussion of power amplifiers has concentrated on BJTs, there are FETs that are fabricated specifically for power applications. Of particular importance is the *n*-channel enhancement MOSFET illustrated in Fig. 9.75a. Compare this with the ordinary *n*-channel enhancement MOSFET shown in Fig. 9.75b. For the latter device, current going from drain to source travels horizontally, whereas for the former, it travels vertically. For this reason, as well as because of the shape of its gate, the "vertical" or "V-shaped" FET depicted in Fig. 9.75a is designated **VMOS.** In previous discussions, it was mentioned that ordinary FETs are symmetrical devices in that source and drain can be interchanged. This, however, is not the case with the VMOS.

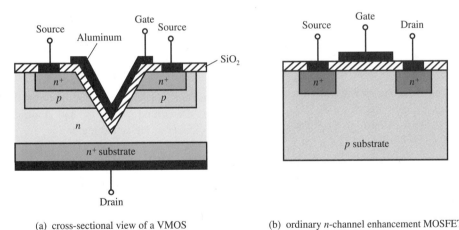

(a) cross-sectional view of a VMOS   (b) ordinary *n*-channel enhancement MOSFET

Fig. 9.75 Two types of *n*-channel enhancement MOSFETs.

Not only is the VMOS capable of handling much more power than an ordinary MOSFET, but VMOS characteristics are much closer to being linear. This property is quite desirable for large-signal applications. Whereas the peak current for an ordinary MOSFET is typically 50 mA, it is not uncommon for a VMOS to have a peak current of 2 A. Although power ratings of ordinary MOSFETs are limited to a few watts, the power ratings for some VMOS are in the tens of watts.

SUMMARY

1. For use as small-signal amplifiers, transistors are biased in the active region.

2. For BJTs and FETs, bias stability is obtained by using a combination of fixed and self-biasing.

3. The ac analysis of transistor amplifiers is simplified by using small-signal models.

4. Analysis with the use of the h-parameter model of a BJT results in a high degree of accuracy.

5. The coupling and bypass capacitors of an amplifier determine its lower cutoff frequency.

6. A transistor's junction capacitances determine an amplifier's upper cutoff frequency.

7. To obtain maximum power output from a class-A amplifier, a transistor's Q point is centered on the ac load line. The maximum power dissipated by the transistor is twice the maximum output signal power.

8. The maximum efficiency of a (transformerless) class-A amplifier is 25 percent.

9. The transistors of a class-B push-pull amplifier are biased at cutoff. The maximum power dissipated by each transistor is only one fifth of the maximum output signal power.

10. The maximum efficiency of a class-B amplifier is 78.5 percent.

Problems

9.1 For the circuit given in Fig. 9.2 on p. 573, suppose that $R_B = 300$ kΩ, $R_C = 1$ kΩ, and $V_{CC} = 10$ V. Find I_{CQ}, V_{CEQ}, and confirm that the BJT is biased in the active region when h_{FE} is (a) 100 and (b) 200.

9.2 For the circuit given in Fig. 9.3 on p. 574, suppose that $h_{FE} = 100$, $R_1 = 90$ kΩ, $R_2 = 10$ kΩ, $R_C = 1$ kΩ, and $V_{CC} = 16$ V. Find I_{CQ}, V_{CEQ}, and confirm that the BJT is biased in the active region.

9.3 For the circuit given in Fig. 9.5 on p. 575, suppose that $h_{FE} = 100$, $R_1 = 90$ kΩ, $R_2 = 10$ kΩ, $R_C = 5$ kΩ, $R_E = 1$ kΩ, and $V_{CC} = 20$ V. Find I_{CQ}, V_{CEQ}, and confirm that the BJT is biased in the active region.

9.4 For the circuit given in Fig. 9.5 on p. 575, suppose that $h_{FE} = 100$, $R_1 = R_2 = 400$ kΩ, $R_C = 0$ Ω, $R_E = 1$ kΩ, and $V_{CC} = 20$ V. Find I_{CQ}, V_{CEQ}, and confirm that the BJT is biased in the active region.

9.5 For the circuit given in Fig. 9.5 on p. 575, suppose that $h_{FE} = 100$, $R_E = 100$ Ω, and $V_{CC} = 20$ V. Given that the parallel combination of R_1 and R_2 is 2.7 kΩ, find R_1, R_2, and R_C such that the BJT is biased in the active region at $I_{CQ} = 10.2$ mA and $V_{CEQ} = 10.2$ V.

9.6 For the circuit given in Fig. 9.5 on p. 575, suppose that $h_{FE} = 100$, $R_1 = 6$ kΩ, $R_2 = 3$ kΩ, and $V_{CC} = 12$ V. Determine R_E and R_C such that the BJT is biased in the active region at $I_{CQ} = 10$ mA and $V_{CEQ} = 6$ V.

9.7 The transistor in the circuit shown in Fig. P9.7 has $h_{FE} = 100$. Determine I_{CQ}, V_{CEQ}, and confirm that the BJT is biased in the active region.

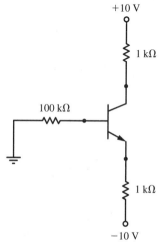

+10 V

1 kΩ

100 kΩ

1 kΩ

−10 V

Fig. P9.7

9.8 The transistor circuit shown in Fig. P9.8 is an example of **collector-feedback bias.** Find I_{CQ}, V_{CEQ}, and confirm that the BJT is biased in the active region given that $h_{FE} = 100$ for the transistor.

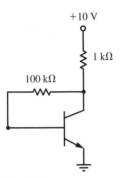

Fig. P9.8

9.9 Both transistors in the circuit shown in Fig. P9.9 have $h_{FE} = 100$. Find I_{CQ} and V_{CEQ} for Q_1 and Q_2, and confirm that the BJTs are biased in the active region.

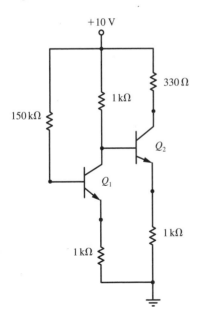

Fig. P9.9

9.10 In the circuit shown in Fig. P9.10, transistors Q_1 and Q_2 are connected in a **Darlington configuration.** Show that when Q_1 and Q_2 are biased in the

active region, and $h_{FE1} \gg 1$ and $h_{FE2} \gg 1$, then

$$i_C \approx h_{FE1} h_{FE2} i_{B1}$$

so that the Darlington connection of Q_1 and Q_2 behaves as a single BJT having $h_{FE} \approx h_{FE1} h_{FE2}$.

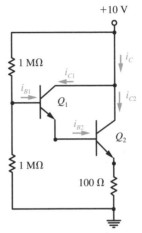

Fig. P9.10

9.11 For the circuit shown in Fig. P9.10, both BJTs have $h_{FE} = 100$. Find I_{CQ} and V_{CEQ} for Q_1 and Q_2, and confirm that the BJTs are biased in the active region.

9.12 The small-signal transistor amplifier given in Fig. 9.7 on p. 577 has $h_{fe} = h_{FE} = 100$, $R_1 = 90$ kΩ, $R_2 = 10$ kΩ, $R_C = 1$ kΩ, $R_E = 0$ Ω, and $V_{CC} = 16$ V. Given that $I_{CQ} = 10$ mA, find (a) g_m, (b) r_e, (c) R_{in}, and (d) A_v.

9.13 The small-signal transistor amplifier given in Fig. 9.7 on p. 577 has $h_{fe} = h_{FE} = 100$, $R_1 = 90$ kΩ, $R_2 = 10$ kΩ, $R_C = 5$ kΩ, $R_E = 1$ kΩ, and $V_{CC} = 20$ V. Given that $I_{CQ} = 1.18$ mA, find (a) g_m, (b) r_e, (c) R_{in}, and (d) A_v.

9.14 For the BJT amplifier given in Fig. 9.7 on p. 577, use the small-signal ac model shown in Fig. 9.10 on p. 579 to derive formulas for (a) R_b and (b) A_v.

9.15 The transistor amplifier given in Fig. 9.14 on p. 583 has $h_{fe} = h_{FE} = 100$, $R_1 = 90$ kΩ, $R_2 = 10$ kΩ, $R_C = 5$ kΩ, $R_E = 1$ kΩ, $R_L = 5$ kΩ, and $V_{CC} = 20$ V. Given that $I_{CQ} = 1.18$ mA, find (a) R_{in}, (b) A_v, and (c) A_i.

9.16 For the **feedback amplifier** shown in Fig. P9.16, $h_{fe} = h_{FE} = 100$. Given that $I_{CQ} = 4.63$ mA, find (a) $A_v = v_c/v_b$, (b) $R_b = v_b/i_b$, (c) $R_{in} = v_b/i_s$, (d) $A_i = i_L/i_b$, and (e) $A_{is} = i_L/i_s$.

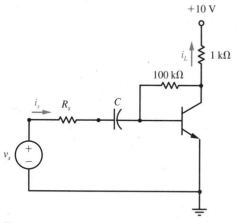

Fig. P9.16

9.17 For the feedback amplifier shown in Fig. P9.16, $h_{fe} = h_{FE} = 100$. Given that $I_{CQ} = 4.63$ mA, find the output resistance R_o that is seen by an external load that is connected to the collector of the BJT.

9.18 The two-stage amplifier shown in Fig. P9.18 has $h_{fe} = h_{FE} = 100$ for both transistors Q_1 and Q_2. Given that $I_{CQ1} = 3.71$ mA and $I_{CQ2} = 5.48$ mA, find (a) $A_v = v_{c2}/v_{b1}$, and (b) $A_i = i_{c2}/i_{b1}$.

9.19 For the emitter follower given in Fig. 9.18 on p. 586, $h_{fe} = h_{FE} = 100$, $R_1 = R_2 = 400$ kΩ, $R_E = R_s = 1$ kΩ, $R_L = 9$ kΩ, and $V_{CC} = 20$ V. Given that $I_{CQ} = 3.09$ mA, find (a) A_v, (b) R_{in}, and (c) $A_i = i_L/i_b$.

9.20 For the emitter follower given in Fig. 9.18 on p. 586, $h_{fe} = h_{FE} = 100$, $R_1 = R_2 = 400$ kΩ, $R_E = R_s = 1$ kΩ, $R_L = 9$ kΩ, and $V_{CC} = 20$ V. Given that $I_{CQ} = 3.09$ mA, find R_o.

9.21 Figure P9.21 shows a **common-base (CB) amplifier.** Given that $h_{fe} = h_{FE} = 100$ and $I_{CQ} = 0.93$ mA, find (a) the voltage gain v_c/v_b, (b) the input resistance R_{in}, and (c) the output resistance R_o that is seen by R_L.

9.22 For the transistor amplifier given in Fig. 9.14 on p. 583, suppose that $R_1 = 90$ kΩ, $R_2 = 10$ kΩ, $R_C = 5$ kΩ, $R_E = 1$ kΩ, $R_L = 5$ kΩ, and the BJT has the h parameters: $h_{ie} = 2.2$ kΩ, $h_{re} = 10^{-4}$, $h_{fe} = 100$, $h_{oe} = 10^{-5}$ ℧. Use the small-signal

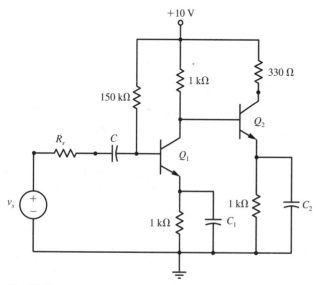

Fig. P9.18

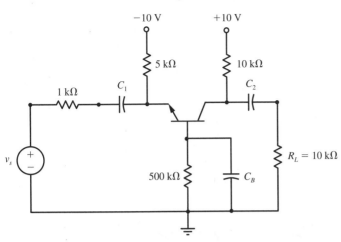

Fig. P9.21

h-parameter model given in Fig. 9.21*b* on p. 590 to determine (a) $A_v = v_e/v_b$, (b) $R_{in} = R_p \parallel R_b$, where $R_b = v_b/i_b$, and (c) $A_i = i_L/i_b$. Compare these results with those obtained for Problem 9.15.

9.23 For the emitter follower given in Fig. 9.18 on p. 586, suppose that $R_1 = R_2 = 400$ kΩ, $R_E = 1$ kΩ, $R_L = 9$ kΩ, and the BJT has the *h* parameters: $h_{ie} = 840$ Ω, $h_{re} = 10^{-4}$, $h_{fe} = 100$, $h_{oe} = 10^{-5}$℧. Use the small-signal *h*-parameter model given in Fig. 9.21*b* on p. 590 to determine (a) $A_v = v_e/v_b$, (b) $R_{in} = R_p \parallel R_b$, where $R_b = v_b/i_b$, and (c) $A_i = i_L/i_b$. Compare these results with those obtained for Problem 9.19.

9.24 Figure P9.24 shows the ac small-signal **hybrid-π model** of a BJT. Use the typical hybrid-π

parameters of $r_b = 100$ Ω, $r_\pi = 2$ kΩ, $r_\mu = 20$ MΩ, $r_o = 200$ kΩ, and $g_m = 45$ m℧ to find the voltage gain $A_v = v_c/v_b$ for the amplifier given in Fig. 9.7 on p. 577, where $R_1 = 90$ kΩ, $R_2 = 10$ kΩ, $R_C = 5$ kΩ, and $R_E = 1$ kΩ.

9.25 For the circuit given in Fig. 9.23 on p. 592, suppose that $I_{DSS} = 12$ mA, $V_p = -4$ V, $R_D = 1$ kΩ, $R_G = 1$ MΩ, $V_{DD} = 12$ V, and $V_{GG} = 2$ V. Find V_{GSQ}, I_{DQ}, V_{DSQ}, and confirm that the JFET is biased in the active region.

9.26 For the circuit given in Fig. 9.25 on p. 593, suppose that $I_{DSS} = 16$ mA, $V_p = -4$ V, $R_D = 1$ kΩ, $R_G = 1$ MΩ, $R_S = 500$ Ω, and $V_{DD} = 10$ V. Find V_{GSQ}, I_{DQ}, V_{DSQ}, and confirm that the JFET is biased in the active region.

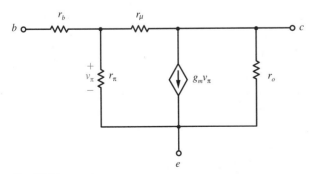

Fig. P9.24

9.27 For the circuit given in Fig. 9.27 on p. 593, suppose that $I_{DSS} = 12.5$ mA, $V_p = -2$ V, $R_D = R_S = 400$ Ω, $R_G = 1$ MΩ, $V_{GG} = 5$ V, and $V_{DD} = 15$ V. Find V_{GSQ}, I_{DQ}, V_{DSQ}, and confirm that the JFET is biased in the active region.

9.28 For the circuit given in Fig. 9.27 on p. 593, suppose that $I_{DSS} = 12$ mA, $V_p = -2$ V, $R_D = 1$ kΩ, $R_G = 1$ MΩ, $V_{GG} = 5$ V, and $V_{DD} = 15$ V. Find I_{DQ}, R_S, and V_{DSQ} such that the JFET is biased in the active region and $V_{GSQ} = -1$ V. Confirm active-region operation.

9.29 For the circuit given in Fig. 9.27 on p. 593, suppose that $I_{DSS} = 16$ mA, $V_p = -2$ V, $R_D = 1.5$ kΩ, $R_G = 1$ MΩ, $V_{GG} = 5$ V, and $V_{DD} = 15$ V. Find V_{GSQ}, R_S, and V_{DSQ} such that the JFET is biased in the active region and $I_{DQ} = 1$ mA. Confirm active-region operation.

9.30 For the circuit given in Fig. 9.27 on p. 593, suppose that $I_{DSS} = 16$ mA, $V_p = -4$ V, $R_G = 1$ MΩ, $V_{GG} = 1$ V, and $V_{DD} = 15$ V. Find V_{GSQ}, R_S, and R_D such that the JFET is biased in the active region at $I_{DQ} = 1$ mA and $V_{DSQ} = 7$ V. Confirm active-region operation.

9.31 For the circuit shown in Fig. 9.33 on p. 597, suppose that $K = 0.25$ mA/V^2, $V_t = 2$ V, $R_D = R_S = 1$ kΩ, $R_G = 1$ MΩ, $V_{DD} = 20$ V, and $V_{GG} = 10$ V. Find V_{GSQ}, I_{DQ}, V_{DSQ}, and confirm that the MOSFET is biased in the active region.

9.32 For the circuit shown in Fig. 9.33 on p. 597, suppose that $K = 0.25$ mA/V^2, $V_t = 2$ V, $R_D = 1$ kΩ, $R_G = 1$ MΩ, $V_{DD} = 12$ V, and $V_{GG} = 6$ V. Find I_{DQ}, R_S, and V_{DSQ} such that the MOSFET is biased in the active region and $V_{GSQ} = -1$ V. Confirm active-region operation.

9.33 For the circuit shown in Fig. 9.33 on p. 597, suppose that $K = 0.25$ mA/V^2, $V_t = 2$ V, $R_G = 1$ MΩ, $V_{DD} = 24$ V, and $V_{GG} = 12.5$ V. Find V_{GSQ}, R_S, and R_D such that the MOSFET is biased in the active region at $I_{DQ} = 9$ mA and $V_{DSQ} = 6$ V. Confirm active-region operation.

9.34 For the circuit given in Fig. 9.34 on p. 598, suppose that $K = 0.25$ mA/V^2, $V_t = 2$ V, $R_D =$

4 kΩ, $R_F = 1$ MΩ, and $V_{DD} = 22$ V. Find V_{GSQ}, V_{DSQ}, and I_{DQ}.

9.35 For the circuit given in Fig. 9.34 on p. 598, suppose that $K = 0.25$ mA/V^2, $V_t = 2$ V, $R_F = 1$ MΩ, and $V_{DD} = 9$ V. Find the value of R_D for which the MOSFET is biased in the active region and $V_{GSQ} = 4$ V.

9.36 For the circuit given in Fig. 9.34 on p. 598, suppose that $K = 0.25$ mA/V^2, $V_t = 2$ V, $R_F = 1$ MΩ, and $V_{DD} = 12$ V. Find the value of R_D for which the MOSFET is biased in the active region and $I_{DQ} = 4$ mA.

9.37 For the circuit shown in Fig. 9.34b on p. 598, connect a resistor R_G between the gate of the MOSFET and the reference. For the resulting circuit, suppose that $K = 0.25$ mA/V^2, $V_t = 2$ V, $R_D = 4$ kΩ, $R_F = R_G = 1$ MΩ, and $V_{DD} = 12$ V. Find V_{GSQ}, V_{DSQ}, and I_{DQ}.

9.38 For the circuit shown in Fig. 9.34b on p. 598, connect a resistor R_G between the gate of the MOSFET and the reference. For the resulting circuit, suppose that $K = 0.25$ mA/V^2, $V_t = 2$ V, $R_F = 1$ MΩ, $R_G = 2$ MΩ, and $V_{DD} = 15$ V. Find the value of R_D for which $I_{DQ} = 4$ mA.

9.39 For the circuit given in Fig. 9.29 on p. 594, remove the capacitor C_S. When $I_{DSS} = 10$ mA, $V_p = -4$ V, $R_1 = 500$ kΩ, $R_2 = 250$ kΩ, $R_D = 1$ kΩ, $R_S = 2.8$ kΩ, and $V_{DD} = 15$ V, then the JFET is biased in the active region at $V_{GSQ} = -2$ V and $I_{DQ} = 2.5$ mA. Use the small-signal model given in Fig. 9.36 on p. 600 to determine the voltage gains (a) v_s/v_g and (b) v_d/v_g.

9.40 For the circuit given in Fig. 9.29 on p. 594, remove the capacitor C_S. Use the small-signal model given in Fig. 9.36 on p. 600 to find expressions for the voltage gains (a) v_s/v_g and (b) v_d/v_g.

9.41 For the common-source amplifier given in Fig. 9.32 on p. 596, when $K = 0.25$ mA/V^2, $V_t = 2$ V, $R_D = R_S = 1$ kΩ, $R_1 = R_2 = 2$ MΩ, and $V_{DD} = 20$ V, then the MOSFET is biased in the active region at $I_{DQ} = 4$ mA and $V_{GSQ} = 6$ V. Use the small-signal model given in Fig. 9.36 on p. 600 to determine (a) v_d/v_g, (b) R_{in}, and (c) the output resistance R_o that is seen by an external load that is connected to the drain of the MOSFET.

9.42 For the common-source amplifier given in Fig. 9.32 on p. 596, when $K = 0.25$ mA$/$V^2, $V_t = 2$ V, $R_D = R_S = 1$ kΩ, $R_1 = R_2 = 2$ MΩ, and $V_{DD} = 20$ V, then the MOSFET is biased in the active region at $I_{DQ} = 4$ mA and $V_{GSQ} = 6$ V. Use the small-signal model given in Fig. 9.37 on p. 602, where $r_d = 10$ kΩ, to determine (a) v_d/v_g, (b) R_{in}, and (c) the output resistance R_o that is seen by an external load that is connected to the drain of the MOSFET.

9.43 For the feedback amplifier given in Fig. 9.43 on p. 608, when $K = 0.25$ mA$/$V^2, $V_t = 2$ V, $R_D = 4$ kΩ, $R_L = 6$ kΩ, $R_F = 1$ MΩ, and $V_{DD} = 22$ V, then the MOSFET is biased in the active region at $I_{DQ} = 4$ mA and $V_{GSQ} = 6$ V. Use the small-signal model given in Fig. 9.36 on p. 600 to determine (a) v_d/v_g and (b) R_{in}.

9.44 For the feedback amplifier given in Fig. 9.43 on p. 608, when $K = 0.25$ mA$/$V^2, $V_t = 2$ V, $R_D = 4$ kΩ, $R_L = 6$ kΩ, $R_F = 1$ MΩ, and $V_{DD} = 22$ V, then the MOSFET is biased in the active region at $I_{DQ} = 4$ mA and $V_{GSQ} = 6$ V. Use the small-signal model given in Fig. 9.37, on p. 602, where $r_d = 12$ kΩ, to determine (a) v_d/v_g and (b) R_{in}.

9.45 For the feedback amplifier given in Fig. 9.43 on p. 608, when $K = 0.25$ mA$/$V^2, $V_t = 2$ V, $R_D = 4$ kΩ, $R_L = 6$ kΩ, $R_F = 1$ MΩ, and $V_{DD} = 22$ V, then the MOSFET is biased in the active region at $I_{DQ} = 4$ mA and $V_{GSQ} = 6$ V. Find the output resistance R_o (seen by the external load R_L) for the case that (a) $r_d = \infty$ and (b) $r_d = 12$ kΩ.

9.46 Show that the output resistance R_o for the MOSFET feedback amplifier of Fig. 9.43 on p. 608 is given by

$$\frac{1}{R_o} = \frac{1}{R_D} + \frac{1 + g_m R_s}{R_F + R_s}$$

9.47 For the ac-equivalent circuit shown in Fig. 9.49 on p. 612, use Thévenin's theorem to replace the combination of v_s, R_s, and R_p by a single voltage source $v = R_p v_s/(R_s + R_p)$ connected in series with

a resistance $R = R_s \parallel R_p = R_s R_p/(R_s + R_p)$. Use the small-signal ac model of a BJT given in Fig. 9.12 on p. 579 to show that the resulting transfer function is

$$\frac{\mathbf{V}_C}{\mathbf{V}_s} =$$

$$\frac{-Kh_{fe}R_T(1 + j\omega R_E C_E)}{R + (1 + h_{fe})(r_e + R_E) + j\omega R_E C_E[R + (1 + h_{fe})r_e]}$$

where $K = R_p/(R_s + R_p)$.

9.48 For the transfer function given in Problem 9.47, there are two critical frequencies—one due to the zero of the transfer function and one due to the pole of the transfer function. Given that the critical frequency of the term $a + j\omega$ is $\omega = a$ rad/s, show that the critical frequencies for the transfer function given in Problem 9.47 are

$$\omega_1 = \frac{1}{R_E C_E} \quad \text{and}$$

$$\omega_2 = \frac{R + (1 + h_{fe})(r_e + R_E)}{R_E C_E[R + (1 + h_{fe})r_e]}$$

9.49 The critical frequency ω_2 given in Problem 9.48 can be written in the form: $\omega_2 = 1/R_{oE}C_E$. Find an expression for R_{oE}. Show that this expression is the same as Eq. 9.31 on p. 613.

9.50 Sketch the amplitude-response portion of the Bode plot for the transfer function given in Problem 9.47 when $C_E = 1$ μF, $h_{fe} = 100$, $R_E = R_s = 1$ kΩ, $R_p = 9$ kΩ, $R_T = 2.5$ kΩ, and $r_e = 22$ Ω.

9.51 For the high-frequency small-signal BJT model given in Fig. 9.57 on p. 619, connect an ac current source $\mathbf{I}_b$ directed from the emitter to the base and place a short circuit between the collector and the emitter. Show that the current transfer function for the resulting circuit is

$$\frac{\mathbf{I}_c}{\mathbf{I}_b} = \frac{g_m - j\omega C_c}{g_m/h_{fe} + j\omega(C_c + C_e)}$$

$$\approx \frac{g_m}{g_m/h_{fe} + j\omega(C_c + C_e)}$$

for $\omega \ll g_m/C_c$.

9.52 For the approximate current transfer function given in Problem 9.51

$$\frac{\mathbf{I}_c}{\mathbf{I}_b} = \frac{g_m}{g_m/h_{fe} + j\omega(C_c + C_e)}$$

use the fact that $h_{fe} \gg 1$ to determine the approximate frequency $f_T = \omega_T/2\pi$ for which $|\mathbf{I}_c/\mathbf{I}_b| = 1$.

9.53 The upper cutoff (3-dB) frequency for the current transfer function of a BJT is called the **bandwidth** f_β of the transistor. (a) Determine an expression for f_β for the approximate current transfer function given in Problem 9.52. (b) Show that the product of the current gain h_{fe} and the bandwidth f_β of the transistor is approximately equal to f_T.

9.54 For the BJT amplifier given in Fig. 9.14 on p. 583, suppose that $C_1 = 1 \mu F$, $C_2 = C_E = 4.7 \mu F$, $C_c = 5$ pF, $f_T = 150$ MHz, $h_{fe} = 100$, $R_1 = 90$ kΩ, $R_2 = 10$ kΩ, $R_C = R_L = 5$ kΩ, $R_E = R_s = 1$ kΩ, and $r_e = 22$ Ω. Find (a) the lower cutoff frequency, (b) the upper cutoff frequency, and (c) the bandwidth.

9.55 For the BJT amplifier given in Fig. 9.14 on p. 583, suppose that $C_1 = C_2 = 1 \mu F$, $C_E = 4.7 \mu F$, $C_c = 5$ pF, $f_T = 120$ MHz, $h_{fe} = 100$, $R_1 = 60$ kΩ, $R_2 = 30$ kΩ, $R_C = R_E = 10$ kΩ, $R_L = 30$ kΩ, $R_s = 1$ kΩ, and $r_e = 26$ Ω. Find (a) the lower cutoff frequency, (b) the upper cutoff frequency, and (c) the bandwidth.

9.56 The MOSFET amplifier given in Fig. 9.51 on p. 615 has $C_1 = C_2 = C_S = 1 \mu F$, $C_{gs} = 2$ pF, $C_{gd} = 3$ pF, $C_{ds} = 1$ pF, $g_m = 3$ m$\mho$, $R_1 = R_2 = 2$ MΩ, $R_D = 1.5$ kΩ, $R_s = 5$ kΩ, $r_d = R_L = 10$ kΩ,

and $R_S = 500$ Ω. Find (a) the lower cutoff frequency, (b) the upper cutoff frequency, and (c) the bandwidth.

9.57 For the MOSFET amplifier given in Fig. 9.51 on p. 615, suppose that $C_1 = C_2 = C_S = 1 \mu F$, $C_{gs} = 4$ pF, $C_{gd} = 2$ pF, $C_{ds} = 1$ pF, $g_m = 0.8$ m$\mho$, $R_1 = R_2 = 1$ MΩ, $R_D = 7.5$ kΩ, $R_L = 5$ kΩ, $r_d = 10$ kΩ, $R_s = 100$ Ω, and $R_S = 2.5$ kΩ. Find (a) the lower cutoff frequency, (b) the upper cutoff frequency, and (c) the bandwidth.

9.58 Show that the circuits given in Fig. P9.58a and b are equivalent by writing an expression relating i and v for both circuits.

9.59 For the feedback amplifier shown in Fig. P9.16, the midband gain is $v_c/v_b = -174$. Find the lower cutoff frequency given that $h_{fe} = 100$, $r_e = 5.62$ Ω, $R_s = 600$ Ω, and $C = 1 \mu F$.

9.60 For the feedback amplifier shown in Fig. P9.16, the midband gain is $v_c/v_b = -174$. Find the upper cutoff frequency given that $g_m = 178$ m$\mho$, $r_\pi = 568$ Ω, $R_s = 600$ Ω, $C_c = 2$ pF, and $C_e = 25$ pF.

9.61 For the feedback amplifier shown in Fig. 9.43 on p. 608, find the lower cutoff frequency given that $R_{in} = 200$ kΩ, $R_o = 3$ kΩ, $R_L = 6$ kΩ, $R_s = 50$ kΩ, $C_1 = 1 \mu F$, and $C_2 = 0.47 \mu F$.

9.62 For the feedback amplifier shown in Fig. 9.43 on p. 608, the midband gain is $v_d/v_g = -4$. Find the upper cutoff frequency given that $r_d = 12$ kΩ, $R_D = 4$ kΩ, $R_L = 6$ kΩ, $R_F = 1$ MΩ, $R_s = 50$ kΩ, $C_{gs} = 2$ pF, $C_{gd} = 4$ pF, and $C_{ds} = 1$ pF.

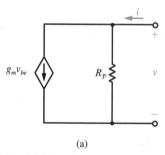

(a)

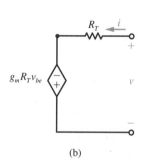

(b)

Fig. P9.58

9.63 For the emitter follower (common-collector BJT amplifier) shown in Fig. P9.63a, use the high-frequency small-signal BJT model given in Fig. 9.57 on p. 619 to obtain the ac-equivalent circuit shown in Fig. P9.63b. Show that the midband voltage gain from base to emitter is

$$A_v = \frac{v_e}{v_b} = \frac{R_E}{R_E + R}$$

where $R = r_e \parallel r_\pi$.

9.64 For the emitter-follower ac-equivalent circuit shown in Fig. P9.63b, given that $C_c = 2$ pF, $C_e = 13$ pF, $g_m = 38.5$ m℧, $R_E = R_s = 1$ kΩ, and $r_\pi = 2.6$ kΩ, find the upper cutoff frequency.

9.65 For the CB amplifier shown in Fig. P9.21, suppose that $C_1 = 1$ μF. Find the lower cutoff frequency for the case that C_2 and C_B are treated as short circuits, $r_e = 28$ Ω, and $h_{fe} = 100$.

9.66 For the CB amplifier shown in Fig. P9.21, suppose that $C_2 = 1$ μF. Find the lower cutoff frequency for the case that C_1 and C_B are treated as short circuits, $r_e = 28$ Ω, and $h_{fe} = 100$.

9.67 For the CB amplifier shown in Fig. P9.21, suppose that $C_B = 10$ μF. Find the lower cutoff frequency for the case that C_1 and C_2 are treated as short circuits, $r_e = 28$ Ω, and $h_{fe} = 100$.

9.68 A CB amplifier typically has a higher upper cutoff frequency than a CE amplifier. For the CB amplifier shown in Fig. P9.21, find the upper cutoff frequency for the case that $C_c = 5$ pF, $C_e = 50$ pF, and $R_{in} = 27.8$ Ω.

9.69 For the class-A amplifier given in Fig. 9.64 on p. 624, suppose that $h_{FE} = 20$, $R_1 = 300$ Ω, $R_2 = 200$ Ω, $R_C = 50$ Ω, $R_E = 75$ Ω, and $V_{CC} = 20$ V. (a) Find the value of R_L that results in a centered Q point. (b) Find the maximum output signal power. (c) Find the minimum value of $P_{D(max)}$ for the BJT.

9.70 For the class-A amplifier shown in Fig. 9.14 on p. 583, suppose that $h_{FE} = 20$, $R_1 = 600$ Ω, $R_2 = 300$ Ω, $R_C = R_E = 100$ Ω, and $V_{CC} = 30$ V. Connect a 300-Ω resistor in series with C_E. (a) Find the value of R_L that results in a centered Q point. (b) Find the maximum output signal power. (c) Find the minimum value of $P_{D(max)}$ for the BJT.

9.71 For the class-A amplifier given in Fig. 9.64 on p. 624, suppose that $h_{FE} = 20$, $R_1 = 300$ Ω, $R_2 = 200$ Ω, $R_E = 75$ Ω, $R_L = \infty$, and $V_{CC} = 20$ V. (a) Find the value of R_C that results in a centered Q point. (b) Find the maximum output signal power. (c) Find the minimum value of $P_{D(max)}$ for the BJT.

9.72 For the class-A amplifier shown in Fig. 9.18 on p. 586, suppose that $h_{FE} = 20$, $R_1 = 300$ Ω, $R_2 = 100$ Ω, $R_E = 75$ Ω, and $V_{CC} = 24$ V. Connect

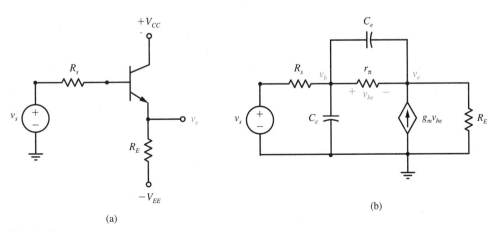

(a)

(b)

Fig. P9.63

a 125-Ω resistor in series with the collector of the transistor. (a) Find the value of R_L that results in a centered Q point. (b) Find the maximum output signal power. (c) Find the minimum value of $P_{D(max)}$ for the BJT.

9.73 For the class-A amplifier given in Fig. 9.64 on p. 624, suppose that $h_{FE} = 25$, $R_1 = 300$ Ω, $R_2 = 150$ Ω, and $V_{CC} = 30$ V. Find R_E, R_C, and R_L such that the BJT is biased on the center of the associated load line at $V_{CEQ} = 10$ V and $I_{CQ} = 50$ mA.

9.74 For the class-A amplifier given in Fig. 9.64 on p. 624, suppose that $R_L = \infty$ and $R_E = 0$ Ω. Show that the resulting amplifier has an efficiency of $\eta = \frac{1}{4}$ when the output signal power is maximum.

9.75 A silicon *npn* transistor has $P_{D(max)} = 1.8$ W at 25°C and a derating factor of 12 mW/°C. Find $P_{D(max)}$ at (a) 50°C, (b) 100°C, and (c) 150°C.

9.76 The circuit shown in Fig. P9.76a is an example of a "transformer-coupled" class-A amplifier. Figure P9.76b shows the dc-equivalent circuit of the amplifier, and Fig. P9.76c gives the ac-equivalent circuit, where $R_p = R_1 \parallel R_2$ and N is the "turns ratio" of the transformer. Suppose that $h_{FE} = 20$, $R_1 = R_2 = 200$ Ω, $R_E = 175$ Ω, $R_L = 8$ Ω, and $V_{CC} = 20$ V. Determine the value of N that results in a centered Q point.

9.77 For the transformer-coupled class-A amplifier given in Fig. P9.76a, remove the bypass capacitor and set $R_E = 0$ Ω. Show that the resulting amplifier has an efficiency of $\eta = \frac{1}{2}$ when the output signal power is maximum.

9.78 For the class-B push-pull emitter follower shown in Fig. 9.71 on p. 632, the BJTs have $h_{FE} = 20$ and $I_S = 1.4 \times 10^{-14}$ A at 300 K, whereas $R_1 = 2.2$ kΩ and $V_{CC} = 30$ V. (a) Find V_{BEQ}. (b) Find I_{CQ}. (c) Justify the use of Eq. 9.54. (d) Find $i_{C(sat)}$. (e) Find the maximum output signal power.

9.79 For the class-B push-pull emitter follower given in Fig. 9.72 on p. 634, the BJTs have $h_{FE} = 25$ and $I_S = 1.9 \times 10^{-14}$ A at 300 K, whereas $R_1 = 2$ kΩ and $V_{CC} = 10$ V. The emitter follower is to be

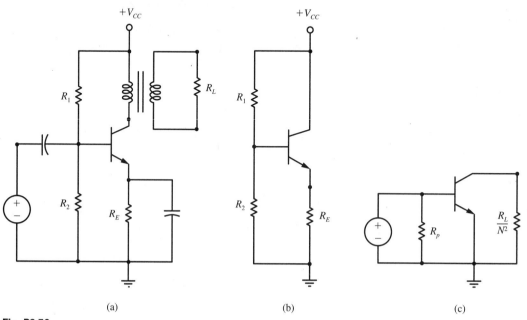

(a)   (b)   (c)

Fig. P9.76

trickle biased at $I_{CQ} = 0.01 i_{C(sat)} = 2$ mA. (a) Find V_{BEQ}. (b) Find R_2. (c) Justify the use of Eq. 9.55. (d) Find R_L. (e) Find the maximum output signal power.

9.80 For the class-B push-pull emitter follower given in Fig. 9.72 on p. 634, the BJTs have $h_{FE} = 25$ at 300 K, whereas $R_2 = 146$ Ω and $V_{CC} = 10$ V. The emitter follower is to be trickle biased at $V_{BEQ} = 0.65$ V and $I_{CQ} = 0.01 i_{C(sat)} = 0.8$ mA. (a) Find R_1. (b) Justify the use of Eq. 9.55. (c) Find R_L. (d) Find I_S. (e) Find the maximum output signal power.

9.81 For the class-B push-pull emitter follower given in Fig. 9.72 on p. 634, the BJTs have $h_{FE} = 25$ and $I_S = 10^{-14}$ A at 300 K, whereas $R_2 = 200$ Ω and $V_{CC} = 15$ V. The emitter follower is to be trickle biased at $V_{BEQ} = 0.64$ V and $I_{CQ} = 0.01 i_{C(sat)}$. (a) Find I_{CQ}. (b) Find R_1. (c) Justify the use of Eq. 9.55. (d) Find R_L. (e) Find the maximum output signal power.

9.82 Show that the integral given in Eq. 9.57 on p. 636 yields Eq. 9.58. (*Hint*: $\sin^2 \theta = \frac{1}{2} - \frac{1}{2} \cos 2\theta$)

9.83 Show that the integral given in Eq. 9.59 on p. 637 yields Eq. 9.60. (*Hint*: $\sin^2 \theta = \frac{1}{2} - \frac{1}{2} \cos 2\theta$)

9.84 For the class-B push-pull CE amplifier given in Fig. 9.73 on p. 635, the BJTs have $h_{FE} = 25$ and $I_S = 1.9 \times 10^{-14}$ A at 300 K, whereas $R_1 = 4$ kΩ, $R_E = 50$ Ω, and $V_{CC} = 20$ V. The amplifier is to be trickle biased at $I_{CQ} = 0.02 i_{C(sat)} = 2$ mA. (a) Find V_{BEQ}. (b) Find R_2—justifying any assumptions, and (c) find R_L.

9.85 For the diode-compensated push-pull emitter follower shown in Fig. P9.85, the BJTs have $h_{FE} = 25$ and $I_S = 1.6 \times 10^{-14}$ A at 300 K. The emitter follower is to be trickle biased at $I_{CQ} = 0.8$ mA. (a) Find V_{BEQ}. (b) Ignore base currents and determine the dc diode current I_{DQ}. (c) Find I_{BQ}. (d) Find $i_{C(sat)}$. (e) Find the maximum output signal power.

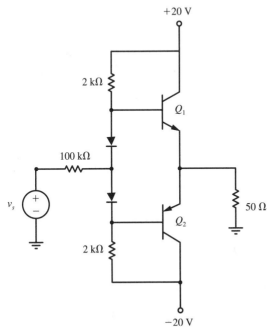

Fig. P9.85

10

Electronic Circuits and Applications

INTRODUCTION

In the preceding chapter, we saw how the coupling and bypass capacitors determine the lower cutoff frequency of an amplifier. If such capacitors can be eliminated, then an amplifier's frequency response can go down to dc (zero frequency). Because its frequency response cannot go all the way down to dc, an amplifier that employs coupling capacitors is known as an **ac amplifier**. In contrast to this, if coupling is accomplished directly without the use of capacitors, we say that the amplifier is **directly coupled** and refer to it as a **direct-coupled (dc) amplifier**. Note that the two ways in which the term "dc" is used are, in essence, synonymous. Provided that a dc (direct-coupled) amplifier has no bypass capacitors, its frequency response will go down to dc (zero frequency). The elimination of coupling and bypass capacitors from amplifier design has an extremely important consequence—amplifiers can be more easily fabricated on IC chips. As a result, amplifiers—in particular, operational amplifiers—can be made small in size and economical in price.

The differential amplifier is a direct-coupled device that is typically the input stage of an operational amplifier (op amp). Because it requires no capacitors for coupling its stages, the op amp lends itself nicely to IC fabrication. It can be produced small in size and high in reliability. Operational amplifiers are useful for numerous linear applications, and nonlinear applications as well. They are versatile and predictable—they are the basic analog IC.

Feedback can be produced in a circuit or system by returning a portion of the output to the input. Because the incorporation of feedback generally will modify the performance and characteristics of a circuit, feedback principles are quite useful in electronic circuit design. In this chapter, we will see how negative feedback can be used to improve amplifier operation, and how positive feedback can be used to produce oscillators.

Oscillators are devices with no external inputs, but which generate repetitive or periodic waveforms such as sinusoids, square waves, and triangular waves. Although an oscillator may seem to be just another electronic device, its importance should not be underestimated. Specifically, even if an electronic system does not contain some type of oscillator, it almost assuredly will be used in conjunction with a system that does. Among the major systems that require oscillators are computers, digital wristwatches, oscilloscopes, radio receivers, television sets, and various types of measuring instruments.

A major aspect of the field of communications is the concept of modulation. By changing or modulating a characteristic of a waveform, it is possible to transmit information efficiently and economically from one place to another. Although the most familiar methods of communication, such as radio and television, typically utilize amplitude modulation (AM) and frequency modulation (FM), there are numerous modulation schemes and a wide variety of applications.

10.1   Integrated-Circuit Amplifiers

We begin our study of IC amplifiers by considering the BJT dc amplifier shown in Fig. 10.1. Here we assume that Q_1 and Q_2 have been fabricated with identical characteristics—a typical assumption for two transistors on the same IC chip. This configuration, known as a **differential pair**, has two ac input voltages v_1 and v_2. For the corresponding dc-equivalent circuit, shown in Fig. 10.2, by symmetry Q_1 and Q_2 have identical currents and voltages. By KVL,

$$R_s i_B + v_{BE} + 2(i_B + i_C)R_E - V_{EE} = 0 \tag{10.1}$$

Since for active-region operation, $i_C = h_{FE}i_B$ and $i_B + i_C = i_B + h_{FE}i_B = (1 + h_{FE})i_B \approx h_{FE}i_B = i_C$, then $i_B = i_C/h_{FE}$ and Eq. 10.1 becomes

$$R_s \frac{i_C}{h_{FE}} + v_{BE} + 2R_E i_C - V_{EE} \approx 0 \quad \Rightarrow \quad i_C \approx \frac{V_{EE} - v_{BE}}{(R_s/h_{FE}) + 2R_E}$$

There is, however, an alternative way to establish bias currents in Q_1 and Q_2. In place of R_E, another transistor Q_3 and its associated resistors R_1, R_2, and R_3 are used

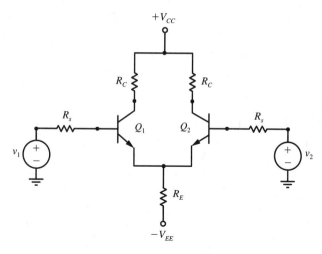

Fig. 10.1 Differential-pair configuration.

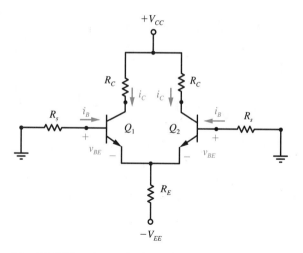

Fig. 10.2 The dc-equivalent circuit.

as illustrated in Fig. 10.3. The reason why this approach is taken in practice will be discussed shortly. First, though, let us concentrate on Q_3 and its associated resistors. By Thévenin's theorem, we can replace the voltage divider consisting of R_1 and R_2 by a voltage source $v_{oc} = -R_1 V_{EE}/(R_1 + R_2) = -KV_{EE}$ connected in series with a resistance $R_p = R_1 \parallel R_2 = R_1 R_2/(R_1 + R_2)$ as shown in Fig. 10.4. Applying KVL, we get

$$-KV_{EE} = R_p i_{B3} + v_{BE3} + R_3(i_{B3} + i_{C3}) - V_{EE}$$

Since for active-region operation $i_{C3} = h_{FE3} i_{B3} \approx i_{B3} + i_{C3}$, then

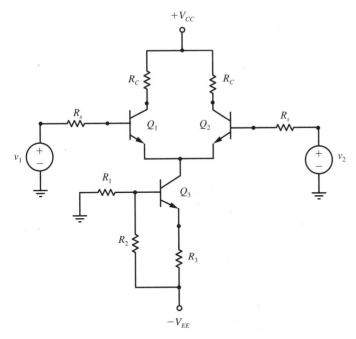

Fig. 10.3 Alternative differential-pair configuration.

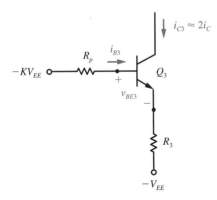

Fig. 10.4 Result of applying Thévenin's theorem to the base circuit of Q_3.

$$V_{EE} - KV_{EE} - v_{BE3} \approx R_p \frac{i_{C3}}{h_{FE3}} + R_3 i_{C3}$$

from which

$$i_{C3} \approx \frac{(1 - K)V_{EE} - v_{BE3}}{(R_p/h_{FE3}) + R_3} = I_O \tag{10.2}$$

where $K = R_1/(R_1 + R_2)$. However, since $i_{C3} \approx 2i_C$, then $i_C \approx i_{C3}/2 = I_O/2$. In other words, the collector currents of Q_1 and Q_2 are determined by Q_3 and its associated components. Thus R_1, R_2, R_3, Q_3, and $-V_{EE}$ combine to act as a constant-current (dc) source I_O. For this reason, the differential pair given in Fig. 10.3 is often drawn as shown in Fig. 10.5.

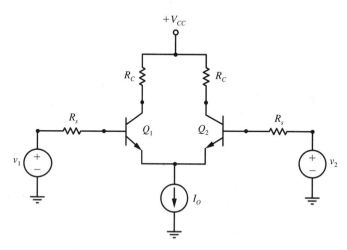

Fig. 10.5 Differential pair with current-source biasing.

Differential-Amplifier AC Analysis

To perform an ac analysis of the differential pair given in Fig. 10.5, we first consider its ac-equivalent circuit shown in Fig. 10.6. Note that because I_O is a dc (constant-current) source, in ac terms it is a current source of value zero (i.e., an open circuit—see Problem 10.1 at the end of this chapter). We proceed by using the principle of superposition.

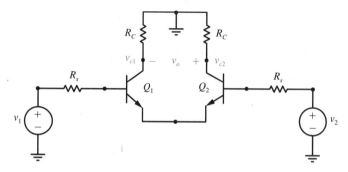

Fig. 10.6 The ac-equivalent circuit of the differential pair with current-source biasing.

To find the response due solely to v_1, set v_2 equal to zero, that is, replace voltage source v_2 with a short circuit. Also, replace Q_1 and Q_2 by the small-signal BJT model given in Fig. 9.12 on p. 579. The resulting circuit is shown in Fig. 10.7. By KVL, we have

$$v_1 = R_s i_{b1} + (1 + h_{fe}) r_e i_{b1} + (1 + h_{fe}) r_e i_{b1} - R_s i_{b2} \tag{10.3}$$

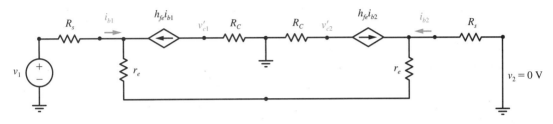

Fig. 10.7 Differential pair ac-equivalent circuit including small-signal BJT model.

Applying KCL at the node where both resistances labeled r_e meet, we get that $i_{b1} + h_{fe} i_{b1} + i_{b2} + h_{fe} i_{b2} = 0$, from which we deduce that $i_{b2} = -i_{b1}$. Substituting this fact into Eq. 10.3, we obtain

$$v_1 = 2R_s i_{b1} + 2(1 + h_{fe}) r_e i_{b1} = 2[R_s + (1 + h_{fe}) r_e] i_{b1} \tag{10.4}$$

However, $v'_{c1} = -R_C h_{fe} i_{b1}$. Thus $i_{b1} = -v'_{c1}/h_{fe} R_C$ and substituting this into Eq. 10.4 yields

$$v'_{c1} = \frac{-h_{fe} R_C}{2[R_s + (1 + h_{fe}) r_e]} v_1 \approx \frac{-h_{fe} R_C}{2(R_s + h_{fe} r_e)} v_1$$

Since $v'_{c2} = -R_C h_{fe} i_{b2}$, then $i_{b1} = -i_{b2} = v'_{c2}/h_{fe} R_C$. Substituting this fact into Eq. 10.4 results in

$$v'_{c2} \approx \frac{h_{fe} R_C}{2(R_s + h_{fe} r_e)} v_1$$

Next consider the case that $v_1 = 0$ V and $v_2 \neq 0$ V. By the symmetry of the differential pair, we deduce that the resulting ac collector voltages are

$$v''_{c1} \approx \frac{h_{fe} R_C}{2(R_s + h_{fe} r_e)} v_2 \quad \text{and} \quad v''_{c2} \approx \frac{-h_{fe} R_C}{2(R_s + h_{fe} r_e)} v_2$$

Hence by the principle of superposition,

$$v_{c1} = v'_{c1} + v''_{c1} \approx \frac{h_{fe}R_C}{2(R_s + h_{fe}r_e)}(v_2 - v_1) = \frac{A}{2}(v_2 - v_1)$$

and

$$v_{c2} = v'_{c2} + v''_{c2} \approx \frac{h_{fe}R_C}{2(R_s + h_{fe}r_e)}(v_1 - v_2) = \frac{A}{2}(v_1 - v_2) \approx -v_{c1}$$

where $A = h_{fe}R_C/(R_s + h_{fe}r_e)$.

If the output v_o of the amplifier is taken between the collectors of Q_1 and Q_2 as indicated in Fig. 10.6—called a **differential output**—then

$$v_o = v_{c2} - v_{c1} = \frac{A}{2}(v_1 - v_2) - \frac{A}{2}(v_2 - v_1) = A(v_1 - v_2) \qquad (10.5)$$

where $A = h_{fe}R_C/(R_s + h_{fe}r_e)$ is called the **differential gain.** Even if the output is chosen to be one of the collector voltages (with respect to the reference), called a **single-ended output,** we see that the output is directly proportional to the difference of the input voltages v_1 and v_2. Because of this, such an amplifier is referred to as a **differential amplifier** or a **difference amplifier.** The input stage of an operational amplifier is typically a differential amplifier.

Note that the differential output of a differential amplifier results in twice as much gain as that for a single-ended output. Also note that a differential amplifier can be used for one single-ended input—we need only connect the other input terminal (base) to the reference, thereby making the second input equal to zero. The output is then directly proportional to the single nonzero input. Specifically, for the case that $v_1 \neq 0$ V and $v_2 = 0$ V, from Eq. 10.5, we have that $v_o = Av_1$. Furthermore, from Eq. 10.4, the resistance R_1 loading the source v_1 is

$$R_1 = \frac{v_1}{i_{b1}} = 2[R_s + (1 + h_{fe})r_e] \approx 2[R_s + h_{fe}r_e] \qquad (10.6)$$

For the case that the signal source resistance $R_s = 0\ \Omega$, the input resistance R_{in} of the differential amplifier is

$$R_{in} = 2(1 + h_{fe})r_e \approx 2h_{fe}r_e \qquad (10.7)$$

In addition to applying single-ended inputs (either one or two at a time) to a differential amplifier, we can also apply a single **differential input** v_d as shown in Fig. 10.8. Since $v_d = v_1 - v_2$, then from Eq. 10.5, we have that $v_o = Av_d$, where

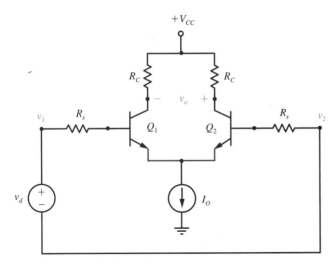

Fig. 10.8 Differential amplifier with a differential-input voltage.

$A = h_{fe}R_C/(R_s + h_{fe}r_e)$. Furthermore, the ac-equivalent circuit of Fig. 10.8 can be obtained from Fig. 10.7 by changing the source v_1 to v_d, erasing the reference symbols at the lower left- and right-hand portions of the circuit, and joining the resulting two loose ends. Proceeding as for the circuit given in Fig. 10.7, we again find that the resistance loading v_d is given by Eq. 10.6, and the input resistance of the differential amplifier (when $R_s = 0\ \Omega$) is given by Eq. 10.7.

Common-Mode Rejection Ratio (CMRR)

When both inputs of a differential amplifier have the same input $v_1 = v_2 = v_{cm}$, called a **common-mode input,** as shown in Fig. 10.9, from Eq. 10.5, the differential-mode output is $v_o = A(v_1 - v_2) = A(v_{cm} - v_{cm}) = 0$ V. This property of a differential amplifier is of great practical importance. Because of environmental or technological factors, quite often a desired signal is subject to interference. Such unwanted signals, referred to as **noise,** often appear in the common mode. As an example, superimposed on the differential output of a differential amplifier can be some ripple (see Section 6.7) due to the amplifier's dc power supply. If the output of the amplifier is connected to the differential input of a second differential amplifier, the ripple will be a common-mode input. Ideally, the second differential amplifier will not amplify the ripple, that is, the second amplifier will reject the ripple.

In practice, a differential amplifier is not perfectly symmetrical. Consequently, when a common-mode input is applied, the output will be nonzero—the smaller the better. The ratio of the gain A_D for a differential-input signal to the gain A_{CM} for a

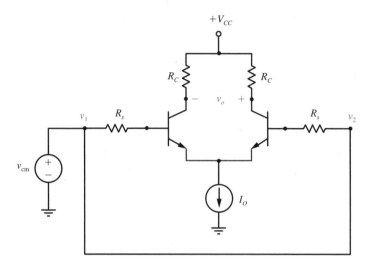

Fig. 10.9 Differential amplifier with a common-mode input.

common-mode input is called the **common-mode rejection ratio** (CMRR) of the amplifier. That is,

$$\text{CMRR} = \left| \frac{A_D}{A_{CM}} \right| \tag{10.8}$$

so the larger the CMRR the better. Normally, the CMRR is expressed in decibels:

$$\text{CMRR}_{\text{dB}} = 20 \log_{10} \left| \frac{A_D}{A_{CM}} \right| \tag{10.9}$$

Previously we mentioned that, in practice, the current-source biasing scheme for a differential amplifier (see Figs. 10.3 and 10.5) was preferred over the use of an emitter resistance R_E (see Fig. 10.1). The main reason for this preference, as we now see, is that the former approach yields a higher CMRR than does the latter. Specifically, consider the differential amplifier given in Fig. 10.1, where $R_s = 0 \ \Omega$. Let us apply a common-mode input v_{cm}. Since the resulting differential output is ideally zero, let us consider the single-ended output v_{c2}. It can be shown (see Problem 10.6 at the end of this chapter) that when $h_{fe} \gg 1$ and $R_E \gg r_e$, then $v_{c2}/v_{cm} \approx -R_C 2R_E$. In addition, if a differential input v_d is applied to the circuit given in Fig. 10.1 with $R_s = 0 \ \Omega$, then it also can be shown (see Problem 10.5 at the end of this chapter) that $v_{c2}/v_d \approx R_C/2r_e$. Thus we deduce that for the differential amplifier given in Fig. 10.1,

$$\text{CMRR} \approx \left| \frac{R_C/2r_e}{-R_C/2R_E} \right| = \frac{R_E}{r_e} \tag{10.10}$$

Therefore, to get a large CMRR, we should try to make R_E large. This, however, reduces the quiescent current and increases r_e.

On the other hand, consider the current-source biasing scheme for a differential amplifier depicted in Fig. 10.3. Although Q_3 and its associated resistors do not form an ideal current source as shown in Fig. 10.5, an accurate model is an ideal current source in parallel with a resistance R_o. Proceeding for the resulting differential amplifier as was done for the amplifier in Fig. 10.1, we deduce that the CMRR for a differential amplifier with current source biasing is

$$\text{CMRR} \approx \frac{R_o}{r_e}$$

where R_o is the output (Norton-equivalent) resistance of the nonideal current source. Since R_o is typically much higher than R_E would have to be to produce the same quiescent current and, hence, the same value of r_e, the CMRR for current-source biasing is generally superior.

Example 10.1

The two outputs of a differential amplifier measure 8 mV rms and 10 mV rms when a common-mode input of 1 V rms is applied. When a differential input of 1 mV rms is applied to the amplifier, the resulting differential output is 75 mV rms. Let us determine the CMRR of this differential amplifier.

The differential output for the common-mode input of 1 V is 10 mV − 8 mV = 2 mV. Thus the common-mode gain has absolute value

$$\left| A_{CM} \right| = \frac{2 \times 10^{-3}}{1} = 2 \times 10^{-3}$$

Furthermore, the absolute value of the differential gain is

$$\left| A_D \right| = \frac{75 \times 10^{-3}}{1 \times 10^{-3}} = 75$$

Thus

$$\text{CMRR} = \left| \frac{A_D}{A_{CM}} \right| = \frac{\left| A_D \right|}{\left| A_{CM} \right|} = \frac{75}{2 \times 10^{-3}} = 37{,}500$$

and

$$\text{CMRR}_{dB} = 20 \log_{10}(37,500) = 91.5 \text{ dB}$$

Drill Exercise 10.1

Given the differential amplifier shown in Fig. 10.1, where $h_{FE} = h_{fe} = 100$, $R_C = R_E = 4.7 \text{ k}\Omega$, $R_s = 0 \text{ }\Omega$, and $V_{CC} = V_{EE} = 10 \text{ V}$, find CMRR_{dB}.

ANSWER 45 dB

Although our discussion has been limited to BJTs, differential amplifiers can also be constructed using FETs (see Problems 10.9 through 10.17 at the end of this chapter). An important property of an FET differential amplifier is its high input resistance.

IC BJT Biasing

In Fig. 10.3 on p. 653, we saw how we could use the current source consisting of R_1, R_2, R_3, and Q_3 to bias a differential pair of transistors. However, on ICs, transistors take up much less space than do resistors. Consequently, it is more advantageous to use fewer resistors at the cost of more transistors. Figure 10.10a shows a more sophisticated current source, known as a **current mirror,** which can be employed to bias a differential amplifier.

For this circuit, it is assumed that Q_1 and Q_2 have identical characteristics. Since $v_{BE1} = v_{BE2}$, then $i_{C1} = i_{C2}$ (see Eq. 7.6), and for the case that Q_1 and Q_2 are biased in the active region, we have that $i_{C1} = h_{FE}i_{B1}$ and $i_{C2} = h_{FE}i_{B2}$. Thus $i_{B1} = i_{C1}/h_{FE}$ and $i_{B2} = i_{C2}/h_{FE} = i_{B1}$, and by KCL,

$$I_{REF} = i_{C1} + 2i_{B1} = i_{C1} + 2\frac{i_{C1}}{h_{FE}} = \left(1 + \frac{2}{h_{FE}}\right)i_{C1} = \left(1 + \frac{2}{h_{FE}}\right)i_{C2}$$

from which

$$i_{C2} = \frac{I_{REF}}{1 + 2/h_{FE}} = I_O$$

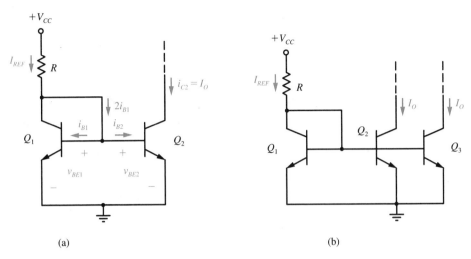

Fig. 10.10 (*a*) A current mirror, and (*b*) a current repeater.

For the typical case that $h_{FE} \gg 2$, then

$$I_O = i_{C2} \approx I_{REF} = \frac{V_{CC} - v_{BE1}}{R}$$

so the biasing current I_O "mirrors" the reference current I_{REF}.

The circuit shown in Fig. 10.10*b*, called a **current repeater,** is an extension of the current mirror shown in Fig. 10.10*a*. For this circuit, all three B-E junctions are connected in parallel so that $v_{BE1} = v_{BE2} = v_{BE3}$, and hence $i_{C1} = i_{C2} = i_{C3}$. For the case that the three BJTs are biased in the active region, then $i_{C1} = h_{FE}i_{B1}$, $i_{C2} = h_{FE}i_{B2}$, and $i_{C3} = h_{FE}i_{B3}$. Thus, $i_{B1} = i_{C1}/h_{FE}$, $i_{B2} = i_{C2}/h_{FE} = i_{B1}$, and $i_{B3} = i_{C3}/h_{FE} = i_{B1}$. Proceeding as above, we obtain

$$I_O = \frac{I_{REF}}{1 + 3/h_{FE}} \approx I_{REF}$$

for $h_{FE} \gg 3$, where $I_{REF} = (V_{CC} - v_{BE1})/R$.

A more intricate current mirror is shown in Fig. 10.11*a*. Again, assume that the BJTs have identical characteristics. Since $v_{BE1} = v_{BE2}$, then $i_{C1} = i_{C2}$, and when Q_1, Q_2, and Q_3 are biased in the active region, then $i_{B1} = i_{C1}/h_{FE}$, $i_{B2} = i_{C2}/h_{FE} = i_{B1}$, and $i_{C3} = h_{FE}i_{B3}$. By KCL,

$$i_{B3} + i_{C3} = i_{B1} + i_{B2} = \frac{2i_{C2}}{h_{FE}} \quad \Rightarrow \quad i_{B3} = \frac{2i_{C2}}{(1 + h_{FE})h_{FE}}$$

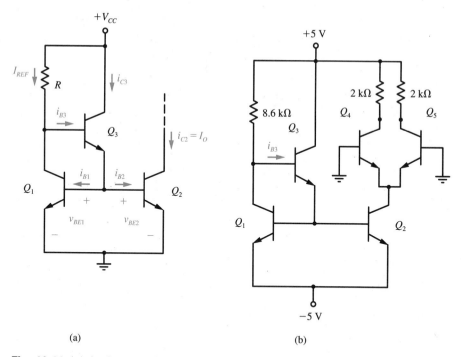

Fig. 10.11 (a) An improved current mirror, and (b) dc-equivalent of a differential amplifier.

But

$$I_{REF} = i_{C1} + i_{B3} = i_{C2} + \frac{2i_{C2}}{(1 + h_{FE})h_{FE}} = \frac{(1 + h_{FE})h_{FE}i_{C2} + 2i_{C2}}{(1 + h_{FE})h_{FE}}$$

from which

$$I_O = i_{C2} = \frac{I_{REF}}{1 + 2/(1 + h_{FE})h_{FE}} \approx \frac{I_{REF}}{1 + 2/h_{FE}^2} \approx I_{REF}$$

since for the typical case that $h_{FE} \gg 1$, then $1 + h_{FE} \approx h_{FE}$ and $1 \gg 2/h_{FE}^2$. For this improved current mirror, $I_{REF} = (V_{CC} - v_{BE1} - v_{BE3})/R$.

Note that by using a third BJT, the current mirror given in Fig. 10.11a yields a better approximation to $I_O \approx I_{REF}$ than does the current mirror depicted in Fig. 10.10a.

Example 10.2

Fig. 10.11b shows the dc-equivalent circuit of a differential amplifier that employs a current mirror similar to the one given in Fig. 10.11a. Let us calculate the power dissipated by this circuit by determining the power supplied by the dc sources labeled +5 V and −5 V.

Assuming that all the BJTs are identical, have $h_{FE} = 50$, and are biased in the active region, then

$$I_{REF} = \frac{5 - v_{C1}}{8600} = \frac{5 - (v_{BE3} + v_{BE1} - 5)}{8600} = \frac{5 - (0.7 + 0.7 - 5)}{8600} = 1 \text{ mA}$$

and from the previous discussion, $i_{C2} = I_O \approx I_{REF} = 1 \text{ mA} \Rightarrow i_{C1} \approx 1 \text{ mA}$. Thus

$$i_{B1} = i_{B2} = \frac{i_{C2}}{h_{FE}} = \frac{1 \times 10^{-3}}{50} = 20 \ \mu\text{A}$$

and $i_{E3} = i_{B1} + i_{B2} = 20 \times 10^{-6} + 20 \times 10^{-6} = 40 \ \mu\text{A}$. But,

$$i_{E3} = i_{B3} + i_{C3} = i_{B3} + h_{FE}i_{B3} = i_{B3} + 50i_{B3} = 51i_{B3} = 40 \ \mu\text{A}$$

from which $i_{B3} = (40 \times 10^{-6})/51 = 0.78 \ \mu\text{A}$ and $i_{C3} = 50i_{B3} = 39 \ \mu\text{A}$. Since

$$i_{B4} + i_{C4} = i_{B4} + 50i_{B4} = 51i_{B4} = \frac{i_{C2}}{2} = \frac{1 \times 10^{-3}}{2} = 0.5 \text{ mA}$$

then $i_{B4} = (0.5 \times 10^{-3})/51 = 9.8 \ \mu\text{A}$ and $i_{C4} = h_{FE}i_{B4} = 50(9.8 \times 10^{-6}) = 0.49 \text{ mA}$. By symmetry, $i_{B5} = i_{B4} = 9.8 \ \mu\text{A}$ and $i_{C5} = i_{C4} = 0.49 \text{ mA}$.

Thus the current directed out of the +5-V power supply is

$$i_{S1} = I_{REF} + i_{C3} + i_{C4} + i_{C5}$$

$$= 1 \times 10^{-3} + 39 \times 10^{-6} + 0.49 \times 10^{-3} + 0.49 \times 10^{-3}$$

$$= 2.02 \text{ mA}$$

so the power supplied by the +5-V source is $P_1 = 5i_{S1} = 5(2.02 \times 10^{-3}) = 10.1 \text{ mW}$.

The current directed into the −5-V power supply is

$$i_{S2} = (i_{B1} + i_{C1}) + (i_{B2} + i_{C2})$$

$$= (20 \times 10^{-6} + 1 \times 10^{-3}) + (20 \times 10^{-6} + 1 \times 10^{-3}) = 2.04 \text{ mA}$$

so the power supplied by the −5-V source is $P_2 = 5i_{S2} = 5(2.04 \times 10^{-3}) = 10.2 \text{ mW}$.

Hence the total power supplied by the two sources is

$$P = P_1 + P_2 = 10.1 \times 10^{-3} + 10.2 \times 10^{-3} = 20.3 \text{ mW}$$

and this is equal to the power that is dissipated by the circuit.

It still remains to confirm that the BJTs are indeed biased in the active region. The details of this are left for the reader as Drill Exercise 10.2.

Drill Exercise 10.2

For the circuit described in Example 10.2, confirm that all the transistors are biased in the active region by finding v_{CE} for each BJT.

ANSWER 1.4 V, 4.3 V, 9.3 V, 4.72 V, 4.72 V

The Widlar Current Source

For the current mirrors just discussed, the output (biasing) current I_O is approximately equal to the reference current I_{REF}. For applications where small biasing currents are required, such circuits require large values for R. However, in ICs, the larger the value of a resistor, the more space is required on a chip. The need of large-valued resistors can be avoided by slightly modifying the current mirror shown in Fig. 10.10a. By connecting a resistor R_E in series with the emitter of Q_2, we get the circuit, known as a **Widlar current source,** shown in Fig. 10.12a.

For an *npn* BJT, we know that for $v_{BE} > 0.2$ V,

$$i_C = I_S(e^{v_{BE}/V_T} - 1) \approx I_S e^{v_{BE}/V_T} \quad \Rightarrow \quad v_{BE} \approx V_T \ln \frac{i_C}{I_S}$$

where $V_T \approx 26$ mV at 300 K (27°C). For the circuit shown in Fig. 10.12a, by KVL,

$$v_{BE1} - v_{BE2} = R_E i_{E2} \tag{10.11}$$

If Q_1 and Q_2 are identical and are in the active region, since typically $h_{FE} \gg 1$ and $i_E = i_B + i_C = i_C/h_{FE} + i_C \approx i_C$, then Eq. 10.11 becomes

$$V_T \ln \frac{i_{C1}}{I_S} - V_T \ln \frac{i_{C2}}{I_S} = R_E i_{C2} \tag{10.12}$$

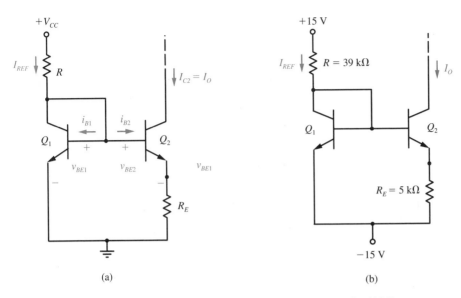

Fig. 10.12 (a) A Widlar current source, and (b) an example of a Widlar current source.

and since $\ln a - \ln b = \ln(a/b)$, then Eq. 10.12 reduces to

$$V_T \ln \frac{i_{C1}/I_S}{i_{C2}/I_S} = V_T \ln \frac{i_{C1}}{i_{C2}} = R_E i_{C2} \tag{10.13}$$

Since $v_{BE2} = v_{BE1} - R_E i_{E2}$, then $v_{BE2} < v_{BE1}$, so $i_{B2} < i_{B1}$. Thus we have that $I_{REF} = i_{B1} + i_{C1} + i_{B2} \approx i_{C1}$ because $i_{C1} \gg i_{B1} > i_{B2}$. Furthermore, by definition, $I_O = i_{C2}$. Hence, from Eq. 10.13,

$$I_O = \frac{V_T}{R_E} \ln \frac{I_{REF}}{I_O} \tag{10.14}$$

where $I_{REF} = (V_{CC} - v_{BE1})/R$.

Suppose we are given values for R, R_E, and V_{CC}. We can also get an approximate value for I_{REF} by using the fact that $v_{BE1} \approx 0.7$ V. Since $V_T \approx 26$ mV (at 300 K), Eq. 10.14 has only one unknown—that being I_O. However, this is a transcendental equation, so there is no closed-form solution. In other words, we cannot obtain a formula for the solution to I_O. Instead, we must determine the value for I_O numerically, as we will now demonstrate.

Example 10.3

Let us find I_O for the Widlar current source shown in Fig. 10.12b. Assuming that Q_1 and Q_2 are identical and in the active region so that $v_{BE1} \approx 0.7$ V, we have that

$$I_{REF} = \frac{15 - (0.7 - 15)}{39 \times 10^3} = 0.751 \text{ mA}$$

Suppose that we guess that the solution to the above transcendental equation is $I_O = 0.1$ mA $= 100$ μA. Substituting this value into the right side of Eq. 10.14 we obtain

$$I_O = \frac{V_T}{R_E} \ln \frac{I_{REF}}{I_O} = \frac{26 \times 10^{-3}}{5 \times 10^3} \ln \frac{0.751 \times 10^{-3}}{0.1 \times 10^{-3}} = 10.5 \text{ μA}$$

Since, under the assumption that $I_O = 100$ μA, we calculated a different value for I_O, it is not the solution to the equation. Therefore, let us now assume the calculated value (10.5 μA) for I_O. Doing so, we obtain

$$I_O = \frac{V_T}{R_E} \ln \frac{I_{REF}}{I_O} = \frac{26 \times 10^{-3}}{5 \times 10^3} \ln \frac{0.751 \times 10^{-3}}{10.5 \times 10^{-6}} = 22.2 \text{ μA}$$

Again, the assumed value of I_O (10.5 μA) is not the same as the calculated value (22.2 μA). But, the difference between the two is much less that it was for the first guess for I_O. Therefore, let us keep calculating new values for I_O until the difference between the assumed and calculated values vanishes. In particular, assuming that $I_O = 22.2$ μA, we calculate

$$I_O = \frac{V_T}{R_E} \ln \frac{I_{REF}}{I_O} = \frac{26 \times 10^{-3}}{5 \times 10^3} \ln \frac{0.751 \times 10^{-3}}{22.2 \times 10^{-6}} = 18.3 \text{ μA}$$

Proceeding in this manner, we calculate the value of I_O sequentially to be 19.3 μA, 19.0 μA, and finally, 19.1 μA. It is the value $I_O = 19.1$ μA that is the solution we seek.

In addition, by KVL we have that

$$v_{BE2} = v_{BE1} - R_E i_{E2} \approx 0.7 - (5 \times 10^3)(19.1 \times 10^{-6}) = 0.605 \text{ V}$$

Our discussion of IC biasing techniques has focused on BJT technology. FET current sources have similar structures, and consequently, this subject matter is covered in the set of problems at the end of this chapter (see Problems 10.23 through 10.25).

10.2 Operational Amplifiers

Perhaps the most important analog IC is the operational amplifier (op amp). As mentioned previously, the first stage of this dc amplifier is typically a differential amplifier. The output of an op amp, however, is usually single ended. By way of a review, the circuit symbol of an op amp is shown in Fig. 10.13a. The input marked with a minus sign (v_1) is the **inverting input,** and the one marked with a plus sign (v_2) is the **noninverting input.** The single-ended output is labeled v_o. Although voltages v_1, v_2, and v_o are measured with respect to the reference so indicated, often the reference is not shown, and, therefore, is implicit. The simplified model of an

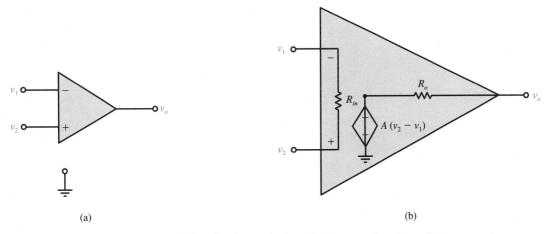

Fig. 10.13 (a) Operational-amplifier circuit symbol, and (b) operational-amplifier model.

op amp is shown in Fig. 10.13b, where A is the gain, R_{in} is the input resistance, and R_o is the output resistance. For an **ideal operational amplifier,** $A = \infty$, $R_{in} = \infty$, and $R_o = 0\ \Omega$.

An actual operational amplifier can be described by the block diagram shown in Fig. 10.14. As already indicated, the first stage of an op amp is a differential amplifier. Since the input resistance of a simple differential amplifier is not necessarily high (e.g., see Eq. 10.7), modified versions of the basic differential amplifier are used, or FETs are used in place of BJTs. Typically, what results is an input resistance of 1 MΩ or more. Following the first stage is a high-gain amplifier that supplies additional gain. The overall gain of an op amp is typically 100,000 (100 dB) or more. After the high-gain amplifier comes a buffer, which is usually an emitter follower, and a level shifter. The level shifter is used so that the output voltage is zero when the input is zero. The last stage of the op amp, the driver, is a large-signal (power) amplifier (this subject was discussed in Section 9.4) with a low output resistance—typically 100 Ω. The driver supplies the output current and voltage.

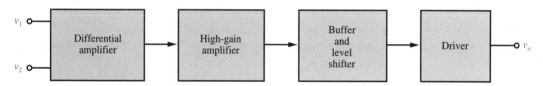

Fig. 10.14 Block diagram of an operational amplifier.

Just as for the case of a differential amplifier, the CMRR of an op amp is the ratio of the differential gain to the common-mode gain (see Eq. 10.8 and Eq. 10.9). The CMRR of an op amp typically ranges between 80 and 100 dB.

The most popular operational amplifier and the industry standard is the 741 op amp shown in Fig. 10.15. This analog (also called linear) IC, which is produced by a number of manufacturers and is available in different types of packages, is comprised of 24 BJTs, 11 resistors, and 1 capacitor. Power is supplied to this device by supply voltages $+V_{CC}$ and $-V_{EE}$. Typically, $V_{CC} = V_{EE} = 15$ V. With such supply voltages, the range of the output voltage v_o is $-10 \le v_o \le 10$ V, or wider. Also, the output of a 741 op amp is **short-circuit protected.** That is, if the output is accidentally "grounded" (i.e., connected to the reference), then the resulting short-circuit output current is limited to 25 mA, and the device will not be destroyed. Typical parameters for the 741 op amp are gain $A = 200{,}000$, input resistance $R_{in} = 2$ MΩ, and output resistance $R_o = 75\ \Omega$.

In addition to gain, input resistance, and output resistance, op amps have other important parameters. We now discuss some of these, citing typical values for the 741 op amp.

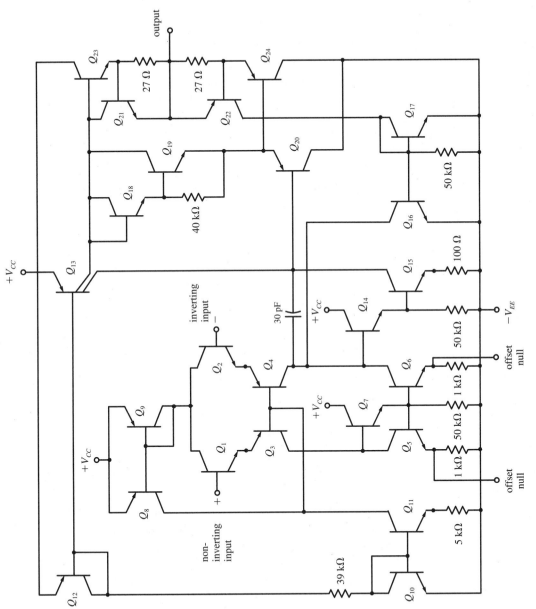

Fig. 10.15 The 741 op amp.

Offset Voltages and Currents

If both input terminals of an op amp are grounded, then $v_1 = v_2 = 0$ V, and ideally, $v_o = 0$ V. However, since an actual op amp is not perfectly symmetrical (or balanced), the output voltage, in general, will be nonzero. This is known as the **output offset voltage** of the op amp. By applying an appropriate voltage V_{os} to the input of the op amp, the output voltage can be made zero. The value $|V_{os}|$ is called the **input offset voltage** of the op amp. For a 741 op amp, the input offset voltage is typically 1.0 mV. By definition, an op amp with a nonzero output offset voltage can be modeled by an op amp with a zero offset voltage and a dc voltage source V_{os} (which can be positive or negative) as shown in Fig. 10.16a. Note that in the circuit diagram of the 741 op amp (Fig. 10.15) there are two terminals (pins) labeled "offset null." These can be used to yield $v_o = 0$ V when $v_1 = v_2 = 0$ V. This is done by connecting a **potentiometer** (an adjustable resistor that is tapped) as shown in Fig. 10.16b. The potentiometer ("pot" for short)—typically 10 kΩ—is adjusted so that $v_o = 0$ V when $v_1 = v_2 = 0$ V.

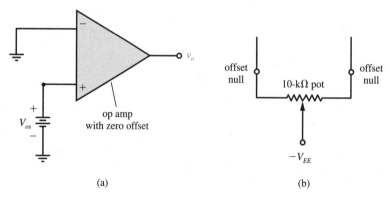

(a)   (b)

Fig. 10.16 (a) Model of an op amp with nonzero offset, and (b) offset-null adjustment.

Since the input stage of an op amp is a differential amplifier, an op amp's inputs are typically the bases of two BJTs. For an op amp to function properly, the input BJTs must be appropriately biased. Thus, there must be a dc path between each base and the reference (ground). (Recall that an ac source acts as a short circuit to dc.) Consequently, in general, there are input (base) bias currents, albeit small currents, in the two input BJTs. The average value of these currents, when the output is zero due to an applied V_{os}, is called the **input bias current** of the op amp. For a 741 op amp, the input bias current is typically 80 nA. The absolute value of the difference between the two input bias currents is known as the **input offset current,** and for a 741 op amp this current is typically 20 nA.

The **slew rate** of an op amp is the maximum rate of change of the output voltage when the input is a large-signal step of voltage. For a 741 op amp, the slew rate is typically 0.5 V/μ. In other words, a change of 20 V in the output voltage due to an ideal step input typically would require $20 \text{ V} \div 0.5 \text{ V}/\mu s = 40 \mu s$.

Example 10.4

If the input voltage for an inverting amplifier (see Fig. 2.15 on p. 75) or a noninverting amplifier (see Fig. 2.16a on p. 75) is set to zero, what results is the op-amp circuit shown in Fig. 10.17a. Let us determine the output offset voltage for this circuit.

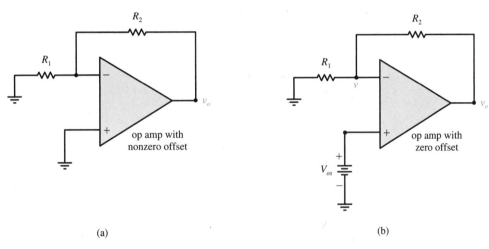

(a)   (b)

Fig. 10.17 (a) Op-amp circuit with zero input, and (b) circuit used to determine the effect of the input offset voltage.

First let us find the output voltage v_o that is the result of the op amp's input offset voltage $|V_{os}|$. As indicated in Fig. 10.16a, let us model the nonideal op amp in Fig. 10.17a by an ideal op amp and a voltage source V_{os} as shown in Fig. 10.17b. (An alternative model is given in Problem 10.26 at the end of this chapter.) Ignoring the effect of the base currents, by KCL, we have that

$$\frac{v}{R_1} + \frac{v - v_o}{R_2} = 0$$

Solving this equation for v_o and using the fact that $v = V_{os}$, we get

$$v_o = \frac{R_1 + R_2}{R_1} v = \frac{R_1 + R_2}{R_1} V_{os} = \left(1 + \frac{R_2}{R_1}\right) V_{os} \qquad (10.15)$$

Next let us determine the effect of the input bias currents $i_{B1} = I_{BQ1}$ and $i_{B2} = I_{BQ2}$ on the output voltage v_o. To do this, we will model the op amp given in

Fig. 10.17a as shown in Fig. 10.18. Since, in essence, $v = 0$ V, there is no current through R_1. This means that the current through R_2 is i_{B1} and

$$v_o = R_2 i_{B1}$$

This output voltage, however, can be reduced by connecting a resistor R_3 to the noninverting input as shown in Fig. 10.19. We proceed by using the principle of superposition.

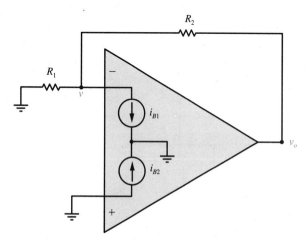

Fig. 10.18 Circuit used to determine the effect of the dc base currents.

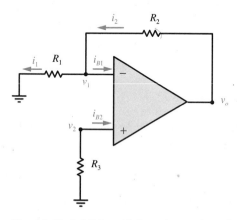

Fig. 10.19 Addition of R_3 reduces the effects of the input bias currents.

Case 1. $i_{B1} \neq 0$ A and $i_{B2} = 0$ A

Since $v_2 = -R_3 i_{B2} = 0$ V, then $v_1 = 0$ V. Thus $i_1 = v_1/R_1 = 0$ A, so $i_2 = i_{B1}$. Therefore, the resulting output voltage is $v_{o1} = R_2 i_2 = R_2 i_{B1}$.

Case 2. $i_{B1} = 0$ A and $i_{B2} \neq 0$ A

Since $v_2 = -R_3 i_{B2} = v_1$, then $i_1 = v_1/R_1 = -R_3 i_{B2}/R_1$. Thus we have that the output voltage v_{o2} is

$$v_{o2} = R_2 i_2 + v_1 = R_2 i_1 + v_1 = -\frac{R_2 R_3}{R_1} i_{B2} - R_3 i_{B2} = -\left(\frac{R_2}{R_1} + 1\right) R_3 i_{B2}$$

Combining the two cases, the output voltage due to the input bias currents is

$$v_o = v_{o1} + v_{o2} = R_2 i_{B1} - \left(\frac{R_2}{R_1} + 1\right) R_3 i_{B2} \tag{10.16}$$

Since i_{B1} and i_{B2} are typically close in value, v_o can be made small by making

$$R_2 = \left(\frac{R_2}{R_1} + 1\right) R_3 \tag{10.17}$$

From this equation, we easily determine that

$$R_3 = \frac{R_1 R_2}{R_1 + R_2} = R_1 \parallel R_2$$

In other words, to minimize the effects of the input bias currents on the output, we select $R_3 = R_1 \parallel R_2$. If we do so, then from Eq. 10.16 and Eq. 10.17, the resulting output voltage is

$$v_o = R_2 i_{B1} - R_2 i_{B2} = R_2(i_{B1} - i_{B2}) = R_2 I_{os} \tag{10.18}$$

where $|I_{os}|$ is the input offset current of the op amp.

Note that even if a resistor R_3 is placed in series with the voltage source V_{os} in Fig. 10.17, Eq. 10.15 will still be valid. Thus, from Eq. 10.15 and Eq. 10.18, the output voltage v_o that results from both the input offset voltage and input offset current of the op amp is (when R_3 is used)

$$v_o = \left(1 + \frac{R_2}{R_1}\right) V_{os} + R_2 I_{os}$$

Since V_{os} and I_{os} each can be positive or negative, the worst-case output offset voltage has an absolute value of

$$|v_o| = \left(1 + \frac{R_2}{R_1}\right) |V_{os}| + R_2 |I_{os}| \tag{10.19}$$

Drill Exercise 10.4

For the circuit given in Fig. 10.19, the op amp has an input offset voltage of 1 mV and input bias currents of 100 nA and 200 nA. Determine R_3 and find the worse-case output offset voltage when $R_1 = 1$ kΩ, and $R_2 = 10$ kΩ.

ANSWER 909 Ω, 12 mV

We have already seen a number of applications for op amps in preceding chapters. They are used to make inverting and noninverting amplifiers (also called scalers), buffers (voltage followers), adders, difference amplifiers, differentiators, integrators, and hence, analog computers, and so on. In subsequent sections of this chapter, we will see additional applications of operational amplifiers.

DC Analysis of the 741 Op Amp

Because of the importance of the 741 op amp, let us analyze the heart of this IC—that being the bias portion and the first two stages of the amplifier. We begin with the dc analysis by replacing the ac input sources with short circuits, indicated in Fig. 10.20 by connecting the bases of both input BJTs (Q_1 and Q_2) to the reference (ground). We also replace the capacitor with an open circuit. In the following analysis, we shall assume that each transistor is biased in the active region and $h_{FE} \gg 1$.

Transistors Q_{10} and Q_{11}, along with the associated 39-kΩ and 5-kΩ resistors, form a Widlar current source. I_O is determined just as was done in Example 10.3 with one slight change. By KVL,

$$15 = v_{EB12} + (39 \times 10^3)I_{REF} + v_{BE10} - 15$$

From this expression, since $v_{EB12} \approx 0.7$ V, we get

$$I_{REF} = \frac{30 - 1.4}{39 \times 10^3} = 0.733 \text{ mA}$$

Proceeding as was done in Example 10.3, we obtain $I_O = 19$ μA.

By symmetry, Q_1 and Q_2 have the same dc collector current I_{CQ}. Since the transistors Q_8 and Q_9 form a current mirror, the collector current for the *pnp* transistor Q_8 is $2I_{CQ}$. Furthermore, the emitter currents for BJTs Q_1 and Q_2 are each

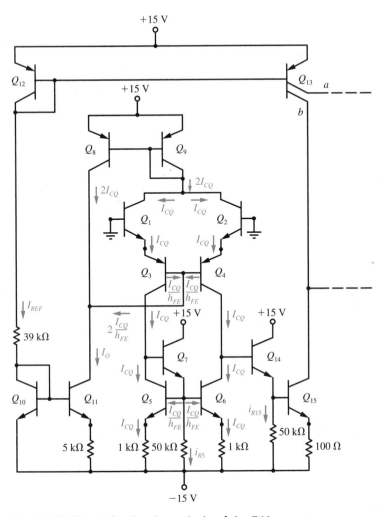

Fig. 10.20 Circuit for the dc analysis of the 741 op amp.

$I_{CQ} + I_{CQ}/h_{FE} \approx I_{CQ}$. Since $i_B + i_C \approx i_C$, the collector current for each of the *pnp* transistors Q_3 and Q_4 is approximately I_{CQ}. The corresponding base currents are I_{CQ}/h_{FE}. Applying KCL to the collector of Q_{11} we get

$$i_{C11} = i_{C8} + i_{B3} + i_{B4}$$

or

$$I_O = 2I_{CQ} + \frac{I_{CQ}}{h_{FE}} + \frac{I_{CQ}}{h_{FE}} = 2\left(I_{CQ} + \frac{I_{CQ}}{h_{FE}}\right) \approx 2I_{CQ}$$

from which we obtain the approximate value

$$I_{CQ} = \frac{I_O}{2} = \frac{19 \times 10^{-6}}{2} = 9.5 \ \mu A$$

If we neglect the base currents of Q_7 and Q_{14} (Q_{16} is part of the short-circuit protection circuitry and is normally OFF), then $i_{C5} = i_{C6} = I_{CQ}$, $i_{B5} = i_{B6} = I_{CQ}/h_{FE}$, and $i_{E5} = i_{E6} \approx I_{CQ}$.

Using the typical saturation current $I_S = 10^{-14}$ A, we get the typical value

$$v_{BE5} = V_T \ln \frac{I_{CQ}}{I_S} = 0.026 \ln \frac{9.5 \times 10^{-6}}{10^{-14}} = 0.537 \ V$$

Thus the current i_{R5} is

$$i_{R5} = \frac{v_{BE5} + 1000 I_{CQ}}{50 \times 10^3} = \frac{0.537 + 1000(9.5 \times 10^{-6})}{50 \times 10^3} = 10.9 \ \mu A$$

Since $i_{B7} + i_{C7} \approx i_{C7}$, by KCL, $i_{C7} = i_{R5} + 2I_{CQ}/h_{FE} = 10.9 \times 10^{-6} + 2(9.5 \times 10^{-6})/h_{FE} \approx 10.9 \ \mu A$. Thus $i_{B7} = i_{C7}/h_{FE} \ll i_{C7} = 10.9 \ \mu A$, and since $I_{CQ} = 9.5 \ \mu A$, then $i_{B7} \ll I_{CQ}$, and ignoring the base current for Q_7 was appropriate.

Note that Q_{12} and Q_{13} also form a current mirror so that the total collector current for Q_{13} is approximately equal to I_{REF}. If Q_{13} is fabricated such that collector a gets 25 percent and collector b gets 75 percent of the total collector current, then

$$i_{C13b} = 0.75 I_{REF} = (0.75)(0.733 \times 10^{-3}) = 0.55 \ mA$$

If we ignore i_{B20}, then $i_{C15} = i_{C13b} = 0.55$ mA. Thus

$$v_{BE15} = V_T \ln \frac{i_{C15}}{I_S} = 0.026 \ln \frac{0.55 \times 10^{-3}}{10^{-14}} = 0.643 \ V$$

and since $i_{B15} + i_{C15} \approx i_{C15}$, then the current i_{R15} is

$$i_{R5} = \frac{v_{BE15} + 100 i_{C15}}{50 \times 10^3} = \frac{0.643 + 100(0.55 \times 10^{-3})}{50 \times 10^3} = 14 \ \mu A$$

For the case that $h_{FE} = 200$, then $i_{B15} = i_{C15}/h_{FE} = (0.55 \ 10^{-3})/200 = 2.75 \ \mu A$. Hence

$$i_{C14} \approx i_{B14} + i_{C14} = i_{B15} + i_{R15} = 2.75 \times 10^{-6} + 14 \times 10^{-6} = 16.8 \ \mu A$$

Since $i_{B14} = i_{C14}/h_{FE} \ll i_{C14} \approx 16.8 \ \mu A$, ignoring the base current of Q_{14} above was appropriate.

AC Analysis of the First Stage

Let us now do the ac analysis for the first stage of a 741 op amp, which, in essence, consists of BJTs Q_1 through Q_{11}.

In the ac-equivalent circuit of the first stage, the $+15$-V supply is replaced by a short circuit. Transistor Q_9 is a diode-connected BJT that can be modeled by its ac emitter resistance r_{e9}. Since the collector of Q_8 is the output of a current mirror (i.e., a dc current source) and the collector of Q_{11} is the output of a Widlar (dc) current source, in the ac-equivalent circuit these collectors are treated as open circuits. The resulting ac-equivalent circuit is shown in Fig. 10.21 for the case that a voltage source v_d is applied as a differential input. Figure 10.22 shows the ac-equivalent circuit that results when each BJT is replaced by the small-signal BJT model depicted in Fig. 9.12 on p. 579.

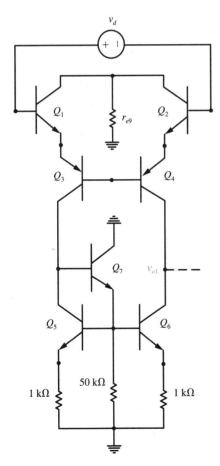

Fig. 10.21 First-stage ac-equivalent circuit of a 741 op amp.

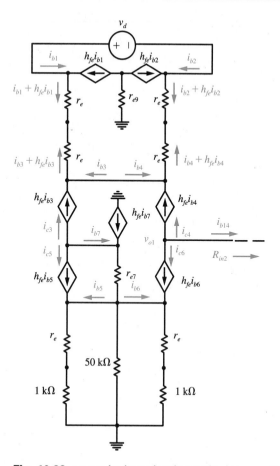

Fig. 10.22 ac-equivalent circuit employing small-signal BJT model.

By KCL, we have that

$$i_{b1} + h_{fe}i_{b1} + i_{b3} + h_{fe}i_{b3} = 0 \quad \Rightarrow \quad i_{b1} = -i_{b3}$$

and

$$i_{b2} + h_{fe}i_{b2} + i_{b4} + h_{fe}i_{b4} = 0 \quad \Rightarrow \quad i_{b2} = -i_{b4}$$

In addition,

$$i_{b3} + i_{b4} = 0 \quad \Rightarrow \quad i_{b4} = -i_{b3} = i_{b1} = -i_{b2}$$

By KVL,

$$v_d = r_e(1 + h_{fe})i_{b1} - r_e(1 + h_{fe})i_{b3} + r_e(1 + h_{fe})i_{b4} - r_e(1 + h_{fe})i_{b2}$$

$$= 4r_e(1 + h_{fe})i_{b1}$$

Thus the ac input resistance of the op amp is

$$R_{in} = \frac{v_d}{i_{b1}} = 4(1 + h_{fe})r_e$$

where $r_e = V_T/I_{CQ}$. Since we have already determined that $I_{CQ} = 9.5$ μA, then for the case that $V_T = 26$ mV and $h_{fe} = 200$, we obtain

$$R_{in} = 4(1 + h_{fe})r_e = 4(1 + 200) \frac{26 \times 10^{-3}}{9.5 \times 10^{-6}} = 2.2 \text{ M}\Omega$$

If we ignore i_{b7} (use reasoning as was done for the dc analysis), then by symmetry,

$$i_{c6} = i_{c5} = -i_{c3}^{*} = -h_{fe}i_{b3} = h_{fe}i_{b1}$$

Thus, the output voltage v_{o1} of the first stage is

$$v_{o1} = R_{in2}i_{b14} = R_{in2}(-i_{c4} - i_{c6}) = R_{in2}(-h_{fe}i_{b4} - h_{fe}i_{b1})$$

$$= -2h_{fe}R_{in2}i_{b1} = -2h_{fe}R_{in2} \frac{v_d}{R_{in}}$$

where R_{in2} is the ac input resistance of the second stage of the op amp. Therefore, since the gain of the first stage can be written as

$$\frac{v_{o1}}{v_d} = -\frac{2h_{fe}R_{in2}}{R_{in}} \tag{10.20}$$

let us now determine R_{in2}.

AC Analysis of the Second Stage

The ac-equivalent of the second stage of the op amp is shown in Fig. 10.23, and replacing each BJT with its small-signal model results in the circuit shown in Fig. 10.24. Here, however, we have not treated the collector of Q_{13} simply as an ideal dc current source, but instead used a more accurate model. In particular, if an h-parameter model with $h_{oe} = 1.25 \times 10^{-5}$ ℧ is used, then the resulting ac output resistance is $1/h_{oe} = 1/(1.25 \times 10^{-5}) = 80$ kΩ, as depicted in Fig. 10.24. We have already determined (see Eq. 9.10 on p. 581) that the resistance looking into the base of a BJT whose emitter is connected in series with a resistor R_E is $R_b = (1 + h_{fe})(r_e + R_E)$. Using the previously calculated values of $I_{CQ14} = 16.8$ μA and $I_{CQ15} = 0.55$ mA, we have that

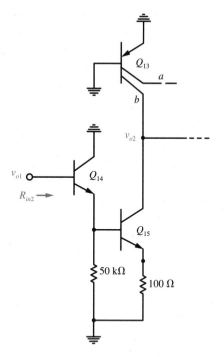

Fig. 10.23 The ac-equivalent circuit of the second stage of a 741 op amp.

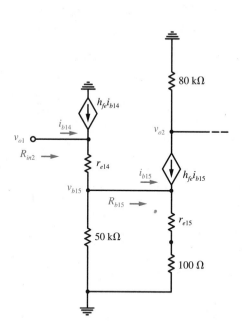

Fig. 10.24 The ac-equivalent circuit incorporating the small-signal BJT model.

$$r_{e14} = \frac{V_T}{I_{CQ14}} = \frac{26 \times 10^{-3}}{16.8 \times 10^{-6}} = 1.55 \text{ k}\Omega$$

and

$$r_{e15} = \frac{V_T}{I_{CQ15}} = \frac{26 \times 10^{-3}}{0.55 \times 10^{-3}} = 47.3 \text{ }\Omega$$

Using the typical value of $h_{fe} = 200$, we get that

$$R_{b15} = (1 + h_{fe})(r_{e15} + R_{E15}) = (1 + 200)(47.3 + 100) = 29.6 \text{ k}\Omega$$

Since this resistance is connected in parallel with 50 kΩ, then $R_{E14} = $ 50 k$\Omega \parallel$ 29.6 kΩ = 18.6 kΩ, and

$$R_{in2} = R_{b14} = (1 + h_{fe})(r_{e14} + R_{E14})$$

$$= (1 + 200)(1.55 \times 10^3 + 18.6 \times 10^3) = 4.05 \text{ M}\Omega$$

Hence from Eq. 10.20, we have that the gain of the first stage of the op amp is

$$\frac{v_{o1}}{v_d} = -\frac{2h_{fe}R_{in2}}{R_{in}} = -\frac{2(200)(4.05 \times 10^6)}{2.2 \times 10^6} = -736$$

Determining the gain of the second stage is a routine matter of circuit analysis for the circuit shown in Fig. 10.24. By KCL, we have that

$$i_{b15} + 200i_{b15} = \frac{v_{b15}}{47.3 + 100} \qquad \Rightarrow \qquad i_{b15} = \frac{v_{b15}}{29.6 \times 10^3}$$

Also, by KCL,

$$\frac{v_{o1} - v_{b15}}{1.55 \times 10^3} = \frac{v_{b15}}{50 \times 10^3} + i_{b15} = \frac{v_{b15}}{50 \times 10^3} + \frac{v_{b15}}{29.6 \times 10^3}$$

Solving this equation for v_{b15} in terms of v_{o1} results in $v_{b15} = 0.923v_{o1}$. Ignoring the load on the second stage, we have that

$$v_{o2} = -(80 \times 10^3)h_{fe}i_{b15} = -(80 \times 10^3)(200)\frac{0.923v_{o1}}{29.6 \times 10^3} = -499v_{o1}$$

so the gain of the second stage is $v_{o2}/v_{o1} = -499 \approx -500$. Hence the gain of the first two stages of the op amp is

$$\frac{v_{o2}}{v_d} = \frac{v_{o1}}{v_d}\frac{v_{o2}}{v_{o1}} = (-736)(-499) = 367,000$$

which is $20 \log_{10} 367,000 = 111$ dB.

10.3 Feedback

When a portion of the output of a circuit or system is returned to the input, the result is known as **feedback.** The consequences of feedback are both useful and important. In this section, we study some of them.

Series-Parallel Feedback

Let us begin by considering the situation depicted in Fig. 10.25. Shown is an amplifier that produces an output voltage v_o across a load R_L. The output voltage is also

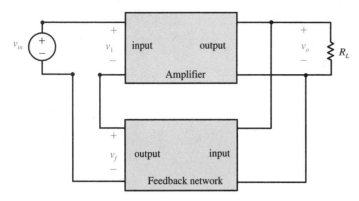

Fig. 10.25 Series-parallel feedback amplifier.

applied to the input of some feedback network (circuit), and the output voltage v_f of the feedback network is returned to the input portion of the amplifier. If the amplifier's voltage gain is $v_o/v_1 = A$, then the voltage gain v_o/v_{in} of the overall circuit, in general, will be some value other than A—what it is depends upon the feedback network.

The overall circuit shown in Fig. 10.25 is an example of a **feedback amplifier,** and it illustrates one of the four basic forms that feedback amplifiers take. In particular, since the output of the feedback network is connected in series with the amplifier input, and the input of the feedback network is connected in parallel with the amplifier output, then this is an example of **series-parallel feedback** (also called **voltage-series feedback**). Since this type of feedback offers the most benefits for voltage-amplification applications (as we shall see), the other forms of feedback will be relegated to the problem section (see Problems 10.36 through 10.44) at the end of this chapter.

An example of a series-parallel feedback amplifier is shown in Fig. 10.26. The amplifier portion is an ideal op amp and the feedback network is a voltage-divider

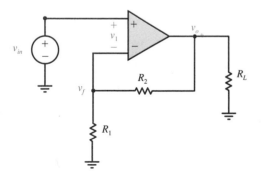

Fig. 10.26 Example of a series-parallel feedback amplifier.

circuit consisting of resistors R_1 and R_2. The voltage that is being amplified by the op amp is $v_1 = v_{in} - v_f$. Since the feedback voltage v_f is subtracted from the overall input voltage v_{in}, this is an example of **negative feedback.**

To find the gain $A_F = v_o/v_{in}$ of the (overall) feedback amplifier, we use the fact that for an ideal op amp, $v_1 = v_{in} - v_f = 0$ V, or $v_{in} = v_f$. Since the inputs of an ideal op amp draw no current, then by voltage division we have

$$v_{in} = v_f = \frac{R_1}{R_1 + R_2} v_o \qquad (10.21)$$

from which

$$A_F = \frac{v_o}{v_{in}} = \frac{R_1 + R_2}{R_1} = 1 + \frac{R_2}{R_1} \qquad (10.22)$$

(Of course, the same result is obtained by a routine application of nodal analysis.) Note that the gain B of the feedback network, which we define as $B = v_f/v_o$, is, by voltage division,

$$B = \frac{v_f}{v_o} = \frac{R_1}{R_1 + R_2} = \frac{1}{A_F} \qquad (10.23)$$

Since the op amp shown in Fig. 10.26 is ideal, the noninverting input draws no current. Therefore, the feedback amplifier (of which the op amp is part) also draws no current and, hence, the feedback amplifier also has an infinite input resistance. The output resistance R_{oF} of the feedback amplifier given in Fig. 10.26 can be determined from the circuit shown in Fig. 10.27. For an ideal op amp, $v_f = v_{in} = 0$ V. Thus, $i_1 = v_f/R_1 = 0$ A, and since the inverting input of the op amp draws no current, then $i_2 = i_1 = 0$ A. Consequently, $v_o = R_1 i_1 + R_2 i_2 = 0$ V and, hence, $R_{oF} = v_o/i_o = 0$ Ω. Therefore, for this example, the feedback network has not altered the input and output resistances.

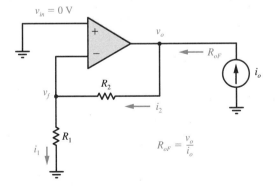

Fig. 10.27 Circuit for determining the output resistance.

Nonideal Op Amp

Now suppose that the op amp shown in Fig. 10.26 is not ideal in that its gain is a finite value A. Let us still assume that its input resistance is $R_{in} = \infty$, and its output resistance $R_o = 0 \, \Omega$ (see Fig. 10.13 on p. 667). Since $v_f = Bv_o$, where $B = R_1/(R_1 + R_2)$ is the gain of the feedback network, then

$$v_o = A(v_{in} - v_f) = Av_{in} - Av_f = AV_{in} - ABv_o$$

Thus

$$v_o + ABv_o = Av_{in}$$

from which the overall gain $A_F = v_o/v_{in}$ of the given feedback amplifier is

$$A_F = \frac{v_o}{v_{in}} = \frac{A}{1 + AB} \tag{10.24}$$

For very large values of A, we have the approximation

$$A_F = \frac{A}{1 + AB} \approx \frac{A}{AB} = \frac{1}{B} = \frac{R_1 + R_2}{R_1}$$

and the gain A_F becomes independent of the op-amp gain. In the limit as $A \to \infty$, this becomes an equality—which is Eq. 10.22, as it should be.

Example 10.5

With the use of Eq. 10.24, we may now demonstrate how negative feedback produces a more stable amplifier gain. For the feedback amplifier given in Fig. 10.26, let $A = 200{,}000$, $R_{in} = \infty$, $R_o = 0 \, \Omega$, $R_1 = 1 \, k\Omega$, and $R_2 = 100 \, k\Omega$. We then have that

$$B = \frac{R_1}{R_1 + R_2} = \frac{1000}{1000 + 100{,}000} = 0.0099$$

From Eq. 10.24, the overall gain of the feedback amplifier is

$$A_F = \frac{200{,}000}{1 + (0.0099)(200{,}000)} = 100.959$$

Now suppose that, because of a change in temperature, the amplifier gain increases 10 percent to $A = 220,000$. The result of this will be that the feedback amplifier gain A_F changes to

$$A_F = \frac{220,000}{1 + (0.0099)(220,000)} = 100.964$$

which is an increase of only 0.005 percent in the gain of the feedback amplifier. The price that we pay for this more stable operation is a reduction of voltage gain from around 200,000 to about 100. Thus in applications where more gain is needed, further stages of amplification would be required.

Drill Exercise 10.5

For the feedback amplifier given in Fig. 10.26, suppose that the op amp has $R_{in} = \infty$, $R_o = 0\ \Omega$, and $A = 100,000$. The gain of this feedback amplifier is to be $A_F = 200$. (a) Determine R_2 when $R_1 = 1\ k\Omega$. (b) Determine R_1 when $R_2 = 50\ k\Omega$.

ANSWER (a) 199 Ω, (b) 251 Ω

Effect of Feedback on Input and Output Resistances

Next, let us investigate what happens to the input resistance R_{iF} of the feedback amplifier given in Fig. 10.26 when the input resistance R_{in} of the op amp is finite. Replacing the op amp with its model (but still assuming that its output resistance is $R_o = 0\ \Omega$), we get the circuit shown in Fig. 10.28. By definition, $R_{iF} = v_{in}/i_1$. However, since $i_1 = v_1/R_{in}$, then

$$R_{iF} = \frac{v_{in}}{i_1} = \frac{v_{in}}{v_1}R_{in} \tag{10.25}$$

From the fact that $v_o = Av_1$, we have that $v_1 = v_o/A$. Substituting this into Eq. 10.25 yields

$$R_{iF} = \frac{v_{in}}{v_o}AR_{in} \tag{10.26}$$

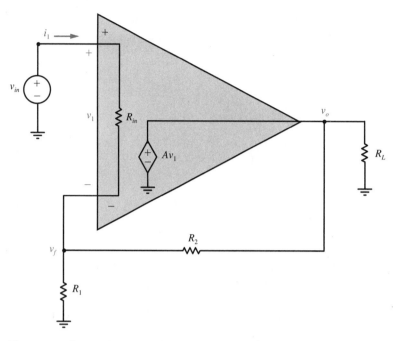

Fig. 10.28 Circuit for determining the input resistance of the feedback amplifier.

For the (typical) case that $R_{in} \gg R_1$, from Eq. 10.24, we have the approximate formula $v_o/v_{in} = A/(1 + AB)$. Substituting this into Eq. 10.26 results in

$$R_{iF} = (1 + AB)R_{in} \qquad (10.27)$$

Thus we see that the input resistance R_{iF} of the feedback amplifier is the product of the input resistance R_{in} of the op amp and the factor $(1 + AB)$. This, therefore, shows how negative feedback can be used to increase the input resistance of an amplifier.

Now let us consider the output resistance R_o of the op amp in Fig. 10.26, and see how it affects the output resistance R_{oF} of the feedback amplifier. To determine R_{oF}, we replace the op amp with its model and set the input voltage to zero. The resulting circuit is shown in Fig. 10.29, where $R_{oF} = v_o/i_o$. Let us assume that $R_{in} \gg R_1$, which is a typical situation. Then, by voltage division, we get the approximation

$$v_f = \frac{R_1}{R_1 + R_2} v_o = Bv_o$$

Since $v_f = -v_1$, then

$$i_o = i_1 + i_2 = \frac{v_o - Av_1}{R_o} + \frac{v_o - v_f}{R_2} = \frac{v_o + Av_f}{R_o} + \frac{v_o - v_f}{R_2}$$

$$= \frac{v_o + ABv_o}{R_o} + \frac{v_o - [R_1/(R_1 + R_2)]v_o}{R_2} = \frac{1 + AB}{R_o} v_o + \frac{1}{R_1 + R_2} v_o$$

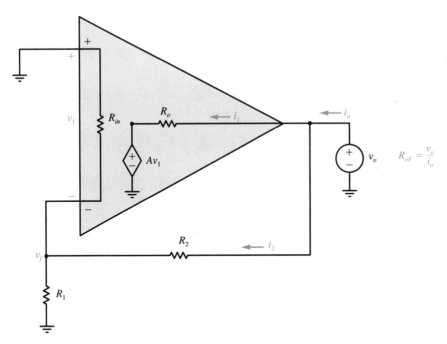

Fig. 10.29 Circuit for determining the output resistance of the feedback amplifier.

Thus

$$\frac{1}{R_{oE}} = \frac{i_o}{v_o} = \frac{1 + AB}{R_o} + \frac{1}{R_1 + R_2}$$

In other words, R_{oF} is equal to the parallel combination of $R_o/(1 + AB)$ and $R_1 + R_2$. That is,

$$R_{oF} = \frac{R_o}{1 + AB} \parallel (R_1 + R_2) \approx \frac{R_o}{1 + AB} \tag{10.28}$$

since, typically, $R_1 + R_2 >> R_o/(1 + AB)$. Whereas negative feedback caused the input resistance of an op amp to be multiplied by the factor $(1 + AB)$, we now see that it causes the output resistance to be divided by that same factor.

Example 10.6

For the feedback amplifier given in Fig. 10.26, suppose that the op amp has $A = 200{,}000$, $R_{in} = 2$ MΩ, and $R_o = 75$ Ω. When $R_1 = 1$ kΩ and $R_2 = 100$ kΩ, then $R_{in} >> R_1$, and

$$B = \frac{R_1}{R_1 + R_2} = \frac{1000}{1000 + 100,000} = 0.0099$$

From Eq. 10.27, the input resistance of the feedback amplifier is

$$R_{iF} = (1 + AB)R_{in} = [1 + (200,000)(0.0099)](2 \times 10^6) = 3960 \text{ M}\Omega$$

while from Eq. 10.28, the output resistance of the feedback amplifier is

$$R_{oF} = \frac{R_o}{1 + AB} = \frac{75}{1 + (200,000)(0.0099)} = 0.038 \ \Omega$$

Drill Exercise 10.6

For the feedback amplifier given in Fig. 10.26, suppose that the op amp has $A = 100,000$, $R_{in} = 1$ MΩ, $R_o = 100$ Ω, $R_1 = 1$ kΩ, and $R_2 = 50$ kΩ. Determine A_F, R_{iF}, and R_{oF}.

ANSWER 51, 1960 MΩ, 0.051 Ω

As mentioned earlier, other forms of feedback are considered in Problems 10.36 through 10.44 at the end of this chapter.

Frequency Response

In our discussion of operational amplifiers prior to this point, we have assumed that a nonideal op amp's gain is some constant A for all frequencies. In actuality, the gain A of an op amp is constant only over a limited range of frequencies. Since it is a dc amplifier, the lower cutoff frequency of an op amp is zero. As frequency increases, the internal capacitances of transistors become more significant. However, the upper cutoff frequency of an op amp can be determined by an appropriately added capacitance. (An example is the 30-pF capacitance shown in the 741 op-amp circuit diagram given in Fig. 10.15 on p. 669.) Furthermore, such a capacitance can produce an op-amp frequency response that is essentially equivalent to a simple low-pass filter (e.g., see Fig. 5.8 on p. 272). Specifically, for such an **internally compensated** op amp, the gain, referred to as the **open-loop gain,** is a function of frequency. Let us denote this function by **A**. Then **A** has the form

$$\mathbf{A} = \frac{A}{1 + j\omega/\omega_H} \tag{10.29}$$

where ω_H is the upper cutoff frequency.

Now suppose that the op amp in the feedback amplifier shown in Fig. 10.26 has an open-loop gain in the form of Eq. 10.29. Thus, from Eq. 10.24, the gain $\mathbf{A}_F$ of the feedback amplifier is also a function of frequency, which is given by

$$\mathbf{A}_F = \frac{\mathbf{A}}{1 + \mathbf{A}B} = \frac{\left[\dfrac{A}{1 + j\omega/\omega_H}\right]}{1 + \left[\dfrac{A}{1 + j\omega/\omega_H}\right]B} \tag{10.30}$$

where $B = R_1/(R_1 + R_2)$. Simplifying Eq. 10.30, we get

$$\mathbf{A}_F = \frac{A}{(1 + AB) + j\dfrac{\omega}{\omega_H}} = \frac{A}{(1 + AB)\left(1 + j\dfrac{\omega}{(1 + AB)\omega_H}\right)}$$

$$= \frac{A}{1 + AB}\frac{1}{1 + j\left[\dfrac{\omega}{(1 + AB)\omega_H}\right]} = \frac{A_F}{1 + j\dfrac{\omega}{\omega_{HF}}}$$

where $A_F = A/(1 + AB)$ is the feedback-amplifier gain for low and middle frequencies (called the **dc gain**), and $\omega_{HF} = (1 + AB)\omega_H$ is the upper cutoff frequency for the feedback amplifier. Therefore, as was indicated in Eq. 10.24, when placed in the feedback configuration shown in Fig. 10.26, the dc gain A of the op amp is divided by the factor $1 + AB$. However, the upper cutoff frequency ω_H (and, hence, the bandwidth) of the op amp is multiplied by the factor $1 + AB$. Thus the use of an op amp as part of a feedback amplifier drastically can increase an amplifier's bandwidth.

Gain-Bandwidth Product

The **gain-bandwidth product** f_T of an op amp is just the product of its dc gain A and its upper cutoff frequency (bandwidth) f_H given in hertz. That is,

$$f_T = Af_H \tag{10.31}$$

Using the 741 op amp typical values of $A = 200,000$ and $f_T = 1$ MHz, we obtain

$$f_H = \frac{f_T}{A} = \frac{1 \times 10^6}{2 \times 10^5} = 5 \text{ Hz} \qquad \Rightarrow \qquad \omega_H = 2\pi f_H = 10\pi = 31.4 \text{ rad/s}$$

If such an op amp is part of the feedback amplifier shown in Fig. 10.26, where $R_1 = 1$ kΩ and $R_2 = 100$ kΩ, then

$$B = \frac{R_1}{R_1 + R_2} = \frac{1000}{1000 + 100,000} = 0.0099$$

and the resulting upper cutoff frequency (bandwidth) is

$$\omega_{HF} = (1 + AB)\omega_H = [1 + (200,000)(0.0099)](10 \pi)$$
$$= 6.22 \times 10^4 \text{ rad/s}$$

or

$$f_{HF} = \frac{\omega_{HF}}{2\pi} = 9.9 \times 10^3 = 9.9 \text{ kHz}$$

By changing the values of R_1 and R_2, we can change the value of the feedback gain $B = R_1/(R_1 + R_2)$, and, hence, the feedback-amplifier dc gain $A_F = A/(1 + AB)$. Although this means that the feedback-amplifier upper cutoff frequency (bandwidth) $\omega_{HF} = (1 + AB)\omega_H = 2\pi f_{HF} = 2\pi(1 + AB)f_H$ will change, the product of the gain and bandwidth of the feedback amplifier is

$$A_F f_{HF} = \frac{A}{1 + AB}(1 + AB)f_H = Af_H = f_T \qquad (10.32)$$

In other words, the gain-bandwidth product of the feedback amplifier is constant and is equal to the gain-bandwidth product of the op amp.

10.4 Sinusoidal Oscillators

In the preceding section, we saw an example of a feedback amplifier (Fig. 10.26) whose gain A_F was given by

$$A_F = \frac{A}{1 + AB}$$

where A is the op-amp gain and B is the feedback-network gain. For the case that the feedback network is a resistive circuit, such as a voltage divider, then $B > 0$ and, hence, $AB > 0$. Therefore, $1 + AB > 1$ and $A_F < A$. When $A_F < A$, we say that the feedback is **negative** (or **degenerative**). On the other hand, when $A_F > A$ due to some other type of feedback network, we call the feedback **positive** (or **regenerative**).

For the feedback amplifier shown in Fig. 10.26, we have seen how negative feedback produced more stability in voltage gain, a higher input resistance, and a lower output resistance. In addition to the usefulness of negative feedback, though, positive feedback can produce desirable results in areas other than voltage amplification.

Consider again the general series-parallel feedback amplifier depicted in Fig. 10.25, where A is the gain of the amplifier and B is the gain of the feedback network. Before an input voltage is applied, the output voltage is $v_o = 0$ V and, hence, the output of the feedback network is $v_f = Bv_o = 0$ V. Therefore, when an input voltage v_{in} is applied, initially, $v_o = A(v_{in} - v_f) = Av_{in}$. This output voltage then gives rise to a nonzero voltage at the output of the feedback network, in particular, $v_f = Bv_o = BAv_{in}$. However, now that v_f is no longer zero, let us suppose that we can immediately make the original input voltage zero. This means that the resulting input to the gain-A amplifier will be $v_1 = v_{in} - v_f = 0 - BAv_{in} = -BAv_{in}$. Therefore, under the condition that $BA = -1$, the gain-A amplifier input will be $v_1 = -BAv_{in} = -(-1)v_{in} = v_{in}$, which is equal to the original input. In other words, the feedback network allows the maintenance of a voltage equal to the original voltage v_{in} at the input of the gain-A amplifier even after the original input voltage is set to zero. Thus also, the same output voltage v_o is maintained in spite of the fact that an external input is no longer applied.

A feedback configuration such as the one just described, which produces a nonzero output when no external input is applied, is called an **oscillator.** We define the **loop gain** of a feedback configuration with no external input to be the product of the individual gains around the loop. In particular, for the configuration given in Fig. 10.25 with the external input v_{in} set to zero, the amplifier gain is $v_o/v_1 = A$ and the feedback-network gain is $v_f/v_o = B$. Note, too, the additional gain $v_1/v_f = -1$. Thus, the loop gain for this case is $-BA$. From the previous discussion, we see that the condition for oscillation is $-BA = 1$, that is, unity loop gain. This condition is known as the **Barkhausen criterion.** Substitution of $BA = -1$ into the feedback-gain formula $A_F = A/(1 + AB)$ indicates that the feedback gain is infinite. This, in turn, justifies the fact that we can have a nonzero output when there is no external input.

Phase-Shift Oscillators

If the amplifier gain A is a constant (not frequency dependent) for the configuration shown in Fig. 10.25, one way to get a unity loop gain is to use a frequency-dependent feedback network having a transfer function **B**, where the magnitude of **B** is $|\mathbf{B}| = 1/A$ and the phase angle (phase shift) of **B** is $\theta = 180$ degrees. If the phase shift of the feedback network is 180 degrees for just a single frequency, then the loop gain will be unity for only one frequency and the oscillator output will be a sinusoid having that particular frequency.

Since a simple RC lead or lag network (which consists of one R and one C) will produce a phase shift of less than 90 degrees for finite frequencies, a minimum of three simple RC stages is required to obtain a 180-degree phase shift. In particular, consider the RC phase-shift network shown in Fig. 10.30a. By KCL, using phasor notation, we have

$$j\omega C(\mathbf{V}_3 - \mathbf{V}_o) + j\omega C(\mathbf{V}_3 - \mathbf{V}_2) + \frac{\mathbf{V}_3}{R} = 0$$

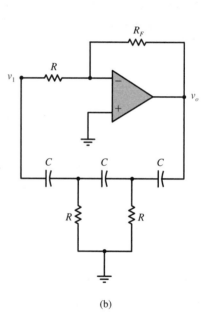

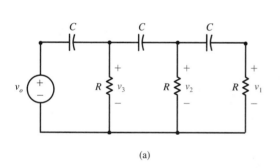

(a)   (b)

Fig. 10.30 (a) An RC phase-shift network, and (b) an op-amp phase-shift oscillator.

as well as

$$j\omega C(\mathbf{V}_2 - \mathbf{V}_3) + j\omega C(\mathbf{V}_2 - \mathbf{V}_1) + \frac{\mathbf{V}_2}{R} = 0$$

and

$$j\omega C(\mathbf{V}_1 - \mathbf{V}_2) + \frac{\mathbf{V}_1}{R} = 0$$

Manipulating these equations eventually results in the voltage transfer function

$$\mathbf{B} = \mathbf{B}(j\omega) = \frac{\mathbf{V}_1}{\mathbf{V}_o} = \frac{\omega^3 R^3 C^3}{\omega RC(\omega^2 R^2 C^2 - 5) + j(1 - 6\omega^2 R^2 C^2)} \tag{10.33}$$

The angle of this transfer function is 180 degrees when $1 - 6\omega^2 R^2 C^2 = 0 \Rightarrow \omega = 1/RC\sqrt{6}$, and substituting the frequency $\omega_0 = 1/RC\sqrt{6}$ into Eq. 10.33, we get

$$B(j\omega_0) = \frac{-1}{29} = \frac{1}{29} \underline{/180°} = |\mathbf{B}| \underline{/\theta}$$

In other words, the gain (the magnitude of the transfer function) of this feedback network is $|\mathbf{B}| = 1/29$ and the phase shift (angle) is $\theta = 180$ degrees when $\omega = 1/RC\sqrt{6} = \omega_0$.

We now can see how the circuit given in Fig. 10.30a can be used as a feedback network to produce an oscillator. Specifically, consider the op-amp circuit shown in Fig. 10.30b. Assuming that the op amp is ideal, the gain v_o/v_1 due to the op amp and resistors R and R_F is $A = -R_F/R$ (see Eq. 2.27 on p. 74). Since the inverting input of the op amp must be at 0 V, then the resistance between node v_1 and the reference is equal to R. In other words, the RC feedback network is loaded with a resistance of value R. Thus the transfer function of the feedback network is given by Eq. 10.33. Hence when

$$\omega = \omega_0 = \frac{1}{RC\sqrt{6}}$$

the loop gain is

$$\mathbf{B}A = -\frac{1}{29}\left(-\frac{R_F}{R}\right) = \frac{R_F}{29R}$$

which is unity when $R_F = 29R$ (i.e., $A = -29$). This, therefore, is the condition for oscillation, and ω_0 is the corresponding frequency of oscillation.

If the loop gain is less than unity, from the previous discussion we know that oscillation will not occur. However, if the loop gain is greater than unity, not only will oscillation occur, but because of positive feedback, the oscillation amplitude will increase with time. Since an actual amplifier is a physical device, its output cannot grow without bound, and eventually the operation of the amplifier will become nonlinear. This, in turn, will limit the oscillation amplitude.

In actual oscillator circuits, because the amplifier portion is not ideal, the theoretical requirement of unity loop gain will not be sufficient to result in oscillation.

Furthermore, an oscillator such as the one shown in Fig. 10.30*b* at no time has an explicit external input. However, when the power supplies are connected, step responses result. When the loop gain is greater than unity, the component at the oscillation frequency of such a response grows in amplitude so as to produce the desired sinusoidal oscillation. Therefore, the loop gain is chosen to be greater than unity—typically by about 5 percent.

In the discussion above, we assumed that the op amp shown in Fig. 10.30*b* was ideal. In particular, we ignored its upper cutoff frequency and, hence, the upper cutoff frequency of the amplifier consisting of the op amp and resistors R and R_F. If the latter cutoff frequency is much greater than the frequency of oscillation, then the same approach can be taken for the case of a nonideal op amp as was taken for an ideal op amp. Yet, higher oscillation frequencies can be obtained if, instead of an op amp, we use a single transistor.

The phase-shift oscillator shown in Fig. 10.31*a* uses the RC phase-shift feedback network described earlier (Fig. 10.30*a*). The amplifier portion of the oscillator consists of a JFET and its associated components. Assuming that $R \gg R_D$ and ignoring

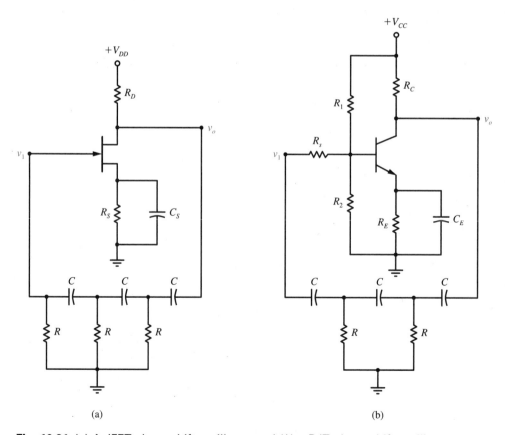

Fig. 10.31 (*a*) A JFET phase-shift oscillator, and (*b*) a BJT phase-shift oscillator.

the ac output resistance r_d of the JFET, the mid-band gain of the amplifier is $A = -g_m R_D$. (To take r_d into consideration, replace R_D by $r_d \parallel R_D$.) Therefore, at the oscillation frequency

$$\omega_0 = \frac{1}{RC\sqrt{6}}$$

setting the loop gain to unity we have

$$\frac{-1}{29}(-g_m R_D) = 1 \quad \Rightarrow \quad g_m = \frac{29}{R_D}$$

Thus, to get oscillation, we must have $g_m \geq 29/R_D$.

In addition to utilizing an FET, a phase-shift oscillator can be constructed by using a BJT as shown in Fig. 10.31b. Unlike an FET, a BJT does not have an (ideally) infinite input resistance. Since the feedback network is to be loaded with a resistance equal to R, we should choose R_s such that the resistance R_{in} between node v_1 and the reference looking into the amplifier is equal to R. In other words, we want to satisfy the equation

$$R_{in} \approx R_s + R_1 \parallel R_2 \parallel h_{fe} r_e = R$$

and since $g_m = 1/r_e$ for a BJT, we wish to have

$$R_s = R - \left(R_1 \parallel R_2 \parallel \frac{h_{fe}}{g_m} \right)$$

If the collector resistor R_C is not ignored, then after a considerable amount of mathematical manipulation, it is found that the oscillation frequency is

$$\omega_0 = \frac{1}{RC\sqrt{6 + 4R_C/R}} .$$

and the minimum value of h_{fe} is $h_{fe} = 23 + 29R_C/R + 4R/R_C$.

The Wien-Bridge Oscillator

A practical op-amp oscillator circuit that sees frequent application is the **Wien-bridge oscillator** shown in Fig. 10.32a. Assuming an ideal op amp, the gain of the amplifier that consists of the op amp and resistors R_1 and R_2, by Eq. 10.22, is

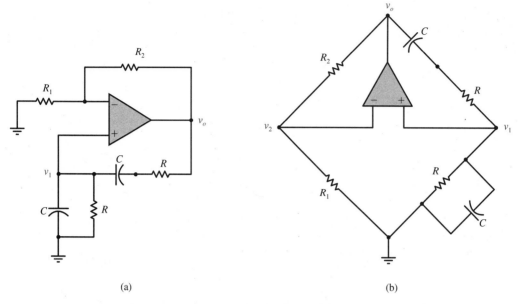

Fig. 10.32 (*a*) Wien-bridge oscillator, and (*b*) equivalent circuit of Wien-bridge oscillator.

$$A = \frac{v_o}{v_1} = \frac{R_1 + R_2}{R_1} = 1 + \frac{R_2}{R_1}$$

For the *RC* feedback network, let us designate the parallel *RC* connection by $\mathbf{Z}_1$ and the series *RC* connection by $\mathbf{Z}_2$. Then

$$\mathbf{Z}_1 = \mathbf{Z}_R \parallel \mathbf{Z}_C = \frac{(R)(1/j\omega C)}{R + 1/j\omega C} = \frac{R}{1 + j\omega RC}$$

and

$$\mathbf{Z}_2 = \mathbf{Z}_R + \mathbf{Z}_C = R + \frac{1}{j\omega C} = \frac{1 + j\omega RC}{j\omega C}$$

Therefore, by voltage division, the voltage transfer function of the feedback network is

$$\mathbf{B} = \mathbf{B}(j\omega) = \frac{\mathbf{V}_1}{\mathbf{V}_o} = \frac{\mathbf{Z}_1}{\mathbf{Z}_1 + \mathbf{Z}_2} = \frac{\dfrac{R}{1 + j\omega RC}}{\dfrac{R}{1 + j\omega RC} + \dfrac{1 + j\omega RC}{j\omega C}}$$

After a few steps of algebraic manipulation, we get

$$\mathbf{B} = \mathbf{B}(j\omega) = \frac{\mathbf{V}_1}{\mathbf{V}_o} = \frac{j\omega RC}{1 - \omega^2 R^2 C^2 + j3\omega RC}$$

Note, however, that when $\omega = \omega_0 = 1/RC$, then $\mathbf{B}(j\omega) = \frac{1}{3}$. Thus for $\omega = \omega_0 = 1/RC$, the loop gain is unity when

$$\mathbf{B}A = \frac{1}{3} A = \frac{1}{3}\left(1 + \frac{R_2}{R_1}\right) = 1$$

that is, when $R_2 = 2R_1$. In summary, the circuit given in Fig. 10.32a oscillates when $R_2 = 2R_1$ (or $R_2 > 2R_1$), and the frequency of oscillation is

$$\omega_0 = 1/RC$$

The circuit given in Fig. 10.32a can be drawn in the equivalent form shown in Fig. 10.32b. From this figure, it is explicitly seen that the resistors and capacitors constitute a configuration known as a **Wien bridge**—hence, the name of the oscillator. Because of the op amp, we must have that $v_1 = v_2$, and for this condition we say that the bridge is **balanced.**

Colpitts and Hartley Oscillators

We may develop other types of oscillators by investigating the general feedback circuit given in Fig. 10.33. This circuit shows an inverting amplifier with input resistance R_{in}, output resistance R_o, and gain A_1. For the sake of simplified calculations, let us assume that the input resistance is infinite—a reasonable assumption when the amplifier is realized with an FET or an op amp.

To determine an expression for the feedback-network transfer function $\mathbf{B} = \mathbf{V}_1/\mathbf{V}_o$, we apply KCL at node v_1. The result is

$$\frac{\mathbf{V}_1}{\mathbf{Z}_1} + \frac{\mathbf{V}_1 - \mathbf{V}_o}{\mathbf{Z}_3} = 0 \quad \Rightarrow \quad \mathbf{B} = \frac{\mathbf{V}_1}{\mathbf{V}_o} = \frac{\mathbf{Z}_1}{\mathbf{Z}_1 + \mathbf{Z}_3}$$

which is the voltage-division formula. Applying KCL at node v_o and using the fact that $\mathbf{V}_1 = \mathbf{B}\mathbf{V}_o$, we get

$$\frac{-A_1\mathbf{V}_1 - \mathbf{V}_o}{R_o} = \frac{\mathbf{V}_o}{\mathbf{Z}_2} + \frac{\mathbf{V}_o - \mathbf{B}\mathbf{V}_o}{\mathbf{Z}_3}$$

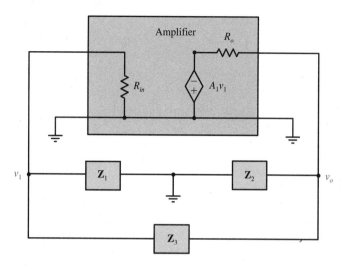

Fig. 10.33 General feedback circuit.

After substituting $\mathbf{B} = \mathbf{Z}_1/(\mathbf{Z}_1 + \mathbf{Z}_3)$ into this equation and doing some manipulation, we obtain the amplifier transfer function

$$\mathbf{A} = \frac{-A_1\mathbf{Z}_2(\mathbf{Z}_1 + \mathbf{Z}_3)}{R_o(\mathbf{Z}_1 + \mathbf{Z}_2 + \mathbf{Z}_3) + \mathbf{Z}_2(\mathbf{Z}_1 + \mathbf{Z}_3)}$$

Hence, the loop gain is

$$\mathbf{AB} = \frac{-A_1\mathbf{Z}_1\mathbf{Z}_2}{R_o(\mathbf{Z}_1 + \mathbf{Z}_2 + \mathbf{Z}_3) + \mathbf{Z}_2(\mathbf{Z}_1 + \mathbf{Z}_3)} \tag{10.34}$$

Suppose we let each impedance $\mathbf{Z}$ be a pure reactance, that is, $\mathbf{Z} = jX$, where $X = \omega L$ for an L-henry inductor or $X = -1/\omega C$ for a C-farad capacitor. Then Eq. 10.34 becomes

$$\mathbf{AB} = \frac{A_1 X_1 X_2}{jR_o(X_1 + X_2 + X_3) - X_2(X_1 + X_3)} \tag{10.35}$$

To obtain oscillation, we require that the loop gain be unity (or greater). To make the imaginary part of the denominator of $\mathbf{AB}$ equal to zero, we set

$$X_1 + X_2 + X_3 = 0 \tag{10.36}$$

From this equation, $X_1 + X_3 = -X_2$, and therefore, the loop gain is

$$\mathbf{AB} = \frac{A_1 X_1 X_2}{-X_2(X_1 + X_3)} = -\frac{A_1 X_1}{X_1 + X_3} = \frac{A_1 X_1}{X_2} \tag{10.37}$$

Setting the loop gain to unity, we find that oscillation occurs when $A_1 = X_2/X_1$ (or more).

For Eq. 10.36 to hold, the reactances can be neither all inductive nor all capacitive. Furthermore, since the loop gain must be positive, and it was assumed implicitly that $A_1 > 0$, then from Eq. 10.37, both X_1 and X_2 must be either capacitive or inductive.

First consider the case that X_1 and X_2 are both capacitive, for example, $\mathbf{Z}_1 = 1/j\omega C_1$ and $\mathbf{Z}_2 = 1/j\omega C_2$. To satisfy Eq. 10.36, let us make X_3 inductive, for example, $\mathbf{Z}_3 = j\omega L$. Thus we have

$$X_1 = -\frac{1}{\omega C_1} \qquad X_2 = -\frac{1}{\omega C_2} \qquad \mathbf{X}_3 = \omega L$$

Therefore, Eq. 10.36 is satisfied when

$$-\frac{1}{\omega C_1} - \frac{1}{\omega C_2} + \omega L = 0$$

or when

$$\omega L = \frac{1}{\omega C_1} + \frac{1}{\omega C_2} = \frac{1}{\omega}\left(\frac{1}{C_1} + \frac{1}{C_2}\right) = \frac{1}{\omega C} \tag{10.38}$$

where we define C by

$$\frac{1}{C} = \frac{1}{C_1} + \frac{1}{C_2} \qquad \Rightarrow \qquad C = \frac{C_2 C_2}{C_1 + C_2}$$

Let us denote the frequency ω that satisfies Eq. 10.38 by ω_0. Then

$$\omega_0 L = \frac{1}{\omega_0 C} \qquad \Rightarrow \qquad \omega_0 = \frac{1}{\sqrt{LC}}$$

where $C = C_1 C_2/(C_1 + C_2)$, and ω_0 is the frequency of oscillation. Note that C is, in effect, the series combination of C_1 and C_2, and ω_0 is the resonance frequency of the parallel combination of L and C. For this case, the circuit given in Fig. 10.33 is known as a **Colpitts oscillator.** An example of a JFET Colpitts oscillator is shown in Fig. 10.34a. Here, the amplifier indicated in Fig. 10.33 consists of a JFET, a gate resistor R_G, a drain resistor R_D, a coupling capacitor C_c, and a power supply voltage V_{DD}.

Suppose that instead of an inductor L, the impedance $\mathbf{Z}_3$ consists of an inductor L connected in series with a capacitor C_3. Then

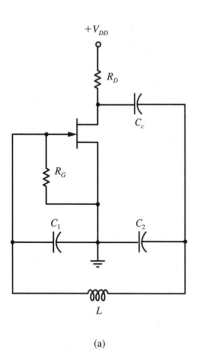 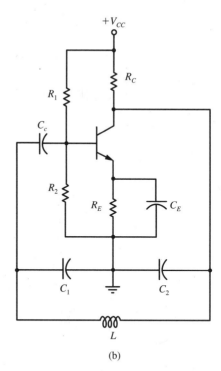

(a)   (b)

Fig. 10.34 (*a*) A JFET Colpitts oscillator, (*b*) a BJT Colpitts oscillator.

$$X_3 = \omega L - \frac{1}{\omega C_3}$$

Substituting this into Eq. 10.36 along with the fact that $X_1 = -1/\omega C_1$ and $X_2 = -1/\omega C_2$, we find that the oscillation frequency is

$$\omega_0 = \frac{1}{\sqrt{LC}}$$

where $1/C = 1/C_1 + 1/C_2 + 1/C_3$. The resulting oscillator is referred to as a **Clapp oscillator.** If $C_1 \gg C_3$ and $C_2 \gg C_3$, then $C \approx C_3$, and the frequency of oscillation is $\omega_0 = 1/\sqrt{LC_3}$. Therefore, by selecting C_1 and C_2 to be much greater than C_3, we can make ω_0 essentially independent of C_1 and C_2. This may be desirable since junction capacitances of transistors change with temperature, and thereby can affect the values of C_1 and C_2. Thus by eliminating the dependence of ω_0 on C_1 and C_2, greater oscillation-frequency stability is achieved.

In the analysis of the general feedback configuration given in Fig. 10.33, we assumed that the input resistance of the amplifier was infinite. This, however, is not a good approximation for a single BJT amplifier. But a similar approach can be taken for the case of a BJT, and the result is that oscillation will occur for sufficient amplifier gain, and the oscillation frequency will be the same as the FET case. An example of a BJT Colpitts oscillator is shown in Fig. 10.34b. Again the oscillation frequency is $\omega_0 = \dfrac{1}{\sqrt{LC}}$, where $C = \dfrac{C_1 C_2}{C_1 + C_2}$.

Finally, let us consider the case that X_1 and X_2 are both inductive, for example, $\mathbf{Z}_1 = j\omega L_1$ and $\mathbf{Z}_2 = j\omega L_2$. Proceeding in a dual manner to the preceding (Colpitts oscillator) approach, we make X_3 capacitive (i.e., $\mathbf{Z}_3 = 1/j\omega C$), and deduce that the frequency of oscillation is $\omega_0 = 1/\sqrt{LC}$, where $L = L_1 + L_2$. This type of oscillator is an example of a **Hartley oscillator.** Figures 10.35a and 10.35b show JFET and BJT Hartley oscillators respectively.

Crystal Oscillators

While RC oscillators can achieve frequency stabilities of about 0.1 percent (e.g., 1000 Hz $\pm$ 1 Hz), stabilities of 0.01 percent are more typical of LC oscillators. Much greater oscillation-frequency stabilities, however, are obtained with oscillators

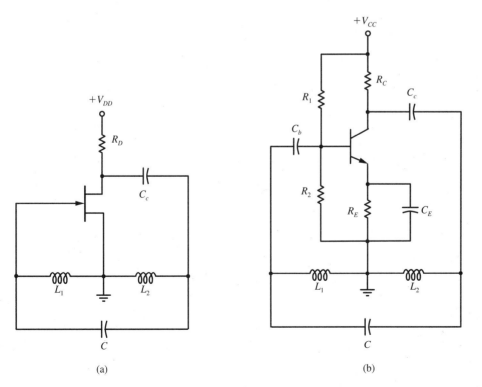

Fig. 10.35 (*a*) A JFET Hartley oscillator, and (*b*) a BJT Hartley oscillator.

that employ crystals—typically quartz. Such **crystal oscillators** can achieve stabilities of 0.001 percent (also denoted as 10 parts per million), and even 0.0001 percent (one part per million). As a consequence, such oscillators are used in digital wristwatches.

Quartz crystals have the property that if they are mechanically vibrated, they produce an ac voltage. Conversely, if an ac voltage is applied across them, they will vibrate at the frequency of the applied voltage. This phenomenon is known as the **piezoelectric effect,** and quartz is an example of a **piezoelectric crystal.**

Although a quartz crystal will vibrate at the frequency of an applied ac voltage, the magnitude of the vibration is a function of frequency. Specifically, the crystal's internal friction, mass, and compliance give rise to a series *RLC* electric equivalent circuit. Further, since a quartz crystal is mounted between two metal plates, what results is a **mounting capacitance** that is connected in parallel with the series *RLC* connection. Shown in Fig. 10.36*a* is the circuit symbol for a quartz crystal (designated XTAL), and Fig. 10.36*b* gives its electric equivalent circuit.

For the circuit shown in Fig. 10.36*b*, the series *LC* combination behaves as a short circuit at its resonance frequency $\omega_0 = 1/\sqrt{LC}$. There is, however, an additional resonance frequency due to the parallel connection of C_m. Since R typically is relatively small, let us ignore it in determining this second resonance frequency. Neglecting R, the admittance of the equivalent circuit shown in Fig. 10.36*b* is

$$
\mathbf{Y} = j\omega C_m + \frac{1}{j\omega L + 1/j\omega C} = \frac{(j\omega C_m)(j\omega L) + (j\omega C_m/j\omega C) + 1}{j\omega L + 1/j\omega C}
$$

$$
= \frac{-\omega^2 L C_m + C_m/C + 1}{j(\omega L - 1/\omega C)} = \frac{j(\omega^2 L C_m - C_m/C - 1)}{\omega L - 1/\omega C}
$$

At the resonance frequency ω_p, we must have that

$$
\omega_p^2 L C_m - \frac{C_m}{C} - 1 = 0
$$

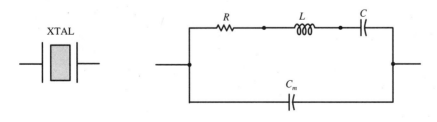

(a) Quartz-crystal symbol　　　　　　　　(b) Equivalent circuit

Fig. 10.36 Quartz crystal (*a*) circuit symbol, and (*b*) equivalent circuit.

from which

$$\omega_p = \sqrt{\frac{1}{L}\left(\frac{1}{C} + \frac{1}{C_m}\right)}$$

Typically, however, for a quartz crystal, $C_m \gg C$, and consequently

$$\frac{1}{C} + \frac{1}{C_m} \approx \frac{1}{C}$$

so that the parallel resonance frequency is

$$\omega_p \approx \sqrt{\frac{1}{LC}} = \frac{1}{\sqrt{LC}} = \omega_s$$

In employing a quartz crystal for an oscillator, we can utilize either its series resonance ω_s or its parallel resonance ω_p—they are approximately equal. In the former case, the crystal behaves as a small (ideally zero) impedance, and in the latter it behaves as a large (ideally infinite) impedance.

One way to employ the series resonance of a quartz crystal is to use it in place of the inductor L in the Colpitts oscillator shown in either Fig. 10.34a or Fig. 10.34b. Another way is indicated by Fig. 10.37a, which shows a JFET **Pierce oscillator.** On

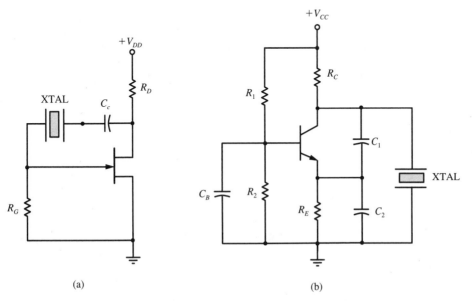

(a)

(b)

Fig. 10.37 (a) A Pierce oscillator, and (b) a modified Colpitts oscillator.

the other hand, the modified BJT Colpitts oscillator shown in Fig. 10.37*b* utilizes the parallel resonance of a quartz crystal.

By electric-circuit standards, the quality factor Q of a quartz crystal is extremely high. For discrete electric-circuit components, a Q of 100 is considered high. However, a quartz crystal with a Q of 10,000 or higher is not uncommon. Even Q values of up to 1,000,000 are possible. These high-Q values for quartz crystals, along with the fact that quartz characteristics are quite stable with respect to temperature and time, mean that crystal oscillators have excellent oscillation-frequency stability.

Quartz crystals with resonance frequencies in the range of 10 kHz to 10 MHz are commonly available. The equivalent circuit, given in Fig. 10.36, of a crystal is actually a simplification of a more complicated model that contains additional series RLC combinations. These give rise to additional resonances, called **overtones,** which are integer multiples of ω_s that is, $2\omega_s$, $3\omega_s$, and so on. By using crystals designed for overtone-mode operation, it is possible to obtain crystal oscillators with oscillation frequencies that go beyond 200 MHz.

10.5 Comparators

In our previous discussions, some type of feedback was always associated with the use of operational amplifiers. There is, however, an important application—albeit nonlinear—for an operational amplifier that does not employ feedback. To demonstrate this, consider the op amp shown in Fig. 10.38*a*, where the supply voltages $+V_{CC}$ and $-V_{CC}$ are explicitly indicated. Let us assume that the gain A of the op amp is very large. (Recall that a typical gain for a 741 op amp is $A = 200{,}000$.) Let us suppose that one of the inputs, the inverting input v_2, is constrained to be some reference voltage $V_r > 0$ V, that is, suppose $v_2 = V_r$. Although the output of the op amp tends to be $v_o = A(v_1 - v_2) = A(v_1 - V_r)$, because of the very large gain, even if v_1 is just slightly larger than V_r, the op-amp output will saturate and (ideally) we

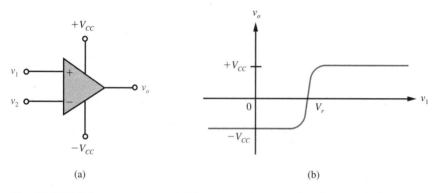

(a)   (b)

Fig. 10.38 (*a*) Comparator, and (*b*) comparator transfer characteristic.

get that $v_o = +V_{CC}$. Similarly, if v_1 is less than V_r, then $v_o = -V_{CC}$. These results are summarized by the transfer characteristic shown in Fig. 10.38b. A nonlinear device such as this is an example of a **comparator.** Specifically, for the comparator shown in Fig. 10.38a, the input voltage v_1 is compared with the reference voltage $v_2 = V_r$; and if $v_1 > V_r$, then $v_o = +V_{CC}$, whereas if $v_1 < V_r$, then $v_o = -V_{CC}$.

For the case that the reference voltage is nonzero ($V_r \neq 0$ V), a comparator is called a **level detector**—if the reference voltage is zero ($V_r = 0$ V) it is referred to as a **zero-crossing detector.** Figure 10.39a shows a zero-crossing detector and Fig. 10.39b shows a sinusoidal input voltage v_1 and the corresponding output voltage v_o. Thus this is an example of a device that converts a sine wave to a square wave.

Although a general-purpose high-gain op amp can be used as a comparator, some op amps are specially designed for such use and are so designated. (These devices have higher speeds than ordinary op amps.) In addition to a single comparator, IC chips that contain two or four independent comparators are on the market.

The Schmitt Trigger

Now suppose that we apply positive feedback to an op amp (comparator) as shown in Fig. 10.40. Such a regenerative comparator is known as a **Schmitt trigger.** When the loop gain is unity or greater, since the feedback is positive (regenerative), the output voltage will saturate at either $+V_{CC}$ or $-V_{CC}$. Which of these two values results depends not only on the value of the input, but also on the value of the output. Let us see why.

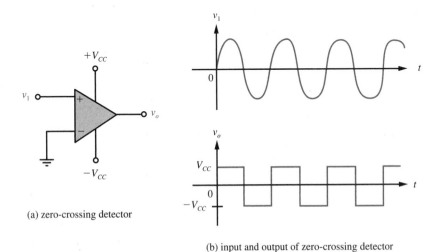

(a) zero-crossing detector

(b) input and output of zero-crossing detector

Fig. 10.39 (*a*) Zero-crossing detector, and (*b*) input and output.

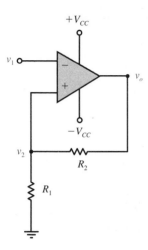

Fig. 10.40 Schmitt trigger.

Assume first that the output voltage is $v_o = +V_{CC}$. Then, by voltage division,

$$v_2 = \frac{R_1}{R_1 + R_2} v_o = BV_{CC}$$

where $B = R_1/(R_1 + R_2)$. If $v_1 < BV_{CC}$, then the op-amp input voltage is $v_2 - v_1 = BV_{CC} - v_1 > 0$ V and the output voltage remains at $+V_{CC}$. Conversely, if $v_1 > BV_{CC}$, then the op-amp input is $v_2 - v_1 = BV_{CC} - v_1 < 0$ V, and the output becomes $-V_{CC}$. The transfer characteristic of the Schmitt trigger for an increasing input voltage v_1 is shown in Fig. 10.41a.

Next assume that the output voltage is $v_o = -V_{CC}$. Then

$$v_2 = \frac{R_1}{R_1 + R_2} v_o = -BV_{CC}$$

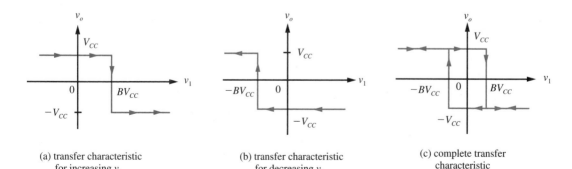

(a) transfer characteristic for increasing v_1

(b) transfer characteristic for decreasing v_1

(c) complete transfer characteristic

Fig. 10.41 Schmitt-trigger transfer characteristic.

If $v_1 > -BV_{CC}$, then the op-amp input is $v_2 - v_1 = -BV_{CC} - v_1 < 0$ V and the output remains at $-V_{CC}$. Conversely, if $v_1 < -BV_{CC}$, then the op-amp input is $v_2 - v_1 = -BV_{CC} - v_1 > 0$ V and the output becomes $+V_{CC}$. The transfer characteristic of the Schmitt trigger for decreasing v_1 is shown in Fig. 10.41b.

Combining the Schmitt-trigger transfer characteristics shown in Figs. 10.41a and b, we get the complete transfer characteristic shown in Fig. 10.41c. Note that an output voltage of $+V_{CC}$ does not change to $-V_{CC}$ until the input voltage increases beyond BV_{CC}, and an output of $-V_{CC}$ does not change to $+V_{CC}$ until the input decreases below $-BV_{CC}$. This phenomenon is known as **hysteresis.** The **width** of the hysteresis is the difference between the two **threshold** (or **triggering**) **voltages** BV_{CC} and $-BV_{CC}$. Thus, in this case, the hysteresis width is $2BV_{CC}$.

The Schmitt trigger is a **bistable** device in the sense that it has two stable states. For one of the stable states, the output is $v_o = +V_{CC}$, whereas for the other stable state, it is $v_o = -V_{CC}$. The circuit will remain in either of its two stable states unless and until it is triggered into its other stable state by the appropriate input voltage— $v_1 > BV_{CC}$ for v_o to change from $+V_{CC}$ to $-V_{CC}$, and $v_1 < -BV_{CC}$ for v_o to change from $-V_{CC}$ to $+V_{CC}$.

Like a zero-crossing detector, a Schmitt trigger can convert a sinusoid (or some other periodic function with evenly spaced zero crossings) into a square wave. In addition to having the ability to "square" a function, a Schmitt trigger has another important use. Specifically, consider the input voltage v_1, which is a sinusoid that is corrupted by noise, given by Fig. 10.42a. If this voltage were applied to a zero-

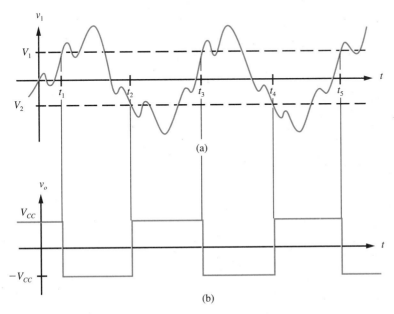

Fig. 10.42 Noise-corrupted input and output voltages for a Schmitt trigger.

crossing detector's input, the output would become $+V_{CC}$ at time $t = 0$ s. However, as a result of the noise, soon thereafter ($0 < t < t_1$) the output would become $-V_{CC}$ and then $+V_{CC}$ again. The same situation could recur later. A Schmitt trigger, on the other hand, offers some noise immunity. For suppose that the voltage v_1 is applied to a Schmitt trigger having an upper threshold of V_1 and a lower threshold of V_2 as indicated in Fig. 10.42a. If the Schmitt-trigger output is $+V_{CC}$, the output will not change to $-V_{CC}$ until v_1 exceeds V_1. As shown, this occurs when $t = t_1$. However, if the output is $-V_{CC}$, it will not change back to $+V_{CC}$ until v_1 becomes less than V_2. Therefore, even though v_1 falls slightly below V_1 just after $t = t_1$, that will in no way cause the output to change from $-V_{CC}$. The output will not change to $+V_{CC}$ until $t = t_2$, at which time v_1 falls below V_2. The Schmitt-trigger output corresponding to the input given in Fig. 10.42a is shown in Fig. 10.42b.

Generalized Schmitt Trigger

The Schmitt trigger given in Fig. 10.40 can be generalized to the one shown in Fig. 10.43. In particular, the former circuit can be obtained from the latter by setting $V_r = 0$ V.

For the circuit shown in Fig. 10.43, by KCL at node v_2,

$$\frac{v_2 - V_r}{R_1} + \frac{v_2 - v_o}{R_2} = 0 \quad \Rightarrow \quad v_2 = \frac{R_1}{R_1 + R_2}v_o + \frac{R_2}{R_1 + R_2}V_r$$

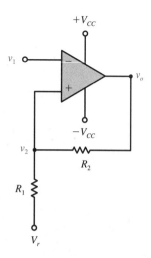

Fig. 10.43 More general Schmitt trigger.

For the case that the output voltage is $v_o = +V_{CC}$, then it will remain at $+V_{CC}$ as long as $v_1 < v_2$, that is, as long as

$$v_1 < v_2 = \frac{R_1}{R_1 + R_2}V_{CC} + \frac{R_2}{R_1 + R_2}V_r = V_1$$

However, the output voltage will change from $+V_{CC}$ to $-V_{CC}$ when $v_1 > v_2$, that is, when

$$v_1 > v_2 = \frac{R_1}{R_1 + R_2}V_{CC} + \frac{R_2}{R_1 + R_2}V_r = V_1$$

Assuming that $0 < V_r < V_{CC}$, then the transfer characteristic describing this behavior is depicted in Fig. 10.44a.

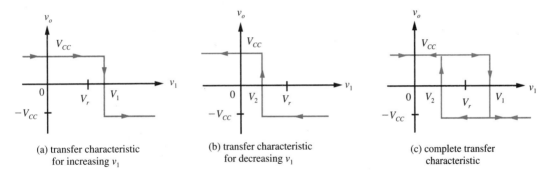

| (a) transfer characteristic for increasing v_1 | (b) transfer characteristic for decreasing v_1 | (c) complete transfer characteristic |

Fig. 10.44 Schmitt-trigger transfer characteristic.

For the case that the output voltage is $v_o = -V_{CC}$, then it will remain at $-V_{CC}$ as long as $v_1 > v_2$, that is, as long as

$$v_1 > v_2 = \frac{R_1}{R_1 + R_2}(-V_{CC}) + \frac{R_2}{R_1 + R_2}V_r = V_2$$

However, the output voltage will change from $-V_{CC}$ to $+V_{CC}$ when $v_1 < v_2$, that is, when

$$v_1 < v_2 = \frac{R_1}{R_2 + R_2}(-V_{CC}) + \frac{R_2}{R_1 + R_2}V_r = V_2$$

Since $V_2 < V_r$ and $V_2 < V_1$, then the transfer characteristic describing this behavior is shown in Fig. 10.44b.

The complete transfer characteristic for the given Schmitt trigger is shown in Fig. 10.44c, where the hysteresis width is

$$V_1 - V_2 = \frac{2R_1}{R_1 + R_2} V_{CC}$$

Since the output v_o is high when the input v_1 is low, and vice versa, then the circuit given in Fig. 10.43 is an example of an **inverting Schmitt trigger.**

Example 10.7

For the Schmitt trigger given in Fig. 10.43, suppose that $R_1 = 1$ kΩ, $R_2 = 9$ kΩ, $V_{CC} = 10$ V, and $V_r = 5$ V. Then from the expressions for V_1 and V_2 above,

$$V_1 = \frac{1000}{1000 + 9000}(10) + \frac{9000}{1000 + 9000}(5) = 5.5 \text{ V}$$

$$V_2 = \frac{1000}{1000 + 9000}(-10) + \frac{9000}{1000 + 9000}(5) = 3.5 \text{ V}$$

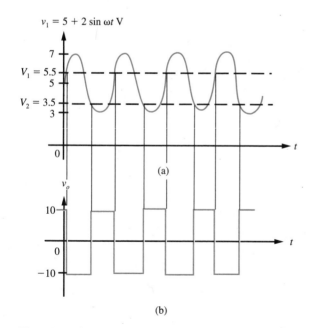

Fig. 10.45 Schmitt-trigger voltages: (a) input, (b) output.

and, as a check, the hysteresis width is

$$V_1 - V_2 = \frac{2R_1}{R_1 + R_2} V_{CC} = \frac{2(1000)}{1000 + 9000}(10) = 2 \text{ V}$$

If the input voltage to the Schmitt trigger is $v_1 = 5 + 2 \sin \omega t$ V as depicted in Fig. 10.45a, then the corresponding output voltage v_o is as shown in Fig. 10.45b.

Drill Exercise 10.7

For the Schmitt trigger given in Fig. 10.43, suppose that $R_1 = 5$ kΩ, $R_2 = 20$ kΩ, $V_{CC} = 12$ V, and $V_r = -5$ V. Find the threshold voltages V_1 and V_2, and the hysteresis width.

ANSWER -1.6 V, -6.4 V, 4.8 V

Relaxation Oscillators

Let us add a resistor R and a capacitor C to the Schmitt trigger in Fig. 10.40, as shown in Fig. 10.46a. Assume that $v_2 > v_1$ and $v_o = +V_{CC}$. By voltage division, we also have that

$$v_2 = \frac{R_1}{R_1 + R_2} v_o = B V_{CC} < V_{CC}$$

where $B = R_1/(R_1 + R_2)$. However, the fact that $v_o = V_{CC}$ means that C will tend to charge up to $+V_{CC}$ with a time constant of $\tau = RC$. Therefore, eventually the voltage v_1 across the capacitor will become larger than $v_2 = B V_{CC}$. When $v_1 > v_2$, then $v_o = -V_{CC}$ and $v_2 = -B V_{CC}$. This, in turn, means that C will then tend to charge to $-V_{CC}$ again with a time constant of $\tau = RC$. Hence eventually v_1 will become more negative than v_2 (i.e., $v_2 > v_1$), and the resulting output will be $v_o = V_{CC}$, putting the circuit back into the originally assumed state. The process repeats, and what results are the waveforms shown in Fig. 10.46b. The device shown in Fig. 10.46a is a nonsinusoidal oscillator whose operation is based on the charging and discharging of a capacitor. We call such a device a **relaxation oscillator.** These types of oscillators are particularly useful at lower frequencies—from 10 Hz to

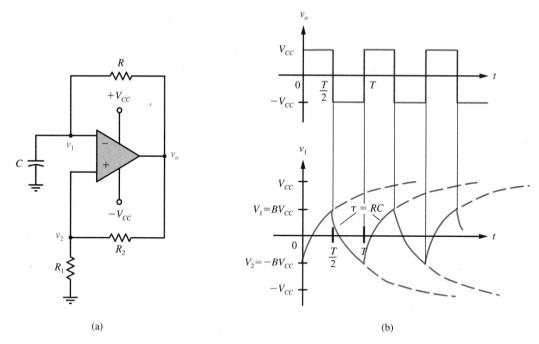

Fig. 10.46 (*a*) Relaxation oscillator, and (*b*) associated waveforms.

10 kHz. At higher frequencies, the square-wave output is far less than ideal due to the slew rate of the op amp (comparator).

As shown in Fig. 10.46*b*, the output voltage of the relaxation oscillator given in Fig. 10.46*a* is (ideally) a square wave, whereas the capacitor voltage is comprised of rising and decaying exponentials. It is from the capacitor voltage that we may determine an expression for the frequency of the oscillator. Specifically, let us consider the half period from $t = 0$ s to $t = T/2$ during which the capacitor charges from $V_2 = -BV_{CC}$ to $V_1 = BV_{CC}$ with a time constant of $\tau = RC$. This corresponds to the application of the voltage V_{CC} to a series RC circuit, where the initial capacitor voltage is $-BV_{CC}$. From Eq. 3.18 on p. 144, the capacitor voltage is

$$v_1(t) = V_{CC} + Ae^{-t/RC} \tag{10.39}$$

Setting $t = 0$ s, we get

$$v_1(0) = V_{CC} + A = -BV_{CC} \quad \Rightarrow \quad A = -BV_{CC} - V_{CC} = -(B + 1)V_{CC}$$

Substituting this value of A into Eq. 10.39, yields

$$v_1(t) = V_{CC} - (B + 1)V_{CC}\, e^{-t/RC} = V_{CC}[1 - (B + 1)e^{-t/RC}]$$

From Fig. 10.46b, we see that $v_1(T/2) = BV_{CC}$. Thus

$$v_1\left(\frac{T}{2}\right) = V_{CC}[1 - (B + 1)e^{-T/2RC}] = BV_{CC} \quad \Rightarrow \quad e^{-T/2RC} = \frac{1 - B}{1 + B}$$

Solving this last equation for the period T, we obtain

$$T = 2RC \ln\frac{1 + B}{1 - B} = 2RC \ln\left(\frac{2R_1}{R_2} + 1\right) \tag{10.40}$$

since $B = R_1/(R_1 + R_2)$. Of course, the frequency (in hertz) of the oscillator is, therefore, $f_0 = 1/T$.

Since the output voltage v_o of the relaxation oscillator is temporarily in one of two stable states ($v_o = +V_{CC}$ and $v_o = V_{CC}$) for half a period and then jumps to the other stable state for the other half period, we say that the device is **astable**. As we have seen, the relaxation oscillator given in Fig. 10.46a, which is also called an **astable multivibrator,** produces a square-wave output voltage and a repetitive exponentially increasing and decreasing capacitor voltage. However, the beginning portion of such an exponential curve approximates a straight-line segment. Thus, by selecting $B = R_1/(R_1 + R_2)$ to be a small quantity, the threshold voltages $V_1 = BV_{CC}$ and $V_2 = -BV_{CC}$ will be relatively small. This, in turn, means that the capacitor voltage will be approximately a triangular waveform. Thus the relaxation oscillator given in Fig. 10.46a can produce an (approximate) triangular wave, as well as a square wave.

A Triangular-Wave Oscillator

There is, however, a way to produce more accurate triangular waveforms. In the circuit shown in Fig. 10.47, the op amp on the left, together with resistors R_1 and R_2, is an example of a **noninverting Schmitt trigger.** Connected to this device is an integrator whose output is fed back to the input of the noninverting Schmitt trigger. (For the sake of simplicity, the power supply voltages $+V_{CC}$ and $-V_{CC}$ are not shown.) Let us see how the entire circuit behaves.

If $v_1 < 0$ V, then $v_2 = -V_{CC}$. Therefore, $i = v_2/R = V_{CC}/R$, and

$$v_o = -\frac{1}{C}\int i\,dt = -\frac{1}{C}\int\frac{-V_{CC}}{R}\,dt = \frac{V_{CC}}{RC}t + K_1 \tag{10.41}$$

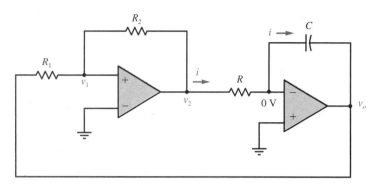

Fig. 10.47 Triangular-wave relaxation oscillator.

Thus we see that the output voltage increases linearly with time. In addition, we have that

$$\frac{v_o - v_1}{R_1} + \frac{v_2 - v_1}{R_2} = 0 \quad \Rightarrow \quad v_1 = \frac{R_2 v_o}{R_1 + R_2} + \frac{R_1 v_2}{R_1 + R_2}$$

Although $v_2 = -V_{CC}$, since v_o is increasing with time, eventually v_1 will become positive, that is, we will get

$$v_1 = \frac{R_2 v_o}{R_1 + R_2} - \frac{R_1 V_{CC}}{R_1 + R_2} > 0 \quad \Rightarrow \quad v_o > \frac{R_1}{R_2} V_{CC}$$

In other words, when $v_o > R_1 V_{CC}/R_2 = V_1$ (the upper threshold voltage), then $v_1 > 0$ V and the Schmitt-trigger output becomes $v_2 = +V_{CC}$. Proceeding as above, the resulting output voltage becomes

$$v_o = -\frac{V_{CC}}{RC} t + K_2$$

which indicates that the output voltage decreases linearly with time. When $v_o < -R_1 V_{CC}/R_2 = V_2$ (the lower threshold voltage), then $v_1 < 0$ V and the process is repeated. The Schmitt-trigger output voltage v_2 and the integrator output voltage v_o are shown in Fig. 10.48.

Having determined the waveforms associated with the oscillator given in Fig. 10.47, it is a simple matter to determine the frequency of oscillation. For the time interval $0 \le t \le T/2$, by Eq. 10.41 the (integrator) output voltage is

$$v_o(t) = -\frac{V_{CC}}{RC} t + V_2 = \frac{V_{CC}}{RC} t - \frac{R_1 V_{CC}}{R_2}$$

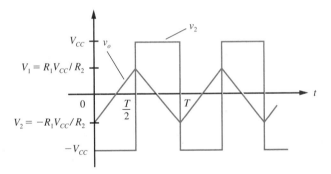

Fig. 10.48 Schmitt-trigger and integrator outputs for triangular-wave oscillator.

Setting $t = T/2$, we get that

$$v_o\left(\frac{T}{2}\right) = -\frac{V_{CC}}{RC}\left(\frac{T}{2}\right) - \frac{R_1 V_{CC}}{R_2} = V_1 = \frac{R_1 V_{CC}}{R_2}$$

Solving this equation for the period T results in

$$T = \frac{4R_1 RC}{R_2} \qquad \Rightarrow \qquad \boxed{\frac{1}{T} = \frac{R_2}{4R_1 RC} = f_0}$$

which is the frequency of oscillation (in hertz).

The 555 IC Timer

Although astable multivibrators (relaxation oscillators) can be constructed using op amps, a popular method for forming such devices is with the use of an IC chip known as a **timer.** The pin connections for the popular 555 timer are shown in Fig. 10.49a. This IC, which contains two comparators, resistors, transistors, and a digital circuit called a "flip-flop," operates as described in the following manner.

The supply voltage v_8 (indicating pin 8) is made $v_8 = V_{CC}$. When an applied **threshold** voltage is $v_6 \geq \frac{2}{3} V_{CC}$, then the **output** voltage v_3 is low ($v_3 \approx 0$ V) and, in addition, a switch (transistor) that is internally connected between **discharge** (pin 7) and ground (pin 1) is turned ON. Conversely, when an applied trigger voltage is $v_2 \leq \frac{1}{3} V_{CC}$, then v_3 is high ($v_3 \approx V_{CC}$) and, in addition, the discharge transistor is turned OFF. When the **reset** voltage v_4 is low, the discharge transistor is ON. However, the reset is disabled when the reset pin is "tied" (directly connected) to the supply voltage.

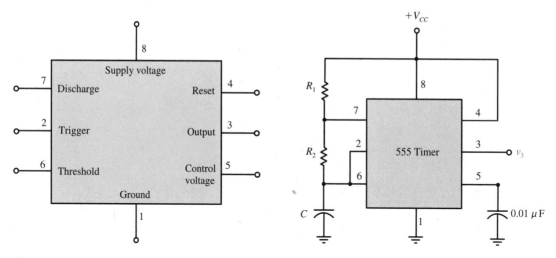

Fig. 10.49 (*a*) The 555 IC timer, and (*b*) an astable multivibrator.

We now demonstrate why the circuit shown in Fig. 10.49*b* is an astable multivibrator. Let us assume that the capacitor C is initially uncharged. Then, since the trigger voltage is $v_2 < \frac{1}{3} V_{CC}$, the output voltage is $v_3 \approx V_{CC}$, and the discharge transistor is OFF. Because the trigger and threshold inputs draw no current (they are connected to comparators), the capacitor C begins to charge with a time constant of $\tau_1 = (R_1 + R_2)C$. When the capacitor voltage reaches the threshold voltage $v_6 = \frac{2}{3} V_{CC}$, then the output voltage becomes $v_3 \approx 0$ V and the discharge transistor turns ON. Thus, C can discharge through R_2 with a time constant of $\tau_2 = R_2C < \tau_1$. After the capacitor discharges to $v_2 = v_6 = \frac{1}{3} V_{CC}$, the output again becomes $v_3 \approx V_{CC}$ and the discharge transistor again goes OFF, allowing the capacitor to recharge. The process repeats and the waveforms shown in Fig. 10.50 result. For the sake of mathematical convenience, as we shall now see, time $t = 0$ s is chosen to be the time when the capacitor initially charges to $\frac{2}{3} V_{CC}$.

In the time interval $0 \le t \le t_1$, from Eq. 3.11 on p. 132, the capacitor voltage $v_C = v_2 = v_6$ is

$$v_C(t) = \tfrac{2}{3} V_{CC} \, e^{-t/R_2 C}$$

At time $t = t_1$, therefore,

$$v_C(t_1) = \tfrac{2}{3} V_{CC} + A e^{-t_1/R_2 C} = \tfrac{1}{3} V_{CC} \quad \Rightarrow \quad t_1 = R_2 C \ln 2$$

However, for $t_1 \le t \le t_2$, as for Eq. 3.18, on p. 144,

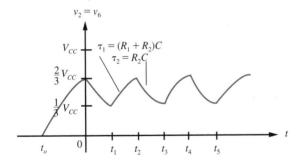

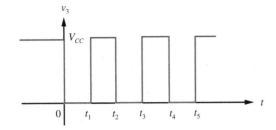

Fig. 10.50 Astable multivibrator capacitor and output voltages.

$$v_C(t) = V_{CC} + Ae^{-(t-t_1)/(R_1+R_2)C}$$

But

$$v_C(t_1) = V_{CC} + A = \tfrac{1}{3} V_{CC} \qquad \Rightarrow \qquad A = -\tfrac{2}{3} V_{CC}$$

so

$$v_C(t) = V_{CC} - \tfrac{2}{3} V_{CC}\, e^{-(t-t_1)/(R_1+R_2)C}$$

At time $t = t_2$, therefore,

$$v_C(t_2) = V_{CC} - \tfrac{2}{3} V_{CC}\, e^{-(t_2-t_1)/(R_1+R_2)C} = \tfrac{2}{3} V_{CC}$$

Solving this equation for t_2, we obtain,

$$t_2 = (R_1 + R_2)C \ln 2 + t_1 = (R_1 + R_2)C \ln 2 + R_2C \ln 2$$

$$= (R_1 + 2R_2)C \ln 2 = 0.693(R_1 + 2R_2)C$$

Since the oscillation period is $T = t_2$, then the frequency of oscillation is

$$f_0 = \frac{1}{T} = \frac{1}{0.693(R_1 + 2R_2)C} = \frac{1.44}{(R_1 + 2R_2)C}$$

The supply voltage for a 555 timer is typically in the range $5 \le V_{CC} \le 16$ V, and the frequency of oscillation usually is between 0.1 Hz and 100 kHz. A 556 IC chip consists of two 555 timers on the same 14-pin package.

A Monostable Multivibrator

Another useful application of the 555 timer is for a circuit known as a **monostable** (or **one-shot**) **multivibrator.** This type of device has one normal (stable) state and one temporary state that is triggered by an appropriate input. After the device leaves its temporary state, it returns to its normal state.

Figure 10.51a shows an example of a monostable multivibrator that employs a 555 timer. The trigger voltage v_2 is the input of the one-shot and v_3 is the output. Given that the input voltage is normally high ($v_3 \approx V_{CC}$) and the output is normally low ($v_3 \approx 0$ V), suppose that the input becomes low for a short period of time. As soon as $v_2 < \frac{1}{3} V_{CC}$, the output goes high ($v_3 \approx V_{CC}$) and the discharge transistor turns OFF. Then C charges with a time constant of $\tau = RC$. If $t = 0$ s is chosen to

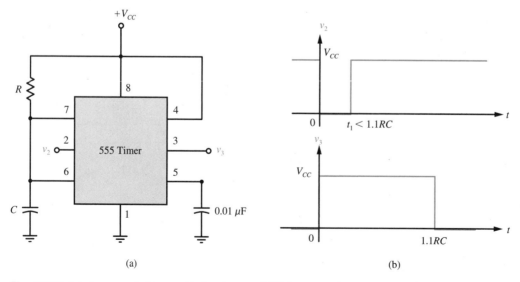

(a) (b)

Fig. 10.51 (a) A monostable multivibrator, and (b) input and output waveforms.

coincide with the time that the input is triggered, then from Eq. 3.18 on p. 144, the capacitor voltage is given by

$$v_6(t) = V_{CC} - V_{CC} \, e^{-t/RC}$$

However, when $v_6 > \frac{2}{3} V_{CC}$, provided that the input is no longer low (otherwise there would be a contradiction), the output goes low, the discharge transistor turns ON, and the capacitor discharges quickly. This means that the output is in its high (temporary) state from time $t = 0$ s until the time when

$$V_{CC} - V_{CC} \, e^{-t/RC} = \frac{2}{3} V_{CC}$$

Solving this equation for t, we get

$$t = RC \ln 3 = 1.099RC \approx 1.1RC$$

Hence the time that the one-shot is in its temporary state is $1.1RC$. Input (v_2) and output (v_3) waveforms for the given monostable multivibrator are shown in Fig. 10.51b.

A Voltage-Controlled Oscillator

In addition to a 555 timer, another IC chip—the 566—is often used as an oscillator. This device produces both a square-wave and a triangular-wave output voltage whose frequency depends on the values of an external resistor R and capacitor C. The pin connections of a 566 IC are shown in Fig. 10.52a. Pin 1 is connected to ground, while pin 8 is connected to the supply voltage $+V_{CC}$. There is no connection (NC) for pin 2, while pin 3 is the square-wave output, and pin 4 is the triangular-wave output. The external resistor R is connected between pin 6 and the supply voltage $+V_{CC}$, and the external capacitor C is connected between pin 7 and ground. Although the frequency of oscillation depends on the values of R and C, it also depends on the voltage v_c, called the **control voltage**, which is applied to pin 5. Specifically, the oscillation frequency f_0 in hertz is given by

$$f_0 = \frac{2}{RC}\left(1 - \frac{v_c}{V_{CC}}\right) \tag{10.42}$$

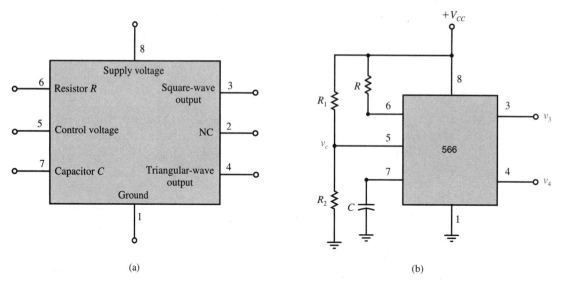

Fig. 10.52 (*a*) The 566 IC chip, and (*b*) an oscillator using a 566 IC chip.

where $v_c < V_{CC}$. Since the frequency of such a device is determined by an applied voltage (in addition to the values of R and C), we refer to it as a **voltage-controlled oscillator** (VCO).

For the oscillator shown in Fig. 10.52*b*, assuming that the control-voltage input draws no current, by voltage division, the control voltage is

$$v_c = \frac{R_2}{R_1 + R_2} V_{CC}$$

and the frequency of oscillation, therefore, is

$$f_0 = \frac{2}{RC}\left(1 - \frac{R_2}{R_1 + R_2}\right) = \frac{2}{RC}\left(\frac{R_1}{R_1 + R_2}\right)$$

Not only can f_0 be changed by changing either R or C; but by varying either R_1 or R_2, the control voltage and hence the frequency, can be varied.

In practical terms, the supply voltage for a 566 IC should be in the range $10 \le V_{CC} \le 24$ V, and the control voltage should be $0.75V_{CC} \le v_c < V_{CC}$. The range of the resistance R should be $2\ \text{k}\Omega \le R \le 20\ \text{k}\Omega$, and the oscillation frequency should be kept below 1 MHz.

Another way of varying the control voltage v_c is to add an ac component to it by connecting an ac signal source to pin 5 via a coupling capacitor. In this way, the frequency of the oscillator will change in accordance with the ac signal. This is an example of a process known as **frequency modulation** (FM). The subject of modulation is discussed in the next section.

10.6   Introduction to Communications

The transmission of information is an extremely important and vital process in a modern technological society. Because of the nature of electromagnetic radiation, however, information in the **audio band** or **spectrum** (20 − 20,000 Hz) cannot be directly transmitted efficiently through the atmosphere or space. Conversely, though, **radio-frequency** (RF) signals (above 20 kHz) can be propagated through space both efficiently and economically. Therefore, a common method for transmitting audio information (and higher-frequency information, too) is to transmit a high-frequency signal, called a **carrier,** which is altered in some way so that it corresponds to the information to be sent. We refer to this as **modulation.** After such a signal is received, the original information is extracted by a process known as **demodulation** or **detection.**

Amplitude Modulation (AM)

Let us now be specific by considering a particular type of modulation that is simple, but still very useful. We begin by selecting the carrier to be a high-frequency sinusoid $v_c(t) = V_c \cos \omega_c t$. For the sake of mathematical convenience, suppose that the audio information to be transmitted is simply a sinusoid $v_m(t) = V_m \cos \omega_m t$, called the **modulating signal,** where we require that the carrier frequency ω_c be much higher than the modulating-signal frequency ω_c (i.e., $\omega_c >> \omega_m$). Suppose, now, that we make the amplitude of the carrier $v_c(t)$ change according to the modulating signal $v_m(t)$. This is known as **amplitude modulation** (AM).

Since the amplitude of a sinusoid is a positive quantity, we cannot simply let the amplitude of the carrier be equal to $V_m \cos \omega_m t$. However, we can let the amplitude of the carrier be

$$V_c + kV_m \cos \omega_m t$$

where k is a positive constant such that $V_c \geq kV_m$. In other words, an AM signal can be written in the form

$$v(t) = (V_c + kV_m \cos \omega_m t) \cos \omega_c t = V_c(1 + \frac{kV_m}{V_c} \cos \omega_m t) \cos \omega_c t$$

$$= V_c(1 + m \cos \omega_m t) \cos \omega_c t \qquad (10.43)$$

where $m = kV_m/V_c$ is called the **modulation index.** Note that when there is no modulating signal $[v_m(t) = V_m \cos \omega_m t = 0]$, then $V_m = 0$ and, hence, $m = 0$. Under this circumstance (zero modulation), the AM signal is simply equal to the carrier, that is, $v(t) = V_c \cos \omega_c t = v_c(t)$. Conversely, since $V_c \geq kV_m$, then $1 \geq kV_m/V_c = m$. Thus the modulation index is bounded by $m \leq 1$. We refer to the case that $m = 1$ as 100 percent modulation. Shown in Fig. 10.53 is an example of a carrier $v_c(t) = V_c \cos \omega_c t$, a modulating signal $v_m(t) = V_m \cos \omega_m t$ (where $\omega_m \ll \omega_c$), and an AM signal $v(t) = V_c(1 + m \cos \omega_m t) \cos \omega_c t$ for the case of roughly 50 percent modulation ($m = 0.5$). The upper boundary for the AM signal is described by $V_c(1 + m \cos \omega_m t)$ and is referred to as the **upper envelope.** Similarly, $-V_c(1 + m \cos \omega_m t)$ is known as the **lower envelope.** The upper and lower envelopes are indicated in Fig. 10.53c by dashed curves.

Although the carrier $v_c(t)$ is an ordinary sinusoid, as is the modulating signal $v_m(t)$, the AM signal $v(t)$ is a sinusoid whose amplitude varies. Yet we can express $v(t)$ in an alternative form as follows. From Eq. 10.43, we have

$$v(t) = V_c \cos \omega_c t + mV_c \cos \omega_c t \cos \omega_m t$$

Using the trigonometric identity $(\cos A)(\cos B) = \frac{1}{2} \cos(A + B) + \frac{1}{2} \cos(A - B)$, we get that

$$v(t) = V_c \cos \omega_c t + \frac{1}{2} mV_c \cos(\omega_c + \omega_m)t + \frac{1}{2} \cos (\omega_c - \omega_m)t \qquad (10.44)$$

Therefore, the AM signal $v(t)$ given by Eq. 10.43 is actually the sum of three (ordinary) sinusoids as demonstrated by Eq. 10.44. One sinusoid is just the carrier $v_c(t) = V_c \cos \omega_c t$ whose amplitude is V_c. Although the other two sinusoids both have amplitudes of $\frac{1}{2} mV_c = \frac{1}{2} kV_m$, one has a frequency of $\omega_c + \omega_m$, and the other a frequency of $\omega_c - \omega_m$. A plot of the amplitudes of the frequency components

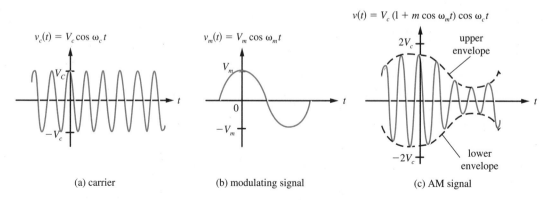

(a) carrier   (b) modulating signal   (c) AM signal

Fig. 10.53 (a) Carrier, (b) modulating-signal, and (c) AM signal.

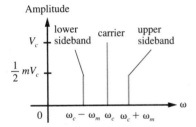

Fig. 10.54 Spectrum of AM signal.

(versus frequency) of the AM signal $v(t)$ is shown in Fig. 10.54a. We refer to such a plot as the **frequency spectrum,** or simply **spectrum,** of $v(t)$. The sinusoid whose frequency is $\omega_c + \omega_m$ is known as the **upper sideband,** and the one whose frequency is $\omega_c - \omega_m$ is the **lower sideband.**

In the preceding discussion, we considered the case that the modulating signal is sinusoidal, that is, $v_m(t) = V_m \cos \omega_m t$. However, now suppose that the modulating signal $v_m(t)$ is nonsinusoidal and has a spectrum that is limited to frequencies between 0 and ω_u as illustrated in Fig. 10.55a, where V_M is the maximum amplitude of the frequency components comprising $v_m(t)$. Then, as in the discussion above, the spectrum of the AM signal is as depicted in Fig. 10.55(b).

For standard AM radio, the bandwidth of the modulating signal is limited to $f_u = \omega_u/2\pi = 5$ kHz. In audio terms, by no means would this be considered "high fidelity." However, because of such a small bandwidth, the carrier frequencies for AM radio stations (which range between 540 and 1600 kHz) are spaced 10 kHz apart.

An example of a single BJT amplitude modulator is shown in Fig. 10.56a. This circuit is, in essence, an amplifier whose input is the relatively high-frequency carrier $v_c(t)$, and whose gain depends upon the relatively low-frequency modulating signal $v_m(t)$. The coupling and bypass capacitors are chosen so that they behave as small impedances (ideally zero) for the carrier frequency and as large impedances (ideally

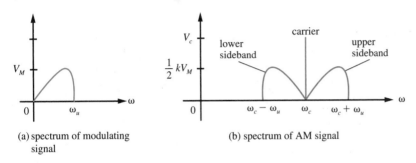

(a) spectrum of modulating
    signal

(b) spectrum of AM signal

Fig. 10.55 (a) Spectrum of modulating signal, and (b) spectrum of AM signal.

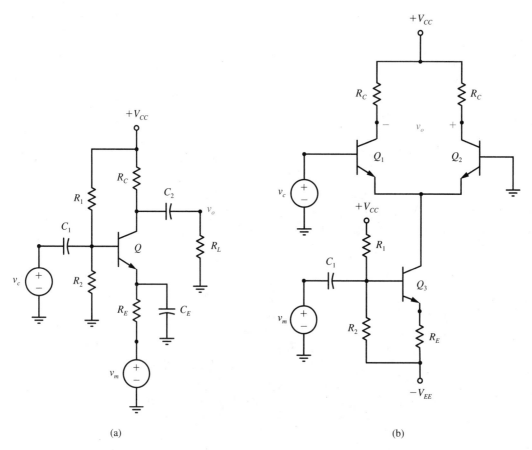

(a)

(b)

Fig. 10.56 (a) A BJT amplitude modulator, and (b) differential-amplifier modulator.

infinite) for the modulating-signal frequencies. Practically speaking, therefore, ω_c should be at least 100 times the value of ω_u. Placing a large modulating signal in series with the emitter will vary r_e and, hence, the gain of the amplifier as $v_m(t)$ varies. The result is an output voltage $v_o(t)$ that is a sinusoid having the carrier frequency ω_c and an amplitude that varies with $v_m(t)$—in other words, an AM signal.

The problem of dealing with amplifier coupling and bypass capacitors, such as the ones shown in Fig. 10.56a, can be eliminated by using the dc-amplifier modulator shown in Fig. 10.56b. Again, the modulating signal will cause the gain of the differential pair to vary accordingly. Although a portion of the modulating signal appears at the collectors of Q_1 and Q_2, because it is in the common mode, this voltage is not present when the output v_o is taken between the two collectors as indicated. The result, as was the case for the circuit given in Fig. 10.56a, is an AM signal.

AM Detection

After an AM signal is transmitted and then received, the original information it contains must be extracted. The process that accomplishes this is called **amplitude demodulation** or **AM detection.** Detecting an AM signal is even less involved than generating it. Specifically, the peak detector given in Fig. 6.56 and repeated in Fig. 10.57*a* is an example of an AM detector, also called an **envelope detector.** If the

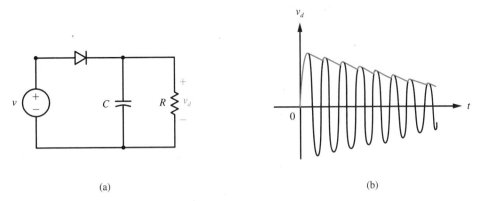

(a)   (b)

Fig. 10.57 (*a*) Envelope detector, and (*b*) output voltage.

input to this circuit is the AM signal $v(t)$ shown in Fig. 10.53*c*, assuming an ideal diode, when $v(t)$ reaches its first peak, the capacitor will charge up to this value. During the time it takes for $v(t)$ to go from its first peak to its second peak, the capacitor voltage will discharge with a time constant of $\tau = RC$. This time constant should be large compared with the period T_c of the carrier. In other words,

$$RC \gg T_c = \frac{1}{f_c} = \frac{2\pi}{\omega_c}$$

However, the time constant $\tau = RC$ should not be too large; otherwise, the capacitor (output) voltage v_d will not be able to follow the successive peaks of the AM signal. In particular, the time constant $\tau = RC$ should be small compared with the period T_m of the modulating signal. In other words,

$$RC \ll T_m = \frac{1}{f_m} = \frac{2\pi}{\omega_m}$$

In the general case where the modulating signal $v_m(t)$ is nonsinusoidal, the time constant should be small compared with the period $T_u = 2\pi/\omega_u$ of the highest frequency component of the modulating signal. Combining the preceding two inequalities, we have that the time constant $\tau = RC$ is chosen such that

$$\frac{2\pi}{\omega_c} << RC << \frac{2\pi}{\omega_m} \tag{10.45}$$

From this constraint, we see that we must have that $\omega_c >> \omega_m$—which was an original condition for amplitude modulation.

When the time constant $\tau = RC$ is chosen according to Inequality 10.45, the output voltage $v_d(t)$ of the peak detector given in Fig. 10.57a is as shown in Fig. 10.57b and will be essentially equal to the upper envelope of the applied AM signal. That is,

$$v_d(t) = V_c(1 + m \cos \omega_m t) = V_c + mV_c \cos \omega_m t$$

$$= V_c + \frac{kV_m}{V_c}V_c \cos \omega_m t = V_c + kV_m \cos \omega_m t$$

Since the output $v_d(t)$ contains the original information $v_m(t) = V_m \cos \omega_m t$, this means that the AM signal has been detected. Connecting the output of the peak detector to a simple RC low-pass filter will eliminate the constant V_c and produce an output that is directly proportional to $V_m \cos \omega_m t$—if such an output is desired.

Let us again consider a band-limited modulating signal $v_m(t)$ having a spectrum, referred to as the **baseband,** as depicted in Fig. 10.58a. If this signal amplitude modulates a carrier that has the frequency $\omega_c >> \omega_u$, then the spectrum of the resulting AM signal is as shown in Fig. 10.58b, where the upper and lower sidebands are symmetrical with respect to the carrier frequency.

If the carrier is removed (or suppressed) from the AM signal, the resulting signal is called a **double-sideband suppressed-carrier** (DSB-SC) signal, and it has the spectrum illustrated in Fig. 10.58c. Since the carrier is not sent along, transmission of a DSB-SC signal requires less power than is needed for the corresponding AM signal. There is a price, however, for this saving of power. To detect a DSB-SC

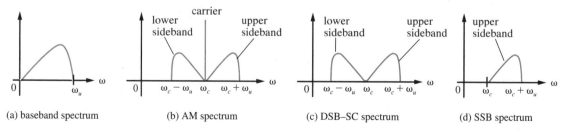

(a) baseband spectrum (b) AM spectrum (c) DSB–SC spectrum (d) SSB spectrum

Fig. 10.58 Some types of amplitude modulation.

signal, an oscillator is required at the receiver so that the carrier can be reinserted. Hence, a DSB-SC detector is more complicated than an ordinary AM detector. This type of modulation is used in conjunction with FM stereo broadcasting.

Since the lower sideband of an AM signal is the mirror image of the upper sideband, it is redundant. By eliminating one of the sidebands, say the lower, and the carrier, we obtain a signal whose spectrum is as shown in Fig. 10.58d. Such a signal, referred to as a **single-sideband** (SSB) signal, requires less power than is needed for either AM or DSB-SC, and has half of the bandwidth. However, because of the missing carrier and sideband, a more complicated detector is needed. Owing to its high efficiency, this type of modulation is popular with radio amateurs.

Yet another form of amplitude modulation is where the carrier, one complete sideband, and only a part or "vestige" of the other sideband are transmitted. The resulting signal is known as a **vestigial-sideband** (VSB) signal. This type of modulation is used for transmitting television video signals. A standard video signal has a maximum frequency component of about $f_u = 4$ MHz [$\omega_u = 2\pi f_u = 2\pi(4 \times 10^6) \approx 25$ Mrad/s]. If such a signal were transmitted by ordinary AM, a bandwidth of 8 MHz would be required. However, since television stations are allocated a bandwidth of only 6 MHz, VSB modulation is used. (Although a smaller bandwidth is needed for SSB, it is much more complicated to detect than is VSB.)

Figure 10.59a shows the baseband for a video signal, where the horizontal axis is labeled in megahertz. If this signal is to be broadcast on channel 3, whose carrier frequency is $f_c = 61.25$ MHz, then the spectrum for channel 3 is as shown in Fig. 10.59b. The entire carrier is transmitted along with the upper sideband. However, only 1.25 MHz of the lower sideband is broadcast. Note, though, that the $4 + 1.25 = 5.25$ MHz required for the VSB video signal does not occupy the allotted 6 MHz band. This is so because part of channel 3's spectrum, which extends from 60 to 66 MHz, is required for the audio signal. Specifically, at a frequency of 4.5 MHz above the video carrier (i.e., at $61.25 + 4.5 = 65.75$ MHz) is the audio carrier. But the audio signal is not AM—it is FM.

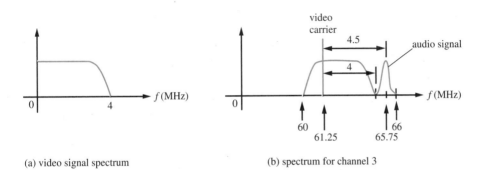

(a) video signal spectrum   (b) spectrum for channel 3

Fig. 10.59 (a) Video baseband, and (b) spectrum for a TV signal.

Frequency Modulation (FM)

Amplitude modulation (AM) is the process by which the amplitude of a waveform—quite often a sinusoid—is varied in accordance with some modulating signal. If the frequency of the waveform (carrier) changes, instead of the amplitude, then the result is known as **frequency modulation** (FM). In particular, if ω_c is the carrier frequency and $v_m(t)$ is the modulating signal, then the expression

$$v(t) = A \cos [\omega_c + k_1 v_m(t)]t$$

where k_1 is a constant, is an example of an FM signal.

Figure 10.60 shows waveforms of an FM carrier $v_c(t) = A \cos \omega_c t$, a modulating signal $v_m(t) = V_m \sin \omega_m t$, where $\omega_m << \omega_c$, and the resulting FM signal $v(t)$. Note that the change in amplitude of the modulating signal causes the change in the frequency of the FM signal.

There is another type of modulation, called **phase modulation** (PM), that is closely related to FM. In this case, the phase angle ϕ is dependent upon the modulating signal $v_m(t)$. Specifically, for a sinusoidal carrier, the expression

$$v(t) = A \cos [\omega_c t + \phi(t)] = A \cos [\omega_c t + k_2 v_m(t)]$$

where k_2 is a constant, is an example of a PM signal. Both FM and PM are generically termed **angle modulation.**

FM signals can be produced in numerous ways. One approach is to use the VCO given in Fig. 10.52*b*. The frequency of this device depends on the control voltage v_c (see Eq. 10.42). By using a coupling capacitor to connect a modulating-signal source to the control-voltage input (pin 5), the frequency of oscillation will vary in accordance with the modulating signal. Although the output of this VCO is triangular or square in shape, by using the appropriate circuits (e.g., a narrow bandpass filter), a frequency-modulated sinusoid can be produced.

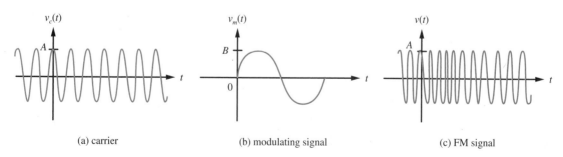

| (a) carrier | (b) modulating signal | (c) FM signal |

Fig. 10.60 Frequency-modulation waveforms.

The bandwidth of an AM signal is determined strictly by the highest frequency component f_u of the modulating signal. In particular, for ordinary AM, the bandwidth is $BW_{AM} = 2f_u$. For an FM signal, however, the bandwidth is determined not only by f_u, but by the amplitude of the modulating signal as well. It is the amplitude of the modulating signal that causes the deviation in the frequency of the carrier, and it is the frequency of the modulating signal that determines the rate of change of this deviation. Although a precise discussion of the bandwidth BW_{FM} of an FM signal is quite involved, an approximate formula for it is given by

$$BW_{FM} \approx 2(\Delta_f + f_u) \tag{10.46}$$

where Δ_f is the maximum frequency deviation of the carrier and f_u is the highest frequency component of the modulating signal. We refer to the case $\Delta_f > f_u$ as **wideband** FM, and $\Delta_f < f_u$ as **narrowband** FM.

The commercial FM broadcasting band extends from 88 to 108 MHz (compare this with 0.54 to 1.6 MHz for AM). Carrier frequencies, which are spaced 200 kHz (0.2 MHz) apart, begin at 88.1 MHz and end at 107.9 MHz. The highest frequency component of an audio modulating signal is 15 kHz (compared with 5 kHz for AM). Thus we see that FM allows for much greater fidelity than does AM. The maximum frequency deviation allowed for broadcast FM is $\Delta_f = 75$ kHz. From Eq. 10.46, we see that the bandwidth of an FM signal with maximum frequency deviation is

$$BW_{FM} \approx 2(75,000 + 15,000) = 180 \text{ kHz}$$

Therefore, we see that the bandwidth of such a wideband FM signal falls within the 200 kHz allocated to a station by the Federal Communications Commission (FCC).

As mentioned in Chapter 5, an FM signal can be demodulated or detected with the use of a phase-locked loop (PLL). The PLL block diagram shown in Fig. 5.36 is repeated in Fig. 10.61. The free-running (zero input voltage) frequency of the VCO is set equal to the FM carrier frequency. The input of the PLL is the FM signal. If the modulating signal is zero, then the input and the VCO will have the same frequency, and the output essentially will be zero. If the input frequency changes,

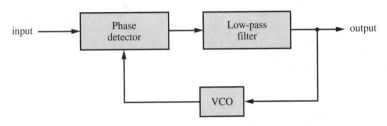

Fig. 10.61 Block diagram of a PLL.

the (low-pass filter) output voltage will change so as to match the frequency of the VCO with that of the input. The result of this is that the output voltage changes, as does the frequency of the input (FM signal). In other words, the output is the detected FM signal. A popular IC chip for commercial broadcast FM detection is the 560 PLL.

As mentioned previously, the audio signal for broadcast TV is FM. The audio carrier is located 4.5 MHz above the video carrier see Fig. 10.59b, for example. The maximum frequency deviation of the audio carrier that is allowed is 25 kHz, and the bandwidth allocated for TV audio is 50 kHz.

In addition to FM radio and TV audio, FM has numerous other applications. Among these are narrowband voice transmissions used by aircraft, fire and police departments, and weather services. Bandwidth requirements for such uses typically are in the range from 10 to 30 kHz.

Electric noise arising from numerous causes (e.g., lightning, automobile ignitions, static electricity discharges) is generally additive to electric signals. As a result of this phenomenon, the amplitude of a signal is affected more by noise than is its frequency. Thus AM is much more susceptible to noise than is FM. Also, the greater the frequency deviation (and, hence, the bandwidth), the greater are the noise-immunity properties of FM. Therefore, in terms of noise, wideband FM is superior to narrowband FM.

FM Stereo

In the preceding discussion of FM, it was assumed that there was a single modulating signal (monaural FM). We are, however, familiar with the concept of FM stereo in which there are two modulating signals—a left and a right. Let us see how FM stereo is produced.

We begin with two signals v_L and v_R, one for the left and one for the right, respectively. We then form the sum $v_L + v_R$ and the difference $v_L - v_R$ of the two signals. To achieve high fidelity, the maximum frequency components of $v_L + v_R$ and $v_L - v_R$ are allowed to be 15 kHz. However, since the spectra of both $v_L + v_R$ and $v_L - v_R$ cover the same frequency band (0–15 kHz) as indicated in Figs. 10.62a and b, we will modulate one of the signals. In particular, for $v_L - v_R$, we use DSB-SC modulation employing a carrier frequency of 38 kHz as indicated in Fig. 10.62c. We now combine (add) the signal for $v_L + v_R$ and the DSB-SC modulated signal of $v_L - v_R$ along with a 19-kHz sinusoid called a **pilot subcarrier.** The spectrum of the resulting signal is shown in Fig. 10.63. (Also shown in Fig. 10.63 is a frequency band labeled "SCA"—ignore this for the time being.) It is this signal that modulates the FM carrier; that is, the frequency spectrum shown in Fig. 10.63 is the baseband for the FM stereo signal. The procedure just described is an example of **frequency multiplexing.**

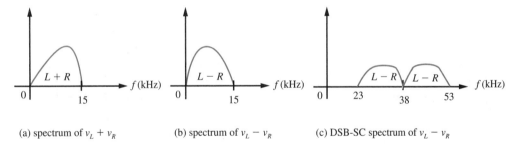

(a) spectrum of $v_L + v_R$ (b) spectrum of $v_L - v_R$ (c) DSB-SC spectrum of $v_L - v_R$

Fig. 10.62 Frequency spectra for $v_L + v_R$, $v_L - v_R$, and modulated form of $v_L - v_R$.

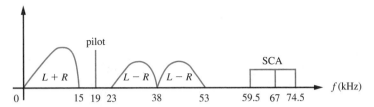

Fig. 10.63 Spectrum of an FM stereo modulating signal.

An ordinary (monaural) FM receiver will detect the $v_L + v_R$ portion of a stereo FM signal, so stereo FM is compatible with ordinary FM. A stereo FM receiver not only detects the $v_L + v_R$ signal, but it also detects the DSB-SC signal and the 19-kHz pilot.[1] The frequency of the pilot is doubled to produce a 38-kHz carrier for the detection of the DSB-SC signal, and DSB-SC demodulation yields the $v_L - v_R$ signal. Taking the sum of $v_L + v_R$ and $v_L - v_R$ gives $2v_L$; taking the difference gives $2v_R$. In this way, the left-channel and the right-channel signals emerge, and stereophonic sound results.

Once an FM stereo signal is detected—for example, by a PLL, as mentioned previously—the job of demodulating the DSB-SC signal and combining the $v_L + v_R$ and $v_L - v_R$ signals appropriately can be accomplished with an IC chip such as the CA3090 produced by RCA.

Since the highest frequency component of an FM stereo signal is much higher than a monophonic signal, to be within the allocated bandwidth, a smaller maximum frequency deviation is required. The result is that FM stereo is narrowband FM and, therefore, does not have the superior noise characteristics of monophonic FM.

Some FM stations also broadcast background music for department stores, elevators, grocery stores, and the like. This type of transmission, known as **subsidiary communication authorization** (SCA), is accomplished with a narrowband FM signal having a carrier[2] frequency of 67 kHz and a bandwidth of 15 kHz. The spectrum

---

[1]It is the detection of the pilot subcarrier that causes a receiver's stereo indicator light to go on.

[2]Since this FM signal is part of an overall FM signal, the term **subcarrier** actually is more accurate.

of an SCA signal is included as part of the FM stereo modulating-signal spectrum given in Fig. 10.63. After such a modulating signal is detected, the SCA signal can then be demodulated. In particular, the circuit shown in Fig. 10.64 will detect SCA signals. The IC chip used in this circuit is a 565 PLL. Connected to the phase-detector input terminals (pins 2 and 3) is a high-pass filter consisting of 500-pF capacitors and 5-kΩ resistors. The resistors also are part of a voltage divider that establishes the appropriate dc bias voltage. The frequency of the VCO squarewave output (pin 4) is compared with the input frequency at pin 5. The free-running frequency of the VCO is determined by the resistance connected to pin 8 and the capacitance connected to pin 9. The cutoff frequency of the PLL's low-pass filter is determined by the capacitor which is connected between pins 7 and 8. Finally, con-nected to the PLL output (pin 7) is a low-pass filter that reduces the high-frequency noise that often occurs with SCA transmissions.

Stereo TV sound is accomplished as is FM stereo. In stereo TV, however, the pilot frequency is the same as the television's horizontal sweep rate, that is, 15.75 kHz. However, instead of an SCA signal, the TV audio signal may include a **second audio program** (SAP) channel. The SAP channel can be used to supply sound in a second language simultaneously. The capability of stereo TV sound and a SAP chan-nel is known as **multichannel television sound** (MTS).

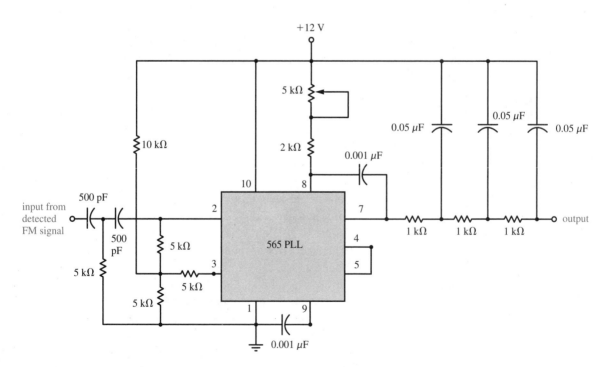

Fig. 10.64 An SCA decoder.

Previously, we stated that a commercial TV broadcast signal consists of an AM (VSB) video signal and an FM audio signal, where the bandwidth of a channel is 6 MHz. In addition to broadcast TV, television signals are transmitted by networks to their affiliates, and pay-TV services send their programming to cable companies throughout the country. A very popular way of program distribution is with the use of communication satellites. Although such satellites have numerous channels, called **transponders,** their bandwidths are not 6 MHz, as is the case for standard broadcast TV. The reason is that satellite-TV transmissions use FM for video as well as audio— a transponder bandwidth of 30 MHz being typical.

SUMMARY

1. Direct-coupled (dc) amplifiers are easily fabricated on IC chips.

2. Differential amplifiers are dc amplifiers that are typically the first stages in operational amplifiers (op amps).

3. Differential amplifiers and, hence, op amps, have the ability to reject common-mode signals (noise). This ability is expressed in terms of the common-mode rejection ratio (CMRR).

4. An ideal operational amplifier has infinite gain, infinite input resistance, and zero output resistance.

5. An op amp typically has a differential input and a single-ended output.

6. The industry standard is the 741 op amp.

7. Negative feedback can be used to improve amplifier stability. It also can be used to improve a voltage amplifier's input and output characteristics. The cost is a reduction of gain.

8. The gain-bandwidth product of a feedback amplifier is a constant.

9. Positive feedback can be used to produce oscillators.

10. The loop gain of an oscillator must be unity (or greater).

11. Phase-shift and Wien-bridge oscillators are examples of RC sinusoidal oscillators.

12. Colpitts, Hartley, and Clapp oscillators are examples of LC sinusoidal oscillators.

13. Crystal oscillators are used when extreme stability is required.

14. An operational amplifier can be used as a comparator. Comparators can be used as level detectors and zero-crossing detectors.

15. The Schmitt trigger is a bistable device that exhibits hysteresis.

16. Comparators and Schmitt triggers can be used to convert sine waves into square waves.

17. A nonsinusoidal oscillator that is based on the charging and discharging of a capacitor is an example of a relaxation oscillator or astable multivibrator.

18. The 555 IC timer is frequently used to construct astable and monostable (one-shot) multivibrators.

19. A voltage-controlled oscillator (VCO), such as the 566 IC chip, can also be used to produce nonsinusoidal, periodic waveforms.

20. Information can be transmitted efficiently with the use of modulation.

PROBLEMS

10.1 Demonstrate that the circuit consisting of R_1, R_2, R_3, Q_3, and V_{EE} given in Fig. 10.3 on p. 653 is an open circuit to ac signals by finding the Norton equivalent (with respect to the collector of Q_3 and the reference) of the corresponding ac-equivalent circuit.

10.2 For the differential amplifier given in Fig. 10.5 on p. 654, suppose that $h_{FE} = h_{fe} = 100$, $I_O = 2.02$ mA, $R_C = 1$ kΩ, and $R_s = 100$ Ω. Find the output voltage $v_o = v_{c2} - v_{c1}$.

10.3 For the differential amplifier given in Fig. 10.1 on p. 652, suppose that $R_s = 0$ Ω. Show that when $h_{fe} \gg 1$ and $R_E \gg r_e$, then the output $v_o = v_{c2} - v_{c1} \approx A(v_1 - v_2)$, where $A = R_C/r_e$.

10.4 For the differential amplifier shown in Fig. 10.1 on p. 652, suppose that $R_s = 0$ Ω and $v_2 = 0$ V. Show that when $h_{fe} \gg 1$ and $R_E \gg r_e$, then the input resistance R_{in} of the differential amplifier is $R_{in} = v_1/i_{b1} \approx 2h_{fe}r_e$.

10.5 For the differential amplifier with a differential input voltage given in Fig. 10.8 on p. 657, set $R_s = 0$ Ω and replace the current source I_O by a resistor R_E connected in series with a supply voltage $-V_{EE}$ (see Fig. 10.1 on p. 652). Show that when $h_{fe} \gg 1$, then for the resulting differential amplifier (a) the input resistance is $R_{in} = v_d/i_{b1} \approx 2h_{fe}r_e$, and (b) $v_{c2}/v_d \approx R_C/2r_e$.

10.6 For the differential amplifier given in Fig. 10.1 on p. 652, let $R_s = 0$ Ω. Apply a common-mode input v_{cm} and show that when $h_{fe} \gg 1$ and $R_E \gg r_e$, then (a) $v_{c2}/v_{cm} \approx -R_C/2R_E$, and (b) the input resistance is $v_{cm}/(i_{b1} + i_{b2}) \approx h_{fe}R_E$.

10.7 For the differential amplifier given in Fig. 10.3 on p. 653, suppose that $h_{FE} = h_{fe} = 100$, $R_1 = 47$ kΩ, $R_2 = 100$ kΩ, $R_3 = 2.7$ kΩ, $R_C = 4.7$ kΩ, $R_s = 0$ Ω, and $V_{CC} = V_{EE} = 10$ V. Find $CMRR_{dB}$ assuming that $R_o = 650$ kΩ.

10.8 A differential amplifier has the inputs $v_1 = 10 \cos 120\pi t + 0.01 \cos 2000\pi t$ V and $v_2 = $

$10 \cos 120\pi t - 0.01 \cos 2000\pi t$ V and output $v_o = 0.2 \cos 120\pi t + 8 \cos 2000\pi t$ V. (a) Find the absolute value of the common-mode gain of the amplifier. (b) Find the absolute value of the differential-mode gain of the amplifier. (c) Find $CMRR_{dB}$.

10.9 In addition to BJTs, differential amplifiers can be fabricated using FETs. The differential amplifier shown in Fig. P10.9 has identical JFETs with $I_{DSS} = 12$ mA and $V_p = -6$ V. In addition, v_1 and v_2 are ac voltage sources. Find V_{GSQ}, I_{DQ}, V_{DSQ}, g_m, and confirm that the JFETs are biased in the active region.

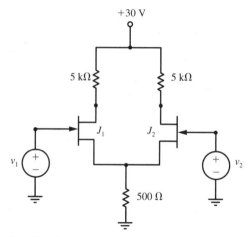

Fig. P10.9

10.10 For the JFET differential amplifier given in Fig. P10.9, suppose that $g_m = 2$ m℧ and $r_d = \infty$. Find the gain v_{d2}/v_1 (where v_{d2} is the voltage at the drain of J_2) for the case that $v_2 = 0$ V.

10.11 For the JFET differential amplifier given in Fig. P10.9, suppose that $g_m = 2$ m℧ and $r_d = \infty$. Find the gain v_{d2}/v_{cm} (where v_{d2} is the voltage at the drain of J_2) for the case that $v_1 = v_2 = v_{cm}$.

10.12 The differential amplifier shown in Fig. P10.12 has $R_D = 5$ kΩ, $I_{DSS} = 12$ mA, and $V_p = -6$ V for identical JFETs J_1 and J_2, while $I_{DSS} = 6$ mA and $V_p = -6$ V for J_3. Find V_{GSQ}, I_{DQ}, V_{DSQ},

g_m, for J_1 and J_2, and confirm that all three JFETs are biased in the active region.

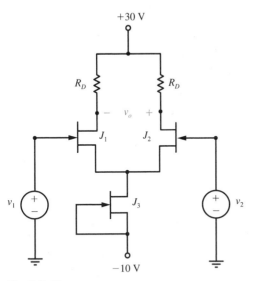

Fig. P10.12

10.13 For the differential amplifier shown in Fig. P10.12, JFETs J_1 and J_2 are identical. Suppose that $g_m = 2$ m℧, $r_d = \infty$, $R_D = 5$ kΩ, and J_3 behaves as an ideal dc current source. Find $v_o = v_{d2} - v_{d1}$.

10.14 For the differential amplifier shown in Fig. P10.12, JFETs J_1 and J_2 are identical. Suppose that $g_m = 2$ m℧, $r_d = \infty$, $R_D = 5$ kΩ, and J_3 behaves as an ideal dc current source that is connected in parallel with $R_o = 50$ kΩ. Determine the differential gain $A_D = v_{d2}/v_d$ (where v_{d2} is the voltage at the drain of J_2) of the amplifier by setting $v_1 = v_d/2$ and $v_2 = -v_d/2$.

10.15 For the differential amplifier shown in Fig. P10.12, JFETs J_1 and J_2 are identical. Suppose that $g_m = 2$ m℧, $r_d = \infty$, $R_D = 5$ kΩ, and J_3 behaves as an ideal dc current source that is connected in parallel with $R_o = 50$ kΩ. Determine (a) the common-mode gain $A_{CM} = v_{d2}/v_{cm}$ (where v_{d2} is the voltage at the drain of J_2) of the amplifier, and (b) CMRR$_{dB}$ for the differential amplifier.

10.16 For the differential amplifier shown in Fig. P10.12, JFETs J_1 and J_2 are identical and have transconductance g_m and $r_d = \infty$. Suppose that J_3 behaves as an ideal dc current source that is connected in parallel with a resistor R_o. Find an expression for the differential gain $A_D = v_{d2}/v_d$ (where v_{d2} is the voltage at the drain of J_2) of the amplifier by setting $v_1 = v_d/2$ and $v_2 = -v_d/2$.

10.17 For the differential amplifier shown in Fig. P10.12, JFETs J_1 and J_2 are identical and have transconductance g_m and $r_d = \infty$. Suppose that J_3 behaves as an ideal dc current source that is connected in parallel with a resistor R_o. Determine expressions for (a) the common-mode gain $A_{CM} = v_{d2}/v_m$ (where v_{d2} is the voltage at the drain of J_2) of the amplifier, and (b) CMRR$_{dB}$ for the differential amplifier.

10.18 For the current mirror shown in Fig. P10.18, Q_1 and Q_2 are identical. Assume that all the BJTs are in the active region and have $h_{FE} = 49$ and $v_{BE} = 0.7$ V. Find (a) I_{REF}, (b) I_O, and (c) I_S for each transistor.

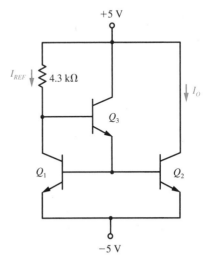

Fig. P10.18

10.19 For the modified current mirror shown in Fig. P10.19, assume that both BJTs are in the active region and have $h_{FE} = 100$ and $v_{BE} = 0.7$ V. Find (a) v_{CE} for each transistor, (b) I_{REF}, and (c) I_O.

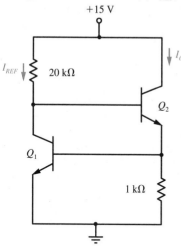

+15 V

I_{REF} 20 kΩ I_O

Q_2

Q_1 1 kΩ

Fig. P10.19

10.20 For the **Wilson current mirror** shown in Fig. P10.20, assume that all the BJTs are in the active region and have value of h_{FE}. Show that

$$I_O = \frac{1 + 2/h_{FE}}{1 + 2/h_{FE} + 2/h_{FE}^2} I_{REF}$$

where $I_{REF} = (V_{CC} - v_{BE2} - v_{BE3})/R$.

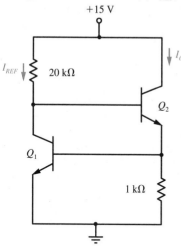

+V_{CC}

I_{REF} R I_O

Q_3

Q_1 Q_2

Fig. P10.20

10.21 For the Widlar current source given in Fig. 10.12*b* on p. 665, suppose that $R = 5$ kΩ and $R_E =$

1 kΩ. Assume that the BJTs are in the active region, $h_{FE} \gg 1$, and $v_{BE1} = 0.7$ V. Find I_O numerically by beginning with the guess that $I_O = 1$ mA.

10.22 For the Widlar current source given in Fig. 10.12*a* on p. 665, suppose that $V_{CC} = 5$ V. Assume that the BJTs are in the active region, $h_{FE} \gg 1$, and $v_{BE1} = 0.7$ V. Find (a) R such that $I_{REF} = 1$ mA, (b) R_E such that $I_O = 10$ μA, (c) v_{BE2}, and (d) I_S.

10.23 For the FET current mirror shown in Fig. P10.23, suppose that for the enhancement MOSFETs, $K = 400$ μA/V^2 for M_1 and M_2, $K = 1.8$ μA/V^2 for M_3, and $V_t = 1$ V for M_1, M_2, and M_3. Find I_{REF} and I_O, and confirm that each MOSFET operates in the active region.

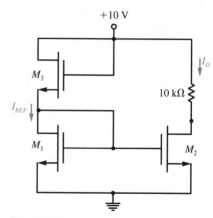

+10 V

M_3 10 kΩ I_O

I_{REF}

M_1 M_2

Fig. P10.23

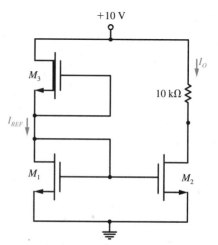

+10 V

M_3 10 kΩ I_O

I_{REF}

M_1 M_2

Fig. P10.24

10.24 For the FET current mirror shown in Fig. P10.24, suppose that for the enhancement MOSFETs, $K = 100 \ \mu\text{A}/\text{V}^2$ and $V_t = 1$ V, whereas for the depletion MOSFET $I_{DSS} = 100 \ \mu\text{A}$ and $V_p = -2$ V. Find I_{REF} and I_O, and confirm that each MOSFET operates in the active region.

10.25 For the FET current mirror shown in Fig. P10.25, suppose that for the enhancement MOSFETs, $K = 200 \ \mu\text{A}/\text{V}^2$ for M_1, M_2, and M_3, $K = 2 \ \mu\text{A}/\text{V}^2$ for M_4, and $V_t = 2$ V for M_1, M_2, M_3, and M_4. Find I_{REF} and I_O, and confirm that each MOSFET operates in the active region.

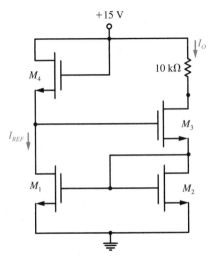

Fig. P10.25

10.26 An alternative circuit for determining the effect of an op amp's input offset voltage is shown in Fig. P10.26. Derive an expression for v_o, and compare it with Eq. 10.15 on p. 671.

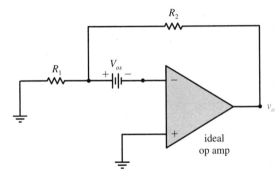

Fig. P10.26

10.27 For the inverting amplifier shown in Fig. P10.27, $R_1 = 100$ kΩ, $R_2 = 900$ kΩ, and the op amp has an input offset voltage of 1 mV, an input offset current of 20 nA, and an input bias current of 80 nA. Find the worst-case output offset voltage of the overall amplifier when R_3 is chosen to minimize the output offset voltage.

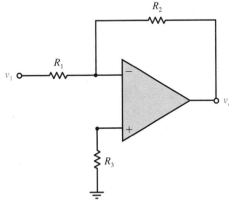

Fig. P10.27

10.28 For the inverting amplifier shown in Fig. P10.27, $R_1 = 100$ kΩ, $R_2 = 900$ kΩ, $R_3 = 0$ Ω, and the op amp has an input offset voltage of 1 mV, an input offset current of 20 nA, and an input bias current of 80 nA. Assume that $i_{B1} \approx i_{B2}$ and determine the worst-case output offset voltage.

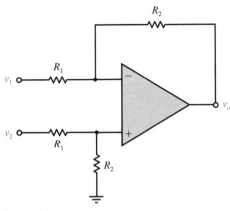

Fig. P10.29

10.29 For the differential amplifier shown in Fig. P10.29, suppose that $R_1 = 100$ kΩ and $R_2 = 900$ kΩ. For the case that the op amp has an input offset voltage of 1 mV, an input offset current of 20 nA, and an input bias current of 80 nA, find the worst-case output offset voltage of the overall amplifier.

10.30 Assume that the op amp in the differential amplifier shown in Fig. P10.29 is ideal. If $R_1 = 100$ kΩ and $R_2 = 900$ kΩ, then the differential gain of the overall amplifier has absolute value $|A_D| = 9$. If the top resistor labeled R_2 changes to 900,100 Ω, then the differential gain is still $|A_D| \approx 9$. Determine the CMRR$_{dB}$ for the resulting slightly unbalanced amplifier.

10.31 For the differential amplifier shown in Fig. P10.29, assume that the op amp is ideal. When $R_1 = 1$ kΩ, the upper resistor labeled $R_2 = 100$ kΩ, and the lower resistor labeled $R_2 = 110$ kΩ, then the differential gain of the overall amplifier has absolute value $|A_D| = 105$. Determine the CMRR$_{dB}$ for the amplifier.

10.32 For the differential amplifier shown in Fig. P10.32, v_1 and v_2 are ac sources. Suppose that the BJTs have $h_{FE} = 100$, $R_E = 10$ kΩ, and $I_O = 20$ μA is a dc source. Find I_{CQ} and V_{CEQ} for each transistor.

10.33 For the differential amplifier shown in Fig. P10.32, set $v_2 = 0$ V. Suppose that the BJTs have $h_{fe} = 100$ and $r_e = 5$ Ω. Find (a) the ac input resistance R_{in} and (b) the voltage gain v_o/v_1.

10.34 For the differential amplifier shown in Fig. P10.32, set $v_2 = 0$ V. Suppose that the BJTs have $h_{fe} = 100$ and $r_e = 5$ Ω. Find the ac output resistance R_o.

10.35 The feedback amplifier shown in Fig. P10.35 illustrates the ideal form of series-parallel feedback. Show that for this feedback amplifier

$$A_F = \frac{v_o}{v_{in}} = \frac{A}{1 + AB} \qquad R_{iF} = \frac{v_{in}}{i_{in}} = (1 + AB)R_{in}$$

$$R_{oF} = \frac{R_o}{1 + AB}$$

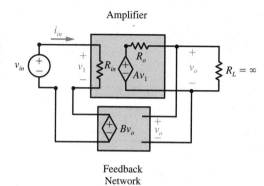

Amplifier

Feedback
Network

Fig. P10.35

10.36 The feedback amplifier shown in Fig. P10.36 illustrates the ideal form of **parallel-parallel**

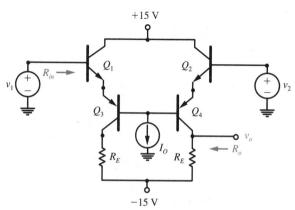

Fig. P10.32

feedback (also called **voltage-shunt feedback**).
Show that for this feedback amplifier

$$A_F = \frac{v_o}{i_{in}} = \frac{A}{1 + AB} \qquad R_{iF} = \frac{v_{in}}{i_{in}} = \frac{R_{in}}{1 + AB}$$

$$R_{oF} = \frac{R_o}{1 + AB}$$

Amplifier

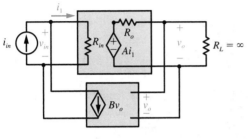

Fig. P10.36

10.37 The op amp circuit shown in Fig. P10.37
is an example of (nonideal) parallel-parallel feedback. Assume that the op amp has $R_{in} = \infty$, $R_o =
0\ \Omega$, and finite gain A. Find $A_F = v_o/i_{in}$ and $R_{iF} =
v_{in}/i_{in}$.

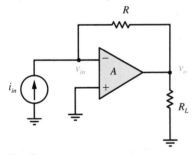

Fig. P10.37

10.38 The feedback amplifier shown in Fig.
P10.38 illustrates the ideal form of **series-series feedback** (also called **current-series feedback**). Show
that for this feedback amplifier

$$A_F = \frac{i_o}{v_{in}} = \frac{A}{1 + AB} \qquad R_{iF} = \frac{v_{in}}{i_{in}} = (1 + AB)R_{in}$$

$$R_{oF} = (1 + AB)R_o$$

Amplifier

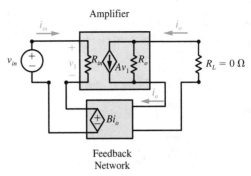

Fig. P10.38

10.39 The op amp circuit shown in Fig. P10.39
is an example of (nonideal) series-series feedback.
Assume that the op amp has $R_{in} = \infty$, $R_o = 0\ \Omega$, and
finite gain A. Find $A_F = i_o/i_{in}$.

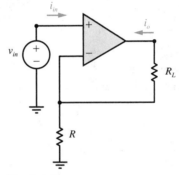

Fig. P10.39

10.40 The op amp circuit shown in Fig. P10.39
is an example of (nonideal) series-series feedback.
Assume that the op amp has finite gain A, finite input
resistance R_{in}, and $R_o = 0\ \Omega$. Show that when
$R_{in} \gg R$, then

$$R_{iF} = \frac{v_{in}}{i_{in}} \approx \left(1 + \frac{AR}{R + R_L}\right)R_{in}$$

10.41 The op amp circuit shown in Fig. P10.39
is an example of (nonideal) series-series feedback.
Assume that the op amp has finite gain A, finite input
resistance R_{in}, and nonzero output resistance R_o.
Show that when $R_{in} \gg R$, then

$$R_{oF} \approx (1 + A)R + R_o$$

10.42 The feedback amplifier shown in Fig.
P10.42 illustrates the ideal form of **parallel-series**

feedback (also called **current-shunt feedback**). Show that for this feedback amplifier

$$A_F = \frac{i_o}{i_{in}} = \frac{A}{1 + AB} \qquad R_{iF} = \frac{v_{in}}{i_{in}} = \frac{R_{in}}{1 + AB}$$

$$R_{oF} = (1 + AB)R_o$$

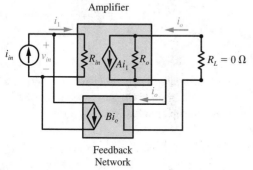

Amplifier

Feedback
Network

Fig. P10.42

10.43 The op amp circuit shown in Fig. P10.43 is an example of (nonideal) parallel-series feedback. Assume that the op amp has $R_{in} = \infty$, $R_o = 0\ \Omega$, and finite gain A. Show that

$$A_F = \frac{i_o}{i_{in}} = \frac{AR_1 + (1 + A)R_2}{R_L + (1 + A)R_2}$$

(*Hint:* Use mesh analysis, where i_{in} and i_o are the mesh currents.)

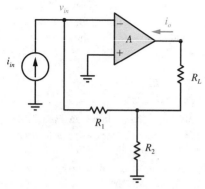

Fig. P10.43

10.44 The op amp circuit shown in Fig. P10.43 is an example of (nonideal) parallel-series feedback.

Assume that the op amp has $R_{in} = \infty$, $R_o = 0\ \Omega$, and finite gain A. Show that

$$R_{iF} = \frac{v_{in}}{i_{in}} = \frac{R_L(R_1 + R_2) + R_1R_2}{R_L + (1 + A)R_2}$$

(*Hint:* Use mesh analysis, where i_{in} and i_o are the mesh currents.)

10.45 For the feedback amplifier given in Fig. 10.26 on p. 682, suppose that the op amp has a gain of 200,000 and a gain-bandwidth product of 1 MHz. Find the bandwidth (in hertz) of the feedback amplifier for the case that $R_1 = 1\ k\Omega$ and (a) $R_2 = 9\ k\Omega$, (b) $R_2 = 1\ M\Omega$.

10.46 For the feedback amplifier shown in Fig. 10.26 on p. 682, suppose that the op amp has a gain of 200,000, a gain-bandwidth product of 1 MHz, and $R_1 = 5\ k\Omega$. Determine the value of R_2 that will result in a bandwidth of 200 kHz for the feedback amplifier.

10.47 For the feedback amplifier shown in Fig. 10.26 on p. 682, suppose that the op amp has a gain of 200,000, a gain-bandwidth product of 1 MHz, and $R_2 = 50\ k\Omega$. Determine the value of R_1 that will result in a bandwidth of 200 kHz for the feedback amplifier.

10.48 By adding a resistance to the feedback amplifier given in Fig. P10.37, we obtain the feedback amplifier shown in Fig. P10.48. Assume that the op amp has $R_{in} = \infty$, $R_o = 0\ \Omega$, and finite gain A. Show that the overall voltage gain is

$$A_F = \frac{v_o}{v_{in}} = \frac{-R_2}{R_1 + (R_1 + R_2)/A}$$

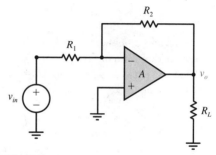

Fig. P10.48

10.49 For the feedback amplifier shown in Fig. P10.48, the op amp has gain A, input resistance R_{in}, and output resistance $R_o = 0 \, \Omega$. Show that the input resistance of the feedback amplifier is

$$R_{iF} = R_1 + R_{in} \parallel \frac{R_2}{1 + A}$$

10.50 For the feedback amplifier shown in Fig. P10.48, the op amp has gain A, output resistance R_o, and input resistance $R_{in} = \infty$. Show that the output resistance of the feedback amplifier is

$$R_{oF} = \frac{R_o(R_1 + R_2)}{R_o + R_1 + R_2 + AR_1}$$

10.51 The voltage gain A_F of the feedback amplifier shown in Fig. P10.48 is given in Problem 10.48, where A is the gain of the op amp. If ω_H is the bandwidth (upper cutoff frequency) of the op amp, show that the bandwidth of the feedback amplifier is

$$\omega_{HF} = \left(\frac{AR_1}{R_1 + R_2} + 1 \right) \omega_H$$

10.52 For the feedback amplifier shown in Fig. P10.48, the op amp has a gain of 200,000 and a gain-bandwidth product of 1 MHz. Use the result given in Problem 10.51 to determine the bandwidth (in hertz) of the feedback amplifier for the case that $R_1 = 1 \, \text{k}\Omega$ and $R_2 = 9 \, \text{k}\Omega$.

10.53 Consider the op-amp phase-shift oscillator shown in Fig. 10.30b on p. 692. (a) Given that $C = 0.001 \, \mu\text{F}$ and the oscillation frequency is to be 10 kHz, determine R and the minimum value of R_F. (b) Given that $R = 1.5 \, \Omega$ and the oscillation frequency is to be 1.2 kHz, determine C and the minimum value of R_F. (c) Given that $C = 0.01 \, \mu\text{F}$ and $R = 10 \, \text{k}\Omega$, determine the minimum value of R_F and the oscillation frequency.

10.54 Consider the JFET phase-shift oscillator shown in Fig. 10.31a on p. 694. (a) Given that $g_m = 5 \, \text{m}\mho$ and $R = 100 \, \text{k}\Omega$, determine the minimum value of R_D and C required for an oscillation frequency of 1 kHz. (b) Given that $g_m = 10 \, \text{m}\mho$ and $C = 0.001 \, \mu\text{F}$, determine the minimum value of R_D

and R required for an oscillation frequency of 1 kHz. (c) Given that $R_D = 5 \, \text{k}\Omega$, $R = 100 \, \Omega$, and $C = 0.001 \, \mu\text{F}$, determine the minimum value of g_m and the frequency of oscillation. (d) Given that $R_D = 5 \, \text{k}\Omega$, $R = 100 \, \Omega$, and the frequency of oscillation is 1 kHz, determine the minimum value of g_m and C.

10.55 Suppose that for the BJT phase-shift oscillator shown in Fig. 10.31b on p. 694, $R_1 = 27 \, \text{k}\Omega$, $R_2 = 3 \, \text{k}\Omega$, $R = 1 \, \text{k}\Omega$, $r_e = 2.16 \, \Omega$, and the oscillation frequency is 1 kHz. (a) Given that $R_C = 860 \, \Omega$, determine the minimum value of h_{fe} and C. Find R_s for the case that $h_{fe} = 100$. (b) Given that $C = 0.047 \, \mu\text{F}$, determine R_C and the minimum value of h_{fe}.

10.56 Consider the Wien-bridge oscillator shown in Fig. 10.32a on p. 696. (a) Given that $R = R_1 = 10 \, \text{k}\Omega$, determine C and the minimum value of R_D required for an oscillation frequency of 10 kHz. (b) Given that $R_2 = 10 \, \text{k}\Omega$ and $C = 0.02 \, \mu\text{F}$, determine R and the maximum value of R_1 required for an oscillation frequency of 1 kHz. (c) Given that $R = R_1 = 5 \, \text{k}\Omega$ and $C = 0.01 \, \mu\text{F}$, determine the frequency of oscillation and the minimum value of R_2.

10.57 For the Wein-bridge oscillator shown in Fig. 10.32a on p. 696, replace the parallel RC combination with a series RC combination, and vice versa. (a) Show that the voltage transfer function for the resulting RC feedback network is

$$\mathbf{B} = \mathbf{B}(j\omega) = \frac{\mathbf{V}_1}{\mathbf{V}_o} = \frac{1 - \omega^2 R^2 C^2 + j2\omega RC}{1 - \omega^2 R^2 C^2 + j3\omega RC}$$

(b) Find the relationship between R_1 and R_2 for which the oscillation frequency is $\omega_0 = 1/RC$.

10.58 Consider the oscillator shown in Fig. P10.58. (a) Show that the voltage transfer function for the RC feedback network is

$$\mathbf{B} = \frac{\mathbf{V}_1}{\mathbf{V}_o} = \frac{1}{3 + j(\omega RC - 1/\omega RC)}$$

(b) Find the oscillation frequency ω_0 and the relationship between R_1 and R_2 for which oscillation occurs.

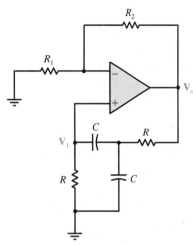

Fig. P10.58

10.59 Consider the oscillator shown in Fig. P10.59. (a) Show that the voltage transfer function for the RC feedback network is that given in Problem 10.58. (b) Find the oscillation frequency ω_0 and the relationship between R_1 and R_2 for which oscillation occurs.

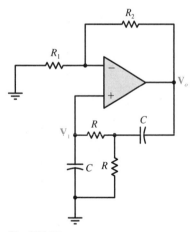

Fig. P10.59

10.60 For the oscillator shown in Fig. P10.59, replace each capacitor C with an inductor L. (a) Show that the voltage transfer function for the resulting RL feedback network is

$$\mathbf{B} = \frac{\mathbf{V}_1}{\mathbf{V}_o} = \frac{1}{3 + j(\omega L/R - R/\omega L)}$$

(b) Find the oscillation frequency ω_0 and the relationship between R_1 and R_2 for which oscillation occurs.

10.61 For the JFET Colpitts oscillator shown in Fig. 10.34a on p. 700, suppose that $C_1 = 1200$ pF, $C_2 = 2400$ pF, and $R_G = 10$ MΩ. (a) Given that $L = 50$ μH and $R_D = 1$ kΩ, find the oscillation frequency and determine the minimum value of g_m required for oscillation. (b) Given that the frequency of oscillation is 500 kHz and $R_D = 500$ Ω, find L and determine the minimum value of g_m required for oscillation.

10.62 For the JFET Hartley oscillator shown in Fig. 10.35a on p. 701, suppose that $L_1 = L_2 = 100$ μH and $R_D = 1$ kΩ. (a) Given that $C = 200$ pF and $R_D = 2$ kΩ, find the oscillation frequency and determine the minimum value of g_m required for oscillation. (b) Given that the frequency of oscillation is 500 kHz and $R_D = 1$ kΩ, find C and determine the minimum value of g_m required for oscillation.

10.63 For the oscillator shown in Fig. P10.63, suppose that the op amp is ideal. (a) Find the frequency of oscillation. (b) Find the minimum value of R for which oscillation occurs.

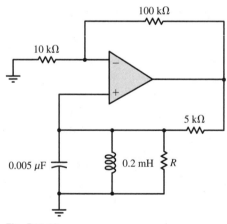

Fig. P10.63

10.64 For the oscillator shown in Fig. P10.64, suppose that the op amp is ideal. (a) Find the frequency of oscillation. (b) Find the minimum value of R for which oscillation occurs.

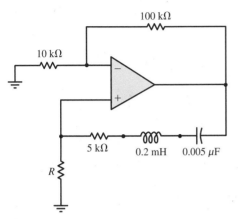

Fig. P10.64

10.65 The electric equivalent circuit of a quartz crystal has the values $R = 1$ kΩ, $L = 5$ H, $C = 0.04$ pF, and $C_m = 12$ pF. Find (a) the series resonance ω_s, (b) the parallel resonance ω_p, and (c) the quality factor Q at the series resonance frequency.

10.66 For the comparator shown in Fig. 10.38a on p. 704, suppose that $V_{CC} = 10$ V and the input voltage is $v_1 = 6 \sin \omega t$ V. Sketch the output voltage v_o for the case that (a) $v_2 = V_r = 3$ V, and (b) $v_2 = V_r = -3$ V.

10.67 For the comparator given in Fig. 10.38a on p. 704, let $v_1 = V_r > 0$ V be the reference voltage and let v_2 be the input voltage. (a) Sketch the transfer characteristic v_o versus v_2. (b) Sketch the output voltage v_o when the input voltage is given by $v_2 = V_r + V_r \sin \omega t$ V.

10.68 For the Schmitt trigger shown in Fig. 10.40 on p. 706, suppose that $R_1 = R_2 = 10$ kΩ and $V_{CC} = 10$ V. Sketch the output voltage v_o when the input voltage is $v_1 = 10 \sin \omega t$ V.

10.69 For the Schmitt trigger given in Fig. 10.43 on p. 708, suppose that $R_1 = 1$ kΩ, $R_2 = 9$ kΩ, and $V_{CC} = 10$ V. Sketch the transfer characteristic for the case that (a) $V_r = 5$ V, (b) $V_r = 10$ V, (c) $V_r = 20$ V, and (d) $V_r = -10$ V.

10.70 For the Schmitt trigger given in Fig. 10.43 on p. 708, suppose that $R_1 = 4$ kΩ, $R_2 = 10$ kΩ, $V_r = -6$ V, and $V_{CC} = 12$ V. Determine the threshold voltages and sketch the transfer characteristic.

10.71 For the noninverting Schmitt trigger shown in Fig. P10.71, verify that

$$v = \frac{R_1}{R_1 + R_2} v_o + \frac{R_2}{R_1 + R_2} v_1$$

and sketch the transfer characteristic for this device when $V_r = 0$ V.

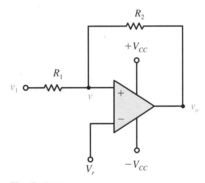

Fig. P10.71

10.72 For the noninverting Schmitt trigger shown in Fig. P10.71, suppose that $R_1 = 10$ kΩ, $R_2 = 25$ kΩ, $V_r = 0$ V, and $V_{CC} = 10$ V. Determine the threshold voltages and sketch the transfer characteristic.

10.73 For the noninverting Schmitt trigger shown in Fig. P10.71, suppose that $R_1 = 10$ kΩ, $R_2 = 20$ kΩ, $V_r = 6$ V, and $V_{CC} = 12$ V. Determine the threshold voltages and sketch the transfer characteristic.

10.74 For the noninverting Schmitt trigger shown in Fig. P10.71, suppose that $R_1 = 10$ kΩ, $R_2 = 20$ kΩ, $V_r = -6$ V, and $V_{CC} = 12$ V. Determine the threshold voltages and sketch the transfer characteristic.

10.75 The BJT circuit shown in Fig. P10.75 is an emitter-coupled Schmitt trigger. When Q_1 is OFF, then Q_2 is ON, and the output v_o is low. When Q_1 is ON, then Q_2 is OFF, and the output v_o is high (5 V). Assume that Q_1 is OFF and Q_2 is ON to find (a) the minimum value of h_{FE2}, (b) the low voltage, and (c) the upper threshold voltage.

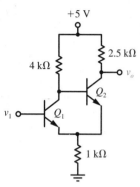

+5 V

2.5 kΩ

4 kΩ

v_o

Q_2

v_1

Q_1

1 kΩ

Fig. P10.75

10.76 The BJT circuit shown in Fig. P10.75 is an emitter-coupled Schmitt trigger. When Q_1 is OFF, then Q_2 is ON, and the output v_o is low. When Q_1 is ON, then Q_2 is OFF, and the output v_o is high (5 V). Assume that Q_1 is ON and Q_2 is OFF to find the lower threshold voltage given that $h_{FE1} = 100$.

10.77 Consider the oscillator shown in Fig. 10.46a on p. 712. (a) Find the oscillation frequency f_0 when $C = 0.05$ μF, $R_1 = R_2 = 10$ kΩ, and $R = 91$ kΩ. (b) Find R when $R_1 = R_2 = 10$ kΩ, $C = 0.1$ μF, and $f_0 = 100$ Hz. (c) Find C when $R = R_1 = R_2 = 22$ kΩ and $f_0 = 100$ Hz. (d) Find R_1 when $R_2 = 10$ kΩ, $R = 20$ kΩ, $C = 0.1$ μF, and $f_0 = 100$ Hz. (e) Find R_2 when $R_1 = 10$ kΩ, $R = 20$ kΩ, $C = 0.1$ μF, and $f_0 = 100$ Hz.

10.78 Consider the oscillator shown in Fig. 10.47 on p. 714. (a) Find the oscillation frequency f_0 when $R = R_1 = R_2 = 10$ kΩ and $C = 0.05$ μF. (b) Find R_2 when $R = R_1 = 10$ kΩ, $C = 0.1$ μF, and $f_0 = 300$ Hz. (c) Find C when $R = R_1 = R_2 = 10$ kΩ and $f_0 = 250$ Hz. (d) Find R_1 when $R = 10$ kΩ, $R_2 = 20$ kΩ, $C = 0.5$ μF, and $f_0 = 100$ Hz. (e) Find

R when $R_1 = R_2 = 10$ kΩ, $C = 0.1$ μF, and $f_0 = 200$ Hz.

10.79 Consider the astable multivibrator shown in Fig. 10.49b on p. 716. (a) Find the oscillation frequency f_0 when $C = 0.1$ μF and $R_1 = R_2 = 5$ kΩ. (b) Find C when $R_1 = R_2 = 5$ kΩ and $f_0 = 1.6$ kHz. (c) Find R_1 when $R_2 = 10$ kΩ, $C = 0.005$ μF, and $f_0 = 10$ kHz. (d) Find R_2 when $R_1 = 10$ kΩ, $C = 0.005$ μF, and $f_0 = 10$ kHz.

10.80 Consider the oscillator shown in Fig. 10.52b on p. 720. (a) Find the oscillation frequency f_0 when $R_1 = 1$ kΩ, $R_2 = 9$ kΩ, $R = 10$ kΩ, and $C = 0.001$ μF. (b) Find R when $R_1 = R_2 = 10$ kΩ, $C = 0.01$ μF, and $f_0 = 10$ kΩ. (c) Find C when $R_1 = 2$ kΩ, $R_2 = 18$ kΩ, $R = 10$ kΩ, and $f_0 = 100$ kHz. (d) Find R_1 when $R = 10$ kΩ, $C = 0.001$ μF, $R_2 = 20$ kΩ, and $f_0 = 40$ kHz. (e) Find R_2 when $R_1 = 3$ kΩ, $C = 0.001$ μF, $R = 10$ kΩ, and $f_0 = 40$ kHz.

10.81 Determine the number of distinct carrier frequencies (i.e., the maximum number of stations for a given location) in the AM broadcast band.

10.82 The envelope detector given in Fig. 10.57a on p. 725 is to be used to demodulate an AM signal whose carrier frequency is 540 kHz, where the highest frequency component of the modulating signal is 5 kHz. (a) Determine the time constant $\tau = RC$ that is geometrically centered for Inequality 10.45 on p. 726, where an inequality $a < x < b$ is geometrically centered when $b/x = x/a$. (b) Given that $R = 1$ kΩ, find the value of C that meets the condition given in part (a).

10.83 Find the frequencies of the video and audio carriers for the following television channels: (a) channel 4 (66 − 72 MHz), (b) channel 5 (76 − 82 MHz), and (c) channel 7 (174 − 180 MHz).

10.84 Determine the number of distinct carrier frequencies (i.e., the maximum number of stations for a given location) in the FM broadcast band.

10.85 For an FM signal, the ratio of the frequency deviation to the modulating frequency is called the

modulation index. The ratio of the maximum frequency deviation Δ_f to the maximum modulating frequency f_u is often called the **deviation ratio.** Assuming maximum frequency deviation (referred to as 100 percent modulation), find the deviation ratio for monophonic FM radio.

10.86 An FM radio station transmits a stereo signal but no SCA signal. Assuming that the station fully occupies its 200-kHz bandwidth, determine the signal's deviation ratio (see Problem 10.85). Is this narrowband or wideband FM?

10.87 An FM radio station transmits a stereo signal plus an SCA signal. Assuming that the station fully occupies its 200-kHz bandwidth, determine the signal's deviation ratio (see Problem 10.85). Is this narrowband or wideband FM?

Part III

Digital Systems

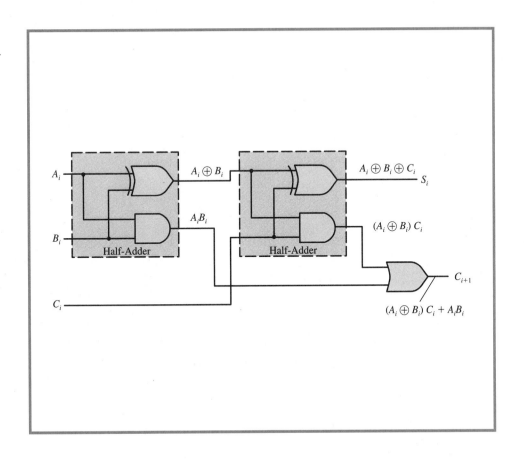

11

Digital Logic

INTRODUCTION

Almost everyone is familiar with digital clocks, digital wristwatches, pocket calculators, video games, and personal computers. Less familiar, perhaps, are general-purpose digital computers and microprocessors. Yet these systems, as well as uncountable others, have a common characteristic—they manipulate information that is in a discrete form. Since this information can be represented by digits, such systems have come to be called **digital systems.**

In our studies in preceding chapters, our preoccupation was with analog signals—that is, signals that could assume any one of an infinite number of values. (For example, at some instant of time the sinusoid $v(t) = V \cos \omega t$ can take on any value between $-V$ and V.) Now, however, we turn our attention to discrete or digital signals and the systems with which they are used.

A mechanically operated switch is a device that is either open or closed. Thus we can associate two discrete values, say 0 and 1, with a switch. Because it can be in either of two possible states, a switch is a **binary** device. Since a transistor can be used as a switch (when it is either OFF or ON), a transistor also can be employed as a binary device. Such binary devices are used to construct digital systems, from the simplest to the most complicated. Before the transistor was invented, electromechanically operated switches (relays) were often used in digital-system design. Although relays were superseded by discrete-component transistor electronics, it was the advances in integrated-circuit (IC) technology that led to the present-day proliferation of digital systems.

Perhaps the most important digital system is the digital computer. As shown in the simplified block diagram in Fig. 11.1, a digital computer has five basic components. A set of instructions, called a **program,** which will be followed, is stored in the computer's memory. Also stored in memory are the data on which the program will be executed. The data and program enter the computer via the **input.** Among the numerous types of input devices are keyboards, disk drives, magnetic-tape readers, paper-tape readers, punched-card readers, display scopes, and various types of sensors. The flow of information from the input to the memory is supervised by the **control unit,** as is the information flow between the other units of the computer. The **processor** or **arithmetic unit** manipulates the data, under the direction of the control unit, as specified by the program's instructions. These data manipulations can be arithmetic operations (e.g., addition, subtraction, multiplication, division) or other data-processing tasks. The results of the computations are available to the user at the computer's **output.** Very often the output is printed on paper or appears in graphical form on either plotters or cathode-ray tube (CRT) displays. Outputs printed on a CRT screen are also quite popular.

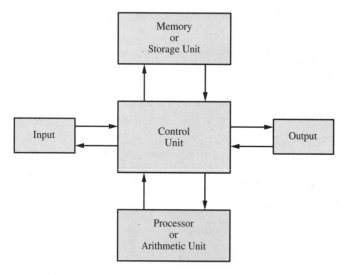

Fig. 11.1 Block diagram of a digital computer.

We now begin our study of the fundamentals of digital computers, as well as of digital systems in general. Since such systems are binary, our first concern is with the representation of information in binary form.

11.1 Binary Numbers

It is said that the number system we ordinarily use is based on the fact that we have 10 fingers. Such a system employs the digits 0, 1, 2, 3, 4, 5, 6, 7, 8, 9 and is said to be **decimal.** For example, the decimal number

1983.64

represents the following sum of the powers of 10:

$$1 \times 10^3 + 9 \times 10^2 + 8 \times 10^1 + 3 \times 10^0 + 6 \times 10^{-1} + 4 \times 10^{-2}$$

In general the decimal number

$$a_n a_{n-1} \ldots a_3 a_2 a_1 a_0 \cdot a_{-1} a_{-2} a_{-3} \ldots a_{-m}$$

represents

$$a_n \times 10^n + a_{n-1} \times 10^{n-1} + \cdots + a_2 \times 10^2 + a_1 \times 10^1$$
$$+ a_0 \times 10^0 + a_{-1} \times 10^{-1} + a_{-2} \times 10^{-2} + a_{-3} \times 10^{-3}$$
$$+ \cdots + a_{-m} \times 10^{-m}$$

where each coefficient a_i is one of the digits 0, 1, 2, 3, 4, 5, 6, 7, 8, 9.

Since it uses powers of 10, we say that the decimal number system is **base** 10, or **radix** 10. By choosing a different base, however, we get a different number system. Specifically, numbers in the base-r system also have the form

$$a_n a_{n-1} \ldots a_3 a_2 a_1 a_0 \cdot a_{-1} a_{-2} a_{-3} \ldots a_{-m}$$

which represents

$$a_n \times r^n + a_{n-1} \times r^{n-1} + \cdots + a_3 \times r^3 + a_2 \times r^2 + a_1 \times r^1$$
$$+ a_0 \times r^0 + a_{-1} \times r^{-1} + a_{-2} \times r^{-2} + a_{-3} \times r^{-3} + \cdots$$
$$+ a_{-m} \times r^{-m}$$

where each coefficient a_i is one of the elements in the set 0, 1, 2, ..., $r - 1$.

For example, the **binary** (base-2) number

1 1 0 1 0 1 . 1 0 1

is

$$1 \times 2^5 + 1 \times 2^4 + 0 \times 2^3 + 1 \times 2^2 + 0 \times 2^1 + 1 \times 2^0 + 1 \times 2^{-1}$$
$$+ 0 \times 2^{-2} + 1 \times 2^{-3}$$

and this is the decimal number 53.625. To be explicit as to the number system, we will place parentheses around a number and use a subscript to indicate the base. For instance, from the discussion above we have that

$$(1\ 1\ 0\ 1\ 0\ 1\ .\ 1\ 0\ 1)_2 = (53.625)_{10}$$

Because of our training in decimal-number arithmetic, converting a binary number into a decimal number—as was just done—is a straightforward, routine procedure. Now let us see how we can convert a decimal number to binary in a simple, systematic manner. To begin with, let the decimal number I be an integer. Expressing I in terms of powers of 2, we have that

$$\begin{aligned} I &= a_n \times 2^n + a_{n-1} \times 2^{n-1} + \cdots + a_3 \times 2^3 + a_2 \times 2^2 \\ &\quad + a_1 \times 2^1 + a_0 \times 2^0 \\ &= 2(a_n \times 2^{n-1} + a_{n-1} \times 2^{n-2} + \cdots + a_3 \times 2^2 + a_2 \times 2^1 \\ &\quad + a_1 \times 2^0) + a_0 \end{aligned}$$

If we divide I by 2, then the remainder will be a_0, where a_0 is 0 or 1, and the quotient is

$$a_n \times 2^{n-1} + a_{n-1} \times 2^{n-2} + \cdots + a_3 \times 2^2 + a_2 \times 2^1 + a_1 \times 2^0$$

If this is divided by 2, then the remainder will be a_1, where a_1 is 0 or 1. The process can be repeated until the quotient is zero, and the successive remainders will be a_2, a_3, . . . , a_{n-1}, a_n.

Example 11.1

Let us determine the binary form of the decimal number 25. By successive divisions by 2, we get the table on p. 753.

Thus we conclude that

$$(25)_{10} = (a_4 a_3 a_2 a_1 a_0)_2 = (11001)_2$$

Decimal Number	Division By 2	Quotient	Remainder
25	$\frac{25}{2}$	12	$1 = a_0$
12	$\frac{12}{2}$	6	$0 = a_1$
6	$\frac{6}{2}$	3	$0 = a_2$
3	$\frac{3}{2}$	1	$1 = a_3$
1	$\frac{1}{2}$	0	$1 = a_4$

Instead of filling out a table like the one just shown, a more convenient way to convert a decimal integer into a binary is to perform repeated divisions from right to left such as this:

	Third remainder	Second remainder	First remainder		
	a_2	a_1	a_0		
$- - - -$	Third quotient	Second quotient	First quotient	I	2

Specifically, for the decimal integer 25, we get

a_4	a_3	a_2	a_1	a_0	
1	1	0	0	1	
0	1	3	6	12	25 (2

Drill Exercise 11.1

Determine the binary form of the following decimal numbers: (a) 11, (b) 261, (c) 3523.

ANSWER   (a) 1011, (b) 100000101, (c) 110111000011

Now suppose the decimal number F is a fraction that is less than one. Expressing F in terms of powers of 2, we have that

$$F = a_{-1} \times 2^{-1} + a_{-2} \times 2^{-2} + a_{-3} \times 2^{-3} + \cdots + a_{-m} \times 2^{-m} + \cdots$$

If we multiply F by 2, then we get

$$2F = a_{-1} + a_{-2} \times 2^{-1} + a_{-3} \times 2^{-3} + \cdots + a_{-m} \times 2^{-m+1} + \cdots$$

which is a number whose integer part is equal to a_{-1}, where a_{-1} is 0 or 1. We then can form

$$2F - a_{-1} = a_{-2} \times 2^{-1} + a_{-3} \times 2^{-2} + \cdots + a_{-m} \times 2^{-m+1} + \cdots$$

Multiplying this by 2 yields a number whose integer part is equal to a_{-2}, where a_{-2} is 0 or 1. By repeating this procedure. we can obtain the remaining coefficients a_{-3}, $a_{-4}, \ldots, a_{-m}, \ldots$.

Example 11.2

Let us determine the binary form of the decimal number 0.684. By successive multiplications by 2, we get the following table.

Decimal Number	Multiplication By 2	Integer Part
0.684	1.368	$1 = a_{-1}$
0.368	0.736	$0 = a_{-2}$
0.736	1.472	$1 = a_{-3}$
0.472	0.944	$0 = a_{-4}$
0.944	1.888	$1 = a_{-5}$
0.888	1.776	$1 = a_{-6}$
0.776	1.552	$1 = a_{-7}$
0.552	1.104	$1 = a_{-8}$
etc.		

Thus we conclude that

$$(0.684)_{10} = (0.a_{-1}a_{-2}a_{-3}a_{-4}a_{-5}a_{-6}a_{-7}a_{-8} \ldots)_2 = (0.10101111 \ldots)_2$$

Drill Exercise 11.2

Determine the binary form of the following decimal numbers: (a) 0.27, (b) 0.625, (c) 0.1875.

ANSWER (1) 0.01000101. . . . , (b) 0.101, (c) 0.0011

Combining the results of the preceding two examples, we have that $(25.684)_{10} =$ $(11001.10101111 . . .)_2$

Other Number Systems

Two other number systems, which are closely related to the binary number system, are sometimes utilized. One of these, the **octal** (base-8) number system, employs the digits 0, 1, 2, 3, 4, 5, 6, 7.

Example 11.3

An example of a number conversion from octal to decimal is

$$(261.64)_8 = 2 \times 8^2 + 6 \times 8^1 + 1 \times 8^0 + 6 \times 8^{-1} + 4 \times 8^{-2}$$
$$= (146.8125)_{10}$$

Converting a decimal number to an octal number can be accomplished as was done for the decimal-to-binary conversion techniques described earlier, with the exception that division or multiplication by 8 is used in place of 2. For instance, for the decimal number $(937)_{10}$, successive divisions by 8 yield

$$
\begin{array}{ccccc}
1 & 6 & 5 & 1 & \\
\overline{(0 \ (\ 1 \ (\ 14 \ (\ 117 \ (\ 937 \ (\ 8}
\end{array}
$$

Thus

$$(937)_{10} = (1651)_8$$

Furthermore, for the decimal number $(0.194)_{10}$, successive multiplications by 8 yield

$$0.194 \times 8 = 1.552 \quad \Rightarrow \quad a_{-1} = 1$$
$$0.552 \times 8 = 4.416 \quad \Rightarrow \quad a_{-2} = 4$$
$$0.416 \times 8 = 3.328 \quad \Rightarrow \quad a_{-3} = 3$$
$$0.328 \times 8 = 2.624 \quad \Rightarrow \quad a_{-4} = 2$$
$$0.624 \times 8 = 4.992 \quad \Rightarrow \quad a_{-5} = 4$$
$$0.992 \times 8 = 7.936 \quad \Rightarrow \quad a_{-6} = 7$$

etc.

Thus

$$(0.194)_{10} = (0.143247\ldots)_8$$

Drill Exercise 11.3

(a) Find the decimal form of $(714.05)_8$. (b) Find the octal form of $(926.3125)_{10}$.

ANSWER (a) $(460.078125)_{10}$, (b) $(1636.24)_8$

The conversion of numbers from binary to octal, and vice versa, is quite simple. Since $2^3 = 8$, the digits of a binary number can be partitioned into groups of three, moving to the left and to the right from the binary point. (If there are insufficient digits to complete a group of three, then a zero or zeros can be added.) Replacing each group of three binary digits by its octal equivalent yields the octal form of the number. For example, with the addition of some zeros, the digits of the binary number 10101011000.1110011 can be partitioned into groups of three as follows:

Additional   Additional
   zero   zeros

010 101 011 000.111 001 100

 2   5   3   0   7   1   4

Substituting the octal equivalent for each group of three binary digits as indicated results in the fact that

$$(10101011000.1110011)_2 = (2530.714)_8$$

Conversely, given an octal number, by replacing each digit by its three-digit binary equivalent, we get the binary form of the number. For example, for the octal number 1704.536, we make the following associations:

 1   7   0   4 . 5   3   6

001 111 000 100   101 011 110

This means that

$$(1704.536)_8 = (1111000100.10101111)_2$$

The other number system that was mentioned as being closely related to the binary number system is the **hexadecimal** (base-16) system. Since such a system requires 16 symbols, the digits 0 through 9 do not suffice. For this reason the first six letters of the alphabet are also used. In other words, the digits for the hexadecimal number system are 0, 1, 2, 3, 4, 5, 6, 7, 8, 9, A, B, C, D, E, F. Converting numbers from binary to hexadecimal, and vice versa, can be accomplished as between binary and octal—the difference being that groups of four binary digits are used instead of groups of three. For example, for the binary number 10101011000.1110011, we have the grouping

$$0101 \quad 0101 \quad 1000 \ . \ 1110 \quad 0110$$
$$5 \qquad 5 \qquad 8 \qquad E \qquad 6$$

so that

$$(10101011000.1110011)_2 = (558.E6)_{16}$$

Table 11.1

Decimal (Base 10)	Binary (Base 2)	Octal (Base 8)	Hexadecimal (Base 16)
0	0	0	0
1	1	1	1
2	10	2	2
3	11	3	3
4	100	4	4
5	101	5	5
6	110	6	6
7	111	7	7
8	1000	10	8
9	1001	11	9
10	1010	12	A
11	1011	13	B
12	1100	14	C
13	1101	15	D
14	1110	16	E
15	1111	17	F

Furthermore, for the hexadecimal number 6A9.0C, we have

$$
\underbrace{6}_{0110} \quad \underbrace{A}_{1010} \quad \underbrace{9}_{1001} \, . \, \underbrace{0}_{0000} \quad \underbrace{C}_{1100}
$$

so that

$$
(6A9.0C)_{16} = (11010101001.000011)_2
$$

A partial summary of number systems is given in Table 11.1 on p. 757, which shows the integers 0–15 in four different bases.

11.2 Binary Arithmetic

Just as we can perform decimal arithmetic operations, so too can we add, subtract, multiply, and divide binary numbers. To perform addition, we use the following rules for adding the binary digits 0 and 1:

$$
\begin{array}{cccc}
0 & 0 & 1 & 1 \\
+\,0 & +\,1 & +\,0 & +\,1 \\
\hline
0 & 1 & 1 & 1\,0
\end{array}
$$

Note that for the first three additions, the sum is a single binary digit. For the fourth addition—which is the binary version of "one plus one equals two"—the sum consists of two binary digits. This means that, just as in ordinary decimal addition, we can have "carries" when we add binary numbers. For example, the binary equivalent of the decimal sum $60 + 42 = 102$ is the following:

```
carries→  1   1
          1  1  1  1  0  0
       +  1  0  1  0  1  0
       ───────────────────
       1  1  0  0  1  1  0
```

Subtraction of binary numbers is accomplished as is decimal subtraction, where it may be necessary to "borrow" from a higher number position. For example, the binary equivalent of the decimal difference $60 - 42 = 18$ is the following:

```
                 borrow
          1  1  1  1  0  0
       -  1  0  1  0  1  0
       ───────────────────
          0  1  0  0  1  0
```

To subtract a positive number from a smaller positive number, simply subtract the smaller number from the larger and place a negative sign in front of the difference.

Multiplication of binary numbers can be performed by using the following product rules for 0 and 1:

$$
\begin{array}{cccc}
0 & 0 & 1 & 1 \\
\times\,0 & \times\,1 & \times\,0 & \times\,1 \\
\hline
0 & 0 & 0 & 1
\end{array}
$$

We then proceed as we do for decimal numbers. For example, the binary equivalent of the decimal product $23 \times 5 = 115$ can be obtained as follows:

$$
\begin{array}{ccccccc}
 & & 1 & 0 & 1 & 1 & 1 \\
 & & & \times\,1 & 0 & 1 \\
\hline
 & & 1 & 0 & 1 & 1 & 1 \\
 & 0 & 0 & 0 & 0 & 0 \\
1 & 0 & 1 & 1 & 1 \\
\hline
1 & 1 & 1 & 0 & 0 & 1 & 1
\end{array}
$$

Division of binary numbers can be accomplished as ordinary long division is for decimal numbers. For example, the decimal division $87 \div 6 = 14.5$ when performed with binary numbers is as follows:

```
                  1 1 1 0 . 1
      1 1 0 ) 1 0 1 0 1 1 1 . 0
              1 1 0
              1 0 0 1
                1 1 0
                1 1 1
                1 1 0
                  1 1   0
                  1 1   0
                        0
```

Subtraction using "carries" may be well suited for hand calculations using pencil and paper, but other techniques are more appropriate for digital-computer calculations. In this vein, we say that the **1's complement** of a binary number is obtained by changing the 0s to 1s and the 1s to 0s. (For a fraction less than unity, the 0 to the left of the binary point is left as is.) For example,

1's complement of 1 0 1 1 0 0 1 0 = 0 1 0 0 1 1 0 1
1's complement of 0 . 1 1 1 0 1 0 0 = 0 . 0 0 0 1 0 1 1
1's complement of 1 0 1 1 . 0 0 1 = 0 1 0 0 . 1 1 0

The difference between two positive binary numbers n and m (i.e., $n - m$) can be obtained using the 1's complement as follows:

1. To n add the 1's complement of m.

2. If the sum has a carry at the left end, add a 1 to the least significant digit (this is called an **end-around carry**) of the sum. The result is $n - m$.

3. If the sum has no carry at the left end, take the 1's complement of the sum and place a negative sign in front of it. The result is $n - m$.

Example 11.4

Given that $n = 1101001$ and $m = 0101110$, to obtain $n - m$, we proceed as follows:

```
          1 1 0 1 0 0 1    n
        + 1 0 1 0 0 0 1    1's complement of m
        1 0 1 1 1 0 1 0    sum
end-around carry
                     → 1
          0 1 1 1 0 1 1    n − m
```

Now to form $m - n$, we have the following:

```
  0 1 0 1 1 1 0    m
+ 0 0 1 0 1 1 0    1's complement of n
  1 0 0 0 1 0 0    sum
− 0 1 1 1 0 1 1    m − n
```

The **10's complement** of a positive decimal number whose integer part has p digits is obtained by subtracting that number from 10^p. For instance, the 10's complement of 47 is $10^2 - 47 = 53$, whereas the 10's complement of 125.84 is $10^3 - 125.84 = 874.16$. The 10's complement of 0 is defined to be 0.

The **2's complement** of a positive binary number whose integer part has p digits is obtained by subtracting that number from 2^p. For instance, the 2's complement of 11010 is $2^5 - 11010 = 100000 - 11010 = 00110$, whereas the 2's complement of 1011.001 is $2^4 - 1011.001 = 10000 - 1011.001 = 0100.111$. The 2's complement of 0 is defined to be 0. The 2's complement of a positive binary number can also be obtained by adding a 1 to the rightmost digit of its 1's complement.

Example 11.5

1's complement of 1 0 1 1 0 0 1 0 = 0 1 0 0 1 1 0 1
$$\phantom{1\text{'s complement of 1 0 1 1 0 0 1 0 = 0 1 0 0 1}} + \ 1$$
2's complement of 1 0 1 1 0 0 1 0 = $\overline{0\ 1\ 0\ 0\ 1\ 1\ 1\ 0}$

1's complement of 0 . 1 1 1 0 1 0 0 = 0 . 0 0 0 1 0 1 1
$$\phantom{1\text{'s complement of 0 . 1 1 1 0 1 0 0 = 0 . 0 0 0 1 0}} + \ 1$$
2's complement of 0 . 1 1 1 0 1 0 0 = $\overline{0\ .\ 0\ 0\ 0\ 1\ 1\ 0\ 0}$

1's complement of 1 0 1 1 . 0 0 1 = 0 1 0 0 . 1 1 0
$$\phantom{1\text{'s complement of 1 0 1 1 . 0 0 1 = 0 1 0 0 . 1}} + \ 1$$
2's complement of 1 0 1 1 . 0 0 1 = $\overline{0\ 1\ 0\ 0\ .\ 1\ 1\ 1}$

Drill Exercise 11.5

Use two different approaches to determine the 2's complement of (a) 11010110, (b) 0.101011, and (c) 1101.011.

ANSWER   (a) 00101010, (b) 0.010101, (c) 0010.101

We are all familiar with ordinary decimal subtraction. There is an alternative technique, however, for performing this type of operation. Instead of subtracting a specific number, we may add its 10's complement and ignore any carry that results. For example, the alternative to $8 - 3 = 5$ is to take the 10's complement of 3, which is 7 (because $10 - 3 = 7$), and form the sum $8 + 7 = 15$. Ignoring the carry, we obtain 5—the desired result. Similarly, the alternative to $27 - 40 = -13$ is to take the 10's complement of 40, which is $10^2 - 40 = 60$, and form the sum

27 + 60 = 87. Since there is no carry, we take the 10's complement of 87 (which is $10^2 - 87 = 13$) and insert a negative sign. The result is -13—as it should be.

The difference between two positive binary numbers n and m (i.e., $n - m$) can be obtained by using the 2's complement as follows:

1. To n add the 2's complement of m.

2. If the sum has a carry at the left end, remove it. The result is $n - m$.

3. If the sum has no carry at the left end, take the 2's complement of the sum and insert a negative sign. The result is $n - m$.

Example 11.6

Given $n = 1101001$ and $m = 0101110$, to obtain $n - m$, we proceed as follows:

	1	0	1	0	0	0	1	1's complement of m
							+ 1	
	1	0	1	0	0	1	0	2's complement of m
	+ 1	1	0	1	0	0	1	n
remove →	1	0	1	1	1	0	1 1	sum
	0	1	1	1	0	1	1	$n - m$

Now to form $m - n$, we have the following:

	0	0	1	0	1	1	0	1's complement of n
							+ 1	
	0	0	1	0	1	1	1	2's complement of n
	+ 0	1	0	1	1	1	0	m
	1	0	0	0	1	0	1	sum
	0	1	1	1	0	1	0	1's complement of sum
insert							+ 1	
negative sign	− 0	1	1	1	0	1	1	$m - n$

Binary Codes

We have already seen how we can express numerical information in binary (base-2) form. Other types of information, such as letters of the alphabet and various kinds of symbols, are also very important. To be handled by digital systems, such as digital computers, information must be in some type of binary form. This can be accomplished by ''encoding'' information with the use of ''binary codes.''

The binary **encoding** of information consists of representing various pieces of information by sequences of the binary digits, called **bits,** 0 and 1. A sequence of 8 bits is referred to as a **byte.** The resulting set of binary sequences is known as a **binary code,** and each sequence is a **code word.** For example, suppose that we had four pieces of information W, X, Y, and Z that we would like to encode in binary form. Since there are four distinct binary pairs (binary sequences with 2 bits), those being 00, 01, 10, and 11, one possibility for a binary encoding of W, X, Y, and Z is to make the associations shown below:

W ↔ 00

X ↔ 01

Y ↔ 10

Z ↔ 11

The resulting set of binary sequences (00, 01, 10, 11) is the binary code.

When dealing with computers, users prefer to use the decimal representations of numbers rather than the binary versions. Because of this, even though real numbers can be expressed in binary (base-2) form, often it is desirable to encode decimal numbers with a binary code. The encoded decimal numbers can then be converted into binary numbers for purposes of arithmetic operations, or alternatively, the arithmetic can be performed directly on the encoded decimal numbers.

Since there are 10 decimal digits (0, 1, 2, 3, 4, 5, 6, 7, 8, 9), a minimum of 4 bits is required to encode these digits into binary. (This is a consequence of the fact that there are only eight distinct binary triples: 000, 001, 010, 011, 100, 101, 110, 111.) A popular scheme for encoding the decimal digits is to use binary quadruples (sequences of 4 bits), where each quadruple is the binary (base-2) representation of the corresponding decimal digit. The resulting code, referred to as **binary-coded decimal** (BCD), is given in Table 11.2. For example, the BCD version of the decimal number 1983 is 0001 1001 1000 0011.

Let us now suppose that eight pieces of information are encoded into the eight distinct binary triples 000, 001, 010, 011, 100, 101, 110, and 111. During the transmission of information, it is possible that electrical noise can cause an error in a code word—that is, either a 0 is changed to 1, or a 1 is changed to 0. If this happens,

Table 11.2

Decimal Digit	BCD
0	0 0 0 0
1	0 0 0 1
2	0 0 1 0
3	0 0 1 1
4	0 1 0 0
5	0 1 0 1
6	0 1 1 0
7	0 1 1 1
8	1 0 0 0
9	1 0 0 1

Table 11.3

Coded Information	Parity Bit	New Code
0 0 0	0	0 0 0 0
0 0 1	1	0 0 1 1
0 1 0	1	0 1 0 1
0 1 1	0	0 1 1 0
1 0 0	1	1 0 0 1
1 0 1	0	1 0 1 0
1 1 0	0	1 1 0 0
1 1 1	1	1 1 1 1

when the triple is decoded, the wrong piece of information will result. To help prevent such an occurrence, an additional bit—called a **parity bit**—can be added to the triples so a new code consisting of binary quadruples will be produced. One possible rule for adding the parity bit is to make each new code word have an even number of 1s. Table 11.3 indicates what parity bit should be chosen and the resulting new code. Since any word in the new code differs from any other word in at least two positions, a single error cannot change one code word into another. If a single error is made in a code word, then the result will be an odd number of 1s in that word. Thus by counting the number of 1s in a code word, a single error can be detected. Two errors in a code word, however, will cause an even number of 1s, and, therefore, two errors cannot be detected. The same statement can be made about four errors. Since three errors in a code word also produce an odd number of 1s, three errors can be detected as well.

Table 11.3 shows an error-detecting code that is obtained by adding a parity bit such that each resulting code word has an even number of 1s. Another error-detecting code can be formed by making the number of 1s in each code word odd (see Problem 11.18 at the end of this chapter). These are normally referred to as **even** and **odd parity,** respectively. Although such codes have error-detection capabilities, they do not have error-correction capabilities. In other words, if a single error is made in one of the code words of the new code given in Table 11.3, the result is a binary quadruple that is not a code word; however, there is no way of knowing for sure what the original code word was. To construct codes with the ability to correct errors, additional parity bits are required. The subject of such **error-correcting codes** is beyond the scope of this book.

Having seen how information can be expressed or represented in binary form by using binary numbers or an appropriate binary encoding scheme, let us turn our attention to the binary operation of electric and electronic circuits.

11.3 Digital Logic Circuits

Consider the simple circuit consisting of a battery V, a switch S, and a lightbulb L shown in Fig. 11.2. Suppose we indicte that the switch is open by $S = 0$ and closed by $S = 1$. In other words, the variable S designates the state of the switch (open or closed) and S takes on one of two values (0 or 1). Since S can take on only two values, we say that it is a **binary variable.** When the switch is open ($S = 0$), there is an open circuit between a and b (the terminals of the switch), and, therefore, the lightbulb is off; however, when the switch is closed ($S = 1$), there is a short circuit between a and b, and the lightbulb is on.

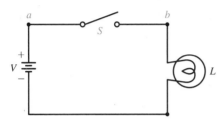

Fig. 11.2 Simple switching circuit.

Table 11.4 Truth Tables for (a) $L = S$ and (b) $L = \bar{S}$.

	(a)		(b)
S	$L = S$	S	$L = \bar{S}$
0	0	0	1
1	1	1	0

Suppose we also designate the state of the lightbulb by $L = 0$ when the lightbulb is off, and $L = 1$ when it is on. Doing this we see that L is also a binary variable, and $S = 0 \Rightarrow L = 0$, while $S = 1 \Rightarrow L = 1$. In other words, $L = S$, and a table indicating the relationship between S and L is given by Table 11.4a. Such a table is called a **truth table.** If the lightbulb being off is designated by $L = 1$ and being on is designated by $L = 0$, then we get the truth table given by Table 11.4b. Under this circumstance we say that L is the **complement** of S and express this mathematically as $L = \bar{S}$. However, unless indicated otherwise, $L = 0$ denotes that the lightbulb is off and $L = 1$ means that the lightbulb is on.

Switching Circuits

A circuit such as the one shown in Fig. 11.2, where a lightbulb L is either off or on depending upon whether a switch (or switches) is open or closed, is referred to as a **switching circuit.**

Let us consider the parallel connection of switches A and B shown in Fig. 11.3, where although the battery V and the lightbulb L of the corresponding switching circuit are not shown, they are assumed to be connected as depicted in Fig. 11.2. When either switch (or both) is closed, there is a short circuit between a and b, and, therefore, the lightbulb is on. In other words, when either $A = 1$ or $B = 1$, then $L = 1$. When both switches are open ($A = 0$ and $B = 0$), then the lightbulb is off

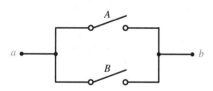

Fig. 11.3 Two switches connected in parallel.

Table 11.5 Truth Table for the OR Operation.

A	B	L = A + B
0	0	0
0	1	1
1	0	1
1	1	1

($L = 0$). This behavior of the two switches connected in parallel is summarized in Table 11.5. This truth table describes the **OR operation** for binary variables. By specifying A and B to be input variables and L to be the output variable, then the output $L = 1$ when either input $A = 1$ OR $B = 1$ (or both)—otherwise $L = 0$. Instead of writing $L = A$ OR B, we use the mathematical notation $L = A + B$.

Figure 11.4 shows the series connection of switches A and B. In this case, the only situation for which $L = 1$ (i.e., there s a short circuit between a and b) is when both $A = 1$ AND $B = 1$ (i.e., both switches are closed). The resulting truth table, given in Table 11.6, describes the **AND operation.** The mathematical notation for this operation is $L = A \cdot B$ or, more simply, $L = AB$.

Table 11.6 Truth Table for the AND Operation.

A	B	L = AB
0	0	0
0	1	0
1	0	0
1	1	1

Fig. 11.4 Two switches connected in series.

For the connection of three switches shown in Fig. 11.5, the resulting truth table is given in Table 11.7. In this case, the truth table requires eight rows instead of four so that all the possible combinations of open and closed switches are included. Since switches B and C are connected in parallel, that connection can be described by $B + C$. This parallel combination, however, is connected in series with switch A. Thus we may write that $L = A(B + C)$. But does the distributive law hold for binary variables? In other words, does $A(B + C) = AB + AC$? The answer to this question can be found by solving Problem 11.19 at the end of this chapter.

Given a switch A and a switch B, if B is closed when A is open (if $B = 1$ when $A = 0$), and if B is open when A is closed (if $B = 0$ when $A = 1$), then from the truth table corresponding to these conditions (Table 11.8), we see that $B = \overline{A}$.

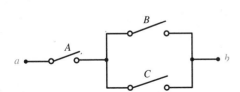

Fig. 11.5 Connection of three switches.

Table 11.7 Truth Table for Three-Switch Connection.

A	B	C	$L = A(B + C)$
0	0	0	0
0	0	1	0
0	1	0	0
0	1	1	0
1	0	0	0
1	0	1	1
1	1	0	1
1	1	1	1

Table 11.8

A	$B = \overline{A}$
0	1
1	0

We can use this concept of a complement in the design of an arbitrary switching circuit—that is, we can connect switches together to realize a given truth table. For example, consider Table 11.9. (This truth table describes the **exclusive-OR operation.**) Since $L = 1$ when $A = 0$ ($\overline{A} = 1$) AND $B = 1$, OR when $A = 1$ AND $B = 0$ ($\overline{B} = 1$), then we have that $L = \overline{A}B + A\overline{B}$. A connection of switches that realizes this expression is shown in Fig. 11.6

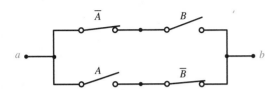

Fig. 11.6 Realization of exclusive-OR operation.

Table 11.9 Truth Table for the Exclusive-OR Operation.

A	B	$L = \overline{A}B + A\overline{B}$
0	0	0
0	1	1
1	0	1
1	1	0

Binary operations such as AND, OR, exclusive-OR, and complement are referred to as **logic operations,** and circuits that perform these operations are called **logic gates.**

Up to this point we have been dealing with switches of the type shown in Fig. 11.7a. Such a switch, called a **single-pole, single-throw** (SPST) switch, either connects a and b or it does not. Another type of switch, a **single-pole, double-throw** (SPDT) switch, is shown in Fig. 11.7b. For this three-terminal device, either a and b are connected or a and c are connected—there is no other switch condition. This means that if a and b are considered to be the terminals of switch A, then terminals a and c comprise switch $\overline{A}$. Of course, by not using either terminal b or terminal c, a SPDT switch can be used as a SPST switch.

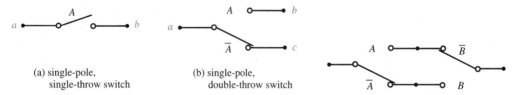

(a) single-pole,
single-throw switch

(b) single-pole,
double-throw switch

Fig. 11.7 Single-pole switches.

Fig. 11.8 Alternative realization of exclusive-OR operation.

By using SPDT switches, it is a simple matter to realize the exclusive-OR operation given in Table 11.9. Specifically, the connection of two SPDT switches as shown in Fig. 11.8 yields $L = \overline{A}B + A\overline{B}$.

The realization of the exclusive-OR operation with switches has a very practical application—a two-way light switch. To see why this is true, again consider Table 11.9. Suppose that a room has two doors, where switch A is next to one door and switch B is next to the other door. We begin with both switches open ($A = B = 0$) and the room light off ($L = 0$). Upon entering door A, the switch is thrown ($A = 1$, $B = 0$) and the light goes on ($L = 1$). Leaving by the same door results in the initial conditions; however, leaving by door B instead and throwing that switch ($A = 1$, $B = 1$) also turns the light off ($L = 0$). If after leaving by door B, door A is entered and the switch is thrown, the result is $A = 0$, $B = 1$, and $L = 1$—that is, the light goes on. Leaving by either door and throwing the corresponding switch again will turn off the light.

The concept of a two-way light switch can be extended to a three-way switch, and so on. Any further discussion of this topic, however, is left as an exercise for the reader (see Problems 11.24 and 11.25 at the end of this chapter).

Logic Gates

We have seen how we can construct circuits corresponding to binary expressions (or functions) by using mechanically actuated switches. However, because transistors can be used as electronic switches, manually operated switches and relays have become passé for purposes of performing logical operations. Instead, as discussed in the electronics portion of this book, various types of electronic circuits are used to realize logic gates. At this point, however, we will not be concerned with the electronic details of how such gates are constructed, but rather will introduce digital-circuit symbols to signify the various types of logic gates.

A digital logic gate whose input is a binary variable, say A, and whose output is also A, is referred to as a **buffer.** The circuit symbol for such a gate is shown in Fig. 11.9. A logic circuit, whose output is $\overline{A}$ (the complement of A) when the input is A, is called a **NOT gate** or **inverter.** The circuit symbol for this gate is obtained from the buffer symbol by placing a small circle at the buffer's output as shown in Fig. 11.10. The truth table for a NOT gate is given in Table 11.8.

Fig. 11.9 Buffer.   **Fig. 11.10** NOT gate.

Another logic circuit is the **OR gate.** Figure 11.11 shows a two-input OR gate whose inputs are A and B, and whose output is $A + B$ (see Table 11.5). Although a NOT gate has only one input, an OR gate can have two or more inputs. (Very shortly we will discuss the situation of more than two inputs in greater detail.) Figure 11.12 shows the circuit symbol of a two-input **AND gate,** that is, a gate that has an output of AB (see Table 11.6) when its inputs are A and B.

Fig. 11.11 Two-input OR gate.   **Fig. 11.12** Two-input AND gate.

By approximately connecting AND, OR, and NOT gates, it is possible to construct a digital logic circuit that realizes any binary function (expression). As an example of this fact, recall the exclusive-OR operation mentioned previously. We use the symbol $\oplus$ to denote this important operation. From Table 11.9, which is repeated by Table 11.10, we have that $A \oplus B = \overline{A}B + A\overline{B}$. The logic circuit shown in Fig. 11.13 utilizes two AND gates, two NOT gates, and one OR gate to realize this operation. Thus we have used AND, OR, and NOT gates to construct an **exclusive-OR gate.** However, since exclusive-OR gates are used frequently, such a gate has its own circuit symbol—shown in Fig. 11.14.

Table 11.10 Truth Table for the Exclusive-OR Operation.

A	B	$A \oplus B = \overline{A}B + A\overline{B}$
0	0	0
0	1	1
1	0	1
1	1	0

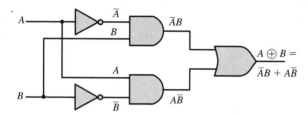

Fig. 11.13 Realization of an exclusive-OR gate.

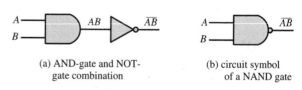

Fig. 11.14 Two-input exclusive-OR gate.

Another very important logic operation is obtained by taking the complement after performing the AND operation. The truth table that describes this situation is given by Table 11.11, and we refer to this operation as **NAND**—a contraction of NOT-AND.[1] Although this operation can be realized with an AND gate followed by a NOT gate as shown in Fig. 11.15a, we give the **NAND gate** its own circuit symbol. This is shown in Fig. 11.15b.

Table 11.11 Truth Table for the NAND Operation.

A	B	AB	$\overline{AB}$
0	0	0	1
0	1	0	1
1	0	0	1
1	1	1	0

(a) AND-gate and NOT-gate combination

(b) circuit symbol of a NAND gate

Fig. 11.15 Two-input NAND gates.

By following an OR gate with a NOT gate as shown in Fig. 11.16a, we obtain a **NOR gate** (NOR is a contraction of NOT-OR). As with the NAND gate, the NOR gate has its own circuit symbol—shown in Fig. 11.16b. The truth table for the NOR operation is given by Table 11.12.

---

[1]The contraction may be misleading because the AND operation is performed prior to the NOT operation.

Table 11.12 Truth Table for the NOR Operation.

A	B	A + B	$\overline{A + B}$
0	0	0	1
0	1	1	0
1	0	1	0
1	1	1	0

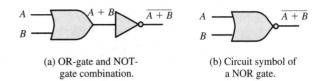

(a) OR-gate and NOT-gate combination.

(b) Circuit symbol of a NOR gate.

Fig. 11.16 Two-input NOR gates.

Example 11.7

For the logic circuit shown in Fig. 11.17, clearly the output is characterized by the binary function $F = \overline{\overline{A + B} + AB}$. By substituting the values of 0 and 1 for the variables A and B, we can obtain the corresponding truth table. This approach is illustrated by Table 11.13.

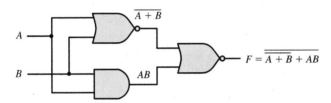

Fig. 11.17 Exclusive-OR gate.

Table 11.13 Truth Table for Example.

A	B	A + B	$\overline{A + B}$	AB	$\overline{A + B} + AB$	$\overline{\overline{A + B} + AB}$
0	0	0	1	0	1	0
0	1	1	0	0	0	1
1	0	1	0	0	0	1
1	1	1	0	1	1	0

Drill Exercise 11.7

For the logic circuit shown in Fig. 11.17, replace each NOR gate with a NAND gate and the AND gate with an OR gate. Determine the resulting function F, and obtain the corresponding truth table.

ANSWER $F = \overline{(\overline{AB})(A + B)}$

From Example 11.7 we see that the truth table for $F = \overline{A + B} + AB$ is identical to that for the exclusive-OR operation. Thus the logic circuit given in Fig. 11.17 realizes an exclusive-OR gate. Furthermore, we must have that

$$F = \overline{A + B} + AB = \overline{A}B + A\overline{B} = A \oplus B \qquad (11.1)$$

Trying to simplify a binary function by constructing its corresponding truth table is a very tedious procedure. Fortunately, though, there is an analytical means for simplifying binary expressions. This is the next topic of discussion.

11.4 Boolean Algebra

We have seen that by connecting logic gates, we can construct a logic circuit whose output is described by some binary function (or expression) F. As indicated by Eq. 11.1, for example, such an expression can have more than one form. We can, however, obtain one form from another by using mathematical manipulations that do not require the construction of truth tables. Let us see how this is accomplished.

We are all familiar with the mathematical system consisting of the set of real numbers together with the operations of multiplication and addition.[2] Another important mathematical system consists of the set of all complex numbers together with the operations of complex-number multiplication and complex-number addition. Fortunately, there is also a structure to the binary elements 0 and 1 along with the operations of AND, OR, and NOT (complement). Such a mathematical system is referred to as **switching algebra** (because of the use of switches in the early days) or **Boolean**[3] **algebra.** We will now describe the rules for this algebra.

First, to recapitulate, suppose that A and B are binary variables. Then the truth tables for the operations AND (denoted by $A \cdot B$ or AB) and OR (denoted $A + B$) are given by Tables 11.14 and 11.15, respectively. Furthermore, for the binary var-

Table 11.14 Truth Table for A AND B.

A	B	AB
0	0	0
0	1	0
1	0	0
1	1	1

Table 11.15 Truth Table for A OR B.

A	B	$A + B$
0	0	0
0	1	1
1	0	1
1	1	1

---

[2]Dividing by a number is equivalent to multiplying by the reciprocal of that number. Subtracting a number is equivalent to adding the negative of that number.

[3]Named for the English mathematician George Boole (1815–1864).

iable A, the truth table for the operation NOT, or complement (denoted $\overline{A}$), is given by Table 11.16. For these operations, it is simply a matter of filling out the corre-

Table 11.16 Truth Table for NOT A (the Complement of A).

A	$\overline{A}$
0	1
1	0

sponding truth tables to show the following equalities:

$$A \cdot 1 = A \qquad A + 0 = A$$
$$A \cdot A = A \qquad A + A = A$$
$$A \cdot 0 = 0 \qquad A + 1 = 1$$
$$A \cdot \overline{A} = 0 \qquad A + \overline{A} = 1$$

A number of other results, which can be confirmed with the use of truth tables, are listed below.

Involution:	$\overline{\overline{A}} = A$
Commutative law for AND:	$AB = BA$
Commutative law for OR:	$A + B = B + A$
Associative law for AND:	$A(BC) = (AB)C$
Associative law for OR:	$A + (B + C) = (A + B) + C$
Distributive law for AND over OR:	$A(B + C) = AB + AC$
Distributive law for OR over AND:	$A + BC = (A + B)(A + C)$

Implicit for these distributive laws is the fact that the AND operation has precedence over the OR operation. Specifically, for the distributive law $A + BC = (A + B)(A + C)$, the AND operation BC is performed prior to the OR operation in the term $A + BC$. Since this distributive law may seem strange (when compared with the distribution of real numbers), let us verify the expression with a truth table. Because the columns labeled $A + BC$ and $(A + B)(A + C)$ are identical, Table 11.17 confirms the distributive law for OR over AND.

Table 11.17 Truth Table for $A + BC = (A + B)(A + C)$.

A	B	C	BC	A + BC	A + B	A + C	(A + B)(A + C)
0	0	0	0	0	0	0	0
0	0	1	0	0	0	1	0
0	1	0	0	0	1	0	0
0	1	1	1	1	1	1	1
1	0	0	0	1	1	1	1
1	0	1	0	1	1	1	1
1	1	0	0	1	1	1	1
1	1	1	1	1	1	1	1

Two absorption laws for Boolean algebra are the following:

$$A + AB = A \qquad \text{and} \qquad A(A + B) = A \qquad \text{absorption laws}$$

Instead of verifying these equalities with truth tables, let us use algebraic manipulations. For example,

$$
\begin{aligned}
A + AB &= A \cdot 1 + AB && \text{since } A \cdot 1 = A \\
&= A(1 + B) && \text{distributive law} \\
&= A \cdot 1 && \text{since } 1 + B = B + 1 \quad \text{and} \quad B + 1 = 1 \\
&= A && \text{since } A \cdot 1 = A
\end{aligned}
$$

Also

$$
\begin{aligned}
A(A + B) &= AA + AB && \text{distributive law} \\
&= A + AB && \text{since } AA = A \\
&= A && \text{preceding absorption law}
\end{aligned}
$$

Two additional types of absorption laws are obtained as follows:

$$
\begin{aligned}
A + \overline{A}B &= (A + \overline{A})(A + B) && \text{distributive law} \\
&= 1(A + B) && \text{since } A + \overline{A} = 1 \\
&= A + B && \text{since } 1 \cdot A = A
\end{aligned}
$$

and

$$A(\overline{A} + B) = A\overline{A} + AB \qquad \text{distributive law}$$
$$= 0 + AB \qquad \text{since } A\overline{A} = 0$$
$$= AB \qquad \text{since } 0 + A = A + 0 \quad \text{and} \quad A + 0 = A$$

In summary, therefore,

$$\boxed{A + \overline{A}B = A + B} \qquad \text{and} \qquad \boxed{A(\overline{A} + B) = AB}$$

Two very important results, known as **De Morgan's theorems**,[4] are

$$\boxed{\overline{A + B} = \overline{A}\,\overline{B}} \qquad \text{and} \qquad \boxed{\overline{AB} = \overline{A} + \overline{B}}$$

The simplest way of establishing these facts is with the use of truth tables. This is done by Tables 11.18 and 11.19, respectively.

Table 11.18 Proof of $\overline{A + B} = \overline{A}\,\overline{B}$.

A	B	A + B	$\overline{A + B}$	$\overline{A}$	$\overline{B}$	$\overline{A}\,\overline{B}$
0	0	0	1	1	1	1
0	1	1	0	1	0	0
1	0	1	0	0	1	0
1	1	1	0	0	0	0

Table 11.19 Proof of $\overline{AB} = \overline{A} + \overline{B}$.

A	B	AB	$\overline{AB}$	$\overline{A}$	$\overline{B}$	$\overline{A} + \overline{B}$
0	0	0	1	1	1	1
0	1	0	1	1	0	1
1	0	0	1	0	1	1
1	1	1	0	0	0	0

---

[4]Named for the English mathematician Augustus De Morgan (1806–1871).

Example 11.8

Let us verify algebraically that $\overline{A + B + AB} = \overline{A}B + A\overline{B}$ (Eq. 11.1).

$$\overline{A + B + AB} = (\overline{A + B})(\overline{AB}) \qquad \text{since } \overline{A + B} = \overline{A}\,\overline{B}$$

$$= (A + B)(\overline{AB}) \qquad \text{involution}$$

$$= (A + B)(\overline{A} + \overline{B}) \qquad \text{since } \overline{AB} = \overline{A} + \overline{B}$$

$$= (A + B)\overline{A} + (A + B)\overline{B} \qquad \text{distributive law}$$

$$= \overline{A}(A + B) + \overline{B}(B + A) \qquad \text{commutative laws}$$

$$= \overline{A}B + \overline{B}A = \overline{A}B + A\overline{B} \qquad \text{since } A(\overline{A} + B) = AB$$

$$\text{and } \overline{B}A = A\overline{B}$$

Drill Exercise 11.8

Verify that $(\overline{AB})(A + B) = \overline{A}B + A\overline{B}$.

A frequently used property of Boolean algebra is the principle of **duality.** Two expressions are said to be **duals** if all the AND and OR operations are interchanged and all the 0s and 1s are interchanged. (The variable names and complements remain unchanged.) The following pairs are some examples of duals

AB and $A + B$		
$A + 0 = A$ and $A \cdot 1 = A$		
$B \cdot 0 = 0$ and $B + 1 = 1$		
$A(BC) = (AB)C$ and $A + (B + C) = (A + B) + C$	associative laws
$A(B + C) = AB + AC$ and $A + BC = (A + B)(A + C)$	distributive laws
$A + AB = A$ and $A(A + B) = A$	absorption laws
$A + \overline{A}B = A + B$ and $A(\overline{A} + B) = AB$	absorption laws
$\overline{A + B} = \overline{A}\,\overline{B}$ and $\overline{AB} = \overline{A} + \overline{B}$	De Morgan's theorems

Once an equality has been demonstrated, its dual is automatically valid and, therefore, does not need to be verified. Thus we get two proofs for the price of one. Just because two expressions are duals does not mean they are equivalent—in general, they are not equivalent.

Example 11.9

The dual of the exclusive-OR operation

$$A \oplus B = \overline{A}B + A\overline{B}$$

is called the **exclusive-NOR** (or **equivalence**) operation, and is denoted $A \odot B$. In other words,

$$A \odot B = (\overline{A} + B)(A + \overline{B})$$

The truth table for this operation is given by Table 11.20, and the circuit symbol for an **exclusive-NOR gate** is shown in Fig. 11.18.

Table 11.20 Truth Table for Exclusive-NOR Operation.

A	B	$\overline{A}$	$\overline{B}$	$\overline{A} + B$	$A + \overline{B}$	$A \odot B$
0	0	1	1	1	1	1
0	1	1	0	1	0	0
1	0	0	1	0	1	0
1	1	0	0	1	1	1

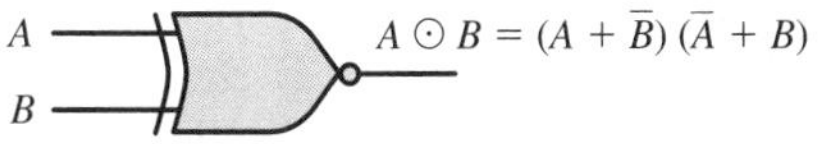

$$A \odot B = (A + \overline{B})(\overline{A} + B)$$

Fig. 11.18 Exclusive-NOR gate.

Drill Exercise 11.9

Find the dual of $(\overline{A} + \overline{B})(A + B)$, and determine the truth table for the resulting expression. What operation does this expression describe?

ANSWER $\overline{A}\overline{B} + AB$, the exclusive-NOR operation

Multiple-Input Gates

The associativity law for the OR operation states that $A + (B + C) = (A + B) + C$. This property, along with the commutative law for the OR operation, $A + B = B + A$, means that the order in which the OR operations are performed on three binary variables does not matter—the results are the same. Therefore, we can write $A + B + C$, and its meaning is unambiguous. A three-input OR gate is a logic gate whose output is $A + B + C$ when its inputs are A, B, and C. The circuit symbol for such a logic gate is shown in Fig. 11.19.

For the case of the AND operation, just as was deduced in the discussion of the OR operation, it is unambiguous to write ABC. The circuit symbol for a three-input AND gate is shown in Fig. 11.20. For the reasons just discussed, the number of inputs for AND gates and OR gates can be extended to four or more.

Fig. 11.19 Three-input OR gate.

Fig. 11.20 Three-input AND gate.

Now consider the NOR operation. Let us denote this operation by the symbol $\downarrow$. Thus $A \downarrow B = \overline{A + B}$. Clearly, commutivity holds for this operation—$A \downarrow B = B \downarrow A$—because the OR operation is commutative. Associativity, however, does not hold; that is,

$$A \downarrow (B \downarrow C) \neq (A \downarrow B) \downarrow C$$

because

$$A \downarrow (B \downarrow C) = \overline{A + \overline{B + C}} = \overline{A + \overline{B}\,\overline{C}} = \overline{A}(\overline{\overline{B}\,\overline{C}}) = \overline{A}(B + C)$$

and

$$(A \downarrow B) \downarrow C = \overline{\overline{A + B} + C} = \overline{\overline{A}\,\overline{B} + C} = (\overline{\overline{A}\,\overline{B}})\overline{C} = (A + B)\overline{C}$$

Therefore, we shall define a three-input NOR gate to be a logic gate whose output is $\overline{A + B + C}$ when its inputs are A, B, and C. The circuit symbol for such a logic gate is shown in Fig. 11.21.

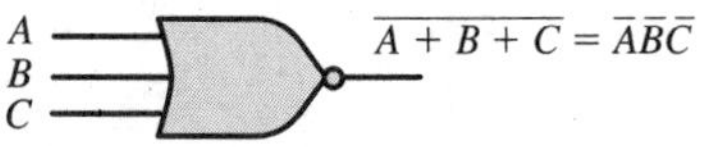

$$\overline{A + B + C} = \overline{A}\,\overline{B}\,\overline{C}$$

Fig. 11.21 Three-input NOR gate.

Let us denote the NAND operation by the symbol $\uparrow$. Thus $A \uparrow B = \overline{AB}$. Similar to the last discussion, the NAND operation is commutative but not associative (see Problem 11.41 at the end of this chapter). Therefore, we define the output of the three-input NAND gate to be $\overline{ABC}$ when the inputs are A, B, and C. The circuit symbol for such a logic gate is shown in Fig. 11.22.

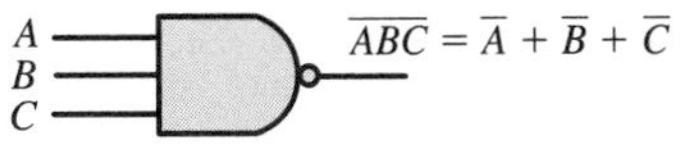

$$\overline{ABC} = \overline{A} + \overline{B} + \overline{C}$$

Fig. 11.22 Three-input NAND gate.

Note that

$$\overline{A + B + C} = \overline{A + (B + C)} = \overline{A}(\overline{B + C}) = \overline{A}(\overline{B}\,\overline{C}) = \overline{A}\,\overline{B}\,\overline{C}$$

and

$$\overline{ABC} = \overline{A(BC)} = \overline{A} + \overline{BC} = \overline{A} + (\overline{B} + \overline{C}) = \overline{A} + \overline{B} + \overline{C}$$

These are just the three-variable versions of De Morgan's theorems, and, therefore, we have alternative expressions for the outputs of the NOR and NAND gates shown in Figs. 11.21 and 11.22, respectively.

Clearly, NOR and NAND gates can have more than three inputs, and De Morgan's theorems can be extended as well. For example, for the case of the four inputs A, B, C, and D, the output of a four-input NOR gate is

$$\overline{A + B + C + D} = \overline{A}\,\overline{B}\,\overline{C}\,\overline{D}$$

and the output of a four-input NAND gate is

$$\overline{ABCD} = \overline{A} + \overline{B} + \overline{C} + \overline{D}$$

11.5 Standard Forms of Boolean Functions

A binary function comprised of binary variables, along with AND, OR, and NOT operations, is referred to as a **Boolean function.** An example of a Boolean function of the two variables A and B is

$$F_1 = \overline{A}B + A\overline{B}$$

As we have already seen, this function describes the exclusive-OR operation, as well as a two-way light switch (Fig. 11.9). In addition to an algebraic expression, a Boolean function can also be specified in terms of a truth table. The truth table that corresponds to $F_1 = \overline{A}B + A\overline{B}$ is given by Table 11.21.

Another example of a Boolean function is $F_2 = (A + B)(\overline{AB})$. The truth table for this function (along with the columns for $A + B$, AB, and $\overline{AB}$, which aid computation) is given in Table 11.22. Since the columns for A, B, and F_1 in Table 11.21 are identical to those for A, B, and F_2 in Table 11.22, respectively, then F_1 and F_2 have the same truth tables. This means that $F_1 = F_2$.

Table 11.21 Truth Table for $F_1 = \overline{A}B + A\overline{B}$.

A	B	F_1
0	0	0
0	1	1
1	0	1
1	1	0

Table 11.22 Truth Table for $F_2 = (A + B)(\overline{AB})$.

A	B	$A + B$	AB	$\overline{AB}$	F_2
0	0	0	0	1	0
0	1	1	0	1	1
1	0	1	0	1	1
1	1	1	1	0	0

Even though $F_1 = \overline{A}B + A\overline{B} = (A + B)(\overline{AB}) = F_2$, it is the expression $F_1 = \overline{A}B + A\overline{B}$ that is evident from the truth table. That is so because $F_1 = 1$ when $A = 0$ $(\overline{A} = 1)$ AND $B = 1$, OR $A = 1$ AND $B = 0$ $(\overline{B} = 1)$.

The Sum of Products

In general, given a function F of n variables, the corresponding truth table has 2^n rows. Typically, the rows of the first n columns consist of the binary numbers ordered from 0 to $2^n - 1$, where the first row is the number 0 and the last row is the number $2^n - 1$. For example, a function F of three variables is described by Table 11.23. In this case, we can proceed to write an expression for F as was just suggested. Specifically,

$$F = \overline{A}\overline{B}C + \overline{A}B\overline{C} + A\overline{B}\overline{C} + ABC$$

Note that F consists of four terms $(\overline{A}\overline{B}C, \overline{A}B\overline{C}, A\overline{B}\overline{C}, \text{and } ABC)$—called **minterms**—which are combined by the OR operation. Because the OR operation is represented as addition is, F is said to be a **sum** of the four minterms. Every minterm consists of the three variables, each being complemented or uncomplemented, which are combined by the AND operation. Because the AND operation is represented as multiplication is, each of the four minterms is said to be a **product** of the three variables. For these reasons, the expression for the Boolean function F given above is said to be in the form of a **sum of products.**

Table 11.23 Truth Table for a Boolean Function F.

A	B	C	F
0	0	0	0
0	0	1	1
0	1	0	1
0	1	1	0
1	0	0	1
1	0	1	0
1	1	0	0
1	1	1	1

Table 11.24 Minterms for the Case of Three Variables A, B, C.

Binary Number			Minterms for Variables A, B, C
0	0	0	$m_0 = \overline{A}\,\overline{B}\,\overline{C}$
0	0	1	$m_1 = \overline{A}\,\overline{B}C$
0	1	0	$m_2 = \overline{A}B\overline{C}$
0	1	1	$m_3 = \overline{A}BC$
1	0	0	$m_4 = A\overline{B}\,\overline{C}$
1	0	1	$m_5 = A\overline{B}C$
1	1	0	$m_6 = AB\overline{C}$
1	1	1	$m_7 = ABC$

For a Boolean function of three variables, there are $2^3 = 8$ minterms—one for each row of the truth table. In other words, there is one minterm for each number between 0 and $2^3 - 1 = 7$. These minterms are designated $m_0, m_1, m_2, m_3, m_4, m_5, m_6, m_7$. (For the case of n variables, there are 2^n minterms, and these are designated $m_0, m_1, m_2, m_3, \ldots, m_{2^n-1}$.) Every minterm consists of a product of the three variables, where each variable is either uncomplemented or complemented. Specifically, if the first bit of the binary number i is 0 (or 1), then the first variable in the minterm m_i is complemented (or uncomplemented), and so on. Table 11.24 shows the binary numbers ranging from 0 to 7 (the case of three variables) and the corresponding minterms.

Given the truth table, an expression for a Boolean function F is obtained by summing the minterms that correspond to the rows in which there is a 1 in the column labeled F. For Table 11.23, we have that

$$F = m_1 + m_2 + m_4 + m_7 = \overline{A}\,\overline{B}C + \overline{A}B\overline{C} + A\overline{B}\,\overline{C} + ABC$$

A logic circuit whose output is described by this Boolean function is shown in Fig. 11.23. For this circuit, which consists of four three-input AND gates and one four-

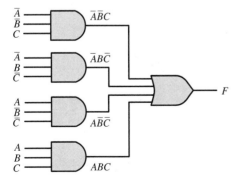

Fig. 11.23 Realization of the Boolean function $F = \overline{A}\,\overline{B}C + \overline{A}B\overline{C} + A\overline{B}\,\overline{C} + ABC$.

input OR gate, it is assumed that each variable and its complement are available. Furthermore, to avoid the pictorial cluttering that results from wires crossing, inputs on different gates are labeled with the same (uncomplemented or complemented) variable.

Given an expression for a Boolean function, it is not necessary to construct a truth table in order to write the function as a sum of minterms.

Example 11.10

Let us express the function

$$F = \overline{A}\overline{C} + (A + \overline{B})C = \overline{A}C + \overline{B}C + AC$$

as a sum of minterms.

To obtain minterms, we can multiply each product in the sum by 1 in an appropriate form. Specifically, since $\overline{A} + A = 1$, then

$$F = \overline{A}(\overline{B} + B)\overline{C} + (\overline{A} + A)\overline{B}C + A(\overline{B} + B)C$$

from which

$$F = \overline{A}\,\overline{B}\,\overline{C} + \overline{A}B\overline{C} + \overline{A}\,\overline{B}C + A\overline{B}C + A\overline{B}C + ABC$$
$$= \overline{A}\,\overline{B}\,\overline{C} + \overline{A}B\overline{C} + \overline{A}\,\overline{B}C + A\overline{B}C + ABC$$
$$= m_0 + m_1 + m_2 + m_5 + m_7$$

Drill Exercise 11.10

Express the Boolean function $F = \overline{A}\overline{B} + B\overline{C}$ as a sum of minterms.

ANSWER $m_0 + m_1 + m_2 + m_6$

As exemplified in Fig. 11.23, a logic circuit whose output is characterized by a Boolean function, which is expressed as a sum of products, can be constructed by using two columns of gates. The first column is comprised of AND gates, and the second column is a single multi-input OR gate. Such a circuit is an example of what is known as **two-level logic.**

In Example 11.10, we demonstrated that the Boolean function $F = m_0 + m_1 + m_2 + m_5 + m_7$ is equivalent to the sum of products

$$F = \overline{A}\,\overline{C} + \overline{B}C + AC$$

A two-level logic circuit that realizes, or implements, this function is shown in Fig. 11.24. Although this realization employs AND gates, an OR gate, and (implicit) NOT gates, it is possible to implement the given function using only NAND gates. Let us see how.

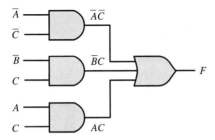

Fig. 11.24 Implementation of $F = \overline{A}\,\overline{C} + \overline{B}C + AC$.

NAND-Gate Realizations

Suppose that the same input A is applied to both inputs of a two-input NAND gate as shown in Fig. 11.25a. Since the output of the gate is $\overline{AA} = \overline{A}$, the gate is an inverter (NOT gate). The equivalent situation, a one-input NAND gate, is symbolized in Fig. 11.25b.

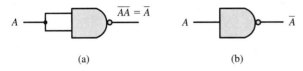

(a) (b)

Fig. 11.25 NAND-gate realization of an inverter (NOT gate).

We will now show how NAND gates can be connected to form an OR gate and how they can be connected to form an AND gate. By De Morgan's theorem

$$A + B = \overline{\overline{A + B}} = \overline{\overline{A}\,\overline{B}}$$

and this expression can be realized by the logic circuit, which, therefore, is an OR gate, shown in Fig. 11.26. Furthermore, since

$$AB = \overline{\overline{AB}}$$

then the NAND-gate circuit shown in Fig. 11.27 realizes an AND gate.

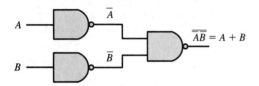

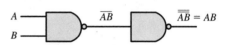

Fig. 11.26 NAND-gate realization of an OR gate.

Fig. 11.27 NAND-gate realization of an AND gate.

Since any Boolean function can be implemented with AND, OR, and NOT gates, it can also be implemented using only NAND gates. This is so because, as we have seen, AND, OR, and NOT gates can be constructed from NAND gates. However, a more direct method for obtaining a NAND-gate realization of a Boolean function is possible through the use of De Morgan's theorem.

Example 11.11

Let us determine a NAND-gate realization of $F = \overline{A}\overline{C} + \overline{B}C + AC$.
Using De Morgan's theorem, we can write

$$F = \overline{A}\overline{C} + \overline{B}C + AC = \overline{\overline{\overline{A}\overline{C} + \overline{B}C + AC}} = (\overline{\overline{A}\overline{C}})(\overline{\overline{B}C})(\overline{AC})$$

and this expression can be implemented by the two-level NAND circuit shown in Fig. 11.28. This implementation, therefore, is an alternative to the one given in Fig. 11.24.

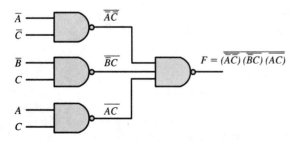

Fig. 11.28 NAND-gate implementation of $F = \overline{A}\overline{C} + \overline{B}C + AC$.

Drill Exercise 11.11

Use De Morgan's theorem to obtain a two-level NAND-gate realization of $F = \overline{A}\overline{B} + B\overline{C}$.

ANSWER $\overline{(\overline{\overline{A}\overline{B}})(\overline{B\overline{C}})}$

Up to this point, we have used an AND gate symbol followed by a small circle (to denote inversion) to signify a NAND gate. This logic-circuit symbol is repeated in Fig. 11.29a for the case of three input variables. We may also use the circuit symbol shown in Fig. 11.29b to indicate a NAND gate. In this case, the small circles preceding the OR gate cause the input variables to be complemented before the OR operation is performed. The resulting output, therefore, is, by De Morgan's theorem,

$$\overline{A} + \overline{B} + \overline{C} = \overline{ABC}$$

and this confirms the NAND operation.

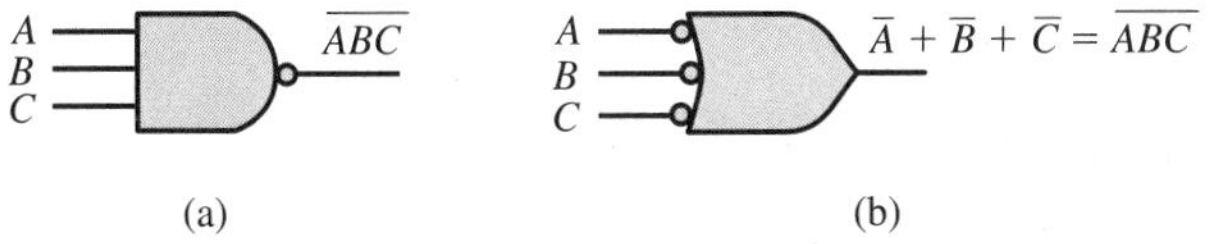

(a)   (b)

Fig. 11.29 Alternative circuit symbols for the NAND gate.

Since an arbitrary Boolean function can be implemented by a logic circuit consisting only of NAND gates, we say that the NAND gate is a **universal gate.** A sum of products is one type of **standard form** of a Boolean function. There is one other type of standard form—a product of sums. We now look at this functional form.

The Product of Sums

Given the truth table for a Boolean function, we have seen how we can express that function as a sum of minterms. Let us describe a procedure for obtaining an alternative, equivalent expression of the function from its truth table. For demonstration purposes, reconsider Table 11.23, which is repeated as Table 11.25. Here a column corresponding to the function $\overline{F}$, the complement of F, has been attached. As described previously, we can express $\overline{F}$ as

$$\overline{F} = m_0 + m_3 + m_5 + m_6 = \overline{A}\,\overline{B}\,\overline{C} + \overline{A}BC + A\overline{B}C + AB\overline{C}$$

Taking the cmplement of both sides of this equation, we get

$$F = \overline{\overline{A}\,\overline{B}\,\overline{C} + \overline{A}BC + A\overline{B}C + AB\overline{C}}$$

By De Morgan's theorems, we have that

$$F = (\overline{\overline{A}\,\overline{B}\,\overline{C}})(\overline{\overline{A}BC})(\overline{A\overline{B}C})(\overline{AB\overline{C}})$$
$$= (A + B + C)(A + \overline{B} + \overline{C})(\overline{A} + B + \overline{C})(\overline{A} + \overline{B} + C)$$

which is a **product of sums.** Each such sum is called a **maxterm.** For a Boolean function of three variables, there are $2^3 = 8$ maxterms—one for each row of the truth table. The maxterms are designated M_0, M_1, M_2, M_3, M_4, M_5, M_6, M_7. Each maxterm consists of the sum of the three variables, where each variable is either uncomplemented or complemented. Specifically, if the first bit of the binary number i is 0 (or 1), then the first variable in the maxterm M_i is uncomplemented (or complemented), and so on. Table 11.26 shows the binary numbers ranging from 0 to 7 and the corresponding maxterms.

Table 11.25 Truth Table for F and $\overline{F}$.

A	B	C	F	$\overline{F}$
0	0	0	0	1
0	0	1	1	0
0	1	0	1	0
0	1	1	0	1
1	0	0	1	0
1	0	1	0	1
1	1	0	0	1
1	1	1	1	0

Table 11.26 Maxterms for the Case of Three Variables A, B, C.

Binary Number			Maxterms for Variables A, B, C
0	0	0	$M_0 = A + B + C$
0	0	1	$M_1 = A + B + \overline{C}$
0	1	0	$M_2 = A + \overline{B} + C$
0	1	1	$M_3 = A + \overline{B} + \overline{C}$
1	0	0	$M_4 = \overline{A} + B + C$
1	0	1	$M_5 = \overline{A} + B + \overline{C}$
1	1	0	$M_6 = \overline{A} + \overline{B} + C$
1	1	1	$M_7 = \overline{A} + \overline{B} + \overline{C}$

Given the truth table for a Boolean function F, an alternative expression to the sum of minterms is obtained by taking the product of the maxterms that correspond to the rows of the table in which there is a 0 in the column labeled F. For Table 11.25, we have that

$$F = M_0 M_3 M_5 M_6 = (A + B + C)(A + \overline{B} + \overline{C})(\overline{A} + B + \overline{C})(\overline{A} + \overline{B} + C)$$

A two-level logic circuit whose output is described by this Boolean function is shown in Fig. 11.30. This circuit is composed of four three-input OR gates and one four-

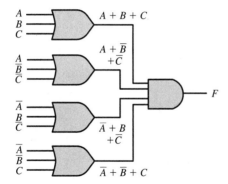

Fig. 11.30 Realization of the Boolean function
$F = (A + B + C)(A + \overline{B} + \overline{C})(\overline{A} + B + \overline{C})(\overline{A} + \overline{B} + C)$.

input AND gate—again it is assumed that each variable and its complement are available.

Given an expression for a Boolean function, it is not necessary to construct a truth table in order to write the function as a product of maxterms.

Example 11.12

Let us express the function $F = \overline{A}\,\overline{C} + (A + \overline{B})C$ as a product of maxterms.
Using the distributive law $A + BC = (A + B)(A + C)$, we can write

$$F = (\overline{A}\,\overline{C} + A + \overline{B})(\overline{A}\,\overline{C} + C) = (A + \overline{A}\,\overline{C} + \overline{B})(C + \overline{C}\overline{A})$$

and since $A + \overline{A}B = A + B$, we have the product of sums

$$F = (A + \overline{C} + \overline{B})(C + \overline{A}) = (A + \overline{B} + \overline{C})(\overline{A} + C)$$

Although $A + \overline{B} + \overline{C}$ is a maxterm, $\overline{A} + C$ is not. To obtain maxterms, to each sum that is not a maxterm we can add 0 in an appropriate form. Specifically, since $B\overline{B} = 0$, then

$$F = (A + \overline{B} + \overline{C})(B\overline{B} + \overline{A} + C)$$

$$= (A + \overline{B} + \overline{C})(B + \overline{A} + C)(\overline{B} + \overline{A} + C)$$

$$= (A + \overline{B} + \overline{C})(\overline{A} + B + C)(\overline{A} + \overline{B} + C)$$

$$= M_3 M_4 M_6$$

For the function F discussed in Example 11.12, $F = (A + \overline{B} + \overline{C})(\overline{A} + C)$ is an equivalent expression. A two-level logic circuit that realizes this function is shown in Fig. 11.31. Although this circuit employs an AND gate, OR gates, and (implicit) NOT gates, it is possible to implement this function, or any Boolean function for that matter, using only NOR gates. Let us see how.

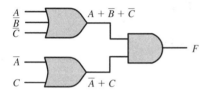

Fig. 11.31 Implementation of $F = (A + \overline{B} + \overline{C})(\overline{A} + C)$.

NOR-Gate Realizations

As was the case for NAND gates, we only need show that AND, OR, and NOT gates can be constructed from NOR gates. By connecting both inputs of a NOR gate together as shown in Fig. 11.32a, we obtain a NOT gate. This is a consequence of the fact that $\overline{A + A} = \overline{A}$. The equivalent situation of the gate given in Fig. 11.32a is the one-input NOR gate symbolized in Fig. 11.32b.

Since $A + B = \overline{\overline{A + B}}$, then the connection of NOR gates shown in Fig. 11.33 realizes an OR gate.

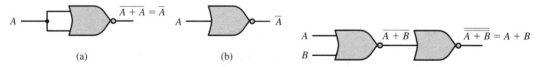

Fig. 11.32 NOR-gate realization of an inverter.　　**Fig. 11.33** NOR-gate realization of an OR gate.

Furthermore, since $AB = \overline{\overline{AB}} = \overline{\overline{A} + \overline{B}}$, then the NOR-gate circuit shown in Fig. 11.34 realizes an AND gate. This, therefore, demonstrates the fact that an arbitrary

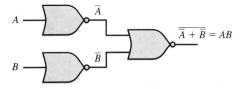

Fig. 11.34 NOR-gate realization of an AND gate.

Boolean function can be implemented using only NOR gates. Like the NAND gate, the NOR gate is a universal gate. Instead of constructing AND, OR, and NOT gates from NOR gates, a more direct approach can be taken. To do this, we use De Morgan's theorem on a Boolean function that is expressed as a product of sums.

Example 11.13

For the function $F = (A + \bar{B} + \bar{C})(\bar{A} + C)$ discussed in Example 11.12, we have that

$$F = (A + \bar{B} + \bar{C})(\bar{A} + C) = \overline{\overline{(A + \bar{B} + \bar{C})(\bar{A} + C)}}$$
$$= \overline{\overline{(A + \bar{B} + \bar{C})} + \overline{(\bar{A} + C)}}$$

and this expression can be implemented by the two-level NOR circuit shown in Fig. 11.35. Such a logic circuit is an alternative to those given in Figs. 11.24, 11.28, and 11.31.

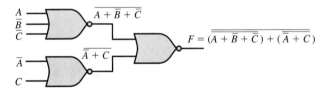

Fig. 11.35 NOR-gate implementation of $F = (A + \bar{B} + \bar{C})(\bar{A} + C)$.

Drill Exercise 11.13

Use De Morgan's theorem to obtain a two-level NOR-gate realization of $F = \overline{A}\overline{B} + B\overline{C}$.

ANSWER $\overline{\overline{(\overline{A} + B)} + \overline{(\overline{A} + \overline{C})} + \overline{(\overline{B} + \overline{C})}}$

As indicated above, a NOR gate is signified by the symbol for an OR gate followed by a small circle. This logic-circuit symbol is repeated in Fig. 11.36a for the case of three input variables. However, a NOR gate can also be represented by the circuit symbol shown in Fig. 11.36b. In this case, the small circles preceding the AND gate indicate the inversion of the input variables prior to the performing of the AND operation. By De Morgan's theorem, the resulting output is given by

$$\overline{A}\,\overline{B}\,\overline{C} = \overline{A + B + C}$$

and this confirms NOR operation.

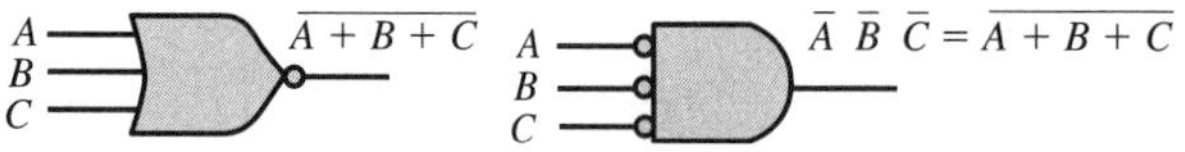

Fig. 11.36 Alternative circuit symbols for the NOR gate.

The realization of Boolean functions using NAND gates or NOR gates is not just an academic exercise. Because they are more easily fabricated, these types of gates see greater use than do AND gates and OR gates. NAND and NOR gates are the basic gates of IC logic.

11.6 Simplification of Boolean Functions

When a Boolean function is expressed either as a sum of minterms or as a product of maxterms, we say that such a function is in **canonical form.** We refer to an uncomplemented variable or a complemented variable as a **literal.** When the output of a logic circuit is characterized by a Boolean function in either canonical form, each literal in the expression corresponds to an input of a gate, and each term (product or sum) in the expression corresponds to a logic gate. If a Boolean function can be simplified so as to consist of fewer literals and terms, the resulting realization of such a function will require less hardware than is needed for the unsimplified form of the function.

One way to simplify Boolean functions is to use algebraic manipulations that employ the relationships given in Section 11.4. For example, consider the following Boolean function expressed as a sum of minterms:

$$F = \overline{A}BC + A\overline{B}\,\overline{C} + A\overline{B}C + AB\overline{C} + ABC$$

It follows that

$$F = \overline{A}BC + A\overline{B}(\overline{C} + C) + AB(\overline{C} + C) = \overline{A}BC + A\overline{B} + AB$$

$$= \overline{A}BC + A(\overline{B} + B) = \overline{A}BC + A = A + BC$$

and the function cannot be simplified further.

Karnaugh Maps

Instead of taking an algebraic approach, Boolean functions can be simplified graphically using maps referred to as **Karnaugh[5] maps.** Specifically, we will discuss only the cases of either three or four variables. We shall avoid the complexity that arises when dealing with more than four variables.

For the case of three variables, we form a map consisting of $2^3 = 8$ boxes as shown in Fig. 11.37. Here the eight boxes form two rows and four columns. The rows, labeled 0 and 1, are associated with variable A, whereas the columns are associated with variables B and C. Note that the columns, going from left to right, are labeled 00, 01, 11, and 10. The labeling is done in this way so that in going from left to right the binary pairs change in only one bit.

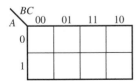

Fig. 11.37 Three-variable Karnaugh map.

Each box corresponds to a distinct binary triple. The first bit of the triple is the box's row designation, whereas the second and third bits are indicated by the box's column designation. For example, the box in the upper left-hand corner corresponds to 000, whereas the box in the lower right-hand corner corresponds to 110. Therefore, with each box we can identify one of the minterms $m_0, m_1, m_2, m_3, m_4, m_5, m_5, m_7$, where m_i is associated with the box corresponding to the binary number i. The minterms associated with the boxes of a three-variable Karnaugh map are as shown in Fig. 11.38.

---

[5]Named for the American electrical engineer Maurice Karnaugh.

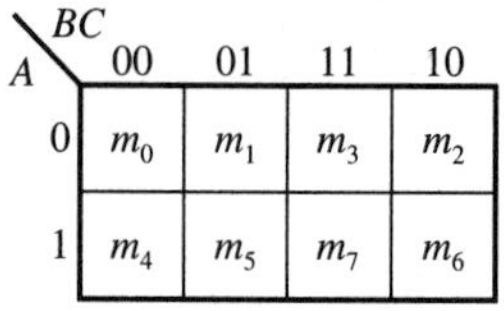

BC A	00	01	11	10
0	m_0	m_1	m_3	m_2
1	m_4	m_5	m_7	m_6

Fig. 11.38 Minterms associated with a three-variable Karnaugh map.

Given a Boolean function in the form of a sum of minterms, place a 1 in each of the boxes that corresponds to a minterm in the sum. Place a 0 in each remaining box, or, to avoid the cluttering of symbols, leave the remaining boxes empty. For example, consider the function

$$F = m_0 + m_1 + m_2 + m_6 = \overline{A}\overline{B}\overline{C} + \overline{A}\overline{B}C + \overline{A}B\overline{C} + AB\overline{C}$$

The corresponding Karnaugh map is shown in Fig. 11.39. The two boxes in the upper left with 1s in them represent the sum $m_0 + m_1 = \overline{A}\overline{B}\overline{C} + \overline{A}\overline{B}C$. Algebraically, we know that this reduces to $\overline{A}\overline{B}(\overline{C} + C) = \overline{A}\overline{B}$. The graphical interpretation is the following: Group the adjacent 1s corresponding to the minterms m_0 and m_1 together as shown in Fig. 11.40. Both 1s are identified with $A = 0$ and $B = 0$. However, $C = 0$ for one box, whereas $C = 1$ for the other. Thus the term that corresponds to the two 1s in the upper left portion of the map is $\overline{A}\overline{B}$.

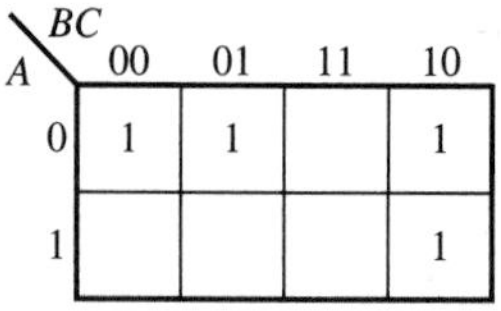

BC A	00	01	11	10
0	1	1		1
1				1

Fig. 11.39 Karnaugh map for $F = m_0 + m_1 + m_2 + m_6$.

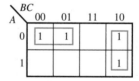

Fig. 11.40 Grouping of minterms.

Using this interpretation, when the 1s on the right are grouped together as shown in Fig. 11.40, since $A = 0$ for one box and $A = 1$ for the other, whereas $B = 1$ and $C = 0$ for both boxes, the corresponding term is $B\overline{C}$. The result of the two groupings shown is the expression

$$F = \overline{A}\overline{B} + B\overline{C}$$

Although this expression is also a sum of products, its form is simpler than the original sum of minterms.

A pair of 1s in a karnaugh map can be adjacent even though they do not appear to be at first glance. For example, for the map shown in Fig. 11.41, the two ones

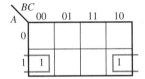

Fig. 11.41 Karnaugh map with adjacent 1s.

indicated are in adjacent boxes, and, therefore, are grouped as depicted. In this case, for both boxes $A = 1$ and $C = 0$, whereas $B = 0$ for one box and $B = 1$ for the other. Thus the term corresponding to this pair of 1s is $A\overline{C}$.

A greater reduction in literals can be made when there are 1s in four adjacent boxes. An example is shown in Fig. 11.42. In this case, each 1 in the group corresponds to $A = 0$, whereas B takes on the values 0 and 1, as does C. Thus, as can be proved algebraically, the term corresponding to this group of four 1s is simply $\overline{A}$.

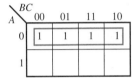

Fig. 11.42 Karnaugh map with four adjacent 1s.

Another group of four 1s is shown in Fig. 11.43. Here the only condition corresponding to all four boxes is $C = 1$. Thus, the term that results is simply C. For the Karnaugh map shown in Fig. 11.44, the group of four adjacent 1s has the common condition $C = 0$. Thus, the corresponding term is $\overline{C}$.

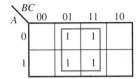

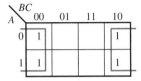

Fig. 11.43 Another group of four adjacent 1s.

Fig. 11.44 Karnaugh map with a group of four adjacent 1s.

Having considered some examples of grouping the 1s in a Karnaugh map, let us now list the steps to take for function simplification.

Boolean-Function Simplification with a Karnaugh Map

1. Given a Boolean function F, express it as a sum of minterms, and place 1s in the corresponding boxes of a Karnaugh map.

2. Group adjacent 1s so as to form the largest possible groups. A particular 1 can be placed in more than one group.

3. A 1 that is not adjacent to any other 1 forms its own group.

4. After all the 1s have been placed in the largest possible groups, without forming more groups than required, take the sum of the terms corresponding to each group—this is the simplified expression for the Boolean function.

Following this procedure results in a sum of products with the minimum number of literals. If alternative groupings are possible, however, then other expressions, also with a minimum number of literals, may be obtained.

Example 11.14

Let us use Karnaugh maps to simplify the following Boolean functions:

(a) $F_1 = m_0 + m_2 + m_3 + m_4 + m_5 + m_6 + m_7$,
(b) $F_2 = m_0 + m_1 + m_2 + m_5 + m_7$, and (c) $F_3 = m_0 + m_1 + m_4 + m_7$.

(a) The map for the Boolean function F_1 is shown in Fig. 11.45. In this case, the bottom four 1s form a group that corresponds to the term A. The two 1s on the lower right can be used again with the two on the upper right to form another group of four 1s. This group corresponds to the term B. Finally, grouping the two 1s on the extreme left with the two 1s on the extreme right, we get the term $\overline{C}$. Thus the function has the simplified form

$$F_1 = A + B + \overline{C}$$

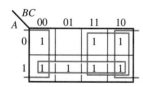

Fig. 11.45 Karnaugh map with groups of four adjacent 1s.

(b) The map for the Boolean function F_2 is shown in Fig. 11.46. In this case, there are no groups consisting of four adjacent 1s. However, as indicated, there are

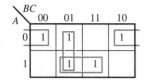

Fig. 11.46 Karnaugh map with groups of two adjacent 1s.

three groups that are comprised of a pair of 1s. The corresponding expression for F_2 is

$$F_2 = AC + \overline{A}\,\overline{C} + \overline{B}C$$

If the 1s are grouped by the alternative scheme shown in Fig. 11.47, then the corresponding equivalent expression is

$$F_2 = AC + \overline{A}\,\overline{C} + \overline{A}\overline{B}$$

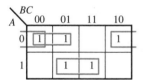

Fig. 11.47 Karnaugh map with alternative grouping.

Even though both of these expressions for F_2 have the minimum number of literals (six) for a sum of products, an expression not in the form of a sum of products may contain less literals. Specifically, from the expressions just obtained, we have that

$$F_2 = (A + \overline{B})C + \overline{A}\,\overline{C} = AC + \overline{A}(\overline{B} + \overline{C})$$

which consists of only five literals.

(c) The map for the Boolean function F_3 is shown in Fig. 11.48. In this case, again there is no group consisting of four adjacent 1s. But, as illustrated, there are two groups, each consisting of a pair of 1s. There is also a 1 that is not adjacent to any other 1. This single 1 corresponds to the minterm ABC, and the Boolean function in simplified sum-of-product form is

$$F_3 = \overline{A}\overline{B} + \overline{B}\,\overline{C} + ABC$$

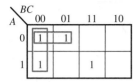

Fig. 11.48 Karnaugh map with a group consisting of a single 1.

Drill Exercise 11.14

Use a Karnaugh map to simplify the Boolean function
$F = m_3 + m_4 + m_5 + m_6 + m_7$.

ANSWER $A + BC$

Four-Variable Maps

A Karnaugh map for a Boolean function of the four variables A, B, C, and D is shown in Fig. 11.49. In this case, there are 16 boxes that correspond to the 16 minterms m_0, m_1, m_2, . . . , m_{15}. Each of the map's boxes is labeled with its associated minterm.

CD \ AB	00	01	11	10
00	m_0	m_1	m_3	m_2
01	m_4	m_5	m_7	m_6
11	m_{12}	m_{13}	m_{15}	m_{14}
10	m_8	m_9	m_{11}	m_{10}

Fig. 11.49 Four-variable map displaying the corresponding minterms.

The simplification procedure is essentially the same as that described for the case of three variables. For four variables, however, it may be possible to form groups of eight adjacent 1s—for three variables, the largest group of adjacent 1s consists of four 1s. (We are ignoring the trivial case that every box has a 1—this corresponds to the Boolean function $F = 1$.)

Example 11.15

Let us use Karnaugh maps to simplify the following Boolean functions:

(a) $F_1 = m_2 + m_3 + m_5 + m_8 + m_{10} + m_{11} + m_{13}$,
(b) $F_2 = m_0 + m_2 + m_4 + m_5 + m_8 + m_{10} + m_{12} + m_{13}$, and
(c) $F_3 = m_1 + m_3 + m_6 + m_8 + m_9 + m_{11} + m_{12}$.

(a) The map for a Boolean function F_1 is shown in Fig. 11.50. The group of two 1s located in approximately the center of the map corresponds to the term $B\bar{C}D$. Since the pair of 1s in the upper right-hand corner is adjacent to the 1s in the lower right-hand corner, these 1s form a group of four. The corresponding term is $\bar{B}C$. Since the remaining 1 in the lower left-hand corner can be grouped with the 1 in the lower right-hand corner, the corresponding term is $A\bar{B}\bar{D}$. Therefore, the simplified sum-of-products form for the given function is

$$F_1 = B\bar{C}D + \bar{B}C + A\bar{B}\bar{D}$$

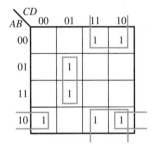

Fig. 11.50 Example of a map for a function with four variables.

(b) The map for the Boolean function F_2 has the Karnaugh map shown in Fig. 11.51. In this case, two groups of four 1s should be immediately obvious. One group

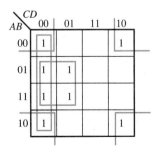

Fig. 11.51 Map with three groups of four 1s.

consists of the four 1s comprising the column on the far left. The corresponding term is $\overline{C}\overline{D}$. Another group consists of the four 1s representing minterms m_4, m_5, m_{12}, and m_{13}. This group corresponds to the term $B\overline{C}$. What remains is the pair of 1s in the upper right-hand corner and the lower right-hand corner of the map. Even though these 1s are adjacent and form a group of two, note that they are also adjacent to the 1s in the upper and lower left-hand corners of the map. Therefore, the 1s in the corners of the map form a group of four 1s as illustrated—the corresponding term being $\overline{B}\overline{D}$. Hence the simplified form of F_2 is

$$F_2 = \overline{C}\overline{D} + B\overline{C} + \overline{B}\overline{D}$$

(c) The map for the Boolean function F_3 has the Karnaugh map shown in Fig. 11.52. The pair of 1s on top can be grouped with the pair of 1s directly under them on the bottom as illustrated. The corresponding term is $\overline{B}D$. The pair of 1s on the extreme left form a group of two 1s whose corresponding term is $A\overline{C}\overline{D}$. The 1 on the extreme right cannot be grouped with any other 1. Since this corresponds to the minterm $m_6 = \overline{A}BC\overline{D}$, the simplified form of the function F_3 associated with the map is

$$F_3 = \overline{B}D + A\overline{C}\overline{D} + \overline{A}BC\overline{D}$$

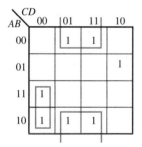

Fig. 11.52 Map with a group consisting of a single 1.

Drill Exercise 11.15

Use a Karnaugh map to simplify the Boolean function
$F = m_1 + m_3 + m_6 + m_8 + m_9 + m_{11} + m_{12} + m_{14}$.

ANSWER $A\overline{C}D + BC\overline{D} + \overline{B}D$

Karnaugh Maps Utilizing Maxterms

Instead of associating the boxes of a Karnaugh map with minterms, we can associate them with maxterms. Specifically, maxterm M_i is associated with the box corresponding to the binary number i. If the first bit of the binary number i is 0 (or 1), then the first variable in the maxterm M_i is uncomplemented (or complemented), and so on. The grouping of the 0s that correspond to maxterms is performed just as was the grouping of the 1s that correspond to minterms. When writing the simplified expression in the form of a product of sums, for the bits designating a group of 0s, a 0 signifies an uncomplemented variable whereas a 1 signifies a complemented variable.

Example 11.16

Let us use Karnaugh maps to simplify the following Boolean functions:

(a) $F_1 = M_0M_1M_4M_6M_7M_9M_{12}M_{14}M_{15}$, (b) $F_2 = M_1M_3M_6M_7M_9M_{11}M_{14}M_{15}$, and (c) $F_3 = M_0M_2M_4M_5M_7M_{10}M_{13}M_{14}M_{15}$.

(a) The map of the Boolean function F_1 is shown in Fig. 11.53, where the 0s are shown explicitly and the 1s are implicit. The four 0s on the right form a group that corresponds to the sum $\overline{B} + \overline{C}$. The term that corresponds to the pair of 0s on the extreme left and the pair of 0s on the extreme right is $\overline{B} + D$. For the pair of 0s on top we get the term $A + B + C$, and for the 0s in the second column we get the term $B + C + \overline{D}$. Thus an expression for F_1 is the simplified product of sums

$$F_1 = (A + B + C)(B + C + \overline{D})(\overline{B} + \overline{C})(\overline{B} + D)$$

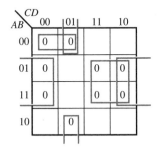

Fig. 11.53 Karnaugh map with 0s shown explicitly.

(b) The map of the Boolean function F_2 is shown in Fig. 11.54 with the 0s shown explicitly. There are two groups, each containing four 0s. The terms that correspond to these groups are $\overline{B} + \overline{C}$ and $B + \overline{D}$. Thus we have that

$$F_2 = (\overline{B} + \overline{C})(B + \overline{D})$$

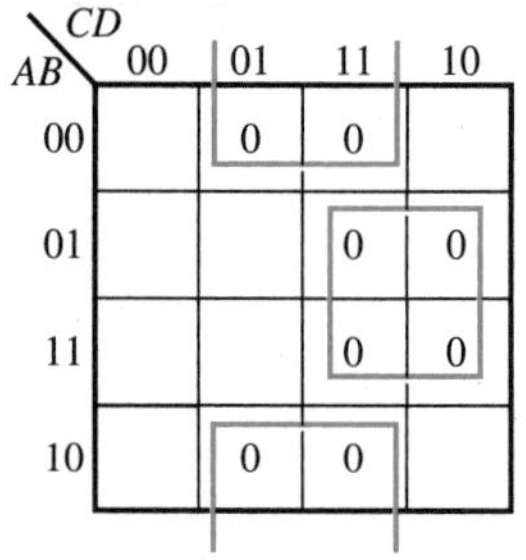

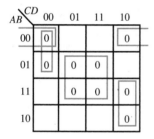

Fig. 11.54 Map with two groups of four 0s.

(c) The map of the Boolean function F_3 is shown in Fig. 11.55. In this case, there is only one group with four 0s, and it corresponds to $\overline{B} + \overline{D}$. The group of two 0s in the upper left corresponds to $A + C + D$, whereas the group in the lower right corresponds to $\overline{A} + \overline{C} + D$. The 0 in the upper right-hand corner can be grouped with either the 0 in the upper left-hand corner or the 0 in the lower right-hand corner. Arbitrarily selecting the former, we get $A + B + D$. The result of these groupings is the simplified form

$$F_3 = (\overline{B} + \overline{D})(A + C + D)(\overline{A} + \overline{C} + D)(A + B + D)$$

Fig. 11.55 Karnaugh map for the Boolean function F_3.

Drill Exercise 11.16

Use a karnaugh map to simplify the Boolean function
$F = M_0 M_2 M_4 M_5 M_7 M_{10} M_{13} M_{15}$.

ANSWERS $(A + C + D)(B + \overline{C} + D)(\overline{B} + \overline{D})$

Even though we demonstrated how to group 0s and write the corresponding product-of-sums functional forms for four-variable Karnaugh maps, it should be obvious that essentially the same technique can be employed for three-variable maps. The details are left as exercises in the following problem set.

SUMMARY

1. The representation of a number depends upon the base of the number system selected.

2. Arithmetic can be performed in the binary number system as well as the decimal number system.

3. Nonnumerical information can be put in binary form by using codes.

4. The position of a switch (i.e., open, closed) in a circuit can be characterized by the binary digits (bits) 0 and 1.

5. AND, OR, NOT (complement), exclusive-OR, NAND, and NOR operations are examples of logic operations.

6. Logic gates are circuits that perform logic operations.

7. Switching or logic circuits are characterized by Boolean functions.

8. Boolean functions can be manipulated by using the rules of Boolean algebra.

9. Boolean functions in standard form can be implemented with two-level logic.

10. NAND and NOR gates are universal gates—either type can be used exclusively to realize arbitrary Boolean functions.

11. Boolean functions can be simplified graphically by using Karnaugh maps.

12. Grouping the 1s on a Karnaugh map produces a simplified sum-of-products expression, whereas grouping the 0s yields a simplified product-of-sums functional form.

PROBLEMS

11.1 Make a list of the binary numbers from zero to fifteen.

11.2 Express the following binary numbers in decimal form: (a) 11011, (b) 101011, (c) 0.11011, (d) 0.10101, (e) 10100.011, (f) 10011.101.

11.3 Express the following decimal numbers in binary form: (a) 43, (b) 27, (c) 0.84375, (d) 0.65625, (e) 19.625, (f) 20.375.

11.4 Express the following octal numbers in decimal form: (a) 33, (b) 53, (c) 0.66, (d) 0.52, (e) 24.3, (f) 23.5.

11.5 Express the following decimal numbers in octal form: (a) 43, (b) 27, (c) 0.84375, (d) 0.65625, (e) 19.625, (f) 20.375.

11.6 Express the following binary numbers in octal form: (a) 110010110.1, (b) 101100101.01, (c) 11101001.111.

11.7 Express the following octal numbers in binary form: (a) 20.375, (b) 413.702, (c) 2610.35.

11.8 Repeat Problem 11.6 converting from binary to hexadecimal numbers.

11.9 Express the following hexadecimal numbers in binary form: (a) 1983.605, (b) 2B2.A07, (c) C3BO.82D2.

11.10 Find the sum $n + m$ for the following binary numbers: (a) $n = 101100$, $m = 10100$; (b) $n = 1110.01$, $m = 1001.11$; (c) $n = 101.011$, $m = 1101.01$; (d) $n = 11001$, $m = 101110$.

11.11 Find the difference $n - m$ for the binary numbers given in Problem 11.10. (Do not use complements.)

11.12 Find the product of $n \times m$ for the following binary numbers: (a) $n = 110.01$, $m = 11000$; (b) $n = 101.11$, $m = 10100$; (c) $n = 11001.1$, $m = 101$; (d) $n = 10010.01$, $m = 110$.

11.13 Divide m by n, that is, find $m \div n$, for the following binary numbers: (a) $m = 10010110$, $n = 11000$; (b) $m = 1110011$, $n = 10100$; (c) $m = 11111111$, $n = 110011$; (d) $m = 11011011$, $n = 110$.

11.14 Use 1's complements fo find the difference $n - m$ for the binary numbers given in Problem 11.10.

11.15 Use 2's complements to find the difference $n - m$ for the binary numbers given in Problem 11.10.

11.16 Express the following decimal numbers in BCD: (a) 1127, (b) 1940, (c) 261.64, (d) 607.83.

11.17 Express the following BCD numbers in decimal form: (a) 100101000011, (b) 0011010100100011, (c) 10011000.0110.

11.18 List the words of the binary code obtained by adding a parity bit to the eight distinct binary triples such that each resulting code word has an odd number of 1s.

11.19 Determine the truth table for the connection of switches shown in Fig. P11.19. (The two distinct switches labeled A are both open when $A = 0$ and both closed when $A = 1$.) Write a mathematical expression for the output variable L in terms of the input variables A, B, and C. How is this expression related to $A(B + C)$?

11.20 Determine the truth table for the connection of switches shown in Fig. P11.20. Write a mathematical expression for L in terms of A, B, and C.

11.21 Repeat Problem 11.19 for the connection of switches shown in Fig. P11.21. How is this expression for L related to the one obtained in Problem 11.20?

11.22 Determine the truth table for the connection of switches shown in Fig. P11.22. Write a mathe-

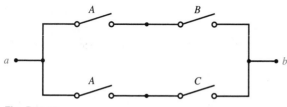

Fig. P11.19

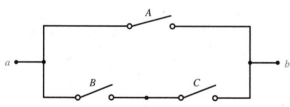

Fig. P11.20

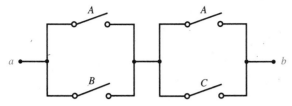

Fig. P11.21

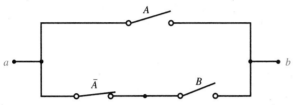

Fig. P11.22

matical expression for L in terms of A, $\bar{A}$, and B. Is this expression for L related either to the OR operation $A + B$ or the AND operation AB?

11.23   Repeat Problem 11.22 for the connection of switches shown in Fig. P11.23.

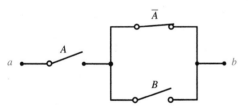

Fig. P11.23

11.24   A room has three doors and three associated switches A, B, and C. By entering any door and throwing the associated switch, the room light is to turn on. Upon leaving through any door and throwing the associated switch, the room light is to turn off. Assume that the light is off when $A = B = C = 0$. (a) Determine the corresponding truth table. (b) Show that $L = (AB + \bar{A}\bar{B})C + (A\bar{B} + \bar{A}B)\bar{C}$.

11.25   Figure P11.25 shows a **double-pole, double-throw** (DPDT) switch. This device is, in essence, two distinct SPDT switches that are operated together. Specifically, when terminals a and b are connected, then a' and b' are connected. In addition,

when a and c are connected, so are a' and c'. Use one SPDT switch and two DPDT switches to construct the three-way light switch described in Problem 11.24.

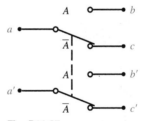

Fig. P11.25

11.26   The truth table given by Table P11.26 describes the exclusive-NOR (or equivalence) operation. This operation is denoted by the symbol $\odot$. Express $A \odot B$ as a function of the binary variables A

Table P11.26

A	B	$A \odot B$
0	0	1
0	1	0
1	0	0
1	1	1

and B, and use this expression to realize an exclusive-NOR gate using AND, OR, and NOT gates.

11.27 Determine the function F that characterizes the output of the logic circuit shown in Fig. P11.27. Construct the corresponding truth table. Does this table describe any operation that was studied previously?

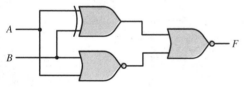

Fig. P11.27

11.28 Determine the function F that characterizes the output of the logic circuit shown in Fig. P11.28. Construct the corresponding truth table. Does this table describe any operation that was studied previously?

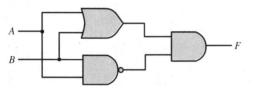

Fig. P11.28

11.29 Determine the function F that characterizes the output of the logic circuit shown in Fig. P11.29. Construct the corresponding truth table. Does this table describe any operation that was studied previously?

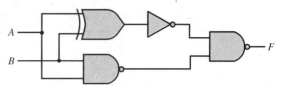

Fig. P11.29

11.30 Determine the function F that characterizes the output of the logic circuit shown in Fig. P11.30. Construct the corresponding truth table. Does this table describe any operation that was studied previously?

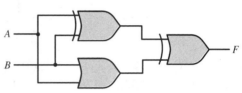

Fig. P11.30

11.31 Determine the function F that characterizes the output of the logic circuit shown in Fig. P11.31. Construct the corresponding truth table. Does this table describe any operation that was studied previously?

11.32 Verify the distributive law for AND over OR by constructing a truth table.

11.33 Use algebraic manipulations to simplify the following:

$$\overline{(A\overline{B} + \overline{A}B)(\overline{A}B)}$$

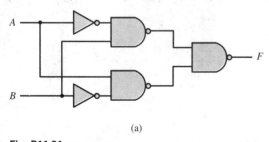

(a)

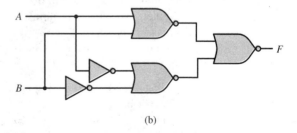

(b)

Fig. P11.31

11.34 Use algebraic manipulations to simplify the following:

$$\overline{(A\overline{B} + \overline{A}B) + \overline{(A + B)}}$$

11.35 Use algebraic manipulations to show that the exclusive-NOR operation has the equivalent forms

$$A \odot B = (A + \overline{B})(\overline{A} + B) = AB + \overline{A}\overline{B}$$

11.36 Use algebraic manipulations to show that the exclusive-OR operation has the equivalent forms

$$A \oplus B = A\overline{B} + \overline{A}B = (A + B)(\overline{AB})$$

11.37 Use algebraic manipulations to show that

$$AB + \overline{A}\overline{C} = (\overline{A} + B)(A + \overline{C})$$

11.38 Find the dual of the expression given in (a) Problem 11.33, (b) Problem 11.34, (c) Problem 11.35, (d) Problem 11.36, and (e) Problem 11.37.

11.39 Use algebraic manipulations to show that the exclusive-OR operation $A \oplus B = A\overline{B} + \overline{A}B$ is (a) commutative, that is, $A \oplus B = B \oplus A$; (b) associative, that is, $A \oplus (B \oplus C) = (A \oplus B) \oplus C$.

11.40 Use algebraic manipulations to show that the exclusive-NOR operation $A \odot B = AB + \overline{A}\overline{B}$ is (a) commutative, that is, $A \odot B = B \odot A$; and (b) associative, that is, $A \odot (B \odot C) = (A \odot B) \odot C$.

11.41 Show that (a) $A \uparrow B = B \uparrow A$, and (b) $A \uparrow (B \uparrow C) \neq (A \uparrow B) \uparrow C$, where $\uparrow$ denotes the NAND operation.

11.42 Construct a truth table for the function $F_1 = A \uparrow (B \uparrow C)$, where $\uparrow$ denotes the NAND operation. Express F_1 as (a) a sum of minterms, and (b) a product of maxterms. Repeat for $F_2 = (A \uparrow B) \uparrow C$.

11.43 Realize an exclusive-OR gate using only (a) NAND gates, and (b) NOR gates. Repeat for an exclusive-NOR gate.

11.44 A **majority gate** is a logic circuit whose output is 1 when a majority (more than half) of its inputs are 1. Construct the truth table for a three-input majority logic gate. Express the Boolean function F that describes the output of the gate as (a) a

sum of minterms, and (b) a product of maxterms. Realize both of these expressions using AND gates and OR gates.

11.45 Show that the Boolean function $F = AB + BC + AC$ describes the output of a three-input majority gate (see Problem 11.44) by constructing the corresponding truth table. Implement this function with a two-level logic circuit employing only NAND gates.

11.46 Show that the Boolean function $F = (A + B)(B + C)(A + C)$ describes the output of a three-input majority gate (see Problem 11.44) by constructing the corresponding truth table. Implement this function with a two-level logic circuit employing only NOR gates.

11.47 Express the Boolean function $F = A + BC$ as (a) a sum of minterms, and (b) a product of maxterms. Realize the expression obtained in part (a), the expression obtained in part (b), and the expression $F = A + BC$ by using AND gates and OR gates. Which realization requires the fewest logic gates?

11.48 Implement the Boolean function $F = A + BC = (A + B)(A + C)$ with a two-level logic circuit employing only (a) NAND gates, and (b) NOR gates.

11.49 Use algebraic manipulations to express the Boolean function $F = AB + BC + AC$ as (a) a sum of minterms, and (b) a product of maxterms.

11.50 Use algebraic manipulations to express the Boolean function $F = \overline{A}B + B\overline{C}$ as (a) a sum of minterms, and (b) a product of maxterms.

11.51 Find the simplified sum-of-products form of the Boolean functions: (a) $\overline{A}\overline{B}C + \overline{A}B\overline{C} + A\overline{B}\overline{C} + ABC$, (b) $\overline{A}\overline{B}\overline{C} + \overline{A}B\overline{C} + \overline{A}BC + A\overline{B}C + ABC$, (c) $\overline{A}BC + A\overline{B}C + AB\overline{C} + ABC$, and (d) $\overline{A}\overline{B}\overline{C} + \overline{A}B\overline{C} + \overline{A}BC + AB\overline{C}$.

11.52 Find the simplified product-of-sums forms of the Boolean functions given in Problem 11.51.

11.53 Find the simplified sum-of-products forms of the Boolean functions that correspond to the Karnaugh maps shown in Fig. P11.53.

(a)

A\\BC	00	01	11	10
0		1	1	1
1	1	1		

(b)

A\\BC	00	01	11	10
0				1
1	1		1	1

(c)

A\\BC	00	01	11	10
0	1	1	1	1
1		1		1

Fig. P11.53

11.54 Find the simplified product-of-sums forms of the Boolean functions corresponding to the Karnaugh maps given in Fig. P11.53.

11.55 Find the simplified product-of-sums form of the Boolean functions:
(a) $(A + B + C)(A + B + \overline{C})(A + \overline{B} + \overline{C})$
$(\overline{A} + B + \overline{C})(\overline{A} + \overline{B} + \overline{C})$
(b) $(A + B + \overline{C})(A + \overline{B} + \overline{C})(A + \overline{B} + C)$
(c) $(A + B + \overline{C})(\overline{A} + B + \overline{C})(\overline{A} + \overline{B} + C)$
$(\overline{A} + \overline{B} + \overline{C})$

11.56 Find the simplified sum-of-products forms of the Boolean functions given in Problem 11.55.

11.57 Find the simplified sum-of-products forms of the Boolean functions that correspond to the Karnaugh maps shown in Fig. P11.57.

11.58 Find the simplified product-of-sums forms of the Boolean functions corresponding to the Karnaugh maps given in Fig. P11.57.

11.59 Find the simplified sum-of-products form of the Boolean functions (a) $\overline{A}(\overline{B} + \overline{C}) + \overline{A}B + ABC$, (b) $A(B + \overline{D}) + \overline{B}(C + AD)$, and (c) $\overline{B}D + \overline{A}B\overline{C} + A(\overline{B}C + B\overline{C})$.

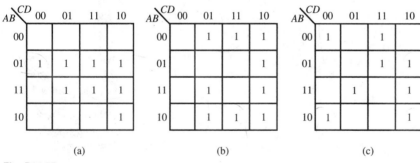

Fig. P11.57

12

Logic Design

INTRODUCTION

In the preceding chapter, we had the opportunity to analyze and design some relatively simple logic circuits. In this chapter, we consider the analysis and design of more complicated logic circuits—with the emphasis on design. In addition to the classical approach of designing logic circuits with individual logic gates that are available as small-scale-integration (SSI) devices, we will also study the design of logic circuits by using medium-scale-integration (MSI) and large-scale integration (LSI) packages. Examples of these are binary and BCD adders, magnitude comparators, encoders, decoders, multiplexers, demultiplexers, read-only memories (ROMs), and programmable logic arrays (PLAs). Devices such as these that see application in numerous types of digital systems, especially digital computers, will be studied in this chapter.

In addition to the devices mentioned above, the logic circuits encountered in the preceding chapter are examples of combinational logic. For such a circuit, the output at a particular time is determined directly from the combination of inputs applied at that time. Although important, a combinational logic circuit is not the only type of logic circuit. In this chapter, we will also discuss sequential logic circuits (or sequential circuits for short). Such a circuit contains storage or memory elements, typically in the form of flip-flops, and the output of the circuit at a particular time depends on the contents of the memory as well as the combination of inputs applied at that time. The chapter concludes with a discussion of flip-flops and the analysis and design of sequential circuits.

807

12.1 Combinational Logic

A logic circuit whose output (or outputs) at any given time is determined directly from the combination of inputs applied at that time is referred to as a **combinational logic circuit** or **combinational logic.** All the logic circuits we have encountered so far are examples of combinational logic. Later we will study noncombinational logic.

Given a combinational logic circuit, the process of **analysis** consists of determining a Boolean function (or functions), constructing a truth table, or formulating a verbal description that characterizes the circuit. As an example of the analysis of a combinational logic circuit, consider the two-level NAND-gate circuit shown in Fig. 12.1. As indicated, the outputs of the two-input NAND gates are, from top to bottom, $\overline{AB}$, $\overline{BC}$, and $\overline{AC}$, respectively. Thus, the Boolean function F describing the output of the circuit is

$$F = \overline{(\overline{AB})(\overline{BC})(\overline{AC})} \tag{12.1}$$

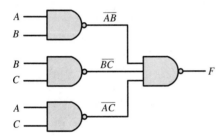

Fig. 12.1 Two-level NAND-gate logic circuit.

From this expression, we can construct a truth table such as Table 12.1. From this table, we have alternative expressions for F—one being the sum of minterms

$$F = m_3 + m_5 + m_6 + m_7 = \overline{A}BC + A\overline{B}C + AB\overline{C} + ABC \tag{12.2}$$

and one being the product of maxterms

$$F = M_0 M_1 M_2 M_4$$
$$= (A + B + C)(A + B + \overline{C})(A + \overline{B} + C)(\overline{A} + B + C)$$

The corresponding Karnaugh maps are given in Figs. 12.2a and b respectively. For the former, the simplified expression for F is

$$F = AB + BC + AC \tag{12.3}$$

Table 12.1 Truth Table for $F = \overline{(\overline{AB})(\overline{BC})(\overline{AC})}$.

A	B	C	AB	$\overline{AB}$	BC	$\overline{BC}$	AC	$\overline{AC}$	$\overline{(\overline{AB})(\overline{BC})(\overline{AC})}$	F
0	0	0	0	1	0	1	0	1	1	0
0	0	1	0	1	0	1	0	1	1	0
0	1	0	0	1	0	1	0	1	1	0
0	1	1	0	1	1	0	0	1	0	1
1	0	0	0	1	0	1	0	1	1	0
1	0	1	0	1	0	1	1	0	0	1
1	1	0	1	0	0	1	0	1	0	1
1	1	1	1	0	1	0	1	0	0	1

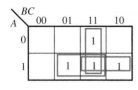

(a) Karnaugh map for
$F = m_3 + m_5 + m_6 + m_7$

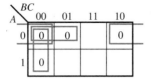

(b) Karnaugh map for
$F = M_0 M_1 M_2 M_4$

Fig. 12.2

and for the latter it is

$$F = (A + B)(B + C)(A + C) \tag{12.4}$$

For this function, however, constructing a truth table and Karnaugh maps can be avoided by using the De Morgan theorem $\overline{ABC} = \overline{A} + \overline{B} + \overline{C}$. Applying this to Eq. 12.1 yields

$$F = \overline{(\overline{AB})(\overline{BC})(\overline{AC})} = \overline{\overline{AB}} + \overline{\overline{BC}} + \overline{\overline{AC}} = AB + BC + AC$$

Furthermore,

$AB + BC + AC$

$\quad = (AB + BC + A)(AB + BC + C) \qquad$ since $A + BC = (A + B)(A + C)$

$\quad = (A + BC)(AB + C) \qquad$ since $A + AB = A$

$\quad = (A + B)(A + C)(A + C)(B + C) \qquad$ since $A + BC = (A + B)(A + C)$

$\quad = (A + B)(B + C)(A + C) \qquad$ since $AA = A$

where commutivity was also employed even though it was not explicitly stated. Therefore, algebraic manipulation is the simplest type of analysis in this case.

Adders

The process of going from a verbal description, a truth table, or a Boolean function (or functions) describing a logic circuit to a circuit diagram is known as **synthesis** or **design.** Analysis is unique in the sense that a given circuit behaves in a particular way. Conversely, synthesis is not unique because, in general, more than one logic circuit can produce a desired behavior. As an example of synthesis, let us synthesize a combinational logic circuit that will add binary numbers; that is, let us design an adder.

One way to describe the addition of two n-bit binary numbers given by $A = A_{n-1}A_{n-2} \ldots A_1A_0$ and $B = B_{n-1}B_{n-2} \ldots B_1B_0$ is with a truth table. Thus, there are $2n$ variables $(A_0, A_1, A_2, \ldots, A_{n-2}A_{n-1}, B_0, B_1, B_2, \ldots, B_{n-1})$ associated with the table. In other words, such a table has 2^{2n} rows. Just to be able to add numbers between 0 and 500 requires that $n = 9$ (because $2^9 = 512$) and, therefore, such a truth table has $2^{2n} = 2^{18} = 262,144$ rows. Hence we should take a different approach.

What we will do, instead, is approach the problem from a bit-by-bit point of view. Specifically, recall the rules for binary addition for single bits. They are

$$
\begin{array}{cccc}
0 & 0 & 1 & 1 \\
+\,0 & +\,1 & +\,0 & +\,1 \\
\hline
0 & 1 & 1 & 10
\end{array}
$$

For the first three cases, addition results in a sum—but no carry—whereas in the fourth case $(1 + 1)$, there is a sum (0) and a carry (1). A truth table describing this process of addition is given by Table 12.2, where A and B are the binary variables to be added, S is the sum, and C is the carry. From this table, we see that the resulting Boolean functions for S and C are

$$S = \overline{A}B + A\overline{B} = A \oplus B \quad \text{and} \quad C = AB$$

Table 12.2 Truth Table for the Addition of 2 Bits.

A	B	S	C
0	0	0	0
0	1	1	0
1	0	1	0
1	1	0	1

We now recognize that the expression for S describes the exclusive-OR operation, whereas C indicates the AND operation. Therefore, these functions can be implemented simply with an exclusive-OR gate and an AND gate, respectively, as shown in Fig. 12.3a. This logic circuit with two inputs A and B, and two outputs S and C, is known as a **half-adder.** A block diagram for a half-adder is shown in Fig. 12.3b.

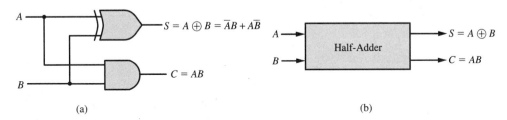

(a)   (b)

Fig. 12.3 (a) Half-adder, and (b) block diagram of half-adder.

Now let us see how we can extend the concept of a half-adder. First consider the example of adding $(93)_{10} = (1011101)_2$ and $(121)_{10} = (1111001)_2$ as shown below.

$$
\begin{array}{rcccccccc}
 & 1 & 1 & 1 & & & 1 & & \leftarrow\text{carries} \\
 & & 1 & 0 & 1 & 1 & 1 & 0 & 1 \\
+ & & 1 & 1 & 1 & 1 & 0 & 0 & 1 \\
\hline
 & 1 & 1 & 0 & 1 & 0 & 1 & 1 & 0 \\
 & & & & & & \uparrow & & \\
 & & & & & & \text{LSB} & &
\end{array}
$$

In determining the **least significant bit** (LSB)—the bit on the extreme right—2 bits are added together. This can be accomplished with a half-adder. To obtain any other bit, however, there must be the capability to deal with carries. In general, given two n-bit binary numbers $A_{n-1}A_{n-2} \ldots A_3A_2A_1A_0$ and $B_{n-1}B_{n-2} \ldots B_3B_2B_1B_0$, the process of addition is described as follows:

$$
\begin{array}{l}
C_nC_{n-1}C_{n-2} \ldots C_3C_2C_1 \\
\quad A_{n-1}A_{n-2} \ldots A_3A_2A_1A_0 \\
+ \quad B_{n-1}B_{n-2} \ldots B_3B_2B_1B_0 \\
\hline
S_n\,S_{n-1}\,S_{n-2} \ldots S_3S_2\,S_1\,S_0
\end{array}
$$

Here S_i is the sum of A_i, B_i, and C_i for $i = 1, 2, 3, \ldots, n - 1$, and C_{i+1} is the resulting carry. This means that the sums $S_1, S_2, S_3, \ldots, S_{n-2}, S_{n-1}$ are formed as are carries $C_2, C_3, \ldots, C_{n-2}, C_{n-1}, C_n$, according to Table 12.3a. Furthermore, Table 12.3b indicates how to obtain S_0 and C_1. (This is a half-adder truth table.) In addition, $S_n = C_n$. The result of adding the two given binary numbers is the binary number $S_nS_{n-1}S_{n-2} \ldots S_3S_2S_1S_0$.

Table 12.3(a) Truth Table for S_i and C_{i+1} for $i = 1, 2, \ldots, n - 1$.

A_i	B_i	C_i	S_i	C_{i+1}
0	0	0	0	0
0	0	1	1	0
0	1	0	1	0
0	1	1	0	1
1	0	0	1	0
1	0	1	0	1
1	1	0	0	1
1	1	1	1	1

Table 12.3(b) Truth Table for S_0 and C_1.

A_0	B_0	S_0	C_1
0	0	0	0
0	1	1	0
1	0	1	0
1	1	0	1

From Table 12.3a, we have that

$$S_i = m_1 + m_2 + m_4 + m_7 = \overline{A_i}\overline{B_i}C_i + \overline{A_i}B_i\overline{C_i} + A_i\overline{B_i}\overline{C_i} + A_iB_iC_i \quad (12.5)$$

and

$$C_{i+1} = m_3 + m_5 + m_6 + m_7 = \overline{A_i}B_iC_i + A_i\overline{B_i}C_i + A_iB_i\overline{C_i} + A_iB_iC_i \quad (12.6)$$

The Karnaugh map for S_i is given in Fig. 12.4. From this map, we see that the expression for S_i cannot be simplified as a sum of products. On the other hand, the expression for C_{i+1} has the form of Eq. 12.2 and the Karnaugh map in Fig. 12.2a and, therefore, can be written as is Eq. 12.3. In other words, we can put C_{i+1} in the form

$$C_{i+1} = A_iB_i + B_iC_i + A_iC_i \quad (12.7)$$

The expressions for S_i and C_i can be implemented by using the two-level AND-OR logic circuits shown in Fig. 12.5. A logic circuit, such as the one given in Fig. 12.5, whose inputs are A_i, B_i, C_i and whose outputs are S_i and C_{i+1} as described by Eq. 12.5 and Eq. 12.7, respectively, is called a **full-adder**. A block diagram for a full-adder is shown in Fig. 12.6.

Fig. 12.4 Karnaugh map for S_i.

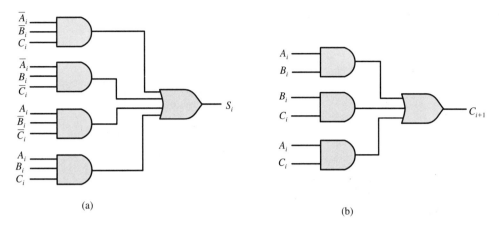

Fig. 12.5 AND-OR implementation of (*a*) S_i and (*b*) C_{i+1}.

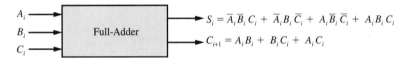

$$S_i = \overline{A}_i\overline{B}_iC_i + \overline{A}_iB_i\overline{C}_i + A_i\overline{B}_i\overline{C}_i + A_iB_iC_i$$

$$C_{i+1} = A_iB_i + B_iC_i + A_iC_i$$

Fig. 12.6 Block diagram for a full-adder.

By connecting a half-adder and $n - 1$ full-adders as shown in Fig. 12.7, we get a combinational logic circuit that will enable us to add any two *n*-bit binary numbers $A_{n-1}A_{n-2} \ldots A_3A_2A_1A_0$ and $B_{n-1}B_{n-2} \ldots B_3B_2B_1B_0$.

An important alternative to the full-adder implementation given in Fig. 12.5 may also be derived. Although Eq. 12.5 is in simplified sum-of-products form, we can manipulate S_i algebraically as follows:

$$S_i = \overline{A}_i\overline{B}_iC_i + \overline{A}_iB_i\overline{C}_i + A_i\overline{B}_i\overline{C}_i + A_iB_iC_i = \overline{A}_i(\overline{B}_iC_i + B_i\overline{C}_i) + A_i(\overline{B}_i\overline{C}_i + B_iC_i)$$

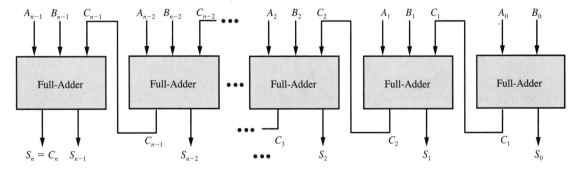

Fig. 12.7 An *n*-bit binary adder.

But note that

$$\overline{B_i}\overline{C_i} + B_i C_i = \overline{\overline{\overline{B_i}C_i} + \overline{B_i\overline{C_i}}} = \overline{(\overline{B_i}C_i + B_i\overline{C_i})}$$

$$= \overline{(B_i + \overline{C_i})(\overline{B_i} + C_i)} = \overline{B_i\overline{B_i} + B_i C_i + \overline{B_i}\overline{C_i} + C_i\overline{C_i}}$$

$$= \overline{B_i C_i + \overline{B_i}\overline{C_i}}$$

If we define $D_i = \overline{B_i}C_i + B_i\overline{C_i}$ then $\overline{D_i} = \overline{\overline{B_i}C_i + B_i\overline{C_i}} = \overline{B_i}\overline{C_i} + B_i C_i$ and we can write

$$S_i = \overline{A_i}D_i + A_i\overline{D_i} = A_i \oplus D_i$$

where

$$D_i = \overline{B_i}C_i + B_i\overline{C_i} = B_i \oplus C_i$$

In other words, we can write S_i as

$$S_i = A_i \oplus (B_i \oplus C_i)$$

However, since the exclusive-OR operation is associative (see Problem 11.39 on p. 805), it is unambiguous to write

$$S_i = A_i \oplus B_i \oplus C_i$$

In addition to this, from Eq. 12.6, we have

$$C_{i+1} = \overline{A_i}B_i C_i + A_i\overline{B_i}C_i + A_i B_i\overline{C_i} + A_i B_i C_i$$

$$= (\overline{A_i}B_i + A_i\overline{B_i})C_i + A_i B_i(\overline{C_i} + C_i) = (A_i \oplus B_i)C_i + A_i B_i$$

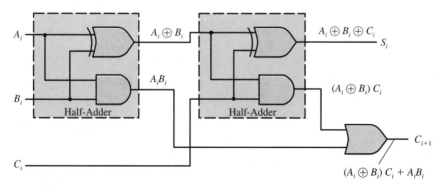

Fig. 12.8 Full-adder consisting of two half-adders and an OR gate.

Using these expressions for S_i and C_{i+1}, we now see that the logic circuit shown in Fig. 12.8, which is comprised of two half-adders and an OR gate, realizes a full-adder.

Subtractions can be accomplished by taking complements and then performing addition (see Section 11.2). Because of this fact, adders can also be used for finding the difference between two binary numbers.

Don't-Care Conditions

Figure 12.9*a* shows either an LCD (liquid-crystal display) or an LED (light-emitting diode) seven-segment display that is used to indicate any of the decimal digits 0, 1, 2, 3, 4, 5, 6, 7, 8, and 9. Figure 12.9*b* shows how the decimal digits can be formed by using the various segments of the display. For example, the decimal digit 1 is comprised of the segments labeled *b* and *c*, whereas the digit 2 consists of the segments labeled *a*, *b*, *d*, *e*, and *g*.

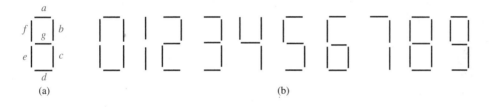

(a)   (b)

Fig. 12.9 (*a*) Seven-segment display, and (*b*) decimal digits obtained with seven-segment display.

Assume that a segment of the display is visible when the bit 1 is applied to it and is invisible when a 0 is applied to it. Given the BCD representation of decimal digits, suppose that we wish to have such a digit shown on the display. This situation is described by the truth table given in Table 12.4. The input variables *A*, *B*, *C*, and *D* give the BCD representation of a digit. The seven output functions *a*, *b*, *c*, *d*, *e*, *f*, and *g* indicate whether or not the corresponding segment is to be visible. For example, the segment labeled *a* is to be visible for the digits 0, 2, 3, 5, 6, 7, 8, and 9. Thus there are 1s in columnn *a* only in the rows corresponding to these digits. Similarly, column *b* has 1s in the rows corresponding to the digits 0, 1, 2, 3, 4, 7, 8, and 9.

In the discussion above, we assumed that information expressed as decimal digits is to be displayed, and such digits are in BCD form. Let us also assume that a binary quadruple that is not the BCD representation of a decimal digit is not produced. Therefore, since the binary quadruples in the last six rows of Table 12.4 (under the variables *A*, *B*, *C*, *D*) cannot occur, then it does not matter whether a 0 or a 1 is

Table 12.4 Truth Table Characterizing the Seven-Segment Display.

A	B	C	D	a	b	c	d	e	f	g
0	0	0	0	1	1	1	1	1	1	0
0	0	0	1	0	1	1	0	0	0	0
0	0	1	0	1	1	0	1	1	0	1
0	0	1	1	1	1	1	1	0	0	1
0	1	0	0	0	1	1	0	0	1	1
0	1	0	1	1	0	1	1	0	1	1
0	1	1	0	1	0	1	1	1	1	1
0	1	1	1	1	1	1	0	0	0	0
1	0	0	0	1	1	1	1	1	1	1
1	0	0	1	1	1	1	1	0	1	1
1	0	1	0	×	×	×	×	×	×	×
1	0	1	1	×	×	×	×	×	×	×
1	1	0	0	×	×	×	×	×	×	×
1	1	0	1	×	×	×	×	×	×	×
1	1	1	0	×	×	×	×	×	×	×
1	1	1	1	×	×	×	×	×	×	×

indicated for each output function. Because of this, the symbol $\times$ is used, and this is referred to as a **don't-care condition.**

The Karnaugh map for the Boolean function a is shown in Fig. 12.10a, where the don't-care conditions are included. When simplifying the expression for a, each $\times$ can be used as a 0 or as a 1—whichever yields the simplest expression. In this case, by making each $\times$ a 1, we can form two groups of eight and two groups of four, as indicated. The result is that

$$a = A + C + BD + \overline{B}\,\overline{D}$$

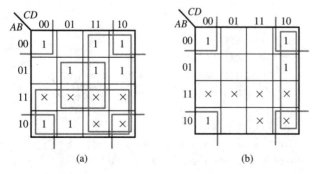

(a)   (b)

Fig. 12.10 (a) Karnaugh map for function a. (b) Karnaugh map for function e.

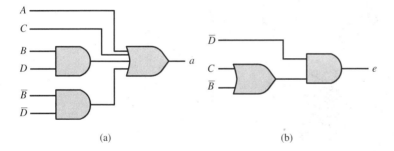

Fig. 12.11 Realization of (a) $a = A + C + BD + \overline{B}\,\overline{D}$, and (b) $e = (\overline{B} + C)\overline{D}$.

and a two-level logic circuit that implements this Boolean function is shown in Fig. 12.11a.

As another example, the Karnaugh map for the Boolean function e is shown in Fig. 12.10b. In this case, two groups of four encompass all the 1s, and this is accomplished by utilizing only two of the six don't-care conditions. As a result, we have that

$$e = \overline{B}\,\overline{D} + C\overline{D} = (\overline{B} + C)\overline{D}$$

and this function can be implemented as shown in Fig. 12.11b.

Under the circumstance that binary quadruples that do not correspond to the BCD representations of decimal digits are allowed to appear, all of the don't-care conditions indicated by an $\times$ in Table 12.4 should be changed to 0s. Doing this will result in an invisible (blank) display when an invalid (non-BCD) quadruple arises.

Integrated-Circuit Gates

Ordinarily, logic gates in integrated-circuit (IC) form are used in logic design. Such ICs typically come in two types of packages—the **dual-in-line package** (DIP) shown in Fig. 12.12a and the **flat package** shown in Fig. 12.12b. Because it is easily installed in circuit boards or sockets, and is low in price, the former IC package is much more popular than the latter. The number of pins on a package can be as few as 8 and can exceed 64—the 14-pin and 16-pin packages being quite common. The dimensions of a typical 14-pin DIP are $20 \times 8 \times 5$ millimeters, whereas a 16-pin flat package can be $20 \times 10 \times 3$ millimeters.

The number of gates on an IC chip indicates what type of device that IC is. If an IC consists of 30 or fewer gates, it is called a **small-scale-integration** (SSI) device. A **medium-scale-integration** (MSI) device contains between 30 and 300 gates, whereas a **large-scale-integration** (LSI) device has between 300 and 3000 gates. An

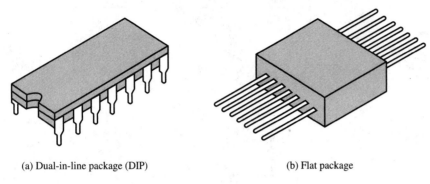

(a) Dual-in-line package (DIP)   (b) Flat package

Fig. 12.12 IC packages.

IC with more that 3000 gates is referred to as a **very-large-scale-integration** (VLSI) device.

Even though there are basically two types of digital logic IC packages, the IC itself can be a member of one of a number of logic-circuit families. The most popular logic families are **transistor-transistor logic** (TTL), **emitter-coupled logic** (ECL), **metal-oxide semiconductor** (MOS), **complementary metal-oxide semiconductor** (CMOS), and **integrated-injection logic** (I^2L).

The labeling on a package indicates to which family that IC belongs. ICs belonging to the 5400 series or the 7400 series are TTL ICs. The ICs in the 5400 series operate over a wider temperature range (-55 to $125°C$) than those in the 7400 series (0 to $70°C$), and, therefore, the 5400 series is designated for military use. On the other hand, ICs in the 7400 series are suitable for industrial applications. The 7400 ICs are designated 7400, 7401, 7402, 7403, and so on. Figure 12.13 shows the functional description of a 7400 TTL IC. This device is a quad 2-input NAND gate; that is, the IC contains four distinct 2-input NAND gates as illustrated. The top view of the IC shows that pins 1 and 2 are the inputs to one NAND gate and pin 3 is the output of that gate. The pin connections for the other NAND gates are also displayed. Pin 7 is labeled GND to indicate that it is to be connected to the reference or ''ground.'' A supply voltage of $V_{CC} = 5$ V connected to pin 14 is required for proper operation. As another example of a TTL IC and its pin connections, Fig. 12.14 shows the 7427 triple 3-input NOR gate.

The 10,000 series is a common ECL designation. An ECL 10107 triple 2-input exclusive-OR/NOR gate is shown in Fig. 12.15. For this particular IC, the typical value for the negative supply-voltage pin is $V_{EE} = 5.2$ V, whereas pins 1 and 16 are connected to ground. Note that this IC is in a 16-pin DIP. Furthermore, each individual gate has two outputs, the one without the small circle indicating the exclusive-OR function and the other (with the small circle) designating the exclusive-NOR function. Figure 12.16 shows a CMOS IC in the 4000 series. This 4002 dual 4-input NOR gate utilizes only 12 of the 14 pins of the DIP. Specifically, pins 6 and 8 are

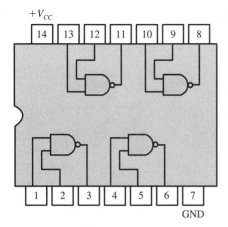

Fig. 12.13 A 7400 quad 2-input NAND gate (TTL).

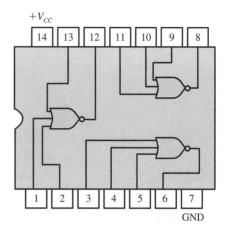

Fig. 12.14 A 7427 triple 3-input NOR gate (TTL).

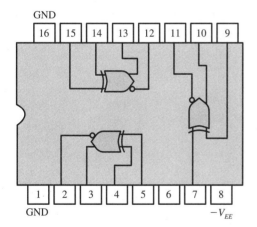

Fig. 12.15 A 10107 triple 2-input exclusive-OR/ NOR gate (ECL).

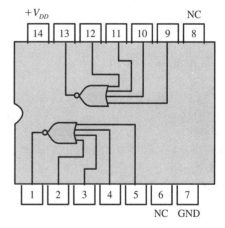

Fig. 12.16 A 4002 dual 4-input NOR gate (CMOS).

labeled NC to indicate "no connection." The supply voltage V_{DD} for this IC can be any value in the range from 3 to 15 V.

Each IC family uses different values for supply voltages, and they have different values of low- and high-voltage levels.[1] Table 12.5 summarizes the typical values for the TTL, ECL, and CMOS families. Not only are there different IC families, but there are also different versions of the same family. In particular, there are a number of forms of TTL, such as standard, high-power standard (designated H), low-power

---

[1]For positive logic, the low voltage is logical 0 and the high voltage is logical 1; conversely for negative logic.

Table 12.5 Comparison of Typical Values Used for IC Families.

Family	Basic Gate	Supply Voltage	Low-Voltage Level	High-Voltage Level
TTL	NAND	5 V	0.2 V	3.4 V
ECL	OR/NOR	−5.2 V	−1.8 V	−0.8 V
CMOS	NOT	3 to 15 V	0 V	3 to 15 V

standard (L), Schottky (S), and low-power Schottky (LS). The specific version of an IC is indicated by placing the appropriate letter designation after its series number. For example, the low-power-Schottky version of the 7400 quad 2-input NAND gate is designated 74LS00, and the high-power-standard version is 74H00. The Schottky version of the 7427 triple 3-input NOR gate is designated 74S27, and the standard-version labeling remains 7427.

Typical characteristics for the basic gate of different versions of TTL as well as ECL and CMOS are listed in Table 12.6. Recall from Chapter 7 that the maximum number of similar output gates that can load a given gate and still have that gate function properly is called the **fan-out** of the gate. Furthermore, the minimum noise voltage at the input of a gate that causes the output to deviate from its proper value is known as the **noise margin** of the gate. The **power dissipation** of a gate is the average power that is delivered to that gate by the supply voltage (power supply). The **propagation delay** is the average time delay for a voltage pulse to propagate through the gate.

The classical approach to logic design, as described earlier, is to write a Boolean function (or functions) and implement it using individual logic gates. As we have just discussed, these gates are available in a wide variety of SSI packages that are fabricated from different IC families. (The CMOS and I^{2}L families generally see LSI and VLSI applications.) There can be, however, many instances in which commercially available MSI or LSI packages can be utilized for the design process. It may

Table 12.6 Characteristics of IC Logic Families.

Logic Family	Fan-out	Noise Margin, V	Power Dissipation, mW	Propagation Delay, ns
Standard TTL	10	0.4	10	10
High-power standard TTL	10	0.4	22	6
Low-power standard TTL	20	0.4	1	33
Schottky TTL	10	0.4	22	3
Low-power Schottky TTL	20	0.4	2	10
ECL	25	0.2	25	2
CMOS	50	3	0.1	25

be that such packages produce the exact implementations that are desired, or that such realizations can be obtained by appropriately connecting ICs together. This approach to design is our next topic of study.

12.2 MSI and LSI Design

In the preceding section, we saw how to design an n-bit binary adder (see Fig. 12.7) using a half-adder and $n - 1$ full-adders. To be able to include the case of a possible previous carry, the half-adder can be replaced by a full-adder. Figure 12.17 shows this more general **binary parallel adder** for the case that $n = 4$. Such a device is available as an MSI package (e.g., the TTL 74283 IC), as are other n-bit ($n \neq 4$) binary parallel adders. By connecting such ICs together appropriately, larger adders can be formed. For example, by connecting two 4-bit adders as shown in Fig. 12.18, we get an 8-bit binary parallel adder.

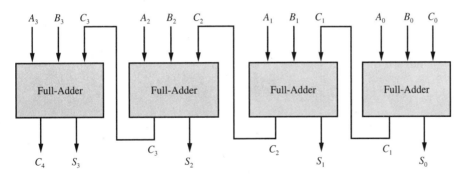

Fig. 12.17 A 4-bit binary parallel adder.

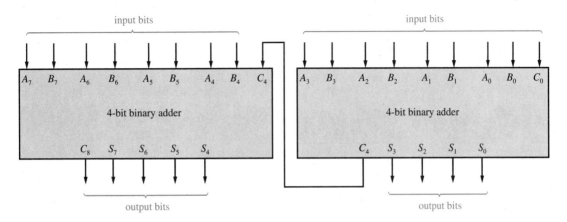

Fig. 12.18 An 8-bit binary parallel adder.

Not only can adders be used to construct larger adders, but they also can be used for other types of applications as well. An example of such a situation is the conversion from BCD to the **excess-3 code.** Table 12.7 shows the BCD representations of the decimal digits and the corresponding words of the excess-3 code. Note that the words of the excess-3 code can be obtained by adding the binary number 3 (0011) to the corresponding BCD representations of the decimal digits. Because of this property, we can use a 4-bit binary parallel adder if we make $B_0 = B_1 = 1$ and $B_2 = B_3 = C_0 = 0$. Doing this will result in the desired addition, and thus produce outputs that are the appropriate excess-3 code words. An MSI implementation of such a code converter is shown in Fig. 12.19.

Table 12.7 Excess-3 Code-Word Correspondence with BCD.

BCD				Excess-3 Code			
A_3	A_2	A_1	A_0	S_3	S_2	S_1	S_0
0	0	0	0	0	0	1	1
0	0	0	1	0	1	0	0
0	0	1	0	0	1	0	1
0	0	1	1	0	1	1	0
0	1	0	0	0	1	1	1
0	1	0	1	1	0	0	0
0	1	1	0	1	0	0	1
0	1	1	1	1	0	1	0
1	0	0	0	1	0	1	1
1	0	0	1	1	1	0	0

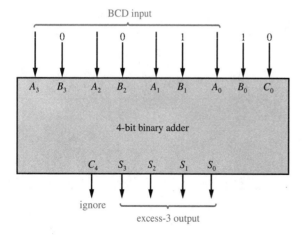

Fig. 12.19 A BCD-to-excess-3-code converter.

BCD Adders

There is an alternative to the binary addition of numbers—specifically, an alternative is decimal addition. By this we mean the addition of decimal digits in BCD form. Suppose we perform binary addition on the BCD representations of two decimal digits. If the sum of the two digits is 9 or less, then performing binary addition results in the appropriate number. If the sum of the digits is greater than 9, however, then what results from the addition is the binary number representation of the sum, not the BCD representation. Table 12.8 lists the sums that result from the binary addition of decimal digits in BCD form, and it also shows the BCD versions of these sums. Even though $9 + 9 = 18$, the table extends to include sums equaling 19 so that carries from a preceding addition can be included.

By applying two decimal digits in BCD form, say $A_3A_2A_1A_0$ and $B_3B_2B_1B_0$, to a 4-bit binary adder (see Fig. 12.17), we obtain a sum $S_4S_3S_2S_1S_0$ in binary form, where $S_4 = C_4$ is the carry bit. If this binary number is 9 or less, then this sum is also in BCD form. If this binary number is greater than 9, however, we must modify the

Table 12.8 Sums of BCD Digits in Binary Form and BCD Form.

Sums in Binary Form					Sums in BCD Form				
S_4	S_3	S_2	S_1	S_0	D_4	D_3	D_2	D_1	D_0
0	0	0	0	0	0	0	0	0	0
0	0	0	0	1	0	0	0	0	1
0	0	0	1	0	0	0	0	1	0
0	0	0	1	1	0	0	0	1	1
0	0	1	0	0	0	0	1	0	0
0	0	1	0	1	0	0	1	0	1
0	0	1	1	0	0	0	1	1	0
0	0	1	1	1	0	0	1	1	1
0	1	0	0	0	0	1	0	0	0
0	1	0	0	1	0	1	0	0	1
0	1	0	1	0	1	0	0	0	0
0	1	0	1	1	1	0	0	0	1
0	1	1	0	0	1	0	0	1	0
0	1	1	0	1	1	0	0	1	1
0	1	1	1	0	1	0	1	0	0
0	1	1	1	1	1	0	1	0	1
1	0	0	0	0	1	0	1	1	0
1	0	0	0	1	1	0	1	1	1
1	0	0	1	0	1	1	0	0	0
1	0	0	1	1	1	1	0	0	1

sum to get its BCD form. From Table 12.8, we see that if we add the binary number 6 (0110) to the sum in binary form, we get the sum in BCD form. But how can we state this in logical terms?

From Table 12.8 we have that the binary number 6 should be added to a sum in binary form when $S_4 = 1$, OR $S_3 = 1$ AND $S_2 = 1$, OR $S_3 = 1$ AND $S_1 = 1$. This description corresponds to the Boolean function

$$F = S_4 + S_3 S_2 + S_3 S_1$$

As a consequence of this discussion we can realize a **BCD adder,** that is, a logic circuit that adds decimal digits in BCD form, by using two 4-bit binary adders as shown in Fig. 12.20. Two decimal digits in BCD form are added together in binary

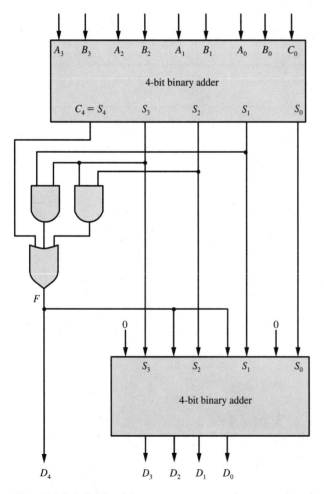

Fig. 12.20 A BCD adder.

by the top 4-bit adder. If the resulting binary number $S_4S_3S_2S_1S_0$ is 9 or less, then $S_4S_3S_2S_1S_0 = S_3S_2S_1S_0$, $F = 0$, and 0 is added to $S_3S_2S_1S_0$ by the bottom 4-bit adder. If the resulting binary number $S_4S_3S_2S_1S_0$ is greater than 9, however, then $F = 1$ and 6 is added to $S_3S_2S_1S_0$ by the bottom 4-bit adder. Because it is a 4-bit adder and S_4 is not applied to its input, the output carry D_4 cannot be obtained for the sums 16, 17, 18, and 19. This is no problem because a carry results when $F = 1$. Therefore, we can use the output of the OR gate to produce D_4. Note, too, that the bottom 4-bit adder employs no carry from a preceding stage and, therefore, no input for such a carry is indicated.

Even though we have spent time discussing how to construct them from two 4-bit adders, BCD adders are available as MSI packages (e.g., the TTL 82S83 IC). Despite this, it is worthwhile studying how to combine MSI circuits that implement one kind of function in order to produce logic circuits that realize another type of function.

Magnitude Comparators

Another very important digital logic function is performed by a **magnitude comparator,** and such a device is also available in MSI form (e.g., the TTL 7485 IC is a 4-bit magnitude comparator). As its name suggests, the function that is performed is the comparison of two numbers A and B, and the determination of whether $A < B$, $A > B$, or $A = B$.

Given two n-bit binary numbers $A = A_{n-1} \ldots A_2A_1A_0$ and $B = B_{n-1} \ldots B_2B_1B_0$, an n-bit magnitude comparator therefore will have $2n$ inputs and 3 outputs C, D, and E, which signify $A < B$, $A > B$, and $A = B$, respectively. (For example, $C = 0$, $D = 1$, $E = 0 \Rightarrow A > B$.) A truth table characterizing magnitude comparison has 2^{2n} rows—therefore, the construction of such a table will be avoided.

Specifically, let us consider the case that $n = 3$. In other words, suppose that we wish to compare the 3-bit binary numbers $A = A_2A_1A_0$ and $B = B_2B_1B_0$. The equality of a pair of bits A_i and B_i can be determined from Table 12.9. From this truth table, we see that $A_i = B_i$ is described by the Boolean expression $F_i = \overline{A_i}\overline{B_i} + A_iB_i = A_i \odot B_i$. We wish to have the output $C = 1$ when $A < B$. This inequality occurs

Table 12.9 Truth Table for the Equality of a Pair of Bits.

A_i	B_i	F_i
0	0	1
0	1	0
1	0	0
1	1	1

when $A_2 = 0$ ($\overline{A}_2 = 1$) AND $B_2 = 1$, OR when $A_1 = 0$ ($\overline{A}_1 = 1$) AND $B_1 = 1$ given that (AND) $A_2 = B_2$ ($F_2 = 1$), OR when $A_0 = 0$ ($\overline{A}_0 = 1$) AND $B_0 = 1$ given that (AND) $A_2 = B_2$ AND $A_1 = B_1$ ($F_2 = 1$ AND $F_1 = 1$). This description can be characterized by the Boolean function

$$C = \overline{A}_2 B_2 + \overline{A}_1 B_1 F_2 + \overline{A}_0 B_0 F_1 F_2$$

where $F_1 = \overline{A}_1 \overline{B}_1 + A_1 B_1 = A_1 \odot B_1$ and $F_2 = \overline{A}_2 \overline{B}_2 + A_2 B_2 = A_2 \odot B_2$ are exclusive-NOR functions.

Proceeding in a similar manner for the case that $A > B$, we get the Boolean function

$$D = A_2 \overline{B}_2 + A_1 \overline{B}_1 F_2 + A_0 \overline{B}_0 F_1 F_2$$

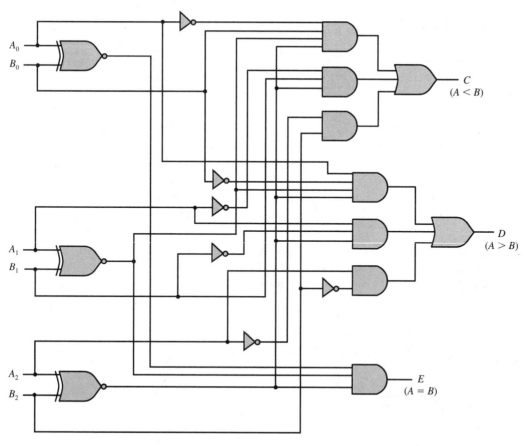

Fig. 12.21 A 3-bit magnitude comparator.

Finally, for the case that $A = B$ (i.e., $A_2 = B_2$ AND $A_1 = B_1$ AND $A_0 = B_0$), we have that

$$E = F_2 F_1 F_0 = (A_2 \odot B_2)(A_1 \odot B_1)(A_0 \odot B_0)$$

A logic circuit for a 3-bit magnitude comparator is shown in Fig. 12.21.

Encoders

In Section 11.2, we mentioned that various pieces of information could be represented by a code, that is, a set of binary sequences. As a simple example, consider the four pieces of information A_1, A_2, A_3, and A_4. To have four distinct binary sequences, each code word must consist of a minimum of 2 bits. Even though we can use sequences of more than 2 bits, for the sake of brevity, let us pick code words with 2 bits as given in Table 12.10. This table indicates that the code words for A_1, A_2, A_3, and A_4 are 00, 01, 10, and 11, respectively.

Table 12.10 A Binary Encoding.

Pieces of Information				Code Words	
A_1	A_2	A_3	A_4	B_1	B_2
1	0	0	0	0	0
0	1	0	0	0	1
0	0	1	0	1	0
0	0	0	1	1	1

Suppose we wish to construct a logic circuit to encode each piece of information. In other words, when $A_1 = 1$, we want 00 to be produced; when $A_2 = 1$, we want 01 to result, and so on. We will also assume that only one piece of information is encoded at a time (e.g., if $A_3 = 1$, then $A_1 = A_2 = A_4 = 0$). Under these conditions, from Table 12.10, we can write the Boolean expressions

$$B_1 = A_3 + A_4 \quad \text{and} \quad B_2 = A_2 + A_4$$

A combinational logic circuit that implements these functions is shown in Fig. 12.22. (Note that input A_1 is not connected to anything.) Such a device is called an **encoder.** In general, an encoder for m pieces of information has m inputs (one for each piece of information) and n outputs, where $2^n \geq m$. This can be called an **m-to-n-line** (designated $m \times n$) **encoder.**

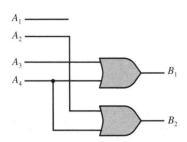

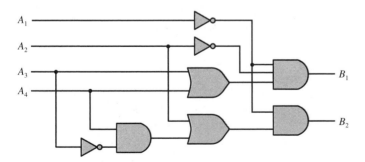

Table 12.11 A Priority Encoding.

	Pieces of Information				Code Words	
A_1	A_2	A_3	A_4		B_1	B_2
1	×	×	×		0	0
0	1	×	×		0	1
0	0	1	×		1	0
0	0	0	1		1	1

Fig. 12.22 Encoder corresponding to Table 12.10.

Now let us consider the situation that two or more inputs are equal to 1 at the same time. For such a case, an encoder's output has no meaning—two pieces of information cannot be encoded at the same time. Therefore, let us assign a priority to the information to be encoded. For the example above, let the subscript of the information indicate its priority, that is, A_1 has the highest priority, A_2 is second, A_3 is third, and A_4 is last. Thus if more than one input is 1, the piece of information with the highest priority is encoded. These conditions are described by Table 12.11. Here the symbol × indicates a don't-care condition. For example, if $A_3 = 1$, it does not matter whether $A_4 = 0$ or $A_4 = 1$; since A_3 has priority over A_4, the resulting encoding is 10. From Table 12.11, we can write the Boolean expressions

$$B_1 = \overline{A}_1\overline{A}_2A_3 + \overline{A}_1\overline{A}_2\overline{A}_3A_4 = \overline{A}_1\overline{A}_2(A_3 + \overline{A}_3A_4) = \overline{A}_1\overline{A}_2(A_3 + A_4)$$

and

$$B_2 = \overline{A}_1A_2 + \overline{A}_1\overline{A}_2\overline{A}_3A_4 = \overline{A}_1(A_2 + \overline{A}_2\overline{A}_3A_4) = \overline{A}_1(A_2 + \overline{A}_3A_4)$$

A logic circuit for such an encoder, known as a **priority encoder,** is shown in Fig. 12.23.

Fig. 12.23 A 4-to-2-line priority encoder.

Because of its simplicity of construction, ordinary (nonpriority) encoders do not come in the form of MSI packages. Priority encoders, however, are available as MSI units. For example, the 74147 IC is a 10-to-4-line decimal-to-BCD priority encoder, and the IC 74148 is an 8-to-3-line octal-to-binary priority encoder.

Decoders

The converse of the process of encoding is called **decoding.** In general, an ***n*-to-*m*-line decoder** has n inputs and m outputs, where $2^n \geq m$. For each n-bit input, there is a unique output that is equal to 1—the remaining outputs being equal to 0.

An example of a 3-to-8-line decoder is a device whose input is a 3-bit binary number between 0 and 7, and whose output is the corresponding octal digit. The truth table describing this decoder is given by Table 12.12, where D_0 through D_7 designate the octal digits 0 through 7 respectively. From this table we see that

$$D_i = m_i \qquad \text{for } i = 0, 1, 2, 3, 4, 5, 6, 7$$

Therefore, each output function is one of the minterms of the input variables. The implementations of these Boolean expressions with logic circuits should be a routine exercise by now, so they are omitted. Do note, however, that each output function can be implemented with one (3-input) AND gate.

Unlike (nonpriority) encoders, decoders are available in MSI form (e.g., the TTL 74138 IC is a 3-to-8-line decoder).

The 3×8 decoder characterized by Table 12.12 is an example of a decoder with n inputs and $m = 2^n$ outputs. The output functions for such a device are all the minterms of the input variables. Since any Boolean function can be expressed as a

Table 12.12 Truth Table for a (3-to-8-Line) Binary-to-Octal Decoder.

Inputs			Outputs							
A	B	C	D_0	D_1	D_2	D_3	D_4	D_5	D_6	D_7
0	0	0	1	0	0	0	0	0	0	0
0	0	1	0	1	0	0	0	0	0	0
0	1	0	0	0	1	0	0	0	0	0
0	1	1	0	0	0	1	0	0	0	0
1	0	0	0	0	0	0	1	0	0	0
1	0	1	0	0	0	0	0	1	0	0
1	1	0	0	0	0	0	0	0	1	0
1	1	1	0	0	0	0	0	0	0	1

sum of minterms, this type of decoder can be connected to an OR gate to implement any Boolean function. For example, from Eq. 12.5 and Eq. 12.6, we know that a full-adder is characterized by the Boolean functions

$$S_i = m_1 + m_2 + m_4 + m_7 \quad \text{and} \quad C_{i+1} = m_3 + m_5 + m_6 + m_7$$

Therefore, we can use OR gates and a 3-to-8-line decoder such as the one described by Table 12.12 to implement a full-adder. Figure 12.24 gives the realization.

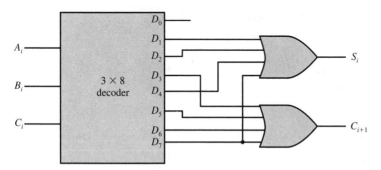

Fig. 12.24 Realization of a full-adder with a 3-to-8-line decoder.

Multiplexers

A **digital multiplexer** is a logic circuit typically with 2^n inputs, one output, and n selection variables or lines. For such a device, the n selection lines are used to make the output equal to one of the inputs. Because of this, a digital multiplexer (designated MUX) is also referred to as a **data selector.**

Figure 12.25 shows a realization of a 4-to-1-line multiplexer (designated 4×1 MUX). For this case ($n = 2$) there are four input lines labeled I_0, I_1, I_2, I_3; two selection lines labeled A, B; and one output line F. Note that when $A = B = 0$, then the output of the top AND gate is $\overline{A}\,\overline{B}I_0 = I_0$ and the output of each remaining AND gate is 0. Thus $F = I_0$. Similarly, when $A = 0$ and $B = 1$, then the output of the second (from the top) AND gate is $\overline{A}BI_1 = I_1$ and each of the other AND gates has an output of 0. Thus $F = I_1$. Continuing in this manner, we see that the output is equal to one of the inputs—which one depends on the selection-line values. These results are summarized by Table 12.13. The MSI representation of a 4-to-1-line multiplexer is shown in Fig. 12.26. Note that a Boolean expression for the output F is given by

$$F = \overline{A}\,\overline{B}I_0 + \overline{A}BI_1 + A\overline{B}I_2 + ABI_3 \tag{12.8}$$

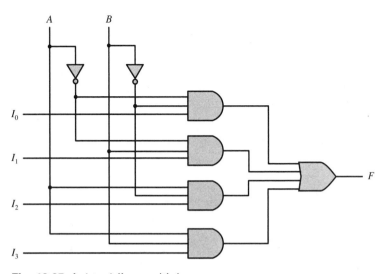

Fig. 12.25 A 4-to-1-line multiplexer.

Table 12.13 Table Characterizing a 4-to-1-Line Multiplexer.

Selection Lines		Output
A	B	F
0	0	I_0
0	1	I_1
1	0	I_2
1	1	I_3

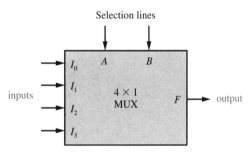

Fig. 12.26 Circuit symbol of a 4-to-1-line multiplexer.

Let us now see how we can use a multiplexer to implement Boolean functions. Suppose that we would like to realize the function F characterized by Table 12.14. From this truth table, we can write

$$F = \overline{A}BC + A\overline{B}\,\overline{C} + AB\overline{C} + ABC = \overline{A}BC + A\overline{B}\,\overline{C} + AB(\overline{C} + C)$$

$$F = \overline{A}\,\overline{B}(0) + \overline{A}BC + A\overline{B}\,\overline{C} + AB(1) \tag{12.9}$$

Therefore, by putting the expression for F (Eq. 12.9) into the form of Eq. 12.8, we can make the associations

$$I_0 = 0 \qquad I_1 = C \qquad I_2 = \overline{C} \qquad I_3 = 1$$

Table 12.14 Truth Table for a Boolean Function F.

A	B	C	F
0	0	0	0
0	0	1	0
0	1	0	0
0	1	1	1
1	0	0	1
1	0	1	0
1	1	0	1
1	1	1	1

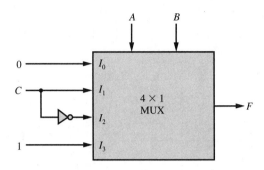

Fig. 12.27 Realization of F with a 4-to-1-line multiplexer.

This means that the 4-to-1-line multiplexer shown in Fig. 12.27 implements the Boolean function F described by Table 12.14.

Demultiplexers

The converse process of multiplexing is **demultiplexing.** Specifically, a **demultiplexer** is a device with one input, n selection lines, and 2^n outputs. For such a device, the n selection lines are used to make one of the outputs equal to the input. Figure 12.28 shows a realization of a 1-to-4-line demultiplexer. For this case ($n = 2$) there is one input line labeled E; two selection lines labeled A, B; and four output lines

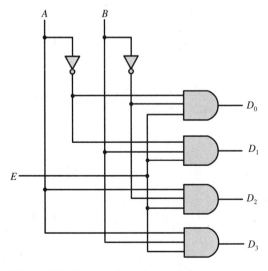

Fig. 12.28 A 1-to-4-line demultiplexer.

labeled D_0, D_1, D_2, D_3. Note that when $A = B = 0$, then the output of the top AND gate is $D_0 = \overline{A}\,\overline{B}E = E$ and the remaining outputs are $D_1 = D_2 = D_3 = 0$. Similarly, when $A = 0$ and $B = 1$, then $D_1 = \overline{A}BE = E$ and $D_0 = D_2 = D_3 = 0$. Continuing in this manner, we see that the input is transferred to one of the outputs—which one depends on the selection-line values. These results are summarized by Table 12.15. The MSI representation of a 1-to-4-line demultiplexer is shown in Fig. 12.29.

Table 12.15 Table Characterizing a 1-to-4-line demultiplexer.

Selection Lines		Outputs			
A	B	D_0	D_1	D_2	D_3
0	0	E	0	0	0
0	1	0	E	0	0
1	0	0	0	E	0
1	1	0	0	0	E

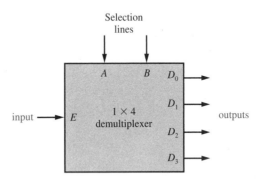

Fig. 12.29 Circuit symbol of a 1-to-4-line demultiplexer.

From Table 12.15, note that if we always have the condition that $E = 1$, then the 1-to-4-line demultiplexer becomes a 2-to-4-line decoder. On the other hand, if we always have $E = 0$, then the device does nothing—all the outputs are 0 regardless of values of the selection lines. Thus we can use a demultiplexer as a decoder; to do this, we just set $E = 1$ and we use the selection lines as the decoder inputs. In this case, E is referred to as the **enable** input—making $E = 1$ enables decoding, whereas $E = 0$ does not. Using the 1-to-4-line demultiplexer given in Fig. 12.29 as a 2-to-4-line decoder yields the MSI representation shown in Fig. 12.30.

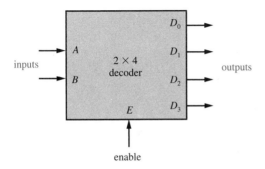

Fig. 12.30 A 2-to-4-line decoder with an enable input.

Read-Only Memories

Previously, we saw how decoders can be used in conjunction with OR gates to implement Boolean functions. A $2^n \times m$ **read-only memory** (ROM) is a device with an n-to-2^n-line decoder and m OR gates, each having 2^n inputs, all on the same IC. Such a device has n inputs and m outputs and is depicted in Fig. 12.31 for the case that $n = 3$ and $m = 2$. In other words, Fig. 12.31 shows a $2^n \times m = 8 \times 2$ ROM. Note that at the inputs of the OR gates there are connections, referred to as **links,** that can be broken if desired. Whether or not a link is broken depends on what the Boolean functions F_1 and F_2 are to be. The decoder outputs are all of the minterms of the input variables, and any or all of these minterms can be applied to each OR gate. If a particular minterm is not to be applied to the input of an OR gate, then the corresponding link is broken. (It is assumed that an open terminal at

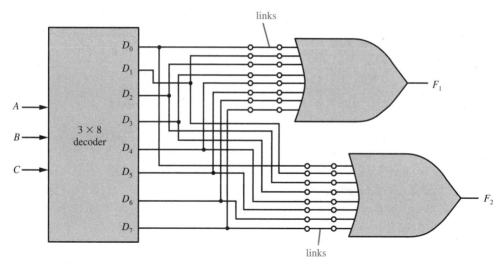

Fig. 12.31 An 8 × 2 ROM.

the input of an OR gate behaves as an input of 0.)

As an example, we have already seen that the output functions of a full-adder are characterized by the Boolean expressions

$$S_i = m_1 + m_2 + m_4 + m_7 \quad \text{and} \quad C_{i+1} = m_3 + m_5 + m_6 + m_7$$

Therefore, for the 8×2 ROM shown in Fig. 12.31 to realize a full-adder, if $F_1 = S_i$ and $F_2 = C_{i+1}$, then for the upper OR gate the links connected to the decoder outputs D_0, D_3, D_5, and D_6 are broken. In addition, for the lower OR gate, the links connected to D_0, D_1, D_2, and D_4 are broken.

Establishing a pattern of connections between the outputs of the decoder and the inputs of the OR gates by breaking links is referred to as **programming** the ROM.

Since complicated Boolean expressions can be realized with a programmed ROM, such functions can be implemented with a single IC. This eliminates the need for interconnecting a number of ICs.

A $2^n \times m$ ROM has n inputs and m outputs. Thus there are 2^n possible n-bit input combinations, and each one is called an **address.** Each m-bit output combination is called a **word** or **output word.** For every address, therefore, an output word corresponds, and such a word is "read" (appears at the output) when the corresponding address (input) is applied. Consequently, it is said that a ROM stores 2^n words— hence the use of the word "memory." Because there are 2^n m-bit words stored in the ROM, we say that it contains $2^n \times m$ bits. For instance, the ROM shown in Fig. 12.31 contains $2^3 \times 2 = 16$ bits, so it is referred to as a 16-bit ROM.

A ROM with 15 inputs and 8 outputs contains $2^{15} \times 8 = 32{,}768 \times 8 = 262{,}144$ bits and, therefore, is a 262,144-bit ROM. Clearly, describing a ROM in this manner is awkward, but by using the letter K to designate the number $2^{10} = 1024$, we can call such a ROM a 256K-bit ROM, or 256K ROM for short. Furthermore, since 1 byte = 8 bits, we may also refer to it as a 32K-byte ROM.

A ROM may be programmed by its manufacturer during its fabrication process— typically the last step. In this case, the user of the ROM is required to specify the characteristics of the particular device by giving the manufacturer the truth table for the ROM. Such a procedure is expensive, but if a large number of the same ROM are to be produced, then such a procedure is economical.

An alternative to a ROM that is programmed by its manufacturer is a **programmable ROM,** designated PROM. For a PROM, all the links are initially intact, and the device is programmed by applying current pulses (through output terminals) to break the appropriate links. If a ROM is programmed in the manufacturing process, its programming cannot be changed by its user. In addition, once a link of a PROM has been broken, it cannot be replaced. It is in this sense that the programming of a ROM or PROM is irreversible. There is, however, another type of ROM—an **erasable PROM,** designated EPROM. For one type of such a device, the programming can be erased by placing it under a certain kind of ultraviolet light for a specified period of time. For another type, erasure can be accomplished electrically; such a device may be referred to as an **electrically alterable ROM,** or EAROM for short.

Programmable Logic Arrays

Suppose that 64 pieces of information are encoded by using a code with 12-bit binary sequences.[2] Furthermore, suppose that these code words are to be converted into

---

[2]Even though there are $2^6 = 64$ distinct 6-bit binary sequences, appropriately using 12 bits for each code word gives the code error-detecting and error-correcting capabilities. This can be very desirable when transmitting information in the presence of noise.

6-bit binary sequences, one for each of the original pieces of information. If a ROM is used for this purpose, the device requires 12 inputs and 6 outputs, that is, a $2^{12} \times 6$ (24K-bit) ROM is needed. Even though such a device has $2^{12} = 4096$ addresses and can accommodate 4096 words, only 64 of these words are utilized. This is considered a wasteful use of a ROM.

Another type of LSI device that is similar to a ROM is the **programmable logic array** (PLA). A PLA does not contain a decoder that generates all possible minterms as does a ROM. Instead of a decoder, a PLA contains a group of AND gates, and the inputs of each AND gate are connected to every input variable and the complement of every variable through links. If the PLA has n inputs, then each AND gate has $2n$ inputs, and $2n$ links are associated with each AND gate. If the PLA has m outputs, then there are m OR gates. If there are p AND gates, then each OR gate has p inputs; the output of every AND gate is connected to each OR gate through links. The output of each OR gate is one output of the PLA. Such a device can be called an $n \times p \times m$ PLA.

An example of a $3 \times 4 \times 2$ PLA is shown in Fig. 12.32. To use a device such as this to realize Boolean functions F_1 and F_2, we must be able to express both F_1

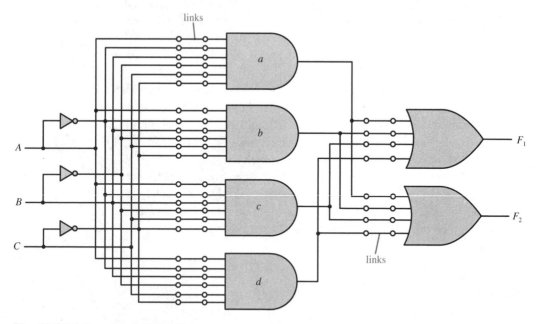

Fig. 12.32 A $3 \times 4 \times 2$ PLA.

and F_2 as a sum of at most four products of the input variables or their complements. These products will be the outputs of the four AND gates, and the outputs of the OR gates will be F_1 and F_2. Specifically, we can realize the Boolean functions given by Table 12.16 using this PLA. From the given truth table, we get the Karnaugh maps shown in Fig. 12.33. From these maps we have that

$$F_1 = \overline{A}\overline{B} + \overline{A}C + AB\overline{C} \qquad \text{and} \qquad F_2 = \overline{A}C + \overline{B}C + AB\overline{C}$$

Table 12.16 Truth Table for Example.

A	B	C	F_1	F_2
0	0	0	1	0
0	0	1	1	1
0	1	0	0	0
0	1	1	1	1
1	0	0	0	0
1	0	1	0	1
1	1	0	1	1
1	1	1	0	0

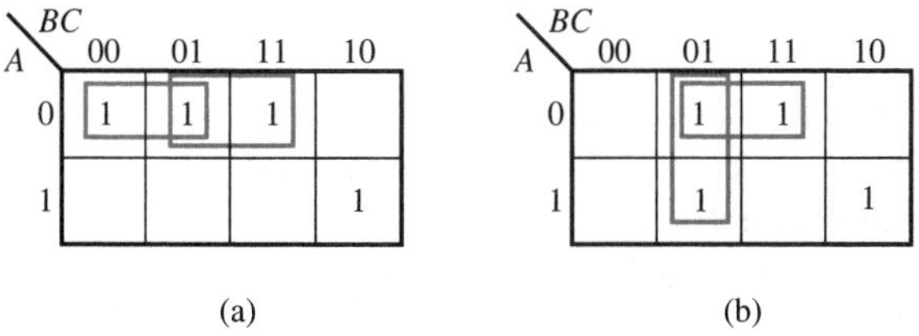

Fig. 12.33 Karnaugh maps for (a) F_1 and (b) F_2.

Thus we see that these two expressions contain four distinct products—$\overline{A}\overline{B}$, $\overline{A}C$, $\overline{B}C$, and $AB\overline{C}$. Therefore, to implement F_1 and F_2 with the given PLA, we need only break the appropriate links. In particular, for the AND gate labeled a, break the links associated with A, B, C, and $\overline{C}$, and the resulting output of AND gate a is $\overline{A}\overline{B}$. For AND gate b, break the links A, B, $\overline{B}$ and $\overline{C}$, and the resulting output is $\overline{A}C$. For AND gate c, break the links A, $\overline{A}$, B, and $\overline{C}$, and the resulting output is $\overline{B}C$. Finally, for AND gate d, break the links $\overline{A}$, $\overline{B}$, and C, and the resulting output is $AB\overline{C}$. (We have assumed here that an open terminal at the input of an AND gate behaves as an input of 1.) Next, for the upper OR gate, break the link connected to the output of AND gate c. Assuming that an open terminal at the input of an OR gate behaves as an input of 0, the output of the OR gate is $F_1 = \overline{A}\overline{B} + \overline{A}C + AB\overline{C}$. For the lower OR gate, break the link connected to the output of AND gate a. Then the output of this OR gate is $F_2 = \overline{A}C + \overline{B}C + AB\overline{C}$. Hence we have the desired implementations.

The above example of the use of a PLA resulted in a relatively small logic circuit. From a practical point of view, implementations of such Boolean functions are better accomplished using the previously studied approaches. The same thing can be said about the example dealing with the use of a ROM. These examples were presented for purposes of demonstration. ROMs and PLAs are useful for the design of large combinational logic circuits—typically the former for the case that most or all of the addresses (possible input combinations) are utilized, and the latter for the case that only a relatively few addresses are needed. As with a ROM, a PLA can be programmed by the manufacturer or the user. For the latter situation, the device is referred to as a **field-programmable logic array** (FPLA). In particular, a typical $n \times p \times m$ PLA, such as the TTL 82S100 IC, has $n = 16$, $p = 48$, and $m = 8$.

12.3 Sequential Logic

Up to this point, our concern has been focused on combinational logic. For this type of circuit, the output (or outputs) at a particular time is determined by a combination of the inputs applied at that time. A more general situation of a logic circuit is to have the output determined not only from a combination of the inputs applied at a particular time, but from the condition or "state" of the circuit at that time as well. We refer to this type of logic as **sequential logic,** and we may represent a general sequential logic circuit by the block diagram shown in Fig. 12.34. Here the information stored in the memory elements at a particular time defines the state of the sequential circuit at that time.

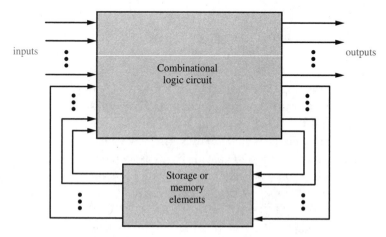

Fig. 12.34 General sequential-logic circuit.

SR Flip-Flops

Typically, sequential-circuit memory elements are **flip-flops.**[3] Such a device stores 1 bit of information, either a 0 or a 1, and has two outputs—one is equal to the bit stored, and the other is equal to its complement.

Consider the connection of the two NOR gates shown in Fig. 12.35. This logic circuit has two inputs labeled R and S, and two outputs labeled Q and $\overline{Q}$. The output Q indicates the state of the circuit—$Q = 1$ ($\overline{Q} = 0$) is called the **set state** and $Q = 0$ ($\overline{Q} = 1$) is called the **reset** or **clear state.** Such a device is known as a **set-reset** (SR) **flip-flop**[4] or an **SR latch.** Let us analyze this sequential circuit.

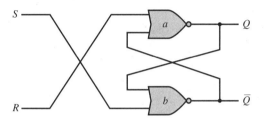

Fig. 12.35 An SR (set-reset) flip-flop.

Suppose at some time the flip-flop is in the reset state $Q = 0$ ($\overline{Q} = 1$). If $R = S = 0$, then the output of NOR gate a is $\overline{R + \overline{Q}} = \overline{0 + 1} = 0$ and the output of NOR gate b is $\overline{Q + S} = \overline{0 + 0} = 1$. Thus when the flip-flop is in the reset state $Q = 0$ and the input $R = S = 0$ is applied, the flip-flop remains in the reset state. Similarly, it can be shown that if the flip-flop is in the set state $Q = 1$ ($\overline{Q} = 0$), then when $R = S = 0$ is applied, the flip-flop remains in the set state.

Now suppose that the flip-flop is in the reset state $Q = 0$ ($\overline{Q} = 1$). If $R = 1$ and $S = 0$, then the output of NOR gate a is $\overline{R + \overline{Q}} = \overline{1 + 1} = 0$ and the output of NOR gate b is $\overline{Q + S} = \overline{0 + 0} = 1$. Thus the flip-flop remains in the reset state. However, if $R = 0$ and $S = 1$, then the output of NOR gate b becomes $\overline{Q + S} = \overline{0 + 1} = 0$—and 0 is the new value for $\overline{Q}$. This, in turn, causes the output of NOR gate a to be $\overline{R + \overline{Q}} = \overline{0 + 0} = 1$—and 1 is the new value for Q. Does this then affect the new output value of NOR gate b? Since $\overline{Q + S} = \overline{1 + 1} = 0$, it does not. Thus the flip-flop changes to the set state. In summary, if the flip-flop is in the reset state ($Q = 0$), then the inputs $R = 1$ and $S = 0$ cause the flip-flop to remain in the reset state ($Q = 0$), whereas the inputs $R = 0$ and $S = 1$ cause the flip-flop to change to the set state ($Q = 1$).

---

[3]A flip-flop is a **bistable multivibrator.**
[4]Also called an RS flip-flop.

Similarly, it can be shown that if the flip-flop is in the set state ($Q = 1$), then the inputs $R = 0$ and $S = 1$ cause the flip-flop to remain in the set state ($Q = 1$), whereas the inputs $R = 1$ and $S = 0$ cause the flip-flop to change to the reset state ($Q = 0$).

Finally, if $R = S = 1$, no matter what the state of the flip-flop, both NAND gate outputs are equal to 0. This would mean that both Q and $\overline{Q}$ would take on the value of 0. Such a condition, however, violates the definition of a flip-flop. Therefore, to use the logic circuit shown in Fig. 12.34 as an SR flip-flop, we will insist that the condition $R = S = 1$ not be allowed to occur.

The above discussion of the operation of the SR flip-flop shown in Fig. 12.35 is summarized by Table 12.17. In this table, Q indicates the state at some time t, and R and S designate the inputs at that time. The next state that results, denoted $Q(t + 1)$, may be the same or different from the present state Q. An input of $S = 1$ sets the next state to 1, and an input of $R = 1$ resets the next state to 0.

Table 12.17 Truth Table for SR Flip-Flop.

Inputs		Present State	Next State
S	R	Q	Q(t + 1)
0	0	0	0
0	0	1	1
0	1	0	0
0	1	1	0
1	0	0	1
1	0	1	1
1	1	NOT ALLOWED	

Typically, the inputs of a flip-flop are connected to the outputs of some logic circuit. It may be that for some of the time these outputs are not equal to their proper values as a result of propagation delays or spurious effects. Because of this, we can modify the SR flip-flop shown in Fig. 12.35 so that the device will ignore the inputs except over some specified time interval when it is known that they will take on their correct values. This can be done by adding two AND gates to the circuit as shown in Fig. 12.36. For this sequential circuit, however, in addition to the set (S) and reset (R) inputs, there is an input for a **clock pulse,** designated C, as shown in Fig. 12.37a. When the clock pulse is equal to 0 (i.e., $C = 0$), then the outputs of the AND gates are 0. Such values will not change the value of Q. But when $C = 1$, then the output of AND gate a is $CS = (1)S = S$ and the output of AND gate b is $CR = (1)R = R$. These values will affect the present state Q as described by Table 12.17. (It is assumed that the clock pulse is sufficiently long that the device

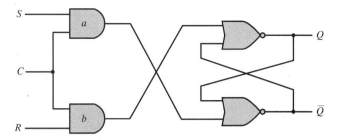

Fig. 12.36 Clocked (gated) SR flip-flop.

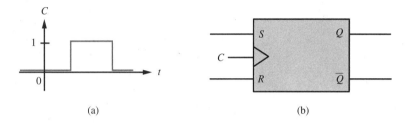

(a) (b)

Fig. 12.37 (*a*) Single clock pulse, and (*b*) circuit symbol for a clocked SR flip-flop.

will be allowed to reach its appropriate next state.) The sequential circuit shown in Fig. 12.36 is known as a **clocked** (or **gated**) **SR flip-flop,** and its block-diagram circuit symbol is shown in Fig. 12.37*b*. A circuit such as this, in which operations are performed in conjunction with repetitive or periodic clock pulses, is called a **synchronous sequential circuit.** Typically, these clock pulses are produced by a timing device known as a **master-clock generator.**

When power is first supplied to a flip-flop, its state is arbitrary. By slightly modifying the SR flip-flop circuit given in Fig. 12.36, the flip-flop can be given the capability of being directly **cleared** (making $Q = 0$) or of being directly **preset** (making $Q = 1$) bfore a clock pulse is applied. This is accomplished by the logic circuit shown in Fig. 12.38. This circuit is the same as the one in Fig. 12.36 with the exception that there is an additional input to each NOR gate. For the upper NOR gate, this input is labeled *CLR* (for clear), and for the lower NOR gate it is labeled *PR* (for preset). When 0s are applied to both of these inputs, the circuit behaves identically to the SR flip-flop given in Fig. 12.36. In the absence of a clock pulse ($C = 0$), the middle inputs of both NOR gates are 0. Making $CLR = 1$ then results in $Q = 0$ and $\overline{Q} = 1$, thereby clearing the flip-flop. On the other hand, making $PR = 1$ causes $\overline{Q} = 0$ and $Q = 1$, thereby presetting the flip-flop. Because these inputs are applied without the need for a clock pulse, they are referred to as **asynchronous** inputs. The block-diagram circuit symbol for a clocked SR flip-flop having direct clear and preset inputs is shown in Fig. 12.39.

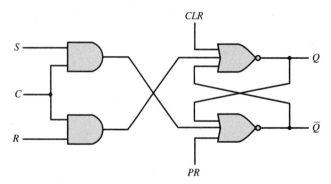

Fig. 12.38 Clocked SR flip-flop with clear and preset capability.

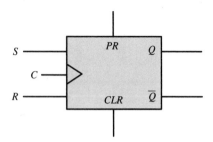

Fig. 12.39 An SR flip-flop with clear and preset inputs.

D Flip-Flops and JK Flip-Flops

Consider the synchronous sequential circuit shown in Fig. 12.40. For this circuit, the inputs to the clocked SR flip-flop are $S = D$ and $R = \overline{D}$. Note that because of this fact, we never can have the undesirable condition for an SR flip-flop that $R = S = 1$. We see from Table 12.17 for the case that the present state is $Q = 0$, when $D = S = 0$ ($R = \overline{D} = 1$) then the next state is $Q(t + 1) = 0$. For the case

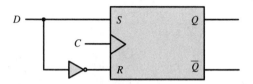

Fig. 12.40 A D flip-flop.

that the present state is $Q = 1$, when $D = S = 0$ ($R = \overline{D} = 1$), then the state is reset to $Q(t + 1) = 0$. Similarly, whether the present state is $Q = 0$ or $Q = 1$, when $D = S = 1$ ($R = \overline{D} = 0$), then the state is set to $Q(t + 1) = 1$. These results are summarized by Table 12.18. Therefore, for the circuit given in Fig. 12.40, an input of 0 results in a next state of 0, and an input of 1 results in a next state of 1. For these reasons, the device shown in Fig. 12.40 is referred to as a (clocked) **delay** or **data** (D) **flip-flop,** and its block diagram circuit symbol is as depicted by Fig. 12.41.

Table 12.18 Truth Table for D Flip-Flop.

Input	Present State	Next State
D	Q	$Q(t + 1)$
0	0	0
0	1	0
1	0	1
1	1	1

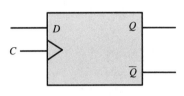

Fig. 12.41 Circuit symbol of a D flip-flop.

Let us now combine two AND gates with a clocked SR flip-flop as shown in Fig. 12.42. We designate the inputs of the resulting sequential circuit by J and K as illustrated. For this circuit, suppose that $J = K = 0$ and $Q = 0$. Then $S = J\overline{Q} = 0(1) = 0$, $R = KQ = 0(0) = 0$, and the next state is $Q(t + 1) = 0$. When $J = K = 0$ and $Q = 1$, then $S = J\overline{Q} = 0$, $R = KQ = 0$, and the next state is $Q(t + 1) = 1$.

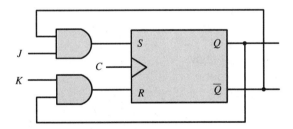

Fig. 12.42 A JK flip-flop.

Next suppose that $J = 0$ and $K = 1$. When $Q = 0$, then $S = J\overline{Q} = 0(1) = 0$, $R = KQ = 1(0) = 0$, and $Q(t + 1) = 0$. When $J = 0$, $K = 1$, and $Q = 1$, then $S = J\overline{Q} = 0(0) = 0$, $R = KQ = 1(1) = 1$, and the state is reset to $Q(t + 1) = 0$. In a similar manner, it can be seen that when $J = 1$, $K = 0$, and $Q = 0$ or $Q = 1$, then $Q(t + 1) = 1$.

The characterization thus far for the sequential circuit shown in Fig. 12.42 is given by the first six rows of Table 12.19. Since these rows are identical to the first six rows of Table 12.17, the circuit shown in Fig. 12.42 behaves as an SR flip-flop except for the case that both inputs are equal to 1. For an SR flip-flop, this condition

Table 12.19 Truth Table for JK Flip-Flop.

Inputs		Present State	Next State
J	K	Q	$Q(t+1)$
0	0	0	0
0	0	1	1
0	1	0	0
0	1	1	0
1	0	0	1
1	0	1	1
1	1	0	1
1	1	1	0

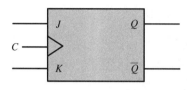

Fig. 12.43 Circuit symbol of a JK flip-flop.

must be avoided—however, it is allowed for the circuit given in Fig. 12.42. Specifically, suppose that $J = K = 1$. If $Q = 0$, then $S = J\overline{Q} = 1(1) = 1$, $R = KQ = 1(0) = 0$, and the state is set to $Q(t+1) = 1$. Furthermore, if $J = K = Q = 1$, then $S = J\overline{Q} = 1(0) = 0$, $R = KQ = 1(1) = 1$, and the state is reset to $Q(t+1) = 0$. These facts yield the last two rows in Table 12.19, and this is the truth table for the circuit shown in Fig. 12.42. Such a device is known as a **JK flip-flop,** and its block-diagram circuit symbol is given by Fig. 12.43. In summary, a JK flip-flop behaves as an SR flip-flop except that a JK flip-flop can have both inputs equal to 1. Connecting the inputs of a JK flip-flop results in what is known as a **toggle** (T) **flip-flop** (see Problem 12.33 at the end of this chapter).

There is a practical problem associated with the JK flip-flop just discussed. As long as the input to the flip-flop is $J = K = 1$, the state **toggles,** i.e., it changes either from 0 to 1, or from 1 to 0. Therefore, if the clock pulse is too long, there will be more than one state transition and the final state of the flip-flop will be indeterminate. To get around this problem, instead of constructing a JK flip-flop from a single SR flip-flop as indicated in Fig. 12.42, let us consider the "master-slave" connection of two SR flip-flops shown in Fig. 12.44. Note that for this sequential circuit, when its clock input is 1 (i.e., $C_2 = 1$), the slave flip-flop simply transfers its input values to its output. This is so because if $S_2 = 0$ and $R_2 = 1$, then when $C_2 = 1$ the flip-flop is reset to $Q_2 = 0$ and $\overline{Q}_2 = 1$. Similarly, if $S_2 = 1$ and $R_2 = 0$, then when $C_2 = 1$ the flip-flop is set to $Q_2 = 1$ and $\overline{Q}_2 = 0$.

Suppose that the circuit shown in Fig. 12.44, called a **JK master-slave flip-flop,** has inputs $J = K = 0$. Then the inputs to the master flip-flop are $S_1 = R_1 = 0$. In the absence of a clock pulse ($C = 0$), the clock input to the slave is $C_2 = \overline{C}_1 = \overline{C} = 1$. Thus the slave state is the same as the master state. When a clock pulse occurs ($C = C_1 = 1$), the master flip-flop does not change state (see Table 12.17). When the clock pulse ends ($C = 0$), then $C_2 = 1$ and the input of the slave flip-

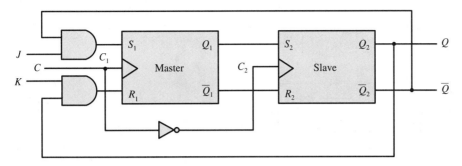

Fig. 12.44 A JK master-slave flip-flop.

flop (output of the master flip-flop) gets transferred to the output of the slave flip-flop. Thus the state of the slave flip-flop remains unchanged.

Now suppose that $J = 0$ and $K = 1$. When $Q = 0$, then $S_1 = J\overline{Q} = 0(1) = 0$ and $R_1 = KQ = 1(0) = 0$. This means that the master state remains unchanged, as does the slave state. However, when $Q = 1$, then $S_1 = J\overline{Q} = 0(0) = 0$ and $R_1 = KQ = 1(1) = 1$. This, in turn, means that the master flip-flop is reset to 0, and the slave flip-flop follows. Proceeding in a similar manner, when $J = 1$ and $K = 0$, the outputs that result are $Q = 1$ and $\overline{Q} = 0$.

Finally, consider the case that $J = K = 1$. When $Q = 1$, then $R_1 = KQ = 1(1) = 1$—which means that the master and slave flip-flops are reset to 0. Conversely, when $Q = 0$, then $S_1 = J\overline{Q} = 1(1) = 1$ and both flip-flops are set to 1. This gives us the toggle operation of the JK flip-flop. Note, however, that the master flip-flop operates when the clock pulse is $C = 1$, whereas the operation of the slave flip-flop follows when $C = 0$. Because the next-state values of the flip-flop occur after the clock pulse goes to 0—and the master flip-flop is inoperative—these values cannot cause the unwanted toggling that can occur for the JK flip-flop given in Fig. 12.42. The fact that the slave flip-flop does not change state until after the clock pulse returns to 0 is why this JK master-slave flip-flop is referred to as a **trailing-edge-triggered flip-flop.**

Sequential-Circuit Analysis

As mentioned previously, the general form of a sequential circuit is as shown in Fig. 12.34, where the storage elements typically are flip-flops. Let us consider a specific example and see what is involved in analyzing a sequential circuit.

For the sake of simplicity, the clock inputs for the two JK flip-flops in the circuit given in Fig. 12.45 are not shown. The contents (i.e., the bits stored) of the flip-flops at a particular time are said to be the **state** of the sequential circuit at that time. Therefore, we can label the state of the circuit with the binary pair Q_1Q_2. For ex-

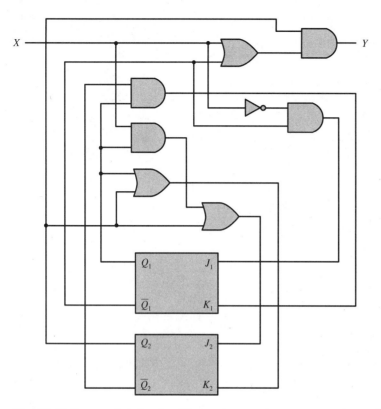

Fig. 12.45 Sequential circuit with one input, one output, and two JK flip-flops.

ample, if at some time $Q_1 = 0$ and $Q_2 = 1$, then we can designate the state as 01. There is one input X to the given circuit and one output Y. Clearly, the expression relating the output Y to the input X and the flip-flop outputs Q_1 and Q_2 is

$$Y = (X + \overline{Q}_1)Q_2$$

Furthermore, we have that

$$J_1 = \overline{X}\overline{Q}_1, \qquad K_1 = Q_1\overline{Q}_2, \qquad J_2 = XQ_1 + Q_2, \qquad K_2 = Q_1 + Q_2$$

Let us now construct a table that includes all possible combinations of the input and the present state. The first three columnns of Table 12.20 enumerate these combinations. Using the expressions for J_1, K_1, J_2, K_2, and Y, we can then fill in the corresponding columns in Table 12.20. Furthermore, knowing the characteristics of a JK flip-flop (Table 12.19), the columns in the table designating the next state can be determined. For example, suppose that the present state of the circuit is $Q_1Q_2 = 01$ and an input of $X = 0$ is applied. This results in $J_1 = \overline{X}\overline{Q}_1 = 1(1) = 1$, $K_1 = Q_1\overline{Q}_2 = 0(0) = 0$, $J_2 = XQ_1 + Q_2 = 0(0) + 1 = 1$, $K_2 = Q_1 + Q_2 = 0 + 1 = 1$, and $Y = (X + \overline{Q}_1)Q_2 = (0 + 1)1 = 1$. The present state of the first flip-flop is

Table 12.20 Transition Table for Given Sequential Circuit.

Input	Present State		Flip-Flop Inputs				Next State		Output
X	Q_1	Q_2	J_1	K_1	J_2	K_2	$Q_1(t+1)$	$Q_2(t+1)$	Y
0	0	0	1	0	0	0	1	0	0
0	0	1	1	0	1	1	1	0	1
0	1	0	0	1	0	1	0	0	0
0	1	1	0	0	1	1	1	0	0
1	0	0	0	0	0	0	0	0	0
1	0	1	0	0	1	1	0	0	1
1	1	0	0	1	1	1	0	1	0
1	1	1	0	0	1	1	1	0	1

$Q_1 = 0$, and an input of $J_1 = 1$, $K_1 = 0$ sets the state to $Q_1(t+1) = 1$. The present state of the second flip-flop is $Q_2 = 1$, and an input of $J_2 = 1$, $K_2 = 1$ toggles the state to $Q_2(t+1) = 0$. Thus we have obtained the information for the second row of the table. Similarly, the remaining rows of the table can be determined. This table is called the **transition table** or the **state table** for the sequential circuit.

The information contained in the transition table (excluding the flip-flop inputs) can be displayed graphically by a **state diagram.** For such a diagram, each state is represented by a circle, called a **node,** that is labeled with its associated state. Thus for the example above, there are four nodes, and they are labeled 00, 01, 10, and 11. If an input of X causes a state to change from Q_1Q_2 to $Q_1'Q_2'$, then there is a line, called an **edge,** directed from Q_1Q_2 to $Q_1'Q_2'$. Furthermore, if the resulting output is Y, then the edge is labeled X/Y. For instance, since an input of 0 changes state 01 to 10, then there is an edge directed from node 01 to node 10, and since the resulting output is 1, then this edge is labeled 0/1. Figure 12.46 shows the state diagram for

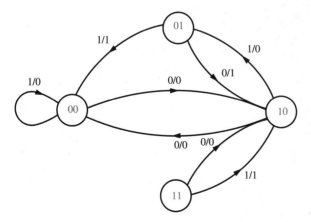

Fig. 12.46 State diagram for given sequential circuit.

the sequential circuit given in Fig. 12.45. From this diagram, we see that once the circuit leaves the state 11, due either to an input of 0 or of 1, there is no possible sequence of inputs that will return the circuit to that state.

A sequential circuit is completely characterized by either its transition table or its state diagram. Knowing either, the sequences of states and outputs of a sequential circuit that result from a sequence of inputs are easily obtained. Therefore, the analysis of a sequential circuit essentially consists of determining either its transition table or its state diagram.

Sequential-Circuit Design

The converse process of analysis is synthesis or design. The design of a sequential circuit consists of starting with a description of the circuit and ending with a logic-circuit diagram of the sequential circuit. Although a sequential circuit can be specified by a word description from which a state diagram may be obtained, the state diagram of a sequential circuit is often given as the initial circuit specification.

Since a sequential circuit employs flip-flops (the type of flip-flop is the choice of the designer), the behavior of different types of flip-flops is a design factor. Instead of directly using the truth tables of the flip-flops studied earlier, it is more convenient to put flip-flop descriptions in alternative forms called **excitation tables.** For example, Table 12.21a repeats the truth table for the SR flip-flop. The information contained in this table is used to construct the excitation table for the SR flip-flop given by Table 12.21b. For such a table, the first column designates the present state Q and the second column designates the next state $Q(t + 1)$. The third and fourth columns, labeled S and R, indicate the inputs required to change the flip-flop's present

Table 12.21 (*a*) Truth Table for the SR Flip-Flop.

Inputs		Present State	Next State
S	R	Q	$Q(t + 1)$
0	0	0	0
0	0	1	1
0	1	0	0
0	1	1	0
1	0	0	1
1	0	1	1
1	1	NOT ALLOWED	

Table 12.21 (*b*) Excitation Table for the SR Flip-Flop.

Present State	Next State	Inputs	
Q	$Q(t + 1)$	S	R
0	0	0	×
0	1	1	0
1	0	0	1
1	1	×	0

state Q to the next state $Q(t + 1)$. Specifically, if the present state of the flip-flop is 0, to get the next state to be 0, we see from Table 12.21a that the inputs should be either $S = 0$, $R = 0$, or $S = 0$, $R = 1$. Therefore, the value of R is irrelevant—as long as $S = 0$ the flip-flop will go from $Q = 0$ to $Q(t + 1) = 0$. This characterization is given by the first row of Table 12.21b. Here R is marked with the don't-care condition $\times$.

If an SR flip-flop is to change from the present state $Q = 0$ to the next state $Q(t + 1) = 1$, we see from Table 12.21a that the inputs should be $S = 1$ and $R = 0$. This condition is given by the second row of Table 12.21b. Proceeding in this manner, the remainder of the excitation table for the SR flip-flop can be obtained. Table 12.21b is this excitation table.

The excitation tables for the D flip-flop and the JK flip-flop are given by Tables 12.22b and 12.23b, respectively. The corresponding truth tables are reviewed by Tables 12.22a and 12.23a, respectively. The confirmation of these results is left as an exercise for the reader.

Table 12.22 (a) Truth Table for the D Flip-Flop.

Input	Present State	Next State
D	Q	$Q(t + 1)$
0	0	0
0	1	0
1	0	1
1	1	1

Table 12.22 (b) Excitation Table for the D Flip-Flop.

Present State	Next State	Input
Q	$Q(t + 1)$	D
0	0	0
0	1	1
1	0	0
1	1	1

Table 12.23 (a) Truth Table for the JK Flip-Flop.

Inputs		Present State	Next State
J	K	Q	$Q(t + 1)$
0	0	0	0
0	0	1	1
0	1	0	0
0	1	1	0
1	0	0	1
1	0	1	1
1	1	0	1
1	1	1	0

Table 12.23 (b) Excitation Table for the JK Flip-Flop.

Present State	Next State	Inputs	
Q	$Q(t + 1)$	J	K
0	0	0	$\times$
0	1	1	$\times$
1	0	$\times$	1
1	1	$\times$	0

We will now see how an excitation table is utilized by going through an example of the design of a sequential circuit. Specifically, let us design a modulo-4 binary up/down counter. Such a device has four states representing the binary numbers 0, 1, 2, and 3. An input of 1 counts up, that is, changes state from 0 to 1, or 1 to 2, or 2 to 3, or 3 to 0. An input of 0 counts down, that is, changes state from 3 to 2, or 2 to 1, or 1 to 0, or 0 to 3. The output of the counter is 0 if the next state is even (0 or 2), and it is 1 if the next state is odd (1 or 3). This description of the sequential circuit is displayed graphically by the state diagram shown in Fig. 12.47.

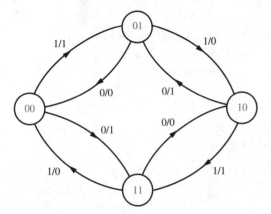

Fig. 12.47 State diagram for given up/down counter.

With the information given, we may now begin to construct the transition table for the sequential circuit. We start with columns for the input X and the present state Q_1Q_2. In these three columns, we put all possible binary triples. This is done in the first three columns of Table 12.24. The next two columns designate the next state $Q_1(t + 1)Q_2(t + 1)$. These columns, as well as the one corresponding to the output Y, can be filled in by referring to the state diagram. For example, the third row of the table represents the condition that an input of 0 is applied when the present state is 10. From the state diagram, we see that the next state is 01 and the output is 1. These values are placed in the appropriate columns in the table.

After the next-state and output columns of the transition table have been completed, the columns for the flip-flop inputs must be filled in. The values that are entered depend on which type of flip-flop is selected. Different flip-flop types will result, in general, in different combinational logic circuits that control the flip-flops. There is no way of knowing ahead of time which type of flip-flop will produce the simplest circuits. For the given problem, let us arbitrarily select SR flip-flops. Doing so, we now employ the excitation table for the SR flip-flop (Table 12.21b). For example, for the third row of the transition table, the state is to change from $Q_1Q_2 = 10$ to $Q_1(t + 1)Q_2(t + 1) = 01$. This means that the first flip-flop is to

Table 12.24 Transition Table for the Up/Down Counter.

Input	Present State		Next State		Flip-Flop Inputs				Output
X	Q_1	Q_2	$Q_1(t+1)$	$Q_2(t+1)$	S_1	R_1	S_2	R_2	Y
0	0	0	1	1	1	0	1	0	1
0	0	1	0	0	0	×	0	1	0
0	1	0	0	1	0	1	1	0	1
0	1	1	1	0	×	0	0	1	0
1	0	0	0	1	0	×	1	0	1
1	0	1	1	0	1	0	0	1	0
1	1	0	1	1	×	0	1	0	1
1	1	1	0	0	0	1	0	1	0

change its state from 1 to 0. From Table 12.21b, the corresponding inputs are $S_1 = 0$ and $R_1 = 1$. The second flip-flop, on the other hand, is to change its state from 0 to 1. Again from Table 12.21b, the corresponding inputs are $S_2 = 1$ and $R_2 = 0$. Hence we have the entries in the third row in the columns for the flip-flop inputs. Proceeding in this manner, the remaining entries can be determined. These are shown in Table 12.24.

Once the columns corresponding to the flip-flop inputs have been obtained, Boolean expressions for the inputs can be found, and then the corresponding logic-circuit implementations may be realized. For the flip-flop input columns and the output column given in Table 12.24, we have the Karnaugh maps shown in Fig. 12.48. From these maps we find that

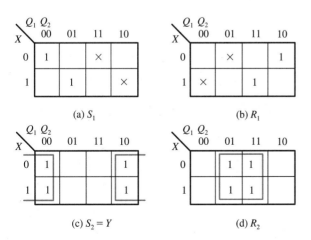

Fig. 12.48 Karnaugh maps for flip-flop inputs.

$$S_1 = \overline{X}\,\overline{Q}_1 Q_2 + X\overline{Q}_1 Q_2 = (\overline{X}\,\overline{Q}_2 + XQ_2)\overline{Q}_1$$

$$R_1 = \overline{X} Q_1 \overline{Q}_2 + X Q_1 Q_2 = (\overline{X}\,\overline{Q}_2 + XQ_2)Q_1$$

$$S_2 = Y = \overline{Q}_2$$

$$R_2 = Q_2$$

Using these expressions, we obtain the sequential circuit shown in Fig. 12.49.

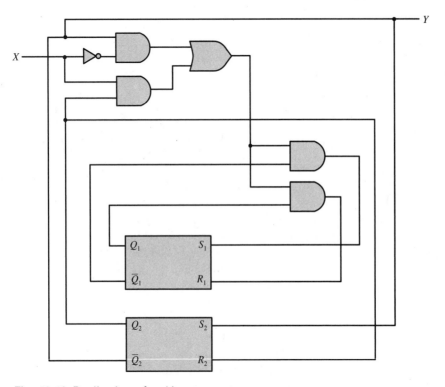

Fig. 12.49 Realization of up/down counter.

S U M M A R Y

1. The output of a combinational logic circuit is determined by the inputs that are applied to the circuit. The output of a sequential logic circuit depends on the state of the circuit as well as the applied inputs.

2. An n-bit binary adder can be constructed with $n - 1$ full-adders and one half-adder or with n full-adders.

3. A truth-table entry that can be either a 0 or a 1 is designated by the don't care condition $\times$.

4. The most popular integrated-circuit (IC) package is the dual-in-line package (DIP).

5. Digital ICs are available as small-scale-integration (SSI), medium-scale-integration (MSI), large-scale-integration (LSI), and very-large-scale-integration (VLSI) devices.

6. Devices such as adders, magnitude comparators, encoders, decoders, multiplexers, demultiplexers, read-only memories (ROMs), and programmable logic arrays (PLAs) are available as MSI and LSI packages.

7. Binary adders can be combined to form larger binary adders. They also can be used to implement BCD adders.

8. A magnitude comparator determines whether one binary number is greater than, less than, or equal to another binary number.

9. An m-to-n-line ($m \times n$) encoder converts m pieces of information to binary n-tuples (sequences of n bits), where $2^n \geq m$.

10. An n-to-m-line ($n \times m$) decoder converts binary n-tuples to m pieces of information, where $2^n \geq m$.

11. A digital multiplexer (MUX) with 2^n inputs and one output employs n selection lines to make the output equal to one of the inputs.

12. A demultiplexer with one input and 2^n outputs uses n selection lines to make one of the outputs equal to the input. A demultiplexer can be used as a decoder.

13. A read-only memory (ROM) consists of a decoder and OR gates. Such devices can be programmed by the manufacturer or the user by breaking appropriate links that connect the decoder outputs to the inputs of the OR gates.

14. A PLA contains a group of AND gates that replaces the decoder in a ROM. PLAs can also be programmed by the manufacturer or the user. PLAs are preferred over ROMs when relatively few input combinations are to be utilized.

15. A flip-flop is a 1-bit memory. Common flip-flops are designated SR (set-reset), JK, D (data or delay), and T (toggle).

16. Flip-flops typically are clocked (synchronous) sequential circuits and often have inputs for direct clear and preset.

17. Flip-flop operation can be improved with a master-slave connection.

18. Flip-flops are characterized by their excitation tables as well as their truth tables.

19. The analysis of a sequential circuit amounts to finding the transition table or the state diagram of the circuit.

20. The design of a sequential circuit typically begins with a word description or a state diagram and culminates with a logic-circuit diagram.

PROBLEMS

12.1 Analyze the NOR-gate logic circuit shown in Fig. P12.1 by (a) determining the Boolean function F, (b) constructing the truth table for F, (c) expressing F as a simplified sum of products, and (d) expressing F as a simplified product of sums.

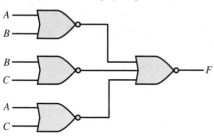

Fig. P12.1

12.2 For the NAND-gate logic circuits shown in Fig. P12.2, express F as a sum of products using only algebraic manipulations.

12.3 Table P12.3 is the truth table for a **half-sub-tractor.** Such a device subtracts 1 bit (designated B)

Table P12.3

A	B	D	R
0	0	0	0
0	1	1	1
1	0	1	0
1	1	0	0

from another (A). The difference is $D = A - B$, where R indicates whether a borrow is required. Write expressions for D and R, and design logic circuits that realize these expressions.

12.4 Table P12.4 is the truth table for a **full-subtractor.** Such a device performs subtraction and takes into consideration that there may have been a previous borrow. The difference between A_i and B_i taking a previous borrow R_i into account is designated D_i, and R_{i+1} indicates whether a borrow is required. Write expressions for D_i and R_{i+1}, and design logic circuits that realize these expressions.

Table P12.4

A_i	B_i	R_i	D_i	R_{i+1}
0	0	0	0	0
0	0	1	1	1
0	1	0	1	1
0	1	1	0	1
1	0	0	1	0
1	0	1	0	0
1	1	0	0	0
1	1	1	1	1

12.5 Show how the full-subtractor described in Problem 12.4 can be realized by using an OR gate

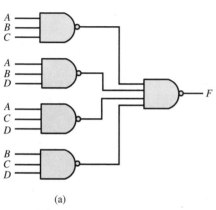

(a)

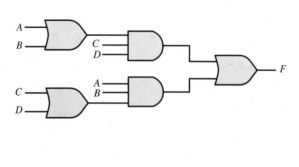

(b)

Fig. P12.2

and two of the half-subtractors (with possible slight modifications) described in Problem 12.3.

12.6 Write a simplified sum-of-products expression for the Boolean function f described in Table 12.4 on p. 816, and implement this function with a two-level AND-OR logic circuit. Repeat for the case that all the don't care conditions are changed to 0s.

12.7 Write a simplified sum-of-products expression for the Boolean function g described in Table 12.4 on p. 816, and implement this function with a two-level AND-OR logic circuit. Repeat for the case that all the don't-care conditions are changed to 0s.

12.8 Table P12.8 shows the truth table that describes the generation of a parity bit that, when attached to its corresponding BCD quadruple, produces

Table 12.8

A	B	C	D	P
0	0	0	0	0
0	0	0	1	1
0	0	1	0	1
0	0	1	1	0
0	1	0	0	1
0	1	0	1	0
0	1	1	0	0
0	1	1	1	1
1	0	0	0	1
1	0	0	1	0
1	0	1	0	×
1	0	1	1	×
1	1	0	0	×
1	1	0	1	×
1	1	1	0	×
1	1	1	1	×

a binary quintuple with an even number of 1s. Write a simplified product-of-sums expression for the Boolean function P, and implement this function with a two-level AND-OR logic circuit.

12.9 Table P12.9 shows the truth table that describes the conversion from BCD to the excess-3 code indicated. (Note that the excess-3 code is obtained by adding the binary number 3 to each BCD quadruple.) Design logic circuits that implement the Boolean functions E, F, G, and H.

Table P12.9

BCD Input				Excess-3 Code Output			
A	B	C	D	E	F	G	H
0	0	0	0	0	0	1	1
0	0	0	1	0	1	0	0
0	0	1	0	0	1	0	1
0	0	1	1	0	1	1	0
0	1	0	0	0	1	1	1
0	1	0	1	1	0	0	0
0	1	1	0	1	0	0	1
0	1	1	1	1	0	1	0
1	0	0	0	1	0	1	1
1	0	0	1	1	1	0	0
1	0	1	0	×	×	×	×
1	0	1	1	×	×	×	×
1	1	0	0	×	×	×	×
1	1	0	1	×	×	×	×
1	1	1	0	×	×	×	×
1	1	1	1	×	×	×	×

12.10 Design a logic circuit whose input is any possible binary quadruple and whose output is the 2's complement of the input.

12.11 Use a 4-bit binary adder to construct an excess-3-to-BCD code converter.

12.12 Design a logic circuit whose output is 1 when two 4-bit binary numbers are equal and 0 when they are not equal.

12.13 Given two 4-bit binary numbers $A = A_3A_2A_1A_0$ and $B = B_3B_2B_1B_0$. A 4-bit magnitude comparator whose inputs are A and B is to have three outputs C, D, and E such that (a) $C = 1$, $D = E = 0$ when $A < B$; (b) $C = 0$, $D = 1$, $E = 0$ when $A > B$; and (c) $C = D = 0$, $E = 1$ when $A = B$. Write Boolean expressions for C, D, and E.

12.14 Design an 8×3 (nonpriority) encoder.

12.15 Write the Boolean expressions for the outputs B_1, B_2, and B_3 of an 8-to-3-line priority encoder having inputs A_1, A_2, A_3, A_4, A_5, A_6, A_7, and A_8, where A_i has priority over A_j if $i < j$.

12.16 (a) The truth table for a half-subtractor is given by Table P12.4. Use a 3-to-8 line decoder to realize a half-subtractor. (b) A majority gate is a logic circuit whose output is 1 when a majority (more than half) of its inputs are 1. Use a 3-to-8-line decoder to implement a 3-input majority gate.

12.17 Table P12.17 shows the truth table that describes the conversion from binary to the **reflected** (or **Gray**) **code** indicated. Note that adjacent words in the reflected code differ by only 1 bit. Design a binary-to-Gray code converter by using a 3-to-8-line decoder.

Table P12.17

Binary			Reflected Code		
A	B	C	D	E	F
0	0	0	0	0	0
0	0	1	0	0	1
0	1	0	0	1	1
0	1	1	0	1	0
1	0	0	1	1	0
1	0	1	1	1	1
1	1	0	1	0	1
1	1	1	1	0	0

12.18 Design a BCD-to-excess-3-code converter by employing a 4-to-16-line decoder.

12.19 Realize a 3-input majority gate (see Problem 12.16*b* using a 4-to-1-line multiplexer.

12.20 Implement the "even-parity-bit" or "three-way-light-switch" truth table given by Table P12.20 with a 4-to-1-line multiplexer.

Table P12.20

A	B	C	F
0	0	0	0
0	0	1	1
0	1	0	1
0	1	1	0
1	0	0	1
1	0	1	0
1	1	0	0
1	1	1	1

12.21 Implement a full-adder with two 4-to-1-line multiplexers.

12.22 Find the Boolean function F that corresponds to the 4-to-1-line multiplexer shown in Fig. P12.22.

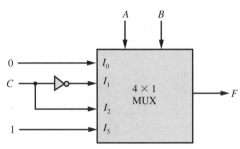

Fig. P12.22

12.23 A Boolean function F of the four variables A, B, C, and D is given by the sum of minterms

$$F = m_1 + m_2 + m_4 + m_7 + m_8$$

(See Table P12.8.) Implement this function by using an 8-to-1-line multiplexer.

12.24 Specify which links must be broken for the ROM given in Fig. 12.31 on p. 834 to implement the full-subtractor characterized by Table P12.24.

12.25 For the PLA shown in Fig. 12.32 on p. 836, add a fifth AND gate (and label it e) so as to form a $3 \times 5 \times 2$ PLA. Specify which links should be broken so that this new PLA will implement the full-subtractor characterized by Table P12.24.

Table P12.24

A	B	C	F_1	F_2
0	0	0	0	0
0	0	1	1	1
0	1	0	1	1
0	1	1	0	1
1	0	0	1	0
1	0	1	0	0
1	1	0	0	0
1	1	1	1	1

12.26 For the logic circuit shown in Fig. P12.26, the inputs are S and R, whereas the outputs are Q and $\overline{Q}$. Construct a truth table including the present state Q and the next state $Q(t + 1)$ for this circuit. Assuming that the input condition $R = S = 1$ does not occur, what type of sequential circuit is this?

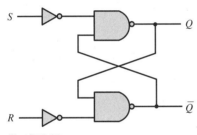

Fig. P12.26

12.27 For the logic circuit shown in Fig. P12.27, the inputs are S and R, whereas the outputs are Q and $\overline{Q}$. Construct a truth table including the present state Q and the next state $Q(t + 1)$ for this circuit.

What happens if the input condition $S = R = 1$ is allowed?

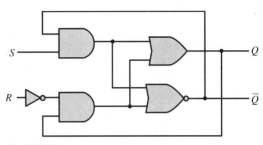

Fig. P12.27

12.28 The logic circuit shown in Fig. P12.28 is an example of a **set-dominate** SR flip-flop. Construct the truth table for this device.

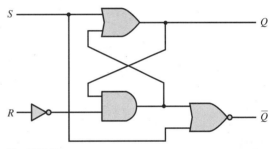

Fig. P12.28

12.29 Use the results of Problem 12.28 to design a set-dominate SR flip-flop using only (a) NAND gates and AND gates, and (b) NOR gates and OR gates.

12.30 The logic circuit shown in Fig. P12.30 is an example of a **reset-dominate** SR flip-flop. Construct the truth table for this device.

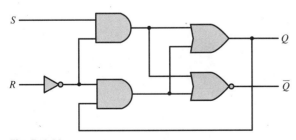

Fig. P12.30

12.31 Use the results of Problem 12.30 to design a reset-dominate SR flip-flop using only (a) NAND gates and AND gates, and (b) NOR gates and OR gates.

12.32 For the clocked sequential NAND-gate circuit shown in Fig. P12.32, the inputs are S and R, whereas the outputs are Q and $\overline{Q}$. Construct a truth table including the present state Q and the next state $Q(t + 1)$ for this circuit. Assuming that the input condition $R = S = 1$ does not occur, what type of sequential circuit is this?

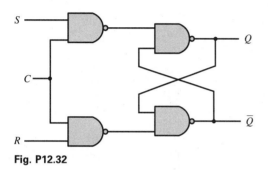

Fig. P12.32

12.33 Connecting the inputs of a JK flip-flop together as shown in Fig. P12.33a results in a toggle or T flip-flop, the circuit symbol of which is shown in Fig. P12.33b. Construct the truth table for a T flip-flop.

12.34 The transition table (excluding the columns corresponding to the flip-flop inputs) for a sequential circuit is given by Table P12.34. Determine the state diagram for this sequential circuit.

Table P12.34

X	Q_1	Q_2	$Q_1(t + 1)$	$Q_2(t + 1)$	Y
0	0	0	0	1	1
0	0	1	1	1	0
0	1	0	1	1	0
0	1	1	1	0	1
1	0	0	0	0	0
1	0	1	0	1	1
1	1	0	0	1	0
1	1	1	0	0	0

12.35 A sequential circuit has one input X, one output Y, and two flip-flops whose states are Q_1 and Q_2. Given that the output and next-state functions are given by

$$Y = XQ_2 + \overline{Q}_1Q_2 \qquad Q_1(t + 1) = \overline{X}\overline{Q}_1 + Q_1Q_2$$
$$Q_2(t + 1) = XQ_1\overline{Q}_2$$

determine the state diagram for the sequential circuit.

12.36 A sequential circuit has one input X, one output Y, and two flip-flops whose states are Q_1 and Q_2. Given that the output and next-state functions are given by

$$Y = \overline{X}\overline{Q}_2 + \overline{Q}_1 \qquad Q_1(t + 1) = XQ_2 + \overline{Q}_1\overline{Q}_2$$
$$Q_2(t + 1) = \overline{X}\overline{Q}_2 + Q_1$$

determine the state diagram for the sequential circuit.

12.37 Find the transition table for the sequential circuit shown in Fig. P12.37.

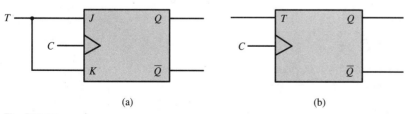

(a)   (b)

Fig. P12.33

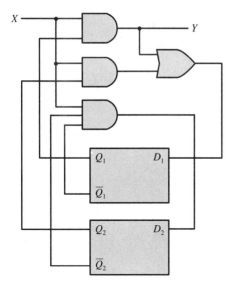

Fig. P12.37

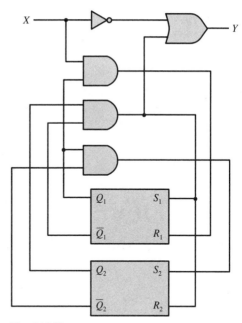

Fig. P12.39

12.38   Find the transition table for the 2-input sequential circuit shown in Fig. P12.38.

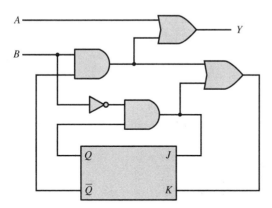

Fig. P12.38

12.39   Find the transition table for the sequential circuit shown in Fig. P12.39.

12.40   Construct the excitation table for the T flip-flop described in Problem 12.33.

12.41   Use JK flip-flops to realize a sequential circuit that is characterized by the state diagram given in Fig. 12.47 on p. 850.

12.42   Use D flip-flops to realize a sequential circuit that is characterized by the state diagram given in Fig. 12.47 on p. 850.

12.43   Use D flip-flops to realize a sequential circuit that is characterized by the state diagram given in Fig. 12.46 on p. 847.

12.44   Use JK flip-flops to realize a sequential circuit that is characterized by the state diagram given in Fig. 12.46 on p. 847. Compare the sequential circuit obtained with the one given in Fig. 12.45 on p. 846. Which circuit is simpler?

13

Digital Devices

INTRODUCTION

In the preceding two chapters, we studied various combinational and sequential logic circuits that can be used in the design of digital systems. In this chapter, we turn our attention to the major components of digital computers, and discuss some of the digital devices that are integral parts of computers.

Counters, as their name suggests, are used to count events in digital systems. In addition to this, though, they can be used to generate timing sequences that control operations in computers. Both synchronous and asynchronous counters are discussed. A counter is one important application of a connection of flip-flops called a register. In this chapter, we also see the usefulness of registers in digital computers.

Information storage in digital computers is accomplished with memory devices, and different types of memories are employed for different purposes. Fast information retrieval is obtained with memories that have random-access capability. In one such memory, the read-only memory (ROM), after its initial programming, information may be retrieved or "read," but not stored or "written." A read-and-write memory (RAM)—also known as a random-access memory—has the capability, however, for both reading and writing information. Both ROMs and RAMs are commonly available as semiconductor IC packages. Although they are not as popular as semiconductor memories, random-access magnetic-core memories are still being used in some digital computers. For mass storage of information, bulk-storage devices such as magnetic tape and magnetic disks are in widespread use. Another type of semi-conductor memory, the charge-coupled device (CCD), is also studied here.

This chapter includes a discussion of the digital devices that make up the typical digital computer. The five major components of a digital computer are the input, the control unit, the arithmetic-logic unit (ALU), the memory, and the output. When the control unit and the ALU are placed on one semiconductor chip, that IC is referred to as a microprocessor.

Although computers process information in digital form, analog information is in-herent in many situations. The conversion of information from analog to digital form, and vice versa, is covered with the discussions on analog-to-digital converters (ADCs) and digital-to-analog converters (DACs), respectively.

13.1   Counters

At the end of the preceding chapter, we saw how to design a modulo-4 up/down counter using SR flip-flops (see Fig. 12.49). Such a device also can be implemented with either D or JK flip-flops (see Problems 12.41 and 12.42 at the end of Chapter 12). A **counter** is a sequential circuit that goes through a prescribed sequence of states when pulses are applied to its input. Most digital systems contain counters because they are an integral part of digital computers. Not only are counters used to keep track of the number of occurrences of an event, but they also are used for controlling operations in digital systems by generating timing sequences.

Binary Counters

A counter that counts in binary is called, quite obviously, a **binary counter.** An n-bit binary counter requires n flip-flops and can have as many as 2^n states. Therefore, it is possible to count from 0 to $2^n - 1$ with a binary counter. The number of states m (where $m \leq 2^n$) of an n-bit binary counter is referred to as its **modulus.** A binary counter with modulus m is called a **modulo-m** (or **mod-m**) **counter.**

The input pulses to a counter, known as **count pulses,** can be either clock pulses or externally generated pulses, and they may be periodically or randomly produced. Let us design a 3-bit binary counter that counts clock pulses modulo 8. The state diagram for such a device is shown in Fig. 13.1. In this case, since the transition from state to state is the result of a clock pulse, there is no indi-cation of an external input in the state diagram. Furthermore, since a display of the state of the counter can be considered the output, no additional external output need be indicated.

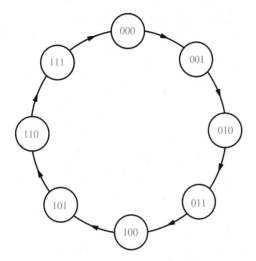

Fig. 13.1 State diagram for a 3-bit modulo-8 binary counter.

The transition table for the counter is given in Table 13.1. As shown in the table, JK flip-flops have been selected for the counter. The columns corresponding to the flip-flop inputs were obtained with the aid of the excitation table for the JK flip-flop

Table 13.1 Transition Table for 3-Bit Modulo-8 Binary Counter.

Present State			Next State			Flip-Flop Inputs					
Q_1	Q_2	Q_3	$Q_1(t+1)$	$Q_2(t+1)$	$Q_3(t+1)$	J_1	K_1	J_2	K_2	J_3	K_3
0	0	0	0	0	1	0	×	0	×	1	×
0	0	1	0	1	0	0	×	1	×	×	1
0	1	0	0	1	1	0	×	×	0	1	×
0	1	1	1	0	0	1	×	×	1	×	1
1	0	0	1	0	1	×	0	0	×	1	×
1	0	1	1	1	0	×	0	1	×	×	1
1	1	0	1	1	1	×	0	×	0	1	×
1	1	1	0	0	0	×	1	×	1	×	1

(Table 12.23*b*). The Karnaugh maps for the flip-flop input variables J_1, K_1, J_2, and K_2 are shown in Fig. 13.2. The resulting Boolean expressions are

$$J_1 = K_1 = Q_2Q_3 \qquad \text{and} \qquad J_2 = K_2 = Q_3$$

Moreover, we do not even have to draw Karnaugh maps for J_3 and K_3 since we can make all the don't-care conditions equal to 1; this results simply in

$$J_3 = K_3 = 1$$

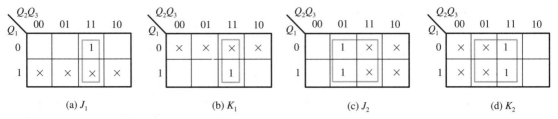

Fig. **13.2** Karnaugh maps for flip-flop input variables.

We may now implement the modulo-8 binary counter with three JK flip-flops as shown in Fig. 13.3. The state of the sequential circuit is $Q_1Q_2Q_3$, and at any given time this is a binary number between 0 and 7.

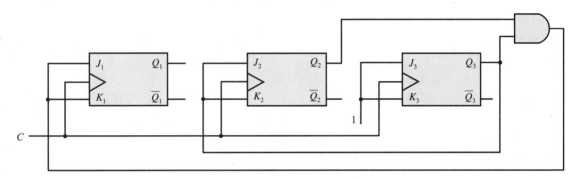

Fig. **13.3** A 3-bit modulo-8 binary counter.

The waveforms associated with the clock pulses and the flip-flop contents comprise its **timing diagram.** Figure 13.4 shows the timing diagram for the modulo-8 binary counter, where it is assumed that the counter is in the state 000 when the first clock pulse arrives. This diagram shows that the flip-flops are triggered on the leading edges of the clock pulses. As indicated by Table 13.1, the first clock pulse changes

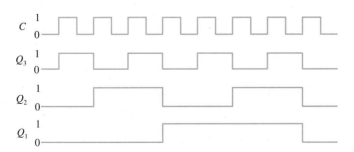

Fig. **13.4** Timing diagram for modulo-8 binary counter.

Q_3 from 0 to 1. The second clock pulse changes Q_3 from 1 to 0 and Q_2 from 0 to 1, and so on. Note that the frequency of the waveform for Q_3 is half the frequency of the waveform for the clock C, the frequency of Q_2 is a fourth of the frequency of C, and the frequency of Q_1 is an eighth of the frequency of C. Because of this, we can refer to a modulo-8 counter as a divide-by-8 counter. More generally, a modulo-m counter can be called a **divide-by-m counter.**

Note that for each flip-flop in Fig. 13.3, $J = K$. A JK flip-flop in which the inputs are connected is a toggle or a T flip-flop. From the truth table for a JK flip-flop (Table 12.19), we see that the truth table for a T flip-flop is given in Table 13.2a, and the excitation table for the T flip-flop is given in Table 13.2b. We leave it as an exercise for the reader to confirm these results (see Problems 12.33 and 12.40 at the end of Chapter 12). Note that an input of 0 will not affect the state of the flip-flop. On the other hand, an input of 1 changes or "toggles" the state from 0 to 1, or from 1 to 0. The circuit symbol for a T flip-flop is shown in Fig. 13.5.

Table 13.2 (*a*) Truth Table for the T Flip-Flop.

Input	Present State	Next State
T	Q	$Q(t + 1)$
0	0	0
0	1	1
1	0	1
1	1	0

Table 13.2 (*b*) Excitation Table for the T Flip-Flop.

Present State	Next State	Input
Q	$Q(t + 1)$	T
0	0	0
0	1	1
1	0	1
1	1	0

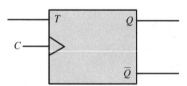

Fig. 13.5 Circuit symbol for the T flip-flop.

Ripple Counters

The binary counters just discussed are examples of **synchronous counters.** Such counters have clock pulses applied to the clock inputs of all the flip-flops. The counter shown in Fig. 13.6 is not a synchronous counter; therefore, it is said to be **asynchronous.** The small circles at the clock-pulse inputs of the T flip-flops signify that they are trailing-edge-triggered devices.

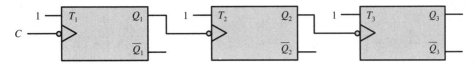

Fig. 13.6 A 3-bit binary ripple counter.

Suppose that the initial state of the counter is 000. Since the flip-flops trigger on the trailing edge of a pulse, Q_1 will not change from 0 to 1 until the first input pulse is completed as shown in Fig. 13.7. Since the transition of Q_1 changing from 0 to 1 is the leading edge of a pulse that is applied to the second flip-flop, Q_2 is not affected by the change of Q_1; however, after the second input pulse, Q_1 changes from 1 to 0. This, in turn, is the trailing edge of a pulse that is applied to the second flip-flop, and, therefore, Q_2 changes from 0 to 1. But Q_2 will not change from 1 to 0 until the next time Q_1 changes from 1 to 0. In a similar manner, we see that Q_3 does not change from 0 to 1 until the first time Q_2 changes from 1 to 0; and the second time Q_2 changes from 1 to 0, Q_3 changes from 1 to 0. The operation of the counter is summarized by the timing diagram shown in Fig. 13.7.

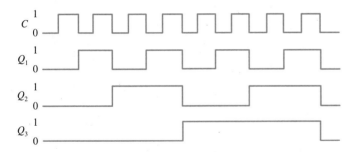

Fig. 13.7 Timing diagram for 3-bit binary ripple counter.

For the counter just described, consider what occurs due to the trailing edge of the fourth input pulse. The value of Q_1 changes from 1 to 0. The resulting trailing edge changes Q_2 from 1 to 0. This, in turn, changes Q_1 from 0 to 1. It is because of this ripple effect that we say that the device shown in Fig. 13.6 is a **ripple counter.** Further note from Fig. 13.7 that this device is a modulo-8 binary counter.

Next consider the ripple counter shown in Fig. 13.8. Suppose that this counter is in the state 000 when the first clock pulse arrives. Since $J_1 = K_1 = \overline{Q}_3 = 1$, $J_3 = Q_1Q_2 = 0(0) = 0$, and $K_3 = Q_3 = 0$, then after the first pulse, $Q_1 = 1$ and $Q_3 = 0$. Because the flip-flops trigger on the trailing edges of pulses, it remains that $Q_2 = 0$.

Since $J_1 = K_1 = \overline{Q}_3 = 1$, $J_3 = Q_1Q_2 = 1(0) = 0$, and $K_3 = Q_3 = 0$, then after the second clock pulse, $Q_1 = 0$ and $Q_3 = 0$. The transition of Q_1 from 1 to 0,

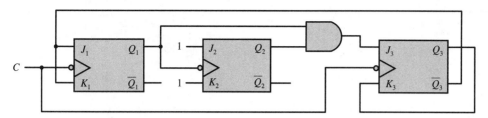

Fig. 13.8 Modulo-5 binary ripple counter.

therefore, results in $Q_2 = 1$. Since $J_1 = K_1 = \overline{Q}_3 = 1$, $J_3 = Q_1Q_2 = 0(1) = 0$, and $K_3 = Q_3 = 0$, then after the third clock pulse, $Q_1 = 1$ and $Q_3 = 0$. The transition of Q_1 from 0 to 1 means that $Q_2 = 1$ remains. Since $J_1 = K_1 = \overline{Q}_3 = 1$, $J_3 = Q_1Q_2 = 1(1) = 1$, and $K_3 = Q_3 = 0$, then after the fourth clock pulse, $Q_1 = 0$ and $Q_3 = 1$. The transition of Q_1 from 1 to 0, therefore, results in $Q_2 = 0$.

Finally, since $J_1 = K_1 = \overline{Q}_3$, $J_3 = Q_1Q_2 = 0(0) = 0$, and $K_3 = Q_3 = 1$, then after the fifth clock pulse, $Q_1 = 0$ and $Q_3 = 0$. Because there is no change in Q_1, it remains that $Q_2 = 0$. Thus after five clock pulses, the counter returns to its initial state 000. In other words, the device shown in Fig. 13.8 is a modulo-5 counter. The timing diagram for this sequential circuit is shown in Fig. 13.9.

Using an analysis similar to the one just described, it can be confirmed that the ripple counter shown in Fig. 13.10 is a modulo-10 or BCD counter. The details of this are left to the reader as an exercise. An alternative way of obtaining a BCD

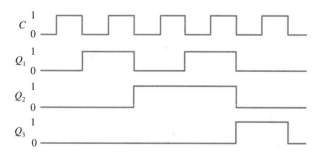

Fig. 13.9 Timing diagram for modulo-5 ripple counter.

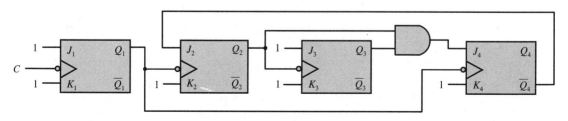

Fig. 13.10 A BCD ripple counter.

counter is to cascade a modulo-2 counter and a modulo-5 counter (see Problem 13.6 at the end of this chapter).

The timing diagrams displayed up to this point have been ideal in that transitions, triggered either on the leading edge or the trailing edge of a pulse, occurred instantaneously. In a physical device, there is some nonzero propagation delay associated with a flip-flop, that is, there is a time differential between when a flip-flop is triggered and when its output responds. Because of the ripple effect, however, in a ripple counter time delays are cumulative. Thus although ripple counters generally have simpler constructions, synchronous counters are faster. Another problem associated with ripple counters is the occurrence of **glitches.** These are voltage spikes that arise as a result of temporary flip-flop states that are caused by propagation delays. This problem may be eliminated by using synchronous counters.

Ring Counters

The synchronous counter shown in Fig. 13.11 is an example of a **ring counter.** Let us assume that the D flip-flops can be directly preset and cleared (inputs not shown), and that the initial state of this counter is $Q_1 = 1$ and $Q_2 = Q_3 = 0$. Since $D_1 = Q_3 = 0$, $D_2 = Q_1 = 1$, and $D_3 = Q_2 = 0$, then after the flip-flops are triggered by the trailing edge of the first clock pulse, the state of the counter is $Q_1 = 0$, $Q_2 = 1$, and $Q_3 = 0$. Similarly, after the second clock pulse, $Q_1 = Q_2 = 0$ and $Q_3 = 1$. Finally, after the third clock pulse, $Q_1 = 1$ and $Q_2 = Q_3 = 0$, so the counter is back in its initial state. This behavior is summarized by Table 13.3. The timing diagram for this 3-bit ring counter is shown in Fig. 13.12. Note that the frequency of the waveform associated with each flip-flop is a third of the frequency of the clock-pulse waveform. Therefore, the device shown in Fig. 13.11 is a divide-by-3 counter.

A ring counter in conjunction with an appropriate encoder can be used to construct a glitch-free binary modulo-m counter—however, m flip-flops are required. For example, the first four columns of Table 13.4 indicate the states of a 4-bit ring counter (assuming that 1000 is the initial state). If $B_1 B_0$ is the binary-number representation indicated by Table 13.4, then we have

$$B_0 = \overline{Q}_1 Q_2 \overline{Q}_3 \overline{Q}_4 + \overline{Q}_1 \overline{Q}_2 \overline{Q}_3 Q_4 \qquad \text{and} \qquad B_1 = \overline{Q}_1 \overline{Q}_2 Q_3 \overline{Q}_4 + \overline{Q}_1 \overline{Q}_2 \overline{Q}_3 Q_4$$

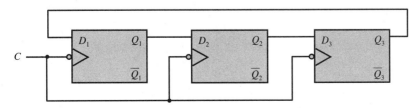

Fig. 13.11 A 3-bit ring counter.

Table 13.3 Behavior of 3-Bit Ring Counter.

Number of Applied Clock Pulses	State of Counter		
	Q_1	Q_2	Q_3
0	1	0	0
1	0	1	0
2	0	0	1
3	1	0	0

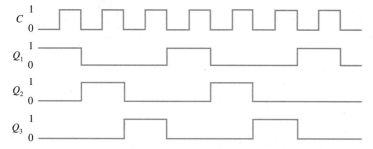

Fig. 13.12 Timing diagram for 3-bit ring counter.

Therefore, the combinational logic circuit (encoder) shown in Fig. 13.13, when connected to a 4-bit ring counter, will yield a binary modulo-4 counter.

An important use of a ring counter is to generate timing signals that control a sequence of operations in a digital system. An m-bit ring counter can generate m timing signals by using m flip-flops. If $m = 2^n$, then 2^n flip-flops are required. An

Table 13.4 Truth Table for Binary Counter Implementation.

Q_1	Q_2	Q_3	Q_4	B_1	B_0
1	0	0	0	0	0
0	1	0	0	0	1
0	0	1	0	1	0
0	0	0	1	1	1

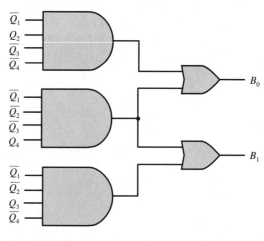

Fig. 13.13 Encoder used for producing a modulo-4 counter.

alternative approach to generating m timing signals is to use an n-bit binary counter (which has 2^n distinct states) in conjunction with an n-to-2^n-line decoder. This, however, requires 2^n AND gates, each of which has n inputs. By using another type of ring counter, we instead can use fewer flip-flops than an ordinary ring counter and not require as complicated a decoder as for a binary counter.

The synchronous counter shown in Fig. 13.14 is known as a **twisted-ring counter** (or **switchtail ring counter** or **Johnson counter**). Let us assume that the initial state of the counter is $Q_1 = Q_2 = Q_3 = 0$. Since $D_1 = \overline{Q}_3 = 1$, $D_2 = Q_1 = 0$, and $D_3 = Q_2 = 0$, then after the first clock pulse $Q_1 = 1$ and $Q_2 = Q_3 = 0$. Since $D_1 = \overline{Q}_3 = 1$, $D_2 = Q_1 = 1$, and $D_3 = Q_2 = 0$, then after the second clock pulse $Q_1 = Q_2 = 1$ and $Q_3 = 0$. Since $D_1 = \overline{Q}_3 = 1$, $D_2 = Q_1 = 1$, and $D_3 = Q_2 = 1$, then after the third clock pulse, $Q_1 = Q_2 = Q_3 = 1$.

Proceeding in this manner, the remaining sequence of states (each being a triple of the form $Q_1Q_2Q_3$) that results is 011, 001, and 000—which is the initial state. This behavior is summarized in Table 13.5. The corresponding timing diagram is shown in Fig. 13.15. Also included are the waveforms obtained by applying various flip-flop outputs to two-input AND gates. Specifically, the waveforms for $Q_1\overline{Q}_2$, $Q_2\overline{Q}_3$, Q_1Q_3, $\overline{Q}_1Q_2$, $\overline{Q}_2Q_3$, and $\overline{Q}_1\overline{Q}_3$ are shown. Therefore, whereas the 3-bit ring counter shown in Fig. 13.11 generates only three timing signals, we see that with

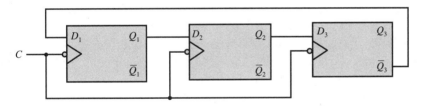

Fig. 13.14 A 3-bit twisted-ring counter.

Table 13.5 Behavior of Twisted-Ring Counter.

Number of Applied Clock Pulses	State of Twisted-Ring Counter		
	Q_1	Q_2	Q_3
0	0	0	0
1	1	0	0
2	1	1	0
3	1	1	1
4	0	1	1
5	0	0	1
6	0	0	0

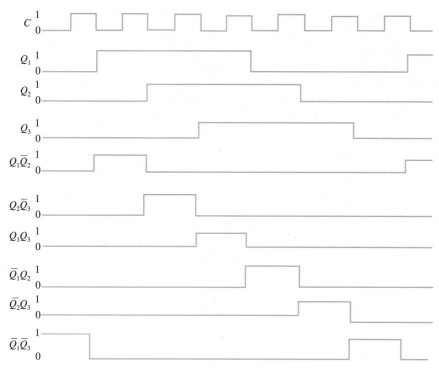

Fig. 13.15 Timing diagram for 3-bit twisted-ring counter.

the aid of six two-input AND gates, the 3-bit twisted-ring counter shown in Fig. 13.14 can produce six timing signals. Finally, note that this device is a divide-by-6 counter.

Various IC counters are available as MSI packages. For example, the 74160 and 74161 ICs are synchronous 4-bit counters, whereas the 74393 consists of two 4-bit binary ripple counters on the same IC. The 7490 and 7493 ICs are BCD ripple counters.

13.2 Registers

We have already seen that a flip-flop is a storage device capable of storing 1 bit of information. A **register** is a collection of flip-flops. Specifically, an n-bit register consists of n flip-flops. Combinational logic gates may or may not be part of a register. Such gates control the register's information transfer. Registers are used typically for the temporary storage of information in binary form, and they are used extensively in digital systems—especially in digital computers. A counter is a type of register whose state changes in some predetermined manner when a sequence of pulses is applied to it.

A simple register consisting of four D flip-flops is shown in Fig. 13.16. For this synchronous sequential circuit, a 4-bit word $A_1A_2A_3A_4$ is applied to the inputs of the flip-flops as indicated. A clock pulse causes this word to be transferred into the register; each bit is stored in its corresponding flip-flop. Transferring information into a register is known as **loading.** The state (information content) of the register can be "read" from the output lines labeled B_1, B_2, B_3, and B_4. The contents of the register remain unchanged until the next clock pulse leads a new word (or the same word) into the register. Since all the bits enter the register at the same time and can be read at the same time, the device in Fig. 13.16 is an example of a **parallel-in, parallel-out register.**

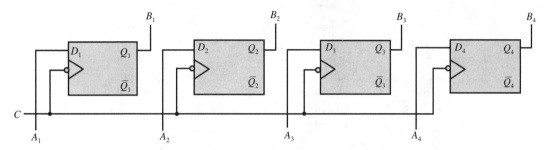

Fig. 13.16 Parallel-in, parallel-out register.

Shift Registers

There is an alternative to the parallel loading of a register as illustrated in Fig. 13.16. An example of this alternative is the simple 4-bit register shown in Fig. 13.17. For this device, information is put into the register serially. Specifically, binary information in the form of a sequence of bits is loaded into this register from left to right by applying an input A while a clock pulse occurs. For example, consider the input sequence A_1, A_2, A_3, A_4, ... Assuming that the initial state of the register is $Q_1 = Q_2 = Q_3 = Q_4 = 0$, before the first clock pulse arrives, $D_1 = A_1$, $D_2 = Q_1$,

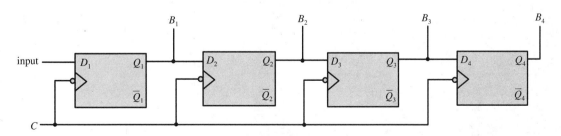

Fig. 13.17 Shift register.

$D_3 = Q_2$, and $D_4 = Q_3$. After the first clock pulse, the state of the register (which can be read from left to right from the output lines) is A_1000. After the second clock pulse, the state of the register is A_2A_100. After the third clock pulse, the state of the register is $A_3A_2A_10$. After the fourth clock pulse, the state of the register is $A_4A_3A_2A_1$, and so on. Note how the input information shifts from left to right. Because of this property, the device shown in Fig. 13.17 is called a **shift register.** In particular, since binary information is loaded serially, and because the contents of the register can be read from the output lines simultaneously, this device is a **series-in, parallel-out shift register.**

Suppose that for the shift register shown in Fig. 13.17 the three output lines labeled B_1, B_2, and B_3 are removed, and only the single output line labeled B_4 remains. For the resulting shift register with an initial state of $Q_1 = Q_2 = Q_3 = Q_4 = 0$ and an input sequence of $A_1, A_2, A_3, A_4, \ldots$, as described above, the first four output bits will be 0 (the first output bit being available before the first clock pulse occurs). After the fourth clock pulse, the output is A_1; after the fifth clock pulse, the output is A_2; after the sixth clock pulse the output is A_3, and so on. Because the output is obtained serially, and because the input is loaded serially, such a device is called a **series-in, series-out shift register.** Note that the ring counter shown in Fig. 13.11 and the twisted-ring counter shown in Fig. 13.14 are series-in, series-out shift registers that have feedback connections.

As an example, for the shift register shown in Fig. 13.17, suppose that the applied input sequence is 1, 1, 0, 1, 0, 0, ... Assuming that the initial state of the device is $Q_1 = Q_2 = Q_3 = Q_4 = 0$, then after the first clock pulse, the state (of the form $Q_1Q_2Q_3Q_4$) of the register is 1000. After the second clock pulse, the state is 1100. The following states are 0110, 1011, 0101, 0010, 0001, 0000, . . . These results are summarized by the timing diagram shown in Fig. 13.18.

A modification of the shift register shown in Fig. 13.17 is required to obtain a **parallel-in, series-out shift register.** One approach that can be taken is shown in

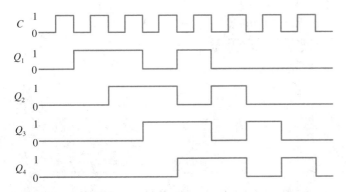

Fig. 13.18 Timing diagram of shift register for the input sequence 1, 1, 0, 1, 0, 0, ...

Fig. 13.19. For this register, there is an enable input E. When $E = 1$, we have that $D_1 = A_1$, $D_2 = A_2$, $D_3 = A_3$, and $D_4 = A_4$. Thus the presence of a clock pulse when $E = 1$ will result in the parallel loading of the register. Conversely, when $E = 0$, we have that $D_1 = 0$, $D_2 = Q_1$, $D_3 = Q_2$, and $D_4 = Q_3$. Under these circumstances, the presence of a clock pulse will shift the contents of the register to the right, loading a 0 into the first flip-flop. Another clock pulse will cause another shift to the right, and so on, until the loaded contents of the register can be completely put out serially.

Just as there are registers that shift to the right, there are also registers that shift to the left. (To see one, take a page with a shift-to-the-right register on it, turn the page over, and hold it up to a light.) However, there are also registers that have the capability of shifting either to the right or to the left—the direction being determined by the application of a control signal. Such a device is referred to as a **bidirectional shift register.** A bidirectional shift register that can be loaded either serially or in parallel, and also has either a serial or parallel output, is known as a **universal shift register.** As was the case for converting a series-in shift register to a parallel-loading shift register (see Fig. 13.19), bidirectional and universal shift registers can be obtained from simpler structures by adding appropriate logic gates and control signals.

A wide variety of shift registers are available as MSI packages. For example, the 7491 IC is an 8-bit series-in, series-out shift register, and the 74165 IC is an 8-bit parallel-in, series-out shift register. An example of a 4-bit universal shift register is the 74194 IC, and the 74198 IC is an 8-bit universal shift register. A series-in, series-out IC shift register has only one input pin and one output pin—regardless of the number of flip-flops. Because of this, very long shift registers can be fabricated by using LSI; an example is the 2401 NMOS IC, which is a 1024-bit shift register.

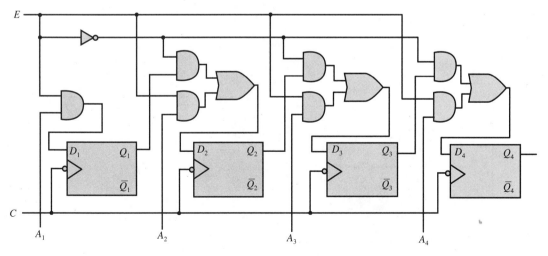

Fig. 13.19 Parallel-in, series-out shift register.

A Serial Adder

A classic application of shift registers is in the construction of binary serial adders. In Chapter 12, we discussed binary parallel adders (e.g., see Figs. 12.7 and 12.17). It was implicitly assumed that in forming the sum $S_n S_{n-1} S_{n-2} \ldots S_2 S_1 S_0$ by adding $A_{n-1} A_{n-2} \ldots A_2 A_1 A_0$ and $B_{n-1} B_{n-2} \ldots B_2 B_1 B_0$, registers for storing these numbers were required. Furthermore, because of the nature of parallel addition, the registers storing the numbers to be added must have parallel outputs, and the register storing the sum must have a parallel loading capability. For serial addition, different types of registers are employed.

The serial addition of two n-bit binary numbers involves n 1-bit binary additions performed one after another (in series). Compare this with the parallel addition of two n-bit binary numbers where n 1-bit binary additions are performed in parallel. Although parallel addition is much faster than serial addition, parallel addition requires n full-adders. Serial addition, on the other hand, needs only one full-adder and one flip-flop. Therefore, serial addition, in general, requires less hardware.

To see how two binary numbers can be added serially, consider Fig. 13.20. Shown are two series-in, series-out shift registers that contain the n-bit binary numbers $A = A_{n-1} A_{n-2} \ldots A_2 A_1 A_0$ and $B = B_{n-1} B_{n-2} \ldots B_2 B_1 B_0$ to be added. Also shown is a full-adder (see Fig. 12.6) and a D flip-flop. After the numbers to be added are shifted into registers A and B, the least significant bit (LSB) of A (which is A_0) is added to the LSB of B (which is B_0) by the full-adder (it is assumed that the initial state of the flip-flop is 0, i.e., $Q = 0$). After the next clock pulse occurs, the sum S_0 and the carry C_1, respectively, enter register S and the flip-flop. The LSB in register A then is A_1 and the LSB in register B then is B_1. The resulting outputs of the full-adder then are S_1 and C_2. The following clock pulse loads these values into register S and the flip-flop, respectively. This process continues until the sum of A and B is loaded

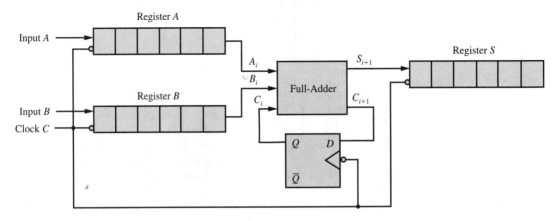

Fig. 13.20 Serial adder.

into register S. As the binary numbers A and B are shifted out of their respective registers, new binary numbers that are to be added can be shifted in.

13.3 Memories

The memory unit of a digital computer (see Fig. 11.1) stores instructions and information in binary form. Information put into a computer by an input device is first stored in the memory unit, and output information comes from the memory unit. The processing of data that takes place in the processor unit employs instructions and data taken from the memory unit. Intermediate and final data-processing results are also stored there. These facts help to demonstrate the usefulness and the importance of the memory unit.

The memory unit of a computer consists of binary storage cells (e.g., flip-flops), each of which is capable of storing 1 bit of information, and associated digital circuits that are used to transfer information into and out of the cells. Typically, memory units are comprised of different types of memory devices, or **memories**—the most common are semiconductor ICs, magnetic cores, and moving surface memories such as magnetic disks, drums, and tapes. Let us begin our discussion of memories by looking at some semiconductor LSI and VLSI memories.

Read-Only Memories (ROMs)

In our previous discussion of read-only memories (ROMs), we presented a logic-circuit depiction of an 8×2 ROM (see Fig. 12.31). Let us now specifically illustrate one type of ROM in which the binary cells are explicitly indicated.

Recall that a $2^n \times m$ ROM has n input lines and m output lines. Each n-bit input combination is called an **address,** and each m-bit output combination is called an **(output) word.** We can form a ROM by using an n-to-2^n-line decoder whose output lines, known as **word lines,** are connected to a matrix array of diodes. Figure 13.21 shows an example for the case that $n = 2$ and $m = 5$. Suppose word line $D_0 = 1$ (and, therefore, $D_1 = D_2 = D_3 = D_4 = 0$); then the diode connected between word line D_0 and output line B_4 is forward biased, resulting in $B_4 = 1$. Similarly, the diode between lines D_0 and B_2 is forward biased, as is the diode between D_0 and B_1. This results in $B_2 = 1$ and $B_1 = 1$, respectively. Because there is no diode between lines D_0 and B_3, or between lines D_0 and B_0 (and because $D_1 = D_2 = D_3 = 0$), it must be true that $B_3 = 0$ and $B_0 = 0$. Hence the resulting output word is $B_4B_3B_2B_1B_0 = 10110$. Table 13.6 summarizes the outputs that result from inputs applied to the ROM shown in Fig. 13.21.

For the ROM shown in Fig. 13.21, a storage cell consists of the presence or absence of a diode. Specifically, if there is a diode connected between word line D_i

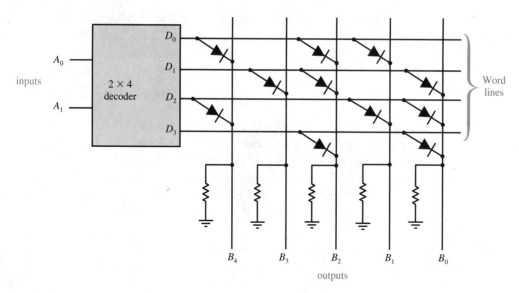

Fig. 13.21 A 4 × 5 (20-bit) ROM.

Table 13.6 Input-Output Description of Given ROM.

Inputs		Decoder	Outputs				
A_1	A_0	Output	B_4	B_3	B_2	B_1	B_0
0	0	D_0	1	0	1	1	0
0	1	D_1	0	1	1	0	1
1	0	D_2	1	0	0	1	1
1	1	D_3	0	0	1	0	1

and output line B_j, then a 1 is stored in the i, j cell. Conversely, if there is no diode between D_i and B_j, then a 0 is stored in the corresponding cell.

A decoder output for a ROM with a diode array, such as the one shown in Fig. 13.21, is required to drive (forward bias) diodes by supplying currents to the diodes and their associated resistive loads. A design improvement is to replace the diodes with transistors. Specifically, let us connect a BJT to word line D_i and output line B_j as shown in Fig. 13.22, where the supply voltage V_{CC} is assumed to be the value of the logical 1 (high) voltage. When $D_i = 0$, the base-emitter junction of the BJT is reverse biased and the transistor is OFF. However, when $D_i = 1$, the BJT turns ON and the output line is driven by the supply voltage and $B_j = 1$.

As with diodes, we see that having a BJT connected between lines D_j and B_j, as shown in Fig. 13.22, corresponds to storing a 1 in that cell, whereas not connecting a transistor corresponds to storing a 0. Alternatively, BJTs can be placed in all the

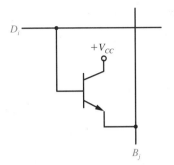

Fig. 13.22 A BJT cell.

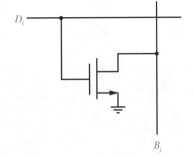

Fig. 13.23 A MOSFET cell.

cells initially and then the connection (or link) between the emitter and the output line for each cell can be left alone or broken. To store a 0, the link is broken; and to store a 1, the connection is left as is. We have already mentioned that this process is called programming the ROM. Moreover, some ROMs can be programmed by their manufacturers (called **mask programming**), and some ROMs (called PROMs) can be programmed by their users.

Instead of using BJT memory cells, ROMs can be fabricated with the use of MOSFETs. In particular, each cell may have the form shown in Fig. 13.23. Instead of having a load resistance connected between each output line and ground, as shown in Fig. 13.21, a MOSFET load is connected between each output line and a supply voltage V_{DD}. For the cell shown in Fig. 13.23, when $D_i = 1$, the enhancement MOSFET is ON and $B_j = 0$. Therefore, connecting the MOSFET as shown corresponds to storing a 0. To store a 1, the MOSFET is made inoperative by fabricating it so that it has a high threshold voltage. The result is a MOSFET that is effectively absent.

For larger ROMs, the structure exemplified by Fig. 13.21 can be inefficient. For example, consider a ROM with ten inputs and four outputs, that is, a $2^{10} \times 4 = 4096 = 4K$-bit ROM. If such a device utilizes a 10×1024 decoder as suggested, the decoder requires 1024 AND gates. Suppose, instead of using a 10×1024 decoder and a 1024×4 cellular matrix, that we use a 7×128 decoder and a 128×32 matrix as shown in Fig. 13.24. Such a matrix also stores $128 \times 32 = 4096$ bits (4K bits), but there are only 128 horizontal (word) lines; however, there are 32 vertical lines. But by using 8-to-1-line (8×1) data selectors or multiplexers (e.g., see Fig. 12.26), we can obtain the output bits, and hence the output words, as shown. In other words, instead of determining an output bit from a column of 1024 cells, we obtain it from eight columns—each consisting of 128 cells—and an 8×1 data selector. The decoder has seven inputs ($A_0, A_1, A_2, A_3, A_4, A_5, A_6$), and each selector uses three additional inputs (A_7, A_8, A_9), for a total of ten inputs to the ROM—as required. Each selector can be implemented with eight AND gates (and one OR gate). Therefore, four selectors require $4 \times 8 = 32$ AND gates (and four

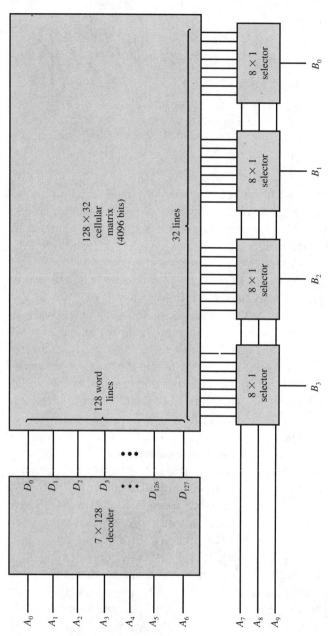

Fig. 13.24 A 4K-bit ROM with two-dimensional addressing.

OR gates). The 7×128 decoder uses 128 AND gates. Thus the ROM shown in Fig. 13.24 employs $32 + 128 = 160$ AND gates. Compare this with the 1024 AND gates required for the ROM structure discussed previously. The memory shown in Fig. 13.24 is an example of a ROM with **two-dimensional addressing.**

Removal of the power supplied to a ROM does not affect the device's physical construction. Reinstating the power, therefore, does not change its programming. Because of this, we say that a ROM is a **nonvolatile memory.** In addition to memory applications, ROMs can be used for code conversions and for realizations of other combinational logic circuits.

Read-and-Write Memories (RAMs)

We know that when an input is applied to a ROM, the word at the corresponding address is read (i.e., the word appears at the output). The time required to select a word and read it is known as the **access time** of the ROM. Since the words in a ROM can be read or accessed in any order, such a device is an example of a **random-access memory.** There is another type of memory for which words not only can be read (retrieved), but can be "written" (stored) at any address in the memory as well. Such a device can be called a **read-and-write memory,** or RAM. (Although "RAM" was originally the acronym for the phrase "random-access memory," it is now used to signify a read-and-write memory.)

Typically, a ROM is used for permanent storage of data and programs that are used frequently. Examples are code converters, look-up tables, and assemblers. A RAM, on the other hand, is used for temporary storage of data and instructions. Such information can be read in accordance with a program that is to be executed, and the results then can be written in the RAM. The time required to complete a read or a write operation is called the **access time** of a RAM.

A logic-equivalent circuit of a RAM cell is shown in Fig. 13.25a. The block diagram of this cell is shown in Fig. 13.25b. For the circuit shown in Fig. 13.25a, there are three inputs labeled data input (A), select (S), and read/write ($R/\overline{W}$), and one output (B).

When the select input is $S = 0$, the output of the cell is $B = 0$ regardless of the value of the data input A or the content Q of the SR flip-flop. Furthermore, $S_c = R_c = 0$, so the state of the flip-flop does not change. When $S = 1$, however, then the cell will be in either the read mode or the write mode, depending on the value of the read/write input $R/\overline{W}$. If $R/\overline{W} = 1$, then $S_c = R_c = 0$ (which does not change Q) and $B = Q$. Thus the output reads the content of the cell. On the other hand, if $R/\overline{W} = 0$, then $S_c = A$ and $R_c = \overline{A}$ (while $B = 0$). When $A = 1$, the flip-flop is set ($Q = 1$); when $A = 0$, the flip-flop is reset ($Q = 0$). In other words, the flip-flop stores (the cell writes) A.

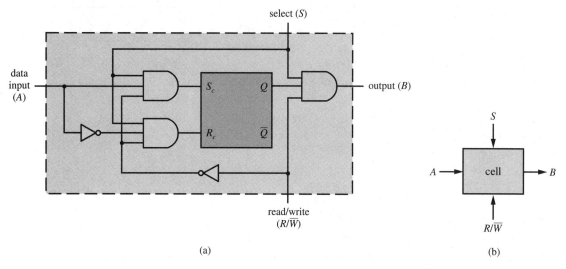

Fig. 13.25 (a) RAM cell logic-equivalent circuit, and (b) block diagram.

A RAM with n inputs and m outputs is a $2^n \times m$ RAM or $2^n \times m$-bit RAM. Figure 13.26 shows a RAM with $n = 1$ and $m = 4$, that is, an 8-bit RAM. As can be seen, the word lines are used to select a row of cells; when $D_0 = 1$, the top row is selected, and when $D_1 = 1$, the bottom row is selected. The $R/\overline{W}$ input places the selected cells in either the read mode ($R/\overline{W} = 1$) or the write mode ($R/\overline{W} = 0$).

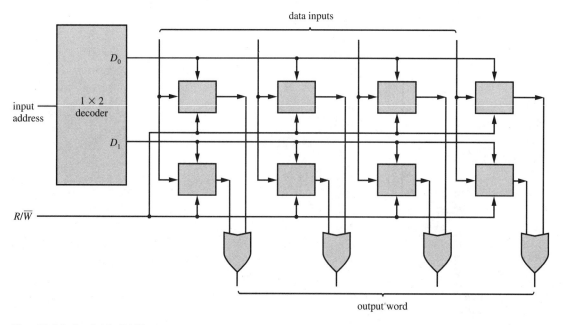

Fig. 13.26 An 8-bit RAM.

When cells are read, their contents appear at the outputs; when input data are written in cells, the outputs are 0.

Just as was the case for ROMs, increased efficiencies can be achieved for RAMs by using two-dimensional addressing. As a specific example, consider a 4096 × 1 (4K-bit) RAM, where instead of having a single column of 4096 cells, the cellular array has 64 rows and 64 columns. Any cell can be selected by designating its row location and its column location. By using six inputs and a 6 × 64 decoder, any row can be selected. Another six inputs applied to a second 6 × 64 decoder can be employed to select any column. Each cell, therefore, must be able to detect signals from both the row and column decoders. Again a cell is in the read mode when $R/\overline{W} = 1$ and in the write mode when $R/\overline{W} = 0$. For the latter case, the single data input A is written into the selected cell, and for the former case, the content of the selected cell is read and appears at the output B. Typically, amplifiers are used to write and to read or sense data. A block diagram representation of this 4K-bit RAM is shown in Fig. 13.27.

Such a RAM has 1-bit output words. Suppose we wish to have a RAM with 4096 4-bit words (i.e., a 16K-bit RAM). We can construct such a device by combining four of the structures shown in Fig. 13.27. In this case, the 12 address inputs (A_0 through A_{11}) are applied in parallel to the four packages. The $R/\overline{W}$ inputs are

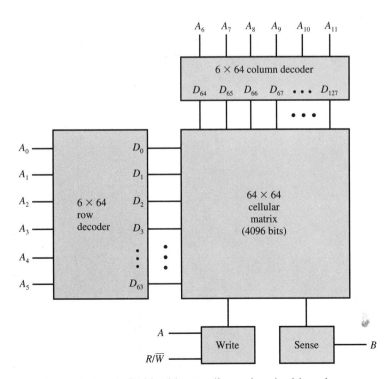

Fig. 13.27 A 4K-bit RAM with two-dimensional addressing.

connected in parallel, the data inputs are entered in parallel, and the four output bits are read in parallel.

The structure of a RAM cell depends not only on the technology to be used (i.e., BJT or MOS), but also on the mode of operation of the components. One type of cell, which contains a transistor flip-flop, maintains its state until a new state is written in (as long as the power is on). A RAM with this cell structure is known as a **static RAM.** Such RAMs usually are fabricated with either BJTs, NMOS, or CMOS.

A second type of cell may be formed with a single NMOS transistor and a capacitance as shown in Fig. 13.28. The gate of the enhancement MOSFET is connected to a word line (a line from the row decoder) and the drain is connected to a bit line (a line from the column decoder). When the capacitor voltage is high, the cell stores a 1; when the capacitor voltage is low, the cell stores a 0. Because the capacitor will lose its charge (voltage) owing to leakage, there must be a **refresh circuit** (not shown) to recharge the capacitor to its appropriate value. A RAM with cells such as this is an example of a **dynamic RAM.** Even with the added complexity of refresh circuits, because of their adaptability to VLSI, dynamic RAMs have become very popular. Unlike the ROM, the RAM—be it static or dynamic—is a **volatile** device. That is, if the power supplied to a RAM is interrupted, the information stored will be lost.

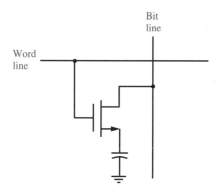

Fig. 13.28 Dynamic RAM cell.

Magnetic-Core Memories

Another type of memory with random access is a **magnetic-core memory.** Such a memory consists of a collection of **magnetic cores,** each of which is magnetic material, typically ferrite, in the shape of a doughnut. As depicted in Fig. 13.29, three conducting lines pass through the hole of each core. By putting appropriate-magnitude currents through both the word line and the bit line, a core can be magnetized

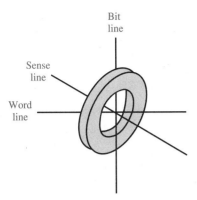

Bit
line

Sense
line

Word
line

Fig. 13.29 Magnetic core.

in one of two directions—say negative (0) or positive (1). A current through only one line is not sufficient to magnetize the core in either direction. Once the core has been magnetized in one direction as the result of word-line and bit-line currents, it cannot be magnetized in the other direction until both word and bit reverse currents are applied. In addition to this, when the magnetization of the core changes from one direction to the other (e.g., from the positive direction to the negative direction), a voltage is induced across the sense line.

A simplified 16×1 (16-bit) magnetic-core memory is shown in Fig. 13.30. Let us consider a specific memory cell—the core in row 2 and column 1. To write into this cell, the row decoder selects word line D_1 and the column decoder selects bit line D_4. Then a control signal sends negative currents through the selected lines; this puts the core into the 0 state (negative magnetization). A single negative current is not sufficient to change a core to the 0 state, so only the core in row 2 and column 1 can be changed to 0. If a 0 is to be written into the cell indicated, as the core is already in the 0 state, no further action need be taken. If a 1 is to be written into the selected cell, a proper control signal sends positive currents through lines D_1 and D_4. This puts the corresponding core into the 1 state. Since a single positive current is not sufficient to change a core to the 1 state, only the core in row 2 and column 1 is changed to 1.

To read the cell in row 2 and column 1, again negative currents are sent through lines D_1 and D_4. As indicated above, this places a 0 in the core. If the initial state of the cell was 1, the change in core magnetization induces a voltage across the sense line and a 1 is entered into the memory buffer register. On the other hand, if the initial cell state was 0, the lack of a change in core magnetization means that there is no voltage induced across the sense line—so a 0 is entered into the memory buffer register. Note, however, that when a cell is read, it is put into the 0 state. This is known as **destructive readout.** Therefore, after a cell is read, its content prior to the reading must be rewritten. If the memory buffer register contains a 0,

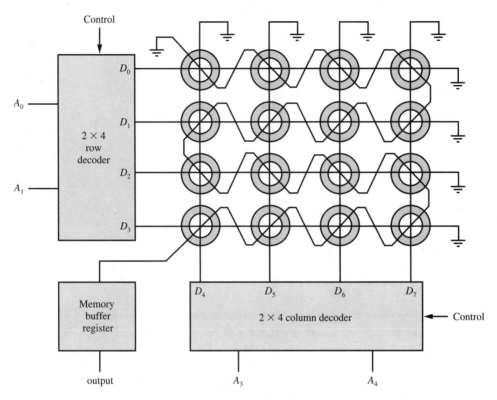

Fig. 13.30 A 16-bit magnetic-core memory.

no further action is required since the cell has been put into the zero state. If the memory buffer register contains a 1, however, positive currents are passed through lines D_1 and D_4 to rewrite a 1 in the cell.

The magnetic-core memory shown in Fig. 13.30 has a 1-bit output. By combining such memories in parallel, as was indicated for the RAM given in Fig. 13.27, magnetic-core memories with n-bit outputs can be formed. Even though a magnetic-core memory has a destructive readout and requires a rewriting cycle, if the power to a magnetic-core memory is turned off, the cores retain their magnetization. Therefore, unlike a RAM, a magnetic-core memory is a nonvolatile memory. Although semiconductor memories are very popular, magnetic-core memories are still in use.

Moving-Surface Memories

The ROM, RAM, and magnetic-core memory are examples of memories in which information can be accessed directly. These types of memories, therefore, afford rapid information access. Where high-speed information retrieval does not have high priority, other types of memories can be utilized. One nonvolatile memory medium that provides bulk or mass storage capability at low cost is magnetic tape. In essence, it is simply plastic tape with a magnetic coating. Magnetic tape is a **serial** or **sequential memory** in that information is available in the same sequence in which it was originally stored. Therefore, the access time for magnetic tape depends on where the desired information is stored and the relative location of that information with respect to the tape read-out.

For a magnetic-tape storage system, the tape moves along a **read/write head** as depicted in Fig. 13.31. The read/write head is basically an iron core (with an air gap) on which a coil of wire is wound. When the head is in the write mode, a current corresponding to the information to be stored is passed through the coil of wire. The resulting magnetic field across the gap magnetizes the magnetic surface coating of the tape as it moves across the head. When the head is in the read mode, and when a magnetized tape moves across the head, the resulting magnetic field in the core induces a voltage across the coil that corresponds to the information stored.

Magnetic tape is an example of a **moving-surface memory.** Another moving surface memory that is used for bulk storage is the **magnetic disk.** The various types of magnetic disks differ in detail, but not in principle. A disk is shaped like a phonograph record and contains numerous concentric circles called **tracks.** As with magnetic tape, information is either read or written in a track by a read/write head.

Magnetic disks can be classified in two categories—**rigid** or **hard disks,** and **flexible disks,** called **floppy disks** or **diskettes.** The read/write head does not come into physical contact with a rigid disk, but instead is cushioned by a thin layer of air. A floppy disk does make mechanical contact with the read/write head as is done with a magnetic-tape memory. The head is movable so that it can be positioned on different tracks. Because of this, a magnetic disk is often called a **direct-access memory.** The access time consists of the time it takes to move the head to the proper track and for the desired portion of the track to reach the head. The access time for

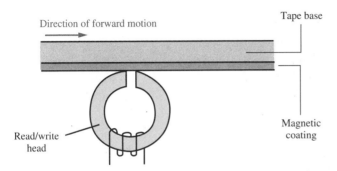

Fig. 13.31 Magnetic tape and read/write head.

a rigid disk typically is of the order of tens of milliseconds, whereas for a floppy disk it is of the order of hundreds of milliseconds.

The rotational speed of a rigid disk typically is 4500 rpm (revolutions per minute), and a floppy disk rotates at 300 rpm. Both $5\frac{1}{4}$-inch and $3\frac{1}{2}$-inch high-density diskettes contain 80 tracks on each of their two sides. The former has 15 sectors per track, whereas the latter has 18 sectors per track, where a sector comprises 512 bytes. The information storage capability is measured in millions of bits for a floppy disk (1.2 Mbytes for a $5\frac{1}{4}$-inch disk and 1.44 Mbytes for a $3\frac{1}{2}$-inch disk) and hundreds of millions of bits for a hard disk.

A $5\frac{1}{4}$-inch floppy disk is placed inside an envelope similar to a jacket for a phonograph record. This protects the disk from such things as dust and scratching. The inside of the protective envelope is made so as to keep the friction between the disk and envelope low, and the disk clean. Figure 13.32 shows such a floppy disk inside its protective envelope. Note that there is a slot in the envelope so that the read/write head has access to the disk. There also is a hole in the protective envelope and a hole in the disk—called an **index hole.** When these holes are aligned, a light passing through them can used to indicate the beginning of a track.

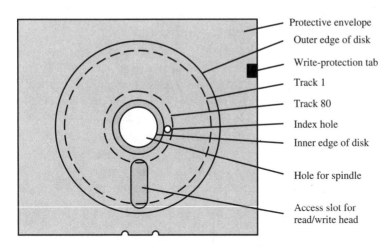

Protective envelope
Outer edge of disk
Write-protection tab
Track 1
Track 80
Index hole
Inner edge of disk
Hole for spindle
Access slot for read/write head

Fig. 13.32 Floppy disk in its protective envelope.

Another example of a moving-surface memory is a **compact disk** (CD), which is a read-only memory (ROM) that is read optically with a laser beam. CD-ROMs have thousands of tracks and can contain hundreds of millions of bytes of information.

Charge-Coupled Devices

Another semiconductor IC memory, unlike a RAM or ROM, is a serial memory. A **charge-coupled device** (CCD) is, in effect, an enhancement MOSFET with numerous gates (see Section 8.2). Figure 13.33 illustrates such a device, where the gates are labeled ϕ_1, ϕ_2, ϕ_3, and ϕ_4. Although the number of gates is large (hundreds or more), only eight are shown. The drain and the source of the MOSFET are not shown in Fig. 13.33.

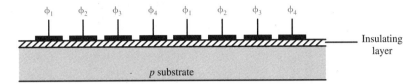

Fig. 13.33 Portion of a CCD.

If a positive voltage (with respect to the substrate) is applied to a gate, then holes (majority carries) are repelled from a region of the p substrate directly under the gate. This region, which is called a **potential well,** is depleted of holes and is capable of attracting and storing electrons. If electrons are supplied to a potential well from the source of the MOSFET, then by applying appropriate voltages to the gates of the device, these electrons can be transferred or shifted from gate to gate along the length of the device. Therefore, a CCD is, in essence, a long shift register. The storage of bits of information in a CCD corresponds to the presence or absence of charge (electrons for the example being discussed) under the gates of the device. Quite clearly, a CCD is a serial memory.

Let us see how charge (information) can be shifted through a CCD. Suppose that a positive voltage is applied to the leftmost gate labeled ϕ_1 in Fig. 13.33. Further suppose that electrons are supplied to this well via the source of the CCD. The resulting situation is depicted in Fig. 13.34a. If gate ϕ_3 is also made positive, then another potential well forms under that gate as shown in Fig. 13.34b, however,

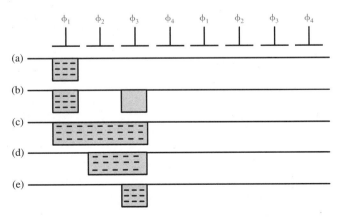

Fig. 13.34 Formation of potential wells in CCD.

electrons are absent from this newly formed well. Now suppose that in addition to ϕ_1 and ϕ_3, ϕ_2 is made positive. Assuming closely spaced gates, what results is the large potential well shown in Fig. 13.34c, where the electrons originally stored in the well under gate ϕ_1 are distributed throughout this larger well. If the positive voltage is removed from gate ϕ_1, its associated potential well is eliminated, leaving the well shown in Fig. 13.34d. Finally, when the positive voltage is also removed from gate ϕ_2, what results is the electron-filled well under gate ϕ_3, as shown in Fig. 13.34e. The net effect of these events is the transfer of charge from (under) gate ϕ_1 to gate ϕ_3. Because of this, a CCD is also referred to as a **charge-transfer device** (CTD).

A summary of the voltages that cause the transfer of charge just described is given in Fig. 13.35. For time interval T_1, the voltage on gate ϕ_1 is positive—this corresponds to the situation depicted by Fig. 13.34a. For time interval T_2, in addition to gate ϕ_1, the voltage on gate ϕ_3 is positive—this corresponds to Fig. 13.34b, and so on. Just as charge was transferred from gate ϕ_1 to gate ϕ_3 by using the clocking voltage waveforms shown in Fig. 13.35, by appropriately extending these waveforms, the charge under gate ϕ_3 can be shifted to the right along the length of the device (see Problem 13.26 at the end of this chapter). Owing to the leakage of charge, however, **refresh amplifiers** are inserted along the length of the structure between groups (ranging from 100 to 1000) of gates so that packets of charge can be refreshed or replenished.

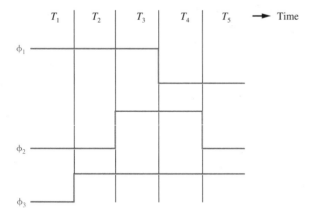

Fig. 13.35 Voltage waveforms used to transfer charge.

13.4   Digital Information Processing

In the electronics section of this book, we studied the fabrication of logic gates using electronic components such as resistors, diodes, BJTs, and FETs (CMOS, NMOS, and PMOS). Logic gates are examples of small digital systems—that is, systems that operate on information in digital form. As demonstrated in Chapter 12, logic gates in SSI form can be combined to make larger digital systems. In addition, fundamental digital building blocks such as adders, magnitude comparators, encoders, decoders, multiplexers, demultiplexers, flip-flops, counters, registers, and memories are available in MSI and LSI packages. These devices are comprised of simpler digital structures, and thus are examples of more complex digital systems. And these more complicated digital devices can be combined to form digital systems that have even greater complexity. Because of the advances in intgrated-circuit technology, however, more complicated structures, such as the components of digital information-processing systems (as well as entire systems such as microprocessors and some microcomputers), are available as LSI or VLSI packages. Let us now discuss the major components of some digital information-processing systems.

Digital-Computer Components

Undoubtedly, the most important digital system is the digital computer. A **digital computer** is a digital system that sequentially executes a set of instructions on data in digital form. Digital computers take many forms as to size and shape, but from a functional point of view, their behavior can be characterized by the simplified block diagram shown in Fig. 13.36.

Information in the form of instructions and data enters a computer via its input. Some of the ways in which binary information is put into computers include punched cards, magnetic tape, paper tape, magnetic disks, teletypewriters, video displays, and keyboards. In addition to the data, the set of instructions (program) to

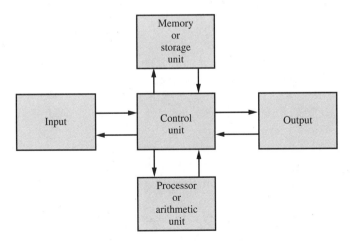

Fig. 13.36 Simplified block diagram of a digital computer.

be executed is stored in a portion of the computer's memory known as **main memory.**

Main memory typically is comprised of memories with random-access capability—specifically, RAMs and ROMs. Although semiconductor IC units predominate, magnetic-core memories still see use. In addition to main memory, many computers have bulk memory devices such as magnetic disks, tapes, and drums for storing large amounts of information. Usually information in bulk storage is not directly available for data processing, but first must be transferred into main memory.

The job of the processor or arithmetic unit is to perform all the arithmetic and logical operations. For this reason, it is also known as the **arithmetic-logic unit** (ALU) of the computer. The arithmetic and logic operations are performed by various logic circuits. Intermediate results of ALU operations may be temporarily stored in ALU registers known as **accumulators.** Overall, however, the ALU receives data from the memory unit, and after the appropriate operations are performed in the ALU, the results are stored in the memory unit.

The control unit is the heart of the computer. After it reads and interprets an instruction, it sends the appropriate timing and control signals to the portions of the computer that perform the required operations. Upon the completion of these operations, the control unit proceeds to the next instruction. Control units contain counters, combinational logic circuits, and registers for the temporary storage of information. The combination of the control unit and the ALU is called the **central processing unit** (CPU). A CPU on a single IC chip is referred to as a **microprocessor.** Information can be put out of a computer by various types of output devices. Examples are teletypewriters, printers, video (CRT) displays, and plotters.

There are basically three classifications of computers. Large computers are known as **mainframe computers** or **maxicomputers.** Intermediate-size computers are referred to as **minicomputers,** and small computers are called **microcomputers.** A computer transfers and manipulates binary information in groups of bits. The size of these groups is called the **word size** of the computer. Typically, the word size for a microcomputer is 8 bits (referred to as 1 byte). The word size for a minicomputer usually ranges from 8 to 32 bits, with 16 bits being quite common; mainframe computers have word sizes ranging from 16 to 64 bits, with 32 bits being typical.

A microcomputer can be formed from a microprocessor, a clock, a ROM, a RAM, and input/output capability. Although the CPU is usually on one chip and the other components on separate chips, it is possible to have all the major components on one LSI chip. Such a ''computer on a chip,'' of course, has greater limitations than does a multiple-chip microcomputer; however, single-chip computers can be gainfully employed in such things as electronic test equipment, microwave ovens, automobile ignition systems, supermarket scales, and video games.

The classical approach to digital-system design is to combine such devices as logic gates, flip-flops, counters, registers, and various combinational logic circuits (e.g., multiplexers, decoders) so as to implement a particular system behavior. An alternative method of digital-system design employs the microcomputer. Instead of interconnecting various digital devices, a microcomputer can be programmed so that a specific behavior is realized. A major advantage of this approach is that if the operating characteristics of the system must be modified, then only the programming of the microcomputer need be changed. In contrast to this, a classically designed digital system could require significant changes in its interconnections and/or components. Furthermore, having a microcomputer available on just one or a few IC chips can result in a digital system that is smaller in size and lower in cost than one that is classically designed.

Digital-to-Analog Converters

Digital computers and microprocessors are digital information-processing systems, but information quite often is in analog form (e.g., speech, music, and video signals). To process this type of information with sophisticated digital techniques, the information first must be converted from analog to digital form. A device that accomplishes this is known as an **analog-to-digital converter** (ADC or A/D converter). Since many types of electronic equipment are inherently analog devices (e.g., stereo amplifiers, radio and television receivers), there are many occasions when it is necessary to transform digital information into analog information. This can be accomplished with a **digital-to-analog converter** (DAC or D/A converter).

The circuit shown in Fig. 13.37 is an example of a simple DAC. The input to the circuit is the 3-bit binary number $A_2 A_1 A_0$, where each of the variables A_0, A_1, and A_2 takes on the value 0 V or 1 V. Summing currents at the inverting input of the operational amplifier (op amp) on the left yields

$$\frac{v_1}{R_F} + \frac{A_0}{R} + \frac{A_1}{R/2} + \frac{A_2}{R/4} = 0$$

from which

$$v_1 = -\frac{R_F}{R}(4A_2 + 2A_1 + A_0)$$

Since the second op-amp stage has a voltage gain of $v_2/v_1 = -R_2/R_1$, the output voltage v_2 is given by

$$v_2 = \frac{R_2 R_F}{R_1 R}(4A_2 + 2A_1 + A_0)$$

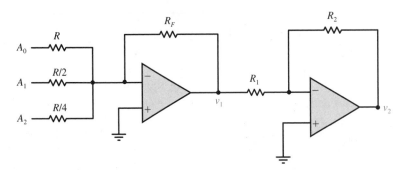

Fig. 13.37 A 3-bit DAC.

Consider the case that $R_1 = R_F$ and $R_2 = R$. Then we have

$$v_2 = 4A_2 + 2A_1 + A_0$$

Under this circumstance, when the input is $A_2A_1A_0 = 000$, the output is $v_2 = 0$ V; whereas when the input is $A_2A_1A_0 = 111$, the output is $v_2 = 7$ V. In general, for these choices of R_1 and R_2, the value of the output voltage is equal to the value of the binary number put in. Different choices for R_1 and R_2 will result in different output voltages (see Problem 13.27 at the end of this chapter). Further, if logical 0 and logical 1 correspond to voltages other than 0 V and 1 V, respectively, different output voltages will again result (see Problem 13.28 at the end of this chapter).

Extending the circuit shown in Fig. 13.37 to implement a DAC with more than three input bits requires additional input resistors $R/8$, $R/16$, $R/32$, and so on (see Problem 13.29 at the end of this chapter). Because of the wide variation of input-resistance values required for DACs with many input bits, a better DAC design is the one shown in Fig. 13.38. This DAC uses a connection of resistors called an **R-2R ladder,** and employs a single op amp.

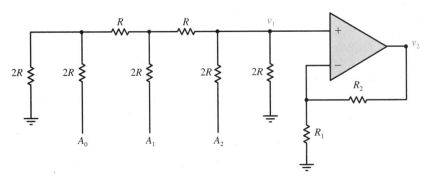

Fig. 13.38 A 3-bit DAC using an *R-2R* ladder.

Since the input of an (ideal) op amp draws no current, we can analyze the ladder network as if the op amp were absent. Let us find the voltage v_1 by using the principle of superposition. If $A_0 = A_1 = 0$, then all the resistors, except the $2R$ resistor connected between A_2 and v_1, can be reduced by parallel and series combinations to an R resistor. Then by voltage division, the voltage at the noninverting input of the op amp is

$$v_{1a} = \frac{R}{R + 2R}A_2 = \tfrac{1}{3}A_2$$

Taking similar approaches, if $A_0 = A_2 = 0$, then $v_{1b} = A_1/6$; if $A_1 = A_2 = 0$, then $v_{1c} = A_0/12$. Therefore, by the principle of superposition,

$$v_1 = v_{1a} + v_{1b} + v_{1c} = \tfrac{1}{3}A_2 + \tfrac{1}{6}A_1 + \tfrac{1}{12}A_0 = \tfrac{1}{12}(4A_2 + 2A_1 + A_0)$$

Since the voltage gain of the op-amp stage is $v_2/v_1 = 1 + R_2/R_1$, the output voltage is

$$v_2 = \left(1 + \frac{R_2}{R_1}\right)\left(\frac{1}{12}\right)(4A_2 + 2A_1 + A_0) = \frac{R_1 + R_2}{12R_1}(4A_2 + 2A_1 + A_0)$$

For the case that $R_2 = 11R_1$, we again have that

$$v_2 = 4A_2 + 2A_1 + A_0$$

As for the 3-bit DAC given in Fig. 13.37, the 3-bit DAC shown in Fig. 13.38 can be extended such that n-bit ($n > 3$) binary information can be converted from digital to analog form (see Problem 13.31 at the end of this chapter).

Analog-to-Digital Converters

There are a number of approaches that can be taken for the construction of an analog-to-digital converter (ADC). One technique incorporates a DAC as shown in Fig. 13.39. For this device, an analog input voltage is applied to one of the inputs of a voltage comparator (see Section 10.5). The output of the DAC is applied to the other input of the comparator. As long as the DAC output voltage is less than the analog input voltage, the comparator output is high (logical 1) and clock pulses are applied to the binary counter. When the count of the binary counter is high enough such that the value of the DAC output exceeds the analog input, however, the comparator output is low (logical 0), so additional clock pulses are not applied to the counter.

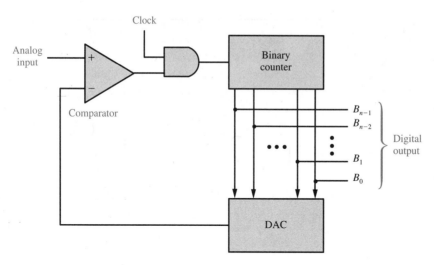

Fig. 13.39 An ADC that employs a DAC.

Therefore, the counter stops counting and the resulting n-bit output is $B_{n-1}B_{n-2} \cdots B_1B_0$. The counter is then reset to zero (provision for this is not indicated in the figure), and the next analog input is applied. Such a device is called a **counting ADC.**

For the ADC shown in Fig. 13.39, it was assumed that the analog input remained constant during the time interval required for the counter to reach its final state. If this is not the case, then the analog signal can be "sampled" by a switch S (which

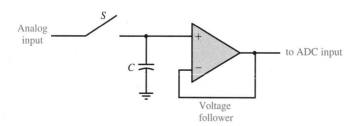

Fig. 13.40 Sample-and-hold circuit.

may be implemented with an FET, for example) as illustrated by Fig. 13.40. The switch S closes momentarily and the capacitor C charges to the value of the analog voltage. Since the input resistance of the voltage-follower stage is very high, with the switch open, the capacitor "holds" its charge, and, hence, holds the value of the sampled analog signal. After the ADC has had sufficient time to reach its final value, the **sample-and-hold circuit** shown in Fig. 13.40 takes the next sample of the analog signal.

For the ADC shown in Fig. 13.39, the binary counter may be required to count up to $2^n - 1$ clock pulses for some analog inputs. A faster method of analog-to-digital conversion is illustrated in Fig. 13.41. For this circuit, a reference (constant) voltage V_r is divided among the comparator inputs as indicated. The analog input is applied to each comparator as shown. When the analog voltage exceeds $V_r/4$, the output of comparator C_1 is 1; otherwise it is 0. Similarly, when the analog voltage exceeds $V_r/2$, the output of C_2 is 1; otherwise it is 0. Finally, when the analog voltage exceeds $3V_r/4$, the output of C_3 is 1; otherwise it is 0. The outputs of the comparators are encoded to give the desired digital output B_1B_0. The truth table for the encoder is given by Table 13.7.

The ADC shown in Fig. 13.41 is a parallel-comparator ADC. In particular, it is a 2-bit ADC. For a 3-bit parallel-comparator ADC, $2^3 - 1 = 7$ comparators are re-

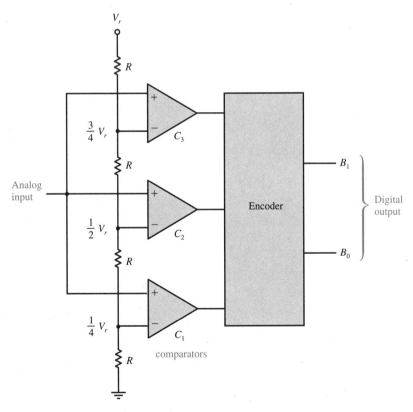

Fig. 13.41 A 2-bit parallel-comparator ADC.

quired. In general, $2^n - 1$ comparators are needed for an n-bit parallel-comparator ADC. This is the fastest type of ADC.

Various types of ADCs and DACs are available in IC form.

Table 13.7 Truth Table for Encoder in ADC.

Endoder Inputs			Encoder Outputs	
C_3	C_2	C_1	B_1	B_0
0	0	0	0	0
0	0	1	0	1
0	1	1	1	0
1	1	1	1	1

SUMMARY

1. Counters are used in digital systems to count events and to generate timing sequences.

2. Modulo-m counters are also known as divide-by-m counters.

3. Ripple counters are examples of asynchronous counters.

4. Synchronous counters are faster and do not have the noise problems associated with ripple counters.

5. Ring counters and twisted-ring counters are used to generate timing signals.

6. Registers may be loaded serially or in parallel, and they may be unloaded serially or in parallel.

7. Shift registers can be used to implement serial adders.

8. A read-only memory (ROM) is a nonvolatile memory. Two-dimensional addressing of ROMs results in more efficient device fabrications.

9. Read-and-write memories (RAMs) are either static or dynamic. In either case, a semiconductor RAM is a volatile memory.

10. Magnetic-core memories are random-access devices that are nonvolatile.

11. magnetic-tape memories are serial memories that are nonvolatile and are used for bulk storage.

12. Magnetic disks are also nonvolatile and used for bulk storage, but they are direct-access devices.

13. A charge-coupled device (CCD) is a serial semiconductor memory. Such a device is, in essence, a MOSFET shift register.

14. The major components of a digital computer are the input, the control unit, the arithmetic-logic unit (ALU), the memory, and the output.

15. The central processing unit (CPU) is comprised of the ALU and the control unit. A microprocessor is a CPU in IC form.

16. Information can be changed from analog to digital form with an analog-to-digital converter (ADC), and from digital to analog form with a digital-to-analog converter (DAC).

PROBLEMS

13.1 Design a modulo-6 binary counter using JK flip-flops.

13.2 Use four JK flip-flops to design a modulo-10 binary counter (also known as a BCD counter).

13.3 Use D flip-flops to design a synchronous counter that is characterized by the state diagram shown in Fig. P13.3.

13.4 The asynchronous counter shown in Fig. P13.4 employs three D flip-flops, each of which triggers on the leading edge of a pulse applied to its clock-pulse input. Draw the timing diagram for this counter. What type of counter is this?

13.5 The 3-bit ripple counter shown in Fig. P13.5 is in example of a binary count-down counter. Draw

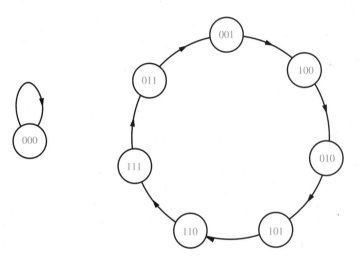

Fig. P13.3

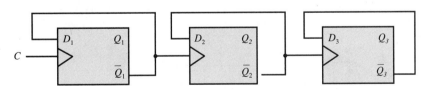

Fig. P13.4

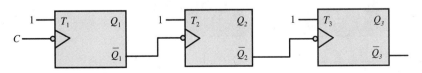

Fig. P13.5

the timing diagram for this counter given that each T flip-flop triggers on the trailing edge of a pulse applied to its clock-pulse input.

13.6 The JK flip-flop shown in Fig. P13.6 is an example of a modulo-2 counter. Cascade this counter with the modulo-5 counter given in Fig. 13.8 on p. 866 by connecting the output Q of this counter to the clock-pulse input of the modulo-5 counter. Draw the circuit diagram and the timing diagram for the resulting BCD ripple counter.

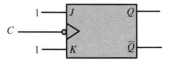

Fig. P13.6

13.7 Design a combinational logic circuit (encoder) to convert a 6-bit ring counter into a binary modulo-6 counter.

13.8 Design a combinational logic circuit to convert a 3-bit twisted-ring counter into a binary modulo-6 counter.

13.9 For the shift register shown in Fig. 13.17 on p. 871, draw the timing diagram for the case that the input is the sequence 1, 0, 1, 1, 0, 0, . . .

13.10 The device shown in Fig. P13.10 is an example of a **feedback shift register.** Find the state diagram for this shift register.

13.11 For the feedback shift register shown in Fig. P13.10, replace the exclusive-OR gate with an

exclusive-NOR gate. Find the state diagram for the resulting shift register.

13.12 (a) For the feedback shift register shown in Fig. P13.10, replace the exclusive-OR gate with an OR gate. Find the state diagram for the resulting shift register. (b) For the feedback shift register shown in Fig. P13.10, replace the exclusive-OR gate with an AND gate. Find the state diagram for the resulting shift register.

13.13 The series-in, series-out shift register and the associated combinational logic circuit shown in Fig. P13.13 are to count clock pulses as indicated by Table P13.13. Such a device can be called a **serial counter.** Design an appropriate combinational logic circuit.

13.14 Design an additional combinational logic circuit which, when combined with the serial counter given in Fig. P13.13, will produce a modulo-8 binary counter.

13.15 The serial counter shown in Fig. P13.13 produces a squence of eight distinct states as indicated by Table P13.13. Use the state diagram to find a different sequence of eight distinct states and design a corresponding combinational logic circuit.

13.16 The feedback shift register shown in Fig. P13.10 can be used to generate the words of a code. A code word is formed by taking the first seven output (Q_3) bits after a binary triple is placed in the register. Determine the eight words in the code.

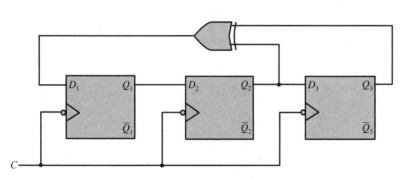

Fig. P13.10

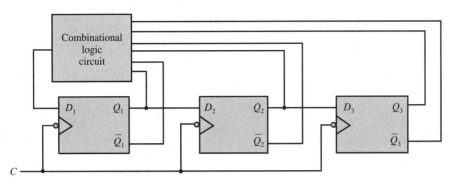

Fig. P13.13

Table P13.13

Number of Clock Pulses	State of Register		
	Q_1	Q_2	Q_3
0	0	0	0
1	1	0	0
2	0	1	0
3	1	0	1
4	1	1	0
5	1	1	1
6	0	1	1
7	0	0	1
8	0	0	0

Table P13.17

Inputs			Outputs			
A_2	A_1	A_0	B_3	B_2	B_1	B_0
0	0	0	0	1	1	0
0	0	1	0	1	0	1
0	1	0	1	0	1	0
0	1	1	0	1	1	1
1	0	0	1	0	1	1
1	0	1	1	1	0	0
1	1	0	0	0	1	1
1	1	1	1	0	0	1

13.17 Draw the circuit diagram for a diode-matrix ROM that is characterized by Table P13.17.

13.18 Draw the circuit diagram for a MOSFET ROM that is characterized by Table P13.17.

13.19 A 4K-bit ROM stores 4096 bits. Find the number of output lines given that the number of input lines is (a) 10, (b) 8, and (c) 6.

13.20 A 2K-bit ROM stores 2048 bits. Find the number of input lines given that the number of output lines is (a) 4, (b) 8, and (c) 16.

13.21 Given a 16 × 8 ROM with inputs A_0, A_1, A_2, A_3 and outputs B_0, B_1, B_2, B_3, B_4, B_5, B_6, B_7. Convert this ROM into a 32 × 4 ROM by adding an additional input line A_4 and using logic gates to form new output lines C_0, C_1, C_2, C_3 such that $C_0 = B_0$, $C_1 = B_1$, $C_2 = B_2$, $C_3 = B_3$ when $A_4 = 0$; while $C_0 = B_4$, $C_1 = B_5$, $C_2 = B_6$, $C_3 = B_7$ when $A_4 = 1$.

13.22 Draw the block diagram for a 256 × 8 (2K-bit) ROM with two-dimensional addressing, where a 64 × 32 cellular matrix is utilized.

13.23 Draw the block diagram for a 4 × 3 (12-bit) RAM that has the form of the RAM shown in Fig. 13.26 on p. 880.

13.24 Determine the number of AND gates required for the address decoder of a 4096 × 1

(4K-bit) RAM that has the form shown in (a) Fig. 13.26 on p. 880 and (b) Fig. 13.27 on p. 881. How many inputs does each AND gate have?

13.25   Each of 40 tracks of a $5\frac{1}{4}$-inch floppy disk is partitioned into 9 sectors—each containing 512 bytes (1 byte = 8 bits). Determine the information-storage content of this floppy disk. What is the information-storage content if the number of tracks is increased to 80?

13.26   Extend the waveforms ϕ_1, ϕ_2, and ϕ_3 shown in Fig. 13.35 on p. 888 and add the waveform for ϕ_4 such that the packet of charge under gate ϕ_3 is shifted to the right to the next gate labeled ϕ_1. (The device that produces waveforms ϕ_1, ϕ_2, ϕ_3, and ϕ_4 is known as a **four-phase clock**.)

13.27   For the 3-bit DAC shown in Fig. 13.37 on p. 892, suppose that $R_1 = 7R_F$, and $R_2 = 10R$. Construct a table indicating the output voltages that correspond to the eight possible 3-bit binary inputs, given that a logical 0 and a logical 1 correspond to 0 V and 1 V, respectively.

13.28   For the 3-bit DAC shown in Fig. 13.37 on p. 892, suppose that $R_1 = R_F$ and $R_2 = R$. Construct a table indicating the output voltages that correspond to the eight possible 3-bit binary inputs, given that a logical 0 has the value 0 V and a logical 1 has the value 1.5 V.

13.29   Based on the DAC shown in Fig. 13.37 on p. 892, design a 4-bit DAC whose maximum output voltage is 10 V, given that a logical 0 and a logical 1 correspond to 0 V and 1 V, respectively.

13.30   For the 3-bit DAC shown in Fig. 13.38, suppose that $R_2 = 17R_1$. Construct a table indicating the output voltages that correspond to the eight possible 3-bit binary inputs, given that a logical 0 and a logical 1 correspond to 0 V and 1 V, respectively.

13.31   Repeat Problem 13.29 for the DAC shown in Fig. 13.38 on p. 892.

13.32   The DAC shown in Fig. P13.32 is comprised of three individual DACs. For the two DACs on the left, when the input is the binary representation of the decimal digit D, the output is D volts. Express R_0 and R_1 in terms of R_F and R such that when the input of the overall DAC is the decimal number $D_1 D_0$, the output that results is $v_2 = D_1.D_0$ volts.

13.33   For the 2-bit parallel-comparator ADC shown in Fig. 13.41 on p. 895, suppose $V_r = 8$ V and the analog input voltage varies between 0 and 8 V. Determine the input-voltage interval that corresponds to a digital output $B_1 B_0$ of (a) 00, (b) 01, (c) 10, and (d) 11.

13.34   Draw a 3-bit parallel-comparator ADC. Construct the truth table relating the encoder inputs and outputs.

13.35   Given an n-bit ADC whose analog input voltage varies over a range of V_r volts, the **resolution** of the ADC is defined to be $V_r/2^n$. In terms of the percentage of V_r, the resolution is $1/2^n$. Determine the resolution of an n-bit ADC for the case that (a) $n = 2$, (b) $n = 3$, (c) $n = 4$, (d) $n = 6$, and (e) $n = 8$.

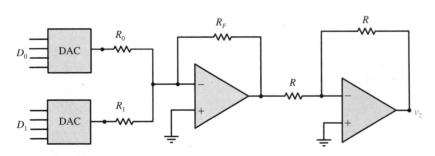

Fig. P13.32

Part IV

Electromechanics

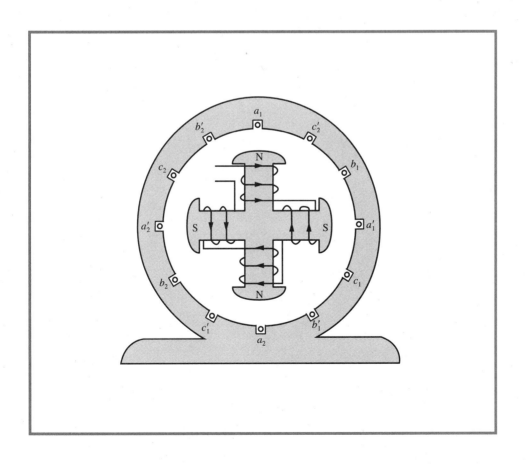

14

Electromagnetics

INTRODUCTION

Most electric power is generated and consumed by devices that employ magnetic fields. In this chapter, we discuss the fundamentals of magnetic circuits and see how electromagnetics gives rise to a device known as a "transformer." Such a device, typically a four-terminal circuit element, is obtained by placing two inductors in physical proximity.

Transformers, which are capable of efficiently transferring electric energy, come in various sizes and have a variety of uses. Relatively small transformers are used in radios and television sets to couple amplifier stages, and larger transformers are used in power-supply sections of these and numerous types of electronic equipment. Large, massive transformers are employed by electric utilities in the distribution of electric power.

14.1 Magnetic Fields

We are all familiar with **permanent magnets** made of steel or iron alloys. A magnet exerts force (attraction and/or repulsion) on other magnets and on moving electric charge (i.e., current) as well. To account for this phenomenon, we visualize a **magnetic field** around a magnet. For example, Fig. 14.1 shows a bar magnet and its associated magnetic field by means of **lines of magnetic force** or **magnetic flux** (or

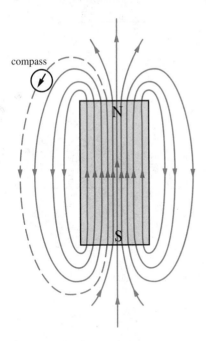

Fig. 14.1 Magnetic field around a magnet.

simply **flux**). The ends of the magnet are its **poles**. The **north pole** (labeled N in Fig. 14.1) is the pole from which the flux emanates, and the **south pole** (labeled S) is the end to which the flux returns. Since lines of magnetic flux always form closed loops, the lines of flux pass through the magnet going from the south pole to the north pole. A compass located in the magnetic field will indicate the direction of the flux line passing through the point at which the compass is placed.

The total flux passing through a given area A is denoted by ϕ, where the unit of flux is the **weber**[1] (Wb). When the flux is uniformly distributed over area A, we say that the **magnetic flux density** B is given by

$$B = \phi/A$$

and the unit of magnetic flux density is the **tesla**[2] (T). Consequently, 1 tesla = 1 weber per square meter (1 T = 1 Wb/m^2).

As flux has direction, so does flux density. The symbol B used above refers to the magnitude of the flux density, which is a vector $\mathbf{B}$. In other words, $|\mathbf{B}| = B$. With reference to Fig. 14.1, note that the flux lines nearer the magnet are more closely spaced. This is so because the strength of the magnetic field decreases with distance; that is, the magnitude of $\mathbf{B}$ decreases with distance.

---

[1]Named for the German physicist Wilhelm Weber (1804–1891).
[2]Named for the American electrician and inventor Nikola Tesla (1856–1943).

As just mentioned, a magnetic field exerts a force on moving electric charge. Suppose a charge of q coulombs has velocity **u** (a vector with magnitude and direction), where $|\mathbf{u}| = u$ is measured in meters per second (m/s). If the location of q has magnetic flux density **B**, then exerted on q is a force **f** (a vector with magnitude and direction), where the unit of $|\mathbf{f}| = f$ is the **newton**[3] (N). The magnitude of the force is given by

$$f = quB \sin \theta \qquad\qquad (14.1)$$

where θ is the angle between the vectors **u** and **B**, and this angle is measured in the plane containing **u** and **B**. The vector **f** is normal to the plane containing **u** and **B**, and the direction of **f** is obtained by the **right-hand screw rule:** If the fingers on the right hand are rotated from **u** to **B** (by an amount less than 180 degrees), then the thumb points in the direction of **f**. This concept is illustrated in Fig. 14.2, where vectors **u** and **B** are in the *x-y* plane. As a consequence of the right-hand screw rule, we see that the vector **f** is on the *z*-axis and is pointed in the positive *z* direction. The magnitude of **f** is given by Eq. 14.1.

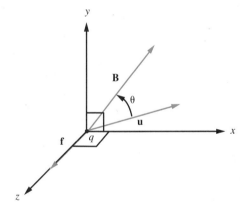

Fig. 14.2 Right-hand screw rule.

Currents and Magnetic Fields

A magnetic field not only is produced by a permanent magnet, but a current going through a conductor also gives rise to a magnetic field. In particular, consider a current i going through a long wire as depicted in the three-dimensional view shown in Fig. 14.3. We see that the lines of flux form concentric circles around the conductor, and the direction of the magnetic field is given by another **right-hand rule:** If the thumb on the right hand points in the direction of the current, then the fingers

---

[3]Named for the English scientist Sir Isaac Newton (1642–1727).

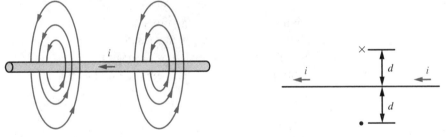

Fig. 14.3 Magnetic field around a conductor.   **Fig. 14.4** Direction of magnetic field.

curl in the direction of the magnetic field. Specifically, for the two-dimensional view shown in Fig. 14.4, a point whose distance is d (measured perpendicularly) from the wire has a flux density of magnitude B given by

$$B = \frac{\mu}{2\pi d} i \tag{14.2}$$

where the term μ is a property of the material that surrounds the conductor, and is called the **permeability** of the material. As 2π is a dimensionless constant, the units of μ are webers per ampere-meter (Wb/A-m). For the situation illustrated in Fig. 14.4, by the right-hand rule, the magnetic field at a point above the wire is perpendicular to and directed into the page (indicated by the cross $\times$), while below the wire, the field is perpendicular to and directed out of the page (indicated by the dot $\cdot$).

We have seen the type of magnetic field that results from a current-carrying straight wire. We can, however, obtain a magnetic field similar to the type produced by the bar magnet in Fig. 14.1 by winding a current-carrying wire into a coil having N turns as shown in Fig. 14.5. When it comes to a coil, the right-hand rule can be

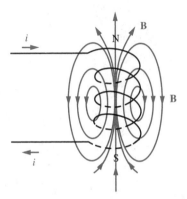

Fig. 14.5 Magnetic field for a coil.

restated for the magnetic field in the core of the coil: If the fingers on the right hand curl in the direction of the current, then the thumb points in the direction of the magnetic field inside the core. A cylindrical coil of wire, whether it has an air core, an iron core, or some other type, is called a **solenoid**. If the radius of the core is r, and the length of the cylinder is l, then when $l \gg r$, it can be shown that the flux density of the interior of the core is approximated by

$$B = \frac{\mu N}{l} i \tag{14.3}$$

Since the cross-sectional area of the core is $A = \pi r^2$, the total flux is given by

$$\phi = BA = \frac{\mu N \pi r^2}{l} i \tag{14.4}$$

From Eq. 14.4, we see that the flux in the core of a solenoid depends on the permeability μ of the core material. The permeability of free space is designated by μ_0 and has the value

$$\mu_0 = 4\pi \times 10^{-7} \ \text{Wb/A-m}$$

We define **relative permeability** μ_r to be the ratio

$$\mu_r = \mu/\mu_0$$

The relative permeability of most materials is approximately unity (examples are air and copper); however, **ferromagnetic** materials (such as iron, nickel, steel, and cobalt) have quite high relative permeabilities. A typical value for a good magnetic material is $\mu_r = 1000$ or greater. Thus we see that the flux, and hence the flux density, of a solenoid with a ferromagnetic core may be a thousand or more times greater than for an air core.

Example 14.1

For the solenoid shown in Fig. 14.5, suppose the core has a radius of 0.01 m = 1 cm, where cm is the abbreviation for centimeter, and a length of 0.2 m (20 cm). Let us determine the number of turns required for a current of 1 A to produce a magnetic flux density of 0.1 T in the core when the core material is (a) air, and (b) iron having a relative permeability of 1200.

From Eq. 14.3 we have that

$$N = \frac{Bl}{\mu i}$$

(a) For an air core, $\mu \approx \mu_0 = 4\pi \times 10^{-7}$ Wb/A-m, so

$$N = \frac{(0.1)(0.2)}{(4\pi \times 10^{-7})(1)} = 15{,}900 \text{ turns}$$

(b) For the iron core, $\mu \approx \mu_r \mu_0$, so

$$N = \frac{(0.1)(0.2)}{(1200)(4\pi \times 10^{-7})(1)} = 13.3 \text{ turns}$$

Drill Exercise 14.1

For the solenoid shown in Fig. 14.5, suppose that the core has a radius of 0.01 m and a length of 0.2 m. Find the magnetic flux in the core when the current through the coil is 0.1 A and the core material is (a) air, and (b) iron having a relative permeability of 1500.

ANSWER (a) 1.97×10^{-7} Wb, (b) 2.96×10^{-4} Wb

Suppose that instead of a cylindrical core, a coil of wire is wound on a doughnut-shaped core as shown in Fig. 14.6. For this **toroid**, by the right-hand rule, for the

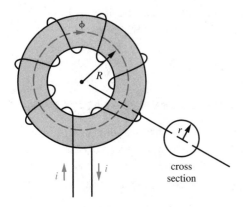

Fig. 14.6 Coil with a toroidal core.

indicated current, the direction of the magnetic flux in the core is clockwise. The expression for this flux ϕ is given by Eq. 14.4, where r is the radius of the cross-sectional area and l is the average length (or mean length) of the core. Since $l = 2\pi R$, then Eq. 14.4 becomes

$$\phi = \frac{\mu N r^2}{2R} i \tag{14.5}$$

and the magnetic flux density is

$$B = \frac{\phi}{A} = \frac{\mu N}{2\pi R} i \tag{14.6}$$

Example 14.2

The toroid shown in Fig. 14.6 consists of 500 turns wound on an iron core having relative permeability $\mu_r = 1500$, average radius $R = 0.1$ m $= 10$ cm, and cross-sectional radius $r = 0.02$ m $= 2$ cm. Let us determine the current that yields a magnetic flux density of magnitude $B = 0.5$ T in the core.

Since $\mu_r = \mu/\mu_0$, the permeability of the core is

$$\mu = \mu_r \mu_0 = 1500(4\pi \times 10^{-7}) = 6\pi \times 10^{-4} \text{ Wb/A-m}$$

Then from Eq. 14.6, we have

$$i = \frac{2\pi R B}{\mu N} = \frac{2\pi(0.1)(0.5)}{(6\pi \times 10^{-4})(500)} = 0.333 \text{ A}$$

Drill Exercise 14.2

For the toroid shown in Fig. 14.6, suppose the iron core has an average radius of 0.1 m, a cross-sectional radius of 0.02 m, and a relative permeability of 1200. Determine the number of turns required to produce a flux of 1 mWb in the core when the coil current is 1 A.

ANSWER 332

Since the cross-sectional area A of the core of a toroid need not necessarily be circular, from Eq. 14.3 and the fact that $\phi = BA$, a more general version of Eq. 14.4 is

$$\phi = \frac{\mu NA}{l} i \qquad (14.7)$$

For this equation, we know that the cross-sectional area A and average length l depend on the geometry of the core, and its permeability μ is a property of the material from which the core is constructed. We see, therefore, that for a given core, the flux ϕ is proportional to the product of the number of coil turns N and the current i. We call this product the **magnetomotive force** (**mmf**), and designate it by the symbol $\mathcal{F}$. In other words,

$$\boxed{\mathcal{F} = Ni} \qquad (14.8)$$

Although N is a dimensionless number, we shall call the units of mmf ampere-turns (A-t) to stress the typical situation of a coil with many turns.

14.2 Magnetic Circuits

Note that Eq. 14.7 can be written in the form

$$\boxed{\phi = \mathcal{F}/\mathcal{R}} \qquad (14.9)$$

where $\mathcal{R} = l/\mu A$, and this is called the **reluctance** of the path of the flux through the core. From Eq. 14.9, we have that

$$\boxed{\mathcal{F} = \mathcal{R}\phi} \qquad (14.10)$$

and from Eq. 14.10, it is apparent that the units of reluctance are ampere-turns per weber (A-t/Wb).

Example 14.3

For the toroid described in Example 14.2, we determined the current required to establish $B = 0.5$ T. That is, we found the current (0.333 A) needed to produce a flux of

$$\phi = BA = (0.5)(\pi)(0.02)^2 = 6.28 \times 10^{-4} \text{ Wb}$$

The reluctance of the path of the flux is

$$\mathcal{R} = \frac{l}{\mu A} = \frac{2\pi R}{\mu \pi r^2} = \frac{2R}{\mu r^2} = \frac{2(0.1)}{(6\pi \times 10^{-4})(0.02)^2} = 2.65 \times 10^5 \text{ A-t}\big/\text{Wb}$$

Thus the mmf is

$$\mathcal{F} = \mathcal{R}\phi = (2.65 \times 10^5)(6.28 \times 10^{-4}) = 166.5 \text{ A-t}$$

and from Eq. 14.8,

$$i = \frac{\mathcal{F}}{N} = \frac{166}{500} = 0.332 \text{ A}$$

and this agrees with our previous calculation for the current.

Drill Exercise 14.3

For the toroid shown in Fig. 14.6, suppose the iron core has an average radius of 0.1 m, a cross-sectional radius of 0.02 m, and a relative permeability of 1200. (a) Find the reluctance of the flux path. (b) Determine the mmf when the flux in the core is 1 mWb. (c) What is the corresponding current in the coil for the case that it has 332 turns?

ANSWER (a) 3.32×10^{-5} A-t$\big/$Wb, (b) 332 A-t, (c) 1 A

Although we obtained the same results, the approach taken in Example 14.3 was slightly different from the one taken in Example 14.2. Example 14.3 employed Eq. 14.10, which may remind you of Ohm's law ($v = Ri$). Drawing such an analogy, we can talk about **magnetic circuits** and employ the analogies summarized in Table 14.1.

Having established electromagnetic analogies, we can use electric circuit principles for analyzing magnetic circuits. The electric analog for the toroid given in Examples 14.2 and 14.3 is shown in Fig. 14.7.

Table 14.1 Analogies for Electric and Magnetic Circuits.

Electric	Magnetic
Current i	Flux ϕ
Voltage v	Mmf $\mathscr{F}$
Resistance R	Reluctance $\mathscr{R}$
$v = Ri$ (Ohm's law)	$\mathscr{F} = \mathscr{R}\phi$

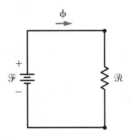

Fig. 14.7 Analogous electric circuit.

Example 14.4

Consider a rectangular iron core, which has a relative permeability of $\mu_r = 1500$, as shown in Fig. 14.8. Let us determine the reluctance and the magnetic flux in this core when a 200-turn coil has a current of 2 A.

Since the mean length of the flux path is $l = 16 + 14 + 16 + 14 = 60$ cm = 0.6 m, and the cross-sectional area is $A = (4 \text{ cm})(3 \text{ cm}) = (0.04)(0.03) = 0.0012$ m^2 = 1.2 mm^2, then the reluctance of the core is

$$\mathscr{R}_c = \frac{l}{\mu A} = \frac{0.6}{(1500)(4\pi \times 10^{-7})(1.2 \times 10^{-3})} = 2.65 \times 10^5 \text{ A-t}/\text{Wb}$$

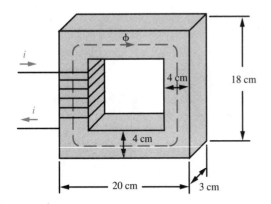

Fig. 14.8 Magnetic circuit with rectangular core.

and the magnetic flux is

$$\phi = \frac{\mathscr{F}}{\mathscr{R}_c} = \frac{Ni}{\mathscr{R}_c} = \frac{(200)(2)}{2.65 \times 10^5} = 1.51 \times 10^{-3} = 1.51 \text{ mWb}$$

The electric analog for this magnetic circuit is again as shown in Fig. 14.7.

Cores With Air Gaps

In the next chapter we will see the importance of iron cores that contain **air gaps**. Therefore, let us now consider an example of such a situation.

Example 14.5

Suppose for the iron core described in Example 14.4, a 1-cm section is removed so as to produce an **air gap** as depicted in Fig. 14.9. Although the flux that passes through the iron core also passes through the air gap, there is a **fringing** of the flux as illustrated in Fig. 14.9. To account for the effect of this fringing in short gaps, traditionally the dimensions of the gap's cross-sectional area are increased by the gap length l_g. In particular, to account for fringing, the effective cross-sectional area is

$$A_g = (4 + 1)(3 + 1) = 20 \text{ cm}^2 = 2 \text{ mm}^2$$

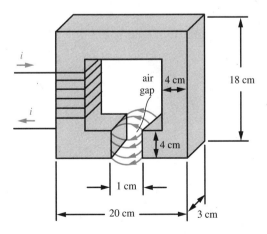

Fig. 14.9 Rectangular core with an air gap.

so the reluctance of the air gap is

$$\mathcal{R}_g = \frac{l_g}{\mu_0 A_g} = \frac{0.01}{(4\pi \times 10^{-7})(2 \times 10^{-3})} = 3.98 \times 10^6 \text{ A-t}/\text{Wb}$$

Since the mean length of the flux path through the iron core material is $l_c = l - 1 = 59$ cm $= 0.59$ m, then the reluctance of the core is

$$\mathcal{R}_c = \frac{l_c}{\mu A} = \frac{0.59}{(1500)(4\pi \times 10^{-7})(1.2 \times 10^{-3})} = 2.61 \times 10^5 \text{ A-t}/\text{Wb}$$

These values illustrate a common occurrence—the reluctance of the air gap is much greater than the reluctance of the iron core. Specifically, in this example, $\mathcal{R}_g$ is more than 15 times larger than $\mathcal{R}_c$.

The electric analog for the magnetic circuit is shown in Fig. 14.10. Since the iron core and the air gap are in series, the total reluctance is

$$\mathcal{R} = \mathcal{R}_c + \mathcal{R}_g = 2.61 \times 10^5 + 3.98 \times 10^6 = 4.24 \times 10^6 \text{ A-t}/\text{Wb}$$

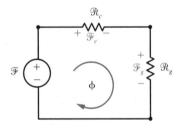

Fig. 14.10 Effect of air gap on analogous electric circuit.

and we see that most of the reluctance is attributable to the air gap. Furthermore,

$$\mathcal{F} = \mathcal{F}_c + \mathcal{F}_g$$

where $\mathcal{F}_c = \mathcal{R}_c \phi$ and $\mathcal{F}_g = \mathcal{R}_g \phi$.

To obtain the same flux as was obtained for the core without an air gap (Fig. 14.8), from

$$\mathcal{F} = Ni = \mathcal{R}\phi$$

we have that

$$i = \frac{\mathcal{R}\phi}{N} = \frac{(4.24 \times 10^6)(1.51 \times 10^{-3})}{200} = 32 \text{ A}$$

In other words, the introduction of the air gap necessitates 16 times the current required to produce the same flux that was obtained without an air gap.

Drill Exercise 14.5

Suppose that the air gap for the rectangular core described in Example 14.5 is reduced from 1 cm to 0.5 cm. Determine the current that results in the same magnetic flux in the core.

ANSWER 21.1 A

We now must point out that our previous magnetic-circuit calculations were by no means exact. One of the reasons for this is the fringing of the flux due to the presence of an air gap. There are, however, additional reasons. More of the flux is concentrated in the inner portion of a core than in the outer portion, as illustrated in the side view of a core depicted in Fig. 14.11. Some of the flux even may take a short-cut through the air and not reach the air gap. Such **leakage flux** is also shown in Fig. 14.11. Furthermore, the permeability, and hence the reluctance, of ferromagnetic core material varies with the flux through it. Because of this inherent nonlinearity, the analysis of magnetic circuits often employs graphical information. Let us see how.

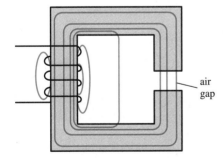

Fig. 14.11 Flux associated with a rectangular core.

Magnetization Curves

As magnetic flux density $B = \phi/A$, and because $\phi = \mathcal{F}/\mathcal{R}$, then since reluctance $\mathcal{R} = l/\mu A$, we have that

$$B = \frac{\mathcal{F}}{\mathcal{R}A} = \frac{\mathcal{F}}{(l/\mu A)A} = \frac{\mu \mathcal{F}}{l}$$

Thus, we can write

$$B = \mu H \tag{14.11}$$

where

$$H = \frac{\mathcal{F}}{l} = \frac{Ni}{l} \tag{14.12}$$

Hence we see that H, called the **magnetic field intensity** or **magnetization force**, has as units ampere-turns per meter (A-t$/$m) and is independent of core material (it depends on the number of turns, the current, and the mean length).

A plot of magnetic flux density B versus magnetic field intensity H is called a **magnetization curve** or **B-H curve**, and graphically describes the average relationship between B and H. (Such curves are typically supplied by manufacturers of magnetic materials.) Figure 14.12 shows magnetization curves for cast iron, cast steel, and sheet steel. Note that for values of B less than 1 T for steel and 0.5 T for cast iron, the curves are approximately straight lines, so that the relationship $B = \mu H$ is essentially a linear one. (Remember that the relationship for nonmagnetic material, e.g., air, is given by $B = \mu_0 H$ and, therefore, is linear.) It is for these reasons that the previously discussed magnetic-circuit analysis is reasonably accurate.

Example 14.6

Suppose the toroid given in Fig. 14.6 on p. 908 is constructed of sheet steel, and suppose the magnetic flux density is to be $B = 1$ T. Let us determine (a) the relative permeability of this core, and (b) the current required to produce the magnetic flux density when the average radius is 0.1 m, the cross-sectional radius is 0.02 m, and the coil has 500 turns.

(a) From the magnetization curve given in Fig. 14.12, we see that the field intensity is $H = 200$ A-t$/$m. From Eq. 14.11,

$$\mu = \frac{B}{H} = \frac{1}{200} = 5 \times 10^{-3} \text{ Wb}/\text{A-m}$$

and, therefore, the relative permeability is

$$\mu_r = \frac{\mu}{\mu_0} = \frac{5 \times 10^{-3}}{4\pi \times 10^{-7}} = 3980$$

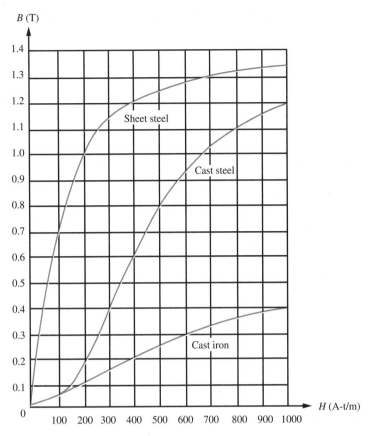

Fig. 14.12 Magnetization curves.

(b) From Eq. 14.12, the mmf is

$$\mathcal{F} = Hl = 200(2\pi)(0.1) = 1256 \text{ A-t}$$

and since $\mathcal{F} = Ni$, the current required to produce the desired magnetic field intensity is

$$i = \frac{\mathcal{F}}{N} = \frac{1256}{500} = 0.252 \text{ A}$$

Drill Exercise 14.6

Suppose the toroid shown in Fig. 14.6 on p. 908 is constructed of cast steel and is to have a magnetic flux density of 0.8 T. (a) Determine the relative permeability of the core. (b) If the core has an average radius of 0.1 m and a cross-sectional radius of 0.02 m, find the current required for a 500-turn coil.

ANSWER (a) 1273, (b) 0.628 A

Having done an example of a core with no air gaps, let us now consider the case of a core with an air gap.

Example 14.7

Suppose the rectangular core given in Fig. 14.9 is cast iron. Let us determine the current required in a 200-turn coil to establish a flux of 0.24 mWb in the iron.
The magnetic flux density in the core is

$$B = \frac{\phi}{A} = \frac{2.4 \times 10^{-4}}{(0.03)(0.04)} = 0.2 \text{ T}$$

From the magnetization curve for cast iron (Fig. 14.12), the magnetic field intensity is $H_c = 400$ A-t/m. Thus the mmf for the iron is

$$\mathcal{F}_c = H_c l_c = 400(0.59) = 236 \text{ A-t}$$

For the air gap, we have already calculated in Example 14.5 that the reluctance is $\mathcal{R}_g = 3.98 \times 10^6$ A-t/Wb. Thus the mmf for the air gap is

$$\mathcal{F}_g = \mathcal{R}_g \phi = (3.98 \times 10^6)(2.4 \times 10^{-4}) = 955 \text{ A-t}$$

Therefore, the total mmf is

$$\mathcal{F} = \mathcal{F}_c + \mathcal{F}_g = 236 + 955 = 1191 \text{ A-t}$$

and the required current is

$$i = \frac{\mathcal{F}}{N} = \frac{1191}{200} = 5.96 \text{ A}$$

Drill Exercise 14.7

Suppose the rectangular core given in Fig. 14.9 on p. 913 is made of cast steel. Determine the number of coil turns required to produce a magnetic flux of 0.48 mWb when the coil current is 7 A.

ANSWER   298

Hysteresis

Not only are magnetic-circuit calculations approximate for the reasons already mentioned, but the magnetic history of magnetic materials is also a factor. Let us see why.

A **demagnetized** magnetic material is one whose magnetic flux density is zero (i.e., $B = 0$ T) when no external magnetizing force is applied (i.e., $H = 0$ A-t$/$m). If a magnetizing force is applied to a demagnetized magnetic material, then the relationship between the magnetic flux density B that results and the magnetizing force H is given by the magnetization curve of the material. A typical B-H curve is shown in Fig. 14.13a. The arrowheads indicate the direction that a point on the curve moves as magnetizing force increases. For smaller values of H, the B-H curve is relatively straight. As H increases, the curve bends and then flattens out so that a further increase in H results in only a slight increase in B. Eventually, an increase in H yields no further increase in B. The material is then said to be **saturated**. If, after the material is saturated, H is reduced to zero, B will decrease, but the original magnetization curve does not describe how B decreases. Instead, the resulting behavior is as illustrated in Fig. 14.13b. The value of the magnetic flux density obtained

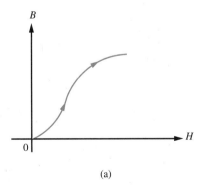

(a)

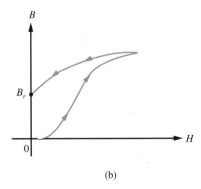

(b)

Fig. 14.13 Typical magnetization behavior.

when H is reduced to zero is $B = B_r$. This value is the **residual magnetism** of the material. (Permanent magnets are made from magnetic materials with very large values of residual magnetism, e.g., 1 T.) If H is increased again, the material eventually will saturate, but the path taken will not be along the one by which point B_r was reached. If H is made increasingly negative, there will be some value $H = -H_c$ for which the flux density is $B = 0$ T (see Fig. 14.14). The value H_c is called the **coercive force** of the material. (Permanent magnets have a large coercive force, e.g., 10^4 A-t/m.) If H is made negative enough, the material will eventually saturate, but this time in the opposite direction. If H is then increased and made large enough, the magnetic material will again saturate in the positive direction as indicated in Fig. 14.14. The resulting loop, shown with solid lines, is called a **hysteresis loop**, and the corresponding phenomenon is known as **hysteresis**. If the magnetizing force applied to the demagnetized material is less than that required to produce saturation, a hysteresis loop such as the dashed loop in Fig. 14.14 may be obtained.

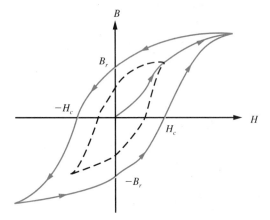

Fig. 14.14 Hysteresis loop.

Certain magnetic materials have hysteresis loops that are approximately rectangular. Such a material, called a **square-loop material**, is depicted in Fig. 14.15. Note that the slopes of the sides of this hysteresis loop are quite large. This means that a small change in H can produce a large change in B. Small cores made from such a material are used as binary memory devices in switching circuits and digital computers.

When magnetic material is periodically magnetized and demagnetized (e.g., due to an alternating current), power, and hence energy, is absorbed by the material and is converted to heat. An empirical formula for this average hysteresis power loss P (in watts) was developed by electrical engineer Charles P. Steinmetz (1865–1923), and is

$$P_h = K_h f B_m^n \tag{14.13}$$

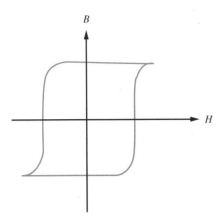

Fig. 14.15 Square hysteresis loop.

where the constant K_h and the value of n (called the **Steinmetz exponent**[4]) depend on the core material, B_m is the maximum flux density, and f is the frequency of the magnetic variations.

14.3 Transformers

Electric energy can be transferred efficiently with the use of a "transformer." Although such a device, which is an example of a magnetic circuit, typically consists of two multiturn coils wound on a common core, more than two windings may also be used. Transformer applications are varied and extremely important.

One major use of transformers is in power distribution. Transformers possess the ability to "step up" and "step down" ac voltages and currents. Because of this, electric power, which is conveniently generated typically from 5 to 50 kV, can be stepped up in voltage and stepped down in current prior to distribution. The transmission of power at high voltages and relatively low currents means less of an RI^2 power loss in the transmission lines and, consequently, greater efficiency. Typically the transmission of power over long distances occurs at 200 to 400 kV, and even may be at voltages up to 1000 kV. Residential-area power lines typically range between 2 and 10 kV. After power is distributed to the load ends, transformers step down voltages typically to 120-V, 240-V levels for ordinary consumption.

In addition to power systems, transformers are commonly used in electronic, control, and communication circuits and systems. Small power transformers are employed for power supplies in various electronic devices and instruments. Transformers are often used for "impedance matching," to accomplish maximum power transfer between source and load. Since they can be employed to couple circuits

---

[4]A typical value of n is 1.6.

(e.g., amplifier stages) without requiring a conducting electrical connection, transformers are also used to insulate one circuit from another, but at the same time they allow for the transfer of electric energy.

Magnetic Induction

It was Michael Faraday who first announced in 1831 that magnetic flux passing through a multiturn coil would induce a voltage, called an **electromotive force** (emf), across the coil, provided the flux was time varying. If the same flux ϕ passes through, or **links**, all the N turns of a coil, then **Faraday's law** says that the emf e induced across the coil is given by

$$e = N\frac{d\phi}{dt}$$ (14.14)

where, or course, the units of e are volts. Let us designate the product of the flux ϕ linking the coil and the number of turns of the coil N by λ, that is, $\lambda = N\phi$. We call λ the number of **flux linkages**, and although N is a dimensionless number, we shall call the units of flux linkages **weber-turns** (Wb-t) to stress the typical situation of a coil with many turns. As a consequence of the definition of λ, we can rewrite Eq. 14.14 as

$$e = \frac{d\lambda}{dt}$$ (14.15)

Suppose an L-henry inductor is made with a coil of wire. We know that the voltage v across it and the current i through it are related by

$$v = L\frac{di}{dt}$$ (14.16)

Thus for the given inductor (coil), from Eq. 14.15 and Eq. 14.16, we have

$$L\frac{di}{dt} = \frac{d\lambda}{dt} \quad \Rightarrow \quad L\frac{di}{dt}\frac{dt}{di} = \frac{d\lambda}{dt}\frac{dt}{di}$$

and by the chain rule of calculus, since $\lambda = N\phi$, then

$$L = \frac{d\lambda}{di} = N\frac{d\phi}{di}$$ (14.17)

For the case that the core of a coil is well below saturation, we can utilize Eq. 14.8 and Eq. 14.10 to deduce that

$$\phi = \frac{Ni}{\mathscr{R}} = Ki \tag{14.18}$$

In other words, the flux is directly proportional to the current (see Eq. 14.4, Eq. 14.5, and Eq. 14.7, also). Taking the derivative with respect to i of both sides of Eq. 14.18 yields $d\phi/di = K$. But also from Eq. 14.18, $K = \phi/i$. Thus $d\phi/di = \phi/i$. Substituting this fact into Eq. 14.17 gives

$$L = \frac{N\phi}{i} = \frac{\lambda}{i} \tag{14.19}$$

Thus inductance is the number of flux linkages per ampere, and we see that 1 H = 1 Wb/A. Previously we saw that the permeability μ of a material was measured in webers/ampere-meter. Now we see that we can alternatively measure permeability in henries/meter (H/m). Actually, H/m is the SI unit for permeability.

Note also from Eq. 14.18 and Eq. 14.19 that

$$L = \frac{N(Ni/\mathscr{R})}{i} \quad \Rightarrow \quad \boxed{L = \frac{N^2}{\mathscr{R}}} \tag{14.20}$$

In other words, inductance is directly proportional to the square of the number of turns.

Example 14.8

Let us determine the inductance of a 500-turn coil wound on the toroidal core shown in Fig. 14.6 on p. 908 when the core has a relative permeability of 1500, an average radius of 0.1 m, and a cross-sectional radius of 0.02 m.

In Example 14.3 it was determined that the reluctance of the core is 2.65×10^5 A-t/Wb. Therefore, the inductance of the coil is

$$L = \frac{N^2}{\mathscr{R}} = \frac{500^2}{2.65 \times 10^5} = 0.943 = 943 \text{ mH}$$

Lenz's Law

Although the value of an induced emf can be obtained by using either Eq. 14.14 or Eq. 14.15, the polarity of this voltage is determined from **Lenz's law**[5] which states:

> **The polarity of the voltage induced by a changing flux tends to oppose the change in flux that produced the induced voltage.**

To demonstrate Lenz's law, consider the core shown in Fig. 14.16. Suppose that a time-varying voltage source v is applied to the left coil and the resulting current is i as illustrated. By the right-hand rule, if at some instant of time i has a positive value, the resulting magnetic flux ϕ is clockwise as shown. If i is increasing with time, then ϕ is increasing with time since $Ni = \mathcal{R}\phi$. Under this circumstance, the coil on the right will tend to have a current in the direction shown as this would produce a counterclockwise flux that opposes the increasing flux ϕ. If nothing is connected to terminals a and b of the coil on the right, then there is no current through this coil. Whether or not there is a current through this coil, there is a voltage v_{ab} induced across it. The polarity of this voltage is easily determined by assuming

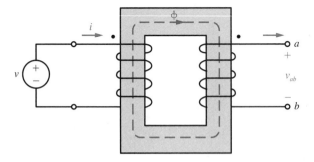

Fig. 14.16 Magnetically coupled coils.

---

[5]Named for the German physicist Heinrich Lenz (1804–1865).

that a resistor is connected to terminals a and b. The resulting current in the coil yields the polarity of v_{ab} (a positive quantity) as shown. Note that when i is increasing with time, since $v = L \, di/dt$, then v is also a positive quantity.

If i is decreasing instead of increasing, then the coil on the right tends to have a current in the direction that is opposite to the one shown, and the polarity of v_{ab} is reversed. Note also that if i is decreasing with time, then v is a negative quantity since $v = L \, di/dt$.

The polarity of the voltage induced across the coil on the right, of course, depends upon how that coil is wound with respect to the coil on the left. One way to indicate such a relationship is with **dots**. The dots at the ends of the coils in Fig. 14.16 indicate that the dotted ends are positive at the same time and, of course, are negative at the same time. Our discussion above confirms this fact. Had we been so inclined, we could have placed both dots at the other ends of their respective coils—their significance remains the same.

Magnetic Coupling

For the rectangular core shown in Fig. 14.16, suppose the coil on the left has N_1 turns and inductance L_1, and the coil on the right has N_2 turns and inductance L_2. Since a time-varying current going through the coil on the left induces a voltage across the coil on the right, we say that the coils are **magnetically coupled**. In particular, when there is no load connected to terminals a and b (as depicted in Fig. 14.16), since from Eq. 14.18,

$$\phi = \frac{N_1 i}{\mathcal{R}}$$

then by using Eq. 14.14, when there is no flux leakage, we find that the induced voltage v_{ab} is

$$v_{ab} = N_2 \frac{d\phi}{dt} = \frac{N_1 N_2}{\mathcal{R}} \frac{di}{dt} = M \frac{di}{dt} \tag{14.21}$$

where $M = N_1 N_2/\mathcal{R}$ is a positive number. Since Eq. 14.21 has the form $v = L \, di/dt$, we deduce that M is an inductance, and we call it a **mutual inductance**. The inductances L_1 and L_2 are referred to as **self-inductances**. As for the case of self-inductance, the unit for mutual inductance is the henry.

From Eq. 14.21, we see that if i is increasing with time, then v_{ab} is positive; if i is decreasing with time, then v_{ab} is negative. In accordance with our previous discussion, these are the correct values of v_{ab} for the polarity indicated in Fig. 14.16.

As a consequence of these facts, we can make a general statement concerning dots: If a current i goes into the dotted end of one coil, then for the other coil, the polarity of the induced voltage $M\,di/dt$ is plus (+) at the dotted end. Conversely, if a current i comes out of the dotted end of one coil, then the polarity of the induced voltage $M\,di/dt$ is minus (−) at the dotted end of the other coil.

The pair of magnetically coupled coils shown in Fig. 14.16 is an example of a **transformer**. One of the coils, say the coil on the left, is designated as the **primary** winding or coil, and the other is the **secondary** winding or coil. A circuit symbol for a transformer is shown in Fig. 14.17a. In general, there may be connections to both windings, so i_1 and i_2 both can be nonzero. As a result, the current i_1 going through the self-inductance L_1 produces a voltage $L_1\,di_1/dt$. The current i_2 going through the secondary, however, induces the voltage $M_{12}\,di_2/dt$ across the primary as well. By the principle of superposition, we have

$$v_1 = L_1\frac{di_1}{dt} + M_{12}\frac{di_2}{dt} \tag{14.22}$$

where the mutual inductance that corresponds to the voltage induced across the primary winding, due to the current in the secondary winding, is M_{12}. In a similar manner, we can obtain

$$v_2 = M_{21}\frac{di_1}{dt} + L_2\frac{di_2}{dt} \tag{14.23}$$

where the mutual inductance that corresponds to the voltage induced across the secondary, due to the current in the primary, is M_{21}.

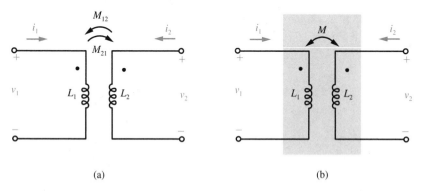

(a)   (b)

Fig. 14.17 Transformer circuit symbol.

The reasoning used to derive Eq. 14.21 suggests that $M = N_1 N_2/\mathcal{R} = M_{12} = M_{21}$. Thus we will denote the mutual inductance of a transformer by M, and we can represent a transformer as shown in Fig. 14.17*b*. Furthermore, Eq. 14.22 and Eq. 14.23 can be rewritten as the following general pair of equations describing a transformer:

$$v_1 = L_1 \frac{di_1}{dt} + M \frac{di_2}{dt} \quad \text{and} \quad v_2 = M \frac{di_1}{dt} + L_2 \frac{di_2}{dt} \tag{14.24}$$

In addition, the energy stored in the transformer can be shown to be

$$w(t) = \frac{1}{2} L_1 i_1^2(t) + \frac{1}{2} L_2 i_2^2(t) + M i_1(t) i_2(t) \tag{14.25}$$

The use of transformers is limited to non-dc applications, since for the dc case, the pair of Eq. 14.24 indicates that the primary and secondary windings behave as short circuits. Just as the time-domain expression $v = L \, di/dt$ for an inductor corresponds to the phasor relationship $\mathbf{V} = j\omega L \mathbf{I}$ for the sinusoidal case, the frequency-domain version of Equation-pair 14.24 is

$$\mathbf{V}_1 = j\omega L_1 \mathbf{I}_1 + j\omega M \mathbf{I}_2 \quad \text{and} \quad \mathbf{V}_2 = j\omega M \mathbf{I}_1 + j\omega L_2 \mathbf{I}_2 \tag{14.26}$$

Example 14.9

Let us find $v_C(t)$ for the ac circuit given in Fig. 14.18 when $v(t) = 36 \cos(3t - 60°)$ V.

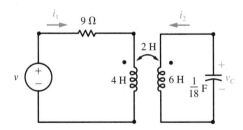

Fig. 14.18 Time-domain transformer circuit.

Shown in Fig. 14.19 is the representation of this circuit in the frequency domain. By KVL, for mesh $\mathbf{I}_1$,

$$\mathbf{V} = 9\mathbf{I}_1 + j12\mathbf{I}_1 + j6\mathbf{I}_2 = (9 + j12)\mathbf{I}_1 + j6\mathbf{I}_2 \tag{14.27}$$

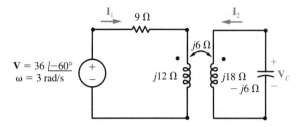

Fig. 14.19 Frequency-domain transformer circuit.

whereas for mesh $\mathbf{I}_2$, by KVL,

$$0 = j6\mathbf{I}_1 + j18\mathbf{I}_2 - j6\mathbf{I}_2 \quad \Rightarrow \quad \mathbf{I}_1 = -2\mathbf{I}_2$$

Substituting this fact in Eq. 14.27 yields

$$\mathbf{V} = -2(9 + j12)\mathbf{I}_2 + j6\mathbf{I}_2 = -18(1 + j)\mathbf{I}_2 \quad \Rightarrow \quad \mathbf{I}_2 = \frac{\mathbf{V}}{-18(1 + j)}$$

Thus

$$\mathbf{V}_C = \mathbf{Z}_C(-\mathbf{I}_2) = (-j6)(-\mathbf{I}_2) = \frac{-j6\mathbf{V}}{18(1 + j)}$$

$$= \frac{(6\underline{/-90°})(36\underline{/-60°})}{(18\underline{/0°})(\sqrt{2}\underline{/45°})} = 6\sqrt{2}\underline{/-195°} = 6\sqrt{2}\underline{/165°} \text{ V}$$

Hence

$$v_C(t) = 6\sqrt{2} \cos(3t + 165°) \text{ V}$$

Drill Exercise 14.9

For the transformer circuit shown in Fig. 14.18, suppose that $v(t) = 36 \cos(3t - 60°)$ V. Find (a) $i_1(t)$, (b) $i_2(t)$, and (c) the voltage $v_1(t)$ across the primary winding of the transformer.

ANSWER (a) $2\sqrt{2} \cos(3t - 105°)$ A, (b) $\sqrt{2} \cos(3t + 75°)$ A, and (c) $18\sqrt{2} \cos(3t - 15°)$ V

Although a transformer is basically a pair of inductors (coils) that are magnetically coupled, when a primary terminal and a secondary terminal are connected as shown in Fig. 14.20a, we can model the transformer by the connection of three (uncoupled) inductors shown in Fig. 14.20b. To verify that the circuit given in Fig. 14.20b is equivalent to the one in Fig. 14.20a, simply apply KVL to the transformer in Fig. 14.20a and its "T-equivalent" shown in Fig. 14.20b. The result is the Equation-pair 14.24.

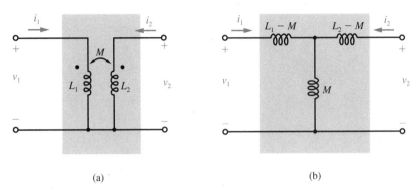

(a) (b)

Fig. 14.20 Transformer and its model.

As with all magnetic circuits, there are certain losses in the cores of transformers. We have already discussed one type of loss—that due to hysteresis. The other type is attributable to eddy currents. Since a magnetic core, such as iron, is a conductor, then a time-varying magnetic flux will induce a voltage that produces circulating currents, called **eddy currents**, in the core. The eddy current depicted in Fig. 14.21a causes an RI^2 power loss that results in core heating. An expression for this eddy-current power loss is

$$P_e = K_e f^2 B_m^2$$

where B_m is the maximum flux density, f is the frequency of the magnetic variations, and the constant K_e depends on the core material. To reduce eddy-current power

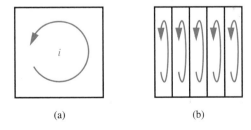

(a) (b)

Fig. 14.21 Eddy currents.

loss, instead of using solid magnetic material, cores are constructed from thin sheets or **laminations** of the material, as illustrated in Fig. 14.21*b*. The laminations,[6] which are typically insulated from each other with thin layers of varnish, have increased resistances and decreased induced voltages. The result is a significant reduction in eddy-current loss.

In addition to transformers, hysteresis and eddy-current losses, collectively known as **core loss**, occur in ac and dc machines. The expression for core loss *P* is given by

$$P = P_h + P_e = K_h f B_m^n + K_e f^2 B_m^2$$

14.4 The Ideal Transformer

The behavior of many iron-core transformers, especially those used in power systems, can be "idealized." This results in simplified voltage-current relationships and circuit models. These, then, can be employed to analyze circuits and obtain very accurate results. Therefore, we now proceed to develop and investigate the concept of an "ideal transformer."

Coupling Coefficient

Since the energy stored in the transformer shown in Fig. 14.17*b* on p. 926 is

$$w = \tfrac{1}{2} L_1 i_1^2 + \tfrac{1}{2} L_2 i_2^2 + M i_1 i_2 = \tfrac{1}{2}[L_1 i_1^2 + L_2 i_2^2] + M i_1 i_2$$

then by completing the square, we get

$$w = \tfrac{1}{2}[(\sqrt{L_1} i_1 + \sqrt{L_2} i_2)^2 - 2\sqrt{L_1 L_2} i_1 i_2] + M i_1 i_2$$

$$= \tfrac{1}{2}(\sqrt{L_1} i_1 + \sqrt{L_2} i_2)^2 + (M - \sqrt{L_1 L_2}) i_1 i_2$$

Consider the case for which

$$\sqrt{L_1} i_1 + \sqrt{L_2} i_2 = 0 \quad \Rightarrow \quad i_2 = -\left(\sqrt{\frac{L_1}{L_2}}\right) i_1$$

Then the resulting energy stored is given by

---

[6]A typical thickness for some types of low-frequency transformers is 0.0356 cm.

$$w = (M - \sqrt{L_1 L_2})i_1\left(-\sqrt{\frac{L_1}{L_2}}\right)i_1 = (\sqrt{L_1 L_2} - M)\left(\sqrt{\frac{L_1}{L_2}}\right)i_1^2$$

Since the energy stored in a transformer is a nonnegative quantity, we must have that

$$0 \le \sqrt{L_1 L_2} - M \qquad \Rightarrow \qquad \boxed{M \le \sqrt{L_1 L_2}} \tag{14.28}$$

Thus we have an upper bound for the mutual inductance M.

If we define the **coefficient of coupling** of a transformer, denoted by k, as

$$\boxed{k = \frac{M}{\sqrt{L_1 L_2}}} \tag{14.29}$$

then by Inequality 14.28 we have that $0 \le k \le 1$.

The coupling coefficient k of a physical transformer is determined by a number of factors, namely, the permeability of the core on which the primary and secondary coils are wound, the number of turns of each coil, and the coils' relative positions and their physical dimensions. If k is close to zero, we say that the coils are **loosely coupled,** and if k is close to one, we say that they are **tightly coupled.** Air-core transformers are typically loosely coupled, and iron-core transformers are usually tightly coupled. As a matter of fact, iron-core transformers can have coupling coefficients approaching unity.

Perfect Coupling

Consider the case of a transformer having coupling coefficient $k = 1$. This is known as **perfect coupling**. We represent such a device as shown in Fig. 14.22. We have that

$$\mathbf{V}_1 = j\omega L_1 \mathbf{I}_1 + j\omega M \mathbf{I}_2 \qquad \Rightarrow \qquad \mathbf{I}_1 = \frac{\mathbf{V}_1 - j\omega M \mathbf{I}_2}{j\omega L_1} \tag{14.30}$$

Also

$$\mathbf{V}_2 = j\omega M \mathbf{I}_1 + j\omega L_2 \mathbf{I}_2 = j\omega M\left(\frac{\mathbf{V}_1 - j\omega M \mathbf{I}_2}{j\omega L_1}\right) + j\omega L_2 \mathbf{I}_2$$

$$= \frac{M \mathbf{V}_1}{L_1} - \frac{j\omega M^2 \mathbf{I}_2}{L_1} + j\omega L_2 \mathbf{I}_2$$

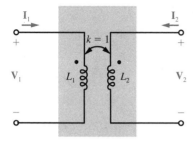

Fig. 14.22 Transformer with perfect coupling ($k = 1$).

But for perfect coupling ($k = 1$), $M = \sqrt{L_1 L_2}$ so that

$$\mathbf{V}_1 = \frac{\sqrt{L_2 L_2}\mathbf{V}_1}{L_1} - \frac{j\omega L_1 L_2 \mathbf{I}_2}{L_1} + j\omega L_2 \mathbf{I}_2 = \left(\sqrt{\frac{L_2}{L_1}}\right)\mathbf{V}_1 = N\mathbf{V}_1$$

where $N = \sqrt{\dfrac{L_2}{L_1}}$

From Eq. 14.20, we know that the inductance of a coil is proportional to the square of the number of turns. Therefore, the primary and secondary self-inductances, respectively, are

$$L_1 = N_1^2/\mathcal{R} \qquad \text{and} \qquad L_2 = N_2^2/\mathcal{R}$$

Since

$$N = \sqrt{\frac{L_2}{L_1}} = \sqrt{\frac{N_2^2/\mathcal{R}}{N_1^2/\mathcal{R}}} = \frac{N_2}{N_1}$$

so we call N the **turns ratio** of the transformer.

For the case of perfect coupling, from Eq. 14.30,

$$\mathbf{I}_1 = \frac{\mathbf{V}_1}{j\omega L_1} - \frac{M\mathbf{I}_2}{L_1} = \frac{\mathbf{V}_1}{j\omega L_1} - \frac{\sqrt{L_1 L_2}\mathbf{I}_2}{L_1} = \frac{\mathbf{V}_1}{j\omega L_1} - \sqrt{\frac{L_2}{L_1}}\mathbf{I}_2 \tag{14.31}$$

A transformer with perfect coupling is said to be **ideal** if L_1 and L_2 approach infinity such that the turns ratio remains constant. For an ideal transformer, Eq. 14.31 becomes

$$\mathbf{I}_1 = -N\mathbf{I}_2 \qquad \text{or} \qquad \mathbf{I}_2 = \frac{-\mathbf{I}_1}{N}$$

Since good coupling is achieved by transformers with iron cores, an ideal transformer is usually represented as shown in Fig. 14.23. Since the relationships between the voltages and the currents are given by the pair

$$\mathbf{V}_2 = N\mathbf{V}_1 \qquad \text{and} \qquad \mathbf{I}_2 = -\mathbf{I}_1 / N$$

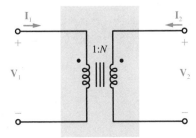

Fig. 14.23 Ideal transformer.

we can model an ideal transformer by either of the two equivalent circuits shown in Figs. 14.24 and 14.25. For the circuit given in Fig. 14.24, clearly, we have that $\mathbf{V}_2 = N\mathbf{V}_1$ and $\mathbf{I}_1 = -N\mathbf{I}_2$, whereas for the circuit depicted in Fig. 14.25, $\mathbf{V}_1 = \mathbf{V}_2 / N$ and $\mathbf{I}_2 = -\mathbf{I}_1 / N$.

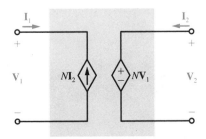

Fig. 14.24 Ideal-transformer model.

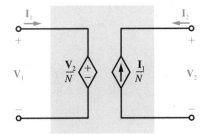

Fig. 14.25 Alternative ideal-transformer model.

The instantaneous power absorbed by the primary of an ideal transformer is $p_1 = v_1 i_1$, and that absorbed by the secondary is $p_2 = v_2 i_2$. Thus the total instantaneous power absorbed by the transformer is

$$p = p_1 + p_2 = v_1 i_1 + v_2 i_2 = v_1 i_1 + (N v_1)\left(\frac{-i_1}{N}\right) = 0 \text{ W}$$

Since the instantaneous power absorbed is zero, so is the average power and the energy stored. This fact can be confirmed from the formula for the energy stored in a transformer (Eq. 14.25) by using the relationships for perfect coupling ($M = \sqrt{L_1 L_2}$) and turns ratio ($N = \sqrt{L_2/L_1}$), along with the ideal transformer formulas. Consequently, we see that the ideal transformer is a lossless device.

Example 14.10

Figure 14.26 depicts the equivalent circuit of a simple "class-A" transistor power amplifier. Let us determine the voltage gain and the power gain for this circuit.
By KVL,

$$\mathbf{V}_g = 750\mathbf{I} + 50(21\mathbf{I}) = 1800\mathbf{I} \quad \Rightarrow \quad \mathbf{I} = \frac{\mathbf{V}_g}{1800}$$

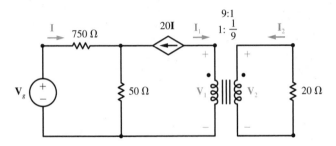

Fig. 14.26 Equivalent circuit of a power amplifier.

Since

$$\mathbf{I}_1 = -20\mathbf{I} = -20\left(\frac{\mathbf{V}_g}{1800}\right) = \frac{-\mathbf{V}_g}{90}$$

then

$$\mathbf{I}_2 = -\frac{\mathbf{I}_1}{1/9} = -9\mathbf{I}_1 = -9\left(-\frac{\mathbf{V}_g}{90}\right) = \frac{\mathbf{V}_g}{10}$$

Thus

$$\mathbf{V}_2 = -20\mathbf{I}_2 = -20\left(\frac{\mathbf{V}_g}{10}\right) = -2\mathbf{V}_g \quad \Rightarrow \quad \frac{\mathbf{V}_2}{\mathbf{V}_g} = -2$$

where V_2/V_g is the voltage transfer function. Hence the voltage gain is $|V_2/V_g| = 2$. The instantaneous power supplied by the independent voltage source is

$$p_g = v_g i = v_g\left(\frac{v_g}{1800}\right) = \frac{v_g^2}{1800}$$

whereas the instantaneous power absorbed by the 20-Ω load resistor is

$$p_2 = \frac{v_2^2}{20} = \frac{(-2v_g)^2}{20} = \frac{v_g^2}{5}$$

Thus the **power gain** p_2/p_g is

$$\frac{p_2}{p_g} = \frac{v_g^2/5}{v_g^2/1800} = 360$$

Drill Exercise 14.10

Consider the transformer circuit given in Fig. 14.26. (a) Find the impedance loading the voltage source. (b) Find the current gain $|I_2/I|$. (c) What is the product of the current gain and the voltage gain?

ANSWER   (a) 1.8 kΩ, (b) 180, (c) 360

Impedance Matching

For the ideal-transformer circuit shown in Fig. 14.27, $V_2 = -Z_L I_2$. Since $V_2 = N V_1$ and $I_2 = -I_1/N$, then

$$NV_1 = -Z_L\left(\frac{-I_1}{N}\right) \quad \Rightarrow \quad V_1 = \left(\frac{Z_L}{N^2}\right)I_1$$

By KVL,

$$V_g = Z_g I_1 + V_1 = Z_g I_1 + \left(\frac{Z_L}{N^2}\right)I_1 = \left[Z_g + \left(\frac{Z_L}{N^2}\right)\right]I_1$$

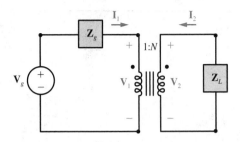

Fig. 14.27 Ideal-transformer circuit.

Thus the impedance $\mathbf{V}_g/\mathbf{V}_1$ loading the source is

$$\frac{\mathbf{V}_g}{\mathbf{I}_1} = \mathbf{Z}_g + \frac{\mathbf{Z}_L}{N^2} \tag{14.32}$$

which is $\mathbf{Z}_g$ in series with the load impedance $\mathbf{Z}_L$ scaled by the factor $1/N^2$. The term $\mathbf{Z}_L/N^2$ is often called the **reflected impedance**. We can also say that this term is the equivalent impedance **referred** to the primary side of the transformer.

From Eq. 14.32,

$$\mathbf{I}_1 = \frac{\mathbf{V}_g}{\mathbf{Z}_g + \mathbf{Z}_L/N^2}$$

Therefore,

$$\mathbf{V}_2 = -\mathbf{Z}_L\mathbf{I}_2 = -\mathbf{Z}_L\left(-\frac{\mathbf{I}_1}{N}\right) = \frac{\mathbf{Z}_L}{N}\left(\frac{\mathbf{V}_g}{\mathbf{Z}_g + \mathbf{Z}_L/N^2}\right) = \frac{N\mathbf{Z}_L}{N^2\mathbf{Z}_g + \mathbf{Z}_L}\mathbf{V}_g$$

As an alternative approach to finding $\mathbf{V}_2$, apply Thévenin's theorem to the circuit given in Fig. 14.27. First remove the load $\mathbf{Z}_L$ and determine $\mathbf{V}_{oc}$ from the circuit shown in Fig. 14.28. Since $\mathbf{I}_2 = 0$ A, then $\mathbf{I}_1 = -N\mathbf{I}_2 = 0$ A, and by KVL,

$$\mathbf{V}_g = \mathbf{Z}_g\mathbf{I}_1 + \mathbf{V}_1 = \mathbf{V}_1 \quad \Rightarrow \quad \mathbf{V}_{oc} = \mathbf{V}_2 = N\mathbf{V}_1 = N\mathbf{V}_g$$

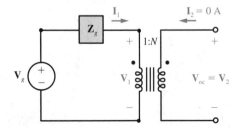

Fig. 14.28 Determination of open-circuit voltage.

To find the Thévenin-equivalent (output) impedance, set $\mathbf{V}_g = 0$ V (i.e., replace $\mathbf{V}_g$ by a short circuit) and take the ratio of $\mathbf{V}_o$ to $\mathbf{I}_o$ as depicted in Fig. 14.29. Since

$$\mathbf{V}_1 = -\mathbf{Z}_g\mathbf{I}_1 = -\mathbf{Z}_g(-N\mathbf{I}_o) = N\mathbf{Z}_g\mathbf{I}_o$$

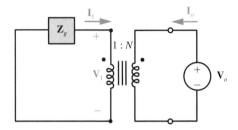

Fig. 14.29 Determination of output impedance.

then

$$\mathbf{V}_o = N\mathbf{V}_1 = N(N\mathbf{Z}_g\mathbf{I}_o) = N^2\mathbf{Z}_g\mathbf{I}_o \quad \Rightarrow \quad \mathbf{Z}_o = \frac{\mathbf{V}_o}{\mathbf{I}_o} = N^2\mathbf{Z}_g$$

is the output impedance.

The resulting Thévenin-equivalent circuit is shown in Fig. 14.30. By voltage division,

$$\mathbf{V}_2 = \frac{\mathbf{Z}_L}{\mathbf{Z}_L + N^2\mathbf{Z}_g}(N\mathbf{V}_g) = \frac{N\mathbf{Z}_L}{N^2\mathbf{Z}_g + \mathbf{Z}_L}\mathbf{V}_g$$

and this is the same result that was obtained above.

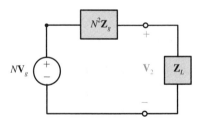

Fig. 14.30 Thévenin-equivalent circuit.

Example 14.11

For the circuit given in Fig. 14.27, suppose $\mathbf{Z}_g = 20$ kΩ and $\mathbf{Z}_L = 8$ Ω. Let us determine the value of the turns ratio N for which the 8-Ω load resistor absorbs maximum power.

Given the Thévenin-equivalent circuit shown in Fig. 14.30, we know that the load absorbs maximum power when the load resistance $\mathbf{Z}_L = R_L = 8 \ \Omega$ is equal to the output (Thévenin-equivalent) resistance $\mathbf{Z}_o = R_o = N^2 \mathbf{Z}_g = 20{,}000N^2$. Thus

$$20{,}000N^2 = 8 \qquad \Rightarrow \qquad N = 1/50$$

and this is the turns ratio for which the 8-Ω load resistor absorbs maximum power.

In this example, we have used a transformer to "match" the 8-Ω load resistor to the 20-kΩ source resistance. When a transformer is used in such an application, it can be referred to as a **matching transformer**. Typically, this type of transformer is not designated by its turns ratio N, but rather by the impedances (resistances) it matches. Specifically, the transformer in this example could be called a "20-kΩ/8-Ω matching transformer."

Drill Exercise 14.11

For the circuit given in Fig. 14.27, suppose $\mathbf{Z}_g = 20$ kΩ, $\mathbf{Z}_L = 8 \ \Omega$, and $N = 1/50$. (a) Find the impedance loading the voltage source. (b) Find the voltage gain $|\mathbf{V}_2/\mathbf{V}_g|$. (c) Find the current gain $|\mathbf{I}_2/\mathbf{I}_1|$. (d) Find the power gain.

ANSWER (a) 40 kΩ, (b) 0.01, (c) 50, (d) 0.5

14.5 Nonideal-Transformer Models

There are many situations for which the assumption that an iron-core transformer is ideal yields quite good results. There may be times, however, when a more accurate model of a practical iron-core transformer is desirable. One such model (there are others) is shown in Fig. 14.31. For this circuit, the resistor R_s accounts for the "copper losses" due to the resistance of the primary winding and the secondary-winding resistance referred to the primary. The reactance X_s accounts for the leakage flux in the primary and the secondary leakage reactance referred to the primary. The resistor R_m accounts for the core losses due to hysteresis and eddy currents. If there is no load connected to the secondary winding of the transformer, then $\mathbf{I}_2 = 0$ A. Yet, for a physical core, even when there is no load, a voltage applied to the primary results in a primary current that will produce a flux in the core. By Eq. 14.19, this indicates an inductance, called the **magnetizing inductance**, and it is accounted for by the reactance X_m. The current $\mathbf{I}_m$ through the magnetizing inductance is known as the **magnetizing current**, and the current $\mathbf{I}_e$ is called the **exciting current**.

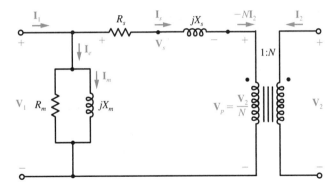

Fig. 14.31 Model of a practical transformer.

We can determine the parameters of the transformer model given in Fig. 14.31 by taking open- and short-circuit measurements.

Example 14.12

A transformer used for a power application is rated 2400:240 V, 60 Hz, 36 kVA. This rating indicates, first, that one winding, the high-voltage winding, can have a maximum rms voltage of 2400 V. The other winding, the low-voltage winding, can have a maximum rms voltage of 240 V. Although this means that the turns ratio is $N = 2400/240 = 10$ when the high-voltage winding is the secondary, either winding may be used as the secondary. The next term in the rating (60 Hz) indicates that operation at this frequency will not result in excessive core loss. Finally, the 36-kVA rating means that, whichever winding is chosen as the secondary, the product of the load's voltage and current can be maintained at this "full-load" value and excessive heating will not result. Note that we can now calculate the current ratings (maximum values) for the windings. For the high-voltage winding, the rated current is 36 kVA/2400 V = 15 A, and for the low-voltage winding, 36 kVA/240 V = 150 A is the rated current.

Suppose the high-voltage winding is the primary. Then the turns ratio is $N = 240/2400 = 1/10$. Let us apply $|\mathbf{V}_1| = 2400$ V (rms) to the primary winding when no load is connected to the secondary. For this **open-circuit test**, suppose we take the voltage, current, and power measurements indicated in Fig. 14.32. Assume that the voltmeter (V), ammeter (A), and wattmeter (W) are ideal, and their respective readings are 2400 V, 0.4 A, and 360 W.

Since there is no load, $\mathbf{I}_2 = 0$ A and there is no current through R_s (i.e., $\mathbf{I}_s = 0$ A). Thus the power absorbed by R_m is 360 W, and thus

$$\frac{|\mathbf{V}_1|^2}{R_m} = 360 \quad \Rightarrow \quad R_m = \frac{|\mathbf{V}_1|^2}{360} = \frac{(2400)^2}{360} = 16 \text{ k}\Omega$$

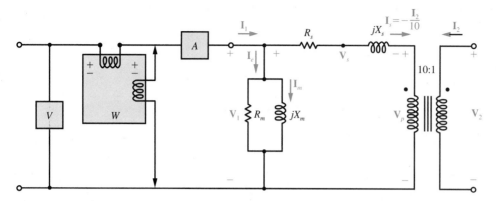

Fig. 14.32 Circuit used for open-circuit and short-circuit testing.

Furthermore, since $|\mathbf{I}_e| = 0.4$ A, then

$$|\mathbf{V}_1|\,|\mathbf{I}_e|\cos\theta_m = 360 \quad \Rightarrow$$

$$\cos\theta_m = \frac{360}{|\mathbf{V}_1|\,|\mathbf{I}_e|} = \frac{360}{(2400)(0.4)} = 0.375$$

Thus $\theta_m = \cos^{-1}(0.375) = 68.0°$. Since this is the angle of the impedance $\mathbf{Z}_m$ consisting of the parallel connection of R_m and X_m, then

$$\mathbf{Z}_m = \frac{(R_m)(jX_m)}{R_m + jX_m} \quad \Rightarrow \quad \theta_m = 90° - \tan^{-1}\left(\frac{X_m}{R_m}\right)$$

from which

$$X_m = R_m \tan(90° - \theta_m) = 16{,}000 \tan(90° - 68.0°) = 6.46 \text{ k}\Omega$$

Since $X_m = \omega L_m = 2\pi f L_m$, then the magnetizing inductance L_m is

$$L_m = \frac{X_m}{2\pi f} = \frac{6460}{2\pi(60)} = 17.1 \text{ H}$$

Now suppose a short circuit is placed across the secondary winding. For this **short-circuit test**, instead of applying 2400 V to the primary, we apply a voltage that yields the rated current of 150 A in the secondary. Suppose the voltmeter has a reading of 50 V, the ammeter has a reading of 15 A, and the wattmeter has a reading of 315 W.

The power absorbed by R_m is

$$\frac{|\mathbf{V}_1|^2}{R_m} = \frac{50^2}{16,000} = 0.156 \text{ W}$$

and so the power absorbed by R_s is approximately 315 W. Since the magnitude of the exciting current is

$$|\mathbf{I}_e| = \frac{|\mathbf{V}_1|}{|\mathbf{Z}_m|} = \frac{|\mathbf{V}_1|}{R_m X_m / \sqrt{R_m^2 + X_m^2}}$$

$$= \frac{50}{(16,000)(6460) / \sqrt{16,000^2 + 6460^2}} = 8.35 \text{ mA}$$

then $|\mathbf{I}_s| \approx 15$ A, and the secondary has the rated current of 150 A. Therefore, the power absorbed by R_s is

$$R_s(15^2) = 315 \quad \Rightarrow \quad R_s = \frac{315}{15^2} = 1.4 \ \Omega$$

When a short circuit is placed across the secondary winding, $\mathbf{V}_2 = 0$ V. Under this circumstance, $\mathbf{V}_p = 0$ V. Thus the voltage across $\mathbf{Z}_s$, the series connection of R_s and jX_s, is $\mathbf{V}_1$. Consequently,

$$|\mathbf{V}_1| |\mathbf{I}_s| \cos \theta_s = 315 \quad \Rightarrow \quad \cos \theta_s = \frac{315}{|\mathbf{V}_1| |\mathbf{I}_s|} = \frac{315}{(50)(15)} = 0.42$$

Thus $\theta_s = 65.2°$. Since this is the angle of the impedance $\mathbf{Z}_s = R_s + jX_s$, then

$$\theta_s = \tan^{-1}\left(\frac{X_s}{R_s}\right) \quad \Rightarrow \quad X_s = R_s(\tan \theta_s)$$

$$= 1.4(\tan 65.2°) = 3.02 \approx 3 \ \Omega$$

Since $X_s = 2\pi f L_s$, then the corresponding inductance is

$$L_s = \frac{X_s}{2\pi f} = \frac{3.02}{2\pi(60)} = 0.00801 \approx 8 \text{ mH}$$

In summary, the open-circuit test yields the parameters R_m and X_m, and the short-circuit test gives R_s and X_s.

Drill Exercise 14.12

A transformer is rated 2400:240 V, 60 Hz, 36 kVA. Suppose the low-voltage winding is the primary. Determine the transformer parameters R_m, X_m, R_s, and X_s when the readings of the meters (as depicted in Fig. 14.32) are 240 V, 3 A, and 360 W for the open-circuit test, and 50 V, 150 A, and 3 kW for the short-circuit test.

ANSWER 160 Ω, 92.4 Ω, 0.133 Ω, 0.304 Ω

Some Transformer Properties

One figure of merit of a device is its **efficiency**, which we define as the ratio (usually expressed in percent) of the output power P_o to the input power P_i, Thus we have that the efficiency (denoted Eff) is

$$\text{Eff} = \frac{P_o}{P_i}$$

Since some of the input power is lost in the device and the remainder reaches the output, if P_l denotes the power loss, we have

$$P_i = P_l + P_o \quad \Rightarrow \quad P_o = P_i - P_l$$

Thus we can write

$$\text{Eff} = \frac{P_o}{P_i} = \frac{P_o}{P_l + P_o} = \frac{P_i - P_l}{P_i} = 1 - \frac{P_l}{P_i} \tag{14.33}$$

In addition to the efficiency of a transformer, we can talk about its **voltage regulation** (also usually expressed in percent), which is the ratio of the change between no-load output (rms) voltage V_{NL} and full-load output (rms) voltage V_{FL} to the full-load output voltage. In other words, the voltage regulation VR is given by

$$\text{VR} = \frac{V_{NL} - V_{FL}}{V_{FL}} \tag{14.34}$$

Example 14.13

Let us determine the voltage regulation and efficiency for the transformer given in Example 14.12 for the case that the transformer load has a lagging power factor (pf) of 0.866.

We know that the full-load (rated) output voltage is $V_{\text{FL}} = 240$ V. Let us take the angle of $\mathbf{V}_2$ to be zero, that is, $\mathbf{V}_2 = 240\underline{/0°}$. Then $\mathbf{V}_p = 2400\underline{/0°}$. Since the full-load output current is 150 A, then $|\mathbf{I}_2| = 150$ A. Also, since the pf is $\cos\theta = 0.866$, then $\theta = 30°$. But as the pf is lagging, then the current lags the voltage by 30 degrees. Thus $-\mathbf{I}_2 = 150\underline{/-30°}$ from which $\mathbf{I}_s = -\mathbf{I}_2/10 = 15\underline{/-30°}$. Therefore,

$$\begin{aligned}
\mathbf{V}_s &= \mathbf{Z}_s\mathbf{I}_s = (R_s + jX_s)\mathbf{I}_s \\
&= (1.4 + j3)(15\underline{/-30°}) = (3.31\underline{/65°})(15\underline{/-30°}) = 49.7\underline{/35°}\ \text{V}
\end{aligned}$$

and

$$\begin{aligned}
\mathbf{V}_1 &= \mathbf{V}_s + \mathbf{V}_p \\
&= (40.7 + j28.5) + 2400 = 2441 + j28.5 = 2441\underline{/0.67°}\ \text{V}
\end{aligned}$$

A phasor diagram (not drawn to scale) is shown in Fig. 14.33.

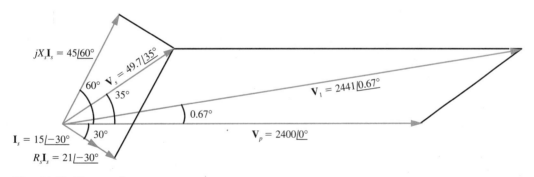

Fig. 14.33 Phasor diagram.

For the case of no load ($\mathbf{I}_2 = 0$ A), we have that $\mathbf{I}_s = 0$ A, and hence $\mathbf{V}_s = 0$ V. Thus $\mathbf{V}_p = \mathbf{V}_1$, and since $\mathbf{V}_2 = \mathbf{V}_p/10 = \mathbf{V}_1/10$, then the no-load output voltage is $V_{\text{NL}} = 244.1$ V. Therefore, by Eq. 14.34, the voltage regulation is

$$\text{VR} = \frac{244.1 - 240}{240} = 0.0171 = 1.71\%$$

To determine the (full-load) efficiency, the copper loss of the windings is the power absorbed by the resistance R_s. This power P_s is

$$P_s = R_s|\mathbf{I}_s|^2 = (1.4)(15^2) = 315 \text{ W}$$

which was the wattmeter reading for the short-circuit test. The core loss (due to leakage flux) is the power absorbed by the resistor R_m. This power P_m is

$$P_m = \frac{|\mathbf{V}_1|^2}{R_m} = \frac{2441^2}{16,000} = 372 \text{ W}$$

Thus the power loss $P_l = P_s + P_m = 315 + 372 = 687$ W.
   The output power is

$$P_o = |\mathbf{V}_2|\,|\mathbf{I}_2|\cos\theta = (240)(150)(0.866) = 31.2 \text{ kW}$$

and from Eq. 14.33, the full-load efficiency is

$$\text{Eff} = \frac{31,200}{687 + 31,200} = 0.987 = 97.8\%$$

Drill Exercise 14.13

Determine the voltage regulation and the efficiency of the transformer described in Drill Exercise 14.12 for the case that the transformer load has a lagging pf of 0.866.

ANSWER   17.5%, 89.9%

Normally, power transformers operate at a fixed frequency (e.g., 60 Hz), and they are designed with this in mind; however, iron-core transformers also see use in situations where a range of frequencies is encountered. One common application is in an audio system in which an ''output transformer'' is used to match the output of an amplifier stage to a loudspeaker load for frequencies ranging, typically, from 30 to 15,000 Hz. The behavior of such a transformer can be accurately modeled by using the circuit parameters discussed previously, along with two capacitors to represent the capacitances of the primary and secondary windings. For power-transformer applications, these capacitors can be neglected, but they become more and more significant as frequency increases.

A model of an output transformer is shown in Fig. 14.34, along with a voltage source $\mathbf{V}_g$, that has an internal resistance R_g, and a resistive load R_L. At low frequen-

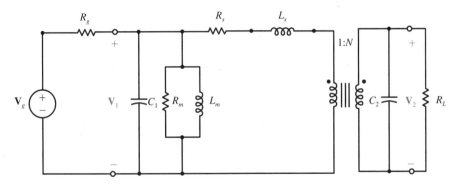

Fig. 14.34 Model of an output transformer.

cies, the impedance of the magnetizing inductance L_m is small. This causes a shunting effect in that a significant amount of current will go through L_m, bypassing the primary coil of the ideal transformer portion of the model. The result, for a fixed magnitude of V_g, is a small magnitude for V_2, which increases as frequency increases. After the frequency reaches around 100 Hz, the impedance of L_m becomes large enough that it can be ignored (treated as an open circuit). From 100 Hz to about 10 kHz, the magnitude of V_2 is essentially constant. In the vicinity of 10 kHz, however, the output voltage magnitude actually increases as a result of the resonance that results from C_2 and the leakage inductance L_s. For frequencies above 10 kHz, not only does the impedance of L_s get larger, but the impedances of the shunting capacitors C_1 and C_2 get smaller with increasing frequency. The result is a sharp rolloff or decrease in the magnitude of V_2 as frequency increases. The frequency dependence of the transformer modeled by the circuit given in Fig. 14.34 is summarized by the amplitude-response curve shown in Fig. 14.35.

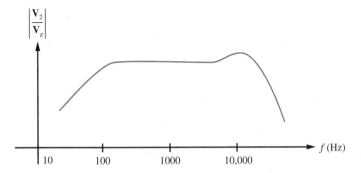

Fig. 14.35 Amplitude-response curve.

A good audio-frequency transformer is one that has a large magnetizing inductance L_m. This property can be achieved by using a core with a high permeability and appropriate geometry. In addition, a good transformer has small values for R_s and L_s. The former can be obtained by forming the primary and secondary coils from wires with large diameters, and the latter by tightly interleaving the primary and secondary windings.

SUMMARY

1. Current going through a conductor produces a magnetic field about the conductor.

2. A current going through a solenoid creates a magnetic field similar to that of a permanent magnet.

3. The relative permeability μ_r of a material is the ratio of the permeability of the material μ to the permeability of free space $\mu_0 = 4\pi \times 10^{-7}$ H/m. That is, $\mu_r = \mu/\mu_0$.

4. For a given core, magnetomotive force (mmf), designated by $\mathscr{F}$, is the product of the number of coil turns N and the current i through the coil. That is, $\mathscr{F} = Ni$.

5. The ratio of mmf $\mathscr{F}$ to magnetic flux ϕ is called reluctance and is designated by $\mathscr{R}$. That is, $\mathscr{R} = \mathscr{F}/\phi$.

6. The magnetic-circuit equation $\mathscr{F} = \mathscr{R}\phi$ is analogous to the electric-circuit equation $v = Ri$ (Ohm's law).

7. Magnetic field intensity H is the ratio of magnetic flux density B to permeability μ. That is, $H = B/\mu$.

8. A plot of magnetic flux density B versus magnetic field intensity H is called a magnetization (or B-H) curve.

9. Power losses in a core are due to hysteresis and eddy currents.

10. A changing magnetic flux passing through a coil induces an electromotive force (emf) across that coil (Faraday's law).

11. Inductance, which is the number of flux linkages per ampere, is directly proportional to the square of the number of turns.

12. The polarity of the voltage induced by a changing flux tends to oppose the change in flux that produced the induced voltage (Lenz's law).

13. The phase relationship between the windings of a transformer is indicated by dots.

14. An ideal transformer has perfect coupling and infinite self-inductances.

15. The energy stored in an ideal transformer is zero.

16. The circuit parameters for a nonideal-transformer model can be calculated from the results of open-circuit and short-circuit tests.

17. A transformer can be rated in terms of its efficiency and voltage regulation.

18. For transformers used over a range of frequencies, the capacitances of the primary and secondary windings are significant at higher frequencies.

PROBLEMS

14.1 Suppose the solenoid shown in Fig. 14.5 on p. 906 consists of 500 turns wound on a core of radius 0.01 m (1 cm) and length 0.2 m (20 cm). Find the current required to produce a magnetic flux density of magnitude $B = 0.5$ T in the core of the solenoid when the core material is (a) air, and (b) iron having a relative permeability of 1500.

14.2 Suppose the solenoid shown in Fig. 14.5 on p. 906 consists of 100 turns wound on an iron core having relative permeability of 1500 and radius of 0.01 m. Determine the length of the solenoid required to produce a flux density of 0.1 T in the core due to a coil current of 0.1 A.

14.3 A toroid, as shown in Fig. 14.6, on p. 908, has a core with an average radius of 0.1 m, a cross-sectional radius of 0.02 m, and a coil current of $\frac{1}{3}$ A. (a) Find the magnetic flux density in the core when the coil has 500 turns and the core material is non-magnetic (i.e., $\mu_r \approx 1$). (b) Determine the number of turns required to produce a flux of 1 mWb in the core when it is iron and has a relative permeability of 1500.

14.4 A toroid, as shown in Fig. 14.6, on p. 908, has an iron core with a relative permeability of 1500, an average radius of 0.01 m, a coil with 100 turns, and a coil current of 1 A. Determine the cross-sectional radius of the core such that the flux in the core is 0.1 mWb.

14.5 A rectangular core has a relative permeability of 1500, and the current through a coil wound around the core is $\frac{1}{3}$ A. Determine the number of turns required to produce a magnetic flux of 1 mWb for the core shown in (a) Fig. 14.8, on p. 912, and (b) Fig. 14.9 on p. 913.

14.6 A rectangular core has a relative permeability of 1500, and a coil with 500 turns is wound around the core. Determine the current through the coil which will produce a magnetic flux of 1 mWb for the core shown in (a) Fig. 14.8, on p. 912, and (b) Fig. 14.9 on p. 913.

14.7 For the rectangular iron core shown in Fig. 14.9, on p. 913, suppose there is a second 1-cm air gap on the top part of the structure. Also suppose that the iron has a relative permeability of 1500. (a) Determine the total reluctance of the flux path. (b) Find the current required to produce a magnetic flux of 1 mWb in the core when the coil has 200 turns. (c) Find the number of coil turns needed to produce a magnetic flux of 1 mWb in the core when the coil current is 2 A. (d) Find the magnetic flux in the core when the coil has 200 turns and a current of 2 A.

14.8 The toroid shown in Fig. 14.6 on p. 908 is constructed of cast iron and is to have a magnetic flux density of 0.2 T. (a) Determine the relative permeability of the core. (b) If the core has an average radius of 0.1 m and a cross-sectional radius of 0.02 m, find the current required for a 500-turn coil.

14.9 The rectangular core shown in Fig. 14.8 on p. 912 is constructed of sheet steel and is to have a magnetic flux density of 0.7 T. (a) Determine the relative permeability of the core. (b) Determine the current required for a 500-turn coil.

14.10 The rectangular core shown in Fig. 14.8 on p. 912 is constructed of cast iron and is to have a magnetic flux of 0.24 mWb. (a) Determine the relative permeability of the core. (b) Determine the current required for a 200-turn coil.

14.11 For the cast-steel core shown in Fig. 14.8, on p. 912, the magnetic flux is to be 0.72 mWb. (a) Determine the relative permeability of the core. (b) Determine the number of turns for the coil when its current is 1.2 A.

14.12 For the cast-steel core shown in Fig. 14.9, on p. 913, the magnetic flux is to be 0.72 mWb. (a) Determine the relative permeability of the core. (b) Determine the current in the coil when it has 200 turns. (c) Determine the magnetic field intensity in the air gap.

14.13 A cast-iron magnetic core has an average length of $l = 0.6$ m and $N = 200$ turns, and the average hysteresis power loss is $P_h = 10$ W when the maximum current is 1.2 A. Use $n = 1.6$ to approximate P_h for the case that the maximum current is (a) 0.9 A and (b) 2.4 A.

14.14 A cast-steel magnetic core has an average length of $l = 0.6$ m and $N = 200$ turns, and the average hysteresis power loss is $P_h = 10$ W when the maximum current is 1.2 A. Use $n = 1.6$ to approximate P_h for the case that the maximum current is (a) 0.9 A and (b) 2.4 A.

14.15 For the case of a core having a relative permeability of 1500, determine the number of turns required to result in an inductance of 50 mH for the coil shown in (a) Fig. 14.6, on p. 908, where the average radius is 0.1 m and the cross-sectional radius is 0.02 m, and (b) Fig. 14.8 on p. 912.

14.16 A 200-turn coil is to have an inductance of 50 mH. Determine the relative permeability of the core that the coil is wound on if the core is the one shown in (a) Fig. 14.6, on p. 908, where the average radius is 0.1 m and the cross-sectional radius is 0.02 m, and (b) Fig. 14.8 on p. 912.

14.17 Show that for the toroidal coil depicted in Fig. 14.6, on p. 908, (a) the inductance is

$$L = \frac{\mu N^2 A}{l}$$

where $A = \pi r^2$ and $l = 2\pi R$, and (b) the energy stored is

$$w = \frac{B^2 A l}{2\mu}$$

where $B = \phi/A$.

14.18 For the core with two coils shown in Fig. P14.18, assume that at some instant of time the current i is a positive quantity. (a) Is the resulting magnetic flux ϕ clockwise or counterclockwise? (b) If ϕ is increasing with time, for the coil on the right, in what direction will current tend to flow? (c) If ϕ is increasing with time, is v_{ab} positive or negative? (d) If ϕ is increasing with time, is v positive or negative?

14.19 For the core with two coils shown in Fig. P14.18, assume that at some instant of time the current i is a positive quantity. (a) Is the resulting magnetic flux ϕ clockwise or counterclockwise? (b) If ϕ is decreasing with time, for the coil on the right, in what direction will current tend to flow? (c) If ϕ is decreasing with time, is v_{ab} positive or negative? (d) If ϕ is decreasing with time, is v positive or negative?

14.20 Place dots appropriately on the transformer shown in Fig. P14.18.

14.21 For the transformers shown in Fig. P14.21, place dots appropriately at the ends of the coils.

14.22 For the transformer shown in Fig. P14.22, (a) write a pair of equations in the time domain relating the voltages v_1 and v_2 to the currents i_1 and i_2, and (b) use Eq. 14.25 on p. 927 to show that the energy stored in this transformer is

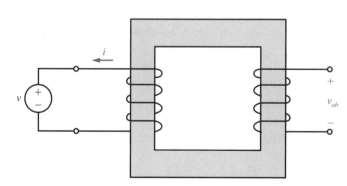

Fig. P14.18

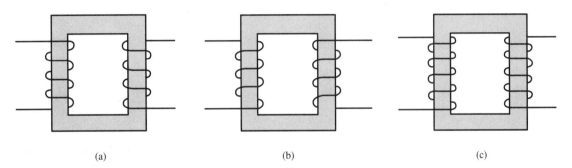

Fig. P14.21

$$w(t) = \tfrac{1}{2} L_1 i_1^2(t) + \tfrac{1}{2} L_2 i_2^2(t) - M i_1(t) i_2(t)$$

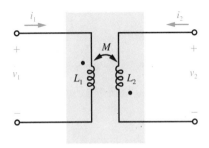

Fig. P14.22

14.23 For the transformer circuit given in Fig. 14.19, on p. 928, find the impedance $\mathbf{V}/\mathbf{I}_1$ loading the voltage source.

14.24 For the transformer circuit given in Fig. 14.18, on p. 927, connect the lower ends of the primary and secondary windings with a short circuit. Use the transformer model given in Fig. 14.20b on p. 929 to determine $i_1(t)$, $i_2(t)$, and $v_C(t)$.

14.25 For the transformer circuit shown in Fig. 14.19, on p. 928, consider the capacitor to be the load. (a) Show that the frequency-domain Thévenin equivalent of the remainder of the circuit is $\mathbf{V}_{oc} = 14.4\underline{/-23.13°}$ V and $\mathbf{Z}_o = 1.44 + j16.08\ \Omega$. (b) Use the results of part (a) to determine $v_C(t)$.

14.26 For the circuit given in Fig. 14.18, on p. 927, suppose $R = 9\ \Omega$, $L_1 = 4$ H, $L_2 = 6$ H, $M = 2$ H, $C = 1/18$ F, and $v(t) = 36\cos\sqrt{3}\,t$ V. Find

(a) $i_1(t)$, (b) $i_2(t)$, (c) $v_C(t)$, and (d) the impedance loading the voltage source.

14.27 For the circuit given in Fig. 14.18, on p. 927, suppose $R = 9\ \Omega$, $L_1 = 3$ H, $L_2 = 4$ H, $M = 3$ H, $C = 1/16$ F, and $v(t) = 36\cos 2t$ V. Find (a) $i_1(t)$, (b) $i_2(t)$, (c) $v_C(t)$, and (d) the impedance loading the voltage source.

14.28 For the circuit given in Fig. 14.18, on p. 927, suppose $R = 9\ \Omega$, $L_1 = 3$ H, $L_2 = 4$ H, $M = 3$ H, $C = 1/16$ F, and $v(t) = 36\cos 4t$ V. Find (a) $i_1(t)$, (b) $i_2(t)$, (c) $v_C(t)$, and (d) the impedance loading the voltage source.

14.29 For the circuit given in Fig. 14.18, on p. 927, move the dot from the upper end of the secondary winding to the lower end, and find $i_1(t)$, $i_2(t)$, and $v_C(t)$.

14.30 Find $v_2(t)$ for the circuit shown in Fig. P14.30. (See p. 950.)

14.31 The transformer given in Fig. P14.31a can be modeled by the T-equivalent circuit shown in Fig. P14.31b. Use the results of Problem 14.22 to determine the values of the inductances labeled with the question marks. (See p. 950.)

14.32 A magnetic circuit with a constant ac voltage amplitude has a core loss totaling 100 W at 120 Hz and totaling 32 W at 60 Hz. Find the core loss at 50 Hz.

14.33 For the ideal-transformer circuit shown in Fig. 14.27, on p. 936, suppose $N = 0.1$ and $\mathbf{Z}_g =$

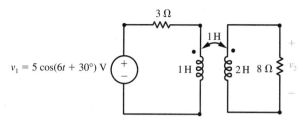

Fig. P14.30

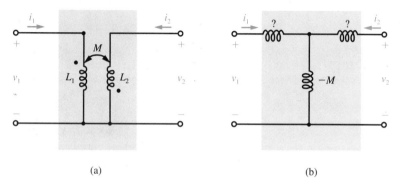

(a)   (b)

Fig. P14.31

10 kΩ. (a) Determine the value of $\mathbf{Z}_L$ such that this load absorbs maximum power. (b) Find the impedance $\mathbf{V}_g/\mathbf{I}_1$ for this value of $\mathbf{Z}_L$. (c) Find the voltage gain $\mathbf{V}_2/\mathbf{V}_g$ for this value of $\mathbf{Z}_L$.

14.34 For the ideal-transformer circuit shown in Fig. 14.27 on p. 936, suppose $\mathbf{Z}_g = 10$ kΩ and $\mathbf{Z}_L = 16$ Ω. (a) Determine N such that $\mathbf{Z}_L$ absorbs maximum power. (b) Find the impedance $\mathbf{V}_g/\mathbf{I}_1$ for this value of N. (c) Find the voltage gain $\mathbf{V}_2/\mathbf{V}_g$ for this value of N.

14.35 For the ideal-transformer circuit shown in Fig. 14.27 on p. 936, suppose $N = 2$, $\mathbf{Z}_g$ is a 2-H

inductor, $\mathbf{Z}_L$ is a $\frac{1}{2}$-F capacitor, and $v_g(t) = 15 \cos 2t$ V. Find (a) $i_1(t)$, (b) $i_2(t)$, (c) $v_1(t)$, and (d) $v_2(t)$.

14.36 For the circuit shown in Fig. 14.18 on p. 927, replace the transformer with an ideal transformer having $N = \frac{1}{3}$. Find (a) $i_1(t)$, (b) $i_2(t)$, and (c) $v_C(t)$ given that $v(t) = 36 \cos(3t - 60°)$ V.

14.37 The circuit shown in Fig. P14.37 is the equivalent circuit of a simple "class-B" transistor power amplifier. (a) Find the voltage gain $\mathbf{V}_2/\mathbf{V}_g$. (b) Assuming that p_g is the power supplied by the voltage source and p_2 is the power absorbed by the 8-Ω load resistor, find the power gain p_2/p_g.

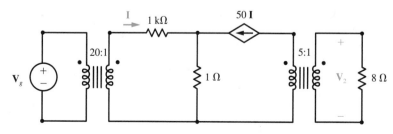

Fig. P14.37

14.38 A power transformer, rated at 2300:230 V, 60 Hz, 46 kVA, is to be modeled as shown in Fig. 14.31 on p. 939. When the high-voltage winding is the primary, the open-circuit test (see Fig. 14.32 on p. 940) results in readings of 2300 V, 0.35 A, and 400 W; and the short-circuit test results in 40 V, 20 A, and 350 W. Find the model parameters R_m, X_m, R_s, and X_s.

14.39 A 2400:240-V, 60-Hz, 48-kVA transformer modeled as shown in Fig. 14.31 on p. 939 has parameters $R_m = 15$ kΩ, $X_m = 7$ kΩ, $R_s = 1$ Ω, and $X_s = 2$ Ω when the high-voltage winding is the primary. Find the voltmeter, ammeter, and wattmeter readings indicated in Fig. 14.32 on p. 940 when (a) the open-circuit test is performed, and (b) the short-circuit test is performed.

14.40 A 2400:240-V, 60-Hz, 48-kVA transformer modeled as shown in Fig. 14.31 on p. 939 has parameters $R_m = 150$ Ω, $X_m = 70$ Ω, $R_s = 0.01$ Ω, and $X_s = 0.02$ Ω when the low-voltage winding is the primary. Find the voltmeter, ammeter, and wattmeter readings indicated in Fig. 14.32 on p. 940 when (a) the open-circuit test is performed, and (b) the short-circuit test is performed.

14.41 A 2400:240-V, 60-Hz, 48-kVA transformer modeled as shown in Fig. 14.31 on p. 939 has parameters $R_m = 15$ kΩ, $X_m = 7$ kΩ, $R_s = 1$ Ω, and $X_s = 2$ Ω when the high-voltage winding is the primary. Find the voltage regulation and full-load efficiency for the case that the load has a leading pf of 0.5.

14.42 A 2400:240-V, 60-Hz, 48-kVA transformer modeled as shown in Fig. 14.31 on p. 939 has parameters $R_m = 13.2$ kΩ, $X_m = 7.57$ kΩ, $R_s = 0.875$ Ω, and $X_s = 1.79$ Ω when the high-voltage winding is the primary. Find the voltage regulation and full-load efficiency for the case that the load has a unity pf.

15

Machines

INTRODUCTION

The subject of electromechanical energy conversion has been, and continues to be, a very important area of study. Because most electric power is generated and consumed by devices that employ magnetic fields, we focus our attention on these types of machines.

The basic principles of electromechanical energy conversion are initially introduced with the use of simple translational transducers—a generator and a motor, as well as a bilateral transducer (a combination of both). In this regard, it is seen how electric and mechanical circuit elements can be used to model machines.

The concept of a transducer is extended to encompass some devices that have moving coils and others that have moving-iron sections. The former, or dynamic transducers, find application in loudspeakers, microphones, and phonograph pickups. The latter see use as relays and solenoid plungers.

Expanding the study of transducers from translation to rotation leads to the d'Arsonval movement. This type of rotational transducer relates current and torque, and, as a consequence, has widespread application in electrical instrumentation. The same thing can be said about another rotational transducer—the electrodynamometer.

When it comes to the conversion of electromechanical energy, there is no question but that, in terms of quantity, the leader is the rotating machine. It is important as both generator and motor. In our study, for the most part, we are concerned with the steady-state behavior of rotating machines that operate under typical conditions. We will limit our discussions to include only the most common types of machines. Emphasis will be placed on simplified forms and models of machines so that any analysis can proceed in a clear and straight-forward manner without becoming obscured by excessive detail.

Most rotating machines operate on the same basic principle—appropriately aligned conductors moving in a magnetic field. The different ways in which this can be accomplished result in the various types of machines. Typically, though, any machine will have a portion that rotates (the rotor) and a portion that is stationary (the stator). In a dc machine, both rotor and stator are supplied with direct current. Basically, there are two types of ac machines. In one, the induction type, alternating current is supplied to both rotor and stator. For the other type, the synchronous machine, rotor and stator receive different excitations—one is supplied with direct current and the other with alternating current. In this chapter, we begin our study of rotating machines with a discussion of both dc and ac machines. Included is material on dc motors and generators, alternators, synchronous motors, and induction motors—both the three-phase and single-phase types.

15.1 Transducers

In the preceding chapter, we discussed Faraday's law, which states that a changing magnetic flux ϕ that passes through an N-turn conducting coil induces an emf of $N \, d\phi/dt$ volts across that conductor. We have considered the case for which the conductor is stationary and the flux is time varying. If, however, the flux is constant and the conductor moves (i.e., its position varies with time), we still achieve a changing magnetic flux relative to the conductor—and hence, an induced emf. (Of course, if both the flux and the conductor's position are time varying, the result again will be an induced emf.)

A Simple Generator

With the above in mind, consider the one-turn loop formed by a movable conductor that is electrically connected to a fixed, conducting track comprised of two rails and a resistor as shown in Fig. 15.1. Suppose the length of the moveable conductor is l, and a uniform magnetic field with a constant magnetic flux density B is directed into the page (indicated by the crosses). If the conductor moves to the right at an average velocity of u m/s (meters per second), then, after dt seconds, it travels $u \, dt$ meters. Since $\phi = BA$, then also after dt seconds, the flux enclosed by the sin-

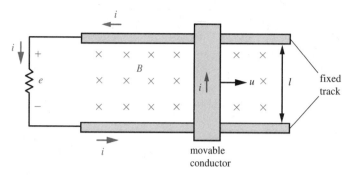

Fig. 15.1 Simple generator.

gle-turn loop comprised of the movable conductor, the track, and the resistance increases by

$$d\phi = d(BA) = B \, dA = Blu \, dt$$

from which we have

$$\frac{d\phi}{dt} = Blu = e \qquad\qquad (15.1)$$

and this is the subsequent induced emf. To obtain the polarity of this voltage, we apply Lenz's law. For the current resulting from the induced emf to establish a magnetic field that opposes the increasing flux, by the right-hand rule, the current must be counterclockwise as depicted. This yields the polarity of e shown in Fig. 15.1. In the derivation of Eq. 15.1, it was assumed that the movable conductor and fixed track had zero resistances, and that the magnetic field resulting from current i was negligible compared with the original magnetic field.

For the setup given in Fig. 15.1, some applied force f, measured in newtons (N), is required to move the length-l conductor with velocity u. Thus we have that for this electromechanical system, the input is an applied force f and the output (response) is a voltage $e = Blu$. A device such as this, which converts mechanical energy into electric energy, is called a **generator**.

A Simple Motor

Let us now modify the system given in Fig. 15.1 by replacing the resistance with a current source as shown in Fig. 15.2. The result is a clockwise current i going through the loop formed by the current source, the fixed track, and the movable conductor.

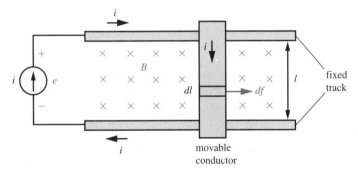

Fig. 15.2 Simple motor.

Suppose that u_q is the average velocity of an increment of charge dq within an incremental section dl of the movable conductor. Thus $u_q = dl/dt$. However, since $f = quB \sin \theta$ (Eq. 14.1) and the direction of the current (charge) is perpendicular to the direction of the magnetic field, then the resulting incremental force df developed on the section dl of the movable conductor is given by

$$df = dq\, u_q B \tag{15.2}$$

and by the right-hand screw rule, this developed force is directed to the right. Since

$$i = \frac{dq}{dt} = \frac{dq}{dl}\frac{dl}{dt} = \frac{dq}{dl} u_q$$

then, by the chain rule of calculus,

$$idl = dq\, u_q$$

Substituting this result into Eq. 15.2 yields

$$df = Bi\, dl$$

and summing df over the entire length l, we deduce that the force f developed on the movable conductor is

$$\boxed{f = Bli} \tag{15.3}$$

and such a force will tend to move the conductor to the right.

Conversely to the generator given in Fig. 15.1, an applied current i is the input to the electromechanical system shown in Fig. 15.2, and the output is a force $f =$

Bli on the movable conductor. A device such as this, which converts electric energy into mechanical energy, is called a **motor**. A device that either converts electric energy to mechanical energy (e.g., a motor) or converts mechanical energy to electric energy (e.g., a generator) is called an **electromechanical transducer**.

A Bilateral Tranducer

For the setup shown in Fig. 15.2, a velocity u directed to the right due to the developed force $f = Bli$ on the movable conductor results in an induced emf $e = Blu$—just as was the case for the system given in Fig. 15.1. In this case, the electric input power supplied by the current source is

$$p_e = ei = Blui$$

and the mechanical output power is

$$p_m = fu = Bliu = p_e$$

so that energy is conserved, and we see that 1 W = 1 N-m/s. For the generator shown in Fig. 15.1, the force on the movable conductor that results from the counterclockwise current has magnitude Bli and is directed to the left. Therefore, to obtain the velocity u directed to the right, the applied force (which is directed to the right) must be $f = Bli$. Consequently, the mechanical input power is

$$p_m = fu = Bliu$$

and the electric output power (absorbed by the resistance) is

$$p_e = ei = Blui = p_m$$

as may have been anticipated.

Now consider the electromechanical system shown in Fig. 15.3, which again consists of a fixed, conducting track and a movable conductor, in addition to a series connection of a voltage source and a resistor. Since

$$i = \frac{V - e}{R}$$

for the case that the movable conductor has velocity u directed to the right, if $e = Blu < V$, then $i > 0$ A and the system behaves as a motor (see Fig. 15.2). Conversely, if $e = Blu > V$, then $i < 0$ A and the system behaves as a generator (see Fig. 15.1).

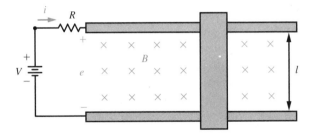

Fig. 15.3 Bilateral electromechanical transducer.

Under this circumstance, the battery will charge (i.e., it will absorb power). Because this transducer can convert either electric energy to mechanical energy or vice versa, we say that it is **bilateral**.

Example 15.1

Two parallel wires, each 3 m in length, have a separation of 4 mm. Let us determine the force exerted on these wires when they conduct a current of 5 A.

Assuming that the permeability of the material between the wires is approximated by $\mu_0 = 4\pi \times 10^{-7}$ Wb/A-m, from Eq. 14.2, the magnetic flux density at a distance of 4 mm from one of the wires is

$$B = \frac{\mu_0}{2\pi d} i = \frac{4\pi \times 10^{-7}}{2\pi(4 \times 10^{-3})}(5) = 2.5 \times 10^{-4} = 0.25 \text{ mT}$$

Therefore, the force on each wire is

$$f = Bli = (2.5 \times 10^{-4})(3)(5) = 3.75 \times 10^{-3} = 3.75 \text{ mN}$$

Drill Exercise 15.1

Two parallel wires, each 2 m in length, conduct a current of 4 A. Assume that the permeability of the material between the wires is approximated by $\mu_0 = 4\pi \times 10^{-7}$ Wb/A-m, and determine the distance between the wires such that the force exerted on each wire is 5 mN.

ANSWER   0.64 mm

Mechancial Circuit Elements and Symbols

In our discussion of a simple generator and motor, we ignored such physical aspects as mass and friction. To describe these quantities, we introduce some mechanical circuit elements. These can be used to model electromechanical devices.

A **mass** M, whose unit is the **kilogram** (kg), with an applied force f and a velocity u, is depicted on a frictionless surface in Fig. 15.4a. With the force and velocity directed to the right, we have that the force f, mass M, and velocity u are related by

$$f = M\frac{du}{dt}$$

(15.4)

where $m\,du/dt$ is the inertia force of the mass. We can represent this situation with the **mechanical circuit element** symbol shown in Fig. 15.4b. Note that a mass for a mechanical circuit is analogous to a capacitor for an electric circuit when force and velocity are analogous to current and voltage, respectively ($i = C\,dv/dt$).

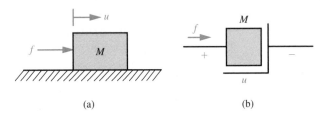

(a)   (b)

Fig. 15.4 Mass and its mechanical circuit symbol.

If the surface indicated in Fig. 15.4a is not frictionless, then we denote by D the **coefficient of sliding friction**. If the mass has velocity u, then the frictional force will be in a direction opposite to the direction if u, and the amount of the force due to friction is Du. A mechanical element that represents friction is the **dash pot** or **damper** shown in Fig. 15.5, and the describing equation is

$$f = Du$$

(15.5)

Since $D = f/u$, then the units for D are N-s/m. Again for the case that force and velocity are analogous to current and voltage, respectively, we see that the dash pot is the mechanical analog of conductance ($i = Gv$).

Another mechanical circuit element is a **spring** that has a **compliance** K, which is the name for the proportionality constant arising from the property that force is proportional to displacement. Since velocity is the derivative of displacement, for a

Fig. 15.5 Symbol for a dash pot or damper.   **Fig. 15.6** Symbol for a spring.

spring with compliance K, as symbolized in Fig. 15.6, the relationship between f and u is given by

$$f = \frac{1}{K}\int u \, dt \qquad \Rightarrow \qquad u = K\frac{df}{dt} \qquad\qquad (15.6)$$

Since $K = u/(df/dt)$, then the units for K are $(m/s)/(N/s) = m/N$. It is interesting to note that a spring and an inductor ($v = L \, di/dt$), being analogs, have the same pictorial representation.

Example 15.2

Figure 15.7 shows a mechanical system consisting of a spring with compliance K, a mass M on a surface with coefficient of friction D, and an applied force f. **D'Alembert's principle**[1] states that the inertia force and the external forces acting on a mass sum to zero. If the initial velocity and displacement are zero, then using Eq. 15.4, Eq. 15.5, and the integral form of Eq. 15.6, application of D'Alembert's principle to the mechanical system yields

$$\frac{1}{K}\int_0^t u \, dt + Du + M\frac{du}{dt} = f \qquad\qquad (15.7)$$

This equation corresponds to the mechanical circuit shown in Fig. 15.8a. The electrical analog of this mechanical circuit is shown in Fig. 15.8b.

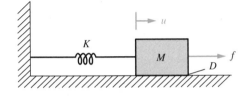

Fig. 15.7 A mechanical system.

---

[1]Named for the French mathematician Jean Le Rond D'Alembert (1717–1783).

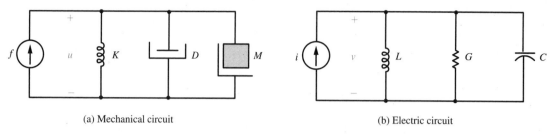

(a) Mechanical circuit   (b) Electric circuit

Fig. 15.8 Analogous mechanical and electric circuits.

Drill Exercise 15.2

For the mechanical system shown in Fig. 15.7, connect a dash pot D_p between the mass M and the vertical wall. Write an equation corresponding to the resulting mechanical system assuming that the initial velocity and displacement are zero. Draw the mechanical circuit for this system.

ANSWER $\dfrac{1}{K}\displaystyle\int_0^t u \, dt + (D + D_p)u + M\dfrac{du}{dt} = f$

The electrical analogs of mechanical circuits that we have just discussed were based on the choice of the force-current and velocity-voltage analogies. We could have instead, however, selected force to be analagous to voltage and velocity to be analogous to current. Doing so would result in alternative electrical analogs (see Problems 15.5 and 15.6 at the end of this chapter).

Model for a Motor

We are now in a position to generalize our previous discussion of simple motors and generators. In particular, reconsider the case of the motor shown in Fig. 15.3. Suppose the battery is replaced by a general voltage source v, and L is the inductance of the moving conductor, and its resistance (and that of the track, too) is lumped together with R. In addition, suppose the moving conductor has mass M, coefficient of friction D, and an external force f (directed to the left) loading it. Then, by KVL, the equation describing the electric portion of the motor is

$$v = Ri + L\frac{di}{dt} + e = Ri + L\frac{di}{dt} + Blu \qquad (15.8)$$

Applying D'Alembert's principle to the mechanical portion of the motor yields

$$f + Du + M\frac{du}{dt} = Bli$$

or

$$f = -Du - M\frac{du}{dt} + Bli \tag{15.9}$$

Thus we can model the motor, described by Eq. 15.8 and Eq. 15.9, by the electro-mechanical circuit shown in Fig. 15.9.

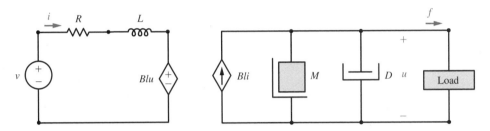

Fig. 15.9 Electromechanical circuit model of a motor.

15.2 Moving-Coil and Moving-Iron Devices

Up to now we have given some thought to simple motors and generators that util-ize movable conductors. These kinds of elementary systems are called **transla-tional transducers**. A **moving-coil** or **dynamic transducer** is a more practical type of translational transducer. Such a device, which can be employed in loud-speakers, microphones, and phonograph pickups, is illustrated in Fig. 15.10. The permanent magnet has an annular air gap with a radial magnetic field (only the flux at the end of the magnet is depicted). A sleeve (shown extended) with an N-turn coil fits over the south pole of the permanent magnet. The sleeve-coil com-bination, called the **moving coil**, is supported so that it can move axially in either direction.

Dynamic Loudspeakers

For the case of a loudspeaker, the moving coil of a dynamic transducer is connected to a cone or diaphragm that vibrates when alternating current passes through the coil. This results in the radiation of acoustic energy.

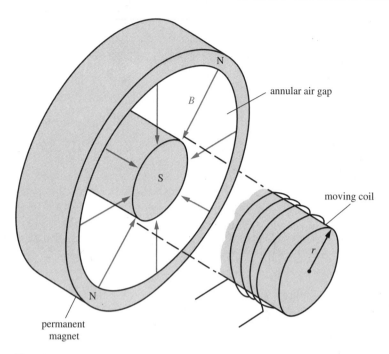

Fig. 15.10 Dynamic (moving-coil) transducer.

Suppose a voltage source v is applied to the coil whose winding resistance is R. Neglecting the inductance of the coil (i.e., $L = 0$ H), when a current i is directed out of the plus terminal of the voltage source, the equation describing the electric portion is, from Eq. 15.8,

$$v = Ri + e = Ri + Blu \tag{15.10}$$

where the length-l coiled conductor is comprised of N turns, each of which has radius r, that is, $l = 2\pi r N$. For the mechanical portion, from Eq. 15.9,

$$f = -Du - M\frac{du}{dt} + Bli \tag{15.11}$$

where D is the coefficient of friction of the moving coil, M is the mass, and again $l = 2\pi r N$. This time, however, we consider that the external force f due to an acoustic load is proportional to the velocity u, that is, $f = D_L u$. Thus we can model the loudspeaker described by Eq. 15.10 and Eq. 15.11 with the electromechanical circuit shown in Fig. 15.11.

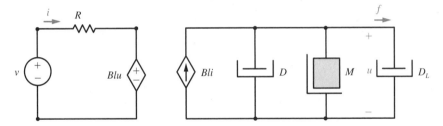

Fig. 15.11 Model of a loudspeaker.

Example 15.3

The magnetic flux density in the annular air gap of a dynamic loudspeaker (see Fig. 15.10) is $B = 0.5$ T. Suppose the moving coil has a radius of $r = 2$ cm and a current of $i(t) = 0.1 \cos \omega t$ A. Let us find the number of turns required to develop a maximum force of $Bli = 0.25$ N.

From the fact that

$$Bli = B2\pi rNi = 0.25 \text{ N}$$

and since the developed force Bli is maximum when i is maximum (denoted i_m), we have that the number of turns is

$$N = \frac{0.25}{B2\pi r i_m} = \frac{0.25}{(0.5)(2\pi)(0.02)(0.1)} \approx 40$$

Now suppose $D = 3$ N-s/m and $D_L = 12$ N-s/m. If we ignore the mass (i.e., $M = 0$ kg), then using current division (as for conductances) for the mechanical portion of the circuit in Fig. 15.11, we find that the force on the load is

$$f = \frac{D_L}{D_L + D} Bli = \frac{12}{12 + 3}(0.25) = 0.2 \text{ N}$$

Drill Exercise 15.3

For the loudspeaker given in Fig. 15.10, suppose $B = 1$ T, $R = 4 \ \Omega$, $M = 1$ g (0.001 kg), $N = 100$, $D = 4$ N-s/m, $D_L = 12$ N-s/m, and $r = 4$ cm.

Determine the velocity $u(t)$ for the case that $v(t) = \cos 16{,}000t$ V. (*Hint*: Use phasors and the fact that for a mass, $\mathbf{U}/\mathbf{F} = 1/j\omega M$.)

ANSWER $36 \cos (16{,}000t - 5.26°)$ mm/s

Other Dynamic Transducers

The idea behind a dynamic microphone is the converse of the dynamic loudspeaker. Although the mechanisms are essentially the same, it is acoustic energy that applies a force f (the input) to the diaphragm of the microphone. This, in turn, translates the moving coil through the magnetic field established by the permanent magnet. The result is an induced emf $e = Blu$ that supplies an electrical signal to a load (e.g., the input of an amplifier). The equations describing the behavior of a dynamic microphone are similar to Eq. 15.10 and Eq. 15.11—the difference being that some signs are reversed since the energy conversion is mechanical to electrical instead of electrical to mechanical. The describing equations are

$$v = -Ri + e = -Ri + Blu \tag{15.12}$$

and

$$f = Du + M\frac{du}{dt} + Bli \tag{15.13}$$

where f is the input, and i and v are the output current and output voltage, respectively. These equations can be modeled in a manner similar to the case of the loudspeaker (see Problem 15.10 at the end of this chapter).

Another use of the moving-coil transducer is as a **dynamic phonograph pickup**, as depicted in Fig. 15.12 by a cross-sectional view. Attached to the moving coil is a **stylus** that produces a vertical displacement $y(t)$ of the moving coil as the stylus tracks a record groove. (The variations in the groove represent the original audio signal.) The motion of the moving coil due to the displacement of the stylus induces an emf across the winding. This voltage can then be applied to an amplifier whose input resistance R_L is the load for the moving coil's winding. Since the behavior of such a phonograph pickup is that of a dynamic microphone, then from Eq. 15.12 and the fact that $i = v/R_L$ (see Problem 15.10), we have that

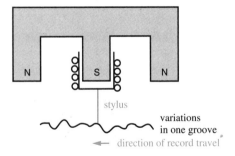

Fig. 15.12 Dynamic phonograph pickup.

$$v = -R\left(\frac{v}{R_L}\right) + Blu$$

from which

$$v = \frac{R_L}{R_L + R} Blu = \frac{R_L Bl}{R_L + R} \frac{dy}{dt}$$

since $u = dy/dt$.

For the case that a sinusoid having a frequency of ω rad/s and a constant amplitude is recorded, the corresponding vertical displacement has the form

$$y(t) = A \cos(\omega t + \theta)$$

and, therefore, the resulting output voltage v is proportional to

$$\frac{dy(t)}{dt} = -A\omega \sin(\omega t + \theta) = A\omega \sin(\omega t + \theta + 180°)$$

$$= A\omega \cos(\omega t + \theta + 90°)$$

Thus the sinusoid undergoes a phase shift of 90 degrees, and its amplitude increases linearly with frequency as well. Because of this phenomenon, phonograph amplifiers contain **equalizing networks** to compensate for such a frequency dependence.

A Relay

Not only does a magnetic field exert force on moving electric charge, but it also will exert a force on magnetic material, such as iron. (Anyone who has ever played with a permanent magnet has experienced this phenomenon.) It is this property that forms the basis for another type of transducer, called a **moving-iron transducer**, which utilizes a magnetic field to attract an initially unmagnetized, movable piece of iron.

Figure 15.13 shows a coil wound around a magnetic core that has an air gap. Such a device is often referred to as an **electromagnet**. In this particular case, part of the core is pivoted and is able to move. The movable portion, called the **armature**, is held in place by a spring. If a current passes through the coil, the resulting force opposes the restraining force of the spring and tends to move the armature so that the gap length gets smaller. This is the case regardless of either the direction of the current or the direction of the resulting flux. Also depicted in Fig. 15.13 is a rigid member extending from the armature. When the coil current is large enough, the armature will move a sufficient distance for the member to close the pair of contacts shown. (These contacts, in turn, may be used as a switch for some other system.) The entire device seen in Fig. 15.13 is a magnetically operated switch called a **relay**.

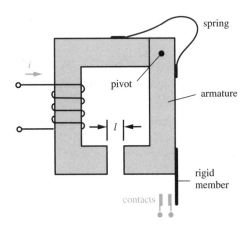

Fig. 15.13 Relay.

But how much force results from an applied current, and how much current is required to close the contacts of such a relay? To answer these questions, we proceed as follows: Suppose as the result of an applied current, the armature moves an infinitesimal horizontal distance dl, called a **virtual displacement**, due to the force f on it. The resulting work, called **virtual work**, done on the armature is $f\,dl$. For the sake of simplicity, let us assume that during the duration dt of the virtual displacement, the current is adjusted so that the flux ϕ remains constant. As a consequence, the induced emf is $e = N\,d\phi/dt = 0$ V and, therefore, the electric energy put into the system is $vi\,dt = Ri^2\,dt$, where R is the winding resistance of the coil.

By Eq. 14.20 on p. 923, the inductance of the device is $L = N^2/\mathscr{R}$, where the reluctance $\mathscr{R} = \mathscr{R}_c + \mathscr{R}_g$ is the sum of the core reluctance $\mathscr{R}_c$ and the gap reluctance $\mathscr{R}_g$. Assuming that the permeability of iron is so high that $\mathscr{R}_g \gg \mathscr{R}_c$, then $\mathscr{R} \approx \mathscr{R}_g$, and the energy stored in the air gap is $w_g = \frac{1}{2} L i^2 = N^2 i^2/2\mathscr{R}$. From Eq. 14.8 and Eq. 14.10 on p. 910, $i = \mathscr{R}\phi/N$, and, thus,

$$w_g = \frac{1}{2}\mathscr{R}\phi^2 = \frac{1}{2}\left(\frac{l}{\mu_0 A}\right)(BA)^2 = \frac{B^2 A l}{2\mu_0}$$

where A is the cross-sectional area of the gap and fringing is ignored. Since the length of the gap decreases by dl, then the energy stored in the gap decreases by $B^2 A\, dl/2\mu_0$. Thus this amount of energy also goes into the system. The motion of the armature is horizontal, and so there is no change in potential energy; and, if we assume that the motion is slow enough, we can neglect kinetic energy.

As mentioned above, one result of the energy put into the system is the virtual work done on the armature. Another is the heat due to the energy absorbed by the winding resistance R—this loss is $R i^2\, dt$. Making the further simplifying assumption of no friction, we can ignore mechanical friction loss. Finally, with a constant flux, there is no loss from hysteresis and eddy currents. Therefore, by the principle of the conservation of energy, we have that

$$R i^2\, dt + \frac{B^2 A\, dl}{2\mu_0} = f\, dl + R i^2\, dt$$

from which the force on the armature is

$$f = \frac{B^2 A}{2\mu_0} \tag{15.14}$$

Since from Eq. 14.8 and Eq. 14.10, we have that $\phi = Ni/\mathscr{R}$, and since $B = \phi/A$, then

$$B = \frac{Ni}{\mathscr{R}A} = \frac{Ni}{(l/\mu_0 A)A} = \frac{\mu_0 Ni}{l}$$

Substituting this expression for B into Eq. 15.14 yields

$$f = \frac{\mu_0 N^2 i^2 A}{2l^2} \tag{15.15}$$

Example 15.4

Suppose the relay given in Fig. 15.13 has a coil with 400 turns, a cross-sectional area of 10 cm^2, and a spring force of 0.5 N. The gap length when the contacts are open is 1 cm, and 0.4 cm when the contacts are closed. Then from Eq. 15.15, the current required to close the contacts, called the **pickup** or **pull-in current**, is

$$i = \sqrt{\frac{2l^2 f}{\mu_0 N^2 A}} = \frac{l}{N}\sqrt{\frac{2f}{\mu_0 A}} \qquad (15.16)$$

$$= \frac{0.01}{400}\sqrt{\frac{2(0.5)}{(4\pi \times 10^{-7})(10)(0.01)^2}} = 0.705 \text{ A}$$

The **dropout** or **hold-in current** is the minimum current required to keep the contacts closed. Some value of current, which must be less than 0.705 A, will result in the contacts opening. Since Eq. 15.16 indicates that current is directly proportional to gap length, we deduce that the dropout current is

$$i = \frac{0.4}{1.0}(0.705) = 0.282 \text{ A}$$

Drill Exercise 15.4

For the relay shown in Fig. 15.13, the core has a cross-sectional area of 10 mm^2 and a spring force of 0.4 N. When the contacts are open, the gap length is 5 mm. Determine the number of turns required for a current of 0.1 A to close the contacts.

ANSWER 399

A Plunger

Another type of moving-iron transducer is shown in Fig. 15.14. Here we have an N-turn coil (a solenoid) with a movable iron core, called a **plunger**. When a current passes through the coil, a force directed to the right (so as to center the plunger in the coil) is exerted on the plunger. The plunger may be connected to some mechanical apparatus such as a valve.

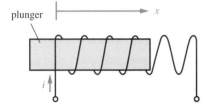

Fig. 15.14 Solenoid-plunger combination.

Suppose as a result of an applied current, the plunger has a virtual displacement of dx meters in dt seconds. Thus the virtual work done on the plunger is $f\,dx$. Also, suppose the coil has zero resistance. Then the voltage induced across the coil is $e = N\,d\phi/dt$, and the electric energy put into the system is

$$ei\,dt = N\frac{d\phi}{dt}i\,dt = Ni\,d\phi = i\,d(N\phi)$$

From Eq. 14.19 on p. 923, $N\phi = Li$. Thus $i\,d(N\phi) = i\,d(Li)$, and if the current is constant, then the electric energy put into the system is

$$ei\,dt = i^2\,dL$$

where dL is the change of inductance. (As the plunger moves to the right, the flux ϕ increases. This, in turn, decreases the reluctance $\mathcal{R} = Ni/\phi$ and, hence, increases the inductance $L = N^2/\mathcal{R}$.) Since $\frac{1}{2}Li^2$ is the energy stored, the increase in energy stored is $\frac{1}{2}i^2\,dL$. As for the case of the relay discussed previously, we will not be concerned with the potential and kinetic energies. Ignoring hysteresis, eddy currents, and friction, by the conservation of energy we have that

$$i^2\,dL = f\,dx + \tfrac{1}{2}i^2\,dL$$

from which the force exerted on the plunger is given by

$$\boxed{f = \tfrac{1}{2}i^2\frac{dL}{dx}} \qquad\qquad (15.17)$$

where the current i is constant.

15.3 Rotating-Coil Devices

Having discussed translational transducers, let us now consider **rotational transducers**, that is, electromechanical transducers that utilize rotation. We begin with a classical example—the **d'Arsonval movement**.[2] The basic components of this mechanism, which employs a rotating coil, are depicted in Fig. 15.15. It consists of a stationary permanent magnet with an air gap in which a stationary iron core is placed. (The air gap between the permanent magnet and the iron core is greatly exaggerated for illustrative purposes.) Around the stationary core is an N-turn coil that is wound on a rectangular form, and this coil is free to rotate around its vertical axis. The coil assembly, which is supported on jeweled bearings, is connected to two springs that not only supply a restraining torque, but also provide an electric connection to the rotating coil. The angular deflection γ of the coil assembly is indicated by a pointer attached to it.

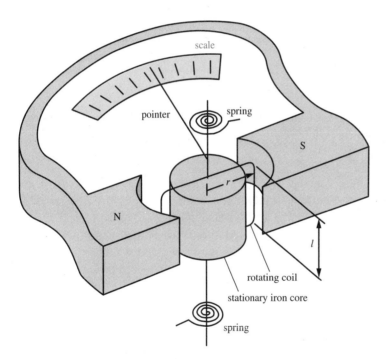

Fig. 15.15 The d'Arsonval movement.

A top view of the air-gap region of the d'Arsonval movement is shown in Fig. 15.16. Inserting a stationary iron core between the poles of the permanent magnet produces a uniform radial magnetic field with flux density B in the resulting air gap. Depicted in Fig. 15.16 is just one turn of the N-turn movable coil. Assuming that the coil's rotation is restricted (typically by mechanical stops) to the uniform field,

---

[2]Named for the French physicist Jacques Arsène d'Arsonval (1851–1940).

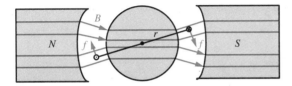

Fig. 15.16 Air-gap region of a d'Arsonval movement.

when a current i goes through the coil, a force will be exerted on it. For the case indicated in which the current direction is into the page for the upper right-hand portion of the coil and out of the page for the lower left-hand portion, the resulting forces have the directions shown. Since one side of a single turn has length l, by Equation 15.3, a force of $f = Bli$ is exerted on this conductor. Consequently, the resulting torque is $fr = Blir$. Because the other side of the turn experiences the same torque, and the coil has N turns, the clockwise torque T on the movable coil is

$$T = 2NBlri \tag{15.18}$$

In opposition to this torque is a counterclockwise torque T_0 attributable to the springs. We thus have that

$$T_0 = \frac{\gamma}{K} \tag{15.19}$$

where K is the rotational compliance of the springs and γ is the angle of deflection of the coil. For the case of equilibrium, from Eq. 15.18 and Eq. 15.19, we obtain

$$\gamma = (2KNBlr)i \tag{15.20}$$

In other words, the deflection angle is directly proportional to the current in the movable coil. For this reason, the scale of the d'Arsonval movement is calibrated linearly, and such an instrument is used as a dc ammeter or dc voltmeter.

Example 15.5

A d'Arsonval movement has a moving coil with dimensions $l = 2$ cm $= 0.02$ m and $r = 1$ cm $= 0.01$ m. The magnetic flux density in the air gap is $B = 0.5$ T, and the spring compliance is $K = 2 \times 10^8$ degrees/N-m. Suppose a full-scale deflection corresponds to $\gamma = 90$ degrees. Let us determine the number of required coil turns if a current of $i = 50$ μA is to cause a full-scale deflection.

From Eq. 15.20, the number of turns required for the moving coil is

$$N = \frac{\gamma}{2KBlri} = \frac{90}{(2)(2 \times 10^8)(0.5)(0.02)(0.01)(50 \times 10^{-6})} = 45$$

Drill Exercise 15.5

For the d'Arsonval movement described in Example 15.5, what value of magnetic flux density is required if the coil has 100 turns?

ANSWER 0.225 T

Ammeters

Clearly, the d'Arsonval movement given in the preceding example can be used as a dc ammeter to measure currents ranging from 0 to 50 μA. It is possible, however, to use this 50-μA movement as a dc ammeter for other current ranges as well.

Example 15.6

Suppose the resistance of the moving coil of a 50-μA d'Arsonval movement is $R_c = 2$ kΩ. To use this movement for a 0- to 100-mA dc ammeter, let us connect

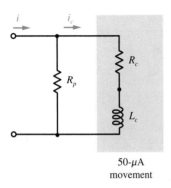

50-μA
movement

Fig. 15.17 A d'Arsonval movement ammeter.

a **shunt** resistance R_p in parallel with the moving coil as illustrated in Fig. 15.17 and determine the appropriate value for R_p.

Since the inductance L_c behaves as a short circuit to dc, by current division, we have that

$$i_c = \frac{R_p}{R_p + R_c} i \quad \Rightarrow \quad R_p = \frac{i_c}{i - i_c} R_c \tag{15.21}$$

For full-scale deflection, $i_c = 50 \ \mu A$ and $i = 100 \ mA$. Thus from Eq. 15.21, the value of the shunt resistance is

$$R_p = \frac{50 \times 10^{-6}}{100 \times 10^{-3} - 50 \times 10^{-6}} (2 \times 10^3) = 1.00 \ \Omega$$

Drill Exercise 15.6

A 2-Ω resistor is connected in parallel with a 50-μA d'Arsonval movement. Given that the coil has a resistance of 2 kΩ, determine the current range for the resulting dc ammeter.

ANSWER 0 to 50 mA

An ideal ammeter has a resistance of zero ohms so that when it is inserted in series with an element whose current is to be measured, the resulting combined resistance remains the same and, thus, the values of the circuit variables do not change. Therefore, since its resistance is 2 kΩ, the 50-μA d'Arsonval movement just mentioned is a reasonable approximation of an ideal 50-μA ammeter when it is used to measure currents through very large resistances ($\gg$ 2 kΩ). However, since the resistance of the 100-mA ammeter discussed in Example 15.6 is 1 Ω, it approximates an ideal ammeter for a much wider variety of situations.

Voltmeters

We have just seen an example of how a 50-μA movement can be used as an ammeter for currents greater than 50 μA. To accomplish this, we utilized a shunt resistance R_p; however, even though a d'Arsonval movement is basically a current-measuring device, we can also use it to measure dc voltages.

Example 15.7

Suppose the resistance of the moving coil of a 50-µA d'Arsonval movement is $R_c = 2$ kΩ. To use this movement for a 0- to 20-V dc voltmeter, let us connect a resistance R_s in series with the moving coil as depicted in Fig. 15.18 and determine the appropriate value for R_s.

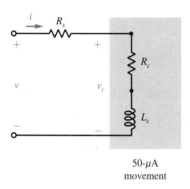

50-µA
movement

Fig. 15.18 A d'Arsonval movement voltmeter.

For full-scale deflection $i = 50$ µA. Since this is to correspond to the case that $v = 20$ V, by Ohm's law, we have that

$$R_s = \frac{v - v_c}{i} = \frac{v - R_c i}{i} = \frac{v}{i} - R_c$$

$$= \frac{20}{50 \times 10^{-6}} - 2 \times 10^3 = 3.98 \times 10^5 = 398 \text{ k}\Omega$$

Thus we see that the voltmeter shown in Fig. 15.18 has a resistance of $R_s + R_c = 398{,}000 + 2000 = 400$ kΩ.

Drill Exercise 15.7

A 238-kΩ resistor is connected in series with a 50-µA d'Arsonval movement. Given that the coil has a resistance of 2 kΩ, determine the voltage range for the resulting dc voltmeter. What is the resistance of this voltmeter?

ANSWER 0 to 12 V, 240 kΩ

An ideal voltmeter has infinite resistance so that, when voltage measurements are made, a circuit will not be disturbed (an ideal voltmeter draws no current). The voltmeter described in Example 15.7 might significantly alter circuit variable values in some types of applications (e.g., see Problem 15.22 at the end of this chapter).

In a voltmeter consisting of a d'Arsonval movement and a series resistor R_s such as the one in Example 15.7, instead of talking about the resistance of the voltmeter (400 kΩ in this case), typically the meter is described in terms of the number of ohms per volt of full-scale deflection. Thus we would say that this meter is rated at

$$\frac{400,000 \ \Omega}{20 \ V} = 20,000 \ \Omega/V = 20 \ k\Omega/V$$

Ohmmeters

In addition to being used in ammeters and voltmeters, d'Arsonval movements can be utilized for measuring the resistances of resistors or combinations of resistors— that is, they can be used to fabricate ohmmeters.

Example 15.8

An elementary method for constructing an ohmmmeter is illustrated in Fig. 15.19. This ohmmeter consists of a d'Arsonval movement connected in series with a resistance R_s and a dc voltage source (battery) V_s. A resistance R is connected to the terminals of the ohmmeter and its value is read on the d'Arsonval's calibrated scale.

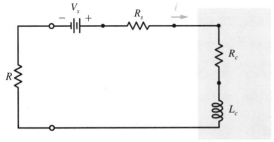

d' Arsonval movement

Fig. 15.19 A d'Arsonval movement ohmmeter.

Suppose a 50-μA movement with a coil resistance of $R_c = 2$ kΩ and a battery of $V_s = 1.5$ V are to be used for the ohmmeter shown in Fig. 15.19. By KVL,

$$V_s = R_s i + R_c i + Ri$$

from which

$$R_s = \frac{V_s - R_c i - Ri}{i} = \frac{V_s}{i} - R_c - R \tag{15.22}$$

To get full-scale deflection (i.e., $i = 50$ μA) when a short circuit (i.e., $R = 0$ Ω) is connected to the terminals of the ohmmeter, from Eq. 15.22 we require that

$$R_s = \frac{1.5}{50 \times 10^{-6}} - 2 \times 10^3 - 0 = 28 \text{ k}\Omega$$

From Eq. 15.22,

$$R = \frac{V_s}{i} - R_c - R_s$$

For half-scale deflection ($i = 25$ μA), we have that the corresponding resistance is, therefore,

$$R = \frac{1.5}{25 \times 10^{-6}} - 28 \times 10^3 - 2 \times 10^3 = 30 \text{ k}\Omega$$

For quarter-scale deflection ($i = 12.5$ μA), the value of the resistance is

$$R = \frac{1.5}{12.5 \times 10^{-6}} - 28 \times 10^3 - 2 \times 10^3 = 90 \text{ k}\Omega$$

Finally, for the case that nothing is connected to the terminals of the ohmmeter (i.e., R is infinite), the current is $i = 0$ A, so there is no meter deflection.

Drill Exercise 15.8

For the ohmmeter shown in Fig. 15.9, suppose that $V_s = 3$ V and the d'Arsonval movement has a coil resistance of 2 kΩ and requires 50 μA for full-scale deflection. (a) Determine the value of R_s required to have full-scale deflection when $R = 0$ Ω. (b) What value of R corresponds to $\frac{3}{4}$-scale deflection? (c) What value of R corresponds to $\frac{1}{2}$-scale deflection? (d) What value of R corresponds to $\frac{1}{4}$-scale deflection?

ANSWER (a) 58 kΩ, (b) 20 kΩ, (c) 60 kΩ, (d) 180 kΩ

Unlike the ammeter and the voltmeter, which indicate values on a linear, increasing scale, as demonstrated in the preceding example, the ohmmeter displays resistance on a decreasing scale that is nonlinear. An actual ohmmeter is more complicated than the one discussed, but the basic notion of measuring resistance is the same.

We have seen that a d'Arsonval movement can be used to measure current, voltage, and resistance. In actuality, the same d'Arsonval movement can be employed to measure a wide variety of current values. Changes in a ammeter scale are easily accomplished by switching to different resistors for the shunt. In an analogous manner, changing the series resistance of a voltmeter or an ohmmeter readily can switch the scales of the respective meters. A meter with a single d'Arsonval movement that can be used to measure multiple ranges of currents and voltages is called a **multimeter**. A common type of multimeter—a combination voltmeter, ohmmeter, and milliammeter—is abbreviated VOM.

The Electrodynamometer

Having studied the d'Arsonval movement—a permanent-magnet, moving-coil mechanism—we now will look at the **electrodynamometer**. The principle of operation of the latter mechanism is the same as for the former, but its construction is quite different. Instead of a permanent magnet, the magnetic field of the electrodynamometer is supplied by a stationary solenoid separated into two sections as illustrated in Fig. 15.20. The core of the solenoid is comprised of nonmagnetic material (thus avoiding eddy-current and hysteresis losses) so that the magnetic flux density B is directly proportional to the current i_s through the solenoid (e.g., see Eq. 14.3). In the solenoid's magnetic field is an N-turn movable coil (only one turn is depicted in

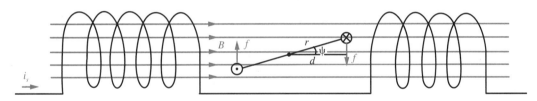

Fig. 15.20 Magnetic-field region for an electrodynamometer.

Fig. 15.20) restrained by springs and equipped with a pointer (not shown), as is done for a d'Arsonval movement. Given that the length of the moving coil is l and the width is $2r$, then when a current i_c goes through it, the force (with the direction shown) on the upper right-hand portion of the coil is $f = Bli_c$. Therefore, the resulting torque is $fd = Bli_c d$. If the plane of the moving coil is at an angle ψ with the direction of B, then $d = r \cos \psi$. Combining this with the fact that the coil has N turns, we see that the resulting clockwise torque on the movable coil is

$$T = 2NBli_c r \cos \psi$$

However, since B is directly proportional to the solenoid current i_s, that is, since $B = K_s i_s$, then

$$T = (2NK_s lr \cos \psi)i_c i_s$$

Equating this torque to the spring restraining torque $T_0 = \gamma/K$, we obtain

$$\gamma = (2NKK_s lr \cos \psi)i_c i_s \tag{15.23}$$

Thus we see that the angle of deflection γ depends on the product of the moving-coil current i_c and the solenoid current i_s.

Although the electrodynamometer can be used in dc ammeters and dc voltmeters, because of its generally stronger magnetic field, the d'Arsonval movement is superior. The electrodynamometer is quite useful, however, for ac measurements.

When both coils carry alternating currents, say $i_c = I_c \cos(\omega t + \theta_1)$ and $i_s = I_s \cos(\omega t + \theta_2)$, the mechanical inertia of the movement causes an averaging of the angle of deflection. Since the average value of $i_c i_s$ is $\frac{1}{2}I_c I_s \cos \theta$, where $\theta = \theta_1 - \theta_2$ (see Problem 15.27 at the end of this chapter), then when the instantaneous deflection-angle expression (Eq. 15.23) is averaged, we find that the angle of deflection has the form

$$\gamma = K_\psi I_c I_s \cos \theta \tag{15.24}$$

where the value of K_ψ varies as the angle ψ varies.

For the case that the solenoid and moving coil are connected in series, the current through them will be the same. Thus when $i_c = i_s = I \cos \omega t$, from Eq. 15.24, we see that the angle of deflection is proportional to I^2. As a consequence, the scale can be graduated nonlinearly to read the rms value $(I/\sqrt{2})$ of the current. Hence the electrodynamometer can be used as an ac ammeter or ac voltmeter. But, because of a practical limitation on the current through the moving coil and because the reac-

tance of the moving coil is significant in ammeter applications, the electrodynamometer's use is much more common for ac voltmeters than for ac ammeters.

The Wattmeter

Perhaps the most important use of the electrodynamometer is for the wattmeter. For
the mechanism to function in this manner, the fixed coil (solenoid) is used for the
load current $I \cos(\omega t + \theta_2)$, and the moving coil connected in series with a large
resistance R_s is used for the load voltage $V \cos(\omega t + \theta_1)$. Assuming that the impedance of the moving coil is negligible compared to R_s, the amplitude of the moving-
coil current is $I_c = V/R_s$. Furthermore, the amplitude of the current in the fixed coil
is $I_s = I$ assuming that $i_s = i$. Thus from Eq. 15.24, we see that the angle of deflection
is proportional to $I_c I_s \cos \theta$, where $\theta = \theta_1 - \theta_2$, and, therefore, it is proportional to
the average power $P = \frac{1}{2} VI \cos \theta$ absorbed by the load. Hence the scale of the
electrodynamometer can be graduated to indicate power.

A typical connection of an electrodynamometer for use as a wattmeter is shown
in Fig. 15.21. For the current directions indicated, if both i_c and i_s are positively
valued, then the pointer of the meter will have an upscale deflection. If both currents
have a negative value, then the deflection again will be up-scale. For this reason, the
appropriate ends of the coils are marked with the symbol $\pm$. If one current has a
positive value and the other has a negative value, then the meter will have a down-
scale deflection.

In the discussion above, it was assumed that the load current is equal to the current
in the fixed coil, that is $i = i_s$. To be more accurate, however, from Fig. 15.21, we

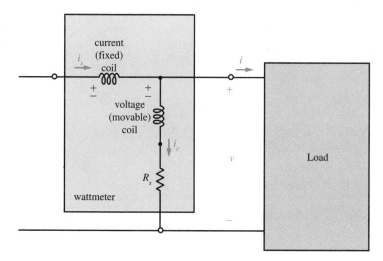

Fig. 15.21 Electrodynamometer wattmeter.

see that $i = i_s - i_c$. Therefore, the load current is less than the current in the fixed coil, and, consequently, the wattmeter reading is larger than the power absorbed by the load. The power indicated by the wattmeter is the sum of the load power and the power absorbed by the series connection of the moving coil and the resistance R_s. If P_L is the power absorbed by the load and P_W is the wattmeter reading, since the amplitude of the load voltage is V, then

$$P_L = P_W - \frac{V^2}{2R_s}$$

There are wattmeters, called **compensated wattmeters**, that directly read P_L. In such a meter, the moving-coil (the voltage coil) current goes through an additional fixed coil located so as to cancel the effect of i_c on the current in the fixed coil.

15.4 DC Machines

Generators and motors are examples of electromechanical **machines**. The simple machines shown in Figs. 15.1, 15.2, and 15.3 are translational devices. Rotating machines, however, are inherently more practical because of their higher velocity, voltage, and power capabilities. Therefore, we will now turn our attention to the subject of rotating machines.

Rotating Machines

Let us begin with the single conducting loop embedded in a uniform magnetic field having a constant flux density B as depicted in Fig. 15.22a. Suppose the loop, which is rectangular with length l and width $2r$, is forced to rotate around the z axis with a constant angular velocity of ω radians per second as indicated. At time t, the angle between the plane of the loop and the y-z plane is ωt, as shown in Fig. 15.22a and the end view in Fig. 15.22b. As indicated in Fig. 15.22a, the ends of the loop are connected to **slip rings** that rotate along with the loop. Touching the slip rings, however, are **brushes** that make electrical connections, but that are fixed. The area of the loop is $A = 2rl$ and the flux ϕ passing through the loop is

$$\phi = B(2r \cos \omega t)l = B2rl \cos \omega t = BA \cos \omega t$$

By Faraday's law, the emf induced across the brushes is

$$e = \frac{d\phi}{dt} = -BA\omega \sin \omega t \qquad (15.25)$$

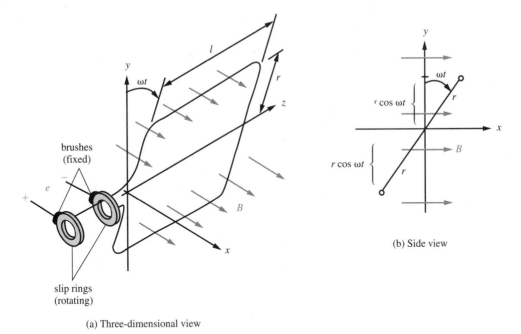

(a) Three-dimensional view

(b) Side view

Fig. 15.22 Simple rotating machine.

and, due to Lenz's law, the polarity of this emf is as indicated. If an *N*-turn coil were used in place of the single loop, the resulting induced emf would be

$$e = -NBA\omega \sin \omega t \qquad (15.26)$$

Since the simple rotating machine shown in Fig. 15.22*a* produces a sinusoidal voltage, it is an example of an ac generator, which is referred to as an **alternator**. Note that, in addition to the number of coil turns, the magnetic field strength, and the area of the coil, the amplitude of the induced voltage also depends on the speed of rotation of the coil.

Now let us modify the simple alternator given in Fig. 15.22*a* as shown in Fig. 15.23*a*. What we have done is to replace the slip rings with a **split-ring commutator**—a side view of which is shown in Fig. 15.23*b*. Furthermore, the brushes have been repositioned on either side of the split-ring commutator. As we have already seen, the voltage induced across the loop is given by Eq. 15.25, and this is the voltage v_{12} between the halves of the split-ring commutator. The brushes, however, are situated such that

$$v = \begin{cases} -v_{12} = -e = BA\omega \sin \omega t & \text{when } 0 < \omega t < \pi \\ v_{12} = e = -BA\omega \sin \omega t & \text{when } \pi < \omega t < 2\pi \end{cases}$$

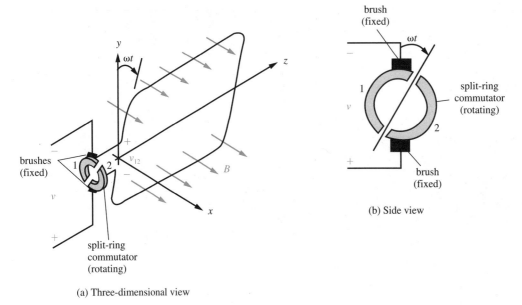

(a) Three-dimensional view

(b) Side view

Fig. 15.23 Simple rotating dc machine.

Therefore, whereas the voltage e between the brushes of the alternator given in Fig. 15.22a is $e = BA\omega \sin \omega t$, the voltage v between the brushes of the rotating machine given in Fig. 15.23a is $v = |BA\omega \sin \omega t|$. These two voltages are shown in Fig. 15.24a and b, respectively. Although the voltage v shown in Fig. 15.24b is not constant, it never takes on a negative value. Because the voltage does not change polarity, the simple machine given in Fig. 15.23a is an example of an elementary dc generator.

DC Generators

As was implied in the preceding discussions of a simple alternator and a simple dc generator, a typical rotating machine is comprised of a stationary portion called the

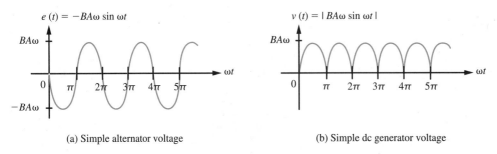

(a) Simple alternator voltage

(b) Simple dc generator voltage

Fig. 15.24 Voltages produced by simple alternator and dc generator.

stator and a rotating portion called the **rotor**. In a dc generator, a magnetic field is produced when a direct current is applied to windings or coils in the stator, and so the stator is referred to as the **field**, and such coils are called **field windings**. The rotor contains the commutator and the conductors across which emfs are induced. This portion of the machine is known as the **armature**.

A cross-sectional view of a simplified dc generator is shown in Fig. 15.25. Although the stator has two poles, one labeled N and one labeled S, an even number, greater than two, of magnetic poles can be used. As indicated, situated on the poles are the field windings. The rotor consists of a cylindrical iron core that contains slots that house the armature conductors and a commutator along with the associated brushes. Shown on the armature are eight conductors that form four coils, and a commutator with four segments. The brushes are depicted as riding on the inside, but they actually make contact with the outside of the commutator. Figure 15.26 gives one possibility for connecting the coils to the commutator. Here, conductors 1 and 1′ form coil 1, and so forth, but not shown are the implicit rear connections between 1 and 1′, 2 and 2′, 3 and 3′, and 4 and 4′.

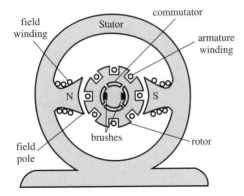

Fig. 15.25 A dc generator.

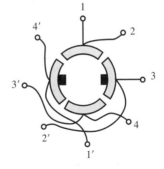

Fig. 15.26 Coil-commutator connections.

Passing (direct) current through the field windings produces essentially uniform magnetic fields in the small air gaps between the poles and the rotor. The emf e induced across a single coil is as shown in Fig. 15.27, where the maximum and

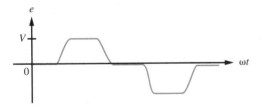

Fig. 15.27 The emf across a single coil.

minimum voltages occur when the coil is in a horizontal position. When the armature windings are connected to the commutator in an appropriate manner, such as the one indicated in Fig. 15.26, the individual induced emfs not only will be rectified, but they will be added together as well. For example, for the connection given in Fig. 15.26, the right-hand brush is connected to conductor 3, which, in turn, is (implicitly) connected to 3′. But 3′ is connected to 4 (via the commutator), which, in turn, is connected to 4′. Since 4′ is connected to the left-hand brush, then the series connection of coils 3 and 4 is connected between the two brushes—so their induced emfs add. Note, too, that coils 1 and 2 are also connected in series for the indicated brush positions. The net effect of the coil connections to the commutator is a brush voltage v such as shown in Fig. 15.28. The small notches in the waveform result from the brushes changing from the different segments of the commutator. In practice, many coils are employed and this **commutator ripple** is quite small.

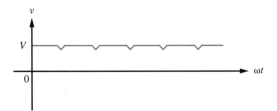

Fig. 15.28 The voltage across the brushes.

Generated Voltage

Let us now establish a formula for the voltage produced by a dc generator having p poles, each with a flux per pole of Φ, where the generator is turning at the rate of n revolutions per minute (rpm).

If the rotor has a speed of n rpm, then it turns at the rate of $n/60$ revolutions per second (rps). Since there are p poles, in one revolution the rotor turns at the rate of $pn/60$ poles per second. Each conductor, however, experiences a flux change of Φ per pole. Therefore, the average rate of change of flux (induced emf) per conductor is $\Phi pn/60$ volts. If there are N armature conductors and a parallel paths between the brushes, then the number of conductors connected in series comprising each parallel path is N/a. Consequently, the generated voltage v_g is given by

$$v_g = \frac{N}{a} p\Phi \frac{n}{60} \tag{15.27}$$

where N is the number of armature conductors, a is the number of parallel paths between the brushes, p is the number of poles, Φ is the magnetic flux per pole, and n is the speed of the rotor in rpm.

Example 15.9

Let us determine the voltage produced by a dc generator such as the one shown in Fig. 15.25, which has two poles and two parallel paths between the brushes, given that each pole has a surface area of $A = 0.01$ m^2, the air-gap magnetic flux density is $B = 1$ T, and each of the four armature coils has 12 turns.

The flux per pole is

$$\Phi = BA = 1(0.01) = 0.01 \text{ Wb}$$

Since each of the four armature coils has 12 turns and each turn contains two conductors, then $N = (4)(12)(2) = 96$.

For a rotor speed of 1500 rpm, the generated voltage is

$$v_g = \frac{N}{a}p\Phi\frac{n}{60} = \frac{96}{2}(2)(0.01)\frac{1500}{60} = 24 \text{ V}$$

Drill Exercise 15.9

A dc generator such as the one shown in Fig. 15.25 has two poles and two parallel paths between the brushes. Determine the rotor speed required to generate 16 V if each of the four armature coils has 10 turns and the flux per pole is 10 mWb.

ANSWER   1200 rpm

Generator with a Load

In the preceding discussion, the generated emf v_g is the voltage across the armature terminals (the brushes) when no load is connected to those terminals. If, however, there is some electrical load connected to the armature, the resulting load voltage, in general, will be different from the no-load voltage. This conclusion is a consequence of the fact that there is a resistance R_a associated with the armature due to such things as the resistances of the windings and the brush contacts. The following example demonstrates the effect of the armature resistance R_a of a dc generator.

Example 15.10

An 8-kW, 200-V dc generator has a full-load current of 40 A at 1200 rpm. Given that the armature resistance is $R_a = 0.5 \; \Omega$, we can model the armature as shown in Fig. 15.29, where R_L is a resistive load. Let us determine the full-load voltage for this generator at 900 rpm.

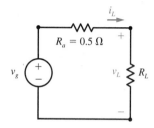

Fig. 15.29 Model of given armature and associated load.

From the fact that

$$v_g = R_a i_L + v_L$$

under full-load conditions (at $n = 1200$ rpm), the generated voltage is

$$v_g = (0.5)(40) + 200 = 220 \; \text{V}$$

and this is the no-load voltage.

From Eq. 15.27, we see that if the flux per pole Φ is kept constant, then the generated voltage is directly proportional to the armature speed n. Thus if the field current is kept constant (and, hence, the flux per pole is kept constant), then at 900 rpm the no-load voltage is

$$v_g = \frac{900}{1200}(220) = 165 \; \text{V}$$

Therefore, under the full-load condition of 40 A, at 900 rpm the full-load voltage is, by KVL,

$$v_L = -R_a i_L + v_g = -(0.5)(40) + 165 = 145 \; \text{V}$$

Drill Exercise 15.10

A 10-kW, 200-V dc generator has a full-load current of 50 A at 1500 rpm. At 1800 rpm, the full-load voltage is 244 V. Find (a) the armature resistance, (b) the no-load voltage at 1500 rpm, and (c) the no-load voltage at 1800 rpm.

ANSWER (a) 0.4 Ω, (b) 220 V, (c) 264 V

Generator Symbol

We can represent a dc generator symbolically as is shown in Fig. 15.30, where i_f indicates the field current and i_a denotes the armature current. Sometimes the symbol for a generator will include an indication of something, called a **prime mover**, that causes the armature to turn; however, we shall consider the prime mover to be implicit.

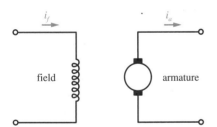

Fig. 15.30 Symbol for a dc generator.

From Equation 15.27, we know that the rotor of a dc generator turning at n rpm induces an emf (the no-load armature voltage) of

$$v_g = K\Phi n \tag{15.28}$$

where $K = Np/60a$ is a constant determined by the construction of the generator. Furthermore, the flux per pole Φ is determined by the field current i_f (see Equations 14.8 and 14.9). In general, the relationship between Φ and i_f is nonlinear. For a given speed n, therefore, let us plot the no-load voltage v_g versus the field current i_f. Such curves, which are shown in Fig. 15.31, can be obtained experimentally from a physical generator by holding the speed constant and measuring the value of v_g that results when different values of i_f are applied. Since v_g is proportional to Φ (Eq. 15.28), which, in turn, is proportional to B, and since i_f is proportional to $\mathcal{F}$ (Eq. 14.12),

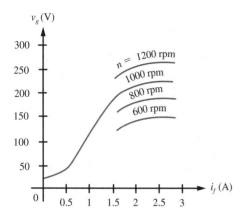

Fig. 15.31 Magnetization curves for a dc generator.

which, in turn, is proportional to H, a plot of v_g versus i_f has the same shape as a B-H curve. Thus the curves shown in Fig. 15.31 are known as **magnetization curves** of the generator. Sometimes, in addition to being labeled in terms of field current, the horizontal axis is also labeled in terms of mmf (ampere-turns) per pole. Once one magnetization curve has been determined (or measured), with the use of Eq. 15.28, it is a simple matter to obtain the magnetization curve corresponding to a different speed. Note that a magnetization curve, in particular, the curve corresponding to $n = 1000$ rpm in Fig. 15.31, does not pass through the origin. This is so because, even when the field current is zero, residual pole magnetism will result in a small generated emf.

Generator Field Excitation

There are four basic ways of establishing the field current for a dc generator. The most obvious approach is to connect its own individual source to the field winding[3] as indicated in Fig. 15.32. This is an example of a **separately excited generator**; however, this requires the use of an additional source V_f. Such a requirement can be eliminated by using the generator's own voltage to produce the field current. One way this can be done is by connecting the field in parallel or in shunt with the armature, as shown in Fig. 15.33a. The resulting **shunt-connected generator** is an example of a **self-excited generator**. In a shunt-connected generator, the field winding typically has many turns, and the relatively high field resistance results in a field current that is only a small fraction of the armature current. The field current can be

---

[3]Although the field winding is represented by the symbol for an ideal inductor, there is an implicit field resistance R_f that is not shown.

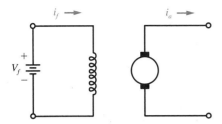

Fig. 15.32 Separately excited generator.

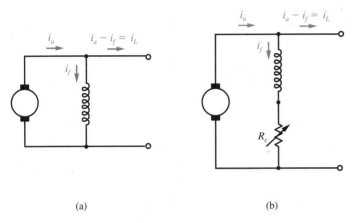

(a)   (b)

Fig. 15.33 Shunt-connected generator.

further adjusted by placing a variable resistance R_c, called a **control rheostat**, in series with the field as shown in Fig. 15.33b.

Another type of self-excitation is obtained for the **series-connected generator** shown in Fig. 15.34a. Since the field current is equal to the armature current, the field winding has few turns and a relatively low resistance. Even so, under load conditions, a large armature current can cause a significant difference between the

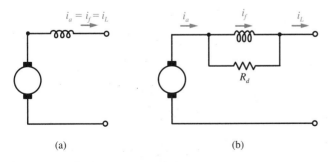

(a)   (b)

Fig. 15.34 Series-connected generator.

generated and load voltages. The relative effect of the series field resistance can be reduced by connecting a low resistance R_d in parallel with the field as shown in Fig. 15.34b. Since such a resistance diverts some of the armature current from the field, R_d is called a **diverter resistance**. Despite the possibility of using diverter resistances, series-connected generators are rarely used.

By combining the concepts of a shunt connection and a series connection, another type of self-excitation can be obtained. In particular, Fig. 15.35a shows a **compound-connected generator**. Again, a control rheostat and a diverter resistance can be employed as shown in Fig. 15.35b. The series portion of the field can be connected so that its mmf either can be added to or subtracted from the mmf of the shunt portion of the field. Typically, it is the former connection, called **cumulative compounding**, that is used, rather than the latter connection, which is known as **differential compounding**.

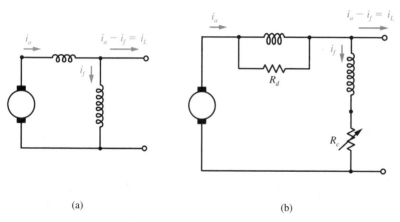

(a)   (b)

Fig. 15.35 Compound-connected generator.

Generator Buildup

For a separately excited generator turning at n rpm, the value of the field current (which can be adjusted by adjusting V_f as shown in Fig. 15.32) determines the value of the generated voltage—the magnetization curve can be used to find the voltage value. Let us consider the shunt-connected generator shown in Fig. 15.36, however, where a control rheostat R_c is included, as is a field switch S. Suppose the magnetization curve for the generator is as given in Fig. 15.37 for the case of a prime mover operating at n rpm. Even before the field switch S is closed, when $i_f = 0$ A, there will be a small generated emf due to residual magnetism. When the switch is closed, there will be a voltage applied across the series connection of the field winding and the control rheostat. This, in turn, produces a nonzero field current and, hence, in-

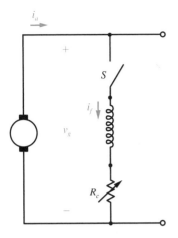

Fig. 15.36 Shunt-connected generator.

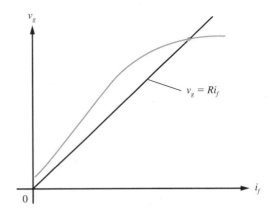

Fig. 15.37 Magnetization curve.

creases the flux, thereby resulting in an increase in the generated emf. This repetitive process of increasing generated voltage and field current is known as generator **buildup**. But how much do the voltage and current build up?

A magnetization curve is a plot of the no-load generated voltage versus the field current. For the case of no load, however, the armature resistance R_a is typically much smaller than R, the sum of the field resistance R_f and the resistance R_c of the control rheostat; that is, $R_a \ll R = R_f + R_c$. Thus in essence, the terminal voltage equals the generated emf under no-load conditions, and, therefore, $v_g = Ri_f$. A plot of this simple Ohm's law equation is also given in Fig. 15.37. Since the voltage and current values for generator buildup correspond to points on the generator's magnetization curve, and since the equation $v_g = Ri_f$ also must be satisfied, the intersection of the magnetization curve and the line $v_g = Ri_f$ gives the point where generator buildup ceases. At this point, the applied field voltage produces the field current necessary to generate that voltage.

Note that by decreasing the value of $R = R_f + R_c$, the intersection of the curve and the line will be to the right of the point shown. That would indicate a greater value for the generated voltage. Conversely, increasing the value of R would decrease the generated voltage. Specifically, if the value of R were increased so much that the line $v_g = Ri_f$ were, for the most part, on the left side of the magnetization curve, there would be very little generator buildup and the generated voltage would be quite small. Even when the value of R is reduced so that the line $v_g = Ri_f$ approximately coincides with the linear portion of the magnetization curve, the generated voltage is not uniquely determined, and it might change from one value to another. Therefore, for the most stable operation, the value of R should be chosen such that the line $v_g = Ri_f$ intersects the nonlinear (saturated) portion of the magnetization curve as illustrated in Fig. 15.37.

External Characteristics

Suppose for a dc generator, the load (or terminal) voltage is denoted by v_L and the load current by i_L. As depicted in Fig. 15.29, for a separately excited generator, the load voltage is related to the load current by

$$v_L = -R_a i_L + v_g \qquad (15.29)$$

where v_g is the generated emf. Thus as the load current increases, the load voltage decreases. A plot of the relationship between the load voltage and the load current (v_L versus i_L) is known as the **external characteristic** of the generator. Although Eq. 15.29 indicates that the external characteristic for a separately excited generator is a straight line, in actuality it is slightly curved, as shown in Fig. 15.38. (The reason for this will be discussed shortly.) Here, the values V_r and I_r are the generator's full-load (rated) voltage and current, respectively.

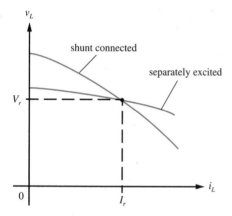

Fig. 15.38 External characteristics for separately excited and shunt-connected generators.

Now consider the case of a shunt-connected generator. As mentioned previously, the field current is small compared with the armature current, so we will assume that the load current is equal to the armature current. Just as was the case for the separately excited generator, therefore, due to the armature resistance, the load voltage will decrease as the load current increases. In addition, however, when the load (terminal) voltage decreases, the field current decreases, thereby reducing the generated voltage even further. As a consequence of this, the load voltage decreases more rapidly with increasing load current than does the separately excited generator. This situation is also depicted in Fig. 15.38.

For some uses, it may be undesirable to have the terminal voltage of a generator decrease with increasing load current, as is illustrated in Fig. 15.38. In a cumulative-compounded generator, the flux from the series-field winding reinforces the flux from the shunt-field winding. By having the series and shunt windings properly proportioned, the generator will produce a full-load (rated) voltage that is equal to the no-load voltage. The external characteristics for such a **flat-compounded generator** is shown in Fig. 15.39. When the series field reinforces the shunt field excessively, the result is an **overcompounded generator**; insufficient series field reinforcement produces an **undercompounded generator**. The external characteristics for these two cases are also depicted in Fig. 15.39.

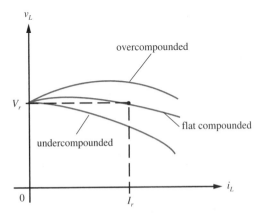

Fig. 15.39 External characteristics for a compound-connected generator.

Previously, we mentioned that an external generator characteristic was not the straight line that it was predicted to be. The cause of this discrepancy is known as **armature reaction**. Current going through an armature will establish a magnetic field that can distort, especially for loaded conditions, the flux distribution of the field. In preceding discussions, we ignored this effect; however, such a distortion can adversely affect the proper commutation for the generator. To help compensate for this condition, **compensating windings**, which are connected in series with the armature, are placed in slots in the pole faces. The result is an opposition to the armature mmf and, therefore, a reduction in the distortion of the flux distribution. Compensating windings are also called **pole-face windings**.

In addition to compensating windings, many dc machines have windings also connected in series with the armature that produce small auxiliary fields that are used to aid current reversals in conductors undergoing commutation so as to alleviate sparking at the brushes. These windings are placed around narrow poles placed midway between the main poles of a machine. These additional poles are known as **commutating poles**, and the resulting fields are referred to as **commutating** or **interpole fields**.

DC Motors

Let us return to the simple rotating machine given in Fig. 15.22 on p. 981. Instead of mechanically forcing this conducting loop to rotate, however, suppose that a current i is applied, as indicated in Fig. 15.40a. (For the sake of simplicity, the slip rings and the brushes have been omitted.) The end view of the loop is shown in Fig. 15.40b, where the current in the upper conductor is directed into the page (indicated by the cross $\times$) and the current in the lower conductor is directed out of the page (indicated by the dot $\cdot$). By the right-hand screw rule, the resulting force on the

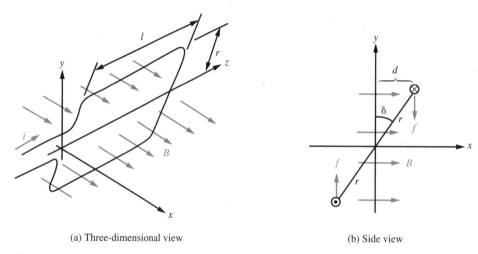

(a) Three-dimensional view   (b) Side view

Fig. 15.40 Simple rotating machine.

upper conductor is $f = Bli$ in the downward direction indicated. (This value of f is also the upward indicated force on the lower conductor.) The torque developed by each conductor is $fd = Blir \sin \delta$, where the angle δ formed by the y-z plane and the plane of the loop is called the **torque angle** or **power angle**. If, instead of a single-turn loop, an N-turn coil is employed, then because such a coil is comprised of $2N$ conductors, the torque developed by the coil is

$$T = 2NBlir \sin \delta = NBAi \sin \delta \qquad (15.30)$$

where $A = 2rl$ is the area of the coil. Note that the developed torque is maximum when $\sin \delta = 1$ ($\delta = 90$ degrees). Since the forces on the conductors are as shown in Fig. 15.40b, the developed torque tends to rotate the loop clockwise, and the simple rotating machine behaves as a motor.

For the simple generator given in Fig. 15.22 on p. 981, at the instant of time shown, a resistive load placed across the brushes has a voltage across it with the polarity indicated. This, in turn, means that the resulting loop current is in the direction opposite to that shown in Fig. 15.40. Consequently, as in the previous discussion, the developed torque tends to rotate the loop in Fig. 15.22 counterclockwise—that is, the developed torque is in opposition to that which forces the loop to rotate clockwise. Note, too, that for the simple motor given in Fig. 15.40, at the instant of time shown, the induced emf due to the loop rotating clockwise tends to produce a current that opposes the applied current.

Armature Model for a DC Motor

A dc motor is constructed basically the same as a dc generator (e.g., see Fig. 15.25 on p. 983). Provided that the field current direction is maintained along with the direction of rotation, when the motor is in use, however, the direction of its armature

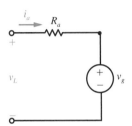

Fig. 15.41 Model of armature portion of a dc motor.

current is opposite to that in the generator. Consequently, we can model the armature portion of a motor by the equivalent electric circuit shown in Fig. 15.41. (Note the direction of the armature current i_a as compared with that in a dc generator.) Again, R_a is the armature resistance, and by Eq. 15.28, $v_g = K\Phi n$, where K is a constant dependent on the construction of the motor, Φ is the flux per pole, and n is the rotor (armature) speed in rpm. In this case, v_L indicates the applied or line voltage of the motor. As was true for the dc generator, the dc motor can be separately or self-excited as depicted in Figs. 15.32 to 15.35, where the directions of i_a and i_L should be reversed.

For a dc motor, from the circuit given in Fig. 15.41, by KVL, we have that

$$v_L = R_a i_a + v_g \tag{15.31}$$

and since $v_g = K\Phi n$, then

$$v_L = R_a i_a + K\Phi n \tag{15.32}$$

From this equation, we have that the speed (in rpm) of the motor is

$$n = \frac{v_L - R_a i_a}{K\Phi} \qquad (15.33)$$

and this is known as the **speed equation** of a dc motor.

As indicated by Eq. 15.30, the torque developed by a dc motor is determined by the armature current and the magnetic flux density, and hence, the field flux (flux per pole). In mathematical terms, we can write that the developed torque T is given by

$$T = K'\Phi i_a \qquad (15.34)$$

where K' is a constant that is determined by the construction of the motor.

DC Shunt Motors

Specifically, let us consider the case of the shunt-connected motor, called a **shunt motor** for short, shown in Fig. 15.42. Suppose we are given that the line voltage v_L is constant. For a fixed setting of the control rheostat R_c, the field flux is constant (ignoring armature reaction). Even though i_a can vary with different loading conditions, since R_a typically is small and $R_a i_a$ can be considered negligible compared with v_L, under these conditions, the shunt motor essentially has a constant speed. For a given load and constant line voltage, however, from Eq. 15.33, we see that we can vary the motor speed by changing Φ. Since Φ depends on i_f and $i_f = v_L/(R_c + R_f)$, where R_f is the field resistance, by adjusting R_c we can vary the speed of the motor.

By combining Eq. 15.33 and Eq. 15.34, we can derive an expression relating speed and torque. In particular, from Eq. 15.34, we have that $i_a = T/K'\Phi$, and substituting this into Eq. 15.33, we get

$$n = \frac{v_L}{K\Phi} - \frac{R_a}{K\Phi}\left(\frac{T}{K'\Phi}\right) = \frac{v_L}{K\Phi} - \frac{R_a}{KK'\Phi^2}T = mT + b$$

where $m = -R_a/KK'\Phi^2$ and $b = v_L/K\Phi$. Therefore, since v_L and (once R_c is fixed) Φ are constant for a shunt motor, a plot of speed n versus torque T is a straight[4] line ($n = mT + b$) having slope m and vertical-axis intercept b. Such a plot, known as the **speed-torque characteristic** of the motor, is shown in Fig. 15.43, where n_r

---

[4]Due to armature reaction, the line is slightly curved.

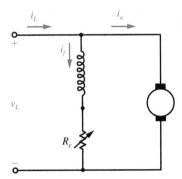

Fig. 15.42 Shunt motor.

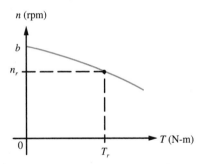

Fig. 15.43 Speed-torque characteristic of a shunt motor.

is the rated speed and T_r is the rated torque at full load. Note that the speed decreases as the developed torque increases. An increasing developed torque is required for an increasing load (i.e., a resisting torque). Thus an increasing load means a decreasing speed. From Eq. 15.33, a decrease in speed means an increase in armature current i_a. Since $v_g = K\Phi n$, a decrease in speed n also means a decrease in v_g.

Example 15.11

A 220-V dc shunt motor has a speed of 1200 rpm and an armature current of 6 A at no load. Given that the armature resistance is 0.5 Ω, let us determine the motor speed when the armature current is 40 A at full load.

From Eq. 15.33,

$$K\Phi = \frac{v_L - R_a i_a}{n}$$

Thus at no load,

$$K\Phi = \frac{220 - (0.5)(6)}{1200} = 0.181$$

Hence at full load,

$$n = \frac{v_L - R_a i_a}{K\Phi} = \frac{220 - (0.5)(40)}{0.181} = 1105 \text{ rpm}$$

> ### Drill Exercise 15.11
>
> A 220-V dc shunt motor has a speed of 1000 rpm at a full-load armature current of 50 A. Given that the armature resistance is 0.4 Ω, determine the no-load speed when the no-load current is 5 A.
>
> **ANSWER** 1090 rpm

As indicated in Example 15.11, the motor speed at full load, denoted n_{FL}, differs from the speed at no load, n_{NL}. A measure of the change in speed is the **speed regulation** SR of the motor, which is defined to be

$$\boxed{SR = \frac{n_{NL} - n_{FL}}{n_{FL}}}$$

For the preceding example, we have that

$$SR = \frac{1200 - 1105}{1105} = 0.086 = 8.6\%$$

The speed regulation for a typical shunt motor usually ranges between 5 percent and 12 percent.

As mentioned above, the speed of a shunt motor can be varied by adjusting the control rheostat R_c. If the armature and field voltages are supplied by separate sources, however, motor speed can be varied by changing the armature voltage while keeping the field voltage (and hence, current and flux) fixed. This conclusion should be apparent from Eq. 15.33. Such a technique offers a wide range of speed control, and is frequently used in industrial applications.

DC Series Motors

If the field of a dc motor is series connected as shown in Fig. 15.44, then the result is a **series motor**. Since the field current is also the armature current, and this current can be large, the field winding typically is comprised of a few turns of heavy wire. Unlike the shunt motor, where we considered Φ to be constant once the control rheostat R_c was set, for the series motor Φ is dependent on the armature current. This is so because the armature current is the field current, and the field flux Φ depends on the field current i_f.

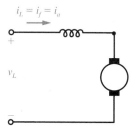

Fig. 15.44 Series motor.

For the sake of analytical convenience, let us assume that a series motor operates on the linear portion of its magnetization curve and

$$\Phi = K_f i_f = K_f i_a$$

where K_f is a constant. Since $v_g = K\Phi n$ and $T = K'\Phi i_a$, we have that

$$v_g = K(K_f i_a)n = KK_f n i_a \tag{15.35}$$

and

$$\boxed{T = K'(K_f i_a)i_a = K'K_f i_a^2} \tag{15.36}$$

Thus we see that the developed torque for a series motor is proportional to the square of the armature current, whereas a shunt motor is proportional to the armature current.

Suppose that for a series motor, R denotes the series combination of the armature resistance R_a and the resistance R_f of the field winding, that is, $R = R_a + R_f$. Then, by KVL and Eq. 15.35, we get

$$v_L = Ri_a + v_g = Ri_a + KK_f n i_a = (R + KK_f n)i_a \quad \Rightarrow$$

$$i_a = \frac{v_L}{R + KK_f n}$$

Substituting this expression for the armature current into Eq. 15.36 yields

$$\boxed{T = \frac{K'K_f v_L^2}{(R + KK_f n)^2}} \tag{15.37}$$

The plot of n versus T for this expression gives us the speed-torque characteristic for a series motor as shown in Fig. 15.45. From this curve, note how rapidly the motor speed increases as the torque gets small. As a result of this fact, to avoid excessive motor speeds and its undesirable consequences, typically a series motor is always mechanically coupled to its load. The nonconstant speed characteristic of a series motor makes it useful in certain types of variable-speed applications such as cranes, hoists, and winches.

From Eq. 15.36 we see that regardless of the direction of the current, the torque T has the same direction. Therefore, for either an ac or a dc line voltage, the developed torque for a series motor is unidirectional. Series motors designed to operate on either ac or dc are known as **universal motors**. Clearly, for ac operation, the developed torque is not constant but rather pulsating; and the mechanical output power for ac operation is less than that for dc operation. Still, universal motors are commonly used in hand power tools, such as electric drills, and small appliances, such as vacuum cleaners.

A compound-connected motor, called a **compound motor** for short, almost always is cumulatively compounded. The speed-torque characteristic for such a machine lies between those for a shunt motor and for a series motor as illustrated in Fig. 15.46. The relative position of the characteristic for the compound motor with respect to the shunt and series motors depends on the relative strengths of the shunt and series fields.

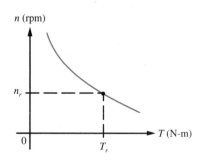

Fig. 15.45 Speed-torque characteristic for a series motor.

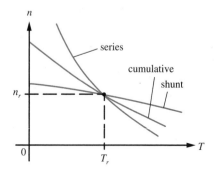

Fig. 15.46 Speed-torque dc motor characteristics.

In the preceding discussions, it was assumed that motor operation was in the steady state. Initially, however, the speed of a motor is zero. As a consequence, from Eq. 15.28, the initial emf of the motor is $v_g = K\Phi n = 0$ V. In the case of a shunt motor, for example, this means that the entire line voltage v_L is initially across the armature resistance. If this voltage were steadily increased from zero to its final value, the motor would proceed smoothly to its rated speed. Because the line voltage is fixed at its given value, however, placing it across the typically small armature resistance can result in a quite excessive armature current. For this reason, when starting a large dc motor, a resistance, which is adjustable continuously or in steps, is placed in series with the armature. Typically, this resistance limits the starting current to a value between 150 percent and 200 percent of the rated current, and it is removed (short-circuited) when the motor reaches its appropriate (steady-state) speed. A device that will perform this duty is known as a **starting box**.

15.5   AC Machines

We have seen that for a dc generator, typically the field (the stator) is stationary and the armature (the rotor) rotates. Conversely, for an ac generator, or alternator, usually the field windings are on the rotor whereas the armature conductors are embedded in slots in the stator. The reason for this is that the armature handles much more power that the field does, and by supplying direct current to the rotating field through slip rings and brushes, there are fewer mechanical and electrical problems than if the armature were to rotate.

The Alternator

A cross-sectional view of a simplified alternator is shown in Fig. 15.47. A constant current is applied to the field winding through stationary brushes riding on slip rings (not shown). This produces a magnetic field, and when the current is in the direction indicated, the resulting north (N) and south (S) poles are as shown. Embedded in slots in the stator are conductors a and a', which are connected in the rear (also not

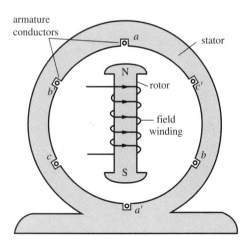

Fig. 15.47 Alternator.

shown) so as to form a loop. As the field rotates, the changing magnetic flux induces an emf across such a coil. By shaping the poles appropriately, such an induced emf can be made sinusoidal. The rotor position shown in Fig. 15.47 corresponds to the case that the induced emf is maximum. Such a voltage completes one cycle for every revolution of the rotor. Therefore, if the rotor speed is n rpm, or equivalently $n/60$ rps, then the induced emf completes $n/60$ cycles per second; that is, the frequency of this voltage is $n/60$ Hz. We now see that the alternator's electrical frequency is synchronized with its mechanical speed and, therefore, the given alternator is known as a **synchronous machine**.

In addition to conductors a and a', also placed in slots in the stator are other conductors. As with a and a', conductors b and b' are connected in the rear, as are c and c'. Note that the axis of the coil formed by b and b' differs by 120 degrees from the axis determined by a and a'. Also, the axis for c and c' differs by 120 degrees from the axis for b and b', and hence, by 240 degrees from a and a'. Thus if the rotor turns clockwise at ω rad/s, then the emfs induced across the three coils are as depicted in Fig. 15.48. By making either a wye (Y) connection or a delta (Δ) connection of the three coils, the given alternator will realize a three-phase voltage source.

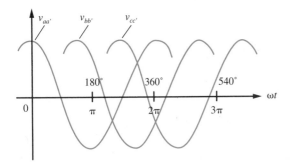

Fig. 15.48 Voltages induced across the armature coils.

The two-pole alternator (also called a "synchronous generator") shown in Fig. 15.47 has poles that are prominent or "salient," and, therefore, it is an example of what is known as a **salient-pole machine**. For such an alternator to produce a 60-Hz sinusoid, according to the previous discussion, we must have that $n/60$ rps = 60 Hz and, hence, the speed of the rotor must be $n = 3600$ rpm. The alternator speed, however, can be reduced by using a rotor that has more than two poles. As an example, consider the four-pole alternator given in Fig. 15.49. For this salient-pole machine with the field winding shown, when direct current is applied in the direction depicted (the slip rings and brushes are not shown), the resulting pole polarities are as indicated. In this case, the armature conductors a_1 and a_1' are connected in the rear (not shown) so as to form a coil, as do a_2 and a_2'. By connecting a_1' and a_2 in front (also not shown), the two coils then become connected in series.

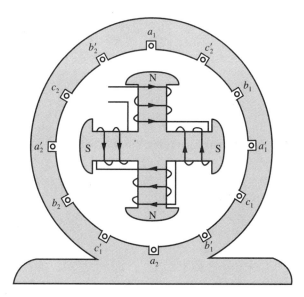

Fig. 15.49 Four-pole, three-phase alternator.

As a consequence of such a construction, one complete revolution of the rotor will produce two complete cycles of the sinusoid induced across the winding consisting of coil a_1-a_1' connected in series with coil a_2-a_2'. The conductors labeled b and c are similarly connected, and again, by making either a Y or Δ connection, a three-phase alternator results.

For the salient-pole alternator given in Fig. 15.49, a generated sinusoid completes two cycles for every revolution of the rotor. Thus a rotor speed of n rpm ($n/60$ rps) corresponds to a sinusoidal frequency of $2n/60 = n/30$ Hz. In general, for an alternator with p poles (where p is even), the frequency f (in hertz) of the generated voltage is

$$f = \frac{p}{2}\frac{n}{60} = \frac{pn}{120} \tag{15.38}$$

where n is the rotor speed in rpm.

From Eq. 15.38, we see that for a fixed frequency, the rotor speed decreases as the number of poles increases. Typically, salient-pole machines have six or more poles. If either two or four poles are employed, then the rotor speed required for a frequency of 60 Hz is 3600 or 1800 rpm, respectively. At such high speeds, however, a salient-pole rotor is more subject to mechanical stresses, windage losses, and noisy operation. To avoid these complications, a **cylindrical** or **nonsalient rotor** is utilized. For such a machine, the field winding is distributed around the rotor and embedded in slots. Alternators driven by steam turbines, also called **turboalternators**, typically operate at 1800 or 3600 rpm and are cylindrical (nonsalient)-rotor machines. Lower-speed alternators, such as those driven by waterwheels, are usually salient-pole machines.

Alternator Armature Model

Just as we modeled the armature of a dc generator with an equivalent electric circuit, we can do the same for an alternator. However, because an alternator generates alternating current instead of direct current, such a model will be slightly more complex. In particular, we shall use the equivalent circuit shown in Fig. 15.50 to model a single phase of the voltage generated by an alternator. As for the case of a dc generator, the armature resistance is designated by R_a. The armature resistance is connected in series with an inductive reactance X_s, known as the **synchronous reactance**, which represents the effects that are a consequence of the armature winding.

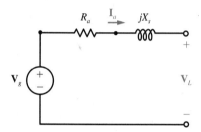

Fig. 15.50 Electric-equivalent circuit of the armature of an alternator.

Typically, the armature resistance is much smaller than the synchronous reactance, and, therefore, quite often R_a is ignored. For a dc generator, the generated voltage is proportional to the field flux and the rotor speed (see Eq. 15.28). This is also the case for an alternator, and, consequently, its steady-state operation can also be described in terms of a magnetization curve. The shape of such a curve, of course, is determined by the construction of the machine. Because an alternator generates ac voltages, the vertical axis of the magnetization curve indicates the rms value of the generated voltage. The field current, however, is dc. The voltage source given in Fig. 15.50 is labeled $\mathbf{V}_g$, which is the phasor representation of the generated voltage and $|\mathbf{V}_g|$ represents the rms value of the generated voltage. The armature (stator) current $\mathbf{I}_a$ is also an rms phasor.

Example 15.12

A six-pole, three-phase 12-kVA alternator is Y connected and has a rated per-phase voltage of 125 V (rms) for a load with a lagging pf of 0.6 at 60 Hz. The synchronous reactance is 2 Ω, and the armature resistance is assumed to be negligible. Let us determine the rotor speed and the generated voltage.

From Eq. 15.38, the rotor speed of the alternator is

$$n = \frac{120f}{p} = \frac{(120)(60)}{6} = 1200 \text{ rpm}$$

Ignoring the armature resistance, the corresponding electric-equivalent per-phase armature circuit is as shown in Fig. 15.44, where $V_L = |\mathbf{V}_L| = 125$ V and $I_a = |\mathbf{I}_a|$ is to be determined.

Since the load operates at a lagging pf of 0.6, then the load current $\mathbf{I}_a$ lags the load voltage $\mathbf{V}_L$ by θ, where $\cos \theta = 0.6$. Thus the pf angle is $\theta = \cos^{-1} 0.6 = 53.1°$. Arbitrarily selecting the angle of $\mathbf{V}_L$ to be $0°$ [ang$(\mathbf{V}_L) = 0°$], we must have that ang$(\mathbf{I}_a) = -53.1°$.

Since the alternator is rated at 12 kVA, the per-phase load current has an rms value of

$$I_a = \frac{(1/3)(12,000)}{V_L} = \frac{4000}{125} = 32 \text{ A}$$

From the circuit given in Fig. 15.51, by KVL, the generated voltage $\mathbf{V}_g$ is found to be

$$\mathbf{V}_g = jX_s\mathbf{I}_a + \mathbf{V}_L = (j2)(32\underline{/-53.1°}) + 125\underline{/0°} = 64\underline{/36.9°} + 125\underline{/0°}$$

$$= 51.2 + j38.4 + 125 = 176 + j38.4 = 180\underline{/12.3°} \text{ V}$$

The corresponding phasor diagram is shown in Fig. 15.52.

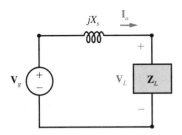

Fig. 15.51 Per-phase alternator-equivalent circuit.

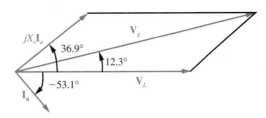

Fig. 15.52 Phasor diagram.

Suppose the per-phase voltage of this alternator is described by the magnetization curve given in Fig. P15.32 on p. 1031. Then from this curve, the (rms) generated voltage of 180 V corresponds to a (dc) field current of 0.8 A.

Drill Exercise 15.12

For the alternator described in Example 15.12, change the load pf to 0.8 leading. (a) Find the resulting generated voltage. (b) Draw the corresponding phasor diagram. (c) Use the magnetization curve given in Fig. P15.32 on p. 1031 to determine the approximate field current.

ANSWER   (a) 101 V, (c) 0.45 A

In Example 5.12, as is typical for the case of a dc generator, the generated voltage of an alternator has a larger magnitude than the load voltage. As demonstrated in Drill Exercise 5.12, however, a change in the pf of the load can result in the converse situation. In such a case, armature reaction results in a reinforcement of the field mmf.

The Rotating Stator Field

Previously, we mentioned that a dc motor has the same construction as a dc generator. So, too, one type of ac motor—a **synchronous motor**—is constructed as is an alternator (e.g., see Fig. 15.47). To describe this type of synchronous machine, as well as the "induction motor" that will be studied in the following section, we need to introduce the concept of a rotating stator field or flux.

To see how this phenomenon occurs, consider the simple three-phase, two-pole machine cross section given in Fig. 15.53, where, for the sake of simplicity, the rotor is not shown. Instead, depicted in the center of the machine are the flux components due to the currents in each of the three stator coils a-a', b-b', and c-c'. The current

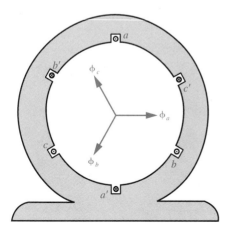

Fig. 15.53 Stator field components of a synchronous motor.

i_a in coil a-a' is coming out of the page in conductor a and going into the page in conductor a' (designated by a dot · and a cross ×, respectively). By the right-hand rule, the direction of the resulting flux ϕ_a is to the right, as illustrated in Fig. 15.53. Similarly, the currents i_b and i_c in coils b-b' and c-c', respectively, yield ϕ_b and ϕ_c. By spacing the coils 120 degrees apart, the directions of ϕ_a, ϕ_b, and ϕ_c are spaced by 120 degrees. But what is the net flux?

To simplify our analysis, let us assume that flux is directly proportional to current, that is, $\phi = ki$. Since the flux produced by each coil has direction, as well as magnitude, let us define the complex number $\mathbf{\Phi} = \phi e^{j\theta}$, where ϕ is flux and θ is the angle between the direction of the flux and the horizontal. Thus

$$\mathbf{\Phi}_a = ki_a e^{j0°} = ki_a \qquad \mathbf{\Phi}_b = ki_b e^{-j120°} \qquad \mathbf{\Phi}_c = ki_c e^{j120°}$$

If the three stator currents are equal ($i_a = i_b = i_c$), since $\mathbf{\Phi} = \mathbf{\Phi}_a + \mathbf{\Phi}_b + \mathbf{\Phi}_c = 0$ Wb, the net flux is zero. However, let us consider the case that i_a, i_b, and i_c are the balanced set of currents

$$i_a = I \cos \omega t \qquad i_b = I \cos(\omega t - 120°)$$

$$i_c = I \cos(\omega t - 240°) = I \cos(\omega t + 120°)$$

which are depicted in Fig. 15.54. Then we get the following set of complex numbers with time-varying magnitudes:

$$\mathbf{\Phi}_a = (kI \cos \omega t)e^{j0°} = kI \cos \omega t \qquad \mathbf{\Phi}_b = [kI \cos(\omega t - 120°)]e^{-j120°}$$

$$\mathbf{\Phi}_c = [kI \cos(\omega t + 120°)]e^{j120°}$$

Using the fact that $e^{j\theta} = \cos \theta + j \sin \theta$ (Euler's formula), we therefore have that $e^{-j120°} = -\dfrac{1}{2} - j\dfrac{\sqrt{3}}{2}$ and $e^{j120°} = -\dfrac{1}{2} + j\dfrac{\sqrt{3}}{2}$. Thus

$$\mathbf{\Phi}_a = kI \cos \omega t$$

$$\mathbf{\Phi}_b = -\frac{1}{2}kI \cos(\omega t - 120°) - j\frac{\sqrt{3}}{2}kI \cos(\omega t - 120°)$$

$$\mathbf{\Phi}_c = -\frac{1}{2}kI \cos(\omega t + 120°) + j\frac{\sqrt{3}}{2}kI \cos(\omega t + 120°)$$

and their sum is

$$\mathbf{\Phi} = \mathbf{\Phi}_a + \mathbf{\Phi}_b + \mathbf{\Phi}_c = kI[\cos \omega t - \tfrac{1}{2}\cos(\omega t - 120°) - \tfrac{1}{2}\cos(\omega t + 120°)]$$

$$+ j(\sqrt{3}/2)kI[\cos(\omega t + 120°) - \cos(\omega t - 120°)]$$

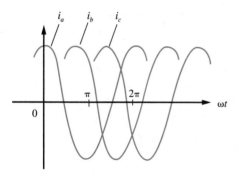

Fig. 15.54 Stator currents.

However, $\cos(\omega t \pm 120°) = (\cos \omega t)(\cos 120°) \mp (\sin \omega t)(\sin 120°) = -\frac{1}{2} \cos \omega t \mp (\sqrt{3}/2) \sin \omega t$. Therefore,

$$\Phi = \Phi_a + \Phi_b + \Phi_c$$

$$= kI\left[\cos \omega t - \frac{1}{2}\left(-\frac{1}{2} \cos \omega t + \frac{\sqrt{3}}{2} \sin \omega t\right) - \frac{1}{2}\left(-\frac{1}{2} \cos \omega t - \frac{\sqrt{3}}{2} \sin \omega t\right)\right]$$

$$+ j\left(\frac{\sqrt{3}}{2}\right)kI\left[\left(-\frac{1}{2} \cos \omega t - \frac{\sqrt{3}}{2} \sin \omega t\right) - \left(-\frac{1}{2} \cos \omega t + \frac{\sqrt{3}}{2} \sin \omega t\right)\right]$$

$$= \frac{3}{2}kI \cos \omega t - j\frac{3}{2}kI \sin \omega t = \frac{3}{2}kI(\cos \omega t - j \sin \omega t)$$

$$= \frac{3}{2}kIe^{-j\omega t}$$

Since $\Phi = \Phi_a + \Phi_b + \Phi_c$ has a constant magnitude $(\frac{3}{2})kI$, but an angle $-\omega t$ that varies with time, this demonstrates that the flux rotates clockwise at a frequency of ω rad/s ($60\omega/2\pi = 30\omega/\pi$ rpm) when the currents i_a, i_b, and i_c given above are applied to coils a-a', b-b', and c-c', respectively.

The Synchronous Motor

Now let us look at the two-pole synchronous motor given in Fig. 15.53, where the rotor is explicitly depicted as in Fig. 15.55. Just as in an alternator, the rotor field is caused by a direct current applied through brushes and slip rings. At some time (e.g., for $t = 0$ s), the stator flux will be as indicated. Suppose the position of the

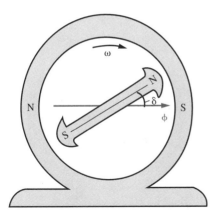

Fig. 15.55 Stator and rotor poles for a synchronous motor.

rotor is as illustrated. As the stator field rotates clockwise, the magnetic attraction between the stator and rotor poles will cause the rotor to turn at the same speed as the stator flux, that is, the rotor and stator fields will be synchronized. For the case of no load on the rotor, the torque angle δ will be small since the developed torque only needs to overcome rotational losses due to friction and windage. (As suggested by Eq. 15.30, if δ is small, the developed torque is small.) Increasing the mechanical load will result in an increase in δ. If the load torque on the rotor becomes greater than can be overcome by the developed torque, then the rotor will slow down and stop. Furthermore, if the rotor- and stator-field axes coincide, that is, if δ = 0°, then no torque will be developed on the rotor. For motor operation, the rotor field must lag the stator field.

Because the construction of a dc motor is the same as for a dc generator, we obtained the circuit model for the former from the latter simply by changing the direction of the armature current. Again, because of the similar construction of synchronous motors and generators (alternators), this approach can also be taken for synchronous machines. Specifically, with reference to the alternator equivalent circuit given in Fig. 15.50, we have that a model of the armature of a synchronous motor is as shown in Fig. 15.56. Again, as in the case of an alternator, the armature resistance R_a is often considered negligible compared with the synchronous reactance X_s.

In an alternator (synchronous generator), the generated voltage $\mathbf{V}_g$ leads the phase voltage $\mathbf{V}_L$ (e.g., see Fig. 15.52). Conversely, for a synchronous motor, $\mathbf{V}_g$ lags $\mathbf{V}_L$, and since the rotor lags the rotating stator field by the torque angle δ, then $\mathbf{V}_g$ lags $\mathbf{V}_L$ by δ. Ignoring the armature resistance, for the circuit given in Fig. 15.56, we have

$$\mathbf{V}_L = jX_s I_a + \mathbf{V}_g \tag{15.39}$$

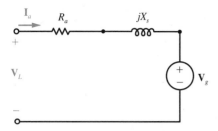

Fig. 15.56 Electric-equivalent circuit of the armature of a synchronous motor.

Assuming that the angle of $\mathbf{V}_L$ is zero, an example of a phasor diagram for a synchronous motor is as shown in Fig. 15.57. For this diagram, let us construct a line

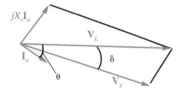

Fig. 15.57 Phasor diagram for a synchronous motor.

labeled l and extend $\mathbf{I}_a$ as shown in Fig. 15.58. If $|\mathbf{I}_a| = I_a$, $|\mathbf{V}_g| = V_g$, and $|\mathbf{V}_L| = V_L$, we have

$$\frac{l}{V_g} = \sin \delta \quad \text{and} \quad \frac{l}{X_s I_a} = \cos \theta \quad \Rightarrow$$
$$l = V_g \sin \delta = X_s I_a \cos \theta \tag{15.40}$$

But the input power (per phase) of a synchronous motor (see Fig. 15.56) is

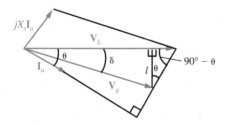

Fig. 15.58 Modified phasor diagram.

$$P = V_L I_a \cos \theta$$

From Eq. 15.40, then

$$P = V_L I_a \frac{V_g \sin \delta}{X_s I_a} = \frac{V_L V_g}{X_s} \sin \delta \qquad (15.41)$$

and we see why the torque angle is also called the power angle. Since an increase in load torque requires an increase in power for a constant speed, the torque angle must increase when the load increases.

Example 15.13

A Y-connected, three-phase synchronous motor has a per-phase voltage of 125 V and a synchronous reactance of 5 Ω. Suppose the motor input power is 6 kW and the torque angle is 27 degrees. Let us determine the pf of the motor.

Since the per-phase input power is $P = 6000/3 = 2000$ W, then from Eq. 15.41 we have that

$$V_g = \frac{P X_s}{V_L \sin \delta} = \frac{(2000)(5)}{(125)(\sin 27°)} = 176 \text{ V}$$

If the motor were characterized by the magnetization curve given in Fig. P15.32 at the end of this chapter, on p. 1031, then the corresponding field current would be approximately 0.78 A. Furthermore, by Eq. 15.39, we have

$$\mathbf{I}_a = \frac{\mathbf{V}_L - \mathbf{V}_g}{jX_s} = \frac{125\underline{/0°} - 176\underline{/-27°}}{5\underline{/90°}} = \frac{86.1\underline{/112°}}{5\underline{/90°}} = 17.2\underline{/22°} \text{ A}$$

and since the current $\mathbf{I}_a$ leads the voltage $\mathbf{V}_L$, the motor operates at a power factor of

$$\text{pf} = \cos 22° = 0.927 \text{ leading}$$

The phasor diagram for the motor is shown in Fig. 15.59. In this case, $V_g > V_L$, and this is an example of an **overexcited** motor. For the situation that $V_g < V_L$, we say that the motor is **underexcited**.

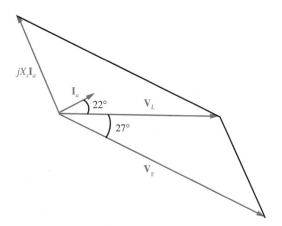

Fig. 15.59 Phasor diagram for example motor.

Drill Exercise 15.13

A Y-connected, three-phase synchronous motor has a per-phase voltage of 125 V, a synchronous reactance of 5 Ω, a generated emf of 150 V, and a torque angle of 30 degrees. Determine the magnitude of the armature current, the pf, and the input power of the motor.

ANSWER   15 A, 0.998 leading, 5613 W

Consider a synchronous motor with a lagging pf, where the associated phasor diagram is as shown in Fig. 15.60a. Suppose the load on the motor remains constant, and the field (rotor) current is increased. This will strengthen (increase) the field flux and produce a larger generated voltage magnitude V_g. The result is a greater magnetic attraction between the rotating (stator) field poles and the rotor poles, and, thus, the

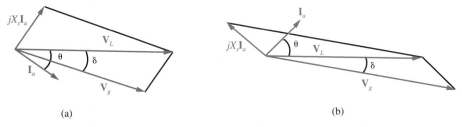

(a)   (b)

Fig. 15.60 Effect of increasing field current on phasor diagram of motor.

torque angle δ is reduced. The effect on the phasor diagram for the motor when the field current is increased is shown in Fig. 15.60b. Hence we see that by increasing the field current sufficiently, the pf of a motor can be changed from lagging to leading (or from lagging to less lagging). It is this property that enables a synchronous motor to be used for power-factor correction, that is, for changing the pf. Typically, a synchronous motor that is used for this purpose is unloaded, so the only input power required is for the motor losses. Since a capacitor has the effect of making a pf more leading, a synchronous motor that is utilized for pf correction is known as a **synchronous condenser.**[5]

Even though our previous discussions explicitly characterized two-pole motors, just as for the case of a synchronous generator (alternator), the relationship between machine speed, ac voltage, frequency, and the number of poles of a synchronous motor is as given by Eq. 15.38. From this equation, we have that the **synchronous speed** n_s of a motor (in rpm) is

$$n_s = \frac{120f}{p} \tag{15.42}$$

For the common situation that $f = 60$ Hz, the maximum synchronous speed is obtained when $p = 2$, and under this circumstance $n_s = 3600$ rpm.

Since the speed of a synchronous motor is exactly the synchronous speed n_s, such a motor has a constant speed—even under different load conditions. If the load torque is made too great, however, the motor will lose synchronism and stop. The torque that causes this is known as the **pull-out torque**.

A variation of a synchronous motor is a motor that does not rotate continuously, but rather turns in steps as the result of an applied digital signal. A simple example of such a **stepper** (or **stepping**) **motor** is shown in Fig. 15.61. For this machine, the rotor is a permanent magnet, and the stator has four poles with their associated windings. When a current goes through a pole winding in the direction indicated, that pole is north (N); when the current has the opposite direction, that pole is south (S). When $i_1 > 0$ A, $i_2 < 0$ A, $i_3 < 0$ A, and $i_4 > 0$ A, then the corresponding poles are N, S, S, and N, respectively. For the situation shown, the resulting rotor angle is $\theta = 0°$. When $i_1 > 0$ A, $i_2 > 0$ A, $i_3 < 0$ A, and $i_4 < 0$ A, then the result is $\theta = 45°$. By applying the appropriate currents to the pole windings, the rotor can be positioned in increments of 45 degrees. In order to get smaller steps, more stator poles are required. For a typical motor with 24 poles, the rotor turns 7.5 degrees per step. Stepper motors can be used for positioning purposes in various types of systems.

---

[5]A capacitor was formerly called a condenser.

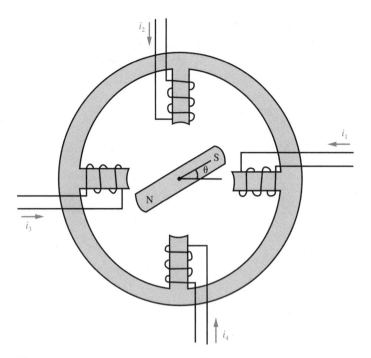

Fig. 15.61 Stepper motor.

In previous discussions, it was assumed that a synchronous motor operates at synchronous speed—for only at this speed does the motor develop a constant torque in the direction of rotation. By itself, a synchronous motor at rest will produce no starting torque. This is a consequence of the fact that a flux rotating around a motionless rotor results in a zero average torque. One way to start a synchronous motor is to modify the rotor in such a way that a starting torque can be obtained. We now discuss the principles that will achieve this end.

The Induction Motor

For a synchronous machine (either a generator or motor), a dc voltage is applied to the rotor (field) and an ac voltage is applied to the stator (armature). The requirement for a dc rotor excitation can be eliminated, however, by constructing a machine that makes use of magnetic induction. **Induction motors**, which operate on this principle, are the most common type of ac motors. Conversely, an induction generator is inferior to a synchronous generator and is considered of little importance.

As in a synchronous motor, when alternating current is applied to the stator of an induction motor, a rotating magnetic flux is produced. For an induction motor, however, the stator is the field and the rotor is the armature. A common type of rotor contains conductors that form a cylinder, and the ends of the conductors are short-circuited together by conducting rings as shown in Fig. 15.62. Since such a configuration resembles a cage that is used to exercise rodents such as hamsters and squirrels, this type of rotor is referred to as a **squirrel-cage rotor**. Unlike a synchronous motor, though, a squirrel-cage induction motor does not require brushes and slip rings for the rotor.

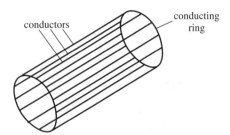

Fig. 15.62 Squirrel-cage configuration.

A cross-sectional view of a simple two-pole, three-phase induction motor is shown in Fig. 15.63. For this machine, a rotating stator flux is established as described previously and turns clockwise at an angular velocity of ω rad/s. Suppose the rotor is initially at rest. As the field rotates, an emf will be induced in coil 1—the coil consisting of conductors 1 and $1'$. With reference to Fig. 15.22, on p. 981, the clockwise rotating coil and fixed field are equivalent to a fixed coil and a counter-clockwise rotating field. The result is an induced emf that produces a current out of the upper conductor. Thus for a fixed coil and a clockwise rotating field, the resulting current is into the upper conductor. This means that the current produced in coil 1 in Fig. 15.63 goes into conductor 1 (i.e., into the page) and comes out of conductor

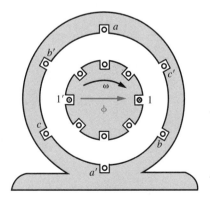

Fig. 15.63 Induction motor.

$1'$. As illustrated in Fig. 15.40, on p. 994, the torque that is developed is clockwise, and so the rotor tends to turn in the direction of the rotating field. The other coils formed by the rotor conductors experience the same type of effect and contribute to the total rotor torque. If this starting torque is sufficient to overcome the load, the rotor will begin to turn and the motor will come up to its steady-state speed of n rpm.

The operating speed n of an induction motor, however, cannot be equal to the synchronous speed n_s. This is so because, if the two speeds were equal, there would be no time rate of change in flux linkages. Then, by Faraday's law, there would be no induced emf and, therefore, no induced current. The result would be no torque. Consequently, the speed n of an induction motor is less than synchronous speed n_s, since this will cause a time rate of change in flux linkages and, hence, will allow motor behavior.

As mentioned in the last section, by itself a synchronous motor at rest will produce no starting torque. But by adding some short-circuited conductors to the rotor of a synchronous motor, as in the case of an induction motor, a starting torque can be obtained. When the machine operates at synchronous speed, however, such squirrel-cage windings will make no contribution to the rotor torque. These windings, called **damping windings**, also serve to damp the tendency for the rotor to ''hunt,'' or oscillate around its final position, when the load on the motor changes suddenly. This is a consequence of the induction-motor torque that is developed by a damping winding when the rotor momentarily operates at a nonsynchronous speed.

The **slip** s of an induction motor having speed n is defined as

$$s = \frac{n_s - n}{n_s} \qquad (15.43)$$

where $n_s = 120f/p$ is synchronous speed. From this definition, we can write that the motor speed is

$$n = n_s - sn_s = (1 - s)n_s \qquad (15.44)$$

A squirrel-cage induction motor at full load typically has a slip that is in the range of 2 to 6 percent.

The difference between synchronous speed n_s and rotor speed n is called the **slip speed**. From Eq. 15.43, we have that the slip speed is

$$n_s - n = sn_s \qquad (15.45)$$

Since this is the relative speed of the rotating stator field with respect to the rotor, from Eq. 15.38, the frequency of the induced emf is

$$\frac{psn_s}{120} = s\left(\frac{pn_s}{120}\right) = sf$$

where f is the frequency of the stator voltage. Such a voltage produces a rotor field that rotates at a speed relative to the rotor of

$$\frac{120(sf)}{p} = s\frac{120f}{p} = sn_s$$

Since the speed of the rotor is n rpm, the rotor field has an absolute speed of

$$n + sn_s = n_s$$

Thus the rotor field has the same speed as the stator field, and, as with a synchronous motor, the developed torque for a given load is constant—despite the fact that the rotor speed n is less than the synchronous speed n_s.

Example 15.14

Suppose we are given that the speed of a three-phase induction motor is 864 rpm at 60 Hz. Let us determine the number of poles for the motor, the slip, and the frequency of the emf induced in the rotor.

Since the synchronous speed is

$$n_s = \frac{120f}{p} = \frac{(120)(60)}{p} = \frac{7200}{p}$$

and since $n = 864$ rpm is slightly below n_s, then p is the closest even integer such that

$$\frac{7200}{p} > 864 \quad \Rightarrow \quad \frac{7200}{864} > 8.33$$

Thus we deduce that $p = 8$ and this value yields $n_s = 7200/8 = 900$ rpm. Therefore, the slip of the motor is

$$s = \frac{n_s - n}{n_s} = \frac{900 - 864}{900} = 0.04 = 4\%$$

and the emf induced in the rotor has a frequency of

$$sf = (0.04)(60) = 2.4 \text{ Hz}$$

> **Drill Exercise 15.14**
>
> A six-pole, three-phase, 220-V, 60-Hz induction motor has a speed of 1170 rpm. Determine the slip and the frequency of the induced emf in the rotor.
>
> **ANSWER** 2.5 percent, 1.5 Hz

An Induction-Motor Model

Suppose the rotor of an induction motor is constrained to be at rest (i.e., $s = 1$). On a per-phase basis, the rotating stator field will induce an emf $\mathbf{V}_2$ having frequency f. If R_2 is the effective resistance of the rotor and X_2 is the reactance, then the resulting current $\mathbf{I}_2$ is given by

$$\mathbf{I}_2 = \frac{\mathbf{V}_2}{R_2 + jX_2} \tag{15.46}$$

Since the emf induced in the rotor is directly proportional to the relative velocity of the rotating field with respect to the rotor, for an induction motor having slip s, the induced emf is $s\mathbf{V}_2$, the frequency is sf, and the reactance is sX_2. Therefore, the resulting rotor current is

$$\mathbf{I}_2 = \frac{s\mathbf{V}_2}{R_2 + jsX_2} = \frac{\mathbf{V}_2}{(R_2/s) + jX_2} \tag{15.47}$$

at slip s. However, since we can write

$$\frac{R_2}{s} = \frac{R_2(1)}{s} = \frac{R_2(s + 1 - s)}{s} = \frac{R_2 s}{s} + \frac{R_2(1 - s)}{s} = R_2 + \frac{1 - s}{s} R_2$$

Eq. 15.47 becomes

$$\mathbf{I}_2 = \frac{\mathbf{V}_2}{R_2 + jX_2 + (1 - s)R_2/s} \tag{15.48}$$

From this expression, we get

$$\mathbf{V}_2 = \left(R_2 + jX_2 + \frac{1 - s}{s} R_2 \right)\mathbf{I}_2 = R_2\mathbf{I}_2 + jX_2\mathbf{I}_2 + \frac{1 - s}{s} R_2\mathbf{I}_2$$

which can be modeled by the circuit shown in Fig. 15.64.

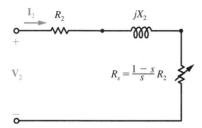

Fig. 15.64 Rotor equivalent circuit for an induction motor.

The power absorbed by the rotor is $V_2 I_2 \cos\theta$, where $V_2 = |\mathbf{V}_2|$, $I_2 = |\mathbf{I}_2|$, and θ is the difference between the angles of $\mathbf{V}_2$ and $\mathbf{I}_2$. Since R_2 is the effective resistance of the rotor, the rotor's ohmic loss $R_2 I_2^2$ is dissipated as heat. The remainder of the power delivered to the rotor, that is, the power absorbed by the speed-dependent resistance $R_s = (1 - s)R_2/s$, represents the electric-to-mechanical power conversion. This quantity is given by $R_s I_2^2$.

An induction motor is basically a transformer with a fixed primary (the stator) and a rotating secondary (the rotor). Consequently, one possibility for modeling an induction motor is to use the equivalent circuit shown in Fig. 15.65, where R_1 and X_1 are the effective resistance and reactance, respectively, of the primary. By KVL, we can write

$$\mathbf{V} = (R_1 + jX_1)\mathbf{I}_1 + \mathbf{V}_1 \tag{15.49}$$

and

$$\mathbf{V}_2 = \left(R_2 + jX_2 + \frac{1-s}{s}R_2\right)\mathbf{I}_2 = \left(\frac{R_2}{s} + jX_2\right)\mathbf{I}_2 \tag{15.50}$$

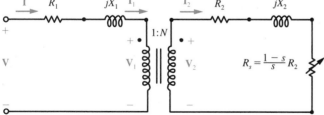

Fig. 15.65 Equivalent circuit of an induction motor.

Using the fact that for the ideal transformer portion of the circuit $\mathbf{V}_2 = N\mathbf{V}_1$ and $\mathbf{I}_2 = \mathbf{I}_1/N$, Eq. 15.49 becomes

$$\mathbf{V} = (R_1 + jX_1)N\mathbf{I}_2 + \frac{\mathbf{V}_2}{N} = (R_1 + jX_1)N\mathbf{I}_2 + \frac{[(R_2/s) + jX_2]\mathbf{I}_2}{N}$$

from which

$$I_2 = \frac{NV}{(R_1 + jX_1)N^2 + (R_2/s) + jX_2} = \frac{V/N}{R_1 + jX_1 + (R_2/sN^2) + jX_2/N^2}$$

Let us define $R'_2 = R_2/N^2$ and $X = X_1 + X_2/N^2$. Then

$$I_2 = \frac{V/N}{R_1 + (R'_2/s) + jX} \quad \Rightarrow \quad I_2 = |I_2| = \frac{V/N}{\sqrt{[R_1 + (R'_2/s)]^2 + X^2}}$$

where $V = |V|$. Thus the power absorbed by $R_s = (1 - s)R_2/s$ is

$$R_s I_2^2 = \frac{V^2/N^2}{[R_1 + (R'_2/s)]^2 + X^2} \frac{(1 - s)R_2}{s} = \frac{1 - s}{s} \frac{V^2 R'_2}{[R_1 + (R'_2/s)]^2 + X^2}$$

and this represents the per-phase electric-to-mechanical power conversion.

Torque-Speed Characteristics

Since mechanical power is the product of torque and angular velocity (i.e., $P = T\omega$), the developed torque per phase is $T = P/\omega = R_s I_2^2/\omega$. From Eq. 15.43, because angular velocity and rpm differ by a constant, it should be obvious that

$$s = \frac{n_s - n}{n_s} = \frac{\omega_s - \omega}{\omega_s}$$

This expression yields $\omega = (1 - s)\omega_s$, and, therefore, the total developed torque for a three-phase induction motor is

$$T = \frac{3R_s I_2^2}{\omega} = \frac{3}{(1 - s)\omega_s} R_s I_2^2 = \frac{3}{(1 - s)\omega_s} \frac{1 - s}{s} \frac{V^2 R'_2}{[R_1 + (R'_2/s)]^2 + X^2}$$

or

$$T = \frac{3}{\omega_s} \frac{V^2(R'_2/s)}{[R_1 + (R'_2/s)]^2 + X^2} \qquad (15.51)$$

Since this expression relates torque T and speed, either $n = (1 - s)n_s$ or $\omega = (1 - s)\omega_s$, it is possible to plot torque versus speed. Such a typical **torque-speed**

curve for a squirrel-cage induction motor is given in Fig. 15.66. Here the torque axis is labeled in terms of percent of the rated torque T_r, and the speed axis is labeled in terms of percent of synchronous speed—either ω_s or n_s. Furthermore, since slip $s = 1 - \omega/\omega_s = 1 - n/n_s$, the horizontal axis can also be labeled in terms of slip as indicated.

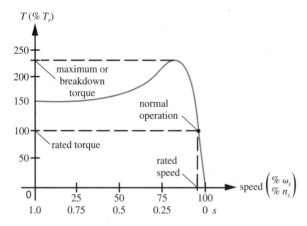

Fig. 15.66 Torque-speed curve for a typical squirrel-cage induction motor.

For normal operation, the slip s is small (between 2 and 6 percent), and in the range of rated load, the induction motor is, in essence, a constant-speed machine. Furthermore, for small values of s, values of R_1 and X are typically small compared with R_2'/s. Under this circumstance, Eq. 15.51 can be approximated by

$$T \approx \frac{3sV^2}{\omega_s R_2'}$$

(normal operation)

and we see that torque is directly proportional to slip; also, torque is inversely proportional to R_2'. This means that for a given slip, a greater torque can be developed by using larger rotor conductors.

At very slow speeds ($s \approx 1$), $R_1 + R_2'/s$ is usually small compared with X and Eq. 15.51 can be approximated by

$$T \approx \frac{3V^2 R_2'}{s\omega_s X^2}$$

(slow speeds)

Thus we see that under this circumstance, torque is inversely proportional to slip s. Further, at rest (or standstill), $s = 1$ and

$$T \approx \frac{3V^2 R_2'}{\omega_s X^2} \qquad \text{(standstill)}$$

This, then, is the starting torque, and we see that it is directly proportional to V^2 as well as R_2'.

The **maximum** or **breakdown torque** can be determined by differentiating Eq. 15.51 with respect to s and setting the result to zero. Doing so will yield the value of the slip that corresponds to the breakdown torque. We leave it to the reader (see Problem 15.59 at the end of this chapter) to show that this slip is

$$s = \frac{R_2'}{\sqrt{R_1^2 + X^2}} \qquad (15.52)$$

and the maximum torque T_m is

$$T_m = \frac{3V^2}{2\omega_s(R_1 + \sqrt{R_1^2 + X^2})} \qquad (15.53)$$

Note that the breakdown torque is independent of R_2'.

Example 15.15

A three-phase induction motor has a per-phase voltage of 120 V and operates at 1152 rpm. This 60-Hz, six-pole machine has the parameters $R_1 = 0.25 \ \Omega$, $R_2' = 0.15 \ \Omega$, and $X = 0.5 \ \Omega$. Let us find the rated, starting, and maximum torques, as well as the rated mechanical output power.

Since $p = 6$, the synchronous speed is

$$n_s = \frac{120f}{p} = \frac{(120)(60)}{6} = 1200 \text{ rpm}$$

and thus the slip is

$$s = \frac{n_s - n}{n_s} = \frac{1200 - 1152}{1200} = 0.04 = 4\%$$

Since one revolution $= 2\pi$ rad, and 1 minute $= 60$ s, then

$$1200 \text{ rpm} = \frac{(1200)(2\pi)}{60} = 40\pi \text{ rad}/s = \omega_s$$

From Eq. 15.51, the rated torque T_r is

$$T_r = \frac{3}{40\pi} \frac{(120)^2(0.15/0.04)}{[0.25 + (0.15/0.04)]^2 + (0.5)^2} = 79.3 \text{ N-m}$$

The starting torque T_{st} corresponds to $s = 1$. From Eq. 15.51,

$$T_{st} = \frac{3}{40\pi} \frac{(120)^2(0.15)}{(0.25 + 0.15)^2 + (0.5)^2} = 1268 \text{ N-m}$$

The maximum torque is, by Eq. 15.53,

$$T_m = \frac{3(120)^2}{2(40\pi)(0.25 + \sqrt{(0.25)^2 + (0.5)^2})} = 212 \text{ N-m}$$

which, from Eq. 15.52, occurs when

$$s = \frac{0.15}{\sqrt{(0.25)^2 + (0.5)^2}} = 0.268 = 26.8\%$$

and the corresponding motor speed is, from Eq. 15.44

$$n = (1 - s)n_s = (1 - 0.268)(1200) = 878 \text{ rpm}$$

Since the rated torque is $T_r = 79.3$ N-m, the rated mechanical output power is

$$T_r\omega = T_r(1 - s)\omega_s = 79.3(1 - 0.04)(40\pi) = 9570 \text{ W}$$

However, 1 **horsepower** (hp) = 746 W. Hence the rated mechanical output power is $9570/746 = 12.8$ hp.

Drill Exercise 15.15

A three-phase induction motor has a per-phase voltage of 254 V and operates at 1710 rpm. For this 60-Hz, six-pole machine, $R_2' = 0.32 \ \Omega$. Find approximate values for the rated torque and output power.

ANSWER 160 N-m, 38.4 hp

The squirrel-cage induction motor has the desirable properties of ruggedness, simplicity of construction, and efficiency. Such a machine is started easily and, for a given load, essentially has a constant speed. Another type of induction motor, however, has the capability of speed control and greater starting torque. In such a machine, referred to as a **wound-rotor motor**, the rotor has a winding similar to that on the stator. With the use of brushes and slip rings, external resistances can be connected in series with the rotor coils. In this way, the resistance of the rotor (R_2, and hence R_2') can be increased. For normal operation, an increase in R_2' means an increase in slip, and, as indicated by a torque-speed curve (e.g., Fig. 15.66), the result is a decrease in speed. Furthermore, even though the maximum developed torque (see Eq. 15.53) does not depend on the value of R_2', the starting torque does. Specifically, increasing R_2' will increase the starting torque (e.g., see Problem 15.60 at the end of this chapter). The price that has to be paid for a wound-rotor induction motor is its higher cost.

Single-Phase Induction Motors

Although our discussion of induction motors thus far has been concerned with three-phase machines (which can have ratings down to a fraction of a horsepower), single-phase motors are the rule rather than the exception for low-power (under 1 hp) applications. Single-phase induction motors are similar to three-phase squirrel-cage induction motors; the main difference is that a single-phase stator winding is used in place of a three-phase winding.

In the discussion of a rotating stator field, we saw how it is produced by a three-phase stator winding. Such a rotating flux results in a unidirectional torque in an induction motor as well as a synchronous motor. For a single-phase stator winding, however, the stator field produced by a sinusoidal stator current is like one of the three components resulting from a three-phase stator winding. For example, a single-phase stator field can have a flux of the form

$$\phi = kI \cos \omega t$$

where ω is the frequency of the stator current (see the discussion associated with Figs. 15.53 and 15.54). By Euler's formula, we can rewrite this expression for ϕ as

$$\phi = \tfrac{1}{2} kI e^{j\omega t} + \tfrac{1}{2} kI e^{-j\omega t}$$

Therefore, we can consider ϕ to consist of two components, each having the same magnitude, but one rotating clockwise at an angular velocity of ω rad/s, and the other rotating counterclockwise at the same speed. This means that the rotor of a single-phase induction motor can turn either clockwise or counterclockwise—once it has started. Because of the opposite rotating stator fields, however, the developed torque will be pulsating. Yet, on a time average, once it has been started, the torque-speed behavior of a single-phase induction motor is similar to that of a three-phase induction motor.

There is more than one way to start a single-phase induction motor. In one method, the stator has two salient poles, and a heavy short-circuited copper coil—known as a **shading coil**—is wound around half of each of these poles. A symbolic representation of such a machine is shown in Fig. 15.67. Currents induced in the shading coils delay the field in the shaded portion. This produces a small rotating-field component, which, in turn, causes a low starting torque. The direction of this torque is from the unshaded half to the shaded half of the poles. This type of single-phase machine is called a **shaded-pole motor**.

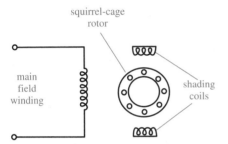

Fig. 15.67 Shaded-pole motor.

Another technique for producing a starting torque is to have an auxiliary winding in addition to the main stator winding. Such a situation is depicted in Fig. 15.68. In this case, the main winding has a higher reactance-to-resistance ratio than the auxiliary winding. Because of this, the current in the main winding reaches its maximum value later than does the current in the auxiliary winding. This results in a rotating stator field that produces a starting torque. When such a motor, called a **split-phase**

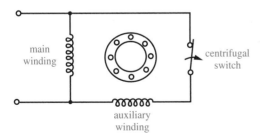

Fig. 15.68 Split-phase motor.

motor, approaches its operating speed, a centrifugal switch that is connected in series with the auxiliary winding opens up and that winding is disconnected.

Another type of split-phase motor is shown in Fig. 15.69. In this case, a capacitor is placed in series with the auxiliary winding to produce the phase shift between the currents in the main and auxiliary windings. Again, when the motor nears operating speed, the centrifugal switch opens and disconnects the auxiliary winding. This machine is an example of one type of **capacitor motor**. Another type can be obtained by replacing the centrifugal switch with a short circuit. Doing so can improve the running performance and pf of the motor if the capacitor value is changed, but the price paid for this is a reduction in the starting torque; however, an alternative approach can be taken. Instead of replacing the centrifugal switch with a short circuit, it is left as is and a second capacitor is placed in parallel with the original capacitor-switch series combination. In essence then, one capacitor is used for starting and one is used for running—and this type of machine may be more desirable.

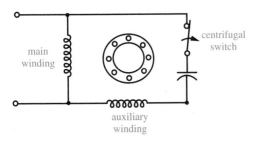

Fig. 15.69 Capacitor motor.

SUMMARY

1. A conductor of length l moving with velocity u at right angles through a magnetic field of flux density B generates an emf $e = Blu$.

2. A conductor of length l carrying a current i at right angles to a magnetic field of flux density B develops a force $f = Bli$.

3. Mechanical circuits can be dealt with in a manner analogous to that for electric circuits.

4. Dynamic transducers are moving-coil devices that can be used for loudspeakers, microphones, and phonograph pickups.

5. An electromagnet is a moving-iron transducer that can be used to fabricate a relay. A solenoid plunger is also a moving-iron transducer.

6. The d'Arsonval movement is a rotational transducer that is widely used for dc ammeters, dc voltmeters, and ohmmeters.

7. The electrodynamometer is a rotational transducer that is used for ac measurements, in particular, voltmeters, and especially wattmeters.

8. A generator converts mechanical energy into electric energy; a motor converts electric energy into mechanical energy.

9. An alternator utilizes brushes and slip rings to generate alternating current. A dc generator requires the use of brushes and a split-ring commutator.

10. For a dc generator, the field is on the stator and the armature is on the rotor. For an alternator,

the field is on the rotor and the armature is on the stator.

11. The voltage generated by a rotating machine is graphically displayed by its magnetization curve.

12. A dc machine can be separately excited or self-excited. Self-excited machines can be shunt connected, series connected, or compound connected.

13. The external characteristic of a dc generator graphically shows the relationship between load voltage and load current.

14. Synchronous machines can have salient-pole rotors or cylindrical (nonsalient) rotors. The former machines typically are used in lower-speed applications (less than 1800 rpm), whereas the latter often see use in higher-speed situations (1800 rpm and above).

15. Three-phase machines are usually analyzed on a per-phase basis.

16. The synchronous speed of a machine depends on the frequency of the applied excitation and the number of poles of the machine.

17. The speed of a dc motor can be easily controlled.

18. The relationship between the speed and the torque of a dc motor is graphically displayed by its speed-torque characteristic.

19. The speed regulation of a dc motor is a measure of the change in speed between no-load and full-load conditions.

20. Bringing a dc motor up to operating speed requires the use of a starting box.

21. The phenomenon of a rotating stator field is employed by an ac motor.

22. A synchronous motor operates at a constant speed over a range of load conditions. Synchronous motors can be used for power-factor correction.

23. An induction motor is basically a transformer with a fixed primary and a rotating secondary.

24. The simplest type of induction motor—the squirrel-cage motor—gets its name from its squirrel-cage rotor.

25. An induction motor typically operates at a speed slightly less than synchronous speed. For a fixed load, it is essentially a constant-speed machine.

26. The slip of an induction motor is a measure of the difference between the operating and synchronous speeds of the motor.

27. The relationship between the torque and the speed (or slip) of an induction motor is graphically given by its torque-speed curve.

28. A wound-rotor induction motor offers speed control and greater starting torque. The price for this type of motor is higher cost.

29. Single-phase induction motors are similar to three-phase squirrel-cage motors, and are used in low-power applications.

30. Shaded-pole motors, split-phase motors, and capacitor motors are all examples of single-phase induction motors.

PROBLEMS

15.1 The electromechanical system shown in Fig. 15.3 on p. 957 is to be operated as a motor. Assume that $B = 0.4$ T and $l = 0.2$ m. (a) Find the current required to develop a force of 0.5 N directed to the right on the moveable conductor. (b) Determine the value of the resistance R that results in a velocity to the right of 25 m/s if the battery has a value of 12 V.

15.2 The electromechanical system shown in Fig. 15.3 on p. 957 is to be operated as a generator. Assume that $B = 0.4$ T and $l = 0.2$ m. (a) Find the

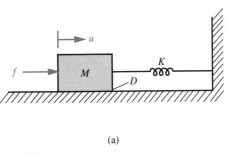

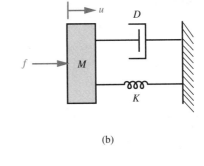

(a)   (b)

Fig. P15.3

current corresponding to a force of 0.5 N directed to the right on the movable conductor. (b) Determine the velocity required to obtain this current for the case that the resistance is 1.28 Ω and the battery has a value of 12 V.

15.3 (a) Write an equation describing the mechanical system shown in Fig. P15.3a. Draw the mechanical circuit and its electric-circuit analog. (b) Write an equation describing the mechanical system shown in Fig. P15.3b. Draw the mechanical circuit and its electric-circuit analog.

15.4 For the mechanical system shown in Fig. 15.7 on p. 959, suppose $D = 0.25$ N-s/m, $K = 0.5$ m/N, and $M = 0.5$ kg. Find the velocity $u(t)$ when the applied force is $f(t) = 3 \cos 2t$ N. (*Hint:* Use phasor analysis.)

15.5 An alternative electrical analogy for a mechanical circuit is obtained when force is analogous to voltage and velocity is analogous to current. (a) What is the electric analog of the mass given in Fig. 15.4? (b) What is the electric analog of the dash pot given in Fig. 15.5? (c) What is the electric analog of the spring given in Fig. 15.6?

15.6 Use the alternative electrical analogy described in Problem 15.5 to determine the analogous electric circuit of the mechanical system shown in Fig. 15.7 on p. 959. Is there any relationship between this electric circuit and the one shown in Fig. 15.8b on p. 960?

15.7 The motor shown in Fig. 15.3 on p. 957 is described by Equations 15.8 and 15.9. Assume the

inductance $L = 0$ H and the movable conductor is initially at rest. Find the velocity $u(t)$ for $t \geq 0$ s for the case that a constant force f is applied at time $t = 0$ s.

15.8 Suppose the electromechanical system shown in Fig. 15.3 on p. 957 is used as a generator when an applied force f is directed to the right. In a manner similar to that for the case of a motor, write the describing equations and model the generator with an appropriate equivalent circuit.

15.9 For the dynamic loudspeaker in Example 15.3 on p. 963, suppose the resistance of the moving coil is $R = 0.4$ Ω. The efficiency of the loudspeaker is the ratio of the output (mechanical) power $p_o = fu$ to the input (electric) power $p_{in} = vi$. Find the efficiency of the loudspeaker.

15.10 Eq. 15.12 and Eq. 15.13 on p. 964 describe the dynamic microphone. Find an equivalent circuit that models such a device for the case that the microphone is loaded with a resistance R_L.

15.11 A force of $f = 0.25$ N is applied to the diaphragm of a dynamic microphone whose 40-turn moving-coil radius is $r = 2$ cm. Ignore the mass (i.e., $M = 0$ kg), assume the coefficient of friction is $D = 3$ N-s/m, and suppose the magnetic flux density in the annular air gap is $B = 0.5$ T. Find the efficiency of the microphone, that is, find the ratio of the output (electric) power $p_o = vi$ to the input (mechanical) power $p_{in} = fu$, for the case that the microphone has a moving-coil resistance of $R = 0.4$ Ω and a load resistance of $R_L = 4$ Ω.

15.12 A dynamic phonograph pickup similar to that shown in Fig. 15.12 on p. 965 has $B = 0.3$ T, $r = 0.2$ cm, $N = 15$, and $R = 0.4\ \Omega$. Given a sinusoidal displacement of amplitude 0.002 cm, determine the rms voltage across a 4-Ω load for the case that the frequency is (a) 1000 Hz and (b) 5000 Hz.

15.13 A relay with 2000 turns has a gap length of 0.8 cm when the contacts are open and 0.2 cm when the contacts are closed. Given that the cross-sectional area is 2 cm^2, determine the pickup and dropout currents for the case that the spring exerts a force of 0.08 N.

15.14 For the relay shown in Fig. 15.13 on p. 966, initially the gap has length l_g. After the armature moves a distance x, the gap length is $l = l_g - x$. Show that the expression for the force on the armature given by Equation 15.15 has the equivalent form

$$f = \frac{1}{2} i^2 \frac{dL}{dx}$$

where $L = N^2/\mathcal{R}$, and the reluctance of the core is negligible compared with the reluctance of the air gap.

15.15 For the solenoid shown in Fig. 15.14 on p. 969, the length of the plunger is 10 cm, as is the length of the cylindrical coil. Suppose that the variation of inductance with displacement is given by

$$L = \tfrac{1}{2} + (1/\pi)\tan^{-1}(20x - 1)$$

where x ranges between $x = 0$ and $x = 10$ cm. Determine the current required to produce a maximum force of 4 N on the plunger.

$$\left[Hint:\ \frac{d}{dx}(\tan^{-1}\theta) = \frac{1}{1 + \theta^2} \frac{d\theta}{dx} \right]$$

15.16 A d'Arsonval movement has a 10-turn moving coil whose length is 2 cm and whose width is 2 cm. The air-gap magnetic flux density is 0.5 T and the spring compliance is 10^8 degrees/N-m. What current will yield a full-scale deflection of 100 degrees?

15.17 A d'Arsonval movement with a moving-coil resistance of $R_c = 2$ kΩ requires 50 μA for full-

scale deflection. Determine the value of the shunt resistance required to use the mechanism for a (a) 0- to 1-mA ammeter, and (b) 0- to 25-mA ammeter.

15.18 For the circuit shown in Fig. P15.18, an ammeter whose resistance is 100 Ω measures the current through the 900-Ω resistor to be 1 mA. What is the actual current through the 900-Ω resistor?

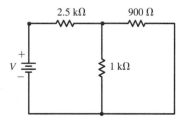

Fig. P15.18

15.19 A 50-μA d'Arsonval movement with a coil resistance of $R_c = 2$ kΩ is to be used for a dc voltmeter. Determine the value of the series resistance required for a (a) 0- to 10-V voltmeter, and (b) 0- to 50-V voltmeter.

15.20 With reference to Problem 15.19, what is the ohms-per-volt rating for the voltmeter in part (a) and in part (b)?

15.21 A 50-μA d'Arsonval movement with moving-coil resistance R_c is to be used for a dc voltmeter whose full-scale deflection is v volts. Show that, as long as $R_c \le v/(50 \times 10^{-6})$, the voltmeter will be rated at 20 kΩ/V.

15.22 For the circuit shown in Fig. P15.22, the voltage across the 300-kΩ resistor is measured with a 0- to 30-V voltmeter to be 8 V. If the voltmeter is rated at 20 kΩ/V, what is the actual voltage across the 300-kΩ resistor?

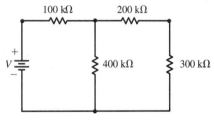

Fig. P15.22

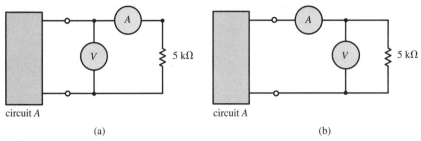

circuit A   circuit A

(a)   (b)

Fig. P15.24

15.23 For the ohmmeter given in Fig. 15.19 on p. 975, suppose $V_s = 1.5$ V, $R_s = 28$ kΩ, $R_c = 2$ kΩ, and the d'Arsonval mechanism is a 50-μA movement. Find the value of the resistance R when the movement has a current of (a) 10 μA, (b) 20 μA, (c) 30 μA, and (d) 40 μA.

15.24 An ammeter and a voltmeter are connected to a 5-kΩ resistor. When the connection is as shown in Fig. P15.24a, the ammeter and voltmeter readings are 9.98 mA and 50 V, respectively. When the connection is as depicted in Fig. P15.24b, the respective readings are 10.03 mA and 49.9 V. Find the resistances of the ammeter and the voltmeter.

15.25 For the meter connections shown in Fig. P15.24, suppose the ammeter resistance is 10 Ω and the voltmeter resistance is 998 kΩ. Determine the ammeter readings and the voltmeter readings for the case that the Thévenin equivalent of circuit A is $v_{oc} = 60$ V and $R_o = 997$ Ω.

15.26 For the meter connections shown in Fig. P15.24, suppose the ammeter resistance is 10 Ω and the voltmeter resistance is 998 kΩ. Determine the ammeter readings and the voltmeter readings for the case that the Thévenin equivalent of circuit A is $v_{oc} = 80$ V and $R_o = 2990$ Ω.

15.27 Show that the average value of the product of $i_c = I_c \cos(\omega t + \theta_1)$ and $i_s = I_s \cos(\omega t + \theta_2)$ is $\frac{1}{2} I_c I_s \cos \theta$, where $\theta = \theta_1 - \theta_2$. Use the result to show that the average value of $I^2 \cos^2 \omega t$ is $\frac{1}{2} I^2$. [*Hint*: $(\cos A)(\cos B) = \frac{1}{2} \cos (A - B) + \frac{1}{2} \cos (A + B)$]

15.28 For the case that $l = 20$ cm, $r = 5$ cm, and $B = 1$ T, determine the voltage generated by a 10-turn coil rotating at 1200 rpm as depicted in (a) Fig. 15.22, on p. 981, and (b) Fig 15.23 on p. 982. What is the rms value of the resulting voltage.

15.29 The dc generator shown in Fig. 15.25 on p. 983 has two poles and two parallel paths between the brushes. Suppose the surface area of each pole is 0.01 m^2, the air-gap magnetic flux density is 1 T, and the generated voltage is to be 40 V. Determine (a) the number of turns required by each of the four coils when the rotor speed is 1000 rpm, and (b) the rotor speed when each of the four coils has 20 turns.

15.30 The magnetic flux per pole of a 10-kW, four-pole, 220-V dc generator is 25 mWb. The armature has 48 coils, each having two turns, and they are connected so that there are two parallel paths between the brushes. Find the rotor speed that achieves the rated voltage at no load.

15.31 An 11-kW dc generator is rated at 220 V, 50 A, and 1500 rpm. If the armature resistance is 0.2 Ω, find (a) the no-load voltage at 1500 rpm, (b) the no-load voltage at 1800 rpm, and (c) the full-load voltage at 1800 rpm given a full-load current of 50 A.

15.32 A separately excited dc generator has the magnetization curve shown in Fig. P15.32. Given that the field current is 1 A, find the no-load gen-

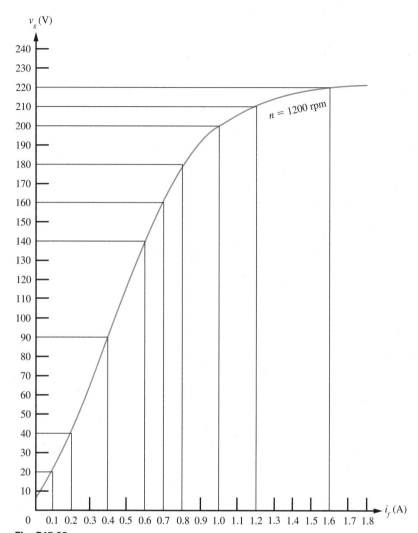

Fig. P15.32

erated voltage when the generator speed is (a) 1200 rpm, (b) 1500 rpm, and (c) 1000 rpm.

15.33 A separately excited dc generator has the magnetization curve shown in Fig. P15.32. Given that the no-load voltage is 180 V, assume that the flux per pole is directly proportional to the field current, and find the field current when the generator speed is (a) 1200 rpm, (b) 1500 rpm, and (c) 1000 rpm.

15.34 A separately excited 10-kW dc generator is rated at 200 V, 50 A, and 1200 rpm. The magnetization curve for this machine is shown in Fig. P15.32. Given that the armature resistance is 0.4 Ω and the field resistance is 100 Ω, (a) find the field current for the rated conditions, and (b) find the no-load voltage when the voltage applied to the field is 120 V.

15.35 A separately excited dc generator has the magnetization curve shown in Fig. P15.32 for the case that $n = 1200$ rpm. Suppose the voltage applied to the field is $V_f = 200$ V and the field resistance is $R_f = 125$ Ω. (a) Determine the no-load generated voltage assuming no control rheostat, that is, $R_c = 0$ Ω. (b) Find the value of R_c that will result in a no-load generated voltage of 210 V.

15.36 A shunt-connected 8-kW dc generator is rated at 200 V, 40 A, and 1200 rpm. The magnetization curve for this machine is as shown in Fig. P15.32. Suppose the armature resistance is 0.5 Ω and the field resistance is 125 Ω. Determine the value R_c of the control rheostat such that (a) the no-load voltage is 200 V, and (b) the full-load voltage is 200 V.

15.37 A 12-kW shunt-connected dc generator is rated at 240 V, 50 A, and 1000 rpm. Under no-load conditions, the generated voltage is 255 V. For full-load conditions, neglect the field current compared with the armature current and determine the armature resistance.

15.38 A shunt-connected dc generator has the magnetization curve shown in Fig. P15.32. Determine the no-load voltage at 1200 rpm when the total resistance of the field circuit is $R = R_c + R_f = 175$ Ω.

15.39 An 8-kW cumulative-compounded dc generator is rated at 200 V, 40 A, and 1200 rpm. The shunt field is separately excited, and ignoring the resistance of the series field, the no-load voltage is 200 V when the field current is 1 A, whereas the full-load voltage is 200 V when the field current is 1.6 A (i.e., a flat-compounded generator). Furthermore, it is given that the shunt winding has 500 turns per pole. (a) Find the mmf in ampere-turns per pole (A-t/pole) of the shunt field at no load and at full load. (b) Determine the mmf that must be supplied by the series field. (c) Find the number of turns per pole for the series winding.

15.40 A 10-turn coil similar to the one shown in Fig. 15.40 has a length of 20 cm and a width of 10 cm, and the magnetic flux density of 1 T. (a) Find the maximum torque that is developed when

the current is 1 A. (b) Determine the current required to develop a maximum torque of 0.5 N-m.

15.41 A 220-V shunt motor has an armature resistance of 0.4 Ω. Given that the motor speed is 1000 rpm when the armature current is 50 A, (a) determine the speed when the armature current is 75 A, and (b) find the armature current when the speed is 1050 rpm.

15.42 A 220-V dc shunt motor has an armature current of 60 A when the speed is 1400 rpm, and an armature current of 40 A when the speed is 1500 rpm. Determine the armature resistance.

15.43 A 220-V dc shunt motor has an armature resistance of 0.5 Ω. When the speed of the motor is 1000 rpm, the armature current is 40 A. Assuming that the resisting torque of the load stays constant, (a) when a 1.5-Ω resistance is placed in series with the armature, find the resulting current and speed; and (b) determine the armature current and speed (without the 1.5-Ω resistance) when the field flux is reduced by 20 percent.

15.44 A 220-V dc shunt motor has an armature resistance of 0.5 Ω. When the motor speed is 1000 rpm, the armature current is 50 A and the developed torque is 100 N-m. Find the developed torque at 1100 rpm.

15.45 A 220-V dc shunt motor has an armature resistance of 0.5 Ω. At full load, the motor speed is 1000 rpm and the armature current is 40 A. At no load, the armature current is 6 A. Determine the speed regulation of the motor.

15.46 A 500-V dc series motor has an armature resistance of 0.3 Ω and a field-winding resistance of 0.2 Ω. Given that the motor is rated at 100 A at 1000 rpm, determine the motor speed when the armature current is 46 A.

15.47 A 500-V dc series motor has an armature resistance of 0.3 Ω, a field-winding resistance of 0.2 Ω, and is rated at 100 A at 1000 rpm. Suppose the rated torque is 430 N-m. For the case that the motor has a speed of 2000 rpm, find (a) the armature current and (b) the torque.

15.48   An alternator is to generate a 60-Hz ac voltage. Find the rotor speed for the case that the alternator has (a) 4 poles, (b) 6 poles, (c) 8 poles, and (d) 18 poles.

15.49   A six-pole, three-phase 12-kVA alternator is Y connected and has a rated per-phase voltage of 125 V (rms) for a load having a unity pf at 60 Hz. The synchronous reactance is 2 Ω, and the armature resistance is assumed to be negligible. (a) Find the resulting generated voltage. (b) Draw the corresponding phasor diagram. (c) Use the magnetization curve given in Fig. P15.32 to determine the field current.

15.50   A single-phase 5-kW alternator is rated at 110 V for a load with a pf of 0.75 lagging. If the armature resistance is 0.1 Ω, and the synchronous reactance is 0.5 Ω, what is the value of the generated voltage?

15.51   A three-phase, 10-kVA alternator is Y connected and has a rated per-phase voltage of 128 V for a load with a pf of 0.8 lagging at 60 Hz. Ignore the armature resistance and determine the synchronous reactance given that the generated voltage is 185 V and it leads the load voltage by 20 degrees.

15.52   A three-phase synchronous motor is Y-connected and has a per-phase voltage of 125 V. Given that the synchronous reactance is 5 Ω and the input power is 6 kW, find the torque angle for the case that the motor operates at a pf of (a) unity, (b) 0.8 lagging, and (c) 0.8 leading.

15.53   A Y-connected, three-phase synchronous motor has a per-phase voltage of 125 V. Given that the input power is 6 kW and the torque angle is 16 degrees, determine the synchronous reactance of the armature for the case that the motor operates at a pf of unity.

15.54   A Y-connected, three-phase synchronous motor has a per-phase voltage of 254 V and a synchronous reactance of 4 Ω. Suppose the generated emf is 200 V and the torque angle is 36.9 degrees.

(a) Find the line (armature) current. (b) Determine the pf at which the motor operates. (c) Find the generated emf and torque angle that result in a unity pf and the same line current magnitude.

15.55   A Y-connected, three-phase synchronous motor has a per-phase voltage of 254 V, a synchronous reactance of 4 Ω, a line (armature) current magnitude of 25 A, and operates at a pf of 0.8 leading. Find the generated emf and determine the torque angle.

15.56   A three-phase synchronous motor operating at a pf of unity draws an armature current of 25 A. (a) Determine the synchronous reactance given that the generated emf is 256 V and the torque angle is 20 degrees. (b) Find the per-phase voltage of the motor. (c) Determine the input power of the motor.

15.57   A three-phase, 60-Hz induction motor operates at 1158 rpm. (a) Determine the number of poles for this machine. (b) Find the slip of the motor. (c) Determine the frequency of the induced rotor emf.

15.58   A three-phase, 60-Hz induction motor has four poles. Find the speed of the motor given that the emf induced in the rotor has a frequency of 3 Hz.

15.59   Show that for the developed-torque expression of an induction motor given by Eq. 15.51 on p. 1020, the breakdown torque is achieved when the slip is given by Eq. 15.52 on p. 1022, and breakdown torque is given by Eq. 15.53.

15.60   Consider the induction motor given in Example 15.15 on p. 1022. (a) Use the approximate formulas given in the text to calculate the rated torque T_r and the starting torque T_{st}. Compare these results with those determined in the example and explain any discrepancies. (b) Determine the value to which R_2' should be changed so that the maximum torque will occur at starting.

15.61   For the induction motor given in Example 15.15 on p. 1022, determine $I = |\mathbf{I}|$ (see Fig. 15.65 on p. 1019) and the power factor of the motor for the condition of (a) rated torque, (b) starting torque, and (c) maximum torque.

15.62 A three-phase induction motor has a per-phase voltage of 120 V. This six-pole, 60-Hz machine has a rated torque of 100 N-m, and has $R_1 = 0.1 \ \Omega$, $R_2' = 0.12 \ \Omega$, and $X = 0.4 \ \Omega$. Find (a) the slip, and (b) the rated speed. Justify any approximation used.

15.63 A 36-hp, 60-Hz induction motor with four poles has a per-phase voltage of 254 V. Given that the motor has a slip of 5 percent at rated speed, find the approximate value for R_2'.

15.64 A four-pole, 60-Hz induction motor has a slip of 5 percent at rated speed. Given that the per-phase voltage is 254 V, and $R_1 = 0.2 \ \Omega$, $R_2' = 0.32 \ \Omega$, and $X = 0.5 \ \Omega$, find the (a) rated speed, (b) rated torque, (c) starting torque, and (d) maximum torque.

16

SPICE

INTRODUCTION

We can perform circuit analysis by writing the equations that characterize a particular circuit and then finding the solutions to these equations. Throughout this book, we have studied techniques to analyze different types of circuits.

The digital computer is a very useful tool when it comes to the analysis of circuits. A computer can be programmed to solve a set of simultaneous algebraic equations, thereby eliminating the drudgery of hand calculations. It can also be used to numerically solve differential equations.

In addition to using a computer to solve equations, there are computer programs available that will perform circuit analysis. The most well-known circuit-analysis software is the **S**imulation **P**rogram with **I**ntegrated-**C**ircuit **E**mphasis, or **SPICE.** SPICE does not require that circuit equations be written. Instead, a circuit is described by an ASCII text file, which is read by the program, and then that circuit is analyzed by SPICE.

SPICE is available for mainframe and minicomputers, and there are also versions for personal computers. The most well-known is called **PSpice,** which is a product of the MicroSim Corporation in Irvine, California. We will now proceed to discuss how to use SPICE—and in particular, PSpice—by specifying how to describe a circuit and how to indicate what types of analyses are to be performed.

A circuit to be analyzed has to be described in a manner that is acceptable to SPICE. Furthermore, a circuit has to be comprised of certain types of electric-circuit elements. Specifically, circuits can consist of resistors, inductors, capacitors, independent sources (both voltage and current), dependent current sources (both current-dependent and voltage-dependent), and dependent voltage sources (both current-dependent and voltage-dependent). In addition, circuits can contain transformers, transmission lines, and the four most common semiconductor devices: diodes, bipolar junction transistors (BJTs), junction field-effect transistors (JFETs), and metal-oxide semiconductor field-effect transistors (MOSFETs).

16.1 PSpice

PSpice can be run on a PC in either of two environments—DOS or Windows. For the former, first create an ASCII text "input" file (the extension for this file must be .CIR) which contains statements that indicate what elements are in the circuit and what types of analyses and printouts to perform. Such statements are called **element statements** and **control statements,** respectively. If no analysis statement is made, PSpice will automatically perform a dc analysis of the given circuit. The first line in the input file is a comment, and the last line of the file is the statement **.END.** The ordering of all other lines is irrelevant. Comment lines can be placed throughout the file provided that the line begins with the symbol *. Furthermore, a PSpice input file is case insensitive—that is, there is no difference between upper case and lower case letters.

The same method can be employed when running the PSpice for Windows. In this case, however, an alternative schematics approach may be used. Since the description of this schematics technique is very involved, we shall limit our discussion of PSpice to the situation where an input file is created, and then PSpice is executed under either DOS or Windows.

To describe a given circuit, first the nodes are labeled 0, 1, 2, 3, . . . , n. The node labeled 0 is always the reference (or ground) node. The remaining node numbers need not be consecutive. After the nodes have been labeled, we specify the nodes of each element. Let us now indicate some of the various element statements for SPICE. In the following discussions, the term NAME represents an alphanumeric string of from zero to seven characters.

Resistors

The general form of a **resistor** specification is

RNAME N1 N2 VALUE

where RNAME is the name of the resistor, N1 and N2 are the nodes of the resistor (node order is irrelevant for resistors), and VALUE is the resistance in ohms—this number can be either positive or negative, but not zero. As an example, the specification

R1 3 5 27

indicatees that resistor R1 is connected between nodes 3 and 5, and it has a value of 27 Ω. Another example is

RL 10 0 1k

and this indicates that resistor RL is connected between node 10 and the reference node, and the value of RL is 1000 Ω. When expressing the values of resistors or other circuit elements, in addition to the ordinary decimal form, we can use either the symbolic form or the exponential form of the value as indicated in Table 16.1.

Table 16.1 SPICE Symbols

Value	Symbolic Form	Exponential Form
10^{-15}	F	1E–15
10^{-12}	P	1E–12
10^{-9}	N	1E–9
10^{-6}	U	1E–6
10^{-3}	M	1E–3
10^{3}	K	1E3
10^{6}	MEG	1E6
10^{9}	G	1E9
10^{12}	T	1E12

Independent Constant Voltage Sources

The specification for an **independent constant (dc) voltage source** has the general form

VNAME N+ N− DC VALUE

where VNAME is the name of the voltage source, N+ is the positive node of the voltage source, N− is the negative node, DC indicates a constant value, and VALUE is the constant voltage in volts. As an example, the specification

Vs 1 0 DC 6

indicates that there is a 6-V independent voltage source (battery) connected between node 1 and the reference node. (Even if the specification "DC" is omitted, the statement "Vs 1 0 6" will yield the same results.)

Example 16.1

Before we look at any more element specifications, let us analyze a dc resistive circuit with PSpice. Consider the circuit shown in Fig. 16.1 (see Fig. 1.41 on p. 29). Here the nodes of the circuit have been labeled 0, 1, 2, 3. Let us call the input file VOLTDIV.CIR. The contents of this file are

```
Input file for Fig. 16.1
Vs 3 0 DC 28
R1 3 1 5
R2 1 0 4
R3 1 2 1
R4 2 0 3
.END
```

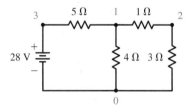

Fig. 16.1 Voltage-divider circuit.

When PSpice is run, since no type of analysis is specified, a dc analysis (called a "small signal bias solution") will automatically be performed. PSpice creates an ASCII text output file named VOLTDIV.OUT. This file contains a copy of the input file and the following small signal bias solution:

NODE VOLTAGE	NODE VOLTAGE	NODE VOLTAGE
(1) 8.0000	(2) 6.0000	(3) 28.0000

which indicates that the voltage at node 1 (with respect to the reference) is 8 V, the voltage at node 2 is 6 V, and the voltage at node 3 is 28 V.

In addition, the output file lists the currents through all of the independent voltage sources in the circuit. Also listed is the total power dissipation, which is the total power absorbed by all of the circuit elements other than the independent voltage sources. Specifically, VOLTDIV.OUT contains

VOLTAGE SOURCE CURRENTS
NAME   CURRENT

Vs   −4.000E+00

and this indicates that the current through the voltage source (directed from + to −) is −4 A.

Also contained in the output file is

TOTAL POWER DISSIPATION   1.12E+02   WATTS

and this means that the total power absorbed by all of the resistors is 112 W.

In general, the total power dissipation is equal to the total power absorbed by all of the circuit elements that are not independent voltage sources.

Drill Exercise 16.1

Determine an input file for the circuit given in Fig. P1.34 on p. 49 for the case that $R = 4\ \Omega$.

ANSWER   Input file for Fig. P1.34
Vs 2 0 DC 12
R1 2 3 4
R2 0 3 2
R3 1 3 2
R4 0 1 2
R5 1 2 4
.END

Independent Constant Current Sources

The specification for an **independent constant current source** has the general form

INAME N+ N− DC VALUE

where INAME is the name of the current source, the current source's arrow is directed from node N+ to node N−, DC indicates a constant value, and VALUE is the constant current in amperes.

Example 16.2

The circuit shown in Fig. 16.2 (see Fig. 2.8 on p. 60) contains a constant current source and is described by the following input file:

Input file for Fig. 16.2
Vs1 1 0 DC 24
Vs2 2 3 DC 4
Is 3 0 DC 3
* The letters "DC" can be left out of the element statements for Vs1, Vs2,
* and Is without affecting the results contained in the output file.
R1 1 2 2
R2 0 2 6
R3 0 3 8
.END

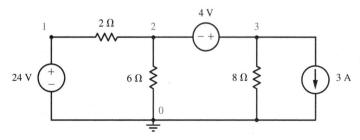

Fig. 16.2 Circuit with a constant current source.

The corresponding output file contains the node voltages:

NODE VOLTAGE	NODE VOLTAGE	NODE VOLTAGE
(1) 24.0000	(2) 12.0000	(3) 8.0000

Drill Exercise 16.2

Determine an input file for the circuit given in Fig. 2.6 on p. 56.

ANSWER Input file for Fig. 2.6
Is1 0 1 DC 6
Is2 2 0 DC 12
R1 0 1 1
R2 0 2 2
R3 1 2 3
.END

Voltage-Dependent Voltage Sources

The general form for a **voltage-dependent** (or **voltage-controlled**) **voltage source** is

ENAME N+ N− NC+ NC− VALUE

where N+ is the positive node of the source, N− is the negative node, NC+ is the positive node of the dependent (or controlling) voltage, NC− is the negative node of the dependent voltage, and VALUE is the ''value'' of the dependent source—in particular, if the dependent source is labeled kv, then k is the value (dimensionless) of the source.

Example 16.3

The circuit shown in Fig. 16.3 (see Fig. 2.9 on p. 64) is described by the following input file:

Input file for Fig. 16.3
Vs 1 0 DC 5
E 0 3 2 0 10
R1 0 2 1
R2 1 2 2
R3 3 4 3
R4 4 0 4
R5 2 4 5
.END

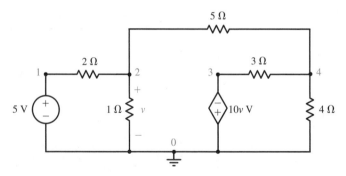

Fig. 16.3 Circuit with a voltage-dependent voltage source.

The corresponding output file contains the node voltages:

NODE VOLTAGE   NODE VOLTAGE   NODE VOLTAGE

(1) 5.0000   (2) 1.0000   (3) −10.0000

NODE VOLTAGE

(4) −4.0000

In addition to the node voltages, the currents through all of the independent voltage sources, and the total power dissipation, by including a **.OP statement** in the input file, the output file will also list the voltage across and the current through each voltage-controlled voltage source. Specifically, for this example, if we add the control statement

.OP

to the input file, then the output file also contains the "operating point information":

VOLTAGE-CONTROLLED VOLTAGE SOURCES

NAME   E
V-SOURCE   1.000E+01
I-SOURCE   −2.000E+00

which indicates that the voltage across the dependent (controlled) voltage source is 10 V with the polarity indicated, and the current through this source (directed from + to −) is −2 A.

Drill Exercise 16.3

Use PSpice to find the voltage across and current through the voltage-dependent voltage source shown in the circuit given in Fig. 2.11 on p. 68.

ANSWER   1.5 V, − 3.5 A

Voltage-Dependent Current Sources

The general form for a **voltage-dependent** (or **voltage-controlled**) **current source** is

GNAME N+ N− NC+ NC− VALUE

where the current source's arrow is directed from node N+ to node N−, NC+ is the positive node of the dependent (or controlling) voltage, NC− is the negative node of the dependent voltage, and VALUE is the "value" of the dependent source—specifically, if the dependent source is labeled kv, then k is the value (in mhos or siemens) of the source.

Example 16.4

Consider the circuit shown in Fig. 16.4 (see Fig. 2.7 on p. 58). The following is an input file for this circuit:

    Input file for Fig. 16.4
    Vs 1 0 DC 3
    G 3 0 2 0 .25
    R1 1 2 2
    R2 0 2 6
    R3 2 3 1
    R4 0 3 8
    .OP
    .END

The resulting output file contains the following results:

NODE VOLTAGE	NODE VOLTAGE	NODE VOLTAGE
(1) 3.0000	(2) 1.5000	(3) 1.0000

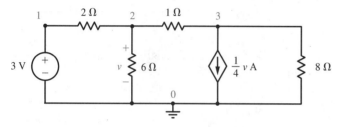

Fig. 16.4 Circuit with a voltage-dependent current source.

VOLTAGE-CONTROLLED CURRENT SOURCES

NAME	G
I-SOURCE	3.750E−01

and this means that the current through the voltage-controlled current source is 0.375 A in the direction indicated in the circuit.

Drill Exercise 16.4

Use PSpice to find the node voltages for the circuit given in Fig. P2.5 on p. 99.

ANSWER −3 V, 6 V, 4V

Current-Dependent Sources

In order to analyze circuits with current-dependent sources, SPICE requires a slight modification of a given circuit. In particular, we must insert an independent voltage source VNAME having a value of zero volts (i.e., a short circuit) in the circuit such that the dependent (or controlling) current goes through this source from the positive side to the negative side.

The general form for a **current-dependent** (or **current-controlled**) **current source** is

FNAME N+ N− VNAME VALUE

where the arrow of the current source is directed from node N+ to node N− and VALUE is the "value" of the dependent source—specifically, if the dependent source is labeled ki, then k is the value (dimensionless) of the dependent source.

Example 16.5

For the circuit shown in Fig. 16.5, a 0-V independent voltage source Vcur is used in conjunction with the current-dependent current source. The following is an input file for this circuit.

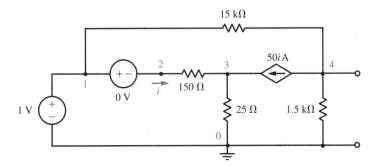

Fig. 16.5 Circuit with a current-dependent current source.

Input file for Fig. 16.5
V1 1 0 DC 1
Vcur 1 2
*Since no value for Vcur is specified, the default value is 0 V.
F 4 3 Vcur 50
R1 2 3 150
R2 0 3 25
R3 0 4 1.5k
R4 1 4 15k
.OP
.END

The resulting output file contains the following results:

NODE VOLTAGE	NODE VOLTAGE	NODE VOLTAGE
(1) 1.0000	(2) 1.0000	(3) .8947

NODE VOLTAGE		
(4) −47.7560		

VOLTAGE SOURCE CURRENTS

NAME	CURRENT
Vs	−3.952E−03
Vcur	7.018E−04

TOTAL POWER DISSIPATION   3.95E−03   WATTS

CURRENT-CONTROLLED CURRENT SOURCES

NAME   F
I-SOURCE   3.509E−02

Drill Exercise 16.5

Use PSpice to find i_e for the circuit given in Fig. P2.15 on p. 101 when $v_1 = 1$ V.

ANSWER 23.14 mA

The general form for a **current-dependent** (or **current-controlled**) **voltage source** is

HNAME N+ N− VNAME VALUE

where N+ is the positive node of the source, N− is the negative node of the source, VNAME is a zero-valued independent voltage source through which the dependent (or controlling) current goes (from the positive side to the negative side), and VALUE is the "value" of the dependent source—specifically, if the dependent source is labeled ki, then k is the value (in ohms) of the dependent source.

Operational Amplifiers

Since an ideal operational amplifier is modeled by a voltage-dependent voltage source (see Fig. 2.13 on p. 72), we can create an element statement for an op amp. Specifically, the general form for an **operational amplifier** will be

ENAME NOUT 0 NIN+ NIN− A

where NOUT is the output node, NIN+ is the noninverting-input node, NIN− is the inverting-input node, and A is the gain of the op amp. For an ideal op amp, A is infinite—therefore, to have a good approximation of an ideal op amp, make the gain a very large positive number.

Example 16.6

For the op-amp circuit shown in Fig. 16.6, we have the following input file:

```
Input file for Fig. 16.6
Vs 1 0 DC .1
R1 1 2 1k
R2 3 0 1k
R3 3 4 10k
R4 2 4 20k
E 4 0 3 2 1G
.END
```

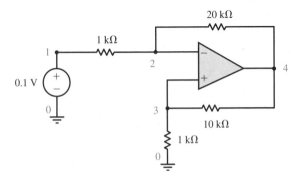

Fig. 16.6 Op-amp circuit.

The resulting output file contains the following:

NODE VOLTAGE	NODE VOLTAGE	NODE VOLTAGE
(1) .1000	(2) .2000	(3) .2000

NODE VOLTAGE
(4) 2.2000

Since an ideal op amp consists of just a voltage-dependent voltage source, there may be a circumstance when you might try to use PSpice to analyze a circuit in which there is only one circuit element that is connected to a particular node. (As an example, for a noninverting amplifier such as that depicted in Fig. 2.16 on p. 75, the node representing the inverting input of the op amp is connected to only one circuit element—that being the independent voltage source.) To avoid such a situation, which is not allowed in PSpice, just connect a very large resistor (ideally, it should be infinite) to the node in question. The other end of the resistor can be connected to a different convenient node.

Drill Exercise 16.6

Determine an input file for the op-amp circuit given in Fig. 2.19 on p. 79 for the case that $v_s = 0.1$ V.

ANSWER Input file for Fig. 2.19
Vs 1 0 DC .1
E 6 0 3 2 1G
R1 1 3 1k
R2 0 2 1k
R3 2 6 100k
R4 0 6 10k
*The next resistor Rin is used so that at least 2
*circuit elements are connected to node 3.
Rin 2 3 1G
.END

Transfer Functions

As mentioned previously, when no type of analysis is specified, PSpice will automatically perform a dc analysis. This type of analysis will determine all of the node voltages for a given circuit. By using the appropriate control statement, we can also have PSpice perform a dc analysis which, in addition to calculating node voltages, will calculate a specific transfer function (gain) as well as the input and output resistances for a circuit. This **.TF statement** has the form

.TF OUTVAR INSRC

where OUTVAR is the output variable and INSRC is the input source. An output variable can be either the voltage between any pair of nodes or the current through an independent voltage source. Specifically, V(N1,N2) is the voltage between nodes N1 and N2 (with the + at node N1 and the − at node N2). If the comma and N2 are omitted, then node N2 is assumed to be the reference node. The notation I(VNAME) signifies the current through voltage source VNAME directed from the positive side to the negative side.

Example 16.7 ▰▰▰▰▰▰▰▰▰▰▰▰▰▰▰▰▰▰▰▰

For the circuit described in Example 16.5, suppose we add the control statement

.TF V(4) Vs

to that input file. The control statement .TF indicates that V(4) is the output variable, and consequently, PSpice will calculate the transfer function V(4)/Vs, the input resistance—that is, the resistance loading the independent current source, and the output (Thévenin-equivalent) resistance (between nodes 4 and 0). The resulting output file contains the following:

SMALL-SIGNAL CHARACTERISTICS

$V(4)/Vs = -4.776E+01$
INPUT RESISTANCE AT Vs = 2.530E+02
OUTPUT RESISTANCE AT V(4) = 1.364E+03

This indicates that the transfer function is $v_4/v_s = -47.76$, the input resistance is $R_{in} = 253 \ \Omega$, and the output resistance is $R_o = 1364 \ \Omega$. (See Problem 2.14 on p. 100.)

Drill Exercise 16.7

Use PSpice to find the transfer function, input resistance, and output resistance for the circuit given in Fig. P2.44 on p. 106.

ANSWER   1.5, 2.667 Ω, 3 Ω

DC Analysis

We have seen that if a circuit contains independent dc sources described by element statements having the form.

SRCNAME N+ N− DC VALUE

where SRCNAME is the name of the independent source, N+ is the positive node, N− is the negative node, and VALUE is the value of the source, then PSpice will automatically perform a dc analysis and determine the node voltages, the independent voltage-source currents, and the total power dissipation. Knowing the node voltages, the voltage between any pair of nodes can be determined readily by using KVL. Furthermore, the currents through resistors can then be found by applying Ohm's law.

It is also possible to directly obtain a printout of the voltages between pairs of nodes and the currents through circuit elements by using a **.DC analysis statement.** The form of this statement is

.DC SRCNAME LIST VALUE

In conjunction with this statement, however, we will be required to use a .PRINT statement that specifies which voltages and currents are to be printed out. The general form of the **.PRINT statement for dc analysis** is

.PRINT DC OV1 <OV2 OV3 . . . >

where OV1, OV2, OV3, . . . are output variables. These can be the voltage at any node (with respect to the reference), the voltage between any pair of nodes, the voltage across any element (where the + is at the first listed node in the element statement), or the current through any element (where the current is directed from the first listed node to the second listed node in the element statement). The brackets < > indicate that the enclosed output variables are optional.

For instance, for the circuit described in Example 16.3, if we add the statements

.DC Vs LIST 5
.PRINT DC I(R2) V(1,2) I(E) V(E) V(4) V(R4) I(Vs)

to the input file, the resulting output file contains

**** DC TRANSFER CURVES   TEMPERATURE = 27.000 DEG C
****

Vs	I(R2)	V(1,2)	I(E)	V(E)
5.000+00	2.000E+00	4.000E+00	−2.000E+00	1.000E+01

**** DC TRANSFER CURVES   TEMPERATURE = 27.000 DEG C
****

Vs	V(4)	V(R4)	I(Vs)
5.000E+00	−4.000E+00	−4.000E+00	−2.000E+00

This indicates that the current through R2 (directed from node 1 to node 2) is 2 A, the voltage between nodes 1 and 2 (with the + at node 1) is 4 V, the current through the voltage-dependent voltage source (directed from node 0 to node 3) is −2 A, the voltage across the voltage-dependent voltage source (with the + at node 0) is 10 V, the voltage at node 4 is −4 V, the voltage across R4 (with the + at node 4) is −4 V, and the current from node 1 to node 0 through the 5-V source is −2 A.

16.2 Transient Analysis

Up to this point, we have considered circuits that contained voltage and current sources (both independent and dependent), op amps, and resistors. Furthermore, all of the independent sources in the circuits that were analyzed had dc (constant) values. As a consequence, all of the voltages and currents that were determined had constant values.

We shall now consider the case that circuits have voltages and currents whose values vary with time. One way to produce such voltages and currents is to use independent sources whose values are described by functions of time. Another way is to employ circuit elements such as inductors and capacitors that have nonzero initial conditions. We shall consider this latter situation first.

Inductors and Capacitors

The general form for an **inductor** is

LNAME N+ N− VALUE <IC=INCOND>

where N+ is the positive node, N− is the negative node, and VALUE is the inductance in henries. The brackets < > indicate that the enclosed term is optional. In this case, the optional term IC =INCOND is the initial condition (current) of the inductor. If none is specified, then the initial condition is set to zero, and the node polarities are irrelevant. If the initial condition is nonzero, then the initial current goes through the inductor from the positive node to the negative node.

The general form for a **capacitor** is

CNAME N+ N− VALUE <IC=INCOND>

where N+ is the positive node, N− is the negative node, and VALUE is the capacitance in farads. In this case, the optional term is the initial condition (voltage) of the capacitor. If none is specified, then the initial condition is set to zero, and the node polarities are irrelevant. If the initial condition is nonzero, then node N+ is the positive plate of the capacitor, and node N− is the negative plate of the capacitor.

Knowing the element statements for inductors and capacitors, we are now in a position to use PSpice to determine natural responses. In order either to print out or plot voltages and currents that vary with time, we use the **transient analysis** feature of PSpice. To perform such analyses, we use the .TRAN control statement. PSpice does not yield analytical expressions, but rather produces printouts or plots (or both) of a circuit variable. Thus along with a .TRAN statement, we will also use either a .PRINT, a .PLOT, or a .PROBE statement.

The form of the **.TRAN statement** is

.TRAN TSTEP TSTOP <TSTART> <UIC>

where TSTEP is the printing or plotting increment and TSTOP is the final time. Specifying the initial time with TSTART is optional—if TSTART is omitted, it is assumed to be zero. (TSTART cannot be a negative number.) If the circuit to be analyzed has nonzero initial conditions, the term UIC is included in the .TRAN statement.

The general form of the **.PRINT statement for transient analysis** is

.PRINT TRAN OV1 <OV2 OV3 . . . >

where OV1, OV2, OV3 . . . are output variables. As before, output variables can be the voltage at a node (with respect to the reference), the voltage between two nodes, and the current through an independent voltage source. In addition, the voltage across or the current through a resistor, an inductor, or a capacitor can be designated as an output variable. For example, V(R1) is the voltage across R1 (where the $+$ is at the first listed node in the element statement for R1), whereas I(R1) is the current through R1 (where the current is directed from the first to the second listed node in the element statement for R1).

The general form of the **.PLOT statement for transient analysis** is

.PLOT TRAN OV1 <OV2 OV3 . . . >

This statement will result in plots of the output variables listed, and in addition, will also result in a printout of the first variable listed.

Example 16.8

Consider the *RC* circuit shown in Fig. 16.7. Suppose the initial condition for this circuit is $v(0) = 6$ V. We can get a printout and a plot of the natural response of the voltage $v(t)$ across the capacitor by using the following input file:

```
Input file for Fig. 16.7
R 1 0 2
C 1 0 .125 IC=6
.TRAN .05 1 UIC
.PLOT TRAN V(1)
.END
```

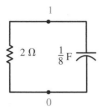

Fig. 16.7 *RC* circuit with nonzero initial condition.

The output variable for the .PLOT statement is the voltage at node 1. The resulting output file contains the printout and plot shown in Fig. 16.8.

```
 TIME          V(1)
 (*)----------      0.0000E+00   2.0000E+00   4.0000E+00   6.0000E+00   8.0000E+00

 0.000E+00    6.000E+00  - - - - - - - - - - - - - - - - - - - - - - * - - - - -
 5.000E-02    4.915E+00  .              .            .          *        .
 1.000E-01    4.024E+00  .              .            .     *             .
 1.500E-01    3.294E+00  .              .         *                      .
 2.000E-01    2.697E+00  .              .    *                           .
 2.500E-01    2.207E+00  .            .*                                 .
 3.000E-01    1.807E+00  .          * .                                  .
 3.500E-01    1.479E+00  .        *                                      .
 4.000E-01    1.211E+00  .      *                                        .
 4.500E-01    9.914E-01  .    *                                          .
 5.000E-01    8.117E-01  .   *                                           .
 5.500E-01    6.644E-01  .  *                                            .
 6.000E-01    5.440E-01  .  *                                            .
 6.500E-01    4.453E-01  . *                                             .
 7.000E-01    3.646E-01  . *                                             .
 7.500E-01    2.984E-01  . *                                             .
 8.000E-01    2.443E-01  . *                                             .
 8.500E-01    2.000E-01  .*                                              .
 9.000E-01    1.637E-01  .*                                              .
 9.500E-01    1.340E-01  .*                                              .
 1.000E+00    1.097E-01  .*                                              .
```

Fig. 16.8 Printout and plot of natural response for *RC* circuit shown in Fig. 16.7.

Drill Exercise 16.8

For the circuit shown in Fig. 16.7, replace the capacitor with a $\frac{1}{2}$-H inductor whose initial current is 3 A directed up. Use PSpice to get a plot of the voltage at node 1.

ANSWER Same plot as in Fig. 16.8.

PROBE

Figure 16.8 demonstrates how crude a plot of voltage versus time is in a typical SPICE output file. Fortunately, PSpice has a companion graphics package known as **PROBE.** In order to use this superb graphics software, a **.PROBE statement** must be added to an input file. When this is done, after PSpice is executed, not only is an output file created, but a PROBE data file is also generated. (If the input file is called FILENAME.CIR, then the output file is FILENAME.OUT and the PROBE data file is FILENAME.DAT.) By invoking PROBE, which is menu driven, a high quality plot can be obtained. Figure 16.9 shows a PROBE plot of the natural response for the circuit described in Example 16.8.

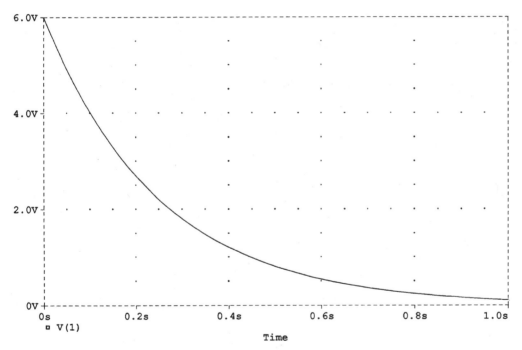

Fig. 16.9 PROBE plot of voltage versus time for the *RC* circuit described in Example 16.8.

As mentioned before, PSpice and PROBE produce printouts and plots of voltage and current responses, but not analytical expressions; however, PROBE does have the capability of plotting analytical expressions. In this way, it is possible to determine whether or not a particular expression describes a certain response. Just have PROBE plot an expression and compare the result with the plot produced by PSpice.

Example 16.9

For the series *RLC* circuit shown in Fig. 16.10*a*, suppose $R = 2\ \Omega$, $L = 1$ H, $C = 0.02$ F, $v(0) = 2$ V, and $i(0) = -0.32$ A. In Example 3.13 on p. 155 it was determined that for $t \geq 0$ s,

$$v(t) = 2e^{-t}(\cos 7t - \sin 7t)\ \text{V} \qquad \text{and}$$

$$i(t) = e^{-t}(-0.32 \cos 7t - 0.24 \sin 7t)\ \text{A}$$

Let us now use PROBE to confirm these results.
We begin with the input file

```
Input file for Fig. 16.10a
R 0 1 2
L 1 2 1 IC=-.32
C 2 0 .02 IC=2
.TRAN .01 3 UIC
.PROBE
.END
```

Executing PROBE after PSpice is run, and plotting the variable V(2), which is the voltage across the capacitor, results in the first plot shown in Fig. 16.10*b*. In this figure there is a second trace that is virtually identical to the trace corresponding to V(2). This second trace is a plot of the above analytical expression for $v(t)$. The syntax used for PROBE for this voltage expression is

$$2*\exp(-\text{time})*(\cos(7*\text{time})-\sin(7*\text{time}))$$

A plot of the current through the inductor—which is also the current through the capacitor and resistor—can be obtained from PROBE by plotting the variable I(L).

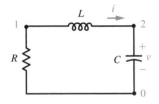

Fig. 16.10 *(a)* Series *RLC* circuit.

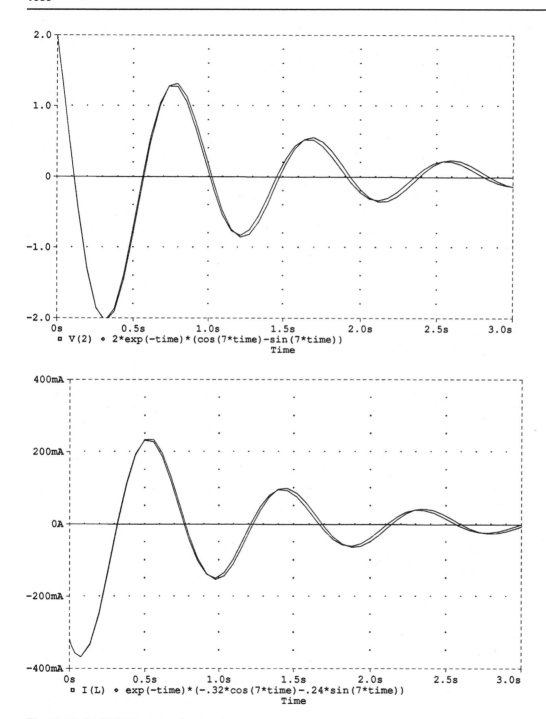

Fig.16.10 *(b)* PROBE plots of natural responses.

This trace is shown in the second plot in Fig. 16.10*b*, as is the virtually identical trace corresponding to the analytical expression for *i(t)* above. The PROBE syntax for this current expression is

$$\exp(-\text{time})*(-.32*\cos(7*\text{time})-.24*\sin(7*\text{time}))$$

Drill Exercise 16.9

For the series *RLC* circuit shown in Fig. 16.10*a*, suppose $R = 5\ \Omega$, $L = \frac{1}{2}$ H, $C = \frac{1}{8}$ F, $v(0) = 2$ V, and $i(0) = 1$ A. Obtain PROBE plots of $v(t)$ and $i(t)$ versus t for $t \geq 0$ s.

ANSWER   See Fig. 3.32 on p. 154.

Nonzero Excitations

Let us now consider the case that a circuit contains an independent voltage or current source whose value is time-dependent. The first independent-source function we will consider is the **PWL (piece-wise linear) function** whose general form is

PWL (T1 V1 <T2 V2 T3 V3 . . . >)

where each pair TN, VN signifies that the value of the source at time TN is VN. The value of the source in the interval TN < t < T(N + 1) is determined by using linear interpolation on the values VN and V(N + 1).

Example 16.10

Suppose we wish to determine the unit step responses for the series *RC* circuit shown in Fig. 16.11*a*. Since the time constant of the circuit is $\tau = RC = (2)(2) = 4$ s, we know that the response will be very close to its final value in about three or four time constants. Therefore, let us plot the step response from 0 to 15 s. Although we cannot produce a step function (because it has a discontinuity at time $t = 0$ s) with PSpice, we can approximate it and still get a very accurate result.

In particular, suppose we approximate a unit step function with a PWL function that is 0 at $t = 0$ s, 1 at $t = 0.001$ s, and remains 1 for $t > 0.001$ s. We can do this with the following input file:

```
Input file for a unit step response
Vs 1 0 PWL(0 0 .001 1 15 1)
R 1 2 2
C 2 0 2
.TRAN .01 15
.PROBE
.END
```

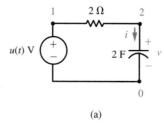

(a)

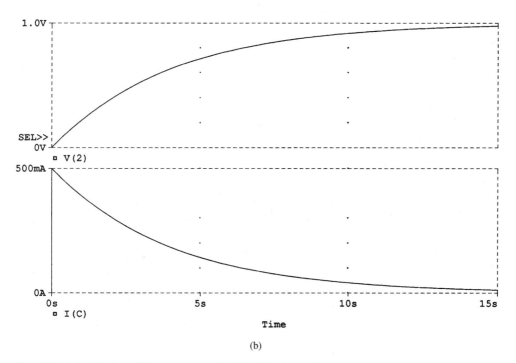

(b)

Fig. 16.11 *(a)* Series *RC* circuit, and *(b)* PROBE plots of unit step responses.

The PROBE plots of the voltage across and the current through the capacitor are shown in Fig. 16.11*b*.

Drill Exercise 16.10

Use PROBE to verify that the voltage step response given in Fig. 16.11*b* is $1 - e^{-t/4}$ V for $t \geq 0$ s and the current step response is $0.5e^{-t/4}$ A for $t \geq 0$ s.

ANSWER   Add a trace to the voltage plot by using $1 - \exp(-\text{time}/4)$.
Add a trace to the current plot by using $.5*\exp(-\text{time}/4)$.

Another way to produce a good approximation to a step function is with the use of a **PULSE function.** The general form of this function is

PULSE(V1 V2 TD TR TF PW PER)

where the parameters are indicated in Fig. 16.12. When using this function, we must specify the **initial value** V1 and the **pulsed value** V2. If any other parameter is omitted, it is given a default value. The **delay time** TD has a default value of 0.

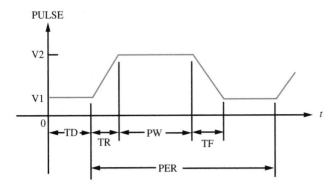

Fig. 16.12 PULSE function.

The **rise time** TR and **fall time** TF have default values of TSTEP, whereas the **pulse width** PW and the **period** PER have default values of TSTOP. (TSTEP and TSTOP are the printing increment and final time, respectively, of the associated .TRAN statement.)

In order to make a pulse function approximate a unit step function, we make V1 zero, V2 unity, TD zero, TR very small, and omit the remaining parameters, thereby incorporating their default values.

Example 16.11

Suppose we wish to determine the inductor-current and the capacitor-voltage unit step responses for the *RLC* circuit shown in Fig. 16.13*a* (see Example 3.18 on p.

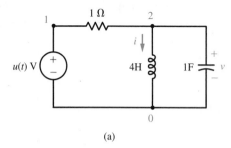

(a)

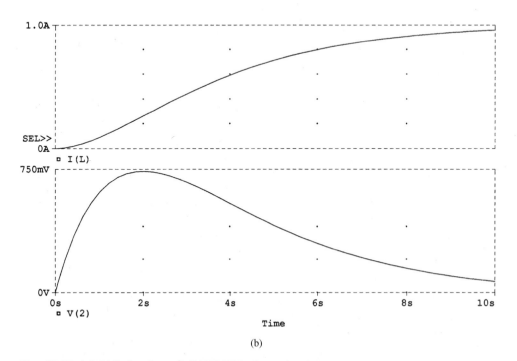

(b)

Fig. 16.13 *(a) RLC circuit, and (b) PROBE plots of unit step responses.*

173). To approximate a unit step function, we have made the rise time TR of the pulse function very small—in this case, TR = 0.001 s. The following is the corresponding input file:

```
Input file for Fig. 16.13a
Vs 1 0 PULSE(0 1 0 .001)
R 1 2 1
L 2 0 4
C 2 0 1
.TRAN .001 10
.PROBE
.END
```

The PROBE plots of the current through the inductor and the voltage across the capacitor (which is also the voltage across the inductor) are shown in Fig. 16.13b.

Drill Exercise 16.11

Use PROBE to verify that the current step response given in Fig. 16.13b is $1 - \frac{1}{2} te^{-t/2} - e^{-t/2}$ A for $t \geq 0$ s and the voltage step response is $te^{-t/2}$ V for $t \geq 0$ s.

ANSWER   Add a trace to the current plot by using
$1 - .5*time*exp(-time/2) - exp(-time/2)$.
Add a trace to the voltage plot by using $time*exp(-time/2)$.

Another independent-source function is the **EXP function.** The general form of this function is

EXP(V1 V2 TD1 TAU1 TD2 TAU2)

where the parameters are indicated in Fig. 16.14. When using this function, we must specify the **initial value** V1 and the **pulsed value** V2. If any other parameter is omitted, it is given a default value. The **first delay time** TD1 has a default value of 0. The **first time constant** TAU1 has a default value of TSTEP, the **second delay time** TD2 has a default value of TD1 + TSTEP, and the **second time constant** TAU2 has a default value of TSTEP. (TSTEP is the printing increment of the associated .TRAN statement.)

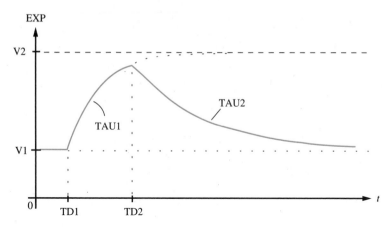

Fig. 16.14 EXP function.

Example 16.12

We can employ the EXP function to obtain a voltage source whose value is the typical decaying exponential expression $Ae^{-t/\tau}$. In particular, if $A = 1$ and the time constant is $\tau = 0.25$ s, then we can use EXP(1 0 0 .25 1). Specifically, consider a

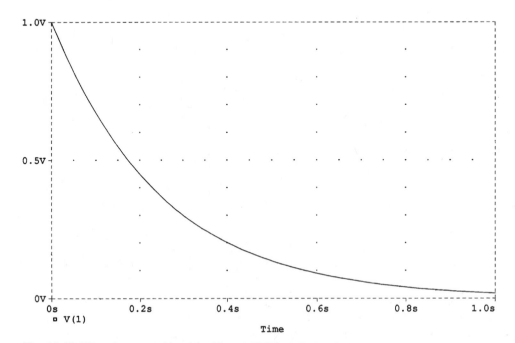

Fig. 16.15 Waveform produced by Vs 1 0 EXP(1 0 0 .25 1).

simple circuit consisting of a voltage source and a 1-Ω resistor connected to the same pair of nodes. The corresponding input file is

Input file for a decaying exponential source
Vs 1 0 EXP(1 0 0 .25 1)
R 1 0 1
.TRAN .01 1
.PROBE
.END

The PROBE plot for the voltage produced by the independent voltage source is shown in Fig. 16.15.

Drill Exercise 16.12

Connect three independent voltage sources, two having EXP functions and one having a PULSE function, in series such that the voltage across the connection is the voltage waveform shown in Fig. 3.30 on p. 150.

ANSWER EXP(1 0 0 1 3), EXP($-$1 0 1 1 3), PULSE(0 1 0 .001 .001 1 3)

16.3 AC Analysis

In addition to dc analysis and transient analysis, SPICE will also perform sinusoidal or ac analysis. The general form of an **independent ac voltage source** is

VNAME N+ N$-$ AC ACMAG <ACPHASE>

where VNAME is the name of the voltage source, N+ is the positive node of the source, N$-$ is the negative node of the source, AC indicates a sinusoidal source, ACMAG is the ac magnitude (i.e., the amplitude of the sinusoid), and ACPHASE is the ac phase (i.e., the phase angle of the sinusoid in degrees). If ACPHASE is omitted, then a value of zero is assumed.

The general form of an **independent ac current source** is

INAME N+ N$-$ AC ACMAG <ACPHASE>

where INAME is the name of the current source, the current source's arrow is directed from node N+ to node N−, and the terms AC, ACMAG, and ACPHASE are as described above.

To use SPICE to analyze an ac circuit for a given frequency, we use the following control statement:

.AC LIN 1 FREQ FREQ

where FREQ is the sinusoidal frequency in hertz. The significance of the terms "LIN" and "1" will be discussed shortly.

Unlike dc analysis, SPICE will not automatically print out the values of the ac node voltages. Instead, we must specify which variables are to be listed with a .PRINT statement. The general form of a **.PRINT statement for ac analysis** is

.PRINT AC OV1 <OV2 OV3 . . . >

where the voltage output variables are designated VM (magnitude), VP (phase), VR (real part), VI (imaginary part), and VDB ($20 \log_{10}$ magnitude). To specify current output variables, replace V by I in these designations.

Example 16.13

Consider the series RC circuit shown in Fig. 16.16 (see Example 4.4 on p. 189).

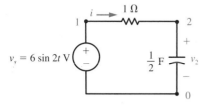

Fig. 16.16 Series RC circuit.

The following input file can be used to determine the capacitor voltage and current:

Input file for Fig. 16.16
Vs 1 0 AC 6 −90
R 1 2 1
C 2 0 .5
*Since the angular frequency is 2 rad/s, then
*the cyclical frequency is $2/2\pi = 1/\pi = 0.31831$ Hz

.AC LIN 1 .31831 .31831
.PRINT AC VM(C) VP(C) IM(C) IP(C)
.END

The resulting output file contains the following:

**** AC ANALYSIS   TEMPERATURE = 27.000 DEG C
****

FREQ	VM(C)	VP(C)	IM(C)	IP(C)
3.183E−01	4.243E+00	−1.350E+02	4.243E+00	−4.500E+01

Thus we have the following phasors

$$\mathbf{V}_2 = 4.243\underline{/-135°}\ \text{V} \quad\text{and}\quad \mathbf{I} = 4.243\underline{/-45°}\ \text{A}$$

and the corresponding functions of time are

$$v_2(t) = 4.243\cos(2t - 135°)\ \text{V} \quad\text{and}\quad i(t) = 4.243\cos(2t - 45°)\ \text{A}$$

Drill Exercise 16.13

Use PSpice to find the clockwise mesh currents for the ac circuit given in Fig. 4.15 on p. 209 when $\mathbf{Z}_L$ is a 48-Ω resistor.

ANSWER $0.08046\cos(4t + 17.01°)$ A, $0.07143\cos(4t + 23.13°)$ A

Transformers

The general form of a (nonideal) **transformer,** that is, magnetically coupled (mutual) inductors, is

    KNAME L1NAME L2NAME VALUE

where L1NAME and L2NAME are the self-inductances of the transformer (each self-inductance is also listed separately with its own element statement), and VALUE is the coefficient of coupling of the transformer. In particular, if a transformer has self-inductances L_1 and L_2, and mutual inductance M, then the coefficient of coupling is $k = M/\sqrt{L_1 L_2}$. The value of the coefficient of coupling (VALUE) is a number that must be greater than zero and less than or equal to one. The "dots" of the transformer correspond to the positive nodes of the self-inductances.

Example 16.14

Let us find $v_2(t)$ for the transformer circuit shown in Fig. 16.17 when $R_1 = 9\ \Omega$, $L_1 = 4$ H, $L_2 = 6$ H, $M = 2$ H, $C = 1/18$ F, and $v_s(t) = 36\cos(3t - 60°)$ V. (See Example 14.9 on p. 927.)

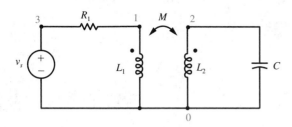

Fig. 16.17 Circuit with a transformer.

Since $C = 1/18 = 55.56$ mF, and

$$k = M/\sqrt{L_1 L_2} = 2/\sqrt{(4)(6)} = 0.4082 \qquad \text{and}$$

$$f = \omega/2\pi = 3/2\pi = 0.4775 \text{ Hz}$$

then we have the following input file:

```
Input file for Fig. 16.17
Vs 3 0 AC 36 −60
R1 1 3 9
C 2 0 55.56m
K L1 L2 .4082
L1 1 0 4
L2 2 0 6
.AC LIN 1 .4775 .4775
.PRINT AC VM(2) VP(2)
.END
```

The resulting output file contains the following:

```
**** AC ANALYSIS              TEMPERATURE = 27.000 DEG C
****************************************************************
      FREQ        VM(2)        VP(2)
    4.775E−01   8.481E+00   1.650E+02
```

and this corresponds to the phasor $\mathbf{V}_2 = 8.481\underline{/165°}$ V. Therefore, we have that $v_2(t) = 8.481 \cos (3t + 165°)$ V.

Drill Exercise 16.14

For the transformer circuit shown in Fig. 16.17, replace the capacitor with an 8-Ω resistor. Find $v_2(t)$ for the case that $R_1 = 3\ \Omega$, $L_1 = 1$ H, $L_2 = 2$ H, $M = 1$ H, and $v_s(t) = 5 \cos (6t + 30°)$ V.

ANSWER $2.828 \cos (6t + 21.86°)$ V

Frequency Response

In addition to single-frequency analysis, PSpice can also perform ac circuit analysis for a frequency range. Frequency can have a decade[1] variation, an octave[2] variation, or a linear variation. The **.AC statement** takes one of three forms:

.AC DEC ND FSTART FSTOP

where DEC indicates a decade variation of frequency, and ND is the number of points per decade.

.AC OCT NO FSTART FSTOP

where OCT indicates an octave variation of frequency, and NO is the number of points per octave.

.AC LIN NP FSTART FSTOP

where LIN indicates a linear variation of frequency, and NP is the number of points.

In all three cases, FSTART (which must be greater than zero) is the starting frequency in hertz, and FSTOP is the final frequency in hertz. To print or plot the results to the output file, respectively, we can use the following control statements:

.PRINT AC OV1 <OV2 OV3 . . . >
.PLOT AC OV1 <OV2 OV3 . . . >

---

[1]A decade is change in frequency by a factor of 10.
[2]An octave is a change in frequency by a factor of 2.

Better yet, we can employ **PROBE** by including the statement .PROBE in the input
file.

Example 16.15

Consider the first-order *RC* high-pass filter shown in Fig. 16.18*a*. We can obtain the
frequency response—consisting of the amplitude and phase responses—for this circuit with the following input file:

Input file for high-pass filter
V1 1 0 AC 1
R 2 0 1k

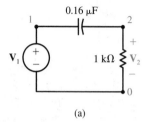

(a)

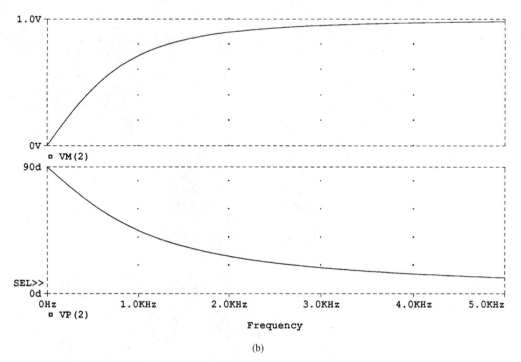

(b)

Fig. 16.18 (a) High-pass filter, and (b) amplitude and phase responses.

```
C 1 2 .16u
.AC LIN 100 1 5k
.PROBE
.END
```

Since the magnitude of V_1 is unity, then the magnitude plot for V_2 (i.e., VM(2) versus frequency) is the same as the amplitude response for the transfer function V_2/V_1. Since the voltage-source element statement does not specify a phase, by default it is 0°. Thus the phase plots for V_2 (i.e., VP(2) versus frequency) is the same as the phase response for V_2/V_1.

After PSpice is executed, we can use PROBE to obtain the amplitude and phase responses shown in Fig. 16.18*b*. Since FSTART must be greater than zero, in order to obtain these plots we are required to make the selection of a linear *x* axis within PROBE.

Drill Exercise 6.15

For the circuit shown in Fig. 5.8 on p. 272, suppose $R = 1$ kΩ. Replace the capacitor with a 160-mH inductor and use PSpice and PROBE to obtain the amplitude and phase responses for V_2/V_1.

ANSWER Same plots as shown in Fig. 16.18*b*.

Bode Plots

By using a logarithmic frequency scale instead of a linear scale, it is possible to use PSpice and PROBE to produce Bode plots. Specifically, we may use either an octave or a decade scale for frequency, a dB scale for the magnitude, and a linear scale for the phase.

Example 16.16

Consider the first-order *RC* low-pass filter shown in Fig. 16.19*a*. We can obtain the Bode plot for this circuit with the following input file:

```
Input file for low-pass filter
V1 1 0 AC 1
R 1 2 1k
```

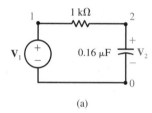

(a)

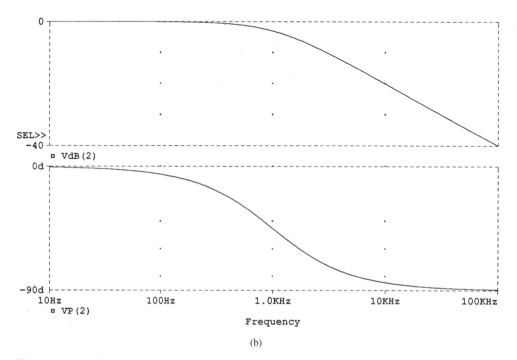

(b)

Fig. 16.19 *(a)* *RC* low-pass filter, and *(b)* its Bode plot.

```
C 2 0 .16u
.AC DEC 50 10 100k
.PROBE
.END
```

For this input file, the frequency variation from 10 to 100 kHz (which is four decades) has 50 points per decade.

For a Bode plot, the amplitude response is a plot of $20 \log_{10} |V_2/V_1|$ versus frequency. Since the magnitude of the source is unity, this is equivalent to a plot of $20 \log_{10} |V_2|$. Such a plot can be obtained with PROBE by selecting the output variable VDB(2). Since the phase of the source is zero, the phase response can be obtained by selecting the output variable to be VP(2). The Bode plot for the low-pass filter given in Fig. 16.19a is shown in Fig. 16.19b.

16.4 Diodes

Even though it is very useful for linear circuits, SPICE was created especially for the analysis of electronic circuits. In this section we begin to use PSpice to analyze circuits that contain diodes.

The general form for a diode is

DNAME N+ N− MODNAME

where N+ is the anode or p side, N− is the cathode or n side, and MODNAME is the model name of the diode. The implication, therefore, is that there is an associated model of the diode which describes the diode's characteristics. Thus in addition to the diode element statement, we are required to use a **.MODEL statement** which establishes the parameters of the diode.

The general form of the .MODEL statement for a diode is

.MODEL MODNAME D (<PNAME1=PVAL1 PNAME2=PVAL2 . . . >)

where PNAME1 is the name of the first diode parameter and PVAL1 is the value of this parameter, etc. Although there are numerous parameters which can be used to model a diode, we shall limit our discussion to those in the following table.

Diode Parameter	Name	Default Value	Units
Is	saturation current	1E−14	A
N	emission coefficient	1	none
BV	breakdown voltage	∞	V

The emission coefficient is N = 1 for Ge and N = 2 for Si.

Example 16.17

Let us determine the diode current and voltage for the circuit shown in Fig. 16.20 for the case that $R = 1$ kΩ, $v_1 = 3$ V, the diode is silicon with a saturation current of 10 nA, and the temperature is 300 K. (See Example 6.7 on p. 369.)

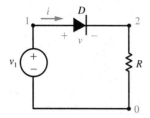

Fig. 16.20 A diode circuit.

Since the default temperature for PSpice is 27°C (300 K), it is not necessary to input temperature information; however, to have a different temperature, use the **.TEMP statement**

.TEMP VALUE

where VALUE is the temperature in degrees Celsius.

An input file for the given circuit is

Input file for Fig. 16.20
V1 1 0 DC 3
D 1 2 DMOD
.MODEL DMOD D (Is=10n N=2)
R 0 2 1k
.OP
.END

The resulting output file contains the following:

**** OPERATING POINT INFORMATION   TEMPERATURE = 27.000 DEG C
****
**** DIODES

NAME   D
MODEL   DMOD
ID   2.36E–03
VD   6.40E–01

and this indicates that $i = 2.36$ mA and $v = 0.640$ V.

> **Drill Exercise 16.17**
>
> Use PSpice to find the currents and the voltages for the silicon diodes in the circuit given in Fig. 6.17 on p. 371 when $I_{S1} = 1$ nA, $I_{S2} = 20$ nA, and the temperature is 300 K (27°C).
>
> **ANSWER**   20 nA, −20 nA, 0.158 V, −5.84 V

The DC Sweep

For a linear circuit, scaling an excitation scales its response by the same factor. This is not the case, however, for a nonlinear circuit or element such as a diode. To see how the response of a nonlinear circuit changes when the excitation changes, we can employ a **DC sweep** by using a **.DC sweep statement.** The form of this statement is

  .DC SRCNAME VSTART VFINAL VINCR

where SRCNAME is the name of the independent source, VSTART is the starting value for the source, VFINAL is the final value for the source, and VINCR is the incrementing value. VSTART may be greater than or less than VFINAL—so the sweep can be in either direction—but VINCR must be a positive number.

  Specifically, by including a .DC statement in an input file, when PSpice is executed, first a dc analysis is performed for the starting value of the source, a second dc analysis is performed for next value (the starting value plus the incrementing value), etc. These individual dc analyses are performed until the independent source achieves its final value. The results of the dc sweep can be printed or plotted in the corresponding output file by respectively using the statements

  .PRINT DC OV1 <OV2 OV3 . . . >

or

  .PLOT DC OV1 <OV2 OV3 . . . >

or can be viewed by using PROBE.

Example 16.18

Let us obtain the *i-v* characteristic for a Si diode having $I_S = 1$ nA at 300 K. To do this, we will connect a voltage source across the diode as shown in Fig. 16.21a, and

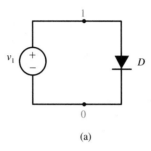

(a)

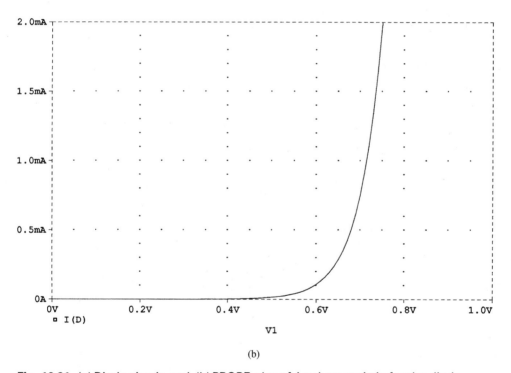

(b)

Fig. 16.21 *(a)* Diode circuit, and *(b)* PROBE plot of *i-v* characteristic for the diode.

then sweep the value of the voltage source from 0 to 1 V in increments of 0.01 V. This can be accomplished with the following input file:

Diode i-v characteristic
V1 1 0

D 1 0 DMOD
.MODEL DMOD D (Is=1n N=2)
.DC V1 0 1 .01
.PROBE
.END

Using PROBE, we can get the diode *i-v* characteristic shown in Fig. 16.21*b* by plotting the current through the diode I(D) versus the voltage across the diode V1.

Drill Exercise 16.18

Consider the circuit shown in Fig. 6.16, on p. 368, where the silicon diode has $I_S = 1$ nA, $R = 500$ Ω, and $v_1 = 1$ V. Use PROBE to determine the intersection of the diode *i-v* characteristic obtained in Example 16.18 and the load line $-V1/500 + 1/500$.

ANSWER 0.69 V, 0.62 mA

The SIN Function

Another very useful independent-source function is the **SIN function.** The general form of this function is

SIN(VO VA FREQ TD DF PHASE)

The corresponding function of time is

$$VO + \sin[(PHASE)(2\pi/360)] \qquad \text{for } 0 \le t \le TD$$

and

$$VO + VA\ e^{-DF(t-TD)} \sin[2\pi(FREQ)(t - TD) + (PHASE)(2\pi/360)] \text{ for } t \ge TD$$

where VO is the **offset,** VA is the **amplitude,** FREQ is the **frequency** in hertz, TD is the **delay time** in seconds, DF is the **damping factor,** and PHASE is the **phase angle** in degrees. The default value for FREQ is 1/TSTOP and the default values for TD, DF, and PHASE are zero. For the special case that TD = 0 s, the corresponding function of time is

$$VO + VA\, e^{-DFt} \sin[2\pi(FREQ)t + (PHASE)(2\pi/360)] \qquad \text{for } t \geq 0 \text{ s}$$

Example 16.19

Let us use PSpice and PROBE to analyze the Zener-diode circuit shown in Fig. 16.22a for the case that the 12-V Si Zener diodes have $I_S = 10$ nA and $v_s(t) = 24 \sin 2\pi t$ V. (See Example 6.16 on p. 398 for the case of ideal Zener diodes.)

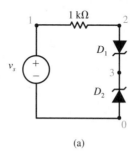

(a)

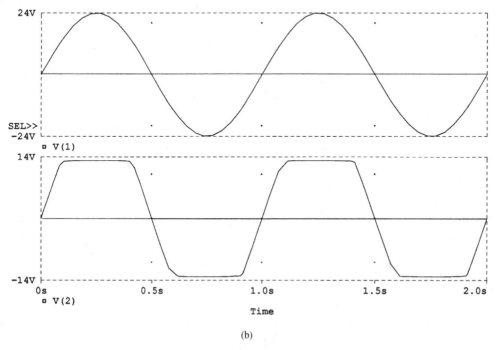

(b)

Fig. 16.22 (a) Zener-diode circuit, and (b) waveforms for $v_s(t)$ and $v_o(t)$.

In order to represent a Zener diode in PSpice, we just use a diode which has a finite breakdown voltage BV. In particular, a 12-V Zener diode would have the parameter BV=12 added to the .MODEL statement for the diode. An input file for the circuit to be analyzed is

Input file for Fig. 16.22a
Vs 1 0 SIN(0 24 1)
R 1 2 1k
D1 2 3 DMOD
D2 0 3 DMOD
.MODEL DMOD D (Is=10n N=2 BV=12)
.TRAN .01 2
.PROBE
.END

The PROBE plots for $v_s(t)$ and the $v_o(t)$ are shown in Fig. 16.22b.

Drill Exercise 16.19

Determine the SIN function that corresponds to the following damped sinusoids: (a) $-8e^{-2t}\sin 2t$, and (b) $5.657e^{-2t}\cos(2t - 45°)$.

ANSWER (a) SIN(0 8 .3183 0 2 180), (b) SIN(0 5.657 .3183 0 2 45)

Device Libraries

PSpice includes a library of models of various electronic devices such as diodes, BJTs, and JFETs. Instead of modeling a device by naming the model and using a corresponding .MODEL statement that specifies its parameters, a device in the library can be used. For example, in the evaluation version of PSpice, there is a library file EVAL.LIB that contains a number of models of electronic devices. One of these, listed as D1N4148, is a switching diode. In order to employ such a diode in a circuit, in the diode element statement of the input file, we do not use a model name, but rather the device model name D1N4148. This statement is then followed, not by a .MODEL statement, but rather the **.LIB statement**

.LIB EVAL.LIB

Example 16.20

Let us obtain a PROBE plot of the output voltage $v_o(t)$ for the peak detector shown in Fig. 16.23a when the diode model is D1N4148 from the device library, $R = 10$ kΩ, $C = 10$ μF, and $v_s(t) = 24 \sin 120\pi t$ V. An input file for this circuit is

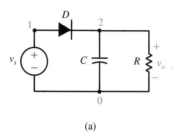

(a)

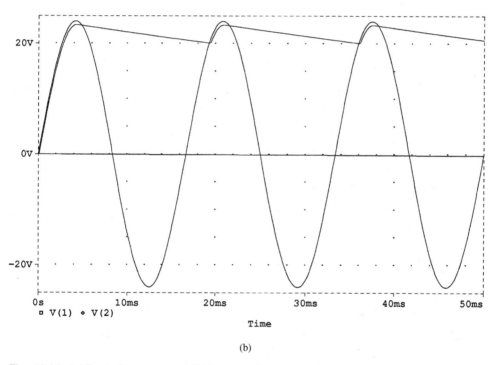

(b)

Fig. 16.23 *(a)* Peak detector, and *(b)* input and output voltages.

Input file for Fig. 16.23a
Vs 1 0 SIN(0 24 60)
D 1 2 D1N4148
.LIB EVAL.LIB

R 2 0 10k
C 2 0 10u
.TRAN 2u 50m
.PROBE
.END

The PROBE plots of both the input voltage $v_s(t)$ and the output voltage $v_o(t)$ are shown in Fig. 16.23b (see Fig. 6.57 on p. 401).

Drill Exercise 16.20

Obtain PROBE plots of the input voltage $v_s(t) = 24 \sin 120\pi t$ V and $v_o(t)$ for the circuit shown in Fig. 6.58 on p. xxx when $R = 10$ kΩ, $C = 100$ μF, and the diode model is D1N4148 from the PSpice device library.

ANSWER See Fig. 6.59 on p. 402.

16.5 Bipolar Junction Transistors (BJTs)

The general form of the element statement for a **bipolar junction transistor** (BJT) is

QNAME NC NB NE MODNAME

where NC is the collector node, NB is the base node, NE is the emitter node, and MODNAME is the model name of the BJT. Similar to the diode, the general form of the .MODEL statement for a *pnp* BJT is

.MODEL MODNAME pnp (<PNAME1=PVAL1 PNAME2=PVAL2 . . . >)

whereas the general form of the .MODEL statement for an *npn* BJT is

.MODEL MODNAME npn (<PNAME1=PVAL1 PNAME2=PVAL2 . . . >)

where PNAME1 is the name of the first BJT parameter and PVAL1 is the value of this parameter, etc. Although there are numerous parameters which can be used to model a BJT, we shall limit our discussion to those in the following table.

BJT Parameter	Name	Default Value	Units
Is	saturation current	1E–16	A
BF	forward beta	100	
BR	reverse beta	1	

Example 16.21

Let us determine the BJT's currents and voltages for the circuit shown in Fig. 16.24 (see Example 7.3 on p. 430) for the case that $I_S = 3.9E{-}15$ A and the temperature is 300 K.

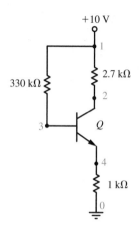

Fig. 16.24 An *npn* BJT circuit.

An input file for the given circuit is

```
Input file for Fig. 16.24
VCC 1 0 DC 10
Q 2 3 4 QMOD
.MODEL QMOD npn (Is=3.9E–15)
RB 1 3 330k
RC 1 2 2.7k
RE 0 4 1k
.OP
.END
```

The resulting output file contains the following:

**** OPERATING POINT INFORMATION   TEMPERATURE = 27.000 DEG C
****
**** BIPOLAR JUNCTION TRANSISTORS

NAME	Q
MODEL	QMOD
IB	2.16E−05
IC	2.16E−03
VBE	6.99E−01
VBC	−1.29E+00
VCE	1.99E+00
BETADC	1.00E+02

Thus we have that

$$i_B = 21.6 \ \mu A, \ i_C = 2.16 \ mA, \ v_{BE} = 0.699 \approx 0.7 \ V, \ v_{BC} = -1.29 \ V,$$

$$v_{CE} = 1.99 \ V$$

Drill Exercise 16.21

Use PSpice to find the currents and the voltages for the BJTs in the circuit given in Example 7.4 on p. 432 when $I_{S1} = 4.7E-15$ A, $I_{S2} = 9E-15$ A, and the temperature is 300 K.

ANSWER   26 μA, 2.6 mA, 0.699 V, 0.324 mV, 0.699 V
49.2 μA, 4.92 mA, 0.699 V, −0.382 V, 1.08 V

Schottky Transistors

Since a Schottky transistor is formed by connecting a Schottky diode between the base and collector of an ordinary BJT (see Fig. 7.36 on p. 470), in order to model a Schottky transistor, we need to have a Schottky-diode element for PSpice. All that is required, however, is a diode with the appropriate parameters. (If a diode with an emission coefficient of N = 1 and a saturation current of Is = 1 nA is used to model a Schottky diode, reasonable numerical results will often be obtained.)

Example 16.22

For the Schottky-BJT circuit shown in Fig. 16.25, suppose $R_B = 20$ kΩ, $R_C = 1$ kΩ, $V_{CC} = 5$ V, $h_{FE} = 50$, $v_1 = 5$ V, $I_S = 8.75E-15$ A for the BJT, and $I_S = 2.3E-11$ A for the diode. Then an input file for this circuit is the following:

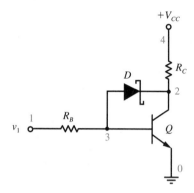

Fig. 16.25 A Schottky-BJT circuit.

```
Input file for Fig. 16.25
V1 1 0 DC 5
VCC 4 0 DC 5
RB 1 3 20k
RC 2 4 1k
D 3 2 DMOD
.MODEL DMOD D (Is=2.3E−11)
Q 2 3 0 QMOD
.MODEL QMOD npn (Is=8.75E−15 BF=50)
.OP
.END
```

The resulting output file contains

```
**** OPERATING POINT INFORMATION   TEMPERATURE = 27.000 DEG C
***************************************************************
**** DIODES

       NAME      D
       MODEL     DMOD
       ID        1.19E−04
       VD        4.00E−01
       REQ       2.18E+02
```

**** BIPOLAR JUNCTION TRANSISTORS

NAME	Q
MODEL	QMOD
IB	9.64E−05
IC	4.82E−03
VBE	6.99E−01
VBC	4.00E−01
VCE	2.99E−01
BETADC	5.00E+01

Thus we see that

$$i_B = 96.4 \ \mu A, \ i_C = 4.82 \ mA, \ i_D = 119 \ \mu A$$

$$v_{BE} = 0.699 \approx 0.7 \ V, \ v_{BC} = 0.4 \ V, \ v_{CE} = 0.299 \approx 0.3 \ V$$

(See Example 7.16 on p. 471.)

Drill Exercise 16.22

For the Schottky-BJT circuit shown in Fig. 16.25, suppose $R_B = 20 \ k\Omega$, $R_C = 500 \ \Omega$, $V_{CC} = 5 \ V$, $h_{FE} = 50$, $v_1 = 5 \ V$, $I_S = 1.7E−14 \ A$ for the BJT, and $I_S = 5E−12 \ A$ for the diode. Find the currents and voltages for the circuit.

ANSWER   189 μA, 9.43 mA, 26.4 μA, 0.699 V, 0.4 V, 0.299 V

Nested DC Sweep

By using a more general form of a DC sweep, called a **nested DC sweep,** we can readily obtain the collector characteristics of a BJT. The form of this more general .DC statement is

.DC SRC1 VSTART1 VFINAL1 VINCR1 SRC2 VSTART2 VFINAL2 VINCR2

In this case, a sweep of the first source (SRC1) is performed for each value of the second source (SRC2), where the value of the second source is swept from VSTART2 to VFINAL2 in increments of VINCR2.

Example 16.23

We can generate the collector characteristics of an *npn* BJT by analyzing the circuit shown in Fig. 16.26a. The collector characteristics of an *npn* BJT are plots of i_C versus v_{CE} for different values of i_B. Therefore, for each value of i_B, we can sweep the value of v_{CE}, and use PSpice to determine the corresponding values of i_C. In particular, we can determine the collector characteristics for a typical 2N3904 *npn* BJT since its model is included in the PSpice evaluation version device library under the model name Q2N3904. An input file for achieving the BJT's collector characteristics is the following:

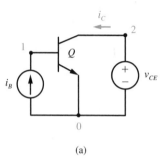

(a)

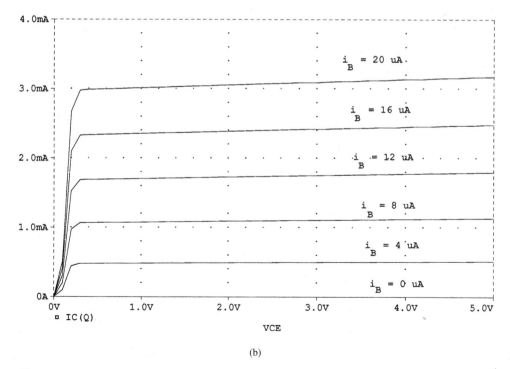

(b)

Fig. 16.26 *(a)* BJT circuit, and *(b)* collector characteristics of 2N3904 *npn* BJT.

Input file for Fig. 16.26a
IB 0 1
VCE 2 0
Q 2 1 0 Q2N3904
.LIB EVAL.LIB
.DC VCE 0 5 .1 IB 0 20u 4u
.PROBE
.END

The resulting PROBE plot of the transistor's collector characteristics is shown in Fig. 16.26b.

Drill Exercise 16.23

Use PSPice and PROBE to obtain the collector characteristics for a 2N3906 *pnp* BJT (the model for this BJT is in the device library under the model name Q2N3906). *Note:* Since PSpice defines i_C to be the current directed ***into*** the collector of a *pnp* transistor, the collector characteristics using PSpice and PROBE are plots of $-i_C$ versus v_{BC} for various values of i_E.

ANSWER   See Fig. 7.6 on p. 424.

Transfer Characteristics

One of the important uses of the DC sweep is for the purpose of obtaining the transfer characteristic of a logic gate.

Example 16.24

Let us obtain the voltage transfer characteristic v_9 versus v_1 for the circuit shown in Fig. 16.27a. In this case we will use 2N3904 BJTs and 1N4148 diodes. This circuit consists of a TTL NAND gate (see Fig. 7.32 on p. 464) and a load that simulates another NAND gate. Since we are interested in the transition of the output voltage, we may use a single-emitter BJT for the input transistor.

The following is an input file for the circuit.

Input file for Fig. 16.27a
V1 1 0
VCC 5 0 DC 5
R1 2 5 5k

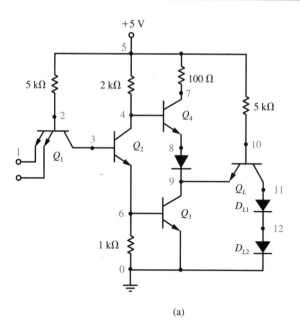

(a)

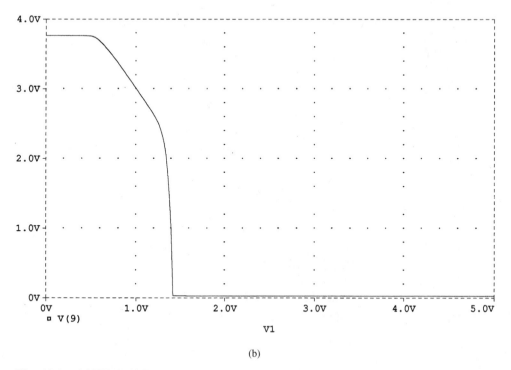

(b)

Fig. 16.27 *(a)* TTL NAND gate with load, and *(b)* voltage transfer characteristic.

R2 4 5 2k
R3 6 0 1k
R4 5 7 100
R5 5 10 5k
Q1 3 2 1 Q2N3904
Q2 4 3 6 Q2N3904
Q3 9 6 0 Q2N3904
Q4 7 4 8 Q2N3904
D 8 9 D1N4148
QL 11 10 9 Q2N3904
DL1 11 12 D1N4148
DL2 12 0 D1N4148
.LIB EVAL.LIB
.DC V1 0 5 .01
.PROBE
.END

The resulting PROBE plot of the transfer characteristic is shown in Fig. 16.27*b*.

Drill Exercise 16.24

Use PSpice and PROBE to obtain the transfer characteristic for the BJT inverter given in Fig. 7.22 on p. 443. Use a 2N3904 transistor.

ANSWER See Fig. 7.23 on p. 445.

16.6 Field-Effect Transistors (FETs)

The general form of the element statement for a **junction field-effect transistor** (JFET) is

JNAME ND NG NS MODNAME

where ND is the drain node, NG is the gate node, NS is the source node, and MODNAME is the model name of the JFET. The general form of the corresponding .MODEL statement for an *n*-channel JFET is

.MODEL MODNAME NJF (<PNAME1=PVAL1 PNAME2=PVAL2 . . . >)

whereas the general form of the .MODEL statement for a *p*-channel JFET is

.MODEL MODNAME PJF (<PNAME1=PVAL1 PNAME2=PVAL2 . . . >)

where PNAME1 is the name of the first JFET parameter and PVAL1 is the value of this parameter, etc. Although there are numerous parameters which can be used to model a JFET, we shall limit our discussion to those in the following table.

JFET Parameter	Name	Default Value	Units
VTO	threshold voltage	-2	V
BETA	transconductance coefficient	1	A/V^2

The parameter VTO is just the pinchoff voltage V_p for the JFET; however, the parameter BETA can be determined from both V_p and the short-circuit drain current I_{DSS}.

The active-region equation used by PSpice for a JFET is

$$i_D = \text{BETA}(v_{GS} - \text{VTO})^2$$

Since the active-region expression (Eq. 8.4 on p. 510) can be rewritten as

$$i_D = I_{DSS}\left(1 - \frac{v_{GS}}{V_p}\right)^2 = \frac{I_{DSS}}{V_p^2}(V_p - v_{GS})^2 = \frac{I_{DSS}}{V_p^2}(v_{GS} - V_p)^2$$

we deduce that $\text{BETA} = I_{DSS}/V_p^2$.

Example 16.25

Let us determine the JFET's current and voltages for the circuit shown in Fig. 16.28 (see Example 8.2 on p. 513) for the case that the *n*-channel JFET has $I_{DSS} = 8$ mA and $V_p = -4$ V.

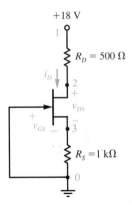

Fig. 16.28 A JFET circuit.

The PSpice parameters are

$$VTO = V_p = -4 \text{ V} \qquad \text{and}$$

$$BETA = I_{DSS}/V_p^2 = (8 \times 10^{-3})/(-4)^2 = 0.5 \text{ mA}/\text{V}^2$$

An input file for the given circuit is

Input file for Fig. 16.28
VDD 1 0 DC 18
J 2 0 3 JMOD
.MODEL JMOD NJF (BETA=.5m VTO=−4)
RD 1 2 500
RS 0 3 1k
.OP
.END

The resulting output file contains the following:

**** OPERATING POINT INFORMATION   TEMPERATURE = 27.000 DEG C
****
**** JFETS

NAME	J
MODEL	JMOD
ID	2.00E−03
VGS	−2.00E+00
VDS	1.50E+01

Thus we have that $i_D = 2$ mA, $v_{GS} = -2$ V, and $v_{DS} = 15$ V.

For the n-channel JFET circuit described in Example 16.25, change the value of the 500-Ω resistor to 7 kΩ and use PSpice to determine the JFET's current and voltages.

ANSWER 2 mA, -2 V, 2V

MOSFETs

The general form of the element statement for a **MOSFET** is

  MNAME ND NG NS NB MODNAME

where ND is the drain node, NG is the gate node, NS is the source node, NB is the substrate node, and MODNAME is the model name of the MOSFET. The general form of the corresponding .MODEL statement for an n-channel MOSFET is

  .MODEL MODNAME NMOS (<PNAME1=PVAL1 PNAME2=PVAL2 . . . >)

whereas the general form of the .MODEL statement for a p-channel MOSFET is

  .MODEL MODNAME PMOS (<PNAME1=PVAL1 PNAME2=PVAL2 . . . >)

where PNAME1 is the name of the first MOSFET parameter and PVAL1 is the value of this parameter, etc. Although there are numerous parameters which can be used to model a MOSFET, we shall limit our discussion to those in the following table.

MOSFET Parameter	Name	Default Value	Units
VTO	threshold voltage	0	V
KP	transconductance coefficient	2E$-$5	A/V^2

Depletion MOSFETs

For a depletion MOSFET, the parameter VTO is just the pinchoff voltage V_p of the MOSFET. However, KP can be determined from the depletion MOSFET parameters V_p and I_{DSS}. The active-region equation used by PSpice for a MOSFET is

$$i_D = \frac{KP}{2} (v_{GS} - VTO)^2$$

Since the active-region expression (Eq. 8.4 on p. 510) for a depletion MOSFET can be rewritten as

$$i_D = I_{DSS}\left(1 - \frac{v_{GS}}{V_p}\right)^2 = \frac{I_{DSS}}{V_p^2} (V_p - v_{GS})^2 = \frac{I_{DSS}}{V_p^2} (v_{GS} - V_p)^2$$

we deduce that $KP = 2I_{DSS}/V_p^2$.

Example 16.26

Let us determine the currents and voltages for the depletion MOSFETs in the circuit shown in Fig. 16.29 (see Example 8.4 on p. 521) for the case that $I_{DSS} = 16$ mA and $V_p = -4$ V.

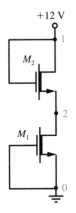

Fig. 16.29 Circuit with two depletion MOSFETs.

The PSpice parameters for the MOSFETs are

$$VTO = V_p = -4 \text{ V} \qquad \text{and}$$

$$KP = 2I_{DSS}/V_p^2 = 2(16 \times 10^{-3})/(-4)^2 = 2 \text{ mA}/V^2$$

An input file for the given circuit is

Input file for Fig. 16.29
VDD 1 0 DC 12
M1 2 0 0 0 MOSMOD
M2 1 1 2 2 MOSMOD
.MODEL MOSMOD NMOS (KP=2m VTO=−4)
.OP
.END

The resulting output file contains the following:

**** OPERATING POINT INFORMATION   TEMPERATURE = 27.000 DEG C
****
**** MOSFETS

NAME	M1	M2
MODEL	MOSMOD	MOSMOD
ID	1.60E−02	1.60E−02
VGS	0.00E+00	1.66E+00
VDS	1.03E+01	1.66E+00
VBS	0.00E+00	0.00E+00

Thus we have that

$$i_{D1} = i_{D2} = 16 \text{ mA}, \; v_{GS1} = 0 \text{ V}, \; v_{GS2} = 1.66 \text{ V}, \; v_{DS1} = 10.3 \text{ V},$$

$$v_{DS2} = 1.66 \text{ V}$$

Drill Exercise 16.26

For the NMOS circuit shown in Fig. 16.29, remove the short circuit between the gate and the source of M_1 and connect it between the gate and the drain of M_1. Use PSpice to determine the MOSFETs' currents and voltages for the case that $I_{DSS} = 16$ mA and $V_p = -4$ V.

ANSWER 84 mA, 84 mA, 6 V, 6 V, 6 V, 6 V

Enhancement MOSFETs

For an enhancement MOSFET, the parameter VTO is just the threshold voltage V_t of the MOSFET. KP can be determined from the enhancement MOSFET parameter K. Again, since the active-region equation used by PSpice for a MOSFET is

$$i_D = \frac{KP}{2}(v_{GS} - VTO)^2$$

and since the active-region expression (Eq. 8.6) for an enhancement MOSFET is

$$i_D = K(v_{GS} - V_t)^2$$

we deduce that $KP = 2K$.

Example 16.27

For the circuit shown in Fig. 16.30, the enhancement MOSFETs have $K = 0.25$ mA$/$V^2 and $V_t = 2$ V. Let us determine the currents and the voltages when $V_{GG} = 12$ V (see Example 8.6 on p. 528).

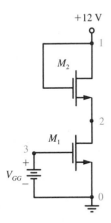

Fig. 16.30 Circuit with two enhancement MOSFETs.

The PSpice parameters for the MOSFETs are

$$VTO = V_t = 2 \text{ V} \qquad \text{and} \qquad KP = 2K = 2(0.25 \times 10^{-3}) = 0.5 \text{ mA}/\text{V}^2$$

An input file for the given circuit is

```
Input file for Fig. 16.30
VDD 1 0 DC 12
VGG 3 0 DC 12
M1 2 3 0 0 MOSMOD
M2 1 1 2 2 MOSMOD
.MODEL MOSMOD NMOS (KP=.5m VTO=2)
.OP
.END
```

The resulting output file contains the following:

```
**** OPERATING POINT INFORMATION   TEMPERATURE = 27.000 DEG C
*****************************************************************
**** MOSFETS
```

NAME	M1	M2
MODEL	MOSMOD	MOSMOD
ID	1.25E−02	1.25E−02
VGS	1.20E+01	9.07E+00
VDS	2.93E+00	9.07E+00
VBS	0.00E+00	0.00E+00

Thus we have that

$$i_{D1} = i_{D2} = 12.5 \text{ mA}, \; v_{GS1} = 12 \text{ V}, \; v_{DS1} = 2.93 \text{ V},$$

$$v_{GS2} = v_{DS2} = 9.07 \text{ V}$$

Drill Exercise 16.27

For the circuit shown in Fig. 16.30, the enhancement MOSFETs have $K = 0.25 \text{ mA/V}^2$ and $V_t = 2$ V. Find i_{D1}, i_{D2}, v_{DS1}, and v_{DS2} when V_{GG} is (a) 4 V, (b) 6 V, and (c) 7 V.

ANSWER (a) 1 mA, 1 mA, 8 V, 4 V; (b) 4 mA, 4 mA, 6 V, 6 V; (c) 6.25 mA, 6.25 mA, 5 V, 7V

FET Drain Characteristics

By using a DC sweep and one of the models of an FET in the PSpice device library, we can readily obtain such FET characteristics as the drain characteristics of an FET.

Example 16.28

Let us generate the drain characteristics of the IRF150 "power" *n*-channel enhancement MOSFET by analyzing the circuit shown in Fig. 16.31*a*. The drain character-

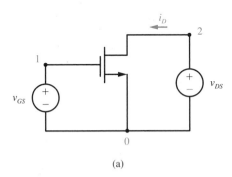

(a)

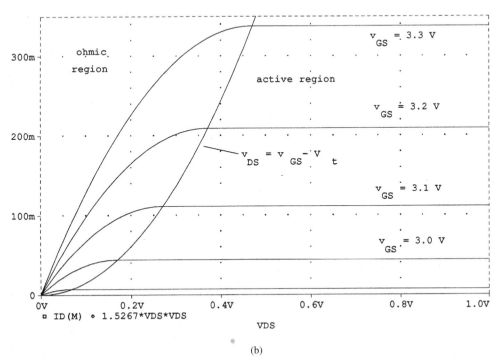

(b)

Fig. 16.31 *(a)* MOSFET circuit, and *(b)* drain characteristics of IRF150 MOSFET.

istics of a MOSFET are plots of i_D versus v_{DS} for different values of v_{GS}. Therefore, for each value of v_{GS}, we can sweep the value of v_{DS}, and use PSpice to determine the corresponding values of i_D. In particular, we can determine the drain characteristics for a typical IRF150 MOSFET since its model is included in the PSpice evaluation version device library under the same model name. Since the threshold voltage for such a device is 2.83 V, an input file for achieving the MOSFET's drain characteristics is the following:

```
Input file for Fig. 16.31a
VGS 1 0
VDS 2 0
M 2 1 0 0 IRF150
.LIB EVAL.LIB
.DC VDS 0 1 .01 VGS 2.8 3.3 .1
.PROBE
.END
```

The resulting PROBE plot of the transistor's drain characteristics is shown in Fig. 16.31*b*.

Drill Exercise 16.28

Use PSpice and PROBE to obtain the drain characteristics for a 2N3819 *n*-channel JFET (the model for this JFET is in the device library under the model name J2N3819).

ANSWER See Fig. 8.7 on p. 509.

MOSFET Inverter Transfer Characteristics

As with the case of BJTs, let us employ the DC sweep for the purpose of obtaining the transfer characteristic of a logic gate—in particular, a MOSFET inverter.

Example 16.29

Let us obtain the voltage transfer characteristic v_2 versus v_1 for the CMOS inverter shown in Fig. 16.32a, where $K = 0.25$ mA$/$V^2 for the MOSFETs, $V_{t1} = 2$ V, and $V_{t2} = -2$ V.

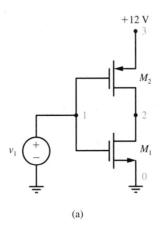

(a)

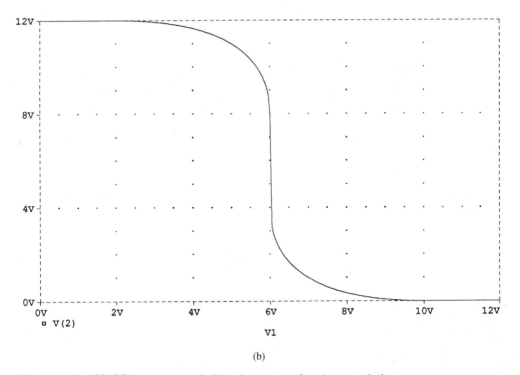

(b)

Fig. 16.32 (a) CMOS inverter, and (b) voltage transfer characteristic.

The following is an input file for the circuit:

```
Input file for Fig. 16.32a
V1 1 0
VDD 3 0 DC 12
M1 2 1 0 0 NMOD
.MODEL NMOD NMOS (KP=.5m VTO=2)
M2 2 1 3 3 PMOD
.MODEL PMOD PMOS (KP=.5m VTO=−2)
.DC V1 0 12 .05
.PROBE
.END
```

The resulting PROBE plot of the transfer characteristic is shown in Fig. 16.32*b*.

Drill Exercuse 16.29

Use PSpice and PROBE to obtain the transfer characteristic for the NMOS inverter given in Fig. 8.29 on p. 538 when M_1 has $K = 0.25 \text{ mA}/\text{V}^2$, $V_t = 2$ V and M_2 has $I_{DSS} = 4$ mA, $V_p = -4$ V.

ANSWER See Fig. 8.30*b* on p. 539.

Suppose we wish to see how a transfer characteristic changes as a transistor parameter changes. Instead of modifying the input file and rerunning PSpice, we can also sweep the transistor parameter as part of the DC statement. A **.DC statement for a model parameter** has the form

.DC MTYPE MODNAME(PNAME) VSTART VFINAL VINCR

where MTYPE is the model type (e.g., NPN, NJF, NMOS), MODNAME is the model name, PNAME is the parameter name (e.g., Is, VTO, KP), VSTART is the starting value, VFINAL is the final value, and VINCR is the increment. VSTART may be greater than or less than VFINAL—so the sweep may be in either direction—but VINCR must be a positive number.

Example 16.30

The MOSFET circuit shown in Fig. 16.30 is an NMOS inverter. Suppose $K_1 = 0.25 \text{ mA}/\text{V}^2$ and $V_{t1} = V_{t2} = 2$ V. Let us determine the transfer characteristic for this inverter for the case that (a) $K_2 = 0.25 \text{ mA}/\text{V}^2$, and (b) $K_2 = 0.05 \text{ mA}/\text{V}^2$.

In terms of PSpice parameters, for M_1, VTO $= V_{t1} = 2$ V and KP $= 2K_1 = 0.5 \text{ mA}/\text{V}^2$; whereas for M_2, VTO $= V_{t2} = 2$ V and (a) KP $= 2K_2 = 0.5 \text{ mA}/\text{V}^2$ and (b) KP $= 2K_2 = 0.1 \text{ mA}/\text{V}^2$. The following input file sweeps not only the applied voltage V_{GG}, but KP for M_2 as well.

```
Input file for Fig. 16.30
VDD 3 0 DC 12
VGG 1 0
M1 2 1 0 0 MOSMOD1
.MODEL MOSMOD1 NMOS (KP=.5m VTO=2)
M2 3 3 2 2 MOSMOD2
.MODEL MOSMOD2 NMOS (KP=.5m VTO=2)
.DC VGG 0 12 .05 NMOS MOSMOD2(KP) .5m .1m .4m
.PROBE
.END
```

The resulting transfer characteristics are shown in Fig. 8.26 on p. 533.

Drill Exercise 16.30

The MOSFET circuit shown in Fig. 8.27a on p. 535 is another NMOS inverter. Suppose $K_1 = 0.25 \text{ mA}/\text{V}^2$, $V_{t1} = V_{t2} = 2$ V, $V_{DD} = 12$ V, and $V_{GG} = 18$. Determine the transfer characteristic for this inverter for the case that (a) $K_2 = 0.25 \text{ mA}/\text{V}^2$, and (b) $K_2 = 0.05 \text{ mA}/\text{V}^2$.

ANSWER See Fig. 8.28b on p. 536.

The .STEP Statement

Another way of varying an element value in a circuit is with the use of a **.STEP statement.** The form of the statement depends upon what is being changed. For example, similar to a DC sweep, for an independent source one form is

.STEP SRCNAME VSTART VFINAL VINCR

where SRCNAME is the name of the independent source, VSTART is the starting value of the source, VFINAL is the final value of the source, and VINCR is the incrementing value. In this case, the circuit analysis is performed for the first value of the source, the entire analysis is performed for the second value of the source, and so on. As with a DC sweep, VSTART may be greater than or less than VFINAL—so the sweep may be in either direction—but VINCR must be a positive number.

Instead of repeated analyses for source values that are stepped linearly from the starting value to the final value, the stepping can be done through a list of values by using the statement

.STEP SRCNAME LIST VALUE1 VALUE2 VALUE3 . . .

Example 16.31

Suppose we wish to determine the node voltages for the MOSFET circuit described in Example 16.27 when V_{GG} is (a) 4 V, (b) 6 V, and (c) 7 V. We can use the following input file:

```
Input file for Fig. 16.30
VDD 1 0 DC 12
VGG 3 0
.STEP VGG LIST 4 6 7
M1 2 3 0 0 MOSMOD
M2 1 1 2 2 MOSMOD
.MODEL MOSMOD NMOS (KP=.5m VTO=2)
.END
```

The resulting output file contains the following:

```
**** SMALL SIGNAL BIAS SOLUTION   TEMPERATURE = 27.000 DEG C
**** CURRENT STEP              VGG = 4
***************************************************************
      NODE VOLTAGE     NODE VOLTAGE         NODE VOLTAGE
      ( 1) 12.0000     ( 2) 8.0000          ( 3) 4.0000

**** SMALL SIGNAL BIAS SOLUTION   TEMPERATURE = 27.000 DEG C
**** CURRENT STEP              VGG = 6
***************************************************************
      NODE VOLTAGE    NODE VOLTAGE          NODE VOLTAGE
      ( 1) 12.0000    ( 2) 6.0000           ( 3) 6.0000
```

```
**** SMALL SIGNAL BIAS SOLUTION   TEMPERATURE = 27.000 DEG C
**** CURRENT STEP                 VGG = 7
***************************************************************
    NODE VOLTAGE        NODE VOLTAGE        NODE VOLTAGE
    ( 1) 12.0000        ( 2) 5.0000         ( 3) 7.0000
```

Drill Exercise 16.31

Use an alternative type of .STEP statement than the one given in Example 16.31 to determine the node voltages for the MOSFET circuit described in Example 16.27 when V_{GG} is (a) 4 V, (b) 5 V, (c) 6 V, and (d) 7 V.

ANSWER   .STEP VGG 4 7 1

16.7 Transistor Amplifiers

PSpice is especially useful when it comes to analyzing transistor amplifiers. In performing hand analysis, first an amplifier's dc-equivalent circuit is analyzed so as to determine the biasing of the amplifier's transistors. After that, the amplifier's ac-equivalent circuit is determined and the transistors are replaced by their small-signal ac models. The resulting circuit is analyzed to determine such parameters as voltage gain, current gain, input resistance, and output resistance. Fortunately, only the original transistor amplifier need be described appropriately in order for PSpice to do a complete analysis—both dc and ac.

Bipolar Junction Transistor Amplifiers

We begin our discussion of transistor-amplifier circuit analysis with PSpice by considering a bipolar junction transistor (BJT) amplifier.

Example 16.32

Let us find the small-signal ac voltage gain $|V_2/V_1|$ and the input resistance R_{in} for the BJT amplifier shown in Fig. 16.33 for the case that the BJT has $h_{FE} = 100$, $I_S = 3.6E-15$ A and v_1 is an ac voltage source. The assumption is, therefore, that the frequency is high enough so that the capacitor approximates a short circuit. We shall select the frequency to be 1000 Hz.

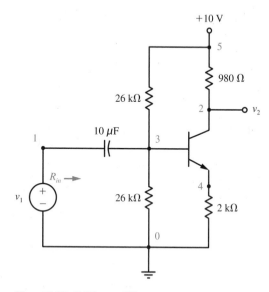

+10 V

5

980 Ω

26 kΩ

2   v_2

10 μF

1   3

R_{in}

v_1 26 kΩ

4

2 kΩ

0

Fig. 16.33 BJT amplifier.

Since $h_{FE} = 100$ is the default value for a BJT, it is not necessary to specify BF in the model statement for the BJT. An input file for this amplifier is

```
Transistor amplifier shown in Fig. 16.33
V1 1 0 AC 1
*Since no angle is specified, the phase of V1 is 0 degrees.
VCC 5 0 DC 10
C 1 3 10u
R1 0 3 26k
R2 3 5 26k
RC 2 5 980
RE 0 4 2k
Q 2 3 4 QMOD
.MODEL QMOD npn (Is=3.6E−15)
.AC LIN 1 1000 1000
.PRINT AC VM(2) VP(2) IM(C) IP(C)
.OP
.END
```

The output file contains the dc node voltages for the amplifier, the dc current through the voltage sources, and the (dc) power dissipated by the amplifier. They appear in the output file as

**** SMALL SIGNAL BIAS SOLUTION   TEMPERATURE = 27.000 DEG C
****

NODE VOLTAGE	NODE VOLTAGE	NODE VOLTAGE
(1) 0.0000	(2) 8.0398	(3) 4.7400

NODE VOLTAGE	NODE VOLTAGE
(4) 4.0405	(5) 10.0000

VOLTAGE SOURCE CURRENTS

NAME	CURRENT
V1	0.000E+00
VCC	−2.203E−03

TOTAL POWER DISSIPATION   2.20E−02 WATTS

In addition, the output file contains the bias currents and voltages as well as the transconductance g_m and the ac beta h_{fe} for the BJT as follows:

**** OPERATING POINT INFORMATION   TEMPERATURE = 27.000 DEG C
****
**** BIPOLAR JUNCTION TRANSISTORS

NAME	Q
MODEL	QMOD
IB	2.00E−05
IC	2.00E−03
VBE	6.99E−01
VBC	−3.30E+00
VCE	4.00E+00
BETADC	1.00E+02
GM	7.73E−02
BETAAC	1.00E+02

The last two lines indicate that $g_m = 77.3$ m℧ and $h_{fe} = 100$.
The results of the small-signal ac analysis are

**** AC ANALYSIS   TEMPERATURE = 27.000 DEG C
****

FREQ	VM(2)	VP(2)	IM(C)	IP(C)
1.000E+03	4.821E−01	−1.799E+02	8.184E−05	7.463E−02

This means that the magnitude of the voltage at node 2 is 0.4821 V. Also, the phase (angle) is $-179.9 \approx -180$ degrees. Since the magnitude of the voltage at node 1 is 1 V, then the ac voltage gain (at 1000 Hz) is $|V_2/V_1| = 0.4821$. Since the phase of the voltage at node 1 is 0 degrees, then $v_2/v_1 = -0.4821$.

In calculating the ac input resistance of an amplifier, it is assumed that all of the coupling and bypass capacitors behave as short circuits. (At 1000 Hz, the impedance of the capacitor is $-j16 \ \Omega$.) Assuming that the capacitor behaves as a short circuit, we can calculate the input resistance of the amplifier to be

$$R_{in} = \text{VM}(1)/\text{IM}(C) = 1/(8.184 \times 10^{-5}) = 12.2 \ \text{k}\Omega$$

Drill Exercise 16.32

For the emitter follower shown in Fig. 9.18 on p. 586, suppose $R_1 = R_2 = 26$ kΩ, $R_E = R_L = 2$ kΩ, $R_s = 0 \ \Omega$, $C_1 = C_2 = 10 \ \mu$F, $V_{CC} = 10$ V, $h_{FE} = 100$, $I_S = 3.6\text{E}{-}15$ A, and v_1 is a sinusoid whose frequency is 1000 Hz. Find v_e/v_s and R_{in}.

ANSWER 0.987, 11.5 kΩ

Determining Output Resistance

Having determined the ac voltage gain and the input resistance for a BJT amplifier, let us now consider the problem of finding the ac output (Thévenin-equivalent) resistance of an amplifier.

There are two different approaches that can be taken. One approach is to set the input independent source to zero, and then determine the ratio of voltage to current at the output when an independent source is applied to the output. Alternatively, we can find the output resistance by taking the ratio of the open-circuit voltage to the short-circuit current at the output.

Example 16.33

Let us find the ac output resistance R_o for the emitter follower shown in Fig. 16.34 when $I_S = 3.6\text{E}{-}15$ A, and the frequency of the independent voltage source is 100 kHz.

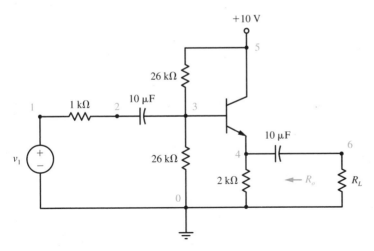

Fig. 16.34 An emitter follower.

In order to find R_o, let us determine the open-circuit voltage and the short-circuit current at the output, and then take the ratio of this voltage to this current. Instead of analyzing two different circuits to get these results, let us analyze the given circuit and vary the value of R_L. Ideally, to get the open-circuit voltage, we should make R_L infinite, and to get the short-circuit current, we should make $R_L = 0 \ \Omega$. Instead, we will make R_L very large and very small, respectively. Although the results will be approximate, they will be quite accurate.

We can vary the value of R_L by using a .STEP statement. In this case, however, we are required to use a .MODEL statement in conjunction with the element statement for the resistor. The element statement for R_L has the form

        RL N+ N− MODNAME VALUE

where MODNAME is the model name and VALUE is the **line value** of the resistor. The corresponding .MODEL statement has the form

        .MODEL MODNAME RES

where RES is the element type—a resistor. The .STEP statement has the form

        .STEP RES MODNAME(R) LIST VALUE1 VALUE2 . . .

By using these three statements, a circuit is analyzed first with the value of R_L equal to the product of VALUE and VALUE1, it is then analyzed with the value of R_L equal to the product of VALUE and VALUE2, and so on.

An input file for the emitter follower is

Input file for Fig. 16.34
V1 1 0 AC 1
VCC 5 0 DC 10
C1 2 3 10u
C2 4 6 10u
R1 0 3 26k
R2 3 5 26k
RE 0 4 2k
RS 1 2 1k
RL 6 0 RMOD 1
.MODEL RMOD RES
.STEP RES RMOD(R) LIST .0001 1E10
Q 2 3 4 QMOD
.MODEL QMOD npn (Is=3.6E−15)
.AC LIN 1 100k 100k
.PRINT AC VM(RL) VP(RL) IM(RL) IP(RL)
.END

The resulting output file contains

```
**** AC ANALYSIS              TEMPERATURE = 27.000 DEG C
**** CURRENT STEP             RMOD R = 100.0000E−06
***********************************************************************
     FREQ        VM(RL)      VP(RL)      IM(RL)      IP(RL)
     1.000E+05   4.221E−06   4.233E−01   4.221E−02   4.223E−01
```

and this indicates that the approximate short-circuit current directed down through R_L is $\mathbf{I}_{sc} = 0.04221\underline{/0.4233°}$ A.

The output file also contains

```
**** AC ANALYSIS              TEMPERATURE = 27.000 DEG C
**** CURRENT STEP             RMOD R = 10.0000E+09
***********************************************************************
     FREQ        VM(RL)      VP(RL)      IM(RL)      IP(RL)
     1.000E+05   9.185E−01   6.899E−04   9.185E−11   6.899E−04
```

and this portion of the output file indicates that the approximate open circuit voltage is $\mathbf{V}_{oc} = 0.9185\underline{/0.0006899°}$ V.

Thus the output impedance is

$$Z_o = \frac{\mathbf{V}_{oc}}{\mathbf{I}_{sc}} = \frac{0.9185\underline{/0.0006899°}}{0.04221\underline{/0.4223°}} = 21.76\underline{/-0.4226°}\ \Omega$$

Since the output impedance has an angle close to 0 degrees, it is in effect a resistance R_o, and therefore, $R_o \approx 21.76\ \Omega$.

Drill Exercise 16.33

For the emitter follower shown in Fig. 16.34, suppose the BJT has $I_S = 3.6\text{E}-15$ A. Determine the output resistance as follows: (i) Replace the independent voltage source with a short circuit. (ii) Replace the load resistance R_L with an independent 100-kHz voltage source $\mathbf{V}_o = 1\underline{/0°}$ V. (a) Use PSpice to determine the current $\mathbf{I}_o$ (directed through the voltage source from $-$ to $+$). (b) Use this result to calculate the approximate output resistance R_o.

ANSWER   (a) $0.04596\underline{/0.4226°}$ A, (b) 21.76 Ω

Field-Effect Transistor Amplifiers

PSpice can be used equally as well to analyze field-effect transistor (FET) amplifiers as it was used to analyze BJT amplifiers.

Example 16.34

Let us determine the ac voltage gain and the input resistance at 1 kHz for the feedback amplifier shown in Fig. 16.35 for the case that MOSFET has $K = 0.25$ mA/V^2 and $V_t = 2$ V.

Since $K = 0.25$ mA/V^2, then KP $= 2K = 0.5$ mA/V^2, and an input file for the feedback amplifier is

```
Input file for Fig. 16.35
Vs 1 0 AC 1
VDD 5 0 DC 10
C1 2 3 10u
C2 4 6 10u
Rs 1 2 1k
```

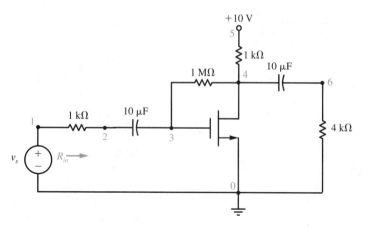

Fig. 16.35 MOSFET feedback amplifier.

```
RF 3 4 1Meg
RD 4 5 1k
RL 6 0 4k
M 4 3 0 0 MOSFET
.MODEL MOSFET NMOS (KP=.5m VTO=2)
.AC LIN 1 1000 1000
.PRINT AC VM(RL) VP(RL) IM(Rs) IP(Rs)
.OP
.END
```

The resulting output file contains

```
**** OPERATING POINT INFORMATION    TEMPERATURE = 27.000 DEG C
******************************************************************
**** MOSFETS

        NAME        M
        MODEL       MOSFET
        ID          4.00E−03
        VGS         6.00E+00
        VDS         6.00E+00
        VBS         0.00E+00
        VTH         2.00E+00
        VDSAT       4.00E+00
        GM          2.00E−03
```

Thus we see that $g_m = 2$ m℧.

Also contained in the output file is

```
**** AC ANALYSIS          TEMPERATURE = 27.000 DEG C
****************************************************************
    FREQ       VM(RL)      VP(RL)      IM(Rs)      IP(Rs)
    1.000E+03  1.594E+00  -1.798E+02  2.591E-06  -2.559E-02
```

Since the voltage across the 4-kΩ load resistor is $1.594\underline{/-179.8°}$ V, then the voltage gain has magnitude 1.594 V and angle $-179.8 \approx -180°$. That is, $v_o/v_s \approx -1.594$. Furthermore, since the current into the amplifier has magnitude 2.591 μA and angle $-0.02559 \approx 0°$, then the input resistance of the amplifier is

$$R_{in} \approx 1/(2.591 \times 10^{-6}) = 386 \text{ k}\Omega$$

Drill Exercise 16.34

Use PSpice to find the output resistance for the MOSFET feedback amplifier described in Example 16.34.

ANSWER 997 Ω

Frequency Response

Finding the lower and upper cutoff frequencies—and hence its bandwidth—and the midband gain of an amplifier can be determined from the amplifier's amplitude response.

Example 16.35

Let us determine the lower and upper cutoff frequencies, the bandwidth, and the midband gain of the amplifier shown in Fig. 16.36a, for the case that the transistor is a 2N2222 *npn* BJT. Since this element is in the device library EVAL.LIB under the model name Q2N2222, the following is an input file for obtaining an amplitude response via PROBE.

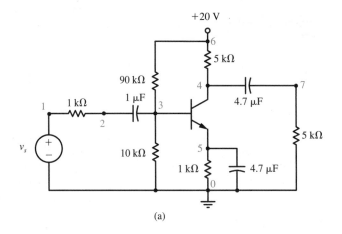

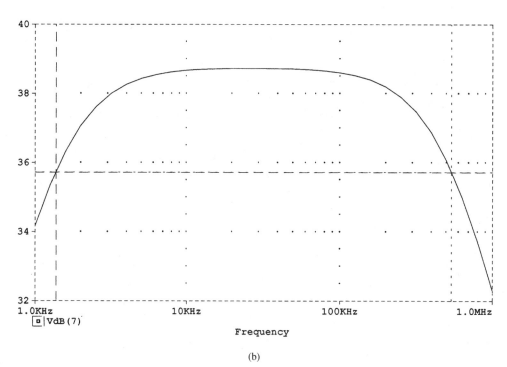

Fig. 16.36 *(a)* BJT amplifier, and *(b)* its amplitude response.

Input file for Fig. 16.36a
Vs 1 0 AC 1
VCC 6 0 DC 20
C1 2 3 1u

C2 4 7 4.7u
CE 5 0 4.7u
R1 0 3 10k
R2 3 6 90k
RE 0 5 1k
RS 1 2 1k
RL 7 0 5k
RC 4 6 5k
Q 4 3 5 Q2N2222
.LIB EVAL.LIB
.AC DEC 10 1k 1Meg
.OP
.END

The resulting PROBE plot is shown in Fig. 16.36*b*. In PROBE, a cursor can be selected so that values on a curve can easily be read. In this case, the midband gain is 38.7 dB. The lower and upper cutoff frequencies are those frequencies for which the gain is approximately $38.7 - 3 = 35.7$ dB. From the PROBE plot, using cursors, these frequencies are approximately 1.37 kHz and 539.5 KHz, respectively. Therefore, the bandwidth of the amplifier is $539.5 - 1.37 \approx 538$ kHz.

Drill Exercise 16.35

Use PSPice and PROBE to find the midband gain, the lower and upper cutoff frequencies, and the bandwidth for the JFET amplifier shown in Fig. 9.29 on p. 594 for the case that $R_s = 1$ kΩ, $R_1 = R_2 = 1$ MΩ, $R_D = R_S = 4$ kΩ, $C = C_S = 1$ μF, $V_{DD} = 12$ V, the FET is a 2N3819 JFET, and the output is the drain voltage.

ANSWER   20.4 dB, 456 Hz, 12.0 MHz, 12.0 MHz

16.8   Operational Amplifiers

Let us now consider the transient analysis of op-amp circuits using PSpice. We will begin by discussing sinusoidal oscillators.

Example 16.36

Let us analyze the *RC* phase-shift oscillator shown in Fig. 16.37*a* for the case that $R = 6.5$ kΩ and $C = 1$ nF. From p. 693, the minimum value of R_F is

$$R_F = 29R = 29(6500) = 188.5 \text{ k}\Omega$$

the frequency of oscillation is

$$f_0 = 1/2\pi\sqrt{6}RC = 1/2\pi\sqrt{6}(6.5 \times 10^3)(1 \times 10^{-9}) = 9996 \approx 10 \text{ kHz}$$

and the period is

$$T = 1/f_0 = 1/9996 \approx 100 \text{ μs}$$

In order to view eight periods, we will set TSTOP to be 800 μs in the TRAN statement. Doing this, however, results in a PROBE plot that lacks accuracy. This is because the default value of the maximum internal time step used by PSpice is TSTOP/50, not TSTEP. Yet there is a way of specifying the maximum value.

The general form of a **.TRAN statement** is

.TRAN TSTEP TSTOP <TSTART <TMAX>> <UIC>

where TMAX is the maximum internal time step. TMAX can be smaller or larger than the printing increment TSTEP—its default value is TSTOP/50.

An input file for the *RC* phase-shift oscillator shown in Fig. 16.37*a* is

```
Input file for Fig. 16.37a
R1 1 5 6.5k
R2 1 2 188.5k
R3 3 0 6.5k
R4 4 0 6.5k
C1 2 3 1n IC=5
C2 3 4 1n
C3 4 5 1n
E 2 0 0 1 1E9
.TRAN 1u 800u 0 1u UIC
.PROBE
.END
```

The resulting PROBE plot of the voltage at the output of the op amp is shown in Fig. 16.37*b*.

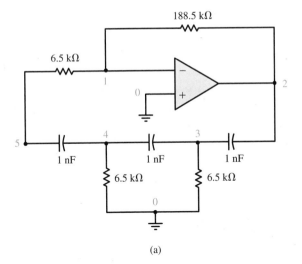

(a)

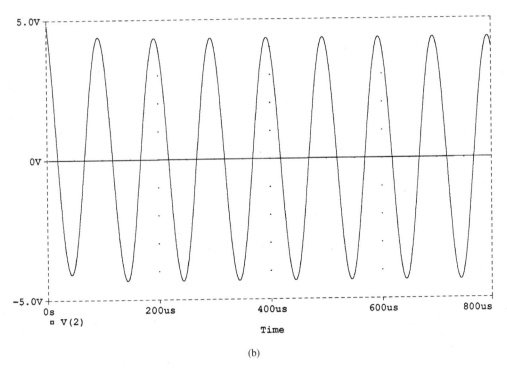

(b)

Fig. 16.37 *(a) RC* phase-shift oscillator, and *(b)* op-amp output voltage.

Since the circuit shown in Fig. 16.37*a* has no applied source, there would be no response if all the initial conditions where zero. Therefore, an initial voltage was placed on one of the capacitors in order to get a nonzero response.

For an actual oscillator, the op amp is not an ideal circuit element, but a practical device requiring power supplies. When these power supplies are connected, transients (e.g., step responses) occur and eventually the output voltage has the form shown in Fig. 16.37*b*.

Drill Exercise 16.36

For the Wien-bridge amplifier shown in Fig. 10.32*a* on p. 696, suppose $R = R_1 = 10$ kΩ, $R_2 = 20$ kΩ, and $C = 1.59$ nF. Assume an initial voltage of 1 V across the horizontally oriented capacitor (+ on the left), and use PROBE to plot the voltage at the output of the op amp from 0 to 800 μs.

ANSWER $3 \sin 20,000\pi t$ V

Nonideal Operational Amplifiers

In the preceding example and drill exercise, the op amp was employed as a linear circuit element. Consequently, we were able to utilize the linear op-amp model consisting of a voltage-dependent voltage source having a very large value. When an op amp is used as a comparator, it is a nonlinear device, and therefore, it is not valid to model it as a linear dependent voltage source.

PSpice does have a model of a practical 741 op amp in its device library EVAL.LIB. This device is a subcircuit which can be included as part of a circuit by appropriately calling it from the library. In order to call a 741 op amp from the library, we use the statements

```
XNAME N+ N- NV+ NV- NOUT uA741
.LIB EVAL.LIB
```

where X indicates a call for a subcircuit, NAME is the name that is selected for the subcircuit, N+ is the noninverting input node, N- is the inverting input node, NV+ is the positive power-supply node, NV- is the negative power-supply node, NOUT is the output node, and uA741 indicates a 741 op amp.

Example 16.37

Let us determine the transfer characteristic for the Schmitt trigger shown in Fig. 16.38a. Since a Schmitt trigger transfer characteristic exhibits hysteresis, we will not attempt to use a DC sweep because such an approach could cause divergence problems in the program due to the sudden change of values during the sweep. Furthermore, both a sweep up and a sweep down are required to determine both threshold voltages. Instead, we will apply the triangular waveform shown in Fig. 16.38b, and perform a transient analysis to obtain v_2. In PROBE, we can then change the x axis from time to v_1, and thereby obtain the transfer characteristic.

An input file for the given Schmitt trigger is

```
Input file for Fig. 16.38a
V1 1 0 PWL(0 0 1 10 2 0)
VCC 4 0 DC 10
VEE 5 0 DC −10
Vr 6 0 DC 5
R1 3 6 1k
R2 2 3 9k
XOPAMP 3 1 4 5 2 uA741
.LIB EVAL.LIB
.TRAN .01 2
.PROBE
.END
```

The resulting PROBE plot of the transfer characteristic is shown in Fig. 16.38c. (See Example 10.7 on p. 710.)

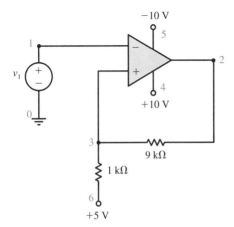

Fig. 16.38 *(a)* Schmitt trigger.

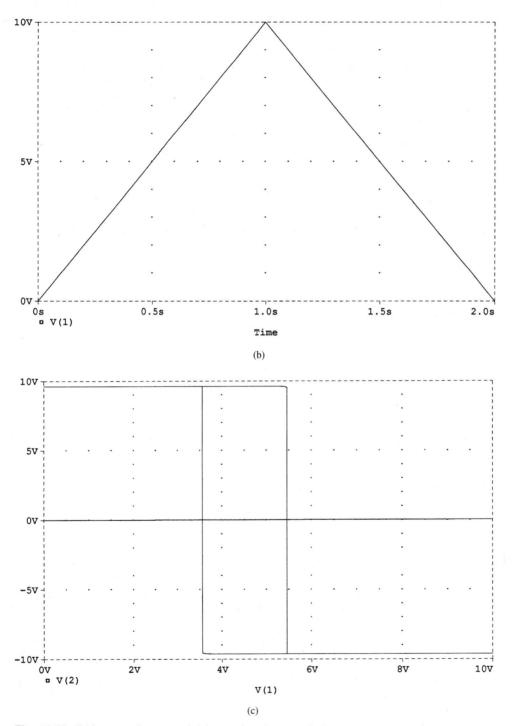

(b)

(c)

Fig. 16.38 *(b)* Input voltage, and *(c)* transfer characteristic.

Drill Exercise 16.37

For the Schmitt trigger described in Example 16.37, suppose $v_1(t) = 5 + 2 \sin 200\pi t$ V. Obtain PROBE plots of $v_1(t)$ and $v_2(t)$ versus time.

ANSWER See Fig. 10.45 on p. 710.

Relaxation Oscillators

By using an op amp as a comparator, it is possible to form relaxation oscillators.

Example 16.38

Let us obtain PROBE plots of the voltages for the relaxation oscillator shown in Fig. 16.39a. From Eq. 10.40 on p. 713, the period of oscillation is

$$T = 2RC \ln(2R_1/R_2 + 1)$$

$$= 2(91 \times 10^3)(50 \times 10^{-9}) \ln (20{,}000/10{,}000 + 1) \approx 10 \text{ ms}$$

Using the initial voltage across the capacitor to be

$$-BV_{CC} = -\frac{R_1}{R_1 + R_2}V_{CC} = -\frac{10{,}000}{10{,}000 + 10{,}000}(10) = -5 \text{ V}$$

an input file for the given relaxation oscillator is

```
Input file for relaxation oscillator in Fig. 16.39a
VCC 4 0 DC 10
VEE 5 0 DC −10
R1 3 0 10k
R2 2 3 10k
R 1 2 91k
C 1 0 .05u IC=−5
XOPAMP 3 1 4 5 2 uA741
.LIB EVAL.LIB
.TRAN .2m 20m UIC
.PROBE
.END
```

PROBE plots of the voltage v_1 across the capacitor and the output voltage v_2 are shown in Fig. 16.39b.

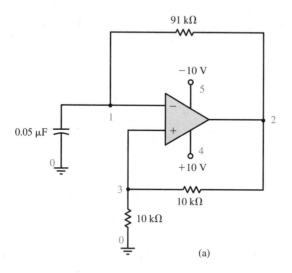

(a)

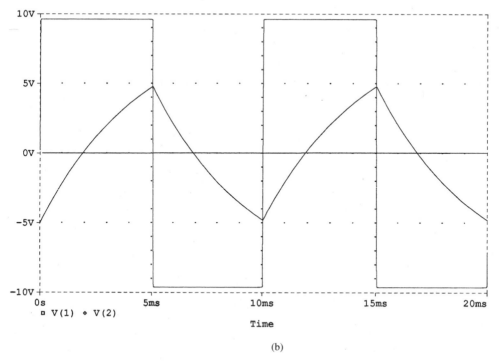

(b)

Fig. 16.39 *(a)* Relaxation oscillator, and *(b)* oscillator voltages.

Drill Exercise 16.38

For the relaxation oscillator shown in Fig. 10.47 on p. 714, the op amp on the right is used as a linear device, so therefore, model it as an ideal op amp. Model the op amp on the left as a uA741 op amp in the device library, and obtain PROBE plots of $v_2(t)$ and $v_o(t)$ for the case that $R_1 = R = 10$ kΩ, $R_2 = 20$ kΩ, and $C = 0.5$ μF.

ANSWER   See Fig. 10.48 on p. 715 where $V_{CC} = 10$ V, $V_1 = 5$ V, and $V_2 = -5$ V.

There is much more to PSpice than what has been covered in this chapter, which has been intended to serve as an introduction to the subject.

S U M M A R Y

SPICE Element	Form of Element Statement
Independent DC Voltage Source	Vname N+ N− DC Value
Independent DC Current Source	Iname N+ N− DC Value
Resistor	Rname N+ N− Value
Capacitor	Cname N+ N− Value <IC=INCOND>
Inductor	Lname N+ N− Value <IC=INCOND>
Voltage-Controlled Voltage Source	Ename N+ N− NC+ NC− Value
Current-Controlled Current Source	Fname N+ N− Vname Value
Voltage-Controlled Current Source	Gname N+ N− NC+ NC− Value
Current-Controlled Voltage Source	Hname N+ N− Vname Value
Operational Amplifier	Ename Nout 0 Nin+ Nin− A
Transformer	Kname Lself1 Lself2 Value
Diode	Dname N+ N− Modname
Diode Model	.Model Modname D (<Pname1=Pvalue1 . . . >)
npn BJT	Qname NC NB NE Modname
npn BJT Model	.Model Modname npn (<Pname1=Pvalue1 . . . >)
pnp BJT	Qname NC NB NE Modname
pnp BJT Model	.Model Modname pnp (<Pname1=Pvalue1 . . . >)
n-channel JFET	Jname ND NG NS Modname
n-channel JFET Model	.Model Modname NJF (<Pname1=Pvalue1 . . . >)
p-channel JFET	Jname ND NG NS Modname
p-channel JFET Model	.Model Modname PJF (<Pname1=PValue1 . . . >)
n-channel MOSFET	Mname ND NG NS NB Modname
n-channel MOSFET Model	.Model Modname NMOS (<Pname1=Pvalue1 . . . >)
p-channel MOSFET	Mname ND NG NS NB Modname
p-channel MOSFET Model	.Model Modname PMOS (<Pname1=Pvalue1 . . . >)

Independent-Source Functions

Exponential Function	EXP(V1 V2 Td1 Tau1 Td2 Tau2)
Piece-Wise Linear Function	PWL(T1 V1 <T2 V2 T3 V3 . . . >)
Pulse Function	PULSE(V1 V2 Td Tr Tf PW Per)
Sin Function	SIN(VO VA FREQ TD DF PHASE)

PSpice Control Statements

AC Analysis	.AC {LIN or DEC or OCT} Npoints Fstart Fstop
DC Analysis	.DC Srcname LIST Value
DC Sweep	.DC Src1 Vstart1 Vfinal1 Vincr1 <Src2 Vstart2 Vfinal2 Vincr2>
DC Parameter Sweep	.DC Mtype Modname(Pname) Vstart Vfinal Vincr
End	.END
Evaluation Library	.LIB EVAL.LIB
Operating Point	.OP
Plot	.PLOT {AC or DC or TRAN} OV1 <OV2 OV3 . . . >
Print	.PRINT {AC or DC or TRAN} OV1 <OV2 OV3 . . . >
Probe	.PROBE
Step Resistance Value	.STEP RES Modname(R) LIST Value1 Value2 . . .
Step Source Value	.STEP Srcname {Vstart Vfinal Vincr or LIST Value1 . . . }
Temperature	.TEMP Value
Transfer Function	.TF OUTVAR INSRC
Transient Analysis	.TRAN Tstep Tstop <Tstart <Tmax>> <UIC>

Problems

16.1 Use PSpice to find the node voltages for the circuit shown in Fig. P2.8 on p. 99.

16.2 Use PSpice to find the node voltages for the circuit shown in Fig. P2.9 on p. 99.

16.3 Use PSpice to find the node voltages for the circuit shown in Fig. P2.10 on p. 100.

16.4 Use PSpice to find the node voltages for the circuit shown in Fig. P2.7 on p. 100.

16.5 Use PSpice to find the i_1 and v_2 for the circuit shown in Fig. P2.14 on p. 101 for the case that $v_1 = 1$ V.

16.6 Use PSpice to find the i_1 and v_2 for the circuit shown in Fig. P2.11 on p. 100 for the case that $h_i = 1$ kΩ, $h_r = 2.5 \times 10^{-4}$, $h_f = 50$, $h_o = 25$ $\mu\mho$, $R_L = 10$ kΩ, and $v_1 = 1$ V.

16.7 Use PSpice to find $G = i/v$ for the circuit shown in Fig. P2.16 on p. 101.

16.8 Use PSpice to find the node voltages for the op-amp circuit shown in Fig. P2.33 on p. 104 for the case that $v_s = 15$ V.

16.9 Use PSpice to find the node voltages for the op-amp circuit shown in Fig. P2.34 on p. 105 for the case that $v_s = 1$ V.

16.10 Use PSpice to find the node voltages for the op-amp circuit shown in Fig. P2.36 on p. 105 for the case that $v_s = 1$ V.

16.11 For the circuit shown in Fig. P2.37 on p. 104, remove the load resistor R_L and use PSpice to determine the Thévenin equivalent of the resulting circuit.

16.12 For the circuit shown in Fig. P2.42 on p. 106, remove the 5-Ω load resistor and use PSpice to determine the Thévenin equivalent of the resulting circuit.

16.13 For the circuit shown in Fig. P2.41 on p. 105, remove the load resistor R_L and use PSpice to determine the Thévenin equivalent of the resulting circuit for the case that $v_1 = 20$ V.

16.14 Use PSpice to determine the Thévenin equivalent of the circuit shown in Fig. P2.48 on p. 107.

16.15 Obtain PROBE plots of $v_R(t)$, $v_L(t)$, and $v(t)$ for the circuit shown in Fig. P3.5 on p. 176 for the case that $i(t)$ is described by the function given in Fig. P3.1b.

16.16 Obtain PROBE plots of $v_R(t)$, $v_L(t)$, and $v(t)$ for the circuit shown in Fig. P3.5 on p. 176 for the case that $i(t)$ is described by the function given in Fig. P3.2.

16.17 Obtain PROBE plots of $i(t)$, $i_R(t)$, $v_R(t)$, $v_s(t)$, and $v_o(t)$ for the circuit shown in Fig. P3.13 on p. 177 for the case that $v(t)$ is described by the function given in Fig. P3.7b.

16.18 Obtain PROBE plots of $i_R(t)$, $v_C(t)$, and $v_o(t)$ for the circuit shown in Fig. P3.18 on p. 179 for the case that $v(t)$ is described by the function given in Fig. P3.16.

16.19 Obtain PROBE plots of $i(t)$ and $v(t)$ for $t \geq 0$ s for the circuit shown in Fig. P3.30 on p. 180 for the case that $i_s(t) = 10$ A for $t < 0$ s and $i_s(t) = 0$ A for $t \geq 0$ s.

16.20 Obtain PROBE plots of $v(t)$ and $i(t)$ for $t \geq 0$ s for the circuit shown in Fig. P3.34 on p. 181 for the case that $v_s(t) = 12$ V for $t < 0$ s and $v_s(t) = 0$ V for $t \geq 0$ s.

16.21 Obtain PROBE plots of $i(t)$ and $v(t)$ for $t \geq 0$ s for the circuit shown in Fig. P3.30 on p. 180 when $i_s(t) = 10u(t)$ A.

16.22 Obtain PROBE plots of $v(t)$ and $i(t)$ for $t \geq 0$ s for the circuit shown in Fig. P3.34 on p. 181 when $v_s(t) = 12u(t)$ V.

16.23 Obtain PROBE plots of $v(t)$ and $i(t)$ for $t \geq 0$ s for the circuit shown in Fig. P3.7a on p. 176

for the case that $v_s(t) = 12e^{-t/2}u(t)$ V. (See Problem 3.51.)

16.24 Obtain PROBE plots of $i(t)$ and $v(t)$ for $t \geq 0$ s for the circuit shown in Fig. P3.1a on p. 176 for the case that $v_s(t) = 12e^{-t}u(t)$ V. (See Problem 3.54.)

16.25 Obtain PROBE plots of $v_1(t)$ and $v_2(t)$ for $t \geq 0$ s for the circuit shown in Fig. P3.66 on p. 184 for the case that $v_s(t) = 6$ V for $t < 0$ s and $v_s(t) = 0$ V for $t \geq 0$ s.

16.26 Obtain PROBE plots of $i(t)$ and $v(t)$ for $t \geq 0$ s for the circuit shown in Fig. P3.67 on p. 184 for the case that $v_s(t) = 6$ V for $t < 0$ s and $v_s(t) = 0$ V for $t \geq 0$ s.

16.27 Obtain a PROBE plot of $v_2(t)$ for $t \geq 0$ s for the circuit shown in Fig. P3.66 on p. 184 for the case that $v_s(t) = 9u(t)$ V. (See Problem 3.75.)

16.28 Obtain PROBE plots of $i(t)$ and $v(t)$ for $t \geq 0$ s for the circuit shown in Fig. P3.67 on p. 184 for the case that $v_s(t) = 6u(t)$ V. (See Problem 3.76.)

16.29 Obtain a PROBE plot of $v_o(t)$ for $t \geq 0$ s for the circuit shown in Fig. P3.77 on p. 185 for the case that $v_s(t) = 8u(t)$ V and $C = \frac{1}{8}$ F. (See Problem 3.78.)

16.30 Obtain a PROBE plot of $v_o(t)$ for $t \geq 0$ s for the circuit shown in Fig. P3.80 on p. 185 for the case that $v_s(t) = 2u(t)$ V and $C = \frac{1}{3}$ F. (See Problem 3.82.)

16.31 For the series RL circuit shown in Fig. P4.6, on p. 255, use PSpice to find $v_o(t)$ when $v_s(t) = 13\cos(2t - 22.6°)$ V.

16.32 For the circuit shown in Fig. P4.9 on p. 256, use PSpice to find $v_o(t)$ and $v_s(t)$ when $i_s(t) = 5\cos 3t$ A.

16.33 Use PSpice to find $v_o(t)$ for the circuit shown in Fig. P4.17 on p. 256 when $v_s(t) = 4\cos(4t - 60°)$ V.

16.34 Use PSpice to find $v_o(t)$ for the circuit shown in Fig. P4.17 on p. 256 when $v_s(t) = 4 \cos(2t - 60°)$ V.

16.35 Use PSpice to find $v_o(t)$ for the circuit shown in Fig. P4.21 on p. 257 when $v_s(t) = 6 \sin 2t$ V.

16.36 Use PSpice to find $v_o(t)$ for the circuit shown in Fig. P4.22 on p. 257 when $v_s(t) = 3 \cos 2t$ V.

16.37 Use PSpice to find $v_o(t)$ for the circuit shown in Fig. P4.23 on p. 258 when $v_s(t) = 4 \cos(2t - 30°)$ V.

16.38 For the circuit shown in Fig. P4.25 on p. 257, use PSpice to find $\mathbf{I}_1$ and $\mathbf{I}_2$ when $\mathbf{V}_{s1} = 250\sqrt{2}\underline{/-30°}$ V, $\mathbf{V}_{s2} = 250\sqrt{2}\underline{/-90°}$ V, $\mathbf{Z} = 26 - j15$ Ω, and the frequency is 60 Hz.

16.39 For the transformer circuit shown in Fig. 16.17, on p. 1066, the primary, secondary, and mutual inductances are 3 H, 4 H, and 3 H, respectively. Furthermore, the capacitor value is 1/16 F. Use PSpice to determine the transformer's currents and voltages when $v_s(t) = 36 \cos 4t$ V and $R_1 = 9$ Ω. (See Problem 14.28.)

16.40 For the circuit in Example 16.14 on p. 1066, replace the transformer with an ideal transformer having a turns ratio of $N = \frac{1}{3}$. Use PSpice to determine the transformer's currents and voltages when $v_s(t) = 36 \cos(3t - 60°)$ V. (See Problem 14.36.) *Hint:* Model the ideal transformer as depicted in Fig. 14.24 on p. 933.

16.41 For the op-amp circuit shown in Fig. P5.3 on p. 333, suppose $R = 1$ Ω and $C = 1$ F. Obtain a PROBE plot of the amplitude response of $\mathbf{V}_2/\mathbf{V}_1$ using a linear frequency scale from 0.001 to 1 Hz. Use the cursor in PROBE to determine the half-power frequency.

16.42 For the circuit shown in Fig. P5.4 on p. 333, suppose $R_1 = R_2 = 1$ Ω, $C_1 = 1$ F, and $C_2 = 0$ F. Obtain a PROBE plot of the amplitude response of $\mathbf{V}_2/\mathbf{V}_1$ using a linear frequency scale from 0.001 to 1 Hz. Use the cursor in PROBE to determine the half-power frequency.

16.43 For the circuit shown in Fig. P5.15 on p. 334, suppose $R = 11$ Ω, $L = 1$ H, and $C = 0.1$ F. Use PROBE to obtain the Bode plot of the amplitude and phase responses of $\mathbf{V}_C/\mathbf{V}_1$. Have the frequency range from 0.01 to 10 Hz. Use the cursor in PROBE to determine the half-power frequency.

16.44 For the circuit shown in Fig. P5.15 on p. 334, suppose $R = 2$ Ω, $L = 1$ H, and $C = 1$ F. Use PROBE to obtain the Bode plot of the amplitude and phase responses of $\mathbf{V}_C/\mathbf{V}_1$. Have the frequency range from 0.01 to 10 Hz. Use the cursor in PROBE to determine the half-power frequency.

16.45 For the circuit shown in Fig. P5.15 on p. 334, suppose $R = 2$ Ω, $L = 1$ H, and $C = 1$ F. Use PROBE to obtain the Bode plot of the amplitude and phase responses of $\mathbf{V}_L/\mathbf{V}_1$. Have the frequency range from 0.005 to 5 Hz. Use the cursor in PROBE to determine the half-power frequency.

16.46 For the circuit shown in Fig. P5.15 on p. 334, suppose $R = 2$ Ω, $L = 1$ H, and $C = 1$ F. Use PROBE to obtain the Bode plot of the amplitude and phase responses of $\mathbf{V}_R/\mathbf{V}_1$. Have the frequency range from 0.002 to 10 Hz. Use the cursor in PROBE to determine the half-power frequencies.

16.47 For the diode circuit shown in Fig. 16.20 on p. 1072, suppose $R = 350$ Ω, $v_1 = 6$ V, and the silicon diode has $I_S = 5$ nA at 300 K. Use PSpice to find i and v.

16.48 For the diode circuit shown in Fig. P6.27 on p. 410, suppose $R = 100$ Ω, $V_s = 1.5$ V, and the silicon diodes have $I_{S1} = 1.5$ nA and $I_{S2} = 1$ nA at 300 K. Use PSpice to find the currents and the voltages for the diodes.

16.49 For the diode circuit shown in Fig. P6.27 on p. 410, suppose $R = 160$ Ω, $V_s = 1.5$ V, and the silicon diodes have $I_{S1} = I_{S2} = 1$ nA at 300 K. Use PSpice to find the currents and the voltages for the diodes.

16.50 For the diode circuit shown in Fig. P6.31 on p. 410, suppose the silicon diodes have $I_{S1} = 5$ nA and $I_{S2} = 10$ nA at 300 K. Use PSpice to find the currents and the voltages for the diodes.

16.51 For the diode circuit shown in Fig. P6.32 on p. 411, suppose $R = 300$ Ω and the silicon diodes have $I_{S1} = 5$ nA and $I_{S2} = 10$ nA at 300 K. Use PSpice to find the currents and the voltages for the diodes.

16.52 Use PSpice to solve Problem 6.35 on p. 411.

16.53 For the circuit shown in Fig. P6.38 on p. 411, suppose that the Ge diode has $I_S = 20$ μA at 300 K and $v_s(t) = 12 \sin 2000\pi t$ V. Obtain PROBE plots of $v_s(t)$ and $v_o(t)$.

16.54 For the circuit shown in Fig. P6.40 on p. 411, suppose the Ge diode has $I_S = 20$ μA at 300 K, $R = 1$ kΩ, and $v_s(t) = 6 \sin 2000\pi t$ V. Obtain PROBE plots of $v_s(t)$ and $v_o(t)$.

16.55 For the circuit shown in Fig. P6.41 on p. 412, suppose the Ge diodes have $I_S = 20$ μA at 300 K and $v_s(t) = 12 \sin 2000\pi t$ V. Obtain PROBE plots of $v_s(t)$ and $v_o(t)$.

16.56 For the circuit shown in Fig. P6.44 on p. 412, suppose the Ge diodes have $I_S = 20$ μA at 300 K, $R = 1$ kΩ, and $v_s(t) = 6 \sin 2000\pi t$ V. Obtain PROBE plots of $v_s(t)$ and $v_o(t)$.

16.57 For the circuit shown in Fig. P6.45 on p. 412, suppose the Ge diodes have $I_S = 20$ μA at 300 K, $R = 1$ kΩ, and $v_s(t) = 6 \sin 2000\pi t$ V. Obtain PROBE plots of $v_s(t)$ and $v_o(t)$.

16.58 Obtain a PROBE plot of the i-v characteristic for a 5-V Si Zener diode having $I_S = 10$ nA at 300 K. Have the voltage range from -6 to 1 V and the current range from -1 to 1 μA.

16.59 Use PSpice to solve Problem 6.63a on p. 414.

16.60 Use PSpice to solve Problem 6.63b on p. 414.

16.61 For the circuit shown in Fig. P6.81 on p. 416, suppose the Ge diodes have $I_S = 20$ μA at 300 K, $A = 12$, $R = 10$ kΩ, $C = 10$ μF, and $v_s(t) = 12 \sin 2000\pi t$ V. Obtain PROBE plots of $v_s(t)$ and $v_o(t)$.

16.62 For the circuit shown in Fig. 7.4 on p. 421, suppose $R_C = R_E = 1$ kΩ, $V_{CC} = V_{EE} = 5$ V, $h_{FE} = 100$, and $I_S = 7.7E-15$ A. Use PSpice to find the BJT's currents and voltages.

16.63 For the circuit shown in Fig. P7.10 on p. 493, suppose $R_1 = 90$ kΩ, $R_2 = 10$ kΩ, $R_C = 5$ kΩ, $R_E = 1$ kΩ, $V_{CC} = 20$ V, $h_{FE} = 100$, and $I_S = 2.1E-15$ A. Use PSpice to find the BJT's currents and voltages.

16.64 For the circuit shown in Fig. P7.24 on p. 494, suppose $R_B = 250$ kΩ, $R_{C1} = 2.2$ kΩ, $R_{C2} = 100$ Ω, $R_E = 0$ Ω, $V_{BB} = V_{CC} = 5$ V, $h_{FE} = 100$, $I_{S1} = 3.1E-15$ A, and $I_{S2} = 4.27E-14$. Use PSpice to find the BJTs' currents and voltages.

16.65 For the circuit shown in Fig. P7.27 on p. 494, suppose $R_B = 65$ kΩ, $R_C = 1$ kΩ, $R_{E1} = 500$ Ω, $R_{E2} = 1$ kΩ, $V_{CC} = 6$ V, $h_{FE} = 100$, $I_{S1} = 8.3E-15$ A, and $I_{S2} = 2.8E-15$. Use PSpice to find the BJTs' currents and voltages.

16.66 Obtain a PROBE plot of the transfer characteristic for the inverter shown in Fig. P7.45 on p. 496 when $R_1 = 15$ kΩ, $R_2 = 50$ kΩ, $R_C = 500$ Ω, and the BJT is a 2N3904 transistor.

16.67 Suppose the DTL NAND gate shown in Fig. 7.26 on p. 451 uses a 2N3904 BJT and 1N4148 diodes. Obtain a PROBE plot of the transfer characteristic for this gate.

16.68 Suppose the DTL NAND gate shown in Fig. 7.28 on p. 457 uses 2N3904 BJTs and 1N4148 diodes. Obtain a PROBE plot of the transfer characteristic for this gate.

16.69 Obtain a PROBE plot of the transfer characteristic for the TTL NAND gate shown in Fig. 7.31 on p. 460 when the BJTs are 2N3904 transistors.

16.70 For the Schottky-BJT circuit shown in Fig. 16.25 on p. 1082, suppose $R_B = 10$ kΩ, $R_C = 1$ kΩ,

$V_{CC} = 5$ V, $h_{FE} = 50$, $v_1 = 5$ V, $I_S = 9\text{E}-15$ A for the BJT, and $I_S = 6.4\text{E}-11$ A for the diode. Find the currents and voltages for the circuit.

16.71 For the ECL OR/NOR gate shown in Fig. 7.43 on p. 478, when $R = 275\ \Omega$ then the low voltage is -1.525 V and the high voltage is -0.7 V. Use 2N3904 transistors and obtain PROBE plots of the transfer characteristics v_o versus v_1 and $\bar{v}_o$ versus v_1.

Note: All problems referred to in Problems 16.72 to 16.87 are found in Chapter 8 of this textbook.

16.72 Use PSpice to solve Problem 8.6 on p. 560.

16.73 Use PSpice to solve Problem 8.7 on p. 560.

16.74 Use PSpice to solve Problem 8.10 on p. 561.

16.75 Use PSpice to solve Problem 8.13 on p. 561.

16.76 Use PSpice to solve Problem 8.14 on p. 561.

16.77 Use PSpice to solve Problem 8.16 on p. 561.

16.78 Use PSpice to solve Problem 8.20 on p. 562.

16.79 Use PSpice to solve Problem 8.22 on p. 562.

16.80 Use PSpice to solve Problem 8.26 on p. 562.

16.81 Use PSpice to solve Problem 8.35 on p. 563.

16.82 Use PSpice to solve Problem 8.36 on p. 563.

16.83 Use PSpice to solve Problem 8.43 on p. 564.

16.84 Use PSpice to solve Problem 8.50 on p. 565.

16.85 Use PSpice to solve Problem 8.56 on p. 565.

16.86 Use PSpice to solve Problem 8.57 on p. 566.

16.87 Use PSpice to solve Problem 8.65 on p. 567.

16.88 For the BJT amplifier shown in Fig. 9.14 on p. 583, suppose $R_s = 0\ \Omega$, $R_1 = 90$ kΩ, $R_2 = 10$ kΩ, $R_C = 5$ kΩ, $R_E = 1$ kΩ, $R_L = 5$ kΩ, $C_1 = C_2 = C_E = 10\ \mu$F, $V_{CC} = 20$ V, $h_{FE} = 100$, and $I_S = 2\text{E}-15$ A. Use SPice to find the ac voltage gain from base to collector and input resistance at 100 kHz.

16.89 For the BJT amplifier shown in Fig. P9.16 on p. 642, suppose $R_s = 0\ \Omega$, $C = 10\ \mu$F, $h_{FE} = 100$, and $I_S = 8.4\text{E}-15$ A. Use PSpice to find the ac voltage gain from base to collector and input resistance at 100 kHz.

16.90 For the BJT amplifier shown in Fig. P9.16 on p. 642, suppose $R_s = 0\ \Omega$, $C = 10\ \mu$F, $h_{FE} = 100$, and $I_S = 8.4\text{E}-15$ A. Use PSpice to find the output resistance (between the collector and the emitter) at 100 kHz.

16.91 For the two-stage amplifier shown in Fig. P9.18 on p. 642, suppose $R_s = 0\ \Omega$, $C = C_1 = C_2 = 10\ \mu$F, $h_{FE1} = h_{FE2} = 100$, $I_{S1} = 6.7\text{E}-15$ A, $I_{S2} = 10\text{E}-15$ A. Use PSpice to find the ac voltage gain v_{c2}/v_s and the input resistance at 100 kHz.

16.92 For the common-base amplifier shown in Fig. P9.21 on p. 643, suppose $C_1 = C_2 = C_B = 10\ \mu$F, $h_{FE} = 100$, and $I_S = 1.7\text{E}-15$ A. Use PSpice to find the ac gain v_c/v_b and the input resistance at 100 kHz.

16.93 For the MOSFET feedback amplifier shown in Fig. 9.43 on p. 608, suppose $R_s = 0\ \Omega$, $R_D = 4$ kΩ, $R_F = 1$ MΩ, $R_L = 6$ kΩ, $C_1 = C_2 = 10\ \mu$F, $V_{DD} = 22$ V, $V_t = 2$ V, and $K = 0.25$ mA/V^2. Use PSpice to find the ac voltage gain v_d/v_g and the input resistance at 100 kHz.

16.94 For the MOSFET feedback amplifier shown in Fig. 9.43 on p. 608, suppose $R_s = 0\ \Omega$, $R_D = 4$ kΩ, $R_F = 1$ MΩ, $R_L = 6$ kΩ, $C_1 = C_2 = 10\ \mu$F, $V_{DD} = 22$ V, $V_t = 2$ V, and $K =$

0.25 mA$/$V^2. Use PSpice to find the output resistance (seen by the external load R_L) at 100 kHz.

16.95 For the BJT amplifier given in Fig. 9.14 on p. 583, suppose $C_1 = C_2 = 1$ μF, $C_E = 4.7$ μF, $R_1 = 60$ kΩ, $R_2 = 30$ kΩ, $R_C = R_E = 10$ kΩ, $R_L = 30$ kΩ, $R_s = 1$ kΩ, $V_{CC} = 33$ V, and the BJT is a 2N3904 transistor. Use PSpice and PROBE to obtain a plot of the amplitude-response portion of the Bode plot for the amplifier. Use the cursor in PROBE to determine the lower cutoff frequency, the upper cutoff frequency, and the bandwidth of the amplifier.

16.96 For the common-base amplifier shown in Fig. P9.21 on p. 643, suppose $C_1 = C_2 = C_B = 10$ μF and the BJT is a 2N3904 transistor. Use PSpice and PROBE to obtain a plot of the amplitude-response portion of the Bode plot for the amplifier. Use the cursor in PROBE to determine the lower cutoff frequency, the upper cutoff frequency, and the bandwidth of the amplifier.

16.97 For the *RC* phase-shift oscillator described in Example 16.36 on p. 1112, use the model for the 741 op amp in the device library and increase the value of R_F to 200 kΩ so that oscillation will occur. Obtain a PROBE plot of the voltage at the output of the op amp.

16.98 For the Wien-bridge oscillator described in Drill Exercise 16.36 on p. 1114, use the model for the 741 op amp in the device library and increase the value of the initial condition to 10 V and the value of R_2 to 20.5 kΩ so that oscillation will occur. Obtain a PROBE plot of the voltage at the output of the op amp.

16.99 For the oscillator shown in Fig. P10.59 on p. 742, suppose $R = R_1 = 10$ kΩ, $R_2 = 20$ kΩ, $C = 1.59$ nF, and the op amp is ideal. Assume the initial voltage across the horizontally oriented capacitor is 1 V (+ on the left). Obtain a PROBE plot of the voltage at the output of the op amp.

16.100 For the oscillator shown in Fig. P10.64 on p. 743, suppose $R = 500$ Ω and the op amp is ideal. Assume the initial voltage across the capacitor is 0.1 V (+ on the left). Obtain a PROBE plot of the voltage at the output of the op amp.

16.101 For the noninverting Schmitt trigger shown in Fig. P10.71 on p. 743, suppose $R_1 = 10$ kΩ, $R_2 = 20$ kΩ, $V_{CC} = 12$ V, and $V_r = -6$ V. Obtain a PROBE plot of the transfer characteristic for this Schmitt trigger.

16.102 For the emitter-coupled Schmitt trigger shown in Fig. P10.75 on p. 744, suppose the BJTs are 2N3904 transistors. Obtain a PROBE plot of the transfer characteristic for this Schmitt trigger.

16.103 For the relaxation oscillator shown in Fig. 10.47 on p. 714, suppose $R = R_1 = 10$ kΩ, $R_2 = 20$ kΩ, $C = 0.5$ μF, and both op amps use ± 10-V power supplies and are modeled by 741 op amps in the device library. Obtain PROBE plots of the voltages at the outputs of the op amps.

BIBLIOGRAPHY

Circuit Analysis

BOBROW, L.S. (1987). *Elementary Linear Circuit Analysis* (2nd ed.). Oxford University Press.

IRWIN, J.D. (1996). *Basic Engineering Circuit Analysis* (5th ed.). Prentice Hall.

KRAUS, A.D. (1991). *Circuit Analysis.* West.

SCOTT, D.E. (1987). *An Introduction to Circuit Analysis: A Systems Approach.* McGraw-Hill.

Electrical Engineering

CARLSON A.B. and D.G. GISSER. (1990). *Electrical Engineering: Concepts and Applications* (2nd ed.). Addison-Wesley.

CAVICCHI, T.J. (1993). *Fundamentals of Electrical Engineering: Principles and Applications.* Prentice Hall.

ROADSTRUM, W.H. and D.H. WOLAVER. (1987). *Electrical Engineering: For All Engineers.* Harper & Row.

Electronics

GHAUSI, M.S. (1985). *Electronic Devices and Circuits: Discrete and Integrated.* Oxford University Press.

SEDRA, A.S. and K.C. SMITH. (1991). *Microelectronic Circuits* (3rd ed.). Oxford University Press.

Signals and Systems

KAMEN, E.W. (1990). *Introduction to Signals and Systems* (2nd ed.). Macmillan.

ZIEMER, R.E., W.H. TRANTER, and D.R. FANNIN. (1989). *Signals and Systems: Continuous and Discrete* (2nd ed.). Macmillan.

Software

ARNEY, D.C. (1992). *The Student Edition of Derive.* Addison-Wesley.

KEOWN, J. (1993). *PSpice and Circuit Analysis* (2nd ed.). Merrill.

Answers to Selected Problems

Chapter 1

1.2 (a) 0 V, (b) 5 V, (c) 0 V, (d) −5 V, (e) 0 V

1.4 (a) $-8e^{-2t}$ A, (b) $3\pi \cos \pi t$ A,
(c) $-12\pi \sin 2\pi t$ A,
(d) $-15e^{-4t} \sin 3t -20e^{-4t} \cos 3t$ A

1.7 (a) 20 V, (b) 8 V, (c) −6 V, (d) −2 V

1.9 (a) −2 A, (b) −4 V, (c) −1 A, (d) 1 V

1.11 (a) 3 V, (b) −12 V, (c) −5.5 V,
(d) −10.d V, (e) 7 V

1.13 (a) 2 A, (b) $-\frac{1}{2}$ A, (c) $\frac{1}{2}$ A, (d) 1 A

1.15 (a) 3 Ω, (b) 6 Ω, (c) 4 Ω, (d) 8 Ω

1.17 (a) 3 A, (b) 0 A, (c) −3 A

1.20 9 A, 3 A, 6 A, −2 A

1.22 −18 A, −6 A, −12 A, −8 A

1.24 (a) $\frac{1}{2}$ V, (b) 0 V, (c) $-\frac{1}{2}$ V

1.26 (a) 6 A, −2 A; (b) 12 V, 0 V;
(c) −24 V, 36 V, 2 A; (d) 15 V, 13 V, −4 A

1.28 (a) 9 V, 4.5 A, −4.5 A; (b) 12 V, 3 A, −6 A;
(c) 13.5 V, 2.25 A; −6.75 A

1.31 2 V, $\frac{2}{3}$ A

1.33 (a) 16 V, (b) 24 V, (c) 32 V,
(d) 8/3 Ω, 8/3 Ω, 8/3 Ω

1.35 3 Ω

1.37 2.4 Ω

1.40 8 V, 0 A, 4 V, 3.6 Ω

1.43 (a) 2 A, (b) 0 A, (c) −2 A

1.45 (a) 1.5 A, (b) −1.5 A

1.47 (a) 2 Ω, (b) v_g, (c) $4.5v_1$

1.49 (a) 8 V, (b) ∞

1.51 (a) 5 A, (b) 0 Ω

1.53 (a) $v_1/1500$, (b) $-50v_1$, (c) $-5 \cos 120\pi t$ V

1.56 −48 W, 192 W, −288 W, 144 W

1.58 (a) −24 W, 24 W, −48 W, 48 W;
(b) −120 W, 24 W, 48 W, 48 W

1.60 (a) −4.5 W, 2.25 W, −9 W, 4.5 W, 6.75 W;
(b) 4.5 W, 2.25 W, −18 W, 4.5 W, 6.75 W

Chapter 2

2.2 (a) 6 V, 8 V, 4 V;
(b) −1 A, −2 A, −1 A, 2 A

2.4 (a) 2 V, −6 V, −2 V,
(b) −1 A, −2 A, −1 A, 2 A

2.6 20 V, $-\frac{2}{3}$ V, 18 V

2.9 10 V, 6 V, 8 V

2.11 (a) −444, (b) 900 Ω

2.15 (a) 30.96, 43.2 Ω

2.17 2.75 ℧

2.20 2 A, 1 A, −1 A

2.22 6 A, 4 A, 2 A

2.24 2.75 ℧

2.26 −3 A, −11 A, 7 A

2.28 $(1 + R_1/R_2)i_s$

2.30 (a) $(1 + R_1/R_2)v_s$, (b) $-R_1R/R_2$

2.32 (a) $0.522v_s$, (b) 5.75 Ω

2.34 (a) $5v_s$, (b) −2.5 Ω

2.36 $-1.8v_s$

2.38 (a) 56/3 V, 4 Ω, (b) 19.4 W, (c) 21.8 W

2.40 (a) 4 V, 2 Ω, (b) 1.92 W, (c) 2 W

2.43 0 A, 3 Ω

2.45 $(1 + R_1/R_2)v_s$, 0 Ω

2.47 $-\frac{1}{4}v_s$, ∞

2.51 4 V

2.53 0.5 A

2.56 (a) 2 V, (b) 4 V, (c) 6 V

2.58 (a) 0.8 V, (b) -3.2 V, (c) -2.4 V

2.60 (a) 2.25 A, 2.81 V, (b) 0.75 A, 0.188 V,
(c) 3 A, 3 V

2.62 (a) 6 A, 120 V, (b) -3 A, -60 V,
(c) 0 A, -30 V, (d) 3 A, 30 V

Chapter 3

3.2 (a) 0 for $t \le 0$ s, -2 for $0 < t \le 1$ s,
2 for $1 < t \le 2$ s, 0 for $t > 2$ s

(b) 0 for $t \le 0$ s, t^2 for $0 < t \le 1$ s,
$(t - 2)^2$ for $1 < t \le 2$ s, 0 for $t > 2$ s

(c) 0 for $t \le 0$ s, $2t^2$ for $0 < t \le 1$ s,
$2(t - 2)^2$ for $1 < t \le 2$ s, 0 for $t > 2$ s

(d) 0 for $t \le 0$ s, $-2t$ for $0 < t \le 1$ s,
$2(t - 2)$ for $1 < t \le 2$ s, 0 for $t > 2$ s

(e) 0 for $t \le 0$ s, $-2t - 2$ for $0 < t \le 1$ s,
$2t - 2$ for $1 < t \le 2$ s, 0 for $t > 2$ s

3.4 (a) 0 for $t \le 0$ s, -2 for $0 < t \le 1$ s,
2 for $1 < t \le 2$ s, 0 for $t > 2$ s

(b) 0 for $t \le 0$ s, t^2 for $0 < t \le 1$ s,
$(t - 2)^2$ for $1 < t \le 2$ s, 0 for $t > 2$ s

(c) 0 for $t \le 0$ s, 2 for $0 < t \le 1$ s,
2 for $1 < t \le 2$ s, 0 for $t > 2$ s

(d) 0 for $t \le 0$ s, -1 for $0 < t \le 1$ s,
1 for $1 < t \le 2$ s, 0 for $t > 2$ s

(e) 0 for $t \le 0$ s, $-1 - t$ for $0 < t \le 1$ s,
$t - 1$ for $1 < t \le 2$ s, 0 for $t > 2$ s

3.6 (a) 0 for $t \le 0$ s, $-2t$ for $0 < t \le 1$ s,
$2t - 4$ for $1 < t \le 2$ s, 0 for $t > 2$ s

(b) 0 for $t \le 0$ s, -2 for $0 < t \le 1$ s,
2 for $1 < t \le 2$ s, 0 for $t > 2$ s

(c) 0 for $t \le 0$ s, $-2t - 2$ for $0 < t \le 1$ s,
$2t - 2$ for $1 < t \le 2$ s, 0 for $t > 2$ s

3.8 (a) 0 for $t \le 0$ s, 2 for $0 < t \le 1$ s,
0 for $t > 1$ s

(b) 0 for $t \le 0$ s, t^2 for $0 < t \le 1$ s,
1 for $t > 1$ s

(c) 0 for $t \le 0$ s, $\frac{1}{2}t^2$ for $0 < t \le 1$ s,
$\frac{1}{2}$ for $t > 1$ s

(d) 0 for $t \le 0$ s, $\frac{1}{2}t$ for $0 < t \le 1$ s,
$\frac{1}{2}$ for $t > 1$ s

(e) 0 for $t \le 0$ s, $\frac{1}{2}t + 2$ for $0 < t \le 1$ s,
$\frac{1}{2}$ for $t > 1$ s

3.10 (a) 0 for $t \le 0$ s, 2 for $0 < t \le 1$ s,
0 for $t > 1$ s

(b) 0 for $t \le 0$ s, $\frac{1}{2}t + 2$ for $0 < t \le 1$ s,
$\frac{1}{2}$ for $t > 1$ s

(c) 0 for $t \le 0$ s, $t + 4$ for $0 < t \le 1$ s,
1 for $t > 1$ s

(d) 0 for $t \le 0$ s, $t + 4$ for $0 < t \le 1$ s,
1 for $t > 1$ s

(e) 0 for $t \le 0$ s, $-t$ for $0 < t \le 1$ s,
-1 for $t > 1$ s

3.12 (a) 0 for $t \le 0$ s, 2 for $0 < t \le 1$ s,
0 for $t > 1$ s

(b) 0 for $t \le 0$ s, $\frac{1}{2}t + 2$ for $0 < t \le 1$ s,
$\frac{1}{2}$ for $t > 1$ s

(c) 0 for $t \le 0$ s, $t + 4$ for $0 < t \le 1$ s,
1 for $t > 1$ s

(d) 0 for $t \le 0$ s, $t + 4$ for $0 < t \le 1$ s,
1 for $t > 1$ s

(e) 0 for $t \le 0$ s, $2t + 4$ for $0 < t \le 1$ s,
2 for $t > 1$ s

3.14 (a) 0 for $t \le 0$ s, 2 for $0 < t \le 1$ s,
0 for $t > 1$ s

(b) 0 for $t \le 0$ s, $\frac{1}{2}t + 2$ for $0 < t \le 1$ s,
$\frac{1}{2}$ for $t > 1$ s

(c) 0 for $t \le 0$ s, $t + 4$ for $0 < t \le 1$ s,
1 for $t > 1$ s

(d) 0 for $t \le 0$ s, $t + 4$ for $0 < t \le 1$ s,
1 for $t > 1$ s

(e) 0 for $t \le 0$ s, $2t + 4$ for $0 < t \le 1$ s,
2 for $t > 1$ s

3.17 (a) 0 for $t \le 0$ s, $\frac{1}{2}t$ for $0 < t \le 1$ s,
$\frac{1}{2}$ for $t > 1$ s

(b) 0 for $t \le 0$ s, $\frac{1}{2}$ for $0 < t \le 1$ s,
0 for $t > 1$ s

(c) 0 for $t \le 0$ s, $\frac{1}{2}t + \frac{1}{2}$ for $0 < t \le 1$ s,
$\frac{1}{2}$ for $t > 1$ s

3.19 (a) 0 for $t \le 0$ s, $\frac{1}{2}t$ for $0 < t \le 1$ s,
$\frac{1}{2}$ for $t > 1$ s

(b) 0 for $t \le 0$ s, 2 for $0 < t \le 1$ s,
0 for $t > 1$ s

(c) 0 for $t \le 0$ s, $\frac{1}{2}t + 2$ for $0 < t \le 1$ s,
$\frac{1}{2}$ for $t > 1$ s

3.21 (a) 0 for $t \le 0$ s, $\frac{1}{2}t$ for $0 < t \le 1$ s,
$\frac{1}{2}$ for $t > 1$ s

(b) 0 for $t \le 0$ s, 2 for $0 < t \le 1$ s,
0 for $t > 1$ s

(c) 0 for $t \le 0$ s, 2 for $0 < t \le 1$ s,
0 for $t > 1$ s

(d) 0 for $t \le 0$ s, $-\frac{1}{2}t$ for $0 < t \le 1$ s,
$-\frac{1}{2}$ for $t > 1$ s

3.23 (a) 0 for $t \le 0$ s, $\frac{1}{2}t$ for $0 < t \le 1$ s,
$\frac{1}{2}$ for $t > 1$ s

(b) 0 for $t \le 0$ s, 2 for $0 < t \le 1$ s,
0 for $t > 1$ s

(c) 0 for $t \le 0$ s, 2 for $0 < t \le 1$ s,
0 for $t > 1$ s

(d) 0 for $t \le 0$ s, $\frac{1}{2}t + 2$ for $0 < t \le 1$ s,
$\frac{1}{2}$ for $t > 1$ s

3.28 12 V for $t < 0$ s, $12e^{-2t}$ V for $t \ge 0$ s;
0 A for $t < 0$ s, $-2.4e^{-2t}$ A for $t \ge 0$ s

3.30 4 A for $t < 0$ s, $4e^{-t}$ A for $t \ge 0$ s;
0 V for $t < 0$ s, $-20e^{-t}$ V for $t \ge 0$ s

3.32 1 A for $t < 0$ s, e^{-9t} A for $t \ge 0$ s;
0 V for $t < 0$ s, $-4.5e^{-9t}$ V for $t \ge 0$ s

3.34 18 V for $t < 0$ s, $18e^{-4t}$ V for $t \ge 0$ s;
0 A for $t < 0$ s, $-3e^{-4t}$ A for $t \ge 0$ s

3.36 1.5 A for $t < 0$ s, $1.5e^{-2t}$ A for $t \ge 0$ s;
0 V for $t < 0$ s, $-48e^{-2t}$ V for $t \ge 0$ s

3.38 6 V for $t < 0$ s, $6e^{-4t}$ V for $t \ge 0$ s;
4 V for $t < 0$ s, $4e^{-2t}$ V for $t \ge 0$ s;
0 A for $t < 0$ s, $-2e^{-4t}$ A for $t \ge 0$ s;
0 A for $t < 0$ s, $-2e^{-2t}$ A for $t \ge 0$ s;
10 V for $t < 0$ s, $6e^{-4t} + 4e^{-2t}$ V for $t \ge 0$ s

3.40 $12(1 - e^{-t/4})u(t)$ V, $6e^{-t/4}u(t)$ A

3.42 $\frac{1}{2}V(1 - e^{-2t/RC})u(t)$, $(V/R)e^{-2t/RC}u(t)$

3.44 $20(1 - e^{-2t})u(t)$ V, $4e^{-2t}u(t)$ A

3.46 $3(1 - e^{-2t})u(t)$ A, $18e^{-2t}u(t)$ V

3.49 $-(VR_2/R_1)(1 - e^{-t/R_2 C})u(t)$,
$[(R_1 + R_2)V/R_1 - (VR_2/R_1)e^{-t/R_2 C}]u(t)$

3.51 $(-12e^{-t/2} + 12e^{-t/4})u(t)$ V,
$(12e^{-t/2} - 6e^{-t/4})u(t)$ A

3.53 $(-6e^{-2t} + 6e^{-t})u(t)$ A, $(24e^{-2t} - 12e^{-t})u(t)$ V

3.56 $6(1 - e^{-4t})u(t) - 6(1 - e^{-4(t - 2)})u(t - 2)$ V,
$e^{-4t}u(t) - e^{-4(t - 2)}u(t - 2)$ A

3.58 0 V for $t < 0$ s, $(5/3)e^{-2t} \sin 4t$ V for $t \ge 0$ s;
4 A for $t < 0$ s,
$e^{-2t}(4 \cos 4t - 2 \sin 4t)$ A for $t \ge 0$ s

3.60 6 A for $t < 0$ s,
$-6e^{-4t} + 12e^{-2t}$ A for $t \ge 0$ s;
0 V for $t < 0$ s, $6e^{-4t} - 6e^{-2t}$ V for $t \ge 0$ s

3.62 6 A for $t < 0$ s, $18te^{-3t} + 6e^{-3t}$ A for $t \ge 0$ s;
0 V for $t < 0$ s, $-12te^{-3t}$ V for $t \ge 0$ s

3.64 10 V for $t < 0$ s,
$e^{-4t}(10 \cos 2t + 20 \sin 2t)$ V for $t \ge 0$ s;
0 A for $t < 0$ s, $-20e^{-4t} \sin 2t$ A for $t \ge 0$ s

3.66 6 V for $t < 0$ s, $-2e^{-8t} + 8e^{-2t}$ V for $t \ge 0$ s;
6 V for $t < 0$ s, $2e^{-8t} + 4e^{-2t}$ V for $t \ge 0$ s

3.68 0 V for $t < 0$ s, $-6e^{-t/2} \sin \frac{1}{2}t$ V for $t \ge 0$ s;
6 A for $t < 0$ s, $6e^{-t/2} \cos \frac{1}{2}t$ A for $t \ge 0$ s

3.70 $(4 - 8te^{-2t} - 4e^{-2t})u(t)$ A, $48te^{-2t}u(t)$ V

3.73 $(2 + e^{-6t} - 3e^{-2t})u(t)$ A,
$(-2e^{-6t} + 2e^{-2t})u(t)$ V

3.76 $[3 + e^{-3t}(-3 \cos t - 3 \sin t)]u(t)$ A,
$[3 + e^{-3t}(-3 \cos t + 3 \sin t)]u(t)$ V

3.78 $[-16 + e^{-t}(16 \cos t + 16 \sin t)]u(t)$ V

3.80 $(3 - 3te^{-t} - 3e^{-t})u(t)$ V

3.84 -2 V for $t <$ s,
$2 + e^{-4t}(-4 \cos 2t - 8 \sin 2t)$ V for $t \ge$ s;
0 A for $t < 0$ s, $8e^{-4t} \sin 2t$ A for $t \ge 0$ s

Chapter 4

4.1 (a) $8.06e^{j60.3°}$, (c) $5.83e^{-j59.0°}$, (e) $4e^{j0°}$,
(g) $7e^{j90°}$

4.2 (b) $-1 + j1.73$, (d) $-3.46 - j2$, (f) $-j1$,
(h) -2

4.3 (a) $j12$, (c) $5 + j8.66$

4.4 (b) $0.375 + j0.65$, (d) $-1.41 - j1.41$

4.5 (a) $4.6 + j4.96$, (c) $1.5 - j2.6$

4.7 $6.37 \cos(2t - 10.8°)$ V, lead network

4.9 $4.47 \cos(3t + 26.6°)$ V, $6.32 \cos(3t - 18.4°)$ V

4.12 (a) leads by $43.8°$, (b) in phase,
(c) lags by $16.7°$

4.14 (a) $0.25 - j0.2\mho$, (b) $0.05\mho$,
(c) $0.01 + j0.28\mho$

4.17 $0.972 \cos(4t - 164°)$ V

4.20 series connection of $3.71 \cos(5t - 15.9°)$-V
voltage source, 2.38-Ω resistor, and
0.295-F capacitor

4.22 $10.6 \cos(2t + 135°)$ V

4.24 $6.8/30°$ A, $6.8/-90°$ A

4.27 1.57 W, 0.749 W

4.29 7.5 W, 0.5 W, 0 W, 0 W, -8.0 W

4.31 0 W, 18 W, 18 W, 0 W, -36 W

4.34 (a) 602 W, 602 W, 602 W;
(b) 601 W, 1202 W

4.36 -0.5 W, 0.5 W, 2.25 W, 1.5 W, -3.75 W

4.38 (b) 1.58, (d) 0.745

4.40 0.707

4.43 (a) 0.5, (b) 209 μF

4.45 0.869 lagging

4.47 0.958 leading, 800 W

4.50 27 VA, 9 VA, 18 VA, $18\sqrt{2}$ VA, no

4.52 (a) $6.01 - j347$ VA, $6.01 - j347$ VA, $6.01 - j347$ VA, (b) $600 - j1040$ VA, 1203 VA

4.54 (a) -347 VAR, -347 VAR, -347 VAR, (b) -1040 VAR, 0 VAR

4.59 826 W, 777 W

4.61 $7.66\underline{/-45°}$ A, $7.66\underline{/-165°}$ A, $7.66\underline{/75°}$ A, 2110 W

4.63 $23\underline{/-53.1°}$ A, $11.5\underline{/-120°}$ A, $8.85\underline{/52.6°}$ A, 3302 W

4.65 $68.9\underline{/-45°}$ A, $68.8\underline{/-165°}$ A, $68.8\underline{/75°}$ A, 18,960 W

4.67 $31.9 + 19.8$ Ω, 37.3 μF

4.69 $j\frac{2}{3}$ Ω, 2 Ω, 1 Ω

4.71 $2 + j2$ Ω, $3 - j3$ Ω, $1 + j1$ Ω

4.73 2 $\mho$, $j\frac{2}{3}$ $\mho$, $-j1.5$ $\mho$

4.75 1666 W, 446 W

4.78 14,950 W, 4,000 W

4.81 (a) 20 Ω, $40\underline{/45°}$ Ω, $j10$ Ω, (b) 2420 W, 856 W, 0 W

4.83 1517 W, 1685 W

4.85 601 W, 1203 W

Chapter 5

5.3 $1/RC$, low-pass filter

5.5 0.5 for all ω, 0° for all ω

5.7 $2/RC$

5.11 a rad/s, low-pass filter

5.20 (a) 2 rad/s, (b) 4, (c) 0.828 rad/s, 4.828 rad/s, (d) 4 rad/s

5.22 $\sqrt{1/LC - R^2/L^2}$

5.24 $1/\sqrt{LC - R^2C^2}$

5.26 $1/\sqrt{LC - L^2/R^2}$

5.28 $\sqrt{1/LC - 1/R^2C^2}$

5.30 (a) $\cos^2(t - 45°)$ J, (b) $\cos^2(t - 90°)$ J, (c) 1.71 J, (d) 1 W, (e) 6.28 J, (f) 1.71

5.33 (a) 10 rad/s, (b) 5, (c) 2 rad/s, (d) 9.05 rad/s, 11.05 rad/s

5.36 (a) 0.1, (b) 9.9 Ω, 19.8 mH, (c) 0.1

5.38 200 kΩ, 50 mH, 0.005 μF, 63.2

5.40 $0.97e^{-6t}\cos(3t - 157°)$ V, $19.5e^{-6t}\cos(3t - 3.6°)$ V, $2.17e^{-6t}\cos(3t + 49.6°)$ V

5.42 $3.42e^{-6t}\cos(3t - 149°)$ V, $22.9e^{-6t}\cos(3t + 4°)$ V, $0.127e^{-6t}\cos(3t + 58°)$ V

5.44 (a) poles: $-0.167 \pm j2.23$, zero: origin, (b) poles: $-0.167 \pm j2.23$, zero: origin, (c) poles: $-0.167 \pm j2.23$, zeros: none

5.46 (a) double pole: $-\frac{1}{2}$, zero: origin, (b) double pole: $-\frac{1}{2}$, zero: origin, (c) double pole: $-\frac{1}{2}$, double zeros: origin, (d) double pole: $-\frac{1}{2}$, zeros: none

5.48 (a) poles: -1.13, -2.87, zeros: none, (b) double pole: -1, zeros: none, (c) poles: $-1 \pm j\sqrt{3}$, zeros: none

5.50 $34.3e^{-6t}\cos(3t - 17.8°)$ V

5.53 $(s^2 + 1)/(s^2 + 7s + 9)$

5.56 $1/(s + 5)$

5.59 (b) $(-4s^2 - 28s + 36)/(s^3 + 7s^2 + 6s)$, (d) $(s - 1)/(s^2 + 6s + 25)$

5.60 (a) $(\beta \cos \phi - [\sin \phi] s)/(s^2 + \beta^2)$, (c) $(\beta \cos \phi - [\sin \phi][s + \alpha])/([s^2 + \alpha^2] + \beta^2)$

5.62 (a) $(-2e^{-3t} + 2\cos 3t + 2\sin 3t)u(t)$, (b) $(1 + 3\cos 2t)u(t)$

5.64 $(6 - 12e^{-t} + 2e^{-6t})u(t)$

5.66 $(20 - 20e^{-2t})u(t)$ V, $4e^{-2t}u(t)$ A

5.68 $3te^{-t/4}u(t)$ V, $(6e^{-t/4} - 1.5te^{-t/4})u(t)$ A

5.70 $(6e^{-t} - 6e^{-2t})u(t)$ A, $(-12e^{-t} + 24e^{-2t})u(t)$ V

5.72 $(18 - 18e^{-4t})u(t)$ V, $3e^{-4t}u(t)$ A

5.74 $(-3 + 3e^{-4t})u(t)$ V

5.76 $-12te^{-4t}u(t)$ V

5.78 $(9e^{-2t} - 6e^{-4t})u(t)$ V

5.80 $(4/3)te^{-2t}u(t)$ V, $4e^{-2t}u(t) - 8te^{-2t}u(t)$ A

5.83 $-6e^{-2t}\sin 2t\, u(t)$ V, $6e^{-2t}\cos 2t\, u(t) + 6e^{-2t}\sin 2t\, u(t)$ A

5.85 $-12te^{-3t}u(t)$ V, $6e^{-3t}u(t) + 18te^{-3t}u(t)$ A

5.87 $-20e^{-4t}\sin 2t\, u(t)$ A, $10e^{-4t}\cos 2t\, u(t) + 20e^{-4t}\sin 2t\, u(t)$ V

5.89 $8e^{-2t}u(t) - 2e^{-8t}u(t)$ V

5.91 $-6e^{-t/2}\sin\frac{1}{2}t\, u(t)$ V, $6e^{-t/2}\cos\frac{1}{2}t\, u(t)$ A

5.93 $48te^{-2t}u(t)$ V, $4u(t) - 4e^{-2t}u(t) - 8te^{-2t}u(t)$ A

5.95 $12u(t) - 12e^{-t}u(t) - 12te^{-t}u(t)$ V, $12te^{-t}u(t)$ A

5.97 $2e^{-2t} \sin 2t\, u(t)$ V,

$2u(t) - 2e^{-2t} \cos 2t\, u(t) - 2e^{-2t} \sin 2t\, u(t)$ A

5.99 $3u(t) - 3e^{-t} \cos t\, u(t) + 3e^{-t} \sin t\, u(t)$ V

5.101 $-16u(t) + 16e^{-t} \cos t\, u(t) + 16e^{-t} \sin t\, u(t)$ V

5.103 $3u(t) - 3e^{-t}u(t) - 3te^{-3t}u(t)$ V

5.105 $2u(t) - 2e^{-t} \cos 2t\, u(t) - e^{-t} \sin 2t\, u(t)$ V

5.107 $8e^{-4t} \sin 2t\, u(t)$ A,

$2u(t) - 4e^{-4t} \cos 2t\, u(t) - 8e^{-4t} \sin 2t\, u(t)$ V

5.109 (a) $(\frac{1}{2} - \frac{1}{2}e^{-2t})u(t)$, (c) $(\frac{1}{2} - e^{-t} + \frac{1}{2}e^{-2t})u(t)$

5.110 (b) $(-e^{-t} + 2e^{-2t})u(t)$, (d) $(e^{-2t} - 2te^{-2t})u(t)$

5.111 (a) $(10 - 9e^{-t})u(t)$, (c) $e^{-10t}u(t)$

5.112 (b) $(s + 1)/(s^2 + 1)$, (d) $1/(s + 1)$

5.113 (a) $(s^2 + 1)/(s^2 + 2s)$,

(c) $(s^2 + 1)/(s^3 + 2s^2 + s)$,

(e) $(s^2 + 1)/(s^3 + 3s^2 + 2s)$

Chapter 6

6.2 (a) 2.55×10^6 A/m^2, (b) 1.89×10^{-4} m/s,
(c) 4.3×10^{-3} m^2/V-s, (d) 5.8×10^7 ℧/m

6.4 (a) 4.32×10^{-4} ℧/m, (b) 23.1 MΩ

6.7 (a) 8 ℧/m, (b) 1.25 kΩ

6.9 (a) 12.8 ℧/m, (b) 781 Ω

6.11 7.9 parts per 10^9

6.13 0.617 V

6.15 0.663 V

6.17 4.96 parts per 10^6

6.19 (a) 3.71 mA, (b) 0.772 V, (c) 349 Ω

6.21 (a) 3.91 nA, (c) 8.55 nA

6.22 (b) 0.781 V

6.23 (b) 6.75 μA, (d) 1.7 μA

6.24 (b) 0.282 V

6.26 (a) 829 Ω, (b) 1.12 kΩ

6.28 (a) 1.08 kΩ, (b) 6.23 Ω

6.30 0.394 V, -5.61 V

6.32 299 Ω

6.34 1.42 A, 0.289 V

6.38 -3 V for $v_s < -4$ V, $\frac{3}{4}v_s$ for $v_s \geq -4$ V

6.40 0 V for $v_s \leq 4$ V, $v_s - 4$ for $v_s > 4$ V

6.42 3 V for $v_s < 4$ V, $\frac{3}{4}v_s$ for $4 \leq v_s \leq 8$ V,
6 V for $v_s > 8$ V

6.44 D_1 ON, D_2 OFF, D_3 ON, D_4 OFF

6.46 (b) A volts

6.48 (a) 0.526 V, (b) 1 V, (c) 10 V

6.50 (a) 0.278 Ω, (b) 0.024 Ω, (c) 0.203 Ω,
(d) 0.018 Ω

6.52 $i = -I_S$ for $v \leq 0$ V, $i = v/R_f - I_S$ for $i > -I_S$

6.54 $i = -I_S$ for $v \leq V_\gamma$,
$i = v/R_f - V_\gamma/R_f - I_S$ for $i > -I_S$

6.56 $i = v/R_Z + V_Z/R_Z$ for $i < 0$ A,
$i = 0$ A for $-V_Z \leq v \leq 0$ V, $v = 0$ V for $i > 0$ A

6.58 $R_1 < 2.83$ kΩ

6.60 0 V for $v_s \leq 0.6$ V, $5v_s/6 - \frac{1}{2}$ for $v_s > 0.6$ V

6.63 (a) -4 V, -2 V, (b) -7 V, -5 V

6.65 (b) 44 V

6.66 (a) 588 Ω, (c) 600 Ω

6.68 $15.3 \leq v_s \leq 45.3$ V

6.70 v_s for $v_s \leq 12$ V, 12 for $v_s > 12$ V

6.75 (b) $15 + v_s$

6.77 (a) $-\frac{1}{2}A + A \sin \omega t$ V

6.79 $-18 + 24 \sin \omega t$ V

6.81 $\frac{1}{2}A$ for $v_s < 0$ V, $\frac{1}{2}A + \frac{1}{2}v_s$ for $v_s \geq 0$ V

6.83 (a) 5.74 μF, (b) 0.729 V

Chapter 7

7.2 (a) 2.15 kΩ, (b) 2 kΩ, (c) 2.75 kΩ

7.4 (a) -5 V, (b) 6.5 V

7.6 (a) 2.7 kΩ, 3.1 kΩ, (b) 1.6 kΩ, 1.86 kΩ

7.8 100 kΩ, 2 kΩ

7.10 (a) 11.8 μA, (b) 1.18 mA, (c) 12.9 V

7.12 30.8 Ω

7.14 554 Ω, 2.19 kΩ

7.16 2.62 kΩ, 2.11 kΩ

7.18 1.98 kΩ, 65 kΩ

7.20 100 kΩ, 1.14 kΩ

7.22 (a) 85, (b) 9 V, (c) 1.45 V

7.24 1.72 mA, 0.7 V, 23.5 mA, 2.65 V

7.26 330 kΩ, 2.4 kΩ, 693 Ω, 300 Ω

7.28 4.6 mA, 3.7 V, 1.59 mA, 2.81 V

7.30 (a) 57.1, (b) 571, (c) 87.5 kΩ, (d) 3.2 V

7.32 (a) 96.7, (b) 157

7.34 (b) 24.6

7.36 41

7.38 23.9

7.40 (a) 0.6 mA, (b) 0.9 mA, (c) 3 V, (d) -1 V,
(e) 0.8 V, (f) 1.5

7.42 (a) 571 Ω, (b) 714 Ω

7.44 (a) 1.8 V, (b) 3.96 V

7.46 (a) 344 Ω, (b) 275 Ω

7.48 12.8 kΩ, 1.96 V, 1.97 V

7.50 956 Ω, 2.25 V, 2.25 V

7.52 (a) 27.4, (b) 1 V, 2.54 V

7.54 22.9

7.56 19.5 mW

7.58 6.3 V, 8.2 V

7.60 127 mW

7.62 (a) 8, (b) 1.11, (c) 8

7.64 1 V, 3.5 V

7.67 0.8 V, 3.3 V

7.70 13

7.72 8.35 mW

7.74 0.2 V, 2.42 V

7.76 25

7.78 122

7.80 4.6, 1.93

7.82 27

7.84 0.325 V, 0.1 V

7.86 (a) 4.3 V, (b) 3.4 V, (c) 300 Ω

7.88 250 Ω, 300 Ω

7.90 (a) 2 V, (b) 16

7.92 (a) 0.2 V, 0.8 V, 0.2 V, 0.2 V,
(b) 0.8 V, 0.2 V, 0.8 V, 0.8 V

7.94 (a) 0.2 V, 0.8 V, 0.8 V, 0.2 V,
(b) 0.8 V, 0.2 V, 0.2 V, 0.8 V

Chapter 8

8.2 (a) 3 mA, (b) 6 V

8.4 (a) 11.7 mA, (b) 3.33 V

8.6 (a) −2 V, (b) 4 mA, (c) 14 V

8.10 (a) −2 V, (b) 1 mA, (c) 6 V

8.13 (a) −2.14 V, (b) 2.57 mA, (c) 1.72 V

8.15 (a) 1 mA, (b) −1 V, (c) 9 V

8.17 2.44 mA, (b) 0.318 V

8.19 722 Ω

8.21 (a) 55.6 Ω, (b) 1.4 kΩ, 100 Ω

8.23 (a) ohmic region, (b) 2.72 V, (c) 2.72 V,
(d) 7.28 mA

8.25 (a) ohmic region, (b) 4.94 V, (c) 4.94 V,
(d) 316 Ω

8.27 (a) 12.5 mA, (b) 2.13 V, (c) 8.87 V

8.29 (a) 16 mA, (b) 4 V, (c) 7 V

8.32 (a) 8 V, (b) 0.6 mA, (c) 6 V, (d) 10 kΩ,
(e) 20 kΩ, (f) 9.9 V

8.35 (a) 6 V, (b) 4 mA, (c) 12 V

8.38 (a) active region, (b) 5 V, (c) 5 V,
(d) 1.56 kΩ

8.40 (a) 2 V, (b) 0.25 mA, (c) 4 kΩ

8.43 $i_D = 1.6$ mA,
$v_{DS} = v_{GS} = 5$ V for all MOSFETs

8.45 9 V

8.47 11.9 V

8.50 3 V

8.52 $-4 < v_1 \leq 1$ V

8.55 0.225 V

8.57 0.1 V

8.59 AND gate

8.61 OR gate

8.63 5 V, 4 mA, 4 mA, 5 V

8.65 9.66 V

8.67 M_1: 0 V, OFF, 0 A, −9 V
M_2: −1 V, OFF, 0 A, −1 V
M_3: 10 V, ON, 0 A, 0 V
M_4: 0 V, OFF, 0 A, 0 V

8.69 M_1: −10 V, ON, 0 A, 0 V
M_2: −10 V, ON, 0 A, 0 V
M_3: 0 V, OFF, 0 A, 10 V
M_4: 0 V, OFF, 0 A, 10 V

8.71 M_1: 0 V, OFF, 0 A, 0 V
M_2: −10 V, ON, 0 A, 0 V
M_3: 10 V, ON, 0 A, 0 V
M_4: 0 V, OFF, 0 A, 10 V

8.73 M_1: −10 V, ON, 0 A, 0 V
M_2: −10 V, ON, 0 A, 0 V
M_3: 0 V, OFF, 0 A, 5 V
M_4: −5 V, OFF, 0 A, 5 V

8.75 (a) V_{DD}, (b) 0 V, (c) $V_{DD}/2$

8.77 (a) $-5 \leq v_1 \leq -3$ V, (b) $-3 < v_1 < 3$ V,
(c) $3 \leq v_1 \leq 5$ V

Chapter 9

9.2 10 mA, 6 V

9.4 3.09 mA, 16.9 V

9.6 307 Ω, 290 Ω

9.8 4.63 mA, 5.32 V

9.12 (a) 385 m℧, (b) 2.6 Ω, (c) 256 Ω, (d) −381

9.14 (a) $r_\pi + (1 + h_{fe})R_E$,
(b) $-h_{fe}R_C/[r_\pi + (1 + h_{fe})R_E]$

9.16 (a) −174, (b) 568 Ω, (c) 285 Ω, (d) 98.8,
(e) −49.6

9.19 (a) 0.991, (b) 62.9 kΩ, (c) 10.1

9.21 (a) 177, (b) 27.8 Ω, (c) 10 kΩ

9.23 (a) 0.991, (b) 63.6 kΩ, (c) 10.3

9.26 −2 V, 4 mA, 4 V

9.28 3 mA, 2 kΩ, 6 V

9.31 6 V, 4 mA, 12 V

9.33 8 V, 500 Ω, 1.5 kΩ

9.35 5 kΩ

9.37 4 V, 8 V, 1 mA

9.40 (a) $R_S/(R_S + 1/g_m)$, (b) $-g_mR_D/(1 + g_mR_S)$

9.42 (a) −1.82, (b) 1 MΩ, (c) 909 Ω

9.44 (a) −3.99, (b) 200 kΩ

9.54 (a) 7.09 krad/s, (b) 2.53 Mrad/s,
(c) 2.52 Mrad/s

9.56 (a) 5.1 krad/s, (b) 13 Mrad/s, (c) 13 Mrad/s

9.59 1.13 krad/s

9.62 1.14 Mrad/s

9.64 434 Mrad/s

9.67 1.35 rad/s

9.69 (a) 69 Ω, (b) 384 mW, (c) 768 mW

9.71 (a) 39 Ω, (b) 424 mW, (c) 847 mW

9.73 175 Ω, 218 Ω, 28.2 Ω

9.76 0.189

9.78 (a) 0.652 V, (b) 1.09 mA, (c) 6.52 mA ≫
54.5 μA, (d) 201 mA, (e) 1.51 W

9.80 (a) 2.1 kΩ, (b) 2.1 mA ≫ 32 μA,
(c) 126 Ω, (d) 1.11×10^{-14} A, (e) 0.4 W

9.84 (a) 0.66 V, (b) 331 Ω, (c) 52 Ω

Chapter 10

10.2 $37(v_1 - v_2)$

10.7 88 dB

10.9 −3 V, 3 mA, 12 V, 2 m℧

10.11 −3.33

10.13 $10(v_1 - v_2)$

10.15 −10/201, (b) 40 dB

10.17 (a) $-g_mR_D/(1 + 2g_mR_o)$, (b) $\frac{1}{2} + g_mR_o$

10.19 (a) 1.4 V, 14.3 V, (b) 0.68 mA, (c) 0.7 mA

10.22 (a) 4.3 kΩ, (b) 12 kΩ, (c) 0.58 kΩ,
(d) 2.05×10^{-15} A

10.25 113 μA, 113 μA

10.28 82 mV

10.30 99.1 dB

10.32 1 mA, 15.7 V, 1 mA, 15.7 V, 1 mA,
−4.3 V, 1 mA, −4.3 V

10.34 10 kΩ

10.37 $-AR/(1 + A), R/(1 + A)$

10.39 $-A/[R_L + (1 + A)R]$

10.45 (a) 100 kHz, (b) 1.00 kHz

10.47 12.5 kΩ

10.53 (a) 6.5 kΩ, 188 kHz, (b) 0.0361 μF, 43.5 kΩ
(c) 290 kΩ, 650 Hz

10.55 (a) 0.0518 μF, 800 Ω, (b) 1.37 kΩ, 65.6

10.61 (a) 796 kHz, 0.5 m℧, (b) 127 μH, 1 m℧

10.64 (a) 159 kHz, (b) 500 Ω

10.67 (a) V_{CC} for $v_2 < V_r$, $-V_{CC}$ for $v_2 > V_r$,
(b) square wave from $-V_{CC}$ to $+V_{CC}$

10.70 0 V, −8 V;
12 V for $v_1 < 0$ V, −12 V for $v_1 > 0$ V;
−12 V for $v_1 > -8$ V, 12 V for $v_1 < -8$ V

10.73 −12 V for $v_1 < 15$ V, 12 V for $v_1 > 15$ V;
12 V for $v_1 > 3$ V, −12 V for $v_1 < 3$ V

10.77 (a) 100 Hz, (b) 45.5 kΩ, (c) 0.207 μF,
(d) 55.9 kΩ, (e) 1.79 kΩ

10.79 (a) 962 Hz, (b) 60.1 nF, (c) 8.86 kΩ,
(d) 9.43 kΩ

10.81 107

10.84 100

10.86 0.887, narrowband FM

Chapter 11

11.2 (a) 27, (c) 0.84375, (e) 20.375

11.3 (a) 101011, (c) 0.11011, (e) 10011.101

11.4 (a) 27, (c) 0.84375, (e) 20.375

11.5 (a) 53, (b) 0.66, (e) 23.5

11.6 (b) 545.2

11.7 (b) 100001011.11100001

11.8 (b) 165.4

11.9 (b) 1010110010.101000000111

11.10 (a) 1000000, (c) 10010.101

11.11 (a) 11000, (c) -111.111
11.12 (a) 10010110.00, (c) 1111111.1
11.13 (a) 110.01, (c) 101
11.16 0001000100100111,
      (c) 001001100001.01100100
11.17 (b) 3523
11.20 $A + BC$
11.22 $A + \overline{A}B$, equals the OR operation
11.26 $\overline{A}B + A\overline{B}$
11.28 $(A \oplus B) + (\overline{A + B})$, AND operation
11.30 $(A \oplus B) \oplus (A + B)$, AND operation
11.33 $\overline{A + \overline{B}}$
11.38 (a) $\overline{(A + \overline{B})(\overline{A} + B)} + \overline{(A + B)}$,
      (c) $A\overline{B} + \overline{A}B = (A + B)(\overline{A} + \overline{B}) = A \oplus B$
11.42 $F_1 = m_0 + m_1 + m_2 + m_3 + m_7 = M_4M_5M_6$,
      $F_2 = m_0 + m_2 + m_4 + m_6 + m_7 = M_1M_3M_5$
11.44 $m_3 + m_5 + m_6 + m_7 = M_0M_1M_2M_4$
11.47 (a) $m_3 + m_4 + m_5 + m_6 + m_7$, (b) $M_0M_1M_2$;
      $A + BC$ requires the fewest gates
11.49 (a) $m_3 + m_5 + m_6 + m_7$, (b) $M_0M_1M_2M_4$
11.51 (a) cannot be simplified, (c) $AC + BC + AB$
11.52 (a) $(A + B + C)(\overline{A} + B + \overline{C})(A + \overline{B} + \overline{C})$
      $(\overline{A} + \overline{B} + C)$
      (b) $(A + B)(A + C)(B + C)$
11.53 (b) $A\overline{C} + AB + B\overline{C}$
11.54 (b) $(A + B)(A + \overline{C})(B + \overline{C})$
11.55 (b) $(A + \overline{B})(A + \overline{C})$
11.56 (b) $A + \overline{B}\overline{C}$
11.57 (b) $\overline{B}D + C\overline{D} + A\overline{C}D$
11.58 (b) $(C + D)(A + \overline{B} + \overline{D})(\overline{B} + \overline{C} + \overline{D})$
11.59 (b) $AB + \overline{B}C + A\overline{C}D$ or $AB + \overline{B}C + A\overline{C}\overline{D}$
11.60 (b) $(A + \overline{B})(A + C)(B + C + \overline{D})$

Chapter 12

12.2 (a) $ABC + ABD + ACD + BCD$,
     (b) $ABC + ABD + ACD + BCD$
12.6 $A + B\overline{C} + B\overline{D} + \overline{C}D$,
     $A\overline{B}\overline{C} + \overline{A}B\overline{C} + \overline{A}B\overline{D} + \overline{A}\,\overline{C}D$
12.8 $A\overline{C}D + B\overline{C}D + BCD + \overline{B}C\overline{D} + \overline{A}\,\overline{B}CD$
12.12 $(A_0 \odot B_0)(A_1 \odot B_1)(A_2 \odot B_2)(A_3 \odot B_3)$
12.14 $B_1 = A_5 + A_6 + A_7 + A_8, B_2 = A_3 + A_4 +$
      $A_7 + A_8, B_3 = A_2 + A_4 + A_6 + A_8$
12.19 $\overline{A}\,\overline{B}I_0 + \overline{A}BI_1 + A\overline{B}I_2 + ABI_3$;
      $I_0 = 0, I_1 = C, I_2 = C, I_3 = 1$

12.22 $AB + AC + B\overline{C}$
12.24 upper OR gate: D_0, D_3, D_5, D_6;
      lower OR gate: D_0, D_4, D_5, D_6
12.26 SR flip-flop
12.29 (a) $\overline{S(RQ)}$, (b) $\overline{S + \overline{R}} + \overline{S + Q}$
12.32 clocked SR flip-flop
12.37

X	Q_1	Q_2	D_1	D_2	$Q_1(t+1)$	$Q_2(t+1)$	Y
0	0	0	0	0	0	0	0
0	0	1	0	0	0	0	0
0	1	0	0	0	0	0	0
0	1	1	0	0	0	0	0
1	0	0	0	1	0	1	0
1	0	1	1	0	1	0	0
1	1	0	1	0	1	0	1
1	1	1	1	0	1	0	1

12.41 $J_1 = K_1 = XQ_2 + \overline{X}\overline{Q}_2 = X \odot Q_2$,
      $J_2 = K_2 = 1$
12.43 $D_1 = \overline{X}\overline{Q}_1 + Q_1Q_2, D_2 = XQ_1\overline{Q}_2$,
      $Y = (X + \overline{Q}_1)Q_2$

Chapter 13

13.1 $J_1 = Q_1Q_2, K_1 = Q_3, J_2 = \overline{Q}_1Q_3, K_2 = Q_3$
13.3 $D_1 = \overline{Q}_2Q_3 + Q_2\overline{Q}_3 = Q_2 \oplus Q_3$,
     $D_2 = Q_1, D_3 = Q_2$
13.7 $m_1 + m_2, m_4 + m_8, m_1 + m_4 + m_{16}$
13.13 $D_1 = \overline{Q}_1\overline{Q}_3 + Q_2\overline{Q}_3 + Q_1\overline{Q}_2Q_3$
13.15 $D_1 = Q_1\overline{Q}_3 + \overline{Q}_2\overline{Q}_3 + \overline{Q}_1Q_2Q_3$
13.19 (b) 16
13.20 (b) 8
13.24 (a) 4096 AND gates, each with 12 inputs
13.28 0 V, 1.5 V, 3 V, 4.5 V, 6 V, 7.5 V, 9 V,
     10.5 V
13.32 $R_0 = 10R_F, R_1 = R_F$
13.35 (a) 25%, (c) 6.25%

Chapter 14

14.1 (a) 159 A, (b) 106 mA
14.3 (a) 333 μWb/m^2, (b) 796 turns
14.5 (a) 795 turns, (b) 12,720 turns
14.7 (a) 8.22×10^6 A-t/Wb, (b) 41.1 A,
     (c) 4110 turns, (d) 48.7 μWb

14.9 (a) 5570, (b) 0.12 A
14.11 (a) 1194, (b) 200 turns
14.13 (a) 6.30 W, (b) 24.4 W
14.15 (a) 115 turns, (b) 115 turns
14.19 (a) counterclockwise, (b) to the right, (c) positive, (d) positive
14.23 $9 + j9\ \Omega$
14.25 (a) $14.4\underline{/-23.1°}$ V, $1.44 + j16.1\ \Omega$, (c) $8.49\cos(3t + 165°)$ V
14.27 (a) 0 A, (b) $6\sin 2t$ A, (c) $48\cos 2t$ V, (d) $\infty\ \Omega$
14.30 $2\sqrt{2}\cos(6t + 22°)$ V
14.32 24.2 W
14.34 (a) 0.04, (b) 20 kΩ, (c) 0.02
14.36 (a) $0.658\cos(3t + 20.5°)$ A, (b) $1.97\cos(3t - 160°)$ A, (c) $11.8\cos(3t - 70°)$ A
14.38 13.2 kΩ, 7.56 kΩ, 0.875 Ω, 1.79 Ω
14.40 (a) 240 V, 3.78 A, 384 W, (b) 4.47 V, 200 A, 400 W
14.42 0.783%, 98.4%

Chapter 15

15.2 (a) −6.25 A, (b) 250 m/s
15.4 $12\cos 2t$ m/s
15.7 $\dfrac{blv - Rf}{(Bl)^2 + RD}(1 - e^{-at})$

where $a = \dfrac{(Bl)^2 + RD}{RM}$

15.9 40.8%
15.11 29.5%
15.13 101 mA, 25.3 mA
15.15 1.12 A
15.17 (a) 105 Ω, (b) 4 Ω
15.20 (a) 20 kΩ/V, (b) 20 kΩ/V
15.22 9.93 V
15.24 (a) 10.0 Ω, (b) 998 kΩ
15.26 (a) 9.98 mA, 50.0 V, (b) 10.03 mA, 49.9 V
15.29 (a) 30 turns, (b) 1500 rpm
15.31 (a) 230 V, (b) 27 V, (c) 266 V
15.33 (a) 0.8 A, (b) 0.64 A, (c) 0.96 A
15.35 (a) 220 V, (b) 42 Ω
15.37 0.3 Ω

15.39 (a) 500 A-t/pole, 800 A-t/pole, (b) 300 A-t/pole, (c) 7.5 A-t/pole
15.41 (a) 950 rpm, (b) 25 A
15.43 (a) 40 A, 700 rpm, (b) 50 A, 1219 rpm
15.45 8.5%
15.47 (a) 52.6 A, (b) 119 N-m
15.50 136 V
15.52 (a) 32.6°, (b) 50.9°, (c) 23.4°
15.54 (a) $38.1\underline{/-38.1°}$ A, (b) 0.787 lagging, (c) 296 $\overline{V}$, 30.9°
15.56 (a) 3.5 Ω, (b) 241 V, (c) 18 kW
15.58 1710 rpm
15.60 (a) 91.7 N-m, 206 N-m, (b) 0.559 Ω
15.62 (a) 3.76%, (b) 1155 rpm
15.64 (a) 1710 rpm, (b) 150 N-m, (c) 631 N-m, (d) 695 N-m

Chapter 16

16.2 10 V, 6 V, 8 V
16.4 6 V, 3 V, −3 V, −9 V
16.6 1.111 mA, −444.44 V
16.9 1 V, 1.4 V, 1 V, 1 V, 5 V
16.11 16 V, 6 Ω
16.14 0 V, 1.333 Ω
16.32 $6.324\cos(3t - 18.43°)$ V, $4.472\cos(3t + 26.56°)$ V
16.35 $13.42\cos(2t - 116.6°)$ V
16.38 $6.8\underline{/29.98°}$ A, $6.8\underline{/-90.02°}$ A
16.40 $0.6576\cos(3t + 20.54°)$ A, $1.973\cos(3t - 159.5°)$ A, $35.51\cos(3t - 69.46°)$ V, $11.84\cos(3t - 69.46°)$ V
16.42 0.225 Hz
16.44 0.103 Hz
16.47 14.9 mA, 0.771 V
16.49 5.14 mA, 0.799 V, 0.763 mA, 0.701 V
16.51 15 mA, 0.771 V, 15 mA, 0.736 V
16.59 −4 V, −2 V
16.62 42.6 μA, 4.26 mA, 0.699 V, −0.742 V, 1.44 V
16.64 17.2 μA, 1.72 mA, 0.699 V, 0.248 mA, 0.699 V
235 μA, 23.5 mA, 0.699 V, −1.95 V, 2.65 V

16.70 329 μA, 0.4 V, 101 μA, 5.03 mA,
0.7 V, 0.4 V, 0.3 V

16.72 3 mA, 6 V

16.75 2.57 mA, −2.14 V, 1.7 V, ohmic region

16.77 8 mA, −1.17 V, 10.8 V

16.79 (a) 5 V, (b) 1.31 V

16.81 4 mA, 6 V, 12 V, active region

16.85 0.1118 V

16.87 9.6569 V

16.90 606 Ω

16.92 179, 27.5 Ω

16.94 3.98 kΩ

16.96 15.7 Hz, 14.1 MHz, 14.1 MHz

Index